ASTM Standards for Welding

2010
2nd Edition

ASTM Stock Number: WELDING10

Editorial Staff

Director:
  Vernice A. Mayer

Editors:
  Sean J. Bailey
  Nicole C. Baldini

Scott Emery
Joe Ermigiotti
Brendan C. Huffman
Emilie Moore
Emilie A. Olcese
Kathleen A. Peters

Jessica L. Rosiak
David A. Terruso
Julie Wright

Publishing Assistant Coordina
  Nicholas Furcola

Library of Congress Cataloging-in-Publication Data

ASTM standards for welding.
  p. cm.
  "ASTM Stock Number: WELDING10."
  ISBN 978-0-8031-8427-5
  1. Gas metal arc welding. I. AWS Committee on Arc Welding and Cutting.

TK4660.A78 2010
671.5'22 021873--dc22

2010017455

Copyright ©2010 ASTM International, West Conshohocken, PA. Prior editions copyrighted 2009 and earlier, by the American Society for Testing and Materials. All rights reserved. This material may not be reproduced or copied, in whole or in part, in any printed, mechanical, electronic, film, or other distribution and storage media, without the written consent of the publisher.

ORDER INFORMATION: Additional copies of this book in print, on CD-ROM, or reprints (single or multiple copies) of individual standards may be obtained by contacting ASTM International at 100 Barr Harbor Drive, PO Box C700, West Conshohocken, PA 19428-2959, or at 610-832-9585 (phone), 610-832-9555 (fax), service@astm.org (email), or through www.astm.org (website).

Photocopy Rights

Authorization to photocopy items for internal, personal, or educational classroom use, or the internal, personal, or educational classroom use of specific clients, is granted by ASTM International provided that the appropriate fee is paid to ASTM International, 100 Barr Harbor Drive, PO Box C700, West Conshohocken, PA 19428–2959, Tel: 610–832–9634; online:http://ww.astm.org/copyright/

Printed in Baltimore, MD
May 2010

Foreword

ASTM International has compiled this valuable companion resource to the American Welding Society (AWS) Structural Welding Code D1.1. All 60 ASTM specifications, practices, and test methods referenced by the AWS Code are found in this publication. The compilation will be a useful, one-source reference to a broad cross-section of users of AWS D1.1, including those quality professionals such as inspectors, supervisors, engineers, and managers.

The content was developed by the following three ASTM International Committees: A01 on Steel, Stainless Steel and Related Alloys; E07 on Nondestructive Testing; and E28 on Mechanical Testing. These committees, which meet twice a year, have jurisdiction over hundreds of standards. You can be part of this important initiative. ASTM members are individuals and corporations who contribute their expertise to influence the standards being set for their industries. The work of ASTM International members makes products and services safer, better, and more cost-effective. In short, ASTM International standards contribute toward product quality, enhancing communication, and overall customer satisfaction.

ASTM International

ASTM International, founded in 1898, is a developer and publisher of technical information designed to promote the understanding and development of technology and to ensure the quality of commodities and services and the safety of products.

ASTM International's primary mission is to develop voluntary full-consensus standards for materials, products, systems, and services. It provides a forum for producers, users, ultimate consumers, and those having general interest (representatives of government and academia) to meet on common ground to write standards that best meet their goals. ASTM International also publishes books containing reports on state-of-the-art testing techniques and their possible applications.

For books and papers related to welding, please visit our website www.astm.org or our digital library www.astm.org/digitallibrary.

Contents

ASTM STANDARDS FOR WELDING

In the serial designations prefixed to the following titles, the number following the dash indicates the year of original issue, or in the case of revision, the year of last revision. Thus, standards adopted or revised during the year 2010 have as their final number 10. A letter following this number indicates more than one revision during that year, that is, 10a indicates the second revision in 2010, 10b the third revision, etc. Standards that have been reapproved without change are indicated by the year of last reapproval in parentheses as part of the designation number, for example, (2010). A superscript epsilon (ϵ) indicates an editorial change since the last revision or reapproval—ϵ^1 for the first change, ϵ^2 for the second change, etc.

Designation	Title
A6/A6M – 09	Specification for General Requirements for Rolled Structural Steel Bars, Plates, Shapes, and Sheet Piling
A29/A29M – 05	Specification for Steel Bars, Carbon and Alloy, Hot-Wrought, General Requirements for
A36/A36M – 08	Specification for Carbon Structural Steel
A53/A53M – 07	Specification for Pipe, Steel, Black and Hot-Dipped, Zinc-Coated, Welded and Seamless
A106/A106M – 08	Specification for Seamless Carbon Steel Pipe for High-Temperature Service
A108 – 07	Specification for Steel Bar, Carbon and Alloy, Cold-Finished
A109/A109M – 08	Specification for Steel, Strip, Carbon (0.25 Maximum Percent), Cold-Rolled
A131/A131M – 08	Specification for Structural Steel for Ships
A139/A139M – 04	Specification for Electric-Fusion (Arc)-Welded Steel Pipe (NPS 4 and Over)
A242/A242M – 04(2009)	Specification for High-Strength Low-Alloy Structural Steel
A252 – 98(2007)	Specification for Welded and Seamless Steel Pipe Piles
A325 – 09a$^{\epsilon1}$	Specification for Structural Bolts, Steel, Heat Treated, 120/105 ksi Minimum Tensile Strength
A333/A333M – 05	Specification for Seamless and Welded Steel Pipe for Low-Temperature Service
A334/A334M – 04a	Specification for Seamless and Welded Carbon and Alloy-Steel Tubes for Low-Temperature Service
A370 – 09a$^{\epsilon1}$	Test Methods and Definitions for Mechanical Testing of Steel Products
A381 – 96(2005)	Specification for Metal-Arc-Welded Steel Pipe for Use With High-Pressure Transmission Systems
A435/A435M – 90(2007)	Specification for Straight-Beam Ultrasonic Examination of Steel Plates
A490 – 09	Specification for Structural Bolts, Alloy Steel, Heat Treated, 150 ksi Minimum Tensile Strength
A496/A496M – 07	Specification for Steel Wire, Deformed, for Concrete Reinforcement
A500/A500M – 09	Specification for Cold-Formed Welded and Seamless Carbon Steel Structural Tubing in Rounds and Shapes
A501 – 07	Specification for Hot-Formed Welded and Seamless Carbon Steel Structural Tubing
A514/A514M – 05(2009)	Specification for High-Yield-Strength, Quenched and Tempered Alloy Steel Plate, Suitable for Welding
A516/A516M – 06	Specification for Pressure Vessel Plates, Carbon Steel, for Moderate- and Lower-Temperature Service
A517/A517M – 06	Specification for Pressure Vessel Plates, Alloy Steel, High-Strength, Quenched and Tempered
A524 – 96(2005)	Specification for Seamless Carbon Steel Pipe for Atmospheric and Lower Temperatures
A529/A529M – 05(2009)	Specification for High-Strength Carbon-Manganese Steel of Structural Quality
A537/A537M – 08	Specification for Pressure Vessel Plates, Heat-Treated, Carbon-Manganese-Silicon Steel
A572/A572M – 07	Specification for High-Strength Low-Alloy Columbium-Vanadium Structural Steel
A573/A573M – 05(2009)	Specification for Structural Carbon Steel Plates of Improved Toughness
A578/A578M – 07	Specification for Straight-Beam Ultrasonic Examination of Rolled Steel Plates for Special Applications
A588/A588M – 05	Specification for High-Strength Low-Alloy Structural Steel, up to 50 ksi [345 MPa] Minimum Yield Point, with Atmospheric Corrosion Resistance
A595/A595M – 06	Specification for Steel Tubes, Low-Carbon or High-Strength Low-Alloy, Tapered for Structural Use
A606/A606M – 09a	Specification for Steel, Sheet and Strip, High-Strength, Low-Alloy, Hot-Rolled and Cold-Rolled, with Improved Atmospheric Corrosion Resistance
A618/A618M – 04	Specification for Hot-Formed Welded and Seamless High-Strength Low-Alloy Structural Tubing
A633/A633M – 01(2006)	Specification for Normalized High-Strength Low-Alloy Structural Steel Plates

CONTENTS

A653/A653M – 09a	Specification for Steel Sheet, Zinc-Coated (Galvanized) or Zinc-Iron Alloy-Coated (Galvannealed) by the Hot-Dip Process
A656/A656M – 05$^{\epsilon 2}$	Specification for Hot-Rolled Structural Steel, High-Strength Low-Alloy Plate with Improved Formability
A671 – 09	Specification for Electric-Fusion-Welded Steel Pipe for Atmospheric and Lower Temperatures
A673/A673M – 07	Specification for Sampling Procedure for Impact Testing of Structural Steel
A678/A678M – 05(2009)	Specification for Quenched-and-Tempered Carbon and High-Strength Low-Alloy Structural Steel Plates
A709/A709M – 09a	Specification for Structural Steel for Bridges
A710/A710M – 02(2007)	Specification for Precipitation–Strengthened Low-Carbon Nickel-Copper-Chromium-Molybdenum-Columbium Alloy Structural Steel Plates
A770/A770M – 03(2007)	Specification for Through-Thickness Tension Testing of Steel Plates for Special Applications
A852/A852M – 03(2007)	Specification for Quenched and Tempered Low-Alloy Structural Steel Plate with 70 ksi [485 MPa] Minimum Yield Strength to 4 in. [100 mm] Thick
A871/A871M – 03(2007)	Specification for High-Strength Low-Alloy Structural Steel Plate With Atmospheric Corrosion Resistance
A913/A913M – 07	Specification for High-Strength Low-Alloy Steel Shapes of Structural Quality, Produced by Quenching and Self-Tempering Process (QST)
A992/A992M – 06a	Specification for Structural Steel Shapes
A1008/A1008M – 09a	Specification for Steel, Sheet, Cold-Rolled, Carbon, Structural, High-Strength Low-Alloy, High-Strength Low-Alloy with Improved Formability, Solution Hardened, and Bake Hardenable
A1011/A1011M – 09b	Specification for Steel, Sheet and Strip, Hot-Rolled, Carbon, Structural, High-Strength Low-Alloy, High-Strength Low-Alloy with Improved Formability, and Ultra-High Strength
A1018/A1018M – 09	Specification for Steel, Sheet and Strip, Heavy-Thickness Coils, Hot-Rolled, Carbon, Commercial, Drawing, Structural, High-Strength Low-Alloy, High-Strength Low-Alloy with Improved Formability, and Ultra-High Strength
A1043/A1043M – 05(2009)	Specification for Structural Steel with Low Yield to Tensile Ratio for Use in Buildings
E23 – 07a$^{\epsilon 1}$	Test Methods for Notched Bar Impact Testing of Metallic Materials
E92 – 82(2003)$^{\epsilon 2}$	Test Method for Vickers Hardness of Metallic Materials
E94 – 04	Guide for Radiographic Examination
E140 – 07	Standard Hardness Conversion Tables for Metals Relationship Among Brinell Hardness, Vickers Hardness, Rockwell Hardness, Superficial Hardness, Knoop Hardness, and Scleroscope Hardness
E165 – 09	Practice for Liquid Penetrant Examination for General Industry
E709 – 08	Guide for Magnetic Particle Testing
E747 – 04	Practice for Design, Manufacture and Material Grouping Classification of Wire Image Quality Indicators (IQI) Used for Radiology
E1025 – 05	Practice for Design, Manufacture, and Material Grouping Classification of Hole-Type Image Quality Indicators (IQI) Used for Radiology
E1032 – 06	Test Method for Radiographic Examination of Weldments

RELATED MATERIAL

Index

Designation: A6/A6M – 09

Standard Specification for
General Requirements for Rolled Structural Steel Bars, Plates, Shapes, and Sheet Piling[1]

This standard is issued under the fixed designation A6/A6M; the number immediately following the designation indicates the year of original adoption or, in the case of revision, the year of last revision. A number in parentheses indicates the year of last reapproval. A superscript epsilon (ε) indicates an editorial change since the last revision or reapproval.

This standard has been approved for use by agencies of the Department of Defense.

1. Scope*

1.1 This general requirements specification[2] covers a group of common requirements that, unless otherwise specified in the applicable product specification, apply to rolled structural steel bars, plates, shapes, and sheet piling covered by each of the following product specifications issued by ASTM:

ASTM Designation[3]	Title of Specification
A36/A36M	Carbon Structural Steel
A131/A131M	Structural Steel for Ships
A242/A242M	High-Strength Low-Alloy Structural Steel
A283/A283M	Low and Intermediate Tensile Strength Carbon Steel Plates
A328/A328M	Steel Sheet Piling
A514/A514M	High-Yield Strength, Quenched and Tempered Alloy Steel Plate Suitable for Welding
A529/A529M	High-Strength Carbon-Manganese Steel of Structural Quality
A572/A572M	High-Strength Low-Alloy Columbium-Vanadium Steel
A573/A573M	Structural Carbon Steel Plates of Improved Toughness
A588/A588M	High-Strength Low-Alloy Structural Steel with 50 ksi (345 MPa) Minimum Yield Point to 4 in. [100 mm] Thick
A633/A633M	Normalized High-Strength Low-Alloy Structural Steel Plates
A656/A656M	Hot-Rolled Structural Steel, High-Strength Low-Alloy Plate with Improved Formability
A678/A678M	Quenched-and-Tempered Carbon and High-Strength Low-Alloy Structural Steel Plates
A690/A690M	High-Strength Low-Alloy Steel H-Piles and Sheet Piling for Use in Marine Environments
A709/A709M	Carbon and High-Strength Low-Alloy Structural Steel Shapes, Plates, and Bars and Quenched-and-Tempered Alloy Structural Steel Plates for Bridges
A710/A710M	Age-Hardening Low-Carbon Nickel-Copper-Chromium-Molybdenum-Columbium Alloy Structural Steel Plates
A769/A769M	Carbon and High-Strength Electric Resistance Welded Steel Structural Shapes
A786/A786M	Rolled Steel Floor Plates
A808/A808M	High-Strength Low-Alloy Carbon, Manganese, Columbium, Vanadium Steel of Structural Quality with Improved Notch Toughness
A827/A827M	Plates, Carbon Steel, for Forging and Similar Applications
A829/A829M	Plates, Alloy Steel, Structural Quality
A830/A830M	Plates, Carbon Steel, Structural Quality, Furnished to Chemical Composition Requirements
A852/A852M	Quenched and Tempered Low-Alloy Structural Steel Plate with 70 ksi [485 Mpa] Minimum Yield Strength to 4 in. [100 mm] Thick
A857/A857M	Steel Sheet Piling, Cold Formed, Light Gage
A871/A871M	High-Strength Low Alloy Structural Steel Plate with Atmospheric Corrosion Resistance
A913/A913M	Specification for High-Strength Low-Alloy Steel Shapes of Structural Quality, Produced by Quenching and Self-Tempering Process (QST)
A945/A945M	Specification for High-Strength Low-Alloy Structural Steel Plate with Low Carbon and Restricted Sulfur for Improved Weldability, Formability, and Toughness
A950/A950M	Specification for Fusion Bonded Epoxy-Coated Structural Steel H-Piles and Sheet Piling
A992/A992M	Specification for Steel for Structural Shapes for Use in Building Framing
A1026	Specification for Alloy Steel Structural Shapes for Use in Building Framing
A1043/A1043M	Specification for Structural Steel with Low Yield to Tensile Ratio for Use in Buildings

1.2 Annex A1 lists permitted variations in dimensions and mass (Note 1) in SI units. The values listed are not exact conversions of the values in Tables 1 to 31 inclusive but are, instead, rounded or rationalized values. Conformance to Annex A1 is mandatory when the "M" specification designation is used.

NOTE 1—The term "weight" is used when inch-pound units are the standard; however, under SI, the preferred term is "mass."

1.3 Annex A2 lists the dimensions of some shape profiles.

1.4 Appendix X1 provides information on coil as a source of structural products.

1.5 Appendix X2 provides information on the variability of tensile properties in plates and structural shapes.

1.6 Appendix X3 provides information on weldability.

1.7 Appendix X4 provides information on cold bending of plates, including suggested minimum inside radii for cold bending.

1.8 This general requirements specification also covers a group of supplementary requirements that are applicable to several of the above product specifications as indicated therein.

---

[1] This specification is under the jurisdiction of ASTM Committee A01 on Steel, Stainless Steel and Related Alloys and is the direct responsibility of Subcommittee A01.02 on Structural Steel for Bridges, Buildings, Rolling Stock and Ships.
Current edition approved April 1, 2009. Published April 2009. Originally approved in 1949. Last previous edition approved in 2008 as A6/A6M – 08a. DOI: 10.1520/A0006_A0006M-09.

[2] For ASME Boiler and Pressure Vessel Code applications, see related Specification SA-6/SA-6M in Section II of that Code.

*A Summary of Changes section appears at the end of this standard.

Such requirements are provided for use where additional testing or additional restrictions are required by the purchaser, and apply only where specified individually in the purchase order.

1.9 In case of any conflict in requirements, the requirements of the applicable product specification prevail over those of this general requirements specification.

1.10 Additional requirements that are specified in the purchase order and accepted by the supplier are permitted, provided that such requirements do not negate any of the requirements of this general requirements specification or the applicable product specification.

1.11 For purposes of determining conformance with this general requirements specification and the applicable product specification, values are to be rounded to the nearest unit in the right-hand place of figures used in expressing the limiting values in accordance with the rounding method of Practice E29.

1.12 The text of this general requirements specification contains notes or footnotes, or both, that provide explanatory material. Such notes and footnotes, excluding those in tables and figures, do not contain any mandatory requirements.

1.13 The values stated in either inch-pound units or SI units are to be regarded separately as standard. Within the text, the SI units are shown in brackets. The values stated in each system are not exact equivalents; therefore, each system is to be used independently of the other, without combining values in any way.

1.14 This general requirements specification and the applicable product specification are expressed in both inch-pound units and SI units; however, unless the order specifies the applicable "M" specification designation (SI units), the structural product is furnished to inch-pound units.

1.15 *This standard does not purport to address all of the safety concerns, if any, associated with its use. It is the responsibility of the user of this standard to establish appropriate safety and health practices and determine the applicability of regulatory limitations prior to use.*

2. Referenced Documents

2.1 *ASTM Standards:*[3]

A131/A131M Specification for Structural Steel for Ships

A370 Test Methods and Definitions for Mechanical Testing of Steel Products

A673/A673M Specification for Sampling Procedure for Impact Testing of Structural Steel

A700 Practices for Packaging, Marking, and Loading Methods for Steel Products for Shipment

A751 Test Methods, Practices, and Terminology for Chemical Analysis of Steel Products

A829/A829M Specification for Alloy Structural Steel Plates

A941 Terminology Relating to Steel, Stainless Steel, Related Alloys, and Ferroalloys

E29 Practice for Using Significant Digits in Test Data to Determine Conformance with Specifications

E112 Test Methods for Determining Average Grain Size

E208 Test Method for Conducting Drop-Weight Test to Determine Nil-Ductility Transition Temperature of Ferritic Steels

2.2 *American Welding Society Standards:*[4]

A5.1/A5.1M Mild Steel Covered Arc-Welding Electrodes

A5.5/A5.5M Low-Alloy Steel Covered Arc-Welding Electrodes

A5.17/A5.17M Specification For Carbon Steel Electrodes And Fluxes For Submerged Arc Welding

A5.18/A5.18M Specification For Carbon Steel Electrodes And Rods For Gas Shielded Arc Welding

A5.20/A5.20M Carbon Steel Electrodes For Flux Cored Arc Welding

A5.23/A5.23M Low Alloy Steel Electrodes And Fluxes For Submerged Arc Welding

A5.28/A5.28M Specification For Low-Alloy Steel Electrodes And Rods For Gas Shielded Arc Welding

A5.29/A5.29M Specification for Low-Alloy Steel Electrodes for Flux Cored Arc Welding

D1.1/D1.1M Structural Welding Code Steel

2.3 *U.S. Military Standards:*[5]

MIL-STD-129 Marking for Shipment and Storage

MIL-STD-163 Steel Mill Products Preparation for Shipment and Storage

2.4 *U.S. Federal Standard:*[5]

Fed. Std. No. 123 Marking for Shipments (Civil Agencies)

2.5 *ASME Boiler Pressure Vessel Code Standard:*[6]

BPVC Section IX Welding and Brazing Qualifications

3. Terminology

3.1 *Definitions of Terms Specific to This Standard:*

3.1.1 *Plates (other than floor plates)*—Flat, hot-rolled steel, ordered to thickness or weight [mass] and typically width and length, commonly classified as follows:

3.1.1.1 *When Ordered to Thickness:*

(1) Over 8 in. [200 mm] in width and 0.230 in. [6 mm] or over in thickness.

(2) Over 48 in. [1200 mm] in width and 0.180 in. [4.5 mm] or over in thickness.

3.1.1.2 *When Ordered to Weight [Mass]:*

(1) Over 8 in. [200 mm] in width and 9.392 lb/ft^2 [47.10 kg/m^2] or heavier.

(2) Over 48 in. [1200 mm] in width and 7.350 lb/ft^2 [35.32 kg/m^2] or heavier.

3.1.1.3 *Discussion*—Steel products are available in various thickness, width, and length combinations depending upon equipment and processing capabilities of various manufacturers and processors. Historic limitations of a product based upon

---

[3] For referenced ASTM standards, visit the ASTM website, www.astm.org, or contact ASTM Customer Service at service@astm.org. For *Annual Book of ASTM Standards* volume information, refer to the standard's Document Summary page on the ASTM website.

[4] Available from American Welding Society (AWS), 550 NW LeJeune Rd., Miami, FL 33126, http://www.aws.org.

[5] Available from Standardization Documents Order Desk, DODSSP, Bldg. 4, Section D, 700 Robbins Ave., Philadelphia, PA 19111-5098, http://www.dodssp.daps.mil.

[6] Available from American Society of Mechanical Engineers (ASME), ASME International Headquarters, Three Park Ave., New York, NY 10016-5990, http://www.asme.org.

dimensions (thickness, width, and length) do not take into account current production and processing capabilities. To qualify any product to a particular product specification requires all appropriate and necessary tests be performed and that the results meet the limits prescribed in that product specification. If the necessary tests required by a product specification cannot be conducted, the product cannot be qualified to that specification. This general requirement standard contains permitted variations for the commonly available sizes. Permitted variations for other sizes are subject to agreement between the customer and the manufacturer or processor, whichever is applicable.

3.1.1.4 Slabs, sheet bars, and skelp, though frequently falling in the foregoing size ranges, are not classed as plates.

3.1.1.5 Coils are excluded from qualification to the applicable product specification until they are decoiled, leveled or straightened, formed (if applicable), cut to length, and, if required, properly tested by the processor in accordance with ASTM specification requirements (see Sections 9-15, 18, and 19 and the applicable product specification).

3.1.2 *Shapes (Flanged Sections)*:

3.1.2.1 *structural-size shapes*—rolled flanged sections having at least one dimension of the cross section 3 in. [75 mm] or greater.

3.1.2.2 *bar-size shapes*—rolled flanged sections having a maximum dimension of the cross section less than 3 in. [75 mm].

3.1.2.3 *"W" shapes*—doubly-symmetric, wide-flange shapes with inside flange surfaces that are substantially parallel.

3.1.2.4 *"HP" shapes*—are wide-flange shapes generally used as bearing piles whose flanges and webs are of the same nominal thickness and whose depth and width are essentially the same.

3.1.2.5 *"S" shapes*—doubly-symmetric beam shapes with inside flange surfaces that have a slope of approximately 16⅔ %.

3.1.2.6 *"M" shapes*—doubly-symmetric shapes that cannot be classified as "W," "S," or "HP" shapes.

3.1.2.7 *"C" shapes*—channels with inside flange surfaces that have a slope of approximately 16⅔ %.

3.1.2.8 *"MC" shapes*—channels that cannot be classified as "C" shapes.

3.1.2.9 *"L" shapes*—shapes having equal-leg and unequal-leg angles.

3.1.3 *sheet piling*—rolled steel sections that are capable of being interlocked, forming a continuous wall when individual pieces are driven side by side.

3.1.4 *bars*—rounds, squares, and hexagons, of all sizes; flats 13/64 in. (0.203 in.) and over [over 5 mm] in specified thickness, not over 6 in. [150 mm] in specified width; and flats 0.230 in. and over [over 6 mm] in specified thickness, over 6 to 8 in. [150 to 200 mm] inclusive, in specified width.

3.1.5 *exclusive*—when used in relation to ranges, as for ranges of thickness in the tables of permissible variations in dimensions, is intended to exclude only the greater value of the range. Thus, a range from 60 to 72 in. [1500 to 1800 mm] exclusive includes 60 in. [1500 mm], but does not include 72 in. [1800 mm].

3.1.6 *rimmed steel*—steel containing sufficient oxygen to give a continuous evolution of carbon monoxide during solidification, resulting in a case or rim of metal virtually free of voids.

3.1.7 *semi-killed steel*—incompletely deoxidized steel containing sufficient oxygen to form enough carbon monoxide during solidification to offset solidification shrinkage.

3.1.8 *capped steel*—rimmed steel in which the rimming action is limited by an early capping operation. Capping is carried out mechanically by using a heavy metal cap on a bottle-top mold or chemically by an addition of aluminum or ferrosilicon to the top of the molten steel in an open-top mold.

3.1.9 *killed steel*—steel deoxidized, either by addition of strong deoxidizing agents or by vacuum treatment, to reduce the oxygen content to such a level that no reaction occurs between carbon and oxygen during solidification.

3.1.10 *mill edge*—the normal edge produced by rolling between horizontal finishing rolls. A mill edge does not conform to any definite contour. Mill edge plates have two mill edges and two trimmed edges.

3.1.11 *universal mill edge*—the normal edge produced by rolling between horizontal and vertical finishing rolls. Universal mill plates, sometimes designated UM Plates, have two universal mill edges and two trimmed edges.

3.1.12 *sheared edge*—the normal edge produced by shearing. Sheared edge plates are trimmed on all edges.

3.1.13 *gas cut edge*—the edge produced by gas flame cutting.

3.1.14 *special cut edge*—usually the edge produced by gas flame cutting involving special practices such as pre-heating or post-heating, or both, in order to minimize stresses, avoid thermal cracking and reduce the hardness of the gas cut edge. In special instances, special cut edge is used to designate an edge produced by machining.

3.1.15 *sketch*—when used to describe a form of plate, denotes a plate other than rectangular, circular, or semi-circular. Sketch plates may be furnished to a radius or with four or more straight sides.

3.1.16 *normalizing*—a heat treating process in which a steel plate is reheated to a uniform temperature above the upper critical temperature and then cooled in air to below the transformation range.

3.1.17 *plate-as-rolled*—when used in relation to the location and number of tests, the term refers to the unit plate rolled from a slab or directly from an ingot. It does not refer to the condition of the plate.

3.1.18 *fine grain practice*—a steelmaking practice that is intended to produce a killed steel that is capable of meeting the requirements for fine austenitic grain size.

3.1.18.1 *Discussion*—It normally involves the addition of one or more austenitic grain refining elements in amounts that have been established by the steel producer as being sufficient. Austenitic grain refining elements include, but are not limited to, aluminum, columbium, titanium, and vanadium.

3.1.19 *structural product*—a hot-rolled steel plate, shape, sheet piling, or bar.

3.1.20 *coil*—hot-rolled steel in coiled form that is intended to be processed into a finished structural product.

3.1.21 *manufacturer*—the organization that directly controls the conversion of steel ingots, slabs, blooms, or billets, by hot-rolling, into an as-rolled structural product or into coil; and for structural products produced from as-rolled structural products, the organization that directly controls, or is responsible for, the operations involved in finishing the structural product.

3.1.21.1 *Discussion*—Such finishing operations include leveling or straightening, hot forming or cold forming (if applicable), welding (if applicable), cutting to length, testing, inspection, conditioning, heat treatment (if applicable), packaging, marking, loading for shipment, and certification.

3.1.22 *processor*—the organization that directly controls, or is responsible for, the operations involved in the processing of coil into a finished structural product. Such processing operations include decoiling, leveling or straightening, hot-forming or cold-forming (if applicable), welding (if applicable), cutting to length, testing, inspection, conditioning, heat treatment (if applicable), packaging, marking, loading for shipment, and certification.

3.1.22.1 *Discussion*—The processing operations need not be done by the organization that did the hot rolling of the coil. If only one organization is involved in the hot rolling and processing operations, that organization is termed the *manufacturer* for the hot rolling operation and the *processor* for the processing operations. If more than one organization is involved in the hot rolling and processing operations, the organization that did the hot rolling is termed the *manufacturer* and an organization that does one or more processing operations is termed a *processor*.

3.2 Refer to Terminology A941 for additional definitions of terms used in this standard.

4. Ordering Information

4.1 Information items to be considered, if appropriate, for inclusion in purchase orders are as follows:

4.1.1 ASTM product specification designation (see 1.1) and year-date;

4.1.2 Name of structural product (plate, shape, bar, or sheet piling);

4.1.3 Shape designation, or size and thickness or diameter;

4.1.4 Grade, class, and type designation, if applicable;

4.1.5 Condition (see Section 6), if other than as-rolled;

4.1.6 Quantity (weight [mass] or number of pieces);

4.1.7 Length;

4.1.8 Exclusion of either structural product produced from coil or structural product produced from an as-rolled structural product (see 5.3 and Appendix X1), if applicable;

4.1.9 Heat treatment requirements (see 6.2 and 6.3), if any;

4.1.10 Testing for fine austenitic grain size (see 8.3.2);

4.1.11 Mechanical property test report requirements (see Section 14), if any;

4.1.12 Special packaging, marking, and loading for shipment requirements (see Section 19), if any;

4.1.13 Supplementary requirements, if any, including any additional requirements called for in the supplementary requirements;

4.1.14 End use, if there are any end-use-specific requirements (see 18.1, 11.3.4, Table 22 or Table A1.22, and Table 24 or Table A1.24);

4.1.15 Special requirements (see 1.10), if any; and

4.1.16 Repair welding requirements (see 9.5), if any.

5. Materials and Manufacture

5.1 The steel shall be made in an open-hearth, basic-oxygen, or electric-arc furnace, possibly followed by additionl refining in a ladle metallurgy furnace (LMF), or secondary melting by vacuum-arc remelting (VAR) or electroslag remelting (ESR).

5.2 The steel shall be strand cast or cast in stationary molds.

5.2.1 *Strand Cast*:

5.2.1.1 When heats of the same nominal chemical composition are consecutively strand cast at one time, the heat number assigned to the cast product need not be changed until all of the steel in the cast product is from the following heat.

5.2.1.2 When two consecutively strand cast heats have different nominal chemical composition ranges, the manufacturer shall remove the transition material by an established procedure that positively separates the grades.

5.3 Structural products shall be produced from an as-rolled structural product or from coil.

5.4 Where part of a heat is rolled into an as-rolled structural product and the balance of the heat is rolled into coil, each part shall be tested separately.

5.5 Structural products produced from coil shall not contain splice welds, unless previously approved by the purchaser.

6. Heat Treatment

6.1 Where the structural product is required to be heat treated, such heat treatment shall be performed by the manufacturer, the processor, or the fabricator, unless otherwise specified in the applicable product specification.

NOTE 2—When no heat treatment is required, the manufacturer or processor has the option of heat treating the structural product by normalizing, stress relieving, or normalizing then stress relieving to meet the applicable product specification.

6.2 Where the heat treatment is to be performed by other than the manufacturer, the order shall so state.

6.2.1 Where the heat treatment is to be performed by other than the manufacturer, the structural products shall be accepted on the basis of tests made on test specimens taken from full thickness test coupons heat treated in accordance with the requirements specified in the applicable product specification or in the purchase order. If the heat-treatment temperatures are not specified, the manufacturer or processor shall heat treat the test coupons under conditions he considers appropriate, provided that the purchaser is informed of the procedure followed in heat treating the test coupons.

6.3 Where the heat treatment is to be performed by the manufacturer or the processor, the structural product shall be heat treated as specified in the applicable product specification, or as specified in the purchase order, provided that the heat treatment specified by the purchaser is not in conflict with the requirements of the applicable product specification.

6.4 Where normalizing is to be performed by the fabricator, the structural product shall be either normalized or heated uniformly for hot forming, provided that the temperature to which the structural product is heated for hot forming does not significantly exceed the normalizing temperature.

6.5 The use of cooling rates that are faster than those obtained by cooling in air to improve the toughness shall be subject to approval by the purchaser, and structural products so treated shall be tempered subsequently in the range from 1100 to 1300°F [595 to 705°C].

7. Chemical Analysis

7.1 *Heat Analysis*:

7.1.1 Sampling for chemical analysis and methods of analysis shall be in accordance with Test Methods, Practices, and Terminology A751.

7.1.2 For each heat, the heat analysis shall include determination of the content of carbon, manganese, phosphorus, sulfur, silicon, nickel, chromium, molybdenum, copper, vanadium, columbium; any other element that is specified or restricted by the applicable product specification for the applicable grade, class, and type; and any austenitic grain refining element whose content is to be used in place of austenitic grain size testing of the heat (see 8.3.2).

7.1.3 Except as allowed by 7.1.4 for primary heats, heat analyses shall conform to the heat analysis requirements of the applicable product specification for the applicable grade, class, and type.

7.1.4 Where vacuum-arc remelting or electroslag remelting is used, a remelted heat is defined as all ingots remelted from a single primary heat. If the heat analysis of the primary heat conforms to the heat analysis requirements of the applicable product specification for the applicable grade, class, and type, the heat analysis for the remelted heat shall be determined from one test sample taken from one remelted ingot, or the product of one remelted ingot, from the primary heat. If the heat analysis of the primary heat does not conform to the heat analysis requirements of the applicable product specification for the applicable grade, type, and class, the heat analysis for the remelted heat shall be determined from one test sample taken from each remelted ingot, or the product of each remelted ingot, from the primary heat.

7.2 *Product Analysis*—For each heat, the purchaser shall have the option of analyzing representative samples taken from the finished structural product. Sampling for chemical analysis and methods of analysis shall be in accordance with Test Methods, Practices, and Terminology A751. The product analyses so determined shall conform to the heat analysis requirements of the applicable product specification for the applicable grade, class, and type, subject to the permitted variations in product analysis given in Table A. If a range is specified, the determinations of any element in a heat shall not vary both above and below the specified range. Rimmed or capped steel is characterized by a lack of homogeneity in its composition, especially for the elements carbon, phosphorus, and sulfur. Therefore, the limitations for these elements shall not be applicable unless misapplication is clearly indicated.

7.3 *Referee Analysis*—For referee purposes, Test Methods, Practices, and Terminology A751 shall be used.

7.4 *Grade Substitution*—Alloy steel grades that meet the chemical requirements of Table 1 of Specification A829/A829M shall not be substituted for carbon steel grades.

8. Metallurgical Structure

8.1 Where austenitic grain size testing is required, such testing shall be in accordance with Test Methods E112 and at least 70 % of the grains in the area examined shall meet the specified grain size requirement.

8.2 *Coarse Austenitic Grain Size*—Where coarse austenitic grain size is specified, one austenitic grain size test per heat shall be made and the austenitic grain size number so determined shall be in the range of 1 to 5 inclusive.

8.3 *Fine Austenitic Grain Size*:

8.3.1 Where fine austenitic grain size is specified, except as allowed in 8.3.2, one austenitic grain size test per heat shall be made and the austenitic grain size number so determined shall be 5 or higher.

NOTE 3—Such austenitic grain size numbers may be achieved with lower contents of austenitic grain refining elements than 8.3.2 requires for austenitic grain size testing to be waived.

8.3.2 Unless testing for fine austenitic grain size is specified in the purchase order, an austenitic grain size test need not be made for any heat that has, by heat analysis, one or more of the following:

8.3.2.1 A total aluminum content of 0.020 % or more.

8.3.2.2 An acid soluble aluminum content of 0.015 % or more.

8.3.2.3 A content for an austenitic grain refining element that exceeds the minimum value agreed to by the purchaser as being sufficient for austenitic grain size testing to be waived, or

8.3.2.4 Contents for the combination of two or more austenitic grain refining elements that exceed the applicable minimum values agreed to by the purchaser as being sufficient for austenitic grain size testing to be waived.

9. Quality

9.1 *General*—Structural products shall be free of injurious defects and shall have a workmanlike finish.

NOTE 4—Unless otherwise specified, structural products are normally furnished in the as-rolled condition and are subjected to visual inspection by the manufacturer or processor. Non-injurious surface or internal imperfections, or both, may be present in the structural product as delivered and the structural product may require conditioning by the purchaser to improve its appearance or in preparation for welding, coating, or other further operations.

More restrictive requirements may be specified by invoking supplementary requirements or by agreement between the purchaser and the supplier.

Structural products that exhibit injurious defects during subsequent fabrication are deemed not to comply with the applicable product specification. (See 17.2.) Fabricators should be aware that cracks may initiate upon bending a sheared or burned edge during the fabrication process; this is not considered to be a fault of the steel but is rather a function of the induced cold-work or the heat-affected zone.

The conditioning requirements in 9.2, 9.3, and 9.4 limit the conditioning allowed to be performed by the manufacturer or processor. Conditioning of imperfections beyond the limits of 9.2, 9.3, and 9.4 may be performed by parties other than the manufacturer or processor at the discretion of the purchaser.

TABLE A Permitted Variations in Product Analysis

Note 1—Where "..." appears in this table, there is no requirement.

Element	Upper Limit, or Maximum Specified Value, %	Permitted Variations, % Under Minimum Limit	Permitted Variations, % Over Maximum Limit
Carbon	to 0.15 incl	0.02	0.03
	over 0.15 to 0.40 incl	0.03	0.04
	over 0.40 to 0.75 incl	0.04	0.05
	over 0.75	0.04	0.06
Manganese[A]	to 0.60 incl	0.05	0.06
	over 0.60 to 0.90 incl	0.06	0.08
	over 0.90 to 1.20 incl	0.08	0.10
	over 1.20 to 1.35 incl	0.09	0.11
	over 1.35 to 1.65 incl	0.09	0.12
	over 1.65 to 1.95 incl	0.11	0.14
	over 1.95	0.12	0.16
Phosphorus	to 0.04 incl	...	0.010
	over 0.04 to 0.15 incl	...	[B]
Sulfur	to 0.06 incl	...	0.010
	over 0.06	[B]	[B]
Silicon	to 0.30 incl	0.02	0.03
	over 0.30 to 0.40 incl	0.05	0.05
	over 0.40 to 2.20 incl	0.06	0.06
Nickel	to 1.00 incl	0.03	0.03
	over 1.00 to 2.00 incl	0.05	0.05
	over 2.00 to 3.75 incl	0.07	0.07
	over 3.75 to 5.30 incl	0.08	0.08
	over 5.30	0.10	0.10
Chromium	to 0.90 incl	0.04	0.04
	over 0.90 to 2.00 incl	0.06	0.06
	over 2.00 to 4.00 incl	0.10	0.10
Molybdenum	to 0.20 incl	0.01	0.01
	over 0.20 to 0.40 incl	0.03	0.03
	over 0.40 to 1.15 incl	0.04	0.04
Copper	0.20 minimum only	0.02	...
	to 1.00 incl	0.03	0.03
	over 1.00 to 2.00 incl	0.05	0.05
Titanium	to 0.15 incl	0.01[C]	0.01
Vanadium	to 0.10 incl	0.01[C]	0.01
	over 0.10 to 0.25 incl	0.02	0.02
	over 0.25	0.02	0.03
	minimum only specified	0.01	...
Boron	any	[B]	[B]
Columbium	to 0.10 incl	0.01[C]	0.01
Zirconium	to 0.15 incl	0.03	0.03
Nitrogen	to 0.030 incl	0.005	0.005

[A] Permitted variations in manganese content for bars and bar size shapes shall be: to 0.90 incl ±0.03; over 0.90 to 2.20 incl ±0.06.
[B] Product analysis not applicable.
[C] 0.005, if the minimum of the range is 0.01 %.

Index to Tables of Permitted Variations

Dimension	Table Inch-Pound Units	Table SI Units
Camber		
Plates, Carbon Steel; Sheared and Gas-Cut	12	A1.12
Plates, Carbon Steel; Universal Mill	11	A1.11
Plates, Other than Carbon Steel; Sheared, Gas-Cut and Universal Mill	11	A1.11
Shapes, Rolled; S, M, C, MC, and L	21	A1.21
Shapes, Rolled; W and HP	24	A1.24
Shapes, Split; L and T	25	A1.25
Cross Section of Shapes and Bars		
Flats	26	A1.26
Hexagons	28	A1.28
Rounds and Squares	27	A1.27
Shapes, Rolled; L, Bulb Angles, and Z	17	A1.17
Shapes, Rolled; W, HP, S, M, C, and MC	16	A1.16
Shapes, Rolled; T	18	A1.18
Shapes, Split; L and T	25	A1.25
Diameter		
Plates, Sheared	6	A1.6
Plates, Other than Alloy Steel, Gas-Cut	7	A1.7
Plates, Alloy Steel, Gas-Cut	10	A1.10
Rounds	27	A1.27
End Out-of-Square		
Shapes, Other than W	20	A1.20
Shapes, W	22	A1.22
Shapes, Milled, Other than W	23	A1.23
Flatness		
Plates, Carbon Steel	13	A1.13
Plates, Other than Carbon Steel	14	A1.14
Plates, Restrictive—Carbon Steel	S27.1	S27.2
Plates, Restrictive—Other than Carbon Steel	S27.3	S27.4
Length		
Bars	30	A1.30
Bars, Recut	31	A1.31
Plates, Sheared and Universal Mill	3	A1.3
Plates, Other than Alloy Steel, Gas-Cut	9	A1.9
Plates, Alloy Steel, Gas-Cut	8	A1.8
Plates, Mill Edge	4	A1.4
Shapes, Rolled; Other than W	19	A1.19
Shapes, Rolled; W and HP	22	A1.22
Shapes, Split; L and T	25	A1.25
Shapes, Milled	23	A1.23
Straightness		
Bars	29	A1.29
Shapes, Other than W	21	A1.21
Sweep		
Shapes, W and HP	24	A1.24
Thickness		
Flats	26	A1.26
Plates, Ordered to Thickness	1	A1.1
Waviness		
Plates	15	A1.15
Weight [Mass]		
Plates, Ordered to Weight [Mass]	2	A1.2
Width		
Flats	26	A1.26
Plates, Sheared	3	A1.3
Plates, Universal Mill	5	A1.5
Plates, Other than Alloy Steel, Gas-Cut	9	A1.9
Plates, Alloy Steel, Gas-Cut	8	A1.8
Plates, Mill Edge	4	A1.4

9.2 *Plate Conditioning*:

9.2.1 The grinding of plates by the manufacturer or processor to remove imperfections on the top or bottom surface shall be subject to the limitations that the area ground is well faired without abrupt changes in contour and the grinding does not reduce the thickness of the plate by (*1*) more than 7 % under the nominal thickness for plates ordered to weight per square foot or mass per square metre, but in no case more than 1/8 in. [3 mm]; or (*2*) below the permissible minimum thickness for plates ordered to thickness in inches or millimetres.

9.2.2 The deposition of weld metal (see 9.5) following the removal of imperfections on the top or bottom surface of plates by chipping, grinding, or arc-air gouging shall be subject to the following limiting conditions:

9.2.2.1 The chipped, ground, or gouged area shall not exceed 2 % of the area of the surface being conditioned.

9.2.2.2 After removal of any imperfections preparatory to welding, the thickness of the plate at any location shall not be reduced by more than 30 % of the nominal thickness of the plate. (Specification A131/A131M restricts the reduction in thickness to 20 % maximum.)

9.2.3 The deposition of weld metal (see 9.5) following the removal of injurious imperfections on the edges of plates by grinding, chipping, or arc-air gouging by the manufacturer or processor shall be subject to the limitation that, prior to welding, the depth of the depression, measured from the plate edge inward, is not more than the thickness of the plate or 1 in. [25 mm], whichever is the lesser.

9.3 *Structural Size Shapes, Bar Size Shapes, and Sheet Piling Conditioning*:

9.3.1 The grinding, or chipping and grinding, of structural size shapes, bar size shapes, and sheet piling by the manufacturer or processor to remove imperfections shall be subject to the limitations that the area ground is well faired without abrupt changes in contour and the depression does not extend below the rolled surface by more than (*1*) 1/32 in. [1 mm], for material less than 3/8 in. [10 mm] in thickness; (*2*) 1/16 in. [2 mm], for material 3/8 to 2 in. [10 to 50 mm] inclusive in thickness; or (*3*) 1/8 in. [3 mm], for material over 2 in. [50 mm] in thickness.

9.3.2 The deposition of weld metal (see 9.5) following removal of imperfections that are greater in depth than the limits listed in 9.3.1 shall be subject to the following limiting conditions:

9.3.2.1 The total area of the chipped or ground surface of any piece prior to welding shall not exceed 2 % of the total surface area of that piece.

9.3.2.2 The reduction of thickness of the material resulting from removal of imperfections prior to welding shall not exceed 30 % of the nominal thickness at the location of the imperfection, nor shall the depth of depression prior to welding exceed 1 1/4 in. [32 mm] in any case except as noted in 9.3.2.3.

9.3.2.3 The deposition of weld metal (see 9.5) following grinding, chipping, or arc-air gouging of the toes of angles, beams, channels, and zees and the stems and toes of tees shall be subject to the limitation that, prior to welding, the depth of the depression, measured from the toe inward, is not more than the thickness of the material at the base of the depression or 1/2 in. [12.5 mm], whichever is the lesser.

9.3.2.4 The deposition of weld metal (see 9.5) and grinding to correct or build up the interlock of any sheet piling section at any location shall be subject to the limitation that the total surface area of the weld not exceed 2 % of the total surface area of the piece.

9.4 *Bar Conditioning*:

9.4.1 The conditioning of bars by the manufacturer or processor to remove imperfections by grinding, chipping, or some other means shall be subject to the limitations that the conditioned area is well faired and the affected sectional area is not reduced by more than the applicable permitted variations (see Section 12).

9.4.2 The deposition of weld metal (see 9.5) following chipping or grinding to remove imperfections that are greater in depth than the limits listed in 9.4.1 shall be subject to the following conditions:

9.4.2.1 The total area of the chipped or ground surface of any piece, prior to welding, shall not exceed 2 % of the total surface area of the piece.

9.4.2.2 The reduction of sectional dimension of a round, square, or hexagon bar, or the reduction in thickness of a flat bar, resulting from removal of an imperfection, prior to welding, shall not exceed 5 % of the nominal dimension or thickness at the location of the imperfection.

9.4.2.3 For the edges of flat bars, the depth of the conditioning depression prior to welding shall be measured from the edge inward and shall be limited to a maximum depth equal to the thickness of the flat bar or 1/2 in. [12.5 mm], whichever is less.

9.5 *Repair by Welding*:

9.5.1 *General Requirements*:

9.5.1.1 Repair by welding shall be in accordance with a welding procedure specification (WPS) using shielded metal arc welding (SMAW), gas metal arc welding (GMAW), flux cored arc welding (FCAW), or submerged arc welding (SAW) processes. Shielding gases used shall be of welding quality.

9.5.1.2 Electrodes and electrode-flux combinations shall be in accordance with the requirements of AWS Specifications A5.1/A5.1M, A5.5/A5.5M, A5.17/A5.17M, A5.18/A5.18M, A5.20/A5.20M, A5.23/A5.23M, A5.28/A5.28M, or A5.29/A5.29M, whichever is applicable. For SMAW, low hydrogen electrodes shall be used.

9.5.1.3 Electrodes and electrode-flux combinations shall be selected so that the tensile strength of the deposited weld metal (after any required heat treatment) is consistent with the tensile strength specified for the base metal being repaired.

9.5.1.4 Welding electrodes and flux materials shall be dry and protected from moisture during storage and use.

9.5.1.5 Prior to repair welding, the surface to be welded shall be inspected to verify that the imperfections intended to be removed have been removed completely. Surfaces to be welded and surfaces adjacent to the weld shall be dry and free of scale, slag, rust, moisture, grease, and other foreign material that would prevent proper welding.

9.5.1.6 Welders and welding operators shall be qualified in accordance with the requirements of AWS D1.1/D1.1M or

BPVC Section IX, except that any complete joint penetration groove weld qualification also qualifies the welder or welding operator to do repair welding.

9.5.1.7 Repair welding of structural products shall be in accordance with a welding procedure specification (WPS) that is in accordance with the requirements of AWS D1.1/D1.1M or BPVC Section IX, with the following exceptions or clarifications:

(1) The WPS shall be qualified by testing a complete joint penetration groove weld or a surface groove weld.

(2) The geometry of the surface groove weld need not be described in other than a general way.

(3) An AWS D1.1/D1.1M prequalified complete joint penetration groove weld WPS is acceptable.

(4) Any material not listed in the prequalified base metal-filler metal combinations of AWS D1.1/D1.1M also is considered to be prequalified if its chemical composition and mechanical properties are comparable to those for one of the prequalified base metals listed in AWS D1.1/D1.1M.

(5) Any material not listed in BPVC Section IX also is considered to be a material with an S-number in BPVC Section IX if its chemical composition and its mechanical properties are comparable to those for one of the materials listed in BPVC Section IX with an S-number.

9.5.1.8 When so specified in the purchase order, the WPS shall include qualification by Charpy V-notch testing, with the test locations, test conditions, and the acceptance criteria meeting the requirements specified for repair welding in the purchase order.

9.5.1.9 When so specified in the purchase order, the welding procedure specification shall be subject to approval by the purchaser prior to repair welding.

9.5.2 *Structural Products with a Specified Minimum Tensile Strength of 100 ksi [690 MPa] or Higher*—Repair welding of structural products with a specified minimum tensile strength of 100 ksi [690 MPa] or higher shall be subject to the following additional requirements:

9.5.2.1 When so specified in the purchase order, prior approval for repair by welding shall be obtained from the purchaser.

9.5.2.2 The surface to be welded shall be inspected using a magnetic particle method or a liquid penetrant method to verify that the imperfections intended to be removed have been completely removed. When magnetic particle inspection is employed, the surface shall be inspected both parallel and perpendicular to the length of the area to be repaired.

9.5.2.3 When weld repairs are to be post-weld heat-treated, special care shall be exercised in the selection of electrodes to avoid those compositions that embrittle as a result of such heat treatment.

9.5.2.4 Repairs on structural products that are subsequently heat-treated at the mill shall be inspected after heat treatment; repairs on structural products that are not subsequently heat-treated at the mill shall be inspected no sooner than 48 h after welding. Such inspection shall use a magnetic particle method or a liquid penetrant method; where magnetic particle inspection is involved, such inspection shall be both parallel to and perpendicular to the length of the repair.

9.5.2.5 The location of the weld repairs shall be marked on the finished piece.

9.5.3 *Repair Quality*—The welds and adjacent heat-affected zone shall be sound and free of cracks, the weld metal being thoroughly fused to all surfaces and edges without undercutting or overlap. Any visible cracks, porosity, lack of fusion, or undercut in any layer shall be removed prior to deposition of the succeeding layer. Weld metal shall project at least 1/16 in. (2 mm) above the rolled surface after welding, and the projecting metal shall be removed by chipping or grinding, or both, to make it flush with the rolled surface, and to produce a workmanlike finish.

9.5.4 *Inspection of Repair*—The manufacturer or processor shall maintain an inspection program to inspect the work to see that:

9.5.4.1 Imperfections have been completely removed.

9.5.4.2 The limitations specified above have not been exceeded.

9.5.4.3 Established welding procedures have been followed, and

9.5.4.4 Any weld deposit is of acceptable quality as defined above.

10. Test Methods

10.1 All tests shall be conducted in accordance with Test Methods and Definitions A370.

10.2 Yield strength shall be determined either by the 0.2 % offset method or by the 0.5 % extension under load method, unless otherwise stated in the material specification.

10.3 *Rounding Procedures*—For purposes of determining conformance with the specification, a calculated value shall be rounded to the nearest 1 ksi [5 MPa] tensile and yield strength, and to the nearest unit in the right-hand place of figures used in expressing the limiting value for other values in accordance with the rounding method given in Practice E29.

10.4 For full-section test specimens of angles, the cross-sectional area used for calculating the yield and tensile strengths shall be a theoretical area calculated on the basis of the weight of the test specimen (see 12.1).

11. Tension Tests

11.1 *Condition*—Test specimens for non-heat-treated structural products shall be taken from test coupons that are representative of the structural products in their delivered condition. Test specimens for heat-treated structural products shall be taken from test coupons that are representative of the structural products in their delivered condition, or from separate pieces of full thickness or full section from the same heat similarly heat treated.

11.1.1 Where the plate is heat treated with a cooling rate faster than still-air cooling from the austenitizing temperature, one of the following shall apply in addition to other requirements specified herein:

11.1.1.1 The gage length of the tension test specimen shall be taken at least $1T$ from any as-heat treated edge where T is the thickness of the plate and shall be at least 1/2 in. [12.5 mm] from flame cut or heat-affected-zone surfaces.

11.1.1.2 A steel thermal buffer pad, $1T$ by $1T$ by at least $3T$, shall be joined to the plate edge by a partial penetration weld completely sealing the buffered edge prior to heat treatment.

11.1.1.3 Thermal insulation or other thermal barriers shall be used during the heat treatment adjacent to the plate edge where specimens are to be removed. It shall be demonstrated that the cooling rate of the tension test specimen is no faster than, and not substantially slower than, that attained by the method described in 11.1.1.2.

11.1.1.4 When test coupons cut from the plate but heat treated separately are used, the coupon dimensions shall be not less than $3T$ by $3T$ by T and each tension specimen cut from it shall meet the requirements of 11.1.1.1.

11.1.1.5 The heat treatment of test specimens separately in the device shall be subject to the limitations that (*1*) cooling rate data for the plate are available; (*2*) cooling rate control devices for the test specimens are available; and, (*3*) the method has received prior approval by the purchaser.

11.2 *Orientation*—For plates wider than 24 in. [600 mm], test specimens shall be taken such that the longitudinal axis of the test specimen is transverse to the final direction of rolling of the plate. Test specimens for all other structural products shall be taken such that the longitudinal axis of the test specimen is parallel to the final direction of rolling.

11.3 *Location*:

11.3.1 *Plates*—Test specimens shall be taken from a corner of the plate.

11.3.2 *W and HP Shapes with Flanges 6 in. [150 mm] or Wider*—Test specimens shall be selected from a point in the flange ⅔ of the way from the flange centerline to the flange toe.

11.3.3 *Shapes Other Than Those in 11.3.2*—Test specimens shall be selected from the webs of beams, channels, and zees; from the stems of rolled tees; and from the legs of angles and bulb angles, except where full-section test specimens for angles are used and the elongation acceptance criteria are increased accordingly. (See 11.6.2.)

11.3.4 *Bars*:

11.3.4.1 Test specimens for bars to be used for pins and rollers shall be taken so that the axis is: midway between the center and the surface for pins and rollers less than 3 in. [75 mm] in diameter; 1 in. [25 mm] from the surface for pins and rollers 3 in. [75 mm] and over in diameter; or as specified in Annex A1 of Test Methods and Definitions A370 if the applicable foregoing requirement is not practicable.

11.3.4.2 Test specimens for bars other than those to be used for pins and rollers shall be taken as specified in Annex A1 of Test Methods and Definitions A370.

11.4 *Test Frequency*:

11.4.1 *Structural Products Produced from an As-Rolled Structural Product*—The minimum number of pieces or plates-as-rolled to be tested for each heat and strength gradation, where applicable, shall be as follows, except that it shall be permissible for any individual test to represent multiple strength gradations:

11.4.1.1 As given in Table B, or

11.4.1.2 One taken from the minimum thickness in the heat and one taken from the maximum thickness in the heat, where thickness means the specified thickness, diameter, or comparable dimension, whichever is appropriate for the applicable structural product rolled.

11.4.2 *Structural Products Produced from Coil and Furnished without Heat Treatment or with Stress Relieving Only*:

11.4.2.1 Except as allowed by 11.4.4, the minimum number of coils to be tested for each heat and strength gradation, where applicable, shall be as given in Table C, except that it shall be permissible for any individual coil to represent multiple strength gradations.

11.4.2.2 Except as required by 11.4.2.3, two tension test specimens shall be taken from each coil tested, with the first being taken immediately prior to the first structural product to be qualified, and the second being taken from the approximate center lap.

11.4.2.3 If, during decoiling, the amount of material decoiled is less than that required to reach the approximate center lap, the second test for the qualification of the decoiled portion of such a coil shall be taken from a location adjacent to the end of the innermost portion decoiled. For qualification of successive portions from such a coil, an additional test shall be taken adjacent to the innermost portion decoiled, until a test is obtained from the approximate center lap.

11.4.3 *Structural Products Produced from Coil and Furnished Heat Treated by other than Stress Relieving*—The minimum number of pieces to be tested for each heat and strength gradation, where applicable, shall be as follows, except that it shall be permissible for any individual test to represent multiple strength gradations:

11.4.3.1 As given in Table B, or

11.4.3.2 One taken from the minimum thickness in the heat and one taken from the maximum thickness in the heat, where thickness means the specified thickness, diameter, or comparable dimension, whichever is appropriate for the applicable structural product rolled.

11.4.4 *Structural Products Produced from Coil and Qualified Using Test Specimens Heat Treated by Other than Stress Relieving*—The minimum number of pieces to be tested for each heat and strength gradation, where applicable, shall be as follows, except that it shall be permissible for any individual test to represent multiple strength gradations:

11.4.4.1 As given in Table B, or

11.4.4.2 One taken from the minimum thickness in the heat, where thickness means the specified thickness, diameter, or comparable dimension, whichever is appropriate for the applicable structural product rolled.

11.5 *Preparation*:

11.5.1 *Plates*:

11.5.1.1 Tension test specimens for plates ¾ in. [20 mm] and under in thickness shall be the full thickness of the plates. The test specimens shall conform to the requirements shown in Fig. 3 of Test Methods and Definitions A370 for either the 1½-in. [40-mm] wide test specimen or the ½-in. [12.5-mm] wide test specimen.

11.5.1.2 For plates up to 4 in. [100 mm] inclusive, in thickness, the use of 1½-in. [40-mm] wide test specimens, full thickness of the plate and conforming to the requirements shown in Fig. 3 of Test Methods and Definitions A370, shall be subject to the limitation that adequate testing machine capacity is available.

11.5.1.3 For plates over ¾ in. [20 mm] in thickness, except as permitted in 11.5.1.2, tension test specimens shall conform to the requirements shown in Fig. 4 of Test Methods and Definitions A370 for the 0.500-in. [12.5-mm] diameter test specimen. The axis of such test specimens shall be located midway between the center of thickness and the top or bottom surface of the plate.

11.5.2 *Shapes*:

11.5.2.1 Except where angles are tested in full section, tension test specimens for shapes ¾ in. [20 mm] and under in thickness shall be the full thickness of the shape. Such test specimen shall conform to the requirements shown in Fig. 3 of Test Methods and Definitions A370 for either the 1½-in. [40-mm] wide test specimen or the ½-in. [12.5-mm] wide test specimen.

11.5.2.2 For shapes up to 5 in. [125 mm] inclusive, in thickness, the use of 1½-in. [40-mm] wide test specimens, full thickness of the shape and conforming to the requirements shown in Fig. 3 of Test Methods and Definitions A370, shall be subject to the limitation that adequate testing machine capacity is available.

TABLE B Minimum Number of Tension Tests Required

Thickness[A] Range Rolled for the Heat	Thickness[A] Difference Between Pieces or Plates-as-rolled in the Thickness[A] Range	Minimum Number of Tension Tests Required
Under ⅜ in. [10 mm]	1/16 in. [2 mm] or less	Two[B] tests per heat, taken from different pieces or plates-as-rolled having any thickness[A] in the thickness[A] range
	More than 1/16 in. [2 mm]	Two[B] tests per heat, one taken from the minimum thickness[A] in the thickness[A] range and one taken from the maximum thickness[A] in the thickness[A] range
⅜ to 2 in. [10 to 50 mm], incl	Less than ⅜ in. [10 mm]	Two[B] tests per heat, taken from different pieces or plates-as-rolled having any thickness[A] in the thickness[A] range
	⅜ in. [10 mm] or more	Two[B] tests per heat, one taken from the minimum thickness[A] in the thickness[A] range and one taken from the maximum thickness[A] in the thickness[A] range
Over 2 in. [50 mm]	Less than 1 in. [25 mm]	Two[B] tests per heat, taken from different pieces or plates-as-rolled having any thickness[A] in the thickness[A] range
	1 in. [25 mm] or more	Two[B] tests per heat, one taken from the minimum thickness[A] in the thickness[A] range and one taken from the maximum thickness[A] in the thickness[A] range

[A] Thickness means the specified thickness, diameter, or comparable dimension, whichever is appropriate for the specific structural product rolled.
[B] One test, if only one piece or plate-as-rolled is to be qualified.

TABLE C Minimum Number of Coils Required to be Tension Tested

NOTE 1—See 11.4.2.2 and 11.4.2.3 for the number of tests to be taken per coil.

Thickness[A] Difference Between Coils in the Heat	Minimum Number of Coils Required to be Tension Tested
Less than 1/16 in. [2 mm]	Two[B] coils per heat, at any thickness[A] in the heat
1/16 in. [2 mm] or more	Two[B] coils per heat, one at the minimum thickness[A] in the heat and one at the maximum thickness[A] in the heat

[A] Thickness means the specified thickness, diameter, or comparable dimension, whichever is appropriate for the specific structural product rolled.
[B] One coil, if the product of only one coil is to be qualified.

11.5.2.3 For shapes over ¾ in. [20 mm] in thickness, except as permitted in 11.5.2.2, tension test specimens shall conform to the requirements shown in Fig. 4 of Test Methods and Definitions A370 for the 0.500–in. [12.5-mm] diameter test specimens. The axis of such test specimens shall be located midway between the center of thickness and the top or bottom surface of the shape.

11.5.3 *Bars*:

11.5.3.1 Except as otherwise provided below, test specimens for bars shall be in accordance with Annex A1 of Test Methods and Definitions A370.

11.5.3.2 Except as provided in 11.5.3.5, test specimens for bars ¾ in. [20 mm] and under in thickness may conform to the requirements shown in Fig. 3 of Test Methods and Definitions A370 for either the 1½-in. [40-mm] wide test specimen or the ½-in. [12.5-mm] wide specimen.

11.5.3.3 Except as provided in 11.5.3.4 and 11.5.3.5, test specimens for bars over ¾ in. [20 mm] in thickness or diameter shall conform either to the requirements for the 1½-in. [40-mm] or ½-in. [12.5-mm] wide test specimen shown in Fig. 3 of Test Methods and Definitions A370, or to the requirements for the 0.500-in. [12.5-mm] diameter test specimen shown in Fig. 4 of Test Methods and Definitions A370.

11.5.3.4 For bars other than those to be used for pins and rollers, the manufacturer or processor shall have the option of using test specimens that are machined to a thickness or diameter of at least ¾ in. [20 mm] for a length of at least 9 in. [230 mm].

11.5.3.5 Test specimens for bars to be used for pins and rollers shall conform to the requirements shown in Fig. 4 of Test Methods and Definitions A370 for the 0.500-in. [12.5-mm] diameter test specimen.

11.6 *Elongation Requirement Adjustments*:

11.6.1 Due to the specimen geometry effect encountered when using the rectangular tension test specimen for testing thin material, adjustments in elongation requirements must be provided for thicknesses under 0.312 in. [8 mm]. Accordingly, the following deductions from the base elongation requirements shall apply:

Nominal Thickness Range, in. [mm]	Elongation Deduction, %
0.299—0.311 [7.60—7.89]	0.5
0.286—0.298 [7.30—7.59]	1.0
0.273—0.285 [7.00—7.29]	1.5
0.259—0.272 [6.60—6.99]	2.0
0.246—0.258 [6.20—6.59]	2.5
0.233—0.245 [5.90—6.19]	3.0
0.219—0.232 [5.50—5.89]	3.5
0.206—0.218 [5.20—5.49]	4.0
0.193—0.205 [4.90—5.19]	4.5
0.180—0.192 [4.60—4.89]	5.0
0.166—0.179 [4.20—4.59]	5.5[A]
0.153—0.165 [3.90—4.19]	6.0[A]
0.140—0.152 [3.60—3.89]	6.5[A]
0.127—0.139 [3.20—3.59]	7.0[A]
< 0.127 [3.20]	7.5[A]

[A] Elongation deductions for thicknesses less than 0.180 in. [4.60 mm] apply to plates and structural shapes only.

11.6.2 Due to the specimen geometry effect encountered when using full-section test specimens for angles, the elongation requirements for structural-size angles shall be increased by six percentage points when full-section test specimens are used.

11.6.3 Due to the inherently lower elongation that is obtainable in thicker structural products, adjustments in elongation requirements shall be provided. For structural products over 3.5 in. [90 mm] in thickness, a deduction of 0.5 percentage point from the specified percentage of elongation in 2 in. [50 mm] shall be made for each 0.5-in. [12.5-mm] increment of thickness over 3.5 in. [90 mm], up to a maximum deduction of 3.0 percentage points. Accordingly, the following deductions from the base elongation requirements shall apply:

Nominal Thickness Range, in. [mm]	Elongation Deduction, %
3.500—3.999 [90.00—102.49]	0.5
4.000—4.499 [102.50—114.99]	1.0
4.500—4.999 [115.00—127.49]	1.5
5.000—5.499 [127.50—139.99]	2.0
5.500—5.999 [140.00—152.49]	2.5
6.000 and thicker [152.50 and thicker]	3.0

11.6.4 The tensile property requirements tables in many of the product specifications covered by this general requirements specification specify elongation requirements in both 8-in. [200-mm] and 2-in. [50-mm] gage lengths. Unless otherwise provided in the applicable product specification, both requirements are not required to be applied simultaneously and the elongation need only be determined in the gage length appropriate for the test specimen used. After selection of the appropriate gage length, the elongation requirement for the alternative gage length shall be deemed not applicable.

11.7 *Yield Strength Application*:

11.7.1 When test specimens do not exhibit a well-defined disproportionate yield point, yield strength shall be determined and substituted for yield point.

11.7.2 The manufacturer or processor shall have the option of substituting yield strength for yield point if the test specimen exhibits a well-defined disproportionate yield point.

11.7.3 Yield strength shall be determined either by the 0.2 % offset method or by the 0.5 % extension-under-load method.

11.8 *Product Tension Tests*—This specification does not provide requirements for product tension testing subsequent to shipment (see 15.1). Therefore, the requirements of 11.1-11.7 inclusive and Section 13 apply only for tests conducted at the place of manufacture prior to shipment.

NOTE 5—Compliance to this specification and the applicable product specification by a manufacturer or processor does not preclude the possibility that product tension test results might vary outside specified ranges. The tensile properties will vary within the same heat or piece, be it as-rolled, control-rolled, or heat-treated. Tension testing according to the requirements of this specification does not provide assurance that all products of a heat will be identical in tensile properties with the products tested. If the purchaser wishes to have more confidence than that provided by this specification testing procedures, additional testing or requirements, such as Supplementary Requirement S4, should be imposed.

11.8.1 Appendix X2 provides additional information on the variability of tensile properties in plates and structural shapes

12. Permitted Variations in Dimensions and Weight [Mass]

12.1 One cubic foot of rolled steel is assumed to weigh 490 lb. One cubic metre of rolled steel is assumed to have a mass of 7850 kg.

12.2 *Plates*—The permitted variations for dimensions and weight [mass] shall not exceed the applicable limits in Tables 1 to 15 [Annex A1, Tables A1.1 to A1.15] inclusive.

12.3 *Shapes*:

12.3.1 Annex A2 lists the designations and dimensions, in both inch-pound and SI units, of shapes that are most commonly available. Radii of fillets and toes of shape profiles vary with individual manufacturers and therefore are not specified.

12.3.2 The permitted variations in dimensions shall not exceed the applicable limits in Tables 16 to 25 [Annex A1, Tables A1.16 to A1.25] inclusive. Permitted variations for special shapes not listed in such tables shall be as agreed upon between the manufacturer and the purchaser.

NOTE 6—Permitted variations are given in Tables 16 to 25 [Annex A1, Tables A1.16 to A1.25] inclusive for some shapes that are not listed in Annex A2 (that is, bulb angles, tees, zees). Addition of such sections to Annex A2 will be considered by Subcommittee A01.02 when and if a need for such listing is shown.

12.3.3 *Shapes Having One Dimension of the Cross Section 3 in. [75 mm] or Greater (Structural-Size Shapes)*—The cross-sectional area or weight [mass] of each shape shall not vary more than 2.5 % from the theoretical or specified amounts except for shapes with a moninal weight of less than 100 lb/ft, in which the variation shall range from −2.5 % to +3.0 % from the theoretical cross-sectional area or the specified nominal weight [mass].

12.4 *Sheet Piling*—The weight [mass] of each steel sheet pile shall not vary more than 2.5 % from the theoretical or specified weight [mass]. The length of each steel sheet pile shall be not less than the specified length, and not more than 5 in. [125 mm] over the specified length.

12.5 *Hot-Rolled Bars*—The permitted variations in dimensions shall not exceed the applicable limits in Tables 26 to 31 [Annex A1, Tables A1.26 to A1.31] inclusive.

12.6 *Conversion of Permitted Variations from Fractions of an Inch to Decimals*—Permitted variations in dimensions for products covered by this specification are generally given as fractions of an inch and these remain the official permitted variations, where so stated. If the material is to be measured by equipment reporting dimensions as decimals, conversion of permitted variations from fractions of an inch to decimals shall be made to three decimal places; using the rounding method prescribed in Practice E29.

13. Retests

13.1 If any test specimen shows defective machining or develops flaws, the manufacturer or processor shall have the option of discarding it and substituting another test specimen.

13.2 If the percentage of elongation of any tension test specimen is less than that specified and any part of the fracture is more than ¾ in. [20 mm] from the center of the gage length of a 2-in. [50-mm] specimen or is outside the middle half of the gage length of an 8-in. [200-mm] specimen, as indicated by scribe scratches marked on the specimen before testing, a retest shall be allowed.

13.3 Except as provided in 13.3.1, if the results from an original tension specimen fails to meet the specified requirements, but are within 2 ksi [14 MPa] of the required tensile strength, within 1 ksi [7 MPa] of the required yield strength or yield point, or within 2 percentage points of the required elongation, a retest shall be permitted to replace the failing test. A retest shall be performed for the failing original test, with the specimen being randomly selected from the heat. If the results of the retest meet the specified requirements, the heat or lot shall be approved.

13.3.1 For structural products that are tested as given in Table C, both tests from each coil tested to qualify a heat are required to meet all mechanical property requirements. Should either test fail to do so, then that coil shall not be used to qualify the heat; however, the portion of that individual coil that is bracketed by acceptable tests (see 11.4.2.3) is considered to be qualified.

13.4 Quenched and tempered steel plates shall be subject to any additional retest requirements contained in the applicable product specification.

13.5 When the full-section option of 11.3.3 is used and the elongation falls below the specified requirement, the manufacturer or processor shall have the option of making another test using a test specimen permitted in 11.5.2.

14. Test Reports

14.1 Test reports for each heat supplied are required and they shall report the following:

14.1.1 The applicable product specification designation, including year-date and whichever of grade, class, and type are specified in the purchase order, to which the structural product is furnished.

14.1.2 The heat number, heat analysis (see 7.1), and nominal sizes.

NOTE 7—If the amount of copper, chromium, nickel, molybdenum, or silicon is less than 0.02 %, the heat analysis for that element may be reported as <0.02 %. If the amount of columbium or vanadium is less than 0.008 %, the heat analysis for that element may be reported as <0.008 %.

14.1.3 For structural products that are tested as given in Table B, two tension test results appropriate to qualify the shipment (see 11.4), except that only one tension test result need be reported if the shipment consists of a single piece or plate-as-rolled.

14.1.3.1 In reporting elongation values, both the percentage increase and the original gage length shall be stated.

14.1.3.2 Yield to tensile ratio when such a requirement is contained in the product specification.

14.1.4 For structural products that are required to be heat treated, either by the applicable product specification or by the purchase order, all heat treatments, including temperature ranges and times at temperature, unless the purchaser and the supplier have agreed to the supply of a heat treatment procedure in place of the actual temperatures and times.

14.1.4.1 Subcritical heat treatment to soften thermally cut edges need not be reported, except for structural products having a specified minimum tensile strength of 95 ksi [655 MPa] or higher, unless such subcritical heating is accomplished at temperatures at least 75°F [40°C] lower than the minimum tempering temperature.

14.1.5 The results of any required austenitic grain size tests (see 8.2 or 8.3, whichever is applicable).

14.1.6 The results of any other test required by the applicable product specification, the applicable supplementary requirements, and the purchase order.

14.2 The thickness of the structural product tested is not necessarily the same as an individual ordered thickness, given that it is the heat that is tested, rather than each ordered item. Tests from specified thicknesses in accordance with 11.4 and encompassing the thicknesses in a shipment shall be sufficient for qualifying the structural product in the shipment. Such test thicknesses are not required to be within previously tested and shipped thicknesses from the same heat.

14.3 For structural products produced from coil that are supplied in the as-rolled condition or have been heat treated by stress relieving only, the test report shall state "Produced from Coil." Both test results shall be reported for each qualifying coil, and the location within the coil for each test shall be stated.

14.4 For structural products produced from coil, both the manufacturer and the processor shall be identified on the test report.

14.5 When full-section test specimens have been used for the qualification of angles, that information shall be stated on the test report.

14.6 A signature is not required on the test report; however, the document shall clearly identify the organization submitting the report. Notwithstanding the absence of a signature, the organization submitting the report is responsible for the content of the report.

14.7 For structural products finished by other than the original manufacturer, the supplier of the structural product shall also provide the purchaser with a copy of the original manufacturer's test report.

14.8 A test report, certificate of inspection, or similar document printed from or used in electronic form from an electronic data interchange (EDI) transmission shall be regarded as having the same validity as a counterpart printed in the certifier's facility. The content of the EDI transmitted document shall meet the requirements of the applicable product specification and shall conform to any existing EDI agreement between the purchaser and the supplier. Notwithstanding the absence of a signature, the organization submitting the EDI transmission shall be responsible for the content of the report.

NOTE 8—The industry definition as invoked here is: EDI is the computer to computer exchange of business information in a standard format such as ANSI ASC X12.

15. Inspection and Testing

15.1 The inspector representing the purchaser shall have free entry, at all times, while work on the contract of the purchaser is being performed, to all parts of the manufacturer's works that concern the manufacture of the structural product ordered. The manufacturer shall afford the inspector all reasonable facilities to be satisfied that the structural product is being furnished in accordance with this general requirements specification, the applicable product specification, and the purchase order. All tests (except product analysis) and inspection shall be made at the place of manufacture prior to shipment, unless otherwise specified, and shall be conducted so as not to interfere with the operation of the manufacturer's works.

15.2 Where structural products are produced from coil, 15.1 shall apply to the processor instead of the manufacturer, and the place of process shall apply instead of the place of manufacture. Where structural products are produced from coil and the processor is different from the manufacturer, the inspector representing the purchaser shall have free entry at all times while work on the contract of the purchaser is being performed to all parts of the manufacturer's works that concern the manufacture of the structural product ordered.

16. Retreatment

16.1 If any heat-treated structural product fails to meet the mechanical property requirements of the applicable product specification, the manufacturer or the processor shall have the option of heat treating the structural product again. All mechanical property tests shall be repeated and the structural product shall be reexamined for surface defects when it is resubmitted for inspection.

17. Rejection

17.1 Any rejection based upon product analysis made in accordance with the applicable product specification shall be reported to the supplier and samples that represent the rejected structural product shall be preserved for two weeks from the date of notification of such rejection. In case of dissatisfaction with the results of the tests, the supplier shall have the option of making claim for a rehearing within that time.

17.2 The purchaser shall have the option of rejecting structural product that exhibits injurious defects subsequent to its acceptance at the manufacturer's or processor's works, and so notifying the manufacturer or processor.

18. Identification of Structural Products

18.1 *Required Plate Markings*:

18.1.1 Except as allowed by 18.1.4.2 and 18.6, plates shall be legibly marked with the following: applicable ASTM designation (see 1.1) (year-date not required); "G" or "MT" if applicable (see 18.1.2); applicable grade; heat number; size and thickness; and name, brand, or trademark of the manufacturer (for plates produced from an as-rolled structural product) or the processor (for plates produced from coil).

18.1.2 Plates that are required to be heat treated, but have not been so heat treated, shall be marked, by the manufacturer or processor, with the letter "G" (denoting green) following the required ASTM designation mark, except that "G" marking is not necessary if such plates are for shipment, for the purpose of obtaining the required heat treatment, to an organization under the manufacturer's control. Such plates shall have been qualified for shipment on the basis of test specimens that have been so heat treated. Plates that are required to be heat treated, and have been so heat treated, shall be marked, by the party that performed the heat treatment, with the letter "MT" (denoting material treated) following the required ASTM designation mark.

18.1.3 Except as allowed by 18.1.4.2 and 18.6, the required markings for plates shall be by steel die stamping, paint marking, or by means of permanently affixed, colorfast, weather-resistant labels or tags. It shall be the responsibility of the supplier that all required markings be intact and fully legible upon receipt by the purchaser.

18.1.4 *Location of Markings*:

18.1.4.1 The required markings for plates shall be in at least one place on each finished plate.

18.1.4.2 For secured lifts of all sizes of plates ⅜ in. [10 mm] (or 5/16 in. [8 mm] for material specified for bridge construction end use) or under in thickness, and for secured lifts of all thicknesses of plates 36 in. [900 mm] or under in width, the manufacturer or processor shall have the option of placing such markings on only the top piece of each lift, or of showing such markings on a substantial tag attached to each lift, unless otherwise specified.

18.2 *Shapes*:

18.2.1 Except as allowed by 18.2.2 and 18.6, shapes shall be marked with the heat number, size of section, length, and mill identification marks on each piece. Shapes with the greatest cross-sectional dimension greater than 6 in. [150 mm] shall have the manufacturer's name, brand, or trademark shown in raised letters at intervals along the length. In addition, shapes shall be identified with the ASTM designation (year-date not required) and grade, either by marking each piece individually, by permanently affixing a colorfast, weather-resistant label or tag, or, if bundled, by attaching a substantial tag to the bundle.

18.2.2 Bundling for shipment of small shapes with the greatest cross-sectional dimension not greater than 6 in. [150 mm] is permissible. Each lift or bundle shall be marked or substantially tagged showing the identification information listed in 18.2.1.

18.2.3 It shall be permissible for the manufacturer to make a full size bundle at the end of a heat by adding product from a consecutively rolled heat of the same nominal chemical composition. The manufacturer shall identify a bundle consisting of product from two heats with the number of the first heat rolled or identify both heats. The manufacturer shall maintain records of the heats contained in each bundle.

18.3 *Steel Sheet Piling*—Steel sheet piling shall be marked with the heat number, size of section, length, and mill identification marks on each piece, either by marking, or by permanently affixing colorfast, weather-resistant label or tag. The manufacturer's name, brand, or trademark shall be shown in raised letters at intervals along the length.

18.4 *Bars*—Bars of all sizes, when loaded for shipment, shall be properly identified with the name or brand of manufacturer, purchaser's name and order number, the ASTM designation number (year-date not required), grade number where appropriate, size and length, weight [mass] of lift, and the heat number for identification. Unless otherwise specified, the method of marking is at the manufacturer's option and shall be made by hot stamping, cold stamping, painting, or marking tags attached to the lifts of bars. Bars are not required to be die-stamped.

18.4.1 It shall be permissible for the manufacturer to make a full size bundle at the end of a heat by adding product from a consecutively rolled heat of the same nominal chemical composition. The manufacturer shall identify a bundle consisting of product from two heats with the number of the first heat rolled or identify both heats. The manufacturer shall maintain records of the heats contained in each bundle.

18.5 *Bar Coding*—In addition to the requirements of 18.1-18.4 inclusive, the manufacturer or processor shall have the option of using bar coding as a supplementary identification method.

NOTE 9—Bar coding should be consistent with AIAG Standards.[7]

18.6 *Subdivided Material*:

18.6.1 Except as allowed by 18.6.2, pieces separated from a master structural product by an organization other than the original manufacturer shall be identified with the ASTM designation (year-date not required), grade, heat number, and the heat treatment identification, if applicable, along with the trademark, brand, or name of the organization subdividing the structural product. The identification methods shall be in accordance with the requirements of 18.1-18.4 inclusive, except that the raised letters method for shapes and steel sheet piling is not required. If the original manufacturer's identification remains intact, the structural product need not be additionally identified by the organization supplying the structural product.

18.6.2 It shall be permissible for pieces from the same heat of structural product to be bundled or placed in secured lifts, with the identification specified in 18.6.1 placed on the top piece of each lift or shown on a substantial tag attached to each bundle or lift.

19. Packaging, Marking, and Loading for Shipment

19.1 Packaging, marking, and loading for shipment shall be in accordance with Practices A700.

19.2 When Level A is specified, and when specified in the contract or order, and for direct procurement by or direct shipment to the U.S. government, preservation, packaging, and packing shall be in accordance with the Level A requirements of MIL-STD-163.

19.3 When specified in the contract or order, and for direct procurement by or direct shipment to the U.S. government, marking for shipment, in addition to requirements specified in the contract or order, shall be in accordance with MIL-STD-129 for military agencies and with Fed. Std. No. 123 for civil agencies.

20. Keywords

20.1 bars; general requirements; plates; rolled; shapes; sheet piling; structural steel

---

[7] Available from Automotive Industry Action Group (AIAG), 26200 Lahser Rd., Suite 200, Southfield, MI 48033, http://www.aiag.org.

TABLE 1 Permitted Variations in Thickness for Rectangular, Carbon, High-Strength, Low-Alloy, and Alloy-Steel Plates, 15 in. and Under in Thickness When Ordered to Thickness

NOTE 1—Tables 1-31 inclusive contain permitted variations in dimensions and weight stated in inch-pound units.
NOTE 2—Permitted variation under specified thickness, 0.01 in.
NOTE 3—Thickness to be measured at 3/8 to 3/4 in. from the longitudinal edge.
NOTE 4—For thicknesses measured at any location other than that specified in Note 3, the permitted variations over specified thickness shall be 1¾ times the amounts in this table, rounded to the nearest 0.01 in.
NOTE 5—Where "..." appears in this table, there is no requirement.

Specified Thickness, in.	Permitted Variations Over Specified Thickness for Widths Given in Inches, in.											
	48 and under	Over 48 to 60, excl	60 to 72, excl	72 to 84, excl	84 to 96, excl	96 to 108, excl	108 to 120, excl	120 to 132, excl	132 to 144, excl	144 to 168, excl	168 to 182, excl	182 and over
To ¼, excl	0.03	0.03	0.03	0.03	0.03	0.03	0.03	0.03	0.04	...	...	...
¼ to 5/16, excl	0.03	0.03	0.03	0.03	0.03	0.03	0.03	0.04	0.04	...	...	...
5/16 to 3/8, excl	0.03	0.03	0.03	0.03	0.03	0.03	0.03	0.04	0.04	0.05	...	...
3/8 to 7/16, excl	0.03	0.03	0.03	0.03	0.03	0.03	0.04	0.04	0.05	0.06	0.06	...
7/16 to ½, excl	0.03	0.03	0.03	0.03	0.03	0.03	0.04	0.04	0.05	0.06	0.06	...
½ to 5/8, excl	0.03	0.03	0.03	0.03	0.03	0.03	0.04	0.04	0.05	0.06	0.07	...
5/8 to ¾, excl	0.03	0.03	0.03	0.03	0.03	0.04	0.04	0.04	0.05	0.06	0.07	0.07
¾ to 1, excl	0.03	0.03	0.03	0.03	0.04	0.04	0.05	0.05	0.06	0.07	0.08	0.09
1 to 2, excl	0.06	0.06	0.06	0.06	0.06	0.07	0.08	0.10	0.10	0.11	0.13	0.16
2 to 3, excl	0.09	0.09	0.09	0.10	0.10	0.11	0.12	0.13	0.14	0.15	0.15	...
3 to 4, excl	0.11	0.11	0.11	0.11	0.11	0.13	0.14	0.14	0.14	0.15	0.17	...
4 to 6, excl	0.15	0.15	0.15	0.15	0.15	0.15	0.15	0.15	0.15	0.20	0.20	...
6 to 10, excl	0.23	0.24	0.24	0.24	0.24	0.24	0.24	0.24	0.24	0.27	0.28	...
10 to 12, excl	0.29	0.29	0.33	0.33	0.33	0.33	0.33	0.33	0.33	0.33	0.35	...
12 to 15, incl	0.29	0.29	0.35	0.35	0.35	0.35	0.35	0.35	0.35	0.35	0.35	...

TABLE 2 Permitted Variations in Weight for Rectangular Sheared Plates and Universal Mill Plates 613.0 lb/ft² and Under When Ordered to Weight

NOTE 1—Permitted variations in overweight for lots of circular and sketch plates shall be 1¼ times the amounts in this table.
NOTE 2—Permitted variations in overweight for single plates shall be 1⅓ times the amounts in this table.
NOTE 3—Permitted variations in overweight for single circular and sketch plates shall be 1⅔ times the amounts in this table.
NOTE 4—The adopted standard density of rolled steel is 490 lb/ft³.
NOTE 5—Where "..." appears in this table, there is no requirement.

Specified Weights, lb/ft²	Permitted Variations in Average Weight of Lots[A] for Widths Given in Inches, Expressed in Percentage of the Specified Weights per Square Foot																	
	48 and under		Over 48 to 60, excl		60 to 72, excl		72 to 84, excl		84 to 96, excl		96 to 108, excl		108 to 120, excl		120 to 132, excl		132 to 144, excl		144 to 168, excl		168 and over	
	Over	Under	Over	Under	Over	Under	Over	Under	Over	Under	Over	Under	Over	Under	Over	Under	Over	Under	Over	Under	Over	Under
To 10, excl	4.0	3.0	4.5	3.0	5.0	3.0	5.5	3.0	6.0	3.0	7.5	3.0	9.0	3.0	11.0	3.0	13.0	3.0	...	...	...	...
10 to 12.5, excl	4.0	3.0	4.5	3.0	4.5	3.0	5.0	3.0	5.5	3.0	6.5	3.0	7.0	3.0	8.0	3.0	9.0	3.0	12.0	3.0	...	...
12.5 to 15.0, excl	4.0	3.0	4.0	3.0	4.5	3.0	4.5	3.0	5.0	3.0	5.5	3.0	6.0	3.0	7.5	3.0	8.0	3.0	11.0	3.0	...	...
15 to 17.5, excl	3.5	3.0	3.5	3.0	4.0	3.0	4.5	3.0	4.5	3.0	5.0	3.0	5.5	3.0	6.0	3.0	7.0	3.0	9.0	3.0	10.0	3.0
17.5 to 20, excl	3.5	2.5	3.5	2.5	3.5	3.0	4.0	3.0	4.5	3.0	4.5	3.0	5.0	3.0	5.5	3.0	6.0	3.0	8.0	3.0	9.0	3.0
20 to 25, excl	3.5	2.5	3.5	2.5	3.5	3.0	3.5	3.0	4.0	3.0	4.0	3.0	4.5	3.0	5.0	3.0	5.5	3.0	7.0	3.0	8.0	3.0
25 to 30, excl	3.0	2.5	3.5	2.5	3.5	2.5	3.5	3.0	3.5	3.0	3.5	3.0	4.0	3.0	4.5	3.0	5.0	3.0	6.5	3.0	7.0	3.0
30 to 40, excl	3.0	2.0	3.0	2.0	3.0	2.0	3.0	2.0	3.5	2.0	3.5	2.5	3.5	2.5	4.0	3.0	4.5	3.0	6.0	3.0	6.5	3.0
40 to 81.7, excl	2.5	2.0	3.0	2.0	3.0	2.0	3.0	2.0	3.5	2.0	3.5	2.0	3.5	2.5	3.5	3.0	4.0	3.0	5.5	3.0	6.0	3.0
81.7 to 122.6, excl	2.5	2.0	3.0	2.0	3.0	2.0	3.0	2.0	3.5	2.0	3.5	2.0	3.5	2.5	3.5	3.0	3.5	3.0	4.0	3.0	4.5	3.0
122.6 to 163.4, excl	2.5	1.5	2.5	1.5	2.5	1.5	2.5	1.5	2.5	2.0	2.5	2.0	2.5	2.0	2.5	2.0	2.5	2.0	3.0	2.0	3.5	2.0
163.4 to 245.1, excl	2.5	1.0	2.5	1.0	2.5	1.0	2.5	1.0	2.5	1.0	2.5	1.0	2.5	1.0	2.5	1.0	2.5	1.0	3.0	1.0	3.5	1.0
245.1 to 409.0, excl	2.5	1.0	2.5	1.0	2.5	1.0	2.5	1.0	2.5	1.0	2.5	1.0	2.5	1.0	2.5	1.0	2.5	1.0	2.5	1.0	3.0	1.0
409.0 to 490.1, excl	2.0	1.0	2.0	1.0	2.5	1.0	2.5	1.0	2.5	1.0	2.5	1.0	2.5	1.0	2.5	1.0	2.5	1.0	2.5	1.0	2.5	1.0
490.1 to 613.0, excl	2.0	1.0	2.0	1.0	2.0	1.0	2.0	1.0	2.5	1.0	2.5	1.0	2.5	1.0	2.5	1.0	2.5	1.0	2.5	1.0	2.5	1.0

[A] The term "lot" means all the plates of each tabular width and weight group represented in each shipment.

TABLE 3 Permitted Variations in Width and Length for Sheared Plates 1½ in. and Under in Thickness; Length Only of Universal Mill Plates 2½ in. and Under in Thickness

Specified Dimensions, in.		Permitted Variations Over Specified Width and Length[A] for Thicknesses Given in Inches or Equivalent Weights Given in Pounds per Square Foot, in.							
		To ⅜, excl		⅜ to ⅝, excl		⅝ to 1, excl		1 to 2, incl[B]	
Length	Width	To 15.3, excl		15.3 to 25.5, excl		25.5 to 40.8, excl		40.8 to 81.7, incl	
		Width	Length	Width	Length	Width	Length	Width	Length
To 120, excl	To 60, excl	⅜	½	7/16	⅝	½	¾	⅝	1
	60 to 84, excl	7/16	⅝	½	11/16	⅝	⅞	¾	1
	84 to 108, excl	½	¾	⅝	⅞	¾	1	1	1⅛
	108 and over	⅝	⅞	¾	1	⅞	1⅛	1⅛	1¼
120 to 240, excl	To 60, excl	⅜	¾	½	⅞	⅝	1	¾	1⅛
	60 to 84, excl	½	¾	⅝	⅞	¾	1	⅞	1¼
	84 to 108, excl	9/16	⅞	11/16	15/16	13/16	1⅛	1	1⅜
	108 and over	⅝	1	¾	1⅛	⅞	1¼	1⅛	1⅜
240 to 360, excl	To 60, excl	⅜	1	½	1⅛	⅝	1¼	¾	1½
	60 to 84, excl	½	1	⅝	1⅛	¾	1¼	⅞	1½
	84 to 108, excl	9/16	1	11/16	1⅛	⅞	1⅜	1	1½
	108 and over	11/16	1⅛	⅞	1¼	1	1⅜	1¼	1¾
360 to 480, excl	To 60, excl	7/16	1⅛	½	1¼	⅝	1⅜	¾	1⅝
	60 to 84, excl	½	1¼	⅝	1⅜	¾	1½	⅞	1⅝
	84 to 108, excl	9/16	1¼	¾	1⅜	⅞	1½	1	1⅞
	108 and over	¾	1⅜	⅞	1½	1	1⅝	1¼	1⅞
480 to 600, excl	To 60, excl	7/16	1¼	½	1½	⅝	1⅝	¾	1⅞
	60 to 84, excl	½	1⅜	⅝	1½	¾	1⅝	⅞	1⅞
	84 to 108, excl	⅝	1⅜	¾	1½	⅞	1⅝	1	1⅞
	108 and over	¾	1½	⅞	1⅝	1	1¾	1¼	1⅞
600 to 720, excl	To 60, excl	½	1¾	⅝	1⅞	¾	1⅞	⅞	2¼
	60 to 84, excl	⅝	1¾	¾	1⅞	⅞	1⅞	1	2¼
	84 to 108, excl	⅝	1¾	¾	1⅞	⅞	1⅞	1⅛	2¼
	108 and over	⅞	1¾	1	2	1⅛	2¼	1¼	2½
720 and over	To 60, excl	9/16	2	¾	2⅛	⅞	2¼	1	2¾
	60 to 84, excl	¾	2	⅞	2⅛	1	2¼	1⅛	2¾
	84 to 108, excl	¾	2	⅞	2⅛	1	2¼	1¼	2¾
	108 and over	1	2	1⅛	2⅜	1¼	2½	1⅜	3

[A] Permitted variation under specified width and length, ¼ in.
[B] Permitted variations in length apply also to Universal Mill plates up to 12 in. in width for thicknesses over 2 to 2½ in., incl, except for alloy steel up to 2 in. thick.

TABLE 4 Permitted Variations in Width for Mill Edge Carbon and High-Strength, Low-Alloy Plates Produced on Strip Mills (Applies to Plates Produced from Coil and to Plates Produced from an As-Rolled Structural Product)

Specified Width, in.	Permitted Variation Over Specified Width, in.[A]
To 14, excl	7/16
14 to 17, excl	½
17 to 19, excl	9/16
19 to 21, excl	⅝
21 to 24, excl	11/16
24 to 26, excl	13/16
26 to 28, excl	15/16
28 to 35, excl	1⅛
35 to 50, excl	1¼
50 to 60, excl	1½
60 to 65, excl	1⅝
65 to 70, excl	1¾
70 to 80, excl	1⅞
80 and over	2

[A] No permitted variation under specified width.

TABLE 5 Permitted Variations in Rolled Width for Universal Mill Plates 15 in. and Under in Thickness

Specified Width, in.	Permitted Variations Over Specified Width[A] for Thicknesses Given in Inches or Equivalent Weights Given in Pounds per Square Foot, in.					
	To ⅜, excl	⅜ to ⅝, excl	⅝ to 1, excl	1 to 2, incl	Over 2 to 10, incl	Over 10 to 15, incl
	To 15.3, excl	15.3 to 25.5, excl	25.5 to 40.8, excl	40.8 to 81.7, incl	81.7 to 409.0, incl	409.0 to 613.0, incl
Over 8 to 20, excl	⅛	⅛	3/16	¼	⅜	½
20 to 36, excl	3/16	¼	5/16	⅜	7/16	9/16
36 and over	5/16	⅜	7/16	½	9/16	⅝

[A] Permitted variation under specified width, ⅛ in.

TABLE 6 Permitted Variations in Diameter for Sheared Circular Plates 1 in. and Under in Thickness

Specified Diameters, in.	Permitted Variations Over Specified Diameter for Thicknesses Given in Inches, in.[A]		
	To 3/8, excl	3/8 to 5/8, excl	5/8 to 1, incl
To 32, excl	1/4	3/8	1/2
32 to 84, excl	5/16	7/16	9/16
84 to 108, excl	3/8	1/2	5/8
108 to 130, excl	7/16	9/16	11/16
130 and over	1/2	5/8	3/4

[A] No permitted variation under specified diameter.

TABLE 7 Permitted Variations in Diameter for Gas-Cut Circular Plates (Not Applicable to Alloy Steel)

Specified Diameter, in.	Permitted Variation Over Specified Diameter for Thicknesses Given in Inches, in.[A]					
	to 1, excl	1 to 2, excl	2 to 4, excl	4 to 6, excl	6 to 8, excl	8 to 15, incl
To 32, excl	3/8	3/8	1/2	1/2	5/8	3/4
32 to 84, excl	3/8	1/2	1/2	5/8	3/4	7/8
84 to 108, excl	1/2	9/16	5/8	3/4	7/8	1
108 to 130, excl	1/2	9/16	11/16	7/8	1	1 1/8
130 and over	5/8	3/4	7/8	1	1 1/8	1 1/4

[A] No permitted variation under specified diameter.

TABLE 8 Permitted Variations in Width and Length for Rectangular Plates When Gas Cuttings is Specified or Required (Applies to Alloy Steel Specifications Only)

NOTE 1—These permitted variations shall be taken all under or divided over and under, if so specified.

NOTE 2—Plates with universal rolled edges will be gas cut to length only.

Specified Thickness, in.	Permitted Variation Over Specified Width and Length, in.
To 2, excl	3/4
2 to 4, excl	1
4 to 6, excl	1 1/8
6 to 8, excl	1 5/16
8 to 15, incl	1 1/2

TABLE 9 Permitted Variations in Width and Length for Rectangular Plates When Gas Cutting is Specified or Required (Not Applicable to Alloy Steel)

NOTE 1—These permitted variations may be taken all under or divided over and under, if so specified.

NOTE 2—Plates with universal rolled edges will be gas cut to length only.

Specified Thickness, in.	Permitted Variation Over Specified Width and Length, in.
To 2, excl	1/2
2 to 4, excl	5/8
4 to 6, excl	3/4
6 to 8, excl	7/8
8 to 15, incl	1

TABLE 10 Permitted Variations in Diameter for Gas-Cut Circular Plates (Applies to Alloy Steel Specifications Only)

Specified Diameter, in.	Permitted Variations Over Specified Diameter for Thicknesses Given in Inches, in.[A]					
	to 1, excl	1 to 2, excl	2 to 4, excl	4 to 6, excl	6 to 8, excl	8 to 15, incl
To 32, excl	1/2	1/2	3/4	3/4	1	1
32 to 84, excl	1/2	5/8	7/8	1	1 1/8	1 1/4
84 to 108, excl	5/8	3/4	1	1 1/8	1 1/4	1 3/8
108 to 130, incl	7/8	1	1 1/8	1 1/4	1 3/8	1 1/2

[A] No permitted variation under specified diameter.

TABLE 11 Permitted Camber[A] for Carbon Steel, High-Strength Low-Alloy Steel, and Alloy Steel Universal Mill Plates and High-Strength Low-Alloy Steel and Alloy Steel Sheared, Special-Cut, or Gas-Cut Rectangular Plates

Specified Thickness, in.	Specified Weight, lb/ft²	Specified Width, in.	Permitted Camber, in.
To 2, incl	to 81.7, incl	all	1/8 × (no. of feet of length/5)
Over 2 to 15, incl	81.7 to 613.0, incl	to 30, incl	3/16 × (no. of feet of length/5)
Over 2 to 15, incl	81.7 to 613.0, incl	over 30	1/4 × (no. of feet of length/5)

[A] Camber as it relates to plates is the horizontal edge curvature in the length, measured over the entire length of the plate in the flat position.

TABLE 12 Permitted Camber[A] for Sheared Plates and Gas-Cut Rectangular Plates, All Thicknesses (Applies to Carbon Steel Only)

Permitted camber, in. = 1/8 × (number of feet of length/5)

[A] Camber as it relates to plates is the horizontal edge curvature in the length, measured over the entire length of the plate in the flat position.

TABLE 13 Permitted Variations From a Flat Surface for Standard Flatness Carbon Steel Plates

NOTE 1—When the longer dimension is under 36 in., the permitted variation from a flat surface shall not exceed 1/4 in. When the longer dimension is from 36 to 72 in., incl, the permitted variation from a flat surface shall not exceed 75 % of the tabular amount for the specified width, but in no case less than 1/4 in.

NOTE 2—These permitted variations apply to plates that have a specified minimum tensile strength of not more than 60 ksi or comparable chemical composition or hardness. The limits in this table are increased 50 % for plates that have a higher specified minimum tensile strength or comparable chemical composition or hardness.

NOTE 3—This table and these notes cover the permitted variations from a flat surface for circular and sketch plates, based upon the maximum dimensions of such plates.

NOTE 4—Where "..." appears in this table, there is no requirement.

NOTE 5—Plates must be in a horizontal position on a flat surface when flatness is measured.

Specified Thickness, in.	Specified Weight, lb/ft²	Permitted Variations from a Flat Surface for Specified Widths Given in Inches, in.[A,B]										
		To 36, excl	36 to 48, excl	48 to 60, excl	60 to 72, excl	72 to 84, excl	84 to 96, excl	96 to 108, excl	108 to 120, excl	120 to 144, excl	144 to 168, excl	168 and Over
To 1/4, excl	To 10.2, excl	9/16	3/4	15/16	1 1/4	1 3/8	1 1/2	1 5/8	1 3/4	1 7/8	...	...
1/4 to 3/8, excl	10.2 to 15.3, excl	1/2	5/8	3/4	15/16	1 1/8	1 1/4	1 3/8	1 1/2	1 5/8	...	...
3/8 to 1/2, excl	15.3 to 20.4, excl	1/2	9/16	5/8	5/8	3/4	7/8	1	1 1/8	1 1/4	1 7/8	2 1/8
1/2 to 3/4, excl	20.4 to 30.6, excl	7/16	1/2	9/16	5/8	5/8	3/4	1	1	1 1/8	1 1/2	2
3/4 to 1, excl	30.6 to 40.8, excl	7/16	1/2	9/16	5/8	5/8	5/8	3/4	7/8	1	1 3/8	1 3/4
1 to 2, excl	40.8 to 81.7, excl	3/8	1/2	1/2	9/16	9/16	5/8	5/8	5/8	11/16	1 1/8	1 1/2
2 to 4, excl	81.7 to 163.4, excl	5/16	3/8	7/16	1/2	1/2	1/2	1/2	9/16	5/8	7/8	1 1/8
4 to 6, excl	163.4 to 245.1, excl	3/8	7/16	1/2	1/2	9/16	9/16	5/8	3/4	7/8	7/8	1
6 to 8, excl	245.1 to 326.8, excl	7/16	1/2	1/2	5/8	11/16	3/4	7/8	7/8	1	1	1
8 to 10, excl	326.8 to 409.0, excl	1/2	1/2	5/8	11/16	3/4	13/16	7/8	15/16	1	1	1
10 to 12, excl	409.0 to 490.1, excl	1/2	5/8	3/4	13/16	7/8	15/16	1	1	1	1	1
12 to 15, excl	490.1 to 613.0, incl	5/8	3/4	13/16	7/8	15/16	1	1	1	1	1	...

[A] *Permitted Variation from a Flat Surface for Length*—The longer dimension specified is considered the length, and the permitted variation from a flat surface along the length shall not exceed the tabular amount for the specified width for plates up to 12 ft in length, or in any 12 ft for longer plates.

[B] *Permitted Variation from a Flat Surface for Width*—The permitted variation from a flat surface across the width shall not exceed the tabular amount for the specified width.

TABLE 14 Permitted Variations From a Flat Surface for Standard Flatness High-Strength Low-Alloy Steel and Alloy Steel Plates, Hot Rolled or Thermally Treated

NOTE 1—When the longer dimension is under 36 in., the permitted variation from a flat surface shall not exceed 3/8 in. When the longer dimension is from 36 to 72 in. incl, the permitted variation from a flat surface shall not exceed 75 % of the tabular amount for the specified width.

NOTE 2—This table and these notes cover the permitted variations from a flat surface for circular and sketch plates, based upon the maximum dimensions of such plates.

NOTE 3—Where "..." appears in this table, there is no requirement.

NOTE 4—Plates must be in a horizontal position on a flat surface when flatness is measured.

Specified Thickness, in.	Specified Weight, lb/ft²	Permitted Variations from a Flat Surface for Specified Widths, in.[A,B]										
		To 36, excl	36 to 48, excl	48 to 60, excl	60 to 72, excl	72 to 84, excl	84 to 96, excl	96 to 108, excl	108 to 120, excl	120 to 144, excl	144 to 168, excl	168 and Over
To 1/4, excl	To 10.2 excl	13/16	1 1/8	1 3/8	1 7/8	2	2 1/4	2 3/8	2 5/8	2 3/4	...	...
1/4 to 3/8, excl	10.2 to 15.3, excl	3/4	15/16	1 1/8	1 3/8	1 3/4	1 7/8	2	2 1/4	2 3/8	...	...
3/8 to 1/2, excl	15.3 to 20.4, excl	3/4	7/8	15/16	15/16	1 1/8	15/16	1 1/2	1 5/8	1 7/8	2 3/4	3 1/8
1/2 to 3/4, excl	20.4 to 30.6, excl	5/8	3/4	13/16	7/8	1	1 1/8	1 1/4	1 3/8	1 5/8	2 1/4	3
3/4 to 1, excl	30.6 to 40.8, excl	5/8	3/4	7/8	7/8	15/16	1	1 1/8	1 5/16	1 1/2	2	2 5/8
1 to 2, excl	40.8 to 81.7, excl	9/16	5/8	3/4	13/16	7/8	15/16	1	1	1	1 5/8	2 1/4
2 to 4, excl	81.7 to 163.4, excl	1/2	9/16	11/16	3/4	3/4	3/4	3/4	7/8	1	1 1/4	1 5/8
4 to 6, excl	163.4 to 245.1, excl	9/16	11/16	3/4	3/4	7/8	7/8	15/16	1 1/8	1 1/4	1 1/4	1 1/2
6 to 8, excl	245.1 to 326.8, excl	5/8	3/4	3/4	15/16	1	1 1/8	1 1/4	1 5/16	1 1/2	1 1/2	1 1/2
8 to 10, excl	326.8 to 409.0, excl	3/4	13/16	15/16	1	1 1/8	1 1/4	15/16	1 3/8	1 1/2	1 1/2	1 1/2
10 to 12, excl	409.0 to 490.1, excl	3/4	15/16	1 1/8	1 1/4	15/16	1 3/8	1 1/2	1 1/2	1 1/2	1 1/2	1 1/2
12 to 15, incl	490.1 to 613.0, incl	7/8	1	1 3/16	15/16	1 3/8	1 1/2	1 1/2	1 1/2	1 1/2	1 1/2	1 1/2

[A] *Permitted Variation from a Flat Surface for Length*—The longer dimension specified is considered the length, and the permitted variation from a flat surface along the length shall not exceed the tabular amount for the specified width in plates up to 12 ft in length, or in any 12 ft for longer plates.

[B] *Permitted Variation from a Flat Surface for Width*—The permitted variation from a flat surface across the width shall not exceed the tabular amount for the specified width.

TABLE 15 Permitted Variations in Waviness for Standard Flatness Plates

Note 1—Waviness denotes the maximum deviation of the surface of the plate from a plane parallel to the surface of the point of measurement and contiguous to the surface of the plate at each of the two adjacent wave peaks, when the plate is resting on a flat horizontal surface, as measured in an increment of less than 12 ft of length. The permitted variation in waviness is a function of the permitted variation from a flat surface as obtained from Table 13 or Table 14, whichever is applicable.

Note 2—Plates must be in a horizontal position on a flat surface when waviness is measured.

Permitted Variation from a Flat Surface (from Table 13 or Table 14), in.	Permitted Variation in Waviness, in., When Number of Waves in 12 ft is						
	1	2	3	4	5	6	7
5/16	5/16	1/4	3/16	1/8	1/8	1/16	1/16
3/8	3/8	5/16	3/16	3/16	1/8	1/16	1/16
7/16	7/16	5/16	1/4	3/16	1/8	1/8	1/16
1/2	1/2	3/8	5/16	3/16	3/16	1/8	1/16
9/16	9/16	7/16	5/16	1/4	3/16	1/8	1/8
5/8	5/8	1/2	3/8	1/4	3/16	1/8	1/8
11/16	11/16	1/2	3/8	5/16	3/16	3/16	1/8
3/4	3/4	9/16	7/16	5/16	1/4	3/16	1/8
13/16	13/16	5/8	7/16	5/16	1/4	3/16	1/8
7/8	7/8	11/16	1/2	3/8	1/4	3/16	1/8
15/16	15/16	11/16	1/2	3/8	5/16	1/4	3/16
1	1	3/4	9/16	7/16	5/16	1/4	3/16
1 1/8	1 1/8	7/8	5/8	1/2	3/8	1/4	3/16
1 1/4	1 1/4	15/16	11/16	1/2	3/8	5/16	1/4
1 3/8	1 3/8	11/16	3/4	9/16	7/16	5/16	1/4
1 1/2	1 1/2	1 1/8	7/8	5/8	1/2	3/8	1/4
1 5/8	1 5/8	1 1/4	15/16	11/16	1/2	3/8	5/16
1 3/4	1 3/4	15/16	1	3/4	9/16	7/16	5/16
1 7/8	1 7/8	1 7/16	1 1/16	13/16	9/16	7/16	5/16
2	2	1 1/2	1 1/8	7/8	5/8	1/2	3/8
2 1/8	2 1/8	1 5/8	1 3/16	7/8	11/16	1/2	3/8
2 1/4	2 1/4	1 11/16	1 1/4	15/16	11/16	9/16	3/8
2 3/8	2 3/8	1 13/16	1 5/16	1	3/4	9/16	7/16
2 1/2	2 1/2	1 7/8	1 7/16	1 1/16	13/16	9/16	7/16
2 5/8	2 5/8	2	1 1/2	1 1/8	13/16	5/8	7/16
2 3/4	2 3/4	2 1/16	1 9/16	1 1/8	7/8	5/8	1/2
2 7/8	2 7/8	2 3/16	1 5/8	1 3/16	15/16	11/16	1/2
3	3	2 1/4	1 11/16	1 1/4	15/16	11/16	9/16
3 1/8	3 1/8	2 3/8	1 3/4	1 5/16	1	3/4	9/16

TABLE 16 Permitted Variations in Cross Section for W, HP, S, M, C, and MC Shapes

NOTE 1—A is measured at center line of web for S, M, and W and HP shapes; at back of web for C and MC shapes. Measurement is overall for C shapes under 3 in. B is measured parallel to flange. C is measured parallel to web.

NOTE 2—Where "..." appears in this table, there is no requirement.

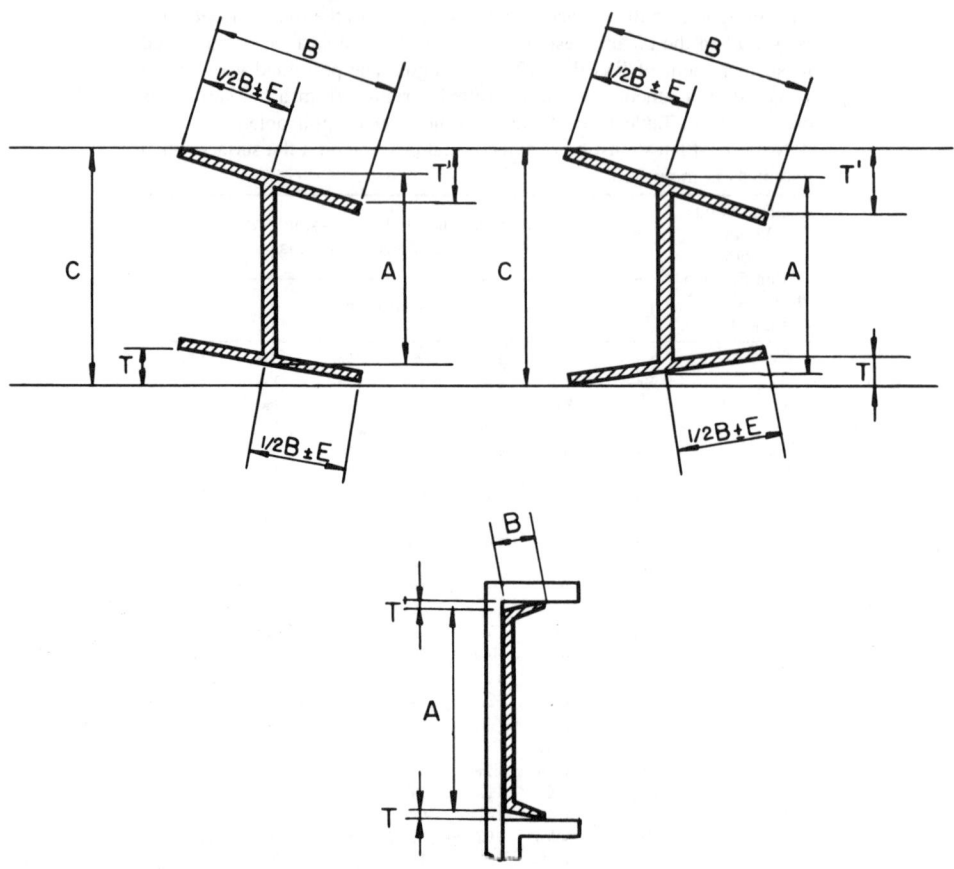

Shape	Section Nominal Sizes, in.	Permitted Variations in Sectional Dimensions Given, in.				$T + T'$ [A] Flanges Out-of-Square[B]	E, Web off Center[C]	C, Maximum Depth at any Cross Section over Theoretical Depth, in.	Permitted Variations Over or Under Theoretical Web Thickness for Thicknesses Given in Inches, in.	
		A, Depth		B, Flange Width						
		Over Theoretical	Under Theoretical	Over Theoretical	Under Theoretical				³⁄₁₆ and under	Over ³⁄₁₆
W and HP	Up to 12, incl	⅛	⅛	¼	³⁄₁₆	¼	³⁄₁₆	¼	...	...
	Over 12	⅛	⅛	¼	³⁄₁₆	⁵⁄₁₆	³⁄₁₆	¼	...	...
S and M	3 to 7, incl	³⁄₃₂	¹⁄₁₆	⅛	⅛	¹⁄₃₂	³⁄₁₆	...	...	...
	Over 7 to 14, incl	⅛	³⁄₃₂	⁵⁄₃₂	⁵⁄₃₂	¹⁄₃₂	³⁄₁₆	...	...	...
	Over 14 to 24, incl	³⁄₁₆	⅛	³⁄₁₆	³⁄₁₆	¹⁄₃₂	³⁄₁₆	...	...	...
C and MC	1½ and under	¹⁄₃₂	¹⁄₃₂	¹⁄₃₂	¹⁄₃₂	¹⁄₃₂	...	...	0.010	0.015
	Over 1½ to 3, excl	¹⁄₁₆	¹⁄₁₆	¹⁄₁₆	¹⁄₁₆	¹⁄₃₂	...	...	0.015	0.020
	3 to 7, incl	³⁄₃₂	¹⁄₁₆	⅛	⅛	¹⁄₃₂	...	...	...	...
	Over 7 to 14, incl	⅛	³⁄₃₂	⅛	⁵⁄₃₂	¹⁄₃₂	...	...	...	...
	Over 14	³⁄₁₆	⅛	⅛	³⁄₁₆	¹⁄₃₂	...	...	...	...

[A] $T + T'$ applies when flanges of channels are toed in or out. For channels ⅝ in. and under in depth, the permitted out-of-square is ³⁄₆₄ in./in. of depth.
[B] Permitted variation is per inch of flange width for S, M, C, and MC shapes.
[C] Permitted variation of ⁵⁄₁₆ in. max for sections over 426 lb/ft.

TABLE 17 Permitted Variations in Cross Section for Angles (L Shapes), Bulb Angles, and Zees

NOTE 1—Where "..." appears in this table, there is no requirement.

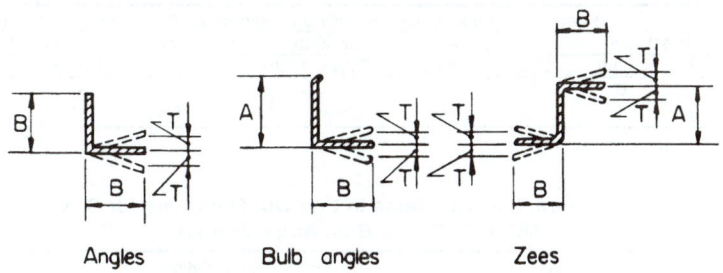

Angles   Bulb angles   Zees

Section	Nominal Size, in.	Permitted Variations in Sectional Dimensions Given, in.					Permitted Variations Over or Under Theoretical Thickness for Thicknesses Given in Inches, in.		
		A, Depth		B, Flange Width or Length of Leg		T, Out-of-Square per Inch of B	3/16 and under	Over 3/16 to 3/8, incl	Over 3/8
		Over Theoretical	Under Theoretical	Over Theoretical	Under Theoretical				
Angles[A] (L Shapes)	1 and under	...	...	1/32	1/32	3/128[B]	0.008	0.010	...
	Over 1 to 2, incl	...	...	3/64	3/64	3/128[B]	0.010	0.010	0.012
	Over 2 to 3, excl	...	...	1/16	1/16	3/128[B]	0.012	0.015	0.015
	3 to 4, incl	...	...	1/8	3/32	3/128[B]	...	...	...
	Over 4 to 6, incl	...	...	1/8	1/8	3/128[B]	...	...	...
	Over 6	...	...	3/16	1/8	3/128[B]	...	...	...
Bulb angles	(Depth) 3 to 4, incl	1/8	1/16	1/8	3/32	3/128[B]	...	...	...
	Over 4 to 6, incl	1/8	1/16	1/8	1/8	3/128[B]	...	...	...
	Over 6	1/8	1/16	3/16	1/8	3/128[B]	...	...	...
Zees	3 to 4, incl	1/8	1/16	1/8	3/32	3/128[B]	...	...	...
	Over 4 to 6, incl	1/8	1/16	1/8	1/8	3/128[B]	...	...	...

[A] For unequal leg angles, longer leg determines classification.
[B] 3/128 in./in. = 1½°.

TABLE 18 Permitted Variations in Sectional Dimensions for Rolled Tees

NOTE 1—*Back of square and center line of stem are to be parallel when measuring "out-of-square."
NOTE 2—Where "..." appears in this table, there is no requirement.

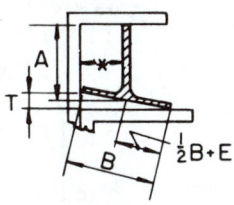

Nominal Size,[A]	Permitted Variations in Sectional Dimensions Given, in. Tees										
	A, Depth[B]		B, Width[B]		T, Out-of-Square per Inch of B	E, Web-off-Center	Stem Out-of-Square[C]	Thickness of Flange		Thickness of Stem	
	Over	Under	Over	Under				Over	Under	Over	Under
1¼ and under	3/64	3/64	3/64	3/64	...	...	1/32	0.010	0.010	0.005	0.020
Over 1¼ to 2, incl	1/16	1/16	1/16	1/16	...	...	1/16	0.012	0.012	0.010	0.020
Over 2 to 3, excl	3/32	3/32	3/32	3/32	...	...	3/32	0.015	0.015	0.015	0.020
3 to 5, incl	3/32	1/16	1/8	1/8	1/32	3/32	...	...	...	...	...
Over 5 to 7, incl	3/32	1/16	1/8	1/8	1/32	1/8	...	...	...	...	...

[A] The longer member of an unequal tee determines the size for permitted variations.
[B] Measurements for both depth and width are overall.
[C] Stem-out-of-square is the permitted variation from its true position of the center line of stem, measured at the point.

TABLE 19 Permitted Variations in Length for S, M, C, MC, L, T, Z, and Bulb Angle Shapes

NOTE 1—Where "..." appears in this table, there is no requirement.

Nominal Size,[A] in.	Permitted Variations from Specified Length for Lengths Given in Feet, in.												
	5 to 10, excl		10 to 20, excl		20 to 30, incl		Over 30 to 40, incl		Over 40 to 50, incl		Over 50 to 65, incl		Over 65 ft	
	Over	Under	Over	Under	Over	Under	Over	Under	Over	Under	Over	Under	Over	Under
Under 3	5/8	0	1	0	1½	0	2	0	2½	0	2½	0	...	...
3 and over	1	0	1½	0	1¾	0	2¼	0	2¾	0	2¾	0	...	...

[A] Greatest cross-sectional dimension.

TABLE 20 Permitted Variations in End Out-Of-Square for S, M, C, MC, L, T, Z, and Bulb Angle Shapes

Shapes	Permitted Variation
S, M, C, and MC	1/64 in. per inch of depth
L[A]	3/128 in. per inch of leg length or 1½ °
Bulb angles	3/128 in. per inch of depth or 1½ °
Rolled Tees[A]	1/64 in. per inch of flange or stem
Zees	3/128 in. per inch of sum of both flange lengths

[A] Permitted variations in end out-of-square are determined on the longer members of the shape.

TABLE 21 Permitted Variations in Straightness for S, M, C, MC, L, T, Z, and Bulb Angle Shapes

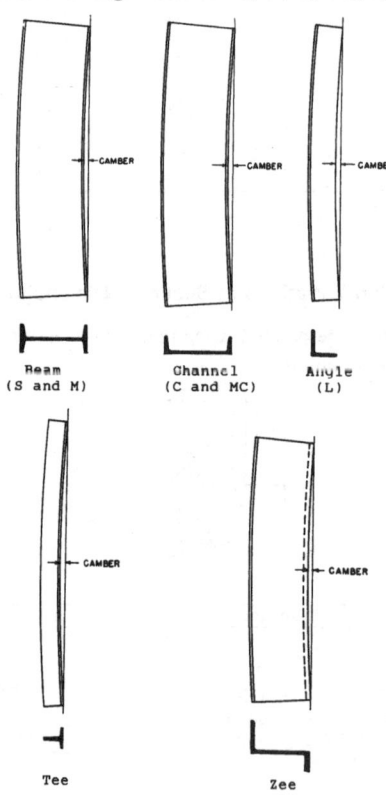

Positions for Measuring Camber of Shapes

Variable	Nominal Size,[A] in.	Permitted Variation, in.
Camber	Under 3	¼ in. in any 5 ft, or ¼ × (number of feet of total length/5)
	3 and over	⅛ × (number of feet of total length/5)
Sweep	All	Due to the extreme variations in flexibility of these shapes, permitted variations for sweep are subject to negotiations between the manufacturer and the purchaser for the individual sections involved.

[A] Greatest cross-sectional dimension.

TABLE 22 Permitted Variations in Length for W and HP Shapes

W and HP Shapes	Permitted Variations from Specified Length for Lengths Given in Feet, in.[A,B]			
	30 and under		Over 30	
	Over	Under	Over	Under
Beams 24 in. and under in nominal depth	3/8	3/8	3/8 plus 1/16 for each additional 5 ft or fraction thereof	3/8
Beams over 24 in. in nominal depth and all columns	1/2	1/2	1/2 plus 1/16 for each additional 5 ft or fraction thereof	1/2

[A] For HP and W shapes specified in the order for use as bearing piles, the permitted variations in length are plus 5 in. and minus 0 in. These permitted variations in length also apply to sheet piles.

[B] The permitted variations in end out-of-square for W and HP shapes shall be 1/64 in. per inch of depth, or per inch of flange width if the flange width is larger than the depth.

TABLE 23 Permitted Variations in Length and End Out-of-Square, Milled Shapes

Nominal Depth, in.	Length, ft[B]	Permitted Variations in Length and End Out-of-Square, in.[A]					
		Milled Both Ends[C]			Milled One-End[C]		End Out-of-Square (for Milled End)
		Length		End Out-of-Square	Length		
		Over	Under		Over	Under	
6 to 36	6 to 70	1/32	1/32	1/32	1/4	1/4	1/32

[A] Length is measured along center line of web. Measurements are made with the steel and tape at the same temperature.

[B] The permitted variations in length and end out-of-square are additive.

[C] End out-of-square is measured by (a) squaring from the center line of the web and (b) squaring from the center line of the flange. The measured variation from true squareness in either plane shall not exceed the total tabular amount.

TABLE 24 Permitted Variations in Straightness for W and HP Shapes

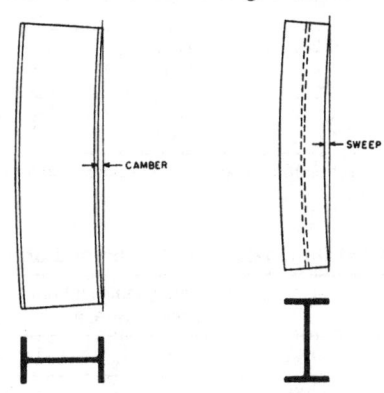

Positions for Measuring Camber and Sweep of W and HP Shapes	
	Permitted Variation in Straightness, in.
Camber and sweep	1/8 × (number of feet of total length/10)[A]
When certain sections[B] with a flange width approximately equal to depth are specified in the order for use as columns:	
Lengths of 45 ft and under	1/8 × (number of feet of total length/10) but not over 3/8
Lengths over 45 ft	3/8 + [1/8 × ([number of feet of total length − 45]/10)]

[A] Sections with a flange width less than 6 in., permitted variation for sweep, in. = 1/8 × (number of feet of total length/5).

[B] Applies only to:
 8-in. deep sections 31 lb/ft and heavier,
 10-in. deep sections 49 lb/ft and heavier,
 12-in. deep sections 65 lb/ft and heavier, and
 14-in. deep sections 90 lb/ft and heavier.
For other sections specified in the order for use as columns, the permitted variation is subject to negotiation with the manufacturer.

TABLE 25 Permitted Variations in Dimensions for Split Tees and Split Angles (L Shapes)[A]

Specified Depth, in.	Permitted Variation Over or Under Specified Depth,[B] in.
To 6, excl (beams and channels)	1/8
6 to 16, excl (beams and channels)	3/16
16 to 20, excl (beams and channels)	1/4
20 to 24, excl (beams)	5/16
24 and over (beams)	3/8

[A] The permitted variations in length for split tees or angles are the same as those applicable to the section from which the tees or angles are split.
[B] The above permitted variations in depth of tees or angles include the permitted variations in depth for the beams or channels before splitting. Permitted variations in dimensions and straightness, as set up for the beams or channels from which these tees or angles are cut, apply, except:
straightness = 1/8 in. × (length in feet/5)

TABLE 26 Permitted Variations in Sectional Dimensions for Square-Edge and Round-Edge Flat Bars

NOTE 1—Where "..." appears in this table, there is no requirement.

Specified Widths, in.	Permitted Variations Over or Under Specified Thickness, for Thicknesses Given in Inches, in.							Permitted Variations From Specified Width, in.	
	0.203 to 0.230, excl	0.230 to 1/4, excl	1/4 to 1/2, incl	Over 1/2 to 1, incl	Over 1 to 2, incl	Over 2 to 3, incl	Over 3	Over	Under
To 1, incl	0.007	0.007	0.008	0.010	...	...	...	1/64	1/64
Over 1 to 2, incl	0.007	0.007	0.012	0.015	1/32	...	...	1/32	1/32
Over 2 to 4, incl	0.008	0.008	0.015	0.020	1/32	3/64	3/64	1/16	1/32
Over 4 to 6, incl	0.009	0.009	0.015	0.020	1/32	3/64	3/64	3/32	1/16
Over 6 to 8, incl	[A]	0.015	0.016	0.025	1/32	3/64	1/16	1/8[B]	3/32[B]

[A] Flats over 6 to 8 in., incl, in width are not available as hot-rolled carbon steel bars in thickness under 0.230 in.
[B] For flats over 6 to 8 in., in width, and to 3 in. incl in thickness.

TABLE 27 Permitted Variations in Sectional Dimensions for Round and Square Bars and Round-Cornered Squares

Specified Size, in.	Permitted Variations from Specified Size, in.		Permitted Out-of-Round or Out-of-Square, in.[A]
	Over	Under	
To 5/16	0.005	0.005	0.008
Over 5/16 to 7/16, incl	0.006	0.006	0.009
Over 7/16 to 5/8, incl	0.007	0.007	0.010
Over 5/8 to 7/8, incl	0.008	0.008	0.012
Over 7/8 to 1, incl	0.009	0.009	0.013
Over 1 to 1 1/8, incl	0.010	0.010	0.015
Over 1 1/8 to 1 1/4, incl	0.011	0.011	0.016
Over 1 1/4 to 1 3/8, incl	0.012	0.012	0.018
Over 1 3/8 to 1 1/2, incl	0.014	0.014	0.021
Over 1 1/2 to 2, incl	1/64	1/64	0.023
Over 2 to 2 1/2, incl	1/32	0	0.023
Over 2 1/2 to 3 1/2, incl	3/64	0	0.035
Over 3 1/2 to 4 1/2, incl	1/16	0	0.046
Over 4 1/2 to 5 1/2, incl	5/64	0	0.058
Over 5 1/2 to 6 1/2, incl	1/8	0	0.070
Over 6 1/2 to 8 1/4, incl	5/32	0	0.085
Over 8 1/4 to 9 1/2, incl	3/16	0	0.100
Over 9 1/2 to 10, incl	1/4	0	0.120

[A] Out-of-round is the difference between the maximum and minimum diameters of the bar, measured at the same transverse cross section. Out-of-square section is the difference in perpendicular distance between opposite faces, measured at the same transverse cross section.

TABLE 28 Permitted Variations in Sectional Dimensions for Hexagons

Specified Sizes Between Opposite Sides, in.	Permitted Variations from Specified Size, in.		Permitted Out-of-Hexagon Section, Three Measurements, in.[A]
	Over	Under	
½ and under	0.007	0.007	0.011
Over ½ to 1, incl	0.010	0.010	0.015
Over 1 to 1½, incl	0.021	0.013	0.025
Over 1½ to 2, incl	1/32	1/64	1/32
Over 2 to 2½, incl	3/64	1/64	3/64
Over 2½ to 3½, incl	1/16	1/64	1/16

[A] Out-of-hexagon section is the greatest difference in distance between any two opposite faces measured at the same transverse cross section.

TABLE 29 Permitted Variations in Straightness for Bars

Permitted Variations in Straightness, in.[A]
¼ in any 5 ft and ¼ × (number of feet of total length/5)

[A] Permitted variations in straightness do not apply to hot-rolled bars if any subsequent heating operation has been performed.

TABLE 30 Permitted Variations in Length for Hot-Cut Steel Bars[A]

NOTE 1—Where "..." appears in this table, there is no requirement.

Specified Sizes of Rounds, Squares, and Hexagons, in.	Specified Sizes of Flats, in.		Permitted Variations Over Specified Length Given in Feet, in. (No Variation Under)				
	Thickness	Width	5 to 10, excl	10 to 20, excl	20 to 30, excl	30 to 40, excl	40 to 60, incl
To 1, incl	To 1, incl	To 3, incl	½	¾	1¼	1¾	2¼
Over 1 to 2, incl	Over 1	To 3, incl	5/8	1	1½	2	2½
Over 1 to 2, incl	To 1, incl	Over 3 to 6, incl	5/8	1	1½	2	2½
Over 2 to 5, incl	Over 1	Over 3 to 6, incl	1	1½	1¾	2¼	2¾
Over 5 to 10, incl	...	...	2	2½	2¾	3	3¼
	0.230 to 1, incl	Over 6 to 8, incl	¾	1¼	1¾	3½	4
	Over 1 to 3, incl	Over 6 to 8, incl	1¼	1¾	2	3½	4
Hot Sawing							
2 to 5, incl[B]	1 and over	3 and over	B	1½	1¾	2¼	2¾
Over 5 to 10, incl	...	...	B	2½	2¾	3	3¼

[A] For flats over 6 to 8 in., incl, in width and over 3 in. in thickness, consult the manufacturer for permitted variations in length.
[B] Smaller sizes and shorter lengths are not commonly hot sawed.

TABLE 31 Permitted Variations in Length for Bars Recut Both Ends After Straightening[A][B]

Sizes of Rounds, Squares, Hexagons, Width of Flats and Maximum Dimension of Other Sections, in.	Permitted Variations from Specified Lengths Given in Feet, in.			
	To 12, incl		Over 12	
	Over	Under	Over	Under
To 3, incl	3/16	1/16	¼	1/16
Over 3 to 6, incl	¼	1/16	3/8	1/16
Over 6 to 8, incl	3/8	1/16	½	1/16
Rounds over 8 to 10, incl	½	1/16	5/8	1/16

[A] For flats over 6 to 8 in., incl, in width, and over 3 in. in thickness, consult the manufacturer or processor for permitted variations in length.
[B] Permitted variations are sometimes required all over or all under the specified length, in which case the sum of the two permitted variations applies.

SUPPLEMENTARY REQUIREMENTS

The following standardized supplementary requirements are for use when desired by the purchaser. Those that are considered suitable for use with each material specification are listed in the specification. Other tests may be performed by agreement between the supplier and the purchaser. These additional requirements shall apply only when specified in the order, in which event the specified tests shall be made by the manufacturer or processor before shipment of the material.

S1. Vacuum Treatment

S1.1 The steel shall be made by a process that includes vacuum degassing while molten. Unless otherwise agreed upon with the purchaser, it is the responsibility of the manufacturer to select suitable process procedures.

S2. Product Analysis

S2.1 Product analyses shall be made for those elements specified or restricted by the applicable product specification for the applicable grade, class, and type. Specimens for analysis shall be taken adjacent to or from the tension test specimen, or from a sample taken from the same relative location as that from which the tension test specimen was taken.

S3. Simulated Post-Weld Heat Treatment of Mechanical Test Coupons

S3.1 Prior to testing, the test specimens representing the structural product for acceptance purposes for mechanical properties shall be thermally treated to simulate a post-weld heat treatment below the critical temperature (Ac_3), using the heat treatment parameters (such as temperature range, time, and cooling rates) specified in the order. The test results for such heat-treated test specimens shall meet the applicable product specification requirements.

S4. Additional Tension Test

S4.1 *Plate*—One tension test shall be made from each unit plate rolled from a slab or directly from an ingot, except that for quenched and tempered plates, a test shall be taken from each unit plate heat treated. The results obtained shall be reported on the mill test reports when such tests are required by the order.

S5. Charpy V-Notch Impact Test

S5.1 Charpy V-notch impact tests shall be conducted in accordance with Specification A673/A673M.

S5.2 The frequency of testing, the test temperature to be used, and the absorbed energy requirements shall be as specified on the order.

S6. Drop-Weight Test (for Material 0.625 in. [16 mm] and Over in Thickness)

S6.1 Drop-weight tests shall be made in accordance with Test Method E208. The specimens shall represent the material in the final condition of heat treatment. Agreement shall be reached between the purchaser and the manufacturer or processor as to the number of pieces to be tested and whether a maximum nil-ductility transition (NDT) temperature is mandatory or if the test results are for information only.

S8. Ultrasonic Examination

S8.1 The material shall be ultrasonically examined in accordance with the requirements specified on the order.

S15. Reduction of Area Measurement

S15.1 The reduction of area, as determined on the 0.500-in. [12.5-mm] diameter round tension test specimen in accordance with Test Methods and Definitions A370, shall not be less than 40 %.

S18. Maximum Tensile Strength

S18.1 Steel having a specified minimum tensile strength of less than 70 ksi [485 MPa] shall not exceed the minimum specified tensile strength by more than 30 ksi [205 MPa].

S18.2 Steel having a minimum specified tensile strength of 70 ksi [485 MPa] or higher shall not exceed the minimum specified tensile strength by more than 25 ksi [170 MPa].

S23. Copper-Bearing Steel (for improved atmospheric corrosion resistance)

S23.1 The copper content shall be a minimum of 0.20 % on heat analysis, 0.18 on product analysis.

S26. Subdivided Material—Marking of Individual Pieces

S26.1 Subdivided pieces shall be individually identified by marking, stenciling, or die stamping the applicable product specification designation (year-date not required), grade, heat number, and the heat treatment identification, if applicable, along with the trademark, brand, or name of the organization that subdivided the structural product. As an alternative, individual subdivided pieces shall be identified by a code traceable to the original required identification, provided that the trademark, name, or brand of the organization that subdivided the structural product is also placed on the structural product and the original required identification, cross referenced on the code, is furnished with the structural product.

S27. Restrictive Plate Flatness

S27.1 As-rolled or normalized carbon steel plates ordered to restrictive flatness shall conform to the permitted variations from a flat surface given in Table S27.1 or Table S27.2, whichever is applicable.

S27.2 As-rolled or normalized high-strength low-alloy steel plates ordered to restrictive flatness shall conform to the permitted variations from a flat surface given in Table S27.3 or Table S27.4, whichever is applicable.

TABLE S27.1 Permitted Variations From a Flat Surface for As-Rolled or Normalized Carbon Steel Plates Ordered to Half-Standard Flatness

NOTE 1—*Permitted Variation From a Flat Surface Along the Length*—The longer dimension specified is considered the length, and the permitted variation from a flat surface along the length shall not exceed the tabular amount for the specified width in plates up to 12 ft in length, or in any 12 ft of longer plates.

NOTE 2—*Permitted Variation From a Flat Surface Across the Width*—The permitted variation from a flat surface across the width shall not exceed the tabular amount for the specified width.

NOTE 3—When the longer dimension is under 36 in., the permitted variation from a flat surface shall not exceed ¼ in. in each direction. When the longer dimension is from 36 to 72 in., incl, the permitted variation from a flat surface shall not exceed 75 % of the tabular amount for the specified width, but in no case less than ¼ in.

NOTE 4—The permitted variations given in this table apply to plates that have a minimum specified tensile strength not over 60 ksi or comparable chemistry or hardness. For plates specified to a higher minimum tensile strength or compatible chemistry or hardness, the permitted variations are 1½ times the amounts in this table.

NOTE 5—This table and these notes cover the permitted variations from a flat surface for circular and sketch plates, based upon the maximum dimensions of such plates.

NOTE 6—Permitted variations in waviness do not apply.

NOTE 7—Plates must be in a horizontal position on a flat surface when flatness is measured.

Specified Thickness, in.	Specified Weights, lb/ft^2	Permitted Variations From a Flat Surface for Specified Widths Given in Inches, in.					
		48 to 60, excl	60 to 72, excl	72 to 84, excl	84 to 96, excl	96 to 108, excl	108 to 120, incl
To ¼, excl	To 10.2, excl	15/32	5/8	11/16	3/4	13/16	7/8
¼ to 3/8, excl	10.2 to 15.3, excl	3/8	15/32	9/16	5/8	11/16	3/4
3/8 to ½, excl	15.3 to 20.4, excl	5/16	5/16	3/8	7/16	1/2	9/16
½ to ¾, excl	20.4 to 30.6, excl	9/32	5/16	5/16	3/8	1/2	1/2
¾ to 1, excl	30.6 to 40.8, excl	9/32	5/16	5/16	5/16	3/8	7/16
1 to 2, incl	40.8 to 51.7, incl	1/4	9/32	9/32	5/16	5/16	5/16

TABLE S27.2 Permitted Variations From a Flat Surface for As-Rolled or Normalized Carbon Steel Plates Ordered to Half-Standard Flatness

NOTE 1—*Permitted Variation From a Flat Surface Along the Length*—The longer dimension specified is considered the length, and the permitted variation from a flat surface along the length shall not exceed the tabular amount for the specified width in plates up to 3700 mm in length, or in any 3700 mm of longer plates.

NOTE 2—*Permitted Variation From a Flat Surface Across the Width*—The permitted variation from a flat surface across the width shall not exceed the tabular amount for the specified width.

NOTE 3—When the longer dimension is under 900 mm, the permitted variation from a flat surface shall not exceed 6 mm in each direction. When the longer dimension is from 900 to 1800 mm, incl., the permitted flatness variation should not exceed 75 % of the tabular amount for the specified width, but in no case less than 6 mm.

NOTE 4—The permitted variations given in this table apply to plates that have a minimum specified tensile strength not over 415 MPa or comparable chemistry or hardness. For plates specified to a higher minimum tensile strength or compatible chemistry or hardness, the permitted variations are 1½ times the amounts in this table.

NOTE 5—This table and these notes cover the permitted variations from a flat surface for circular and sketch plates, based upon the maximum dimensions of such plates.

NOTE 6—Permitted variations in waviness do not apply.

NOTE 7—Plates must be in a horizontal position on a flat surface when flatness is measured.

Specified Thickness, mm	Specified Weights, kg/m^2	Permitted Variations From a Flat Surface for Specified Widths Given in Millimetres, mm					
		1200 to 1500, excl	1500 to 1800, excl	1800 to 2100, excl	2100 to 2400, excl	2400 to 2700, excl	2700 to 3000, incl
To 6, excl	To 47.1 excl	12	16	17	19	20	22
6 to 10, excl	47.1 to 78.5, excl.	9	12	14	16	17	19
10 to 12, excl	78.5 to 94.2, excl	8	8	9	11	12	14
12 to 20, excl	94.2 to 157.0, excl	7	8	8	9	12	12
20 to 25, excl	157.0 to 196.2, excl	7	8	8	8	9	11
25 to 50, incl	196.2 to 392.5, incl	6	7	7	8	8	8

TABLE S27.3 Permitted Variations From a Flat Surface for As-Rolled or Normalized High-Strength Low-Alloy Steel Plates Ordered to Half-Standard Flatness

NOTE 1—*Permitted Variation From a Flat Surface Along the Length*—The longer dimension specified is considered the length, and the permitted variation from a flat surface along the length shall not exceed the tabular amount for the specified width in plates up to 12 ft in length, or in any 12 ft of longer plates.

NOTE 2—*Permitted Variation From a Flat Surface Across the Width*—The permitted variation from a flat surface across the width shall not exceed the tabular amount for the specified width.

NOTE 3—When the longer dimension is under 36 in., the permitted variation from a flat surface shall not exceed ⅜ in. in each direction. When the larger dimension is from 36 to 72 in., incl, the permitted variation from a flat surface shall not exceed 75 % of the tabular amount for the specified width, but in no case less than ⅜ in.

NOTE 4—This table and these notes cover the permitted variations from a flat surface for circular and sketch plates, based upon the maximum dimensions of those plates.

NOTE 5—Permitted variations in waviness do not apply.

NOTE 6—Plates must be in a horizontal position on a flat surface when flatness is measured.

Specified Thickness, in.	Specified Weights, lb/ft²	Permitted Variations From a Flat Surface for Specified Widths Given in Inches, in.					
		48 to 60, excl	60 to 72, excl	72 to 84, excl	84 to 96, excl	96 to 108, excl	108 to 120, incl
To ¼, excl	To 10.2 excl	11/16	15/16	1	1 ⅛	1 3/16	1 5/16
¼ to ⅜, excl	10.2 to 15.3, excl	9/16	11/16	⅞	15/16	1	1 ⅛
⅜ to ½, excl	15.3 to 20.4, excl	15/32	15/32	9/16	21/32	¾	13/16
½ to ¾, excl	20.4 to 30.6, excl	13/32	7/16	½	9/16	⅝	11/16
¾ to 1, excl	30.6 to 40.8, excl	7/16	7/16	15/32	½	9/16	21/32
1 to 2, incl	40.8 to 51.7, incl	⅜	13/32	7/16	15/32	½	½

TABLE S27.4 Permitted Variations From a Flat Surface for As-Rolled or Normalized High-Strength Low-Alloy Steel Plates Ordered to Half-Standard Flatness

NOTE 1— *Permitted Variation From a Flat Surface Along the Length*—The longer dimension specified is considered the length, and the permitted variation from a flat surface along the length shall not exceed the tabular amount for the specified width in plates up to 3700 mm in length, or in any 3700 mm of longer plates.

NOTE 2—*Permitted Variation From a Flat Surface Across the Width*—The permitted variation from a flat surface across the width shall not exceed the tabular amount for the specified width.

NOTE 3—When the longer dimension is under 900 mm, the permitted variation from a flat surface shall not exceed 10 mm in each direction. When the larger dimension is from 900 to 1800 mm, incl., the permitted variation from a flat surface shall not exceed 75 % of the tabular amount for the specified width but in no case less than 10 mm.

NOTE 4—This table and these notes cover the permitted variations from a flat surface for circular and sketch plates, based upon the maximum dimensions of such plates.

NOTE 5—Permitted variations in waviness do not apply.

NOTE 6—Plates must be in a horizontal position on a flat surface when flatness is measured.

Specified Thickness, mm	Specified Weights, kg/m²	Permitted Variations From a Flat Surface for Specified Widths Given in Millimetres, mm					
		1200 to 1500, excl	1500 to 1800, excl	1800 to 2100, excl	2100 to 2400, excl	2400 to 2700, excl	2700 to 3000, incl
To 6, excl	To 47.1 excl	17	24	25	28	30	33
6 to 10, excl	47.1 to 78.5, excl	14	17	22	24	25	28
10 to 12, excl	78.5 to 94.2, excl	12	12	14	16	19	20
12 to 20, excl	94.2 to 157.0, excl	11	11	12	14	16	17
20 to 25, excl	157.0 to 196.2, excl	11	11	12	12	14	16
25 to 50, incl	196.2 to 392.5, incl	9	10	11	12	12	12

S28. Fine Grain Practice

S28.1 The steel shall be made to fine grain practice.

S29. Fine Austenitic Grain Size

S29.1 The requirements for fine austenitic grain size (see 8.1 and 8.3) shall be met.

S30. Charpy V-Notch Impact Test for Structural Shapes: Alternate Core Location

S30.1 For shapes with a flange thickness equal to or greater than 1½ in. [38.1 mm] that are specified in the purchase order to be tested in accordance with this supplementary requirement, Charpy V-notch impact tests shall be conducted in accordance with Specification A673/A673M, using specimens taken from the alternate core location. Unless otherwise specified in the purchase order, the minimum average absorbed energy for each test shall be 20 ft·lbf [27 J] and the test temperature shall be 70°F [21°C].

S30.2 The frequency of testing shall be Frequency (H), except that, for rolled shapes produced from ingots, the frequency shall be Frequency (P) and the specimens shall be taken from a location representing the top of an ingot or part of an ingot used to produce the product represented by such specimens.

S31. Maximum Carbon Equivalent for Weldability

S31.1 Plates and shapes shall be supplied with a specific maximum carbon equivalent value as specified by the purchaser. This value shall be based upon heat analysis. The required chemical analysis as well as the carbon equivalent shall be reported.

S31.2 The carbon equivalent shall be calculated using the following formula:

$$CE = C + Mn/6 + (Cr + Mo + V)/5 + (Ni + Cu)/15$$

S31.3 For additional information on the weldability of steel, see Appendix X3.

S32. Single Heat Bundles

S32.1 Bundles containing shapes or bars shall be from a single heat of steel.

ANNEXES

(Mandatory Information)

A1. PERMITTED VARIATIONS IN DIMENSIONS AND MASS IN SI UNITS

A1.1 Tables A1.1-A1.31 inclusive contain permitted variations in dimensions and mass stated in SI Units.

TABLE A1.1 Permitted Variations in Thickness for Rectangular Carbon, High-Strength Low Alloy, and Alloy Steel Plates, 300 mm and Under in Thickness When Ordered to Thickness

NOTE 1—Permitted variation under specified thickness, 0.3 mm.
NOTE 2—Thickness to be measured at 10 to 20 mm from the longitudinal edge.
NOTE 3—For specified thicknesses not listed in this table, the permitted variations in thickness shall be as given for the next higher value of specified thickness that is listed in this table.
NOTE 4—For thickness measured at any location other than that specified in Note 2, the permitted variations over specified thickness shall be 1¾ times the amounts in this table, rounded to the nearest 0.1 mm.
NOTE 5—Where "..." appears in this table, there is no requirement.

Specified Thickness, mm	Permitted Variations Over Specified Thickness for Widths Given in Millimetres, mm										
	1200 and Under	Over 1200 to 1500, excl	1500 to 1800, excl	1800 to 2100, excl	2100 to 2400, excl	2400 to 2700, excl	2700 to 3000, excl	3000 to 3300, excl	3300 to 3600, excl	3600 to 4200, excl	4200 and Over
5.0	0.8	0.8	0.8	0.8	0.8	0.8	0.8	0.9	1.0	...	...
5.5	0.8	0.8	0.8	0.8	0.8	0.8	0.8	0.9	1.0	...	...
6.0	0.8	0.8	0.8	0.8	0.8	0.8	0.9	1.0	1.1	...	...
7.0	0.8	0.8	0.8	0.8	0.8	0.8	0.9	1.0	1.2	1.4	...
8.0	0.8	0.8	0.8	0.8	0.8	0.8	0.9	1.0	1.2	1.4	...
9.0	0.8	0.8	0.8	0.8	0.8	0.8	1.0	1.0	1.3	1.5	...
10.0	0.8	0.8	0.8	0.8	0.8	0.8	1.0	1.0	1.3	1.5	1.7
11.0	0.8	0.8	0.8	0.8	0.8	0.8	1.0	1.0	1.3	1.5	1.7
12.0	0.8	0.8	0.8	0.8	0.8	0.9	1.0	1.0	1.3	1.5	1.8
14.0	0.8	0.8	0.8	0.8	0.9	0.9	1.0	1.1	1.3	1.5	1.8
16.0	0.8	0.8	0.8	0.8	0.9	0.9	1.0	1.1	1.3	1.5	1.8
18.0	0.8	0.8	0.8	0.8	0.9	1.0	1.1	1.2	1.4	1.6	2.0
20.0	0.8	0.8	0.8	0.8	0.9	1.0	1.2	1.2	1.4	1.6	2.0
22.0	0.8	0.9	0.9	0.9	1.0	1.1	1.3	1.3	1.5	1.8	2.0
25.0	0.9	0.9	1.0	1.0	1.0	1.2	1.3	1.5	1.5	1.8	2.2
28.0	1.0	1.0	1.1	1.1	1.1	1.3	1.4	1.8	1.8	2.0	2.2
30.0	1.1	1.1	1.2	1.2	1.2	1.4	1.5	1.8	1.8	2.1	2.4
32.0	1.2	1.2	1.3	1.3	1.3	1.5	1.6	2.0	2.0	2.3	2.6
35.0	1.3	1.3	1.4	1.4	1.4	1.6	1.7	2.3	2.3	2.5	2.8
38.0	1.4	1.4	1.5	1.5	1.5	1.7	1.8	2.3	2.3	2.7	3.0
40.0	1.5	1.5	1.6	1.6	1.6	1.8	2.0	2.5	2.5	2.8	3.3
45.0	1.6	1.6	1.7	1.8	1.8	2.0	2.3	2.8	2.8	3.0	3.5
50.0	1.8	1.8	1.8	2.0	2.0	2.3	2.5	3.0	3.0	3.3	3.8
55.0	2.0	2.0	2.0	2.2	2.2	2.5	2.8	3.3	3.3	3.5	3.8
60.0	2.3	2.3	2.3	2.4	2.4	2.8	3.0	3.4	3.4	3.8	4.0
70.0	2.5	2.5	2.5	2.6	2.6	3.0	3.3	3.5	3.6	4.0	4.0
80.0	2.8	2.8	2.8	2.8	2.8	3.3	3.5	3.5	3.6	4.0	4.0
90.0	3.0	3.0	3.0	3.0	3.0	3.5	3.5	3.5	3.6	4.0	4.4
100.0	3.3	3.3	3.3	3.3	3.5	3.8	3.8	3.8	3.8	4.4	4.4
110.0	3.5	3.5	3.5	3.5	3.5	3.8	3.8	3.8	3.8	4.4	4.4
120.0	3.8	3.8	3.8	3.8	3.8	3.8	3.8	3.8	3.8	4.8	4.8
130.0	4.0	4.0	4.0	4.0	4.0	4.0	4.0	4.0	4.0	5.2	5.2
140.0	4.3	4.3	4.3	4.3	4.3	4.3	4.3	4.3	4.3	5.6	5.6
150.0	4.5	4.5	4.5	4.5	4.5	4.5	4.5	4.5	4.5	5.6	5.6
160.0	4.8	4.8	4.8	4.8	4.8	4.8	4.8	4.8	4.8	5.6	5.6
180.0	5.4	5.4	5.4	5.4	5.4	5.4	5.4	5.4	5.4	6.3	6.3
200.0	5.8	5.8	6.0	6.0	6.0	6.0	6.0	6.0	6.0	7.0	7.0
250.0	7.5	7.5	7.5	7.5	7.5	7.5	7.5	7.5	7.5	7.5	8.8
300.0	7.5	7.5	9.0	9.0	9.0	9.0	9.0	9.0	9.0	9.0	9.0

TABLE A1.2 Permitted Variations in Mass for Rectangular Sheared Plates and Universal Mill Plates 2983 kg/m² and Under When Ordered to Mass

NOTE 1—Permitted variations in excess mass for lots of circular and sketch plates shall be 1¼ times the amounts in this table.
NOTE 2—Permitted variations in excess mass for single plates shall be 1⅓ times the amounts in this table.
NOTE 3—Permitted variations in excess mass for single circular and sketch plates shall be 1⅔ times the amounts in this table.
NOTE 4—The adopted standard density for rolled steel is 7850 kg/m³.
NOTE 5—Where " ..." appears in this table, there is no requirement.

Specified Mass, kg/m²	Permitted Variations in Average Mass of Lots[A] for Widths Given in Millimetres, Expressed in Percentage of the Specified Masses per Square Metre																			
	1200 and Under		Over 1200 to 1500, excl		1500 to 1800, excl		1800 to 2100, excl		2100 to 2400, excl		2400 to 2700, excl		2700 to 3000, excl		3000 to 3300, excl		3300 to 3600, excl		3600 to 4200, excl		4200 and Over	
	Over	Under	Over	Under	Over	Under	Over	Under	Over	Under	Over	Under	Over	Under	Over	Under	Over	Under	Over	Under	Over	Under
To 51.02, excl	4.0	3.0	4.5	3.0	5.0	3.0	5.5	3.0	6.0	3.0	7.5	3.0	9.0	3.0	...	...	...	...	...	...	...	...
51.02 to 62.80, excl	4.0	3.0	4.5	3.0	5.0	3.0	5.5	3.0	6.0	3.0	6.5	3.0	7.0	3.0	8.0	3.0	9.0	3.0	...	...	...	...
62.80 to 74.58, excl	4.0	3.0	4.0	3.0	4.5	3.0	5.0	3.0	5.5	3.0	5.5	3.0	6.0	3.0	7.5	3.0	8.0	3.0	11	3.0	...	...
74.58 to 86.35, excl	3.5	3.0	3.5	3.0	4.0	3.0	4.5	3.0	5.0	3.0	5.0	3.0	5.5	3.0	6.0	3.0	7.0	3.0	9.0	3.0	10	3.0
86.35 to 102.0, excl	3.5	2.5	3.5	2.5	3.5	3.0	4.0	3.0	4.5	3.0	4.5	3.0	5.0	3.0	5.5	3.0	6.0	3.0	8.0	3.0	9.0	3.0
102.0 to 125.6, excl	3.5	2.5	3.5	2.5	3.5	3.0	3.5	3.0	4.0	3.0	4.0	3.0	4.5	3.0	5.0	3.0	5.5	3.0	7.0	3.0	8.0	3.0
125.6 to 149.2, excl	3.0	2.5	3.5	2.5	3.5	2.5	3.5	3.0	3.5	3.0	3.5	3.0	4.0	3.0	4.5	3.0	5.0	3.0	6.5	3.0	7.0	3.0
149.2 to 196.2, excl	3.0	2.0	3.0	2.0	3.0	2.0	3.0	2.0	3.5	2.0	3.5	2.5	3.5	2.5	4.0	3.0	4.5	3.0	6.0	3.0	6.5	3.0
196.2 to 392.5, excl	2.5	2.0	3.0	2.0	3.0	2.0	3.0	2.0	3.5	2.0	3.5	2.0	3.5	2.5	3.5	3.0	4.0	3.0	5.5	3.0	6.0	3.0
392.5 to 588.8, excl	2.5	2.0	3.0	2.0	3.0	2.0	3.0	2.0	3.5	2.0	3.5	2.0	3.5	2.5	3.5	3.0	3.5	3.0	4.0	3.0	4.5	3.0
588.8 to 785.0, excl	2.5	1.5	2.5	1.5	2.5	1.5	2.5	1.5	2.5	2.0	2.5	2.0	2.5	2.0	2.5	2.0	2.5	2.0	3.0	2.0	3.5	2.0
785.0 to 1178, excl	2.5	1.0	2.5	1.0	2.5	1.0	2.5	1.0	2.5	1.0	2.5	1.0	2.5	1.0	2.5	1.0	2.5	1.0	3.0	1.0	3.5	1.0
1178 to 1962, excl	2.5	1.0	2.5	1.0	2.5	1.0	2.5	1.0	2.5	1.0	2.5	1.0	2.5	1.0	2.5	1.0	2.5	1.0	2.5	1.0	3.0	1.0
1962 to 2355, excl	2.0	1.0	2.0	1.0	2.5	1.0	2.5	1.0	2.5	1.0	2.5	1.0	2.5	1.0	2.5	1.0	2.5	1.0	2.5	1.0	2.5	1.0
2355 to 2983, incl	2.0	1.0	2.0	1.0	2.0	1.0	2.0	1.0	2.5	1.0	2.5	1.0	2.5	1.0	2.5	1.0	2.5	1.0	2.5	1.0	2.5	1.0

[A] The term "lot" means all the plates of each tabular width and mass group represented in each shipment.

TABLE A1.3 Permitted Variations in Width and Length for Sheared Plates 40 mm and Under in Thickness; Length Only of Universal Mill Plates 65 mm and Under in Thickness

Specified Dimensions, mm		Permitted Variations Over Specified Width and Length[A] for Thicknesses Given in Millimetres and Equivalent Masses Given in Kilograms per Square Metre, mm							
		To 10.5, excl		10.5 to 16, excl		16 to 25, excl		25 to 50, incl[B]	
Length	Width	To 78.50, excl		78.50 to 125.6, excl		125.6 to 196.2, excl		196.2 to 392.5, excl	
		Width	Length	Width	Length	Width	Length	Width	Length
To 3000, excl	To 1500, excl	10	13	11	16	13	19	16	25
	1500 to 2100, excl	11	16	13	18	16	22	19	25
	2100 to 2700, excl	13	19	16	22	19	25	25	29
	2700 and over	16	22	19	25	22	29	29	32
3000 to 6000, excl	To 1500, excl	10	19	13	22	16	25	19	29
	1500 to 2100, excl	13	19	16	22	19	25	22	32
	2100 to 2700, excl	14	22	18	24	21	29	25	35
	2700 and over	16	25	19	29	22	32	29	35
6000 to 9000, excl	To 1500, excl	10	25	13	29	16	32	19	38
	1500 to 2100, excl	13	25	16	29	19	32	22	38
	2100 to 2700, excl	14	25	18	32	22	35	25	38
	2700 and over	18	29	22	32	25	35	32	44
9000 to 12 000, excl	To 1500, excl	11	29	13	32	16	35	19	41
	1500 to 2100, excl	13	32	16	35	19	38	22	41
	2100 to 2700, excl	14	32	19	35	22	38	25	48
	2700 and over	19	35	22	38	25	41	32	48
12 000 to 15 000, excl	To 1500, excl	11	32	13	38	16	41	19	48
	1500 to 2100, excl	13	35	16	38	19	41	22	48
	2100 to 2700, excl	16	35	19	38	22	41	25	48
	2700 and over	19	38	22	41	25	44	32	48
15 000 to 18 000, excl	To 1500, excl	13	44	16	48	19	48	22	57
	1500 to 2100, excl	16	44	19	48	22	48	25	57
	2100 to 2700, excl	16	44	19	48	22	48	29	57
	2700 and over	22	44	25	51	29	57	32	64
18 000 and over	To 1500, excl	14	51	19	54	22	57	25	70
	1500 to 2100, excl	19	51	22	54	25	57	29	70
	2100 to 2700, excl	19	51	22	54	25	57	32	70
	2700 and over	25	51	29	60	32	64	35	76

[A] Permitted variations under specified width and length, 6 mm.
[B] Permitted variations in length apply also to Universal Mill plates up to 300 mm in width for thicknesses over 50 to 65 mm, incl, except for alloy steel up to 50 mm thick.

TABLE A1.4 Permitted Variations in Width for Mill Edge Carbon and High Strength Low-Alloy Plates Produced on Strip Mills (Applies to Plates Produced from Coil and to Plates Produced from an As-Rolled Structural Product)

Specified Width, mm	Permitted Variation Over Specified Width, mm[A]
To 360, excl	11
360 to 430, excl	13
430 to 480, excl	14
480 to 530, excl	16
530 to 610, excl	17
610 to 660, excl	21
660 to 710, excl	24
710 to 890, excl	29
890 to 1270, excl	32
1270 to 1520, excl	38
1520 to 1650, excl	41
1650 to 1780, excl	44
1780 to 2030, excl	47
2030 and over	51

[A] No permitted variation under specified width.

TABLE A1.5 Permitted Variations in Rolled Width for Universal Mill Plates 380 mm and Under in Thickness

Specified Width, mm	Permitted Variations Over Specified Width[A] for Thickness Given in Millimetres or Equivalent Masses Given in Kilograms per Square Metre, mm					
	To 10, excl	10 to 16, excl	16 to 25, incl	25 to 50, incl	Over 50 to 250, incl	Over 250 to 400, incl
	To 78.50, excl	78.50 to 125.6, excl	125.6 to 196.2, excl	196.2 to 392.5, incl	Over 392.5 to 1962, incl	Over 1962 to 3140, incl
Over 200 to 500, excl	3	3	5	6	10	13
500 to 900, excl	5	6	8	10	11	14
900 and over	8	10	11	13	14	16

[A] Permitted variation under specified width, 3 mm.

TABLE A1.6 Permitted Variations in Diameter for Sheared Circular Plates 25 mm and Under in Thickness

Specified Diameters, mm	Permitted Variations Over Specified Diameter for Thicknesses Given in Millimetres, mm[A]		
	To 10, excl	10 to 16, excl	16 to 25, incl
To 800, excl	6	10	13
800 to 2100, excl	8	11	14
2100 to 2700, excl	10	13	16
2700 to 3300, excl	11	14	17
3300 and over	13	16	19

[A] No permitted variation under specified diameter.

TABLE A1.7 Permitted Variations in Diameter for Gas-Cut Circular Plates (Not Applicable to Alloy Steel)

Specified Diameters, mm	Permitted Variation Over Specified Diameter for Thicknesses Given, mm[A]					
	To 25, excl	25 to 50, excl	50 to 100, excl	100 to 150, excl	150 to 200, excl	200 to 400, incl
To 800, excl	10	10	13	13	16	19
800 to 2100, excl	10	13	13	16	19	22
2100 to 2700, excl	13	14	16	19	22	25
2700 to 3300, excl	13	14	17	22	25	29
3300 and over	16	19	22	25	29	32

[A] No permitted variations under specified diameter.

TABLE A1.8 Permitted Variations in Width and Length for Rectangular Plates When Gas Cutting is Specified or Required (Applies to Alloy Steel Specifications Only)

NOTE 1—Plates with universal rolled edges will be gas cut to length only.
NOTE 2—These permitted variations shall be taken all under or divided over and under, if so specified.

Specified Thickness, mm	Permitted Variation Over Specified Width and Length, mm
To 50, excl	19
50 to 100, excl	25
100 to 150, excl	29
150 to 200, excl	33
200 to 400, excl	38

TABLE A1.9 Permitted Variations in Width and Length for Rectangular Plates When Gas Cutting is Specified or Required (Not Applicable to Alloy Steel)

NOTE 1—Plates with universal rolled edges will be gas cut to length only.

Specified Thickness, mm	Permitted Variation Over Specified Width and Length, mm[A]
To 50, excl	13
50 to 100, excl	16
100 to 150, excl	19
150 to 200, excl	22
200 to 400, incl	25

[A] These permitted variations shall be taken all under or divided over and under, if so specified.

TABLE A1.10 Permitted Variations in Diameter for Gas-Cut Circular Plates (Applies to Alloy Steel Specifications Only)

Specified Diameter, mm	Permitted Variations Over Specified Diameter for Specified Thicknesses Given in Millimetres, mm[A]					
	To 25, excl	25 to 50, excl	50 to 100, excl	100 to 150, excl	150 to 200, excl	200 to 400, incl
To 800, excl	13	13	19	19	25	25
800 to 2100, excl	13	16	22	25	29	32
2100 to 2700, excl	16	19	25	29	32	35
2700 to 3300, incl	22	25	29	32	35	38

[A] No permitted variations under specified diameter.

TABLE A1.11 Permitted Camber[A] for Carbon Steel, High-Strength Low-Alloy Steel, and Alloy Steel Universal Mill Plates and High-Strength Low-Alloy Steel and Alloy Steel Sheared or Gas-Cut Rectangular Plates

Specified Width, mm	Permitted Camber, mm
To 750, incl	Length in millimetres/300
Over 750 to 1500	Length in millimetres/250

[A] Camber as it relates to plates is the horizontal edge curvature in the length, measured over the entire length of the plate in the flat position.

TABLE A1.12 Permitted Camber[A] for Sheared Plates and Gas-Cut Rectangular Plates, All Thicknesses (Applies to Carbon Steel Only)

Permitted camber, mm = length in millimetres/500

[A] Camber as it relates to plates is the horizontal edge curvature in the length, measured over the entire length of the plate in the flat position.

TABLE A1.13 Permitted Variations From a Flat Surface for Standard Flatness Carbon Steel Plates

NOTE 1—When the longer dimension is under 900 mm, the permitted variation from a flat surface shall not exceed 6 mm. When the longer dimension is from 900 to 1800 mm, incl, the permitted variation from a flat surface shall not exceed 75 % of the tabular amount for the specified width, but in no case less than 6 mm.

NOTE 2—These permitted variations apply to plates that have a specified minimum tensile strength of not more than 415 MPa or comparable chemical composition or hardness. The limits in this table are increased 50 % for plates that have a higher specified minimum tensile strength or comparable chemical composition or hardness.

NOTE 3—This table and these notes cover the permitted variations from a flat surface for circular and sketch plates, based upon the maximum dimensions of such plates.

NOTE 4—Where "..." appears in this table, there is no requirement.

NOTE 5—Plates must be in a horizontal position on a flat surface when flatness is measured.

Specified Thickness, mm	Specified Mass, kg/m²	Permitted Variations From a Flat Surface for Specified Widths Given in Millimetres, mm[A,B]									
		To 900, excl	900 to 1200, excl	1200 to 1500, excl	1500 to 1800, excl	1800 to 2100, excl	2100 to 2400, excl	2400 to 2700, excl	2700 to 3000, excl	3000 to 3600, excl	3600 to 4200, excl	4200 and over
To 6, excl	To 47.1, excl	14	19	24	32	35	38	41	44	48	...	...
6 to 10, excl	47.1 to 78.5, excl	13	16	19	24	29	32	35	38	41	...	...
10 to 12, excl	78.5 to 94.2, excl	13	14	16	16	19	22	25	29	32	48	54
12 to 20, excl	94.2 to 157.0, excl	11	13	14	16	16	19	25	25	29	38	51
20 to 25, excl	157.0 to 196.2, excl	11	13	14	16	16	16	19	22	25	35	44
25 to 50, excl	196.2 to 392.5, excl	10	13	13	14	14	16	16	16	18	29	38
50 to 100, excl	392.5 to 785.0, excl	8	10	11	13	13	13	13	14	16	22	29
100 to 150, excl	785.0 to 1178, excl	10	11	13	13	14	14	16	19	22	22	25
150 to 200, excl	1178 to 1570, excl	11	13	13	16	18	19	22	22	25	25	25
200 to 250, excl	1570 to 1962, excl	13	13	16	18	19	21	22	24	25	25	25
250 to 300, excl	1962 to 2355, excl	13	16	19	21	22	24	25	25	25	25	25
300 to 400, incl	2355 to 3140, incl	16	19	21	22	24	25	25	25	25	25	...

[A] *Permitted Variation From a Flat Surface Along the Length*—The longer dimension specified is considered the length, and the permitted variation from a flat surface along the length shall not exceed the tabular amount for the specified width for plates up to 4000 mm in length, or in any 4000 mm for longer plates.

[B] *Permitted Variation From a Flat Surface Across the Width*—The permitted variation from a flat surface across the width shall not exceed the tabular amount for the specified width.

TABLE A1.14 Permitted Variations From a Flat Surface for Standard Flatness High-Strength Low-Alloy Steel and Alloy Steel Plates, Hot Rolled or Thermally Treated

NOTE 1—When the longer dimension is under 900 mm, the permitted variation from a flat surface shall not exceed 10 mm. When the longer dimension is from 900 to 1800 mm, incl, the permitted variation from a flat surface shall not exceed 75 % of the tabular amount for the specified width.
NOTE 2—This table and these notes cover the permitted variations from a flat surface for circular and sketch plates, based upon the maximum dimensions of such plates.
NOTE 3—Where "..." appears in this table, there is no requirement.
NOTE 4—Plates must be in a horizontal position on a flat surface when flatness is measured.

Specified Thickness, mm	Specified Mass, kg/m²	Permitted Variations from a Flat Surface for Specified Widths Given in Millimetres, mm[A,B]										
		To 900, excl	900 to 1200, excl	1200 to 1500, excl	1500 to 1800, excl	1800 to 2100, excl	2100 to 2400, excl	2400 to 2700, excl	2700 to 3000, excl	3000 to 3600, excl	3600 to 4200, excl	4200 and over
To 6, excl	To 47.1, excl	21	29	35	48	51	57	60	67	70	...	...
6 to 10, excl	47.1 to 78.5, excl	19	24	29	35	44	48	51	57	60	...	...
10 to 12, excl	78.5 to 94.2, excl	19	22	24	24	29	33	38	41	48	70	79
12 to 20, excl	94.2 to 157.0, excl	16	19	22	22	25	29	32	35	41	57	76
20 to 25, excl	157.0 to 196.2, excl	16	19	22	22	24	25	29	33	38	51	67
25 to 50, excl	196.2 to 392.5, excl	14	16	19	21	22	24	25	25	25	41	57
50 to 100, excl	392.5 to 785.0, excl	13	14	18	19	19	19	19	22	25	32	41
100 to 150, excl	785.0 to 1178, excl	14	18	19	19	22	22	24	29	32	32	38
150 to 200, excl	1178 to 1570, excl	16	19	19	24	25	29	32	33	38	38	38
200 to 250, excl	1570 to 1962, excl	19	21	24	25	29	32	33	35	38	38	38
250 to 300, excl	1962 to 2355, excl	19	24	29	32	33	35	38	38	38	38	38
300 to 400, incl	2355 to 3140, incl	22	25	30	33	35	38	38	38	38	38	38

[A] *Permitted Variation From a Flat Surface Along the Length*—The longer dimension specified is considered the length, and the permitted variation from a flat surface along the length shall not exceed the tabular amount for the specified width in plates up to 4000 mm in length, or in any 4000 mm for longer plates.
[B] *Permitted Variation From a Flat Surface Across the Width*—The permitted variation from a flat surface across the width shall not exceed the tabular amount for the specified width.

A6/A6M – 09

TABLE A1.15 Permitted Variations in Waviness for Standard Flatness Plates

NOTE 1—Waviness denotes the maximum deviation of the surface of the plate from a plane parallel to the surface of the point of measurement and contiguous to the surface of the place at each of the two adjacent wave peaks, when the plate is resting on a flat horizontal surface, as measured in an increment of less than 4000 mm of length. The permitted variation in waviness is a function of the permitted variation from a flat surface as obtained from Table A1.13 or Table A1.14, whichever is applicable.

NOTE 2—Plates must be in a horizontal position on a flat surface when waviness is measured.

Permitted Variation from a Flat Surface (from Table Table A1.13 or Table A1.14), mm	Permitted Variations in Waviness, mm, When Number of Waves in 4000 mm is						
	1	2	3	4	5	6	7
8	8	6	5	3	3	2	2
10	10	8	5	5	3	2	2
11	11	8	6	5	3	3	2
13	13	10	8	5	5	3	2
14	14	11	8	6	5	3	2
16	16	13	10	6	5	3	2
17	17	13	10	8	5	5	2
19	19	14	11	8	6	5	2
21	21	16	11	8	6	5	2
22	22	17	13	10	6	5	2
24	24	17	13	10	8	6	5
25	25	19	14	11	8	6	5
29	29	22	16	13	10	6	5
32	32	24	17	13	10	8	6
35	35	27	19	14	11	8	6
38	38	29	22	16	13	10	6
41	41	32	24	17	13	10	8
44	44	33	25	19	14	11	8
48	48	37	27	21	14	11	8
51	51	38	29	22	16	13	10
54	54	41	30	22	17	13	10
57	57	43	32	24	17	14	10
60	60	46	33	25	19	14	11
64	64	48	37	27	21	14	11
67	67	51	38	29	21	16	11
70	70	52	40	29	22	16	13
73	73	56	41	30	24	17	13
76	76	57	43	32	24	17	14
79	79	60	44	33	25	19	14

TABLE A1.16 Permitted Variations in Cross Section for W, HP, S, M, C, and MC Shapes

NOTE 1—*A* is measured at center lines of web for S, M, W, and HP shapes; at back of web for C and MC shapes. Measurement is overall for C shapes under 75 mm. *B* is measured parallel to flange. *C* is measured parallel to web.

NOTE 2—Where "..." appears in this table, there is no requirement.

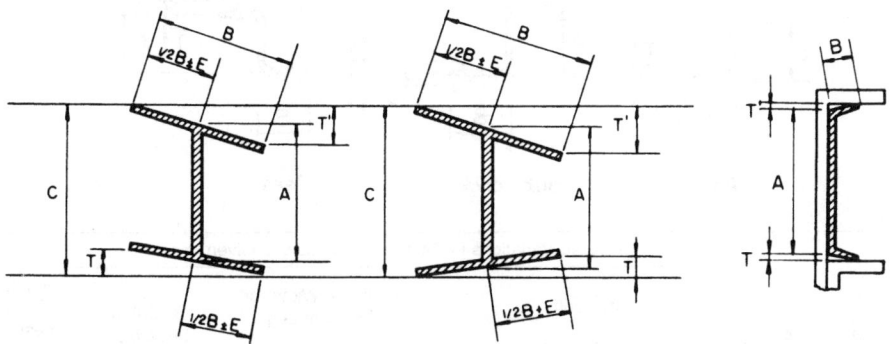

| Shape | Section Nominal Size, mm | Permitted Variations in Sectional Dimensions Given, mm ||||||| Permitted Variations Over or Under Theoretical Web Thickness for Thicknesses Given in Millimetres, mm ||
| | | A, Depth || B, Flange Width || $T + T'^A$ Flanges Out-of-Square[B] | E, Web off Center[C] | C, Maximum Depth at any Cross Section over Theoretical Depth | | |
		Over Theoretical	Under Theoretical	Over Theoretical	Under Theoretical				5 and Under	Over 5
W and HP	up to 310, incl	4	3	6	5	6	5	6	...	...
	over 310	4	3	6	5	8	5	6	...	...
S and M	75 to 180, incl	2	2	3	3	0.03	5	...	...	...
	over 180 to 360, incl	3	2	4	4	0.03	5	...	...	...
	over 360 to 610, incl	5	3	5	5	0.03	5	...	...	...
C and MC	40 and under	1	1	1	1	0.03	...	...	0.2	0.4
	over 40 to 75, excl	2	2	2	2	0.03	...	...	0.4	0.5
	75 to 180, incl	3	2	3	3	0.03	...	...	...	...
	over 180 to 360, incl	3	3	3	4	0.03	...	...	...	...
	over 360	5	4	3	5	0.03	...	...	...	...

[A] $T + T'$ applies when flanges of channels are toed in or out. For channels 16 mm and under in depth, the permitted out-of-square is 0.05 mm/mm of depth. The permitted variation shall be rounded to the nearest millimetre after calculation.
[B] Permitted variation is per millimetre of flange width for S, M, C, and MC shapes.
[C] Permitted variation of 8 mm max for sections over 634 kg/m.

TABLE A1.17 Permitted Variations in Cross Section for Angles (L Shapes), Bulb Angles, and Zees

NOTE 1—Where "..." appears in this table, there is no requirement.

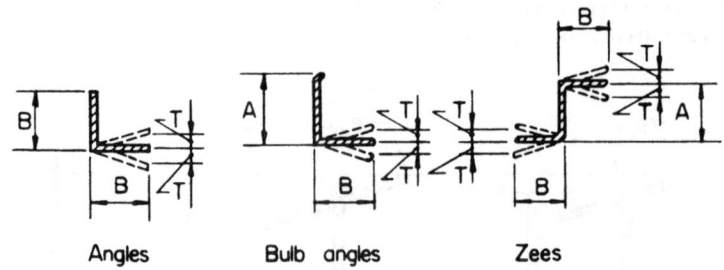

Angles   Bulb angles   Zees

Section	Nominal Size, mm	Permitted Variations in Sectional Dimensions Given, mm					Permitted Variations Over or Under Theoretical Thickness for Thicknesses Given in Millimetres, mm		
		A, Depth		B, Flange Width, or Length of Leg		T, Out-of-Square per Millimetre of B	5 and Under	Over 5 to 10	Over 10
		Over Theoretical	Under Theoretical	Over Theoretical	Under Theoretical				
Angles[A] (L shapes)	25 and under	...	...	1	1	0.026[B]	0.2	0.2	...
	over 25 to 50, incl	...	...	1	1	0.026[B]	0.2	0.2	0.3
	over 50 to 75, excl	...	...	2	2	0.026[B]	0.3	0.4	0.4
	75 to 100, incl	...	...	3	2	0.026[B]	...	...	...
	over 100 to 150 incl	...	...	3	3	0.026[B]	...	...	...
	over 150	...	...	5	3	0.026[B]	...	...	...
Bulb angles	(depth) 75 to 100, incl	3	2	4	2	0.026[B]	...	...	...
	over 100 to 150, incl	3	2	4	3	0.026[B]	...	...	...
	over 150	3	2	5	3	0.026[B]	...	...	...
Zees	75 to 100, incl	3	2	4	2	0.026[B]	...	...	...
	over 100 to 150, incl	3	2	4	3	0.026[B]	...	...	...

[A] For unequal leg angles, longer leg determines classification.
[B] 0.026 mm/mm = 1½ °. The permitted variation shall be rounded to the nearest millimetre after calculation.

TABLE A1.18 Permitted Variations in Sectional Dimensions for Rolled Tees

NOTE 1—*Back of square and center line of stem are to be parallel when measuring "out-of-square."
NOTE 2—Where "..." appears in this table, there is no requirement.

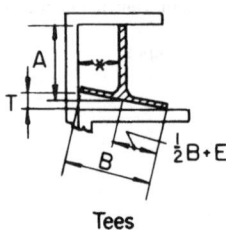

Tees

Nominal Size[A]	Permitted Variations in Sectional Dimensions Given, mm										
	A, Depth[B]		B, Width[B]		T, Out-of-Square per Millimetre of B	E, Web Off-Center, max	Stem Out-of-Square[C]	Thickness of Flange		Thickness of Stem	
	Over	Under	Over	Under				Over	Under	Over	Under
30 and under	1	1	1	1	...	...	1	0.2	0.2	0.1	0.5
Over 30 to 50, incl	2	2	2	2	...	...	2	0.3	0.3	0.2	0.5
Over 50 to 75, excl	2	2	2	2	...	...	2	0.4	0.4	0.4	0.5
75 to 125, incl	2	2	3	3	0.03	2	...	...	...	...	...
Over 125 to 180, incl	2	2	3	3	0.03	3	...	...	...	...	...

[A] The longer member of an unequal tee determines the size for permitted variations.
[B] Measurements for both depth and width are overall.
[C] Stem out-of-square is the permitted variation from its true position of the center line of stem, measured at the point.

TABLE A1.19 Permitted Variations in Length for S, M, C, MC, L, T, Z, and Bulb Angle Shapes

Note 1—Where "..." appears in this table, there is no requirement.

Nominal Size,[A] mm	Permitted Variations From Specified Length for Lengths Given in Metres, mm													
	1.5 to 3, excl		3 to 6, excl		6 to 9, incl		Over 9 to 12, incl		Over 12 to 15, incl		Over 15 to 20, incl		Over 20 m	
	Over	Under	Over	Under	Over	Under	Over	Under	Over	Under	Over	Under	Over	Under
Under 75	16	0	25	0	38	0	51	0	64	0	64	0	...	...
75 and over	25	0	38	0	45	0	57	0	70	0	70	0	...	...

[A] Greatest cross-sectional dimension.

TABLE A1.20 Permitted Variations in End Out-of-Square for S, M, C, MC, L, T, Z, and Bulb Angle Shapes

Shapes	Permitted Variation
S, M, C, and MC	0.017 mm per millimetre of depth
L[A]	0.026 mm per millimetre of leg length or 1½ °
Bulb angles	0.026 mm per millimetre of depth or 1½ °
Rolled tees[A]	0.017 mm per millimetre of flange or stem
Zees	0.026 mm per millimetre of sum of both flange lengths

[A] Permitted variations in ends out-of-square are determined on the longer members of the shape.

TABLE A1.21 Permitted Variations in Straightness for S, M, C, MC, L, T, Z, and Bulb Angle Shapes

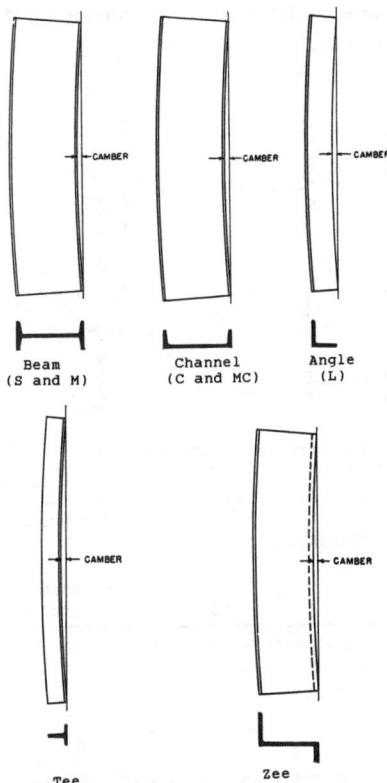

Positions for Measuring Camber of Shapes

Variable	Nominal Size,[A] mm	Permitted Variation, mm
Camber	Under 75	4 × number of metres of total length
	75 and over	2 × number of metres of total length
Sweep	All	Due to the extreme variations in flexibility of these shapes, permitted variations for sweep are subject to negotiations between the manufacturer and the purchaser for the individual sections involved.

[A] Greatest cross-sectional dimension.

TABLE A1.22 Permitted Variations in Length for W and HP Shapes

W Shapes	Permitted Variations From Specified Length for Lengths Given in Metres, mm[A,B]			
	9 and Under		Over 9	
	Over	Under	Over	Under
Beams 610 mm and under in nominal depth	10	10	10 plus 1 for each additional 1 m or fraction thereof	10
Beams over 610 mm in nominal depth and all columns	13	13	13 plus 1 for each additional 1 m or fraction thereof	13

[A] For HP and W shapes specified in the order for use as bearing piles, the permitted variations in length are plus 125 and minus 0 mm. These permitted variations in length also apply to sheet piles.

[B] The permitted variations in end out-of-square for W and HP shapes shall be 0.016 mm per millimetre of depth, or per millimetre of flange width if the flange width is larger than the depth. The permitted variations shall be rounded to the nearest millimetre after calculation.

TABLE A1.23 Permitted Variations for Length and End Out-of-Square, Milled Shapes

Nominal Depth, mm	Length,[B] m	Permitted Variations in Length and End Out-of-Square, mm[A]					
		Milled Both Ends[C]			Milled One End[C]		
		Length		End Out-of-Square	Length		End Out-of-Square-(for Milled End)
		Over	Under		Over	Under	
150 to 920	2 to 21	1	1	1	6	6	1

[A] The permitted variations in length and end out-of-square are additive.

[B] Length is measured along center line of web. Measurements are made with the steel and tape at the same temperature.

[C] End out-of-square is measured by (a) squaring from the center line of the web and (b) squaring from the center line of the flange. The measured variation from true squareness in either plane shall not exceed the total tabular amount.

TABLE A1.24 Permitted Variations in Straightness for W and HP Shapes

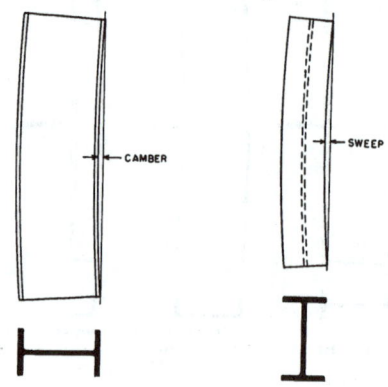

Positions for Measuring Camber and Sweep of W and HP Shapes

	Permitted Variation in Straightness, mm
Camber and sweep	1 × number of metres of total length[A]
When certain sections[B] with a flange width approximately equal to depth are specified in the order for use as columns:	
Lengths of 14 m and under	1 × number of metres of total length, but not over 10
Lengths over 14 m	10 + [1 × (number of metres of total length − 14 m)]

[A] Sections with a flange width less than 150 mm, permitted variation for sweep, mm = 2 × number of metres of total length.

[B] Applies only to:
200-mm deep sections—46.1 kg/m and heavier,
250-mm deep sections—73 kg/m and heavier,
310-mm deep sections—97 kg/m and heavier, and
360-mm deep sections—116 kg/m and heavier.
For other sections specified in the order for use as columns, the permitted variation is subject to negotiation with the manufacturer.

TABLE A1.25 Permitted Variations in Dimensions for Split Tees and Split Angles (L Shapes)[A]

Specified Depth, mm	Permitted Variation Over or Under Specified Depth,[B] mm
To 150, excl (beams and channels)	3
150 to 410, excl (beams and channels)	5
410 to 510, excl (beams and channels)	6
510 to 610, excl (beams)	8
610 and over (beams)	10

[A] The permitted variations in length for split tees or angles are the same as those applicable to the section from which the tees or angles are split.
[B] The above permitted variations in depth of tees or angles include the permitted variations in depth for the beams or channels before splitting. Permitted variations in dimensions and straightness, as set up for the beams or channels from which these tees or angles are cut, apply, except
straightness = 2 mm × length in metres

TABLE A1.26 Permitted Variations in Sectional Dimensions for Square-Edge and Round-Edge Flat Bars

NOTE 1—Where "..." appears in this table, there is no requirement.

Specified Widths, mm	Permitted Variations Over or Under Specified Thickness, for Thicknesses Given in Millimetres, mm						Permitted Variations from Specified Width, mm	
	Over 5 to 6, incl	Over 6 to 12, incl	Over 12 to 25, incl	Over 25 to 50, incl	Over 50 to 75	Over 75	Over	Under
To 25, incl	0.18	0.20	0.25	...	...	...	0.5	0.5
Over 25 to 50, incl	0.18	0.30	0.40	0.8	...	...	1.0	1.0
Over 50 to 100, incl	0.20	0.40	0.50	0.8	1.2	1.2	1.5	1.0
Over 100 to 150, incl	0.25	0.40	0.50	0.8	1.2	1.2	2.5	1.5
Over 150 to 200, incl	[A]	0.40	0.65	0.8	1.2	1.6	3.0	2.5

[A] Flats over 150 to 200 mm, incl, in width are not available as hot-rolled bars in thickness 6 mm and under.

TABLE A1.27 Permitted Variations in Sectional Dimensions for Round and Square Bars and Round-Cornered Squares

NOTE 1—Where "..." appears in this table, there is no requirement.

Specified Sizes, mm	Permitted Variation Over or Under Specified Size		Permitted Out-of-Round or Out-of-Square Section[A]	
	mm	%	mm	%
Up to 7.0, incl	0.13	...	0.20	...
Over 7.0 to 11.0, incl	0.15	...	0.22	...
Over 11.0 to 15.0, incl	0.18	...	0.27	...
Over 15.0 to 19.0, incl	0.20	...	0.30	...
Over 19.0 to 250, incl	...	1[B]	...	1½[B]

[A] Out-of-round is the difference between the maximum and minimum diameters of the bar, measured at the same transverse cross section. Out-of-square section is the difference in perpendicular distance between opposite faces, measured at the same transverse cross section.
[B] The permitted variation shall be rounded to the nearest tenth of a millimetre after calculation.

TABLE A1.28 Permitted Variations in Sectional Dimensions for Hexagons

Specified Sizes Between Opposite Sides, mm	Permitted Variations from Specified Size, mm		Out-of-Hexagon Section, mm[A]
	Over	Under	
To 13 incl	0.18	0.18	0.3
Over 13 to 25 incl	0.25	0.25	0.4
Over 25 to 40 incl	0.55	0.35	0.6
Over 40 to 50 incl	0.8	0.40	0.8
Over 50 to 65 incl	1.2	0.40	1.2
Over 65 to 80 incl	1.6	1.6	

[A] Out-of-hexagon section is the greatest difference in distance between any two opposite faces, measured at the same transverse cross section.

TABLE A1.29 Permitted Variations in Straightness for Bars

Maximum Permitted Variation in Straightness, mm[A]
6 mm in any 1500 mm and (length in millimetres/250)[B]

[A] Permitted variations in straightness do not apply to hot-rolled bars if any subsequent heating operation has been performed.
[B] Round to the nearest whole millimetre.

TABLE A1.30 Permitted Variations in Length for Hot-Cut Steel Bars[A]

NOTE 1—Where "..." appears in this table, there is no requirement.

Specified Sizes of Rounds, Squares, and Hexagons, mm	Specified Sizes of Flats, mm		Permitted Variations Over Specified Lengths Given in Metres, mm (No Variation Under)				
	Thickness	Width	1.5 to 3, excl	3 to 6, excl	6 to 9, excl	9 to 12, excl	12 to 18, incl
To 25, incl	to 25, incl	to 75, incl	15	20	35	45	60
Over 25 to 50, incl	over 25	to 75, incl	15	25	40	50	65
	to 25, incl	over 75 to 150, incl	15	25	40	50	65
Over 50 to 125, incl	over 25	over 75 to 150, incl	25	40	45	60	70
Over 125 to 250, incl	...	...	50	65	70	75	85
	over 6 to 25, incl	over 150 to 200, incl	20	30	45	90	100
	over 25 to 75, incl	over 150 to 200, incl	30	45	50	90	100
Bar size sections	...	...	15	25	40	50	65
Hot Sawing							
50 to 125, incl	25 and over	75 and over	[B]	40	45	60	70
Over 125 to 250, incl	...	...	[B]	65	70	75	85

[A] For flats over 150 to 200 mm, incl, in width and over 75 mm in thickness, consult the manufacturer for permitted variations in length.
[B] Smaller sizes and shorter lengths are not commonly hot sawed.

TABLE A1.31 Permitted Variations in Length for Bars Recut Both Ends After Straightening[A,B]

Sizes of Rounds, Squares, Hexagons, Widths of Flats and Maximum Dimensions of Other Sections, mm	Permitted Variations Over Specified Length Given in Metres, mm (No Variation Under)	
	to 3.7, incl	over 3.7
To 75, incl	6	8
Over 75 to 150, incl	8	11
Over 150 to 200, incl	11	14
Rounds over 200 to 250, incl	14	18

[A] For flats over 150 to 200 mm, incl, in width, and over 75 mm in thickness, consult the manufacturer or the processor for permitted variations in length.
[B] Permitted variations are sometimes required all over or all under the specified length, in which case the sum of the two permitted variations applies.

A2. DIMENSIONS OF STANDARD SHAPE PROFILES

A2.1 Listed herein are dimensions and weight [mass] of some standard shape profiles. The values stated in inch-pound units are independent of the values stated in SI units, and the values from the two systems are not to be combined in any way. Unless the order specifies the applicable "M" specification designation (SI units), the material shall be furnished to inch-pound units.

TABLE A2.1 "W" Shapes

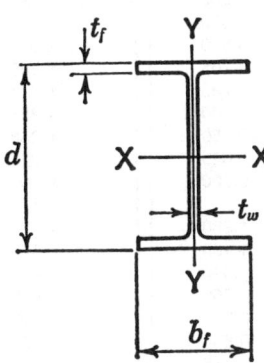

Designation (Nominal Depth in Inches and Weight in Pounds per Linear Foot)	Area A, in.²	Depth d, in.	Flange		Web Thickness t_w, in.[A]	Designation [Nominal Depth in Millimetres and Mass in Kilograms per Metre]	Area A, mm²	Depth d, mm	Flange		Web Thickness t_w mm[A]
			Width b_f, in.	Thickness t_f, in.[A]					Width b_f, mm	Thickness, t_f, mm[A]	
W44 X 335	98.7	44.02	15.945	1.770	1.025	W1100 X 499	63 500	1 118	405	45.0	26.0
X 290	85.8	43.62	15.825	1.575	0.865	X 433	55 100	1 108	402	40.0	22.0
X 262	77.2	43.31	15.750	1.415	0.785	X 390	49 700	1 100	400	36.0	20.0
X 230	67.9	42.91	15.750	1.220	0.710	X 343	43 600	1 090	400	31.0	18.0
W40 X 593	174.4	42.99	16.690	3.230	1.790	W1000 X 883	112 500	1 092	424	82.0	45.5
X 503	147.8	42.05	16.415	2.755	1.535	X 748	95 300	1 068	417	70.0	39.0
X 431	126.7	41.26	16.220	2.360	1.340	X 642	81 800	1 048	412	60.0	34.0
X 397	117.0	40.95	16.120	2.200	1.220	X 591	75 300	1 040	409	55.9	31.0
X 372	109.4	40.63	16.065	2.045	1.160	X 554	70 600	1 032	408	52.0	29.5
X 362	107.0	40.55	16.020	2.010	1.120	X 539	68 700	1 030	407	51.1	28.4
X 324	95.3	40.16	15.910	1.810	1.000	X 483	61 500	1 020	404	46.0	25.4
X 297	87.4	39.84	15.825	1.650	0.930	X 443	56 400	1 012	402	41.9	23.6
X 277	81.3	39.69	15.830	1.575	0.830	X 412	52 500	1 008	402	40.0	21.1
X 249	73.3	39.38	15.750	1.420	0.750	X 371	47 300	1 000	400	36.1	19.0
X 215	63.3	38.98	15.750	1.220	0.650	X 321	40 800	990	400	31.0	16.5
X 199	58.4	38.67	15.750	1.065	0.650	X 296	37 700	982	400	27.1	16.5
W40 X 392	115.3	41.57	12.360	2.520	1.415	W1000 X 584	74 400	1 056	314	64.0	36.0
X 331	97.5	40.79	12.165	2.125	1.220	X 494	62 900	1 036	309	54.0	31.0
X 327	95.9	40.79	12.130	2.130	1.180	X 486	61 900	1 036	308	54.1	30.0
X 294	86.2	40.39	12.010	1.930	1.060	X 438	55 600	1 026	305	49.0	26.9
X 278	81.9	40.16	11.970	1.810	1.025	X 415	52 800	1 020	304	46.0	26.0
X 264	77.6	40.00	11.930	1.730	0.960	X 393	50 100	1 016	303	43.9	24.4
X 235	68.9	39.69	11.890	1.575	0.830	X 350	44 600	1 008	302	40.0	21.1
X 211	62.0	39.37	11.810	1.415	0.750	X 314	40 000	1 000	300	35.9	19.1
X 183	53.7	38.98	11.810	1.200	0.650	X 272	34 600	990	300	31.0	16.5
X 167	49.1	38.59	11.810	1.025	0.650	X 249	31 700	980	300	26.0	16.5
X 149	43.8	38.20	11.810	0.830	0.630	X 222	28 200	970	300	21.1	16.0
W36 X 652	191.7	41.05	17.575	3.540	1.970	W920 X 970	123 700	1 043	446	89.9	50.0
X 529	155.6	39.79	17.220	2.910	1.610	X 787	100 400	1 011	437	73.9	40.9
X 487	143.2	39.33	17.105	2.680	1.500	X 725	92 400	999	434	68.1	38.1
X 441	129.7	38.85	16.965	2.440	1.360	X 656	83 700	987	431	62.0	34.5
X 395	116.2	38.37	16.830	2.200	1.220	X 588	75 000	975	427	55.9	31.0
X 361	106.1	37.99	16.730	2.010	1.120	X 537	68 500	965	425	51.1	28.4
X 330	97.0	37.67	16.630	1.850	1.020	X 491	62 600	957	422	47.0	25.9
X 302	88.8	37.33	16.655	1.680	0.945	X 449	57 600	948	423	42.7	24.0
X 282	82.9	37.11	16.595	1.570	0.885	X 420	53 500	943	422	39.9	22.5

TABLE A2.1 *Continued*

Designation (Nominal Depth in Inches and Weight in Pounds per Linear Foot)	Area A, in.²	Depth d, in.	Flange Width b_f, in.	Flange Thickness t_f, in.[A]	Web Thickness t_w, in.[A]	Designation [Nominal Depth in Millimetres and Mass in Kilograms per Metre]	Area A, mm²	Depth d, mm	Flange Width b_f, mm	Flange Thickness t_f, mm[A]	Web Thickness t_w, mm[A]
X 262	77.0	36.85	16.550	1.440	0.840	X 390	49 700	936	420	36.6	21.3
X 247	72.5	36.67	16.510	1.350	0.800	X 368	46 800	931	419	34.3	20.3
X 231	68.0	36.49	16.470	1.260	0.760	X 344	43 900	927	418	32.0	19.3
W36 X 256	75.4	37.43	12.215	1.730	0.960	W920 X 381	48 600	951	310	43.9	24.4
X 232	68.1	37.12	12.120	1.570	0.870	X 345	44 000	943	308	39.9	22.1
X 210	61.8	36.69	12.180	1.360	0.830	X 313	39 900	932	309	34.5	21.1
X 194	57.0	36.49	12.115	1.260	0.765	X 289	36 800	927	308	32.0	19.4
X 182	53.6	36.33	12.075	1.180	0.725	X 271	34 600	923	307	30.0	18.4
X 170	50.0	36.17	12.030	1.100	0.680	X 253	32 300	919	306	27.9	17.3
X 160	47.0	36.01	12.000	1.020	0.650	X 238	30 300	915	305	25.9	16.5
X 150	44.2	35.85	11.975	0.940	0.625	X 223	28 500	911	304	23.9	15.9
X 135	39.7	35.55	11.950	0.790	0.600	X 201	25 600	903	304	20.1	15.2
W33 X 387	114.0	35.95	16.200	2.280	1.260	W840 X 576	73 500	913	411	57.9	32.0
X 354	104.1	35.55	16.100	2.090	1.160	X 527	67 200	903	409	53.1	29.5
X 318	93.5	35.16	15.985	1.890	1.040	X 473	60 300	893	406	48.0	26.4
X 291	85.6	34.84	15.905	1.730	0.960	X 433	55 200	885	404	43.9	24.4
X 263	77.4	34.53	15.805	1.570	0.870	X 392	49 900	877	401	39.9	22.1
X 241	70.9	34.18	15.860	1.400	0.830	X 359	45 700	868	403	35.6	21.1
X 221	65.0	33.93	15.805	1.275	0.775	X 329	41 900	862	401	32.4	19.7
X 201	59.1	33.68	15.745	1.150	0.715	X 299	38 100	855	400	29.2	18.2
W33 X 169	49.5	33.82	11.500	1.220	0.670	W840 X 251	31 900	859	292	31.0	17.0
X 152	44.7	33.49	11.565	1.055	0.635	X 226	28 800	851	294	26.8	16.1
X 141	41.6	33.30	11.535	0.960	0.605	X 210	26 800	846	293	24.4	15.4
X 130	38.3	33.09	11.510	0.855	0.580	X 193	24 700	840	292	21.7	14.7
X 118	34.7	32.86	11.480	0.740	0.550	X 176	22 400	835	292	18.8	14.0
W30 X 391	115.0	33.19	15.590	2.440	1.360	W760 X 582	74 200	843	396	62.0	34.5
X 357	104.8	32.80	15.470	2.240	1.240	X 531	67 600	833	393	56.9	31.5
X 326	95.7	32.40	15.370	2.050	1.140	X 484	61 700	823	390	52.1	29.0
X 292	85.7	32.01	15.255	1.850	1.020	X 434	55 300	813	387	47.0	25.9
X 261	76.7	31.61	15.155	1.650	0.930	X 389	49 500	803	385	41.9	23.6
X 235	69.0	31.30	15.055	1.500	0.830	X 350	44 500	795	382	38.1	21.1
X 211	62.0	30.94	15.105	1.315	0.775	X 314	40 000	786	384	33.4	19.7
X 191	56.1	30.68	15.040	1.185	0.710	X 284	36 200	779	382	30.1	18.0
X 173	50.8	30.44	14.985	1.065	0.655	X 257	32 800	773	381	27.1	16.6
W30 X 148	43.5	30.67	10.480	1.180	0.650	W760 X 220	28 100	779	266	30.0	16.5
X 132	38.9	30.31	10.545	1.000	0.615	X 196	25 100	770	268	25.4	15.6
X 124	36.5	30.17	10.515	0.930	0.585	X 185	23 500	766	267	23.6	14.9
X 116	34.2	30.01	10.495	0.850	0.565	X 173	22 100	762	267	21.6	14.4
X 108	31.7	29.83	10.475	0.760	0.545	X 161	20 500	758	266	19.3	13.8
X 99	29.1	29.65	10.450	0.670	0.520	X 147	18 800	753	265	17.0	13.2
X 90	26.4	29.53	10.400	0.610	0.470	X 134	17 000	750	264	15.5	11.9
W27 X 539	158.4	32.52	15.255	3.540	1.970	W690 X 802	102 200	826	387	89.9	50.0
X 368	108.1	30.39	14.665	2.480	1.380	X 548	69 800	772	372	63.0	35.1
X 336	98.7	30.0	14.550	2.280	1.260	X 500	63 700	762	369	57.9	32.0
X 307	90.2	29.61	14.445	2.090	1.160	X 457	58 200	752	367	53.1	29.5
X 281	82.6	29.29	14.350	1.930	1.060	X 419	53 300	744	364	49.0	26.9
X 258	75.7	28.98	14.270	1.770	0.980	X 384	48 900	736	362	45.0	24.9
X 235	69.1	28.66	14.190	1.610	0.910	X 350	44 600	728	360	40.9	23.1
X 217	63.8	28.43	14.115	1.500	0.830	X 323	41 100	722	359	38.1	21.1
X 194	57.0	28.11	14.035	1.340	0.750	X 289	36 800	714	356	34.0	19.0
X 178	52.3	27.81	14.085	1.190	0.725	X 265	33 700	706	358	30.2	18.4
X 161	47.4	27.59	14.020	1.080	0.660	X 240	30 600	701	356	27.4	16.8
X 146	42.9	27.38	13.965	0.975	0.605	X 217	27 700	695	355	24.8	15.4
W27 X 129	37.8	27.63	10.010	1.100	0.610	W690 X 192	24 400	702	254	27.9	15.5
X 114	33.5	27.29	10.070	0.930	0.570	X 170	21 600	693	256	23.6	14.5
X 102	30.0	27.09	10.015	0.830	0.515	X 152	19 400	688	254	21.1	13.1
X 94	27.7	26.92	9.990	0.745	0.490	X 140	17 900	684	254	18.9	12.4
X 84	24.8	26.71	9.960	0.640	0.460	X 125	16 000	678	253	16.3	11.7
W24 X 370	108.0	27.99	13.660	2.720	1.520	W610 X 551	70 200	711	347	69.1	38.6
X 335	98.4	27.52	13.520	2.480	1.380	X 498	63 500	699	343	63.0	35.1
X 306	89.8	27.13	13.405	2.280	1.260	X 455	57 900	689	340	57.9	32.0

44

TABLE A2.1 Continued

Designation (Nominal Depth in Inches and Weight in Pounds per Linear Foot)	Area A, in.²	Depth d, in.	Flange Width b_f, in.	Flange Thickness t_f, in.[A]	Web Thickness t_w, in.[A]	Designation [Nominal Depth in Millimetres and Mass in Kilograms per Metre]	Area A, mm²	Depth d, mm	Flange Width b_f, mm	Flange Thickness, t_f, mm[A]	Web Thickness t_w, mm[A]
X 279	82.0	26.73	13.305	2.090	1.160	X 415	52 900	679	338	53.1	29.5
X 250	73.5	26.34	13.185	1.890	1.040	X 372	47 400	669	335	48.0	26.4
X 229	67.2	26.02	13.110	1.730	0.960	X 341	43 400	661	333	43.9	24.4
X 207	60.7	25.71	13.010	1.570	0.870	X 307	39 100	653	330	39.9	22.1
X 192	56.3	25.47	12.950	1.460	0.810	X 285	36 100	647	329	37.1	20.6
X 176	51.7	25.24	12.890	1.340	0.750	X 262	33 300	641	327	34.0	19.0
X 162	47.7	25.00	12.955	1.220	0.705	X 241	30 800	635	329	31.0	17.9
X 146	43.0	24.74	12.900	1.090	0.650	X 217	27 700	628	328	27.7	16.5
X 131	38.5	24.48	12.855	0.960	0.605	X 195	24 800	622	327	24.4	15.4
X 117	34.4	24.26	12.800	0.850	0.550	X 174	22 200	616	325	21.6	14.0
X 104	30.6	24.06	12.750	0.750	0.500	X 155	19 700	611	324	19.0	12.7
W24 X 103	30.3	24.53	9.000	0.980	0.550	W610 X 153	19 600	623	229	24.9	14.0
X 94	27.7	24.31	9.065	0.875	0.515	X 140	17 900	617	230	22.2	13.1
X 84	24.7	24.10	9.020	0.770	0.470	X 125	15 900	612	229	19.6	11.9
X 76	22.4	23.92	8.990	0.680	0.440	X 113	14 500	608	228	17.3	11.2
X 68	20.1	23.73	8.965	0.585	0.415	X 101	13 000	603	228	14.9	10.5
W24 X 62	18.2	23.74	7.040	0.590	0.430	W610 X 92	11 700	603	179	15.0	10.9
X 55	16.2	23.57	7.005	0.505	0.395	X 82	10 500	599	178	12.8	10.0
W21 X 201	59.2	23.03	12.575	1.630	0.910	W530 X 300	38 200	585	319	41.4	23.1
X 182	53.7	22.72	12.500	1.480	0.830	X 272	34 600	577	317	37.6	21.1
X 166	48.9	22.48	12.420	1.360	0.750	X 248	31 500	571	315	34.5	19.0
X 147	43.2	22.06	12.510	1.150	0.720	X 219	27 900	560	318	29.2	18.3
X 132	38.8	21.83	12.440	1.035	0.650	X 196	25 000	554	316	26.3	16.5
X 122	35.9	21.68	12.390	0.960	0.600	X 182	23 200	551	315	24.4	15.2
X 111	32.7	21.51	12.340	0.875	0.550	X 165	21 100	546	313	22.2	14.0
X 101	29.8	21.36	12.290	0.800	0.500	X 150	19 200	543	312	20.3	12.7
W21 X 93	27.3	21.62	8.420	0.930	0.580	W530 X 138	17 600	549	214	23.6	14.7
X 83	24.3	21.43	8.355	0.835	0.515	X 123	15 700	544	212	21.2	13.1
X 73	21.5	21.24	8.295	0.740	0.455	X 109	13 900	539	211	18.8	11.6
X 68	20.0	21.13	8.270	0.685	0.430	X 101	12 900	537	210	17.4	10.9
X 62	18.3	20.99	8.240	0.615	0.400	X 92	11 800	533	209	15.6	10.2
X 55	16.2	20.80	8.220	0.522	0.375	X 82	10 500	528	209	13.3	9.50
X 48	14.1	20.62	8.140	0.430	0.350	X 72	9 180	524	207	10.9	9.00
W21 X 57	16.7	21.06	6.555	0.650	0.405	W530 X 85	10 800	535	166	16.5	10.3
X 50	14.7	20.83	6.530	0.535	0.380	X 74	9 480	529	166	13.6	9.7
X 44	13.0	20.66	6.500	0.450	0.350	X 66	8 390	525	165	11.4	8.9
W18 X 311	91.5	22.32	12.005	2.740	1.520	W460 X 464	59 100	567	305	69.6	38.6
X 283	83.2	21.85	11.890	2.500	1.400	X 421	53 700	555	302	63.5	35.6
X 258	75.9	21.46	11.770	2.300	1.280	X 384	49 000	545	299	58.4	32.5
X 234	68.8	21.06	11.650	2.110	1.160	X 349	44 400	535	296	53.6	29.5
X 211	62.1	20.67	11.555	1.910	1.060	X 315	40 100	525	293	48.5	26.9
X 192	56.4	20.35	11.455	1.750	0.960	X 286	36 400	517	291	44.4	24.4
X 175	51.3	20.04	11.375	1.590	0.890	X 260	33 100	509	289	40.4	22.6
X 158	46.3	19.72	11.300	1.440	0.810	X 235	29 900	501	287	36.6	20.6
X 143	42.1	19.49	11.220	1.320	0.730	X 213	27 100	495	285	33.5	18.5
X 130	38.2	19.25	11.160	1.200	0.670	X 193	24 700	489	283	30.5	17.0
X 119	35.1	18.97	11.265	1.060	0.655	X 177	22 600	482	286	26.9	16.6
X 106	31.1	18.73	11.200	0.940	0.590	X 158	20 100	476	284	23.9	15.0
X 97	28.5	18.59	11.145	0.870	0.535	X 144	18 400	472	283	22.1	13.6
X 86	25.3	18.39	11.090	0.770	0.480	X 128	16 300	467	282	19.6	12.2
X 76	22.3	18.21	11.035	0.680	0.425	X 113	14 400	463	280	17.3	10.8
W18 X 71	20.8	18.47	7.635	0.810	0.495	W460 X 106	13 400	469	194	20.6	12.6
X 65	19.1	18.35	7.590	0.750	0.450	X 97	12 300	466	193	19.0	11.4
X 60	17.6	18.24	7.555	0.695	0.415	X 89	11 400	463	192	17.7	10.5
X 55	16.2	18.11	7.530	0.630	0.390	X 82	10 500	460	191	16.0	9.9
X 50	14.7	17.99	7.495	0.570	0.355	X 74	9 480	457	190	14.5	9.0
W18 X 46	13.5	18.06	6.060	0.605	0.360	W460 X 68	8 710	459	154	15.4	9.1
X 40	11.8	17.90	6.015	0.525	0.315	X 60	7 610	455	153	13.3	8.0
X 35	10.3	17.70	6.000	0.425	0.300	X 52	6 650	450	152	10.8	7.6
W16 X 100	29.4	16.97	10.425	0.985	0.585	W410 X 149	19 000	431	265	25.0	14.9

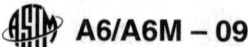

TABLE A2.1 Continued

Designation (Nominal Depth in Inches and Weight in Pounds per Linear Foot)	Area A, in.²	Depth d, in.	Flange		Web Thickness t_w, in.[A]	Designation [Nominal Depth in Millimetres and Mass in Kilograms per Metre]	Area A, mm²	Depth d, mm	Flange		Web Thickness t_w mm[A]
			Width b_f, in.	Thickness t_f, in.[A]					Width b_f, mm	Thickness, t_f, mm[A]	
X 89	26.2	16.75	10.365	0.875	0.525	X 132	16 900	425	263	22.2	13.3
X 77	22.6	16.52	10.295	0.760	0.455	X 114	14 600	420	261	19.3	11.6
X 67	19.7	16.33	10.235	0.665	0.395	X 100	12 700	415	260	16.9	10.0
W16 X 57	16.8	16.43	7.120	0.715	0.430	W410 X 85	10 800	417	181	18.2	10.9
X 50	14.7	16.26	7.070	0.630	0.380	X 75	9 480	413	180	16.0	9.7
X 45	13.3	16.13	7.035	0.565	0.345	X 67	8 580	410	179	14.4	8.8
X 40	11.8	16.01	6.995	0.505	0.305	X 60	7 610	407	178	12.8	7.7
X 36	10.6	15.86	6.985	0.430	0.295	X 53	6 840	403	177	10.9	7.5
W16 X 31	9.12	15.88	5.525	0.440	0.275	W410 X 46.1	5 880	403	140	11.2	7.0
X 26	7.68	15.69	5.500	0.345	0.250	X 38.8	4 950	399	140	8.8	6.4
W14X 730	215.0	22.42	17.890	4.910	3.070	W360 X 1086	139 000	569	454	125.0	78.0
X 665	196.0	21.64	17.650	4.520	2.830	X 990	126 000	550	448	115.0	71.9
X 605	178.0	20.92	17.415	4.160	2.595	X 900	115 000	531	442	106.0	65.9
X 550	162.0	20.24	17.200	3.820	2.380	X 818	105 000	514	437	97.0	60.5
X 500	147.0	19.60	17.010	3.500	2.190	X 744	94 800	498	432	88.9	55.6
X 455	134.0	19.02	16.835	3.210	2.015	X 677	86 500	483	428	81.5	51.2
X 426	125.0	18.67	16.695	3.035	1.875	X 634	80 600	474	424	77.1	47.6
X 398	117.0	18.29	16.590	2.845	1.770	X 592	75 500	465	421	72.3	45.0
X 370	109.0	17.92	16.475	2.660	1.655	X 551	70 300	455	418	67.6	42.0
X 342	101.0	17.54	16.360	2.470	1.540	X 509	65 200	446	416	62.7	39.1
X 311	91.4	17.12	16.230	2.260	1.410	X 463	59 000	435	412	57.4	35.8
X 283	83.3	16.74	16.110	2.070	1.290	X 421	53 700	425	409	52.6	32.8
X 257	75.6	16.38	15.995	1.890	1.175	X 382	48 800	416	406	48.0	29.8
X 233	68.5	16.04	15.890	1.720	1.070	X 347	44 200	407	404	43.7	27.2
X 211	62.0	15.72	15.800	1.560	0.980	X 314	40 000	399	401	39.6	24.9
X 193	56.8	15.48	15.710	1.440	0.890	X 287	36 600	393	399	36.6	22.6
X 176	51.8	15.22	15.650	1.310	0.830	X 262	33 400	387	398	33.3	21.1
X 159	46.7	14.98	15.565	1.190	0.745	X 237	30 100	380	395	30.2	18.9
X 145	42.7	14.78	15.500	1.090	0.680	X 216	27 500	375	394	27.7	17.3
W14 X 132	38.8	14.66	14.725	1.030	0.645	W360 X 196	25 000	372	374	26.2	16.4
X 120	35.3	14.48	14.670	0.940	0.590	X 179	22 800	368	373	23.9	15.0
X 109	32.0	14.32	14.605	0.860	0.525	X 162	20 600	364	371	21.8	13.3
X 99	29.1	14.16	14.565	0.780	0.485	X 147	18 800	360	370	19.8	12.3
X 90	26.5	14.02	14.520	0.710	0.440	X 134	17 100	356	369	18.0	11.2
W14 X 82	24.1	14.31	10.130	0.855	0.510	W360 X 122	15 500	363	257	21.7	13.0
X 74	21.8	14.17	10.070	0.785	0.450	X 110	14 100	360	256	19.9	11.4
X 68	20.0	14.04	10.035	0.720	0.415	X 101	12 900	357	255	18.3	10.5
X 61	17.9	13.89	9.995	0.645	0.375	X 91	11 500	353	254	16.4	9.5
W14 X 53	15.6	13.92	8.060	0.660	0.370	W360 X 79	10 100	354	205	16.8	9.4
X 48	14.1	13.79	8.030	0.595	0.340	X 72	9 100	350	204	15.1	8.6
X 43	12.6	13.66	7.995	0.530	0.305	X 64	8 130	347	203	13.5	7.7
W14 X 38	11.2	14.10	6.770	0.515	0.310	W360 X 58	7 230	358	172	13.1	7.9
X 34	10.0	13.98	6.745	0.455	0.285	X 51	6 450	355	171	11.6	7.2
X 30	8.85	13.84	6.730	0.385	0.270	X 44.6	5 710	352	171	9.8	6.9
W14 X 26	7.69	13.91	5.025	0.420	0.255	W360 X 39.0	4 960	353	128	10.7	6.5
X 22	6.49	13.74	5.000	0.335	0.230	X 32.9	4 190	349	127	8.5	5.8
W12 X 336	98.8	16.82	13.385	2.955	1.775	W310 X 500	63 700	427	340	75.1	45.1
X 305	89.6	16.32	13.235	2.705	1.625	X 454	57 800	415	336	68.7	41.3
X 279	81.9	15.85	13.140	2.470	1.530	X 415	52 800	403	334	62.7	38.9
X 252	74.1	15.41	13.005	2.250	1.395	X 375	47 800	391	330	57.2	35.4
X 230	67.7	15.05	12.895	2.070	1.285	X 342	43 700	382	328	52.6	32.6
X 210	61.8	14.71	12.790	1.900	1.180	X 313	39 900	374	325	48.3	30.0
X 190	55.8	14.38	12.670	1.735	1.060	X 283	36 000	365	322	44.1	26.9
X 170	50.0	14.03	12.570	1.560	0.960	X 253	32 300	356	319	39.6	24.4
X 152	44.7	13.71	12.480	1.400	0.870	X 226	28 800	348	317	35.6	22.1
X 136	39.9	13.41	12.400	1.250	0.790	X 202	25 700	341	315	31.8	20.1
X 120	35.3	13.12	12.320	1.105	0.710	X 179	22 800	333	313	28.1	18.0
X 106	31.2	12.89	12.220	0.990	0.610	X 158	20 100	327	310	25.1	15.5
X 96	28.2	12.71	12.160	0.900	0.550	X 143	18 200	323	309	22.9	14.0
X 87	25.6	12.53	12.125	0.810	0.515	X 129	16 500	318	308	20.6	13.1

TABLE A2.1 *Continued*

Designation (Nominal Depth in Inches and Weight in Pounds per Linear Foot)	Area A, in.²	Depth d, in.	Flange Width b_f in.	Flange Thickness t_f in.[A]	Web Thickness t_w, in.[A]	Designation [Nominal Depth in Millimetres and Mass in Kilograms per Metre]	Area A, mm²	Depth d, mm	Flange Width b_f mm	Flange Thickness, t_f mm[A]	Web Thickness t_w mm[A]
X 79	23.2	12.38	12.080	0.735	0.470	X 117	15 000	314	307	18.7	11.9
X 72	21.1	12.25	12.040	0.670	0.430	X 107	13 600	311	306	17.0	10.9
X 65	19.1	12.12	12.000	0.605	0.390	X 97	12 300	308	305	15.4	9.9
W12 X 58	17.0	12.19	10.010	0.640	0.360	W310 X 86	11 000	310	254	16.3	9.1
X 53	15.6	12.06	9.995	0.575	0.345	X 79	10 100	306	254	14.6	8.8
W12 X 50	14.7	12.19	8.080	0.640	0.370	W310 X 74	9 480	310	205	16.3	9.4
X 45	13.2	12.06	8.045	0.575	0.335	X 67	8 520	306	204	14.6	8.5
X 40	11.8	11.94	8.005	0.515	0.295	X 60	7 610	303	203	13.1	7.5
W12 X 35	10.3	12.50	6.560	0.520	0.300	W310 X 52	6 650	317	167	13.2	7.6
X 30	8.79	12.34	6.520	0.440	0.260	X 44.5	5 670	313	166	11.2	6.6
X 26	7.65	12.22	6.490	0.380	0.230	X 38.7	4 940	310	165	9.7	5.8
W12 X 22	6.48	12.31	4.030	0.425	0.260	W310 X 32.7	4 180	313	102	10.8	6.6
X 19	5.57	12.16	4.005	0.350	0.235	X 28.3	3 590	309	102	8.9	6.0
X 16	4.71	11.99	3.990	0.265	0.220	X 23.8	3 040	305	101	6.7	5.6
X 14	4.16	11.91	3.970	0.225	0.200	X 21.0	2 680	303	101	5.7	5.1
W10 X 112	32.9	11.36	10.415	1.250	0.755	W250 X 167	21 200	289	265	31.8	19.2
X 100	29.4	11.10	10.340	1.120	0.680	X 149	19 000	282	263	28.4	17.3
X 88	25.9	10.84	10.265	0.990	0.605	X 131	16 700	275	261	25.1	15.4
X 77	22.6	10.60	10.190	0.870	0.530	X 115	14 600	269	259	22.1	13.5
X 68	20.0	10.40	10.130	0.770	0.470	X 101	12 900	264	257	19.6	11.9
X 60	17.6	10.22	10.080	0.680	0.420	X 89	11 400	260	256	17.3	10.7
X 54	15.8	10.09	10.030	0.615	0.370	X 80	10 200	256	255	15.6	9.4
X 49	14.4	9.98	10.000	0.560	0.340	X 73	9 290	253	254	14.2	8.6
W10 X 45	13.3	10.10	8.020	0.620	0.350	W250 X 67	8 580	257	204	15.7	8.9
X 39	11.5	9.92	7.985	0.530	0.315	X 58	7 420	252	203	13.5	8.0
X 33	9.71	9.73	7.960	0.435	0.290	X 49.1	6 260	247	202	11.0	7.4
W10 X 30	8.84	10.47	5.810	0.510	0.300	W250 X 44.8	5 700	266	148	13.0	7.6
X 26	7.61	10.33	5.770	0.440	0.260	X 38.5	4 910	262	147	11.2	6.6
X 22	6.49	10.17	5.750	0.360	0.240	X 32.7	4 190	258	146	9.1	6.1
W10 X 19	5.62	10.24	4.020	0.395	0.250	W250 X 28.4	3 630	260	102	10.0	6.4
X 17	4.99	10.11	4.010	0.330	0.240	X 25.3	3 220	257	102	8.4	6.1
X 15	4.41	9.99	4.000	0.270	0.230	X 22.3	2 850	254	102	6.9	5.8
X 12	3.54	9.87	3.960	0.210	0.190	X 17.9	2 280	251	101	5.3	4.8
W8 X 67	19.7	9.00	8.280	0.935	0.570	W200 X 100	12 700	229	210	23.7	14.5
X 58	17.1	8.75	8.220	0.810	0.510	X 86	11 000	222	209	20.6	13.0
X 48	14.1	8.50	8.110	0.685	0.400	X 71	9 100	216	206	17.4	10.2
X 40	11.7	8.25	8.070	0.560	0.360	X 59	7 550	210	205	14.2	9.1
X 35	10.3	8.12	8.020	0.495	0.310	X 52	6 650	206	204	12.6	7.9
X 31	9.13	8.00	7.995	0.435	0.285	X 46.1	5 890	203	203	11.0	7.2
W8 X 28	8.25	8.06	6.535	0.465	0.285	W200 X 41.7	5 320	205	166	11.8	7.2
X 24	7.08	7.93	6.495	0.400	0.245	X 35.9	4 570	201	165	10.2	6.2
W8 X 21	6.16	8.28	5.270	0.400	0.250	W200 X 31.3	3 970	210	134	10.2	6.4
X 18	5.26	8.14	5.250	0.330	0.230	X 26.6	3 390	207	133	8.4	5.8
W8 X 15	4.44	8.11	4.015	0.315	0.245	W200 X 22.5	2 860	206	102	8.0	6.2
X 13	3.84	7.99	4.000	0.255	0.230	X 19.3	2 480	203	102	6.5	5.8
X 10	2.96	7.89	3.940	0.205	0.170	X 15.0	1 910	200	100	5.2	4.3
W6 X 25	7.34	6.38	6.080	0.455	0.320	W150 X 37.1	4 740	162	154	11.6	8.1
X 20	5.87	6.20	6.020	0.365	0.260	X 29.8	3 790	157	153	9.3	6.6
X 15	4.43	5.99	5.990	0.260	0.230	X 22.5	2 860	152	152	6.6	5.8
W6 X 16	4.74	6.28	4.030	0.405	0.260	W150 X 24.0	3 060	160	102	10.3	6.6
X 12	3.55	6.03	4.000	0.280	0.230	X 18.0	2 290	153	102	7.1	5.8
X 9	2.68	5.90	3.940	0.215	0.170	X 13.5	1 730	150	100	5.5	4.3
X 8.5	2.52	5.83	3.940	0.195	0.170	X 13.0	1 630	148	100	4.9	4.3
W5 X 19	5.54	5.15	5.030	0.430	0.270	W130 X 28.1	3 590	131	128	10.9	6.9

TABLE A2.1 Continued

Designation (Nominal Depth in Inches and Weight in Pounds per Linear Foot)	Area A, in.²	Depth d, in.	Flange Width b_f in.	Flange Thickness t_f in.[A]	Web Thickness t_w, in.[A]	Designation [Nominal Depth in Millimetres and Mass in Kilograms per Metre]	Area A, mm²	Depth d, mm	Flange Width b_f mm	Flange Thickness, t_f mm[A]	Web Thickness t_w mm[A]
X 16	4.68	5.01	5.000	0.360	0.240	X 23.8	3 040	127	127	9.1	6.1
W4 X 13	3.83	4.16	4.060	0.345	0.280	W100 X 19.3	2 470	106	103	8.8	7.1

[A] Actual flange and web thicknesses vary due to mill rolling practices; however, permitted variations for such dimensions are not addressed.

TABLE A2.2 "S" Shapes

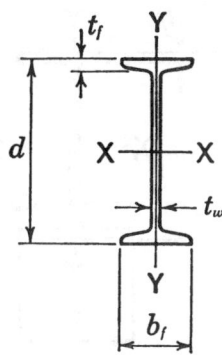

Designation (Nominal Depth in Inches and Weight in Pounds per Linear Foot)	Area A, in.²	Depth d, in.	Flange		Web Thickness t_w, in.[A]	Designation [Nominal Depth in Millimetres and Mass in Kilograms per Metre]	Area a, mm²	Depth d, mm	Flange		Web Thickness t_w mm[A]
			Width b_f, in.	Thickness t_f, in.[A]					Width b_f, mm	Thickness, t_f, mm[A]	
S 24 X 121	35.6	24.50	8.050	1.090	0.800	S 610 X 180	23 000	622	204	27.7	20.3
X 106	31.2	24.50	7.870	1.090	0.620	X 158	20 100	622	200	27.7	15.7
S 24 X 100	29.3	24.00	7.245	0.870	0.745	S 610 X 149	18 900	610	184	22.1	18.9
X 90	26.5	24.00	7.125	0.870	0.625	X 134	17 100	610	181	22.1	15.9
X 80	23.5	24.00	7.000	0.870	0.500	X 119	15 200	610	178	22.1	12.7
S 20 X 96	28.2	20.30	7.200	0.920	0.800	S 510 X 143	18 200	516	183	23.4	20.3
X 86	25.3	20.30	7.060	0.920	0.660	X 128	16 300	516	179	23.4	16.8
S 20 X 75	22.0	20.00	6.385	0.795	0.635	S 510 X 112	14 200	508	162	20.2	16.1
X 66	19.4	20.00	6.255	0.795	0.505	X 98	12 500	508	159	20.2	12.8
S 18 X 70	20.6	18.00	6.251	0.691	0.711	S 460 X 104	13 300	457	159	17.6	18.1
X 54.7	16.1	18.00	6.001	0.691	0.461	X 81.4	10 400	457	152	17.6	11.7
S 15 X 50	14.7	15.00	5.640	0.622	0.550	S 380 X 74	9 480	381	143	15.8	14.0
X 42.9	12.6	15.00	5.501	0.622	0.411	X 64	8 130	381	140	15.8	10.4
S 12 X 50	14.7	12.00	5.477	0.659	0.687	S 310 X 74	9 480	305	139	16.7	17.4
X 40.8	12.0	12.00	5.252	0.659	0.462	X 60.7	7 740	305	133	16.7	11.7
S 12 X 35	10.3	12.00	5.078	0.544	0.428	S 310 X 52	6 650	305	129	13.8	10.9
X 31.8	9.35	12.00	5.000	0.544	0.350	X 47.3	6 030	305	127	13.8	8.9
S 10 X 35	10.3	10.00	4.944	0.491	0.594	S 250 X 52	6 650	254	126	12.5	15.1
X 25.4	7.46	10.00	4.661	0.491	0.311	X 37.8	4 810	254	118	12.5	7.9
S 8 X 23	6.77	8.00	4.171	0.425	0.441	S 200 X 34	4 370	203	106	10.8	11.2
X 18.4	5.41	8.00	4.001	0.425	0.271	X 27.4	3 480	203	102	10.8	6.9
S 6 X 17.25	5.07	6.00	3.565	0.359	0.465	S 150 X 25.7	3 270	152	91	9.1	11.8
X 12.5	3.67	6.00	3.332	0.359	0.232	X 18.6	2 360	152	85	9.1	5.9
S 5 X 10	2.94	5.00	3.004	0.326	0.214	S 130 X 15	1 880	127	76	8.3	5.4
S 4 X 9.5	2.79	4.00	2.796	0.293	0.326	S 100 X 14.1	1 800	102	71	7.4	8.3
X 7.7	2.26	4.00	2.663	0.293	0.193	X 11.5	1 450	102	68	7.4	4.9
S 3 X 7.5	2.21	3.00	2.509	0.260	0.349	S 75 X 11.2	1 430	76	64	6.6	8.9
X 5.7	1.67	3.00	2.330	0.260	0.170	X 8.5	1 080	76	59	6.6	4.3

[A] Actual flange and web thicknesses vary due to mill rolling practices; however, permitted variations for such dimensions are not addressed.

TABLE A2.3 "M" Shapes

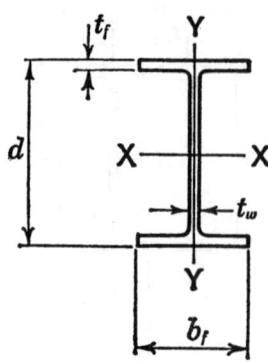

Designation (Nominal Depth in Inches and Weight in Pounds per Linear Foot)	Area A, in.²	Depth d, in.	Flange Width b_f, in.	Flange Thickness t_f, in.[A]	Web Thickness t_w, in.[A]	Designation [Nominal Depth in Millimetres and Mass in Kilograms per Metre]	Area A, mm²	Depth d, mm	Flange Width b_f, mm	Flange Thickness, t_f, mm[A]	Web Thickness t_w, mm[A]
M 12.5X12.4	3.66	12.534	3.750	0.228	0.155	M318 X18.5	2 361	318	95	5.8	3.9
X11.6	3.43	12.500	3.500	0.211	0.155	M318X17.3	2 213	317	89	5.4	3.9
M 12 X11.8	3.47	12.00	3.065	0.225	0.177	M 310 X17.6	2 240	305	78	5.7	4.5
X10.8	3.18	11.97	3.065	0.210	0.160	M 310 X16.1	2 050	304	78	5.3	4.1
X10.0	2.94	11.97	3.250	0.180	0.149	M 310 X14.9	1 900	304	83	4.6	3.8
M 10 X9.0	2.65	10.00	2.690	0.206	0.157	M 250 X13.4	1 710	254	68	4.6	3.6
X8.0	2.35	9.95	2.690	0.182	0.141	M 250 X11.9	1 520	253	68	5.2	4.0
M 10 X7.5	2.21	9.99	2.688	0.173	0.130	M 250 X11.2	1 430	253	68	4.4	3.3
M 8 X6.5	1.92	8.00	2.281	0.189	0.135	M 200 X9.7	1 240	203	57	4.8	3.4
X6.2	1.81	8.00	2.281	0.177	0.129	M 200 X9.2	1 170	203	58	4.5	3.3
M 6 X4.4	1.29	6.00	1.844	0.171	0.114	M 150 X6.6	832	152	47	4.3	2.9
X3.7	1.09	5.92	2.000	0.129	0.098	M 150 X5.5	703	150	51	3.3	2.5
M 5 X18.9	5.55	5.00	5.003	0.416	0.316	M 130 X28.1	3 580	127	127	10.6	8.0
M 4 X6.0	1.78	3.80	3.80	0.160	0.130	M 100 X8.9	1 150	97	97	4.1	3.3
X 4.08	1.20	4.00	2.250	0.170	0.115	M 100 X 6.1	775	102	57	4.3	2.9
X 3.45	1.029	4.00	2.250	0.130	0.092	M 100 X 5.1	665	102	57	3.3	2.8
X 3.2	0.94	4.00	2.250	0.130	0.092	M 100 X 4.8	610	102	57	3.3	2.3
M 3X 2.9	0.853	3.00	2.250	0.130	0.090	M 75 X 4.3	550	76	57	3.3	2.3

[A] Actual flange and web thicknesses vary due to mill rolling practices; however, permitted variations for such dimensions are not addressed.

TABLE A2.4 "HP" Shapes

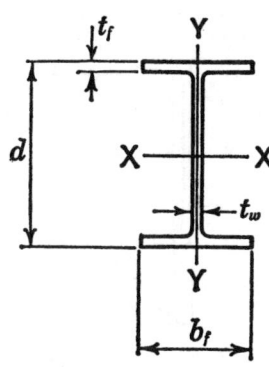

Designation (Nominal Depth in Inches and Weight in Pounds per Linear Foot)	Area A, in.²	Depth d, in.	Flange		Web Thickness tw, in.[A]	Designation [Nominal Depth in Millimetres and Mass in Kilograms per Metre]	Area A, mm²	Depth d, mm	Flange		Web Thickness t_w, mm[A]
			Width b_f, in.	Thickness t_f, in.[A]					Width b_f, mm	Thickness, t_f, mm[A]	
HP14 X 117	34.4	14.21	14.885	0.805	0.805	HP360 X 174	22 200	361	378	20.4	20.4
X 102	30.0	14.01	14.785	0.705	0.705	X 152	19 400	356	376	17.9	17.9
X 89	26.1	13.83	14.695	0.615	0.615	X 132	16 800	351	373	15.6	15.6
X 73	21.4	13.61	14.585	0.505	0.505	X 108	13 800	346	370	12.8	12.8
HP12 X 84	24.6	12.28	12.295	0.685	0.685	HP310 X 125	15 900	312	312	17.4	17.4
X 74	21.8	12.13	12.215	0.610	0.605	X 110	14 100	308	310	15.5	15.4
X 63	18.4	11.94	12.125	0.515	0.515	X 93	11 900	303	308	13.1	13.1
X 53	15.5	11.78	12.045	0.435	0.435	X 79	10 000	299	306	11.0	11.0
HP10 X 57	16.8	9.99	10.225	0.565	0.565	HP250X 85	10 800	254	260	14.4	14.4
X 42	12.4	9.70	10.075	0.420	0.415	X 62	8 000	246	256	10.7	10.5
HP8 X 36	10.6	8.02	8.155	0.445	0.445	HP200 X 53	6 840	204	207	11.3	11.3

[A] Actual flange and web thicknesses vary due to mill rolling practices; however, permitted variations for such dimensions are not addressed.

TABLE A2.5 "C" Shapes

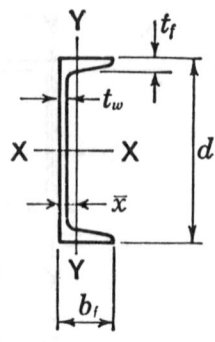

Designation (Nominal Depth in Inches and Weight in Pounds per Linear Foot)	Area A, in.^{2}	Depth d, in.	Flange		Web Thickness t_w, in.[A]	Designation [Nominal Depth in Millimetres in Mass in Kilograms per Metre]	Area A, mm^2	Depth d, mm	Flange		Web Thickness t_w, mm[A]
			Width b_f, in.	Thickness t_f, in.[A]					Width b_f, mm	Thickness t_f, mm[A]	
C15 X 50	14.7	15.00	3.716	0.650	0.716	C380 X 74	9 480	381	94	16.5	18.2
X 40	11.8	15.00	3.520	0.650	0.520	X 60	7 610	381	89	16.5	13.2
X 33.9	9.96	15.00	3.400	0.650	0.400	X 50.4	6 430	381	86	16.5	10.2
C12 X 30	8.82	12.00	3.170	0.501	0.510	C310 X 45	5 690	305	80	12.7	13.0
X 25	7.35	12.00	3.047	0.501	0.387	X 37	4 740	305	77	12.7	9.8
X 20.7	6.09	12.00	2.942	0.501	0.282	X 30.8	3 930	305	74	12.7	7.2
C10 X 30	8.82	10.00	3.033	0.436	0.673	C250 X 45	5 690	254	76	11.1	17.1
X 25	7.35	10.00	2.886	0.436	0.526	X 37	4 740	254	73	11.1	13.4
X 20	5.88	10.00	2.739	0.436	0.379	X 30	3 790	254	69	11.1	9.6
X 15.3	4.49	10.00	2.600	0.436	0.240	X 22.8	2 900	254	65	11.1	6.1
C9 X 20	5.88	9.00	2.648	0.413	0.448	C230 X 30	3 790	229	67	10.5	11.4
X 15	4.41	9.00	2.485	0.413	0.285	X 22	2 850	229	63	10.5	7.2
X 13.4	3.94	9.00	2.433	0.413	0.233	X 19.9	2 540	229	61	10.5	5.9
C8 X 18.75	5.51	8.00	2.527	0.390	0.487	C200 X 27.9	3 550	203	64	9.9	12.4
X 13.75	4.04	8.00	2.343	0.390	0.303	X 20.5	2 610	203	59	9.9	7.7
X 11.5	3.38	8.00	2.260	0.390	0.220	X 17.1	2 180	203	57	9.9	5.6
C7 X 14.75	4.33	7.00	2.299	0.366	0.419	C180 X 22	2 790	178	58	9.3	10.6
X 12.25	3.60	7.00	2.194	0.366	0.314	X 18.2	2 320	178	55	9.3	8.0
X 9.8	2.87	7.00	2.090	0.366	0.210	X 14.6	1 850	178	53	9.3	5.3
C6 X 13	3.83	6.00	2.157	0.343	0.437	C150 X 19.3	2 470	152	54	8.7	11.1
X 10.5	3.09	6.00	2.034	0.343	0.314	X 15.6	1 990	152	51	8.7	8.0
X 8.2	2.40	6.00	1.920	0.343	0.200	X 12.2	1 550	152	48	8.7	5.1
C5 X 9	2.64	5.00	1.885	0.320	0.325	C130 X 13	1 700	127	47	8.1	8.3
X 6.7	1.97	5.00	1.750	0.320	0.190	X 10.4	1 270	127	44	8.1	4.8
C4 X 7.25	2.13	4.00	1.721	0.296	0.321	C100 X 10.8	1 370	102	43	7.5	8.2
X 6.25	1.84	4.00	1.647	0.272	0.247	X 9.3	1 187	102	42	6.9	6.3
X 5.4	1.59	4.00	1.584	0.296	0.184	X 8	1 030	102	40	7.5	4.7
X 4.5	1.32	4.00	1.584	0.296	0.125	X 6.7	852	102	40	7.5	3.2
C3 X 6	1.76	3.00	1.596	0.273	0.356	C75 X 8.9	1 130	76	40	6.9	9.0
X 5	1.47	3.00	1.498	0.273	0.258	X 7.4	948	76	37	6.9	6.6
X 4.1	1.21	3.00	1.410	0.273	0.170	X 6.1	781	76	35	6.9	4.3
X 3.5	1.03	3.00	1.372	0.273	0.132	X 5.2	665	76	35	6.9	3.4

[A] Actual flange and web thicknesses vary due to mill rolling practices; however, permitted variations for such dimensions are not addressed.

TABLE A2.6 "MC" Shapes

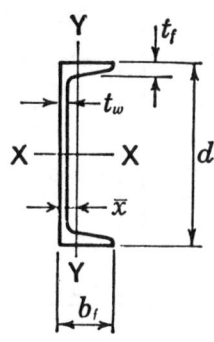

Designation (Nominal Depth in Inches and Weight in Pounds per Linear Foot)	Area A, in.²	Depth d, in.	Flange		Web Thickness tw, in.[A]	Designation [Nominal Depth in Millimetres and Mass in Kilograms per Metre]	Area A, mm²	Depth d, mm	Flange		Web Thickness t_w, mm[A]
			Width b_f, in.	Thickness t_f, in.[A]					Width b_f, mm	Thickness t_f, mm	
MC 18 X 58	17.1	18.00	4.200	0.625	0.700	MC 460 X 86	11 000	457	107	15.9	17.8
X 51.9	15.3	18.00	4.100	0.625	0.600	X 77.2	9 870	457	104	15.9	15.2
X 45.8	13.5	18.00	4.000	0.625	0.500	X 68.2	8 710	457	102	15.9	12.7
X 42.7	12.6	18.00	3.950	0.625	0.450	X 63.5	8 130	457	100	15.9	11.4
MC 13 X 50	14.7	13.00	4.412	0.610	0.787	MC 330 X 74	9 480	330	112	15.5	20.0
X 40	11.8	13.00	4.185	0.610	0.560	X 60	7 610	330	106	15.5	14.2
X 35	10.3	13.00	4.072	0.610	0.447	X 52	6 640	330	103	15.5	11.4
X 31.8	9.35	13.00	4.000	0.610	0.375	X 47.3	6 030	330	102	15.5	9.5
MC 12 X 50	14.7	12.00	4.135	0.700	0.835	MC 310 X 74	9 480	305	105	17.8	21.2
X 45	13.2	12.00	4.010	0.700	0.710	X 67	8 502	305	102	17.8	18.0
X 40	11.8	12.00	3.890	0.700	0.590	X 60	7 610	305	98	17.8	15.0
X 35	10.3	12.00	3.765	0.700	0.465	X 52	6 620	305	96	17.8	11.8
X 31	9.12	12.00	3.670	0.700	0.370	X 46	5 890	305	93	17.8	9.4
MC 12 X 14.3	4.19	12.00	2.125	0.313	0.250	MC 310 X 21.3	2 700	305	54	8.0	6.4
X 10.6	3.10	12.00	1.500	0.309	0.190	X 15.8	2 000	305	38	7.8	4.8
MC 10 X 41.1	12.1	10.00	4.321	0.575	0.796	MC 250 X 61.2	7 810	254	110	14.6	20.2
X 33.6	9.87	10.00	4.100	0.575	0.575	X 50	6 370	254	104	14.6	14.6
X 28.5	8.37	10.00	3.950	0.575	0.425	X 42.4	5 400	254	100	14.6	10.8
MC 10 X 25	7.35	10.00	3.405	0.575	0.380	MC 250 X 37	4 740	254	86	14.6	9.7
X 22	6.45	10.00	3.315	0.575	0.290	X 33	4 160	254	84	14.6	7.4
MC 10 X 8.4	2.46	10.00	1.500	0.280	0.170	MC 250 X 12.5	1 590	254	38	7.1	4.3
X 6.5	1.91	10.00	1.17	0.202	0.152	X 9.7	1 240	254	28	5.1	3.9
MC 9 X 25.4	7.47	9.00	3.500	0.550	0.450	MC 230 X 37.8	4 820	229	88	14.0	11.4
X 23.9	7.02	9.00	3.450	0.550	0.400	X 35.6	4 530	229	87	14.0	10.2
MC 8 X 22.8	6.70	8.00	3.502	0.525	0.427	MC 200 X 33.9	4 320	203	88	13.3	10.8
X 21.4	6.28	8.00	3.450	0.525	0.375	X 31.8	4 050	203	87	13.3	9.5
MC 8 X 20	5.88	8.00	3.025	0.500	0.400	MC 200 X 29.8	3 790	203	76	12.7	10.2
X 18.7	5.50	8.00	2.978	0.500	0.353	X 27.8	3 550	203	75	12.7	9.0
MC 8 X 8.5	2.50	8.00	1.874	0.311	0.179	MC 200 X 12.6	1 610	203	47	7.9	4.5
MC 7 X 22.7	6.67	7.00	3.603	0.500	0.503	MC 180 X 33.8	4 300	178	91	12.7	12.8
X 19.1	5.61	7.00	3.452	0.500	0.352	X 28.4	3 620	178	87	12.7	8.9
MC 6 X 18	5.29	6.00	3.504	0.475	0.379	MC 150 X 26.8	3 410	152	88	12.1	9.6
X 15.3	4.50	6.00	3.500	0.385	0.340	X 22.8	2 900	152	88	9.8	8.6
MC 6 X 16.3	4.79	6.00	3.000	0.475	0.375	MC 150 X 24.3	3 090	152	76	12.1	9.5
X 15.1	4.44	6.00	2.941	0.475	0.316	X 22.5	2 860	152	74	12.1	8.0
MC 6 X 12	3.53	6.00	2.497	0.375	0.310	MC 150 X 17.9	2 280	152	63	9.5	7.9

TABLE A2.6 *Continued*

Designation (Nominal Depth in Inches and Weight in Pounds per Linear Foot)	Area A, in.^{2}	Depth d, in.	Flange Width b_f, in.	Flange Thickness t_f, in.[A]	Web Thickness t_w, in.[A]	Designation [Nominal Depth in Millimetres and Mass in Kilograms per Metre]	Area A, mm^2	Depth d, mm	Flange Width b_f, mm	Flange Thickness t_f, mm	Web Thickness t_w, mm[A]
MC 6 X 7.0	2.07	6.00	1.875	0.291	0.179	MC 150 X 10.4	1 341	152	48	7.4	4.5
X6.5	1.93	6.00	1.850	0.291	0.155	X 9.7	1 250	152	47	7.4	3.9
MC 4 X 13.8	4.02	4.00	2.500	0.500	0.500	MC 100 X20.5	2 594	102	64	13	13
MC 3 x 7.1	2.09	3.00	1.938	0.351	0.312	MC 75 X 10.6	1 348	76	49	8.9	7.9

TABLE A2.7 "L" Shapes (Equal Legs)

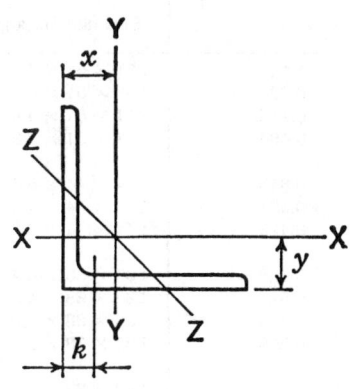

Size and Thickness, in.	Weight per Foot, lb	Area, in.²	Size and Thickness, mm	Mass per Metre, kg	Area, mm²
L8 × 8 × 1⅛	56.9	16.7	L203 × 203 × 28.6	84.7	10 800
L8 × 8 × 1	51.0	15.0	L203 × 203 × 25.4	75.9	9 680
L8 × 8 × ⅞	45.0	13.2	L203 × 203 × 22.2	67.0	8 500
L8 × 8 × ¾	38.9	11.4	L203 × 203 × 19.0	57.9	7 360
L8 × 8 × ⅝	32.7	9.61	L203 × 203 × 15.9	48.7	6 200
L8 × 8 × 9⁄16	29.6	8.68	L203 × 203 × 14.3	44.0	5 600
L8 × 8 × ½	26.4	7.75	L203 × 203 × 12.7	39.3	5 000
L6 × 6 × 1	37.4	11.0	L152 × 152 × 25.4	55.7	7 100
L6 × 6 × ⅞	33.1	9.73	L152 × 152 × 22.2	49.3	6 280
L6 × 6 × ¾	28.7	8.44	L152 × 152 × 19.0	42.7	5 450
L6 × 6 × ⅝	24.2	7.11	L152 × 152 × 15.9	36.0	4 590
L6 × 6 × 9⁄16	21.9	6.43	L152 × 152 × 14.3	32.6	4 150
L6 × 6 × ½	19.6	5.75	L152 × 152 × 12.7	29.2	3 710
L6 × 6 × 7⁄16	17.2	5.06	L152 × 152 × 11.1	25.6	3 270
L6 × 6 × ⅜	14.9	4.36	L152 × 152 × 9.5	22.2	2 810
L6 × 6 × 5⁄16	12.4	3.65	L152 × 152 × 7.9	18.5	2 360
L5 × 5 × ⅞	27.2	7.98	L127 × 127 × 22.2	40.5	5 150
L5 × 5 × ¾	23.6	6.94	L127 × 127 × 19.0	35.1	4 480
L5 × 5 × ⅝	20.0	5.86	L127 × 127 × 15.9	29.8	3 780
L5 × 5 × ½	16.2	4.75	L127 × 127 × 12.7	24.1	3 070
L5 × 5 × 7⁄16	14.3	4.18	L127 × 127 × 11.1	21.3	2 700
L5 × 5 × ⅜	12.3	3.61	L127 × 127 × 9.5	18.3	2 330
L5 × 5 × 5⁄16	10.3	3.03	L127 × 127 × 7.9	15.3	1 960
L4 × 4 × ¾	18.5	5.44	L102 × 102 × 19.0	27.5	3 510
L4 × 4 × ⅝	15.7	4.61	L102 × 102 × 15.9	23.4	2 970
L4 × 4 × ½	12.8	3.75	L102 × 102 × 12.7	19.0	2 420
L4 × 4 × 7⁄16	11.3	3.31	L102 × 102 × 11.1	16.8	2 140
L4 × 4 × ⅜	9.80	2.86	L102 × 102 × 9.5	14.6	1 850
L4 × 4 × 5⁄16	8.20	2.40	L102 × 102 × 7.9	12.2	1 550
L4 × 4 × ¼	6.60	1.94	L102 × 102 × 6.4	9.8	1 250
L3½ × 3½ × ½	11.1	3.25	L89 × 89 × 12.7	16.5	2 100
L3½ × 3½ × 7⁄16	9.80	2.87	L89 × 89 × 11.1	14.6	1 850
L3½ × 3½ × ⅜	8.50	2.48	L89 × 89 × 9.5	12.6	1 600
L3½ × 3½ × 5⁄16	7.20	2.09	L89 × 89 × 7.9	10.7	1 350
L3½ × 3½ × ¼	5.80	1.69	L89 × 89 × 6.4	8.6	1 090
L3 × 3 × ½	9.40	2.75	L76 × 76 × 12.7	14.0	1 770
L3 × 3 × 7⁄16	8.30	2.43	L76 × 76 × 11.1	12.4	1 570
L3 × 3 × ⅜	7.20	2.11	L76 × 76 × 9.5	10.7	1 360
L3 × 3 × 5⁄16	6.10	1.78	L76 × 76 × 7.9	9.1	1 150
L3 × 3 × ¼	4.90	1.44	L76 × 76 × 6.4	7.3	929
L3 × 3 × 3⁄16	3.71	1.09	L76 × 76 × 4.8	5.5	703
L2½ × 2½ × ½	7.70	2.25	L64 × 64 × 12.7	11.4	1 450
L2½ × 2½ × ⅜	5.90	1.73	L64 × 64 × 9.5	8.7	1 120
L2½ × 2½ × 5⁄16	5.00	1.46	L64 × 64 × 7.9	7.4	942
L2½ × 2½ × ¼	4.10	1.19	L64 × 64 × 6.4	6.1	768
L2½ × 2½ × 3⁄16	3.07	0.90	L64 × 64 × 4.8	4.6	581
L2 × 2 × ⅜	4.70	1.36	L51 × 51 × 9.5	7.0	877

TABLE A2.7 Continued

Size and Thickness, in.	Weight per Foot, lb	Area, in.²	Size and Thickness, mm	Mass per Metre, kg	Area, mm²
L2 × 2 × 5/16	3.92	1.15	L51 × 51 × 7.9	5.8	742
L2 × 2 × 1/4	3.19	0.938	L51 × 51 × 6.4	4.7	605
L2 × 2 × 3/16	2.44	0.715	L51 × 51 × 4.8	3.6	461
L2 × 2 × 1/8	1.65	0.484	L51 × 51 × 3.2	2.4	312
L1¾ × 1¾ × 1/4	2.77	0.813	L44 × 44 × 6.4	4.1	525
L1¾ × 1¾ × 3/16	2.12	0.621	L44 × 44 × 4.8	3.1	401
L1¾ × 1¾ × 1/8	1.44	0.422	L44 × 44 × 3.2	2.1	272
L1½ × 1½ × 1/4	2.34	0.688	L38 × 38 × 6.4	3.4	444
L1½ × 1½ × 3/16	1.80	0.527	L38 × 38 × 4.8	2.7	340
L1½ × 1½ × 5/32	1.52	0.444	L38 × 38 × 4.0	2.2	286
L1½ × 1½ × 1/8	1.23	0.359	L38 × 38 × 3.2	1.8	232
L1¼ × 1¼ × 1/4	1.92	0.563	L32 × 32 × 6.4	2.8	363
L1¼ × 1¼ × 3/16	1.48	0.434	L32 × 32 × 4.8	2.2	280
L1¼ × 1¼ × 1/8	1.01	0.297	L32 × 32 × 3.2	1.5	192
L1 × 1 × 1/4	1.49	0.438	L25 × 25 × 6.4	2.2	283
L1 × 1 × 3/16	1.16	0.340	L25 × 25 × 4.8	1.8	219
L1 × 1 × 1/8	0.80	0.234	L25 × 25 × 3.2	1.2	151
L¾ × ¾ × 1/8	0.59	0.172	L19 × 19 × 3.2	0.9	111

TABLE A2.8 "L" Shapes (Unequal Legs)

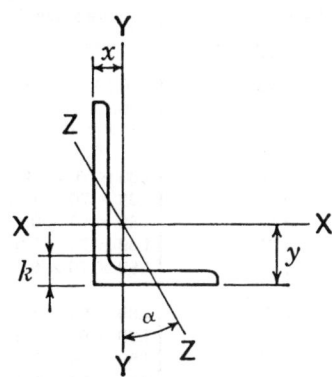

Size and Thickness, in.	Weight per Foot, lb	Area, in.²	Size and Thickness, mm	Mass per Metre, kg	Area, mm²
L8 × 6 × 1	44.2	13.0	L203 × 152 × 25.4	65.5	8 390
L8 × 6 × ⅞	39.1	11.5	L203 × 152 × 22.2	57.9	7 420
L8 × 6 × ¾	33.8	9.94	L203 × 152 × 19.0	50.1	6 410
L8 × 6 × ⅝	28.5	8.36	L203 × 152 × 15.9	42.2	5 390
L8 × 6 × 9/16	25.7	7.56	L203 × 152 × 14.3	38.1	4 880
L8 × 6 × ½	23.0	6.75	L203 × 152 × 12.7	34.1	4 350
L8 × 6 × 7/16	20.2	5.93	L203 × 152 × 11.1	29.9	3 830
L8 × 4 × 1	37.4	11.0	L203 × 102 × 25.4	55.4	7 100
L8 × 4 × ⅞	33.1	9.73	L203 × 102 × 22.2	49.3	6 280
L8 × 4 × ¾	28.7	8.44	L203 × 102 × 19.0	42.5	5 450
L8 × 4 × ⅝	24.2	7.11	L203 × 102 × 15.9	36.0	4 590
L8 × 4 × 9/16	21.9	6.43	L203 × 102 × 14.3	32.4	4 150
L8 × 4 × ½	19.6	5.75	L203 × 102 × 12.7	29.0	3 710
L8 × 4 × 7/16	17.2	5.06	L203 × 102 × 11.1	25.6	3 260
L7 × 4 × ¾	26.2	7.69	L178 × 102 × 19.0	38.8	4 960
L7 × 4 × ⅝	22.1	6.48	L178 × 102 × 15.9	32.7	4 180
L7 × 4 × ½	17.9	5.25	L178 × 102 × 12.7	26.5	3 390
L7 × 4 × 7/16	15.7	4.62	L178 × 102 × 11.1	23.4	2 980
L7 × 4 × ⅜	13.6	3.98	L178 × 102 × 9.5	20.2	2 570
L6 × 4 × ⅞	27.2	7.98	L152 × 102 × 22.2	40.3	5 150
L6 × 4 × ¾	23.6	6.94	L152 × 102 × 19.0	35.0	4 480
L6 × 4 × ⅝	20.0	5.86	L152 × 102 × 15.9	29.6	3 780
L6 × 4 × 9/16	18.1	5.31	L152 × 102 × 14.3	26.8	3 430
L6 × 4 × ½	16.2	4.75	L152 × 102 × 12.7	24.0	3 060
L6 × 4 × 7/16	14.3	4.18	L152 × 102 × 11.1	21.2	2 700
L6 × 4 × ⅜	12.3	3.61	L152 × 102 × 9.5	18.2	2 330
L6 × 4 × 5/16	10.3	3.03	L152 × 102 × 7.9	15.3	1 950
L6 × 3½ × ½	15.3	4.50	L152 × 89 × 12.7	22.7	2 900
L6 × 3½ × ⅜	11.7	3.42	L152 × 89 × 9.5	17.3	2 210
L6 × 3½ × 5/16	9.80	2.87	L152 × 89 × 7.9	14.5	1 850
L5 × 3½ × ¾	19.8	5.81	L127 × 89 × 19.0	29.3	3 750
L5 × 3½ × ⅝	16.8	4.92	L127 × 89 × 15.9	24.9	3 170
L5 × 3½ × ½	13.6	4.00	L127 × 89 × 12.7	20.2	2 580
L5 × 3½ × ⅜	10.4	3.05	L127 × 89 × 9.5	15.4	1 970
L5 × 3½ × 5/16	8.70	2.56	L127 × 89 × 7.9	12.9	1 650
L5 × 3½ × ¼	7.00	2.06	L127 × 89 × 6.4	10.4	1 330
L5 × 3 × ½	12.8	3.75	L127 × 76 × 12.7	19.0	2 420
L5 × 3 × 7/16	11.3	3.31	L127 × 76 × 11.1	16.7	2 140
L5 × 3 × ⅜	9.80	2.86	L127 × 76 × 9.5	14.5	1 850
L5 × 3 × 5/16	8.20	2.40	L127 × 76 × 7.9	12.1	1 550
L5 × 3 × ¼	6.60	1.94	L127 × 76 × 6.4	9.8	1 250
L4 × 3½ × ½	11.9	3.50	L102 × 89 × 12.7	17.6	2 260
L4 × 3½ × ⅜	9.10	2.67	L102 × 89 × 9.5	13.5	1 720
L4 × 3½ × 5/16	7.70	2.25	L102 × 89 × 7.9	11.4	1 450
L4 × 3½ × ¼	6.20	1.81	L102 × 89 × 6.4	9.2	1 170
L4 × 3 × ⅝	13.6	3.98	L102 × 76 × 15.9	20.2	2 570

TABLE A2.8 Continued

Size and Thickness, in.	Weight per Foot, lb	Area, in.²	Size and Thickness, mm	Mass per Metre, kg	Area, mm²
L4 × 3 × ½	11.1	3.25	L102 × 76 × 12.7	16.4	2 100
L4 × 3 × ⅜	8.50	2.48	L102 × 76 × 9.5	12.6	1 600
L4 × 3 × 5/16	7.20	2.09	L102 × 76 × 7.9	10.7	1 350
L4 × 3 × ¼	5.80	1.69	L102 × 76 × 6.4	8.6	1 090
L3½ × 3 × ½	10.2	3.00	L89 × 76 × 12.7	15.1	1 940
L3½ × 3 × 7/16	9.10	2.65	L89 × 76 × 11.1	13.5	1 710
L3½ × 3 × ⅜	7.90	2.30	L89 × 76 × 9.5	11.7	1 480
L3½ × 3 × 5/16	6.60	1.93	L89 × 76 × 7.9	9.8	1 250
L3½ × 3 × ¼	5.40	1.56	L89 × 76 × 6.4	8.0	1 010
L3½ × 2½ × ½	9.40	2.75	L89 × 64 × 12.7	13.9	1 770
L3½ × 2½ × ⅜	7.20	2.11	L89 × 64 × 9.5	10.7	1 360
L3½ × 2½ × 5/16	6.10	1.78	L89 × 64 × 7.9	9.0	1 150
L3½ × 2½ × ¼	4.90	1.44	L89 × 64 × 6.4	7.3	929
L3 × 2½ × ½	8.50	2.50	L76 × 64 × 12.7	12.6	1 610
L3 × 2½ × 7/16	7.60	2.21	L76 × 64 × 11.1	11.3	1 430
L3 × 2½ × ⅜	6.60	1.92	L76 × 64 × 9.5	9.8	1 240
L3 × 2½ × 5/16	5.60	1.62	L76 × 64 × 7.9	8.3	1 050
L3 × 2½ × ¼	4.50	1.31	L76 × 64 × 6.4	6.7	845
L3 × 2½ × 3/16	3.39	0.996	L76 × 64 × 4.8	5.1	643
L3 × 2 × ½	7.70	2.25	L76 × 51 × 12.7	11.5	1 450
L3 × 2 × ⅜	5.90	1.73	L76 × 51 × 9.5	8.8	1 120
L3 × 2 × 5/16	5.00	1.46	L76 × 51 × 7.9	7.4	942
L3 × 2 × ¼	4.10	1.19	L76 × 51 × 6.4	6.1	768
L3 × 2 × 3/16	3.07	0.902	L76 × 51 × 4.8	4.6	582
L2½ × 2 × ⅜	5.30	1.55	L64 × 51 × 9.5	7.9	1 000
L2½ × 2 × 5/16	4.50	1.31	L64 × 51 × 7.9	6.7	845
L2½ × 2 × ¼	3.62	1.06	L64 × 51 × 6.4	5.4	684
L2½ × 2 × 3/16	2.75	0.809	L64 × 51 × 4.8	4.2	522
L2½ × 1½ × ¼	3.19	0.938	L64 × 38 × 6.4	4.8	605
L2½ × 1½ × 3/16	2.44	0.715	L64 × 38 × 4.8	3.6	461
L2 × 1½ × ¼	2.77	0.813	L51 × 38 × 6.4	4.2	525
L2 × 1½ × 3/16	2.12	0.621	L51 × 38 × 4.8	3.1	401
L2 × 1½ × ⅛	1.44	0.422	L51 × 38 × 3.2	2.1	272

APPENDIXES

(Nonmandatory Information)

X1. COIL AS A SOURCE OF STRUCTURAL PRODUCTS

X1.1 Continuous wide hot strip rolling mills are normally equipped with coilers. Regardless of the different types of systems employed during or following the rolling operations, it is common for the steel to be reeled into the coiler at temperatures in the stress-relieving range. In general, such temperatures are higher as the steel thickness increases. The coils subsequently cool to ambient temperature with outer and inner laps cooling more rapidly than central laps. The difference in cooling rate can result in measurable differences in the mechanical properties throughout a coil. Data confirm reduced yield and tensile strength with increased percent elongation for the steel with slower cooling rates from the coiling temperature to ambient. Such differences are in addition to the effects on mechanical properties caused by differences in heat analysis and chemical segregation.

X2. VARIATION OF TENSILE PROPERTIES IN PLATES AND SHAPES

X2.1 The tension testing requirements of this specification are intended only to characterize the tensile properties of a heat of steel for determination of conformance to the requirements of the applicable product specification. Such testing procedures are not intended to define the upper or lower limits of tensile properties at all possible test locations within a heat of steel. It is well known and documented that tensile properties will vary within a heat or individual piece of steel as a function of chemical composition, processing, testing procedure and other factors. It is, therefore, incumbent on designers and engineers to use sound engineering judgement when using tension test results shown on mill test reports. The testing procedures of this specification have been found to provide structural products adequate for normal structural design criteria.

X2.2 A survey of the variation to be expected in tensile properties obtained from plates and structural shapes was conducted by the American Iron and Steel Institute (AISI).[8] The results of this survey are contained in a *Contributions to the Metallurgy of Steel* entitled "The Variation of Product Analysis and Tensile Properties—Carbon Steel Plates and Wide Flange Shapes" (SU/18, SU/19 and SU/20), published in September 1974. The data are presented in tables of probability that tensile properties at other than the official location may differ from those of the reported test location. Another survey sponsored by the AISI entitled "Statistical Analysis of Structural Plate Mechanical Properties" was published in January 2003. That survey analyzed the results of variability testing on more modern as-rolled steels that were generally of higher minimum yield strength steels and also compared those results statistically to the previous surveys.[9]

X2.3 This specification contains no requirements applicable to product tension tests; conformance to the applicable product specification is determined on the basis of tests performed at the place of manufacture or processing prior to shipment, unless otherwise specified.

X2.4 A task group of ASTM Subcommittee A01.02 has determined, based on review of the earlier AISI data,[8] that the variation in tensile properties of plates and structural shapes can be expressed as a function of specified requirements: one standard deviation equals approximately 4 % of required tensile strength, 8 % of required yield strength, and 3 percentage points of required elongation. The January 2003 survey resulted in similar findings.

X2.5 Acceptance criteria for product testing based upon these values, either below the minimum or above the maximum allowed by the applicable product specification, are generally acceptable to manufacturers. Such tolerances could be considered by users of structural products as a reasonable basis for acceptance of structural products that, due to their inherent variability, deviate from the applicable product specification requirements when subjected to product tension testing.

---

[8] Originally published by the American Iron and Steel Institute (AISI), 1140 Connecticut Ave., NW, Suite 705, Washington, DC 20036, http://www.steel.org. Available from ASTM Headquarters as PCN: 29-000390-02.

[9] Available from American Iron and Steel Institute (AISI) directly at http://www.steel.org/infrastructure/bridges/index.html.

X3. WELDABILITY OF STEEL

X3.1 *Weldability* is a term that usually refers to the relative ease with which a metal can be welded using conventional practice. Difficulties arise in steel when the cooling rates associated with weld thermal cycles produce microstructures (for example, martensite) that are susceptible to brittle fracture or, more commonly, hydrogen-induced (or cold) cracking.[10] (Solidification or hot cracking is a relatively rare phenomenon that will not be addressed here. See Randall[11] for further information.)

X3.2 The relative sensitivity of steels to forming cold cracking microstructures is called hardenability and can be measured in a number of ways. Perhaps the most popular method of assessing this is by the carbon equivalent (CE) formula, which attempts to equate the relative hardening contributions of a steel's constituent elements (for example, manganese, vanadium) to an equivalent amount of carbon, which is the most significant hardening agent. The most popular formula is the International Institute of Welding (IIW) equation presented in S31.2, which has been found suitable for predicting hardenability in a wide range of commonly used carbon-manganese and low alloy steels.[12]

X3.3 It should be noted, however, that for the current generation of low carbon (<0.10 %) low alloy steels that derive strength from a combination of microalloys and thermal processing methods the use of other formulae may more accurately assess hardenability and cold cracking sensitivity.[13]

X3.4 For a vast number of common structural applications it is unnecessary to specify the use of CE limits. However, in order to obtain a higher level of confidence in avoiding cold cracking, the chemistry controls in S31 are available. A purchaser who specifies the use of S31 should be aware that there are several factors involved in the judicious selection of a maximum CE value, such as the following:

X3.4.1 Actual production joint restraint/base metal thickness(es),

X3.4.2 Filler metal and base metal strength compatibility,

X3.4.3 Deposited weld metal diffusible hydrogen content,

X3.4.4 Preheat and interpass temperatures,

X3.4.5 Filler metal and base metal cleanliness, and

X3.4.6 Heat input.

X3.5 Though it is widely believed that low CE steels are immune to weld cracking problems, failure to consider these factors and others have resulted in weld or base metal HAZ (heat affected zone) cracks in such steels.[11]

X3.6 It is important to note that carbon equivalence is only a qualitative assessment of potential welding problems, and should never be solely relied on to ensure weld integrity. The proper use of welding specifications, coupled with the knowledge of actual construction conditions, must also be used.

---

[10] Graville, B. A., *The Principles of Cold Cracking Control in Welds*, Dominion Bridge Company, 1975.
[11] Randall, M. D., "Welding Procedure Factors Affecting Weldability for Service," *Weldability of Steels*, by Stout and Doty, Welding Research Council.
[12] Bailey, N., "The Development and Use of Carbon Equivalent in Britain," *Hardenability of Steels*, Abington Publishing, 1990.

[13] International Institute of Welding, "Guide to the Metallurgy of Welding and Weldability of Low Carbon Microalloyed Hot Rolled Steels," Document IIS/IIW-843-87.

X4. RADIUS FOR COLD BENDING

X4.1 Suggested minimum inside bend radii for cold forming are referenced to Group Designations A to F inclusive as defined in Table X4.1. The suggested radii listed in Table X4.2 should be used as minimums in typical shop fabrication. Material that does not form satisfactorily when fabricated in accordance with Table X4.2 may be subject to rejection pending negotiation with the steel supplier. When tighter bends are required, the manufacturer should be consulted.

X4.2 The bend radius and the radius of the male die should be as liberal as the finished part will permit. The width across the shoulders of the female die should be at least eight times the plate thickness. Higher strength steels require larger die openings. The surface of the dies in the area of radius should be smooth.

X4.2.1 Since cracks in cold bending commonly originate from the outside edges, shear burrs and gas cut edges should be removed by grinding. Sharp corners on edges and on punched or gas cut holes should be removed by chamfering or grinding to a radius.

X4.2.2 If possible, parts should be formed such that the bend line is perpendicular to the direction of final rolling. If it is necessary to bend with the bend line parallel to the direction of final rolling, a more generous radius is suggested (1½ times applicable value given in Table X4.2 for bend lines perpendicular to the direction of rolling).

TABLE X4.1 Group Designations for Cold Bending

Specification	Grade	Group Designation[A]
A36/A36M	[B]	B
A131/A131M	A, B, D, CS and E	B
	A, B, D, CS and E (all cold flanging)	B
	AH32, DH32, EH32 and FH 32	C
	AH36, DH36, EH36 andFH36	C
	AH40, DH40, EH40 and FH40	C
A242/A242M	[B]	C
A283/A283M	A or B	A
	C or D	B
A514/A514M	any	F
A529/A529M	50 [345] or 55 [380]	C
A572/A572M	42 [290]	B
	50 [345]	C
	55 [380]	D
	60 [415] or 65 [450]	E
A573/A573M	58 [400] or 65 [450]	B
	70 [485]	C
A588/A588M	any	C
A633/A633M	any	B
A656/A656M	50 [345]	B
	60 [415]	D
	70 [485]	E
	80 [550]	F
A678/A678M	A or B	C
	C or D	D
A709/A709M	36 [250]	B
	50 [345], 50W [345W] or HPS 50W [HPS 345W]	C
	HPS70W [HPS485W]	D
	HPS 100W [HPS 690W]	F
A710/A710M	A	F
A808/A808M	[B]	C
A852/A852M	[B]	D
A871/A871M	60 [415] or 65 [450]	E
A945/A945M	50 [345] or 65 [450]	B

[A] Steels having a ratio of specified minimum tensile strength to specified minimum yield strength of 1.15 or less are in Group F; other steels are in Groups A to E inclusive, which are grouped on the basis of their having similar specified values for minimum elongation in 2 in. [50 mm].
[B] Grade designations are not applicable for this specification.

TABLE X4.2 Suggested Minimum Inside Radii for Cold Bending[A]

Group Designation[B]	Thickness (t), in. [mm]			
	Up to ¾ in. [20 mm]	Over ¾ in. [20 mm] To 1 [25 mm, incl]	Over 1 in. [25 mm] To 2 in. [50 mm], incl	Over 2 in. [50 mm]
A	1.5t	1.5t	1.5t	1.5t
B	1.5t	1.5t	1.5t	2.0t
C	1.5t	1.5t	2.0t	2.5t
D	1.5t	1.5t	2.5t	3.0t
E	1.5t	1.5t	3.0t	3.5t
F	1.75t	2.25t	4.5t	5.5t

[A] Values are for bend lines perpendicular to the direction of final rolling. These radii apply when the precautions listed in X4.2 are followed. If bend lines are parallel to the direction of final rolling, multiply values by 1.5.
[B] Steel specifications included in the group designations may not include the entire thickness range shown in this table.

X4.3 References

X4.3.1 Holt, G. E., et al, "Minimum Cold Bend Radii Project—Final Report," Concurrent Technologies Corporation, January 27, 1997.[14]

X4.3.2 Brockenbrough, R. L., "Fabrication Guidelines for Cold Bending," R. L. Brockenbrough & Associates, June 28, 1998.[14]

---

[14] Available from American Iron and Steel Institute (AISI), 1140 Connecticut Ave., NW, Suite 705, Washington, DC 20036, http://www.steel.org.

SUMMARY OF CHANGES

Committee A01 has identified the location of the following changes to this standard since the last issue (A6/A6M – 08a) that may impact the use of this standard. (Approved April 1, 2009.)

(1) Revised Table X4.1.

Committee A01 has identified the location of the following changes to this standard since the last issue (A6/A6M – 08) that may impact the use of this standard. (Approved November 1, 2008.)

(1) Added 12.6 on conversions from factions to decimals.

Committee A01 has identified the location of the following changes to this standard since the last issue (A6/A6M – 07) that may impact the use of this standard. (Approved March 1, 2008.)

(1) Added 2.5.

ASTM International takes no position respecting the validity of any patent rights asserted in connection with any item mentioned in this standard. Users of this standard are expressly advised that determination of the validity of any such patent rights, and the risk of infringement of such rights, are entirely their own responsibility.

This standard is subject to revision at any time by the responsible technical committee and must be reviewed every five years and if not revised, either reapproved or withdrawn. Your comments are invited either for revision of this standard or for additional standards and should be addressed to ASTM International Headquarters. Your comments will receive careful consideration at a meeting of the responsible technical committee, which you may attend. If you feel that your comments have not received a fair hearing you should make your views known to the ASTM Committee on Standards, at the address shown below.

This standard is copyrighted by ASTM International, 100 Barr Harbor Drive, PO Box C700, West Conshohocken, PA 19428-2959, United States. Individual reprints (single or multiple copies) of this standard may be obtained by contacting ASTM at the above address or at 610-832-9585 (phone), 610-832-9555 (fax), or service@astm.org (e-mail); or through the ASTM website (www.astm.org).

Designation: A29/A29M – 05

Standard Specification for
Steel Bars, Carbon and Alloy, Hot-Wrought, General Requirements for[1]

This standard is issued under the fixed designation A29/A29M; the number immediately following the designation indicates the year of original adoption or, in the case of revision, the year of last revision. A number in parentheses indicates the year of last reapproval. A superscript epsilon (ε) indicates an editorial change since the last revision or reapproval.

This standard has been approved for use by agencies of the Department of Defense.

1. Scope*

1.1 This specification[2] covers a group of common requirements which, unless otherwise specified in the purchase order or in an individual specification, shall apply to carbon and alloy steel bars under each of the following ASTM specifications (or under any other ASTM specification which invokes this specification or portions thereof):

Title of Specification	ASTM Designation[A]
Hot-Rolled Carbon Steel Bars:	
Steel Bars, Carbon, Quenched and Tempered	A321
Steel Bars and Shapes, Carbon Rolled from "T" Rails	A499
Steel Bars, Carbon, Merchant Quality, M-Grades	A575
Steel Bars, Carbon, Hot-Wrought, Special Quality	A576
Steel Bars, Carbon, Merchant Quality, Mechanical Properties	A663/A663M
Steel Bars, Carbon, Hot-Wrought, Special Quality, Mechanical Properties	A675/A675M
Steel Bars for Springs, Carbon and Alloy	A689
Cold-Finished Carbon Steel Bars:	
Steel Bars, Carbon and Alloy, Cold-Finished	A108
Cold-Drawn Stress-Relieved Carbon Steel Bars Subject to Mechanical Property Requirements	A311/A311M
Hot-Rolled Alloy Steel Bars:	
Steel Bars, Alloy, Standard Grades	A322
Carbon and Alloy Steel Bars Subject to End-Quench Hardenability Requirements	A304
Steel Bars, Alloy, Hot-Wrought or Cold-Finished, Quenched and Tempered	A434
Steel Bars, Alloy, Hot-Wrought, for Elevated Temperature or Pressure-Containing Parts, or Both	A739
Cold-Finished Alloy Steel Bars:	
Steel Bars, Alloy, Hot-Rolled or Cold-Finished, Quenched and Tempered	A434
Steel Bars, Carbon, Hot-Wrought or Cold-Finished, Special Quality, for Pressure Piping Components	A696

[A] These designations refer to the latest issue of the respective specifications, which appear either in the *Annual Book of ASTM Standards*, Vol 01.05, or as reprints obtainable from ASTM.

1.2 In case of any conflict in requirements, the requirements of the purchase order, the individual material specification, and this general specification shall prevail in the sequence named.

1.3 The values stated in inch-pound units or SI units are to be regarded as the standard. Within the text, the SI units are shown in brackets. The values stated in each system are not exact equivalents; therefore, each system must be used independently of the other. Combining values from the two systems may result in nonconformance with the specification.

1.4 For purposes of determining conformance to this specification and the various material specifications referenced in 1.1, dimensional values shall be rounded to the nearest unit in the right-hand place of figures used in expressing the limiting values in accordance with the rounding method of Practice E29.

NOTE 1—Specification A29 previously listed dimensional tolerances for cold-finished bars; these are now found in Specification A108.

2. Referenced Documents

2.1 *ASTM Standards:*[3]

A108 Specification for Steel Bar, Carbon and Alloy, Cold-Finished

A304 Specification for Carbon and Alloy Steel Bars Subject to End-Quench Hardenability Requirements

A311/A311M Specification for Cold-Drawn, Stress-Relieved Carbon Steel Bars Subject to Mechanical Property Requirements

A321 Specification for Steel Bars, Carbon, Quenched and Tempered[4]

A322 Specification for Steel Bars, Alloy, Standard Grades

A370 Test Methods and Definitions for Mechanical Testing of Steel Products

A434 Specification for Steel Bars, Alloy, Hot-Wrought or Cold-Finished, Quenched and Tempered

A499 Specification for Steel Bars and Shapes, Carbon Rolled from "T" Rails

A575 Specification for Steel Bars, Carbon, Merchant Quality, M-Grades

---

[1] This specification is under the jurisdiction of ASTM Committee A01 on Steel, Stainless Steel and Related Alloys and is the direct responsibility of Subcommittee A01.15 on Bars.
Current edition approved May 1, 2005. Published May 2005. Originally approved in 1957. Last previous edition approved in 2004 as A29/A29M–04ε1. DOI: 10.1520/A0029_A0029M-05.
[2] For ASME Boiler and Pressure Vessel Code applications see related Specification SA-29/SA-29M in Section II of that Code.

[3] For referenced ASTM standards, visit the ASTM website, www.astm.org, or contact ASTM Customer Service at service@astm.org. For Annual Book of ASTM Standards volume information, refer to the standard's Document Summary page on the ASTM website.
[4] Withdrawn. The last approved version of this historical standard is referenced on www.astm.org.

A Summary of Changes section appears at the end of this standard.

Copyright © ASTM International, 100 Barr Harbor Drive, PO Box C700, West Conshohocken, PA 19428-2959, United States.

A576 Specification for Steel Bars, Carbon, Hot-Wrought, Special Quality

A663/A663M Specification for Steel Bars, Carbon, Merchant Quality, Mechanical Properties

A675/A675M Specification for Steel Bars, Carbon, Hot-Wrought, Special Quality, Mechanical Properties

A689 Specification for Carbon and Alloy Steel Bars for Springs

A696 Specification for Steel Bars, Carbon, Hot-Wrought or Cold-Finished, Special Quality, for Pressure Piping Components

A700 Practices for Packaging, Marking, and Loading Methods for Steel Products for Shipment

A739 Specification for Steel Bars, Alloy, Hot-Wrought, for Elevated Temperature or Pressure-Containing Parts, or Both

A751 Test Methods, Practices, and Terminology for Chemical Analysis of Steel Products

E29 Practice for Using Significant Digits in Test Data to Determine Conformance with Specifications

E112 Test Methods for Determining Average Grain Size

2.2 *Federal Standards:*[5]

Fed. Std. No. 123 Marking for Shipment (Civil Agencies)

Fed. Std. No. 183 Continuous Identification Marking of Iron and Steel Products

2.3 *Military Standard:*[5]

MIL-STD-163 Steel Mill Products—Preparation for Shipment and Storage

2.4 *Other Standards:*

AIAG B-1 Bar Code Symbology Standard for 3-of-9 Bar Codes[6]

AIAG B-5 02.00 Primary Metals Tag Application Standard[6]

3. Terminology

3.1 *Definitions of Terms Specific to This Standard:*

3.1.1 *Hot-Wrought Steel Bars*—Steel bars produced by hot forming ingots, blooms, billets, or other semifinished forms to yield straight lengths (or coils, depending upon size, section, and mill equipment) in sections that are uniform throughout their length, and in the following sections and sizes:

3.1.1.1 *Rounds*—7/32 to 10.0 in. [5.5 to 250 mm], inclusive,

3.1.1.2 *Squares*—7/32 to 6.0 in. [6 to 160 mm], inclusive,

3.1.1.3 *Round-Cornered Squares*—7/32 to 8.0 in. [6 to 200 mm], inclusive,

3.1.1.4 *Flats*—1/4 to 8 in. inclusive, in width: 13/64 in. in minimum thickness up to 6 in. in width; and 0.230 in. in minimum thickness for over 6 to 8 in. in width, inclusive [over 5 mm in thickness up to 150 mm in width; and over 6 mm in thickness for over 150 mm through 200 mm in width]. Maximum thickness for all widths is 4 in. [100 mm].

3.1.1.5 *Hexagons and Octagons*—1/4 to 4 1/16 in. [6 to 103 mm], inclusive, between parallel surfaces,

3.1.1.6 *Bar Size Shapes*—Angles, channels, tees, zees, when their greatest cross-sectional dimension is under 3 in. [75 mm], and

3.1.1.7 *Special Bar Sections*—Half-rounds, ovals, half-ovals, other special bar size sections.

3.1.2 *Cold-Finished Steel Bars*—Steel bars produced by cold finishing previously hot-wrought bars by means of cold drawing, cold forming, turning, grinding, or polishing (singly or in combination) to yield straight lengths or coils in sections that are uniform throughout their length and in the following sections and sizes:

3.1.2.1 *Rounds*—9 in. [230 mm] and under in diameter,

3.1.2.2 *Squares*—6 in. [150 mm] and under between parallel surfaces,

3.1.2.3 *Hexagons*—4 in. [100 mm] and under between parallel surfaces,

3.1.2.4 *Flats*—1/8 in. [3 mm] and over in thickness and not over 12 in. [300 mm] in width, and

3.1.2.5 *Special Bar Sections.—*

3.1.3 *Lot*—Unless otherwise specified in the contract or order, a lot shall consist of all bars submitted for inspection at the same time of the same heat, condition, finish, size, or shape. For bars specified in the quenched and tempered condition, when heat treated in batch-type furnaces, a lot shall consist of all bars from the same heat, of the same prior condition, the same size, and subjected to the same heat treatment in one tempering charge. For bars specified in the quenched and tempered condition, when heat treated without interruption in a continuous-type furnace, a lot shall consist of all bars from the same heat, of the same prior condition, of the same size, and subjected to the same heat treatment.

4. Chemical Composition

4.1 *Limits*:

4.1.1 The chemical composition shall conform to the requirements specified in the purchase order or the individual product specifications. For convenience the grades commonly specified for carbon steel bars are shown in Tables 1 and 2. Bars may be ordered to these grade designations and when so ordered shall conform to the specified limits by heat analysis.

4.1.2 When compositions other than those shown in Tables 1 and 2 are required, the composition limits shall be prepared using the ranges and limits shown in Table 3 for carbon steel and Table 4 for alloy steel.

4.2 *Heat or Cast Analysis*:

4.2.1 The chemical composition of each heat or cast shall be determined by the manufacturer in accordance with Test Methods, Practices, and Terminology A751.

4.2.2 The heat or cast analysis shall conform to the requirements specified in the product specification or purchase order. These can be the heat chemical range and limit for a grade designated in Tables 1 and 2, or another range and limit in accordance with 4.1.2, or with requirements of the product specification.

NOTE 2—Heat analysis for lead is not determinable since lead is added to the ladle stream while each ingot is poured. When specified as an added element to a standard steel, the percentage of lead is reported as 0.15 to 0.35 incl, which is the range commonly specified for this element.

---

[5] Copies of military specifications, military standards, and federal standards required by contractors in connection with specific procurement functions should be obtained from the procuring activity or as directed by the contracting officer, or from the Standardization Documents Order Desk, Bldg. 4 Section D, 700 Robbins Ave., Philadelphia, PA 19111-5094, Attn: NPODS.

[6] Available from Automotive Industry Action Group, North Park Plaza, Ste. 830, 17117 W. Nine Mile Rd., Southfield, MI 48075.

TABLE 1 Grade Designations and Chemical Compositions of Carbon Steel Bars

Grade Designation	Heat Chemical Ranges and Limits, %			
	Carbon	Manganese	Phosphorus, max	Sulfur, max[A]
Nonresulfurized Carbon Steels[B,C,D,E,F]				
1005	0.06 max	0.35 max	0.040	0.050
1006	0.08 max	0.25–0.40	0.040	0.050
1008	0.10 max	0.30–0.50	0.040	0.050
1010	0.08–0.13	0.30–0.60	0.040	0.050
1011	0.08–0.13	0.60–0.90	0.040	0.050
1012	0.10–0.15	0.30–0.60	0.040	0.050
1013	0.11–0.16	0.50–0.80	0.040	0.050
1015	0.13–0.18	0.30–0.60	0.040	0.050
1016	0.13–0.18	0.60–0.90	0.040	0.050
1017	0.15–0.20	0.30–0.60	0.040	0.050
1018	0.15–0.20	0.60–0.90	0.040	0.050
1019	0.15–0.20	0.70–1.00	0.040	0.050
1020	0.18–0.23	0.30–0.60	0.040	0.050
1021	0.18–0.23	0.60–0.90	0.040	0.050
1022	0.18–0.23	0.70–1.00	0.040	0.050
1023	0.20–0.25	0.30–0.60	0.040	0.050
1025	0.22–0.28	0.30–0.60	0.040	0.050
1026	0.22–0.28	0.60–0.90	0.040	0.050
1029	0.25–0.31	0.60–0.90	0.040	0.050
1030	0.28–0.34	0.60–0.90	0.040	0.050
1034	0.32–0.38	0.50–0.80	0.040	0.050
1035	0.32–0.38	0.60–0.90	0.040	0.050
1037	0.32–0.38	0.70–1.00	0.040	0.050
1038	0.35–0.42	0.60–0.90	0.040	0.050
1039	0.37–0.44	0.70–1.00	0.040	0.050
1040	0.37–0.44	0.60–0.90	0.040	0.050
1042	0.40–0.47	0.60–0.90	0.040	0.050
1043	0.40–0.47	0.70–1.00	0.040	0.050
1044	0.43–0.50	0.30–0.60	0.040	0.050
1045	0.43–0.50	0.60–0.90	0.040	0.050
1046	0.43–0.50	0.70–1.00	0.040	0.050
1049	0.46–0.53	0.60–0.90	0.040	0.050
1050	0.48–0.55	0.60–0.90	0.040	0.050
1053	0.48–0.55	0.70–1.00	0.040	0.050
1055	0.50–0.60	0.60–0.90	0.040	0.050
1059	0.55–0.65	0.50–0.80	0.040	0.050
1060	0.55–0.65	0.60–0.90	0.040	0.050
1064	0.60–0.70	0.50–0.80	0.040	0.050
1065	0.60–0.70	0.60–0.90	0.040	0.050
1069	0.65–0.75	0.40–0.70	0.040	0.050
1070	0.65–0.75	0.60–0.90	0.040	0.050
1071	0.65–0.70	0.75–1.05	0.040	0.050
1074	0.70–0.80	0.50–0.80	0.040	0.050
1075	0.70–0.80	0.40–0.70	0.040	0.050
1078	0.72–0.85	0.30–0.60	0.040	0.050
1080	0.75–0.88	0.60–0.90	0.040	0.050
1084	0.80–0.93	0.60–0.90	0.040	0.050
1086	0.80–0.93	0.30–0.50	0.040	0.050
1090	0.85–0.98	0.60–0.90	0.040	0.050
1095	0.90–1.03	0.30–0.50	0.040	0.050
Resulfurized Carbon Steels[B,D,F]				
1108	0.08–0.13	0.60–0.80	0.040	0.08–0.13
1109	0.08–0.13	0.60–0.90	0.040	0.08–0.13
1110	0.08–0.13	0.30–0.60	0.040	0.08–0.13
1116	0.14–0.20	1.10–1.40	0.040	0.16–0.23
1117	0.14–0.20	1.00–1.30	0.040	0.08–0.13
1118	0.14–0.20	1.30–1.60	0.040	0.08–0.13
1119	0.14–0.20	1.00–1.30	0.040	0.24–0.33
1132	0.27–0.34	1.35–1.65	0.040	0.08–0.13
1137	0.32–0.39	1.35–1.65	0.040	0.08–0.13
1139	0.35–0.43	1.35–1.65	0.040	0.13–0.20
1140	0.37–0.44	0.70–1.00	0.040	0.08–0.13
1141	0.37–0.45	1.35–1.65	0.040	0.08–0.13
1144	0.40–0.48	1.35–1.65	0.040	0.24–0.33
1145	0.42–0.49	0.70–1.00	0.040	0.04–0.07
1146	0.42–0.49	0.70–1.00	0.040	0.08–0.13
1151	0.48–0.55	0.70–1.00	0.040	0.08–0.13

ASTM A29/A29M – 05

Rephosphorized and Resulfurized Carbon Steels[D,F]

Grade Designation	Carbon	Manganese	Phosphorous	Sulfur	Lead
1211	0.13 max	0.60–0.90	0.07–0.12	0.10–0.15	...
1212	0.13 max	0.70–1.00	0.07–0.12	0.16–0.23	...
1213	0.13 max	0.70–1.00	0.07–0.12	0.24–0.33	...
1215	0.09 max	0.75–1.05	0.04–0.09	0.26–0.35	...
12L13	0.13 max	0.70–1.00	0.07–0.12	0.24–0.33	0.15–0.35
12L14	0.15 max	0.85–1.15	0.04–0.09	0.26–0.35	0.15–0.35
12L15	0.09 max	0.75–1.05	0.04–0.09	0.26–0.35	0.15–0.35

High-Manganese Carbon Steels[B,C,D,E,F]

Grade Designation	Former Designation	Carbon	Manganese	Phosphorous, max	Sulfur, max
1513	...	0.10–0.16	1.10–1.40	0.040	0.050
1518	...	0.15–0.21	1.10–1.40	0.040	0.050
1522	...	0.18–0.24	1.10–1.40	0.040	0.050
1524	1024	0.19–0.25	1.35–1.65	0.040	0.050
1525	...	0.23–0.29	0.80–1.10	0.040	0.050
1526	...	0.22–0.29	1.10–1.40	0.040	0.050
1527	1027	0.22–0.29	1.20–1.50	0.040	0.050
1536	1036	0.30–0.37	1.20–1.50	0.040	0.050
1541	1041	0.36–0.44	1.35–1.65	0.040	0.050
1547	...	0.43–0.51	1.35–1.65	0.040	0.050
1548	1048	0.44–0.52	1.10–1.40	0.040	0.050
1551	1051	0.45–0.56	0.85–1.15	0.040	0.050
1552	1052	0.47–0.55	1.20–1.50	0.040	0.050
1561	1061	0.55–0.65	0.75–1.05	0.040	0.050
1566	1066	0.60–0.71	0.85–1.15	0.040	0.050
1572	1072	0.65–0.76	1.00–1.30	0.040	0.050

Heat Chemical Ranges and Limits, percent
Merchant Quality M Series Carbon Steel Bars

Grade Designation	Carbon	Manganese[G]	Phosphorous, max	Sulfur, max
M 1008	0.10 max	0.25–0.60	0.04	0.05
M 1010	0.07–0.14	0.25–0.60	0.04	0.05
M 1012	0.09–0.16	0.25–0.60	0.04	0.05
M 1015	0.12–0.19	0.25–0.60	0.04	0.05
M 1017	0.14–0.21	0.25–0.60	0.04	0.05
M 1020	0.17–0.24	0.25–0.60	0.04	0.05
M 1023	0.19–0.27	0.25–0.60	0.04	0.05
M 1025	0.20–0.30	0.25–0.60	0.04	0.05
M 1031	0.26–0.36	0.25–0.60	0.04	0.05
M 1044	0.40–0.50	0.25–0.60	0.04	0.05

[A] Maximum unless otherwise indicated.
[B] When silicon is required, the following ranges and limits are commonly specified: 0.10 %, max, 0.10 % to 0.20 %, 0.15 % to 0.35 %, 0.20 % to 0.40 %, or 0.30 % to 0.60 %.
[C] Copper can be specified when required as 0.20 % minimum.
[D] When lead is required as an added element to a standard steel, a range of 0.15 to 0.35 % inclusive is specified. Such a steel is identified by inserting the letter "L" between the second and third numerals of the grade designation, for example, 10 L 45. A cast or heat analysis is not determinable when lead is added to the ladle stream.
[E] When boron treatment for killed steels is specified, the steels can be expected to contain 0.0005 to 0.003 % boron. If the usual titanium additive is not permitted, the steels can be expected to contain up to 0.005 % boron.
[F] The elements bismuth, calcium, selenium, or tellurium may be added as agreed upon between purchaser and supplier.
[G] Unless prohibited by the purchaser, the manganese content may exceed 0.60 % on heat analysis to a maximum of 0.75 %, provided the carbon range on heat analysis has the minimum and maximum reduced by 0.01 % for each 0.05 % manganese over 0.60 %.

4.2.3 If requested or required, the heat analysis shall be reported to the purchaser or his representative.

4.2.4 Reporting of significant figures and rounding shall be in accordance with Test Methods, Practices, and Terminology A751.

4.3 *Product Analysis*:

4.3.1 Merchant quality carbon bar steel is not subject to rejection for product analysis unless misapplication of a heat is clearly indicated.

4.3.2 Analyses may be made by the purchaser from finished bars other than merchant quality representing each heat of open-hearth, basic-oxygen, or electric-furnace steel. The chemical composition thus determined shall not vary from the limits specified in the applicable specification by more than the amounts prescribed in Table 5 and Table 6, but the several determinations of any element, excluding lead, in a heat may not vary both above and below the specified range. Rimmed or capped steel is characterized by a lack of homogeneity in its composition, especially for the elements carbon, phosphorus, and sulfur; therefore, when rimmed or capped steel is specified or required, the limitations for these elements shall not be applicable. Because of the degree to which phosphorus and sulfur segregate, the limitations for these elements shall not be applicable to rephosphorized or resulfurized steels.

4.3.3 Samples for product analysis shall be taken by one of the following methods:

4.3.3.1 Applicable to small sections whose cross-sectional area does not exceed 0.75 in.^{2} [500 mm^2] such as rounds,

TABLE 2 Grade Designations and Chemical Compositions of Alloy Steel Bars

NOTE 1—Small quantities of certain elements are present in alloy steels, which are not specified or required. These elements are considered as incidental and may be present to the following maximum amounts: copper, 0.35 %; nickel, 0.25 %; chromium, 0.20 % and molybdenum, 0.06 %.
NOTE 2—Where minimum and maximum sulfur content is shown it is indicative of resulfurized steel.
NOTE 3—The chemical ranges and limits shown in Table 2 are produced to product analysis tolerances shown in Table 6.
NOTE 4—Standard alloy steels can be produced with a lead range of 0.15–0.35 %. Such steels are identified by inserting the letter "L" between the second and third numerals of the AISI number, for example, 41 L 40. A cast or heat analysis is not determinable when lead is added to the ladle stream.

Grade Designation	Heat Chemical Ranges and Limits, %							
	Carbon	Manganese	Phosphorus, max	Sulfur, max	Silicon[A]	Nickel	Chromium	Molybdenum
1330	0.28–0.33	1.60–1.90	0.035	0.040	0.15 to 0.35	...	...	...
1335	0.33–0.38	1.60–1.90	0.035	0.040	0.15 to 0.35	...	...	...
1340	0.38–0.43	1.60–1.90	0.035	0.040	0.15 to 0.35	...	...	...
1345	0.43–0.48	1.60–1.90	0.035	0.040	0.15 to 0.35	...	...	...
4012	0.09–0.14	0.75–1.00	0.035	0.040	0.15 to 0.35	...	...	0.15–0.25
4023	0.20–0.25	0.70–0.90	0.035	0.040	0.15 to 0.35	...	...	0.20–0.30
4024	0.20–0.25	0.70–0.90	0.035	0.035–0.050	0.15 to 0.35	...	...	0.20–0.30
4027	0.25–0.30	0.70–0.90	0.035	0.040	0.15 to 0.35	...	...	0.20–0.30
4028	0.25–0.30	0.70–0.90	0.035	0.035–0.050	0.15 to 0.35	...	...	0.20–0.30
4032	0.30–0.35	0.70–0.90	0.035	0.040	0.15 to 0.35	...	...	0.20–0.30
4037	0.35–0.40	0.70–0.90	0.035	0.040	0.15 to 0.35	...	...	0.20–0.30
4042	0.40–0.45	0.70–0.90	0.035	0.040	0.15 to 0.35	...	...	0.20–0.30
4047	0.45–0.50	0.70–0.90	0.035	0.040	0.15 to 0.35	...	...	0.20–0.30
4118	0.18–0.23	0.70–0.90	0.035	0.040	0.15 to 0.35	...	0.40–0.60	0.08–0.15
4120	0.18–0.23	0.90–1.20	0.035	0.040	0.15 to 0.35	...	0.40–0.60	0.13–0.20
4121	0.18–0.23	0.75–1.00	0.035	0.040	0.15 to 0.35	...	0.45–0.65	0.20–0.30
4130	0.28–0.33	0.40–0.60	0.035	0.040	0.15 to 0.35	...	0.80–1.10	0.15–0.25
4135	0.33–0.38	0.70–0.90	0.035	0.040	0.15 to 0.35	...	0.80–1.10	0.15–0.25
4137	0.35–0.40	0.70–0.90	0.035	0.040	0.15 to 0.35	...	0.80–1.10	0.15–0.25
4140	0.38–0.43	0.75–1.00	0.035	0.040	0.15 to 0.35	...	0.80–1.10	0.15–0.25
4142	0.40–0.45	0.75–1.00	0.035	0.040	0.15 to 0.35	...	0.80–1.10	0.15–0.25
4145	0.43–0.48	0.75–1.00	0.035	0.040	0.15 to 0.35	...	0.80–1.10	0.15–0.25
4147	0.45–0.50	0.75–1.00	0.035	0.040	0.15 to 0.35	...	0.80–1.10	0.15–0.25
4150	0.48–0.53	0.75–1.00	0.035	0.040	0.15 to 0.35	...	0.80–1.10	0.15–0.25
4161	0.56–0.64	0.75–1.00	0.035	0.040	0.15 to 0.35	...	0.70–0.90	0.25–0.35
4320	0.17–0.22	0.45–0.65	0.035	0.040	0.15 to 0.35	1.65–2.00	0.40–0.60	0.20–0.30
4340	0.38–0.43	0.60–0.80	0.035	0.040	0.15 to 0.35	1.65–2.00	0.70–0.90	0.20–0.30
E4340	0.38–0.43	0.65–0.85	0.025	0.025	0.15 to 0.35	1.65–2.00	0.70–0.90	0.20–0.30
4419	0.18–0.23	0.45–0.65	0.035	0.040	0.15 to 0.35	...	...	0.45–0.60
4422	0.20–0.25	0.70–0.90	0.035	0.040	0.15 to 0.35	...	...	0.35–0.45
4427	0.24–0.29	0.70–0.90	0.035	0.040	0.15 to 0.35	...	...	0.35–0.45
4615	0.13–0.18	0.45–0.65	0.035	0.040	0.15 to 0.35	1.65–2.00	...	0.20–0.30
4620	0.17–0.22	0.45–0.65	0.035	0.040	0.15 to 0.35	1.65–2.00	...	0.20–0.30
4621	0.18–0.23	0.70–0.90	0.035	0.040	0.15 to 0.35	1.65–2.00	...	0.20–0.30
4626	0.24–0.29	0.45–0.65	0.035	0.040	0.15 to 0.35	0.70–1.00	...	0.15–0.25
4715	0.13–0.18	0.70–0.90	0.035	0.040	0.15 to 0.35	0.70–1.00	0.45–0.65	0.45–0.60
4718	0.16–0.21	0.70–0.90	0.035	0.040	0.15 to 0.35	0.90–1.20	0.35–0.55	0.30–0.40
4720	0.17–0.22	0.50–0.70	0.035	0.040	0.15 to 0.35	0.90–1.20	0.35–0.55	0.15–0.25
4815	0.13–0.18	0.40–0.60	0.035	0.040	0.15 to 0.35	3.25–3.75	...	0.20–0.30
4817	0.15–0.20	0.40–0.60	0.035	0.040	0.15 to 0.35	3.25–3.75	...	0.20–0.30
4820	0.18–0.23	0.50–0.70	0.035	0.040	0.15 to 0.35	3.25–3.75	...	0.20–0.30
5015	0.12–0.17	0.30–0.50	0.035	0.040	0.15 to 0.35	...	0.30–0.50	...
5046	0.43–0.48	0.75–1.00	0.035	0.040	0.15 to 0.35	...	0.20–0.35	...
5115	0.13–0.18	0.70–0.90	0.035	0.040	0.15 to 0.35	...	0.70–0.90	...
5120	0.17–0.22	0.70–0.90	0.035	0.040	0.15 to 0.35	...	0.70–0.90	...
5130	0.28–0.33	0.70–0.90	0.035	0.040	0.15 to 0.35	...	0.80–1.10	...
5132	0.30–0.35	0.60–0.80	0.035	0.040	0.15 to 0.35	...	0.75–1.00	...
5135	0.33–0.38	0.60–0.80	0.035	0.040	0.15 to 0.35	...	0.80–1.05	...
5140	0.38–0.43	0.70–0.90	0.035	0.040	0.15 to 0.35	...	0.70–0.90	...
5145	0.43–0.48	0.70–0.90	0.035	0.040	0.15 to 0.35	...	0.70–0.90	...
5147	0.46–0.51	0.70–0.95	0.035	0.040	0.15 to 0.35	...	0.85–1.15	...
5150	0.48–0.53	0.70–0.90	0.035	0.040	0.15 to 0.35	...	0.70–0.90	...
5155	0.51–0.59	0.70–0.90	0.035	0.040	0.15 to 0.35	...	0.70–0.90	...
5160	0.56–0.61	0.75–1.00	0.035	0.040	0.15 to 0.35	...	0.70–0.90	...
E50100	0.98–1.10	0.25–0.45	0.025	0.025	0.15 to 0.35	...	0.40–0.60	...
E51100	0.98–1.10	0.25–0.45	0.025	0.025	0.15 to 0.35	...	0.90–1.15	...
E52100	0.98–1.10	0.25–0.45	0.025	0.025	0.15 to 0.35	...	1.30–1.60	...
52100[B]	0.93–1.05	0.25–0.45	0.025	0.015	0.15 to 0.35	...	1.35–1.60	...
6118	0.16–0.21	0.50–0.70	0.035	0.040	0.15 to 0.35	...	0.50–0.70	(0.10–0.15 V)
6150	0.48–0.53	0.70–0.90	0.035	0.040	0.15 to 0.35	...	0.80–1.10	(0.15 min V)
8115	0.13–0.18	0.70–0.90	0.035	0.040	0.15 to 0.35	0.20–0.40	0.30–0.50	0.08–0.15
8615	0.13–0.18	0.70–0.90	0.035	0.040	0.15 to 0.35	0.40–0.70	0.40–0.60	0.15–0.25

TABLE 2 Continued

Grade Designation	Heat Chemical Ranges and Limits, %							
	Carbon	Manganese	Phosphorus, max	Sulfur, max	Silicon[A]	Nickel	Chromium	Molybdenum
8617	0.15–0.20	0.70–0.90	0.035	0.040	0.15 to 0.35	0.40–0.70	0.40–0.60	0.15–0.25
8620	0.18–0.23	0.70–0.90	0.035	0.040	0.15 to 0.35	0.40–0.70	0.40–0.60	0.15–0.25
8622	0.20–0.25	0.70–0.90	0.035	0.040	0.15 to 0.35	0.40–0.70	0.40–0.60	0.15–0.25
8625	0.23–0.28	0.70–0.90	0.035	0.040	0.15 to 0.35	0.40–0.70	0.40–0.60	0.15–0.25
8627	0.25–0.30	0.70–0.90	0.035	0.040	0.15 to 0.35	0.40–0.70	0.40–0.60	0.15–0.25
8630	0.28–0.33	0.70–0.90	0.035	0.040	0.15 to 0.35	0.40–0.70	0.40–0.60	0.15–0.25
8637	0.35–0.40	0.75–1.00	0.035	0.040	0.15 to 0.35	0.40–0.70	0.40–0.60	0.15–0.25
8640	0.38–0.43	0.75–1.00	0.035	0.040	0.15 to 0.35	0.40–0.70	0.40–0.60	0.15–0.25
8642	0.40–0.45	0.75–1.00	0.035	0.040	0.15 to 0.35	0.40–0.70	0.40–0.60	0.15–0.25
8645	0.43–0.48	0.75–1.00	0.035	0.040	0.15 to 0.35	0.40–0.70	0.40–0.60	0.15–0.25
8650	0.48–0.53	0.75–1.00	0.035	0.040	0.15 to 0.35	0.40–0.70	0.40–0.60	0.15–0.25
8655	0.51–0.59	0.75–1.00	0.035	0.040	0.15 to 0.35	0.40–0.70	0.40–0.60	0.15–0.25
8660	0.56–0.64	0.75–1.00	0.035	0.040	0.15 to 0.35	0.40–0.70	0.40–0.60	0.15–0.25
8720	0.18–0.23	0.70–0.90	0.035	0.040	0.15 to 0.35	0.40–0.7	0.40–0.60	0.20–0.30
8740	0.38–0.43	0.75–1.00	0.035	0.040	0.15 to 0.35	0.40–0.70	0.40–0.60	0.20–0.30
8822	0.20–0.25	0.75–1.00	0.035	0.040	0.15 to 0.35	0.40–0.70	0.40–0.60	0.30–0.40
9254	0.51–0.59	0.60–0.80	0.035	0.040	1.20–1.60	...	0.60–0.80	...
9255	0.51–0.59	0.70–0.95	0.035	0.040	1.80–2.20	...	...	...
9259	0.56–0.64	0.75–1.00	0.035	0.040	0.70–1.10	...	0.45–0.65	...
9260	0.56–0.64	0.75–1.00	0.035	0.040	1.80–2.20	...	...	...
E9310	0.08–0.13	0.45–0.65	0.025	0.025	0.15 to 0.30	3.00–3.50	1.00–1.40	0.08–0.15
Standard Boron Steels[C]								
50B44	0.43–0.48	0.75–1.00	0.035	0.040	0.15–0.35	...	0.20–0.60	...
50B46	0.44–0.49	0.75–1.00	0.035	0.040	0.15–0.35	...	0.20–0.35	...
50B50	0.48–0.53	0.75–1.00	0.035	0.040	0.15–0.35	...	0.40–0.60	...
50B60	0.56–0.64	0.75–1.00	0.035	0.040	0.15–0.35	...	0.40–0.60	...
51B60	0.56–0.64	0.75–1.00	0.035	0.040	0.15–0.35	...	0.70–0.90	...
81B45	0.43–0.48	0.75–1.00	0.035	0.040	0.15–0.35	0.20–0.40	0.35–0.55	0.08–0.15
94B17	0.15–0.20	0.75–1.00	0.035	0.040	0.15–0.35	0.30–0.60	0.30–0.50	0.80–0.15
94B30	0.28–0.33	0.75–1.00	0.035	0.040	0.15–0.35	0.30–0.60	0.30–0.50	0.08–0.15

[A] Silicon may be specified by the purchaser as 0.10 % maximum. The need for 0.10% maximum generally relates to severe cold-formed parts.
[B] The purchaser may also require the following maximums: copper 0.30 %; aluminum 0.050 %; oxygen 0.0015 %.
[C] These steels can be expected to contain 0.0005 to 0.003 % boron. If the usual titanium additive is not permitted, the steels can be expected to contain up to 0.005 % boron.

squares, hexagons, and the like. Chips are taken by milling or machining the full cross section of the piece. Drilling is not a feasible method for sampling sizes 0.75 in.² and smaller.

4.3.3.2 Applicable to products where the width of the cross section greatly exceeds the thickness, such as bar size shapes and light flat bars. Chips are taken by drilling entirely through the steel at a point midway between the edge and the middle of the section, or by milling or machining the entire cross section.

4.3.3.3 Applicable to large rounds, squares semifinished, etc. Chips are taken at any point midway between the outside and the center of the piece by drilling parallel to the axis or by milling or machining the full cross section. In cases where these methods are not practicable, the piece may be drilled on the side, but chips are not taken until they represent the portion midway between the outside and the center.

4.3.3.4 When the steel is subject to tension test requirements, the tension test specimen can also be used for product analysis. In that case, chips for product analysis can be taken by drilling entirely through the tension test specimens or by the method described in 4.3.3.1.

4.3.4 When chips are taken by drilling, the diameter of the drill used shall conform to the following:

Area of Sample Cross Section, in.²(cm²)	Approximate Drill Diameter, in. (mm)
16 [100] or less	½ [12.5]
Over 16 [100]	1 [25.0]

4.3.5 The minimum number of samples to be taken from material representing the same heat or lot before rejection by the purchaser shall be as follows:

	Minimum Number of Samples
15 tons [15 Mg] and under	4
Over 15 tons [15 Mg]	6

4.3.6 In case the number of pieces in a heat is less than the number of samples required, one sample from each piece shall be considered sufficient.

4.3.7 In the event that product analysis determinations are outside the permissible limits as prescribed in 4.3.2, additional samples shall be analyzed and the acceptability of the heat negotiated between the purchaser and the producer.

4.4 *Referee Analysis*—In case a referee analysis is required and agreed upon to resolve a dispute concerning the results of a chemical analysis, the referee analysis shall be performed in accordance with the latest issue of Test Methods, Practices, and

TABLE 3 Heat Analysis Chemical Ranges and Limits of Carbon Steel Bars

Element	Chemical Ranges and Limits, %		
	When Maximum of Specified Elements is:	Range	Lowest Maximum
Carbon[A]	...	...	0.06
	to 0.12, incl	...	...
	over 0.12 to 0.25, incl	0.05	...
	over 0.25 to 0.40, incl	0.06	...
	over 0.40 to 0.55, incl	0.07	...
	over 0.55 to 0.80, incl	0.10	...
	over 0.80	0.13	...
Manganese	...	...	0.35
	to 0.40, incl	0.15	...
	over 0.40 to 0.50, incl	0.20	...
	over 0.50 to 1.65, incl	0.30	...
Phosphorus	to 0.040, incl	...	0.040[B]
	over 0.040 to 0.08, incl	0.03	...
	over 0.08 to 0.13, incl	0.05	...
Sulfur	to 0.050, incl	...	0.050[B]
	over 0.050 to 0.09, incl	0.03	...
	over 0.09 to 0.15, incl	0.05	...
	over 0.15 to 0.23, incl	0.07	...
	over 0.23 to 0.50, incl	0.09	...
Silicon[C]	...	...	0.10
	to 0.10, incl	...	...
	over 0.10 to 0.15, incl	0.08	...
	over 0.15 to 0.20, incl	0.10	...
	over 0.20 to 0.30, incl	0.15	...
	over 0.30 to 0.60, incl	0.20	...
Copper	When copper is required 0.20 min is generally used		
Lead[D]	When lead is required, a range of 0.15 to 0.35 is specified		
Bismuth[E] Calcium[E] Selenium[E] Tellurium[E]			

[A] The carbon ranges shown in the column headed "Range" apply when the specified maximum limit for manganese does not exceed 1.10 %. When the maximum manganese limit exceeds 1.10 %, add 0.01 to the carbon ranges shown above.

[B] For steels produced in merchant quality the phosphorus maximum is 0.04 % and the sulfur maximum is 0.05 %.

[C] It is not common practice to produce a rephosphorized and resulfurized carbon steel to specified limits for silicon because of its adverse effect on machinability.

[D] A cast or heat analysis is not determinable when lead is added to the ladle stream.

[E] Element specification range as agreed upon between purchaser and supplier.

Terminology A751, unless otherwise agreed upon between the manufacturer and the purchaser.

5. Grain Size Requirement

5.1 *Austenitic Grain Size*:

5.1.1 When a coarse austenitic grain size is specified, the steel shall have a grain size number of 1 to 5 exclusive as determined in accordance with Test Methods E112. Conformance to this grain size of 70 % of the grains in the area examined shall constitute the basis of acceptance. One test per heat shall be made.

5.1.2 When a fine austenitic grain size is specified, the steel shall have a grain size number of 5 or higher as determined in accordance with Test Methods E112. Conformance to this grain size of 70 % of the area examined shall constitute the basis of acceptance. One test per heat shall be made unless the provisions of 5.1.2.1 or 5.1.2.2 are exercised.

5.1.2.1 When aluminum is used as the grain refining element, the fine austenitic grain size requirement shall be deemed to be fulfilled if, on heat analysis, the aluminum content is not less than 0.020 % total aluminum or, alternately, 0.015 % acid soluble aluminum. The aluminum content shall be reported. The grain size test specified in 5.1.2 shall be the referee test.

5.1.2.2 By agreement between purchaser and supplier, columbium or vanadium or both may be used for grain refining instead of or with aluminum. When columbium or vanadium is used as a grain refining element, the fine austenitic grain size requirement shall be deemed to be fulfilled if, on heat analysis, the columbium or vanadium content is as follows (the content of the elements shall be reported with the heat analysis):

Steels having 0.25 % carbon or less:
Cb   0.025 min
V   0.05 min

Steels having over 0.25 % carbon:
Cb   0.015 min
V   0.02 min

The maximum contents shall be:
Cb   0.05 max
V   0.08 max
Cb + V   0.06 max

5.1.2.3 When provisions of 5.1.2.1 or 5.1.2.2 are exercised, a grain size test is not required unless specified by the purchaser. Unless otherwise specified, fine austenitic grain size shall be certified using the analysis of grain refining element(s).

5.1.2.4 *Referee Test*—In the event that the chemical analysis of columbium or vanadium does not meet the requirements of 5.1.2.2, the grain size test shown in 5.1.2 shall be the referee test unless an alternative test method is agreed upon between the manufacturer and the purchaser.

TABLE 4 Heat Analysis Chemical Ranges and Limits of Alloy Steel Bars

Note 1—Boron steels can be expected to have 0.0005 % minimum boron content.
Note 2—Alloy steels can be produced with a lead range of 0.15–0.35 %. A cast or heat analysis is not determinable when lead is added to the ladle stream.

Element	When Maximum of Specified Element is:	Chemical Ranges and Limits, %		
		Open-Hearth or Basic-Oxygen Steel	Electric Furnace Steel	Maximum Limit, %[A]
Carbon	To 0.55, incl	0.05	0.05	
	Over 0.55–0.70, incl	0.08	0.07	
	Over 0.70 to 0.80, incl	0.10	0.09	
	Over 0.80–0.95, incl	0.12	0.11	
	Over 0.95–1.35, incl	0.13	0.12	
Manganese	To 0.60, incl	0.20	0.15	
	Over 0.60–0.90, incl	0.20	0.20	
	Over 0.90–1.05, incl	0.25	0.25	
	Over 1.05–1.90, incl	0.30	0.30	
	Over 1.90–2.10, incl	0.40	0.35	
Phosphorus	Basic open-hearth or basic-oxygen steel			0.035
	Acid open-hearth steel			0.050
	Basic electric-furnace steel			0.025
	Acid electric-furnace steel			0.050
Sulfur	To 0.050, incl	0.015	0.015	
	Over 0.050–0.07, incl	0.02	0.02	
	Over 0.07–0.10, incl	0.04	0.04	
	Over 0.10–0.14, incl	0.05	0.05	
	Basic open-hearth or basic-oxygen steel			0.040
	Acid open-hearth steel			0.050
	Basic electric-furnace steel			0.025
	Acid electric-furnace steel			0.050
Silicon	To 0.20, incl	0.08	0.08	
	Over 0.20–0.30, incl	0.15	0.15	
	Over 0.30–0.60, incl	0.20	0.20	
	Over 0.60–1.00, incl	0.30	0.30	
	Over 1.00–2.20, incl	0.40	0.35	
	Acid steels[B]			
Nickel	To 0.50, incl	0.20	0.20	
	Over 0.50–1.50, incl	0.30	0.30	
	Over 1.50–2.00, incl	0.35	0.35	
	Over 2.00–3.00, incl	0.40	0.40	
	Over 3.00–5.30, incl	0.50	0.50	
	Over 5.30–10.00, incl	1.00	1.00	
Chromium	To 0.40, incl	0.15	0.15	
	Over 0.40–0.90, incl	0.20	0.20	
	Over 0.90–1.05, incl	0.25	0.25	
	Over 1.05–1.60, incl	0.30	0.30	
	Over 1.60–1.75, incl	[C]	0.35	
	Over 1.75–2.10, incl	[C]	0.40	
	Over 2.10–3.99, incl	[C]	0.50	
Molybdenum	To 0.10, incl	0.05	0.05	
	Over 0.10–0.20, incl	0.07	0.07	
	Over 0.20–0.50, incl	0.10	0.10	
	Over 0.50–0.80, incl	0.15	0.15	
	Over 0.80–1.15, incl	0.20	0.20	
Tungsten	To 0.50, incl	0.20	0.20	
	Over 0.50–1.00, incl	0.30	0.30	
	Over 1.00–2.00, incl	0.50	0.50	
	Over 2.00–4.00, incl	0.60	0.60	
Vanadium	To 0.25, incl	0.05	0.05	
	Over 0.25–0.50, incl	0.10	0.10	
Aluminum	Up to 0.10, incl	0.05	0.05	
	Over 0.10–0.20, incl	0.10	0.10	
	Over 0.20–0.30, incl	0.15	0.15	
	Over 0.30–0.80, incl	0.25	0.25	
	Over 0.80–1.30, incl	0.35	0.35	
	Over 1.30–1.80, incl	0.45	0.45	
Copper	To 0.60, incl	0.20	0.20	
	Over 0.60–1.50, incl	0.30	0.30	
	Over 1.50–2.00, incl	0.35	0.35	

[A] Applies to only nonrephosphorized and nonresulfurized steels.
[B] Minimum silicon limit for acid open-hearth or acid electric-furnace alloy steels is 0.15 %.
[C] Not normally produced in open-hearth.

TABLE 5 Permissible Variations for Product Analysis of Carbon Steel

Element	Limit, or Maximum of Specified Range, %	Over Maximum Limit, %	Under Minimum Limit, %
Carbon[A]	0.25 and under	0.02	0.02
	over 0.25 to 0.55, incl	0.03	0.03
	over 0.55	0.04	0.04
Manganese	0.90 and under	0.03	0.03
	over 0.90 to 1.65, incl	0.06	0.06
Phosphorus[A,B]	basic steels	0.008	...
	acid bessemer steel	0.01	0.01
Sulfur[A,B]		0.008	...
Silicon	0.35 and under	0.02	0.02
	over 0.35 to 0.60, incl	0.05	0.05
Copper	under minimum only	...	0.02
Lead[C]	0.15 to 0.35, incl	0.03	0.03

[A] Rimmed and capped steels are not subject to rejection on product analysis unless misapplication is clearly indicated.

[B] Resulfurized or rephosphorized steels are not subject to rejection on product analysis for these elements unless misapplication is clearly indicated.

[C] Product analysis tolerance for lead applies both over and under to a specified range of 0.15 to 0.35 %.

TABLE 6 Permissible Variations for Product Analysis of Alloy Steel

Elements	Limit, or Maximum of Specified Range, %	Permissible Variations Over Maximum Limit or Under Minimum Limit, %
Carbon	0.30 and under	0.01
	over 0.30 to 0.75, incl	0.02
	over 0.75	0.03
Manganese	0.90 and under	0.03
	over 0.90 to 2.10, incl	0.04
Phosphorus	over maximum only	0.005
Sulfur	0.060 and under	0.005
Silicon	0.40 and under	0.02
	over 0.40 to 2.20, incl	0.05
Nickel	1.00 and under	0.03
	over 1.00 to 2.00, incl	0.05
	over 2.00 to 5.30, incl	0.07
	over 5.30 to 10.00, incl	0.10
Chromium	0.90 and under	0.03
	over 0.90 to 2.10, incl	0.05
	over 2.10 to 3.99, incl	0.10
Molybdenum	0.20 and under	0.01
	over 0.20 to 0.40, incl	0.02
	over 0.40 to 1.15, incl	0.03
Vanadium	0.10 and under	0.01
	over 0.10 to 0.25, incl	0.02
	over 0.25 to 0.50, incl	0.03
	minimum value specified, under minimum limit only	0.01
Tungsten	1.00 and under	0.04
	over 1.00 to 4.00, incl	0.08
Aluminum	0.10 and under	0.03
	over 0.10 to 0.20, incl	0.04
	over 0.20 to 0.30, incl	0.05
	over 0.30 to 0.80, incl	0.07
	over 0.80 to 1.80, incl	0.10
Lead[A]	0.15 to 0.35, incl	0.03
Copper	to 1.00 incl	0.03
	over 1.00 to 2.00, incl	0.05

[A] Product analysis tolerance for lead applies both over and under to a specified range of 0.15 to 0.35 %.

6. Mechanical Property Requirements

6.1 *Test Specimens*:

6.1.1 *Selection*—Test specimens shall be selected in accordance with the requirements of the applicable product specification or in accordance with Supplement I of the latest issue of Test Methods and Definitions A370, in the sequence named.

6.1.2 *Preparation*—Unless otherwise specified in the applicable product specification, test specimens shall be prepared in accordance with the latest issue of Test Methods and Definitions A370, and especially Supplement I thereof.

6.2 *Methods of Mechanical Testing*—All mechanical tests shall be conducted in accordance with the latest issue of Test Methods and Definitions A370, and especially Supplement I thereof, on steel bar products.

6.3 *Retests*:

6.3.1 If any test specimen shows defective machining or develops flaws, the specimen may be discarded and another substituted.

6.3.2 If the percentage elongation of any tension specimen is less than that specified and any part of the fracture is more than ¾ in. [20 mm] from the center of a 2-in. [50-mm] specimen, or is outside the middle half of the gage length of an 8-in. [200-mm] specimen as indicated by scribe scratches marked on the specimen before testing, a retest shall be allowed.

6.3.3 For "as-wrought" material, if the results for any original tension specimen are within 2000 psi [14 MPa] of the required tensile strength, within 1000 psi [7 MPa] of the required yield point, or within 2 % of the required elongation, retesting shall be permitted. If the original testing required only one test, the retest shall consist of two random tests from the heat or lot involved. If the original testing required two tests of which one failed by the amounts listed in this paragraph, the retest shall be made on one random test from the heat or lot. If the results on the retest specimen or specimens meet the specified requirements, the heat or test lot will be accepted. If the results of one retest specimen do not meet the specified requirements, the material is subject to rejection.

6.3.4 For thermally treated bars, if the results of the mechanical tests do not conform to the requirements specified, two more tests may be selected for each bar failing, and each of these retests shall conform to the requirements of the product specification.

6.3.5 If a bend specimen fails, due to conditions of bending more severe than required by the specification, a retest shall be permitted from the heat or test lot involved for which one random specimen for each original specimen showing failure shall be used. If the results on the retest specimen meet the requirements of the specification, the heat or test lot will be accepted.

7. Dimensions, Mass, and Permissible Variations

7.1 *Hot-Wrought Bars*—The permissible variations for dimensions of hot-wrought carbon and alloy steel bars shall not exceed the applicable limits stated in Annex A1 for inch-pound values and Annex A2 for metric values.

8. Workmanship, Finish, and Appearance

8.1 The material shall be free of injurious defects and shall have a workmanlike finish.

9. Rework and Retreatment

9.1 For thermally treated bars only, the manufacturer may retreat a lot one or more times, and retests shall be made in the same manner as the original tests. Each such retest shall conform to the requirements specified.

10. Inspection

10.1 The inspector representing the purchaser shall have entry, at all times while work on the contract of the purchaser is being performed, to all parts of the manufacturer's works that concern the manufacture of the material ordered. The manufacturer shall afford the inspector all reasonable facilities to satisfy him that the material is being furnished in accordance with this specification. All tests (except product analysis) and inspection shall be made at the place of manufacture prior to shipment, unless otherwise specified, and shall be so conducted as not to interfere unnecessarily with the operation of the works.

10.2 All required tests and inspection shall be made by the manufacturer prior to shipment.

11. Rejection

11.1 Unless otherwise specified, any rejection because of noncompliance to the requirements of the specification shall be reported by the purchaser to the manufacturer within 30 working days after receipt of samples.

11.2 Material that shows imperfections capable of adversely affecting processibility subsequent to its acceptance at the purchaser's works will be rejected, and the manufacturer shall be notified.

12. Rehearing

12.1 Samples that represent rejected material shall be preserved for two weeks from the date rejection is reported to the manufacturer. In case of dissatisfaction with the results of the tests, the manufacturer may make claim for a rehearing within that time.

13. Product Marking

13.1 *Civilian Procurement*—Bars of all sizes, when loaded for shipment, shall be properly identified with the name or brand of manufacturer, purchaser's name and order number, the ASTM designation (year date is not required), grade number where appropriate, size and length, weight of lift, and the heat number for identification. Unless otherwise specified, the method of marking is at the manufacturer's option and may be made by hot stamping, cold stamping, painting, or marking tags attached to the lifts of bars.

13.1.1 Bar code marking may be used as an auxiliary method of identification. Such bar-code markings shall be of the 3-of-9 type and shall conform to AIAG B1. When barcoded tags are used, they shall conform to AIAG B5.

13.2 *Government Procurement*:

13.2.1 Marking for shipment shall be in accordance with the requirements specified in the contract or order and shall be in accordance with MIL-STD-163 for military agencies and in accordance with Fed. Std. No. 123 for civil agencies.

13.2.2 For government procurement by the Defense Supply Agency, the bars shall be continuously marked for identification in accordance with Fed. Std. No. 183.

14. Packaging

14.1 *Civilian Procurement*—Unless otherwise specified, the bars shall be packaged and loaded in accordance with Practices A700.

14.2 *Government Procurement*—MIL-STD-163 shall apply when packaging is specified in the contract or order, or when Level A for preservation, packaging, and packing is specified for direct procurement by or direct shipment to the government.

15. Keywords

15.1 alloy steel bars; carbon steel bars; cold finished steel bars; general delivery requirements; hot wrought steel bars; steel bars

---

SUPPLEMENTARY REQUIREMENTS

The following supplementary requirements shall apply only when specified by the purchaser in the contract or order.

S1. Flat Bar Thickness Tolerances

S1.1 When flat bars are specified in metric units to a thickness under tolerance of 0.3 mm, the thickness tolerance of Table S1.1 shall apply.

TABLE S1.1 Thickness and Width Tolerances for Hot-Wrought Square-Edge and Round-Edge Flat Bars Ordered to 0.3 mm Under Tolerance[A]

NOTE—Tolerance under specified thickness 0.3 mm.

Specified Width, mm	Tolerances over Specified Thickness for Thickness Given, mm					Tolerance from Specified Width, mm	
	Over 6 to 12, incl	Over 12 to 25, incl	Over 25 to 50, incl	Over 50 to 75, incl	Over 75	Over	Under
To 25, incl	...	...	...	...	...	0.5	0.5
Over 25 to 50, incl	...	0.5	1.3	...	...	1.0	1.0
Over 50 to 100, incl	0.5	0.7	1.3	2.1	2.1	1.5	1.0
Over 100 to 150, incl	0.5	0.7	1.3	2.1	2.1	2.5	1.5
Over 150 to 200, incl	0.5	1.0	1.3	2.1	2.9	3.0	2.5

[A] When a square is held against a face and an edge of a square-edge flat bar, the edge shall not deviate by more than 3° or 5 % of the thickness.

ANNEXES

(Mandatory Information)

A1. PERMISSIBLE VARIATIONS IN DIMENSIONS, ETC.—INCH-POUND UNITS

A1.1 Listed below are permissible variations in dimensions expressed in inch-pound units of measurement.

TABLE A1.1 Permissible Variations in Cross Section for Hot-Wrought Round, Square, and Round-Cornered Square Bars of Steel

Specified Size, in.	Permissible Variation from Specified Size, in.[A]		Out-of-Round or Out-of-Square, in.[B]
	Over	Under	
To 5/16, incl	0.005	0.005	0.008
Over 5/16 to 7/16, incl	0.006	0.006	0.009
Over 7/16 to 5/8, incl	0.007	0.007	0.010
Over 5/8 to 7/8, incl	0.008	0.008	0.012
Over 7/8 to 1, incl	0.009	0.009	0.013
Over 1 to 1 1/8, incl	0.010	0.010	0.015
Over 1 1/8 to 1 1/4, incl	0.011	0.011	0.016
Over 1 1/4 to 1 3/8, incl	0.012	0.012	0.018
Over 1 3/8 to 1 1/2, incl	0.014	0.014	0.021
Over 1 1/2 to 2, incl	1/64	1/64	0.023
Over 2 to 2 1/2, incl	1/32	0	0.023
Over 2 1/2 to 3 1/2, incl	3/64	0	0.035
Over 3 1/2 to 4 1/2, incl	1/16	0	0.046
Over 4 1/2 to 5 1/2, incl	5/64	0	0.058
Over 5 1/2 to 6 1/2, incl	1/8	0	0.070
Over 6 1/2 to 8 1/4, incl	5/32	0	0.085
Over 8 1/4 to 9 1/2, incl	3/16	0	0.100
Over 9 1/2 to 10, incl	1/4	0	0.120

[A] Steel bars are regularly cut to length by shearing or hot sawing, which can cause end distortion resulting in those portions of the bar being outside the applicable size tolerance. When this end condition is objectionable, a machine cut end should be considered.

[B] Out-of-round is the difference between the maximum and minimum diameters of the bar, measured at the same cross section. Out-of-square is the difference in the two dimensions at the same cross section of a square bar between opposite faces.

TABLE A1.2 Permissible Variations in Cross Section for Hot-Wrought Hexagonal Bars of Steel

Specified Sizes Between Opposite Sides, in.	Permissible Variations from Specified Size, in.[A]		Out-of-Hexagon (Carbon Steel and Alloy Steel) or Out-of-Octagon (Alloy Steel), in.[B]
	Over	Under	
To ½ , incl	0.007	0.007	0.011
Over ½ to 1, incl	0.010	0.010	0.015
Over 1 to 1½ , incl	0.021	0.013	0.025
Over 1½ to 2, incl	1/32	1/64	1/32
Over 2 to 2½ , incl	3/64	1/64	3/64
Over 2½ to 3½ , incl	1/16	1/64	1/16
Over 3½ to 4 1/16 , incl	5/64	1/64	5/64

[A] Steel bars are regularly cut to length by shearing or hot sawing, which can cause end distortion resulting in those portions of the bar being outside the applicable size tolerance. When this end condition is objectionable, a machine cut end should be considered.

[B] Out-of-hexagon or out-of-octagon is the greatest difference between any two dimensions at the same cross section between opposite faces.

TABLE A1.3 Permissible Variations in Thickness and Width for Hot-Wrought Square Edge and Round Edge Flat Bars[A]

Specified Width, in.	Permissible Variations in Thickness, for Thickness Given, Over and Under, in.[B]							Permissible Variations in Width, in.	
	0.203 to 0.230, excl	0.230 to ¼ , excl	¼ to ½ , incl	Over ½ to 1, incl	Over 1 to 2, incl	Over 2 to 3, incl	Over 3	Over	Under
To 1, incl	0.007	0.007	0.008	0.010	...	...	...	1/64	1/64
Over 1 to 2, incl	0.007	0.007	0.012	0.015	1/32	...	...	1/32	1/32
Over 2 to 4, incl	0.008	0.008	0.015	0.020	1/32	3/64	3/64	1/16	1/32
Over 4 to 6, incl	0.009	0.009	0.015	0.020	1/32	3/64	3/64	3/32	1/16
Over 6 to 8, incl	[C]	0.015	0.016	0.025	1/32	3/64	1/16	1/8	3/32

[A] When a square is held against a face and an edge of a square edge flat bar, the edge shall not deviate by more than 3° or 5 % of the thickness.

[B] Steel bars are regularly cut to length by shearing or hot sawing, which can cause end distortion resulting in those portions of the bar being outside the applicable size tolerance. When this end condition is objectionable, a machine cut end should be considered.

[C] Flats over 6 to 8 in., incl, in width, are not available as hot-wrought steel bars in thickness under 0.230 in.

TABLE A1.4 Permissible Variations in Thickness, Length, and Out-of-Square for Hot-Wrought Bar Size Angles of Carbon Steel

Specified Length of Leg, in.[A]	Permissible Variations in Thickness, for Thicknesses Given, Over and Under, in.			Permissible Variations for Length of Leg, Over and Under, in.
	To 3/16 , incl	Over 3/16 to 3/8 , incl	Over 3/8	
To 1, incl	0.008	0.010	...	1/32
Over 1 to 2, incl	0.010	0.010	0.012	3/64
Over 2 to 3, excl	0.012	0.015	0.015	1/16

[A] The longer leg of an unequal angle determines the size for tolerance. The out-of-square tolerance in either direction is 1½ °.

TABLE A1.5 Permissible Variations in Dimensions for Hot-Wrought Bar Size Channels of Carbon Steel

Specified Size of Channel, in.	Permissible Variations in Size, Over and Under, in.				Out-of-Square[A] if Either Flange, in./in. of Flange Width
	Depth of Section[B]	Width of Flanges[B]	Thickness of Web for Thickness Given		
			To 3/16 , incl	Over 3/16	
To 1½ , incl	1/32	1/32	0.010	0.015	1/32
Over 1½ to 3, excl	1/16	1/16	0.015	0.020	1/32

[A] For channels 5/8 in. and under in depth, the out-of-square tolerance is 3/64 in./in. of depth.

[B] Measurements for depth of section and width of flanges are overall.

TABLE A1.6 Permissible Variations in Dimensions for Hot-Wrought Bar Size Tees of Carbon Steel

Specified Size of Tee, in.[A]	Permissible Variations in Size, in.						Stem out-of-Square[C]
	Width or Depth[B]		Thickness of Flange		Thickness of Stem		
	Over	Under	Over	Under	Over	Under	
To 1¼ , incl	3/64	3/64	0.010	0.010	0.005	0.020	1/32
Over 1¼ to 2, incl	1/16	1/16	0.012	0.012	0.010	0.020	1/16
Over 2 to 3, excl	3/32	3/32	0.015	0.015	0.015	0.020	3/32

[A] The longer member of the unequal tee determines the size for tolerances.
[B] Measurements for both width and depth are overall.
[C] Stem out-of-square is the variation from its true position of the center line of the stem measured at the point.

TABLE A1.7 Permissible Variations in Dimensions for Half-Rounds, Ovals, Half-Ovals, and Other Special Bar Size Sections

Due to mill facilities, tolerances on half-rounds, ovals, half-ovals, and other special bar size sections vary among the manufacturers and such tolerances should be negotiated between the manufacturer and the purchaser.

TABLE A1.8 Permissible Variations in Length for Hot-Wrought Rounds, Squares, Hexagons, Flats, and Bar Size Sections of Steel

Specified Size of Rounds, Squares, and Hexagons, in.	Specified Size of Flats, in.		Permissible Variations Over Specified Length, in.[A]				
	Thickness	Width	5 to 10 ft, excl	10 to 20 ft, excl	20 to 30 ft, excl	30 to 40 ft, excl	40 to 60 ft, excl
			Mill Shearing				
To 1, incl	to 1, incl	to 3, incl	½	¾	1¼	1¾	2¼
Over 1 to 2, incl	over 1	to 3, incl	5/8	1	1½	2	2½
	to 1, incl	over 3 to 6, incl	5/8	1	1½	2	2½
Over 2 to 5, incl	over 1	over 3 to 6, incl	1	1½	1¾	2¼	2¾
Over 5 to 10, incl	...	...	2	2½	2¾	3	3¼
	0.230 to 1, incl	over 6 to 8, incl	¾	1¼	1¾	3½	4
	over 1 to 3, incl	over 6 to 8, incl	1¼	1¾	2	3½	4
Bar Size Sections	...	...	5/8	1	1½	2	2½
			Hot Sawing				
2 to 3½ , incl	1 and over	3 and over	[B]	1½	1¾	2¼	2¾
Over 3½ to 5, incl				2	2¼	2 5/8	3
Over 5 to 10, incl	...	...	[B]	2½	2¾	3	3¼

[A] No permissible variations under.
[B] Smaller sizes and shorter lengths are not hot sawed.

TABLE A1.9 Permissible Variations in Length for Recutting of Bars Meeting Special Straightness Tolerances

Sizes of Rounds, Squares, Hexagons, Width of Flats and Maximum Dimension of Other Sections, in.[A]	Tolerances Over Specified Length, in.[A]	
	To 12 ft, incl	Over 12 ft
To 3, incl	¼	5/16
Over 3 to 6, incl	5/16	7/16
Over 6 to 8, incl	7/16	9/16
Rounds over 8 to 10, incl.	9/16	11/16

[A] No tolerance under.

TABLE A1.10 Permissible Variations in Straightness for Hot-Wrought Bars and Bar Size Sections of Steel[A]

Standard tolerances	¼ in. in any 5 ft and (¼ in. × length in ft)/5
Special tolerances	⅛ in. in any 5 ft and (⅛ in. × length in ft)/5

[A] Because of warpage, straightness tolerances do not apply to bars if any subsequent heating operation or controlled cooling has been performed.

A2. DIMENSIONAL TOLERANCES—SI UNITS

A2.1 Listed below are permissible variations in dimensions expressed in SI units of measurement.

TABLE A2.1 Tolerances in Sectional Dimensions for Round and Square Bars and Round-Cornered Square Bars

Size, mm	Tolerance from Specified Size, Over and Under, mm or %[A]	Out-of-Round, or Out-of-Square Section,[B] mm or %[A]
To 7, incl	0.13 mm	0.20 mm
Over 7 to 11, incl	0.15 mm	0.22 mm
Over 11 to 15, incl	0.18 mm	0.27 mm
Over 15 to 19, incl	0.20 mm	0.30 mm
Over 19 to 250, incl	1 %	1.5 %

[A] The tolerance shall be rounded to the nearest tenth of a millimetre after calculation.
[B] Out-of-round is the difference between the maximum and the minimum diameters of the bar, measured at the same cross section. Out-of-square is the difference in the two dimensions at the same cross section of a square bar between opposite faces.

TABLE A2.2 Tolerances in Cross Section for Hot-Wrought Hexagonal and Octagonal Steel Bars

Specified Size Between Opposite Sides, mm	Tolerance from Specified Size, mm		Out of Hexagon or Out of Octagon, mm[A]
	Over	Under	
To 13, incl	0.18	0.18	0.3
Over 13 to 25, incl	0.25	0.25	0.4
Over 25 to 40, incl	0.55	0.35	0.6
Over 40 to 50, incl	0.8	0.40	0.8
Over 50 to 65, incl	1.2	0.40	1.2
Over 65 to 80, incl	1.6	0.40	1.6
Over 80 to 100, incl	2.0	0.40	2.0

[A] Out of hexagon or out of octagon is the greatest difference between any two dimensions at the cross section between opposite faces.

TABLE A2.3 Thickness and Width Tolerances for Hot-Wrought Square-Edge and Round-Edge Flat Bars[A,B]

Specified Width, mm	Tolerances from Specified Thickness for Thickness Given Over and Under, mm						Tolerances from Specified Width, mm	
	Over 5 to 6, incl	Over 6 to 12, incl	Over 12, to 25, incl	Over 25 to 50, incl	Over 50 to 75	Over 75	Over	Under
To 25, incl	0.18	0.20	0.25	...	...	...	0.5	0.5
Over 25 to 50, incl	0.18	0.30	0.40	0.8	...	...	1.0	1.0
Over 50 to 100, incl	0.20	0.40	0.50	0.8	1.2	1.2	1.5	1.0
Over 100 to 150, incl	0.25	0.40	0.50	0.8	1.2	1.2	2.5	1.5
Over 150 to 200, incl	[A]	0.40	0.65	0.8	1.2	1.6	3.0	2.5

[A] When a square is held against a face and an edge of a square edge flat bar, the edge shall not deviate by more than 3° or 5 % of the thickness.
[B] Flats over 150 to 200 mm, incl in width are not available as hot-wrought bars in thickness 6 mm and under.

TABLE A2.4 Thickness, Length, and Out-of-Square Tolerances for Hot-Wrought Bar Size Angles

Specified Length of Leg, mm[A,B]	Tolerances in Thickness for Thickness Given, Over and Under, mm			Tolerances for Length of Leg Over and Under, mm
	To 5, incl	Over 5 to 10, incl	Over 10	
To 50, incl	0.2	0.2	0.3	1
Over 50 to 75, excl	0.3	0.4	0.4	2

[A] The longer leg of an unequal angle determines the size for tolerance.
[B] Out of square tolerances in either direction is 1½ ° = 0.026 mm/mm.

TABLE A2.5 Dimensional Tolerances for Hot-Wrought Bar Size Channels

Specified Size of Channel, mm	Tolerances in Size, Over and Under, mm				Out of Square of Either Flange per mm of Flange Width,[B] mm
	Depth of Section[A]	Width of Flanges[A]	Thickness of Web		
			To 5, incl	Over 5	
To 40, incl	1	1	0.2	0.4	0.03
Over 40 to 75, excl	2	2	0.4	0.5	0.03

[A] Measurements for depth of section and width of flanges are overall.
[B] For channels 16 mm and under in depth, out of square tolerance is 0.05 mm/mm.

TABLE A2.6 Dimensional Tolerances for Hot-Wrought Bar Size Tees

Specified Size of Tee,[A] mm	Tolerances in Size, mm						Stem Out of Square[C]
	Width or Depth,[B]		Thickness of Flange		Thickness of Stem		
	Over	Under	Over	Under	Over	Under	
To 30, incl	1	1	0.2	0.2	0.1	0.5	1
Over 30 to 50, incl	2	2	0.3	0.3	0.2	0.5	2
Over 50 to 75, excl	2	2	0.4	0.4	0.4	0.5	2

[A] The longer member of the unequal tee determines the size for tolerances.
[B] Measurements for width and depth are over all.
[C] Stem out of square is the tolerance from its true position of the center line of the stem measured at the point.

TABLE A2.7 Permissible Variations in Dimensions for Half-Rounds, Ovals, Half-Ovals, and Other Special Bar Size Sections

Due to mill facilities, tolerances on half-rounds, ovals, and other special bar size sections vary among the manufacturers and such tolerances should be negotiated between the manufacturer and the purchaser.

TABLE A2.8 Length Tolerances for Hot-Wrought Rounds, Squares, Hexagons, Octagons, Flats, and Bar Size Sections

Specified Size of Rounds, Squares, Hexagons and Octagons, mm	Specified Size of Flats, mm		Tolerances over Specified Length, mm[A]				
	Thickness	Width	1500 to 3000, excl	3000 to 6000, excl	6000 to 9000, excl	9000 to 12 000, excl	12 000 to 18 000, excl
		Hot Shearing					
To 25, incl	to 25, incl	to 75, incl	15	20	35	45	60
Over 25 to 50, incl	over 25	to 75, incl	15	25	40	50	65
	to 25, incl	over 75 to 150, incl	15	25	40	50	65
Over 50 to 125, incl	over 25	over 75 to 150, incl	25	40	45	60	70
Over 125 to 250, incl	...	...	50	65	70	75	85
Bar Size Sections	over 6 to 25, incl	over 150 to 200, incl	20	30	45	90	100
	over 25 to 75, incl	over 150 to 200, incl	30	45	50	90	100
	...	...	15	25	40	50	65
		Hot Sawing					
50 to 90, incl	25 and over	75 and over	[B]	40	45	60	70
Over 90 to 125, incl				50	60	65	75
Over 125 to 250, incl	...	...	[B]	65	70	75	85

[A] No tolerance under.
[B] Smaller sizes and shorter lengths are not hot sawed.

TABLE A2.9 Length Tolerances for Recutting of Bars Meeting Special Straightness Tolerances

Sizes of Rounds, Squares, Hexagons, Octagons, Widths of Flats and Maximum Dimensions of Other Sections, mm	Tolerances over Specified Length, mm[A]	
	To 3700 mm, incl	Over 3700 mm
To 75, incl	6	8
Over 75 to 150, incl	8	11
Over 150 to 200, incl	11	14
Rounds over 200 to 250, incl	14	18

[A] No tolerance under.

TABLE A2.10 Straightness Tolerances for Hot-Wrought Bars and Bar Size Sections[A]

Standard Tolerances	6 mm in any 1500 mm and (length in mm/250)[B]
Special Tolerances	3 mm in any 1500 mm and (length in mm/500)[B]

[A] Because of warpage, straightness tolerances do not apply to bars if any subsequent heating operation or controlled cooling has been performed.
[B] Round to the nearest whole millimetre.

SUMMARY OF CHANGES

Committee A01 has identified the location of selected changes since A29/A29M-04[ε1] that may impact the use of this standard. (Approved May 1, 2005.)

(1) Changed the hot sawing tolerances in Table A1.8 and Table A2.8.

Committee A01 has identified the location of selected changes since A29/A29M-03 that may impact the use of this standard. (Approved March 1, 2004.)

(1) Removed "and Cold-Finished" from Title
(2) Removed A331 and A695 (dropped) from list of specifications in Scope
(3) Added Note to end of Scope.
(4) Removed 7.2
(5) Removed Tables A1.11, 12, 13, 14 and A2.11, 12, 13, 14
(6) Corrected Cr on grade 52100 in Table 2

ASTM International takes no position respecting the validity of any patent rights asserted in connection with any item mentioned in this standard. Users of this standard are expressly advised that determination of the validity of any such patent rights, and the risk of infringement of such rights, are entirely their own responsibility.

This standard is subject to revision at any time by the responsible technical committee and must be reviewed every five years and if not revised, either reapproved or withdrawn. Your comments are invited either for revision of this standard or for additional standards and should be addressed to ASTM International Headquarters. Your comments will receive careful consideration at a meeting of the responsible technical committee, which you may attend. If you feel that your comments have not received a fair hearing you should make your views known to the ASTM Committee on Standards, at the address shown below.

This standard is copyrighted by ASTM International, 100 Barr Harbor Drive, PO Box C700, West Conshohocken, PA 19428-2959, United States. Individual reprints (single or multiple copies) of this standard may be obtained by contacting ASTM at the above address or at 610-832-9585 (phone), 610-832-9555 (fax), or service@astm.org (e-mail); or through the ASTM website (www.astm.org).

Designation: A36/A36M − 08

Standard Specification for Carbon Structural Steel[1]

This standard is issued under the fixed designation A36/A36M; the number immediately following the designation indicates the year of original adoption or, in the case of revision, the year of last revision. A number in parentheses indicates the year of last reapproval. A superscript epsilon (ε) indicates an editorial change since the last revision or reapproval.

This standard has been approved for use by agencies of the Department of Defense.

1. Scope*

1.1 This specification[2] covers carbon steel shapes, plates, and bars of structural quality for use in riveted, bolted, or welded construction of bridges and buildings, and for general structural purposes.

1.2 Supplementary requirements are provided for use where additional testing or additional restrictions are required by the purchaser. Such requirements apply only when specified in the purchase order.

1.3 When the steel is to be welded, a welding procedure suitable for the grade of steel and intended use or service is to be utilized. See Appendix X3 of Specification A6/A6M for information on weldability.

1.4 The values stated in either inch-pound units or SI units are to be regarded separately as standard. Within the text, the SI units are shown in brackets. The values stated in each system are not exact equivalents; therefore, each system is to be used independently of the other, without combining values in any way.

1.5 The text of this specification contains notes or footnotes, or both, that provide explanatory material. Such notes and footnotes, excluding those in tables and figures, do not contain any mandatory requirements.

1.6 For structural products produced from coil and furnished without heat treatment or with stress relieving only, the additional requirements, including additional testing requirements and the reporting of additional test results, of A6/A6M apply.

2. Referenced Documents

2.1 *ASTM Standards:*[3]

A6/A6M Specification for General Requirements for Rolled Structural Steel Bars, Plates, Shapes, and Sheet Piling

A27/A27M Specification for Steel Castings, Carbon, for General Application

A307 Specification for Carbon Steel Bolts and Studs, 60 000 PSI Tensile Strength

A325 Specification for Structural Bolts, Steel, Heat Treated, 120/105 ksi Minimum Tensile Strength

A325M Specification for Structural Bolts, Steel, Heat Treated 830 MPa Minimum Tensile Strength (Metric)

A500 Specification for Cold-Formed Welded and Seamless Carbon Steel Structural Tubing in Rounds and Shapes

A501 Specification for Hot-Formed Welded and Seamless Carbon Steel Structural Tubing

A502 Specification for Rivets, Steel, Structural

A563 Specification for Carbon and Alloy Steel Nuts

A563M Specification for Carbon and Alloy Steel Nuts (Metric)

A668/A668M Specification for Steel Forgings, Carbon and Alloy, for General Industrial Use

A1011/A1011M Specification for Steel, Sheet and Strip, Hot-Rolled, Carbon, Structural, High-Strength Low-Alloy, High-Strength Low-Alloy with Improved Formability, and Ultra-High Strength

A1018/A1018M Specification for Steel, Sheet and Strip, Heavy-Thickness Coils, Hot-Rolled, Carbon, Commercial, Drawing, Structural, High-Strength Low-Alloy, High-Strength Low-Alloy with Improved Formability, and Ultra-High Strength

F568M Specification for Carbon and Alloy Steel Externally

---

[1] This specification is under the jurisdiction of ASTM Committee A01 on Steel, Stainless Steel and Related Alloys and is the direct responsibility of Subcommittee A01.02 on Structural Steel for Bridges, Buildings, Rolling Stock and Ships.
Current edition approved May 15, 2008. Published June 2008. Originally approved in 1960. Last previous edition approved in 2005 as A36/A36M – 05. DOI: 10.1520/A0036_A0036M-08.

[2] For ASME Boiler and Pressure Vessel Code Applications, see related Specifications SA-36 in Section II of that Code.

[3] For referenced ASTM standards, visit the ASTM website, www.astm.org, or contact ASTM Customer Service at service@astm.org. For *Annual Book of ASTM Standards* volume information, refer to the standard's Document Summary page on the ASTM website.

*A Summary of Changes section appears at the end of this standard.

Copyright © ASTM International, 100 Barr Harbor Drive, PO Box C700, West Conshohocken, PA 19428-2959, United States.

Threaded Metric Fasteners (Metric)
F1554 Specification for Anchor Bolts, Steel, 36, 55, and 105-ksi Yield Strength

3. Appurtenant Materials

3.1 When components of a steel structure are identified with this ASTM designation but the product form is not listed in the scope of this specification, the material shall conform to one of the standards listed in Table 1 unless otherwise specified by the purchaser.

4. General Requirements for Delivery

4.1 Structural products furnished under this specification shall conform to the requirements of the current edition of Specification A6/A6M, for the specific structural product ordered, unless a conflict exists in which case this specification shall prevail.

4.2 Coils are excluded from qualification to this specification until they are processed into a finished structural product. Structural products produced from coil means structural products that have been cut to individual lengths from a coil. The processor directly controls, or is responsible for, the operations involved in the processing of a coil into a finished structural product. Such operations include decoiling, leveling or straightening, hot-forming or cold-forming (if applicable), cutting to length, testing, inspection, conditioning, heat treatment (if applicable), packaging, marking, loading for shipment, and certification.

NOTE 1—For structural products produced from coil and furnished without heat treatment or with stress relieving only, two test results are to be reported for each qualifying coil. Additional requirements regarding structural products produced from coil are described in Specification A6/A6M.

5. Bearing Plates

5.1 Unless otherwise specified, plates used as bearing plates for bridges shall be subjected to mechanical tests and shall conform to the tensile requirements of Section 8.

TABLE 1 Appurtenant Material Specifications

NOTE 1—The specifier should be satisfied of the suitability of these materials for the intended application. Chemical composition or mechanical properties, or both, may be different than specified in A36/A36M.

Material	ASTM Designation
Steel rivets	A502, Grade 1
Bolts	A307, Grade A or F568M, Class 4.6
High-strength bolts	A325 or A325M
Steel nuts	A563 or A563M
Cast steel	A27/A27M, Grade 65–35 [450–240]
Forgings (carbon steel)	A668/A668M, Class D
Hot-rolled sheets and strip	A1011/A1011M, SS Grade 36 [250] Type 1 or Type 2 or A1018/A1018M, SS Grade 36 [250]
Cold-formed tubing	A500, Grade B
Hot-formed tubing	A501
Anchor bolts	F1554, Grade 36

TABLE 2 Tensile Requirements[A]

Plates, Shapes,[B] and Bars:	
Tensile strength, ksi [MPa]	58–80 [400–550]
Yield point, min, ksi [MPa]	36 [250][C]
Plates and Bars:[D,E]	
Elongation in 8 in. [200 mm], min, %	20
Elongation in 2 in. [50 mm], min, %	23
Shapes:	
Elongation in 8 in. [200 mm], min, %	20
Elongation in 2 in. [50 mm], min, %	21[B]

[A] See the Orientation subsection in the Tension Tests section of Specification A6/A6M.
[B] For wide flange shapes with flange thickness over 3 in. [75 mm], the 80 ksi [550 MPa] maximum tensile strength does not apply and a minimum elongation in 2 in. [50 mm] of 19 % applies.
[C] Yield point 32 ksi [220 MPa] for plates over 8 in. [200 mm] in thickness.
[D] Elongation not required to be determined for floor plate.
[E] For plates wider than 24 in. [600 mm], the elongation requirement is reduced two percentage points. See the Elongation Requirement Adjustments subsection under the Tension Tests section of Specification A6/A6M.

5.2 Unless otherwise specified, mechanical tests shall not be required for plates over 1½ in. [40 mm] in thickness used as bearing plates in structures other than bridges, subject to the requirement that they shall contain 0.20 to 0.33 % carbon by heat analysis, that the chemical composition shall conform to the requirements of Table 3 in phosphorus and sulfur content, and that a sufficient discard shall be made to secure sound plates.

6. Materials and Manufacture

6.1 The steel for plates and bars over ½ in. [12.5 mm] in thickness and shapes with flange or leg thicknesses over 1 in. [25 mm] shall be semi-killed or killed.

7. Chemical Composition

7.1 The heat analysis shall conform to the requirements prescribed in Table 3, except as specified in 5.2.

7.2 The steel shall conform on product analysis to the requirements prescribed in Table 3, subject to the product analysis tolerances in Specification A6/A6M.

8. Tension Test

8.1 The material as represented by the test specimen, except as specified in 5.2 and 8.2, shall conform to the requirements as to the tensile properties prescribed in Table 2.

8.2 Shapes less than 1 in.^{2} [645 mm^2] in cross section and bars, other than flats, less than ½ in. [12.5 mm] in thickness or diameter need not be subjected to tension tests by the manufacturer, provided that the chemical composition used is appropriate for obtaining the tensile properties in Table 2.

9. Keywords

9.1 bars; bolted construction; bridges; buildings; carbon; plates; riveted construction; shapes; steel; structural steel; welded construction

TABLE 3 Chemical Requirements

NOTE 1—Where "..." appears in this table, there is no requirement. The heat analysis for manganese shall be determined and reported as described in the heat analysis section of Specification A6/A6M.

Product	Shapes[A]	Plates[B]					Bars[B]			
Thickness, in. [mm]	All	To ¾ [20], incl	Over ¾ to 1½ [20 to 40], incl	Over 1½ to 2½ [40 to 65], incl	Over 2½ to 4 [65 to 100], incl	Over 4 [100]	To ¾ [20], incl	Over ¾ to 1½ [20 to 40], incl	Over 1½ to 4 [100], incl	Over 4 [100]
Carbon, max, %	0.26	0.25	0.25	0.26	0.27	0.29	0.26	0.27	0.28	0.29
Manganese, %	...	...	0.80–1.20	0.80–1.20	0.85–1.20	0.85–1.20	...	0.60–0.90	0.60–0.90	0.60–0.90
Phosphorus, max, %	0.04	0.04	0.04	0.04	0.04	0.04	0.04	0.04	0.04	0.04
Sulfur, max, %	0.05	0.05	0.05	0.05	0.05	0.05	0.05	0.05	0.05	0.05
Silicon, %	0.40 max	0.40 max	0.40 max	0.15–0.40	0.15–0.40	0.15–0.40	0.40 max	0.40 max	0.40 max	0.40 max
Copper, min, % when copper steel is specified	0.20	0.20	0.20	0.20	0.20	0.20	0.20	0.20	0.20	0.20

[A] Manganese content of 0.85–1.35 % and silicon content of 0.15–0.40 % is required for shapes with flange thickness over 3 in. [75 mm].
[B] For each reduction of 0.01 percentage point below the specified carbon maximum, an increase of 0.06 percentage point manganese above the specified maximum will be permitted, up to the maximum of 1.35 %.

SUPPLEMENTARY REQUIREMENTS

These requirements shall not apply unless specified in the order.

Standardized supplementary requirements for use at the option of the purchaser are listed in Specification A6/A6M. Those that are considered suitable for use with this specification are listed by title:

S5. Charpy V-Notch Impact Test.

S30. Charpy V-Notch Impact Test for Structural Shapes: Alternate Core Location

S32. Single Heat Bundles

S32.1 Bundles containing shapes or bars shall be from a single heat of steel.

In addition, the following optional supplementary requirement is also suitable for use with this specification:

S97. Limitation on Rimmed or Capped Steel

S97.1 The steel shall be other than rimmed or capped.

SUMMARY OF CHANGES

Committee A01 has identified the location of selected changes to this standard since the last issue (A36/A36M – 05) that may impact the use of this standard. (Approved May 15, 2008.)

(1) Added information to Table 1 on forgings (carbon steel) and anchor bolts.

ASTM International takes no position respecting the validity of any patent rights asserted in connection with any item mentioned in this standard. Users of this standard are expressly advised that determination of the validity of any such patent rights, and the risk of infringement of such rights, are entirely their own responsibility.

This standard is subject to revision at any time by the responsible technical committee and must be reviewed every five years and if not revised, either reapproved or withdrawn. Your comments are invited either for revision of this standard or for additional standards and should be addressed to ASTM International Headquarters. Your comments will receive careful consideration at a meeting of the responsible technical committee, which you may attend. If you feel that your comments have not received a fair hearing you should make your views known to the ASTM Committee on Standards, at the address shown below.

This standard is copyrighted by ASTM International, 100 Barr Harbor Drive, PO Box C700, West Conshohocken, PA 19428-2959, United States. Individual reprints (single or multiple copies) of this standard may be obtained by contacting ASTM at the above address or at 610-832-9585 (phone), 610-832-9555 (fax), or service@astm.org (e-mail); or through the ASTM website (www.astm.org).

Designation: A53/A53M – 07

Standard Specification for
Pipe, Steel, Black and Hot-Dipped, Zinc-Coated, Welded and Seamless[1]

This standard is issued under the fixed designation A53/A53M; the number immediately following the designation indicates the year of original adoption or, in the case of revision, the year of last revision. A number in parentheses indicates the year of last reapproval. A superscript epsilon (ε) indicates an editorial change since the last revision or reapproval.

This standard has been approved for use by agencies of the Department of Defense.

1. Scope*

1.1 This specification[2] covers seamless and welded black and hot-dipped galvanized steel pipe in NPS ⅛ to NPS 26 [DN 6 to DN 650] (Note 1), inclusive, with nominal wall thickness (Note 2) as given in Table X2.2 and Table X2.3. It shall be permissible to furnish pipe having other dimensions provided that such pipe complies with all other requirements of this specification. Supplementary requirements of an optional nature are provided and shall apply only when specified by the purchaser.

NOTE 1—The dimensionless designators NPS (nominal pipe size) [DN (diameter nominal)] have been substituted in this specification for such traditional terms as "nominal diameter," "size," and "nominal size."

NOTE 2—The term nominal wall thickness has been assigned for the purpose of convenient designation, existing in name only, and is used to distinguish it from the actual wall thickness, which may vary over or under the nominal wall thickness.

1.2 This specification covers the following types and grades:

1.2.1 *Type F*—Furnace-butt-welded, continuous welded Grade A,

1.2.2 *Type E*—Electric-resistance-welded, Grades A and B, and

1.2.3 *Type S*—Seamless, Grades A and B.

NOTE 3—See Appendix X1 for definitions of types of pipe.

1.3 Pipe ordered under this specification is intended for mechanical and pressure applications and is also acceptable for ordinary uses in steam, water, gas, and air lines. It is suitable for welding, and suitable for forming operations involving coiling, bending, and flanging, subject to the following qualifications:

1.3.1 Type F is not intended for flanging.

1.3.2 If Type S or Type E is required for close coiling or cold bending, Grade A is the preferred grade; however, this is not intended to prohibit the cold bending of Grade B pipe.

1.3.3 Type E is furnished either nonexpanded or cold expanded at the option of the manufacturer.

1.4 The values stated in either SI units or inch-pound units are to be regarded separately as standard. The values stated in each system may not be exact equivalents; therefore, each system is to be used independently of the other.

1.5 The following precautionary caveat pertains only to the test method portion, Sections 7, 8, 9, 13, 14, and 15 of this specification: *This standard does not purport to address all of the safety concerns, if any, associated with its use. It is the responsibility of the user of this standard to establish appropriate safety and health practices and determine the applicability of regulatory requirements prior to use.*

1.6 The text of this specification contains notes or footnotes, or both, that provide explanatory material. Such notes and footnotes, excluding those in tables and figures, do not contain any mandatory requirements.

2. Referenced Documents

2.1 *ASTM Standards:*[3]

A90/A90M Test Method for Weight [Mass] of Coating on Iron and Steel Articles with Zinc or Zinc-Alloy Coatings

A370 Test Methods and Definitions for Mechanical Testing of Steel Products

A530/A530M Specification for General Requirements for Specialized Carbon and Alloy Steel Pipe

A700 Practices for Packaging, Marking, and Loading Methods for Steel Products for Shipment

A751 Test Methods, Practices, and Terminology for Chemical Analysis of Steel Products

A865 Specification for Threaded Couplings, Steel, Black or Zinc-Coated (Galvanized) Welded or Seamless, for Use in Steel Pipe Joints

---

[1] This specification is under the jurisdiction of ASTM Committee A01 on Steel, Stainless Steel and Related Alloys and is the direct responsibility of Subcommittee A01.09 on Carbon Steel Tubular Products.

Current edition approved Sept. 1, 2007. Published October 2007. Originally approved in 1915. Last previous edition approved in 2006 as A53/A53M – 06a. DOI: 10.1520/A0053_A0053M-07.

[2] For ASME Boiler and Pressure Vessel Code applications, see related Specification SA-53 in Section II of that code.

[3] For referenced ASTM standards, visit the ASTM website, www.astm.org, or contact ASTM Customer Service at service@astm.org. For *Annual Book of ASTM Standards* volume information, refer to the standard's Document Summary page on the ASTM website.

---

*A Summary of Changes section appears at the end of this standard.

B6 Specification for Zinc

E29 Practice for Using Significant Digits in Test Data to Determine Conformance with Specifications

E213 Practice for Ultrasonic Testing of Metal Pipe and Tubing

E273 Practice for Ultrasonic Examination of the Weld Zone of Welded Pipe and Tubing

E309 Practice for Eddy-Current Examination of Steel Tubular Products Using Magnetic Saturation

E570 Practice for Flux Leakage Examination of Ferromagnetic Steel Tubular Products

E1806 Practice for Sampling Steel and Iron for Determination of Chemical Composition

2.2 *ANSI Standards:*
ASC X12[4]
B1.20.1 Pipe Threads, General Purpose[4]

2.3 *ASME Standard:*
B36.10M Welded and Seamless Wrought Steel Pipe[5]

2.4 *Military Standards:*
MIL-STD-129 Marking for Shipment and Storage[6]
MIL-STD-163 Steel Mill Products Preparation for Shipment and Storage[6]

2.5 *Federal Standards:*
Fed. Std. No. 123 Marking for Shipment (Civil Agencies)[7]
Fed. Std. No 183 Continuous Identification Marking of Iron and Steel Products[7]

2.6 *API Standard:*
5B Specification for Threading, Gauging, and Thread Inspection of Casing, Tubing, and Line Pipe Threads[8]

3. Ordering Information

3.1 Information items to be considered, if appropriate, for inclusion in the purchase order are as follows:

3.1.1 Specification designation (A53 or A53M, including year-date),

3.1.2 Quantity (feet, metres, or number of lengths),

3.1.3 Grade (A or B),

3.1.4 Type (F, E, or S; see 1.2),

3.1.5 Finish (black or galvanized),

3.1.6 Size (either nominal (NPS) [DN] and weight class or schedule number, or both; or outside diameter and wall thickness, see Table X2.2 and Table X2.3),

3.1.7 Length (specific or random, see Section 16),

3.1.8 End finish (plain end or threaded, Section 11),

3.1.8.1 Threaded and coupled, if desired,

3.1.8.2 Threads only (no couplings), if desired,

3.1.8.3 Plain end, if desired,

3.1.8.4 Couplings power tight, if desired,

3.1.8.5 Taper-tapped couplings for NPS 2 [DN 50] and smaller, if desired,

3.1.9 Close coiling, if desired (see 7.2.2),

3.1.10 Nondestructive electric test for seamless pipe (see 9.2),

3.1.11 Certification (see Section 20),

3.1.12 Report of the length of the end effect, if desired (see 9.2.7),

3.1.13 Marking (see Section 21),

3.1.14 End use of pipe,

3.1.15 Special requirements,

3.1.16 Supplementary requirements, if any,

3.1.17 Selection of applicable level of preservation and packaging and level of packing required, if other than as specified or if MIL-STD-163 applies (see 22.1), and

3.1.18 Packaging and package marking, if desired (see 23.1).

4. Materials and Manufacture

4.1 The steel for both seamless and welded pipe shall be made by one or more of the following processes: open-hearth, electric-furnace, or basic-oxygen.

4.2 If steels of different grades are sequentially strand cast, identification of the resultant transition material is required. The steel producer shall remove the transition material by any established procedure that positively separates the grades.

4.3 The weld seam of electric-resistance welded pipe in Grade B shall be heat treated after welding to a minimum of 1000 °F [540 °C] so that no untempered martensite remains, or otherwise processed in such a manner that no untempered martensite remains.

4.4 When pipe is cold expanded, the amount of expansion shall not exceed 1½ % of the specified outside diameter of the pipe.

5. Chemical Composition

5.1 The steel shall conform to the requirements as to chemical composition given in Table 1 and the chemical

---

[4] Available from American National Standards Institute (ANSI), 25 W. 43rd St., 4th Floor, New York, NY 10036, http://www.ansi.org.

[5] Available from American Society of Mechanical Engineers (ASME), ASME International Headquarters, Three Park Ave., New York, NY 10016-5990, http://www.asme.org.

[6] Available from Standardization Documents Order Desk, DODSSP, Bldg. 4, Section D, 700 Robbins Ave., Philadelphia, PA 19111-5098

[7] Available from General Services Administration, Washington, DC 20405.

[8] Available from American Petroleum Institute (API), 1220 L. St., NW, Washington, DC 20005-4070, http://api-ec.api.org.

TABLE 1 Chemical Requirements

	Composition, max, %								
	Carbon	Manganese	Phosphorus	Sulfur	Copper[A]	Nickel[A]	Chromium[A]	Molybdenum[A]	Vanadium[A]
Type S (seamless pipe)									
Grade A	0.25	0.95	0.05	0.045	0.40	0.40	0.40	0.15	0.08
Grade B	0.30	1.20	0.05	0.045	0.40	0.40	0.40	0.15	0.08
Type E (electric-resistance-welded)									
Grade A	0.25	0.95	0.05	0.045	0.50	0.40	0.40	0.15	0.08
Grade B	0.30	1.20	0.05	0.045	0.50	0.40	0.40	0.15	0.08
Type F (furnace-welded pipe)									
Grade A	0.30	1.20	0.05	0.045	0.40	0.40	0.40	0.15	0.08

[A] The total composition for these five elements shall not exceed 1.00 %.

analysis shall be in accordance with Test Methods, Practices, and Terminology A751.

6. Product Analysis

6.1 The purchaser is permitted to perform an analysis of two pipes from each lot of 500 lengths, or fraction thereof. Samples for chemical analysis, except for spectrographic analysis, shall be taken in accordance with Practice E1806. The chemical composition thus determined shall conform to the requirements given in Table 1.

6.2 If the analysis of either pipe does not conform to the requirements given in Table 1, analyses shall be made on additional pipes of double the original number from the same lot, each of which shall conform to the specified requirements.

7. Mechanical Properties

7.1 *Tension Test*:

7.1.1 For tension tests other than transverse weld tension tests, the yield strength corresponding to a permanent offset of 0.2 % of the gage length or to an extension of 0.5 % of the gage length under load, the tensile strength, and the elongation in 2 in. or 50 mm shall be determined, and the tension test results shall conform to the applicable tensile property requirements given in Table 2.

7.1.2 For transverse weld tension tests, the tensile strength shall be determined, and the tension test results shall conform to the applicable tensile strength requirement given in Table 2.

7.1.3 Electric-resistance-welded pipe NPS 8 [DN 200] or larger shall be tested using two transverse test specimens, one taken across the weld and one taken opposite the weld.

7.1.4 Transverse tension test specimens shall be approximately 1½ in. [38 mm] wide in the gage length and shall represent the full wall thickness of the pipe from which the test specimens were cut.

7.2 *Bend Test*:

7.2.1 For pipe NPS 2 [DN 50] or smaller, a sufficient length of pipe shall be capable of being bent cold through 90° around a cylindrical mandrel, the diameter of which is twelve times the specified outside diameter of the pipe, without developing cracks at any portion and without opening the weld.

7.2.2 If ordered for close coiling, the pipe shall stand being bent cold through 180° around a cylindrical mandrel, the diameter of which is eight times the specified outside diameter of the pipe, without failure.

7.2.3 Double-extra-strong pipe over NPS 1¼ [DN 32] need not be subjected to the bend test.

7.3 *Flattening Test*:

7.3.1 The flattening test shall be made on welded pipe over NPS 2 [DN 50] in extra-strong weight or lighter.

7.3.2 *Seamless Pipe*:

7.3.2.1 Although testing is not required, pipe shall be capable of meeting the flattening test requirements of Supplementary Requirement S1, if tested.

7.3.3 *Electric-Resistance-Welded Pipe*:

7.3.3.1 A test specimen at least 4 in. [100 mm] in length shall be flattened cold between parallel plates in three steps, with the weld located either 0° or 90° from the line of direction of force as required by 7.3.3.2 or 7.3.3.3, whichever is applicable. During the first step, which is a test for ductility of the weld, except as allowed by 7.3.5, 7.3.6, and 7.3.7, no cracks or breaks on the inside or outside surface at the weld shall be present before the distance between the plates is less than two thirds of the specified outside diameter of the pipe. As a second step, the flattening shall be continued as a test for ductility away from the weld. During the second step, except as allowed by 7.3.6 and 7.3.7, no cracks or breaks on the inside or outside surface away from the weld shall be present before the distance between the plates is less than one third of the specified outside diameter of the pipe but is not less than five times the specified wall thickness of the pipe. During the third step, which is a test for soundness, the flattening shall be continued until the test specimen breaks or the opposite walls of the test specimen meet. Evidence of laminated or unsound material or of incomplete weld that is revealed by the flattening test shall be cause for rejection.

7.3.3.2 For pipe produced in single lengths, the flattening test specified in 7.3.3.1 shall be made using a test specimen taken from each end of each length of pipe. The tests from each end shall be made alternately with the weld at 0° and at 90° from the line of direction of force.

7.3.3.3 For pipe produced in multiple lengths, the flattening test specified in 7.3.3.1 shall be made as follows:

(1) Test specimens taken from, and representative of, the front end of the first pipe intended to be supplied from each coil, the back end of the last pipe intended to be supplied from each coil, and each side of any intermediate weld stop location shall be flattened with the weld located at 90° from the line of direction of force.

(2) Test specimens taken from pipe at any two locations intermediate to the front end of the first pipe and the back end of the last pipe intended to be supplied from each coil shall be flattened with the weld located at 0° from the line of direction of force.

7.3.3.4 For pipe that is to be subsequently reheated throughout its cross section and hot formed by a reducing process, the manufacturer shall have the option of obtaining the flattening test specimens required by 7.3.3.2 or 7.3.3.3, whichever is applicable, either prior to or after such hot reducing.

TABLE 2 Tensile Requirements

	Grade A	Grade B
Tensile strength, min, psi [MPa]	48 000 [330]	60 000 [415]
Yield strength, min, psi [MPa]	30 000 [205]	35 000 [240]
Elongation in 2 in. or 50 mm	A,B	A,B

[A] The minimum elongation in 2 in. [50 mm] shall be that determined by the following equation:

$$e = 625\,000\,[1940]\,A^{0.2}/U^{0.9}$$

where:
- e = minimum elongation in 2 in. or 50 mm in percent, rounded to the nearest percent,
- A = the lesser of 0.75 in.^{2} [500 mm^2] and the cross-sectional area of the tension test specimen, calculated using the specified outside diameter of the pipe, or the nominal width of the tension test specimen and the specified wall thickness of the pipe, with the calculated value rounded to the nearest 0.01 in.^{2} [1 mm^2], and
- U = specified minimum tensile strength, psi [MPa].

[B] See Table X4.1 or Table X4.2, whichever is applicable, for the minimum elongation values that are required for various combinations of tension test specimen size and specified minimum tensile strength.

7.3.4 *Continuous-Welded Pipe*—A test specimen at least 4 in. [100 mm] in length shall be flattened cold between parallel plates in three steps. The weld shall be located at 90° from the line of direction of force. During the first step, which is a test for ductility of the weld, except as allowed by 7.3.5, 7.3.6, and 7.3.7, no cracks or breaks on the inside, outside, or end surfaces at the weld shall be present before the distance between the plates is less than three fourths of the specified outside diameter of the pipe. As a second step, the flattening shall be continued as a test for ductility away from the weld. During the second step, except as allowed by 7.3.6 and 7.3.7, no cracks or breaks on the inside, outside, or end surfaces away from the weld shall be present before the distance between the plates is less than 60 % of the specified outside diameter of the pipe. During the third step, which is a test for soundness, the flattening shall be continued until the test specimen breaks or the opposite walls of the test specimen meet. Evidence of laminated or unsound material or of incomplete weld that is revealed by the flattening test shall be cause for rejection.

7.3.5 Surface imperfections in the test specimen before flattening, but revealed during the first step of the flattening test, shall be judged in accordance with the finish requirements in Section 12.

7.3.6 Superficial ruptures as a result of surface imperfections shall not be cause for rejection.

7.3.7 For pipe with a D-to-t ratio less than 10, because the strain imposed due to geometry is unreasonably high on the inside surface at the 6 and 12 o'clock locations, cracks at such locations shall not be cause for rejection.

8. Hydrostatic Test

8.1 The hydrostatic test shall be applied, without leakage through the weld seam or the pipe body.

8.2 Plain-end pipe shall be hydrostatically tested to the applicable pressure given in Table X2.2, and threaded-and-coupled pipe shall be hydrostatically tested to the applicable pressure given in Table X2.3. It shall be permissible, at the discretion of the manufacturer, to perform the hydrostatic test on pipe with plain ends, with threads only, or with threads and couplings; and it shall also be permissible to test pipe in either single lengths or multiple lengths.

Note 4—The hydrostatic test pressures given herein are inspection test pressures, are not intended as a basis for design, and do not have any direct relationship to working pressures.

8.3 The minimum hydrostatic test pressure required to satisfy the requirements specified in 8.2 need not exceed 2500 psi [17 200 kPa] for pipe NPS 3 [DN 80] or smaller, or 2800 psi [19 300 kPa] for pipe larger than NPS 3 [DN 80]; however, the manufacturer has the option of using higher test pressures. For all sizes of seamless pipe and electric-resistance-welded pipe, the hydrostatic test pressure shall be maintained for at least 5 s.

9. Nondestructive Electric Test

9.1 *Type E Pipe*:

9.1.1 Except for pipe produced on a hot-stretch reducing mill, the weld seam of each length of electric-resistance-welded pipe NPS 2 [DN 50] or larger shall be tested with a nondestructive electric test in accordance with Practices E213, E273, E309, or E570. Each length of electric-resistance-welded pipe NPS 2 [DN 50] or larger and produced on a hot-stretch-reducing mill shall be tested with a nondestructive electric test that inpsects the full volume of the pipe in accordance with Practices E213, E309, or E570.

9.1.2 *Ultrasonic and Electromagnetic Inspection*—Any equipment utilizing the ultrasonic or electromagnetic principles and capable of continuous and uninterrupted inspection of the weld seam shall be used. The equipment shall be checked with an applicable reference standard as described in 9.1.3 at least once every working turn or not more than 8 h to demonstrate its effectiveness and the inspection procedures. The equipment shall be adjusted to produce well-defined indications when the reference standard is scanned by the inspection unit in a manner simulating the inspection of the product.

9.1.3 *Reference Standards*—The length of the reference standards shall be determined by the pipe manufacturer, and they shall have the same specified diameter and thickness as the product being inspected. Reference standards shall contain machined notches, one on the inside surface and one on the outside surface, or a drilled hole, as shown in Fig. 1, at the option of the pipe manufacturer. The notches shall be parallel to the weld seam, and shall be separated by a distance sufficient to produce two separate and distinguishable signals. The 1/8-in. [3.2-mm] hole shall be drilled through the wall and perpendicular to the surface of the reference standard as shown in Fig. 1. Care shall be taken in the preparation of the reference standard to ensure freedom from fins or other edge roughness, or distortion of the pipe.

Note 5—The calibration standards shown in Fig. 1 are convenient standards for calibration of nondestructive testing equipment. The dimensions of such standards are not to be construed as the minimum sizes of imperfections detectable by such equipment.

9.1.4 *Acceptance Limits*—Table 3 gives the height of acceptance limit signals in percent of the height of signals produced by reference standards. Imperfections in the weld seam that produce a signal greater than the acceptance limit signal given in Table 3 shall be considered a defect unless the pipe manufacturer can demonstrate that the imperfection does not reduce the effective wall thickness beyond 12.5 % of the specified wall thickness.

9.2 *Type S Pipe*—As an alternative to the hydrostatic test at the option of the manufacturer or if specified in the purchase order, the full body of each seamless pipe shall be tested with a nondestructive electric test in accordance with Practice E213, E309, or E570. In such cases, each length so furnished shall include the mandatory marking of the letters "NDE." Except as allowed by 9.2.6.2, it is the intent of this nondestructive electric test to reject pipe with imperfections that produce test signals equal to or greater than those produced by the applicable calibration standards.

9.2.1 If the nondestructive electric test has been performed, the lengths shall be marked with the letters "NDE." The certification, if required, shall state Nondestructive Electric Tested and shall indicate which of the tests was applied. Also, the letters NDE shall be appended to the product specification number and grade shown on the certification.

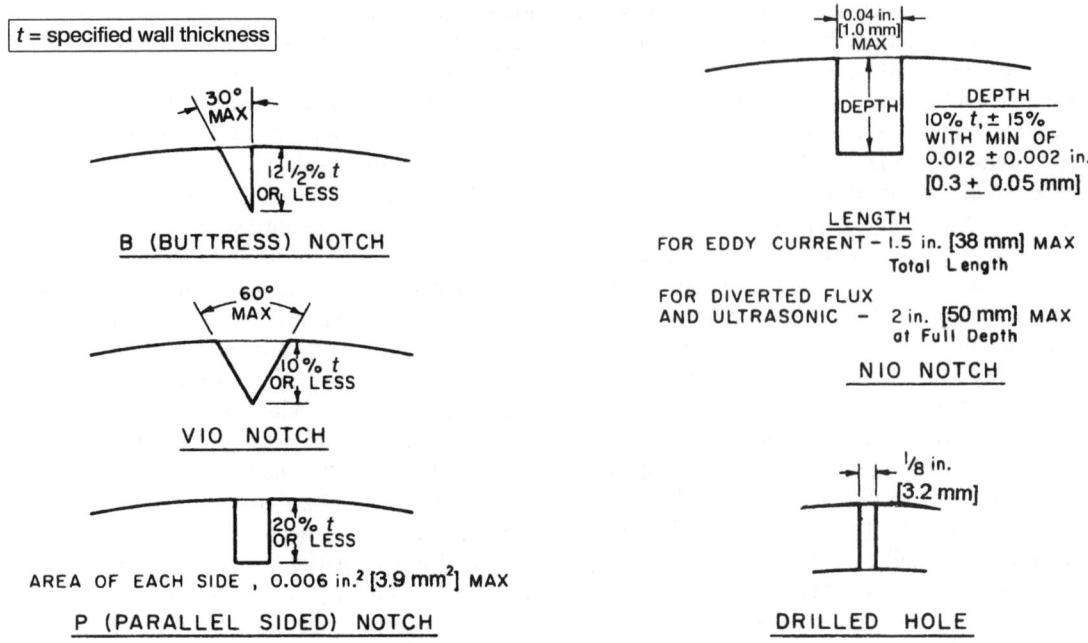

FIG. 1 Calibration Standards

TABLE 3 Acceptance Limits

Type Notch	Size of Hole		Acceptance Limit Signal, %
	in.	mm	
N10, V10	1/8	3.2	100
B, P	...	...	80

9.2.2 The following information is intended to facilitate the use of this specification:

9.2.2.1 The calibration standards defined in 9.2.3 through 9.2.5 are convenient standards for calibration of nondestructive testing equipment. The dimensions of such standards are not to be construed as the minimum sizes of imperfections detectable by such equipment.

9.2.2.2 The ultrasonic testing referred to in this specification is capable of detecting the presence and location of significant longitudinally or circumferentially oriented imperfections; however, different techniques need to be employed for the detection of differently oriented imperfections. Ultrasonic testing is not necessarily capable of detecting short, deep imperfections.

9.2.2.3 The eddy current examination referenced in this specification has the capability of detecting significant discontinuities, especially of the short abrupt type.

9.2.2.4 The flux leakage examination referred to in this specification is capable of detecting the presence and location of significant longitudinally or transversely oriented discontinuities. The provisions of this specification only require longitudinal calibration for flux leakage. Different techniques need to be employed for the detection of differently oriented imperfections.

9.2.2.5 The hydrostatic test referred to in 8.2 has the capability of finding imperfections of a size permitting the test fluid to leak through the tube wall and may be either visually seen or detected by a loss of pressure. Hydrostatic testing is not necessarily capable of detecting very tight through-the-wall imperfections or imperfections that extend an appreciable distance into the wall without complete penetration.

9.2.2.6 A purchaser interested in ascertaining the nature (type, size, location, and orientation) of imperfections that are capable of being detected in the specific application of these examinations is directed to discuss this with the manufacturer of the tubular product.

9.2.3 For ultrasonic testing, the calibration reference notches shall be at the option of the manufacturer, and shall be any one of the three common notch shapes shown in Practice E213. The depth of notch shall not exceed 12.5 % of the specified wall thickness of the pipe or 0.004 in. [0.1 mm], whichever is the greater.

9.2.4 For eddy current testing, the calibration pipe shall contain, at the option of the manufacturer, any one of the following calibration standards to establish a minimum sensitivity level for rejection.

9.2.4.1 *Drilled Hole*—The calibration pipe shall contain three holes spaced 120° apart or four holes spaced 90° apart, sufficiently separated longitudinally to ensure separately distinguishable responses. The holes shall be drilled radially and completely through the pipe wall, care being taken to avoid distortion of the pipe while drilling. Dependent upon the nominal pipe size, the calibration pipe shall contain the following hole:

NPS	DN	Diameter of Drilled Hole
≤ 1/2	≤ 15	0.039 in. [1.0 mm]
> 1/2 ≤ 1 1/4	> 15 ≤ 32	0.055 in. [1.4 mm]
> 1 1/4 ≤ 2	> 32 ≤ 50	0.071 in. [1.8 mm]
> 2 ≤ 5	> 50 ≤ 125	0.087 in. [2.2 mm]
> 5	> 125	0.106 in. [2.7 mm]

9.2.4.2 *Transverse Tangential Notch*—Using a round tool or file with a 1/4 in. [6 mm] diameter, a notch shall be filed or milled tangential to the surface and transverse to the longitudinal axis of the pipe. The notch shall have a depth not exceeding 12.5 % of the specified wall thickness of the pipe or 0.012 in. [0.3 mm], whichever is the greater.

9.2.4.3 *Longitudinal Notch*—A notch 0.031 in. [0.8 mm] or less in width shall be machined in a radial plane parallel to the pipe axis on the outside surface of the pipe, to a depth not exceeding 12.5 % of the specified wall thickness of the pipe or 0.012 in. [0.3 mm], whichever is the greater. The length of the notch shall be compatible with the testing method.

9.2.4.4 *Compatibility*—The calibration standards in the calibration pipe shall be compatible with the testing equipment and the method being used.

9.2.5 For flux leakage testing, the longitudinal calibration reference notches shall be straight-sided notches machined in a radial plane parallel to the pipe axis. For specified wall thicknesses less than 0.500 in. [12.7 mm], outside and inside notches shall be used. For specified wall thicknesses equal to or greater than 0.500 in. [12.7 mm], only an outside notch shall be used. The notch depth shall not exceed 12.5 % of the specified wall thickness, or 0.012 in. [0.3 mm], whichever is the greater. The notch length shall not exceed 1 in. [25 mm], and the notch width shall not exceed the notch depth. Outside diameter and inside diameter notches shall be located sufficiently apart to allow separation and identification of the signals.

9.2.6 Pipe containing one or more imperfections that produce a signal equal to or greater than the signal produced by the calibration standard shall be rejected or the area producing the signal shall be rejected.

9.2.6.1 Test signals produced by imperfections that cannot be identified, or produced by cracks or crack-like imperfections, shall result in rejection of the pipe, unless it is repaired and retested. To be accepted, the pipe shall pass the same specification test to which it was originally subjected and the remaining wall thickness shall not have been decreased below that permitted by the specification. It shall be permissible to reduce the outside diameter at the point of grinding by the amount so removed.

9.2.6.2 It shall be permissible to evaluate test signals produced by visual imperfections in accordance with the provisions of Section 12. A few examples of such imperfections are straightener marks, cutting chips, scratches, steel die stamps, stop marks, or pipe reducer ripple.

9.2.7 The test methods described in Section 9 are not necessarily capable of inspecting the end portion of pipes. This condition is referred to as end effect. The length of the end effect shall be determined by the manufacturer and, if specified in the purchase order, reported to the purchaser.

10. Permissible Variations in Weight (Mass) and Dimensions

10.1 *Weight (Mass)*—The weight (mass) of the pipe shall not vary more than ± 10 % from its specified weight (mass), as derived by multiplying its measured length by its specified weight (mass) per unit length, as given in Table X2.2 or Table X2.3, or as calculated using the relevant equation in ASME B36.10M.

Note 6—For pipe NPS 4 [DN 100] or smaller, the weight (mass) tolerance is applicable to the weights (masses) of the customary lifts of pipe as produced for shipment by the mill. For pipe larger than NPS 4 [DN 100], where individual lengths are weighed, the weight (mass) tolerance is applicable to the individual lengths.

10.2 *Diameter*—For pipe NPS 1½ [DN 40] or smaller, the outside diameter at any point shall not vary more than ± 1/64 in. [0.4 mm] from the specified outside diameter. For pipe NPS 2 [DN 50] or larger, the outside diameter shall not vary more than ± 1 % from the specified outside diameter.

10.3 *Thickness*—The minimum wall thickness at any point shall be not more than 12.5 % under the specified wall thickness. The minimum wall thickness on inspection shall conform to the requirements given in Table X2.4.

11. End Finish

11.1 If ordered with plain ends, the pipe shall be furnished to the following practice, unless otherwise specified.

11.1.1 *NPS 1½ [DN 40] or Smaller*—Unless otherwise specified in the purchase order, end finish shall be at the option of the manufacturer.

11.1.2 *Larger than NPS 1½ [DN 40]*:

11.1.2.1 Pipe of standard-weight or extra-strong weight, or in wall thickness less than 0.500 in. [12.7 mm], other than double extra-strong weight pipe, shall be plain-end beveled with ends beveled to an angle of 30°, +5°, -0°, measured from a line drawn perpendicular to the axis of the pipe, and with a root face of 1/16 in. ± 1/32 in. [1.6 mm ± 0.8 mm].

11.1.2.2 Pipe with a specified wall thickness greater than 0.500 in. [12.7 mm], and all double extra-strong weight pipe, shall be plain-end square cut.

11.2 If ordered with threaded ends, the pipe ends shall be provided with a thread in accordance with the gaging practice and tolerances of ANSI B1.20.1. For standard-weight pipe NPS 6 [DN 150] or smaller, refer to Table X3.1 for threading data. For standard-weight pipe NPS 8 [DN 200] or larger and all sizes of extra-strong weight pipe and double extra-strong weight pipe, refer to Table X3.2 for threading data. Threaded pipe NPS 4 [DN 100] or larger shall have thread protectors on the ends not protected by a coupling.

11.3 If ordered with couplings, one end of each length of pipe shall be provided with a coupling manufactured in accordance with Specification A865. The coupling threads shall be in accordance with the gaging practice of ANSI B1.20.1. The coupling shall be applied handling-tight, unless power-tight is specified in the purchase order. Couplings are to be made of steel. Taper-tapped couplings shall be furnished on all threaded pipe NPS 2½ [DN 65] or larger. For pipe smaller than NPS 2½ [DN 65], it is regular practice to furnish straight-tapped couplings for standard-weight pipe and taper-tapped couplings for extra-strong and double extra-strong weight pipe. If taper-tapped couplings are required for standard-weight pipe smaller than NPS 2½ [DN 65], it is recommended that line pipe threads in accordance with API Specification 5B be ordered. The taper-tapped couplings provided on line pipe in such sizes may be used on mill-threaded standard-weight pipe of the same size.

12. Workmanship, Finish, and Appearance

12.1 The pipe manufacturer shall explore a sufficient number of visual surface imperfections to provide reasonable assurance that they have been properly evaluated with respect to depth.

12.2 Surface imperfections that penetrate more than 12.5 % of the specified wall thickness or encroach on the minimum wall thickness shall be considered defects. Pipe with defects shall be given one or more of the following dispositions:

12.2.1 The defect shall be removed by grinding, provided that the remaining wall thickness is within specified limits,

12.2.2 Type S pipe and the parent metal of Type E pipe, except within ½ in. [13 mm] of the fusion line of the electric-resistance-weld seam, are permitted to be repaired in accordance with the welding provisions of 12.5. Repair welding of Type F pipe and the weld seam of Type E pipe is prohibited.

12.2.3 The section of pipe containing the defect shall be cut off within the limits of requirement on length, or

12.2.4 Rejected.

12.3 At the purchaser's discretion, pipe shall be subjected to rejection if surface defects repaired in accordance with 12.2 are not scattered, but appear over a large area in excess of what is considered a workmanlike finish. Disposition of such pipe shall be a matter of agreement between the manufacturer and the purchaser.

12.4 For the removal of imperfections and defects by grinding, a smooth curved surface shall be maintained, and the wall thickness shall not be decreased below that permitted by this specification. It shall be permissible to reduce the outside diameter at the point of grinding by the amount so removed.

12.4.1 Wall thickness measurements shall be made with a mechanical caliper or with a properly calibrated nondestructive testing device of appropriate accuracy. In the case of a dispute, the measurement determined by use of the mechanical caliper shall govern.

12.5 Weld repair shall only be permitted with the approval of the purchaser and in accordance with Specification A530/A530M.

12.6 The finished pipe shall be reasonably straight.

12.7 The pipe shall contain no dents greater than 10 % of the pipe diameter or ¼ in. [6 mm], whichever is smaller, measured as the gap between the lowest point of the dent and a prolongation of the original contour of the pipe. Cold-formed dents deeper than ⅛ in. [3 mm] shall be free of sharp-bottomed gouges; it shall be permissible to remove the gouges by grinding, provided that the remaining wall thickness is within specified limits. The length of the dent in any direction shall not exceed one half the specified outside diameter of the pipe.

13. Number of Tests

13.1 Except as required by 13.2, one of each of the tests specified in Section 7 shall be made on test specimens taken from one length of pipe from each lot of each pipe size. For continuous-welded pipe, each lot shall contain no more than 25 tons [23 Mg] of pipe for pipe sizes NPS 1½ [DN 40] and smaller, and no more than 50 tons [45 Mg] of pipe for pipe sizes larger than NPS 1½ [DN 40]. For seamless and electric-resistance-welded pipe, a lot shall contain no more than one heat, and at the option of the manufacturer shall contain no more than 500 lengths of pipe (as initially cut after the final pipe-forming operation, prior to any further cutting to the required ordered lengths) or 50 tons [45 Mg] of pipe.

13.2 The number of flattening tests for electric-resistance-welded pipe shall be in accordance with 7.3.3.2 or 7.3.3.3, whichever is applicable.

13.3 Except as allowed by 9.2, each length of pipe shall be subjected to the hydrostatic test (see Section 8).

14. Retests

14.1 Except for flattening tests of electric-resistance-welded pipe, if the results of a mechanical test for a lot fail to conform to the applicable requirements specified in Section 7, the lot shall be rejected unless tests of additional pipe from the affected lot of double the number originally tested are subsequently made and each such test conforms to the specified requirements. Only one retest of any lot will be permitted. Any individual length of pipe that conforms to the test requirements is acceptable. Any individual length of pipe that does not conform to the test requirements may be resubmitted for test and will be considered acceptable if tests taken from each pipe end conform to the specified requirements.

14.2 *Electric-Resistance-Welded Pipe Produced in Single Lengths*—If any flattening test result fails to conform to the requirements specified in 7.3.3, the affected single length shall be rejected unless the failed end is subsequently retested using the same weld orientation as the failed test and a satisfactory test result is obtained before the pipe's length is reduced by such testing to less than 80 % of its length after the initial cropping.

14.3 *Electric-Resistance-Welded Pipe Produced in Multiple Lengths*—If any flattening test result fails to conform to the requirements specified in 7.3.3, the affected multiple length shall be rejected or flattening tests shall be made using a test specimen taken from each end of each individual length in the failed multiple length. For each pipe end, such tests shall be made with the weld alternately at 0° and 90° from the line of direction of force. Individual lengths are considered acceptable if the test results for both pipe ends conform to the specified requirements.

15. Test Methods

15.1 The test specimens and the tests required by this specification shall conform to those described in the latest issue of Test Methods and Definitions A370.

15.2 Each longitudinal tension test specimen shall be taken from a pipe end and shall not be flattened between the gage marks.

15.3 Test specimens for bend tests and flattening tests shall be taken from pipe. Test specimens for flattening tests shall be smooth on the ends and free from burrs.

15.4 Tests shall be conducted at room temperature.

16. Lengths

16.1 Unless otherwise specified, pipe lengths shall be in accordance with the following regular practices:

16.1.1 Except as allowed by 16.1.2 and 16.1.4, pipe lighter than extra-strong weight shall be in single-random lengths of 16 to 22 ft [4.88 to 6.71 m], with not more than 5 % of the total number of threaded lengths furnished being jointers (two pieces coupled together).

16.1.2 For plain-end pipe lighter than extra-strong weight, it shall be permissible for not more than 5 % of the total number of pipe to be in lengths of 12 to 16 ft [3.66 to 4.88 m].

16.1.3 Pipe of extra-strong weight or heavier shall be in random lengths of 12 to 22 ft [3.66 to 6.71 m], except that it shall be permissible for not more than 5 % of the total of pipe to be in lengths of 6 to 12 ft [1.83 to 3.66 m].

16.1.4 For extra-strong weight or lighter pipe ordered in double-random lengths, the minimum lengths shall be not less than 22 ft [6.71 m] and the minimum average length for the order shall be not less than 35 ft [10.67 m].

16.1.5 For pipe heavier than extra-strong weight ordered in lengths longer than single random, the lengths shall be as agreed upon between the manufacturer and the purchaser.

16.1.6 If pipe is furnished threaded and coupled, the length shall be measured to the outer face of the coupling.

17. Galvanized Pipe

17.1 Galvanized pipe ordered under this specification shall be coated with zinc inside and outside by the hot-dip process. The zinc used for the coating shall be any grade of zinc conforming to Specification B6.

17.2 *Weight (Mass) per Unit Area of Coating*—The weight (mass) per unit area of zinc coating shall be not less than 1.8 oz/ft^2 [0.55 kg/m^2] as determined from the average results of the two specimens taken for test in the manner prescribed in 17.5 and not less than 1.6 oz/ft^2 [0.49 kg/m^2] for each of these specimens. The weight (mass) per unit area of coating, expressed in ounces per square foot [kilograms per square metre] shall be calculated by dividing the total weight (mass) of zinc, inside plus outside, by the total area, inside plus outside, of the surface coated. Each specimen shall have not less than 1.3 oz/ft^2 [0.40 kg/m^2] of zinc coating on each surface, calculated by dividing the total weight (mass) of zinc on the given surface (outside or inside) by the area of the surface coated (outside or inside).

17.3 *Weight (Mass) per Unit Area of Coating Test*—The weight (mass) per unit area of zinc coating shall be determined by stripping tests in accordance with Test Method A90/A90M.

17.4 *Test Specimens*—Test specimens for determination of weight (mass) per unit area of coating shall be cut approximately 4 in. [100 mm] in length.

17.5 *Number of Tests*—Two test specimens for the determination of weight (mass) per unit area of coating shall be taken, one from each end of one length of galvanized pipe selected at random from each lot of 500 lengths, or fraction thereof, of each size.

17.6 *Retests*—If the weight (mass) per unit area of coating of any lot does not conform to the requirements specified in 17.2, retests of two additional pipes from the same lot shall be made, each of which shall conform to the specified requirements.

17.7 If pipe ordered under this specification is to be galvanized, the tension, flattening, and bend tests shall be made on the base material before galvanizing, if practicable. If specified, results of the mechanical tests on the base material shall be reported to the purchaser. If it is impracticable to make the mechanical tests on the base material before galvanizing, it shall be permissible to make such tests on galvanized samples, and any flaking or cracking of the zinc coating shall not be considered cause for rejection. If galvanized pipe is bent or otherwise fabricated to a degree that causes the zinc coating to stretch or compress beyond the limit of elasticity, some flaking of the coating is acceptable.

18. Inspection

18.1 The inspector representing the purchaser shall have entry, at all times while work on the contract of the purchaser is being performed, to all parts of the manufacturer's works that concern the manufacture of the pipe ordered. The manufacturer shall afford the inspector all reasonable facilities to be satisfied that the pipe is being furnished in accordance with this specification. All tests (except product analysis) and inspection shall be made at the place of manufacture prior to shipment, unless otherwise specified, and shall be so conducted as not to interfere unnecessarily with the operation of the works.

19. Rejection

19.1 The purchaser is permitted to inspect each length of pipe received from the manufacturer and, if it does not meet the requirements of this specification based upon the inspection and test method as outlined in the specification, the length shall be rejected and the manufacturer shall be notified. Disposition of rejected pipe shall be a matter of agreement between the manufacturer and the purchaser.

19.2 Pipe found in fabrication or in installation to be unsuitable for the intended use, under the scope and requirements of this specification, shall be set aside and the manufacturer notified. Such pipe shall be subject to mutual investigation as to the nature and severity of the deficiency and the forming or installation, or both, conditions involved. Disposition shall be a matter for agreement.

20. Certification

20.1 The manufacturer or supplier shall, upon request, furnish to the purchaser a certificate of compliance stating that the material has been manufactured, sampled, tested, and inspected in accordance with this specification (including year-date), and has been found to meet the requirements.

20.2 *Test Report*—For Types E and S, the manufacturer or supplier shall furnish to the purchaser a chemical analysis report for the elements given in Table 1.

20.3 *EDI*—A certificate of compliance or test report printed from, or used in, electronic form from an electronic data interchange (EDI) transmission shall be regarded as having the same validity as a counterpart printed in the certifier's facility. The use and format of the EDI document are subject to agreement between the purchaser and the manufacturer or supplier.

NOTE 7—EDI is the computer to computer exchange of business information in a standard format such as ANSI ASC X12.

20.4 Notwithstanding the absence of a signature, the organization submitting the certificate of compliance or test report is responsible for its content.

21. Product Marking

21.1 Except as allowed by 21.5 and 21.6, each length of pipe shall be legibly marked in the following sequence to show:

21.1.1 Manufacturer's name or mark,

21.1.2 Specification number (year-date not required),

NOTE 8—Pipe that complies with multiple compatible specifications may be marked with the appropriate designation for each specification.

21.1.3 Size (NPS and weight class, schedule number, or specified wall thickness; or specified outside diameter and specified wall thickness),

21.1.4 Grade (A or B),

21.1.5 Type of pipe (F, E, or S),

21.1.6 Test pressure, seamless pipe only (if applicable, in accordance with Table 4),

21.1.7 Nondestructive electric test, seamless pipe only (if applicable, in accordance with Table 4),

21.2 Unless another marking format is specified in the purchase order, length shall be marked in feet and tenths of a foot, or metres to two decimal places, dependent upon the units to which the pipe was ordered. The location of such marking shall be at the option of the manufacturer.

21.3 Heat number, lot number, run number, or a combination thereof shall be marked at the option of the manufacturer, unless specific marking is specified in the purchase order. The location of such marking shall be at the option of the manufacturer.

21.4 Any additional information desired by the manufacturer or specified in the purchase order.

21.5 For pipe NPS 1½ [DN 40] and smaller that is bundled, it shall be permissible to mark the required information on a tag securely attached to each bundle.

21.6 If pipe sections are cut into shorter lengths by a processor for resale as pipe, the processor shall transfer the complete identification, including the name or brand of the manufacturer, to each unmarked cut length, or to metal tags securely attached to unmarked pipe bundled in accordance with the requirements of 21.5. The same material designation shall be included with the information transferred, and the processor's name, trademark, or brand shall be added.

21.7 *Bar Coding*—In addition to the requirements in 21.1, 21.5, and 21.6, bar coding is acceptable as a supplementary identification method. It is recommended that bar coding be consistent with the Automotive Industry Action Group (AIAG) standard prepared by the Primary Metals Subcommittee of the AIAG Bar Code Project Team.

22. Government Procurement

22.1 If specified in the contract, the pipe shall be preserved, packaged, and packed in accordance with the requirements of MIL-STD-163. The applicable levels shall be as specified in the contract. Marking for shipment of such pipe shall be in accordance with Fed. Std. No. 123 for civil agencies and MIL-STD-129 or Federal Std. No. 183 if continuous marking is required, for military agencies.

22.2 *Inspection*—Unless otherwise specified in the contract, the manufacturer is responsible for the performance of all inspection and test requirements specified herein. Except as otherwise specified in the contract, the manufacturer shall use its own or any other suitable facilities for performing the inspection and test requirements specified herein, unless otherwise disapproved by the purchaser in the contract or purchase order. The purchaser shall have the right to perform any of the inspections and tests set forth in this specification where deemed necessary to ensure that the pipe conforms to the specified requirements.

23. Packaging and Package Marking

23.1 If specified in the purchase order, packaging, marking, and loading for shipment shall be in accordance with those procedures recommended by Practices A700.

24. Keywords

24.1 black steel pipe; seamless steel pipe; steel pipe; welded steel pipe; zinc coated steel pipe

TABLE 4 Marking of Seamless Pipe

Hydro	NDE	Marking
Yes	No	Test pressure
No	Yes	NDE
Yes	Yes	Test Pressure/NDE

SUPPLEMENTARY REQUIREMENTS

The following supplementary requirements shall apply only when specified in the purchase order. The purchaser may specify a different frequency of test than is provided in the supplementary requirement. Subject to agreement between the purchaser and manufacturer, retest and retreatment provisions of these supplementary requirements may also be modified.

S1. Flattening Test, Seamless Pipe

S1.1 A test specimen at least 2½ in. [60 mm] in length shall be flattened cold between parallel plates in two steps. During the first step, which is a test for ductility, except as allowed by S1.3, S1.4, and S1.5, no cracks or breaks on the inside, outside, or end surfaces shall be present before the distance between the plates is less than the value of H calculated as follows:

$$H = (1 + e)t/(e + t/D)$$

where:
H = distance between flattening plates, in. [mm],
e = deformation per unit length (constant for a given grade of steel, 0.09 for Grade A, and 0.07 for Grade B),
t = specified wall thickness, in. [mm], and
D = specified outside diameter, in. [mm]

The H values have been calculated for standard-weight and extra-heavy weight pipe from NPS 2½ to NPS 24 [DN 65 to DN 600], inclusive, and are given in Table X2.1.

S1.2 During the second step, which is a test for soundness, the flattening shall be continued until the test specimen breaks or the opposite sides of the test specimen meet. Evidence of laminated or unsound material that is revealed during the entire flattening test shall be cause for rejection.

S1.3 Surface imperfections in the test specimen before flattening, but revealed during the first step of the flattening test, shall be judged in accordance with the finish requirements in Section 12.

S1.4 Superficial ruptures as a result of surface imperfections shall not be cause for rejection.

S1.5 For pipe with a D-to-t ratio less than 10, because the strain imposed due to geometry is unreasonably high on the inside surface at the 6 and 12 o'clock locations, cracks at such locations shall not be cause for rejection.

S1.6 One test shall be made on test specimens taken from one length of pipe from each lot of each pipe size. A lot shall contain no more than one heat, and at the option of the manufacturer shall contain no more than 500 lengths of pipe (as initially cut after the final pipe-forming operation, prior to any further cutting to the required ordered lengths) or 50 tons [45 Mg] of pipe.

S1.7 If the results of a test for a lot fail to conform to the applicable requirements, the lot shall be rejected unless tests of additional pipe from the affected lot of double the number originally tested are subsequently made and each such test conforms to the specified requirements. Only one retest of any lot will be permitted. Any individual length of pipe that conforms to the test requirements is acceptable. Any individual length of pipe that does not conform to the test requirements may be resubmitted for test and will be considered acceptable if tests taken from each pipe end conform to the specified requirements.

APPENDIXES

(Nonmandatory Information)

X1. DEFINITIONS OF TYPES OF PIPE

X1.1 *Type F, Furnace-Butt-Welded Pipe, Continuous-Welded Pipe*—Pipe produced in multiple lengths from coiled skelp and subsequently cut into individual lengths, having its longitudinal butt joint forge welded by the mechanical pressure developed in rolling the hot-formed skelp through a set of round pass welding rolls.

X1.2 *Type E, Electric-Resistance-Welded Pipe*—Pipe produced in single lengths, or in multiple lengths from coiled skelp and subsequently cut into individual lengths, having a longitudinal butt joint wherein coalescence is produced by the heat obtained from resistance of the pipe to the flow of electric current in a circuit of which the pipe is a part, and by the application of pressure.

X1.3 *Type S, Seamless Pipe*—Pipe made without a welded seam. It is manufactured by hot working steel and, if necessary, by subsequently cold finishing the hot-worked tubular product to produce the desired shape, dimensions, and properties.

X2. TABLES FOR DIMENSIONAL AND CERTAIN MECHANICAL REQUIREMENTS

X2.1 Tables X2.1-X2.4 address dimensional and certain mechanical requirements.

TABLE X2.1 Calculated H Values for Seamless Pipe

NPS Designator	DN Designator	Specified Outside Diameter, in. [mm]	Specified Wall Thickness, in. [mm]	Distance, in. [mm], Between Plates "H" by Formula: $H = (1 + e)t/(e + t/D)$	
				Grade A	Grade B
2½	65	2.875 [73.0]	0.203 [5.16]	1.378 [35.0]	1.545 [39.2]
			0.276 [7.01]	1.618 [41.1]	1.779 [45.2]
3	80	3.500 [88.9]	0.216 [5.49]	1.552 [39.4]	1.755 [44.6]
			0.300 [7.62]	1.861 [47.3]	2.062 [52.4]
3½	90	4.000 [101.6]	0.226 [5.74]	1.682 [42.7]	1.912 [48.6]
			0.318 [8.08]	2.045 [51.9]	2.276 [57.8]
4	100	4.500 [114.3]	0.237 [6.02]	1.811 [46.0]	2.067 [52.5]
			0.337 [8.56]	2.228 [56.6]	2.489 [63.2]
5	125	5.563 [141.3]	0.258 [6.55]	2.062 [52.4]	2.372 [60.2]
			0.375 [9.52]	2.597 [66.0]	2.920 [74.2]
6	150	6.625 [168.3]	0.280 [7.11]	2.308 [58.6]	2.669 [67.8]
			0.432 [10.97]	3.034 [77.1]	3.419 [86.8]
8	200	8.625 [219.1]	0.277 [7.04]	2.473 [62.8]	2.902 [73.7]
			0.322 [8.18]	2.757 [70.0]	3.210 [81.5]
			0.500 [12.70]	3.683 [93.5]	4.181 [106.2]
10	250	10.750 [273.0]	0.279 [7.09][A]	2.623 [66.6]	3.111 [79.0]
			0.307 [7.80]	2.823 [71.7]	3.333 [84.7]
			0.365 [9.27]	3.210 [81.5]	3.757 [95.4]
			0.500 [12.70]	3.993 [101.4]	4.592 [116.6]
12	300	12.750 [323.8]	0.300 [7.62]	3.105 [78.9]	3.683 [93.5]
			0.375 [9.52]	3.423 [86.9]	4.037 [102.5]
			0.500 [12.70]	4.218 [107.1]	4.899 [124.4]
14	350	14.000 [355.6]	0.375 [9.52]	3.500 [88.9]	4.146 [105.3]
			0.500 [12.70]	4.336 [110.1]	5.061 [128.5]
16	400	16.000 [406.4]	0.375 [9.52]	3.603 [91.5]	4.294 [109.1]
			0.500 [12.70]	4.494 [114.1]	5.284 [134.2]
18	450	18.000 [457]	0.375 [9.52]	3.688 [93.7]	4.417 [112.2]
			0.500 [12.70]	4.628 [117.6]	5.472 [139.0]
20	500	20.000 [508]	0.375 [9.52]	3.758 [95.5]	4.521 [114.8]
			0.500 [12.70]	4.740 [120.4]	5.632 [143.1]
24	600	24.000 [610]	0.375 [9.52]	3.869 [98.3]	4.686 [119.0]
			0.500 [12.70]	4.918 [124.9]	5.890 [149.6]

[A] Special order only.

TABLE X2.2 Dimensions, Weights (Masses) per Unit Length, and Test Pressures for Plain-End Pipe

NPS Designator	DN Designator	Specified Outside Diameter, in. [mm]	Specified Wall Thickness, in. [mm]	Nominal Weight (Mass) per Unit Length, Plain End, lb/ft [kg/m]	Weight Class	Schedule No.	Test Pressure,[A] psi [kPa] Grade A	Test Pressure,[A] psi [kPa] Grade B
⅛	6	0.405 [10.3]	0.068 [1.73]	0.24 [0.37]	STD	40	700 [4800]	700 [4800]
			0.095 [2.41]	0.31 [0.47]	XS	80	850 [5900]	850 [5900]
¼	8	0.540 [13.7]	0.088 [2.24]	0.43 [0.63]	STD	40	700 [4800]	700 [4800]
			0.119 [3.02]	0.54 [0.80]	XS	80	850 [5900]	850 [5900]
⅜	10	0.675 [17.1]	0.091 [2.31]	0.57 [0.84]	STD	40	700 [4800]	700 [4800]
			0.126 [3.20]	0.74 [1.10]	XS	80	850 [5900]	850 [5900]
½	15	0.840 [21.3]	0.109 [2.77]	0.85 [1.27]	STD	40	700 [4800]	700 [4800]
			0.147 [3.73]	1.09 [1.62]	XS	80	850 [5900]	850 [5900]
			0.188 [4.78]	1.31 [1.95]	...	160	900 [6200]	900 [6200]
			0.294 [7.47]	1.72 [2.55]	XXS	...	1000 [6900]	1000 [6900]
¾	20	1.050 [26.7]	0.113 [2.87]	1.13 [1.69]	STD	40	700 [4800]	700 [4800]
			0.154 [3.91]	1.48 [2.20]	XS	80	850 [5900]	850 [5900]
			0.219 [5.56]	1.95 [2.90]	...	160	950 [6500]	950 [6500]
			0.308 [7.82]	2.44 [3.64]	XXS	...	1000 [6900]	1000 [6900]
1	25	1.315 [33.4]	0.133 [3.38]	1.68 [2.50]	STD	40	700 [4800]	700 [4800]
			0.179 [4.55]	2.17 [3.24]	XS	80	850 [5900]	850 [5900]
			0.250 [6.35]	2.85 [4.24]	...	160	950 [6500]	950 [6500]
			0.358 [9.09]	3.66 [5.45]	XXS	...	1000 [6900]	1000 [6900]
1¼	32	1.660 [42.2]	0.140 [3.56]	2.27 [3.39]	STD	40	1200 [8300]	1300 [9000]
			0.191 [4.85]	3.00 [4.47]	XS	80	1800 [12 400]	1900 [13 100]
			0.250 [6.35]	3.77 [5.61]	...	160	1900 [13 100]	2000 [13 800]
			0.382 [9.70]	5.22 [7.77]	XXS	...	2200 [15 200]	2300 [15 900]
1½	40	1.900 [48.3]	0.145 [3.68]	2.72 [4.05]	STD	40	1200 [8300]	1300 [9000]
			0.200 [5.08]	3.63 [5.41]	XS	80	1800 [12 400]	1900 [13 100]
			0.281 [7.14]	4.86 [7.25]	...	160	1950 [13 400]	2050 [14 100]
			0.400 [10.16]	6.41 [9.56]	XXS	...	2200 [15 200]	2300 [15 900]
2	50	2.375 [60.3]	0.154 [3.91]	3.66 [5.44]	STD	40	2300 [15 900]	2500 [17 200]
			0.218 [5.54]	5.03 [7.48]	XS	80	2500 [17 200]	2500 [17 200]
			0.344 [8.74]	7.47 [11.11]	...	160	2500 [17 200]	2500 [17 200]
			0.436 [11.07]	9.04 [13.44]	XXS	...	2500 [17 200]	2500 [17 200]
2½	65	2.875 [73.0]	0.203 [5.16]	5.80 [8.63]	STD	40	2500 [17 200]	2500 [17 200]
			0.276 [7.01]	7.67 [11.41]	XS	80	2500 [17 200]	2500 [17 200]
			0.375 [9.52]	10.02 [14.90]	...	160	2500 [17 200]	2500 [17 200]
			0.552 [14.02]	13.71 [20.39]	XXS	...	2500 [17 200]	2500 [17 200]
3	80	3.500 [88.9]	0.125 [3.18]	4.51 [6.72]	...	...	1290 [8900]	1500 [1000]
			0.156 [3.96]	5.58 [8.29]	...	...	1600 [11 000]	1870 [12 900]
			0.188 [4.78]	6.66 [9.92]	...	...	1930 [13 330]	2260 [15 600]
			0.216 [5.49]	7.58 [11.29]	STD	40	2220 [15 300]	2500 [17 200]
			0.250 [6.35]	8.69 [12.93]	...	...	2500 [17 200]	2500 [17 200]
			0.281 [7.14]	9.67 [14.40]	...	...	2500 [17 200]	2500 [17 200]
			0.300 [7.62]	10.26 [15.27]	XS	80	2500 [17 200]	2500 [17 200]
			0.438 [11.13]	14.34 [21.35]	...	160	2500 [17 200]	2500 [17 200]
			0.600 [15.24]	18.60 [27.68]	XXS	...	2500 [17 200]	2500 [17 200]
3½	90	4.000 [101.6]	0.125 [3.18]	5.18 [7.72]	...	...	1120 [7700]	1310 [19 000]
			0.156 [3.96]	6.41 [9.53]	...	...	1400 [6700]	1640 [11 300]
			0.188 [4.78]	7.66 [11.41]	...	...	1690 [11 700]	1970 [13 600]
			0.226 [5.74]	9.12 [13.57]	STD	40	2030 [14 000]	2370 [16 300]
			0.250 [6.35]	10.02 [14.92]	...	...	2250 [15 500]	2500 [17 200]
			0.281 [7.14]	11.17 [16.63]	...	...	2500 [17 200]	2500 [17 200]
			0.318 [8.08]	12.52 [18.63]	XS	80	2800 [19 300]	2800 [19 300]
4	100	4.500 [114.3]	0.125 [3.18]	5.85 [8.71]	...	...	1000 [6900]	1170 [8100]
			0.156 [3.96]	7.24 [10.78]	...	...	1250 [8600]	1460 [10 100]
			0.188 [4.78]	8.67 [12.91]	...	...	1500 [10 300]	1750 [12 100]
			0.219 [5.56]	10.02 [14.91]	...	...	1750 [12 100]	2040 [14 100]
			0.237 [6.02]	10.80 [16.07]	STD	40	1900 [13 100]	2210 [15 200]
			0.250 [6.35]	11.36 [16.90]	...	...	2000 [13 800]	2330 [16 100]
			0.281 [7.14]	12.67 [18.87]	...	...	2250 [15 100]	2620 [18 100]

TABLE X2.2 Continued

NPS Designator	DN Designator	Specified Outside Diameter, in. [mm]	Specified Wall Thickness, in. [mm]	Nominal Weight (Mass) per Unit Length, Plain End, lb/ft [kg/m]	Weight Class	Schedule No.	Test Pressure,[A] psi [kPa] Grade A	Grade B
			0.312 [7.92]	13.97 [20.78]	...	...	2500 [17 200]	2800 [19 300]
			0.337 [8.56]	15.00 [22.32]	XS	80	2700 [18 600]	2800 [19 300]
			0.438 [11.13]	19.02 [28.32]	...	120	2800 [19 300]	2800 [19 300]
			0.531 [13.49]	22.53 [33.54]	...	160	2800 [19 300]	2800 [19 300]
			0.674 [17.12]	27.57 [41.03]	XXS	...	2800 [19 300]	2800 [19 300]
5	125	5.563 [141.3]	0.156 [3.96]	9.02 [13.41]	...	...	1010 [7000]	1180 [8100]
			0.188 [4.78]	10.80 [16.09]	...	...	1220 [8400]	1420 [9800]
			0.219 [5.56]	12.51 [18.61]	...	...	1420 [9800]	1650 [11 400]
			0.258 [6.55]	14.63 [21.77]	STD	40	1670 [11 500]	1950 [13 400]
			0.281 [7.14]	15.87 [23.62]	...	...	1820 [12 500]	2120 [14 600]
			0.312 [7.92]	17.51 [26.05]	...	...	2020 [13 900]	2360 [16 300]
			0.344 [8.74]	19.19 [28.57]	...	...	2230 [15 400]	2600 [17 900]
			0.375 [9.52]	20.80 [30.94]	XS	80	2430 [16 800]	2800 [19 300]
			0.500 [12.70]	27.06 [40.28]	...	120	2800 [19 300]	2800 [19 300]
			0.625 [15.88]	32.99 [49.11]	...	160	2800 [19 300]	2800 [19 300]
			0.750 [19.05]	38.59 [57.43]	XXS	...	2800 [19 300]	2800 [19 300]
6	150	6.625 [168.3]	0.188 [4.78]	12.94 [19.27]	...	...	1020 [7000]	1190 [8200]
			0.219 [5.56]	15.00 [22.31]	...	...	1190 [8200]	1390 [9600]
			0.250 [6.35]	17.04 [25.36]	...	...	1360 [9400]	1580 [10 900]
			0.280 [7.11]	18.99 [28.26]	STD	40	1520 [10 500]	1780 [12 300]
			0.312 [7.92]	21.06 [31.32]	...	...	1700 [11 700]	1980 [13 700]
			0.344 [8.74]	23.10 [34.39]	...	...	1870 [12 900]	2180 [15 000]
			0.375 [9.52]	25.05 [37.28]	...	...	2040 [14 100]	2380 [16 400]
			0.432 [10.97]	28.60 [42.56]	XS	80	2350 [16 200]	2740 [18 900]
			0.562 [14.27]	36.43 [54.20]	...	120	2800 [19 300]	2800 [19 300]
			0.719 [18.26]	45.39 [67.56]	...	160	2800 [19 300]	2800 [19 300]
			0.864 [21.95]	53.21 [79.22]	XXS	...	2800 [19 300]	2800 [19 300]
8	200	8.625 [219.1]	0.188 [4.78]	16.96 [25.26]	...	...	780 [5400]	920 [6300]
			0.203 [5.16]	18.28 [27.22]	...	...	850 [5900]	1000 [6900]
			0.219 [5.56]	19.68 [29.28]	...	...	910 [6300]	1070 [7400]
			0.250 [6.35]	22.38 [33.31]	...	20	1040 [7200]	1220 [8400]
			0.277 [7.04]	24.72 [36.31]	...	30	1160 [7800]	1350 [9300]
			0.312 [7.92]	27.73 [41.24]	...	...	1300 [9000]	1520 [10 500]
			0.322 [8.18]	28.58 [42.55]	STD	40	1340 [9200]	1570 [10 800]
			0.344 [8.74]	30.45 [45.34]	...	...	1440 [9900]	1680 [11 600]
			0.375 [9.52]	33.07 [49.20]	...	...	1570 [10 800]	1830 [12 600]
			0.406 [10.31]	35.67 [53.08]	...	60	1700 [11 700]	2000 [13 800]
			0.438 [11.13]	38.33 [57.08]	...	...	1830 [12 600]	2130 [14 700]
			0.500 [12.70]	43.43 [64.64]	XS	80	2090 [14 400]	2430 [16 800]
			0.594 [15.09]	51.00 [75.92]	...	100	2500 [17 200]	2800 [19 300]
			0.719 [18.26]	60.77 [90.44]	...	120	2800 [19 300]	2800 [19 300]
			0.812 [20.62]	67.82 [100.92]	...	140	2800 [19 300]	2800 [19 300]
			0.875 [22.22]	72.49 [107.88]	XXS	...	2800 [19 300]	2800 [19 300]
			0.906 [23.01]	74.76 [111.27]	...	160	2800 [19 300]	2800 [19 300]
10	250	10.750 [273.0]	0.188 [4.78]	21.23 [31.62]	...	...	630 [4300]	730 [5000]
			0.203 [5.16]	22.89 [34.08]	...	...	680 [4700]	800 [5500]
			0.219 [5.56]	24.65 [36.67]	...	...	730 [5000]	860 [5900]
			0.250 [6.35]	28.06 [41.75]	...	20	840 [5800]	980 [6800]
			0.279 [7.09]	31.23 [46.49]	...	...	930 [6400]	1090 [7500]
			0.307 [7.80]	34.27 [51.01]	...	30	1030 [7100]	1200 [8300]
			0.344 [8.74]	38.27 [56.96]	...	...	1150 [7900]	1340 [9200]
			0.365 [9.27]	40.52 [60.29]	STD	40	1220 [8400]	1430 [9900]
			0.438 [11.13]	48.28 [71.87]	...	...	1470 [10 100]	1710 [11 800]
			0.500 [12.70]	54.79 [81.52]	XS	60	1670 [11 500]	1950 [13 400]
			0.594 [15.09]	64.49 [95.97]	...	80	1990 [13 700]	2320 [16 000]
			0.719 [18.26]	77.10 [114.70]	...	100	2410 [16 600]	2800 [19 300]
			0.844 [21.44]	89.38 [133.00]	...	120	2800 [19 300]	2800 [19 300]
			1.000 [25.40]	104.23 [155.09]	XXS	140	2800 [19 300]	2800 [19 300]
			1.125 [28.57]	115.75 [172.21]	...	160	2800 [19 300]	2800 [19 300]
12	300	12.750 [323.8]	0.203 [5.16]	27.23 [40.55]	...	...	570 [3900]	670 [4600]
			0.219 [5.56]	29.34 [43.63]	...	...	620 [4300]	720 [5000]
			0.250 [6.35]	33.41 [49.71]	...	20	710 [4900]	820 [5700]
			0.281 [7.14]	37.46 [55.75]	...	...	790 [5400]	930 [6400]
			0.312 [7.92]	41.48 [61.69]	...	...	880 [6100]	1030 [7100]
			0.330 [8.38]	43.81 [65.18]	...	30	930 [6400]	1090 [7500]

TABLE X2.2 Continued

NPS Designator	DN Designator	Specified Outside Diameter, in. [mm]	Specified Wall Thickness, in. [mm]	Nominal Weight (Mass) per Unit Length, Plain End, lb/ft [kg/m]	Weight Class	Schedule No.	Test Pressure,[A] psi [kPa] Grade A	Test Pressure,[A] psi [kPa] Grade B
			0.344 [8.74]	45.62 [67.90]	...	...	970 [6700]	1130 [7800]
			0.375 [9.52]	49.61 [73.78]	STD	...	1060 [7300]	1240 [8500]
			0.406 [10.31]	53.57 [79.70]	...	40	1150 [7900]	1340 [9200]
			0.438 [11.13]	57.65 [85.82]	...	...	1240 [8500]	1440 [9900]
			0.500 [12.70]	65.48 [97.43]	XS	...	1410 [9700]	1650 [11 400]
			0.562 [14.27]	73.22 [108.92]	...	60	1590 [11 000]	1850 [12 800]
			0.688 [17.48]	88.71 [132.04]	...	80	1940 [13 400]	2270 [15 700]
			0.844 [21.44]	107.42 [159.86]	...	100	2390 [16 500]	2780 [19 200]
			1.000 [25.40]	125.61 [186.91]	XXS	120	2800 [19 300]	2800 [19 300]
			1.125 [28.57]	139.81 [208.00]	...	140	2800 [19 300]	2800 [19 300]
			1.312 [33.32]	160.42 [238.68]	...	160	2800 [19 300]	2800 [19 300]
14	350	14.000 [355.6]	0.210 [5.33]	30.96 [46.04]	...	...	540 [3700]	630 [4300]
			0.219 [5.56]	32.26 [47.99]	...	...	560 [3900]	660 [4500]
			0.250 [6.35]	36.75 [54.69]	...	10	640 [4400]	750 [5200]
			0.281 [7.14]	41.21 [61.35]	...	...	720 [5000]	840 [5800]
			0.312 [7.92]	45.65 [67.90]	...	20	800 [5500]	940 [6500]
			0.344 [8.74]	50.22 [74.76]	...	...	880 [6100]	1030 [7100]
			0.375 [9.52]	54.62 [81.25]	STD	30	960 [6600]	1120 [7700]
			0.438 [11.13]	63.50 [94.55]	...	40	1130 [7800]	1310 [9000]
			0.469 [11.91]	67.84 [100.94]	...	...	1210 [8300]	1410 [9700]
			0.500 [12.70]	72.16 [107.39]	XS	...	1290 [8900]	1500 [10 300]
			0.594 [15.09]	85.13 [126.71]	...	60	1530 [10 500]	1790 [12 300]
			0.750 [19.05]	106.23 [158.10]	...	80	1930 [13 300]	2250 [15 500]
			0.938 [23.83]	130.98 [194.96]	...	100	2410 [16 600]	2800 [19 300]
			1.094 [27.79]	150.93 [224.65]	...	120	2800 [19 300]	2800 [19 300]
			1.250 [31.75]	170.37 [253.56]	...	140	2800 [19 300]	2800 [19 300]
			1.406 [35.71]	189.29 [281.70]	...	160	2800 [19 300]	2800 [19 300]
			2.000 [50.80]	256.56 [381.83]	...	...	2800 [19 300]	2800 [19 300]
			2.125 [53.97]	269.76 [401.44]	...	...	2800 [19 300]	2800 [19 300]
			2.200 [55.88]	277.51 [413.01]	...	...	2800 [19 300]	2800 [19 300]
			2.500 [63.50]	307.34 [457.40]	...	...	2800 [19 300]	2800 [19 300]
16	400	16.000 [406.4]	0.219 [5.56]	36.95 [54.96]	...	...	490 [3400]	570 [3900]
			0.250 [6.35]	42.09 [62.64]	...	10	560 [3900]	660 [4500]
			0.281 [7.14]	47.22 [70.30]	...	...	630 [4300]	740 [5100]
			0.312 [7.92]	52.32 [77.83]	...	20	700 [4800]	820 [5700]
			0.344 [8.74]	57.57 [85.71]	...	...	770 [5300]	900 [6200]
			0.375 [9.52]	62.64 [93.17]	STD	30	840 [5800]	980 [6800]
			0.438 [11.13]	72.86 [108.49]	...	...	990 [6800]	1150 [7900]
			0.469 [11.91]	77.87 [115.86]	...	...	1060 [7300]	1230 [8500]
			0.500 [12.70]	82.85 [123.30]	XS	40	1120 [7700]	1310 [9000]
			0.656 [16.66]	107.60 [160.12]	...	60	1480 [10 200]	1720 [11 900]
			0.844 [21.44]	136.74 [203.53]	...	80	1900 [13 100]	2220 [15 300]
			1.031 [26.19]	164.98 [245.56]	...	100	2320 [16 000]	2710 [18 700]
			1.219 [30.96]	192.61 [286.64]	...	120	2740 [18 900]	2800 [19 300]
			1.438 [36.53]	223.85 [333.19]	...	140	2800 [19 300]	2800 [19 300]
			1.594 [40.49]	245.48 [365.35]	...	160	2800 [19 300]	2800 [19 300]
18	450	18.000 [457]	0.250 [6.35]	47.44 [70.60]	...	10	500 [3400]	580 [4000]
			0.281 [7.14]	53.23 [79.24]	...	...	560 [3900]	660 [4500]
			0.312 [7.92]	58.99 [87.75]	...	20	620 [4300]	730 [5000]
			0.344 [8.74]	64.93 [96.66]	...	...	690 [4800]	800 [5500]
			0.375 [9.52]	70.65 [105.10]	STD	...	750 [5200]	880 [6100]
			0.406 [10.31]	76.36 [113.62]	...	...	810 [5600]	950 [6500]
			0.438 [11.13]	82.23 [122.43]	...	30	880 [6100]	1020 [7000]
			0.469 [11.91]	87.89 [130.78]	...	...	940 [6500]	1090 [7500]
			0.500 [12.70]	93.54 [139.20]	XS	...	1000 [6900]	1170 [8100]
			0.562 [14.27]	104.76 [155.87]	...	40	1120 [7700]	1310 [9000]
			0.750 [19.05]	138.30 [205.83]	...	60	1500 [10 300]	1750 [12 100]
			0.938 [23.83]	171.08 [254.67]	...	80	1880 [13 000]	2190 [15 100]
			1.156 [29.36]	208.15 [309.76]	...	100	2310 [15 900]	2700 [18 600]
			1.375 [34.92]	244.37 [363.64]	...	120	2750 [19 000]	2800 [19 300]
			1.562 [39.67]	274.48 [408.45]	...	140	2800 [19 300]	2800 [19 300]
			1.781 [45.24]	308.79 [459.59]	...	160	2800 [19 300]	2800 [19 300]
20	500	20.000 [508]	0.250 [6.35]	52.78 [78.55]	...	10	450 [3100]	520 [3600]
			0.281 [7.14]	59.23 [88.19]	...	...	510 [3500]	590 [4100]
			0.312 [7.92]	65.66 [97.67]	...	...	560 [3900]	660 [4500]
			0.344 [8.74]	72.28 [107.60]	...	...	620 [4300]	720 [5000]

TABLE X2.2 *Continued*

NPS Designator	DN Designator	Specified Outside Diameter, in. [mm]	Specified Wall Thickness, in. [mm]	Nominal Weight (Mass) per Unit Length, Plain End, lb/ft [kg/m]	Weight Class	Schedule No.	Test Pressure,[A] psi [kPa] Grade A	Test Pressure,[A] psi [kPa] Grade B
			0.375 [9.52]	78.67 [117.02]	STD	20	680 [4700]	790 [5400]
			0.406 [10.31]	84.04 [126.53]	...	...	730 [5000]	850 [5900]
			0.438 [11.13]	91.59 [136.37]	...	...	790 [5400]	920 [6300]
			0.469 [11.91]	97.92 [145.70]	...	...	850 [5900]	950 [6500]
			0.500 [12.70]	104.23 [155.12]	XS	30	900 [6200]	1050 [7200]
			0.594 [15.09]	123.23 [183.42]	...	40	1170 [8100]	1250 [8600]
			0.812 [20.62]	166.56 [247.83]	...	60	1460 [10 100]	1710 [11 800]
			1.031 [26.19]	209.06 [311.17]	...	80	1860 [12 800]	2170 [15 000]
			1.281 [32.54]	256.34 [381.53]	...	100	2310 [15 900]	2690 [18 500]
			1.500 [38.10]	296.65 [441.49]	...	120	2700 [18 600]	2800 [19 300]
			1.750 [44.45]	341.41 [508.11]	...	140	2800 [19 300]	2800 [19 300]
			1.969 [50.01]	379.53 [564.81]	...	160	2800 [19 300]	2800 [19 300]
24	600	24.000 [610]	0.250 [6.35]	63.47 [94.46]	...	10	380 [2600]	440 [3000]
			0.281 [7.14]	71.25 [106.08]	...	...	420 [2900]	490 [3400]
			0.312 [7.92]	79.01 [117.51]	...	...	470 [3200]	550 [3800]
			0.344 [8.74]	86.99 [129.50]	...	...	520 [3600]	600 [4100]
			0.375 [9.52]	94.71 [140.88]	STD	20	560 [3900]	660 [4500]
			0.406 [10.31]	102.40 [152.37]	...	...	610 [4200]	710 [4900]
			0.438 [11.13]	110.32 [164.26]	...	...	660 [4500]	770 [5300]
			0.469 [11.91]	117.98 [175.54]	...	...	700 [4800]	820 [5700]
			0.500 [12.70]	125.61 [186.94]	XS	...	750 [5200]	880 [6100]
			0.562 [14.27]	140.81 [209.50]	...	30	840 [5800]	980 [6800]
			0.688 [17.48]	171.45 [255.24]	...	40	1030 [7100]	1200 [8300]
			0.938 [23.83]	231.25 [344.23]	...	...	1410 [9700]	1640 [11 300]
			0.969 [24.61]	238.57 [355.02]	...	60	1450 [10 000]	1700 [11 700]
			1.219 [30.96]	296.86 [441.78]	...	80	1830 [12 600]	2130 [14 700]
			1.531 [38.89]	367.74 [547.33]	...	100	2300 [15 900]	2680 [18 500]
			1.812 [46.02]	429.79 [639.58]	...	120	2720 [18 800]	2800 [19 300]
			2.062 [52.37]	483.57 [719.63]	...	140	2800 [19 300]	2800 [19 300]
			2.344 [59.54]	542.64 [807.63]	...	160	2800 [19 300]	2800 [19 300]
26	650	26.000 [660]	0.250 [6.35]	68.82 [102.42]	...	...	350 [2400]	400 [2800]
			0.281 [7.14]	77.26 [115.02]	...	...	390 [2700]	450 [3100]
			0.312 [7.92]	85.68 [127.43]	...	10	430 [3000]	500 [3400]
			0.344 [8.74]	94.35 [140.45]	...	...	480 [3300]	560 [3900]
			0.375 [9.52]	102.72 [152.80]	STD	...	520 [3600]	610 [4200]
			0.406 [10.31]	111.08 [165.28]	...	...	560 [3900]	660 [4500]
			0.438 [11.13]	119.69 [178.20]	...	...	610 [4200]	710 [4900]
			0.469 [11.91]	128.00 [190.46]	...	...	650 [4500]	760 [5200]
			0.500 [12.70]	136.30 [202.85]	XS	20	690 [4800]	810 [5600]
			0.562 [14.27]	152.83 [227.37]	...	...	780 [5400]	910 [6300]

[A] The minimum test pressure for outside diameters and wall thicknesses not listed shall be computed by the formula given below. The computed test pressure shall be used in all cases, except as follows:

(1) For specified wall thicknesses greater than the heaviest specified wall thickness listed in this table for the applicable specified outside diameter, the test pressure shall be the highest value listed for the applicable specified outside diameter and grade.

(2) For pipe smaller than NPS 2 [DN 50] with a specified wall thickness less than the lightest specified wall thickness listed in this table for the applicable specified outside diameter and grade.

(3) For all sizes of Grade A and B pipe smaller than NPS 2 [DN 50], the test pressures were assigned arbitrarily. Test pressures for intermediate specified outside diameters need not exceed those given in this table for the next larger listed size.

$P = 2St/D$

where:
P = minimum hydrostatic test pressure, psi [kPa],
S = 0.60 times the specified minimum yield strength, psi [kPa],
t = specified wall thickness, in. [mm], and
D = specified outside diameter, in. [mm].

TABLE X2.3 Dimensions, Weights (Masses) per Unit Length, and Test Pressures for Threaded and Coupled Pipe

NPS Designator	DN Designator	Specified Outside Diameter, in. [mm]	Specified Wall Thickness, in. [mm]	Nominal Weight (Mass) per Unit Length, Threaded and Coupled, lb/ft [kg/m]	Weight Class	Schedule No.	Test Pressure, psi [kPa]	
							Grade A	Grade B
⅛	6	0.405 [10.3]	0.068 [1.73] 0.095 [2.41]	0.25 [0.37] 0.32 [0.46]	STD XS	40 80	700 [4800] 850 [5900]	700 [4800] 850 [5900]
¼	8	0.540 [13.7]	0.088 [2.24] 0.119 [3.02]	0.43 [0.63] 0.54 [0.80]	STD XS	40 80	700 [4800] 850 [5900]	700 [4800] 850 [5900]
⅜	10	0.675 [17.1]	0.091 [2.31] 0.126 [3.20]	0.57 [0.84] 0.74 [1.10]	STD XS	40 80	700 [4800] 850 [5900]	700 [4800] 850 [5900]
½	15	0.840 [21.3]	0.109 [2.77] 0.147 [3.73] 0.294 [7.47]	0.86 [1.27] 1.09 [1.62] 1.72 [2.54]	STD XS XXS	40 80 ...	700 [4800] 850 [5900] 1000 [6900]	700 [4800] 850 [5900] 1000 [6900]
¾	20	1.050 [26.7]	0.113 [2.87] 0.154 [3.91] 0.308 [7.82]	1.14 [1.69] 1.48 [2.21] 2.45 [3.64]	STD XS XXS	40 80 ...	700 [4800] 850 [5900] 1000 [6900]	700 [4800] 850 [5900] 1000 [6900]
1	25	1.315 [33.4]	0.133 [3.38] 0.179 [4.55] 0.358 [9.09]	1.69 [2.50] 2.19 [3.25] 3.66 [5.45]	STD XS XXS	40 80 ...	700 [4800] 850 [5900] 1000 [6900]	700 [4800] 850 [5900] 1000 [6900]
1¼	32	1.660 [42.2]	0.140 [3.56] 0.191 [4.85] 0.382 [9.70]	2.28 [3.40] 3.03 [4.49] 5.23 [7.76]	STD XS XXS	40 80 ...	1000 [6900] 1500 [10 300] 1800 [12 400]	1100 [7600] 1600 [11 000] 1900 [13 100]
1½	40	1.900 [48.3]	0.145 [3.68] 0.200 [5.08] 0.400 [10.16]	2.74 [4.04] 3.65 [5.39] 6.41 [9.56]	STD XS XXS	40 80 ...	1000 [6900] 1500 [10 300] 1800 [12 400]	1100 [7600] 1600 [11 000] 1900 [13 100]
2	50	2.375 [60.3]	0.154 [3.91] 0.218 [5.54] 0.436 [11.07]	3.68 [5.46] 5.08 [7.55] 9.06 [13.44]	STD XS XXS	40 80 ...	2300 [15 900] 2500 [17 200] 2500 [17 200]	2500 [17 200] 2500 [17 200] 2500 [17 200]
2½	65	2.875 [73.0]	0.203 [5.16] 0.276 [7.01] 0.552 [14.02]	5.85 [8.67] 7.75 [11.52] 13.72 [20.39]	STD XS XXS	40 80 ...	2500 [17 200] 2500 [17 200] 2500 [17 200]	2500 [17 200] 2500 [17 200] 2500 [17 200]
3	80	3.500 [88.9]	0.216 [5.49] 0.300 [7.62] 0.600 [15.24]	7.68 [11.35] 10.35 [15.39] 18.60 [27.66]	STD XS XXS	40 80 ...	2200 [15 200] 2500 [17 200] 2500 [17 200]	2500 [17 200] 2500 [17 200] 2500 [17 200]
3½	90	4.000 [101.6]	0.226 [5.74] 0.318 [8.08]	9.27 [13.71] 12.67 [18.82]	STD XS	40 80	2000 [13 800] 2800 [19 300]	2400 [16 500] 2800 [19 300]
4	100	4.500 [114.3]	0.237 [6.02] 0.337 [8.56] 0.674 [17.12]	10.92 [16.23] 15.20 [22.60] 27.62 [41.09]	STD XS XXS	40 80 ...	1900 [13 100] 2700 [18 600] 2800 [19 300]	2200 [15 200] 2800 [19 300] 2800 [19 300]
5	125	5.563 [141.3]	0.258 [6.55] 0.375 [9.52] 0.750 [19.05]	14.90 [22.07] 21.04 [31.42] 38.63 [57.53]	STD XS XXS	40 80 ...	1700 [11 700] 2400 [16 500] 2800 [19 300]	1900 [13 100] 2800 [19 300] 2800 [19 300]
6	150	6.625 [168.3]	0.280 [7.11] 0.432 [10.97] 0.864 [21.95]	19.34 [28.58] 28.88 [43.05] 53.19 [79.18]	STD XS XXS	40 80 ...	1500 [10 300] 2300 [15 900] 2800 [19 300]	1800 [12 400] 2700 [18 600] 2800 [19 300]
8	200	8.625 [219.1]	0.277 [7.04] 0.322 [8.18] 0.500 [12.70] 0.875 [22.22]	25.53 [38.07] 29.35 [43.73] 44.00 [65.41] 72.69 [107.94]	... STD XS XXS	30 40 80 ...	1200 [8300] 1300 [9000] 2100 [14 500] 2800 [19 300]	1300 [9000] 1600 [11 000] 2400 [16 500] 2800 [19 300]
10	250	10.750 [273.0]	0.279 [7.09] 0.307 [7.80] 0.365 [9.27] 0.500 [12.70]	32.33 [48.80] 35.33 [53.27] 41.49 [63.36] 55.55 [83.17]	... ... STD XS	... 30 40 60	950 [6500] 1000 [6900] 1200 [8300] 1700 [11 700]	1100 [7600] 1200 [8300] 1400 [9700] 2000 [13 800]
12	300	12.750 [323.8]	0.330 [8.38] 0.375 [9.52] 0.500 [12.70]	45.47 [67.72] 51.28 [76.21] 66.91 [99.4]	... STD XS	30	950 [6500] 1100 [7600] 1400 [9700]	1100 [7600] 1200 [8300] 1600 [11 000]

TABLE X2.4 Table of Minimum Permissible Wall Thicknesses on Inspection for Pipe Specified Wall Thicknesses

NOTE 1—The following equation, upon which this table is based, shall be applied to calculate minimum permissible wall thickness from specified wall thickness:

$$t_s \times 0.875 = t_m$$

where:
t_s = specified wall thickness, in. [mm], and
t_m = minimum permissible wall thickness, in. [mm].

The wall thickness is expressed to three [two] decimal places, the fourth [third] decimal place being carried forward or dropped in accordance with Practice E29.

NOTE 2—This table is a master table covering wall thicknesses available in the purchase of different classifications of pipe, but it is not meant to imply that all of the walls listed therein are obtainable under this specification.

Specified Wall Thickness (t_s), in. [mm]	Minimum Permissible Wall Thickness on Inspection (t_m), in. [mm]	Specified Wall Thickness (t_s), in. [mm]	Minimum Permissible Wall Thickness on Inspection (t_m), in. [mm]	Specified Wall Thickness (t_s), in. [mm]	Minimum Permissible Wall Thickness on Inspection (t_m), in. [mm]
0.068 [1.73]	0.060 [1.52]	0.294 [7.47]	0.257 [6.53]	0.750 [19.05]	0.656 [16.66]
0.088 [2.24]	0.077 [1.96]	0.300 [7.62]	0.262 [6.65]	0.812 [20.62]	0.710 [18.03]
0.091 [2.31]	0.080 [2.03]	0.307 [7.80]	0.269 [6.83]	0.844 [21.44]	0.739 [18.77]
0.095 [2.41]	0.083 [2.11]	0.308 [7.82]	0.270 [6.86]	0.864 [21.94]	0.756 [19.20]
0.109 [2.77]	0.095 [2.41]	0.312 [7.92]	0.273 [6.93]	0.875 [22.22]	0.766 [19.46]
0.113 [2.87]	0.099 [2.51]	0.318 [8.08]	0.278 [7.06]	0.906 [23.01]	0.793 [20.14]
0.119 [3.02]	0.104 [2.64]	0.322 [8.18]	0.282 [7.16]	0.938 [23.82]	0.821 [20.85]
0.125 [3.18]	0.109 [2.77]	0.330 [8.38]	0.289 [7.34]	0.968 [24.59]	0.847 [21.51]
0.126 [3.20]	0.110 [2.79]	0.337 [8.56]	0.295 [7.49]	1.000 [25.40]	0.875 [22.22]
0.133 [3.38]	0.116 [2.95]	0.343 [8.71]	0.300 [7.62]	1.031 [26.19]	0.902 [22.91]
0.140 [3.56]	0.122 [3.10]	0.344 [8.74]	0.301 [7.65]	1.062 [26.97]	0.929 [26.30]
0.145 [3.68]	0.127 [3.23]	0.358 [9.09]	0.313 [7.95]	1.094 [27.79]	0.957 [24.31]
0.147 [3.73]	0.129 [3.28]	0.365 [9.27]	0.319 [8.10]	1.125 [28.58]	0.984 [24.99]
0.154 [3.91]	0.135 [3.43]	0.375 [9.52]	0.328 [8.33]	1.156 [29.36]	1.012 [25.70]
0.156 [3.96]	0.136 [3.45]	0.382 [9.70]	0.334 [8.48]	1.219 [30.96]	1.067 [27.08]
0.179 [4.55]	0.157 [3.99]	0.400 [10.16]	0.350 [8.89]	1.250 [31.75]	1.094 [27.79]
0.187 [4.75]	0.164 [4.17]	0.406 [10.31]	0.355 [9.02]	1.281 [32.54]	1.121 [28.47]
0.188 [4.78]	0.164 [4.17]	0.432 [10.97]	0.378 [9.60]	1.312 [33.32]	1.148 [29.16]
0.191 [4.85]	0.167 [4.24]	0.436 [11.07]	0.382 [9.70]	1.343 [34.11]	1.175 [29.85]
0.200 [5.08]	0.175 [4.44]	0.437 [11.10]	0.382 [9.70]	1.375 [34.92]	1.203 [30.56]
0.203 [5.16]	0.178 [4.52]	0.438 [11.13]	0.383 [9.73]	1.406 [35.71]	1.230 [31.24]
0.216 [5.49]	0.189 [4.80]	0.500 [12.70]	0.438 [11.13]	1.438 [36.53]	1.258 [31.95]
0.218 [5.54]	0.191 [4.85]	0.531 [13.49]	0.465 [11.81]	1.500 [38.10]	1.312 [33.32]
0.219 [5.56]	0.192 [4.88]	0.552 [14.02]	0.483 [12.27]	1.531 [38.89]	1.340 [34.04]
0.226 [5.74]	0.198 [5.03]	0.562 [14.27]	0.492 [12.50]	1.562 [39.67]	1.367 [34.72]
0.237 [6.02]	0.207 [5.26]	0.594 [15.09]	0.520 [13.21]	1.594 [40.49]	1.395 [35.43]
0.250 [6.35]	0.219 [5.56]	0.600 [15.24]	0.525 [13.34]	1.750 [44.45]	1.531 [38.89]
0.258 [6.55]	0.226 [5.74]	0.625 [15.88]	0.547 [13.89]	1.781 [45.24]	1.558 [39.57]
0.276 [7.01]	0.242 [6.15]	0.656 [16.66]	0.574 [14.58]	1.812 [46.02]	1.586 [40.28]
0.277 [7.04]	0.242 [6.15]	0.674 [17.12]	0.590 [14.99]	1.968 [49.99]	1.722 [43.74]
0.279 [7.09]	0.244 [6.20]	0.688 [17.48]	0.602 [15.29]	2.062 [52.37]	1.804 [45.82]
0.280 [7.11]	0.245 [6.22]	0.719 [18.26]	0.629 [15.98]	2.344 [59.54]	2.051 [52.10]
0.281 [7.14]	0.246 [6.25]				

X3. BASIC THREADING DATA

X3.1 Fig. X3.1 is to be used with Table X3.1. Fig. X3.2 is to be used with Table X3.2.

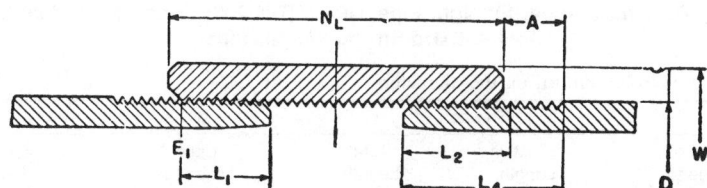

FIG. X3.1 Dimensions of Hand Tight Assembly for Use with Table X3.1

TABLE X3.1 Basic Threading Data for Standard-Weight Pipe, NPS 6 [DN 150] or Smaller

NOTE 1—All dimensions in this table are nominal and subject to mill tolerances.
NOTE 2—The taper of threads is ¾ in./ft [62.5 mm/m] on the diameter.

Pipe			Threads					Coupling		
NPS Designator	DN Designator	Specified Outside Diameter, in. [mm] D	Number per inch	End of Pipe to Hand Tight Plane, in. [mm] L_1	Effective Length, in. [mm] L_2	Total Length, in. [mm] L_4	Pitch Diameter at Hand Tight Plane, in. [mm] E_1	Specified Outside Diameter, in. [mm] W	Length, min., in. [mm] N_L	Hand Tight Stand-Off (Number of Threads) A
⅛	6	0.405 [10.3]	27	0.1615 [4.1021]	0.2638 [6.7005]	0.3924 [9.9670]	0.37360 [9.48944]	0.563 [14.3]	¾ [19]	4
¼	8	0.540 [13.7]	18	0.2278 [5.7861]	0.4018 [10.2057]	0.5946 [15.1028]	0.49163 [12.48740]	0.719 [18.3]	1⅛ [29]	5½
⅜	10	0.675 [17.1]	18	0.240 [6.096]	0.4078 [10.3581]	0.6006 [15.2552]	0.62701 [15.92605]	0.875 [22.2]	1⅛ [29]	5
½	15	0.840 [21.3]	14	0.320 [8.128]	0.5337 [13.5560]	0.7815 [19.8501]	0.77843 [19.77212]	1.063 [27.0]	1½ [38]	5
¾	20	1.050 [26.7]	14	0.339 [8.611]	0.5457 [13.8608]	0.7935 [20.1549]	0.98887 [25.11730]	1.313 [33.4]	1⁹⁄₁₆ [40]	5
1	25	1.315 [33.4]	11½	0.400 [10.160]	0.6828 [17.3431]	0.9845 [25.0063]	1.23863 [31.46120]	1.576 [40.0]	1¹⁵⁄₁₆ [49]	5
1¼	32	1.660 [42.2]	11½	0.420 [10.668]	0.7068 [17.9527]	1.0085 [25.6159]	1.58338 [40.21785]	1.900 [48.3]	2 [50]	5
1½	40	1.900 [48.3]	11½	0.420 [10.668]	0.7235 [18.3769]	1.0252 [26.0401]	1.82234 [46.28744]	2.200 [55.9]	2 [50]	5½
2	50	2.375 [60.3]	11½	0.436 [11.074]	0.7565 [19.2151]	1.0582 [26.8783]	2.29627 [58.32526]	2.750 [69.8]	2¹⁄₁₆ [52]	5½
2½	65	2.875 [73.0]	8	0.682 [17.323]	1.1376 [28.8950]	1.5712 [39.9085]	2.76216 [70.15886]	3.250 [82.5]	3¹⁄₁₆ [78]	5½
3	80	3.500 [88.9]	8	0.766 [19.456]	1.2000 [30.4800]	1.6337 [41.4960]	3.38850 [86.06790]	4.000 [101.6]	3³⁄₁₆ [81]	5½
3½	90	4.000 [101.6]	8	0.821 [20.853]	1.2500 [31.7500]	1.6837 [42.7660]	3.88881 [98.77577]	4.625 [117.5]	3⁵⁄₁₆ [84]	5½
4	100	4.500 [114.3]	8	0.844 [21.438]	1.3000 [33.0200]	1.7337 [44.0360]	4.38713 [111.43310]	5.000 [127.0]	3⁷⁄₁₆ [87]	5
5	125	5.563 [141.3]	8	0.937 [23.800]	1.4063 [35.7200]	1.8400 [46.7360]	5.44929 [138.41200]	6.296 [159.9]	3¹¹⁄₁₆ [94]	5
6	150	6.625 [168.3]	8	0.958 [24.333]	1.5125 [38.4175]	1.9462 [49.4335]	6.50597 [165.25164]	7.390 [187.7]	3¹⁵⁄₁₆ [100]	6

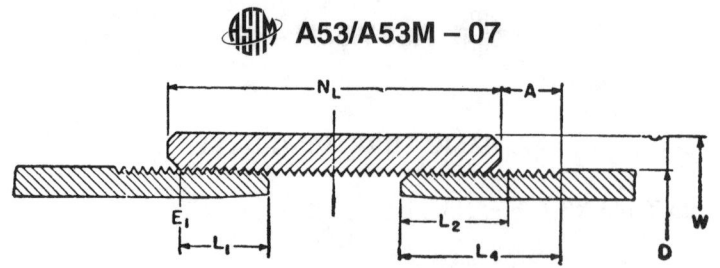

FIG. X3.2 Dimensions of Hand Tight Assembly for Use with Table X3.2

TABLE X3.2 Basic Threading Data for Standard-Weight Pipe, NPS 8 [DN 200] or Larger, and all Sizes of Extra-Strong and Double-Extra-Strong Weight Pipe

NOTE 1—The taper of threads is ¾ in./ft [62.5 mm/m] on the diameter.

Pipe				Threads								Coupling				
NPS Designator	DN Designator	Specified Outside Diameter, in. D	[mm]	Number per Inch	End of Pipe to Hand Tight Plane, in. L_1	[mm]	Effective Length, in. L_2	[mm]	Total Length, in. L_4	[mm]	Pitch Diameter at Hand Tight Plane, in. E_1	[mm]	Specified Outside Diameter, in. W	[mm]	Length, min, in. N_L	[mm]	Hand Tight Stand-Off (Number of Threads)
⅛	6	0.405	[10.3]	27	0.1615	[4.1021]	0.2638	[6.7005]	0.3924	[9.9670]	0.37360	[9.48944]	0.563	[14.3]	1¹⁄₁₆	[27]	3
¼	8	0.540	[13.7]	18	0.2278	[5.7861]	0.4018	[10.2057]	0.5946	[15.1028]	0.49163	[12.48740]	0.719	[18.3]	1⅝	[41]	3
⅜	10	0.675	[17.1]	18	0.240	[6.096]	0.4078	[10.3581]	0.6006	[15.2552]	0.62701	[15.92605]	0.875	[22.2]	1⅝	[41]	3
½	15	0.840	[21.3]	14	0.320	[8.128]	0.5337	[13.5560]	0.7815	[19.8501]	0.77843	[19.77212]	1.063	[27.0]	2⅛	[54]	3
¾	20	1.050	[26.7]	14½	0.339	[8.611]	0.5457	[13.8608]	0.7935	[20.1549]	0.98887	[25.11730]	1.313	[33.4]	2⅛	[54]	3
1	25	1.315	[33.4]	11	0.400	[10.160]	0.6828	[17.3431]	0.9505	[25.0063]	1.23863	[31.46120]	1.576	[40.0]	2⅝	[67]	3
1¼	32	1.660	[42.2]	11½	0.420	[10.668]	0.7068	[17.9527]	1.0085	[25.6159]	1.58338	[40.21785]	2.054	[52.2]	2¾	[70]	3
1½	40	1.900	[48.3]	11½	0.420	[10.668]	0.7235	[18.3769]	1.0252	[26.0401]	1.82234	[46.28744]	2.200	[55.9]	2¾	[70]	3
2	50	2.375	[60.3]	11½	0.436	[11.074]	0.7565	[19.2151]	1.0582	[26.8783]	2.29627	[58.32526]	2.875	[73.0]	2⅞	[73]	3
2½	65	2.875	[73.0]	8	0.682	[17.323]	1.1375	[28.8950]	1.5712	[39.9085]	2.76216	[70.15886]	3.375	[85.7]	4⅛	[105]	2
3	80	3.500	[88.9]	8	0.766	[19.456]	1.2000	[30.4800]	1.6337	[41.4960]	3.38850	[86.06790]	4.000	[101.6]	4¼	[108]	2
3½	90	4.000	[101.6]	8	0.821	[20.853]	1.2500	[31.7500]	1.6837	[42.7660]	3.88881	[98.77577]	4.625	[117.5]	4⅜	[111]	2
4	100	4.500	[114.3]	8	0.844	[21.438]	1.3000	[33.0200]	1.7337	[44.0360]	4.38713	[111.43310]	5.200	[132.1]	4½	[114]	2
5	125	5.563	[141.3]	8	0.937	[23.800]	1.4063	[35.7200]	1.8400	[46.7360]	5.44929	[138.41200]	6.296	[159.9]	4⅝	[117]	2
6	150	6.625	[168.3]	8	0.958	[24.333]	1.5125	[38.4175]	1.9462	[49.4335]	6.50597	[165.25164]	7.390	[187.7]	4⅞	[124]	2
8	200	8.625	[219.1]	8	1.063	[27.000]	1.7125	[43.4975]	2.1462	[54.5135]	8.50003	[215.90076]	9.625	[244.5]	5¼	[133]	2
10	250	10.750	[273.0]	8	1.210	[30.734]	1.9250	[48.8950]	2.3587	[59.9110]	10.62094	[269.77188]	11.750	[298.4]	5¾	[146]	2
12	300	12.750	[323.8]	8	1.360	[34.544]	2.1250	[53.9750]	2.5587	[64.9910]	12.61781	[320.49237]	14.000	[355.6]	6⅛	[156]	2
14	350	14.000	[355.6]	8	1.562	[39.675]	2.2500	[57.1500]	2.6837	[68.1660]	13.87263	[352.36480]	15.000	[381.0]	6⅜	[162]	2
16	400	16.000	[406.4]	8	1.812	[46.025]	2.4500	[62.2300]	2.8837	[73.2460]	15.87575	[403.24405]	17.000	[432]	6¾	[171]	2
18	450	18.000	[457]	8	2.000	[50.800]	2.6500	[67.3100]	3.0837	[78.3260]	17.87500	[454.02500]	19.000	[483]	7⅛	[181]	2
20	500	20.000	[508]	8	2.125	[53.975]	2.8500	[72.3900]	3.2837	[83.4060]	19.87031	[504.70587]	21.000	[533]	7⅝	[194]	2

X4. ELONGATION VALUES

X4.1 Tabulated in Table X4.1 are the minimum elongation values in inch-pound units, calculated using the equation given in Table 2.

TABLE X4.1 Elongation Values

Area, A, in.²	Specified Wall Thickness, in.			Elongation in 2 in., min, %	
	Tension Test Specimen			Specified Minimum Tensile Strength, psi	
	¾-in. Specimen	1-in. Specimen	1½-in. Specimen	48 000	60 000
0.75 and greater	0.994 and greater	0.746 and greater	0.497 and greater	36	30
0.74	0.980–0.993	0.735–0.745	0.490–0.496	36	29
0.73	0.967–0.979	0.726–0.734	0.484–0.489	36	29
0.72	0.954–0.966	0.715–0.725	0.477–0.483	36	29
0.71	0.941–0.953	0.706–0.714	0.471–0.476	36	29
0.70	0.927–0.940	0.695–0.705	0.464–0.470	36	29
0.69	0.914–0.926	0.686–0.694	0.457–0.463	36	29
0.68	0.900–0.913	0.675–0.685	0.450–0.456	35	29
0.67	0.887–0.899	0.666–0.674	0.444–0.449	35	29
0.66	0.874–0.886	0.655–0.665	0.437–0.443	35	29

A53/A53M – 07

TABLE X4.1 *Continued*

Area, A, in.²	Specified Wall Thickness, in.			Elongation in 2 in., min, %	
	Tension Test Specimen			Specified Minimum Tensile Strength, psi	
	¾-in. Specimen	1-in. Specimen	1½-in. Specimen	48 000	60 000
0.65	0.861–0.873	0.646–0.654	0.431–0.436	35	29
0.64	0.847–0.860	0.635–0.645	0.424–0.430	35	29
0.63	0.834–0.846	0.626–0.634	0.417–0.423	35	29
0.62	0.820–0.833	0.615–0.625	0.410–0.416	35	28
0.61	0.807–0.819	0.606–0.614	0.404–0.409	35	28
0.60	0.794–0.806	0.595–0.605	0.397–0.403	35	28
0.59	0.781–0.793	0.586–0.594	0.391–0.396	34	28
0.58	0.767–0.780	0.575–0.585	0.384–0.390	34	28
0.57	0.754–0.766	0.566–0.574	0.377–0.383	34	28
0.56	0.740–0.753	0.555–0.565	0.370–0.376	34	28
0.55	0.727–0.739	0.546–0.554	0.364–0.369	34	28
0.54	0.714–0.726	0.535–0.545	0.357–0.363	34	28
0.53	0.701–0.713	0.526–0.534	0.351–0.356	34	28
0.52	0.687–0.700	0.515–0.525	0.344–0.350	34	27
0.51	0.674–0.686	0.506–0.514	0.337–0.343	33	27
0.50	0.660–0.673	0.495–0.505	0.330–0.336	33	27
0.49	0.647–0.659	0.486–0.494	0.324–0.329	33	27
0.48	0.634–0.646	0.475–0.485	0.317–0.323	33	27
0.47	0.621–0.633	0.466–0.474	0.311–0.316	33	27
0.46	0.607–0.620	0.455–0.465	0.304–0.310	33	27
0.45	0.594–0.606	0.446–0.454	0.297–0.303	33	27
0.44	0.580–0.593	0.435–0.445	0.290–0.296	32	27
0.43	0.567–0.579	0.426–0.434	0.284–0.289	32	26
0.42	0.554–0.566	0.415–0.425	0.277–0.283	32	26
0.41	0.541–0.553	0.406–0.414	0.271–0.276	32	26
0.40	0.527–0.540	0.395–0.405	0.264–0.270	32	26
0.39	0.514–0.526	0.386–0.394	0.257–0.263	32	26
0.38	0.500–0.513	0.375–0.385	0.250–0.256	32	26
0.37	0.487–0.499	0.366–0.374	0.244–0.249	31	26
0.36	0.474–0.486	0.355–0.365	0.237–0.243	31	26
0.35	0.461–0.473	0.346–0.354	0.231–0.236	31	25
0.34	0.447–0.460	0.335–0.345	0.224–0.230	31	25
0.33	0.434–0.446	0.326–0.334	0.217–0.223	31	25
0.32	0.420–0.433	0.315–0.325	0.210–0.216	30	25
0.31	0.407–0.419	0.306–0.314	0.204–0.209	30	25
0.30	0.394–0.406	0.295–0.305	0.197–0.203	30	25
0.29	0.381–0.393	0.286–0.294	0.191–0.196	30	24
0.28	0.367–0.380	0.275–0.285	0.184–0.190	30	24
0.27	0.354–0.366	0.266–0.274	0.177–0.183	29	24
0.26	0.340–0.353	0.255–0.265	0.170–0.176	29	24
0.25	0.327–0.339	0.246–0.254	0.164–0.169	29	24
0.24	0.314–0.326	0.235–0.245	0.157–0.163	29	24
0.23	0.301–0.313	0.226–0.234	0.151–0.156	29	23
0.22	0.287–0.300	0.215–0.225	0.144–0.150	28	23
0.21	0.274–0.286	0.260–0.214	0.137–0.143	28	23
0.20	0.260–0.273	0.195–0.205	0.130–0.136	28	23
0.19	0.247–0.259	0.186–0.194	0.124–0.129	27	22
0.18	0.234–0.246	0.175–0.185	0.117–0.123	27	22
0.17	0.221–0.233	0.166–0.174	0.111–0.116	27	22
0.16	0.207–0.220	0.155–0.165	0.104–0.110	27	22
0.15	0.194–0.206	0.146–0.154	0.097–0.103	26	21
0.14	0.180–0.193	0.135–0.145	0.091–0.096	26	21
0.13	0.167–0.179	0.126–0.134	0.084–0.090	25	21
0.12	0.154–0.166	0.115–0.125	0.077–0.083	25	20
0.11	0.141–0.153	0.106–0.114	0.071–0.076	25	20
0.10	0.127–0.140	0.095–0.105	0.064–0.070	24	20
0.09	0.114–0.126	0.086–0.094	0.057–0.063	24	19
0.08	0.100–0.113	0.075–0.085	0.050–0.056	23	19
0.07	0.087–0.099	0.066–0.074	0.044–0.049	22	18
0.06	0.074–0.086	0.055–0.065	0.037–0.043	22	18
0.05	0.061–0.073	0.046–0.054	0.031–0.036	21	17
0.04	0.047–0.060	0.035–0.045	0.024–0.030	20	16
0.03	0.034–0.046	0.026–0.034	0.017–0.023	19	16
0.02	0.020–0.033	0.015–0.025	0.010–0.016	17	14
0.01 and less	0.019 and less	0.014 and less	0.009 and less	15	12

X4.2 Tabulated in Table X4.2 are the minimum elongation values in SI units, calculated using the equation given in Table 2.

TABLE X4.2 Elongation Values

Area, A, mm²	Specified Wall Thickness, mm			Elongation in 50 mm, min, %	
	Tension Test Specimen			Specified Minimum Tensile Strength, MPa	
	19-mm Specimen	25-mm Specimen	38-mm Specimen	330	415
500 and greater	26.3 and greater	20.0 and greater	13.2 and greater	36	30
480-499	25.3-26.2	19.2-19.9	12.7-13.1	36	30
460-479	24.2-25.2	18.4-19.1	12.1-12.6	36	29
440-459	23.2-24.1	17.6-18.3	11.6-12.0	36	29
420-439	22.1-23.1	16.8-17.5	11.1-11.5	35	29
400-419	21.1-22.0	16.0-16.7	10.6-11.0	35	29
380-399	20.0-21.0	15.2-15.9	10.0-10.5	35	28
360-379	19.0-19.9	14.4-15.0	9.5-9.9	34	28
340-359	17.9-18.9	13.6-14.3	9.0-9.4	34	28
320-339	16.9-17.8	12.8-13.5	8.5-8.9	34	27
300-319	15.8-16.8	12.0-12.7	7.9-8.4	33	27
280-299	14.8-15.7	11.2-11.9	7.4-7.8	33	27
260-279	13.7-14.7	10.4-11.1	6.9-7.3	32	26
240-259	12.7-13.6	9.6-10.3	6.4-6.8	32	26
220-239	11.6-12.6	8.8-9.5	5.8-6.3	31	26
200-219	10.5-11.5	8.0-8.7	5.3-5.7	31	25
190-199	10.0-10.4	7.6-7.9	5.0-5.2	30	25
180-189	9.5-9.9	7.2-7.5	4.8-4.9	30	24
170-179	9.0-9.4	6.8-7.1	4.5-4.7	30	24
160-169	8.4-8.9	6.4-6.7	4.2-4.4	29	24
150-159	7.9-8.3	6.0-6.3	4.0-4.1	29	24
140-149	7.4-7.8	5.6-5.9	3.7-3.9	29	23
130-139	6.9-7.3	5.2-5.5	3.5-3.6	28	23
120-129	6.3-6.8	4.8-5.1	3.2-3.4	28	23
110-119	5.8-6.2	4.4-4.7	2.9-3.1	27	22
100-109	5.3-5.7	4.0-4.3	2.7-2.8	27	22
90-99	4.8-5.2	3.6-3.9	2.4-2.6	26	21
80-89	4.2-4.7	3.2-3.5	2.1-2.3	26	21
70-79	3.7-4.1	2.8-3.1	1.9-2.0	25	21
60-69	3.2-3.6	2.4-2.7	1.6-1.8	24	20
50-59	2.7-3.1	2.0-2.3	...	24	19
40-49	2.1-2.6	1.6-1.9	...	23	19
30-39	1.6-2.0	...	...	22	18

SUMMARY OF CHANGES

Committee A01 has identified the location of selected changes to this specification since the last issue, A53/A53M – 06a, that may impact the use of this specification. (Approved September 1, 2007)

(*1*) Revised 9.1.1 to require the use of full-volumetric NDE on Type E pipe produced on a hot-stretch reducing mill.

Committee A01 has identified the location of selected changes to this specification since the last issue, A53/A53M – 06, that may impact the use of this specification. (Approved October 1, 2006)

(*1*) Revised 1.1 to address supplementary requirements.
(*2*) Added new 3.1.16 and renumbered subsequent paragraphs.
(*3*) Revised 7.3.1.
(*4*) Revised 7.3.2.
(*5*) Deleted Note 4 and renumbered subsequent notes.
(*6*) Added Supplementary Requirement S1.

Committee A01 has identified the location of selected changes to this specification since the last issue, A53/A53M – 05, that may impact the use of this specification. (Approved May 1, 2006)

(*1*) Revised the minimum coupling length for NPS 6 in Table X3.1.

(*2*) Editorially corrected the minimum coupling length for NPS ¾ in Table X3.1 and the DN designation for NPS 6 in the title for Table X3.1.

ASTM International takes no position respecting the validity of any patent rights asserted in connection with any item mentioned in this standard. Users of this standard are expressly advised that determination of the validity of any such patent rights, and the risk of infringement of such rights, are entirely their own responsibility.

This standard is subject to revision at any time by the responsible technical committee and must be reviewed every five years and if not revised, either reapproved or withdrawn. Your comments are invited either for revision of this standard or for additional standards and should be addressed to ASTM International Headquarters. Your comments will receive careful consideration at a meeting of the responsible technical committee, which you may attend. If you feel that your comments have not received a fair hearing you should make your views known to the ASTM Committee on Standards, at the address shown below.

This standard is copyrighted by ASTM International, 100 Barr Harbor Drive, PO Box C700, West Conshohocken, PA 19428-2959, United States. Individual reprints (single or multiple copies) of this standard may be obtained by contacting ASTM at the above address or at 610-832-9585 (phone), 610-832-9555 (fax), or service@astm.org (e-mail); or through the ASTM website (www.astm.org).

Designation: A106/A106M – 08

Used in USDOE-NE standards

Standard Specification for
Seamless Carbon Steel Pipe for High-Temperature Service[1]

This standard is issued under the fixed designation A106/A106M; the number immediately following the designation indicates the year of original adoption or, in the case of revision, the year of last revision. A number in parentheses indicates the year of last reapproval. A superscript epsilon (ε) indicates an editorial change since the last revision or reapproval.

This standard has been approved for use by agencies of the Department of Defense.

1. Scope*

1.1 This specification[2] covers seamless carbon steel pipe for high-temperature service (Note 1) in NPS ⅛ to NPS 48 [DN 6 to DN 1200] (Note 2) inclusive, with nominal (average) wall thickness as given in ASME B 36.10M. It shall be permissible to furnish pipe having other dimensions provided such pipe complies with all other requirements of this specification. Pipe ordered under this specification shall be suitable for bending, flanging, and similar forming operations, and for welding. When the steel is to be welded, it is presupposed that a welding procedure suitable to the grade of steel and intended use or service will be utilized.

NOTE 1—It is suggested, consideration be given to possible graphitization.

NOTE 2—The dimensionless designator NPS (nominal pipe size) [DN (diameter nominal)] has been substituted in this standard for such traditional terms as "nominal diameter," "size," and "nominal size."

1.2 Supplementary requirements of an optional nature are provided for seamless pipe intended for use in applications where a superior grade of pipe is required. These supplementary requirements call for additional tests to be made and when desired shall be so stated in the order.

1.3 The values stated in either SI units or inch-pound units are to be regarded separately as standard. The values stated in each system may not be exact equivalents; therefore, each system shall be used independently of the other. Combining values from the two systems may result in non-conformance with the standard.

1.4 The following precautionary caveat pertains only to the test method portion, Sections 11, 12, and 13 of this specification: *This standard does not purport to address all of the safety concerns, if any, associated with its use. It is the responsibility of the user of this standard to establish appropriate safety and health practices and determine the applicability of regulatory limitations prior to use.*

2. Referenced Documents

2.1 *ASTM Standards:*[3]
A530/A530M Specification for General Requirements for Specialized Carbon and Alloy Steel Pipe
E213 Practice for Ultrasonic Testing of Metal Pipe and Tubing
E309 Practice for Eddy-Current Examination of Steel Tubular Products Using Magnetic Saturation
E381 Method of Macroetch Testing Steel Bars, Billets, Blooms, and Forgings
E570 Practice for Flux Leakage Examination of Ferromagnetic Steel Tubular Products

2.2 *ASME Standard:*
ASME B 36.10M Welded and Seamless Wrought Steel Pipe[4]

2.3 *Military Standards:*
MIL-STD-129 Marking for Shipment and Storage[5]
MIL-STD-163 Steel Mill Products, Preparation for Shipment and Storage[5]

2.4 *Federal Standard:*
Fed. Std. No. 123 Marking for Shipments (Civil Agencies)[5]
Fed. Std. No. 183 Continuous Identification Marking of Iron and Steel Products[5]

2.5 *Other Standards:*
SSPC-SP 6 Surface Preparation Specification No. 6[6]

3. Ordering Information

3.1 The inclusion of the following, as required will describe the desired material adequately, when ordered under this specification:

3.1.1 Quantity (feet, metres, or number of lengths),
3.1.2 Name of material (seamless carbon steel pipe),
3.1.3 Grade (Table 1),

---

[1] This specification is under the jurisdiction of Committee A01 on Steel, Stainless Steel and Related Alloys and is the direct responsibility of Subcommittee A01.09 on Carbon Steel Tubular Products.
Current edition approved July 15, 2008. Published August 2008. Originally approved in 1926. Last previous edition in 2006 as A106/A106M– 06a. DOI: 10.1520/A0106_A0106M-08.

[2] For ASME Boiler and Pressure Vessel Code applications see related Specifications SA-106 in Section II of that Code.

[3] For referenced ASTM standards, visit the ASTM website, www.astm.org, or contact ASTM Customer Service at service@astm.org. For *Annual Book of ASTM Standards* volume information, refer to the standard's Document Summary page on the ASTM website.

[4] Available from American Society of Mechanical Engineers (ASME), ASME International Headquarters, Three Park Ave., New York, NY 10016-5990, http://www.asme.org.

[5] Available from Standardization Documents Order Desk, DODSSP, Bldg. 4, Section D, 700 Robbins Ave., Philadelphia, PA 19111-5098.

[6] Available from Society for Protective Coatings (SSPC), 40 24th St., 6th Floor, Pittsburgh, PA 15222-4656, http://www.sspc.org.

---

*A Summary of Changes section appears at the end of this standard.

Copyright © ASTM International, 100 Barr Harbor Drive, PO Box C700, West Conshohocken, PA 19428-2959, United States.

TABLE 1 Chemical Requirements

	Composition, %		
	Grade A	Grade B	Grade C
Carbon, max[A]	0.25	0.30	0.35
Manganese	0.27–0.93	0.29–1.06	0.29–1.06
Phosphorus, max	0.035	0.035	0.035
Sulfur, max	0.035	0.035	0.035
Silicon, min	0.10	0.10	0.10
Chrome, max[B]	0.40	0.40	0.40
Copper, max[B]	0.40	0.40	0.40
Molybdenum, max[B]	0.15	0.15	0.15
Nickel, max[B]	0.40	0.40	0.40
Vanadium, max[B]	0.08	0.08	0.08

[A] For each reduction of 0.01 % below the specified carbon maximum, an increase of 0.06 % manganese above the specified maximum will be permitted up to a maximum of 1.35 %.

[B] These five elements combined shall not exceed 1 %.

3.1.4 Manufacture (hot-finished or cold-drawn),

3.1.5 Size (NPS [DN] and weight class or schedule number, or both; outside diameter and nominal wall thickness; or inside diameter and nominal wall thickness),

3.1.6 Special outside diameter tolerance pipe (16.2.2),

3.1.7 Inside diameter tolerance pipe, over 10 in. [250 mm] ID (16.2.3),

3.1.8 Length (specific or random, Section 17),

3.1.9 Optional requirements (Section 9 and S1 to S8),

3.1.10 Test report required (Section on Certification of Specification A530/A530M),

3.1.11 Specification designation (A106 or A106M, including year-date),

3.1.12 End use of material,

3.1.13 Hydrostatic test in accordance with Specification A530/A530M or 13.3 of this specification, or NDE in accordance with Section 14 of this specification.

3.1.14 Special requirements.

4. Process

4.1 The steel shall be killed steel, with the primary melting process being open-hearth, basic-oxygen, or electric-furnace, possibly combined with separate degassing or refining. If secondary melting, using electroslag remelting or vacuum-arc remelting is subsequently employed, the heat shall be defined as all of the ingots remelted from a single primary heat.

4.2 Steel cast in ingots or strand cast is permissible. When steels of different grades are sequentially strand cast, identification of the resultant transition material is required. The producer shall remove the transition material by any established procedure that positively separates the grades.

4.3 For pipe NPS 1½ [DN 40] and under, it shall be permissible to furnish hot finished or cold drawn.

4.4 Unless otherwise specified, pipe NPS 2 [DN 50] and over shall be furnished hot finished. When agreed upon between the manufacturer and the purchaser, it is permissible to furnish cold-drawn pipe.

5. Heat Treatment

5.1 Hot-finished pipe need not be heat treated. Cold-drawn pipe shall be heat treated after the final cold draw pass at a temperature of 1200 °F (650 °C) or higher.

6. General Requirements

6.1 Material furnished to this specification shall conform to the applicable requirements of the current edition of Specification A530/A530M unless otherwise provided herein.

7. Chemical Composition

7.1 The steel shall conform to the requirements as to chemical composition prescribed in Table 1.

8. Heat Analysis

8.1 An analysis of each heat of steel shall be made by the steel manufacturer to determine the percentages of the elements specified in Section 7. If the secondary melting processes of 5.1 are employed, the heat analysis shall be obtained from one remelted ingot or the product of one remelted ingot of each primary melt. The chemical composition thus determined, or that determined from a product analysis made by the manufacturer, if the latter has not manufactured the steel, shall be reported to the purchaser or the purchaser's representative, and shall conform to the requirements specified in Section 7.

9. Product Analysis

9.1 At the request of the purchaser, analyses of two pipes from each lot (see 20.1) shall be made by the manufacturer from the finished pipe. The results of these analyses shall be reported to the purchaser or the purchaser's representative and shall conform to the requirements specified in Section 7.

9.2 If the analysis of one of the tests specified in 9.1 does not conform to the requirements specified in Section 7, analyses shall be made on additional pipes of double the original number from the same lot, each of which shall conform to requirements specified.

10. Tensile Requirements

10.1 The material shall conform to the requirements as to tensile properties given in Table 2.

11. Bending Requirements

11.1 For pipe NPS 2 [DN 50] and under, a sufficient length of pipe shall stand being bent cold through 90° around a cylindrical mandrel, the diameter of which is twelve times the outside diameter (as shown in ASME B 36.10M) of the pipe, without developing cracks. When ordered for close coiling, the pipe shall stand being bent cold through 180° around a cylindrical mandrel, the diameter of which is eight times the outside diameter (as shown in ASME B 36.10M) of the pipe, without failure.

11.2 For pipe whose diameter exceeds 25 in. [635 mm] and whose diameter to wall thickness ratio, where the diameter to wall thickness ratio is the specified outside diameter divided by the nominal wall thickness, is 7.0 or less, the bend test shall be conducted. The bend test specimens shall be bent at room temperature through 180° with the inside diameter of the bend being 1 in. [25 mm] without cracking on the outside portion of the bent portion.

Example: For 28 in. [711 mm] diameter 5.000 in. [127 mm] thick pipe the diameter to wall thickness ratio = 28/5 = 5.6 [711/127 = 5.6].

TABLE 2 Tensile Requirements

	Grade A		Grade B		Grade C	
Tensile strength, min, psi [MPa]	48 000 [330]		60 000 [415]		70 000 [485]	
Yield strength, min, psi [MPa]	30 000 [205]		35 000 [240]		40 000 [275]	
	Longitudinal	Transverse	Longitudinal	Transverse	Longitudinal	Transverse
Elongation in 2 in. [50 mm], min, %:						
Basic minimum elongation transverse strip tests, and for all small sizes tested in full section	35	25	30	16.5	30	16.5
When standard round 2-in. [50-mm] gage length test specimen is used	28	20	22	12	20	12
For longitudinal strip tests	A		A		A	
For transverse strip tests, a deduction for each 1/32-in. [0.8-mm] decrease in wall thickness below 5/16 in. [7.9 mm] from the basic minimum elongation of the following percentage shall be made		1.25		1.00		1.00

[A] The minimum elongation in 2 in. [50 mm] shall be determined by the following equation:

$$e = 625\,000 A^{0.2} / U^{0.9}$$

for inch-pound units, and

$$e = 1\,940 A^{0.2} / U^{0.9}$$

for SI units,

where:
e = minimum elongation in 2 in. [50 mm], %, rounded to the nearest 0.5 %,
A = cross-sectional area of the tension test specimen, in.² [mm²], based upon specified outside diameter or nominal specimen width and specified wall thickness, rounded to the nearest 0.01 in.² [1 mm²]. (If the area thus calculated is equal to or greater than 0.75 in.² [500 mm²], then the value 0.75 in.² [500 mm²] shall be used.), and
U = specified tensile strength, psi [MPa].

12. Flattening Tests

12.1 Although testing is not required, pipe shall be capable of meeting the flattening test requirements of Supplementary Requirement S3, if tested.

13. Hydrostatic Test

13.1 Except as allowed by 13.2, 13.3, and 13.4, each length of pipe shall be subjected to the hydrostatic test without leakage through the pipe wall.

13.2 As an alternative to the hydrostatic test at the option of the manufacturer or where specified in the purchase order, it shall be permissible for the full body of each pipe to be tested with a nondestructive electric test described in Section 14.

13.3 Where specified in the purchase order, it shall be permissible for pipe to be furnished without the hydrostatic test and without the nondestructive electric test in Section 14; in this case, each length so furnished shall include the mandatory marking of the letters "NH." It shall be permissible for pipe meeting the requirements of 13.1 or 13.2 to be furnished where pipe without either the hydrostatic or nondestructive electric test has been specified in the purchase order; in this case, such pipe need not be marked with the letters "NH." Pipe that has failed either the hydrostatic test of 13.1 or the nondestructive electric test of 13.2 shall not be furnished as "NH" pipe.

13.4 Where the hydrostatic test and the nondestructive electric test are omitted and the lengths marked with the letters "NH," the certification, where required, shall clearly state "Not Hydrostatically Tested," and the letters "NH" shall be appended to the product specification number and material grade shown on the certification.

14. Nondestructive Electric Test

14.1 As an alternative to the hydrostatic test at the option of the manufacturer or where specified in the purchase order as an alternative or addition to the hydrostatic test, the full body of each pipe shall be tested with a nondestructive electric test in accordance with Practice E213, E309, or E570. In such cases, the marking of each length of pipe so furnished shall include the letters "NDE." It is the intent of this nondestructive electric test to reject pipe with imperfections that produce test signals equal to or greater than that produced by the applicable calibration standard.

14.2 Where the nondestructive electric test is performed, the lengths shall be marked with the letters "NDE." The certification, where required, shall state "Nondestructive Electric Tested" and shall indicate which of the tests was applied. Also, the letters "NDE" shall be appended to the product specification number and material grade shown on the certification.

14.3 The following information is for the benefit of the user of this specification:

14.3.1 The reference standards defined in 14.4 through 14.6 are convenient standards for calibration of nondestructive testing equipment. The dimensions of such standards are not to be construed as the minimum sizes of imperfections detectable by such equipment.

14.3.2 The ultrasonic testing referred to in this specification is capable of detecting the presence and location of significant longitudinally or circumferentially oriented imperfections: however, different techniques need to be employed for the detection of such differently oriented imperfections. Ultrasonic testing is not necessarily capable of detecting short, deep imperfections.

14.3.3 The eddy current examination referenced in this specification has the capability of detecting significant imperfections, especially of the short abrupt type.

14.3.4 The flux leakage examination referred to in this specification is capable of detecting the presence and location of significant longitudinally or transversely oriented imperfections: however, different techniques need to be employed for the detection of such differently oriented imperfections.

14.3.5 The hydrostatic test referred to in Section 13 has the capability of finding defects of a size permitting the test fluid to leak through the tube wall and may be either visually seen or detected by a loss of pressure. Hydrostatic testing is not necessarily capable of detecting very tight, through-the-wall imperfections or imperfections that extend an appreciable distance into the wall without complete penetration.

14.3.6 A purchaser interested in ascertaining the nature (type, size, location, and orientation) of discontinuities that can be detected in the specific applications of these examinations is directed to discuss this with the manufacturer of the tubular product.

14.4 For ultrasonic testing, the calibration reference notches shall be, at the option of the producer, any one of the three common notch shapes shown in Practice E213. The depth of notch shall not exceed 12½ % of the specified wall thickness of the pipe or 0.004 in. [0.1 mm], whichever is greater.

14.5 For eddy current testing, the calibration pipe shall contain, at the option of the producer, any one of the following discontinuities to establish a minimum sensitivity level for rejection:

14.5.1 *Drilled Hole*—The calibration pipe shall contain depending upon the pipe diameter three holes spaced 120° apart or four holes spaced 90° apart and sufficiently separated longitudinally to ensure separately distinguishable responses. The holes shall be drilled radially and completely through the pipe wall, care being taken to avoid distortion of the pipe while drilling. Depending upon the pipe diameter the calibration pipe shall contain the following hole:

NPS	DN	Diameter of Drilled Hole
≤ ½	≤ 15	0.039 in. [1 mm]
> ½ ≤ 1¼	> 15 ≤ 32	0.055 in. [1.4 mm]
> 1¼ ≤ 2	> 32 ≤ 50	0.071 in. [1.8 mm]
> 2 ≤ 5	> 50 ≤ 125	0.087 in. [2.2 mm]
>5	> 125	0.106 in. [2.7 mm]

14.5.2 *Transverse Tangential Notch*—Using a round tool or file with a ¼-in. [6-mm] diameter, a notch shall be filed or milled tangential to the surface and transverse to the longitudinal axis of the pipe. The notch shall have a depth not exceeding 12 ½ % of the specified wall thickness of the pipe or 0.004 in. [0.1 mm], whichever is greater.

14.5.3 *Longitudinal Notch*—A notch 0.031 in. [0.8 mm] or less in width shall be machined in a radial plane parallel to the tube axis on the outside surface of the pipe, to have a depth not exceeding 12 ½ % of the specified wall thickness of the tube or 0.004 in. [0.1 mm], whichever is greater. The length of the notch shall be compatible with the testing method.

14.5.4 *Compatibility*—The discontinuity in the calibration pipe shall be compatible with the testing equipment and the method being used.

14.6 For flux leakage testing, the longitudinal calibration reference notches shall be straight-sided notches machined in a radial plane parallel to the pipe axis. For wall thicknesses under ½ in. [12.7 mm], outside and inside notches shall be used; for wall thicknesses equal to and above ½ in. [12.7 mm], only an outside notch shall be used. Notch depth shall not exceed 12½ % of the specified wall thickness, or 0.004 in. [0.1 mm], whichever is greater. Notch length shall not exceed 1 in. [25 mm], and the width shall not exceed the depth. Outside diameter and inside diameter notches shall be located sufficiently apart to allow separation and identification of the signals.

14.7 Pipe containing one or more imperfections that produce a signal equal to or greater than the signal produced by the calibration standard shall be rejected or the area producing the signal shall be reexamined.

14.7.1 Test signals produced by imperfections which cannot be identified, or produced by cracks or crack-like imperfections shall result in rejection of the pipe, unless it is repaired and retested. To be accepted, the pipe must pass the same specification test to which it was originally subjected, provided that the remaining wall thickness is not decreased below that permitted by this specification. The OD at the point of grinding may be reduced by the amount so reduced.

14.7.2 Test signals produced by visual imperfections such as those listed below may be evaluated in accordance with the provisions of Section 18:

14.7.2.1 Dinges,
14.7.2.2 Straightener marks,
14.7.2.3 Cutting chips,
14.7.2.4 Scratches,
14.7.2.5 Steel die stamps,
14.7.2.6 Stop marks, or
14.7.2.7 Pipe reducer ripple.

14.8 The test methods described in this section are not necessarily capable of inspecting the end portion of pipes, a condition referred to as "end effect." The length of such end effect shall be determined by the manufacturer and, when specified in the purchase order, reported to the purchaser.

15. Nipples

15.1 Nipples shall be cut from pipe of the same dimensions and quality described in this specification.

16. Dimensions, Mass, and Permissible Variations

16.1 *Mass*—The mass of any length of pipe shall not vary more than 10 % over and 3.5 % under that specified. Unless otherwise agreed upon between the manufacturer and the purchaser, pipe in NPS 4 [DN 100] and smaller may be weighed in convenient lots; pipe larger than NPS 4 [DN 100] shall be weighed separately.

16.2 *Diameter*—Except as provided for thin-wall pipe in paragraph 11.2 of Specification A530/A530M, the tolerances for diameter shall be in accordance with the following:

16.2.1 Except for pipe ordered as special outside diameter tolerance pipe or as inside diameter tolerance pipe, variations in outside diameter shall not exceed those given in Table 3.

16.2.2 For pipe over 10 in. [250 mm] OD ordered as special outside diameter tolerance pipe, the outside diameter shall not vary more than 1 % over or 1 % under the specified outside diameter.

16.2.3 For pipe over 10 in. [250 mm] ID ordered as inside diameter tolerance pipe, the inside diameter shall not vary more than 1 % over or 1 % under the specified inside diameter.

16.3 *Thickness*—The minimum wall thickness at any point shall not be more than 12.5 % under the specified wall thickness.

17. Lengths

17.1 Pipe lengths shall be in accordance with the following regular practice:

17.1.1 The lengths required shall be specified in the order, and

TABLE 3 Variations in Outside Diameter

NPS [DN Designator]	Permissible Variations in Outside Diameter			
	Over		Under	
	in.	mm	in.	mm
1/8 to 1½ [6 to 40], incl	1/64 (0.015)	0.4	1/64 (0.015)	0.4
Over 1½ to 4 [40 to 100], incl	1/32 (0.031)	0.8	1/32 (0.031)	0.8
Over 4 to 8 [100 to 200], incl	1/16 (0.062)	1.6	1/32 (0.031)	0.8
Over 8 to 18 [200 to 450], incl	3/32 (0.093)	2.4	1/32 (0.031)	0.8
Over 18 to 26 [450 to 650], incl	1/8 (0.125)	3.2	1/32 (0.031)	0.8
Over 26 to 34 [650 to 850], incl	5/32 (0.156)	4.0	1/32 (0.031)	0.8
Over 34 to 48 [850 to 1200], incl	3/16 (0.187)	4.8	1/32 (0.031)	0.8

17.1.2 No jointers are permitted unless otherwise specified.

17.1.3 If definite lengths are not required, pipe may be ordered in single random lengths of 16 to 22 ft [4.8 to 6.7 m] with 5 % 12 to 16 ft [3.7 to 4.8 m], or in double random lengths with a minimum average of 35 ft [10.7 m] and a minimum length of 22 ft [6.7 m] with 5 % 16 to 22 ft [4.8 to 6.7 m].

18. Workmanship, Finish and Appearance

18.1 The pipe manufacturer shall explore a sufficient number of visual surface imperfections to provide reasonable assurance that they have been properly evaluated with respect to depth. Exploration of all surface imperfections is not required but consideration should be given to the necessity of exploring all surface imperfections to assure compliance with 18.2.

18.2 Surface imperfections that penetrate more than 12½ % of the nominal wall thickness or encroach on the minimum wall thickness shall be considered defects. Pipe with such defects shall be given one of the following dispositions:

18.2.1 The defect shall be removed by grinding, provided that the remaining wall thickness is within the limits specified in 16.3.

18.2.2 Repaired in accordance with the repair welding provisions of 18.6.

18.2.3 The section of pipe containing the defect may be cut off within the limits of requirements on length.

18.2.4 Rejected.

18.3 To provide a workmanlike finish and basis for evaluating conformance with 18.2 the pipe manufacturer shall remove by grinding the following noninjurious imperfections:

18.3.1 Mechanical marks and abrasions—such as cable marks, dinges, guide marks, roll marks, ball scratches, scores, and die marks—and pits, any of which imperfections are deeper than 1/16 in. [1.6 mm].

18.3.2 Visual imperfections commonly referred to as scabs, seams, laps, tears, or slivers found by exploration in accordance with 18.1 to be deeper than 5 % of the nominal wall thickness.

18.4 At the purchaser's discretion, pipe shall be subjected to rejection if surface imperfections acceptable under 18.2 are not scattered, but appear over a large area in excess of what is considered a workmanlike finish. Disposition of such pipe shall be a matter of agreement between the manufacturer and the purchaser.

18.5 When imperfections or defects are removed by grinding, a smooth curved surface shall be maintained, and the wall thickness shall not be decreased below that permitted by this specification. The outside diameter at the point of grinding is permitted to be reduced by the amount so removed.

18.5.1 Wall thickness measurements shall be made with a mechanical caliper or with a properly calibrated nondestructive testing device of appropriate accuracy. In case of dispute, the measurement determined by use of the mechanical caliper shall govern.

18.6 Weld repair shall be permitted only subject to the approval of the purchaser and in accordance with Specification A530/A530M.

18.7 The finished pipe shall be reasonably straight.

19. End Finish

19.1 The Pipe shall be furnished to the following practice, unless otherwise specified.

19.1.1 *NPS 1½ [DN 40] and Smaller*—All walls shall be either plain-end square cut, or plain-end beveled at the option of the manufacturer.

19.1.2 *NPS 2 [DN 50] and Larger*—Walls through extra strong weights, shall be plain-end-beveled.

19.1.3 *NPS 2 [DN 50] and Larger*—Walls over extra strong weights, shall be plain-end square cut.

19.2 Plain-end beveled pipe shall be plain-end pipe having a bevel angle of 30°, +5° or -0°, as measured from a line drawn perpendicular to the axis of the pipe with a root face of ¹⁄₁₆ ± ¹⁄₃₂ in. [1.6 ± 0.8 mm]. Other bevel angles may be specified by agreement between the purchaser and the manufacturer.

20. Sampling

20.1 For product analysis (see 9.1) and tensile tests (see 21.1), a lot is the number of lengths of the same size and wall thickness from any one heat of steel; of 400 lengths or fraction thereof, of each size up to, but not including, NPS 6 [DN 150]; and of 200 lengths or fraction thereof of each size NPS 6 [DN 150] and over.

20.2 For bend tests (see 21.2), a lot is the number of lengths of the same size and wall thickness from any one heat of steel, of 400 lengths or fraction thereof, of each size.

20.3 For flattening tests, a lot is the number of lengths of the same size and wall thickness from any one heat of steel, of 400 lengths or fraction thereof of each size over NPS 2 [DN 50], up to but not including NPS 6 [DN 150], and of 200 lengths or fraction thereof, of each size NPS 6 [DN 150] and over.

21. Number of Tests

21.1 The tensile requirements specified in Section 10 shall be determined on one length of pipe from each lot (see 20.1).

21.2 For pipe NPS 2 [DN 50] and under, the bend test specified in 11.1 shall be made on one pipe from each lot (see 20.2). The bend test, where used as required by 11.2, shall be made on one end of 5 % of the pipe from each lot. For small lots, at least one pipe shall be tested.

21.3 If any test specimen shows flaws or defective machining, it shall be permissible to discard it and substitute another test specimen.

22. Retests

22.1 If the percentage of elongation of any tension test specimen is less than that given in Table 1 and any part of the fracture is more than ¾ in. [19 mm] from the center of the gage length of a 2-in. [50-mm] specimen as indicated by scribe scratches marked on the specimen before testing, a retest shall be allowed. If a specimen breaks in an inside or outside surface flaw, a retest shall be allowed.

23. Test Specimens and Test Methods

23.1 On NPS 8 [DN 200] and larger, specimens cut either longitudinally or transversely shall be acceptable for the tension test. On sizes smaller than NPS 8 [DN 200], the longitudinal test only shall be used.

23.2 When round tension test specimens are used for pipe wall thicknesses over 1.0 in. [25.4 mm], the mid–length of the longitudinal axis of such test specimens shall be from a location midway between the inside and outside surfaces of the pipe.

23.3 Test specimens for the bend test specified in Section 11 and for the flattening tests shall consist of sections cut from a pipe. Specimens for flattening tests shall be smooth on the ends and free from burrs, except when made on crop ends.

23.4 Test specimens for the bend test specified in 11.2 shall be cut from one end of the pipe and, unless otherwise specified, shall be taken in a transverse direction. One test specimen shall be taken as close to the outer surface as possible and another from as close to the inner surface as possible. The specimens shall be either ½ by ½ in. [12.5 by 12.5 mm] in section or 1 by ½ in. [25 by 12.5 mm] in section with the corners rounded to a radius not over ¹⁄₁₆ in. [1.6 mm] and need not exceed 6 in. [150 mm] in length. The side of the samples placed in tension during the bend shall be the side closest to the inner and outer surface of the pipe respectively.

23.5 All routine check tests shall be made at room temperature.

24. Certification

24.1 When test reports are requested, in addition to the requirements of Specification A530/A530M, the producer or supplier shall furnish to the purchaser a chemical analysis report for the elements specified in Table 1.

25. Product Marking

25.1 In addition to the marking prescribed in Specification A530/A530M, the marking shall include heat number, the information as per Table 4, an additional symbol "S" if one or

TABLE 4 Marking

Hydro	NDE	Marking
Yes	No	Test Pressure
No	Yes	NDE
No	No	NH
Yes	Yes	Test Pressure/NDE

more of the supplementary requirements apply; the length, OD 1 %, if ordered as special outside diameter tolerance pipe; ID 1 %, if ordered as special inside diameter tolerance pipe; the schedule number, weight class, or nominal wall thickness; and, for sizes larger than NPS 4 [DN 100], the weight. Length shall be marked in feet and tenths of a foot [metres to two decimal places], depending on the units to which the material was ordered, or other marking subject to agreement. For sizes NPS 1½, 1¼, 1, and ¾ [DN 40, 32, 25, and 20], each length shall be marked as prescribed in Specification A530/A530M. These sizes shall be bundled in accordance with standard mill practice and the total bundle footage marked on the bundle tag; individual lengths of pipe need not be marked with footage. For sizes less than NPS ¾ [DN 20], all the required markings shall be on the bundle tag or on each length of pipe and shall include the total footage; individual lengths of pipe need not be marked with footage. If not marked on the bundle tag, all required marking shall be on each length.

25.2 When pipe sections are cut into shorter lengths by a subsequent processor for resale as material, the processor shall transfer complete identifying information, including the name or brand of the manufacturer to each unmarked cut length, or to metal tags securely attached to bundles of unmarked small diameter pipe. The same material designation shall be included with the information transferred, and the processor's name, trademark, or brand shall be added.

25.3 *Bar Coding*—In addition to the requirements in 25.1 and 25.2, bar coding is acceptable as a supplementary identification method. The purchaser may specify in the order a specific bar coding system to be used.

26. Government Procurement

26.1 When specified in the contract, material shall be preserved, packaged, and packed in accordance with the requirements of MIL-STD-163. The applicable levels shall be as specified in the contract. Marking for the shipment of such material shall be in accordance with Fed. Std. No. 123 for civil agencies and MIL-STD-129 or Fed. Std. No. 183 if continuous marking is required for military agencies.

26.2 *Inspection*—Unless otherwise specified in the contract, the producer is responsible for the performance of all inspection and test requirements specified herein. Except as otherwise specified in the contract, the producer shall use his own, or any other suitable facilities for the performance of the inspection and test requirements specified herein, unless disapproved by the purchaser. The purchaser shall have the right to perform any of the inspections and tests set forth in this specification where such inspections are deemed necessary to ensure that the material conforms to the prescribed requirements.

27. Keywords

27.1 carbon steel pipe; seamless steel pipe; steel pipe

SUPPLEMENTARY REQUIREMENTS

One or more of the following supplementary requirements shall apply only when specified in the purchase order. The purchaser may specify a different frequency of test or analysis than is provided in the supplementary requirement. Subject to agreement between the purchaser and manufacturer, retest and retreatment provisions of these supplementary requirements may also be modified.

S1. Product Analysis

S1.1 Product analysis shall be made on each length of pipe. Individual lengths failing to conform to the chemical composition requirements shall be rejected.

S2. Transverse Tension Test

S2.1 A transverse tension test shall be made on a specimen from one end or both ends of each pipe NPS 8 [DN 200] and over. If this supplementary requirement is specified, the number of tests per pipe shall also be specified. If a specimen from any length fails to meet the required tensile properties (tensile, yield, and elongation), that length shall be rejected subject to retreatment in accordance with Specification A530/A530M and satisfactory retest.

S3. Flattening Test, Standard

S3.1 For pipe over NPS 2 [DN 50], a section of pipe not less than 2½ in. [63.5 mm] in length shall be flattened cold between parallel plates until the opposite walls of the pipe meet. Flattening tests shall be in accordance with Specification A530/A530M, except that in the formula used to calculate the "H" value, the following "e" constants shall be used:

0.08 for Grade A
0.07 for Grades B and C

S3.2 When low D-to-t ratio tubulars are tested, because the strain imposed due to geometry is unreasonably high on the inside surface at the six and twelve o'clock locations, cracks at these locations shall not be cause for rejection if the D-to-t ratio is less than ten.

S3.3 The flattening test shall be made on one length of pipe from each lot of 400 lengths or fraction thereof of each size over NPS 2 [DN 50], up to but not including NPS 6 [DN 150], and from each lot of 200 lengths or fraction thereof, of each size NPS 6 [DN 150] and over.

S3.4 Should a crop end of a finished pipe fail in the flattening test, one retest is permitted to be made from the failed end. Pipe shall be normalized either before or after the first test, but pipe shall be subjected to only two normalizing treatments.

S4. Flattening Test, Enhanced

S4.1 The flattening test of Specification A530/A530M shall be made on a specimen from one end or both ends of each pipe. Crop ends may be used. If this supplementary requirement is specified, the number of tests per pipe shall also be specified.

If a specimen from any length fails because of lack of ductility prior to satisfactory completion of the first step of the flattening test requirement, that pipe shall be rejected subject to retreatment in accordance with Specification A530/A530M and satisfactory retest. If a specimen from any length of pipe fails because of a lack of soundness, that length shall be rejected, unless subsequent retesting indicates that the remaining length is sound.

S5. Metal Structure and Etching Test

S5.1 The steel shall be homogeneous as shown by etching tests conducted in accordance with the appropriate sections of Method E381. Etching tests shall be made on a cross section from one end or both ends of each pipe and shall show sound and reasonably uniform material free from injurious laminations, cracks, and similar objectionable defects. If this supplementary requirement is specified, the number of tests per pipe required shall also be specified. If a specimen from any length shows objectionable defects, the length shall be rejected, subject to removal of the defective end and subsequent retests indicating the remainder of the length to be sound and reasonably uniform material.

S6. Carbon Equivalent

S6.1 The steel shall conform to a carbon equivalent (CE) of 0.50 maximum as determined by the following formula:

$$CE = \%C + \frac{\%Mn}{6} + \frac{\%Cr + \%Mo + \%V}{5} + \frac{\%Ni + \%Cu}{15}$$

S6.2 A lower CE maximum may be agreed upon between the purchaser and the producer.

S6.3 The CE shall be reported on the test report.

S7. Heat Treated Test Specimens

S7.1 At the request of the purchaser, one tensile test shall be performed by the manufacturer on a test specimen from each heat of steel furnished which has been either stress relieved at 1250 °F or normalized at 1650 °F, as specified by the purchaser. Other stress relief or annealing temperatures, as appropriate to the analysis, may be specified by agreement between the purchaser and the manufacturer. The results of this test shall meet the requirements of Table 1.

S8. Internal Cleanliness–Government Orders

S8.1 The internal surface of hot finished ferritic steel pipe and tube shall be manufactured to a free of scale condition equivalent to the visual standard listed in SSPC-SP 6. Cleaning shall be performed in accordance with a written procedure that has been shown to be effective. This procedure shall be available for audit.

S9. Requirements for Carbon Steel Pipe for Hydrofluoric Acid Alkylation Service

S9.1 Pipe shall be provided in the normalized heat-treated condition.

S9.2 The carbon equivalent (CE), based upon heat analysis, shall not exceed 0.43 % if the specified wall thickness is equal to or less than 1 in. [25.4 mm] or 0.45 % if the specified wall thickness is greater than 1 in. [25.4 mm].

S9.3 The carbon equivalent (CE) shall be determined using the following formula:

$$CE = C + Mn/6 + (Cr + Mo + V)/5 + (Ni + Cu)/15$$

S9.4 Based upon heat analysis in mass percent, the vanadium content shall not exceed 0.02 %, the niobium content shall not exceed 0.02 %, and the sum of the vanadium and niobium contents shall not exceed 0.03 %.

S9.5 Based upon heat analysis in mass percent, the sum of the nickel and copper contents shall not exceed 0.15 %.

S9.6 Based upon heat analysis in mass percent, the carbon content shall not be less than 0.18 %.

S9.7 Welding consumables of repair welds shall be of low hydrogen type. E60XX electrodes shall not be used and the resultant weld chemical composition shall meet the chemical composition requirements specified for the pipe.

S9.8 The designation "HF-N" shall be stamped or marked on each pipe to signify that the pipe complies with this supplementary requirement.

SUMMARY OF CHANGES

Committee A01 has identified the location of selected changes to this specification since the last issue, A106/A106M – 06a, that may impact the use of this specification. (Approved July 15, 2008)

(1) Revised 16.2 to permit OD tolerance for thin-wall pipe to default to Specification A530/A530M.

ASTM International takes no position respecting the validity of any patent rights asserted in connection with any item mentioned in this standard. Users of this standard are expressly advised that determination of the validity of any such patent rights, and the risk of infringement of such rights, are entirely their own responsibility.

This standard is subject to revision at any time by the responsible technical committee and must be reviewed every five years and if not revised, either reapproved or withdrawn. Your comments are invited either for revision of this standard or for additional standards and should be addressed to ASTM International Headquarters. Your comments will receive careful consideration at a meeting of the responsible technical committee, which you may attend. If you feel that your comments have not received a fair hearing you should make your views known to the ASTM Committee on Standards, at the address shown below.

This standard is copyrighted by ASTM International, 100 Barr Harbor Drive, PO Box C700, West Conshohocken, PA 19428-2959, United States. Individual reprints (single or multiple copies) of this standard may be obtained by contacting ASTM at the above address or at 610-832-9585 (phone), 610-832-9555 (fax), or service@astm.org (e-mail); or through the ASTM website (www.astm.org).

Designation: A108 − 07

Standard Specification for
Steel Bar, Carbon and Alloy, Cold-Finished[1]

This standard is issued under the fixed designation A108; the number immediately following the designation indicates the year of original adoption or, in the case of revision, the year of last revision. A number in parentheses indicates the year of last reapproval. A superscript epsilon (ε) indicates an editorial change since the last revision or reapproval.

This standard has been approved for use by agencies of the Department of Defense.

1. Scope*

1.1 This specification covers cold-finished carbon and alloy steel bars produced in straight length and coil to chemical compositions. Cold-finished bars are suitable for heat treatment, for machining into components, or for use in the as-finished condition as shafting, or in constructional applications, or for other similar purposes (Note 1). Grades of steel are identified by grade numbers or by chemical composition.

NOTE 1—A guide for the selection of steel bars is contained in Practice A400.

1.2 Some end uses may require one or more of the available designations shown under Supplementary Requirements. Supplementary requirements shall apply only when specified individually by the purchaser.

1.3 The values stated in inch-pound units are to be regarded as the standard. The values given in parentheses are for information only.

2. Referenced Documents

2.1 *ASTM Standards:*[2]

A29/A29M Specification for Steel Bars, Carbon and Alloy, Hot-Wrought, General Requirements for

A304 Specification for Carbon and Alloy Steel Bars Subject to End-Quench Hardenability Requirements

A322 Specification for Steel Bars, Alloy, Standard Grades

A370 Test Methods and Definitions for Mechanical Testing of Steel Products

A400 Practice for Steel Bars, Selection Guide, Composition, and Mechanical Properties

A510 Specification for General Requirements for Wire Rods and Coarse Round Wire, Carbon Steel

A576 Specification for Steel Bars, Carbon, Hot-Wrought, Special Quality

2.2 *Other Documents:*

SAE Handbook and SAE J1086 Recommended Practice for Numbering Metals and Alloys (UNS)[3]

ISS Steel Bar Product Guidelines [4]

3. Terminology

3.1 *Definition:*

3.1.1 *product tolerance levels*—cold-finished steel bar is produced with up to four (4) increasingly tight tolerance levels, for the individual product characteristics, dependent on the method of manufacture necessary to meet purchaser-ordered specification requirements. (Product Tolerance Level 1 is selected, unless otherwise specified by purchaser.)

4. Ordering Information

4.1 Orders for cold-finished steel bar to this specification should include the following items to adequately describe the material:

4.1.1 Name of material,

4.1.2 ASTM specification number and date of issue,

4.1.3 Chemical composition, grade designation or limits,

4.1.4 Silicon level, if required,

4.1.5 Additional machinability-enhancing elements (see Footnote F to Table 1 of Specification A29/A29M),

4.1.6 Condition (Surface Roughness tolerances listed in Table A1.7),

4.1.7 Tolerance Levels (Reference tolerances listed in Table A1.1 through Table A1.9),

4.1.8 Shape (round, hexagon, square, flat, etc.), size, and length,

4.1.9 Report of heat analysis, if required,

4.1.10 End use,

4.1.11 Additions to the specification and special or supplementary requirements, if required, and

4.1.12 For coiled product, the coil weights, inside diameter, outside diameter, and coil height limitations, when required.

NOTE 2—A typical ordering description is as follows: Steel Bar; ASTM A108, dated _____; SAE 1117; Coarse Grain; Cold Drawn; 1.500-in.

---

[1] This specification is under the jurisdiction of ASTM Committee A01 on Steel, Stainless Steel and Related Alloys and is the direct responsibility of Subcommittee A01.15 on Bars.

Current edition approved Sept. 1, 2007. Published September 2007. Originally approved in 1926. Last previous edition approved in 2003 as A108 – 03[e1]. DOI: 10.1520/A0108-07.

[2] For referenced ASTM standards, visit the ASTM website, www.astm.org, or contact ASTM Customer Service at service@astm.org. For *Annual Book of ASTM Standards* volume information, refer to the standard's Document Summary page on the ASTM website.

[3] Available from Society of Automotive Engineers (SAE), 400 Commonwealth Dr., Warrendale, PA 15096-0001.

[4] Available from ISS Customer Service, 410 Commonwealth Drive, Warrendale, PA, 15086.

A Summary of Changes section appears at the end of this standard.

(38.10 mm) diameter round; 12 ft (3657.61 mm) long; Heat Analysis Required; Precision Machined Parts.

NOTE 3—A more complex ordering description is as follows: Steel Bar; ASTM A108, dated ____; SAE 1045; Fine Grain; Cold Drawn, Turned, Ground and Polished; chamfer both ends; Mechanical Property Test Results; Hardness test; 2.000-in. (50.80 mm) diameter round; 12 ft (3657.61 mm) long; Heat Analysis Required; Precision Machined Parts. Product codes allow you to abbreviate, yet identify a complex ordering description in the following simplified description: Steel Bar: ASTM A108, dated ____; SAE 1045; Fine Grain; 2.000-in. (50.80 mm) diameter round; 12 ft (3657.61 mm) long; Heat Analysis Required; Precision Machined Parts.

5. General Requirements

5.1 Material furnished under this specification shall conform to the applicable requirements of the current edition of Specification A29/A29M.

6. Materials and Manufacture

6.1 *Feedstock*—Cold-finished steel bar shall be produced from hot-wrought carbon or alloy steel bar (Specification A29/A29M), or from hot-wrought rod designated for cold-finished bar (Specification A510).

6.2 *Condition*—The product shall be furnished in one of the following conditions as specified by the purchaser.

6.2.1 *Rounds*:
6.2.1.1 Cold drawn,
6.2.1.2 Cold drawn, turned, and polished,
6.2.1.3 Cold drawn, ground, and polished,
6.2.1.4 Cold drawn, turned, ground, and polished,
6.2.1.5 Cold drawn, turned, and ground,
6.2.1.6 Hot wrought, turned, and polished,
6.2.1.7 Hot wrought, turned, ground and polished,
6.2.1.8 Hot wrought, turned, and ground, and
6.2.1.9 Hot wrought, rough turned.
6.2.2 *Squares, Hexagons*:
6.2.2.1 Cold drawn, and
6.2.2.2 Cold rolled.
6.2.3 *Flats*:
6.2.3.1 Cold drawn, and
6.2.3.2 Cold rolled.
6.2.4 *Special Bar Sections*:
6.2.4.1 Cold drawn, and
6.2.4.2 Cold rolled.

6.3 *Heat Treatment*:
6.3.1 Unless otherwise specified, the bars shall be furnished as cold-finished. Plain Carbon Steels with a maximum carbon over 0.55 % and Alloy Steels with a maximum carbon over 0.38 % shall be annealed prior to cold finishing.
6.3.2 The following heat-treatment processes may be performed singularly or in combination:
6.3.2.1 Annealed,
6.3.2.2 Normalized,
6.3.2.3 Stress relieved, and
6.3.2.4 Quenched and tempered.

7. Chemical Composition

7.1 *Chemical Composition*:

7.1.1 The chemical analysis of the steel shall conform to that specified in Specification A29/A29M for the steel grade ordered, or to such other limits as may be specified using the standard ranges in Specification A29/A29M.

7.1.2 Steels may be selected from: Specifications A29/A29M, A304, A322, A510, and A576; the SAE Handbook; or the ISS Steel Bar Product Guideline for Bar Steel.

7.1.3 When a steel's composition cannot be identified by a standard grade number in accordance with 7.1.1 and 7.1.2, the limits for each required element may be specified using the chemical ranges shown in the table (Heat Analysis Chemical Ranges and Limits of Carbon Steel Bars) of Specification A29/A29M.

8. Tolerance Levels

8.1 *Cold-Finished Bars*—The permissible dimensional variations for cold-finished carbon and alloy steel bar shall not exceed the applicable tolerance levels or limits stated in Annex A1 for inch-pound values.

9. Workmanship, Finish, and Product Presentation

9.1 *Workmanship*:
9.1.1 Within the limits of good manufacturing and inspection practices, the bars shall be free of injurious imperfections which, due to their nature, degree, or extent, will interfere with the use of the material in machining or fabrication of suitable parts.
9.1.2 Table A1.8 contains the recommended minimum stock removal to ensure removal of surface discontinuities in cold finished bars.

9.2 *Finish*:
9.2.1 Unless otherwise specified, the bars shall have a commercial bright smooth surface finish, obtained by conventional cold-finishing operations such as cold drawing or cold rolling.
9.2.2 When a superior bar surface finish is required, bars may be obtained as: turned and polished, ground and polished, or turned, ground, and polished. (Reference Table A1.7)
9.2.3 Bars that are thermally treated after cold finishing may exhibit a discolored or oxidized surface.

9.3 *Product Presentation*:
9.3.1 The bars shall be given a surface coating of oil or other rust inhibitor to protect against corrosion during shipment.
9.3.2 The bar bundles shall be identified, packaged and loaded to preserve the physical appearance, product tolerance and identity of the cold-finished product, as agreed upon between the purchaser and supplier.

10. Certification

10.1 Upon request of the purchaser in the contract or order, a manufacturer's certification that the material was manufactured and tested in accordance with this specification together with a report of the test results shall be furnished at the time of shipment.

11. Keywords

11.1 alloy steel; carbon steel; cold-finished; steel bar

SUPPLEMENTARY REQUIREMENTS

One or more of the following supplementary requirements shall be applied only when specified by the purchaser in the inquiry, contract, or order. Details of these supplementary requirements shall be agreed upon in writing, by the manufacturer and the purchaser. Supplementary requirements shall in no way negate any requirement of the specification itself.

S1. Hot Rolling Reduction Ratio

S1.1 When required, purchaser may require the supplier to report the reduction ratio of the initial Bloom/Billet cross sectional area to finished hot rolled cross sectional area.

S2. Steel Melting Process

S2.1 When required, purchaser may require the supplier to report the steel melting process (Basic Oxygen Furnace, Electric Arc Furnace, etc.) for each initial heat/lot number supplied to the purchaser.

S3. Steel Refinement Process

S3.1 When required, purchaser may require the supplier to report the steel refinement processes performed after melting and before casting (Vacuumed Degassed, etc.) on the heat/lot number supplied to the purchaser.

S4. Continuous Casting Process

S4.1 When required, purchaser may require the supplier to report the casting process (Bloom, Billet, etc.) for each heat/lot number supplied to the purchaser.

S5. Country or Countries of Origin

S5.1 When required, purchaser may require the supplier to report the country of origin where the steel was melted for each heat/lot number supplied to the purchaser.

S5.2 When required, purchaser may require the supplier to report the country of origin where the steel was hot rolled for each heat/lot number supplied to the purchaser.

S5.3 When required, purchaser may require the supplier to report the country of origin where the steel was cold finished for each heat/lot number supplied to the purchaser.

S6. Mechanical Properties

S6.1 When required, purchaser may require the supplier to report the cold-finished steel bar mechanical properties for each heat/lot number supplied to the purchaser. Mechanical properties shall be evaluated in accordance with Test Methods and Definitions in Test Methods A370.

S7. Surface Inspection

S7.1 When required, purchaser may require the supplier to inspect the cold finish steel bar surface within an electromagnetic surface inspection process to detect and sort surface discontinuities that exceed the maximum allowed depth tolerances listed in Table A1.8 or other tolerances agreed upon between the purchaser and supplier.

S8. Bar Marking

S8.1 When required, bar marking specification requirements shall be agreed upon between the purchaser and supplier.

ASTM A108 – 07

ANNEX

(Mandatory Information)

A1. PERMISSIBLE VARIATIONS IN QUALITY CHARACTERISTICS—INCH-POUND AND METRIC UNITS

TABLE A1.1 Size Tolerances for Level One Cold-Finished Carbon Steel Bars, Cold Drawn or Turned and Polished

Size, in. (mm)[A]	Maximum of Carbon Range 0.28 % or less	Maximum of Carbon Range over 0.28 % to 0.55 % incl	Maximum of Carbon Range to 0.55 % incl, Stress Relieved or Annealed after Cold Finishing	Maximum of Carbon Range over 0.55 % or All Grades Quenched and Tempered or Normalized and Tempered before Cold Finishing
	All tolerances are in inches (mm) and are minus[B]			
Rounds—Cold Drawn[C] to 6 in.(152.4 mm) or Turned and Polished				
To 1½ (38.1) incl, in coils, or cut lengths	0.002 (.051)	0.003 (.076)	0.004 (.102)	0.005 (.127)
Over 1½ (38.10) to 2½ (63.50) incl	0.003 (.076)	0.004 (.102)	0.005 (.127)	0.006 (.152)
Over 2½ (63.50) to 4 (101.60) incl	0.004 (.102)	0.005 (.127)	0.006 (.152)	0.007 (.178)
Over 4 (101.60) to 6 (152.40) incl	0.005 (.127)	0.006 (.152)	0.007 (.178)	0.008 (.203)
Over 6 (152.40) to 8 (203.20) incl	0.006 (.152)	0.007 (.178)	0.008 (.203)	0.009 (.229)
Over 8 (203.20) to 9 (228.60) incl	0.007 (.178)	0.008 (.203)	0.009 (.229)	0.010 (.254)
Hexagons				
To ¾ (19.05) incl	0.002 (.051)	0.003 (.076)	0.004 (.102)	0.006 (.152)
Over ¾ (19.05) to 1½ (38.10) incl	0.003 (.076)	0.004 (.102)	0.005 (.127)	0.007 (.178)
Over 1½ (38.10) to 2½ (63.50) incl	0.004 (.102)	0.005 (.127)	0.006 (.152)	0.008 (.203)
Over 2½ (63.50) to 3⅛ (79.38) incl	0.005 (.127)	0.006 (.152)	0.007 (.178)	0.009 (.229)
Over 3⅛ (79.38) to 4 (101.60) incl	0.005 (.127)	0.006 (.152)	...	...
Squares				
To ¾ (19.05) incl	0.002 (.051)	0.004 (.102)	0.005 (.127)	0.007 (.178)
Over ¾ (19.05) to 1½ (38.10) incl	0.003 (.076)	0.005 (.127)	0.006 (.152)	0.008 (.203)
Over 1½ (38.10) to 2½ (63.50) incl	0.004 (.102)	0.006 (.152)	0.007 (.178)	0.009 (.229)
Over 2½ (63.50) to 4 (101.60) incl	0.006 (.152)	0.008 (.203)	0.009 (.229)	0.011 (.279)
Over 4 (101.60) to 5 (127.00) incl	0.010 (.254)	...	...	...
Over 5 (127.00) to 6 (152.4) incl	0.014 (.356)	...	...	...
Flats[D]				
Width:				
To ¾ (19.05) incl	0.003 (.076)	0.004 (.102)	0.006 (.152)	0.008 (.203)
Over ¾ (19.05) to 1 ½ (38.10) incl	0.004 (.102)	0.005 (.127)	0.008 (.203)	0.010 (.254)
Over 1½ (38.10) to 3 (76.2) incl	0.005 (.127)	0.006 (.152)	0.010 (.254)	0.012 (.305)
Over 3 (76.2) to 4 (101.60) incl	0.006 (.152)	0.008 (.203)	0.011 (.279)	0.016 (.410)
Over 4 (101.60) to 6 (152.40) incl	0.008 (.203)	0.010 (.254)	0.012 (.305)	0.020 (.508)
Over 6 (152.40)	0.013 (.330)	0.015 (.381)	...	...

[A] Standard manufacturing practice is shear cut for cold drawn bars (size limits vary by producer) which can cause end distortion resulting in those portions of the bar being outside the applicable size tolerance. When this end condition is undesirable, a saw cut end to remove end distortion should be considered.

[B] While size tolerances are usually specified as minus, tolerances may be ordered all plus, or distributed plus and minus, with the sum being equivalent to the tolerances listed.

[C] Maximum allowable deviation in roundness around the circumference of the same cross-section of a round cold drawn bar is ½ the size tolerance range.

[D] Width governs the tolerances for both width and thickness of flats. For example, when the maximum of carbon range is 0.28 % or less, for a flat 2 in. (50.80 mm) wide and 1 in. (25.40 mm) thick, the width tolerance is 0.005 in. (.127 mm) and the thickness tolerance is the same, namely, 0.005 in. (.127 mm).

TABLE A1.2 Size Tolerances for Level One Cold-Finished Alloy Steel Bars, Cold Drawn or Turned and Polished

Size, in. (mm)[A]	Maximum of Carbon Range 0.28 % or less	Maximum of Carbon Range over 0.28 % to 0.55 % incl	Maximum of Carbon Range to 0.55 % incl, Stress Relieved or Annealed after Cold Finishing	Maximum of Carbon Range over 0.55 % with or without Stress Relieving or Annealing after Cold Finishing. Also, all Carbons, Quenched and tempered (heat treated), or Normalized and Tempered, before Cold Finishing
	All tolerances are in inches (mm) and are minus[B]			
Rounds-Cold Drawn[C] to 6 in. (152.40 mm) or Turned and Polished				
To 1 (25.4) incl, in coils	0.002 (.051)	0.003 (.076)	0.004 (.102)	0.005 (.127)
Cut Lengths:				
To 1½ (38.10) incl	0.003 (.076)	0.004 (.102)	0.005 (.127)	0.006 (.152)
Over 1½ (38.10) to 2½ (63.50) incl	0.004 (.102)	0.005 (.127)	0.006 (.152)	0.007 (.178)
Over 2½ (63.50) to 4 (101.60) incl	0.005 (.127)	0.006 (.152)	0.007 (.178)	0.008 (.203)
Over 4 (101.60) to 6 (152.40) incl	0.006 (.152)	0.007 (.178)	0.008 (.203)	0.009 (.229)
Over 6 (152.40) to 8 (203.20) incl	0.007 (.178)	0.008 (.203)	0.009 (.229)	0.010 (.254)
Over 8 (203.20) to 9 (228.60) incl	0.008 (.203)	0.009 (.229)	0.010 (.254)	0.011 (.279)
Hexagons				
To ¾ (19.05) incl	0.003 (.076)	0.004 (.102)	0.005 (.127)	0.007 (.178)
Over ¾ (19.05) to 1½ 38.10) incl	0.004 (.102)	0.005 (.127)	0.006 (.152)	0.008 (.203)
Over 1½ (38.10) to 2½ (63.50) incl	0.005 (.127)	0.006 (.152)	0.007 (.178)	0.009 (.229)
Over 2½ (63.50) to 3⅛ (79.38) incl	0.006 (.152)	0.007 (.178)	0.008 (.203)	0.010 (.254)
Over 3⅛ (79.38) to 4 (101.60) incl	0.006 (.152)	...	...	...
Squares				
To ¾ (19.05) incl	0.003 (.076)	0.005 (.127)	0.006 (.152)	0.008 (.203)
Over ¾ (19.05) to 1½ (38.10) incl	0.004 (.102)	0.006 (.152)	0.007 (.178)	0.009 (.229)
Over 1½ (38.10) to 2½ (63.50) incl	0.005 (.127)	0.007 (.178)	0.008 (.203)	0.010 (.254)
Over 2½ (63.50) to 4 (101.60) incl	0.007 (.178)	0.009 (.229)	0.010 (.254)	0.012 (.305)
Over 4 (101.60) to 5 (127.00) incl	0.011 (.279)	...	...	...
Flats[D]				
To ¾ (19.05) incl	0.004 (.102)	0.005 (.127)	0.007 (.178)	0.009 (.229)
Over ¾ (19.05) to 1½ (38.10) incl	0.005 (.127)	0.006 (.152)	0.009 (.229)	0.011 (.279)
Over 1½ (38.10) to 3 (76.2) incl	0.006 (.152)	0.007 (.178)	0.011 (.279)	0.013 (.330)
Over 3 (76.2) to 4 (101.60) incl	0.007 (.178)	0.009 (.229)	0.012 (.305)	0.017 (.432)
Over 4 (101.60) to 6 (152.40) incl	0.009 (.229)	0.011 (.279)	0.013 (.330)	0.021 (.533)
Over 6 (152.40)	0.014 (.356)	...	...	...

[A] Standard manufacturing practice is shear cut for cold drawn bars (size limits vary by producer) which can cause end distortion resulting in those portions of the bar being outside the applicable size tolerance. When this end condition is undesirable, a saw cut end to remove end distortion should be considered.
[B] While size tolerances are usually specified as minus, tolerances may be ordered all plus, or distributed plus and minus, with the sum being equivalent to the tolerances listed.
[C] Maximum allowable deviation in roundness around the circumference of the same cross-section of a round cold drawn bar is ½ the size tolerance range.
[D] Width governs the tolerances for both width and thickness of flats. For example, when the maximum of carbon range is 0.28 % or less, for a flat 2 in. (50.80 mm) wide and 1 in. (25.40 mm) thick, the width tolerance is 0.006 in. (.152 mm) and the thickness tolerance is the same, namely, 0.006 in. (.152mm).

TABLE A1.3 Size Tolerances for Level Two and Level Three Cold Finished Round Bars Cold Drawn, Ground and Polished, or Turned, Ground and Polished

Size, in. (mm) Cold Drawn, Ground and Polished[A]	Size, in. (mm) Turned, Ground and Polished[A]	Tolerances from Specified Size, Minus Only, in. (mm)	
		Level 2	Level 3
To 1½ (38.10) incl	To 1½ (38.10) incl	0.001 (.0254)	0.0008 (.0203)
Over 1½ (38.10) to 2½ (63.50) excl	Over 1½ (38.10) to 2½ (63.50) excl	0.0015 (.0381)	0.0013 (.033)
2½ (63.50) to 3 (76.20) incl	2½ (63.50) to 3 (76.20) incl	0.002 (.0508)	0.001 5(.0381)
Over 3 (76.20) to 4 (101.60) incl	Over 3 (76.20) to 4 (101.60) incl	0.003 (.0762)	0.0025 (.0635)
...	Over 4 (101.60) to 6 (152.40) incl	0.004 (.1016)[B]	0.003 (.0762)[B]
...	Over 6 (152.40)	0.005 (.127)[B]	0.004 (.1016)[B]

[A] Maximum allowable deviation of roundness or ovality tolerances are agreed upon between purchaser and supplier.
[B] For nonresulfurized steels (steels specified to maximum sulfur limits under 0.08 %), or for steels thermally treated, the tolerance is increased by 0.001 in. (.025 mm).

TABLE A1.4 Straightness Tolerances for Level One Cold Finished Bars[A,B]

NOTE—All grades quenched and tempered or normalized and tempered to Brinell 302 max before cold finishing; and all grades stress relieved or annealed after cold finishing. Straightness tolerances are not applicable to bars having Brinell hardness exceeding 302.

Size, in. (mm)	Length, ft (mm)	Straightness Tolerances, in. (mm) (Maximum Deviation) from Straightness in any 10-ft Portion of the Bar			
		Maximum of Carbon Range, 0.28 % or Less		Maximum of Carbon Range, over 0.28 % and All Grades Thermally Treated	
		Rounds	Squares, Hexagons, and Octagons	Rounds	Squares, Hexagons, and Octagons
Less than 5/8 (15.88)	Less than 15 (4572)	1/8 (3.17)	3/16 (4.76)	3/16 (4.76)	1/4 (6.35)
Less than 5/8 (15.88)	15 (4572) and over	1/8 (3.17)	5/16 (7.94)	5/16 (7.94)	3/8 (9.53)
5/8 (15.88) and over	Less than 15 (4572)	1/16 (1.59)	1/8 (3.17)	1/8 (3.17)	3/16 (4.76)
5/8 (15.88) and over	15 (4572) and over	1/8 (3.17)	3/16 (4.76)	3/16 (4.76)	1/4 (6.35)

[A] The foregoing tolerances are based on the following method of measuring straightness: Departure from straightness is measured by placing the bar on a level table so that the arc or departure from straightness is horizontal, and the depth of the arc is measured with a feeler gage and a straightedge.

[B] It should be recognized that straightness is a perishable quality and may be altered by mishandling. The preservation of straightness in cold-finished bars requires the utmost care in subsequent handling. Specific straightness tolerances are sometimes required for carbon and alloy steels in which case the purchaser should inform the manufacturer of the straightness tolerances and the methods to be used in checking the straightness.

TABLE A1.5 Length Tolerances for Cold Finished Steel Bar

Product Tolerance Level	Tolerances, inches (mm) Plus Allowable Deviation above Specified Uniform Length		
	Cutting Process	Minimum	Maximum
Level 1	Shear Cut	0.000	2.000 (50.80)
Level 2	In-Line Saw Cut	0.000	1.000 (25.40)
Level 3	Off-Line Saw Cut	0.000	0.500 (12.70)

TABLE A1.6 Across-Corner Tolerances for Hexagon and Square Cold Drawn Steel Bar[A]

Product Tolerance Level	Tolerance Range Applied to Across Corner Calculations	
	Hexagon, Inches (mm), Minus	Square, Inches (mm), Minus
Level 1	0.025 (.64)	0.030 (.76)
Level 2	0.020 (.51)	0.025 (.64)
Level 3	0.015 (.38)	0.020 (.51)
	Sharp Corner Hexagon Calculation = $(1.1547 \times D)$	Sharp Corner Square Calculation = $(1.4142 \times D)$
	Round Corner Hexagon Calculation = $[1.1547 \times (D - 2r)] + 2r$	Round Corner Square Calculation = $[1.4142 \times (D - 2r)] + 2r$

[A] When required, type of corner must be specified at time of order inquiry.

TABLE A1.7 Surface Roughness Average[A] (Ra) Tolerances for Cold-Finished Steel Bar

Product Tolerance Level	Allowable Maximum Deviation of Surface Roughness Average (Ra) Measurement	
	Turned and Polished Maximum, (µin.) (Ra)	Ground and Polished Maximum, (µin.) (Ra)
Level 1	Not Required	40
Level 2[B]	60	30
Level 3[B]	40	20

[A] RMS (root mean square calculation) is no longer applied to measure surface roughness. Roughness average (Ra) is current technology measurement output data.

[B] Special surface Ra restrictions must be agreed upon at time of order inquiry, between purchaser and supplier. Lower Ra values are available with additional bar passes and/or special processing conditions.

TABLE A1.8 Surface Discontinuity Tolerances for Cold-Finished Steel Bar[A]

Product Tolerance Level	Maximum Allowable Surface Discontinuity Depth					
	Carbon and Alloy Non-resulfurized		Carbon and Alloy Resulfurized (0.08 thru 0.19 % Sulfur)		Carbon and Alloy Resulfurized (0.20 thru 0.35 % Sulfur)	
	Maximum Depth ¼ (6.35 mm) thru ⅝ (15.88 mm) max. inches (mm)	Maximum Depth (% of Size) over ⅝ (15.88 mm) thru 6 (152.40 mm) (max. percentage)	Maximum Depth ¼ (6.35 mm) thru ⅝ (15.88 mm) max. inches (mm)	Maximum Depth (% of Size) over ⅝ (15.88 mm) thru 6 (152.40 mm) max. percentage)	Maximum Depth ¼ (6.35 mm) thru ⅝ (15.88 mm) max. inches (mm)	Maximum Depth (% of Size) ⅝ (15.88 mm) thru 6 (152.40 mm) (max., percentage)
Level 1	0.008 (.20)	1.6 %	0.010 (.25)	2.0 %	0.012 (.30)	2.4 %
Level 2	0.006 (.15)	1.0 %	0.008 (.20)	1.3 %	0.010 (.25)	1.6 %
Level 3	0.006 (.15)	0.75 %	0.006 (.15)	1.0 %	0.008 (.20)	1.3 %
Level 4[B]	Nil	Nil	Nil	Nil	Nil	Nil

[A] The information in the chart is the expected maximum surface discontinuity depth within the limits of good manufacturing practice. Occasional bars in a shipment may have surface discontinuities that exceed these limits. For critical applications, the purchaser may require the cold finish steel bar supplier to eddy current test the bars prior to shipment.
[B] Level 4 requires metal removal by turning or multiple grinding passes for small bars.

TABLE A1.9 Surface Decarburization Tolerances for Cold-Finished Steel Bar

Product Tolerance Level	Maximum Affected Depth All Carbon or Alloy Steel Grades	
	Maximum Inches of Decarburization per Side of Bar ¼ (6.35 mm) thru ⅝ (15.88 mm) Sizes, All Shapes, Max. inches (mm)	Maximum Percentage of Decarburization per Side Based on Percentage of Size over ⅝ (15.88 mm) thru 6 (152.40 mm) Sizes, All Shapes, (max., %)
Level 1	0.010 in.(.25)	1.6 %
Level 2	0.006 in.(.15)	1.0 %
Level 3[A]	Nil	Nil

[A] Level 3 requires metal removal by turning or multiple grinding for small bars.

SUMMARY OF CHANGES

Committee A01 has identified the location of selected changes to this standard since the last issue (A108 – 03$^{\varepsilon 1}$) that may impact the use of this standard. (Approved Sept. 1, 2007.)

(1) Changes made to 9.1.1.

(2) Added 9.1.2.

ASTM International takes no position respecting the validity of any patent rights asserted in connection with any item mentioned in this standard. Users of this standard are expressly advised that determination of the validity of any such patent rights, and the risk of infringement of such rights, are entirely their own responsibility.

This standard is subject to revision at any time by the responsible technical committee and must be reviewed every five years and if not revised, either reapproved or withdrawn. Your comments are invited either for revision of this standard or for additional standards and should be addressed to ASTM International Headquarters. Your comments will receive careful consideration at a meeting of the responsible technical committee, which you may attend. If you feel that your comments have not received a fair hearing you should make your views known to the ASTM Committee on Standards, at the address shown below.

This standard is copyrighted by ASTM International, 100 Barr Harbor Drive, PO Box C700, West Conshohocken, PA 19428-2959, United States. Individual reprints (single or multiple copies) of this standard may be obtained by contacting ASTM at the above address or at 610-832-9585 (phone), 610-832-9555 (fax), or service@astm.org (e-mail); or through the ASTM website (www.astm.org).

Designation: A109/A109M − 08

Standard Specification for
Steel, Strip, Carbon (0.25 Maximum Percent), Cold-Rolled[1]

This standard is issued under the fixed designation A109/A109M; the number immediately following the designation indicates the year of original adoption or, in the case of revision, the year of last revision. A number in parentheses indicates the year of last reapproval. A superscript epsilon (ε) indicates an editorial change since the last revision or reapproval.

This standard has been approved for use by agencies of the Department of Defense.

1. Scope*

1.1 This specification covers cold-rolled carbon steel strip in cut lengths or coils, furnished to closer tolerances than cold-rolled carbon steel sheet, with specific temper, with specific edge or specific finish, and in sizes as follows:

Width, in.	Thickness, in.
Over ½ to 23¹⁵⁄₁₆	0.300 and under
Over 12.5 to 600 mm	7.6 mm and under

1.2 Cold-rolled strip is produced with a maximum specified carbon not exceeding 0.25 percent.

1.3 Strip tolerance products may be available in widths wider than 23¹⁵⁄₁₆ in. [600 mm] by agreement between purchaser and supplier. However, such products are technically classified as cold rolled sheet. The tolerances, finishes, tempers, edges, and available widths and thicknesses differentiate cold rolled strip from the product known as cold rolled sheet which is defined by Specification A568/A568M and from cold rolled high carbon strip which is defined by Specification A682/A682M.

1.4 For the purpose of determining conformance with this specification, values shall be rounded to the nearest unit in the right hand place of figures used in expressing the limiting values in accordance with the rounding method of Practice E29.

1.5 The SI portions of the tables contained herein list permissible variations in dimensions and mass (see Note 1) in SI (metric) units. The values listed are not exact conversions of the values listed in the inch-pound tables, but instead are rounded or rationalized values. Conformance to SI tolerances is mandatory when the "M" specification is used.

NOTE 1—The term *weight* is used when inch-pound units are the standard. However, under SI the preferred term is *mass*.

1.6 The values stated in either SI units or inch-pound units are to be regarded separately as standard. The values stated in each system may not be exact equivalents; therefore, each system shall be used independently of the other. Combining values from the two systems may result in non-conformance with the standard.

1.7 This specification is expressed in both inch-pound units and SI units. However, unless the order specifies the applicable "M" specification designation (SI units), the material shall be furnished to inch-pound units.

2. Referenced Documents

2.1 *ASTM Standards:*[2]

A370 Test Methods and Definitions for Mechanical Testing of Steel Products

A568/A568M Specification for Steel, Sheet, Carbon, Structural, and High-Strength, Low-Alloy, Hot-Rolled and Cold-Rolled, General Requirements for

A682/A682M Specification for Steel, Strip, High-Carbon, Cold-Rolled, General Requirements For

A700 Practices for Packaging, Marking, and Loading Methods for Steel Products for Shipment

A751 Test Methods, Practices, and Terminology for Chemical Analysis of Steel Products

A941 Terminology Relating to Steel, Stainless Steel, Related Alloys, and Ferroalloys

E8 Test Methods for Tension Testing of Metallic Materials

E29 Practice for Using Significant Digits in Test Data to Determine Conformance with Specifications

E430 Test Methods for Measurement of Gloss of High-Gloss Surfaces by Abridged Goniophotometry[3]

2.2 *Military Standard:*

MIL-STD-129 Marking for Shipment and Storage[4]

2.3 *Federal Standard:*

123 Marking for Shipments (Civil Agencies)[4]

183 Continuous Identification Marking of Iron and Steel Products[4]

---

[1] This specification is under the jurisdiction of ASTM Committee A01 on Steel, Stainless Steel and Related Alloys and is the direct responsibility of Subcommittee A01.19 on Steel Sheet and Strip.
Current edition approved Oct. 1, 2008. Published October 2008. Originally approved in 1926. Last previous edition approved in 2003 as A109/A109M – 03. DOI: 10.1520/A0109_A0109M-08.

[2] For referenced ASTM standards, visit the ASTM website, www.astm.org, or contact ASTM Customer Service at service@astm.org. For *Annual Book of ASTM Standards* volume information, refer to the standard's Document Summary page on the ASTM website.

[3] Withdrawn. The last approved version of this historical standard is referenced on www.astm.org.

[4] Available from Standardization Documents Order Desk, DODSSP, Bldg. 4, Section D, 700 Robbins Ave., Philadelphia, PA 19111-5098, http://www.dodssp.daps.mil.

*A Summary of Changes section appears at the end of this standard.

3. Terminology

3.1 *Definitions of Terms Specific to This Standard:*

3.1.1 *annealing*—the process of heating to and holding at a suitable temperature and then cooling at a suitable rate, for such purposes as reducing hardness, facilitating cold working, producing a desired microstructure, or obtaining desired mechanical, physical, or other properties.

3.1.1.1 *box annealing*—involves annealing in a sealed container under conditions that minimize oxidation. The strip is usually heated slowly to a temperature below the transformation range, but sometimes above or within it, and is then cooled slowly.

3.1.1.2 *continuous annealing*—involves heating the strip in continuous strands through a furnace having a controlled atmosphere followed by a controlled cooling.

3.1.2 *carbon steel*—the designation for steel when no minimum content is specified or required for aluminum, chromium, cobalt, columbium, molybdenum, nickel, titanium, tungsten, vanadium, zirconium or any other element added to obtain a desired alloying effect; when the specified minimum for copper does not exceed 0.40 % or when the maximum content specified for any of the following elements does not exceed the percentage noted: manganese 1.65, silicon 0.60, or copper 0.60.

3.1.2.1 *Discussion*—In all carbon steels small quantities of certain residual elements unavoidably retained from raw materials are sometimes found which are not specified or required, such as copper, nickel, molybdenum, chromium, etc. These elements are considered as incidental and are not normally reported.

3.1.3 *cold reduction*—the process of reducing the thickness of the strip at room temperature. The amount of reduction is greater than that used in skin-rolling (see 3.1.7).

3.1.4 *dead soft*—the temper of strip produced without definite control of stretcher straining or fluting. It is intended for deep drawing applications where such surface disturbances are not objectionable.

3.1.5 *finish*—the degree of smoothness or luster of the strip. The production of specific finishes requires special preparation and control of the roll surfaces employed.

3.1.6 *normalizing*—heating to a suitable temperature above the transformation range and then cooling in air to a temperature substantially below the transformation range. In bright normalizing the furnace atmosphere is controlled to prevent oxidizing of the strip surface.

3.1.7 *skin-rolled*—a term denoting a relatively light cold rolling operation following annealing. It serves to reduce the tendency of the steel to flute or stretcher strain during fabrication. It is also used to impart surface finish, or affect hardness or other mechanical properties, or to improve flatness.

3.1.8 *temper*— a designation by number to indicate the hardness as a minimum, as a maximum, or as a range. The tempers are obtained by the selection and control of chemical composition, by amounts of cold reduction, by thermal treatment, and by skin-rolling.

3.2 Refer to Terminology A941 for additional definitions of terms used in this Specification.

4. Ordering Information

4.1 Orders for material to this specification shall include the following information, as necessary, to describe adequately the desired product:

4.1.1 Quantity,
4.1.2 Name of material (cold-rolled carbon steel strip),
4.1.3 Condition (oiled or not oiled),
4.1.4 Temper (Section 7),
4.1.5 Edge (Section 8),
4.1.6 Dimensions (Section 9),
4.1.7 Workmanship, Finish, and Appearance (Section 10),
4.1.8 Coil size requirements (15.2),
4.1.9 ASTM designation and year of issue,
4.1.10 Copper-bearing steel, if required,
4.1.11 Application (part identification or description),
4.1.12 Cast or heat analysis (request, if required), and
4.1.13 Special requirements, if required.

Note 2—A typical ordering description is as follows: 20 000 lb Cold-Rolled Strip, Oiled, Temper 4, Edge 3, Finish 3, 0.035 by 9 in. by coil, 5000 lb max, 16-in. ID ASTM A 109-XX, for Toaster Shells.

5. Materials and Manufacture

5.1 The steel shall be made by the open-hearth, basic-oxygen, or electric-furnace process.

5.2 Cold-rolled carbon steel strip is normally manufactured from continuously cast steel with aluminum used as the deoxidizer. However, some applications are specified as silicon killed. Ingot cast rimmed, capped and semi-killed steels are subject to limited availability.

5.3 Cold-rolled carbon steel strip is manufactured from hot-rolled descaled coils by cold reducing to the desired thickness on a single stand mill or on a tandem mill consisting of several single stands in series. Sometimes an anneal is used at some intermediate thickness to facilitate further cold reduction or to obtain desired temper and mechanical properties in the finished strip. An anneal and skin pass is typically used as the final step for Temper 4 and 5.

6. Chemical Composition

6.1 *Heat Analysis*— An analysis for each heat of steel shall be made by the manufacturer to determine the percentage of elements shown in Table 1. This analysis shall conform to the requirements shown in Table 1. When requested, heat analysis shall be reported to purchaser or his representative.

6.2 *Product, Check, or Verification Analysis* may be made by the purchaser on the finished material.

6.2.1 Capped or rimmed steels are not technologically suited to product analysis due to the nonuniform character of their chemical composition and therefore, the tolerances in Table 2 do not apply. Product analysis is appropriate on these types of steel only when misapplication is apparent or for copper when copper steel is specified.

6.2.2 For steels other than rimmed or capped, when product analysis is made by the purchaser, the chemical analysis shall not vary from the limits specified by more than the amounts in Table 2. The several determinations of any element shall not vary both above and below the specified range.

TABLE 1 Heat Analysis [A]

Element	Composition– Wt %	
	Temper No. 1, 2, 3	Temper No. 4, 5
Carbon, max	0.25	0.15
Manganese, max	0.90	0.60
Phosphorous, max	0.025	0.025
Sulfur, max	0.025	0.025
Silicon[A,B]	...	...
Aluminum[A,B]	...	...
Copper, when copper steel is specified, min	0.20	0.20
Copper, max[C]	0.20	0.20
Nickel, max[C]	0.20	0.20
Chromium, max[C,D]	0.15	0.15
Molybdenum, max[C]	0.06	0.06
Vanadium[E]	...	...
Columbium[E]	...	...
Titanium[E]	...	...

[A] Where an ellipsis (...) appears in this table, there is no requirement, but the analysis shall be reported unless otherwise specified in this specification.
[B] The analysis shall be reported. When killed steel is specified and aluminum is the deoxidizing element, the minimum is 0.02, and the analysis shall be reported.
[C] The sum of copper, nickel, chromium, and molybdenum shall not exceed 0.50 % on heat analysis. When one or more of these elements is specified, the sum does not apply; in which case, only the individual limits on the remaining elements will apply.
[D] Chromium is permitted, at the producer's option, to 0.25 % maximum when the carbon is less than or equal to 0.05 %. In such case, the limit on the sum of the four elements in Footnote C does not apply.
[E] Reporting shall be required when the level for any of these elements exceeds 0.008 wt%.

TABLE 2 Tolerances for Product Analysis

Element	Limit or Maximum of Specified Element %	Tolerance	
		Under Minimum Limit	Over Maximum Limit
Carbon	to 0.15, incl	0.02	0.03
	over 0.15 to 0.25, incl	0.03	0.04
Manganese	to 0.60, incl	0.03	0.03
Phosphorus		...	0.01
Sulfur		...	0.01
Copper		0.02	...

6.3 For referee purposes, if required, Test Methods, Practices, and Terminology A751 shall be used.

6.4 For applications where cold-rolled strip is to be welded, care must be exercised in selection of chemical composition, as well as mechanical properties, for compatibility with the welding process and its effect on altering the properties.

7. Temper and Bend Test Requirement

7.1 Cold-rolled carbon strip specified to temper numbers shall conform to the Rockwell hardness requirements shown in Table 3.

7.1.1 When a temper number is not specified, rockwell hardness requirements are established by agreement.

7.2 It is recommended that hardness values be specified in the same scale as that which will be used in testing the strip.

7.3 Bend tests shall be conducted at room temperature and test specimens shall be capable of being bent to the requirements shown in Table 4.

7.4 All mechanical tests are to be conducted in accordance with Test Methods and Definitions A370.

TABLE 3 Hardness Requirements

	INCH-POUND UNITS			
Temper	Thickness, in.		Rockwell Hardness	
	Under	Through	Minimum	Maximum (approx.)
No. 1 (hard)	0.025	...	15T90	...
	0.040	0.025	30T76	...
	0.070	0.040	B90.0	...
	0.300	0.070	B84.0	...
No. 2[A] (half-hard)	0.025	...	15T83.5	15T88.5
	0.040	0.025	30T63.5	30T73.5
	0.300	0.040	B70.0	B85
No. 3[A] (quarter-hard)	0.025	...	15T80	15T85
	0.040	0.025	30T56.5	30T67
	0.300	0.040	B60	B75
No. 4[A,B] (skin-rolled)	0.025	...	...	15T82
	0.040	0.025	...	30T60
	0.300	0.040	...	B65
No. 5[A,B] (dead-soft)	0.025	...	...	15T78.5
	0.040	0.025	...	30T53
	0.300	0.040	...	B55
	SI UNITS			
Temper	Thickness, mm		Rockwell Hardness	
	Under	Through	Minimum	Maximun (approx.)
No. 1 (hard)	0.6	...	15T90	...
	1.0	0.6	30T76	...
	1.8	1.0	B90.0	...
	7.6	1.8	B84.0	...
No. 2[A] (half-hard)	0.6	...	15T83.5	15T88.5
	1.0	0.6	30T63.5	30T73.5
	7.6	1.0	B70.0	B85
No. 3[A] (quarter-hard)	0.6	...	15T80	15T85
	1.0	0.6	30T56	30T67
	7.6	1.0	B60	B75
No. 4[A,B] (skin-rolled)	0.6	...	...	15T82
	1.0	0.6	...	30T60
	7.6	1.0	...	B65
No. 5[A,B] (dead-soft)	0.6	...	...	15T78.5
	1.0	0.6	...	30T53
	7.6	1.0	...	B55

[A] Rockwell hardness values apply at time of shipment. Aging may cause slightly higher values when tested at a later date.
[B] Where No. 4 and 5 tempers are ordered with a carbon range of 0.15 to 0.25 %, the maximum hardness requirement is established by agreement.

8. Edge

8.1 The desired edge number shall be specified as follows:

8.1.1 *Number 1 Edge* is a prepared edge of a specified contour (round or square), which is produced when a very accurate width is required or when an edge condition suitable for electroplating is required, or both.

8.1.2 *Number 2 Edge* is a natural mill edge carried through the cold rolling from the hot-rolled strip without additional processing of the edge.

8.1.3 *Number 3 Edge* is an approximately square edge, produced by slitting, on which the burr is not eliminated. Normal coiling or piling does not necessarily provide a definite positioning of the slitting burr.

8.1.4 *Number 4 Edge* is a rounded edge produced by edge rolling either the natural edge of hot-rolled strip or slit-edge

TABLE 4 Bend Test Requirement

NOTE 1—Test specimens shall be capable of being bent as specified above without cracking on the outside of the bent portion. (See applicable figure in Test Methods and Definitions A370.)

Temper	Bend Test Requirement
No. 1 (hard)	Not required to make bends in any direction.
No. 2 (half-hard)	Bend 90° transverse around a radius equal to that of the thickness.
No. 3 (quarter-hard)	Bend 180° transverse over one thickness of the strip and 90° longitudinal around a radius equal to the thickness.
No. 4 (skin-rolled)	Bend flat upon itself in any direction.
No. 5 (dead-soft)	Bend flat upon itself in any direction.

strip. This edge is produced when the width tolerance and edge condition are not as exacting as for No. 1 edge.

8.1.5 *Number 5 Edge* is an approximately square edge produced from slit-edge material on which the burr is eliminated usually by rolling or filing.

8.1.6 *Number 6 Edge* is a square edge produced by edge rolling the natural edge of hot-rolled strip or slit-edge strip. This edge is produced when the width tolerance and edge condition are not as exacting as for No. 1 edge.

8.1.7 *Skived Edges* are custom shaped edges produced by mechanical edge shaving with special tooling.

9. Dimensional Tolerances

9.1 The dimensional tolerances shall be in accordance with Tables 5-11 as follows:

Tolerances for	Table Number
Thickness, in.	5
Width, in.	6,7,8
Length, in.	9
Camber, in.	10
Flatness, in.	11

10. Workmanship, Finish, and Appearance

10.1 Cut lengths shall have a workmanlike appearance and shall not have imperfections of a nature or degree for the product, the grade, and the description ordered that will be detrimental to the fabrication of the finished part.

10.2 Coils may contain some abnormal imperfections which render a portion of the coil unusable since the inspection of coils does not afford opportunity to remove portions containing imperfections as in the case with cut lengths.

10.3 Cold-rolled strip steel finishes are usually specified to one of the following finishes. Typical surface roughness (Ra) ranges for each are included in Table 10.

10.3.1 *Number 1 or Matte (Dull) Finish* is a finish without luster produced by rolling on rolls roughened by mechanical or other means. This finish is especially suitable for paint adhesion and may aid in drawing by reducing friction between die

TABLE 5 Thickness Tolerances of Cold-Rolled Carbon Steel Strip[A,B,C]

Cold-Rolled Carbon Strip Steel Including High-Carbon Strip Steel			
Inch-Pound Units (in.)			
Thickness Tolerances (Plus or Minus, in.)			
Nominal Gage (in.)	Over ½ to less than 12 wide	12 to less than 18	18 to 23$^{15}/_{16}$
0.251 - 0.300	0.0030	0.0035	0.0040
0.160 - 0.250	0.0025	0.0032	0.0036
0.125 - 0.1599	0.0022	0.0028	0.0032
0.070 - 0.1249	0.0018	0.0022	0.0028
0.040 - 0.0699	0.0014	0.0018	0.0024
0.030 - 0.0399	0.0012	0.0015	0.0020
0.020 - 0..0299	0.0010	0.0013	0.0015
0.015 - 0.0199	0.0008	0.0010	0.0012
0.010 - 0.0149	0.0005	0.0008	0.0010
<0.010	0.0003	0.0006	0.0008
SI Units (mm)			
Thickness Tolerances (Plus and Minus, mm)			
Nominal Gage (mm)	Over 12.7 to less than 300	300 to less than 450	450 to 600
6.40 - 7.50	0.080	0.090	0.100
4.00 - 6.39	0.065	0.080	0.090
3.20 - 3.99	0.055	0.070	0.080
1.80 - 3.19	0.045	0.055	0.070
1.00 - 1.79	0.035	0.045	0.060
0.75 - 0.99	0.030	0.035	0.050
0.50 - 0.74	0.025	0.030	0.040
0.38 - 0.49	0.020	0.025	0.030
0.25 - 0.37	0.013	0.020	0.025
<0.25	0.007	0.015	0.020

[A] Measured ⅜ in. or more in from edge; and on narrower than 1 in., at any place between edges.
[B] Measured 10 mm or more in from edge; and on narrower than 25 mm, at any place between edges.
[C] Number 3 edge strip with thickness tolerance guaranteed at less than ⅜ in. [10 mm] from the slit edge is available by agreement between the consumer and the strip manufacturer.

TABLE 6 Width Tolerances of Edge Numbers 1, 4, 5, and 6 of Cold-Rolled Carbon-Steel Strip

INCH - POUND UNITS

Edge Number	Specified Width, in.[A] Over	Specified Width, in.[A] Through	Specified Thickness, in.[B] min	Specified Thickness, in.[B] max	Width Tolerance, Plus and Minus, in.[C]
1	½	¾	...	0.0938	0.005
1	¾	5	...	0.125	0.005
4	½	1	0.025	0.1875	0.015
4	1	2	0.025	0.2499	0.025
4	2	4	0.035	0.2499	0.047
4	4	6	0.047	0.2499	0.047
5	½	¾	...	0.0938	0.005
5	¾	5	...	0.125	0.005
5	5	9	0.008	0.125	0.010
5	9	20	0.015	0.105	0.010
5	20	23¹⁵⁄₁₆	0.023	0.080	0.015
6	½	1	0.025	0.1875	0.015
6	1	2	0.025	0.2499	0.025
6	2	4	0.035	0.2499	0.047
6	4	6	0.047	0.2499	0.047

SI UNITS

Edge No.	Specified Width, mm[A] Over	Specified Width, mm[A] Through	Specific Thickness, mm[B] min	Specific Thickness, mm[B] max	Width Tolerance, Plus and Minus, in.[C]
1	12.5	200	...	3.0	0.13
4	...	25	0.6	5.0	0.38
4	25	50	0.6	6.0	0.65
4	50	150	1.0	6.0	1.20
5	...	100	...	3.0	0.13
5	100	500	0.4	3.0	0.25
5	500	600	0.6	2.0	0.38
6	...	25	0.6	5.0	0.38
6	25	50	0.6	6.0	0.65
6	50	150	1.0	6.0	1.20

[A] Specified width must be within ranges stated for specified edge number.
[B] Specified thickness must be within ranges stated for specified width.
[C] When edge, width and thickness are not defined by this table, tolerances are by agreement between producer and supplier.

TABLE 7 Width Tolerances of Edge Number 2 of Cold-Rolled Carbon Steel Strip

INCH - POUND UNITS

Specified Width, in. Over	Specified Width, in. Through	Width Tolerance, Plus and Minus, in.
½	2	¹⁄₃₂
2	5	³⁄₆₄
5	10	⁵⁄₆₄
10	15	³⁄₃₂
15	20	⅛
20	23¹⁵⁄₁₆	⁵⁄₃₂

SI UNITS

Specified Width, mm Over	Specified Width, mm Through	Width Tolerance, mm Plus and Minus
12.5	50	0.8
50	100	1.2
100	200	1.6
200	400	2.5
400	500	3.0
500	600	4.0

and steel surface. The user and the producer should agree on the permissible surface roughness range, based on the intended end-use.

10.3.2 *Number 2 or Regular Bright Finish* is produced by rolling on moderately smooth rolls. It is suitable for many requirements, but not generally applicable to bright plating.

10.3.3 *Number 2½ or Better Bright Finish* is a smooth finish suitable for those plating applications where high luster is not required.

10.3.4 *Number 3 or Best Bright Finish* is generally of high luster produced by special rolling practices, including the use of specially prepared rolls. It is the highest quality finish commonly produced and is particularly suited for bright plating. The production of this finish requires extreme care in processing and extensive inspection. Paper interleaving is

TABLE 8 Width Tolerances for Edge Number 3 (Slit), Cold-Rolled Carbon Steel Strip

INCH-POUND UNITS						
Specified Thickness, in.		Width Tolerance, Plus and Minus, in. For Specified Width, in.[A]				
Over	Through	Over ½ Through 6	Over 6 Through 9	Over 9 Through 12	Over 12 Through 20	Over 20 Through 23¹⁵⁄₁₆
...	0.016	0.005	0.005	0.010	0.016	0.020
0.016	0.068	0.005	0.005	0.010	0.016	0.020
0.068	0.099	0.008	0.010	0.010	0.016	0.020
0.099	0.160	0.010	0.016	0.016	0.020	0.020
0.160	0.300	0.016	0.020	0.020	0.031	0.031
S.I. UNITS						
Specified Thickness mm		Width Tolerance, Plus and Minus, mm For Specified Width, mm[A]				
Over	Through	Through 100	Over 100 Through 200	Over 200 Through 300	Over 300 Through 400	Over 450 Through 600
...	1.5	0.13	0.13	0.25	0.40	0.50
1.5	2.5	0.20	0.25	0.25	0.40	0.50
2.5	4.5	0.25	0.40	0.40	0.50	0.50
4.5	7.5	0.40	0.50	0.50	0.80	0.80

[A] Width is measured from the shear surface of the slit edge and not from the break.

TABLE 9 Length Tolerances of Cold-Rolled Carbon Steel Strip

INCH-POUND UNITS				
Specified Width, in.		Length Tolerance, Plus Only, in. for Specified Length, in.		
Over	Through	From 24 Through 60	Over 60 Through 120	Over 120 Through 240
½	12	¼	½	¾
12	23¹⁵⁄₁₆	½	¾	1
SI UNITS				
Specified Width, mm		Length Tolerance, Plus Only, mm for Specified Length, mm		
Over	Through	From 600 Through 1500	Over 1500 Through 3000	Over 3000
...	300	10	15	25
300	600	15	20	25

frequently used for protection. In addition to the surface roughness values in Table 12, the user and producer may agree on goniophotometric measurement values (Rs/DI) in accordance with Test Methods E430.

11. Inspection

11.1 When purchaser's order stipulates that inspection and tests (except product analysis) for acceptance on the steel be made prior to shipment from the mill, the manufacturer shall afford the purchaser's inspector all reasonable facilities to satisfy him that the steel is being manufactured and furnished in accordance with the specification. Mill inspection by the purchaser shall not interfere unnecessarily with the manufacturer's operation. All tests and inspection (except product analysis) shall be made at the place of manufacture unless otherwise agreed.

12. Rejection and Rehearing

12.1 Unless otherwise specified, any rejection shall be reported to the producer within a reasonable time after receipt of material by the purchaser.

12.2 Material that is reported to be defective subsequent to the acceptance at the purchaser's works shall be set aside, adequately protected, and correctly identified. The producer shall be notified as soon as possible so that an investigation may be initiated.

12.3 Samples that are representative of the rejected material shall be made available to the producer. In the event that the producer is dissatisfied with the rejection, he may request a rehearing.

13. Test Reports and Certification

13.1 When test reports are required by the purchaser, the supplier shall report the results of all tests required by this specification and any additional tests required by this specification and/or the purchase order.

13.2 When certification is required by the purchase order, the supplier shall furnish a certification that the material has been manufactured and tested in accordance with the requirements of this specification.

13.3 A signature is not required on test reports. However, the document shall clearly identify the organization submitting the document. Notwithstanding the absence of a signature, the organization submitting the document is responsible for the content of the document.

13.4 When test reports are required, it is acceptable for the supplier to report test data from the original manufacturer, provided such data is not rendered invalid by the stripmaking process.

13.5 A Material Test Report, Certificate of Inspection, or similar document printed from or used in electronic form from

TABLE 10 Camber Tolerances of Cold-Rolled Carbon Steel Strip

INCH-POUND UNITS

Note 1—Camber is the greatest deviation of a side edge from a straight line, the measurement being taken on the concave side with a straight edge.
Note 2—Camber tolerances as shown in the table are for any 8 ft. of length. For strip length under 8 ft., camber tolerance shall be subject to negotiation.
Note 3—When the camber tolerances shown in Table 10 are suitable for a particular purpose, cold-rolled strip is sometimes machine straightened.

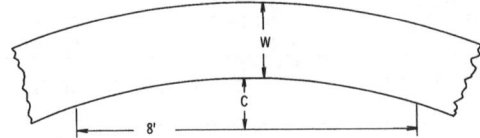

W = Width of strip, in.
C = Camber, in.

Specified Width, in.		Camber Tolerance, in.
Over	Through	
½	1½	½
1½	23¹⁵⁄₁₆	¼

SI UNITS

Note 1—Camber is the greatest deviation of a side edge from a straight line, the measurement being taken on the concave side with a straight edge.
Note 2—Camber tolerances as shown in the table are for any 2000 mm length. For strip length under 2000 mm, camber tolerance shall be subject to negotiation.
Note 3—When the camber tolerances shown in Table 10 are suitable for a particular purpose, cold-rolled strip is sometimes machine straightened.

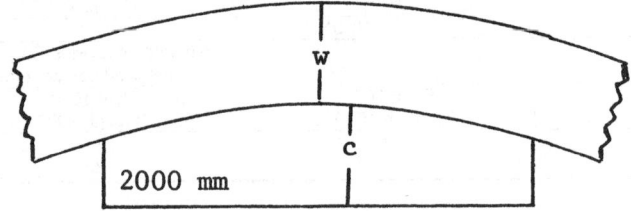

W = Width of strip, mm
C = Camber, mm

Width, in.		Standard Camber Tolerance, mm
Over	Through	
...	50	10
50	600	5

TABLE 11 Flatness Tolerances of Cold-Rolled Carbon Steel Strip

It has not been practical to formulate flatness tolerances for cold-rolled carbon steel strip to represent the wide range of widths and thicknesses and variety of tempers produced.

TABLE 12 Typical Surface Roughness Ranges[A]

Number 1 or Matte (Dull)[B]	Ra 20-80 µin.
Number 2 or Regular Bright[C]	Ra 20 µin. max
Number 2½ or Better Bright[C]	Ra 10 µin. max
Number 3 or Best Bright[C]	Ra 4 µin. max

[A] Due to vagaries in measuring surface roughness, as well as the inherent variability in such rolled surfaces, these values are only typical, and values outside these ranges would not be considered unexpected.
[B] Measured either parallel with or across the rolling direction.
[C] Measured across the rolling direction.

an electronic data interchange (EDI) transmission shall be regarded as having the same validity as a counterpart printed in the certifier's facility. The content of the EDI transmitted document must meet the requirements of the invoked ASTM standard(s) and conform to any existing EDI agreement between the purchaser and the supplier. Notwithstanding the absence of a signature, the organization submitting the EDI transmission is responsible for the context of the report.

NOTE 3—The industry definition as invoked here is: EDI is the computer to computer exchange of business information in an agreed upon standard format such as ANSI ASC X12.

14. Product Marking

14.1 As a minimum requirement, the material shall be identified by having the manufacturer's name, ASTM designation, weight, purchaser's order number, and material identification legibly stenciled on the top of each lift or shown on a tag attached to the coils or shipping units.

14.2 Bar coding is acceptable as a supplementary identification method. Bar coding should be consistent with the Automotive Industry Action Group (AIAG) standard prepared by the primary metals subcommittee of the AIAG bar code project team.

15. Packaging and Package Marking

15.1 Unless otherwise specified, the strip shall be packed and loaded in accordance with Practices A700.

15.2 When coils are ordered it is customary to specify a minimum or range of inside diameter, maximum outside diameter, and a maximum coil weight, if required. The ability of manufacturers to meet the maximum coil weights depends upon individual mill equipment. When required, minimum coil weights are subject to negotiation.

16. Keywords

16.1 carbon steel, strip; cold rolled steel strip; steel strip

APPENDIX

(Nonmandatory Information)

X1. GENERAL INFORMATION AND METALLURGICAL ASPECTS

X1.1 Mechanical Properties

X1.1.1 Table X1.1 shows the approximate mechanical properties corresponding to the five commercial tempers of cold-rolled carbon steel strip. This table is presented as a matter of general information. The limits of tensile strength, etc., are not intended as criteria for acceptance or rejection unless specifically agreed to by the manufacturer when accepting the order. The exact processing by different manufacturers will naturally vary slightly, so that absolute identity cannot be expected in their commercial tempers of cold-rolled strip.

X1.2 Identified Part

X1.2.1 Cold-rolled carbon steel strip can be furnished in the various tempers to make an identified part provided the fabrication of the part is compatible with the grade and temper of the steel specified. Proper identification of parts may include visual examination, prints or descriptions, or a combination of these. It is the general experience that most identified parts can be satisfactorily produced from one of the tempers. There are applications or requirements that necessitate additional controls or limit the choice of processing methods. For most end part application only one kind of mechanical test requirement is normally employed. This test requirement is generally the Rockwell hardness test.

X1.3 Rockwell Scales and Loads

X1.3.1 Various scales and loads are employed in Rockwell testing, depending on the hardness and thickness of the strip to be tested. It is common practice to make the Rockwell hardness test at a point midway between the side edges on a single thickness only. There is some overlapping among the different scales, but the best scale to use in any given case is the one which will give the maximum penetration, without showing undue evidence of impression on the undersurface and without exceeding B100 or its equivalent on the dial. The use of a lighter load results in a loss of sensitivity, while a heavier load leads to a loss in accuracy. If the Rockwell ball is flattened by using it on a hard sample, it should be replaced, otherwise the subsequent readings will be affected. A tolerance for check testing, of two Rockwell points on the B scale below the minimum and above the maximum of the range specified, is commonly allowed to compensate for normal differences in equipment. It is recommended that hardness numbers be specified to the same scale as that to be used during testing.

X1.4 Aging Phenomenon

X1.4.1 Although the maximum ductility is obtained in steel strip in its dead soft (annealed last) condition, such strip is unsuited for many forming operations due to its tendency to stretcher strain or flute. A small amount of cold rolling (skin-rolling) will prevent this tendency, but the effect is only temporary due to a phenomenon called aging. The phenomenon of aging is accompanied by a loss of ductility with an increase in hardness, yield point, and tensile strength. For those uses in which stretcher straining, fluting, or breakage due to aging of the steel is likely to occur, the steel should be fabricated as promptly as possible after skin-rolling. When the above aging characteristics are undesirable, special killed (generally aluminum killed) steel is used.

TABLE X1.1 Approximate Mechanical Properties for Various Tempers of Cold-Rolled Carbon Strip

NOTE 1—These values are given as information only and are not intended as criteria for acceptance or rejection. S.I. units appear in brackets.

Temper	Tensile Strength,[A] † psi [MPa]	Elongation in 2 in. (50 mm) for 0.050 in. (1.27 mm) Thickness of Strip,[B] %	Remarks
No. 1 (hard)	90 000 ± 10 000 [620 ± 70]	...	A very stiff, cold-rolled strip intended for flat blanking only, and not requiring ability to withstand cold forming.
No. 2 (half-hard)	65 000 ± 10 000 [450 ± 70]	10 ± 6	A moderately stiff cold-rolled strip intended for limited bending.
No. 3 (quarter-hard)	55 000 ± 10 000 [380 ± 70]	20 ± 7	A medium soft cold-rolled strip intended for limited bending, shallow drawing and stamping.
No. 4 (skin-rolled)	48 000 ± 6 000 [330 ± 40]	32 ± 8	A soft ductile cold-rolled strip intended for deep drawing where no surface strain or fluting is permissible.[C]
No. 5 (dead-soft)	44 000 ± 6 000 [300 ± 40]	39 ± 6	A soft ductile cold-rolled strip intended for deep drawing where stretcher strains or fluting are permissible.[C] Also for extrusions.

[A] Tensile properties are based on the standard tension-test specimen for sheet metals, see appropriate figure in Test Methods and Definitions A370.
[B] Elongation in 2 in. (50 mm) varies with thickness of strip. For Temper No. 5, dead-soft temper, the percentage of elongation = $41 + 10 \log t$ (t = thickness, in. (mm)). Other tempers vary in a similar way.
[C] See X1.4 for Aging Phenomenon.
† Editorially changed from ksi to psi.

A109/A109M – 08

SUMMARY OF CHANGES

Committee A01 has identified the location of selected changes to this standard since the last issue (A109/A109M – 03) that may impact the use of this standard. (Approved October 1, 2008.)

(1) Clarified wording in Scope.
(2) Removed references to MIL STD-163.
(3) Added Section 3.2.

ASTM International takes no position respecting the validity of any patent rights asserted in connection with any item mentioned in this standard. Users of this standard are expressly advised that determination of the validity of any such patent rights, and the risk of infringement of such rights, are entirely their own responsibility.

This standard is subject to revision at any time by the responsible technical committee and must be reviewed every five years and if not revised, either reapproved or withdrawn. Your comments are invited either for revision of this standard or for additional standards and should be addressed to ASTM International Headquarters. Your comments will receive careful consideration at a meeting of the responsible technical committee, which you may attend. If you feel that your comments have not received a fair hearing you should make your views known to the ASTM Committee on Standards, at the address shown below.

This standard is copyrighted by ASTM International, 100 Barr Harbor Drive, PO Box C700, West Conshohocken, PA 19428-2959, United States. Individual reprints (single or multiple copies) of this standard may be obtained by contacting ASTM at the above address or at 610-832-9585 (phone), 610-832-9555 (fax), or service@astm.org (e-mail); or through the ASTM website (www.astm.org).

Designation: A131/A131M – 08

Standard Specification for
Structural Steel for Ships[1]

This standard is issued under the fixed designation A131/A131M; the number immediately following the designation indicates the year of original adoption or, in the case of revision, the year of last revision. A number in parentheses indicates the year of last reapproval. A superscript epsilon (ε) indicates an editorial change since the last revision or reapproval.

This standard has been approved for use by agencies of the Department of Defense.

1. Scope*

1.1 This specification covers structural steel plates, shapes, bars, and rivets intended primarily for use in ship construction.

1.2 Material under this specification is available in the following categories:

1.2.1 *Ordinary Strength*—Grades A, B, D, and E with a specified minimum yield point of 34 ksi [235 MPa], and

1.2.2 *Higher Strength*—Grades AH, DH, EH, and FH with a specified minimum yield point of 46 ksi [315 MPa], 51 ksi [350 MPa], or 57 ksi [390 MPa].

1.3 Shapes and bars are normally available as Grades A, AH32, and AH36. Other grades may be furnished by agreement between the purchaser and the manufacturer.

1.4 The maximum thickness of products furnished under this specification is 4 in. [100 mm] for plates and 2 in. [50 mm] for shapes and bars.

1.5 When the steel is to be welded, it is presupposed that a welding procedure suitable for the grade of steel and intended use or service will be utilized. See Appendix X3 of Specification A6/A6M for information on weldability.

1.6 The values stated in either inch-pound units or SI units are to be regarded separately as the standard. Within the text, the SI units are shown in brackets. The values stated in each system are not exact equivalents; therefore, each system must be used independently of the other. Combining values from the two systems may result in nonconformance with this specification.

2. Referenced Documents

2.1 *ASTM Standards:*[2]

A6/A6M Specification for General Requirements for Rolled Structural Steel Bars, Plates, Shapes, and Sheet Piling

A370 Test Methods and Definitions for Mechanical Testing of Steel Products

E112 Test Methods for Determining Average Grain Size

3. Terminology

3.1 *Definitions of Terms Specific to This Standard:*

3.1.1 *control rolling, n*—a steel treatment that consists of final rolling within the range used for normalizing heat treatments so that the austenite completely recrystallizes.

3.1.2 *thermo-mechanical controlled processing, n*—a steel treatment that consists of strict control of the steel temperature and the rolling reduction. A high proportion of the rolling reduction is to be carried out close to or below the Ar_3 transformation temperature and may involve rolling towards the lower end of the temperature range of the intercritical dual-phase region, thus permitting little if any recrystallization of the austenite. The process may involve accelerated cooling on completion of rolling.

4. Ordering Information

4.1 Specification A6/A6M establishes the rules for the ordering information that should be complied with when purchasing material to this specification.

4.2 Additional ordering considerations specific to this specification are:

---

[1] This specification is under the jurisdiction of ASTM Committee A01 on Steel, Stainless Steel and Related Alloys and is the direct responsibility of Subcommittee A01.02 on Structural Steel for Bridges, Buildings, Rolling Stock and Ships.
Current edition approved March 1, 2008. Published March 2008. Originally approved in 1931. Last previous edition approved in 2007 as A131/A131M – 07. DOI: 10.1520/A0131_A0131M-08.

[2] For referenced ASTM standards, visit the ASTM website, www.astm.org, or contact ASTM Customer Service at service@astm.org. For *Annual Book of ASTM Standards* volume information, refer to the standard's Document Summary page on the ASTM website.

*A Summary of Changes section appears at the end of this standard.

4.2.1 Condition (control rolled or thermo-mechanical control processed, if applicable).

5. Materials and Manufacture

5.1 Rimmed steels shall not be applied.

5.2 Except for Grades A and B steel, semi-killed steels shall not be applied.

5.3 Grades D, E, AH32, AH36, AH40, DH32, DH36, DH40, EH32, EH36, EH40, FH32, FH36, and FH40 shall be made using a fine grain practice. For ordinary-strength grades, aluminum shall be used to obtain grain refinement. For higher-strength grades, aluminum, vanadium, or columbium (niobium) may be used for grain refinement.

5.4 Plates in all thicknesses ordered to Grade E shall be normalized, or thermo-mechanical control processed. Plates over 1⅜ in. [35 mm] in thickness ordered to Grade D shall be normalized, control rolled, or thermo-mechanical control processed. See Table 1.

5.5 Plates in all thicknesses ordered to Grades EH32 and EH36 shall be normalized, or thermo-mechanical control processed. Plates in all thicknesses ordered to Grade EH40, FH32, FH36, and FH40 shall be normalized, thermo-mechanical control processed, or quenched and tempered. Plates ordered to Grades AH32, AH36, AH40, DH32, DH36, and DH40 shall be normalized, control rolled, thermo-mechanical control processed, or quenched and tempered when so specified. See Table 2.

5.6 In the case of shapes, the thicknesses referred to are those of the flange. Heat treatment and rolling requirements for shapes and bars are given in Table 1 and Table 2.

6. Chemical Requirements

6.1 The heat analysis shall conform to the requirements for chemical composition given in Table 3 and Table 4.

6.1.1 When specified, the steel shall conform on product analysis to the requirements given in Table 3 and Table 4, subject to the product analysis tolerances in Specification A6/A6M.

6.2 For thermo-mechanical control process steel, the carbon equivalent shall be determined from the heat analysis and shall conform to the requirements given in Table 5.

7. Metallurgical Structure

7.1 The steel grades indicated in 5.3 shall be made to fine grain practice, and the requirements for fine austenitic grain size in Specification A6/A6M shall be met.

7.2 Where the use of fine grain practice using columbium, vanadium, or combinations is permitted in 5.3, one or more of the following shall be met:

7.2.1 Minimum columbium (niobium) content of 0.020 % or minimum vanadium content of 0.050 % for each heat, or

7.2.2 When vanadium and aluminum are used in combination, minimum vanadium content of 0.030 % and minimum acid-soluble aluminum content of 0.010 %, or minimum total aluminum content of 0.015 %.

7.2.3 When columbium (niobium) and aluminum are used in combination, minimum columbium (niobium) content of 0.010 % and minimum acid-soluble aluminum content of 0.010 %, or minimum total aluminum content of 0.015 %.

7.2.4 A McQuaid-Ehn austenitic grain size of 5 or finer in accordance with Test Methods E112 for each ladle of each heat.

8. Mechanical Requirements

8.1 *Tension Test*:

8.1.1 Except as specified in the following paragraphs, the material as represented by the test specimens shall conform to the tensile requirements prescribed in Table 6.

8.1.1.1 Shapes less than 1 in.² [645 mm²] in cross section, and bars, other than flats, less than ½ in. [12.5 mm] in thickness or diameter need not be subjected to tension tests by the manufacturer, but chemistry consistent with the required tensile properties must be applied.

8.1.1.2 The elongation requirement of Table 6 does not apply to floor plates with a raised pattern. However, for floor plates over ½ in. [12.5 mm] in thickness, test specimens shall be bent cold with the raised pattern on the inside of the

TABLE 1 Condition of Supply and Frequency of Impact Tests for Ordinary-Strength Structural Steel

Grade	Deoxidation	Product[A]	Condition of Supply[B] (Frequency of Impact Test[C])			
			Thickness (t), in. [mm]			
			t > 0.25 [6.4] t ≤ 1.0 [25]	t > 1.0 [25] t ≤ 1.375 [35]	t > 1.375 [35] t ≤ 2.0 [50]	t > 2.0 [50] t ≤ 4.0 [100]
A	Semi-Killed	All	A (–)			NA[D]
	Killed	P				N (–)[E], TM (–), CR (50 [45]), AR (50 [45])
		S				NA[D]
B	Semi-Killed	All				NA[D]
	Killed	P	A (–)	A (50 [45])		N (50 [45]), TM (50 [45]), CR (25 [23]), AR (25 [23])
		S				NA[D]
D	Killed, Fine Grain Practice	P	A (50 [45]), N (50 [45])		N (50 [45]), TM (50 [45]), CR (50 [45])	N (50 [45]), TM (50 [45]), CR (25 [23])
		S				NA[D]
E	Killed, Fine Grain Practice	P	N (P), TM (P)			N (P), TM (P)
		S	N (25 [23]), TM (25 [23]), CR (15 [14])			NA[D]

[A] Product: P = plate; S = shapes and bars
[B] Condition of Supply: A = any condition; AR = as-rolled; N = normalized; CR = control rolled; TM = thermo-mechanical controlled processing
[C] Frequency of Impact Test: (impact test lot size in tons [Mg] from each heat); (–) = no impact test required; (P) = each plate-as-rolled
[D] Condition of supply is not applicable
[E] Impact tests for Grade A are not required if material is produced using a fine grain practice and normalized

TABLE 2 Condition of Supply and Frequency of Impact Tests for Higher-Strength Structural Steel

Grade	Deoxidation	Grain Refining Element	Product[A]	Condition of Supply[B] (Frequency of Impact Test[C])					
				Thickness (t), in. [mm]					
				t >0.25 [6.4] t ≤0.5 [12.5]	t >0.5 [12.5] t ≤0.80 [20]	t >0.80 [20] t ≤1.0 [25]	t >1.0 [25] t ≤1.375 [35]	t >1.375 [35] t ≤2.0 [50]	t >2.0 [50] t ≤4.0 [100]
AH32 AH36	Killed, Fine Grain Practice	Cb V	P	A (50 [45])	N (50 [45]), TM (50 [45]), CR (50 [45])				N (50 [45]), TM (50 [45]), CR (25 [23])
			S	A (50 [45])	N (50 [45]), TM (50 [45]), CR (50 [45]), AR (25 [23])				NA[D]
		Al	P	A (50 [45])		AR (25 [23]), N (50 [45]), TM (50 [45]), CR (50 [45])	N (50 [45]), TM (50 [45]), CR (50 [45])		N (50 [45],) TM (50 [45]), CR (25 [23])
		Al + Ti	S	A (50 [45])		AR (25 [23]), N (50 [45]), TM (50 [45]), CR (50 [45])	N (50 [45]), TM (50 [45]), CR (50 [45]), AR (25 [23])		NA[D]
DH32 DH36		Cb	P	A (50 [45])	N (50 [45]), TM (50 [45]), CR (50 [45])				N (50 [45]), TM (50 [45]), CR (25 [23])
		V	S	A (50 [45])	N (50 [45]), TM (50 [45]), CR (50 [45])				NA[D]
		Al	P	A (50 [45])		AR (25 [23]), N (50 [45]), TM (50 [45]), CR (50 [45])	N (50 [45]), TM (50 [45]), CR (50 [45])		N (50 [45]), TM (50 [45]), CR (25 [23])
		Al + Ti	S	A (50 [45])		AR (25 [23]), N (50 [45]), TM (50 [45]), CR (50 [45])	N (50 [45]), TM (50 [45]), CR (50 [45])		NA[D]
EH32 EH36		Any	P	N (P), TM (P)					N (P), TM (P)
			S	N (25 [23]), TM (25 [23]), CR (15 [14])					NA[D]
FH32 FH36		Any	P	N (P), TM (P), QT (P)					N (P), TM (P), QT
			S	N (25 [23]), TM (25 [23]), QT (25 [23])					NA[D]
AH40		Any	P	A (50 [45])	N (50 [45]), TM (50 [45]), CR (50 [45])				N (50 [45]), TM (50 [45]), QT (P)
			S	A (50 [45])	N (50 [45]), TM (50 [45]), CR (50 [45])				NA[D]
DH40		Any	P		N (50 [45]), TM (50 [45]), CR (50 [45])				N (50 [45]), TM (50 [45]), QT (P)
			S		N (50 [45]), TM (50 [45]), CR (50 [45])				NA[D]
EH40		Any	P	N (P), TM (P), QT (P)					N (P), TM (P), QT (P)
			S	N (25 [23]), TM (25 [23]), CR (25 [23])					NA[D]
FH40		Any	P	N (P), TM (P), QQT (P)					N (P), TM (P), QT (P)
			S	N (25 [23]), TM (25), CR (25 [23])					NA[D]

[A] Product: P = plate; S = shapes and bars
[B] Condition of Supply: A = any condition; AR = as-rolled; TM = thermo-mechanical controlled processing; CR = control rolled; QT = quenched and tempered; N = normalized
[C] Frequency of Impact Test: (impact test lot size in tons [Mg] from each heat); (P) = each plate-as-rolled
[D] Condition of supply is not applicable

specimen through an angle of 180° without cracking when subjected to a bend test in which the inside diameter is three times plate thickness. Sampling for bend testing shall be as specified for the tension tests in 8.1.2.

8.1.2 One tension test shall be made from each of two different plates, shapes, or bars from each heat of structural steel unless the finished product from a heat is less than 50 tons [45 Mg], in which case one tension test is sufficient. If, however, product from one heat differs ⅜ in. [10 mm] or more in thickness or diameter, one tension test shall be made from both the thickest and the thinnest structural product rolled, regardless of the weight [mass] represented.

8.1.3 For quenched and tempered steel, including Grades EH40, FH32, FH36, and FH40, one tension test shall be made on each plate as quenched and tempered.

8.2 *Toughness Tests*:

8.2.1 Charpy V-notch tests shall be made on Grade A material over 2 in. [50 mm] in thickness, on Grade B material over 1 in. [25 mm] in thickness and on material over ¼ in. [6.4 mm] in thickness of Grades D, E, AH32, AH36, AH40, DH32, DH36, DH40, EH32, EH36, EH40, FH32, FH36, and FH40, as required by Table 1 and Table 2. The frequency of Charpy V-notch impact tests shall be as given in Table 1 and Table 2. The test results shall conform to the requirements given in Table 7.

8.2.2 For Grades EH32, EH36, EH40, FH32, FH36, and FH40 plate material, one set of three impact specimens shall be made from each plate-as-rolled.

8.2.3 For Grade A, B, D, AH32, AH36, AH40, DH32, DH36, and DH40 plate material, and for all shape material, and all bar material, one set of three impact specimens shall be made from the thickest material in each test lot size of each heat, as required by Table 1 and Table 2. If heat testing is required, a set of three specimens shall be tested for each test lot size indicated in Table 1 and Table 2, of the same type of product produced on the same mill from each heat of steel. The set of impact specimens shall be taken from different as-rolled or heat-treated pieces of the heaviest gage produced. An

TABLE 3 Chemical Requirements for Ordinary-Strength Structural Steel

Element	Chemical Composition (heat analysis), % max unless otherwise specified[A]			
	Grade A	Grade B	Grade D	Grade E
	Deoxidation and Thickness (*t*), in. [mm]			
	Killed or Semi-Killed $t \leq 2.0$ in. [50 mm] Killed $t >2.0$ in. [50 mm]	Killed or Semi-Killed $t \leq 2.0$ in. [50 mm] Killed $t >2.0$ in. [50 mm]	Killed, Fine Grain Practice[B]	Killed, Fine Grain Practice[B]
C	0.21[C]	0.21	0.21	0.18
Mn, min	2.5 × C	0.60	0.60	0.70
Si	0.50	0.35	0.10–0.35[D]	0.10–0.35[D]
P	0.035	0.035	0.035	0.035
S	0.035	0.035	0.035	0.035
Ni	E	E	E	E
Cr	E	E	E	E
Mo	E	E	E	E
Cu	E	E	E	E
C + Mn/6	0.40	0.40	0.40	0.40

[A] Intentionally added elements are to be determined and reported.
[B] Grade D steel over 1.0 in. [25 mm] and Grade E steel are to contain at least one of the grain refining elements in sufficient amount to meet the fine grain practice requirements (see Section 7).
[C] A maximum carbon content of 0.23 % is acceptable for Grade A shapes and bars.
[D] Where the content of acid soluble aluminum is not less than 0.015 %, the minimum required silicon content does not apply.
[E] The contents of nickel, chromium, molybdenum, and copper are to be determined and reported. When the amount does not exceed 0.02 %, these elements may be reported as ≤0.02 %.

TABLE 4 Chemical Requirements for Higher-Strength Structural Steel

Element	Chemical Composition[A] (heat analysis), % max unless otherwise specified	
	Grades AH/DH/EH32, AH/DH/EH36, and AH/DH/EH40	Grades FH32/36/40
	Deoxidation	
	Killed, Fine Grain Practice[B]	Killed, Fine Grain Practice[B]
C	0.18	0.16
Mn	0.90–1.60[C]	0.90–1.60
Si	0.10–0.50[D]	0.10–0.50[D]
P	0.035	0.025
S	0.035	0.025
Al (acid soluble), min[E,F]	0.015	0.015
Cb[F]	0.02–0.05	0.02–0.05
V[F]	0.05–0.10	0.05–0.10
Ti	0.02	0.02
Cu	0.35	0.35
Cr	0.20	0.20
Ni	0.40	0.40
Mo	0.08	0.08
N	...	0.009[G]

[A] The contents of any other element intentionally added is to be determined and reported.
[B] The steel is to contain at least one of the grain refining elements in sufficient amount to meet the fine grain practice requirement (see Section 7).
[C] Grade AH 0.5 in. [12.5 mm] and under in thickness may have a minimum manganese content of 0.70 %.
[D] If the content of soluble aluminum is not less than 0.015 %, the minimum required silicon content does not apply.
[E] The total aluminum content may be used instead of acid soluble content, in accordance with 7.1.
[F] The indicated amount of aluminum, columbium, and vanadium applies if any such element is used singly. If used in combination, the minimum content in 7.2.2 and 7.2.3, as appropriate, will apply.
[G] 0.012 if aluminum is present.

as-rolled piece refers to the product rolled from a slab, billet, bloom, or directly from an ingot. Where the maximum thickness or diameter of various sections differs by 3/8 in. [10 mm] or more, one set of impacts shall be made from both the thickest and the thinnest material rolled regardless of the weight represented.

8.2.4 The specimens for plates shall be taken from a corner of the material and the specimens from shapes shall be taken from the end of a shape at a point one third the distance from the outer edge of the flange or leg to the web or heel of the shape. Specimens for bars shall be in accordance with Specification A6/A6M.

8.2.5 The largest size specimens possible for the material thickness are to be machined. The longitudinal axis of each specimen shall be located midway between the surface and the center of the structural product thickness, and the length of the notch shall be perpendicular to the rolled surface of the structural product.

TABLE 5 Carbon Equivalent for Higher-Strength Structural Steel Produced by TMCP

Grade	Carbon Equivalent[A], max, %	
	Thickness (t), in. [mm]	
	$t \leq 2.0$ in. [50 mm]	$t > 2.0$ in. [50 mm] $t \leq 4.0$ in. [100 mm]
AH32, DH32, EH32, FH32	0.36	0.38
AH36, DH36, EH36, FH36	0.38	0.40
AH40, DH40, EH40, FH40	0.40	0.42

[A] The following carbon equivalent formula shall be used to calculate the carbon equivalent, C_{eq}:

$$C_{eq} = C + \frac{Mn}{6} + \frac{Cr + Mo + V}{5} + \frac{Ni + Cu}{15} (\%)$$

8.2.6 Unless a specific orientation is called for on the purchase order, the longitudinal axis of the specimens may be parallel or transverse to the final direction of rolling of the structural product at the option of the steel manufacturer.

8.2.7 The impact test shall be made in accordance with the Charpy Impact Testing section in Test Methods and Definitions A370.

8.2.8 Each impact test shall constitute the average value of three specimens taken from a single test location. The average value shall meet the specified minimum average with not more than one value below the specified minimum average but in no case below 70 % of the specified minimum average.

8.2.8.1 If the results fail to meet the preceding requirements but 8.2.8.1 (*2*) and (*3*) are complied with, three additional specimens may be taken from the location as close to the initial specimens as possible and their test results added to those previously obtained to form a new average. The structural product represented may be accepted if for the six specimens 8.2.8.1 (*1*), (*2*), and (*3*) are met.

(*1*) The average is not less than the required minimum average.

(*2*) No more than two individual values are below the required minimum average.

(*3*) No more than one individual value is below 70 % of the required minimum average.

8.2.8.2 If the required energy values are not obtained upon retest, the material may be heat treated at the option of the producer in the case of as-rolled material or reheat treated in the case of heat-treated material.

8.2.8.3 After heat treatment or reheat treatment, a set of three specimens shall be tested and evaluated in the same manner as for the original material.

8.2.8.4 If the impact test result fails to meet the requirement for the thickest product tested when heat testing, that material shall be rejected and the next thickest material may be tested to qualify the balance of the heat in accordance with 8.2.8. At the option of the producer, retests may be made on each piece of the rejected material, in which case each piece shall stand on the results of its own test. It shall also be the option of the producer to heat treat the product prior to retesting if desired.

9. General Requirements for Delivery

9.1 Material furnished under this specification shall conform to the requirements of the current edition of Specification A6/A6M, for the specific structural product ordered, unless a conflict exists in which case this specification shall prevail.

10. Plate Conditioning

10.1 After removal of any imperfection preparatory to welding the thickness of the plate at any location must not be reduced by more than 20 % of the nominal thickness of the plate.

11. Test Reports

11.1 When test reports are required by the purchase order, the report shall show the results of each test required by Sections 7 and 8, except that the results of only one set of tests need be reported when the amount of material from a heat in a shipment is less than 10 tons [9 Mg] and when the thickness variations described in Section 8 are not exceeded.

11.2 The thickness of the product tested may not necessarily be the same as an individual ordered thickness since it is the heat that is tested rather than each ordered item.

12. Marking

12.1 Plates produced to a normalized heat treatment condition shall be marked with the suffix **N** to indicate that the plates have been normalized.

12.2 Plates produced to a control rolled condition shall be marked with the suffix **CR** to indicate that the plates have been control rolled.

12.3 Plates produced to a thermo-mechanical control processed condition shall be marked with the suffix **TM** to indicate that the plates have been thermo-mechanical control processed.

12.4 Plates produced to a quenched and tempered heat treatment condition shall be marked with the suffix **QT** to indicate that the plates have been quenched and tempered.

13. Keywords

13.1 bars; higher strength; ordinary strength; plates; rivets; shapes; ship construction; steel; structural steel

TABLE 6 Tensile Requirements for Ordinary-Strength and Higher-Strength Structural Steel

Grade	Tensile Strength, ksi [MPa]	Yield Point, min, ksi [MPa]	Elongation in 8 in. [200 mm][A,B], min, %	Elongation in 2 in. [50 mm][B,C], min %
Ordinary strength:				
A, B, D, E	58 to 75 [400 to 520][D]	34 [235]	21	24
Higher strength:				
AH32, DH32, EH32, FH32	64 to 85 [440 to 590]	46 [315]	19	22
AH36, DH36, EH36, FH36	71 to 90 [490 to 620]	51 [355]	19	22
AH40, DH40, EH40, FH40	74 to 94 [510 to 650]	57 [390]	19	22

[A] For nominal thickness or diameter under 5/16 in. [8 mm], a deduction from the specified percentage of elongation in 8 in. [200 mm] shall be made. See elongation requirement adjustments under the Tension Tests section of Specification A6/A6M for deduction values.
[B] Elongation is not required for floor plate.
[C] For nominal thickness or diameter over 3.5 in. [90 mm], a deduction from the specified percentage of elongation in 2 in. [50 mm] shall be made. See elongation requirement adjustments under the Tension Tests section of Specification A6/A6M for deduction values.
[D] For Grade A shapes and bars, the upper limit of tensile strength may be 80 ksi [550 MPa].

TABLE 7 Charpy V-Notch Impact Requirements for Ordinary-Strength and Higher-Strength Structural Steel

Grade[A,B]	Test Temperature, °F [°C]	Average Absorbed Energy[C], min, ft·lbf [J]					
		Thickness (t), in. [mm]					
		$t \leq 2.0$ in. [50 mm]		$t > 2.0$ in. [50 mm] $t \leq 2.8$ in. [70 mm]		$t > 2.8$ in. [70 mm] $t \leq 4.0$ in. [100 mm]	
		Charpy V-notch Impact Specimen Orientation[D]					
		Longitudinal	Transverse	Longitudinal	Transverse	Longitudinal	Transverse
A[E]	68 [20]	...	...	25 [34]	17 [24]	30 [41]	20 [27]
B	32 [0]	20 [27]	14 [20]	25 [34]	17 [24]	30 [41]	20 [27]
AH32	32 [0]	23 [31]	16 [22]	28 [38]	19 [26]	34 [46]	23 [31]
AH36	32 [0]	25 [34]	17 [24]	30 [41]	20 [27]	37 [50]	25 [34]
AH40	32 [0]	29 [39]	19 [26]	34 [46]	23 [31]	41 [55]	27 [37]
D	−4 [−20]	20 [27]	14 [20]	25 [34]	17 [24]	30 [41]	20 [27]
DH32	−4 [−20]	23 [31]	16 [22]	28 [38]	19 [26]	34 [46]	23 [31]
DH36	−4 [−20]	25 [34]	17 [24]	30 [41]	20 [27]	37 [50]	25 [34]
DH40	−4 [−20]	29 [39]	19 [26]	34 [46]	23 [31]	41 [55]	27 [37]
E	−40 [−40]	20 [27]	14 [20]	25 [34]	17 [24]	30 [41]	20 [27]
EH32	−40 [−40]	23 [31]	16 [22]	28 [38]	19 [26]	34 [46]	23 [31]
EH36	−40 [−40]	25 [34]	17 [24]	30 [41]	20 [27]	37 [50]	25 [34]
EH40	−40 [−40]	29 [39]	19 [26]	34 [46]	23 [31]	41 [55]	27 [37]
FH32	−76 [−60]	23 [31]	16 [22]	28 [38]	19 [26]	34 [46]	23 [31]
FH36	−76 [−60]	25 [34]	17 [24]	30 [41]	20 [27]	37 [50]	25 [34]
FH40	−76 [−60]	29 [39]	19 [26]	34 [46]	23 [31]	41 [55]	27 [37]

[A] Charpy V-notch impact test requirements for ordinary-strength structural steel grades apply where such test is required by Table 1.
[B] Charpy V-notch impact test requirements for higher-strength structural steel grades apply where such test is required by Table 2.
[C] The energy shown is minimum for full-sized (0.394 by 0.394-in. [10 × 10-mm]) specimen. For sub-sized specimens, the energy shall be as follows:
  Specimen Size, in. [mm]   0.394 × 0.295 [10 × 7.5]   0.394 × 0.197 [10 × 5.0]   0.394 × 0.098 [10 × 2.5]
  Required Energy   5E/6   2E/3   E/2
  E—energy required for full-sized specimen.
[D] Either direction is acceptable.
[E] Impact tests for Grade A are not required when the material is produced using fine grain practice and normalized.

SUPPLEMENTARY REQUIREMENTS

The following supplementary requirements shall apply only when specified in the order:

S32. Single Heat Bundles

S32.1 Bundles containing shapes or bars shall be from a single heat of steel.

S85. Product Chemical Analysis

S85.1 The chemical composition shall be determined for plates, shapes, or bars in accordance with 6.1.1. The number of pieces to be tested shall be stated on the order.

S86. Orientation of Impact Specimens

S86.1 The orientation of the impact test specimens shall be as specifically stated on the order. (The purchaser shall state whether the tests are to be longitudinal or transverse.)

S87. Heat-Treatment of Grade DH

S87.1 Grade DH aluminum-treated steel over ¾ in. [19 mm] in thickness shall be normalized.

S88. Additional Tension Tests

S88.1 At least one tension test shall be made from each 50 tons [45 Mg] or fraction thereof from each heat. If the material differs by 0.375 in. [10 mm] or more in nominal thickness or diameter, one tension test shall be made from both the thickest and thinnest material in each 50 tons.

A131/A131M – 08

SUMMARY OF CHANGES

Committee A01 has identified the location of selected changes to this standard since the last issue (A131/A131M – 07) that may impact the use of this standard. (Approved March 1, 2008.)

(1) Added 1.4 and renumbered 1.5 and 1.6.
(2) Revised 5.5.
(3) Revised Table 1, Table 2, Table 3, Table 6, and Table 7.
(4) Deleted all references to Grade CS throughout.

Committee A01 has identified the location of selected changes to this standard since the last issue (A131/A131M – 04$^{\varepsilon 1}$) that may impact the use of this standard. (Approved March 1, 2007.)

(1) Revised 5.3.
(2) Deleted paragraph 5.3.1.
(3) Revised Table 1, Table 2, Table 3, Table 5, and Table 7.

ASTM International takes no position respecting the validity of any patent rights asserted in connection with any item mentioned in this standard. Users of this standard are expressly advised that determination of the validity of any such patent rights, and the risk of infringement of such rights, are entirely their own responsibility.

This standard is subject to revision at any time by the responsible technical committee and must be reviewed every five years and if not revised, either reapproved or withdrawn. Your comments are invited either for revision of this standard or for additional standards and should be addressed to ASTM International Headquarters. Your comments will receive careful consideration at a meeting of the responsible technical committee, which you may attend. If you feel that your comments have not received a fair hearing you should make your views known to the ASTM Committee on Standards, at the address shown below.

This standard is copyrighted by ASTM International, 100 Barr Harbor Drive, PO Box C700, West Conshohocken, PA 19428-2959, United States. Individual reprints (single or multiple copies) of this standard may be obtained by contacting ASTM at the above address or at 610-832-9585 (phone), 610-832-9555 (fax), or service@astm.org (e-mail); or through the ASTM website (www.astm.org).

Designation: A139/A139M – 04

Standard Specification for Electric-Fusion (Arc)-Welded Steel Pipe (NPS 4 and Over)[1]

This standard is issued under the fixed designation A139/A139M; the number immediately following the designation indicates the year of original adoption or, in the case of revision, the year of last revision. A number in parentheses indicates the year of last reapproval. A superscript epsilon (ε) indicates an editorial change since the last revision or reapproval.

This standard has been approved for use by agencies of the Department of Defense.

1. Scope*

1.1 This specification covers five grades of electric-fusion (arc)-welded straight-seam or helical-seam steel pipe. Pipe of NPS 4 (Note 1) and larger with nominal (average) wall thickness of 1.0 in. [25.4 mm] and less are covered. Listing of standardized dimensions are for reference (Note 2). The grades of steel are pipe mill grades having mechanical properties which differ from standard plate grades. The pipe is intended for conveying liquid, gas, or vapor.

NOTE 1—The dimensionless designator NPS (nominal pipe size) has been substituted in this standard for such traditional terms as "nominal diameter,"" size," and "nominal size."
NOTE 2—A comprehensive listing of standardized pipe dimensions is contained in ASME B36.10M[2].
NOTE 3—The suitability of pipe for various purposes is somewhat dependent on its dimensions, properties, and conditions of service. For example, for high-temperature service see applicable codes and Specification A691.

1.2 The values stated in either inch-pound units or in SI units are to be regarded separately as standard. Within the text, the SI units are shown in brackets. The values in each system are not exact equivalents; therefore, each system is to be used independently of the other.

2. Referenced Documents

2.1 *ASTM Standards:*[3]
A370 Test Methods and Definitions for Mechanical Testing of Steel Products
A691 Specification for Carbon and Alloy Steel Pipe, Electric-Fusion-Welded for High-Pressure Service at High Temperatures
A751 Test Methods, Practices, and Terminology for Chemical Analysis of Steel Products
E59 Practice for Sampling Steel and Iron for Determination of Chemical Composition[4]

2.2 *American Welding Society Standard:*[5]
AWS B2.1 Standard for Welding Procedure and Performance Qualifications Welding Handbook, Vol 1, 8th ed

2.3 *ASME Standards:*[6]
ASME B36.10M Welded and Seamless Wrought Steel Pipe
ASME B36.19M Stainless Steel Pipe
ASME Boiler and Pressure Vessel Code: Section IX, Welding Qualifications

3. Ordering Information

3.1 Orders for material under this specification should include the following, as required, to describe the desired material adequately:

3.1.1 Quantity (feet, metres, or number of lengths),
3.1.2 Name of material (electric-fusion-(arc) welded steel pipe),
3.1.3 Grade (Table 1),
3.1.4 Size (NPS, or outside diameter, and nominal wall thickness, or schedule number),
3.1.5 Lengths (specific or random, Section 17),
3.1.6 End finish (Section 18),
3.1.7 Hydrostatic test pressure (Section 16, Note 8, and Note 9),
3.1.8 ASTM specification designation, and
3.1.9 End use of material.

4. Process

4.1 The steel shall be made by one or more of the following processes: open-hearth, basic-oxygen, or electric-furnace.
4.2 Steel may be cast in ingots or may be strand cast. When steels of different grades are sequentially strand cast, identification of the resultant transition material is required. The producer shall remove the transition material by any established procedure that positively separates the grades.

NOTE 4—The term "basic-oxygen steelmaking" is used generically to

---

[1] This specification is under the jurisdiction of ASTM Committee A01 on Steel, Stainless Steel and Related Alloys, and is the direct responsibility of Subcommittee A01.09 on Carbon Steel Tubular Products.
Current edition approved March 1, 2004. Published April 2004. Originally approved in 1932. Last previous edition approved in 2000 as A139 – 00. DOI: 10.1520/A0139_A0139M-04.
[2] *Annual Book of ASTM Standards*, Vol 01.01.
[3] For referenced ASTM standards, visit the ASTM website, www.astm.org, or contact ASTM Customer Service at service@astm.org. For *Annual Book of ASTM Standards* volume information, refer to the standard's Document Summary page on the ASTM website.

[4] Withdrawn.
[5] Available from American Welding Society, 550 NW LeJeune Rd., Miami, FL 33135.
[6] Available from American Society of Mechanical Engineers, Three Park Ave., New York, NY 10016-5990.

---

*A Summary of Changes section appears at the end of this standard.

Copyright © ASTM International, 100 Barr Harbor Drive, PO Box C700, West Conshohocken, PA 19428-2959, United States.

TABLE 1 Chemical Requirements

Element	Composition, max, %				
	Grade A	Grade B	Grade C	Grade D	Grade E
Carbon	0.25	0.26	0.28	0.30	0.30
Manganese	1.00	1.00	1.20	1.30	1.40
Phosphorus	0.035	0.035	0.035	0.035	0.035
Sulfur	0.035	0.035	0.035	0.035	0.035

describe processes in which molten iron is refined to steel under a basic slag in a cylindrical furnace lined with basic refractories, by directing a jet of high-purity gaseous oxygen onto the surface of the hot metal bath.

5. Manufacture

5.1 The longitudinal edges of the steel shall be shaped to give the most satisfactory results by the particular welding process employed. The weld shall be made by automatic (Note 5) means (except tack welds if used) and shall be of reasonably uniform width and height for the entire length of the pipe.

NOTE 5—Upon agreement between the purchaser and the manufacturer, manual welding by qualified procedure and welders may be used as an equal alternative under these specifications.

5.2 All weld seams made in manufacturing pipe shall be made using complete joint penetration groove welds.

6. Chemical Composition

6.1 The steel shall conform to the chemical requirements prescribed in Table 1 and the chemical analysis shall be in accordance with Test Methods, Practices, and Terminology A751.

7. Tensile Requirements for the Steel

7.1 Longitudinal tension test specimens taken from the steel shall conform to the requirements as to tensile properties prescribed in Table 2. At the manufacturer's option, the tension test specimen for sizes 8⅝ in. [219.1 mm] in outside diameter and larger may be taken transversely as described in 19.4.

7.2 The yield point shall be determined by the drop of the beam, by the halt in the gage of the testing machine, by the use of dividers, or by other approved methods. The yield strength corresponding to a permanent offset of 0.2 % of the gage length of the specimen, or to a total extension of 0.5 % of the gage length under load shall be determined.

8. Tensile Requirements of Production Welds

8.1 Reduced-section tension test specimens taken perpendicularly across the weld in the pipe, with the weld reinforcement removed, shall show a tensile strength not less than 95 %

TABLE 2 Tensile Requirements

	Grade A	Grade B	Grade C	Grade D	Grade E
Tensile strength, min, ksi [MPa]	48 [330]	60 [415]	60 [415]	60 [415]	66 [455]
Yield strength, min, ksi [MPa]	30 [205]	35 [240]	42 [290]	46 [315]	52 [360]
Elongation in 2 in. or 50 mm, min, %:					
Basic minimum elongation for walls 5/16 in. [7.9 mm] and over in thickness, longitudinal strip tests	35	30	25	23	22
For longitudinal strips tests, a deduction for each 1/32-in. [0.8-mm] decrease in wall thickness below 5/16 in. [7.9 mm] from the basic minimum elongation of the following percentage[A]	1.75[A]	1.50[A]	1.25	1.50	2.0
Elongation in 8 in. or 200 mm, min, %[B,C]	Inch Pound Units, *1500/specified minimum tensile strength (ksi)*				
	SI Units, *10 300/specified minimum tensile strength [MPa]*				

[A] The table below gives the computed minimum values.
[B] For wall thicknesses ½ in. [12.7 mm] and greater, the elongation may be taken in 8 in. or 200 mm.
[C] The elongation in 8 in. or 200 mm need not exceed 30 %.

Wall Thickness		Elongation in 2 in. or 50 mm, min, %	
in.	mm	Grade A	Grade B
5/16 (0.312)	7.9	35.00	30.00
9/32 (0.281)	7.1	33.25	28.50
1/4 (0.250)	6.4	31.50	27.00
7/32 (0.219)	5.6	29.75	25.50
3/16 (0.188)	4.8	28.00	24.00
5/32 (0.156)	4.0	26.25	22.50
1/8 (0.125)	3.7	24.50	21.00
3/32 (0.094)	2.4	22.75	19.50
1/16 (0.062)	1.6	21.00	18.00

Note—The above table gives the computed minimum elongation values for each 1/32-in. [0.8-mm] decrease in wall thickness. Where the wall thickness lies between two values shown above, the minimum elongation value shall be determined by the following equation:

Grade	Equation Inch-Pound Units	Equation SI Units
A	$E = 56t + 17.50$	$E = 2.20t + 17.50$
B	$E = 48t + 15.00$	$E = 1.89t + 15.00$
C	$E = 40t + 12.50$	$E = 1.57t + 12.50$
D	$E = 48t + 8$	$E = 1.89t + 8$
E	$E = 64t + 2$	$E = 2.52t + 2$

where:
E = elongation in 2 in. or 50 mm, %, and
t = actual thickness of specimen, in. [mm]

of the minimum specified in Section 7. At the manufacturer's option, the test may be made without removing the weld reinforcement, in which case the tensile strength shall be not less than that specified in Section 7.

9. Heat Analysis

9.1 An analysis of each heat of steel shall be made by the manufacturer to determine the percentages of the elements specified in Section 6. This analysis shall be made from a test ingot taken during the pouring of the heat. When requested by the purchaser, the chemical composition thus determined shall be reported to the purchaser or his representative, and shall conform to the requirements specified in Section 6.

10. Product Analysis

10.1 An analysis may be made by the purchaser on samples of pipe selected at random and shall conform to the requirements specified in Section 6. Samples for chemical analysis, except for spectrochemical analysis, shall be taken in accordance with Method E59. The number of samples shall be determined as follows:

NPS	Number of Samples Selected
Under 14	2 for each lot of 200 pipes or fraction thereof
14 to 36, incl	2 for each lot of 100 pipes or fraction thereof
Over 36	2 for each 3000 ft or fraction thereof

10.2 *Retests*—If the analysis of either length of pipe or length of skelp does not conform to the requirements specified in Section 6, analyses of two additional lengths from the same lot shall be made, each of which shall conform to the requirements specified.

11. Dimensions, Mass, and Permissible Variations

11.1 *Mass*—The specified mass per unit length shall be calculated using the following equation:

$$M = C(D - t)t \qquad (1)$$

where:
C = 10.69 [0.02466],
M = mass per unit length, lb/ft [kg/m],
D = outside diameter, in. [mm], specified or calculated (from inside diameter and wall thickness), and
t = specified wall thickness, in. (to 3 decimal places) [mm] (to 2 decimal places)

NOTE 6—The mass per unit length given in ASME B36.10M and ASME B36.19M and the calculated mass given by the equation of 11.1 are for carbon steel pipe. The mass per unit length of pipe made of ferritic stainless steels may be about 5 % less, and that made of austenitic stainless steel about 2 % greater than the values given. The specified mass of an individual pipe length shall be calculated as its specified mass per unit length times its length.

11.1.1 The mass of any length of pipe shall not vary more than 10 % over its specified mass.

11.1.2 The mass of any length of pipe shall not vary more than 5 % under the specified mass if the specified wall thickness is 0.188 in. [4.78 mm] or less or more than 5.5 % under if the specified wall thickness is greater than 0.188 in. [4.78 mm].

11.1.3 The mass of a carload lot shall not vary more than 1.75 % under the specified mass. A carload lot is considered to be a minimum of 40 000 lb [18 Mg] shipped on a conveyance.

11.2 *Thickness*—The minimum wall thickness at any point shall be not more than 12.5 % under the nominal wall thickness specified.

11.3 *Circumference*—The pipe shall be substantially round. The outside circumference of the pipe shall not vary more than ±1.0 %, but not exceeding ±¾ in. [19.0 mm], from the nominal outside circumference based upon the diameter specified, except that the circumference at ends shall be sized, if necessary, to meet the requirements of Section 18.

11.4 *Straightness*—Finished pipe shall be commercially straight. When specific straightness requirements are desired, the order should so state, and the tolerance shall be a matter of agreement between the purchaser and the manufacturer.

11.5 *Ovality (Out-of-Roundness)*—The pipe diameter, within 4.0 in. [100 mm] of ends, shall not vary more than 1 % from the specified diameter as measured across any single plane with a bar gage, caliper, or other instrument capable of measuring actual diameter.

12. Finish

12.1 *Repair by Welding*—The manual, or automatic arc, welding of injurious defects in the pipe wall, provided their depth does not exceed one third the specified wall thickness, will be permitted. Defects in the welds, such as sweats or leaks, shall be repaired or the piece rejected at the option of the manufacturer. Repairs of this nature shall be made by completely removing the defect, cleaning the cavity, and then welding.

12.2 All repaired pipe shall be tested hydrostatically in accordance with Section 16.

13. Retests

13.1 If any specimen tested under Sections 8 or 15 fails to meet the requirements, retests of two additional specimens from the same lot of pipe shall be made, all of which shall meet the specified requirements. If any of the retests fail to conform to the requirements, test specimens may be taken from each untested pipe length, at the manufacturer's option, and each specimen shall meet the requirements specified, or that pipe shall be rejected.

14. Number of Production Test Specimens

14.1 One longitudinal tension test specimen specified in 19.2 shall be made from the steel of each heat, or fraction thereof, used in the manufacture of the pipe.

14.2 One reduced-section production weld test specimen specified in 19.5 shall be taken from a length of pipe from each lot of 3000 ft (914 m) of pipe, or fraction thereof, of each size and wall thickness.

14.3 If any test specimen shows defective machining or develops flaws not associated with the quality of the steel or the welding, it may be discarded and another specimen substituted.

14.4 Each length of pipe shall be subjected to the hydrostatic test specified in Section 16.

15. Qualification of Welding Procedure

15.1 Welding procedures shall be qualified in accordance with the requirements of AWS B2.1; ASME Boiler and Pressure Vessel Code, Section IX; or other qualification procedures as noted in the American Welding Society Welding Handbook. Tests and test values shall be as specified in 15.2 and 15.3.

15.2 Two reduced-section tension specimens made in accordance with Fig. 1, with the weld reinforcement removed, shall show a tensile strength not less than 100 % of the minimum specified tensile strength of the grade of steel used.

15.3 Bend test specimens (two face-bend and two root-bend or four side-bend as designated by the welding procedure according to thickness) shall be prepared in accordance with Fig. 2 and shall withstand being bent 180° in a jig substantially in accordance with Fig. 3. The bend test shall be acceptable if no cracks or other defects exceeding 1/8 in. [3.2 mm] in any direction are present in the weld metal or between the weld and the pipe metal after bending. Cracks that originate along the edges of the specimens during testing, and that are less than 1/4 in. [6.4 mm] in any direction shall not be considered. (If necessary, the specimen shall be broken apart to permit examination of the fracture.)

16. Hydrostatic Test (Note 7)

16.1 Each length of pipe shall be tested by the manufacturer to a hydrostatic pressure that will produce in the pipe wall a stress of not less than 60 % of the specified minimum yield strength at room temperature. The pressure shall be determined by the following equation:

$$P = 2St/D \qquad (2)$$

where:
P = hydrostatic test pressure, psi [MPa] (not to exceed 2800 psi [19.3 MPa] in any case) (Note 8),
S = 0.60 to 0.85 times the specified minimum yield strength of the grade of steel used in psi [MPa],
t = specified wall thickness, in. [mm], and
D = specified outside diameter, in.[mm]

NOTE 7—A hydrostatic sizing operation is not to be considered a hydrostatic test or a substitute for it.

NOTE 8—When the diameter and wall thickness of pipe are such that the capacity limits of testing equipment are exceeded by these requirements, the test pressures may be reduced by agreement between the purchaser and the manufacturer.

NOTE 9—Where specified in the purchase order, the pipe may be tested: (1) to 1.5 times the specified working pressure, provided the test pressure does not exceed 2800 psi [19.3 MPa] or produce a fiber stress in excess of 85 % of the specified minimum yield strength for the applicable pipe grade, or (2) to a fiber stress of 85 % or less of the specified minimum yield strength for the applicable pipe grade, provided that the test pressure

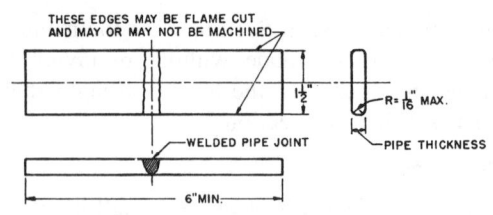

Metric Equivalents

in.	1/16	1 1/2	6
mm	1.6	38	150

NOTE 1—Weld reinforcement may or may not be removed flush with the surface of the specimen.

NOTE 2—Shown in Fig. 2 is a root- or face-bend specimen. Side-bend specimens shall have a thickness (T) of 3/8 in. (9.5 mm) and a width equal to the pipe wall thickness.

FIG. 2 Guided-Bend Test Specimen

does not exceed 2800 psi [19.3 MPa].

16.2 Test pressure shall be held for not less than 5 s, or for a longer time as agreed upon between the purchaser and the manufacturer.

17. Lengths

17.1 Pipe lengths shall be supplied in accordance with the following regular practice:

17.1.1 Specific lengths shall be as specified on the order with a tolerance of ±1/2 in. [12.7 mm], except that the shorter lengths from which test coupons have been cut shall also be shipped.

17.1.2 Unless otherwise specified random lengths shall be furnished in lengths averaging 29 ft [8.9 m] or over, with a minimum length of 20 ft [6.1 m], but not more than 5 % may be under 25 ft [7.6 m].

17.1.3 Pipe lengths containing circumferentially welded joints (Note 6) shall be permitted by agreement between the purchaser and the manufacturer. Tests of these welded joints shall be made in accordance with the production weld tests described in Section 8. The number of production weld tests shall be one for each lot of 100 joints or fraction thereof, but not less than one for each welder or welding operator.

NOTE 10—Circumferentially welded joints are defined for the purpose of these specifications as a welded seam lying in one plane, used to join lengths of straight pipe.

18. Ends

18.1 Pipe shall be furnished with plain right-angle cut or beveled ends as specified. All burrs at the ends of pipe shall be removed.

18.2 When pipe is specified to have the ends prepared for field welding of circumferential joints, the ends shall be beveled on the outside to an angle of 35°, measured from a line drawn perpendicular to the axis of the pipe, with a tolerance of ±2 1/2 ° and with a width of root face (or flat at the end of the pipe) of 1/16 ± 1/32 in. [1.6 ± 0.8 mm]. Unless otherwise specified, the outside circumference of pipe ends for a distance of not less than 4 in. [101.6 mm] shall not vary more than ±60 % of the nominal wall thickness of the pipe from the

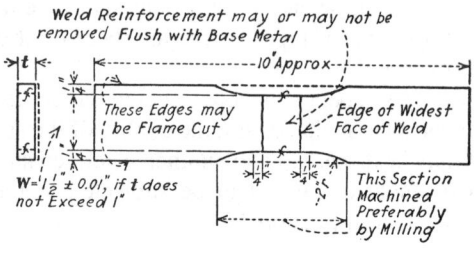

Metric Equivalents

in.	0.01	1/4	1 1/2	10
mm	0.3	6.4	38	250

FIG. 1 Reduced-Section Tension Test Specimen

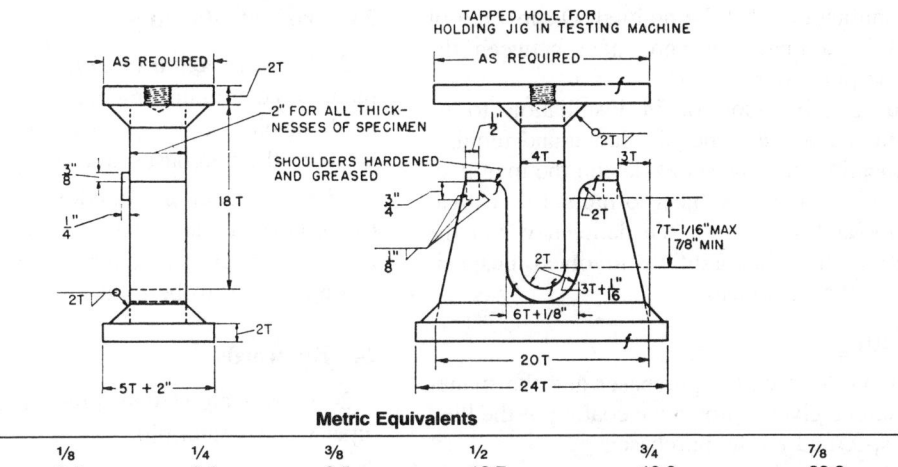

in.	1/16	1/8	1/4	3/8	1/2	3/4	7/8	2
mm	1.6	3.2	6.4	9.5	12.7	19.0	22.2	50.8

FIG. 3 Jig for Guided-Bend Test

nominal outside circumference based on the diameter specified, except that the tolerance shall be not less than ±3/16 in. [4.8 mm].

18.3 Pipe ends for use with mechanical couplings shall have tolerances within the limits required by the manufacturer of the type of coupling to be used.

18.4 Upon agreement between the purchaser and the manufacturer, the ends of the pipe may be sized within agreed-upon tolerances, if necessary to meet the requirements of special installations.

19. Production Test Specimens and Methods of Testing

19.1 The test specimens and the tests required by these specifications shall conform to those described in Test Methods and Definitions A370.

19.2 The longitudinal tension tests specimen of the steel shall be taken from the end of the pipe in accordance with Fig. 4, or by agreement between the purchaser and the manufacturer, or may be taken from the skelp or plate, at a point which will be approximately 90° of arc from the weld in the finished pipe.

19.3 If the tension test specimen is taken transversely, the specimen shall be taken in accordance with Fig. 5.

19.4 The specimens for the reduced-section tension test of production welds shall be taken perpendicularly across the weld at the end of the pipe. The test specimens shall have the weld approximately in the middle of the specimen. The specimens shall be straightened and tested at room temperature.

FIG. 4 Location from Which Longitudinal Tension Test Specimens Are To Be Cut from Large Diameter Tubing

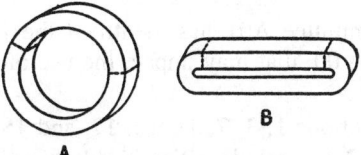

FIG. 5 Location of Transverse Tension Test Specimen in Ring Cut from Tubular Steel Products

19.5 Reduced-section tension test specimens shall be prepared in accordance with Fig. 1.

20. Inspection

20.1 The inspector representing the purchaser shall have entry, at all times while work on the contract of the purchaser is being performed, to all parts of the manufacturer's works that concern the manufacture of the material ordered. The manufacturer shall afford the inspector all reasonable facilities to satisfy him that the material is being furnished in accordance with this specification. All tests and inspection shall be made at the place of manufacture prior to shipment and, unless otherwise specified, shall be so conducted as not to interfere unnecessarily with the operation of the works. If agreed upon, the manufacturer shall notify the purchaser in time so that he may have his inspector present to witness any part of the manufacture or tests that may be desired.

20.2 *Certification*—Upon request of the purchaser in the contract or order, a manufacturer's certification that the material was manufactured and tested in accordance with this specification together with a report of the chemical and tensile tests shall be furnished.

21. Rejection

21.1 Each length of pipe received from the manufacturer may be inspected by the purchaser and, if it does not meet the requirements of this specification based on the inspection and test method as outlined in the specification, the length may be rejected and the manufacturer shall be notified. Disposition of rejected pipe shall be a matter of agreement between the manufacturer and the purchaser.

21.2 Pipe found in fabrication or in installation to be unsuitable for the intended use, under the scope and requirements of this specification, may be set aside and the manufacturer notified. Such pipe shall be subject to mutual investigation as to the nature and severity of the deficiency and the forming or installation, or both, conditions involved. Disposition shall be a matter for agreement.

22. Protective Coating

22.1 If agreed upon between the purchaser and the manufacturer, the pipe shall be given a protective coating of the kind and in the manner specified by the purchaser.

23. Product Marking

23.1 Each section of pipe shall be marked with the manufacturer's distinguishing marking, the specification number, the grade of pipe, and other marking if required and agreed upon between the purchaser and the manufacturer.

23.2 *Bar Coding*—In addition to the requirements in 23.1, bar coding is acceptable as a supplemental identification method. The purchaser may specify in the order a specific bar coding system to be used.

24. Keywords

24.1 arc welded steel pipe; fusion welded steel pipe; steel pipe; welded steel pipe

SUMMARY OF CHANGES

Committee A01 has identified the location of selected changes to this specification since the last issue, A139 – 00, that may impact the use of this specification. (Approved March 1, 2004)

(*1*) Revised Sections 1, 3, 7, 11, 16, 17, and 18, Table 2, and Figures 1 and 2 to include rationalized SI units, creating a combined standard.

ASTM International takes no position respecting the validity of any patent rights asserted in connection with any item mentioned in this standard. Users of this standard are expressly advised that determination of the validity of any such patent rights, and the risk of infringement of such rights, are entirely their own responsibility.

This standard is subject to revision at any time by the responsible technical committee and must be reviewed every five years and if not revised, either reapproved or withdrawn. Your comments are invited either for revision of this standard or for additional standards and should be addressed to ASTM International Headquarters. Your comments will receive careful consideration at a meeting of the responsible technical committee, which you may attend. If you feel that your comments have not received a fair hearing you should make your views known to the ASTM Committee on Standards, at the address shown below.

This standard is copyrighted by ASTM International, 100 Barr Harbor Drive, PO Box C700, West Conshohocken, PA 19428-2959, United States. Individual reprints (single or multiple copies) of this standard may be obtained by contacting ASTM at the above address or at 610-832-9585 (phone), 610-832-9555 (fax), or service@astm.org (e-mail); or through the ASTM website (www.astm.org).

Designation: A242/A242M – 04 (Reapproved 2009)

American Association State Highway and Transportation Officials Standard AASHTO No.: M 161

Standard Specification for High-Strength Low-Alloy Structural Steel[1]

This standard is issued under the fixed designation A242/A242M; the number immediately following the designation indicates the year of original adoption or, in the case of revision, the year of last revision. A number in parentheses indicates the year of last reapproval. A superscript epsilon (ε) indicates an editorial change since the last revision or reapproval.

This standard has been approved for use by agencies of the Department of Defense.

1. Scope

1.1 This specification covers high-strength low-alloy structural steel shapes, plates, and bars for welded, riveted, or bolted construction intended primarily for use as structural members where savings in weight [mass] or added durability are important. The atmospheric corrosion resistance of the steel in most environments is substantially better than that of carbon structural steels with or without copper addition. When properly exposed to the atmosphere, this steel can be used bare (unpainted) for many applications (see Note 1). This specification is limited to material up to 4 in. [100 mm], inclusive, in thickness.

NOTE 1—For methods of estimating the atmospheric corrosion resistance of low-alloy steels, see Guide G101.

1.2 When the steel is to be welded, it is presupposed that a welding procedure suitable for the grade of steel and intended use or service will be utilized. See Appendix X3 of Specification A6/A6M for information on weldability.

1.3 The values stated in either inch-pound units or SI units are to be regarded as standard. Within the text, the SI units are shown in brackets. The values stated in each system are not exact equivalents; therefore, each system must be used independently of the other. Combining values from the two systems may result in nonconformance with the specification.

1.4 For structural products produced from coil and furnished without heat treatment or with stress relieving only, the additional requirements, including additional testing requirements and the reporting of additional test results, of Specification A6/A6M apply.

2. Referenced Documents

2.1 *ASTM Standards:*[2]

A6/A6M Specification for General Requirements for Rolled Structural Steel Bars, Plates, Shapes, and Sheet Piling

G101 Guide for Estimating the Atmospheric Corrosion Resistance of Low-Alloy Steels

3. General Requirements for Delivery

3.1 Structural products furnished under this specification shall conform to the requirements of the current edition of Specification A6/A6M, for the specific structural product ordered, unless a conflict exists, in which case this specification shall prevail.

3.2 Coils are excluded from qualification to this specification until they are processed into a finished structural product. Structural products produced from coil means structural products that have been cut to individual lengths from a coil. The processor directly controls, or is responsible for, the operations involved in the processing of a coil into a finished structural product. Such operations include decoiling, leveling or straightening, hot-forming or cold-forming (if applicable), cutting to length, testing, inspection, conditioning, heat treatment (if applicable), packaging, marking, loading for shipment, and certification.

NOTE 2—For structural products produced from coil and furnished without heat treatment or with stress relieving only, two test results are to be reported for each qualifying coil. Additional requirements regarding structural products produced from coil are described in Specification A6/A6M.

4. Materials and Manufacture

4.1 The steel shall be semi-killed or killed.

---

[1] This specification is under the jurisdiction of ASTM Committee A01 on Steel, Stainless Steel and Related Alloys and is the direct responsibility of Subcommittee A01.02 on Structural Steel for Bridges, Buildings, Rolling Stock and Ships.
Current edition approved April 1, 2009. Published May 2009. Originally approved in 1941. Last previous edition approved in 2004 as A242/A242M–04ε1. DOI: 10.1520/A0242_A0242M-04R09.

[2] For referenced ASTM standards, visit the ASTM website, www.astm.org, or contact ASTM Customer Service at service@astm.org. For *Annual Book of ASTM Standards* volume information, refer to the standard's Document Summary page on the ASTM website.

TABLE 1 Tensile Requirements

	Plates and Bars[A]			Structural Shapes		
	For thicknesses ¾ in.[20 mm], and under	For thicknesses over ¾ to 1½ in. [20 to 40 mm], incl	For thicknesses over 1½ in. to 4 in. [40 to 100 mm], incl	For flange or leg thicknesses 1.5 in. [40 mm] and under	For flange thicknesses over 1.5 in. [40 mm] to 2 in. [50 mm], incl	For flange thicknesses over 2 in. [50 mm]
Tensile strength, min, ksi [MPa]	70 [480]	67 [460]	63 [435]	70 [485]	67 [460]	63 [435]
Yield point, min, ksi [MPa]	50 [345]	46 [315]	42 [290]	50 [345]	46 [315]	42 [290]
Elongation in 8 in. [200 mm], min, %	18[B,C]	18[B,C]	18[B,C]	18[C]	18	18
Elongation in 2 in. [50 mm], min, %	21[C]	21[C]	21[C]	21	21	21[D]

[A] See the Orientation subsection in the Tension Tests section of Specification A6/A6M.
[B] Elongation not required to be determined for floor plate.
[C] For plates wider than 24 in. [600 mm] the elongation requirement is reduced two percentage points. See the Elongation Requirement Adjustments subsection in the Tension Tests section of Specification A6/A6M.
[D] For wide flange shapes over 426 lb/ft [634 kg/m], elongation in 2 in. [50 mm] of 18 % minimum applies.

5. Chemical Composition

5.1 The heat analysis shall conform to the requirements prescribed in Table 2.

5.2 The steel shall conform on product analysis to the requirements prescribed in Table 2, subject to the product analysis tolerances in Specification A6/A6M.

5.3 Choice and use of alloying elements, combined with carbon, manganese, phosphorus, sulfur, and copper within the limits prescribed in 5.1 to give the mechanical properties prescribed in Section 6 and to provide the atmospheric corrosion resistance of 1.1, shall be made by the manufacturer and included and reported in the heat analysis to identify the type of steel applied. Elements commonly added include: chromium, nickel, silicon, vanadium, titanium, and zirconium.

5.4 The atmospheric corrosion-resistance index, calculated on the basis of the heat analysis of the steel, as described in Guide G101–Predictive Method Based on the Data of Larabee and Coburn, shall be 6.0 or higher.

NOTE 3—The user is cautioned that the Guide G101 predictive equation (Predictive Method Based on the Data of Larabee and Coburn) for calculation of an atmospheric corrosion-resistance index has only been verified for the composition limits stated in the guide.

5.5 When required, the manufacturer shall supply evidence of corrosion resistance satisfactory to the purchaser.

6. Tension Test

6.1 The material as represented by the test specimens shall conform to the requirements as to tensile properties prescribed in Table 1.

7. Keywords

7.1 atmospheric corrosion resistance; bars; bolted construction; durability; high-strength; low-alloy; plates; riveted construction; shapes; steel; structural steel; weight; welded construction

TABLE 2 Chemical Requirements (Heat Analysis)

Element	Composition, %
	Type 1
Carbon, max	0.15
Manganese, max	1.00
Phosphorus, max	0.15
Sulfur, max	0.05
Copper, min	0.20

SUPPLEMENTARY REQUIREMENTS

Standardized supplementary requirements for use at the option of the purchaser are listed in Specification A6/A6M. Those that are considered suitable for use with this specification are listed by title:

S2. Product Analysis,
S3. Simulated Post-Weld Heat Treatment of Mechanical Test Coupons,
S5. Charpy V-Notch Impact Test,
S6. Drop Weight Test (for Material 0.625 in. [16 mm] and over in Thickness),
S8. Ultrasonic Examination, and
S15. Reduction of Area Measurement.
S32. Single Heat Bundles.

ASTM International takes no position respecting the validity of any patent rights asserted in connection with any item mentioned in this standard. Users of this standard are expressly advised that determination of the validity of any such patent rights, and the risk of infringement of such rights, are entirely their own responsibility.

This standard is subject to revision at any time by the responsible technical committee and must be reviewed every five years and if not revised, either reapproved or withdrawn. Your comments are invited either for revision of this standard or for additional standards and should be addressed to ASTM International Headquarters. Your comments will receive careful consideration at a meeting of the responsible technical committee, which you may attend. If you feel that your comments have not received a fair hearing you should make your views known to the ASTM Committee on Standards, at the address shown below.

This standard is copyrighted by ASTM International, 100 Barr Harbor Drive, PO Box C700, West Conshohocken, PA 19428-2959, United States. Individual reprints (single or multiple copies) of this standard may be obtained by contacting ASTM at the above address or at 610-832-9585 (phone), 610-832-9555 (fax), or service@astm.org (e-mail); or through the ASTM website (www.astm.org).

Designation: A252 − 98 (Reapproved 2007)

Standard Specification for
Welded and Seamless Steel Pipe Piles[1]

This standard is issued under the fixed designation A252; the number immediately following the designation indicates the year of original adoption or, in the case of revision, the year of last revision. A number in parentheses indicates the year of last reapproval. A superscript epsilon (ε) indicates an editorial change since the last revision or reapproval.

This standard has been approved for use by agencies of the Department of Defense.

1. Scope

1.1 This specification covers nominal (average) wall steel pipe piles of cylindrical shape and applies to pipe piles in which the steel cylinder acts as a permanent load-carrying member, or as a shell to form cast-in-place concrete piles.

1.2 The values stated in inch-pound units are to be regarded as standard. The values given in parentheses are mathematical conversions of the values in inch-pound units to values in SI units.

1.3 The text of this specification contains notes and footnotes that provide explanatory material. Such notes and footnotes, excluding those in tables and figures, do not contain any mandatory requirements.

1.4 The following precautionary caveat pertains only to the test method portion, Section 16 of this specification. *This standard does not purport to address all of the safety problems, if any, associated with its use. It is the responsibility of the user of this standard to establish appropriate safety and health practices and determine the applicability of regulatory limitations prior to use.*

2. Referenced Documents

2.1 *ASTM Standards:*[2]
A370 Test Methods and Definitions for Mechanical Testing of Steel Products
A751 Test Methods, Practices, and Terminology for Chemical Analysis of Steel Products
A941 Terminology Relating to Steel, Stainless Steel, Related Alloys, and Ferroalloys
E29 Practice for Using Significant Digits in Test Data to Determine Conformance with Specifications

3. Terminology

3.1 *Definitions*—Definitions of terms used in this specification shall be in accordance with Terminology A941.

3.1.1 *defect*—an imperfection of sufficient size or magnitude to be cause for rejection.

3.1.2 *imperfection*—any discontinuity or irregularity found in the pipe.

4. Ordering Information

4.1 Orders for material under this specification shall contain information concerning as many of the following items as are required to describe the desired material adequately:

4.1.1 Quantity (feet or number of lengths),
4.1.2 Name of material (steel pipe piles),
4.1.3 Method of manufacture (seamless or welded),
4.1.4 Grade (Tables 1 and 2),
4.1.5 Size (outside diameter and nominal wall thickness),
4.1.6 Lengths (single random, double random, or uniform) (see Section 13),
4.1.7 End finish (Section 15), and
4.1.8 ASTM specification designation and year of issue,
4.1.9 Location of purchaser's inspection (see 19.1), and
4.1.10 Bar coding (see 22.2).

5. Materials and Manufacture

5.1 The piles shall be made by the seamless, electric resistance welded, flash welded, or fusion welded process. The seams of welded pipe piles shall be longitudinal, helical-butt, or helical-lap.

NOTE 1—For welded pipe piles, the weld should not fail when the product is properly fabricated and installed and subjected to its intended end use.

6. Process

6.1 The steel shall be made by one or more of the following processes: open-hearth, basic-oxygen, or electric-furnace.

---

[1] This specification is under the jurisdiction of ASTM Committee A01 on Steel, Stainless Steel and Related Alloys and is the direct responsibility of Subcommittee A01.09 on Carbon Steel Tubular Products.
Current edition approved Nov. 1, 2007. Published January 2008. Originally approved in 1944. Last previous edition approved in 2002 as A252 – 98(2002). DOI: 10.1520/A0252-98R07.
[2] For referenced ASTM standards, visit the ASTM website, www.astm.org, or contact ASTM Customer Service at service@astm.org. For *Annual Book of ASTM Standards* volume information, refer to the standard's Document Summary page on the ASTM website.

TABLE 1 Tensile Requirements

NOTE—Where an ellipsis (...) appears in this table, there is no requirement.

	Grade 1	Grade 2	Grade 3
Tensile strength, min, psi (MPa)	50 000 (345)	60 000 (415)	66 000 (455)
Yield point or yield strength, min, psi (MPa)	30 000 (205)	35 000 (240)	45 000 (310)
Basic minimum elongation for nominal wall thicknesses 5/16 in. (7.9 mm) or more:			
Elongation in 8 in. (203.2 mm), min, %	18	14	...
Elongation in 2 in. (50.8 mm), min, %	30	25	20
For nominal wall thicknesses less than 5/16 in. (7.9 mm), the deduction from the basic minimum elongation in 2 in. (50.08 mm) for each 1/32 - in. (0.8 mm) decrease in nominal wall thickness below 5/16 in. (7.9 mm), in percentage points	1.50[A]	1.25[A]	1.0[A]

[A] Table 2 gives the computed minimum values:

TABLE 2 Calculated Minimum Elongation Values[A]

Nominal Wall Thickness		Elongation in 2 in. (50.8 mm), min, %		
in.	mm	Grade 1	Grade 2	Grade 3
5/16 or 0.312	7.9	30.00	25.00	20.00
9/32 or 0.281	7.1	28.50	23.75	19.00
1/4 or 0.250	6.4	27.00	22.50	18.00
7/32 or 0.219	5.6	25.50	21.25	17.00
3/16 or 0.188	4.8	24.00	20.00	16.00
11/64 or 0.172	4.4	23.25	19.50	15.50
5/32 or 0.156	4.0	22.50	18.75	15.00
9/64 or 0.141	3.6	21.75	18.25	14.50
1/8 or 0.125	3.2	21.00	17.50	14.00
7/64 or 0.109	2.8	20.25	16.75	13.50

[A] The above table gives the calculated minimum elongation values for various nominal wall thicknesses. Where the specified nominal wall thickness is intermediate to those shown above, the minimum elongation value shall be determined as follows:

Grade
1     $E = 48t + 15.00$
2     $E = 40t + 12.50$
3     $E = 32t + 10.00$

where:
E = elongation in 2 in., %, and
t = specified nominal wall thickness, in.

7. Chemical Composition

7.1 The steel shall contain no more than 0.050 % phosphorous.

8. Heat Analysis

8.1 Each heat analysis shall conform to the requirement specified in 7.1. When requested by the purchaser, the applicable heat analyses shall be reported to the purchaser ro the purchaser's representative.

9. Product Analysis

9.1 Chemical analysis shall be in accordance with Test Methods, Practices, and Terminology A751.

9.2 It shall be permissible for the purchaser to make product analyses using samples from lots of pipe piles as follows:

Pipe Size Outside Diameter, in. (mm)	Number of Samples and Size of Lot
Under 14 (355.6)	2 from 200 pipe or fraction thereof
14 to 36, incl (355.6 to 914)	2 from 100 pipe or fraction thereof
Over 36 (914)	2 from 3000 ft (914 m) or fraction thereof

The product analyses shall conform to the requirement in 7.1.

9.3 If the chemical compositions of both of the samples representing a lot fail to conform to the specified requirement, the lot shall be rejected or analyses of four additional samples selected from the lot shall be made, and each shall conform to the specified requirement. If the chemical composition of only one of the samples representing a lot fails to conform to the specified requirement, the lot shall be rejected or analyses of two additional samples selected from the lot shall be made, and each shall conform to the specified requirement.

10. *Tensile Requirements* Tensile Requirements

10.1 The material shall conform to the requirements as to tensile properties prescribed in Tables 1 and 2.

10.2 The yield point shall be determined by the drop of the beam, by the halt in the gage of the testing machine, by the use of dividers, or by other approved methods. When a definite yield point is not exhibited, the yield strength corresponding to a permanent offset of 0.2 % of the gage length of the specimen, or to a total extension of 0.5 % of the gage length under load shall be determined.

11. Weights Per Unit Length

11.1 The weights per unit length for various sizes of pipe piles are listed in Table 3.

11.2 For pipe pile sizes not listed in Table 3, the weight per unit length shall be calculated as follows:

$$W = 10.69(D - t)t \quad (1)$$

where:
- W = weight per unit length, lb/ft,
- D = specified outside diameter, in., and
- t = specified nominal wall thickness, in.

12. Permissible Variations in Weights and Dimensions

12.1 *Weight*—Each length of pipe pile shall be weighed separately and its weight shall not vary more than 15 % over or 5 % under its theoretical weight, calculated using its length and its weight per unit length (see Section 11).

12.2 *Outside Diameter*—The outside diameter of pipe piles shall not vary more than ±1 % from the specified outside diameter.

12.3 *Wall Thickness*—The wall thickness at any point shall not be more than 12.5 % under the specified nominal wall thickness.

Note 2—The minimum permissible wall thickness on inspection is shown in Table X1.1 (see Appendix) for various nominal wall thicknesses.

13. Lengths

13.1 Pipe piles shall be furnished in single random lengths, double random lengths, or in uniform lengths as specified in the purchase order, in accordance with the following limits:

Single random lengths	16 to 25 ft (4.88 to 7.62 mm), incl
Double random lengths	over 25 ft (7.62 m) with a minimum average of 35 ft (10.67 m)
Uniform lengths	length as specified with a permissible variation of ±1 in.

13.2 Lengths that have been spliced at the mill by welding shall be acceptable as the equivalent of unspliced lengths provided tension test specimens cut from sample splices conform to the tensile strength requirements prescribed in Tables 1 and 2. The welding bead shall not be removed for this test. Such specimens shall be made in accordance with the provisions specified in Sections 16-18.

TABLE 3 Common Sizes and Weights Per Unit Length[A]

Outside Diameter, in.	Nominal Wall Thickness, in.[B]	Weight per Unit Lengths, lb/ft[C]	Outside Diameter, in.[B]	Nominal Wall Thickness, in.[B]	Weight per Unit Lengths, lb/ft[C]
6	0.134	8.40	12	0.134	17.00
	0.141	8.83		0.141	17.87
	0.156	9.75		0.150	19.00
	0.164	10.23		0.164	20.75
	0.172	10.72		0.172	21.75
				0.179	22.62
8	0.141	11.85		0.188	23.74
	0.172	14.39		0.203	25.60
				0.219	27.58
8⅝	0.109	9.92		0.230	28.94
	0.141	12.79		0.250	31.40
	0.172	15.54		0.281	35.20
	0.188	16.96		0.312	38.98
	0.203	18.28			
	0.219	19.68	12¾	0.109	14.73
	0.250	22.38		0.134	18.07
	0.277	24.72		0.141	19.01
	0.312	27.73		0.150	20.20
	0.322	28.58		0.164	22.07
	0.344	30.45		0.172	23.13
	0.375	33.07		0.179	24.05
	0.438	38.33		0.188	25.25
	0.500	43.43		0.203	27.23
				0.219	29.34
10	0.109	11.53		0.230	30.78
	0.120	12.67		0.250	33.41
	0.134	14.13		0.281	37.46
	0.141	14.86		0.312	41.48
	0.150	15.79		0.330	43.81
	0.164	17.24		0.344	45.62
	0.172	18.07		0.375	49.61
	0.179	18.79		0.438	57.65
	0.188	19.72		0.500	65.48
	0.203	21.26			
	0.219	22.90	14	0.134	19.86
	0.230	24.02		0.141	20.89
	0.250	26.06		0.150	22.21
				0.164	24.26
10¾	0.109	12.40		0.172	25.43
	0.120	13.64		0.179	26.45
	0.134	15.21		0.188	27.76
	0.141	15.99		0.203	29.94
	0.150	17.00		0.219	32.26
	0.164	18.56		0.230	33.86
	0.172	19.45		0.250	36.75
	0.179	20.23		0.281	41.21
	0.188	21.23		0.312	45.65

TABLE 3 Continued

Outside Diameter, in.	Nominal Wall Thickness, in.[B]	Weight per Unit Lengths, lb/ft[C]	Outside Diameter, in.[B]	Nominal Wall Thickness, in.[B]	Weight per Unit Lengths, lb/ft[C]
	0.203	22.89		0.344	50.22
	0.219	24.65		0.375	54.62
	0.230	25.87		0.438	63.50
	0.250	28.06		0.469	67.84
	0.279	31.23		0.500	72.16
	0.307	34.27			
	0.344	38.27	16	0.134	22.73
	0.141	23.90			
	0.150	25.42			
	0.164	27.76			
16	0.172	29.10	20	0.188	31.78
	0.179	30.27		0.219	46.31
				0.250	52.78
	0.188	30.61		0.281	59.23
	0.203	34.28		0.312	65.66
	0.219	36.95		0.344	72.28
	0.230	38.77		0.375	78.67
	0.250	42.09		0.438	91.59
	0.281	47.22		0.469	97.92
	0.312	52.32		0.500	104.23
	0.344	57.57			
	0.375	62.64	22	0.172	40.13
	0.438	72.86		0.188	43.84
	0.469	77.87		0.219	50.99
	0.500	82.85		0.250	58.13
				0.281	65.24
18	0.141	26.92		0.312	72.34
	0.172	32.78		0.375	86.69
	0.188	35.80		0.438	100.66
	0.219	41.63		0.469	107.95
	0.230	43.69		0.500	114.92
	0.250	47.44			
	0.281	53.23	24	0.172	43.81
	0.312	58.99		0.188	47.86
	0.344	64.93		0.219	55.67
	0.375	70.65		0.250	63.47
	0.438	82.23		0.281	71.25
	0.469	87.89		0.312	79.01
	0.500	93.54		0.375	94.71
				0.438	110.32
20	0.141	29.93		0.469	117.98
	0.172	36.46		0.500	125.62

[A] Subject to agreement between the manufacturer and the purchaser, sizes and weights per unit length other than those listed shall be permitted.
[B] 1 in. = 25.4 mm
[C] 1 lb/ft = 1.49 kg/m.

14. Workmanship, Finish, and Appearance

14.1 The finished pipe piles shall be reasonably straight and shall not contain imperfections in such number or of such character as to render the pipe unsuitable for pipe piles.

14.2 Surface imperfections having a depth not in excess of 25 % of the specified nominal wall thickness shall be acceptable. It shall be permissible to establish the depth of such imperfections by grinding or filing.

14.3 Surface imperfections having a depth in excess of 25 % of the specified nominal wall thickness shall be considered to be defects. It shall be permissible for defects not deeper than 33⅓ % of the specified nominal wall thickness to be repaired by welding, provided that the defect is completely removed prior to welding.

15. Ends

15.1 Pipe piles shall be furnished with plain ends. Unless otherwise specified, pipe piles shall have either flame–cut or machine–cut ends, with the burrs at the ends removed. Where ends are specified to be beveled, they shall be beveled to an angle of 30 +5, −0°, measured from a line drawn perpendicular to the axis of the pipe pile.

16. Number of Tests

16.1 One tension test shall be made on one length or fraction thereof of each size, or one piece of skelp representing each lot of 200 lengths or fraction thereof of each size.

16.2 A retest shall be allowed if the percentage of elongation of any test tension specimen is less than that prescribed in Tables 1 and 2 and any part of the fracture is more than ¾ in. (19 mm) from the center of the gage length for test specimens having a 2–in. (50 mm) gage length, or is outside of the middle third of the gage length for test specimens having an 8–in. (200 mm) gage length, as indicated by scribe scratches marked on the specimen before testing. A retest shall also be allowed if any part of the fracture is in an inside or outside surface imperfection.

16.3 It shall be permissible to discard any test specimen that shows defective machining or develops imperfections and substitute another test specimen.

17. Retests

17.1 If the results of the tension test representing any lot fail to conform to the applicable requirements prescribed in Tables 1 and 2, the lot shall be rejected or retested using two additional lengths from the lot, with each such test being required to conform to such specified requirements.

18. Test Specimens and Test Methods

18.1 The tension test specimens and test methods shall be in accordance with Test Methods and Definitions A370, especially Annex A2.

18.2 At the option of the manufacturer, the tension test specimens shall be longitudinal or transverse strip test specimens, with a gage length of 2 in. (50 mm) or 8 in. (200 mm), taken from the pipe or the skelp. Within their gage length, longitudinal strip test specimens shall be nominally 1½ in. (38 mm) wide, non-flattened, and with parallel sides.

18.3 For welded pipe piles, the tension test specimens shall be taken as follows:

18.3.1 For longitudinal–seam pipe piles, any longitudinal strip test specimens shall be taken from the pipe parallel to the pipe axis and 90° from the weld, or from the skelp at a corresponding location and orientation, and any transverse strip test specimens shall be taken from the pipe 90° to the pipe axis and 180° from the weld, or from the skelp at a corresponding location and orientation.

18.3.2 For helical-seam pipe piles, any longitudinal strip test specimens shall be taken from the pipe parallel to the pipe axis and at such a location that the center of the specimen is located at least a quarter of the distance between adjacent weld convolutions, or from the skelp at a corresponding location and orientation; and transverse specimens shall be taken from the pipe 90° to the pipe axis and at such a location that the center of the specimen is located approximately half the distance between adjacent weld convolutions, or from the skelp at a corresponding location and orientation.

18.4 Specimens shall be tested at room temperature.

19. Inspection

19.1 The inspector representing the purchaser shall have entry, at all times while work on the contract of the purchaser is being performed, to all parts of the manufacturer's works that concern the manufacture of the material ordered. The manufacturer shall afford the inspector all reasonable facilities to satisfy the inspector that the material is being furnished in accordance with the requirements of this specification and any other requirements specified in the purchase order. All tests and inspections shall be made at the place of manufacture prior to shipment, unless otherwise specified in the purchase order, and shall be so conducted as not to interfere unnecessarily with the operation of the works.

20. Rejection

20.1 It shall be permissible for the purchaser inspect the pipe piles received from the manufacturer and reject any pipe pile that does not meet the requirements of this specification and the purchase order, based upon the applicable inspection and test methods. The purchaser shall notify the manufacturer of any pipe pile that has been rejected, and the disposition of such pipe piles shall be subject to agreement between the manufacturer and the purchaser.

20.2 It shall be permissible for the purchaser to set aside any pipe pile that is found in fabrication or installation within the scope of this specification to be unsuitable for the intended end use, based on the requirements of this specification. The purchaser shall notify the manufacturer of any pipe pile that has been set aside. Such pipe piles shall be subject to mutual investigation as to the nature and severity of the deficiency and the forming or installation, or both, conditions involved. The disposition of such pipe piles shall be subject to agreement between the manufacturer and the purchaser.

21. Certification

21.1 Where specified in the purchase order, the manufacturer shall furnish a certificate of compliance stating that the pipe pile was manufactured, tested, and inspected in accordance with the requirements of this specification (including year date) and any requirements specified in the purchase order, and was found to meet such requirements, and shall furnish a test report containing the results of the applicable heat analyses, product analyses, and tension tests.

22. Product Marking

22.1 Each length of pipe pile shall be legibly marked by stenciling, stamping, or rolling to show: the name or brand of the manufacturer; the heat number; the process of manufacture (seamless, flash welded, fusion welded, or electric resistance welded), the type of helical seam (helical-lap or helical-butt), if applicable; the outside diameter, nominal wall thickness, length, and weight per unit length; the specification designation (year date not required); and the grade.

22.2 *Bar Coding*—In addition to the requirements in 22.1, it shall be permissible for bar coding to be used as a supplementary identification method; when a specific bar coding system is specified in the purchase order, that system shall be used.

23. Keywords

23.1 seamless steel pipe; steel piles; steel pipe; welded steel pipe

ASTM A252 − 98 (2007)

APPENDIX

(Nonmandatory Information)

X1. Minimum Permissible Pipe Wall Thicknesses on Inspection

X1.1 See Table X1.1 for minimum wall thicknesses.

TABLE X1.1 Table of Minimum Wall Thicknesses on Inspection for Nominal (Average) Pipe Wall Thicknesses

NOTE 1—The following equation, upon which this table is based, may be applied to calculate minimum wall thickness from nominal (average) wall thickness:

$$t_n \times 0.875 = t_m$$

where:
t_n = nominal wall thickness, in., and
t_m = minimum permissible wall thickness, in.

The wall thickness is expressed to three decimal places, with rounding being in accordance with Practice E29.

NOTE 2—This table is a master table covering some of the nominal wall thicknesses available in the purchase of different classifications of pipe, but it is not meant to imply that all of these nominal wall thicknesses are necessarily obtainable.

Nominal Wall Thickness (t_n), in.[A]	Minimum Permissible Wall Thickness on Inspection (t_m), in.[A]	Nominal Wall Thickness (t_n), in.[A]	Minimum Permissible Wall Thickness on Inspection (t_m), in.[A]	Nominal Wall Thickness (t_n), in.[A]	Minimum Permissible Wall Thickness on Inspection (t_m), in.[A]
0.068	0.060	0.276	0.242	0.674	0.590
0.088	0.077	0.277	0.242	0.687	0.601
0.091	0.080	0.279	0.244	0.719	0.629
0.095	0.083	0.280	0.245	0.750	0.656
0.109	0.095	0.281	0.246	0.812	0.710
0.113	0.099	0.294	0.257	0.843	0.738
0.119	0.104	0.300	0.262	0.864	0.756
0.120	0.105	0.307	0.269	0.875	0.766
0.125	0.109	0.308	0.270	0.906	0.793
0.126	0.110	0.312	0.273	0.937	0.820
0.133	0.116	0.318	0.278	0.968	0.847
0.134	0.117	0.322	0.282	1.000	0.875
0.140	0.122	0.330	0.289	1.031	0.902
0.141	0.123	0.337	0.295	1.062	0.929
0.145	0.127	0.343	0.300	1.093	0.956
0.147	0.129	0.344	0.301	1.125	0.984
0.150	0.131	0.358	0.313	1.156	1.012
0.154	0.135	0.365	0.319	1.218	1.066
0.156	0.136	0.375	0.328	1.250	1.094
0.164	0.143	0.382	0.334	1.281	1.121
0.172	0.150	0.400	0.350	1.312	1.148
0.179	0.157	0.406	0.355	1.343	1.175
0.187	0.164	0.432	0.378	1.375	1.203
0.188	0.164	0.436	0.382	1.406	1.230
0.191	0.167	0.437	0.382	1.438	1.258
0.200	0.175	0.438	0.383	1.500	1.312
0.203	0.178	0.469	0.410	1.531	1.340
0.216	0.189	0.500	0.438	1.562	1.367
0.218	0.191	0.531	0.465	1.593	1.394
0.219	0.192	0.552	0.483	1.750	1.531
0.226	0.198	0.562	0.492	1.781	1.558
0.230	0.201	0.593	0.519	1.812	1.586
0.237	0.207	0.600	0.525	1.968	1.722
0.250	0.219	0.625	0.547	2.062	1.804
0.258	0.226	0.656	0.574	2.343	2.050

[A] 1 in. = 25.4 mm

ASTM International takes no position respecting the validity of any patent rights asserted in connection with any item mentioned in this standard. Users of this standard are expressly advised that determination of the validity of any such patent rights, and the risk of infringement of such rights, are entirely their own responsibility.

This standard is subject to revision at any time by the responsible technical committee and must be reviewed every five years and if not revised, either reapproved or withdrawn. Your comments are invited either for revision of this standard or for additional standards and should be addressed to ASTM International Headquarters. Your comments will receive careful consideration at a meeting of the responsible technical committee, which you may attend. If you feel that your comments have not received a fair hearing you should make your views known to the ASTM Committee on Standards, at the address shown below.

This standard is copyrighted by ASTM International, 100 Barr Harbor Drive, PO Box C700, West Conshohocken, PA 19428-2959, United States. Individual reprints (single or multiple copies) of this standard may be obtained by contacting ASTM at the above address or at 610-832-9585 (phone), 610-832-9555 (fax), or service@astm.org (e-mail); or through the ASTM website (www.astm.org).

Designation: A325 − 09a$^{\varepsilon 1}$

American Association State Highway and Transportation Officials Standard AASHTO No.: M 164

Standard Specification for Structural Bolts, Steel, Heat Treated, 120/105 ksi Minimum Tensile Strength[1]

This standard is issued under the fixed designation A325; the number immediately following the designation indicates the year of original adoption or, in the case of revision, the year of last revision. A number in parentheses indicates the year of last reapproval. A superscript epsilon (ε) indicates an editorial change since the last revision or reapproval.

This standard has been approved for use by agencies of the Department of Defense.

ε^1 NOTE—Table 5 was editorially corrected in March 2010.

1. Scope*

1.1 This specification[2] covers two types of quenched and tempered steel heavy hex structural bolts having a minimum tensile strength of 120 ksi for sizes 1.0 in. and less and 105 ksi for sizes over 1.0 to 1½ in., inclusive.

1.2 The bolts are intended for use in structural connections. These connections are covered under the requirements of the Specification for Structural Joints Using ASTM A325 or A490 Bolts, approved by the Research Council on Structural Connections, endorsed by the American Institute of Steel Construction and by the Industrial Fastener Institute.[3]

1.3 The bolts are furnished in sizes ½ to 1½ in., inclusive. They are designated by type, denoting chemical composition as follows:

Type	Description
Type 1	Medium carbon, carbon boron, or medium carbon alloy steel.
Type 2	Withdrawn in November 1991.
Type 3	Weathering steel.

NOTE 1—Bolts for general applications, including anchor bolts, are covered by Specification A449. Also refer to Specification A449 for quenched and tempered steel bolts and studs with diameters greater than 1½ in. but with similar mechanical properties.

1.4 The values stated in inch-pound units are to be regarded as standard. No other units of measurement are included in this standard.

NOTE 2—A complete metric companion to Specification A325 has been developed—Specification A325M; therefore, no metric equivalents are presented in this specification.

1.5 This specification is applicable to heavy hex structural bolts only. For bolts of other configurations and thread lengths with similar mechanical properties, see Specification A449.

1.6 Terms used in this specification are defined in Terminology F1789.

1.7 The following safety hazard caveat pertains only to the test methods portion, Section 10, of this specification: *This standard does not purport to address all of the safety concerns, if any, associated with its use. It is the responsibility of the user of this standard to establish appropriate safety and health practices and determine the applicability of regulatory limitations prior to use.*

2. Referenced Documents

2.1 *ASTM Standards:*[4]

A194/A194M Specification for Carbon and Alloy Steel Nuts for Bolts for High Pressure or High Temperature Service, or Both

A449 Specification for Hex Cap Screws, Bolts and Studs, Steel, Heat Treated, 120/105/90 ksi Minimum Tensile Strength, General Use

A490 Specification for Structural Bolts, Alloy Steel, Heat Treated, 150 ksi Minimum Tensile Strength

A563 Specification for Carbon and Alloy Steel Nuts

A751 Test Methods, Practices, and Terminology for Chemical Analysis of Steel Products

B695 Specification for Coatings of Zinc Mechanically Deposited on Iron and Steel

D3951 Practice for Commercial Packaging

F436 Specification for Hardened Steel Washers

F606 Test Methods for Determining the Mechanical Properties of Externally and Internally Threaded Fasteners, Washers, Direct Tension Indicators, and Rivets

F788/F788M Specification for Surface Discontinuities of

---

[1] This specification is under the jurisdiction of ASTM Committee F16 on Fasteners and is the direct responsibility of Subcommittee F16.02 on Steel Bolts, Nuts, Rivets and Washers.

Current edition approved Dec. 1, 2009. Published December 2009. Originally approved in 1964. Last previous edition approved in 2009 as A325 – 09. DOI: 10.1520/A0325-09.

[2] For *ASME Boiler and Pressure Vessel Code* applications see related Specification SA-325 in Section II of that Code.

[3] Published by American Institute of Steel Construction (AISC), One E. Wacker Dr., Suite 700, Chicago, IL 60601-2001, http://www.aisc.org.

[4] For referenced ASTM standards, visit the ASTM website, www.astm.org, or contact ASTM Customer Service at service@astm.org. For *Annual Book of ASTM Standards* volume information, refer to the standard's Document Summary page on the ASTM website.

A Summary of Changes section appears at the end of this standard.

Copyright © ASTM International, 100 Barr Harbor Drive, PO Box C700, West Conshohocken, PA 19428-2959, United States.

Bolts, Screws, and Studs, Inch and Metric Series
F959 Specification for Compressible-Washer-Type Direct Tension Indicators for Use with Structural Fasteners
F1136 Specification for Zinc/Aluminum Corrosion Protective Coatings for Fasteners
F1470 Practice for Fastener Sampling for Specified Mechanical Properties and Performance Inspection
F1789 Terminology for F16 Mechanical Fasteners
F2329 Specification for Zinc Coating, Hot-Dip, Requirements for Application to Carbon and Alloy Steel Bolts, Screws, Washers, Nuts, and Special Threaded Fasteners
G101 Guide for Estimating the Atmospheric Corrosion Resistance of Low-Alloy Steels

2.2 *ASME Standards:*[5]
B 1.1 Unified Screw Threads
B 18.2.6 Fasteners for Use in Structural Applications
B 18.24 Part Identification Number (PIN) Code System Standard for B18 Fastener Products

3. Ordering Information

3.1 Orders for heavy hex structural bolts under this specification shall include the following:

3.1.1 Quantity (number of pieces of bolts and accessories).

3.1.2 Size, including nominal bolt diameter, thread pitch, and bolt length.

3.1.3 Name of product, heavy hex structural bolts.

3.1.4 When bolts threaded full length are required, Supplementary Requirement S1 shall be specified.

3.1.5 Type of bolt: Type 1 or 3. When type is not specified, either Type 1 or Type 3 shall be furnished at the supplier's option.

3.1.6 ASTM designation and year of issue.

3.1.7 Other components such as nuts, washers, and compressible washer-type direct-tension indicators, if required.

3.1.7.1 When such other components are specified to be furnished, also state "Nuts, washers, and direct tension indicators, or combination thereof, shall be furnished by lot number."

3.1.8 *Zinc Coating*—Specify the zinc coating process required, for example, hot dip, mechanically deposited, Zinc/Aluminum Corrosion Protective Coating or no preference (see 4.3).

3.1.9 *Other Finishes*—Specify other protective finish, if required.

3.1.10 Test reports, if required (see Section 13).

3.1.11 Supplementary or special requirements, if required.

3.1.12 For establishment of a part identifying system, see ASME B 18.24.

NOTE 3—A typical ordering description follows: 1000 pieces 1⅛-7 UNC in. dia × 4 in. long heavy hex structural bolt, *Type 1 ASTM A325-02*, each with one hardened washer, ASTM F436 Type 1, and one heavy hex nut, ASTM A563 Grade DH. Each component hot-dip zinc-coated. Nuts lubricated.

3.2 *Recommended Nuts*:

3.2.1 Nuts conforming to the requirements of Specification A563 are the recommended nuts for use with Specification A325 heavy hex structural bolts. The nuts shall be of the class and have a surface finish for each type of bolt as follows:

Bolt Type and Finish	Nut Class and Finish
1, plain (noncoated)	A563-C, C3, D, DH, DH3, plain
1, zinc coated	A563-DH, zinc coated
1, coated in accordance with Specification F1136, Grade 3	A563–DH coated in accordance with Specification F1136, Grade 5
3, plain	A563-C3, DH3, plain

3.2.2 Alternatively, nuts conforming to Specification A194/A194M Gr. 2H are considered a suitable substitute for use with Specification A325 Type 1 heavy hex structural bolts.

3.2.3 When Specification A194/A194M Gr. 2H zinc-coated nuts are supplied, the zinc coating, overtapping, lubrication, and rotational capacity testing shall be in accordance with Specification A563.

3.3 *Recommended Washers*:

3.3.1 Washers conforming to Specification F436 are the recommended washers for use with Specification A325 heavy hex structural bolts. The washers shall have a surface finish for each type of bolt as follows:

Bolt Type and Finish	Washer Finish
1, plain (uncoated)	plain (uncoated)
1, zinc coated	zinc coated
1, coated in accordance with Specification F1136, Grade 3	coated in accordance with Specification F1136, Grade 3
3, plain	weathering steel, plain

3.4 *Other Accessories*:

3.4.1 When compressible washer type direct tension indicators are specified to be used with these bolts, they shall conform to Specification F959, Type 325.

4. Materials and Manufacture

4.1 *Heat Treatment*:

4.1.1 Type 1 bolts produced from medium carbon steel shall be quenched in a liquid medium from the austenitizing temperature. Type 1 bolts produced from medium carbon steel to which chromium, nickel, molybdenum, or boron were intentionally added shall be quenched only in oil from the austenitizing temperature.

4.1.2 Type 3 bolts shall be quenched only in oil from the austenitizing temperature.

4.1.3 Type 1 bolts, regardless of the steel used, and Type 3 bolts shall be tempered by reheating to not less than 800°F.

4.2 *Threading*—Threads shall be cut or rolled.

4.3 *Zinc Coatings, Hot-Dip and Mechanically Deposited, Zinc/Aluminum Corrosion Protective Coating*:

4.3.1 When zinc-coated fasteners are required, the purchaser shall specify the zinc coating process, for example, hot dip, mechanically deposited, Zinc/Aluminum Corrosion Protective Coating, or no preference.

4.3.2 When hot-dip is specified, the fasteners shall be zinc-coated by the hot-dip process and the coating shall

---

[5] Available from American Society of Mechanical Engineers (ASME), ASME International Headquarters, Three Park Ave., New York, NY 10016-5990, http://www.asme.org.

conform to the coating weight/thickness and performance requirements of Specification F2329.

4.3.3 When mechanically deposited is specified, the fasteners shall be zinc-coated by the mechanical deposition process and the coating shall conform to the coating weight/thickness and performance requirements of Class 55 of Specification B695.

4.3.4 When Zinc/Aluminum Corrosion Protective Coating is specified, the coating shall conform to the coating weight/thickness and performance requirements of Grade 3 of Specification F1136.

4.3.5 When no preference is specified, the supplier shall furnish either a hot-dip zinc coating in accordance with Specification F2329, a mechanically deposited zinc coating in accordance with Specification B695, Class 55, or a Zinc/Aluminum Corrosion Protective Coating in accordance with Specification F1136, Grade 3. Threaded components (bolts and nuts) shall be coated by the same zinc-coating process and the supplier's option is limited to one process per item with no mixed processes in a lot.

4.4 *Lubrication*—When zinc-coated nuts are ordered with the bolts, the nuts shall be lubricated in accordance with Specification A563, Supplementary Requirement S1, to minimize galling.

4.5 *Secondary Processing*:

4.5.1 If any processing, which can affect the mechanical properties or performance of the bolts, is performed after the initial testing, the bolts shall be retested for all specified mechanical properties and performance requirements affected by the reprocessing.

4.5.2 When the secondary process is heat treatment, the bolts shall be tested for all specified mechanical properties. Hot dip zinc-coated bolts shall be tested for all specified mechanical properties and rotational capacity. If zinc-coated nuts are relubricated after the initial rotational capacity tests, the assemblies shall be retested for rotational capacity.

5. Chemical Composition

5.1 Type 1 bolts shall be plain carbon steel, carbon boron steel, alloy steel or alloy boron steel at the manufacturer's option, conforming to the chemical composition specified in Table 1.

5.2 Type 3 bolts shall be weathering steel and shall conform to one of the chemical compositions specified in Table 2. The selection of the chemical composition, A, B, C, D, E, or F, shall be at the option of the bolt manufacturer. See Guide G101 for methods of estimating the atmospheric corrosion resistance of low alloy steels.

5.3 Product analyses made on finished bolts representing each lot shall conform to the product analysis requirements specified in Tables 1 and 2, as applicable.

5.4 Heats of steel to which bismuth, selenium, tellurium, or lead has been intentionally added shall not be permitted for bolts.

5.5 Compliance with 5.4 shall be based on certification that heats of steel having any of the listed elements intentionally added were not used to produce the bolts.

5.6 Chemical analyses shall be performed in accordance with Test Methods, Practices, and Terminology A751.

TABLE 1 Chemical Requirements for Type 1 Bolts

Element	Carbon Steel	
	Heat Analysis	Product Analysis
Carbon	0.30–0.52	0.28–0.55
Manganese, min	0.60	0.57
Phosphorus, max	0.040	0.048
Sulfur, max	0.050	0.058
Silicon	0.15–0.30	0.13–0.32

Element	Carbon Boron Steel	
	Heat Analysis	Product Analysis
Carbon	0.30–0.52	0.28–0.55
Manganese, min	0.60	0.57
Phosphorus, max	0.040	0.048
Sulfur, max	0.050	0.058
Silicon	0.10–0.30	0.08–0.32
Boron	0.0005–0.003	0.0005–0.003

Element	Alloy Steel	
	Heat Analysis	Product Analysis
Carbon	0.30–0.52	0.28–0.55
Manganese, min	0.60	0.57
Phosphorus, max	0.035	0.040
Sulfur, max	0.040	0.045
Silicon	0.15–0.35	0.13–0.37
Alloying Elements	[A]	[A]

	Alloy Boron Steel	
	Heat Analysis	Product Analysis
Carbon	0.30–0.52	0.28–0.55
Manganese, min	0.60	0.57
Phosphorus, max	0.035	0.040
Sulfur, max	0.040	0.045
Silicon	0.15–0.35	0.13–0.37
Boron	0.0005–0.003	0.0005–0.003
Alloying Elements	[A]	[A]

[A] Steel, as defined by the American Iron and Steel Institute, shall be considered to be alloy when the maximum of the range given for the content of alloying elements exceeds one or more of the following limits: Manganese, 1.65 %; silicon, 0.60 %; copper, 0.60 % or in which a definite range or a definite minimum quantity of any of the following elements is specified or required within the limits of the recognized field of constructional alloy steels: aluminum, chromium up to 3.99 %, cobalt, columbium, molybdenum, nickel, titanium, tungsten, vanadium, zirconium, or any other alloying elements added to obtain a desired alloying effect.

6. Mechanical Properties

6.1 *Hardness*—The bolts shall conform to the hardness specified in Table 3.

6.2 *Tensile Properties*:

6.2.1 Except as permitted in 6.2.1.1 for long bolts and 6.2.1.2 for short bolts, sizes 1.00 in. and smaller having a length of 2¼ D and longer, and sizes larger than 1.00 in. having a length of 3D and longer, shall be wedge tested full size and shall conform to the minimum wedge tensile load and proof load or alternative proof load specified in Table 4. The load achieved during proof load testing shall be equal to or greater than the specified proof load.

6.2.1.1 When the length of the bolt makes full-size testing impractical, machined specimens shall be tested and shall conform to the requirements specified in Table 5. When bolts

TABLE 2 Chemical Requirements for Type 3 Heavy Hex Structural Bolts[A]

	Composition, %					
	Type 3 Bolts[A]					
Element	A	B	C	D	E	F
Carbon:						
Heat analysis	0.33–0.40	0.38–0.48	0.15–0.25	0.15–0.25	0.20–0.25	0.20–0.25
Product analysis	0.31–0.42	0.36–0.50	0.14–0.26	0.14–0.26	0.18–0.27	0.19–0.26
Manganese:						
Heat analysis	0.90–1.20	0.70–0.90	0.80–1.35	0.40–1.20	0.60–1.00	0.90–1.20
Product analysis	0.86–1.24	0.67–0.93	0.76–1.39	0.36–1.24	0.56–1.04	0.86–1.24
Phosphorus:						
Heat analysis	0.035 max	0.06–0.12	0.035 max	0.035 max	0.035 max	0.035 max
Product analysis	0.040 max	0.06–0.125	0.040 max	0.040 max	0.040 max	0.040 max
Sulfur:						
Heat analysis	0.040 max	0.040 max	0.040 max	0.040 max	0.040 max	0.040 max
Product analysis	0.045 max	0.045 max	0.045 max	0.045 max	0.045 max	0.045 max
Silicon:						
Heat analysis	0.15–0.35	0.30–0.50	0.15–0.35	0.25–0.50	0.15–0.35	0.15–0.35
Product analysis	0.13–0.37	0.25–0.55	0.13–0.37	0.20–0.55	0.13–0.37	0.13–0.37
Copper:						
Heat analysis	0.25–0.45	0.20–0.40	0.20–0.50	0.30–0.50	0.30–0.60	0.20–0.40
Product analysis	0.22–0.48	0.17–0.43	0.17–0.53	0.27–0.53	0.27–0.63	0.17–0.43
Nickel:						
Heat analysis	0.25–0.45	0.50–0.80	0.25–0.50	0.50–0.80	0.30–0.60	0.20–0.40
Product analysis	0.22–0.48	0.47–0.83	0.22–0.53	0.47–0.83	0.27–0.63	0.17–0.43
Chromium:						
Heat analysis	0.45–0.65	0.50–0.75	0.30–0.50	0.50–1.00	0.60–0.90	0.45–0.65
Product analysis	0.42–0.68	0.47–0.83	0.27–0.53	0.45–1.05	0.55–0.95	0.42–0.68
Vanadium:						
Heat analysis	[B]	[B]	0.020 min	[B]	[B]	[B]
Product analysis	[B]	[B]	0.010 min	[B]	[B]	[B]
Molybdenum:						
Heat analysis	[B]	0.06 max	[B]	0.10 max	[B]	[B]
Product analysis	[B]	0.07 max	[B]	0.11 max	[B]	[B]
Titanium:						
Heat analysis	[B]	[B]	[B]	0.05 max	[B]	[B]
Product analysis	[B]	[B]	[B]	0.06 max	[B]	[B]

[A] A, B, C, D, E, and F are classes of material used for Type 3 bolts. Selection of a class shall be at the option of the bolt manufacturer.
[B] These elements are not specified or required.

TABLE 3 Hardness Requirements for Bolts

Bolt Size, in.	Bolt Length, in.	Brinell Min	Brinell Max	Rockwell C Min	Rockwell C Max
½ to 1, incl	Less than 2D	253	319	25	34
	2D and over	...	319	...	34
1⅛ to 1½, incl	Less than 3D	223	286	19	30
	3D and over	...	286	...	30

D = Nominal diameter or thread size.

are tested by both full-size and machined specimen methods, the full-size test shall take precedence.

6.2.1.2 Sizes 1.00 in. and smaller having a length shorter than 2¼ D down to 2D, inclusive, that cannot be wedge tensile tested shall be axially tension tested full size and shall conform to the minimum tensile load and proof load or alternate proof load specified in Table 4. Sizes 1.00 in. and smaller having a length shorter than 2D and sizes larger than 1.00 in. with lengths shorter than 3D that cannot be axially tensile tested shall be qualified on the basis of hardness.

6.2.2 For bolts on which both hardness and tension tests are performed, acceptance based on tensile requirements shall take precedence in the event of low hardness readings.

6.3 *Rotational Capacity Test:*

6.3.1 *Definition*—The rotational capacity test is intended to evaluate the presence of a lubricant, the efficiency of the lubricant, and the compatibility of assemblies as represented by the components selected for testing.

6.3.2 *Requirement*—Zinc-coated bolts, zinc-coated washers, and zinc-coated and lubricated nuts tested full size in an assembled joint or tension measuring device, in accordance with 10.2, shall not show signs of failure when subjected to the nut rotation in Table 6. The test shall be performed by the responsible party (see Section 14) prior to shipment after zinc coating and lubrication of nuts (see 10.2 and Note 4).

TABLE 4 Tensile Load Requirements for Bolts Tested Full-Size

Bolt Size, Threads per Inch, and Series Designation	Stress Area,[A] in.^{2}	Tensile Load,[B] min, lbf	Proof Load,[B] Length Measurement Method	Alternative Proof Load,[B] Yield Strength Method
Column 1	Column 2	Column 3	Column 4	Column 5
½ –13 UNC	0.142	17 050	12 050	13 050
⅝ –11 UNC	0.226	27 100	19 200	20 800
¾ –10 UNC	0.334	40 100	28 400	30 700
⅞ –9 UNC	0.462	55 450	39 250	42 500
1–8 UNC	0.606	72 700	51 500	55 750
1⅛ –7 UNC	0.763	80 100	56 450	61 800
1¼ –7 UNC	0.969	101 700	71 700	78 500
1⅜ –6 UNC	1.155	121 300	85 450	93 550
1½ –6 UNC	1.405	147 500	104 000	113 800

[A] The stress area is calculated as follows:

$$As = 0.7854\,[D - (0.9743/n)]^2$$

where:
As = stress area, in.^{2},
D = nominal bolt size, and
n = threads per inch.

[B] Loads tabulated are based on the following:

Bolt Size, in.	Column 3	Column 4	Column 5
½ to 1, incl	120 000 psi	85 000 psi	92 000 psi
1⅛ to 1½, incl	105 000 psi	74 000 psi	81 000 psi

6.3.3 *Acceptance Criterion*—The bolt and nut assembly shall be considered as non-conforming if the assembly fails to pass any one of the following specified requirements:

6.3.3.1 Inability to install the assembly to the nut rotation in Table 6.

6.3.3.2 Inability to remove the nut after installing to the rotation specified in Table 6.

6.3.3.3 Shear failure of the threads as determined by visual examination of bolt and nut threads following removal.

6.3.3.4 Torsional or torsional/tension failure of the bolt. Elongation of the bolt, in the threads between the nut and bolt head, is to be expected at the required rotation and is not to be classified as a failure.

7. Dimensions

7.1 *Head and Body*:

7.1.1 The bolts shall conform to the dimensions for heavy hex structural bolts specified in ASME B 18.2.6.

7.1.2 The thread length shall not be changed except as provided in Supplementary Requirement S1. Bolts with thread lengths other than those required by this specification shall be ordered under Specification A449.

7.2 *Threads*:

7.2.1 *Uncoated*—Threads shall be the Unified Coarse Thread Series as specified in ASME B 1.1, and shall have Class 2A tolerances.

7.2.2 *Coated*—Unless otherwise specified, zinc-coated bolts to be used with zinc-coated nuts or tapped holes that are tapped oversize, in accordance with Specification A563, shall have Class 2A threads before hot-dip or mechanically deposited zinc coating. After zinc coating, the maximum limits of pitch and major diameter shall not exceed the Class 2A limit by more than the following amounts:

Nominal Bolt Diameter (in.)	Oversize Limit, in.[A] Hot-Dip Zinc	Mechanical Zinc
½	0.018	0.012
⁹⁄₁₆, ⅝, ¾	0.020	0.013
⅞	0.022	0.015
1 to 1¼	0.024	0.016
1⅜, 1½	0.027	0.018

[A] Hot-dip zinc nuts are tapped oversize after coating, and mechanical zinc-coated nuts are tapped oversize before coating.

7.2.3 The gaging limit for bolts shall be verified during manufacture. In case of dispute, a calibrated thread ring gage of the same size as the oversize limit in 7.2.2 (Class X tolerance, gage tolerance plus) shall be used to verify compliance. The gage shall assemble with hand effort following application of light machine oil to prevent galling and damage to the gage. These inspections, when performed to resolve controversy, shall be conducted at the frequency specified in the quality assurance provisions of ASME B 18.2.6.

8. Workmanship

8.1 The allowable limits, inspection, and evaluation of the surface discontinuities, quench cracks, forging cracks, head bursts, shear bursts, seams, folds, thread laps, voids, tool marks, nicks, and gouges shall be in accordance with Specification F788/F788M.

9. Number of Tests and Retests

9.1 *Testing Responsibility*:

9.1.1 Each lot shall be tested by the manufacturer prior to shipment in accordance with the lot identification control quality assurance plan in 9.2 through 9.5.

9.1.2 When bolts are furnished by a source other than the manufacturer, the Responsible Party as defined in 14 shall be responsible for assuring all tests have been performed and the bolts comply with the requirements of this specification (see 4.5).

9.2 *Purpose of Lot Inspection*—The purpose of a lot inspection program is to ensure that each lot conforms to the requirements of this specification. For such a plan to be fully effective it is essential that secondary processors, distributors, and purchasers maintain the identification and integrity of each lot until the product is installed.

9.3 *Lot Method*—All bolts shall be processed in accordance with a lot identification-control quality assurance plan. The manufacturer, secondary processors, and distributors shall identify and maintain the integrity of each production lot of bolts from raw-material selection through all processing operations and treatments to final packing and shipment. Each lot shall be assigned its own lot-identification number, each lot shall be tested, and the inspection test reports for each lot shall be retained.

9.4 *Lot Definition*—A lot shall be a quantity of uniquely identified heavy hex structural bolts of the same nominal size and length produced consecutively at the initial operation from

TABLE 5 Tensile Strength Requirements for Specimens Machined from Bolts

Bolt Diameter, in.	Tensile Strength, min, psi (MPa)	Yield Strength, min, psi (MPa)	Elongation, in 4D, min, %	Reduction of Area, min, %
½ to 1, incl.	120 000 (825)	92 000 (635)	14	35
Over 1 to 1½	105 000 (725)	81 000 (560)	14	35[†]

[†] Table alignment was editorially corrected in March 2010

TABLE 6 Rotational Capacity Test for Zinc-Coated Bolts

Bolt Length, in.	Nominal Nut Rotation, degrees (turn)
Up to and including 4 × dia	240 (⅔)
Over 4 × dia, but not exceeding 8 × dia	360 (1)
Over 8 × dia, but not exceeding 12 × dia	420 (1⅙)
Over 12 × dia.	Test not applicable

a single mill heat of material and processed at one time, by the same process, in the same manner so that statistical sampling is valid. The identity of the lot and lot integrity shall be maintained throughout all subsequent operations and packaging.

9.5 *Number of Tests*—The minimum number of tests from each lot for the tests specified below shall be as follows:

Tests	Number of Tests in Accordance With
Hardness, tensile strength, proof load, and rotational capacity	Practice F1470
Coating weight/thickness	The referenced coating specification[A]
Surface discontinuities	Specification F788/F788M
Dimensions and thread fit	ASME B 18.2.6

[A] Practice F1470 applies if the coating specification does not specify a testing frequency.

10. Test Methods

10.1 *Tensile, Proof Load, and Hardness*:

10.1.1 Tensile, proof load, and hardness tests shall be conducted in accordance with Test Methods F606.

10.1.2 Tensile strength shall be determined using the Wedge or Axial Tension Testing Method of Full Size Product Method or the Machined Test Specimens Method depending on size and length as specified in 6.2.1-6.2.2. Fracture on full-size tests shall be in the body or threads of the bolt without a fracture at the junction of the head and body.

10.1.3 Proof load shall be determined using Method 1, Length Measurement, or Method 2, Yield Strength, at the option of the manufacturer.

10.2 *Rotational Capacity*—The zinc-coated bolt shall be placed in a steel joint or tension measuring device and assembled with a zinc-coated washer and a zinc-coated and lubricated nut with which the bolt is intended to be used (see Note 4). The nut shall have been provided with the lubricant described in the last paragraph of the Manufacturing Processes section of Specification A563. The joint shall be one or more flat structural steel plates or fixture stack up with a total thickness, including the washer, such that 3 to 5 full threads of the bolt are located between the bearing surfaces of the bolt head and nut. The hole in the joint shall have the same nominal diameter as the hole in the washer. The initial tightening of the nut shall produce a load in the bolt not less than 10 % of the specified proof load. After initial tightening, the nut position shall be marked relative to the bolt, and the rotation shown in Table 6 shall be applied. During rotation, the bolt head shall be restrained from turning. After the tightening rotation has been applied, the assembly shall be taken apart and examined for compliance with 6.3.3.

Note 4—Rotational capacity tests shall apply only to matched assembly lots that contain one A325 bolt, one A563 lubricated nut, and one F436 washer that have been zinc coated in accordance with either Specifications F2329 or B695. Both the bolt and nut components of the matched assembly shall be zinc coated using the same process.

11. Inspection

11.1 If the inspection described in 11.2 is required by the purchaser, it shall be specified in the inquiry and contract or order.

11.2 The purchaser's representative shall have free entry to all parts of the manufacturer's works, or supplier's place of business, that concern the manufacture or supply of the material ordered. The manufacturer or supplier shall afford the purchaser's representative all reasonable facilities to satisfy him that the material is being furnished in accordance with this specification. All tests and inspections required by the specification that are requested by the purchaser's representative shall be made before shipment, and shall be conducted as not to interfere unnecessarily with the operation of the manufacturer's works or supplier's place of business.

12. Rejection and Rehearing

12.1 Disposition of nonconforming bolts shall be in accordance with the Practice F1470 section titled "Disposition of Nonconforming Lots."

13. Certification

13.1 When specified on the purchase order, the manufacturer or supplier, whichever is the responsible party as defined in Section 14, shall furnish the purchaser a test reports that includes the following:

13.1.1 Heat analysis, heat number, and a statement certifying that heats having the elements listed in 5.4 intentionally added were not used to produce the bolts,

13.1.2 Results of hardness, tensile, and proof load tests,

13.1.3 Results of rotational capacity tests. This shall include the test method used (solid plate or tension measuring device); and the statement " Nuts lubricated" for zinc-coated nuts when shipped with zinc-coated bolts,

13.1.4 Zinc coating measured coating weight/thickness for coated bolts,

13.1.5 Statement of compliance of visual inspection for surface discontinuities (Section 8),

13.1.6 Statement of compliance with dimensional and thread fit requirements,

13.1.7 Lot number and purchase order number,

13.1.8 Complete mailing address of responsible party, and

13.1.9 Title and signature of the individual assigned certification responsibility by the company officers.

13.2 Failure to include all the required information on the test report shall be cause for rejection.

14. Responsibility

14.1 The party responsible for the fastener shall be the organization that supplies the fastener to the purchaser.

15. Product Marking

15.1 *Manufacturer's Identification*—All Type 1 and 3 bolts shall be marked by the manufacturer with a unique identifier to identify the manufacturer or private label distributor, as appropriate.

15.2 *Grade Identification*:

15.2.1 Type 1 bolts shall be marked "A325."

15.2.2 Type 3 bolts shall be marked "A325" with the "A325" underlined. The use of additional distinguishing marks to indicate that the bolts are weathering steel shall be at the manufacturer's option.

15.3 *Marking Location and Methods*—All marking shall be located on the top of the bolt head and shall be either raised or depressed at the manufacturer's option.

15.4 *Acceptance Criteria*—Bolts which are not marked in accordance with these provisions shall be considered nonconforming and subject to rejection.

15.5 Type and manufacturer's or private label distributor's identification shall be separate and distinct. The two identifications shall preferably be in different locations and, when on the same level, shall be separated by at least two spaces.

16. Packaging and Package Marking

16.1 *Packaging*:

16.1.1 Unless otherwise specified, packaging shall be in accordance with Practice D3951.

16.1.2 When zinc coated nuts are included on the same order as zinc coated bolts, the bolts and nuts shall be shipped in the same container.

16.1.3 When special packaging requirements are required, they shall be defined at the time of the inquiry and order.

16.2 *Package Marking*:

16.2.1 Each shipping unit shall include or be plainly marked with the following information:

16.2.1.1 ASTM designation and type,

16.2.1.2 Size,

16.2.1.3 Name and brand or trademark of the manufacturer,

16.2.1.4 Number of pieces,

16.2.1.5 Lot number; when nuts, washers or direct tension indicators, or combination thereof, are ordered with A325 heavy hex structural bolts, the shipping unit shall be marked with the lot number in addition to the marking required by the applicable product specification,

16.2.1.6 Purchase order number, and

16.2.1.7 Country of origin.

17. Keywords

17.1 bolts; carbon steel; steel; structural; weathering steel

---

SUPPLEMENTARY REQUIREMENTS

The following supplementary requirements shall apply only when specified by the purchaser in the contract or order. Details of these supplementary requirements shall be agreed upon in writing between the manufacturer and purchaser. Supplementary requirements shall in no way negate any requirement of the specification itself.

S1. Bolts Threaded Full Length

S1.1 Bolts with nominal lengths equal to or shorter than four times the nominal bolt diameter shall be threaded full length. Bolts need not have a shoulder, and the distance from the underhead bearing surface to the first complete (full form) thread, as measured with a GO thread ring gage, assembled by hand as far as the thread will permit, shall not exceed the length of 2½ threads for bolt sizes 1 in. and smaller, and 3½ threads for bolt sizes larger than 1 in.

S1.2 Bolts shall be marked in accordance with Section 15, except that the symbol shall be " A325 T" instead of "A325."

A325 – 09aᵉ¹

SUMMARY OF CHANGES

Committee F16 has identified the location of selected changes to this standard since the last issue (A325–07a) that may impact the use of this standard. (Approved Dec. 1, 2009)

(*1*) 6.2 and Table 3 and Table 4 were revised to clarify testing requirements for large diameter bolts with short lengths.

Committee F16 has identified the location of selected changes to this standard since the last issue (A325–07a) that may impact the use of this standard. (Approved Jan. 1, 2009)

(*1*) *Revised*—4.3 to add provision for specifying Zinc/Aluminum Corrosion Protective Coating conforming to Specification F1136.

(*2*) In 8.1 deleted Note 4 in regard to non injurious bursts.

ASTM International takes no position respecting the validity of any patent rights asserted in connection with any item mentioned in this standard. Users of this standard are expressly advised that determination of the validity of any such patent rights, and the risk of infringement of such rights, are entirely their own responsibility.

This standard is subject to revision at any time by the responsible technical committee and must be reviewed every five years and if not revised, either reapproved or withdrawn. Your comments are invited either for revision of this standard or for additional standards and should be addressed to ASTM International Headquarters. Your comments will receive careful consideration at a meeting of the responsible technical committee, which you may attend. If you feel that your comments have not received a fair hearing you should make your views known to the ASTM Committee on Standards, at the address shown below.

This standard is copyrighted by ASTM International, 100 Barr Harbor Drive, PO Box C700, West Conshohocken, PA 19428-2959, United States. Individual reprints (single or multiple copies) of this standard may be obtained by contacting ASTM at the above address or at 610-832-9585 (phone), 610-832-9555 (fax), or service@astm.org (e-mail); or through the ASTM website (www.astm.org). Permission rights to photocopy the standard may also be secured from the ASTM website (www.astm.org/COPYRIGHT/).

Designation: A333/A333M – 05

Standard Specification for
Seamless and Welded Steel Pipe for Low-Temperature Service[1]

This standard is issued under the fixed designation A333/A333M; the number immediately following the designation indicates the year of original adoption or, in the case of revision, the year of last revision. A number in parentheses indicates the year of last reapproval. A superscript epsilon (ε) indicates an editorial change since the last revision or reapproval.

This standard has been approved for use by agencies of the Department of Defense.

1. Scope*

1.1 This specification[2] covers nominal (average) wall seamless and welded carbon and alloy steel pipe intended for use at low temperatures. Several grades of ferritic steel are included as listed in Table 1. Some product sizes may not be available under this specification because heavier wall thicknesses have an adverse affect on low-temperature impact properties.

1.2 Supplementary Requirement S1 of an optional nature is provided. This shall apply only when specified by the purchaser.

1.3 The values stated in either inch-pound units or SI units are to be regarded separately as standard. Within the text, the SI units are shown in brackets. The values stated in each system are not exact equivalents; therefore, each system must be used independently of the other. Combining values from the two systems may result in nonconformance with the specification. The inch-pound units shall apply unless the "M" designation of this specification is specified in the order.

NOTE 1—The dimensionless designator NPS (nominal pipe size) has been substituted in this standard for such traditional terms as "nominal diameter," "size," and "nominal size."

2. Referenced Documents

2.1 *ASTM Standards:*[3]
A370 Test Methods and Definitions for Mechanical Testing of Steel Products

A999/A999M Specification for General Requirements for Alloy and Stainless Steel Pipe
A671 Specification for Electric-Fusion-Welded Steel Pipe for Atmospheric and Lower Temperatures
E23 Test Methods for Notched Bar Impact Testing of Metallic Materials

3. Ordering Information

3.1 Orders for material under this specification should include the following, as required, to describe the material adequately:

3.1.1 Quantity (feet, centimetres, or number of lengths),
3.1.2 Name of material (seamless or welded pipe),
3.1.3 Grade (Table 1),
3.1.4 Size (NPS or outside diameter and schedule number of average wall thickness),
3.1.5 Lengths (specific or random) (Section 9), (see the Permissible Variations in Length section of Specification A999/A999M),
3.1.6 End finish (see the Ends section of Specification A999/A999M),
3.1.7 Optional requirements, (see the Heat Analysis requirement in the Chemical Composition section of A999/A999M, the Repair by Welding section, and the section on Nondestructive Test Requirements),
3.1.8 Test report required, (see the Certification section of Specification A999/A999M),
3.1.9 Specification designation, and
3.1.10 Special requirements or exceptions to this specification.

4. Materials and Manufacture

4.1 *Manufacture*—The pipe shall be made by the seamless or welding process with the addition of no filler metal in the welding operation. Grade 4 shall be made by the seamless process.

NOTE 2—For electric-fusion-welded pipe, with filler metal added, see Specification A671.

---

[1] This specification is under the jurisdiction of ASTM Committee A01 on Steel, Stainless Steel and Related Alloys and is the direct responsibility of Subcommittee A01.10 on Stainless and Alloy Steel Tubular Products.
Current edition approved March 1, 2005. Published March 2005. Originally approved in 1950. Last previous edition approved in 2004 as A333/A333M – 04a. DOI: 10.1520/A0333_A0333M-05.
[2] For ASME Boiler and Pressure Vessel Code applications see related Specification SA-333 in Section II of that Code.
[3] For referenced ASTM standards, visit the ASTM website, www.astm.org, or contact ASTM Customer Service at service@astm.org. For *Annual Book of ASTM Standards* volume information, refer to the standard's Document Summary page on the ASTM website.

*A Summary of Changes section appears at the end of this standard.

Copyright © ASTM International, 100 Barr Harbor Drive, PO Box C700, West Conshohocken, PA 19428-2959, United States.

A333/A333M – 05

TABLE 1 Chemical Requirements

Element	Composition, %								
	Grade 1[A]	Grade 3	Grade 4	Grade 6[A]	Grade 7	Grade 8	Grade 9	Grade 10	Grade 11
Carbon, max	0.30	0.19	0.12	0.30	0.19	0.13	0.20	0.20	0.10
Manganese	0.40–1.06	0.31–0.64	0.50–1.05	0.29–1.06	0.90 max	0.90 max	0.40–1.06	1.15–1.50	0.60 max
Phosphorus, max	0.025	0.025	0.025	0.025	0.025	0.025	0.025	0.035	0.025
Sulfur, max	0.025	0.025	0.025	0.025	0.025	0.025	0.025	0.015	0.025
Silicon	...	0.18–0.37	0.08–0.37	0.10 min	0.13–0.32	0.13–0.32	...	0.10–0.35	0.35 max
Nickel	...	3.18–3.82	0.47–0.98	...	2.03–2.57	8.40–9.60	1.60–2.24	0.25 max	35.0–37.0
Chromium	...	...	0.44–1.01	...	...	...	...	0.15 max	0.50 max
Copper	...	...	0.40–0.75	...	...	...	0.75–1.25	0.15 max	...
Aluminum	...	...	0.04–0.30	...	...	...	...	0.06 max	...
Vanadium, max	...	...	...	...	...	...	...	0.12	...
Columbium, max	...	...	...	...	...	...	...	0.05	...
Molybdenum, max	...	...	...	...	...	...	...	0.05	0.50 max
Cobalt	...	...	...	...	...	...	...	...	0.50 max

[A] For each reduction of 0.01 % carbon below 0.30 %, an increase of 0.05 % manganese above 1.06 % would be permitted to a maximum of 1.35 % manganese.

4.2 *Heat Treatment*:

4.2.1 All seamless and welded pipe, other than Grades 8 and 11, shall be treated to control their microstructure in accordance with one of the following methods:

4.2.1.1 Normalize by heating to a uniform temperature of not less than 1500 °F [815 °C] and cool in air or in the cooling chamber of an atmosphere controlled furnace.

4.2.1.2 Normalize as in 4.2.1.1, and, at the discretion of the manufacturer, reheat to a suitable tempering temperature.

4.2.1.3 For the seamless process only, reheat and control hot working and the temperature of the hot-finishing operation to a finishing temperature range from 1550 to 1750 °F [845 to 945 °C] and cool in air or in a controlled atmosphere furnace from an initial temperature of not less than 1550 °F [845 °C].

4.2.1.4 Treat as in 4.2.1.3 and, at the discretion of the manufacturer, reheat to a suitable tempering temperature.

4.2.1.5 Seamless pipe of Grades 1, 6, and 10 may be heat treated by heating to a uniform temperature of not less than 1500 °F [815 °C], followed by quenching in liquid and reheating to a suitable tempering temperature, in place of any of the other heat treatments provided for in 4.2.1.

4.2.2 Grade 8 pipe shall be heat treated by the manufacturer by either of the following methods:

4.2.2.1 *Quenched and Tempered*—Heat to a uniform temperature of 1475 ± 25 °F [800 ± 15 °C]; hold at this temperature for a minimum time in the ratio of 1 h/in. [2 min/mm] of thickness, but in no case less than 15 min; quench by immersion in circulating water. Reheat until the pipe attains a uniform temperature within the range from 1050 to 1125 °F [565 to 605 °C]; hold at this temperature for a minimum time in the ratio of 1 h/in. [2 min/mm] of thickness, but in no case less than 15 min; cool in air or water quench at a rate no less than 300 °F [165 °C]/h.

4.2.2.2 *Double Normalized and Tempered*—Heat to a uniform temperature of 1650 ± 25 °F [900 ± 15 °C]; hold at this temperature for a minimum time in the ratio of 1 h/in. [2 min/mm] of thickness, but in no case less than 15 min; cool in air. Reheat until the pipe attains a uniform temperature of 1450 ± 25 °F [790 ± 15 °C]; hold at this temperature for a minimum time in the ratio of 1 h/in. [2 min/mm] of thickness, but in no case less than 15 min; cool in air. Reheat to a uniform temperature within the range of 1050 to 1125 °F [565 to 605 °C]; hold at this temperature for a minimum time of 1 h/in. [2 min/mm] of thickness but in no case less than 15 min; cool in air or water quench at a rate not less than 300 °F [165 °C]/h.

4.2.3 Whether to anneal Grade 11 pipe is per agreement between purchaser and supplier. When Grade 11 pipe is annealed, it shall be normalized in the range of 1400 to 1600 °F [760 to 870 °C].

4.2.4 Material from which test specimens are obtained shall be in the same condition of heat treatment as the pipe furnished. Material from which specimens are to be taken shall be heat treated prior to preparation of the specimens.

4.2.5 When specified in the order the test specimens shall be taken from full thickness test pieces which have been stress relieved after having been removed from the heat-treated pipe. The test pieces shall be gradually and uniformly heated to the prescribed temperature, held at that temperature for a period of time in accordance with Table 2, and then furnace cooled at a temperature not exceeding 600 °F [315 °C]. Grade 8 shall be cooled at a minimum rate of 300 °F [165 °C]/h in air or water to a temperature not exceeding 600 °F [315 °C].

5. Chemical Composition

5.1 The steel shall conform to the requirements as to chemical composition prescribed in Table 1.

5.2 When Grades 1, 6, or 10 are ordered under this specification, supplying an alloy grade that specifically requires the addition of any element other than those listed for the ordered grade in Table 1 is not permitted. However, the addition of elements required for the deoxidation of the steel is permitted.

TABLE 2 Stress Relieving of Test Pieces

Metal Temperature[A,B]				Minimum Holding Time, h/in. [min/mm] of Thickness
Grades 1, 3, 6, 7, and 10		Grade 4[C]		
°F	°C	°F	°C	
1100	600	1150	620	1 [2.4]
1050	565	1100	600	2 [4.7]
1000	540	1050	565	3 [7.1]

[A] For intermediate temperatures, the holding time shall be determined by straight-line interpolation.
[B] Grade 8 shall be stress relieved at 1025 to 1085 °F, [550 to 585 °C], held for a minimum time of 2 h for thickness up to 1.0 in. [25.4 mm], plus a minimum of 1 h for each additional inch [25.4 mm] of thickness and cooled at a minimum rate of 300 °F [165 °C]/h in air or water to a temperature not exceeding 600 °F [315 °C].
[C] Unless otherwise specified, Grade 4 shall be stress relieved at 1150 °F [620 °C].

6. Product Analysis

6.1 At the request of the purchaser, an analysis of one billet or two samples of flat-rolled stock from each heat or of two pipes from each lot shall be made by the manufacturer. A lot of pipe shall consist of the following:

NPS Designator	Length of Pipe in Lot
Under 2	400 or fraction thereof
2 to 6	200 or fraction thereof
Over 6	100 or fraction thereof

6.2 The results of these analyses shall be reported to the purchaser or the purchaser's representative and shall conform to the requirements specified.

6.3 If the analysis of one of the tests specified in 6.1 does not conform to the requirements specified, an analysis of each billet or pipe from the same heat or lot may be made, and all billets or pipe conforming to the requirements shall be accepted.

7. Tensile Requirements

7.1 The material shall conform to the requirements as to tensile properties prescribed in Table 3.

8. Impact Requirements

8.1 For Grades 1, 3, 4, 6, 7, 9, and 10, the notched-bar impact properties of each set of three impact specimens, including specimens for the welded joint in welded pipe with wall thicknesses of 0.120 in. [3 mm] and larger, when tested at temperatures in conformance with 14.1 shall be not less than the values prescribed in Table 4. The impact test is not required for Grade 11.

8.1.1 If the impact value of one specimen is below the minimum value, or the impact values of two specimens are less than the minimum average value but not below the minimum value permitted on a single specimen, a retest shall be allowed. The retest shall consist of breaking three additional specimens and each specimen must equal or exceed the required average value. When an erratic result is caused by a defective specimen, or there is uncertainty in test procedures, a retest will be allowed.

8.2 For Grade 8 each of the notched bar impact specimens shall display a lateral expansion opposite the notch of not less than 0.015 in. [0.38 mm].

8.2.1 When the average lateral expansion value for the three impact specimens equals or exceeds 0.015 in. [0.38 mm] and the value for one specimen is below 0.015 in. [0.38 mm] but not below 0.010 in. [0.25 mm], a retest of three additional specimens may be made. The lateral expansion of each of the retest specimens must equal or exceed 0.015 in. [0.38 mm].

8.2.2 Lateral expansion values shall be determined by the procedure in Test Methods and Definitions A370.

8.2.3 The values of absorbed energy in foot-pounds and the fracture appearance in percentage shear shall be recorded for information. A record of these values shall be retained for a period of at least 2 years.

9. Lengths

9.1 If definite lengths are not required, pipe may be ordered in single random lengths of 16 to 22 ft (Note 3) with 5 % 12 to 16 ft (Note 3), or in double random lengths with a minimum average of 35 ft (Note 3) and a minimum length of 22 ft (Note 3) with 5 % 16 to 22 ft (Note 3).

NOTE 3—This value(s) applies when the inch-pound designation of this specification is the basis of purchase. When the "M" designation of this specification is the basis of purchase, the corresponding metric value(s) shall be agreed upon between the manufacturer and purchaser.

10. Workmanship, Finish and Appearance

10.1 The pipe manufacturer shall explore a sufficient number of visual surface imperfections to provide reasonable assurance that they have been properly evaluated with respect to depth. Exploration of all surface imperfections is not required but may be necessary to ensure compliance with 10.2.

10.2 Surface imperfections that penetrate more than 12½ % of the nominal wall thickness or encroach on the minimum wall thickness shall be considered defects. Pipe with such defects shall be given one of the following dispositions:

10.2.1 The defect may be removed by grinding provided that the remaining wall thickness is within specified limits.

10.2.2 Repaired in accordance with the repair welding provisions of 10.5.

10.2.3 The section of pipe containing the defect may be cut off within the limits of requirements on length.

10.2.4 The defective pipe may be rejected.

10.3 To provide a workmanlike finish and basis for evaluating conformance with 10.2, the pipe manufacturer shall remove by grinding the following:

10.3.1 Mechanical marks, abrasions and pits, any of which imperfections are deeper than 1/16 in. [1.6 mm], and

10.3.2 Visual imperfections commonly referred to as scabs, seams, laps, tears, or slivers found by exploration in accordance with 10.1 to be deeper than 5 % of the nominal wall thickness.

10.4 At the purchaser's discretion, pipe shall be subject to rejection if surface imperfections acceptable under 10.2 are not scattered, but appear over a large area in excess of what is considered a workmanlike finish. Disposition of such pipe shall be a matter of agreement between the manufacturer and the purchaser.

10.5 When imperfections or defects are removed by grinding, a smooth curved surface shall be maintained, and the wall thickness shall not be decreased below that permitted by this specification. The outside diameter at the point of grinding may be reduced by the amount so removed.

10.5.1 Wall thickness measurements shall be made with a mechanical caliper or with a properly calibrated nondestructive testing device of appropriate accuracy. In case of dispute, the measurement determined by use of the mechanical caliper shall govern.

10.6 Weld repair shall be permitted only subject to the approval of the purchaser and in accordance with Specification A999/A999M.

10.7 The finished pipe shall be reasonably straight.

11. General Requirements

11.1 Material furnished to this specification shall conform to the applicable requirements of the current edition of Specification A999/A999M unless otherwise provided herein.

TABLE 3 Tensile Requirements

	Grade 1		Grade 3		Grade 4		Grade 6		Grade 7		Grade 8		Grade 9		Grade 10		Grade 11	
	psi	MPa	psi	MPa	psi	MPa	psi	MPa	psi	MPa	psi	MPa	psi	MPa	psi	MPa	psi	MPa
Tensile strength, min	55 000	380	65 000	450	60 000	415	60 000	415	65 000	450	100 000	690	63 000	435	80 000	550	65 000	450
Yield strength, min	30 000	205	35 000	240	35 000	240	35 000	240	35 000	240	75 000	515	46 000	315	65 000	450	35 000	240
	Longi-tudinal	Trans-verse	Longi-tudinal	Trans-verse	Longi-tudinal	Trans-verse	Longi-tudinal	Trans-verse	Longi-tudinal	Trans-verse	Longi-tudinal	Trans-verse	Longi-tudinal	Trans-verse	Longi-tudinal	Trans-verse	Longi-tudinal	
Elongation in 2 in. or 50 mm, (or 4D), min, %: Basic minimum elongation for walls 5/16 in. [8 mm] and over in thickness, strip tests, and for all small sizes tested in full section	35	25	30	20	30	16.5	30	16.5	30	22	22	...	28	...	22	...	18[A]	
When standard round 2-in. or 50-mm gage length or proportionally smaller size test specimen with the gage length equal to 4D (4 times the diameter) is used	28	20	22	14	22	12	22	12	22	14	16	...	...	...	16	...	...	
For strip tests, a deduction for each 1/32 in. [0.8 mm] decrease in wall thickness below 5/16 in. [8 mm] from the basic minimum elongation of the following percentage	1.75[B]	1.25[B]	1.50[B]	1.00[B]	1.50[B]	1.00[B]	1.50[B]	1.00[B]	1.50[B]	1.00[B]	1.25[B]	...	1.50[B]	...	1.25[B]	...	...	

Wall Thickness		Elongation in 2 in. or 50 mm, min, %[C]														
		Grade 1		Grade 3		Grade 4		Grade 6		Grade 7		Grade 8		Grade 9		Grade 10	
in.	mm	Longi-tudinal	Trans-verse	Longi-tudinal	Trans-verse	Longi-tudinal	Trans-verse	Longi-tudinal	Trans-verse	Longi-tudinal	Trans-verse	Longi-tudinal	Trans-verse	Longi-tudinal	Trans-verse	Longi-tudinal	Trans-verse
5/16 (0.312)	8	35	25	30	20	30	16	30	16	30	22	22	...	28	...	22	...
9/32 (0.281)	7.2	33	24	28	19	28	15	28	15	28	21	21	...	26	...	21	...
1/4 (0.250)	6.4	32	23	27	18	27	15	27	15	27	20	20	...	25	...	20	...
7/32 (0.219)	5.6	30	...	26	...	26	...	26	...	26	...	18	...	24	...	18	...
3/16 (0.188)	4.8	28	...	24	...	24	...	24	...	24	...	17	...	22	...	17	...
5/32 (0.156)	4	26	...	22	...	22	...	22	...	22	...	16	...	20	...	16	...
1/8 (0.125)	3.2	25	...	21	...	21	...	21	...	21	...	15	...	19	...	15	...
3/32 (0.094)	2.4	23	...	20	...	20	...	20	...	20	...	13	...	18	...	13	...
1/16 (0.062)	1.6	21	...	18	...	18	...	18	...	18	...	12	...	16	...	12	...

[A] Elongation of Grade 11 is for all walls and small sizes tested in full section.
[B] The following table gives the calculated minimum values.
[C] Calculated elongation requirements shall be rounded to the nearest whole number.

Note—The preceding table gives the computed minimum elongation values for each 1/32-in. [0.80-mm] decrease in wall thickness. Where the wall thickness lies between two values shown above, the minimum elongation value is determined by the following equation:

Grade	Direction of Test	Equation
1	Longitudinal	$E = 56t + 17.50$ [$E = 2.19t + 17.50$]
	Transverse	$E = 40t + 12.50$ [$E = 1.56t + 12.50$]
3	Longitudinal	$E = 48t + 15.00$ [$E = 1.87t + 15.00$]
	Transverse	$E = 32t + 10.00$ [$E = 1.25t + 10.00$]
4	Longitudinal	$E = 48t + 15.00$ [$E = 1.87t + 15.00$]
	Transverse	$E = 32t + 6.50$ [$E = 1.25t + 6.50$]
6	Longitudinal	$E = 48t + 15.00$ [$E = 1.87t + 15.00$]
	Transverse	$E = 32t + 6.50$ [$E = 1.25t + 6.50$]
7	Longitudinal	$E = 48t + 15.00$ [$E = 1.87t + 15.00$]
	Transverse	$E = 32t + 11.00$ [$E = 1.25t + 11.00$]
8 and 10	Longitudinal	$E = 40t + 9.50$ [$E = 1.56t + 9.50$]
9	Longitudinal	$E = 48t + 13.00$ [$E = 1.87t + 13.00$]

where:
E = elongation in 2 in. or 50 mm, in %, and
t = actual thickness of specimen, in. [mm].

12. Mechanical Testing

12.1 *Sampling*—For mechanical testing, the term "lot" applies to all pipe of the same nominal size and wall thickness (or schedule) that is produced from the same heat of steel and subjected to the same finishing treatment in a continuous

TABLE 4 Impact Requirements for Grades 1, 3, 4, 6, 7, 9, and 10

Size of Specimen, mm	Minimum Average Notched Bar Impact Value of Each Set of Three Specimens[A]		Minimum Notched Bar Impact Value of One Specimen Only of a Set[A]	
	ft·lbf	J	ft·lbf	J
10 by 10	13	18	10	14
10 by 7.5	10	14	8	11
10 by 6.67	9	12	7	9
10 by 5	7	9	5	7
10 by 3.33	5	7	3	4
10 by 2.5	4	5	3	4

[A] Straight line interpolation for intermediate values is permitted.

furnace. If the final heat treatment is in a batch-type furnace, the lot shall include only those pipes that are heat treated in the same furnace charge.

12.2 *Transverse or Longitudinal Tensile Test and Flattening Test*—For material heat treated in a batch-type furnace, tests shall be made on 5 % of the pipe from each lot. If heat treated by the continuous process, tests shall be made on a sufficient number of pipe to constitute 5 % of the lot, but in no case less than 2 pipes.

12.3 *Impact Test*—One notched bar impact test, consisting of breaking three specimens, shall be made from each heat represented in a heat-treatment load on specimens taken from the finished pipe. This test shall represent only pipe from the same heat and the same heat-treatment load, the wall thicknesses of which do not exceed by more than ¼ in. [6.3 mm] the wall thicknesses of the pipe from which the test specimens are taken. If heat treatment is performed in continuous or batch-type furnaces controlled within a 50 °F [30 °C] range and equipped with recording pyrometers so that complete records of heat treatment are available, then one test from each heat in a continuous run only shall be required instead of one test from each heat in each heat-treatment load.

12.4 *Impact Tests (Welded Pipe)*—On welded pipe, additional impact tests of the same number as required in 12.3 shall be made to test the weld.

12.5 Specimens showing defects while being machined or prior to testing may be discarded and replacements shall be considered as original specimens.

12.6 Results obtained from these tests shall be reported to the purchaser or his representative.

13. Specimens for Impact Test

13.1 Notched bar impact specimens shall be of the simple beam, Charpy-type, in accordance with Test Methods E23, Type A with a V notch. Standard specimens 10 by 10 mm in cross section shall be used unless the material to be tested is of insufficient thickness, in which case the largest obtainable subsize specimens shall be used. Charpy specimens of width along the notch larger than 0.394 in. [10 mm] or smaller than 0.099 in. [2.5 mm] are not provided for in this specification.

13.2 Test specimens shall be obtained so that the longitudinal axis of the specimen is parallel to the longitudinal axis of the pipe while the axis of the notch shall be perpendicular to the surface. On wall thicknesses of 1 in. [25 mm] or less, the specimens shall be obtained with their axial plane located at the midpoint; on wall thicknesses over 1 in. [25 mm], the specimens shall be obtained with their axial plane located ½ in. [12.5 mm] from the outer surface.

13.3 When testing welds the specimen shall be, whenever diameter and thickness permit, transverse to the longitudinal axis of the pipe with the notch of the specimen in the welded joint and perpendicular to the surface. When diameter and thickness do not permit obtaining transverse specimens, longitudinal specimens in accordance with 13.2 shall be obtained; the bottom of the notch shall be located at the weld joint.

14. Impact Test

14.1 Except when the size of the finished pipe is insufficient to permit obtaining subsize impact specimens, all material furnished to this specification and marked in accordance with Section 16 shall be tested for impact resistance at the minimum temperature for the respective grades as shown in Table 5.

14.1.1 Special impact tests on individual lots of material may be made at other temperatures as agreed upon between the manufacturer and the purchaser.

14.1.2 When subsize Charpy impact specimens are used and the width along the notch is less than 80 % of the actual wall thickness of the original material, the specified Charpy impact test temperature for Grades 1, 3, 4, 6, 7, 9, and 10 shall be lower than the minimum temperature shown in Table 5 for the respective grade. Under these circumstances the temperature reduction values shall be by an amount equal to the difference (as shown in Table 6) between the temperature reduction corresponding to the actual material thickness and the temperature reduction corresponding to the Charpy specimen width actually tested. Appendix X1 shows some examples of how the temperature reductions are determined.

14.2 The notched bar impact test shall be made in accordance with the procedure for the simple beam, Charpy-type test of Test Methods E23.

14.3 Impact tests specified for temperatures lower than 70 °F [20 °C] should be made with the following precautions. The impact test specimens as well as the handling tongs shall be cooled a sufficient time in a suitable container so that both reach the desired temperature. The temperature shall be measured with thermocouples, thermometers, or any other suitable devices and shall be controlled within 3 °F [2 °C]. The specimens shall be quickly transferred from the cooling device to the anvil of the Charpy impact testing machine and broken with a time lapse of not more than 5 s.

15. Hydrostatic or Nondestructive Electric Test

15.1 Each pipe shall be subjected to the nondestructive electric test or the hydrostatic test. The type of test to be used

TABLE 5 Impact Temperature

Grade	Minimum Impact Test Temperature	
	°F	°C
1	−50	−45
3	−150	−100
4	−150	−100
6	−50	−45
7	−100	−75
8	−320	−195
9	−100	−75
10	−75	−60

TABLE 6 Impact Temperature Reduction

Specimen Width Along Notch or Actual Material Thickness		Temperature Reduction, Degrees Colder[A]	
in.	mm	°F	°C
0.394	10 (standard size)	0	0
0.354	9	0	0
0.315	8	0	0
0.295	7.5 (¾ std. size)	5	3
0.276	7	8	4
0.262	6.67 (⅔ std. size)	10	5
0.236	6	15	8
0.197	5 (½ std. size)	20	11
0.158	4	30	17
0.131	3.33 (⅓ std. size)	35	19
0.118	3	40	22
0.099	2.5 (¼ std. size)	50	28

[A] Straight line interpolation for intermediate values is permitted.

shall be at the option of the manufacturer, unless otherwise specified in the purchase order.

15.2 The hydrostatic test shall be in accordance with Specification A999/A999M.

15.3 *Nondestructive Electric Test*—Nondestructive electric tests shall be in accordance with Specification A999/A999M, with the following addition:

15.3.1 If the test signals were produced by visual imperfections (listed in 15.3.2), the pipe may be accepted based on visual examination, provided the imperfection is less than 0.004 in. (0.1 mm) or 12½ % of the specified wall thickness (whichever is greater).

15.3.2 *Visual Imperfections*:

15.3.2.1 Scratches,

15.3.2.2 Surface roughness,

15.3.2.3 Dings,

15.3.2.4 Straightener marks,

15.3.2.5 Cutting chips,

15.3.2.6 Steel die stamps,

15.3.2.7 Stop marks, or

15.3.2.8 Pipe reducer ripple.

16. Product Marking

16.1 Except as modified in 16.1.1, in addition to the marking prescribed in Specification A999/A999M, the marking shall include whether hot finished, cold drawn, seamless or welded, the schedule number and the letters "LT" followed by the temperature at which the impact tests were made, except when a lower test temperature is required because of reduced specimen size, in which case, the higher impact test temperature applicable to a full-size specimen should be marked.

16.1.1 When the size of the finished pipe is insufficient to obtain subsize impact specimens, the marking shall not include the letters "LT" followed by an indicated test temperature unless Supplementary Requirement S1 is specified.

16.1.2 When the pipe is furnished in the quenched and tempered condition, the marking shall include the letters "QT," and the heat treatment condition shall be reported to the purchaser or his representative.

17. Keywords

17.1 low; low temperature service; seamless steel pipe; stainless steel pipe; steel pipe; temperature service applications

---

SUPPLEMENTARY REQUIREMENTS

The following supplementary requirement shall apply only when specified by the purchaser in the contract or order.

S1. Subsize Impact Specimens

S1.1 When the size of the finished pipe is insufficient to permit obtaining subsize impact specimens, testing shall be a matter of agreement between the manufacturer and the purchaser.

S2. Requirements for Pipe for Hydrofluoric Acid Alkylation Service

S2.1 Pipe shall be provided in the normalized heat-treated condition.

S2.2 The carbon equivalent (CE), based on heat analysis, shall not exceed 0.43 % if the specified wall thickness is equal to or less than 1 in. [25.4 mm] or 0.45 % if the specified wall thickness is greater than 1 in. [25.4 mm].

S2.3 The carbon equivalent shall be determined using the following formula:

$$CE = C + Mn/6 + (Cr + Mo + V)/5 + (Ni + Cu)/15$$

S2.4 Based upon heat analysis in mass percent, the vanadium content shall not exceed 0.02 %, the niobium content shall not exceed 0.02 % and the sum of the vanadium and niobium contents shall not exceed 0.03 %.

S2.5 Based upon heat analysis in mass percent, the sum of the nickel and copper contents shall not exceed 0.15 %.

S2.6 Based upon heat analysis in mass percent, the carbon content shall not be less than 0.18 %.

S2.7 Welding consumables for repair welds shall be of low hydrogen type. E60XX electrodes shall not be used, and the resultant weld chemistry shall meet the chemical composition requirements specified for the pipe.

S2.8 The designation "HF-N" shall be stamped or marked on each pipe to signify that the pipe complies with this supplementary requirement.

APPENDIX

(Nonmandatory Information)

X1. DETERMINATION OF TEMPERATURE REDUCTIONS

X1.1 Under the circumstances stated in 14.1.2, the impact test temperatures specified in Table 5 must be lowered. The following examples are offered to describe the application of the provisions of 14.1.2.

X1.1.1 When subsize specimens are used (see 10.1) and the width along the notch of the subsize specimen in 80 % or greater of the actual wall thickness of the original material, the provisions of 14.1.2 do not apply.

X1.1.1.1 For example, if the actual wall thickness of pipe was 0.200 in. [5.0 mm] and the width along the notch of the largest subsize specimen obtainable is 0.160 in. [4 mm] or greater, no reduction in test temperature is required.

X1.1.2 When the width along the subsize specimen notch is less than 80 % of the actual wall thickness of the pipe, the required reduction in test temperature is computed by taking the difference between the temperature reduction values shown in Table 6 for the actual pipe thickness and the specimen width used.

X1.1.2.1 For example, if the pipe were 0.262 in. [6.67 mm] thick and the width along the Charpy specimen notch was 3.33 mm (1/3 standard size), the test temperature would have to be lowered by 25 °F [14 °C]. That is, the temperature reduction corresponding to the subsize specimen is 35 °F [19 °C]; the temperature reduction corresponding to the actual pipe thickness is 10 °F [5 °C]; the difference between these two values is the required reduction in test temperature.

SUMMARY OF CHANGES

Committee A01 has identified the location of selected changes to this specification since the last issue, A333/A333M – 04a, that may impact the use of this specification. (Approved March 1, 2005)

(1) Removed old paragraph 12.3 and renumbered subsequent paragraphs.

(2) Added Supplementary Requirement S2 for HF acid alkylation service.

Committee A01 has identified the location of selected changes to this specification since the last issue, A333/A333M – 04, that may impact the use of this specification. (Approved May 1, 2004)

(1) Replaced all references to Note 4 in 9.1 with Note 3.

(2) Revised Section 13 to incorporate Note 4 into the text.

Committee A01 has identified the location of selected changes to this specification since the last issue, A333/A333M – 99, that may impact the use of this specification. (Approved March 1, 2004)

(1) Replaced Specification A530/A530M with Specification A999/A999M in the Referenced Documents.
(2) Replaced Specification A530/A530M with Specification A999/A999M in Sections 3, 11, 15, and 16.
(3) Extensively revised Section 15.

ASTM International takes no position respecting the validity of any patent rights asserted in connection with any item mentioned in this standard. Users of this standard are expressly advised that determination of the validity of any such patent rights, and the risk of infringement of such rights, are entirely their own responsibility.

This standard is subject to revision at any time by the responsible technical committee and must be reviewed every five years and if not revised, either reapproved or withdrawn. Your comments are invited either for revision of this standard or for additional standards and should be addressed to ASTM International Headquarters. Your comments will receive careful consideration at a meeting of the responsible technical committee, which you may attend. If you feel that your comments have not received a fair hearing you should make your views known to the ASTM Committee on Standards, at the address shown below.

This standard is copyrighted by ASTM International, 100 Barr Harbor Drive, PO Box C700, West Conshohocken, PA 19428-2959, United States. Individual reprints (single or multiple copies) of this standard may be obtained by contacting ASTM at the above address or at 610-832-9585 (phone), 610-832-9555 (fax), or service@astm.org (e-mail); or through the ASTM website (www.astm.org).

Designation: A334/A334M – 04a

Standard Specification for
Seamless and Welded Carbon and Alloy-Steel Tubes for Low-Temperature Service[1]

This standard is issued under the fixed designation A334/A334M; the number immediately following the designation indicates the year of original adoption or, in the case of revision, the year of last revision. A number in parentheses indicates the year of last reapproval. A superscript epsilon (ε) indicates an editorial change since the last revision or reapproval.

1. Scope*

1.1 This specification[2] covers several grades of minimum-wall-thickness, seamless and welded, carbon and alloy-steel tubes intended for use at low temperatures. Some product sizes may not be available under this specification because heavier wall thicknesses have an adverse affect on low-temperature impact properties.

1.2 Supplementary Requirement S1 of an optional nature is provided. This shall apply only when specified by the purchaser.

NOTE 1—For tubing smaller than 1/2 in. [12.7 mm] in outside diameter, the elongation values given for strip specimens in Table 1 shall apply. Mechanical property requirements do not apply to tubing smaller than 1/8 in. [3.2 mm] in outside diameter and with a wall thickness under 0.015 in. [0.4 mm].

1.3 The values stated in either inch-pound units or SI units are to be regarded separately as standard. Within the text, the SI units are shown in brackets. The values stated in each system are not exact equivalents; therefore, each system must be used independently of the other. Combining values from the two systems may result in nonconformance with the specification. The inch-pound units shall apply unless the "M" designation of this specification is specified in the order.

2. Referenced Documents

2.1 *ASTM Standards:*[3]

A370 Test Methods and Definitions for Mechanical Testing of Steel Products

A1016/A1016M Specification for General Requirements for Ferritic Alloy Steel, Austenitic Alloy Steel, and Stainless Steel Tubes

E23 Test Methods for Notched Bar Impact Testing of Metallic Materials

3. Ordering Information

3.1 Orders for material under this specification should include the following, as required to describe the desired material adequately:

3.1.1 Quantity (feet, metres, or number of lengths),

3.1.2 Name of material (seamless or welded tubes),

3.1.3 Grade (Table 1),

3.1.4 Size (outside diameter and minimum wall thickness),

3.1.5 Length (specific or random),

3.1.6 Optional requirements (other temperatures, Section 14; hydrostatic or electric test, Section 16),

3.1.7 Test report required, (Certification Section of Specification A1016/A1016M),

3.1.8 Specification designation, and

3.1.9 Special requirements and any supplementary requirements selected.

4. General Requirements

4.1 Material furnished under this specification shall conform to the applicable requirements of the current edition of Specification A1016/A1016M, unless otherwise provided herein.

5. Materials and Manufacture

5.1 The tubes shall be made by the seamless or automatic welding process with no addition of filler metal in the welding operation.

6. Heat Treatment

6.1 All seamless and welded tubes, other than Grades 8 and 11, shall be treated to control their microstructure in accordance with one of the following methods:

6.1.1 Normalize by heating to a uniform temperature of not less than 1550 °F [845 °C] and cool in air or in the cooling chamber of an atmosphere controlled furnace.

---

[1] This specification is under the jurisdiction of ASTM Committee A01 on Steel, Stainless Steel and Related Alloys and is the direct responsibility of Subcommittee A01.10 on Stainless and Alloy Steel Tubular Products.
Current edition approved May 1, 2004. Published June 2004. Originally approved in 1951. Last previous edition approved in 2004 as A334/A334M – 04. DOI: 10.1520/A0334_A0334M-04A.

[2] For ASME Boiler and Pressure Vessel Code applications see related Specification SA-334 in Section II of that Code.

[3] For referenced ASTM standards, visit the ASTM website, www.astm.org, or contact ASTM Customer Service at service@astm.org. For *Annual Book of ASTM Standards* volume information, refer to the standard's Document Summary page on the ASTM website.

*A Summary of Changes section appears at the end of this standard.

TABLE 1 Chemical Requirements

Element	Composition, %						
	Grade 1[A]	Grade 3	Grade 6[A]	Grade 7	Grade 8	Grade 9	Grade 11
Carbon, max	0.30	0.19	0.30	0.19	0.13	0.20	0.10
Manganese	0.40–1.06	0.31–0.64	0.29–1.06	0.90 max	0.90 max	0.40–1.06	0.60 max
Phosphorus, max	0.025	0.025	0.025	0.025	0.025	0.025	0.025
Sulfur, max	0.025	0.025	0.025	0.025	0.025	0.025	0.025
Silicon	...	0.18–0.37	0.10 min	0.13–0.32	0.13–0.32	...	0.35 max
Nickel	...	3.18–3.82	...	2.03–2.57	8.40–9.60	1.60–2.24	35.0–37.0
Chromium	...	...	...	...	...	...	0.50 max
Copper	...	...	...	...	...	0.75–1.25	...
Cobalt	...	...	...	...	...	...	0.50 max
Molybdenum	...	...	...	...	...	...	0.50 max

[A] For each reduction of 0.01 % carbon below 0.30 %, an increase of 0.05 % manganese above 1.06 % will be permitted to a maximum of 1.35 % manganese.

6.1.2 Normalize as in 10.1.1, and, at the discretion of the manufacturer, reheat to a suitable tempering temperature.

6.1.3 For the seamless process only, reheat and control hot working and the temperature of the hot-finishing operation to a finishing temperature range from 1550 to 1750 °F [845 to 955 °C] and cool in a controlled atmosphere furnace from an initial temperature of not less than 1550 °F [845 °C].

6.1.4 Treat as in 6.1.3 and, at the discretion of the manufacturer, reheat to a suitable tempering temperature.

6.2 Grade 8 tubes shall be heat treated by the manufacturer by either of the following methods.

6.2.1 *Quenched and Tempered*—Heat to a uniform temperature of 1475 ± 25 °F [800 ± 15 °C]; hold at this temperature for a minimum time in the ratio of 1 h/in. [2 min/mm] of thickness, but in no case less than 15 min; quench by immersion in circulating water. Reheat until the pipe attains a uniform temperature within the range from 1050 to 1125 °F [565 to 605 °C]; hold at this temperature for a minimum time in the ratio of 1 h/in. [2 min/mm] of thickness, but in no case less than 15 min; cool in air or water quench at a rate no less than 300 °F [165 °C]/h.

6.2.2 *Double Normalized and Tempered*— Heat to a uniform temperature of 1650 ± 25 °F [900 ± 15 °C]; hold at this temperature for a minimum time in the ratio of 1 h/in. [2 min/mm] of thickness, but in no case less than 15 min; cool in air. Reheat until the pipe attains a uniform temperature of 1450 ± 25 °F [790 ± 15 °C]; hold at this temperature for a minimum time in the ratio of 1 h/in. [2 min/mm] of thickness, but in no case less than 15 min; cool in air. Reheat to a uniform temperature within the range from 1050 to 1125 °F [565 to 605 °C]; hold at this temperature for a minimum time of 1 h/in. [2 min/mm] of thickness but in no case less than 15 min; cool in air or water quench at a rate not less than 300 °F [165 °C]/h.

6.3 Material from which impact specimens are obtained shall be in the same condition of heat treatment as the finished tubes.

6.4 Whether to anneal Grade 11 tubes is per agreement between purchaser and supplier. When Grade 11 tubes are annealed they shall be normalized in the range of 1400 to 1600 °F [760 to 870 °C].

7. Chemical Composition

7.1 The steel shall conform to the requirements as to chemical composition prescribed in Table 1.

7.2 When Grades 1 or 6 are ordered under this specification, supplying an alloy grade that specifically requires the addition of any element other than those listed for the ordered grade in Table 1 is not permitted. However, the addition of elements required for the deoxidation of the steel is permitted.

8. Product Analysis

8.1 An analysis of either one billet or one length of flat-rolled stock or one tube shall be made for each heat. The chemical composition thus determined shall conform to the requirements specified.

8.2 If the original test for product analysis fails, retests of two additional billets, lengths of flat-rolled stock, or tubes shall be made. Both retests, for the elements in question, shall meet the requirements of the specification; otherwise all remaining material in the heat or lot shall be rejected or, at the option of the manufacturer, each billet, length of flat-rolled stock, or tube may be individually tested for acceptance. Billets, lengths of flat-rolled stock, or tubes which do not meet the requirements of the specification shall be rejected.

9. Sampling

9.1 For flattening, flare, and flange requirements, the term *lot* applies to all tubes prior to cutting of the same nominal size and wall thickness which are produced from the same heat of steel. When final heat treatment is in a batch-type furnace, a lot shall include only those tubes of the same size and from the same heat which are heat treated in the same furnace charge. When the final heat treatment is in a continuous furnace, the number of tubes of the same size and from the same heat in a lot shall be determined from the size of the tubes as prescribed in Table 2.

TABLE 2 Heat-Treatment Lot

Size of Tube	Size of Lot
2 in. [50.8 mm] and over in outside diameter and 0.200 in. [5.1 mm] and over in wall thickness	not more than 50 tubes
Under 2 in. [50.8 mm] but over 1 in. [25.4 mm] in outside diameter, or over 1 in. [25.4 mm] in outside diameter and under 0.200 in. [5.1 mm] in thickness	not more than 75 tubes
1 in. [25.4 mm] or under in outside diameter	not more than 125 tubes

9.2 For tensile and hardness test requirements, the term *lot* applies to all tubes prior to cutting, of the same nominal diameter and wall thickness which are produced from the same heat of steel. When final heat treatment is in a batch-type furnace, a lot shall include only those tubes of the same size and the same heat which are heat treated in the same furnace charge. When the final heat treatment is in a continuous furnace, a lot shall include all tubes of the same size and heat, heat treated in the same furnace at the same temperature, time at heat and furnace speed.

10. Tensile Requirements

10.1 The material shall conform to the requirements as to tensile properties prescribed in Table 3.

11. Hardness Requirements

11.1 The tubes shall have a hardness number not exceeding those prescribed in Table 4.

12. Impact Requirements

12.1 For Grades 1, 3, 6, 7 and 9, the notched-bar impact properties of each set of three impact specimens, including specimens for the welded joint in welded pipe with wall thicknesses of 0.120 in. [3 mm] and larger, when tested at temperatures in conformance with 14.1 shall be not less than the values prescribed in Table 5. The impact test is not required for Grade 11.

12.1.1 If the impact value of one specimen is below the minimum value, or the impact values of two specimens are less

TABLE 3 Tensile Requirements

	Grade 1		Grade 3		Grade 6		Grade 7		Grade 8		Grade 9		Grade 11	
	ksi	MPa	ksi	MPa	ksi	MPa	ksi	MPa	ksi	MPa	ksi	MPa	ksi	MPa
Tensile Strength, min	55	380	65	450	60	415	65	450	100	690	63	435	65	450
Yield Strength, min	30	205	35	240	35	240	35	240	75	520	46	315	35	240
Elongation in 2 in. or 50 mm (or 4*D*), min, %:														
Basic minimum elongation for walls 5/16 in. [8 mm] and over in thickness, strip tests, and for all small sizes tested in full section	35		30		30		30		22		28		18[A]	
When standard round 2-in. or 50 mm gage length or proportionally smaller size specimen with the gage length equal to 4*D* (4 times the diameter) is used	28		22		22		22		16		...		...	
For strip tests, a deduction for each 1/32 in. [0.8 mm] decrease in wall thickness below 5/16 in. [8 mm] from the basic minimum elongation of the following percentage points	1.75[B]		1.50[B]		1.50[B]		1.50[B]		1.25[B]		1.50[B]		...	

[A] Elongation of Grade 11 is for all walls and for small sizes tested in full section.
[B] The following table gives the calculated minimum values:

Wall Thickness		Elongation in 2 in. or 50 mm, min %[A]					
in.	mm	Grade 1	Grade 3	Grade 6	Grade 7	Grade 8	Grade 9
5/16 (0.312)	8	35	30	30	30	22	28
9/32 (0.281)	7.2	33	28	28	28	21	26
1/4 (0.250)	6.4	32	27	27	27	20	25
7/32 (0.219)	5.6	30	26	26	26	18	24
3/16 (0.188)	4.8	28	24	24	24	17	22
5/32 (0.156)	4	26	22	22	22	16	20
1/8 (0.125)	3.2	25	21	21	21	15	19
3/32 (0.094)	2.4	23	20	20	20	13	18
1/16 (0.062)	1.6	21	18	18	18	12	16

[A] Calculated elongation requirements shall be rounded to the nearest whole number.

Note—The above table gives the computed minimum elongation values for each 1/32-in. [0.8-mm] decrease in wall thickness. Where the wall thickness lies between two values shown above, the minimum elongation value is determined by the following equations:

Grade	Equation[A]
1	$E = 56t + 17.50$ [$E = 2.19t + 17.50$]
3	$E = 48t + 15.00$ [$E = 1.87t + 15.00$]
6	$E = 48t + 15.00$ [$E = 1.87t + 15.00$]
7	$E = 48t + 15.00$ [$E = 1.87t + 15.00$]
8	$E = 40t + 9.50$ [$E = 1.56t + 9.50$]
9	$E = 48t + 13.00$ [$E = 1.87t + 13.00$]

[A] where:
E = elongation in 2 in. or 50 mm, %, and
t = actual thickness of specimen, in. [mm].

TABLE 4 Maximum Hardness Number

Grade	Rockwell	Brinell
1	B 85	163
3	B 90	190
6	B 90	190
7	B 90	190
8	...	...
11	B 90	190

TABLE 5 Impact Requirements for Grades 1, 3, 6, 7, and 9

Size of Specimen, mm	Minimum Average Notched Bar Impact Value of Each Set of Three Specimens[A]		Minimum Notched Bar Impact Value of One Specimen Only of a Set[A]	
	ft·lbf	J	ft·lbf	J
10 by 10	13	18	10	14
10 by 7.5	10	14	8	11
10 by 6.67	9	12	7	9
10 by 5	7	9	5	7
10 by 3.33	5	7	3	4
10 by 2.5	4	5	3	4

[A] Straight line interpolation for intermediate values is permitted.

than the minimum average value but not below the minimum value permitted on a single specimen, a retest shall be allowed. The retest shall consist of breaking three additional specimens and each specimen must equal or exceed the required average value. When an erratic result is caused by a defective specimen, or there is uncertainty in test procedures, a retest will be allowed.

12.2 For Grade 8 each of the notched bar impact specimens shall display a lateral expansion opposite the notch not less than 0.015 in. [0.38 mm].

12.2.1 When the average lateral expansion value for the three impact specimens equals or exceeds 0.015 in. [0.38 mm] and the value for one specimen is below 0.015 in. [0.38 mm] but not below 0.010 in. [0.25 mm], a retest of three additional specimens may be made. The lateral expansion of each of the retest specimens must equal or exceed 0.015 in. [0.38 mm].

12.2.2 Lateral expansion values shall be determined in accordance with Test Methods and Definitions A370.

12.2.3 The values of absorbed energy in foot-pounds and the fracture appearance in percentage shear shall be recorded for information. A record of these values shall be retained for a period of at least 2 years.

13. Mechanical Tests

13.1 *Tension Test*—One tension test shall be made on a specimen for lots of not more than 50 tubes. Tension tests shall be made on specimens from two tubes for lots of more than 50 tubes.

13.2 *Flattening Test*—One flattening test shall be made on specimens from each end of one finished tube of each lot but not the one used for the flare or flange test.

13.3 *Flare Test (Seamless Tubes)*—One flare test shall be made on specimens from each end of one finished tube of each lot, but not the one used for the flattening test.

13.4 *Flange Test (Welded Tubes)*—One flange test shall be made on specimens from each end of one finished tube of each lot, but not the one used for the flattening test.

13.5 *Reverse Flattening Test*—For welded tubes, one reverse flattening test shall be made on a specimen from each 1500 ft [460 m] of finished tubing.

13.6 *Hardness Test*—Brinell or Rockwell hardness tests shall be made on specimens from two tubes from each lot.

13.7 *Impact Tests*—One notched-bar impact test, consisting of breaking three specimens, shall be made from each heat represented in a heat-treatment load on specimens taken from the finished tube. This test shall represent only tubes from the same heat, which have wall thicknesses not exceeding by more than 1/4 in. [6.3 mm] the wall thicknesses of the tube from which the test specimens are taken. If heat treatment is performed in continuous or batch-type furnaces controlled within a 50 °F [30 °C] range and equipped with recording pyrometers which yield complete heat-treatment records, then one test from each heat in a continuous run only shall be required instead of one test from each heat in each heat-treatment load.

13.8 *Impact Tests (Welded Tubes)*—On welded tube, additional impact tests of the same number as required in 13.7 shall be made to test the weld.

13.9 Specimens showing defects while being machined or prior to testing may be discarded and replacements shall be considered as original specimens.

14. Specimens for Impact Test

14.1 Notched-bar impact specimens shall be of the simple beam, Charpy-type, in accordance with Test Methods E23, Type A, with a V notch. Standard specimens 10 by 10 mm in cross section shall be used unless the material to be tested is of insufficient thickness, in which case the largest obtainable subsize specimens shall be used. Charpy specimens of width along the notch larger than 0.394 in. [10 mm] or smaller than 0.099 in. [2.5 mm] are not provided for in this specification.

14.2 Test specimens shall be obtained so that the longitudinal axis of the specimen is parallel to the longitudinal axis of the tube while the axis of the notch shall be perpendicular to the surface. On wall thicknesses of 1 in. [25 mm] or less, the specimens shall be obtained with their axial plane located at the midpoint; on wall thicknesses over 1 in. [25 mm], the specimens shall be obtained with their axial plane located 1/2 in. [12.5 mm] from the outer surface.

14.3 When testing welds the specimen shall be, whenever diameter and thickness permits, transverse to the longitudinal axis of the tube with the notch of the specimen in the welded joint and perpendicular to the surface. When diameter and thickness does not permit obtaining transverse specimens, longitudinal specimens in accordance with 14.2 shall be obtained. The bottom of the notch shall be located at the weld joint.

15. Impact Test

15.1 Except when the size of the finished tube is insufficient to permit obtaining subsize impact specimens, all material furnished under this specification and marked in accordance with Section 17 shall be tested for impact resistance at the temperature for the respective grades as prescribed in Table 6.

TABLE 6 Impact Temperature

Grade	Impact Test Temperature	
	°F	°C
1	−50	−45
3	−150	−100
6	−50	−45
7	−100	−75
8	−320	−195
9	−100	−75

TABLE 7 Impact Temperature Reduction

Specimen Width Along Notch or Actual Material Thickness[A]		Temperature Reduction, Degrees Colder	
Inches	Millimetres	°F	°C
0.394	10 (standard size)	0	0
0.354	9	0	0
0.315	8	0	0
0.295	7.5 (3 / 4 standard size)	5	3
0.276	7	8	4
0.262	6.67 (2 / 3 standard size)	10	5
0.236	6	15	8
0.197	5 (1 / 2 standard size)	20	11
0.158	4	30	17
0.131	3.33 (1 / 3 standard size)	35	19
0.118	3	40	22
0.099	2.5 (1 / 4 standard size)	50	28

[A]Straight line interpolation for intermediate values is permitted.

15.1.1 Special impact tests on individual lots of material may be made at other temperatures if agreed upon between the manufacturer and the purchaser.

15.2 The notched-bar impact test shall be made in accordance with the procedure for the simple beam, Charpy-type of test of Test Methods E23.

15.3 Impact tests specified for temperatures lower than +70 °F [20 °C] should be made with the following precautions. The impact test specimens as well as the handling tongs shall be cooled a sufficient time in a suitable container so that both reach the desired temperature. The temperature shall be measured with thermocouples, thermometers, or any other suitable devices and shall be controlled within ±3 °F [2 °C]. The specimens shall be quickly transferred from the cooling device to the anvil of the Charpy impact testing machine and broken with a time lapse of not more than 5 s.

15.4 When subsize Charpy impact specimens are used and the width along the notch is less than 80 % of the actual wall thickness of the original material, the specified Charpy impact test temperature for Grades 1, 3, 6, 7, and 9 shall be lower than the minimum temperature shown in Table 6 for the respective grade. Under these circumstances the temperature reduction values shall be by an amount equal to the difference (as shown in Table 7) between the temperature reduction corresponding to the actual material thickness and the temperature reduction corresponding to Charpy specimen width actually tested. The appendix shows some examples of how the temperature reductions are determined.

16. Hydrostatic or Nondestructive Electric Test

16.1 Each tube shall be subjected to the nondestructive electric test or the hydrostatic test in accordance with Specification A1016/A1016M. The type of test to be used shall be at the option of the manufacturer, unless otherwise specified in the purchase order.

17. Product Marking

17.1 Except as modified in 16.1.1, in addition to the marking prescribed in Specification A1016/A1016M, the marking shall include whether hot-finished, cold-drawn, seamless, or welded, and the letters "LT" followed by the temperature at which the impact tests were made, except when a lower test temperature is required because of reduced specimen size, in which case, the higher impact test temperature applicable to a full-size specimen should be marked.

17.1.1 When the size of the finished tube is insufficient to obtain subsize impact specimens, the marking shall not include the letters LT followed by an indicated test temperature unless Supplementary Requirement S 1 is specified.

SUPPLEMENTARY REQUIREMENTS

The following supplementary requirement shall apply only when specified by the purchaser in the inquiry, contract, or order.

S1. Nonstandard Test Specimens

S1.1 When the size of the finished tube is insufficient to permit obtaining subsize impact specimens, testing shall be a matter of agreement between the manufacturer and the purchaser.

APPENDIX

(Nonmandatory Information)

X1. DETERMINATION OF TEMPERATURE REDUCTIONS

X1.1 Under the circumstances stated in 15.4, the impact test temperatures specified in Table 6 must be lowered. The following examples are offered to describe the application of the provisions of 15.4.

X1.1.1 When subsize specimens are used (see 14.1) and the width along the notch of the subsize specimen is 80% or greater of the actual wall thickness of the original material, the provisions of 15.4 do not apply.

X1.1.1.1 For example, if the actual wall thickness of pipe was 0.200 in. [5.0 mm] and the width along the notch of the largest subsize specimen obtainable is 0.160 in. [4 mm] or greater, no reduction in test temperature is required.

X1.1.2 When the width along the subsize specimen notch is less than 80 % of the actual wall thickness of the pipe, the required reduction in test temperature is computed by taking the difference between the temperature reduction values shown in Table 7 for the actual pipe thickness and the specimen width used.

X1.1.2.1 For example, if the pipe were 0.262 in. [6.67 mm] thick and the width along the Charpy specimen notch was 3.33 mm (1/3 standard size), the test temperature would have to be lowered by 25 °F [14 °C] (that is, the temperature reduction corresponding to the subsize specimen is 35 °F [19 °C], the temperature reduction corresponding to the actual pipe thickness is 10 °F [5 °C]; the difference between these two values is the required reduction in test temperature).

SUMMARY OF CHANGES

Committee A01 has identified the location of selected changes to this specification since the last issue, A334/A334M – 04, that may impact the use of this specification. (Approved May 1, 2004)

(1) Moved Notes 2 and 3 into new Section 9 defining the sampling requirements.

(2) Renumbered subsequent sections and deleted all references to Notes 2 and 3 throughout.

Committee A01 has identified the location of selected changes to this specification since the last issue, A334/A334M – 99, that may impact the use of this specification. (Approved March 1, 2004)

(1) Replaced Specification A450/A450M with Specification A1016/A1016M in sections 2, 3, 4, 15, and 16.

Designation: A370 − 09aᵋ¹

Standard Test Methods and Definitions for Mechanical Testing of Steel Products[1]

This standard is issued under the fixed designation A370; the number immediately following the designation indicates the year of original adoption or, in the case of revision, the year of last revision. A number in parentheses indicates the year of last reapproval. A superscript epsilon (ε) indicates an editorial change since the last revision or reapproval.

This standard has been approved for use by agencies of the Department of Defense.

ε¹ NOTE—Sections 20 and 22.2.1 were editorially corrected in August 2009.

1. Scope*

1.1 These test methods[2] cover procedures and definitions for the mechanical testing of steels, stainless steels, and related alloys. The various mechanical tests herein described are used to determine properties required in the product specifications. Variations in testing methods are to be avoided, and standard methods of testing are to be followed to obtain reproducible and comparable results. In those cases in which the testing requirements for certain products are unique or at variance with these general procedures, the product specification testing requirements shall control.

1.2 The following mechanical tests are described:

	Sections
Tension	5 to 13
Bend	14
Hardness	15
Brinell	16
Rockwell	17
Portable	18
Impact	19 to 28
Keywords	29

1.3 Annexes covering details peculiar to certain products are appended to these test methods as follows:

	Annex
Bar Products	A1.1
Tubular Products	Annex A2
Fasteners	Annex A3
Round Wire Products	Annex A4
Significance of Notched-Bar Impact Testing	Annex A5
Converting Percentage Elongation of Round Specimens to Equivalents for Flat Specimens	Annex A6
Testing Multi-Wire Strand	Annex A7
Rounding of Test Data	Annex A8
Methods for Testing Steel Reinforcing Bars	Annex A9
Procedure for Use and Control of Heat-Cycle Simulation	Annex A10

1.4 The values stated in inch-pound units are to be regarded as the standard.

1.5 When this document is referenced in a metric product specification, the yield and tensile values may be determined in inch-pound (ksi) units then converted into SI (MPa) units. The elongation determined in inch-pound gauge lengths of 2 or 8 in. may be reported in SI unit gauge lengths of 50 or 200 mm, respectively, as applicable. Conversely, when this document is referenced in an inch-pound product specification, the yield and tensile values may be determined in SI units then converted into inch-pound units. The elongation determined in SI unit gauge lengths of 50 or 200 mm may be reported in inch-pound gauge lengths of 2 or 8 in., respectively, as applicable.

1.6 Attention is directed to ISO/IEC 17025 when there may be a need for information on criteria for evaluation of testing laboratories.

1.7 *This standard does not purport to address all of the safety concerns, if any, associated with its use. It is the responsibility of the user of this standard to establish appropriate safety and health practices and determine the applicability of regulatory limitations prior to use.*

2. Referenced Documents

2.1 *ASTM Standards:*[3]

A703/A703M Specification for Steel Castings, General Requirements, for Pressure-Containing Parts

A781/A781M Specification for Castings, Steel and Alloy, Common Requirements, for General Industrial Use

A833 Practice for Indentation Hardness of Metallic Materials by Comparison Hardness Testers

E4 Practices for Force Verification of Testing Machines

E6 Terminology Relating to Methods of Mechanical Testing

E8/E8M Test Methods for Tension Testing of Metallic Materials

E10 Test Method for Brinell Hardness of Metallic Materials

E18 Test Methods for Rockwell Hardness of Metallic Materials

---

[1] These test methods and definitions are under the jurisdiction of ASTM Committee A01 on Steel, Stainless Steel and Related Alloys and are the direct responsibility of Subcommittee A01.13 on Mechanical and Chemical Testing and Processing Methods of Steel Products and Processes.
Current edition approved June 1, 2009. Published August 2009. Originally approved in 1953. Last previous edition approved in 2009 as A370 – 09. DOI: 10.1520/A0370-09AE01.

[2] For *ASME Boiler and Pressure Vessel Code* applications see related Specification SA-370 in Section II of that Code.

[3] For referenced ASTM standards, visit the ASTM website, www.astm.org, or contact ASTM Customer Service at service@astm.org. For *Annual Book of ASTM Standards* volume information, refer to the standard's Document Summary page on the ASTM website.

A Summary of Changes section appears at the end of this standard.

Copyright © ASTM International, 100 Barr Harbor Drive, PO Box C700, West Conshohocken, PA 19428-2959, United States.

E23 Test Methods for Notched Bar Impact Testing of Metallic Materials

E29 Practice for Using Significant Digits in Test Data to Determine Conformance with Specifications

E83 Practice for Verification and Classification of Extensometer Systems

E110 Test Method for Indentation Hardness of Metallic Materials by Portable Hardness Testers

E190 Test Method for Guided Bend Test for Ductility of Welds

E290 Test Methods for Bend Testing of Material for Ductility

2.2 *ASME Document:*[4]
ASME Boiler and Pressure Vessel Code, Section VIII, Division I, Part UG-8

2.3 *ISO Standard:*[5]
ISO/IEC 17025 General Requirements for the Competence of Testing and Calibration Laboratories

3. General Precautions

3.1 Certain methods of fabrication, such as bending, forming, and welding, or operations involving heating, may affect the properties of the material under test. Therefore, the product specifications cover the stage of manufacture at which mechanical testing is to be performed. The properties shown by testing prior to fabrication may not necessarily be representative of the product after it has been completely fabricated.

3.2 Improper machining or preparation of test specimens may give erroneous results. Care should be exercised to assure good workmanship in machining. Improperly machined specimens should be discarded and other specimens substituted.

3.3 Flaws in the specimen may also affect results. If any test specimen develops flaws, the retest provision of the applicable product specification shall govern.

3.4 If any test specimen fails because of mechanical reasons such as failure of testing equipment or improper specimen preparation, it may be discarded and another specimen taken.

4. Orientation of Test Specimens

4.1 The terms "longitudinal test" and "transverse test" are used only in material specifications for wrought products and are not applicable to castings. When such reference is made to a test coupon or test specimen, the following definitions apply:

4.1.1 *Longitudinal Test*, unless specifically defined otherwise, signifies that the lengthwise axis of the specimen is parallel to the direction of the greatest extension of the steel during rolling or forging. The stress applied to a longitudinal tension test specimen is in the direction of the greatest extension, and the axis of the fold of a longitudinal bend test specimen is at right angles to the direction of greatest extension (Fig. 1, Fig. 2a, and 2b).

4.1.2 *Transverse Test*, unless specifically defined otherwise, signifies that the lengthwise axis of the specimen is at right angles to the direction of the greatest extension of the steel

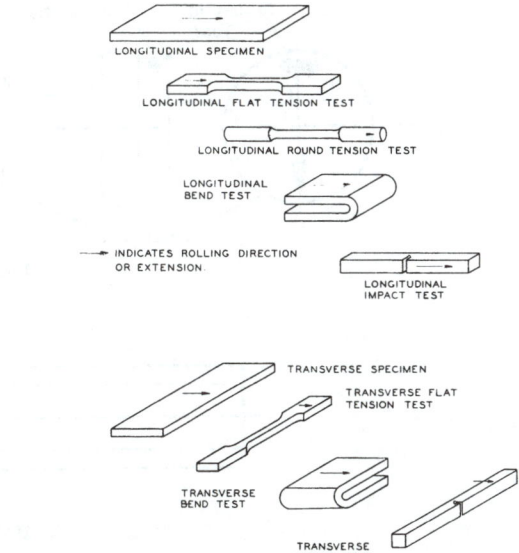

FIG. 1 The Relation of Test Coupons and Test Specimens to Rolling Direction or Extension (Applicable to General Wrought Products)

during rolling or forging. The stress applied to a transverse tension test specimen is at right angles to the greatest extension, and the axis of the fold of a transverse bend test specimen is parallel to the greatest extension (Fig. 1).

4.2 The terms "radial test" and "tangential test" are used in material specifications for some wrought circular products and are not applicable to castings. When such reference is made to a test coupon or test specimen, the following definitions apply:

4.2.1 *Radial Test*, unless specifically defined otherwise, signifies that the lengthwise axis of the specimen is perpendicular to the axis of the product and coincident with one of the radii of a circle drawn with a point on the axis of the product as a center (Fig. 2a).

4.2.2 *Tangential Test*, unless specifically defined otherwise, signifies that the lengthwise axis of the specimen is perpendicular to a plane containing the axis of the product and tangent to a circle drawn with a point on the axis of the product as a center (Fig. 2a, 2b, 2c, and 2d).

TENSION TEST

5. Description

5.1 The tension test related to the mechanical testing of steel products subjects a machined or full-section specimen of the material under examination to a measured load sufficient to cause rupture. The resulting properties sought are defined in Terminology E6.

5.2 In general, the testing equipment and methods are given in Test Methods E8/E8M. However, there are certain exceptions to Test Methods E8/E8M practices in the testing of steel, and these are covered in these test methods.

6. Terminology

6.1 For definitions of terms pertaining to tension testing, including tensile strength, yield point, yield strength, elongation, and reduction of area, reference should be made to Terminology E6.

---

[4] Available from American Society of Mechanical Engineers (ASME), ASME International Headquarters, Three Park Ave., New York, NY 10016-5990.

[5] Available from American National Standards Institute (ANSI), 25 W. 43rd St., 4th Floor, New York, NY 10036, http://www.ansi.org.

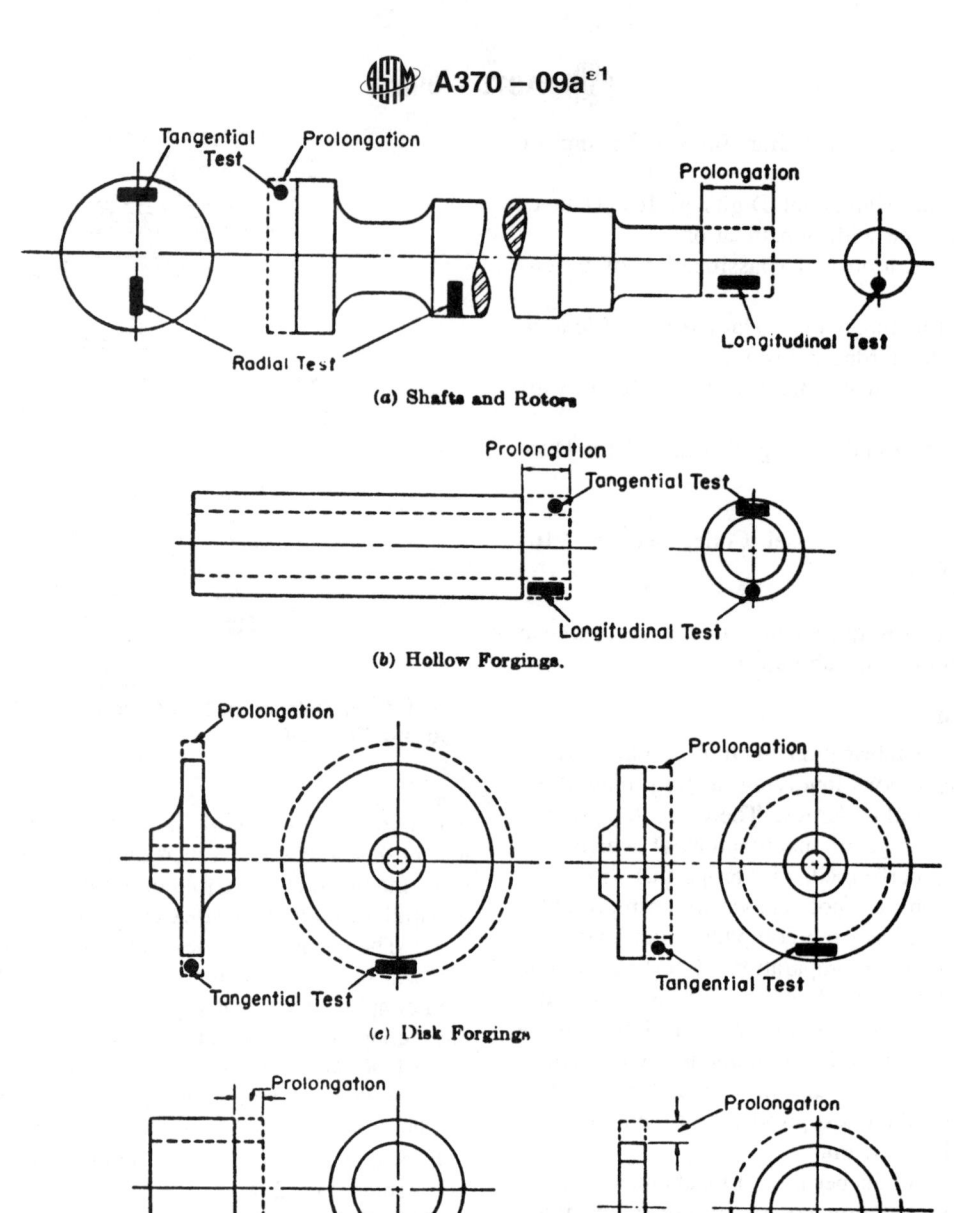

FIG. 2 Location of Longitudinal Tension Test Specimens in Rings Cut from Tubular Products

7. Testing Apparatus and Operations

7.1 *Loading Systems*—There are two general types of loading systems, mechanical (screw power) and hydraulic. These differ chiefly in the variability of the rate of load application. The older screw power machines are limited to a small number of fixed free running crosshead speeds. Some modern screw power machines, and all hydraulic machines permit stepless variation throughout the range of speeds.

7.2 The tension testing machine shall be maintained in good operating condition, used only in the proper loading range, and calibrated periodically in accordance with the latest revision of Practices E4.

Note 1—Many machines are equipped with stress-strain recorders for autographic plotting of stress-strain curves. It should be noted that some recorders have a load measuring component entirely separate from the load indicator of the testing machine. Such recorders are calibrated separately.

7.3 *Loading*—It is the function of the gripping or holding device of the testing machine to transmit the load from the heads of the machine to the specimen under test. The essential requirement is that the load shall be transmitted axially. This implies that the centers of the action of the grips shall be in alignment, insofar as practicable, with the axis of the specimen at the beginning and during the test and that bending or twisting be held to a minimum. For specimens with a reduced section, gripping of the specimen shall be restricted to the grip section. In the case of certain sections tested in full size, nonaxial loading is unavoidable and in such cases shall be permissible.

7.4 *Speed of Testing*—The speed of testing shall not be greater than that at which load and strain readings can be made accurately. In production testing, speed of testing is commonly expressed: (*1*) in terms of free running crosshead speed (rate of movement of the crosshead of the testing machine when not under load), (*2*) in terms of rate of separation of the two heads of the testing machine under load, (*3*) in terms of rate of stressing the specimen, or (*4*) in terms of rate of straining the specimen. The following limitations on the speed of testing are recommended as adequate for most steel products:

NOTE 2—Tension tests using closed-loop machines (with feedback control of rate) should not be performed using load control, as this mode of testing will result in acceleration of the crosshead upon yielding and elevation of the measured yield strength.

7.4.1 Any convenient speed of testing may be used up to one half the specified yield point or yield strength. When this point is reached, the free-running rate of separation of the crossheads shall be adjusted so as not to exceed 1/16 in. per min per inch of reduced section, or the distance between the grips for test specimens not having reduced sections. This speed shall be maintained through the yield point or yield strength. In determining the tensile strength, the free-running rate of separation of the heads shall not exceed 1/2 in. per min per inch of reduced section, or the distance between the grips for test specimens not having reduced sections. In any event, the minimum speed of testing shall not be less than 1/10 the specified maximum rates for determining yield point or yield strength and tensile strength.

7.4.2 It shall be permissible to set the speed of the testing machine by adjusting the free running crosshead speed to the above specified values, inasmuch as the rate of separation of heads under load at these machine settings is less than the specified values of free running crosshead speed.

7.4.3 As an alternative, if the machine is equipped with a device to indicate the rate of loading, the speed of the machine from half the specified yield point or yield strength through the yield point or yield strength may be adjusted so that the rate of stressing does not exceed 100 000 psi (690 MPa)/min. However, the minimum rate of stressing shall not be less than 10 000 psi (70 MPa)/min.

8. Test Specimen Parameters

8.1 *Selection*—Test coupons shall be selected in accordance with the applicable product specifications.

8.1.1 *Wrought Steels*—Wrought steel products are usually tested in the longitudinal direction, but in some cases, where size permits and the service justifies it, testing is in the transverse, radial, or tangential directions (see Fig. 1 and Fig. 2).

8.1.2 *Forged Steels*—For open die forgings, the metal for tension testing is usually provided by allowing extensions or prolongations on one or both ends of the forgings, either on all or a representative number as provided by the applicable product specifications. Test specimens are normally taken at mid-radius. Certain product specifications permit the use of a representative bar or the destruction of a production part for test purposes. For ring or disk-like forgings test metal is provided by increasing the diameter, thickness, or length of the forging. Upset disk or ring forgings, which are worked or extended by forging in a direction perpendicular to the axis of the forging, usually have their principal extension along concentric circles and for such forgings tangential tension specimens are obtained from extra metal on the periphery or end of the forging. For some forgings, such as rotors, radial tension tests are required. In such cases the specimens are cut or trepanned from specified locations.

8.2 *Size and Tolerances*—Test specimens shall be the full thickness or section of material as-rolled, or may be machined to the form and dimensions shown in Figs. 3-6, inclusive. The selection of size and type of specimen is prescribed by the applicable product specification. Full section specimens shall be tested in 8-in. (200-mm) gauge length unless otherwise specified in the product specification.

8.3 *Procurement of Test Specimens*—Specimens shall be sheared, blanked, sawed, trepanned, or oxygen-cut from portions of the material. They are usually machined so as to have a reduced cross section at mid-length in order to obtain uniform distribution of the stress over the cross section and to localize the zone of fracture. When test coupons are sheared, blanked, sawed, or oxygen-cut, care shall be taken to remove by machining all distorted, cold-worked, or heat-affected areas from the edges of the section used in evaluating the test.

8.4 *Aging of Test Specimens*—Unless otherwise specified, it shall be permissible to age tension test specimens. The time-temperature cycle employed must be such that the effects of previous processing will not be materially changed. It may be accomplished by aging at room temperature 24 to 48 h, or in shorter time at moderately elevated temperatures by boiling in water, heating in oil or in an oven.

8.5 *Measurement of Dimensions of Test Specimens*:

8.5.1 *Standard Rectangular Tension Test Specimens*—These forms of specimens are shown in Fig. 3. To determine the cross-sectional area, the center width dimension shall be measured to the nearest 0.005 in. (0.13 mm) for the 8-in. (200-mm) gauge length specimen and 0.001 in. (0.025 mm) for the 2-in. (50-mm) gauge length specimen in Fig. 3. The center thickness dimension shall be measured to the nearest 0.001 in. for both specimens.

8.5.2 *Standard Round Tension Test Specimens*—These forms of specimens are shown in Fig. 4 and Fig. 5. To determine the cross-sectional area, the diameter shall be measured at the center of the gauge length to the nearest 0.001 in. (0.025 mm) (see Table 1).

8.6 *General*—Test specimens shall be either substantially full size or machined, as prescribed in the product specifications for the material being tested.

8.6.1 Improperly prepared test specimens often cause unsatisfactory test results. It is important, therefore, that care be exercised in the preparation of specimens, particularly in the machining, to assure good workmanship.

8.6.2 It is desirable to have the cross-sectional area of the specimen smallest at the center of the gauge length to ensure fracture within the gauge length. This is provided for by the

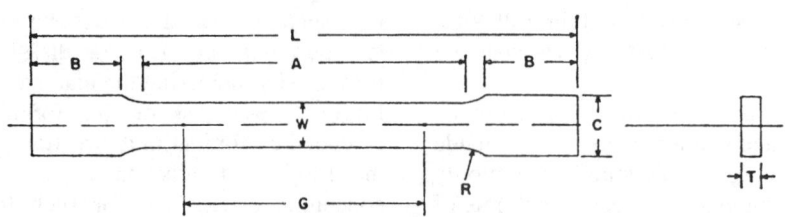

DIMENSIONS

	Standard Specimens						Subsize Specimen	
	Plate-Type, 1½-in. (40-mm) Wide							
	8-in. (200-mm) Gauge Length		2-in. (50-mm) Gauge Length		Sheet-Type, ½ in. (12.5-mm) Wide		¼-in. (6-mm) Wide	
	in.	mm	in.	mm	in.	mm	in.	mm
G—Gauge length (Notes 1 and 2)	8.00 ± 0.01	200 ± 0.25	2.000 ± 0.005	50.0 ± 0.10	2.000 ± 0.005	50.0 ± 0.010	1.000 ± 0.003	25.0 ± 0.08
W—Width (Notes 3, 5, and 6)	1½ + ⅛ − ¼	40 + 3 − 6	1½ + ⅛ − ¼	40 + 3 − 6	0.500 ± 0.010	12.5 ± 0.25	0.250 ± 0.002	6.25 ± 0.05
T—Thickness (Note 7)				Thickness of Material				
R—Radius of fillet, min (Note 4)	½	13	½	13	½	13	¼	6
L—Overall length, min (Notes 2 and 8)	18	450	8	200	8	200	4	100
A—Length of reduced section, min	9	225	2¼	60	2¼	60	1¼	32
B—Length of grip section, min (Note 9)	3	75	2	50	2	50	1¼	32
C—Width of grip section, approximate (Notes 4, 10, and 11)	2	50	2	50	¾	20	⅜	10

Note 1—For the 1½-in. (40-mm) wide specimens, punch marks for measuring elongation after fracture shall be made on the flat or on the edge of the specimen and within the reduced section. For the 8-in. (200-mm) gauge length specimen, a set of nine or more punch marks 1 in. (25 mm) apart, or one or more pairs of punch marks 8 in. (200 mm) apart may be used. For the 2-in. (50-mm) gauge length specimen, a set of three or more punch marks 1 in. (25 mm) apart, or one or more pairs of punch marks 2 in. (50 mm) apart may be used.

Note 2—For the ½-in. (12.5-mm) wide specimen, punch marks for measuring the elongation after fracture shall be made on the flat or on the edge of the specimen and within the reduced section. Either a set of three or more punch marks 1 in. (25 mm) apart or one or more pairs of punch marks 2 in. (50 mm) apart may be used.

Note 3—For the four sizes of specimens, the ends of the reduced section shall not differ in width by more than 0.004, 0.004, 0.002, or 0.001 in. (0.10, 0.10, 0.05, or 0.025 mm), respectively. Also, there may be a gradual decrease in width from the ends to the center, but the width at either end shall not be more than 0.015 in., 0.015 in., 0.005 in., or 0.003 in. (0.40, 0.40, 0.10 or 0.08 mm), respectively, larger than the width at the center.

Note 4—For each specimen type, the radii of all fillets shall be equal to each other with a tolerance of 0.05 in. (1.25 mm), and the centers of curvature of the two fillets at a particular end shall be located across from each other (on a line perpendicular to the centerline) within a tolerance of 0.10 in. (2.5 mm).

Note 5—For each of the four sizes of specimens, narrower widths (W and C) may be used when necessary. In such cases, the width of the reduced section should be as large as the width of the material being tested permits; however, unless stated specifically, the requirements for elongation in a product specification shall not apply when these narrower specimens are used. If the width of the material is less than W, the sides may be parallel throughout the length of the specimen.

Note 6—The specimen may be modified by making the sides parallel throughout the length of the specimen, the width and tolerances being the same as those specified above. When necessary, a narrower specimen may be used, in which case the width should be as great as the width of the material being tested permits. If the width is 1½ in. (38 mm) or less, the sides may be parallel throughout the length of the specimen.

Note 7—The dimension T is the thickness of the test specimen as provided for in the applicable product specification. Minimum nominal thickness of 1½-in. (40-mm) wide specimens shall be 3⁄16 in. (5 mm), except as permitted by the product specification. Maximum nominal thickness of ½-in. (12.5-mm) and ¼-in. (6-mm) wide specimens shall be ¾ in. (19 mm) and ¼ in. (6 mm), respectively.

Note 8—To aid in obtaining axial loading during testing of ¼-in. (6-mm) wide specimens, the overall length should be as large as the material will permit.

Note 9—It is desirable, if possible, to make the length of the grip section large enough to allow the specimen to extend into the grips a distance equal to two thirds or more of the length of the grips. If the thickness of ½-in. (13-mm) wide specimens is over ⅜ in. (10 mm), longer grips and correspondingly longer grip sections of the specimen may be necessary to prevent failure in the grip section.

Note 10—For standard sheet-type specimens and subsize specimens, the ends of the specimen shall be symmetrical with the center line of the reduced section within 0.01 and 0.005 in. (0.25 and 0.13 mm), respectively, except that for steel if the ends of the ½-in. (12.5-mm) wide specimen are symmetrical within 0.05 in. (1.0 mm), a specimen may be considered satisfactory for all but referee testing.

Note 11—For standard plate-type specimens, the ends of the specimen shall be symmetrical with the center line of the reduced section within 0.25 in. (6.35 mm), except for referee testing in which case the ends of the specimen shall be symmetrical with the center line of the reduced section within 0.10 in. (2.5 mm).

FIG. 3 Rectangular Tension Test Specimens

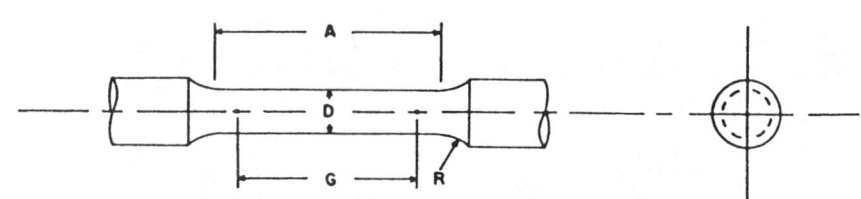

Nominal Diameter	Standard Specimen				Small-Size Specimens Proportional to Standard					
	in.	mm	in.	mm	in.	mm	in.	mm	in.	mm
	0.500	12.5	0.350	8.75	0.250	6.25	0.160	4.00	0.113	2.50
G—Gauge length	2.00± 0.005	50.0 ± 0.10	1.400± 0.005	35.0 ± 0.10	1.000± 0.005	25.0 ± 0.10	0.640± 0.005	16.0 ± 0.10	0.450± 0.005	10.0 ± 0.10
D—Diameter (Note 1)	0.500± 0.010	12.5± 0.25	0.350± 0.007	8.75 ± 0.18	0.250± 0.005	6.25 ± 0.12	0.160± 0.003	4.00 ± 0.08	0.113± 0.002	2.50 ± 0.05
R—Radius of fillet, min	3/8	10	1/4	6	3/16	5	5/32	4	3/32	2
A—Length of reduced section, min (Note 2)	2¼	60	1¾	45	1¼	32	¾	20	5/8	16

Note 1—The reduced section may have a gradual taper from the ends toward the center, with the ends not more than 1 percent larger in diameter than the center (controlling dimension).
Note 2—If desired, the length of the reduced section may be increased to accommodate an extensometer of any convenient gauge length. Reference marks for the measurement of elongation should, nevertheless, be spaced at the indicated gauge length.
Note 3—The gauge length and fillets shall be as shown, but the ends may be of any form to fit the holders of the testing machine in such a way that the load shall be axial (see Fig. 9). If the ends are to be held in wedge grips it is desirable, if possible, to make the length of the grip section great enough to allow the specimen to extend into the grips a distance equal to two thirds or more of the length of the grips.
Note 4—On the round specimens in Fig. 5 and Fig. 6, the gauge lengths are equal to four times the nominal diameter. In some product specifications other specimens may be provided for, but unless the 4-to-1 ratio is maintained within dimensional tolerances, the elongation values may not be comparable with those obtained from the standard test specimen.
Note 5—The use of specimens smaller than 0.250-in. (6.25-mm) diameter shall be restricted to cases when the material to be tested is of insufficient size to obtain larger specimens or when all parties agree to their use for acceptance testing. Smaller specimens require suitable equipment and greater skill in both machining and testing.
Note 6—Five sizes of specimens often used have diameters of approximately 0.505, 0.357, 0.252, 0.160, and 0.113 in., the reason being to permit easy calculations of stress from loads, since the corresponding cross sectional areas are equal or close to 0.200, 0.100, 0.0500, 0.0200, and 0.0100 in.^{2}, respectively. Thus, when the actual diameters agree with these values, the stresses (or strengths) may be computed using the simple multiplying factors 5, 10, 20, 50, and 100, respectively. (The metric equivalents of these fixed diameters do not result in correspondingly convenient cross sectional area and multiplying factors.)

FIG. 4 Standard 0.500-in. (12.5-mm) Round Tension Test Specimen with 2-in. (50-mm) Gauge Length and Examples of Small-Size Specimens Proportional to the Standard Specimens

taper in the gauge length permitted for each of the specimens described in the following sections.

8.6.3 For brittle materials it is desirable to have fillets of large radius at the ends of the gauge length.

9. Plate-Type Specimens

9.1 The standard plate-type test specimens are shown in Fig. 3. Such specimens are used for testing metallic materials in the form of plate, structural and bar-size shapes, and flat material having a nominal thickness of 3/16 in. (5 mm) or over. When product specifications so permit, other types of specimens may be used.

Note 3—When called for in the product specification, the 8-in. (200-mm) gauge length specimen of Fig. 3 may be used for sheet and strip material.

10. Sheet-Type Specimen

10.1 The standard sheet-type test specimen is shown in Fig. 3. This specimen is used for testing metallic materials in the form of sheet, plate, flat wire, strip, band, and hoop ranging in nominal thickness from 0.005 to 1 in. (0.13 to 25 mm). When product specifications so permit, other types of specimens may be used, as provided in Section 9 (see Note 3).

11. Round Specimens

11.1 The standard 0.500-in. (12.5-mm) diameter round test specimen shown in Fig. 4 is frequently used for testing metallic materials.

11.2 Fig. 4 also shows small size specimens proportional to the standard specimen. These may be used when it is necessary to test material from which the standard specimen or specimens shown in Fig. 3 cannot be prepared. Other sizes of small round specimens may be used. In any such small size specimen it is important that the gauge length for measurement of elongation be four times the diameter of the specimen (see Note 4, Fig. 4).

11.3 The type of specimen ends outside of the gauge length shall accommodate the shape of the product tested, and shall properly fit the holders or grips of the testing machine so that axial loads are applied with a minimum of load eccentricity and slippage. Fig. 5 shows specimens with various types of ends that have given satisfactory results.

12. Gauge Marks

12.1 The specimens shown in Figs. 3-6 shall be gauge marked with a center punch, scribe marks, multiple device, or drawn with ink. The purpose of these gauge marks is to determine the percent elongation. Punch marks shall be light,

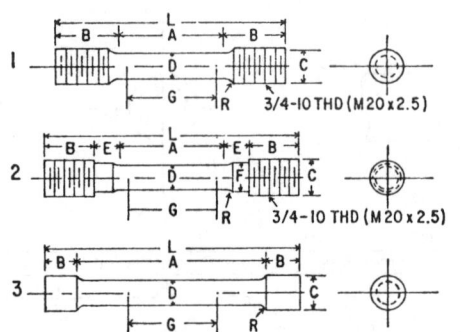
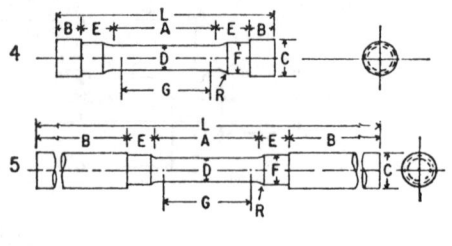

DIMENSIONS

	Specimen 1		Specimen 2		Specimen 3		Specimen 4		Specimen 5	
	in.	mm	in.	mm	in.	mm	in.	mm	in.	mm
G—Gauge length	2.000±0.005	50.0±0.10	2.000±0.005	50.0±0.10	2.000±0.005	50.0±0.10	2.000±0.005	50.0±0.10	2.00±0.005	50.0±0.10
D—Diameter (Note 1)	0.500±0.010	12.5±0.25	0.500±0.010	12.5±0.25	0.500±0.010	12.5±0.25	0.500±0.010	12.5±0.25	0.500±0.010	12.5±0.25
R—Radius of fillet, min	⅜	10	⅜	10	1/16	2	⅜	10	⅜	10
A—Length of reduced section	2¼, min	60, min	2¼, min	60, min	4, approximately	100, approximately	2¼, min	60, min	2¼, min	60, min
L—Overall length, approximate	5	125	5½	140	5½	140	4¾	120	9½	240
B—Grip section (Note 2)	1⅜, approximately	35, approximately	1, approximately	25, approximately	¾, approximately	20, approximately	½, approximately	13, approximately	3, min	75, min
C—Diameter of end section	¾	20	¾	20	23/32	18	⅞	22	¾	20
E—Length of shoulder and fillet section, approximate	...	...	⅝	16	...	...	¾	20	⅝	16
F—Diameter of shoulder	...	...	⅝	16	...	...	⅝	16	19/32	15

NOTE 1—The reduced section may have a gradual taper from the ends toward the center with the ends not more than 0.005 in. (0.10 mm) larger in diameter than the center.

NOTE 2—On Specimen 5 it is desirable, if possible, to make the length of the grip section great enough to allow the specimen to extend into the grips a distance equal to two thirds or more of the length of the grips.

NOTE 3—The types of ends shown are applicable for the standard 0.500-in. round tension test specimen; similar types can be used for subsize specimens. The use of UNF series of threads (¾ by 16, ½ by 20, ⅜ by 24, and ¼ by 28) is suggested for high-strength brittle materials to avoid fracture in the thread portion.

FIG. 5 Suggested Types of Ends for Standard Round Tension Test Specimens

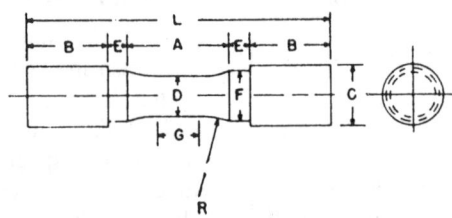

DIMENSIONS

	Specimen 1		Specimen 2		Specimen 3	
	in.	mm	in.	mm	in.	mm
G—Length of parallel	Shall be equal to or greater than diameter D					
D—Diameter	0.500±0.010	12.5±0.25	0.750±0.015	20.0±0.40	1.25±0.025	30.0±0.60
R—Radius of fillet, min	1	25	1	25	2	50
A—Length of reduced section, min	1¼	32	1½	38	2¼	60
L—Over-all length, min	3¾	95	4	100	6⅜	160
B—Grip section, approximate	1	25	1	25	1¾	45
C—Diameter of end section, approximate	¾	20	1⅛	30	1⅞	48
E—Length of shoulder, min	¼	6	¼	6	5/16	8
F—Diameter of shoulder	⅝ ± 1/64	16.0±0.40	15/16 ± 1/64	24.0±0.40	1 7/16 ± 1/64	36.5±0.40

NOTE 1—The reduced section and shoulders (dimensions A, D, E, F, G, and R) shall be shown, but the ends may be of any form to fit the holders of the testing machine in such a way that the load shall be axial. Commonly the ends are threaded and have the dimensions B and C given above.

FIG. 6 Standard Tension Test Specimens for Cast Iron

TABLE 1 Multiplying Factors to Be Used for Various Diameters of Round Test Specimens

Standard Specimen			Small Size Specimens Proportional to Standard					
0.500 in. Round			0.350 in. Round			0.250 in. Round		
Actual Diameter, in.	Area, in.²	Multiplying Factor	Actual Diameter, in.	Area, in.²	Multiplying Factor	Actual Diameter, in.	Area, in.²	Multiplying Factor
0.490	0.1886	5.30	0.343	0.0924	10.82	0.245	0.0471	21.21
0.491	0.1893	5.28	0.344	0.0929	10.76	0.246	0.0475	21.04
0.492	0.1901	5.26	0.345	0.0935	10.70	0.247	0.0479	20.87
0.493	0.1909	5.24	0.346	0.0940	10.64	0.248	0.0483	20.70
0.494	0.1917	5.22	0.347	0.0946	10.57	0.249	0.0487	20.54
0.495	0.1924	5.20	0.348	0.0951	10.51	0.250	0.0491	20.37
0.496	0.1932	5.18	0.349	0.0957	10.45	0.251	0.0495 (0.05)[A]	20.21 (20.0)[A]
0.497	0.1940	5.15	0.350	0.0962	10.39	0.252	0.0499 (0.05)[A]	20.05 (20.0)[A]
0.498	0.1948	5.13	0.351	0.0968	10.33	0.253	0.0503 (0.05)[A]	19.89 (20.0)[A]
0.499	0.1956	5.11	0.352	0.0973	10.28	0.254	0.0507	19.74
0.500	0.1963	5.09	0.353	0.0979	10.22	0.255	0.0511	19.58
0.501	0.1971	5.07	0.354	0.0984	10.16	...	...	...
0.502	0.1979	5.05	0.355	0.0990	10.10	...	...	...
0.503	0.1987	5.03	0.356	0.0995 (0.1)[A]	10.05 (10.0)[A]	...	...	...
0.504	0.1995	5.01	0.357	0.1001 (0.1)[A]	9.99 (10.0)[A]	...	...	...
0.505	0.2003 (0.2)[A]	4.99 (5.0)[A]	...	...	...	...	...	...
0.506	0.2011 (0.2)[A]	4.97 (5.0)[A]	...	...	...	...	...	...
0.507	0.2019	4.95	...	...	...	...	...	...
0.508	0.2027	4.93	...	...	...	...	...	...
0.509	0.2035	4.91	...	...	...	...	...	...
0.510	0.2043	4.90	...	...	...	...	...	...

[A] The values in parentheses may be used for ease in calculation of stresses, in pounds per square inch, as permitted in 5 of Fig. 4.

sharp, and accurately spaced. The localization of stress at the marks makes a hard specimen susceptible to starting fracture at the punch marks. The gauge marks for measuring elongation after fracture shall be made on the flat or on the edge of the flat tension test specimen and within the parallel section; for the 8-in. gauge length specimen, Fig. 3, one or more sets of 8-in. gauge marks may be used, intermediate marks within the gauge length being optional. Rectangular 2-in. gauge length specimens, Fig. 3, and round specimens, Fig. 4, are gauge marked with a double-pointed center punch or scribe marks. One or more sets of gauge marks may be used; however, one set must be approximately centered in the reduced section. These same precautions shall be observed when the test specimen is full section.

13. Determination of Tensile Properties

13.1 *Yield Point*—Yield point is the first stress in a material, less than the maximum obtainable stress, at which an increase in strain occurs without an increase in stress. Yield point is intended for application only for materials that may exhibit the unique characteristic of showing an increase in strain without an increase in stress. The stress-strain diagram is characterized by a sharp knee or discontinuity. Determine yield point by one of the following methods:

13.1.1 *Drop of the Beam or Halt of the Pointer Method*—In this method, apply an increasing load to the specimen at a uniform rate. When a lever and poise machine is used, keep the beam in balance by running out the poise at approximately a steady rate. When the yield point of the material is reached, the increase of the load will stop, but run the poise a trifle beyond the balance position, and the beam of the machine will drop for a brief but appreciable interval of time. When a machine equipped with a load-indicating dial is used there is a halt or hesitation of the load-indicating pointer corresponding to the drop of the beam. Note the load at the "drop of the beam" or the "halt of the pointer" and record the corresponding stress as the yield point.

13.1.2 *Autographic Diagram Method*—When a sharp-kneed stress-strain diagram is obtained by an autographic recording device, take the stress corresponding to the top of the knee (Fig. 7), or the stress at which the curve drops as the yield point.

13.1.3 *Total Extension Under Load Method*—When testing material for yield point and the test specimens may not exhibit a well-defined disproportionate deformation that characterizes a yield point as measured by the drop of the beam, halt of the pointer, or autographic diagram methods described in 13.1.1 and 13.1.2, a value equivalent to the yield point in its practical significance may be determined by the following method and may be recorded as yield point: Attach a Class C or better extensometer (Note 4 and Note 5) to the specimen. When the load producing a specified extension (Note 6) is reached record the stress corresponding to the load as the yield point (Fig. 8).

NOTE 4—Automatic devices are available that determine the load at the specified total extension without plotting a stress-strain curve. Such devices may be used if their accuracy has been demonstrated. Multiplying calipers and other such devices are acceptable for use provided their accuracy has been demonstrated as equivalent to a Class C extensometer.

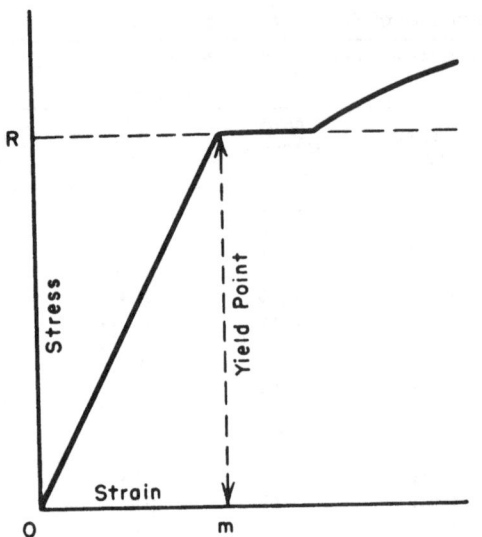

FIG. 7 Stress-Strain Diagram Showing Yield Point Corresponding with Top of Knee

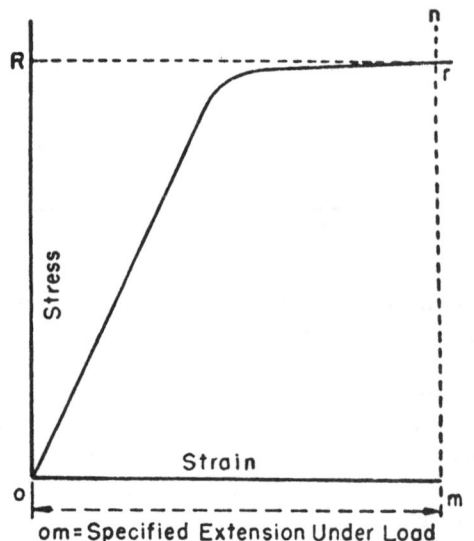

FIG. 8 Stress-Strain Diagram Showing Yield Point or Yield Strength by Extension Under Load Method

NOTE 5—Reference should be made to Practice E83.

NOTE 6—For steel with a yield point specified not over 80 000 psi (550 MPa), an appropriate value is 0.005 in./in. of gauge length. For values above 80 000 psi, this method is not valid unless the limiting total extension is increased.

NOTE 7—The shape of the initial portion of an autographically determined stress-strain (or a load-elongation) curve may be influenced by numerous factors such as the seating of the specimen in the grips, the straightening of a specimen bent due to residual stresses, and the rapid loading permitted in 7.4.1. Generally, the aberrations in this portion of the curve should be ignored when fitting a modulus line, such as that used to determine the extension-under-load yield, to the curve.

13.2 *Yield Strength*—Yield strength is the stress at which a material exhibits a specified limiting deviation from the proportionality of stress to strain. The deviation is expressed in terms of strain, percent offset, total extension under load, etc. Determine yield strength by one of the following methods:

13.2.1 *Offset Method*—To determine the yield strength by the "offset method," it is necessary to secure data (autographic or numerical) from which a stress-strain diagram with a distinct modulus characteristic of the material being tested may be drawn. Then on the stress-strain diagram (Fig. 9) lay off *Om* equal to the specified value of the offset, draw *mn* parallel to *OA*, and thus locate *r*, the intersection of *mn* with the stress-strain curve corresponding to load *R*, which is the yield-strength load. In recording values of yield strength obtained by this method, the value of offset specified or used, or both, shall be stated in parentheses after the term yield strength, for example:

$$\text{Yield strength } (0.2 \% \text{ offset}) = 52\,000 \text{ psi } (360 \text{ MPa}) \tag{1}$$

When the offset is 0.2 % or larger, the extensometer used shall qualify as a Class B2 device over a strain range of 0.05 to 1.0 %. If a smaller offset is specified, it may be necessary to specify a more accurate device (that is, a Class B1 device) or reduce the lower limit of the strain range (for example, to 0.01 %) or both. See also Note 9 for automatic devices.

NOTE 8—For stress-strain diagrams not containing a distinct modulus, such as for some cold-worked materials, it is recommended that the extension under load method be utilized. If the offset method is used for materials without a distinct modulus, a modulus value appropriate for the material being tested should be used: 30 000 000 psi (207 000 MPa) for carbon steel; 29 000 000 psi (200 000 MPa) for ferritic stainless steel; 28 000 000 psi (193 000 MPa) for austenitic stainless steel. For special alloys, the producer should be contacted to discuss appropriate modulus values.

13.2.2 *Extension Under Load Method*—For tests to determine the acceptance or rejection of material whose stress-strain characteristics are well known from previous tests of similar material in which stress-strain diagrams were plotted, the total strain corresponding to the stress at which the specified offset (see Note 9 and Note 10) occurs will be known within

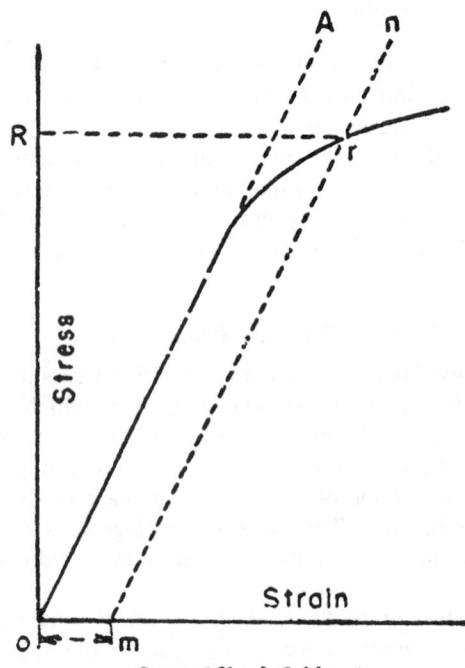

FIG. 9 Stress-Strain Diagram for Determination of Yield Strength by the Offset Method

satisfactory limits. The stress on the specimen, when this total strain is reached, is the value of the yield strength. In recording values of yield strength obtained by this method, the value of "extension" specified or used, or both, shall be stated in parentheses after the term yield strength, for example:

$$\text{Yield strength } (0.5 \% \ EUL) = 52\,000 \text{ psi } (360 \text{ MPa}) \quad (2)$$

The total strain can be obtained satisfactorily by use of a Class B1 extensometer (Note 4, Note 5, and Note 7).

NOTE 9—Automatic devices are available that determine offset yield strength without plotting a stress-strain curve. Such devices may be used if their accuracy has been demonstrated.

NOTE 10—The appropriate magnitude of the extension under load will obviously vary with the strength range of the particular steel under test. In general, the value of extension under load applicable to steel at any strength level may be determined from the sum of the proportional strain and the plastic strain expected at the specified yield strength. The following equation is used:

$$\text{Extension under load, in./in. of gauge length} = (YS/E) + r \quad (3)$$

where:
YS = specified yield strength, psi or MPa,
E = modulus of elasticity, psi or MPa, and
r = limiting plastic strain, in./in.

13.3 *Tensile Strength*— Calculate the tensile strength by dividing the maximum load the specimen sustains during a tension test by the original cross-sectional area of the specimen.

13.4 *Elongation*:

13.4.1 Fit the ends of the fractured specimen together carefully and measure the distance between the gauge marks to the nearest 0.01 in. (0.25 mm) for gauge lengths of 2 in. and under, and to the nearest 0.5 % of the gauge length for gauge lengths over 2 in. A percentage scale reading to 0.5 % of the gauge length may be used. The elongation is the increase in length of the gauge length, expressed as a percentage of the original gauge length. In recording elongation values, give both the percentage increase and the original gauge length.

13.4.2 If any part of the fracture takes place outside of the middle half of the gauge length or in a punched or scribed mark within the reduced section, the elongation value obtained may not be representative of the material. If the elongation so measured meets the minimum requirements specified, no further testing is indicated, but if the elongation is less than the minimum requirements, discard the test and retest.

13.4.3 Automated tensile testing methods using extensometers allow for the measurement of elongation in a method described below. Elongation may be measured and reported either this way, or as in the method described above, fitting the broken ends together. Either result is valid.

13.4.4 Elongation at fracture is defined as the elongation measured just prior to the sudden decrease in force associated with fracture. For many ductile materials not exhibiting a sudden decrease in force, the elongation at fracture can be taken as the strain measured just prior to when the force falls below 10 % of the maximum force encountered during the test.

13.4.4.1 Elongation at fracture shall include elastic and plastic elongation and may be determined with autographic or automated methods using extensometers verified over the strain range of interest. Use a class B2 or better extensometer for materials having less than 5 % elongation; a class C or better extensometer for materials having elongation greater than or equal to 5 % but less than 50 %; and a class D or better extensometer for materials having 50 % or greater elongation. In all cases, the extensometer gauge length shall be the nominal gauge length required for the specimen being tested. Due to the lack of precision in fitting fractured ends together, the elongation after fracture using the manual methods of the preceding paragraphs may differ from the elongation at fracture determined with extensometers.

13.4.4.2 Percent elongation at fracture may be calculated directly from elongation at fracture data and be reported instead of percent elongation as calculated in 13.4.1. However, these two parameters are not interchangeable. Use of the elongation at fracture method generally provides more repeatable results.

13.5 *Reduction of Area*—Fit the ends of the fractured specimen together and measure the mean diameter or the width and thickness at the smallest cross section to the same accuracy as the original dimensions. The difference between the area thus found and the area of the original cross section expressed as a percentage of the original area is the reduction of area.

BEND TEST

14. Description

14.1 The bend test is one method for evaluating ductility, but it cannot be considered as a quantitative means of predicting service performance in all bending operations. The severity of the bend test is primarily a function of the angle of bend of the inside diameter to which the specimen is bent, and of the cross section of the specimen. These conditions are varied according to location and orientation of the test specimen and the chemical composition, tensile properties, hardness, type, and quality of the steel specified. Test Method E190 and Test Method E290 may be consulted for methods of performing the test.

14.2 Unless otherwise specified, it shall be permissible to age bend test specimens. The time-temperature cycle employed must be such that the effects of previous processing will not be materially changed. It may be accomplished by aging at room temperature 24 to 48 h, or in shorter time at moderately elevated temperatures by boiling in water or by heating in oil or in an oven.

14.3 Bend the test specimen at room temperature to an inside diameter, as designated by the applicable product specifications, to the extent specified without major cracking on the outside of the bent portion. The speed of bending is ordinarily not an important factor.

HARDNESS TEST

15. General

15.1 A hardness test is a means of determining resistance to penetration and is occasionally employed to obtain a quick approximation of tensile strength. Table 2, Table 3, Table 4, and Table 5 are for the conversion of hardness measurements from one scale to another or to approximate tensile strength. These conversion values have been obtained from computer-generated curves and are presented to the nearest 0.1 point to

TABLE 2 Approximate Hardness Conversion Numbers for Nonaustenitic Steels[A] **(Rockwell C to Other Hardness Numbers)**

Rockwell C Scale, 150-kgf Load, Diamond Penetrator	Vickers Hardness Number	Brinell Hardness, 3000-kgf Load, 10-mm Ball	Knoop Hardness, 500-gf Load and Over	Rockwell A Scale, 60-kgf Load, Diamond Penetrator	Rockwell Superficial Hardness			Approximate Tensile Strength, ksi (MPa)
					15N Scale, 15-kgf Load, Diamond Penetrator	30N Scale, 30-kgf Load, Diamond Penetrator	45N Scale, 45-kgf Load, Diamond Penetrator	
68	940	...	920	85.6	93.2	84.4	75.4	...
67	900	...	895	85.0	92.9	83.6	74.2	...
66	865	...	870	84.5	92.5	82.8	73.3	...
65	832	739	846	83.9	92.2	81.9	72.0	...
64	800	722	822	83.4	91.8	81.1	71.0	...
63	772	706	799	82.8	91.4	80.1	69.9	...
62	746	688	776	82.3	91.1	79.3	68.8	...
61	720	670	754	81.8	90.7	78.4	67.7	...
60	697	654	732	81.2	90.2	77.5	66.6	...
59	674	634	710	80.7	89.8	76.6	65.5	351 (2420)
58	653	615	690	80.1	89.3	75.7	64.3	338 (2330)
57	633	595	670	79.6	88.9	74.8	63.2	325 (2240)
56	613	577	650	79.0	88.3	73.9	62.0	313 (2160)
55	595	560	630	78.5	87.9	73.0	60.9	301 (2070)
54	577	543	612	78.0	87.4	72.0	59.8	292 (2010)
53	560	525	594	77.4	86.9	71.2	58.6	283 (1950)
52	544	512	576	76.8	86.4	70.2	57.4	273 (1880)
51	528	496	558	76.3	85.9	69.4	56.1	264 (1820)
50	513	482	542	75.9	85.5	68.5	55.0	255 (1760)
49	498	468	526	75.2	85.0	67.6	53.8	246 (1700)
48	484	455	510	74.7	84.5	66.7	52.5	238 (1640)
47	471	442	495	74.1	83.9	65.8	51.4	229 (1580)
46	458	432	480	73.6	83.5	64.8	50.3	221 (1520)
45	446	421	466	73.1	83.0	64.0	49.0	215 (1480)
44	434	409	452	72.5	82.5	63.1	47.8	208 (1430)
43	423	400	438	72.0	82.0	62.2	46.7	201 (1390)
42	412	390	426	71.5	81.5	61.3	45.5	194 (1340)
41	402	381	414	70.9	80.9	60.4	44.3	188 (1300)
40	392	371	402	70.4	80.4	59.5	43.1	182 (1250)
39	382	362	391	69.9	79.9	58.6	41.9	177 (1220)
38	372	353	380	69.4	79.4	57.7	40.8	171 (1180)
37	363	344	370	68.9	78.8	56.8	39.6	166 (1140)
36	354	336	360	68.4	78.3	55.9	38.4	161 (1110)
35	345	327	351	67.9	77.7	55.0	37.2	156 (1080)
34	336	319	342	67.4	77.2	54.2	36.1	152 (1050)
33	327	311	334	66.8	76.6	53.3	34.9	149 (1030)
32	318	301	326	66.3	76.1	52.1	33.7	146 (1010)
31	310	294	318	65.8	75.6	51.3	32.5	141 (970)
30	302	286	311	65.3	75.0	50.4	31.3	138 (950)
29	294	279	304	64.6	74.5	49.5	30.1	135 (930)
28	286	271	297	64.3	73.9	48.6	28.9	131 (900)
27	279	264	290	63.8	73.3	47.7	27.8	128 (880)
26	272	258	284	63.3	72.8	46.8	26.7	125 (860)
25	266	253	278	62.8	72.2	45.9	25.5	123 (850)
24	260	247	272	62.4	71.6	45.0	24.3	119 (820)
23	254	243	266	62.0	71.0	44.0	23.1	117 (810)
22	248	237	261	61.5	70.5	43.2	22.0	115 (790)
21	243	231	256	61.0	69.9	42.3	20.7	112 (770)
20	238	226	251	60.5	69.4	41.5	19.6	110 (760)

[A] This table gives the approximate interrelationships of hardness values and approximate tensile strength of steels. It is possible that steels of various compositions and processing histories will deviate in hardness-tensile strength relationship from the data presented in this table. The data in this table should not be used for austenitic stainless steels, but have been shown to be applicable for ferritic and martensitic stainless steels. The data in this table should not be used to establish a relationship between hardness values and tensile strength of hard drawn wire. Where more precise conversions are required, they should be developed specially for each steel composition, heat treatment, and part. Caution should be exercised if conversions from this table are used for the acceptance or rejection of product. The approximate interrelationships may affect acceptance or rejection.

TABLE 3 Approximate Hardness Conversion Numbers for Nonaustenitic Steels[A] (Rockwell B to Other Hardness Numbers)

Rockwell B Scale, 100-kgf Load 1/16-in. (1.588-mm) Ball	Vickers Hardness Number	Brinell Hardness, 3000-kgf Load, 10-mm Ball	Knoop Hardness, 500-gf Load and Over	Rockwell A Scale, 60-kgf Load, Diamond Penetrator	Rockwell F Scale, 60-kgf Load, 1/16-in. (1.588-mm) Ball	Rockwell Superficial Hardness			Approximate Tensile Strength ksi (MPa)
						15T Scale, 15-kgf Load, 1/16-in. (1.588-mm) Ball	30T Scale, 30-kgf Load, 1/16-in. (1.588-mm) Ball	45T Scale, 45-kgf Load, 1/16-in. (1.588-mm) Ball	
100	240	240	251	61.5	...	93.1	83.1	72.9	116 (800)
99	234	234	246	60.9	...	92.8	82.5	71.9	114 (785)
98	228	228	241	60.2	...	92.5	81.8	70.9	109 (750)
97	222	222	236	59.5	...	92.1	81.1	69.9	104 (715)
96	216	216	231	58.9	...	91.8	80.4	68.9	102 (705)
95	210	210	226	58.3	...	91.5	79.8	67.9	100 (690)
94	205	205	221	57.6	...	91.2	79.1	66.9	98 (675)
93	200	200	216	57.0	...	90.8	78.4	65.9	94 (650)
92	195	195	211	56.4	...	90.5	77.8	64.8	92 (635)
91	190	190	206	55.8	...	90.2	77.1	63.8	90 (620)
90	185	185	201	55.2	...	89.9	76.4	62.8	89 (615)
89	180	180	196	54.6	...	89.5	75.8	61.8	88 (605)
88	176	176	192	54.0	...	89.2	75.1	60.8	86 (590)
87	172	172	188	53.4	...	88.9	74.4	59.8	84 (580)
86	169	169	184	52.8	...	88.6	73.8	58.8	83 (570)
85	165	165	180	52.3	...	88.2	73.1	57.8	82 (565)
84	162	162	176	51.7	...	87.9	72.4	56.8	81 (560)
83	159	159	173	51.1	...	87.6	71.8	55.8	80 (550)
82	156	156	170	50.6	...	87.3	71.1	54.8	77 (530)
81	153	153	167	50.0	...	86.9	70.4	53.8	73 (505)
80	150	150	164	49.5	...	86.6	69.7	52.8	72 (495)
79	147	147	161	48.9	...	86.3	69.1	51.8	70 (485)
78	144	144	158	48.4	...	86.0	68.4	50.8	69 (475)
77	141	141	155	47.9	...	85.6	67.7	49.8	68 (470)
76	139	139	152	47.3	...	85.3	67.1	48.8	67 (460)
75	137	137	150	46.8	99.6	85.0	66.4	47.8	66 (455)
74	135	135	147	46.3	99.1	84.7	65.7	46.8	65 (450)
73	132	132	145	45.8	98.5	84.3	65.1	45.8	64 (440)
72	130	130	143	45.3	98.0	84.0	64.4	44.8	63 (435)
71	127	127	141	44.8	97.4	83.7	63.7	43.8	62 (425)
70	125	125	139	44.3	96.8	83.4	63.1	42.8	61 (420)
69	123	123	137	43.8	96.2	83.0	62.4	41.8	60 (415)
68	121	121	135	43.3	95.6	82.7	61.7	40.8	59 (405)
67	119	119	133	42.8	95.1	82.4	61.0	39.8	58 (400)
66	117	117	131	42.3	94.5	82.1	60.4	38.7	57 (395)
65	116	116	129	41.8	93.9	81.8	59.7	37.7	56 (385)
64	114	114	127	41.4	93.4	81.4	59.0	36.7	...
63	112	112	125	40.9	92.8	81.1	58.4	35.7	...
62	110	110	124	40.4	92.2	80.8	57.7	34.7	...
61	108	108	122	40.0	91.7	80.5	57.0	33.7	...
60	107	107	120	39.5	91.1	80.1	56.4	32.7	...
59	106	106	118	39.0	90.5	79.8	55.7	31.7	...
58	104	104	117	38.6	90.0	79.5	55.0	30.7	...
57	103	103	115	38.1	89.4	79.2	54.4	29.7	...
56	101	101	114	37.7	88.8	78.8	53.7	28.7	...
55	100	100	112	37.2	88.2	78.5	53.0	27.7	...
54	...	...	111	36.8	87.7	78.2	52.4	26.7	...
53	...	...	110	36.3	87.1	77.9	51.7	25.7	...
52	...	...	109	35.9	86.5	77.5	51.0	24.7	...
51	...	...	108	35.5	86.0	77.2	50.3	23.7	...
50	...	...	107	35.0	85.4	76.9	49.7	22.7	...
49	...	...	106	34.6	84.8	76.6	49.0	21.7	...
48	...	...	105	34.1	84.3	76.2	48.3	20.7	...
47	...	...	104	33.7	83.7	75.9	47.7	19.7	...
46	...	...	103	33.3	83.1	75.6	47.0	18.7	...
45	...	...	102	32.9	82.6	75.3	46.3	17.7	...
44	...	...	101	32.4	82.0	74.9	45.7	16.7	...
43	...	...	100	32.0	81.4	74.6	45.0	15.7	...
42	...	...	99	31.6	80.8	74.3	44.3	14.7	...
41	...	...	98	31.2	80.3	74.0	43.7	13.6	...
40	...	...	97	30.7	79.7	73.6	43.0	12.6	...
39	...	...	96	30.3	79.1	73.3	42.3	11.6	...
38	...	...	95	29.9	78.6	73.0	41.6	10.6	...
37	...	...	94	29.5	78.0	72.7	41.0	9.6	...
36	...	...	93	29.1	77.4	72.3	40.3	8.6	...
35	...	...	92	28.7	76.9	72.0	39.6	7.6	...
34	...	...	91	28.2	76.3	71.7	39.0	6.6	...
33	...	...	90	27.8	75.7	71.4	38.3	5.6	...

TABLE 3 *Continued*

Rockwell B Scale, 100-kgf Load 1/16-in. (1.588-mm) Ball	Vickers Hardness Number	Brinell Hardness, 3000-kgf Load, 10-mm Ball	Knoop Hardness, 500-gf Load and Over	Rockwell A Scale, 60-kgf Load, Diamond Penetrator	Rockwell F Scale, 60-kgf Load, 1/16-in. (1.588-mm) Ball	Rockwell Superficial Hardness			Approximate Tensile Strength ksi (MPa)
						15T Scale, 15-kgf Load, 1/16-in. (1.588-mm) Ball	30T Scale, 30-kgf Load, 1/16-in. (1.588-mm) Ball	45T Scale, 45-kgf Load, 1/16-in. (1.588-mm) Ball	
32	...	...	89	27.4	75.2	71.0	37.6	4.6	...
31	...	...	88	27.0	74.6	70.7	37.0	3.6	...
30	...	...	87	26.6	74.0	70.4	36.3	2.6	...

[A] This table gives the approximate interrelationships of hardness values and approximate tensile strength of steels. It is possible that steels of various compositions and processing histories will deviate in hardness-tensile strength relationship from the data presented in this table. The data in this table should not be used for austenitic stainless steels, but have been shown to be applicable for ferritic and martensitic stainless steels. The data in this table should not be used to establish a relationship between hardness values and tensile strength of hard drawn wire. Where more precise conversions are required, they should be developed specially for each steel composition, heat treatment, and part.

TABLE 4 Approximate Hardness Conversion Numbers for Austenitic Steels (Rockwell C to other Hardness Numbers)

Rockwell C Scale, 150-kgf Load, Diamond Penetrator	Rockwell A Scale, 60-kgf Load, Diamond Penetrator	Rockwell Superficial Hardness		
		15N Scale, 15-kgf Load, Diamond Penetrator	30N Scale, 30-kgf Load, Diamond Penetrator	45N Scale, 45-kgf Load, Diamond Penetrator
48	74.4	84.1	66.2	52.1
47	73.9	83.6	65.3	50.9
46	73.4	83.1	64.5	49.8
45	72.9	82.6	63.6	48.7
44	72.4	82.1	62.7	47.5
43	71.9	81.6	61.8	46.4
42	71.4	81.0	61.0	45.2
41	70.9	80.5	60.1	44.1
40	70.4	80.0	59.2	43.0
39	69.9	79.5	58.4	41.8
38	69.3	79.0	57.5	40.7
37	68.8	78.5	56.6	39.6
36	68.3	78.0	55.7	38.4
35	67.8	77.5	54.9	37.3
34	67.3	77.0	54.0	36.1
33	66.8	76.5	53.1	35.0
32	66.3	75.9	52.3	33.9
31	65.8	75.4	51.4	32.7
30	65.3	74.9	50.5	31.6
29	64.8	74.4	49.6	30.4
28	64.3	73.9	48.8	29.3
27	63.8	73.4	47.9	28.2
26	63.3	72.9	47.0	27.0
25	62.8	72.4	46.2	25.9
24	62.3	71.9	45.3	24.8
23	61.8	71.3	44.4	23.6
22	61.3	70.8	43.5	22.5
21	60.8	70.3	42.7	21.3
20	60.3	69.8	41.8	20.2

permit accurate reproduction of those curves. Since all converted hardness values must be considered approximate, however, all converted Rockwell hardness numbers shall be rounded to the nearest whole number.

15.2 *Hardness Testing*:

15.2.1 If the product specification permits alternative hardness testing to determine conformance to a specified hardness requirement, the conversions listed in Table 2, Table 3, Table 4, and Table 5 shall be used.

15.2.2 When recording converted hardness numbers, the measured hardness and test scale shall be indicated in parentheses, for example: 353 HBW (38 HRC). This means that a hardness value of 38 was obtained using the Rockwell C scale and converted to a Brinell hardness of 353.

16. Brinell Test

16.1 *Description*:

16.1.1 A specified load is applied to a flat surface of the specimen to be tested, through a tungsten carbide ball of specified diameter. The average diameter of the indentation is used as a basis for calculation of the Brinell hardness number. The quotient of the applied load divided by the area of the surface of the indentation, which is assumed to be spherical, is

TABLE 5 Approximate Hardness Conversion Numbers for Austenitic Steels (Rockwell B to other Hardness Numbers)

Rockwell B Scale, 100-kgf Load, 1/16-in. (1.588-mm) Ball	Brinell Indentation Diameter, mm	Brinell Hardness, 3000-kgf Load, 10-mm Ball	Rockwell A Scale, 60-kgf Load, Diamond Penetrator	Rockwell Superficial Hardness		
				15T Scale, 15-kgf Load, 1/16-in. (1.588-mm) Ball	30T Scale, 30-kgf Load, 1/16-in. (1.588-mm) Ball	45T Scale, 45-kgf Load, 1/16-in. (1.588-mm) Ball
100	3.79	256	61.5	91.5	80.4	70.2
99	3.85	248	60.9	91.2	79.7	69.2
98	3.91	240	60.3	90.8	79.0	68.2
97	3.96	233	59.7	90.4	78.3	67.2
96	4.02	226	59.1	90.1	77.7	66.1
95	4.08	219	58.5	89.7	77.0	65.1
94	4.14	213	58.0	89.3	76.3	64.1
93	4.20	207	57.4	88.9	75.6	63.1
92	4.24	202	56.8	88.6	74.9	62.1
91	4.30	197	56.2	88.2	74.2	61.1
90	4.35	192	55.6	87.8	73.5	60.1
89	4.40	187	55.0	87.5	72.8	59.0
88	4.45	183	54.5	87.1	72.1	58.0
87	4.51	178	53.9	86.7	71.4	57.0
86	4.55	174	53.3	86.4	70.7	56.0
85	4.60	170	52.7	86.0	70.0	55.0
84	4.65	167	52.1	85.6	69.3	54.0
83	4.70	163	51.5	85.2	68.6	52.9
82	4.74	160	50.9	84.9	67.9	51.9
81	4.79	156	50.4	84.5	67.2	50.9
80	4.84	153	49.8	84.1	66.5	49.9

termed the Brinell hardness number (HBW) in accordance with the following equation:

$$HBW = P/[(\pi D/2)(D - \sqrt{D^2 - d^2})] \quad (4)$$

where:
HBW = Brinell hardness number,
P = applied load, kgf,
D = diameter of the tungsten carbide ball, mm, and
d = average diameter of the indentation, mm.

NOTE 11—The Brinell hardness number is more conveniently secured from standard tables such as Table 6, which show numbers corresponding to the various indentation diameters, usually in increments of 0.05 mm.

NOTE 12—In Test Method E10 the values are stated in SI units, whereas in this section kg/m units are used.

16.1.2 The standard Brinell test using a 10-mm tungsten carbide ball employs a 3000-kgf load for hard materials and a 1500 or 500-kgf load for thin sections or soft materials (see Annex A2 on Steel Tubular Products). Other loads and different size indentors may be used when specified. In recording hardness values, the diameter of the ball and the load must be stated except when a 10-mm ball and 3000-kgf load are used.

16.1.3 A range of hardness can properly be specified only for quenched and tempered or normalized and tempered material. For annealed material a maximum figure only should be specified. For normalized material a minimum or a maximum hardness may be specified by agreement. In general, no hardness requirements should be applied to untreated material.

16.1.4 Brinell hardness may be required when tensile properties are not specified.

16.2 *Apparatus*—Equipment shall meet the following requirements:

16.2.1 *Testing Machine*—A Brinell hardness testing machine is acceptable for use over a loading range within which its load measuring device is accurate to ±1 %.

16.2.2 *Measuring Microscope*—The divisions of the micrometer scale of the microscope or other measuring devices used for the measurement of the diameter of the indentations shall be such as to permit the direct measurement of the diameter to 0.1 mm and the estimation of the diameter to 0.05 mm.

NOTE 13—This requirement applies to the construction of the microscope only and is not a requirement for measurement of the indentation, see 16.4.3.

16.2.3 *Standard Ball*—The standard tungsten carbide ball for Brinell hardness testing is 10 mm (0.3937 in.) in diameter with a deviation from this value of not more than 0.005 mm (0.0004 in.) in any diameter. A tungsten carbide ball suitable for use must not show a permanent change in diameter greater than 0.01 mm (0.0004 in.) when pressed with a force of 3000 kgf against the test specimen. Steel ball indentors are no longer permitted for use in Brinell hardness testing in accordance with these test methods.

16.3 *Test Specimen*—Brinell hardness tests are made on prepared areas and sufficient metal must be removed from the surface to eliminate decarburized metal and other surface irregularities. The thickness of the piece tested must be such that no bulge or other marking showing the effect of the load appears on the side of the piece opposite the indentation.

16.4 *Procedure*:

16.4.1 It is essential that the applicable product specifications state clearly the position at which Brinell hardness indentations are to be made and the number of such indentations required. The distance of the center of the indentation from the edge of the specimen or edge of another indentation must be at least two and one-half times the diameter of the indentation.

16.4.2 Apply the load for 10 to 15 s.

TABLE 6 Brinell Hardness Numbers[A]
(Ball 10 mm in Diameter, Applied Loads of 500, 1500, and 3000 kgf)

Diameter of Indentation, mm	Brinell Hardness Number			Diameter of Indentation, mm	Brinell Hardness Number			Diameter of Indentation, mm	Brinell Hardness Number			Diameter of Indentation, mm	Brinell Hardness Number		
	500-kgf Load	1500-kgf Load	3000-kgf Load		500-kgf Load	1500-kgf Load	3000-kgf Load		500-kgf Load	1500-kgf Load	3000-kgf Load		500-kgf Load	1500-kgf Load	3000-kgf Load
2.00	158	473	945	2.60	92.6	278	555	3.20	60.5	182	363	3.80	42.4	127	255
2.01	156	468	936	2.61	91.8	276	551	3.21	60.1	180	361	3.81	42.2	127	253
2.02	154	463	926	2.62	91.1	273	547	3.22	59.8	179	359	3.82	42.0	126	252
2.03	153	459	917	2.63	90.4	271	543	3.23	59.4	178	356	3.83	41.7	125	250
2.04	151	454	908	2.64	89.7	269	538	3.24	59.0	177	354	3.84	41.5	125	249
2.05	150	450	899	2.65	89.0	267	534	3.25	58.6	176	352	3.85	41.3	124	248
2.06	148	445	890	2.66	88.4	265	530	3.26	58.3	175	350	3.86	41.1	123	246
2.07	147	441	882	2.67	87.7	263	526	3.27	57.9	174	347	3.87	40.9	123	245
2.08	146	437	873	2.68	87.0	261	522	3.28	57.5	173	345	3.88	40.6	122	244
2.09	144	432	865	2.69	86.4	259	518	3.29	57.2	172	343	3.89	40.4	121	242
2.10	143	428	856	2.70	85.7	257	514	3.30	56.8	170	341	3.90	40.2	121	241
2.11	141	424	848	2.71	85.1	255	510	3.31	56.5	169	339	3.91	40.0	120	240
2.12	140	420	840	2.72	84.4	253	507	3.32	56.1	168	337	3.92	39.8	119	239
2.13	139	416	832	2.73	83.8	251	503	3.33	55.8	167	335	3.93	39.6	119	237
2.14	137	412	824	2.74	83.2	250	499	3.34	55.4	166	333	3.94	39.4	118	236
2.15	136	408	817	2.75	82.6	248	495	3.35	55.1	165	331	3.95	39.1	117	235
2.16	135	404	809	2.76	81.9	246	492	3.36	54.8	164	329	3.96	38.9	117	234
2.17	134	401	802	2.77	81.3	244	488	3.37	54.4	163	326	3.97	38.7	116	232
2.18	132	397	794	2.78	80.8	242	485	3.38	54.1	162	325	3.98	38.5	116	231
2.19	131	393	787	2.79	80.2	240	481	3.39	53.8	161	323	3.99	38.3	115	230
2.20	130	390	780	2.80	79.6	239	477	3.40	53.4	160	321	4.00	38.1	114	229
2.21	129	386	772	2.81	79.0	237	474	3.41	53.1	159	319	4.01	37.9	114	228
2.22	128	383	765	2.82	78.4	235	471	3.42	52.8	158	317	4.02	37.7	113	226
2.23	126	379	758	2.83	77.9	234	467	3.43	52.5	157	315	4.03	37.5	113	225
2.24	125	376	752	2.84	77.3	232	464	3.44	52.2	156	313	4.04	37.3	112	224
2.25	124	372	745	2.85	76.8	230	461	3.45	51.8	156	311	4.05	37.1	111	223
2.26	123	369	738	2.86	76.2	229	457	3.46	51.5	155	309	4.06	37.0	111	222
2.27	122	366	732	2.87	75.7	227	454	3.47	51.2	154	307	4.07	36.8	110	221
2.28	121	363	725	2.88	75.1	225	451	3.48	50.9	153	306	4.08	36.6	110	219
2.29	120	359	719	2.89	74.6	224	448	3.49	50.6	152	304	4.09	36.4	109	218
2.30	119	356	712	2.90	74.1	222	444	3.50	50.3	151	302	4.10	36.2	109	217
2.31	118	353	706	2.91	73.6	221	441	3.51	50.0	150	300	4.11	36.0	108	216
2.32	117	350	700	2.92	73.0	219	438	3.52	49.7	149	298	4.12	35.8	108	215
2.33	116	347	694	2.93	72.5	218	435	3.53	49.4	148	297	4.13	35.7	107	214
2.34	115	344	688	2.94	72.0	216	432	3.54	49.2	147	295	4.14	35.5	106	213
2.35	114	341	682	2.95	71.5	215	429	3.55	48.9	147	293	4.15	35.3	106	212
2.36	113	338	676	2.96	71.0	213	426	3.56	48.6	146	292	4.16	35.1	105	211
2.37	112	335	670	2.97	70.5	212	423	3.57	48.3	145	290	4.17	34.9	105	210
2.38	111	332	665	2.98	70.1	210	420	3.58	48.0	144	288	4.18	34.8	104	209
2.39	110	330	659	2.99	69.6	209	417	3.59	47.7	143	286	4.19	34.6	104	208
2.40	109	327	653	3.00	69.1	207	415	3.60	47.5	142	285	4.20	34.4	103	207
2.41	108	324	648	3.01	68.6	206	412	3.61	47.2	142	283	4.21	34.2	103	205
2.42	107	322	643	3.02	68.2	205	409	3.62	46.9	141	282	4.22	34.1	102	204
2.43	106	319	637	3.03	67.7	203	406	3.63	46.7	140	280	4.23	33.9	102	203
2.44	105	316	632	3.04	67.3	202	404	3.64	46.4	139	278	4.24	33.7	101	202
2.45	104	313	627	3.05	66.8	200	401	3.65	46.1	138	277	4.25	33.6	101	201
2.46	104	311	621	3.06	66.4	199	398	3.66	45.9	138	275	4.26	33.4	100	200
2.47	103	308	616	3.07	65.9	198	395	3.67	45.6	137	274	4.27	33.2	99.7	199
2.48	102	306	611	3.08	65.5	196	393	3.68	45.4	136	272	4.28	33.1	99.2	198
2.49	101	303	606	3.09	65.0	195	390	3.69	45.1	135	271	4.29	32.9	98.8	198
2.50	100	301	601	3.10	64.6	194	388	3.70	44.9	135	269	4.30	32.8	98.3	197
2.51	99.4	298	597	3.11	64.2	193	385	3.71	44.6	134	268	4.31	32.6	97.8	196
2.52	98.6	296	592	3.12	63.8	191	383	3.72	44.4	133	266	4.32	32.4	97.3	195
2.53	97.8	294	587	3.13	63.3	190	380	3.73	44.1	132	265	4.33	32.3	96.8	194
2.54	97.1	291	582	3.14	62.9	189	378	3.74	43.9	132	263	4.34	32.1	96.4	193
2.55	96.3	289	578	3.15	62.5	188	375	3.75	43.6	131	262	4.35	32.0	95.9	192
2.56	95.5	287	573	3.16	62.1	186	373	3.76	43.4	130	260	4.36	31.8	95.5	191
2.57	94.8	284	569	3.17	61.7	185	370	3.77	43.1	129	259	4.37	31.7	95.0	190
2.58	94.0	282	564	3.18	61.3	184	368	3.78	42.9	129	257	4.38	31.5	94.5	189
2.59	93.3	280	560	3.19	60.9	183	366	3.79	42.7	128	256	4.39	31.4	94.1	188
4.40	31.2	93.6	187	5.05	23.3	69.8	140	5.70	17.8	53.5	107	6.35	14.0	42.0	84.0
4.41	31.1	93.2	186	5.06	23.2	69.5	139	5.71	17.8	53.3	107	6.36	13.9	41.8	83.7
4.42	30.9	92.7	185	5.07	23.1	69.2	138	5.72	17.7	53.1	106	6.37	13.9	41.7	83.4
4.43	30.8	92.3	185	5.08	23.0	68.9	138	5.73	17.6	52.9	106	6.38	13.8	41.5	83.1
4.44	30.6	91.8	184	5.09	22.9	68.6	137	5.74	17.6	52.7	105	6.39	13.8	41.4	82.8
4.45	30.5	91.4	183	5.10	22.8	68.3	137	5.75	17.5	52.5	105	6.40	13.7	41.2	82.5
4.46	30.3	91.0	182	5.11	22.7	68.0	136	5.76	17.4	52.3	105	6.41	13.7	41.1	82.2
4.47	30.2	90.5	181	5.12	22.6	67.7	135	5.77	17.4	52.1	104	6.42	13.6	40.9	81.9
4.48	30.0	90.1	180	5.13	22.5	67.4	135	5.78	17.3	51.9	104	6.43	13.6	40.8	81.6

TABLE 6 *Continued*

Diameter of Indentation, mm	Brinell Hardness Number			Diameter of Indentation, mm	Brinell Hardness Number			Diameter of Indentation, mm	Brinell Hardness Number			Diameter of Indentation, mm	Brinell Hardness Number		
	500-kgf Load	1500-kgf Load	3000-kgf Load		500-kgf Load	1500-kgf Load	3000-kgf Load		500-kgf Load	1500-kgf Load	3000-kgf Load		500-kgf Load	1500-kgf Load	3000-kgf Load
4.49	29.9	89.7	179	5.14	22.4	67.1	134	5.79	17.2	51.7	103	6.44	13.5	40.6	81.3
4.50	29.8	89.3	179	5.15	22.3	66.9	134	5.80	17.2	51.5	103	6.45	13.5	40.5	81.0
4.51	29.6	88.8	178	5.16	22.2	66.6	133	5.81	17.1	51.3	103	6.46	13.4	40.4	80.7
4.52	29.5	88.4	177	5.17	22.1	66.3	133	5.82	17.0	51.1	102	6.47	13.4	40.2	80.4
4.53	29.3	88.0	176	5.18	22.0	66.0	132	5.83	17.0	50.9	102	6.48	13.4	40.1	80.1
4.54	29.2	87.6	175	5.19	21.9	65.8	132	5.84	16.9	50.7	101	6.49	13.3	39.9	79.8
4.55	29.1	87.2	174	5.20	21.8	65.5	131	5.85	16.8	50.5	101	6.50	13.3	39.8	79.6
4.56	28.9	86.8	174	5.21	21.7	65.2	130	5.86	16.8	50.3	101	6.51	13.2	39.6	79.3
4.57	28.8	86.4	173	5.22	21.6	64.9	130	5.87	16.7	50.2	100	6.52	13.2	39.5	79.0
4.58	28.7	86.0	172	5.23	21.6	64.7	129	5.88	16.7	50.0	99.9	6.53	13.1	39.4	78.7
4.59	28.5	85.6	171	5.24	21.5	64.4	129	5.89	16.6	49.8	99.5	6.54	13.1	39.2	78.4
4.60	28.4	85.4	170	5.25	21.4	64.1	128	5.90	16.5	49.6	99.2	6.55	13.0	39.1	78.2
4.61	28.3	84.8	170	5.26	21.3	63.9	128	5.91	16.5	49.4	98.8	6.56	13.0	38.9	78.0
4.62	28.1	84.4	169	5.27	21.2	63.6	127	5.92	16.4	49.2	98.4	6.57	12.9	38.8	77.6
4.63	28.0	84.0	168	5.28	21.1	63.3	127	5.93	16.3	49.0	98.0	6.58	12.9	38.7	77.3
4.64	27.9	83.6	167	5.29	21.0	63.1	126	5.94	16.3	48.8	97.7	6.59	12.8	38.5	77.1
4.65	27.8	83.3	167	5.30	20.9	62.8	126	5.95	16.2	48.7	97.3	6.60	12.8	38.4	76.8
4.66	27.6	82.9	166	5.31	20.9	62.6	125	5.96	16.2	48.5	96.9	6.61	12.8	38.3	76.5
4.67	27.5	82.5	165	5.32	20.8	62.3	125	5.97	16.1	48.3	96.6	6.62	12.7	38.1	76.2
4.68	27.4	82.1	164	5.33	20.7	62.1	124	5.98	16.0	48.1	96.2	6.63	12.7	38.0	76.0
4.69	27.3	81.8	164	5.34	20.6	61.8	124	5.99	16.0	47.9	95.9	6.64	12.6	37.9	75.7
4.70	27.1	81.4	163	5.35	20.5	61.5	123	6.00	15.9	47.7	95.5	6.65	12.6	37.7	75.4
4.71	27.0	81.0	162	5.36	20.4	61.3	123	6.01	15.9	47.6	95.1	6.66	12.5	37.6	75.2
4.72	26.9	80.7	161	5.37	20.3	61.0	122	6.02	15.8	47.4	94.8	6.67	12.5	37.5	74.9
4.73	26.8	80.3	161	5.38	20.3	60.8	122	6.03	15.7	47.2	94.4	6.68	12.4	37.3	74.7
4.74	26.6	79.9	160	5.39	20.2	60.6	121	6.04	15.7	47.0	94.1	6.69	12.4	37.2	74.4
4.75	26.5	79.6	159	5.40	20.1	60.3	121	6.05	15.6	46.8	93.7	6.70	12.4	37.1	74.1
4.76	26.4	79.2	158	5.41	20.0	60.1	120	6.06	15.6	46.7	93.4	6.71	12.3	36.9	73.9
4.77	26.3	78.9	158	5.42	19.9	59.8	120	6.07	15.5	46.5	93.0	6.72	12.3	36.8	73.6
4.78	26.2	78.5	157	5.43	19.9	59.6	119	6.08	15.4	46.3	92.7	6.73	12.2	36.7	73.4
4.79	26.1	78.2	156	5.44	19.8	59.3	119	6.09	15.4	46.2	92.3	6.74	12.2	36.6	73.1
4.80	25.9	77.8	156	5.45	19.7	59.1	118	6.10	15.3	46.0	92.0	6.75	12.1	36.4	72.8
4.81	25.8	77.5	155	5.46	19.6	58.9	118	6.11	15.3	45.8	91.7	6.76	12.1	36.3	72.6
4.82	25.7	77.1	154	5.47	19.5	58.6	117	6.12	15.2	45.7	91.3	6.77	12.1	36.2	72.3
4.83	25.6	76.8	154	5.48	19.5	58.4	117	6.13	15.2	45.5	91.0	6.78	12.0	36.0	72.1
4.84	25.5	76.4	153	5.49	19.4	58.2	116	6.14	15.1	45.3	90.6	6.79	12.0	35.9	71.8
4.85	25.4	76.1	152	5.50	19.3	57.9	116	6.15	15.1	45.2	90.3	6.80	11.9	35.8	71.6
4.86	25.3	75.8	152	5.51	19.2	57.7	115	6.16	15.0	45.0	90.0	6.81	11.9	35.7	71.3
4.87	25.1	75.4	151	5.52	19.2	57.5	115	6.17	14.9	44.8	89.6	6.82	11.8	35.5	71.1
4.88	25.0	75.1	150	5.53	19.1	57.2	114	6.18	14.9	44.7	89.3	6.83	11.8	35.4	70.8
4.89	24.9	74.8	150	5.54	19.0	57.0	114	6.19	14.8	44.5	89.0	6.84	11.8	35.3	70.6
4.90	24.8	74.4	149	5.55	18.9	56.8	114	6.20	14.7	44.3	88.7	6.85	11.7	35.2	70.4
4.91	24.7	74.1	148	5.56	18.9	56.6	113	6.21	14.7	44.2	88.3	6.86	11.7	35.1	70.1
4.92	24.6	73.8	148	5.57	18.8	56.3	113	6.22	14.7	44.0	88.0	6.87	11.6	34.9	69.9
4.93	24.5	73.5	147	5.58	18.7	56.1	112	6.23	14.6	43.8	87.7	6.88	11.6	34.8	69.6
4.94	24.4	73.2	146	5.59	18.6	55.9	112	6.24	14.6	43.7	87.4	6.89	11.6	34.7	69.4
4.95	24.3	72.8	146	5.60	18.6	55.7	111	6.25	14.5	43.5	87.1	6.90	11.5	34.6	69.2
4.96	24.2	72.5	145	5.61	18.5	55.5	111	6.26	14.5	43.4	86.7	6.91	11.5	34.5	68.9
4.97	24.1	72.2	144	5.62	18.4	55.2	110	6.27	14.4	43.2	86.4	6.92	11.4	34.3	68.7
4.98	24.0	71.9	144	5.63	18.3	55.0	110	6.28	14.4	43.1	86.1	6.93	11.4	34.2	68.4
4.99	23.9	71.6	143	5.64	18.3	54.8	110	6.29	14.3	42.9	85.8	6.94	11.4	34.1	68.2
5.00	23.8	71.3	143	5.65	18.2	54.6	109	6.30	14.2	42.7	85.5	6.95	11.3	34.0	68.0
5.01	23.7	71.0	142	5.66	18.1	54.4	109	6.31	14.2	42.6	85.2	6.96	11.3	33.9	67.7
5.02	23.6	70.7	141	5.67	18.1	54.2	108	6.32	14.1	42.4	84.9	6.97	11.3	33.8	67.5
5.03	23.5	70.4	141	5.68	18.0	54.0	108	6.33	14.1	42.3	84.6	6.98	11.2	33.6	67.3
5.04	23.4	70.1	140	5.69	17.9	53.7	107	6.34	14.0	42.1	84.3	6.99	11.2	33.5	67.0

[A] Prepared by the Engineering Mechanics Section, Institute for Standards Technology.

16.4.3 Measure two diameters of the indentation at right angles to the nearest 0.1 mm, estimate to the nearest 0.05 mm, and average to the nearest 0.05 mm. If the two diameters differ by more than 0.1 mm, discard the readings and make a new indentation.

16.4.4 The Brinell hardness test is not recommended for materials above 650 HBW.

16.4.4.1 If a ball is used in a test of a specimen which shows a Brinell hardness number greater than the limit for the ball as detailed in 16.4.4, the ball shall be either discarded and replaced with a new ball or remeasured to ensure conformance with the requirements of Test Method E10.

16.5 *Brinell Hardness Values:*

16.5.1 Brinell hardness values shall not be designated by a number alone because it is necessary to indicate which indenter and which force has been employed in making the test. Brinell hardness numbers shall be followed by the symbol HBW, and be supplemented by an index indicating the test conditions in the following order:

16.5.1.1 Diameter of the ball, mm,

16.5.1.2 A value representing the applied load, kgf, and,

16.5.1.3 The applied force dwell time, s, if other than 10 s to 15 s.

16.5.1.4 The only exception to the above requirement is for the HBW 10/3000 scale when a 10 s to 15 s dwell time is used. Only in the case of this one Brinell hardness scale may the designation be reported simply as HBW.

16.5.1.5 *Examples*:

220 HBW = Brinell hardness of 220 determined with a ball of 10 mm diameter and with a test force of 3000 kgf applied for 10 s to 15 s; 350 HBW 5/1500 = Brinell hardness of 350 determined with a ball of 5 mm diameter and with a test force of 1500 kgf applied for 10 s to 15 s.

16.6 *Detailed Procedure*—For detailed requirements of this test, reference shall be made to the latest revision of Test Method E10.

17. Rockwell Test

17.1 *Description*:

17.1.1 In this test a hardness value is obtained by determining the depth of penetration of a diamond point or a steel ball into the specimen under certain arbitrarily fixed conditions. A minor load of 10 kgf is first applied which causes an initial penetration, sets the penetrator on the material and holds it in position. A major load which depends on the scale being used is applied increasing the depth of indentation. The major load is removed and, with the minor load still acting, the Rockwell number, which is proportional to the difference in penetration between the major and minor loads is determined; this is usually done by the machine and shows on a dial, digital display, printer, or other device. This is an arbitrary number which increases with increasing hardness. The scales most frequently used are as follows:

Scale Symbol	Penetrator	Major Load, kgf	Minor Load, kgf
B	1/16-in. steel ball	100	10
C	Diamond brale	150	10

17.1.2 Rockwell superficial hardness machines are used for the testing of very thin steel or thin surface layers. Loads of 15, 30, or 45 kgf are applied on a hardened steel ball or diamond penetrator, to cover the same range of hardness values as for the heavier loads. The superficial hardness scales are as follows:

Scale Symbol	Penetrator	Major Load, kgf	Minor Load, kgf
15T	1/16-in. steel ball	15	3
30T	1/16-in. steel ball	30	3
45T	1/16-in. steel ball	45	3
15N	Diamond brale	15	3
30N	Diamond brale	30	3
45N	Diamond brale	45	3

17.2 *Reporting Hardness*—In recording hardness values, the hardness number shall always precede the scale symbol, for example: 96 HRB, 40 HRC, 75 HR15N, or 77 HR30T.

17.3 *Test Blocks*—Machines should be checked to make certain they are in good order by means of standardized Rockwell test blocks.

17.4 *Detailed Procedure*—For detailed requirements of this test, reference shall be made to the latest revision of Test Methods E18.

18. Portable Hardness Test

18.1 Although the use of the standard, stationary Brinell or Rockwell hardness tester is generally preferred, it is not always possible to perform the hardness test using such equipment due to the part size or location. In this event, hardness testing using portable equipment as described in Practice A833 or Test Method E110 shall be used.

CHARPY IMPACT TESTING

19. Summary

19.1 A Charpy V-notch impact test is a dynamic test in which a notched specimen is struck and broken by a single blow in a specially designed testing machine. The measured test values may be the energy absorbed, the percentage shear fracture, the lateral expansion opposite the notch, or a combination thereof.

19.2 Testing temperatures other than room (ambient) temperature often are specified in product or general requirement specifications (hereinafter referred to as the specification). Although the testing temperature is sometimes related to the expected service temperature, the two temperatures need not be identical.

20. Significance and Use

20.1 *Ductile vs. Brittle Behavior*—Body-centered-cubic or ferritic alloys exhibit a significant transition in behavior when impact tested over a range of temperatures. At temperatures above transition, impact specimens fracture by a ductile (usually microvoid coalescence) mechanism, absorbing relatively large amounts of energy. At lower temperatures, they fracture in a brittle (usually cleavage) manner absorbing appreciably less energy. Within the transition range, the fracture will generally be a mixture of areas of ductile fracture and brittle fracture.

20.2 The temperature range of the transition from one type of behavior to the other varies according to the material being tested. This transition behavior may be defined in various ways for specification purposes.

20.2.1 The specification may require a minimum test result for absorbed energy, fracture appearance, lateral expansion, or a combination thereof, at a specified test temperature.

20.2.2 The specification may require the determination of the transition temperature at which either the absorbed energy or fracture appearance attains a specified level when testing is performed over a range of temperatures. Alternatively the specification may require the determination of the fracture appearance transition temperature (FATTn) as the temperature at which the required minimum percentage of shear fracture (n) is obtained.

20.3 Further information on the significance of impact testing appears in Annex A5.

21. Apparatus

21.1 *Testing Machines*:

21.1.1 A Charpy impact machine is one in which a notched specimen is broken by a single blow of a freely swinging pendulum. The pendulum is released from a fixed height. Since the height to which the pendulum is raised prior to its swing, and the mass of the pendulum are known, the energy of the blow is predetermined. A means is provided to indicate the energy absorbed in breaking the specimen.

21.1.2 The other principal feature of the machine is a fixture (See Fig. 10) designed to support a test specimen as a simple beam at a precise location. The fixture is arranged so that the notched face of the specimen is vertical. The pendulum strikes the other vertical face directly opposite the notch. The dimensions of the specimen supports and striking edge shall conform to Fig. 10.

21.1.3 Charpy machines used for testing steel generally have capacities in the 220 to 300 ft·lbf (300 to 400 J) energy range. Sometimes machines of lesser capacity are used; however, the capacity of the machine should be substantially in excess of the absorbed energy of the specimens (see Test Methods E23). The linear velocity at the point of impact should be in the range of 16 to 19 ft/s (4.9 to 5.8 m/s).

NOTE 14—An investigation of striker radius effect is available.[6]

21.2 *Temperature Media*:

21.2.1 For testing at other than room temperature, it is necessary to condition the Charpy specimens in media at controlled temperatures.

21.2.2 Low temperature media usually are chilled fluids (such as water, ice plus water, dry ice plus organic solvents, or liquid nitrogen) or chilled gases.

21.2.3 Elevated temperature media are usually heated liquids such as mineral or silicone oils. Circulating air ovens may be used.

21.3 *Handling Equipment*—Tongs, especially adapted to fit the notch in the impact specimen, normally are used for removing the specimens from the medium and placing them on the anvil (refer to Test Methods E23). In cases where the machine fixture does not provide for automatic centering of the test specimen, the tongs may be precision machined to provide centering.

22. Sampling and Number of Specimens

22.1 *Sampling*:

22.1.1 Test location and orientation should be addressed by the specifications. If not, for wrought products, the test location shall be the same as that for the tensile specimen and the orientation shall be longitudinal with the notch perpendicular to the major surface of the product being tested.

22.1.2 *Number of Specimens*.

22.1.2.1 A Charpy impact test consists of all specimens taken from a single test coupon or test location.

22.1.2.2 When the specification calls for a minimum average test result, three specimens shall be tested.

22.1.2.3 When the specification requires determination of a transition temperature, eight to twelve specimens are usually needed.

22.2 *Type and Size*:

22.2.1 Use a standard full size Charpy V-notch specimen as shown in Fig. 11, except as allowed in 22.2.2.

22.2.2 *Subsized Specimens*.

22.2.2.1 For flat material less than 7/16 in. (11 mm) thick, or when the absorbed energy is expected to exceed 80 % of full scale, use standard subsize test specimens.

22.2.2.2 For tubular materials tested in the transverse direction, where the relationship between diameter and wall thickness does not permit a standard full size specimen, use standard subsize test specimens or standard size specimens containing outer diameter (OD) curvature as follows:

(1) Standard size specimens and subsize specimens may contain the original OD surface of the tubular product as shown in Fig. 12. All other dimensions shall comply with the requirements of Fig. 11.

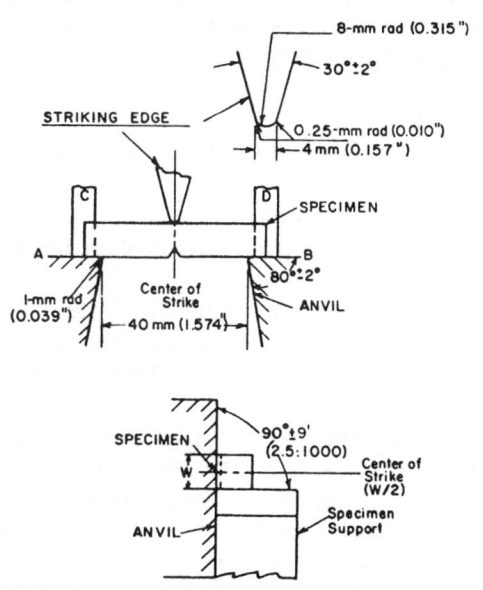

All dimensional tolerances shall be ±0.05 mm (0.002 in.) unless otherwise specified.

NOTE 1—A shall be parallel to B within 2:1000 and coplanar with B within 0.05 mm (0.002 in.).

NOTE 2—C shall be parallel to D within 20:1000 and coplanar with D within 0.125 mm (0.005 in.).

NOTE 3—Finish on unmarked parts shall be 4 μm (125 μin.).

FIG. 10 Charpy (Simple-Beam) Impact Test

---

[6] Supporting data have been filed at ASTM International Headquarters and may be obtained by requesting Research Report RR:A01-1001.

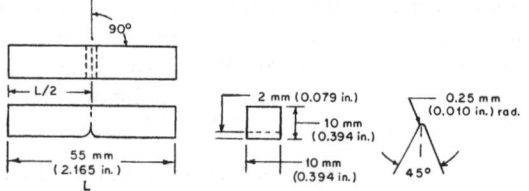

NOTE 1—Permissible variations shall be as follows:

Notch length to edge	90 ±2°
Adjacent sides shall be at	90° ± 10 min
Cross-section dimensions	±0.075 mm (±0.003 in.)
Length of specimen (L)	+ 0, – 2.5 mm (+ 0, – 0.100 in.)
Centering of notch (L/2)	±1 mm (±0.039 in.)
Angle of notch	±1°
Radius of notch	±0.025 mm (±0.001 in.)
Notch depth	±0.025 mm (±0.001 in.)
Finish requirements	2 µm (63 µin.) on notched surface and opposite face; 4 µm (125 µin.) on other two surfaces

(a) **Standard Full Size Specimen**

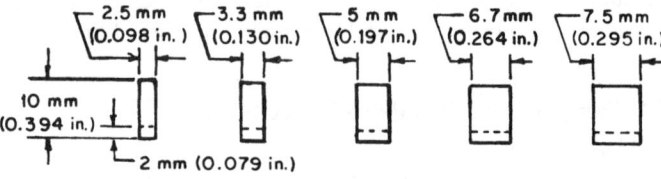

NOTE 2—On subsize specimens, all dimensions and tolerances of the standard specimen remain constant with the exception of the width, which varies as shown above and for which the tolerance shall be ±1 %.

(b) **Standard Subsize Specimens**

FIG. 11 Charpy (Simple Beam) Impact Test Specimens

NOTE 15—For materials with toughness levels in excess of about 50 ft-lbs, specimens containing the original OD surface may yield values in excess of those resulting from the use of conventional Charpy specimens.

22.2.2.3 If a standard full-size specimen cannot be prepared, the largest feasible standard subsize specimen shall be prepared. The specimens shall be machined so that the specimen does not include material nearer to the surface than 0.020 in. (0.5 mm).

22.2.2.4 Tolerances for standard subsize specimens are shown in Fig. 11. Standard subsize test specimen sizes are: 10 × 7.5 mm, 10 × 6.7 mm, 10 × 5 mm, 10 × 3.3 mm, and 10 × 2.5 mm.

22.2.2.5 Notch the narrow face of the standard subsize specimens so that the notch is perpendicular to the 10 mm wide face.

22.3 *Notch Preparation*—The machining of the notch is critical, as it has been demonstrated that extremely minor variations in notch radius and profile, or tool marks at the bottom of the notch may result in erratic test data. (See Annex A5).

23. Calibration

23.1 *Accuracy and Sensitivity*—Calibrate and adjust Charpy impact machines in accordance with the requirements of Test Methods E23.

24. Conditioning—Temperature Control

24.1 When a specific test temperature is required by the specification or purchaser, control the temperature of the heating or cooling medium within ±2°F (1°C).

NOTE 16—For some steels there may not be a need for this restricted temperature, for example, austenitic steels.

NOTE 17—Because the temperature of a testing laboratory often varies from 60 to 90°F (15 to 32°C) a test conducted at "room temperature" might be conducted at any temperature in this range.

25. Procedure

25.1 *Temperature*:

25.1.1 Condition the specimens to be broken by holding them in the medium at test temperature for at least 5 min in liquid media and 30 min in gaseous media.

25.1.2 Prior to each test, maintain the tongs for handling test specimens at the same temperature as the specimen so as not to affect the temperature at the notch.

25.2 *Positioning and Breaking Specimens*:

25.2.1 Carefully center the test specimen in the anvil and release the pendulum to break the specimen.

25.2.2 If the pendulum is not released within 5 s after removing the specimen from the conditioning medium, do not break the specimen. Return the specimen to the conditioning medium for the period required in 25.1.1.

25.3 *Recovering Specimens*—In the event that fracture appearance or lateral expansion must be determined, recover the matched pieces of each broken specimen before breaking the next specimen.

25.4 *Individual Test Values*:

25.4.1 *Impact energy*— Record the impact energy absorbed to the nearest ft·lbf (J).

25.4.2 *Fracture Appearance*:

25.4.2.1 Determine the percentage of shear fracture area by any of the following methods:

(1) Measure the length and width of the brittle portion of the fracture surface, as shown in Fig. 13 and determine the percent shear area from either Table 7 or Table 8 depending on the units of measurement.

(2) Compare the appearance of the fracture of the specimen with a fracture appearance chart as shown in Fig. 14.

(3) Magnify the fracture surface and compare it to a precalibrated overlay chart or measure the percent shear fracture area by means of a planimeter.

(4) Photograph the fractured surface at a suitable magnification and measure the percent shear fracture area by means of a planimeter.

25.4.2.2 Determine the individual fracture appearance values to the nearest 5 % shear fracture and record the value.

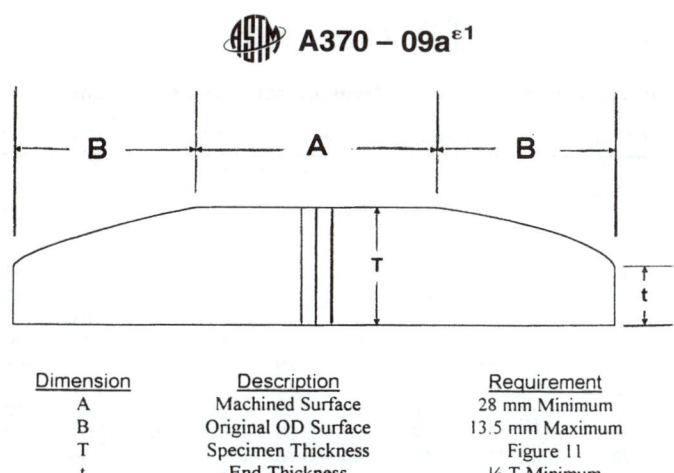

Dimension	Description	Requirement
A	Machined Surface	28 mm Minimum
B	Original OD Surface	13.5 mm Maximum
T	Specimen Thickness	Figure 11
t	End Thickness	½ T Minimum

FIG. 12 Tubular Impact Specimen Containing Original OD Surface

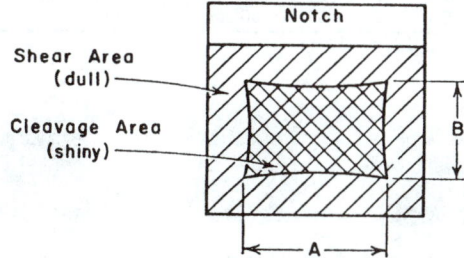

Note 1—Measure average dimensions A and B to the nearest 0.02 in. or 0.5 mm.
Note 2—Determine the percent shear fracture using Table 7 or Table 8.

FIG. 13 Determination of Percent Shear Fracture

TABLE 7 Percent Shear for Measurements Made in Inches

Note 1—Since this table is set up for finite measurements or dimensions A and B, 100% shear is to be reported when either A or B is zero.

Dimension B, in.	Dimension A, in.																
	0.05	0.10	0.12	0.14	0.16	0.18	0.20	0.22	0.24	0.26	0.28	0.30	0.32	0.34	0.36	0.38	0.40
0.05	98	96	95	94	94	93	92	91	90	90	89	88	87	86	85	85	84
0.10	96	92	90	89	87	85	84	82	81	79	77	76	74	73	71	69	68
0.12	95	90	88	86	85	83	81	79	77	75	73	71	69	67	65	63	61
0.14	94	89	86	84	82	80	77	75	73	71	68	66	64	62	59	57	55
0.16	94	87	85	82	79	77	74	72	69	67	64	61	59	56	53	51	48
0.18	93	85	83	80	77	74	72	68	65	62	59	56	54	51	48	45	42
0.20	92	84	81	77	74	72	68	65	61	58	55	52	48	45	42	39	36
0.22	91	82	79	75	72	68	65	61	57	54	50	47	43	40	36	33	29
0.24	90	81	77	73	69	65	61	57	54	50	46	42	38	34	30	27	23
0.26	90	79	75	71	67	62	58	54	50	46	41	37	33	29	25	20	16
0.28	89	77	73	68	64	59	55	50	46	41	37	32	28	23	18	14	10
0.30	88	76	71	66	61	56	52	47	42	37	32	27	23	18	13	9	3
0.31	88	75	70	65	60	55	50	45	40	35	30	25	20	18	10	5	0

25.4.3 *Lateral Expansion*:

25.4.3.1 Lateral expansion is the increase in specimen width, measured in thousandths of an inch (mils), on the compression side, opposite the notch of the fractured Charpy V-notch specimen as shown in Fig. 15.

25.4.3.2 Examine each specimen half to ascertain that the protrusions have not been damaged by contacting the anvil, machine mounting surface, and so forth. Discard such samples since they may cause erroneous readings.

25.4.3.3 Check the sides of the specimens perpendicular to the notch to ensure that no burrs were formed on the sides during impact testing. If burrs exist, remove them carefully by rubbing on emery cloth or similar abrasive surface, making sure that the protrusions being measured are not rubbed during the removal of the burr.

25.4.3.4 Measure the amount of expansion on each side of each half relative to the plane defined by the undeformed portion of the side of the specimen using a gauge similar to that shown in Fig. 16 and Fig. 17.

25.4.3.5 Since the fracture path seldom bisects the point of maximum expansion on both sides of a specimen, the sum of the larger values measured for each side is the value of the test. Arrange the halves of one specimen so that compression sides are facing each other. Using the gauge, measure the protrusion on each half specimen, ensuring that the same side of the

TABLE 8 Percent Shear for Measurements Made in Millimetres

NOTE 1—Since this table is set up for finite measurements or dimensions A and B, 100% shear is to be reported when either A or B is zero.

Dimension B, mm	Dimension A, mm																	
	1.0	1.5	2.0	2.5	3.0	3.5	4.0	4.5	5.0	5.5	6.0	6.5	7.0	7.5	8.0	8.5	9.0	9.5	10
1.0	99	98	98	97	96	96	95	94	94	93	92	92	91	91	90	89	89	88	88
1.5	98	97	96	95	94	93	92	92	91	90	89	88	87	86	85	84	83	82	81
2.0	98	96	95	94	92	91	90	89	88	86	85	84	82	81	80	79	77	76	75
2.5	97	95	94	92	91	89	88	86	84	83	81	80	78	77	75	73	72	70	69
3.0	96	94	92	91	89	87	85	83	81	79	77	76	74	72	70	68	66	64	62
3.5	96	93	91	89	87	85	82	80	78	76	74	72	69	67	65	63	61	58	56
4.0	95	92	90	88	85	82	80	77	75	72	70	67	65	62	60	57	55	52	50
4.5	94	92	89	86	83	80	77	75	72	69	66	63	61	58	55	52	49	46	44
5.0	94	91	88	85	81	78	75	72	69	66	62	59	56	53	50	47	44	41	37
5.5	93	90	86	83	79	76	72	69	66	62	59	55	52	48	45	42	38	35	31
6.0	92	89	85	81	77	74	70	66	62	59	55	51	47	44	40	36	33	29	25
6.5	92	88	84	80	76	72	67	63	59	55	51	47	43	39	35	31	27	23	19
7.0	91	87	82	78	74	69	65	61	56	52	47	43	39	34	30	26	21	17	12
7.5	91	86	81	77	72	67	62	58	53	48	44	39	34	30	25	20	16	11	6
8.0	90	85	80	75	70	65	60	55	50	45	40	35	30	25	20	15	10	5	0

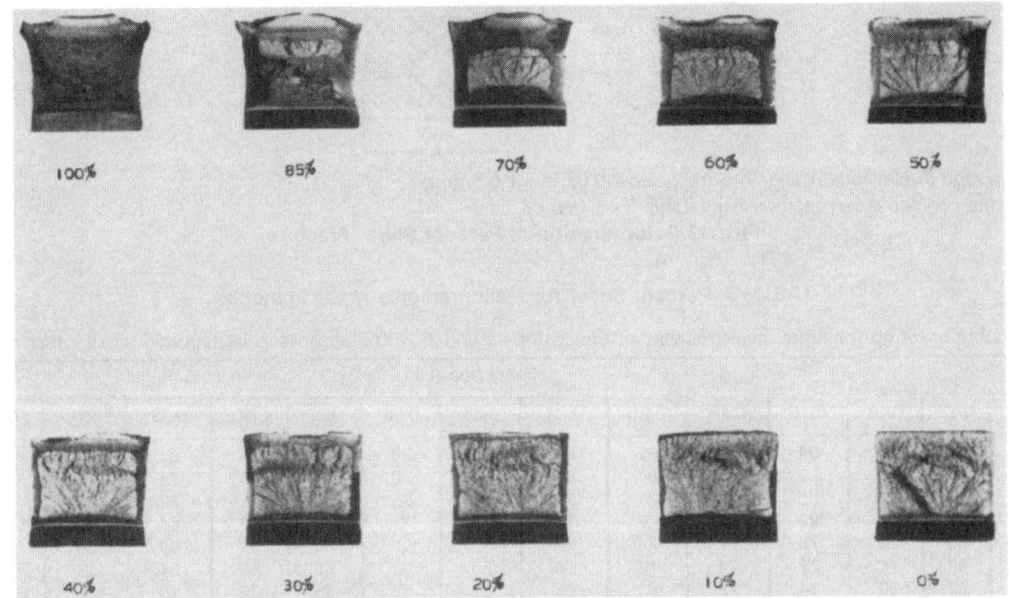

FIG. 14 Fracture Appearance Charts and Percent Shear Fracture Comparator

specimen is measured. Measure the two broken halves individually. Repeat the procedure to measure the protrusions on the opposite side of the specimen halves. The larger of the two values for each side is the expansion of that side of the specimen.

25.4.3.6 Measure the individual lateral expansion values to the nearest mil (0.025 mm) and record the values.

25.4.3.7 With the exception described as follows, any specimen that does not separate into two pieces when struck by a single blow shall be reported as unbroken. If the specimen can be separated by force applied by bare hands, the specimen may be considered as having been separated by the blow.

26. Interpretation of Test Result

26.1 When the acceptance criterion of any impact test is specified to be a minimum average value at a given temperature, the test result shall be the average (arithmetic mean) of the individual test values of three specimens from one test location.

26.1.1 When a minimum average test result is specified:

26.1.1.1 The test result is acceptable when all of the below are met:

(1) The test result equals or exceeds the specified minimum average (given in the specification),

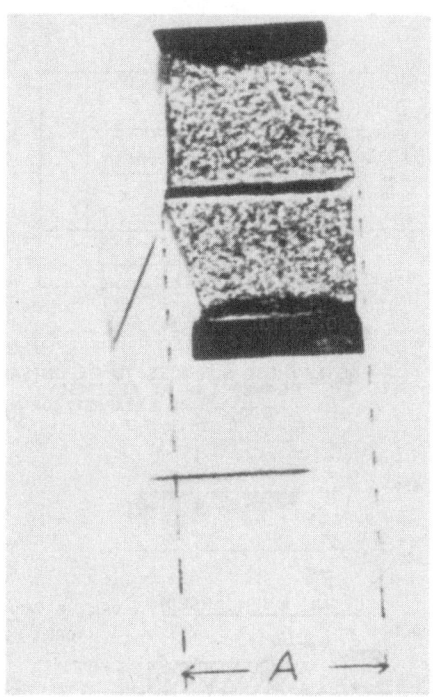

FIG. 15 Halves of Broken Charpy V-Notch Impact Specimen Joined for the Measurement of Lateral Expansion, Dimension A

FIG. 16 Lateral Expansion Gauge for Charpy Impact Specimens

(2) The individual test value for not more than one specimen measures less than the specified minimum average, and

(3) The individual test value for any specimen measures not less than two-thirds of the specified minimum average.

26.1.1.2 If the acceptance requirements of 26.1.1.1 are not met, perform one retest of three additional specimens from the same test location. Each individual test value of the retested specimens shall be equal to or greater than the specified minimum average value.

26.2 *Test Specifying a Minimum Transition Temperature*:

26.2.1 *Definition of Transition Temperature*—For specification purposes, the transition temperature is the temperature at which the designated material test value equals or exceeds a specified minimum test value.

26.2.2 *Determination of Transition Temperature*:

26.2.2.1 Break one specimen at each of a series of temperatures above and below the anticipated transition temperature using the procedures in Section 25. Record each test temperature to the nearest 1°F (0.5°C).

26.2.2.2 Plot the individual test results (ft·lbf or percent shear) as the ordinate versus the corresponding test temperature as the abscissa and construct a best-fit curve through the plotted data points.

26.2.2.3 If transition temperature is specified as the temperature at which a test value is achieved, determine the

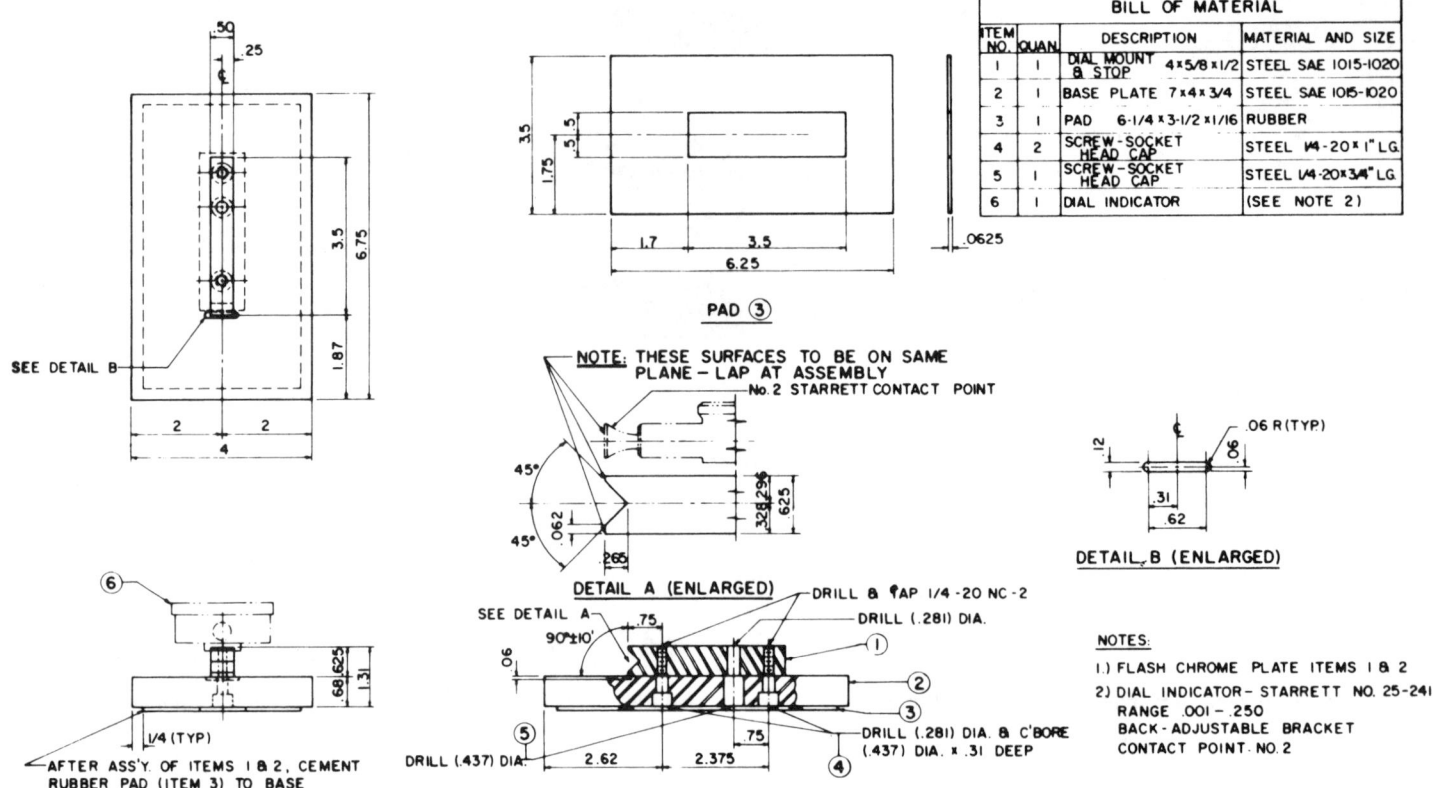

FIG. 17 Assembly and Details for Lateral Expansion Gauge

temperature at which the plotted curve intersects the specified test value by graphical interpolation (extrapolation is not permitted). Record this transition temperature to the nearest 5°F (3°C). If the tabulated test results clearly indicate a transition temperature lower than specified, it is not necessary to plot the data. Report the lowest test temperature for which test value exceeds the specified value.

26.2.2.4 Accept the test result if the determined transition temperature is equal to or lower than the specified value.

26.2.2.5 If the determined transition temperature is higher than the specified value, but not more than 20°F (12°C) higher than the specified value, test sufficient samples in accordance with Section 25 to plot two additional curves. Accept the test results if the temperatures determined from both additional tests are equal to or lower than the specified value.

26.3 When subsize specimens are permitted or necessary, or both, modify the specified test requirement according to Table 9 or test temperature according to ASME Boiler and Pressure Vessel Code, Table UG-84.2, or both. Greater energies or lower test temperatures may be agreed upon by purchaser and supplier.

27. Records

27.1 The test record should contain the following information as appropriate:

27.1.1 Full description of material tested (that is, specification number, grade, class or type, size, heat number).

27.1.2 Specimen orientation with respect to the material axis.

27.1.3 Specimen size.

TABLE 9 Charpy V-Notch Test Acceptance Criteria for Various Sub-Size Specimens

Full Size, 10 by 10 mm		¾ Size, 10 by 7.5 mm		⅔ Size, 10 by 6.7 mm		½ Size, 10 by 5 mm		⅓ Size, 10 by 3.3 mm		¼ Size, 10 by 2.5 mm	
ft·lbf	[J]	ft·lbf	[J]	ft·lbf	[J]	ft·lbf	[J]	ft·lbf	[J]	ft·lbf	[J]
40	[54]	30	[41]	27	[37]	20	[27]	13	[18]	10	[14]
35	[48]	26	[35]	23	[31]	18	[24]	12	[16]	9	[12]
30	[41]	22	[30]	20	[27]	15	[20]	10	[14]	8	[11]
25	[34]	19	[26]	17	[23]	12	[16]	8	[11]	6	[8]
20	[27]	15	[20]	13	[18]	10	[14]	7	[10]	5	[7]
16	[22]	12	[16]	11	[15]	8	[11]	5	[7]	4	[5]
15	[20]	11	[15]	10	[14]	8	[11]	5	[7]	4	[5]
13	[18]	10	[14]	9	[12]	6	[8]	4	[5]	3	[4]
12	[16]	9	[12]	8	[11]	6	[8]	4	[5]	3	[4]
10	[14]	8	[11]	7	[10]	5	[7]	3	[4]	2	[3]
7	[10]	5	[7]	5	[7]	4	[5]	2	[3]	2	[3]

27.1.4 Test temperature and individual test value for each specimen broken, including initial tests and retests.

27.1.5 Test results.

27.1.6 Transition temperature and criterion for its determination, including initial tests and retests.

28. Report

28.1 The specification should designate the information to be reported.

29. Keywords

29.1 bend test; Brinell hardness; Charpy impact test; elongation; FATT (Fracture Appearance Transition Temperature); hardness test; portable hardness; reduction of area; Rockwell hardness; tensile strength; tension test; yield strength

ANNEXES

(Mandatory Information)

A1. STEEL BAR PRODUCTS

A1.1 Scope

A1.1.1 This annex contains testing requirements for Steel Bar Products that are specific to the product. The requirements contained in this annex are supplementary to those found in the general section of this specification. In the case of conflict between requirements provided in this annex and those found in the general section of this specification, the requirements of this annex shall prevail. In the case of conflict between requirements provided in this annex and requirements found in product specifications, the requirements found in the product specification shall prevail.

A1.2 Orientation of Test Specimens

A1.2.1 Carbon and alloy steel bars and bar-size shapes, due to their relatively small cross-sectional dimensions, are customarily tested in the longitudinal direction. In special cases where size permits and the fabrication or service of a part justifies testing in a transverse direction, the selection and location of test or tests are a matter of agreement between the manufacturer and the purchaser.

A1.3 Tension Test

A1.3.1 *Carbon Steel Bars*—Carbon steel bars are not commonly specified to tensile requirements in the as-rolled condition for sizes of rounds, squares, hexagons, and octagons under ½ in. (13 mm) in diameter or distance between parallel faces nor for other bar-size sections, other than flats, less than 1 in.^{2} (645 mm^2) in cross-sectional area.

A1.3.2 *Alloy Steel Bars*—Alloy steel bars are usually not tested in the as-rolled condition.

A1.3.3 When tension tests are specified, the practice for selecting test specimens for hot-rolled and cold-finished steel bars of various sizes shall be in accordance with Table A1.1, unless otherwise specified in the product specification.

A1.4 Bend Test

A1.4.1 When bend tests are specified, the recommended practice for hot-rolled and cold-finished steel bars shall be in accordance with Table A1.2.

A1.5 Hardness Test

A1.5.1 *Hardness Tests on Bar Products*—flats, rounds, squares, hexagons and octagons—is conducted on the surface after a minimum removal of 0.015 in. to provide for accurate hardness penetration.

TABLE A1.1 Practices for Selecting Tension Test Specimens for Steel Bar Products

NOTE 1—For bar sections where it is difficult to determine the cross-sectional area by simple measurement, the area in square inches may be calculated by dividing the weight per linear inch of specimen in pounds by 0.2833 (weight of 1 in.^{3} of steel) or by dividing the weight per linear foot of specimen by 3.4 (weight of steel 1 in. square and 1 ft long).

Thickness, in. (mm)	Width, in. (mm)	Hot-Rolled Bars	Cold-Finished Bars
		Flats	
Under ⅝ (16)	Up to 1½ (38), incl	Full section by 8-in. (200-mm) gauge length (Fig. 3).	Mill reduced section to 2-in. (50-mm) gauge length and approximately 25% less than test specimen width.
	Over 1½ (38)	Full section, or mill to 1½ in. (38 mm) wide by 8-in. (200-mm) gauge length (Fig. 3).	Mill reduced section to 2-in. gauge length and 1½ in. wide.
⅝ to 1½ (16 to 38), excl	Up to 1½ (38), incl	Full section by 8-in. gauge length or machine standard ½ by 2-in. (13 by 50-mm) gauge length specimen from center of section (Fig. 4).	Mill reduced section to 2-in. (50-mm) gauge length and approximately 25% less than test specimen width or machine standard ½ by 2-in. (13 by 50-mm) gauge length specimen from center of section (Fig. 4).
	Over 1½ (38)	Full section, or mill 1½ in. (38 mm) width by 8-in. (200-mm) gauge length (Fig. 3) or machine standard ½ by 2-in. gauge (13 by 50-mm) gauge length specimen from midway between edge and center of section (Fig. 4).	Mill reduced section to 2-in. gauge length and 1½ in. wide or machine standard ½ by 2-in. gauge length specimen from midway between edge and center of section (Fig. 4).
1½ (38) and over		Full section by 8-in. (200-mm) gauge length, or machine standard ½ by 2-in. (13 by 50-mm) gauge length specimen from midway between surface and center (Fig. 4).	Machine standard ½ by 2-in. (13 by 50-mm) gauge length specimen from midway between surface and center (Fig. 4).
		Rounds, Squares, Hexagons, and Octagons	
Diameter or Distance Between Parallel Faces, in. (mm)		Hot-Rolled Bars	Cold-Finished Bars
Under ⅝		Full section by 8-in. (200-mm) gauge length or machine to subsize specimen (Fig. 4).	Machine to sub-size specimen (Fig. 4).
⅝ to 1½ (16 to 38), excl		Full section by 8-in. (200-mm) gauge length or machine standard ½ in. by 2-in. (13 by 50-mm) gauge length specimen from center of section (Fig. 4).	Machine standard ½ in. by 2-in. gauge length specimen from center of section (Fig. 4).
1½ (38) and over		Full section by 8-in. (200-mm) gauge length or machine standard ½ in. by 2-in. (13 by 50-mm) gauge length specimen from midway between surface and center of section (Fig. 4).	Machine standard ½ by 2-in. (13 by 50-mm gauge length specimen from midway between surface and center of section (Fig. 4)).
		Other Bar-Size Sections	
All sizes		Full section by 8-in. (200-mm) gauge length or prepare test specimen 1½ in. (38 mm) wide (if possible) by 8-in. (200-mm) gauge length.	Mill reduced section to 2-in. (50-mm) gauge length and approximately 25% less than test specimen width.

TABLE A1.2 Recommended Practice for Selecting Bend Test Specimens for Steel Bar Products

NOTE 1—The length of all specimens is to be not less than 6 in. (150 mm).
NOTE 2—The edges of the specimen may be rounded to a radius not exceeding ¹⁄₁₆ in. (1.6 mm).

	Flats	
Thickness, in. (mm)	Width, in. (mm)	Recommended Size
Up to ½ (13), incl	Up to ¾ (19), incl	Full section.
	Over ¾ (19)	Full section or machine to not less than ¾ in. (19 mm) in width by thickness of specimen.
Over ½ (13)	All	Full section or machine to 1 by ½ in. (25 by 13 mm) specimen from midway between center and surface.
	Rounds, Squares, Hexagons, and Octagons	
Diameter or Distance Between Parallel Faces, in. (mm)		Recommended Size
Up to 1½ (38), incl		Full section.
Over 1½ (38)		Machine to 1 by ½-in. (25 by 13-mm) specimen from midway between center and surface.

A2. STEEL TUBULAR PRODUCTS

A2.1 Scope

A2.1.1 This annex contains testing requirements for Steel Tubular Products that are specific to the product. The requirements contained in this annex are supplementary to those found in the general section of this specification. In the case of conflict between requirements provided in this annex and those found in the general section of this specification, the requirements of this annex shall prevail. In the case of conflict between requirements provided in this annex and requirements found in product specifications, the requirements found in the product specification shall prevail.

A2.1.2 Tubular shapes covered by this specification include, round, square, rectangular, and special shapes.

A2.2 Tension Test

A2.2.1 *Full-Size Longitudinal Test Specimens*:

A2.2.1.1 As an alternative to the use of longitudinal strip test specimens or longitudinal round test specimens, tension test specimens of full-size tubular sections are used, provided that the testing equipment has sufficient capacity. Snug-fitting metal plugs should be inserted far enough in the end of such tubular specimens to permit the testing machine jaws to grip the specimens properly without crushing. A design that may be used for such plugs is shown in Fig. A2.1. The plugs shall not extend into that part of the specimen on which the elongation is measured (Fig. A2.1). Care should be exercised to see that insofar as practicable, the load in such cases is applied axially. The length of the full-section specimen depends on the gauge length prescribed for measuring the elongation.

A2.2.1.2 Unless otherwise required by the product specification, the gauge length is 2 in. or 50 mm, except that for tubing having an outside diameter of ⅜ in. (9.5 mm) or less, it is customary for a gauge length equal to four times the outside diameter to be used when elongation comparable to that obtainable with larger test specimens is required.

A2.2.1.3 To determine the cross-sectional area of the full-section specimen, measurements shall be recorded as the average or mean between the greatest and least measurements of the outside diameter and the average or mean wall thickness, to the nearest 0.001 in. (0.025 mm) and the cross-sectional area is determined by the following equation:

$$A = 3.1416 t (D - t) \quad (A2.1)$$

where:
A = sectional area, in.^{2}
D = outside diameter, in., and
t = thickness of tube wall, in.

NOTE A2.1—There exist other methods of cross-sectional area determination, such as by weighing of the specimens, which are equally accurate or appropriate for the purpose.

A2.2.2 *Longitudinal Strip Test Specimens*:

A2.2.2.1 As an alternative to the use of full-size longitudinal test specimens or longitudinal round test specimens, longitudinal strip test specimens, obtained from strips cut from the tubular product as shown in Fig. A2.2 and machined to the dimensions shown in Fig. A2.3 are used. For welded structural tubing, such test specimens shall be from a location at least 90° from the weld; for other welded tubular products, such test specimens shall be from a location approximately 90° from the weld. Unless otherwise required by the product specification, the gauge length is 2 in. or 50 mm. The test specimens shall be tested using grips that are flat or have a surface contour corresponding to the curvature of the tubular product, or the ends of the test specimens shall be flattened without heating prior to the test specimens being tested using flat grips. The test specimen shown as specimen no. 4 in Fig. 3 shall be used, unless the capacity of the testing equipment or the dimensions and nature of the tubular product to be tested makes the use of specimen nos. 1, 2, or 3 necessary.

NOTE A2.2—An exact formula for calculating the cross-sectional area of specimens of the type shown in Fig. A2.3 taken from a circular tube is given in Test Methods E8/E8M.

A2.2.2.2 The width should be measured at each end of the gauge length to determine parallelism and also at the center. The thickness should be measured at the center and used with

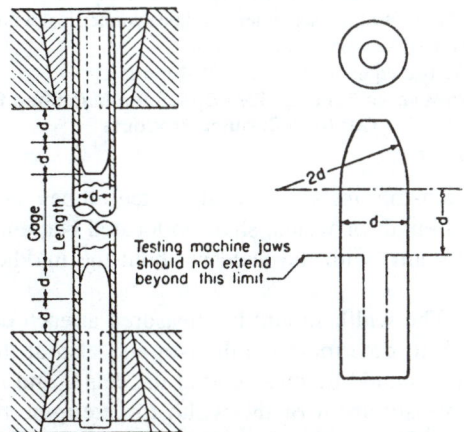

FIG. A2.1 Metal Plugs for Testing Tubular Specimens, Proper Location of Plugs in Specimen and of Specimen in Heads of Testing Machine

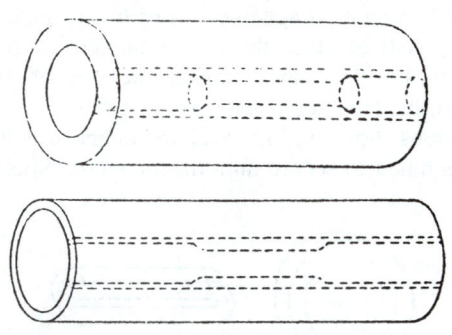

NOTE 1—The edges of the blank for the specimen shall be cut parallel to each other.

FIG. A2.2 Location of Longitudinal Tension–Test Specimens in Rings Cut from Tubular Products

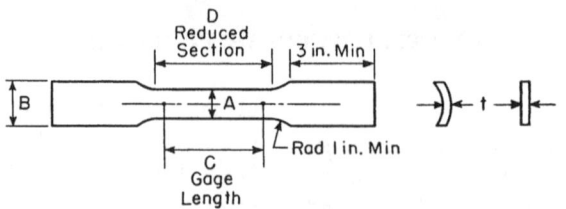

DIMENSIONS

Specimen No.	Dimensions, in.			
	A	B	C	D
1	½ ± 0.015	¹¹⁄₁₆ approximately	2 ± 0.005	2¼ min
2	¾ ± 0.031	1 approximately	2 ± 0.005	2¼ min
			4 ± 0.005	4½ min
3	1 ± 0.062	1½ approximately	2 ± 0.005	2¼ min
			4 ± 0.005	4½ min
4	1½ ± ⅛	2 approximately	2 ± 0.010	2¼ min
			4 ± 0.015	4½ min
			8 ± 0.020	9 min

Note 1—Cross-sectional area may be calculated by multiplying A and t.
Note 2—The dimension t is the thickness of the test specimen as provided for in the applicable material specifications.
Note 3—The reduced section shall be parallel within 0.010 in. and may have a gradual taper in width from the ends toward the center, with the ends not more than 0.010 in. wider than the center.
Note 4—The ends of the specimen shall be symmetrical with the center line of the reduced section within 0.10 in.
Note 5—Metric equivalent: 1 in. = 25.4 mm.
Note 6—Specimens with sides parallel throughout their length are permitted, except for referee testing, provided: (a) the above tolerances are used; (b) an adequate number of marks are provided for determination of elongation; and (c) when yield strength is determined, a suitable extensometer is used. If the fracture occurs at a distance of less than 2A from the edge of the gripping device, the tensile properties determined may not be representative of the material. If the properties meet the minimum requirements specified, no further testing is required, but if they are less than the minimum requirements, discard the test and retest.

FIG. A2.3 Dimensions and Tolerances for Longitudinal Strip Tension Test Specimens for Tubular Products

the center measurement of the width to determine the cross-sectional area. The center width dimension should be recorded to the nearest 0.005 in. (0.127 mm), and the thickness measurement to the nearest 0.001 in.

A2.2.3 *Transverse Strip Test Specimens*:

A2.2.3.1 In general, transverse tension tests are not recommended for tubular products, in sizes smaller than 8 in. in nominal diameter. When required, transverse tension test specimens may be taken from rings cut from ends of tubes or pipe as shown in Fig. A2.4. Flattening of the specimen may be done either after separating it from the tube as in Fig. A2.4 (a), or before separating it as in Fig. A2.4 (b), and may be done hot or cold; but if the flattening is done cold, the specimen may subsequently be normalized. Specimens from tubes or pipe for which heat treatment is specified, after being flattened either hot or cold, shall be given the same treatment as the tubes or pipe. For tubes or pipe having a wall thickness of less than ¾ in. (19 mm), the transverse test specimen shall be of the form and dimensions shown in Fig. A2.5 and either or both surfaces may be machined to secure uniform thickness. Specimens for

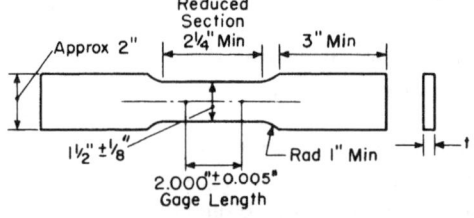

Note 1—The dimension t is the thickness of the test specimen as provided for in the applicable material specifications.
Note 2—The reduced section shall be parallel within 0.010 in. and may have a gradual taper in width from the ends toward the center, with the ends not more than 0.010 in. wider than the center.
Note 3—The ends of the specimen shall be symmetrical with the center line of the reduced section within 0.10 in.
Note 4—Metric equivalent: 1 in. = 25.4 mm.

FIG. A2.5 Transverse Tension Test Specimen Machined from Ring Cut from Tubular Products

transverse tension tests on welded steel tubes or pipe to determine strength of welds, shall be located perpendicular to the welded seams with the weld at about the middle of their length.

A2.2.3.2 The width should be measured at each end of the gauge length to determine parallelism and also at the center. The thickness should be measured at the center and used with the center measurement of the width to determine the cross-sectional area. The center width dimension should be recorded to the nearest 0.005 in. (0.127 mm), and the thickness measurement to the nearest 0.001 in. (0.025 mm).

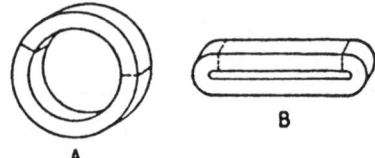

FIG. A2.4 Location of Transverse Tension Test Specimens in Ring Cut from Tubular Products.

A2.2.4 *Round Test Specimens*:

A2.2.4.1 When provided for in the product specification, the round test specimen shown in Fig. 4 may be used.

A2.2.4.2 The diameter of the round test specimen is measured at the center of the specimen to the nearest 0.001 in. (0.025 mm).

A2.2.4.3 Small-size specimens proportional to standard, as shown in Fig. 4, may be used when it is necessary to test material from which the standard specimen cannot be prepared. Other sizes of small-size specimens may be used. In any such small-size specimen, it is important that the gauge length for measurement of elongation be four times the diameter of the specimen (see Note 4, Fig. 4). The elongation requirements for the round specimen 2-in. gauge length in the product specification shall apply to the small-size specimens.

A2.2.4.4 For transverse specimens, the section from which the specimen is taken shall not be flattened or otherwise deformed.

A2.2.4.5 Longitudinal test specimens are obtained from strips cut from the tubular product as shown in Fig. A2.2.

A2.3 Determination of Transverse Yield Strength, Hydraulic Ring-Expansion Method

A2.3.1 Hardness tests are made on the outside surface, inside surface, or wall cross-section depending upon product-specification limitation. Surface preparation may be necessary to obtain accurate hardness values.

A2.3.2 A testing machine and method for determining the transverse yield strength from an annular ring specimen, have been developed and described in A2.3.3-8.1.2.

A2.3.3 A diagrammatic vertical cross-sectional sketch of the testing machine is shown in Fig. A2.6.

A2.3.4 In determining the transverse yield strength on this machine, a short ring (commonly 3 in. (76 mm) in length) test specimen is used. After the large circular nut is removed from the machine, the wall thickness of the ring specimen is determined and the specimen is telescoped over the oil resistant rubber gasket. The nut is then replaced, but is not turned down tight against the specimen. A slight clearance is left between the nut and specimen for the purpose of permitting free radial movement of the specimen as it is being tested. Oil under pressure is then admitted to the interior of the rubber gasket through the pressure line under the control of a suitable valve. An accurately calibrated pressure gauge serves to measure oil pressure. Any air in the system is removed through the bleeder line. As the oil pressure is increased, the rubber gasket expands which in turn stresses the specimen circumferentially. As the pressure builds up, the lips of the rubber gasket act as a seal to prevent oil leakage. With continued increase in pressure, the ring specimen is subjected to a tension stress and elongates accordingly. The entire outside circumference of the ring specimen is considered as the gauge length and the strain is measured with a suitable extensometer which will be described later. When the desired total strain or extension under load is reached on the extensometer, the oil pressure in pounds per square inch is read and by employing Barlow's formula, the unit yield strength is calculated. The yield strength, thus determined, is a true result since the test specimen has not been cold worked by flattening and closely approximates the same condition as the tubular section from which it is cut. Further, the test closely simulates service conditions in pipe lines. One testing machine unit may be used for several different sizes of pipe by the use of suitable rubber gaskets and adapters.

NOTE A2.3—Barlow's formula may be stated two ways:

$$(1) \quad P = 2St/D \quad (A2.2)$$
$$(2) \quad S = PD/2t \quad (A2.3)$$

where:
P = internal hydrostatic pressure, psi,
S = unit circumferential stress in the wall of the tube produced by the internal hydrostatic pressure, psi,
t = thickness of the tube wall, in., and
D = outside diameter of the tube, in.

A2.3.5 A roller chain type extensometer which has been found satisfactory for measuring the elongation of the ring specimen is shown in Fig. A2.7 and Fig. A2.8. Fig. A2.7 shows the extensometer in position, but unclamped, on a ring specimen. A small pin, through which the strain is transmitted to and

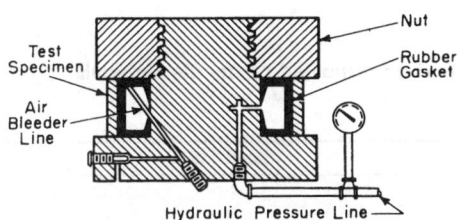

FIG. A2.6 Testing Machine for Determination of Transverse Yield Strength from Annular Ring Specimens

FIG. A2.7 Roller Chain Type Extensometer, Unclamped

FIG. A2.8 Roller Chain Type Extensometer, Clamped

measured by the dial gauge, extends through the hollow threaded stud. When the extensometer is clamped, as shown in Fig. A2.8, the desired tension which is necessary to hold the instrument in place and to remove any slack, is exerted on the roller chain by the spring. Tension on the spring may be regulated as desired by the knurled thumb screw. By removing or adding rollers, the roller chain may be adapted for different sizes of tubular sections.

A2.4 Hardness Tests

A2.4.1 Hardness tests are made either on the outside or the inside surfaces on the end of the tube as appropriate.

A2.4.2 The standard 3000-kgf Brinell load may cause too much deformation in a thin-walled tubular specimen. In this case the 500-kgf load shall be applied, or inside stiffening by means of an internal anvil should be used. Brinell testing shall not be applicable to tubular products less than 2 in. (51 mm) in outside diameter, or less than 0.200 in. (5.1 mm) in wall thickness.

A2.4.3 The Rockwell hardness tests are normally made on the inside surface, a flat on the outside surface, or on the wall cross-section depending upon the product limitation. Rockwell hardness tests are not performed on tubes smaller than 5/16 in. (7.9 mm) in outside diameter, nor are they performed on the inside surface of tubes with less than 1/4 in. (6.4 mm) inside diameter. Rockwell hardness tests are not performed on annealed tubes with walls less than 0.065 in. (1.65 mm) thick or cold worked or heat treated tubes with walls less than 0.049 in. (1.24 mm) thick. For tubes with wall thicknesses less than those permitting the regular Rockwell hardness test, the Superficial Rockwell test is sometimes substituted. Transverse Rockwell hardness readings can be made on tubes with a wall thickness of 0.187 in. (4.75 mm) or greater. The curvature and the wall thickness of the specimen impose limitations on the Rockwell hardness test. When a comparison is made between Rockwell determinations made on the outside surface and determinations made on the inside surface, adjustment of the readings will be required to compensate for the effect of curvature. The Rockwell B scale is used on all materials having an expected hardness range of B0 to B100. The Rockwell C scale is used on material having an expected hardness range of C20 to C68.

A2.4.4 Superficial Rockwell hardness tests are normally performed on the outside surface whenever possible and whenever excessive spring back is not encountered. Otherwise, the tests may be performed on the inside. Superficial Rockwell hardness tests shall not be performed on tubes with an inside diameter of less than 1/4 in. (6.4 mm). The wall thickness limitations for the Superficial Rockwell hardness test are given in Table A2.1 and Table A2.2.

A2.4.5 When the outside diameter, inside diameter, or wall thickness precludes the obtaining of accurate hardness values, tubular products shall be specified to tensile properties and so tested.

A2.5 Manipulating Tests

A2.5.1 The following tests are made to prove ductility of certain tubular products:

A2.5.1.1 *Flattening Test*—The flattening test as commonly made on specimens cut from tubular products is conducted by subjecting rings from the tube or pipe to a prescribed degree of flattening between parallel plates (Fig. A2.4). The severity of the flattening test is measured by the distance between the parallel plates and is varied according to the dimensions of the tube or pipe. The flattening test specimen should not be less than 2½ in. (63.5 mm) in length and should be flattened cold to the extent required by the applicable material specifications.

TABLE A2.1 Wall Thickness Limitations of Superficial Hardness Test on Annealed or Ductile Materials for Steel Tubular Products[A]

("T" Scale (1/16-in. Ball))

Wall Thickness, in. (mm)	Load, kgf
Over 0.050 (1.27)	45
Over 0.035 (0.89)	30
0.020 and over (0.51)	15

[A] The heaviest load recommended for a given wall thickness is generally used.

TABLE A2.2 Wall Thickness Limitations of Superficial Hardness Test on Cold Worked or Heat Treated Material for Steel Tubular Products[A]

("N" Scale (Diamond Penetrator))

Wall Thickness, in. (mm)	Load, kgf
Over 0.035 (0.89)	45
Over 0.025 (0.51)	30
0.015 and over (0.38)	15

[A] The heaviest load recommended for a given wall thickness is generally used.

A2.5.1.2 *Reverse Flattening Test*—The reverse flattening test is designed primarily for application to electric-welded tubing for the detection of lack of penetration or overlaps resulting from flash removal in the weld. The specimen consists of a length of tubing approximately 4 in. (102 mm) long which is split longitudinally 90° on each side of the weld. The sample is then opened and flattened with the weld at the point of maximum bend (Fig. A2.9).

A2.5.1.3 *Crush Test*—The crush test, sometimes referred to as an upsetting test, is usually made on boiler and other pressure tubes, for evaluating ductility (Fig. A2.10). The specimen is a ring cut from the tube, usually about 2½ in. (63.5 mm) long. It is placed on end and crushed endwise by hammer or press to the distance prescribed by the applicable material specifications.

A2.5.1.4 *Flange Test*—The flange test is intended to determine the ductility of boiler tubes and their ability to withstand the operation of bending into a tube sheet. The test is made on a ring cut from a tube, usually not less than 4 in. (100 mm) long and consists of having a flange turned over at right angles to the body of the tube to the width required by the applicable material specifications. The flaring tool and die block shown in Fig. A2.11 are recommended for use in making this test.

A2.5.1.5 *Flaring Test*—For certain types of pressure tubes, an alternate to the flange test is made. This test consists of driving a tapered mandrel having a slope of 1 in 10 as shown in Fig. A2.12 (*a*) or a 60° included angle as shown in Fig. A2.12 (*b*) into a section cut from the tube, approximately 4 in. (100 mm) in length, and thus expanding the specimen until the inside diameter has been increased to the extent required by the applicable material specifications.

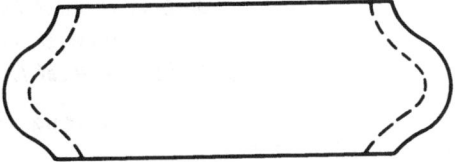

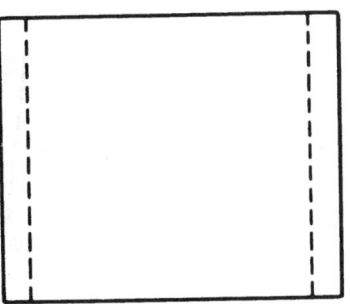

FIG. A2.10 Crush Test Specimen

A2.5.1.6 *Bend Test*—For pipe used for coiling in sizes 2 in. and under a bend test is made to determine its ductility and the soundness of weld. In this test a sufficient length of full-size pipe is bent cold through 90° around a cylindrical mandrel having a diameter 12 times the nominal diameter of the pipe. For close coiling, the pipe is bent cold through 180° around a mandrel having a diameter 8 times the nominal diameter of the pipe.

A2.5.1.7 *Transverse Guided Bend Test of Welds*—This bend test is used to determine the ductility of fusion welds. The specimens used are approximately 1½ in. (38 mm) wide, at least 6 in. (152 mm) in length with the weld at the center, and are machined in accordance with Fig. A2.13 for face and root bend tests and in accordance with Fig. A2.14 for side bend tests. The dimensions of the plunger shall be as shown in Fig. A2.15 and the other dimensions of the bending jig shall be substantially as given in this same figure. A test shall consist of a face bend specimen and a root bend specimen or two side bend specimens. A face bend test requires bending with the inside surface of the pipe against the plunger; a root bend test requires bending with the outside surface of the pipe against the plunger; and a side bend test requires bending so that one of the side surfaces becomes the convex surface of the bend specimen.

(*a*) Failure of the bend test depends upon the appearance of cracks in the area of the bend, of the nature and extent described in the product specifications.

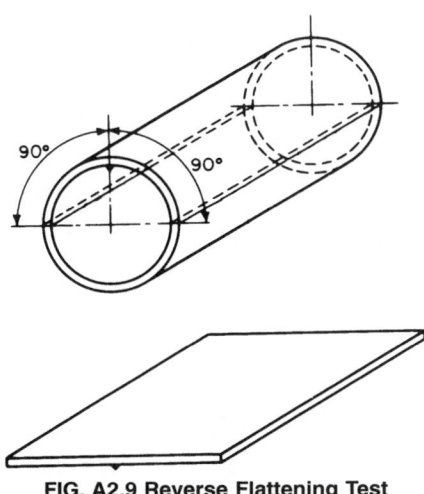

FIG. A2.9 Reverse Flattening Test

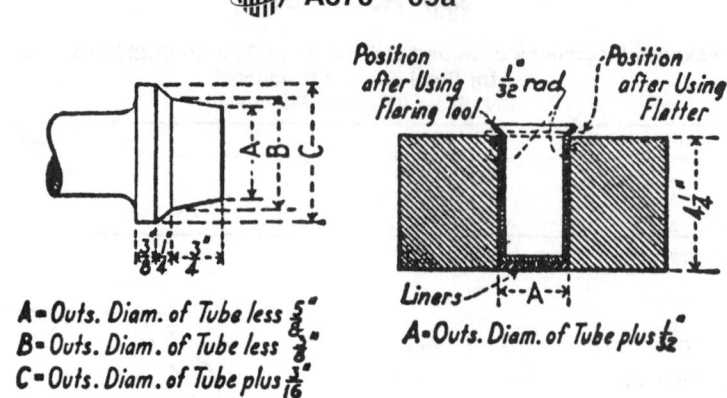

NOTE 1—Metric equivalent: 1 in. = 25.4 mm.
FIG. A2.11 Flaring Tool and Die Block for Flange Test

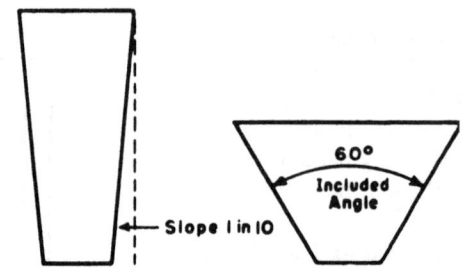

FIG. A2.12 Tapered Mandrels for Flaring Test

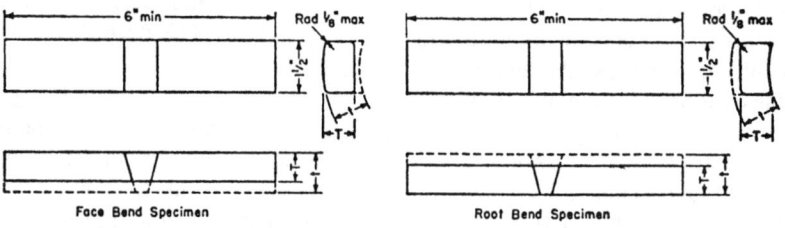

NOTE 1—Metric equivalent: 1 in. = 25.4 mm.

Pipe Wall Thickness (t), in.	Test Specimen Thickness, in.
Up to ⅜, incl	t
Over ⅜	⅜

FIG. A2.13 Transverse Face- and Root-Bend Test Specimens

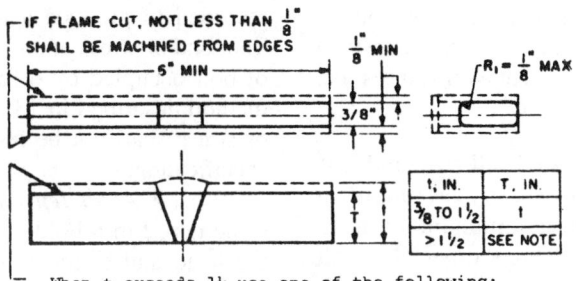

NOTE 1—Metric equivalent: 1 in. = 25.4 mm.
FIG. A2.14 Side-Bend Specimen for Ferrous Materials

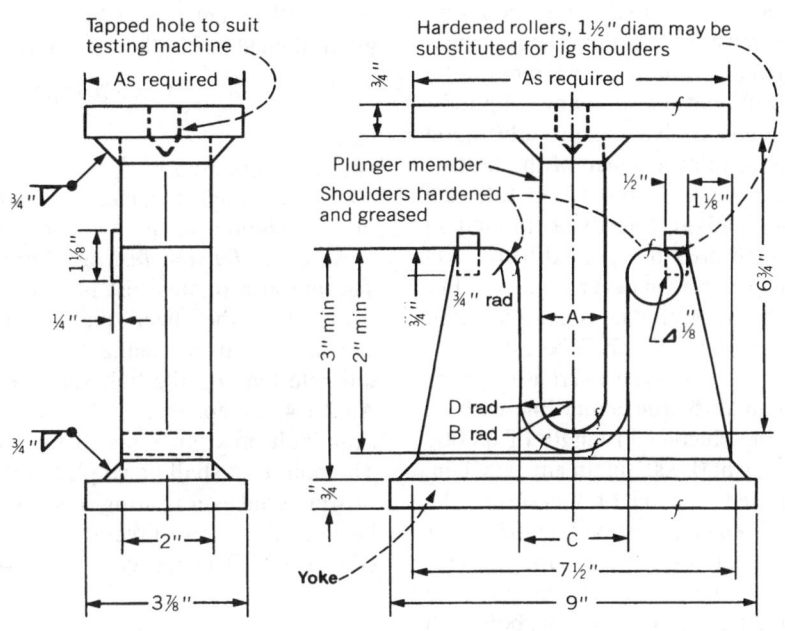

NOTE 1—Metric equivalent: 1 in. = 25.4 mm.

Test Specimen Thickness, in.	A	B	C	D	
3/8	1½	3/4	2 3/8	1 3/16	
t	4t	2t	6t + 1/8	3t + 1/16	
					Material
3/8	2½	1¼	3 3/8	1 11/16	Materials with a specified minimum tensile strength of 95 ksi or greater.
t	6⅔ t	3⅓ t	8⅔ t + 1/8	4½ t + 1/16	

FIG. A2.15 Guided-Bend Test Jig

A3. STEEL FASTENERS

A3.1 Scope

A3.1.1 This annex contains testing requirements for Steel Fasteners that are specific to the product. The requirements contained in this annex are supplementary to those found in the general section of this specification. In the case of conflict between requirements provided in this annex and those found in the general section of this specification, the requirements of this annex shall prevail. In the case of conflict between requirements provided in this annex and requirements found in product specifications, the requirements found in the product specification shall prevail.

A3.1.2 These tests are set up to facilitate production control testing and acceptance testing with certain more precise tests to be used for arbitration in case of disagreement over test results.

A3.2 Tension Tests

A3.2.1 It is preferred that bolts be tested full size, and it is customary, when so testing bolts to specify a minimum ultimate load in pounds, rather than a minimum ultimate strength in pounds per square inch. Three times the bolt nominal diameter has been established as the minimum bolt length subject to the tests described in the remainder of this section. Sections A3.2.1.1-A3.2.1.3 apply when testing bolts full size. Section A3.2.1.4 shall apply where the individual product specifications permit the use of machined specimens.

A3.2.1.1 *Proof Load*— Due to particular uses of certain classes of bolts it is desirable to be able to stress them, while in use, to a specified value without obtaining any permanent set. To be certain of obtaining this quality the proof load is specified. The proof load test consists of stressing the bolt with a specified load which the bolt must withstand without permanent set. An alternate test which determines yield strength of a full size bolt is also allowed. Either of the following Methods, 1 or 2, may be used but Method 1 shall be the arbitration method in case of any dispute as to acceptance of the bolts.

A3.2.1.2 *Proof Load Testing Long Bolts*—When full size tests are required, proof load Method 1 is to be limited in application to bolts whose length does not exceed 8 in. (203 mm) or 8 times the nominal diameter, whichever is greater. For bolts longer than 8 in. or 8 times the nominal diameter, whichever is greater, proof load Method 2 shall be used.

(a) *Method 1, Length Measurement*—The overall length of a straight bolt shall be measured at its true center line with an instrument capable of measuring changes in length of 0.0001 in. (0.0025 mm) with an accuracy of 0.0001 in. in any 0.001-in. (0.025-mm) range. The preferred method of measuring the length shall be between conical centers machined on the center line of the bolt, with mating centers on the measuring anvils. The head or body of the bolt shall be marked so that it can be placed in the same position for all measurements. The bolt shall be assembled in the testing equipment as outlined in A3.2.1.4, and the proof load specified in the product specification shall be applied. Upon release of this load the length of the bolt shall be again measured and shall show no permanent elongation. A tolerance of ±0.0005 in. (0.0127 mm) shall be allowed between the measurement made before loading and that made after loading. Variables, such as straightness and thread alignment (plus measurement error), may result in apparent elongation of the fasteners when the proof load is initially applied. In such cases, the fastener may be retested using a 3 percent greater load, and may be considered satisfactory if the length after this loading is the same as before this loading (within the 0.0005-in. tolerance for measurement error).

A3.2.1.3 *Proof Load-Time of Loading*—The proof load is to be maintained for a period of 10 s before release of load, when using Method 1.

(*1*) *Method 2, Yield Strength*—The bolt shall be assembled in the testing equipment as outlined in A3.2.1.4. As the load is applied, the total elongation of the bolt or any part of the bolt which includes the exposed six threads shall be measured and recorded to produce a load-strain or a stress-strain diagram. The load or stress at an offset equal to 0.2 percent of the length of bolt occupied by 6 full threads shall be determined by the method described in 13.2.1 of these methods, A370. This load or stress shall not be less than that prescribed in the product specification.

A3.2.1.4 *Axial Tension Testing of Full Size Bolts*—Bolts are to be tested in a holder with the load axially applied between the head and a nut or suitable fixture (Fig. A3.1), either of which shall have sufficient thread engagement to develop the full strength of the bolt. The nut or fixture shall be assembled on the bolt leaving six complete bolt threads unengaged between the grips, except for heavy hexagon structural bolts which shall have four complete threads unengaged between the grips. To meet the requirements of this test there shall be a tensile failure in the body or threaded section with no failure at the junction of the body, and head. If it is necessary to record or report the tensile strength of bolts as psi values the stress area shall be calculated from the mean of the mean root and pitch diameters of Class 3 external threads as follows:

$$A_s = 0.7854\,[D - (0.9743/n)]^2 \qquad (A3.1)$$

where:
A_s = stress area, in.^{2},
D = nominal diameter, in., and
n = number of threads per inch.

A3.2.1.5 *Tension Testing of Full-Size Bolts with a Wedge*— The purpose of this test is to obtain the tensile strength and demonstrate the "head quality" and ductility of a bolt with a standard head by subjecting it to eccentric loading. The ultimate load on the bolt shall be determined as described in A3.2.1.4, except that a 10° wedge shall be placed under the same bolt previously tested for the proof load (see A3.2.1.1). The bolt head shall be so placed that no corner of the hexagon or square takes a bearing load, that is, a flat of the head shall be aligned with the direction of uniform thickness of the wedge (Fig. A3.2). The wedge shall have an included angle of 10°

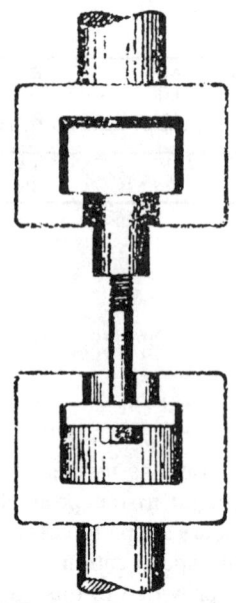

FIG. A3.1 Tension Testing Full-Size Bolt

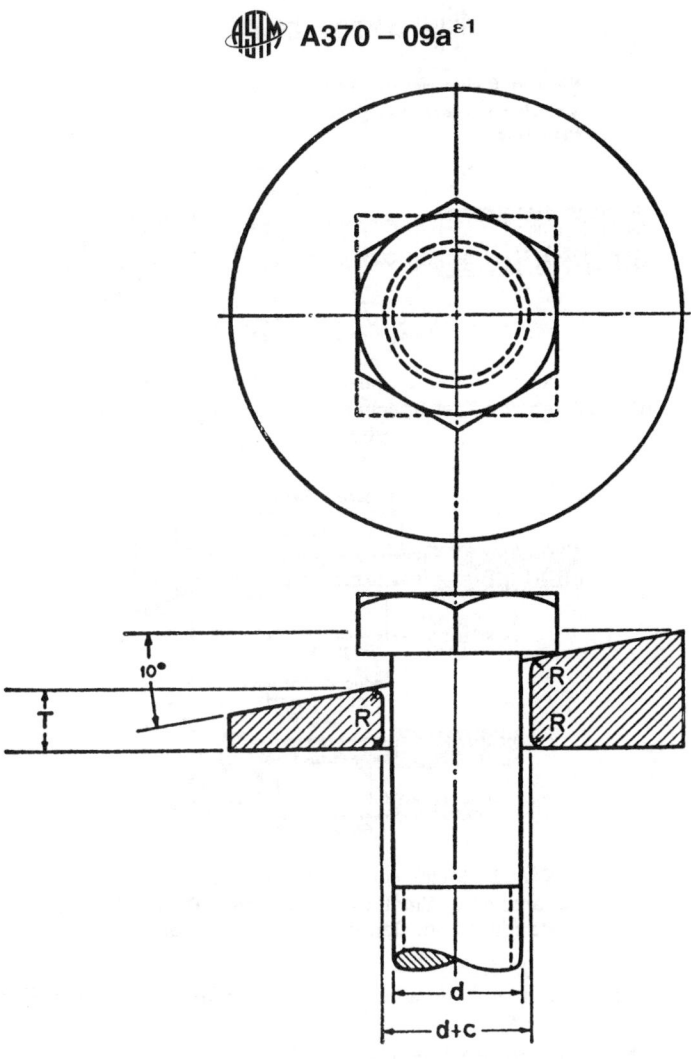

c = Clearance of wedge hole
d = Diameter of bolt
R = Radius
T = Thickness of wedge at short side of hole equal to one-half diameter of bolt

FIG. A3.2 Wedge Test Detail

between its faces and shall have a thickness of one-half of the nominal bolt diameter at the short side of the hole. The hole in the wedge shall have the following clearance over the nominal size of the bolt, and its edges, top and bottom, shall be rounded to the following radius:

Nominal Bolt Size, in.	Clearance in Hole, in. (mm)	Radius on Corners of Hole, in. (mm)
¼ to ½	0.030 (0.76)	0.030 (0.76)
⁹⁄₁₆ to ¾	0.050 (1.3)	0.060 (1.5)
⅞ to 1	0.063 (1.5)	0.060 (1.5)
1⅛ to 1¼	0.063 (1.5)	0.125 (3.2)
1⅜ to 1½	0.094 (2.4)	0.125 (3.2)

A3.2.1.6 *Wedge Testing of HT Bolts Threaded to Head*—For heat-treated bolts over 100 000 psi (690 MPa) minimum tensile strength and that are threaded 1 diameter and closer to the underside of the head, the wedge angle shall be 6° for sizes ¼ through ¾ in. (6.35 to 19.0 mm) and 4° for sizes over ¾ in.

A3.2.1.7 *Tension Testing of Bolts Machined to Round Test Specimens*:

(1) Bolts under 1½ in. (38 mm) in diameter which require machined tests shall preferably use a standard ½-in., (13-mm) round 2-in. (50-mm) gauge length test specimen (Fig. 4); however, bolts of small cross-section that will not permit the taking of this standard test specimen shall use one of the small-size-specimens-proportional-to-standard (Fig. 4) and the specimen shall have a reduced section as large as possible. In all cases, the longitudinal axis of the specimen shall be concentric with the axis of the bolt; the head and threaded section of the bolt may be left intact, as in Fig. A3.3 and Fig. A3.4, or shaped to fit the holders or grips of the testing machine so that the load is applied axially. The gauge length for measuring the elongation shall be four times the diameter of the specimen.

(2) For bolts 1½ in. and over in diameter, a standard ½-in. round 2-in. gauge length test specimen shall be turned from the

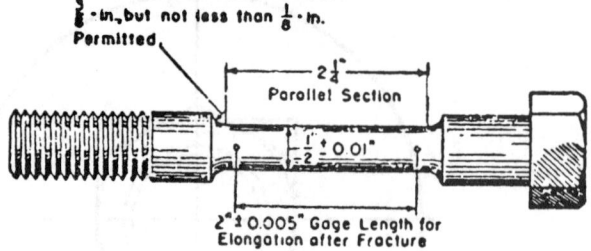

NOTE 1—Metric equivalent: 1 in. = 25.4 mm.
FIG. A3.3 Tension Test Specimen for Bolt with Turned-Down Shank

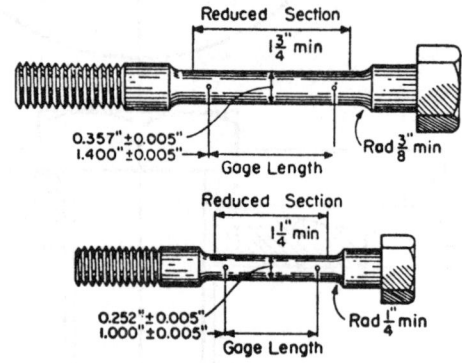

NOTE 1—Metric equivalent: 1 in. = 25.4 mm.
FIG. A3.4 Examples of Small Size Specimens Proportional to Standard 2-in. Gauge Length Specimen

bolt, having its axis midway between the center and outside surface of the body of the bolt as shown in Fig. A3.5.

(3) Machined specimens are to be tested in tension to determine the properties prescribed by the product specifications. The methods of testing and determination of properties shall be in accordance with Section 13 of these test methods.

A3.3 Hardness Tests for Externally Threaded Fasteners

A3.3.1 When specified, externally threaded fasteners shall be hardness tested. Fasteners with hexagonal or square heads shall be Brinell or Rockwell hardness tested on the side or top of the head. Externally threaded fasteners with other type of

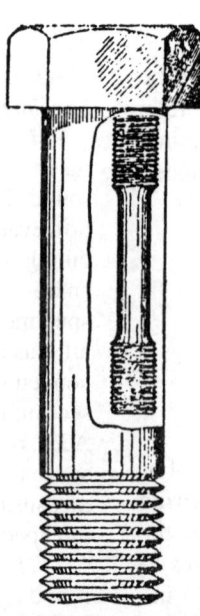

FIG. A3.5 Location of Standard Round 2-in. Gauge Length Tension Test Specimen When Turned from Large Size Bolt

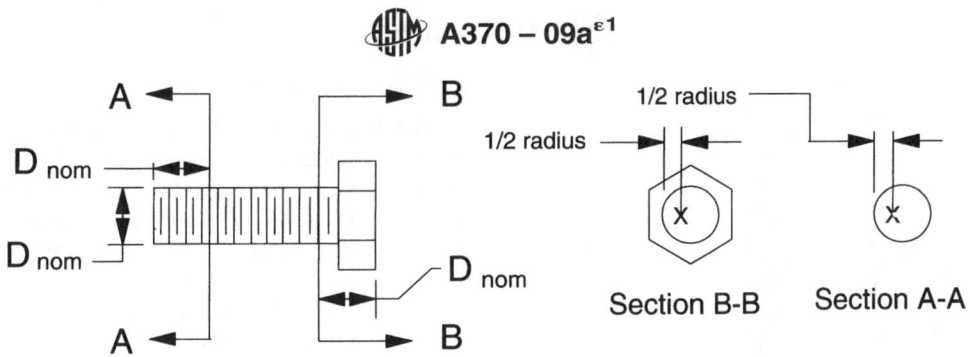

X=Location of Hardness Impressions
FIG. A3.6 Hardness Test Locations for Bolts in a Dispute

heads and those without heads shall be Brinell or Rockwell hardness tested on one end. Due to possible distortion from the Brinell load, care should be taken that this test meets the requirements of Section 16 of these test methods. Where the Brinell hardness test is impractical, the Rockwell hardness test shall be substituted. Rockwell hardness test procedures shall conform to Section 18 of these test methods.

A3.3.2 In cases where a dispute exists between buyer and seller as to whether externally threaded fasteners meet or exceed the hardness limit of the product specification, for purposes of arbitration, hardness may be taken on two transverse sections through a representative sample fastener selected at random. Hardness readings shall be taken at the locations shown in Fig. A3.6. All hardness values must conform with the hardness limit of the product specification in order for the fasteners represented by the sample to be considered in compliance. This provision for arbitration of a dispute shall not be used to accept clearly rejectable fasteners.

A3.4 Testing of Nuts

A3.4.1 *Hardness Test*— Rockwell hardness of nuts shall be determined on the top or bottom face of the nut. Brinell hardness shall be determined on the side of the nuts. Either method may be used at the option of the manufacturer, taking into account the size and grade of the nuts under test. When the standard Brinell hardness test results in deforming the nut it will be necessary to use a minor load or substitute a Rockwell hardness test.

A4. STEEL ROUND WIRE PRODUCTS

A4.1 Scope

A4.1.1 This annex contains testing requirements for Round Wire Products that are specific to the product. The requirements contained in this annex are supplementary to those found in the general section of this specification. In the case of conflict between requirements provided in this annex and those found in the general section of this specification, the requirements of this annex shall prevail. In the case of conflict between requirements provided in this annex and requirements found in product specifications, the requirements found in the product specification shall prevail.

A4.2 Apparatus

A4.2.1 *Gripping Devices*—Grips of either the wedge or snubbing types as shown in Fig. A4.1 and Fig. A4.2 shall be used (Note A4.1). When using grips of either type, care shall be taken that the axis of the test specimen is located approximately at the center line of the head of the testing machine (Note A4.2). When using wedge grips the liners used behind the grips shall be of the proper thickness.

NOTE A4.1—Testing machines usually are equipped with wedge grips. These wedge grips, irrespective of the type of testing machine, may be referred to as the "usual type" of wedge grips. The use of fine (180 or 240) grit abrasive cloth in the "usual" wedge type grips, with the abrasive contacting the wire specimen, can be helpful in reducing specimen slipping and breakage at the grip edges at tensile loads up to about 1000 pounds. For tests of specimens of wire which are liable to be cut at the edges by the "usual type" of wedge grips, the snubbing type gripping device has proved satisfactory.

For testing round wire, the use of cylindrical seat in the wedge gripping device is optional.

NOTE A4.2—Any defect in a testing machine which may cause non-axial application of load should be corrected.

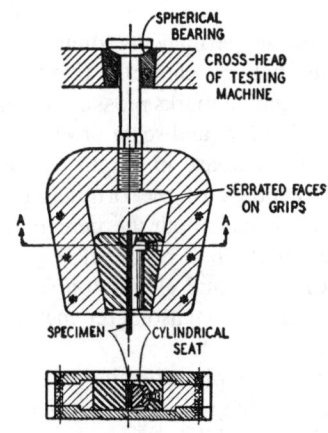

FIG. A4.1 Wedge-Type Gripping Device

A4.2.2 *Pointed Micrometer*—A micrometer with a pointed spindle and anvil suitable for reading the dimensions of the wire specimen at the fractured ends to the nearest 0.001 in. (0.025 mm) after breaking the specimen in the testing machine shall be used.

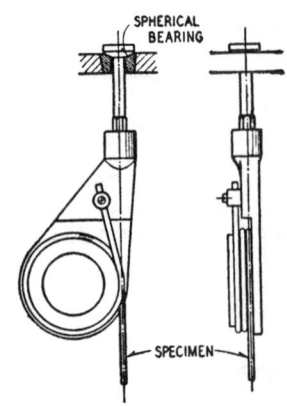

FIG. A4.2 Snubbing-Type Gripping Device

A4.3 Test Specimens

A4.3.1 Test specimens having the full cross-sectional area of the wire they represent shall be used. The standard gauge length of the specimens shall be 10 in. (254 mm). However, if the determination of elongation values is not required, any convenient gauge length is permissible. The total length of the specimens shall be at least equal to the gauge length (10 in.) plus twice the length of wire required for the full use of the grip employed. For example, depending upon the type of testing machine and grips used, the minimum total length of specimen may vary from 14 to 24 in. (360 to 610 mm) for a 10-in. gauge length specimen.

A4.3.2 Any specimen breaking in the grips shall be discarded and a new specimen tested.

A4.4 Elongation

A4.4.1 In determining permanent elongation, the ends of the fractured specimen shall be carefully fitted together and the distance between the gauge marks measured to the nearest 0.01 in. (0.25 mm) with dividers and scale or other suitable device. The elongation is the increase in length of the gauge length, expressed as a percentage of the original gauge length. In recording elongation values, both the percentage increase and the original gauge length shall be given.

A4.4.2 In determining total elongation (elastic plus plastic extension) autographic or extensometer methods may be employed.

A4.4.3 If fracture takes place outside of the middle third of the gauge length, the elongation value obtained may not be representative of the material.

A4.5 Reduction of Area

A4.5.1 The ends of the fractured specimen shall be carefully fitted together and the dimensions of the smallest cross section measured to the nearest 0.001 in. (0.025 mm) with a pointed micrometer. The difference between the area thus found and the area of the original cross section, expressed as a percentage of the original area, is the reduction of area.

A4.5.2 The reduction of area test is not recommended in wire diameters less than 0.092 in. (2.34 mm) due to the difficulties of measuring the reduced cross sections.

A4.6 Rockwell Hardness Test

A4.6.1 On heat–treated wire of diameter 0.100 in. (2.54 mm) and larger, the specimen shall be flattened on two parallel sides by grinding before testing. The hardness test is not recommended for any diameter of hard drawn wire or heat-treated wire less than 0.100 in. (2.54 mm) in diameter. For round wire, the tensile strength test is greatly preferred over the hardness test.

A4.7 Wrap Test

A4.7.1 This test is used as a means for testing the ductility of certain kinds of wire.

A4.7.2 The test consists of coiling the wire in a closely spaced helix tightly against a mandrel of a specified diameter for a required number of turns. (Unless other specified, the required number of turns shall be five.) The wrapping may be done by hand or a power device. The wrapping rate may not exceed 15 turns per min. The mandrel diameter shall be specified in the relevant wire product specification.

A4.7.3 The wire tested shall be considered to have failed if the wire fractures or if any longitudinal or transverse cracks develop which can be seen by the unaided eye after the first complete turn. Wire which fails in the first turn shall be retested, as such fractures may be caused by bending the wire to a radius less than specified when the test starts.

A4.8 Coiling Test

A4.8.1 This test is used to determine if imperfections are present to the extent that they may cause cracking or splitting during spring coiling and spring extension. A coil of specified length is closed wound on an arbor of a specified diameter. The closed coil is then stretched to a specified permanent increase in length and examined for uniformity of pitch with no splits or fractures. The required arbor diameter, closed coil length, and permanent coil extended length increase may vary with wire diameter, properties, and type.

A5. NOTES ON SIGNIFICANCE OF NOTCHED-BAR IMPACT TESTING

A5.1 Notch Behavior

A5.1.1 The Charpy and Izod type tests bring out notch behavior (brittleness versus ductility) by applying a single overload of stress. The energy values determined are quantitative comparisons on a selected specimen but cannot be converted into energy values that would serve for engineering design calculations. The notch behavior indicated in an individual test applies only to the specimen size, notch geometry, and testing conditions involved and cannot be generalized to other sizes of specimens and conditions.

A5.1.2 The notch behavior of the face-centered cubic metals and alloys, a large group of nonferrous materials and the austenitic steels can be judged from their common tensile properties. If they are brittle in tension they will be brittle when notched, while if they are ductile in tension, they will be ductile when notched, except for unusually sharp or deep notches (much more severe than the standard Charpy or Izod specimens). Even low temperatures do not alter this characteristic of these materials. In contrast, the behavior of the ferritic steels under notch conditions cannot be predicted from their properties as revealed by the tension test. For the study of these materials the Charpy and Izod type tests are accordingly very useful. Some metals that display normal ductility in the tension test may nevertheless break in brittle fashion when tested or when used in the notched condition. Notched conditions include restraints to deformation in directions perpendicular to the major stress, or multiaxial stresses, and stress concentrations. It is in this field that the Charpy and Izod tests prove useful for determining the susceptibility of a steel to notch-brittle behavior though they cannot be directly used to appraise the serviceability of a structure.

A5.1.3 The testing machine itself must be sufficiently rigid or tests on high-strength low-energy materials will result in excessive elastic energy losses either upward through the pendulum shaft or downward through the base of the machine. If the anvil supports, the pendulum striking edge, or the machine foundation bolts are not securely fastened, tests on ductile materials in the range of 80 ft·lbf (108 J) may actually indicate values in excess of 90 to 100 ft·lbf (122 to 136 J).

A5.2 Notch Effect

A5.2.1 The notch results in a combination of multiaxial stresses associated with restraints to deformation in directions perpendicular to the major stress, and a stress concentration at the base of the notch. A severely notched condition is generally not desirable, and it becomes of real concern in those cases in which it initiates a sudden and complete failure of the brittle type. Some metals can be deformed in a ductile manner even down to the low temperatures of liquid air, while others may crack. This difference in behavior can be best understood by considering the cohesive strength of a material (or the property that holds it together) and its relation to the yield point. In cases of brittle fracture, the cohesive strength is exceeded before significant plastic deformation occurs and the fracture appears crystalline. In cases of the ductile or shear type of failure, considerable deformation precedes the final fracture and the broken surface appears fibrous instead of crystalline. In intermediate cases the fracture comes after a moderate amount of deformation and is part crystalline and part fibrous in appearance.

A5.2.2 When a notched bar is loaded, there is a normal stress across the base of the notch which tends to initiate fracture. The property that keeps it from cleaving, or holds it together, is the "cohesive strength." The bar fractures when the normal stress exceeds the cohesive strength. When this occurs without the bar deforming it is the condition for brittle fracture.

A5.2.3 In testing, though not in service because of side effects, it happens more commonly that plastic deformation precedes fracture. In addition to the normal stress, the applied load also sets up shear stresses which are about 45° to the normal stress. The elastic behavior terminates as soon as the shear stress exceeds the shear strength of the material and deformation or plastic yielding sets in. This is the condition for ductile failure.

A5.2.4 This behavior, whether brittle or ductile, depends on whether the normal stress exceeds the cohesive strength before the shear stress exceeds the shear strength. Several important facts of notch behavior follow from this. If the notch is made sharper or more drastic, the normal stress at the root of the notch will be increased in relation to the shear stress and the bar will be more prone to brittle fracture (see Table A5.1). Also, as the speed of deformation increases, the shear strength increases and the likelihood of brittle fracture increases. On the other hand, by raising the temperature, leaving the notch and the speed of deformation the same, the shear strength is lowered and ductile behavior is promoted, leading to shear failure.

A5.2.5 Variations in notch dimensions will seriously affect the results of the tests. Tests on E4340 steel specimens[7] have shown the effect of dimensional variations on Charpy results (see Table A5.1).

A5.3 Size Effect

A5.3.1 Increasing either the width or the depth of the specimen tends to increase the volume of metal subject to distortion, and by this factor tends to increase the energy absorption when breaking the specimen. However, any increase in size, particularly in width, also tends to increase the degree of restraint and by tending to induce brittle fracture, may decrease the amount of energy absorbed. Where a standard-size specimen is on the verge of brittle fracture, this is particularly true, and a double-width specimen may actually require less energy for rupture than one of standard width.

A5.3.2 In studies of such effects where the size of the material precludes the use of the standard specimen, as for example when the material is ¼-in. plate, subsize specimens

---

[7] Fahey, N. H., "Effects of Variables in Charpy Impact Testing," *Materials Research & Standards*, Vol 1, No. 11, November, 1961, p. 872.

TABLE A5.1 Effect of Varying Notch Dimensions on Standard Specimens

	High-Energy Specimens, ft·lbf (J)	High-Energy Specimens, ft·lbf (J)	Low-Energy Specimens, ft·lbf (J)
Specimen with standard dimensions	76.0 ± 3.8 (103.0 ± 5.2)	44.5 ± 2.2 (60.3 ± 3.0)	12.5 ± 1.0 (16.9 ± 1.4)
Depth of notch, 0.084 in. (2.13 mm)[A]	72.2 (97.9)	41.3 (56.0)	11.4 (15.5)
Depth of notch, 0.0805 in. (2.04 mm)[A]	75.1 (101.8)	42.2 (57.2)	12.4 (16.8)
Depth of notch, 0.0775 in. (1.77 mm)[A]	76.8 (104.1)	45.3 (61.4)	12.7 (17.2)
Depth of notch, 0.074 in. (1.57 mm)[A]	79.6 (107.9)	46.0 (62.4)	12.8 (17.3)
Radius at base of notch, 0.005 in. (0.127 mm)[B]	72.3 (98.0)	41.7 (56.5)	10.8 (14.6)
Radius at base of notch, 0.015 in. (0.381 mm)[B]	80.0 (108.5)	47.4 (64.3)	15.8 (21.4)

[A] Standard 0.079 ± 0.002 in. (2.00 ± 0.05 mm).
[B] Standard 0.010 ± 0.001 in. (0.25 ± 0.025 mm).

are necessarily used. Such specimens (see Fig. 6 of Test Methods E23) are based on the Type A specimen of Fig. 4 of Test Methods E23.

A5.3.3 General correlation between the energy values obtained with specimens of different size or shape is not feasible, but limited correlations may be established for specification purposes on the basis of special studies of particular materials and particular specimens. On the other hand, in a study of the relative effect of process variations, evaluation by use of some arbitrarily selected specimen with some chosen notch will in most instances place the methods in their proper order.

A5.4 Effects of Testing Conditions

A5.4.1 The testing conditions also affect the notch behavior. So pronounced is the effect of temperature on the behavior of steel when notched that comparisons are frequently made by examining specimen fractures and by plotting energy value and fracture appearance versus temperature from tests of notched bars at a series of temperatures. When the test temperature has been carried low enough to start cleavage fracture, there may be an extremely sharp drop in impact value or there may be a relatively gradual falling off toward the lower temperatures. This drop in energy value starts when a specimen begins to exhibit some crystalline appearance in the fracture. The transition temperature at which this embrittling effect takes place varies considerably with the size of the part or test specimen and with the notch geometry.

A5.4.2 Some of the many definitions of transition temperature currently being used are: (*1*) the lowest temperature at which the specimen exhibits 100 % fibrous fracture, (*2*) the temperature where the fracture shows a 50 % crystalline and a 50 % fibrous appearance, (*3*) the temperature corresponding to the energy value 50 % of the difference between values obtained at 100 % and 0 % fibrous fracture, and (*4*) the temperature corresponding to a specific energy value.

A5.4.3 A problem peculiar to Charpy-type tests occurs when high-strength, low-energy specimens are tested at low temperatures. These specimens may not leave the machine in the direction of the pendulum swing but rather in a sidewise direction. To ensure that the broken halves of the specimens do not rebound off some component of the machine and contact the pendulum before it completes its swing, modifications may be necessary in older model machines. These modifications differ with machine design. Nevertheless the basic problem is the same in that provisions must be made to prevent rebounding of the fractured specimens into any part of the swinging pendulum. Where design permits, the broken specimens may be deflected out of the sides of the machine and yet in other designs it may be necessary to contain the broken specimens within a certain area until the pendulum passes through the anvils. Some low-energy high-strength steel specimens leave impact machines at speeds in excess of 50 ft (15.3 m)/s although they were struck by a pendulum traveling at speeds approximately 17 ft (5.2 m)/s. If the force exerted on the pendulum by the broken specimens is sufficient, the pendulum will slow down and erroneously high energy values will be recorded. This problem accounts for many of the inconsistencies in Charpy results reported by various investigators within the 10 to 25-ft·lbf (14 to 34 J) range. The Apparatus Section (the paragraph regarding Specimen Clearance) of Test Methods E23 discusses the two basic machine designs and a modification found to be satisfactory in minimizing jamming.

A5.5 Velocity of Straining

A5.5.1 Velocity of straining is likewise a variable that affects the notch behavior of steel. The impact test shows somewhat higher energy absorption values than the static tests above the transition temperature and yet, in some instances, the reverse is true below the transition temperature.

A5.6 Correlation with Service

A5.6.1 While Charpy or Izod tests may not directly predict the ductile or brittle behavior of steel as commonly used in large masses or as components of large structures, these tests can be used as acceptance tests of identity for different lots of the same steel or in choosing between different steels, when correlation with reliable service behavior has been established. It may be necessary to make the tests at properly chosen temperatures other than room temperature. In this, the service temperature or the transition temperature of full-scale specimens does not give the desired transition temperatures for Charpy or Izod tests since the size and notch geometry may be so different. Chemical analysis, tension, and hardness tests may not indicate the influence of some of the important processing factors that affect susceptibility to brittle fracture nor do they comprehend the effect of low temperatures in inducing brittle behavior.

A6. PROCEDURE FOR CONVERTING PERCENTAGE ELONGATION OF A STANDARD ROUND TENSION TEST SPECIMEN TO EQUIVALENT PERCENTAGE ELONGATION OF A STANDARD FLAT SPECIMEN

A6.1 Scope

A6.1.1 This method specifies a procedure for converting percentage elongation after fracture obtained in a standard 0.500-in. (12.7-mm) diameter by 2-in. (51-mm) gauge length test specimen to standard flat test specimens ½ in. by 2 in. and 1½ in. by 8 in. (38.1 by 203 mm).

A6.2 Basic Equation

A6.2.1 The conversion data in this method are based on an equation by Bertella,[8] and used by Oliver[9] and others. The relationship between elongations in the standard 0.500-in. diameter by 2.0-in. test specimen and other standard specimens can be calculated as follows:

$$e = e_o [4.47 (\sqrt{A})/L]^a \quad (A6.1)$$

where:
- e_o = percentage elongation after fracture on a standard test specimen having a 2-in. gauge length and 0.500-in. diameter,
- e = percentage elongation after fracture on a standard test specimen having a gauge length L and a cross-sectional area A, and
- a = constant characteristic of the test material.

A6.3 Application

A6.3.1 In applying the above equation the constant a is characteristic of the test material. The value $a = 0.4$ has been found to give satisfactory conversions for carbon, carbon-manganese, molybdenum, and chromium-molybdenum steels within the tensile strength range of 40 000 to 85 000 psi (275 to 585 MPa) and in the hot-rolled, in the hot-rolled and normalized, or in the annealed condition, with or without tempering. Note that the cold reduced and quenched and tempered states are excluded. For annealed austenitic stainless steels, the value $a = 0.127$ has been found to give satisfactory conversions.

A6.3.2 Table A6.1 has been calculated taking $a = 0.4$, with the standard 0.500-in. (12.7-mm) diameter by 2-in. (51-mm) gauge length test specimen as the reference specimen. In the case of the subsize specimens 0.350 in. (8.89 mm) in diameter by 1.4-in. (35.6-mm) gauge length, and 0.250-in. (6.35- mm) diameter by 1.0-in. (25.4-mm) gauge length the factor in the equation is 4.51 instead of 4.47. The small error introduced by using Table A6.1 for the subsized specimens may be neglected. Table A6.2 for annealed austenitic steels has been calculated taking $a = 0.127$, with the standard 0.500-in. diameter by 2-in. gauge length test specimen as the reference specimen.

A6.3.3 Elongation given for a standard 0.500-in. diameter by 2-in. gauge length specimen may be converted to elongation for ½ in. by 2 in. or 1½ in. by 8-in. (38.1 by 203-mm) flat specimens by multiplying by the indicated factor in Table A6.1 and Table A6.2.

A6.3.4 These elongation conversions shall not be used where the width to thickness ratio of the test piece exceeds 20, as in sheet specimens under 0.025 in. (0.635 mm) in thickness.

TABLE A6.1 Carbon and Alloy Steels—Material Constant $a = 0.4$. Multiplication Factors for Converting Percent Elongation from ½-in. Diameter by 2-in. Gauge Length Standard Tension Test Specimen to Standard ½ by 2-in. and 1½ by 8-in. Flat Specimens

Thickness, in.	½ by 2-in. Specimen	1½ by 8-in. Specimen	Thickness in.	1½ by 8-in. Specimen
0.025	0.574	...	0.800	0.822
0.030	0.596	...	0.850	0.832
0.035	0.614	...	0.900	0.841
0.040	0.631	...	0.950	0.850
0.045	0.646	...	1.000	0.859
0.050	0.660	...	1.125	0.880
0.055	0.672	...	1.250	0.898
0.060	0.684	...	1.375	0.916
0.065	0.695	...	1.500	0.932
0.070	0.706	...	1.625	0.947
0.075	0.715	...	1.750	0.961
0.080	0.725	...	1.875	0.974
0.085	0.733	...	2.000	0.987
0.090	0.742	0.531	2.125	0.999
0.100	0.758	0.542	2.250	1.010
0.110	0.772	0.553	2.375	1.021
0.120	0.786	0.562	2.500	1.032
0.130	0.799	0.571	2.625	1.042
0.140	0.810	0.580	2.750	1.052
0.150	0.821	0.588	2.875	1.061
0.160	0.832	0.596	3.000	1.070
0.170	0.843	0.603	3.125	1.079
0.180	0.852	0.610	3.250	1.088
0.190	0.862	0.616	3.375	1.096
0.200	0.870	0.623	3.500	1.104
0.225	0.891	0.638	3.625	1.112
0.250	0.910	0.651	3.750	1.119
0.275	0.928	0.664	3.875	1.127
0.300	0.944	0.675	4.000	1.134
0.325	0.959	0.686	...	...
0.350	0.973	0.696	...	...
0.375	0.987	0.706	...	...
0.400	1.000	0.715	...	...
0.425	1.012	0.724	...	...
0.450	1.024	0.732	...	...
0.475	1.035	0.740	...	...
0.500	1.045	0.748	...	...
0.525	1.056	0.755	...	...
0.550	1.066	0.762	...	...
0.575	1.075	0.770	...	...
0.600	1.084	0.776	...	...
0.625	1.093	0.782	...	...
0.650	1.101	0.788	...	...
0.675	1.110	...	...	...
0.700	1.118	0.800	...	...
0.725	1.126	...	...	...
0.750	1.134	0.811	...	...

A6.3.5 While the conversions are considered to be reliable within the stated limitations and may generally be used in specification writing where it is desirable to show equivalent elongation requirements for the several standard ASTM tension specimens covered in Test Methods A370, consideration must be given to the metallurgical effects dependent on the thickness of the material as processed.

---

[8] Bertella, C. A., *Giornale del Genio Civile*, Vol 60, 1922, p. 343.
[9] Oliver, D. A., *Proceedings of the Institution of Mechanical Engineers*, 1928, p. 827.

TABLE A6.2 Annealed Austenitic Stainless Steels—Material Constant a = 0.127. Multiplication Factors for Converting Percent Elongation from ½-in. Diameter by 2-in. Gauge Length Standard Tension Test Specimen to Standard ½ by 2-in. and 1½ by 8-in. Flat Specimens

Thickness, in.	½ by 2-in. Specimen	1½ by 8-in. Specimen	Thickness, in.	1½ by 8-in. Specimen
0.025	0.839	...	0.800	0.940
0.030	0.848	...	0.850	0.943
0.035	0.857	...	0.900	0.947
0.040	0.864	...	0.950	0.950
0.045	0.870	...	1.000	0.953
0.050	0.876	...	1.125	0.960
0.055	0.882	...	1.250	0.966
0.060	0.886	...	1.375	0.972
0.065	0.891	...	1.500	0.978
0.070	0.895	...	1.625	0.983
0.075	0.899	...	1.750	0.987
0.080	0.903	...	1.875	0.992
0.085	0.906	...	2.000	0.996
0.090	0.909	0.818	2.125	1.000
0.095	0.913	0.821	2.250	1.003
0.100	0.916	0.823	2.375	1.007
0.110	0.921	0.828	2.500	1.010
0.120	0.926	0.833	2.625	1.013
0.130	0.931	0.837	2.750	1.016
0.140	0.935	0.841	2.875	1.019
0.150	0.940	0.845	3.000	1.022
0.160	0.943	0.848	3.125	1.024
0.170	0.947	0.852	3.250	1.027
0.180	0.950	0.855	3.375	1.029
0.190	0.954	0.858	3.500	1.032
0.200	0.957	0.860	3.625	1.034
0.225	0.964	0.867	3.750	1.036
0.250	0.970	0.873	3.875	1.038
0.275	0.976	0.878	4.000	1.041
0.300	0.982	0.883	...	...
0.325	0.987	0.887	...	...
0.350	0.991	0.892	...	...
0.375	0.996	0.895	...	...
0.400	1.000	0.899	...	...
0.425	1.004	0.903	...	...
0.450	1.007	0.906	...	...
0.475	1.011	0.909	...	...
0.500	1.014	0.912	...	...
0.525	1.017	0.915	...	...
0.550	1.020	0.917	...	...
0.575	1.023	0.920	...	...
0.600	1.026	0.922	...	...
0.625	1.029	0.925	...	...
0.650	1.031	0.927	...	...
0.675	1.034	...	...	...
0.700	1.036	0.932	...	...
0.725	1.038	...	...	...
0.750	1.041	0.936	...	...

A7. METHOD OF TESTING MULTI-WIRE STRAND FOR PRESTRESSED CONCRETE

A7.1 Scope

A7.1.1 This method provides procedures for the tension testing of multi-wire strand for prestressed concrete. This method is intended for use in evaluating the strand properties prescribed in specifications for" prestressing steel strands."

A7.2 General Precautions

A7.2.1 Premature failure of the test specimens may result if there is any appreciable notching, cutting, or bending of the specimen by the gripping devices of the testing machine.

A7.2.2 Errors in testing may result if the seven wires constituting the strand are not loaded uniformly.

A7.2.3 The mechanical properties of the strand may be materially affected by excessive heating during specimen preparation.

A7.2.4 These difficulties may be minimized by following the suggested methods of gripping described in A7.4.

A7.3 Gripping Devices

A7.3.1 The true mechanical properties of the strand are determined by a test in which fracture of the specimen occurs in the free span between the jaws of the testing machine. Therefore, it is desirable to establish a test procedure with suitable apparatus which will consistently produce such results. Due to inherent physical characteristics of individual machines, it is not practical to recommend a universal gripping procedure that is suitable for all testing machines. Therefore, it is necessary to determine which of the methods of gripping described in A7.3.2 to A7.3.8 is most suitable for the testing equipment available.

A7.3.2 *Standard V-Grips with Serrated Teeth (Note A7.1)*.

A7.3.3 *Standard V-Grips with Serrated Teeth (Note A7.1), Using Cushioning Material*—In this method, some material is placed between the grips and the specimen to minimize the notching effect of the teeth. Among the materials which have been used are lead foil, aluminum foil, carborundum cloth, bra shims, etc. The type and thickness of material required is dependent on the shape, condition, and coarseness of the teeth.

A7.3.4 *Standard V-Grips with Serrated Teeth (Note A7.1), Using Special Preparation of the Gripped Portions of the Specimen*—One of the methods used is tinning, in which the gripped portions are cleaned, fluxed, and coated by multiple dips in molten tin alloy held just above the melting point. Another method of preparation is encasing the gripped portions in metal tubing or flexible conduit, using epoxy resin as the bonding agent. The encased portion should be approximately twice the length of lay of the strand.

A7.3.5 *Special Grips with Smooth, Semi-Cylindrical Grooves (Note A7.2)*—The grooves and the gripped portions of the specimen are coated with an abrasive slurry which holds the specimen in the smooth grooves, preventing slippage. The slurry consists of abrasive such as Grade 3-F aluminum oxide and a carrier such as water or glycerin.

A7.3.6 *Standard Sockets of the Type Used for Wire Rope*—The gripped portions of the specimen are anchored in the sockets with zinc. The special procedures for socketing usually employed in the wire rope industry must be followed.

A7.3.7 *Dead-End Eye Splices*—These devices are available in sizes designed to fit each size of strand to be tested.

A7.3.8 *Chucking Devices*—Use of chucking devices of the type generally employed for applying tension to strands in casting beds is not recommended for testing purposes.

NOTE A7.1—The number of teeth should be approximately 15 to 30 per in., and the minimum effective gripping length should be approximately 4 in. (102 mm).

NOTE A7.2—The radius of curvature of the grooves is approximately the same as the radius of the strand being tested, and is located 1/32 in. (0.79 mm) above the flat face of the grip. This prevents the two grips from closing tightly when the specimen is in place.

A7.4 Specimen Preparation

A7.4.1 If the molten-metal temperatures employed during hot-dip tinning or socketing with metallic material are too high, over approximately 700°F (370°C), the specimen may be heat affected with a subsequent loss of strength and ductility. Careful temperature controls should be maintained if such methods of specimen preparation are used.

A7.5 Procedure

A7.5.1 *Yield Strength*— For determining the yield strength use a Class B-1 extensometer (Note A7.3) as described in Practice E83. Apply an initial load of 10 % of the expected minimum breaking strength to the specimen, then attach the extensometer and adjust it to a reading of 0.001 in./in. of gauge length. Then increase the load until the extensometer indicates an extension of 1 %. Record the load for this extension as the yield strength. The extensometer may be removed from the specimen after the yield strength has been determined.

A7.5.2 *Elongation*— For determining the elongation use a Class D extensometer (Note A7.3), as described in Practice E83, having a gauge length of not less than 24 in. (610 mm) (Note A7.4). Apply an initial load of 10 % of the required minimum breaking strength to the specimen, then attach the extensometer (Note A7.3) and adjust it to a zero reading. The extensometer may be removed from the specimen prior to rupture after the specified minimum elongation has been exceeded. It is not necessary to determine the final elongation value.

A7.5.3 *Breaking Strength*—Determine the maximum load at which one or more wires of the strand are fractured. Record this load as the breaking strength of the strand.

NOTE A7.3—The yield-strength extensometer and the elongation extensometer may be the same instrument or two separate instruments. Two separate instruments are advisable since the more sensitive yield-strength extensometer, which could be damaged when the strand fractures, may be removed following the determination of yield strength. The elongation extensometer may be constructed with less sensitive parts or be constructed in such a way that little damage would result if fracture occurs while the extensometer is attached to the specimen.

NOTE A7.4—Specimens that break outside the extensometer or in the jaws and yet meet the minimum specified values are considered as meeting the mechanical property requirements of the product specification, regardless of what procedure of gripping has been used. Specimens that break outside of the extensometer or in the jaws and do not meet the minimum specified values are subject to retest. Specimens that break between the jaws and the extensometer and do not meet the minimum specified values are subject to retest as provided in the applicable specification.

A8. ROUNDING OF TEST DATA

A8.1 Rounding

A8.1.1 An observed value or a calculated value shall be rounded off in accordance with the applicable product specification. In the absence of a specified procedure, the rounding-off method of Practice E29 shall be used.

A8.1.1.1 Values shall be rounded up or rounded down as determined by the rules of Practice E29.

A8.1.1.2 In the special case of rounding the number "5" when no additional numbers other than "0" follow the "5," rounding shall be done in the direction of the specification limits if following Practice E29 would cause rejection of material.

A8.1.2 Recommended levels for rounding reported values of test data are given in Table A8.1. These values are designed to provide uniformity in reporting and data storage, and should be used in all cases except where they conflict with specific requirements of a product specification.

NOTE A8.1—To minimize cumulative errors, whenever possible, values should be carried to at least one figure beyond that of the final (rounded) value during intervening calculations (such as calculation of stress from load and area measurements) with rounding occurring as the final operation. The precision may be less than that implied by the number of significant figures.

TABLE A8.1 Recommended Values for Rounding Test Data

Test Quantity	Test Data Range	Rounded Value[A]
Yield Point, Yield Strength, Tensile Strength	up to 50 000 psi, excl (up to 50 ksi)	100 psi (0.1 ksi)
	50 000 to 100 000 psi, excl (50 to 100 ksi)	500 psi (0.5 ksi)
	100 000 psi and above (100 ksi and above)	1000 psi (1.0 ksi)
	up to 500 MPa, excl	1 MPa
	500 to 1000 MPa, excl	5 MPa
	1000 MPa and above	10 MPa
Elongation	0 to 10 %, excl	0.5 %
	10 % and above	1 %
Reduction of Area	0 to 10 %, excl	0.5 %
	10 % and above	1 %
Impact Energy	0 to 240 ft·lbf (or 0 to 325 J)	1 ft·lbf (or 1 J)[B]
Brinell Hardness	all values	tabular value[C]
Rockwell Hardness	all scales	1 Rockwell Number

[A] Round test data to the nearest integral multiple of the values in this column. If the data value is exactly midway between two rounded values, round in accordance with A8.1.1.2.

[B] These units are not equivalent but the rounding occurs in the same numerical ranges for each. (1 ft·lbf = 1.356 J.)

[C] Round the mean diameter of the Brinell impression to the nearest 0.05 mm and report the corresponding Brinell hardness number read from the table without further rounding.

A9. METHODS FOR TESTING STEEL REINFORCING BARS

A9.1 Scope

A9.1.1 This annex contains testing requirements for Steel Reinforcing Bar that are specific to the product. The requirements contained in this annex are supplementary to those found in the general section of this specification. In the case of conflict between requirements provided in this annex and those found in the general section of this specification, the requirements of this annex shall prevail. In the case of conflict between requirements provided in this annex and requirements found in product specifications, the requirements found in the product specification shall prevail.

A9.2 Test Specimens

A9.2.1 All test specimens shall be the full section of the bar as rolled.

A9.3 Tension Testing

A9.3.1 *Test Specimen*— Specimens for tension tests shall be long enough to provide for an 8-in. (200-mm) gauge length, a distance of at least two bar diameters between each gauge mark and the grips, plus sufficient additional length to fill the grips completely leaving some excess length protruding beyond each grip.

A9.3.2 *Gripping Device*— The grips shall be shimmed so that no more than ½ in. (13 mm) of a grip protrudes from the head of the testing machine.

A9.3.3 *Gauge Marks*— The 8-in. (200-mm) gauge length shall be marked on the specimen using a preset 8-in. (200-mm) punch or, alternately, may be punch marked every 2 in. (50 mm) along the 8-in. (200-mm) gauge length, on one of the longitudinal ribs, if present, or in clear spaces of the deformation pattern. The punch marks shall not be put on a transverse deformation. Light punch marks are desirable because deep marks severely indent the bar and may affect the results. A bullet-nose punch is desirable.

A9.3.4 The yield strength or yield point shall be determined by one of the following methods:

A9.3.4.1 Extension under load using an autographic diagram method or an extensometer as described in 13.1.2 and 13.1.3,

A9.3.4.2 By the drop of the beam or halt in the gauge of the testing machine as described in 13.1.1 where the steel tested as a sharp-kneed or well-defined type of yield point.

A9.3.5 The unit stress determinations for yield and tensile strength on full-size specimens shall be based on the nominal bar area.

A9.4 Bend Testing

A9.4.1 Bend tests shall be made on specimens of sufficient length to ensure free bending and with apparatus which provides:

A9.4.1.1 Continuous and uniform application of force throughout the duration of the bending operation,

A9.4.1.2 Unrestricted movement of the specimen at points of contact with the apparatus and bending around a pin free to rotate, and

A9.4.1.3 Close wrapping of the specimen around the pin during the bending operation.

A9.4.2 Other acceptable more severe methods of bend testing, such as placing a specimen across two pins free to rotate and applying the bending force with a fix pin, may be used.

A9.4.3 When retesting is permitted by the product specification, the following shall apply:

A9.4.3.1 Sections of bar containing identifying roll marking shall not be used.

A9.4.3.2 Bars shall be so placed that longitudinal ribs lie in a plane at right angles to the plane of bending.

A10. PROCEDURE FOR USE AND CONTROL OF HEAT-CYCLE SIMULATION

A10.1 Purpose

A10.1.1 To ensure consistent and reproducible heat treatments of production forgings and the test specimens that represent them when the practice of heat-cycle simulation is used.

A10.2 Scope

A10.2.1 Generation and documentation of actual production time—temperature curves (MASTER CHARTS).

A10.2.2 Controls for duplicating the master cycle during heat treatment of production forgings. (Heat treating within the essential variables established during A1.2.1).

A10.2.3 Preparation of program charts for the simulator unit.

A10.2.4 Monitoring and inspection of the simulated cycle within the limits established by the ASME Code.

A10.2.5 Documentation and storage of all controls, inspections, charts, and curves.

A10.3 Referenced Documents

A10.3.1 *ASME Standards:*[4]

ASME Boiler and Pressure Vessel Code Section III, latest edition.

ASME Boiler and Pressure Vessel Code Section VIII, Division 2, latest edition.

A10.4 Terminology

A10.4.1 *Definitions*:

A10.4.1.1 *master chart*—a record of the heat treatment received from a forging essentially identical to the production forgings that it will represent. It is a chart of time and temperature showing the output from thermocouples imbedded in the forging at the designated test immersion and test location or locations.

A10.4.1.2 *program chart*—the metallized sheet used to program the simulator unit. Time-temperature data from the master chart are manually transferred to the program chart.

A10.4.1.3 *simulator chart*—a record of the heat treatment that a test specimen had received in the simulator unit. It is a chart of time and temperature and can be compared directly to the master chart for accuracy of duplication.

A10.4.1.4 *simulator cycle*—one continuous heat treatment of a set of specimens in the simulator unit. The cycle includes heating from ambient, holding at temperature, and cooling. For example, a simulated austenitize and quench of a set of specimens would be one cycle; a simulated temper of the same specimens would be another cycle.

A10.5 Procedure

A10.5.1 *Production Master Charts*:

A10.5.1.1 Thermocouples shall be imbedded in each forging from which a master chart is obtained. Temperature shall be monitored by a recorder with resolution sufficient to clearly define all aspects of the heating, holding, and cooling process. All charts are to be clearly identified with all pertinent information and identification required for maintaining permanent records.

A10.5.1.2 Thermocouples shall be imbedded 180° apart if the material specification requires test locations 180° apart.

A10.5.1.3 One master chart (or two if required in accordance with A10.5.3.1) shall be produced to represent essentially identical forgings (same size and shape). Any change in size or geometry (exceeding rough machining tolerances) of a forging will necessitate that a new master cooling curve be developed.

A10.5.1.4 If more than one curve is required per master forging (180° apart) and a difference in cooling rate is achieved, then the most conservative curve shall be used as the master curve.

A10.5.2 *Reproducibility of Heat Treatment Parameters on Production Forgings*:

A10.5.2.1 All information pertaining to the quench and temper of the master forging shall be recorded on an appropriate permanent record, similar to the one shown in Table A10.1.

A10.5.2.2 All information pertaining to the quench and temper of the production forgings shall be appropriately recorded, preferably on a form similar to that used in A10.5.2.1. Quench records of production forgings shall be retained for future reference. The quench and temper record of the master forging shall be retained as a permanent record.

A10.5.2.3 A copy of the master forging record shall be stored with the heat treatment record of the production forging.

A10.5.2.4 The essential variables, as set forth on the heat treat record, shall be controlled within the given parameters on the production forging.

A10.5.2.5 The temperature of the quenching medium prior to quenching each production forging shall be equal to or lower than the temperature of the quenching medium prior to quenching the master forging.

A10.5.2.6 The time elapsed from opening the furnace door to quench for the production forging shall not exceed that elapsed for the master forging.

A10.5.2.7 If the time parameter is exceeded in opening the furnace door to beginning of quench, the forging shall be placed back into the furnace and brought back up to equalization temperature.

TABLE A10.1 Heat-Treat Record-Essential Variables

	Master Forging	Production Forging 1	Production Forging 2	Production Forging 3	Production Forging 4	Production Forging 5
Program chart number						
Time at temperature and actual temperature of heat treatment						
Method of cooling						
Forging thickness						
Thermocouple immersion						
Beneath buffer (yes/no)						
Forging number						
Product						
Material						
Thermocouple location—0 deg						
Thermocouple location—180 deg						
Quench tank No.						
Date of heat treatment						
Furnace number						
Cycle number						
Heat treater						
Starting quench medium temperature						
Time from furnace to quench						
Heating rate above 1000°F (538°C)						
Temperature upon removal from quench after 5 min						
Orientation of forging in quench						

A10.5.2.8 All forgings represented by the same master forging shall be quenched with like orientation to the surface of the quench bath.

A10.5.2.9 All production forgings shall be quenched in the same quench tank, with the same agitation as the master forging.

A10.5.2.10 *Uniformity of Heat Treat Parameters*—*(1)* The difference in actual heat treating temperature between production forgings and the master forging used to establish the simulator cycle for them shall not exceed ±25°F (±14°C) for the quench cycle. *(2)* The tempering temperature of the production forgings shall not fall below the actual tempering temperature of the master forging. *(3)* At least one contact surface thermocouple shall be placed on each forging in a production load. Temperature shall be recorded for all surface thermocouples on a Time Temperature Recorder and such records shall be retained as permanent documentation.

A10.5.3 *Heat-Cycle Simulation*:

A10.5.3.1 Program charts shall be made from the data recorded on the master chart. All test specimens shall be given the same heating rate above, the AC1, the same holding time and the same cooling rate as the production forgings.

A10.5.3.2 The heating cycle above the AC1, a portion of the holding cycle, and the cooling portion of the master chart shall be duplicated and the allowable limits on temperature and time, as specified in (a)–(c), shall be established for verification of the adequacy of the simulated heat treatment.

(a) Heat Cycle Simulation of Test Coupon Heat Treatment for Quenched and Tempered Forgings and Bars—If cooling rate data for the forgings and bars and cooling rate control devices for the test specimens are available, the test specimens may be heat-treated in the device.

(b) The test coupons shall be heated to substantially the same maximum temperature as the forgings or bars. The test coupons shall be cooled at a rate similar to and no faster than the cooling rate representative of the test locations and shall be within 25°F (14°C) and 20 s at all temperatures after cooling begins. The test coupons shall be subsequently heat treated in accordance with the thermal treatments below the critical temperature including tempering and simulated post weld heat treatment.

(c) Simulated Post Weld Heat Treatment of Test Specimens (for ferritic steel forgings and bars)—Except for carbon steel (P Number 1, Section IX of the Code) forgings and bars with a nominal thickness or diameter of 2 in. (51 mm) or less, the test specimens shall be given a heat treatment to simulate any thermal treatments below the critical temperature that the forgings and bars may receive during fabrication. The simulated heat treatment shall utilize temperatures, times, and cooling rates as specified on the order. The total time at temperature(s) for the test material shall be at least 80 % of the total time at temperature(s) to which the forgings and bars are subjected during postweld heat treatment. The total time at temperature(s) for the test specimens may be performed in a single cycle.

A10.5.3.3 Prior to heat treatment in the simulator unit, test specimens shall be machined to standard sizes that have been determined to allow adequately for subsequent removal of decarb and oxidation.

A10.5.3.4 At least one thermocouple per specimen shall be used for continuous recording of temperature on an independent external temperature-monitoring source. Due to the sensitivity and design peculiarities of the heating chamber of certain equipment, it is mandatory that the hot junctions of control and monitoring thermocouples always be placed in the same relative position with respect to the heating source (generally infrared lamps).

A10.5.3.5 Each individual specimen shall be identified, and such identification shall be clearly shown on the simulator chart and simulator cycle record.

A10.5.3.6 The simulator chart shall be compared to the master chart for accurate reproduction of simulated quench in accordance with A10.5.3.2*(a)*. If any one specimen is not heat treated within the acceptable limits of temperature and time, such specimen shall be discarded and replaced by a newly machined specimen. Documentation of such action and reasons for deviation from the master chart shall be shown on the simulator chart, and on the corresponding nonconformance report.

A10.5.4 *Reheat Treatment and Retesting*:

A10.5.4.1 In the event of a test failure, retesting shall be handled in accordance with rules set forth by the material specification.

A10.5.4.2 If retesting is permissible, a new test specimen shall be heat treated the same as previously. The production forging that it represents will have received the same heat treatment. If the test passes, the forging shall be acceptable. If it fails, the forging shall be rejected or shall be subject to reheat treatment if permissible.

A10.5.4.3 If reheat treatment is permissible, proceed as follows: *(1)* Reheat treatment same as original heat treatment (time, temperature, cooling rate): Using new test specimens from an area as close as possible to the original specimens, repeat the austenitize and quench cycles twice, followed by the tempering cycle (double quench and temper). The production forging shall be given the identical double quench and temper as its test specimens above. *(2)* Reheat treatment using a new heat treatment practice. Any change in time, temperature, or cooling rate shall constitute a new heat treatment practice. A new master curve shall be produced and the simulation and testing shall proceed as originally set forth.

A10.5.4.4 In summation, each test specimen and its corresponding forging shall receive identical heat treatment or heat treatment; otherwise the testing shall be invalid.

A10.5.5 *Storage, Recall, and Documentation of Heat-Cycle Simulation Data*—All records pertaining to heat-cycle simulation shall be maintained and held for a period of 10 years or as designed by the customer. Information shall be so organized that all practices can be verified by adequate documented records.

A370 − 09aᵋ¹

SUMMARY OF CHANGES

Committee A01 has identified the location of selected changes to this standard since the last issue (A370 – 09) that may impact the use of this standard. (Approved June 1, 2009.)

(*1*) Clarifications made to the following sections: 1.1, 2.1, 11.1, 11.2, 14.1, and 24.1.

(*2*) Deleted previous Section 8.1.3.

(*3*) Revised A1.1.1, A2.1.1, A3.1.1, A4.1.1, and A9.1.1.

Committee A01 has identified the location of selected changes to this standard since the last issue (A370 – 08a) that may impact the use of this standard. (Approved January 1, 2009.)

(*1*) Changed Brinell notation to HBW.

Committee A01 has identified the location of selected changes to this standard since the last issue (A370 – 08a) that may impact the use of this standard. (Approved October 15, 2008.)

(*1*) Appended Table 2, Note A.

(*2*) Revised 20.1, 20.2.2, and 22.2.1 to cover FATT.

Committee A01 has identified the location of selected changes to this standard since the last issue (A370 – 08) that may impact the use of this standard. (Approved April 15, 2008.)

(*1*) Changed thickness range in Section 10.1.

Committee A01 has identified the location of selected changes to this standard since the last issue (A370 – 07b) that may impact the use of this standard. (Approved April 1, 2008.)

(*1*) Added Note 14 with reference to a research report on striker radius effect in Section 21.

Committee A01 has identified the location of selected changes to this standard since the last issue (A370 – 07a) that may impact the use of this standard. (Approved November 1, 2007.)

(*1*) Replaced referenced document for laboratory evaluation.

(*2*) Dropped proof load testing of nuts.

ASTM International takes no position respecting the validity of any patent rights asserted in connection with any item mentioned in this standard. Users of this standard are expressly advised that determination of the validity of any such patent rights, and the risk of infringement of such rights, are entirely their own responsibility.

This standard is subject to revision at any time by the responsible technical committee and must be reviewed every five years and if not revised, either reapproved or withdrawn. Your comments are invited either for revision of this standard or for additional standards and should be addressed to ASTM International Headquarters. Your comments will receive careful consideration at a meeting of the responsible technical committee, which you may attend. If you feel that your comments have not received a fair hearing you should make your views known to the ASTM Committee on Standards, at the address shown below.

This standard is copyrighted by ASTM International, 100 Barr Harbor Drive, PO Box C700, West Conshohocken, PA 19428-2959, United States. Individual reprints (single or multiple copies) of this standard may be obtained by contacting ASTM at the above address or at 610-832-9585 (phone), 610-832-9555 (fax), or service@astm.org (e-mail); or through the ASTM website (www.astm.org). Permission rights to photocopy the standard may also be secured from the ASTM website (www.astm.org/COPYRIGHT/).

Designation: A381 − 96 (Reapproved 2005)

Standard Specification for Metal-Arc-Welded Steel Pipe for Use With High-Pressure Transmission Systems[1]

This standard is issued under the fixed designation A381; the number immediately following the designation indicates the year of original adoption or, in the case of revision, the year of last revision. A number in parentheses indicates the year of last reapproval. A superscript epsilon (ε) indicates an editorial change since the last revision or reapproval.

1. Scope

1.1 This specification covers straight seam, double-submerged-arc-welded steel pipe (Note 1) suitable for high-pressure service, 16 in. (406 mm) and larger in outside diameter, with wall thicknesses from 5/16 to 1½ in. (7.9 to 38 mm). The pipe is intended for fabrication of fittings and accessories for compressor or pump-station piping. Pipe ordered to this specification shall be suitable for bending, flanging (vastoning), corrugating, and similar operations.

NOTE 1—A comprehensive listing of standardized pipe dimensions is contained in ANSI B36.10.

NOTE 2—The term "double welded" is commonly used in the gas and oil transmission industry, for which this pipe is primarily intended, to indicate welding with at least two weld passes, of which one is on the outside of the pipe and one on the inside. For some sizes of the pipe covered by this specification, it becomes expedient to use manual welding, in which case the provisions of Note 3 shall be followed.

1.2 Nine classes of pipe, based on minimum yield point requirements, are covered as indicated in Table 1.

1.3 The values stated in inch-pound units are to be regarded as standard. The values given in parentheses are mathematical conversions to SI units that are provided for information only and are not considered standard.

1.4 The following caveat applies to the test methods portion, Sections 9 and 10, only. *This standard does not purport to address all of the safety concerns, if any, associated with its use. It is the responsibility of the user of this standard to establish appropriate safety and health practices and determine the applicability of regulatory limitations prior to use.*

2. Referenced Documents

2.1 *ASTM Standards:*[2]

A370 Test Methods and Definitions for Mechanical Testing of Steel Products

TABLE 1 Tensile Requirements

Class	Yield Strength, min, psi (MPa)	Tensile Strength, min, psi (MPa)	Elongation in 2 in. (50.8 mm), min, %
Y 35	35 000 (240)	60 000 (415)	26
Y 42	42 000 (290)	60 000 (415)	25
Y 46	46 000 (316)	63 000 (435)	23
Y 48	48 000 (330)	62 000 (430)	21
Y 50	50 000 (345)	64 000 (440)	21
Y 52	52 000 (360)	66 000 (455)	20
Y 56	56 000 (385)	71 000 (490)	20
Y 60	60 000 (415)	75 000 (515)	20
Y 65	65 000 (450)	77 000 (535)	20

A530/A530M Specification for General Requirements for Specialized Carbon and Alloy Steel Pipe

E30 Test Methods for Chemical Analysis of Steel, Cast Iron, Open-Hearth Iron, and Wrought Iron[3]

2.2 *ASME Boiler and Pressure Vessel Code:*[4]

Section VIII Pressure Vessels

Section IX Welding Qualifications

2.3 *ANSI Standard:*[5]

ANSI B36.10 Welded and Seamless Wrought Steel Pipe

3. Ordering Information

3.1 Orders for material to this specification should include the following, as required, to describe the desired material adequately:

3.1.1 Quantity (feet, centimetres, or number of lengths),

3.1.2 Name of material (metal-arc welded pipe),

3.1.3 Class (Table 1),

3.1.4 Material (carbon or alloy steel, Section 5),

3.1.5 Size (outside diameter and wall thickness),

3.1.6 Length (specific or random) (Section 13),

3.1.7 Ends (Section 14),

3.1.8 Heat treatment (stress-relieved or normalized) (see 5.6),

3.1.9 Optional requirements (see 5.2 (Note 3), Sections 11 and 15),

---

[1] This specification is under the jurisdiction of ASTM Committee A01 on Steel, Stainless Steel and Related Alloys and is the direct responsibility of Subcommittee A01.09 on Carbon Steel Tubular Products.
Current edition approved Oct. 1, 2005. Published October 2005. Originally approved in 1954. Last previous edition approved in 2001 as A381 – 96 (2001). DOI: 10.1520/A0381-96R05.

[2] For referenced ASTM standards, visit the ASTM website, www.astm.org, or contact ASTM Customer Service at service@astm.org. For *Annual Book of ASTM Standards* volume information, refer to the standard's Document Summary page on the ASTM website.

[3] Withdrawn.

[4] Available from American Society of Mechanical Engineers, 345 E. 47th St., New York, NY 10017.

[5] Available from American National Standards Institute (ANSI), 25 W. 43rd St., 4th Floor, New York, NY 10036, http://www.ansi.org.

3.1.10 Specification number, and

3.1.11 Special requirements or exceptions to this specification.

4. General Requirements

4.1 Material furnished to this specification shall conform to the applicable requirements of the current edition of Specification A530/A530M, unless otherwise provided herein.

5. Materials and Manufacture

5.1 The steel plate used in the manufacture of the pipe shall be of suitable welding quality carbon steel, or of suitable welding quality high-strength, low-alloy steel, as agreed upon between the manufacturer and purchaser.

5.2 The longitudinal edges of the plate shall be shaped to give the most satisfactory results by the particular welding process employed. The plate shall be properly formed and may be tacked preparatory to welding. The weld (except tack welds) shall be made preferably by the automatic submerged-arc-welding process (Note 3) and shall be of reasonably uniform width and height for the entire length of the pipe.

NOTE 3—By agreement between the manufacturer and the purchaser, manual welding by qualified welders using a qualified procedure may be used as an equal alternate to this specification.

5.3 Both longitudinal and circumferential (if any) joints shall be double welded, full penetration welds being made in accordance with procedures and by welders or welding operators qualified in accordance with the ASME Boiler and Pressure Vessel Code, Section IX.

5.4 The contour of the reinforcement shall be smooth, with no valley or groove along the edge or in the center of the weld, and the deposited metal shall be fused smoothly and uniformly into the plate surface. The finish of the welded joint shall be reasonably smooth and free from irregularities, grooves, or depressions.

5.5 All pipe, after welding, shall be heat treated at a temperature of 1100°F (593°C) or higher.

5.6 When specified in the purchase order, all pipe after welding shall be heated at 1650 to 1750°F (899 to 954°C) and air cooled.

6. Chemical Composition

6.1 The carbon steels shall conform to the requirements as to chemical composition specified in Table 2.

6.2 The high-strength low-alloy steels shall be of specified chemical composition in order to ensure weldability and specified minimum tensile properties including elongation.

6.3 Mill test reports, as provided by the manufacturer of the plate, shall be furnished representing the chemical analysis of each heat of steel from which the plates are rolled. This chemical analysis shall conform to the requirements of 5.1, 6.1, or 6.2.

6.4 For referee purposes, Test Methods E30 shall be used.

7. Tensile Requirements

7.1 The tensile properties of transverse body-test specimens taken from the finished pipe shall conform to the requirements prescribed in Table 1. The tensile strength of the transverse weld-test specimens shall conform to that specified in Table 1.

7.2 Transverse body-test specimens shall be taken approximately opposite the weld; transverse weld-test specimens shall be taken with the weld at the center of the specimen. For pipe wall thicknesses up to ¾ in. (19 mm), incl, all transverse test specimens shall be approximately 1½ in. (38 mm) wide in the gauge length and shall represent the full wall thickness of the pipe from which the specimen was cut (see Fig. 23, Test Methods and Definitions A370). For pipe with wall thicknesses over ¾ in. (19 mm), the standard 0.505-in. (12.83-mm) round tension test specimen with 2-in. (50.8-mm) gauge length shall be used (see Fig. 5, Test Methods and Definitions A370).

7.3 If the tension test specimen from any lot of pipe fails to conform to the requirements for the particular grade of pipe ordered, the manufacturer may elect to make retests on two additional lengths of pipe from the same lot, each of which shall conform to the requirements prescribed in Table 2. If one or both of the retests fail to conform to the requirements, the manufacturer may elect to test each of the remaining lengths of pipe in the lot. Retests are required only for the particular test with which the pipe specimen did not comply originally.

7.4 All test specimens which are flattened cold may be reheat treated before machining.

8. Transverse Guided-Bend Tests Weld

8.1 Transverse weld test specimens shall be subject to face and root guided-bend tests. The specimens shall be approximately 1½ in. (38.1 mm) wide, at least 6 in. (152 mm) in length with the weld at the center, and shall be machined in accordance with Fig. 1. One specimen shall be bent with the inside surface of the pipe against the plunger, and the other specimen with the outside surface against the plunger. The dimensions of the plunger for the bending jig shall be in accordance with Fig. 2 and the other dimensions shall be substantially as shown in Fig. 2.

8.2 The bend test shall be acceptable if no cracks or other defects exceeding ⅛ in. (3.17 mm) in any direction are present in the weld metal or between the weld and pipe metal after bending. Cracks which originate along the edges of the specimen during testing, and that are less than ¼ in. (6.35 mm), measured in any direction, shall not be considered.

9. Hydrostatic Test

9.1 Each length of pipe with wall thickness of ½ in. (12.7 mm) and less shall be tested to a hydrostatic pressure which will produce in the pipe wall a stress of not less than 85 % of the minimum specified yield point. This pressure shall be determined by the following equation:

$$P = 2St/D$$

TABLE 2 Chemical Requirements for Carbon Steels on Product Analysis

Element	Composition, %, max	
	Ladle	Check
Carbon	0.26	0.30
Manganese	1.40	1.50
Phosphorus	0.025	0.030
Sulfur	0.025	0.025

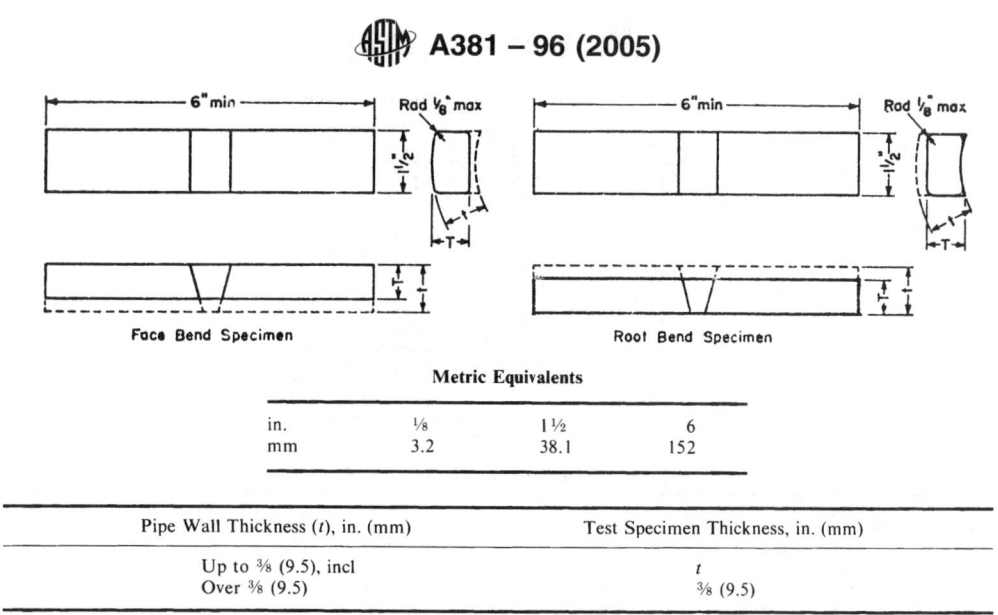

FIG. 1 Transverse Face- and Root-Bend Test Specimens

where:
P = hydrostatic test pressure, psi,
S = 85 % of the specified minimum yield strength of Table 1,
t = specified wall thickness, in., and
D = specified outside diameter, in.

9.2 Each length of pipe with a wall thickness over ½ in. (12.7 mm) shall be tested to a hydrostatic pressure calculated as in 9.1 except that the stress S shall be 70 % of the specified yield point, and that a 3000-psi (20.6-MPa) maximum test pressure shall apply.

9.3 When specified in the order, pipe may be furnished without hydrostatic testing, and each length so furnished shall include with the mandatory marking the letters "NH."

9.4 When certification is required by the purchaser and the hydrostatic test has been omitted, the certification shall clearly state "Not Hydrostatically Tested," and the specification number and class, as shown on the certification, shall be followed by the letters "NH."

10. Mechanical Tests Required

10.1 *Transverse Body Tension Test*—One test shall be made on one length of pipe from each lot of 100 lengths or less, of each size and heat, to determine the yield strength, tensile strength, and percent of elongation in 2 in. (50.8 mm).

10.2 *Transverse Weld Tension Test*—One test shall be made on one length of pipe from each lot of 100 lengths or less, of each size, for tensile strength only.

10.3 *Transverse Guided-Bend Weld Test:*

10.3.1 Two weld bend test specimens as described in 8.1 shall be cut from a length of pipe from each lot of 50 lengths or less, of each size. Bend test specimens shall be cut from pipe ends which have not been repaired.

10.3.2 If either test fails to conform to specified requirements, the manufacturer may elect to make retests on two additional lengths of pipe from the same lot, each of which shall conform to the requirements specified in 8.2. If any of the retests fail to conform to the requirements, the manufacturer may elect to test each of the remaining lengths of pipe in the lot.

10.4 *Hydrostatic Test*—Each length of pipe shall be subjected to the hydrostatic test.

11. Radiographic Examination

11.1 The manufacturer shall employ radiography as a production control on the welding employed in the manufacture of pipe to this specification. At least 5 % of the total linear footage of welding shall be subjected to radiographic examination to ensure that the welding equipment is consistently producing the required quality. The selection of the sections to be so examined shall be at the discretion of the manufacturer's inspector. The purchaser's inspector shall have access to the radiographic films and records of current production.

11.2 When so specified on the purchase order, all welding performed under these specifications shall be fully radiographed. The procedures and requirements shall conform to Paragraph UW-51 of the ASME Boiler and Pressure Vessel Code, Section VIII (latest edition).

12. Permissible Variations in Dimensions

12.1 Permissible variations in dimensions shall not exceed the following:

12.1.1 *Outside Diameter*—±0.5 % of the specified outside diameter for the outside diameter based on circumferential measurement, except that in sizes 24 in. (610 mm) and smaller this tolerance shall be ±⅛ in. (3.2 mm).

12.1.2 *Out-of-Roundness*—1 %, that is, the difference between the major and minor outside diameter.

12.1.3 *Thickness*—The minimum wall thickness shall not be more than 0.01 in. (0.25 mm) under the specified thickness. Localized (isolated and noncontinuous) reductions in wall thickness caused by noninjurious surface defects may be permitted up to a depth not exceeding 6½ % the specified pipe wall thickness.

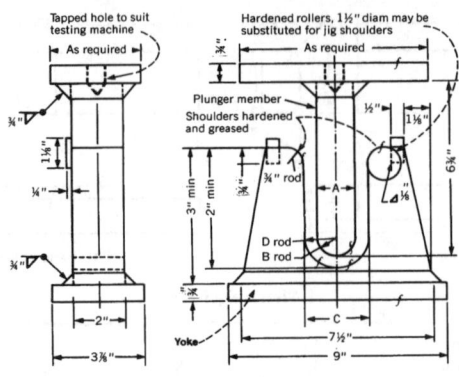

Metric Equivalents

in.	mm	in.	mm	in.	mm
1/16	1.6	1 11/16	42.9	3 1/8	79.4
1/8	3.2	1 3/4	44.4	3 3/8	85.7
1/4	6.4	1 7/8	47.6	3 1/2	88.9
3/8	9.5	1 15/16	49.2	3 3/4	95.2
1/2	12.7	2	50.8	3 7/8	98.4
3/4	19.0	2 1/8	54.0	4 1/4	108.0
15/16	23.8	2 1/4	57.2	4 5/8	117.4
1 1/8	28.6	2 5/16	58.7	5 1/2	139.7
1 5/16	33.3	2 5/8	66.6	6 3/4	171.4
1 3/8	34.9	2 3/4	69.8	7 1/2	190.5
1 1/2	38.1	3	76.2	9	228.6
1 9/16	39.7				

Class of Steel	Y35		Y42		Y46		Y48, Y50, and Y52		Y56 and Y60		Y65	
Thickness of Specimen, in.	3/8	t	3/8	t	3/8	t	3/8	t	3/8	t	3/8	t
"A" dimension	1 7/8	$5t$	2 1/4	$6t$	2 5/8	$7t$	3	$8t$	3 3/8	$9t$	3 3/4	$10t$
"B" dimension	15/15	$(tt/2)$	1 1/8	$3t$	15/16	$(7t/2)$	1 1/2	$4t$	1 11/16	$(9t/2)$	1 7/8	$5t$
"C" dimension	2 3/4	$7t + 1/8$	3 1/8	$8t + 1/8$	3 1/2	$9t + 1/8$	3 7/8	$10t + 1/8$	4 1/4	$11t + 1/8$	4 5/8	$12t + 1/8$
"D" dimension	1 3/8	$(7t/2) + 1/16$	1 9/16	$4t + 1/16$	1 3/4	$(9t/2) + 1/16$	1 15/16	$5t + 1/16$	2 1/8	$5 1/2 + 1/16$	2 5/16	$6t + 1/16$

NOTE 1—"t" equals wall thickness of pipe.
NOTE 2—The dimensions in the above table are based on the following ratio of diameter of bend to thickness of specimen:

Class	Ratio
Y35	5
Y42	6
Y46	7
Y48, Y50, and Y52	8
Y56 and Y60	9
Y65	10

FIG. 2 Guided Bend Test Jig

13. Lengths

13.1 Unless otherwise specified, pipe shall be furnished in approximately 20-ft (6.1-m) lengths.

13.2 Where longer lengths are required, circumferentially welded joints shall be permitted.

13.3 Shorter lengths, when required, shall be specified in the order.

14. Ends

14.1 Pipe ends shall be furnished beveled as specified in the order. The width of the end shall be 1/16 in. (1.6 mm) with a tolerance of ±1/32 in. (0.8 mm).

14.2 The end of the pipe shall not be out of square more than 1/16 in. (1.6 mm).

15. Workmanship, Finish, and Appearance

15.1 The finished pipe shall be free of injurious defects and shall have a workmanlike finish.

15.2 *Repair of Plate Defects by Machining or Grinding*—Pipe showing moderate slivers may be machined or ground inside or outside to a depth which shall ensure the removal of all included scale and slivers, providing the wall thickness is not reduced below the specified minimum wall thickness.

15.3 *Repair of Plate Defects by Welding*—Repair of plate defects by welding shall be permitted. Welding of injurious defects shall not be permitted when the depth of defect exceeds 33 1/3 % of the specified pipe wall thickness or the length of repair exceeds 25 % of the specified diameter of the pipe.

Defects must be thoroughly removed and the welding performed by a welder qualified in accordance with the requirements of the ASME Boiler and Pressure Vessel Code, Section IX. Such repair welding shall be ground or machined flush with the surface of the pipe. All repair welding shall be done before final heat treatment.

16. Coating

16.1 Unless otherwise specified in the purchase order, the pipe shall be furnished uncoated.

17. Inspection

17.1 The inspector representing the purchaser shall have entry, at all times while work on the contract of the purchaser is being performed, to all parts of the manufacturer's works that concern the manufacture of the material ordered. All reasonable facilities shall be afforded the inspector, to satisfy him that the material is being furnished in accordance with this specification. All tests called for by this specification and inspection shall be made at the place of manufacture prior to shipment unless otherwise specified, and shall be so conducted as not to interfere unnecessarily with the operation of the works.

18. Product Marking

18.1 In addition to the marking prescribed in Specification A530/A530M, the marking shall include the hydrostatic test pressure. Marking shall be by stenciling along the welded seam.

18.2 *Bar Coding*—In addition to the requirements in 18.1, bar coding is acceptable as a supplementary identification method. Bar coding should be consistent with the Automotive Industry Action Group (AIAG) standard prepared by the Primary Metals Subcommittee of the AIAG Bar Code Project Team.

19. Keywords

19.1 arc welded steel pipe; steel pipe

Designation: A435/A435M − 90 (Reapproved 2007)

Standard Specification for
Straight-Beam Ultrasonic Examination of Steel Plates[1]

This standard is issued under the fixed designation A435/A435M; the number immediately following the designation indicates the year of original adoption or, in the case of revision, the year of last revision. A number in parentheses indicates the year of last reapproval. A superscript epsilon (ε) indicates an editorial change since the last revision or reapproval.

1. Scope

1.1 This specification[2] covers the procedure and acceptance standards for straight-beam, pulse-echo, ultrasonic examination of rolled fully killed carbon and alloy steel plates, ½ in. [12.5 mm] and over in thickness. It was developed to assure delivery of steel plates free of gross internal discontinuities such as pipe, ruptures, or laminations and is to be used whenever the inquiry, contract, order, or specification states that the plates are to be subjected to ultrasonic examination.

1.2 Individuals performing examinations in accordance with this specification shall be qualified and certified in accordance with the requirements of the latest edition of ASNT SNT-TC-1A or an equivalent accepted standard. An equivalent standard is one which covers the qualification and certification of ultrasonic nondestructive examination candidates and which is acceptable to the purchaser.

1.3 The values stated in either inch-pound units or SI units are to be regarded separately as standard. Within the text, the SI units are shown in brackets. The values stated in each system are not exact equivalents, therefore, each system must be used independently of the other. Combining values from the two systems may result in nonconformance with the specification.

2. Referenced Documents

2.1 *ASNT Documents:*[3]
ASNT SNT-TC-1A Recommended Practice for Personnel Qualification and Certification in Nondestructive Testing

3. Apparatus

3.1 The manufacturer shall furnish suitable ultrasonic equipment and qualified personnel necessary for performing the test. The equipment shall be of the pulse-echo straight beam type. The transducer is normally 1 to 1⅛ in. [25 to 30 mm] in diameter or 1 in. [25 mm] square; however, any transducer having a minimum active area of 0.7 in.[2] [450 mm^2] may be used. The test shall be performed by one of the following methods: direct contact, immersion, or liquid column coupling.

3.2 Other search units may be used for evaluating and pinpointing indications.

4. Test Conditions

4.1 Conduct the examination in an area free of operations that interfere with proper functioning of the equipment.

4.2 Clean and smooth the plate surface sufficiently to maintain a reference back reflection from the opposite side of the plate at least 50 % of the full scale during scanning.

4.3 The surface of plates inspected by this method may be expected to contain a residue of oil or rust or both. Any specified identification which is removed when grinding to achieve proper surface smoothness shall be restored.

5. Procedure

5.1 Ultrasonic examination shall be made on either major surface of the plate. Acceptance of defects in close proximity may require inspection from the second major surface. Plates ordered in the quenched and tempered condition shall be tested following heat treatment.

5.2 A nominal test frequency of 2¼ MHz is recommended. Thickness, grain size, or microstructure of the material and nature of the equipment or method may require a higher or lower test frequency. However, frequencies less than 1 MHz may be used only on agreement with the purchaser. A clear, easily interpreted trace pattern should be produced during the examination.

5.3 Conduct the examination with a test frequency and instrument adjustment that will produce a minimum 50 to a maximum 75 % of full scale reference back reflection from the opposite side of a sound area of the plate. While calibrating the instrument, sweep the crystal along the plate surface for a distance of at least 1*T* or 6 in. [150 mm], whichever is the greater, and note the position of the back reflection. A shift in location of the back reflection during calibration shall be cause for recalibration of the instrument.

---

[1] This specification is under the jurisdiction of ASTM Committee A01 on Steel, Stainless Steel and Related Alloys and is the direct responsibility of Subcommittee A01.11 on Steel Plates for Boilers and Pressure Vessels.
Current edition approved March 1, 2007. Published March 2007. Originally approved in 1959. Last previous edition approved in 2001 as A435/A435M – 90 (2001). DOI: 10.1520/A0435_A0435M-90R07.

[2] For ASME Boiler and Pressure Vessel Code applications, see related Specifications SA-435/SA-435M in Section II of that Code. Available from American Society of Mechanical Engineers (ASME), ASME International Headquarters, Three Park Ave., New York, NY 10016-5990, http://www.asme.org.

[3] Available from American Society for Nondestructive Testing (ASNT), P.O. Box 28518, 1711 Arlingate Ln., Columbus, OH 43228-0518, http://www.asnt.org.

5.4 Scanning shall be continuous along perpendicular grid lines on nominal 9-in. [225-mm] centers, or at the manufacturer's option, shall be continuous along parallel paths, transverse to the major plate axis, on nominal 4-in. [100-mm] centers, or shall be continuous along parallel paths parallel to the major plate axis, on 3-in. [75-mm] or smaller centers. A suitable couplant such as water, soluble oil, or glycerin, shall be used.

5.5 Scanning lines shall be measured from the center or one corner of the plate. An additional path shall be scanned within 2 in. [50 mm] of all edges of the plate on the scanning surface.

5.6 Where grid scanning is performed and complete loss of back reflection accompanied by continuous indications is detected along a given grid line, the entire surface area of the squares adjacent to this indication shall be scanned continuously. Where parallel path scanning is performed and complete loss of back reflection accompanied by continuous indications is detected, the entire surface area of a 9 by 9-in. [225 by 225-mm] square centered on this indication shall be scanned continuously. The true boundaries where this condition exists shall be established in either method by the following technique: Move the transducer away from the center of the discontinuity until the heights of the back reflection and discontinuity indications are equal. Mark the plate at a point equivalent to the center of the transducer. Repeat the operation to establish the boundary.

6. Acceptance Standards

6.1 Any discontinuity indication causing a total loss of back reflection which cannot be contained within a circle, the diameter of which is 3 in. [75 mm] or one half of the plate thickness, whichever is greater, is unacceptable.

6.2 The manufacturer reserves the right to discuss rejectable ultrasonically tested plates with the purchaser with the object of possible repair of the ultrasonically indicated defect before rejection of the plate.

6.3 The purchaser's representative may witness the test.

7. Marking

7.1 Plates accepted in accordance with this specification shall be identified by stamping or stenciling UT 435 adjacent to marking required by the material specification.

SUPPLEMENTARY REQUIREMENTS

The following shall apply only if specified in the order:

S1. Instead of the scanning procedure specified by 5.4 and 5.5, and as agreed upon between manufacturer and purchaser, 100 % of one major plate surface shall be scanned. Scanning shall be continuous along parallel paths, transverse or parallel to the major plate axis, with not less than 10 % overlap between each path.

ASTM International takes no position respecting the validity of any patent rights asserted in connection with any item mentioned in this standard. Users of this standard are expressly advised that determination of the validity of any such patent rights, and the risk of infringement of such rights, are entirely their own responsibility.

This standard is subject to revision at any time by the responsible technical committee and must be reviewed every five years and if not revised, either reapproved or withdrawn. Your comments are invited either for revision of this standard or for additional standards and should be addressed to ASTM International Headquarters. Your comments will receive careful consideration at a meeting of the responsible technical committee, which you may attend. If you feel that your comments have not received a fair hearing you should make your views known to the ASTM Committee on Standards, at the address shown below.

This standard is copyrighted by ASTM International, 100 Barr Harbor Drive, PO Box C700, West Conshohocken, PA 19428-2959, United States. Individual reprints (single or multiple copies) of this standard may be obtained by contacting ASTM at the above address or at 610-832-9585 (phone), 610-832-9555 (fax), or service@astm.org (e-mail); or through the ASTM website (www.astm.org).

Designation: A490 − 09

American Association State
Highway and Transportation
Officials Standard
AASHTO No.: M 253

Standard Specification for
Structural Bolts, Alloy Steel, Heat Treated, 150 ksi Minimum Tensile Strength[1]

This standard is issued under the fixed designation A490; the number immediately following the designation indicates the year of original adoption or, in the case of revision, the year of last revision. A number in parentheses indicates the year of last reapproval. A superscript epsilon (ε) indicates an editorial change since the last revision or reapproval.

This standard has been approved for use by agencies of the Department of Defense.

1. Scope*

1.1 This specification covers two types of quenched and tempered, alloy steel, heavy hex structural bolts having a tensile strength of 150 to 173 ksi.

1.2 These bolts are intended for use in structural connections. These connections are covered under the requirements of the Specification for Structural Joints Using Specification A325 or A490 bolts, approved by the Research Council on Structural Connections; endorsed by the American Institute of Steel Construction and by the Industrial Fastener Institute.[2]

1.3 The bolts are furnished in sizes ½ to 1½ in., inclusive. They are designated by type denoting chemical composition as follows:

Type	Description
Type 1	Medium carbon alloy steel
Type 2	Withdrawn in 2002
Type 3	Weathering steel

1.4 This specification provides that heavy hex structural bolts shall be furnished unless other dimensional requirements are specified on the purchase order.

1.5 Terms used in this specification are defined in Terminology F1789 unless otherwise defined herein.

1.6 For metric bolts, see Specification A490M Classes 10.9 and 10.9.3

1.7 The values stated in inch-pound units are to be regarded as standard. No other units of measurement are included in this standard.

1.8 The following safety hazards caveat pertains only to the Test Methods portion, Section 12 of this specification: *This standard does not purport to address all of the safety concerns, if any, associated with its use. It is the responsibility of the user of this standard to establish appropriate safety and health practices and determine the applicability of regulatory limitations prior to use.*

2. Referenced Documents

2.1 *ASTM Standards:*[3]

A194/A194M Specification for Carbon and Alloy Steel Nuts for Bolts for High Pressure or High Temperature Service, or Both

A325 Specification for Structural Bolts, Steel, Heat Treated, 120/105 ksi Minimum Tensile Strength

A354 Specification for Quenched and Tempered Alloy Steel Bolts, Studs, and Other Externally Threaded Fasteners

A490M Specification for High-Strength Steel Bolts, Classes 10.9 and 10.9.3, for Structural Steel Joints (Metric)

A563 Specification for Carbon and Alloy Steel Nuts

A751 Test Methods, Practices, and Terminology for Chemical Analysis of Steel Products

D3951 Practice for Commercial Packaging

E384 Test Method for Microindentation Hardness of Materials

E709 Guide for Magnetic Particle Testing

E1444 Practice for Magnetic Particle Testing

F436 Specification for Hardened Steel Washers

F606 Test Methods for Determining the Mechanical Properties of Externally and Internally Threaded Fasteners, Washers, Direct Tension Indicators, and Rivets

F788/F788M Specification for Surface Discontinuities of Bolts, Screws, and Studs, Inch and Metric Series

F959 Specification for Compressible-Washer-Type Direct Tension Indicators for Use with Structural Fasteners

F1136 Specification for Zinc/Aluminum Corrosion Protective Coatings for Fasteners

F1470 Practice for Fastener Sampling for Specified Mechanical Properties and Performance Inspection

---

[1] This specification is under the jurisdiction of ASTM Committee F16 on Fasteners and is the direct responsibility of Subcommittee F16.02 on Steel Bolts, Nuts, Rivets and Washers.
Current edition approved Nov. 1, 2009. Published November 2009. Originally approved in 1964. Last previous edition approved in 2008 as A490 – 08b. DOI: 10.1520/A0490-09.

[2] Available from American Institute of Steel Construction (AISC), One E. Wacker Dr., Suite 700, Chicago, IL 60601-2001, http://www.aisc.org.

[3] For referenced ASTM standards, visit the ASTM website, www.astm.org, or contact ASTM Customer Service at service@astm.org. For *Annual Book of ASTM Standards* volume information, refer to the standard's Document Summary page on the ASTM website.

*A Summary of Changes section appears at the end of this standard.

F1789 Terminology for F16 Mechanical Fasteners

F2328 Test Method for Determining Decarburization and Carburization in Hardened and Tempered Threaded Steel Bolts, Screws and Studs

G101 Guide for Estimating the Atmospheric Corrosion Resistance of Low-Alloy Steels

2.2 *ASME Standards:*[4]

B1.1 Unified Screw Threads

B18.2.6 Fasteners for Use in Structural Applications

B18.24 Part Identification Number (PIN) Code System Standard for B18 Fastener Products

2.3 *IFI Standard:*[5]

IFI 144 Test Evaluation Procedures for Coating Qualification Intended for Use on High-Strength Bolts

3. Ordering Information

3.1 Orders for heavy hex structural bolts under this specification shall include the following:

3.1.1 Quantity (number of pieces of bolts and accessories);

3.1.2 Size, including nominal bolt diameter, thread pitch, and bolt length. The thread length shall not be changed;

3.1.3 Name of product: heavy hex structural bolts, or other such bolts as specified;

3.1.4 Type of bolt (Type 1 or 3). When type is not specified, either Type 1 or Type 3 shall be furnished at the supplier's option;

3.1.5 ASTM designation and year of issue,

3.1.6 Other components such as nuts, washers, and washer-type direct tension indicators, if required;

3.1.7 Test Reports, if required (see Section 15); and

3.1.8 Protective coating per Specification F1136, Grade 3, if required. See 4.3.

3.1.9 Special requirements.

3.1.10 For establishment of a part identifying system, see ASME B18.24.

NOTE 1—A typical ordering description follows: 1000 pieces 1-8 in. dia × 4 in. long heavy hex structural bolt, Type 1, *ASTM A490M – 02;* each with two hardened washers, ASTM F436 Type 1; and one heavy hex nut, ASTM A563 Grade DH.

3.2 *Recommended Nuts:*

3.2.1 Nuts conforming to the requirements of Specification A563 are the recommended nuts for use with Specification A490M heavy hex structural bolts. The nuts shall be of the class and have a surface finish for each type of bolt as follows:

Bolt Type and Finish	Nut Class and Finish
1, plain (uncoated)	A563—DH, DH3 plain (uncoated)
1, coated in accordance with Specification F1136, Grade 3	A563—coated in accordance with Specification F1136, Grade 5.
3, weathering steel	A563—DH3, weathering steel

3.2.2 Alternatively, nuts conforming to Specification A194/A194M Gr. 2H plain (uncoated) are considered a suitable substitute for use with Specification A490M Type 1 heavy hex structural bolts.

3.3 *Recommended Washers*—Washers conforming to Specification F436 are the recommended washers for use with Specification A490M heavy hex structural bolts. The washers shall have a surface finish for each type of bolt as follows:

Bolt Type and Finish	Washer Finish
1, plain (uncoated)	plain (uncoated)
1, coated in accordance with F1136, Grade 3	plain, coated in accordance with F1136, Grade 3
3, weathering steel	weathering steel

3.4 *Other Accessories*—When compressible washer type direct tension indicators are specified to be used with these bolts, they shall conform to Specification F959 Type 490.

4. Materials and Manufacture

4.1 *Heat Treatment*—Type 1 and Type 3 bolts shall be heat treated by quenching in oil from the austenitic temperature and then tempered by reheating to a temperature of not less than 800°F.

4.2 *Threading*—The threads shall be cut or rolled.

4.3 *Protective Coatings:*

4.3.1 When a protective coating is required and specified, the bolts shall be coated with Zinc/Aluminum Corrosion Protective Coatings in accordance with Specification F1136, Grade 3. This coating has been qualified based on the findings of an investigation founded on IFI 144. [6]

4.3.2 No other metallic coatings are permitted unless authorized by Committee F16. Future consideration of any coating will be based on results of testing performed in accordance with the procedures in IFI 144, and submitted to Committee F16 for review (See note 2).

NOTE 2—For more detail see the H. E. Townsend Report "Effects of Zinc Coatings on Stress Corrosion Cracking and Hydrogen Embrittlement of Low Alloy Steel," published in Metallurgical Transactions, Vol. 6, April 1975.

5. Chemical Composition

5.1 Type 1 bolts shall be alloy steel conforming to the chemical composition specified in Table 1. The steel shall

---

[4] Available from American Society of Mechanical Engineers (ASME), ASME International Headquarters, Three Park Ave., New York, NY 10016-5990, http://www.asme.org.

[5] Available from Industrial Fastener Institute, (IFI), 6363 Oak Tree Boulevard, Independence, OH 44131. http://www.industrial-fasteners.org.

[6] Supporting data have been filed at ASTM International Headquarters and may be obtained by requesting Research Report F16-1001.

TABLE 1 Chemical Requirements for Type 1 Bolts

Element	Heat Analysis, %	Product Analysis, %
Carbon		
For sizes through 1⅜ in.	0.30–0.48	0.28–0.50
For size 1½ in.	0.35–0.53	0.33–0.55
Phosphorus, max	0.040	0.045
Sulfur, max	0.040	0.045
Alloying Elements	→ See 5.1 ←	

contain sufficient alloying elements to qualify it as an alloy steel (see Note 3).

NOTE 3—Steel is considered to be alloy by the American Iron and Steel Institute when the maximum of the range given for the content of alloying elements exceeds one or more of the following limits: manganese, 1.65 %; silicon, 0.60 %; copper, 0.60 %; or in which a definite range or a definite minimum quantity of any of the following elements is specified or required within the limits of the recognized field of constructional alloy steels: aluminum, chromium up to 3.99 %, cobalt, columbium, molybdenum, nickel, titanium, tungsten, vanadium, zirconium, or any other alloying elements added to obtain a desired alloying effect.

5.2 Type 3 bolts shall be weathering steel conforming to the chemical composition requirements in Table 2. See Guide G101 for methods of estimating the atmospheric corrosion resistance of low alloy steel.

5.3 Product analyses made on finished bolts representing each lot shall conform to the product analysis requirements specified in Tables 1 and 2, as applicable.

5.4 Heats of steel to which bismuth, selenium, tellurium, or lead has been intentionally added shall not be used for bolts furnished to this specification. Compliance with this requirement shall be based on certification that steels having these elements intentionally added were not used.

5.5 Chemical analyses shall be performed in accordance with Test Methods, Practices, and Terminology A751.

6. Mechanical Properties

6.1 *Hardness*—The bolts shall conform to the hardness specified in Table 3.

6.2 *Tensile Properties*:

6.2.1 Except as permitted in 6.2.2 for long bolts and 6.2.3 for short bolts, sizes 1.00 in. and smaller having a length of 2¼ D and longer and sizes larger than 1.00 in. having a length of 3D and longer shall be wedge tested full size and shall conform to the minimum and maximum wedge tensile load, and proof load or alternative proof load specified in Table 4. The load achieved during proof load testing shall be equal to or greater than the specified proof load.

6.2.2 When the length of the bolt makes full-size testing impractical, machined specimens shall be tested and shall conform to the requirements specified in Table 5. When bolts are tested by both full-size and machined specimen methods, the full-size test shall take precedence.

6.2.3 Sizes 1.00 in. and smaller having a length shorter than 2¼ D down to 2D, inclusive, that cannot be wedge tensile tested shall be axially tension tested full size and shall conform to the minimum tensile load and proof load or alternate proof load specified in Table 4. Sizes 1.00 in. and smaller having a length shorter than 2D that cannot be axially tensile tested shall be qualified on the basis of hardness.

6.2.4 For bolts on which hardness and tension tests are performed, acceptance based on tensile requirements shall take precedence in the event of low hardness readings.

7. Carburization/Decarburization

7.1 This test is intended to evaluate the presence or absence of carburization and decarburization as determined by the difference in microhardness near the surface and core.

7.2 *Requirements*:

7.2.1 *Carburization*—The bolts shall show no evidence of a carburized surface when evaluated in accordance with 12.2.

7.2.2 *Decarburization*—Hardness value differences shall not exceed the requirements set forth for decarburization in Test Method F2328 materials when evaluated in accordance with 12.2.

8. Dimensions

8.1 *Head and Body*:

8.1.1 Unless otherwise specified, bolts shall conform to the dimensions for heavy hex structural bolts specified in ASME B18.2.6.

8.1.2 The thread length shall not be changed from that specified in ASME B18.2.6 for heavy hex structural bolts. Bolts requiring thread lengths other than those required by this specification shall be ordered under Specification A354 Gr. BD.

8.2 *Threads*—Threads shall be the Unified Coarse Thread Series as specified in ASME B1.1 and shall have Class 2A tolerances.

9. Workmanship

9.1 The allowable limits, inspection, and evaluation of the surface discontinuities, quench cracks, forging cracks, head bursts, shear bursts, seams, folds, thread laps, voids, tool marks, nicks, and gouges shall be in accordance with Specification F788/F788M.

10. Magnetic Particle Inspection for Longitudinal Discontinuities and Transverse Cracks

10.1 *Requirements*:

10.1.1 Each sample representative of the lot shall be magnetic particle inspected for longitudinal discontinuities and transverse cracks.

10.1.2 The lot, as represented by the sample, shall be free from nonconforming bolts, as defined in Specification F788/F788M, when inspected in accordance with 10.2.1-10.2.3.

10.2 *Inspection Procedure*:

10.2.1 The inspection sample shall be selected at random from each lot in accordance with Practice F1470 and examined for longitudinal discontinuities and transverse cracks.

10.2.2 Magnetic particle inspection shall be conducted in accordance with Guide E709 or Practice E1444. Guide E709 shall be used for referee purposes. If any nonconforming bolt is found during the manufacturer's examination of the lot selected in 10.2.1, the lot shall be 100 % magnetic particle

TABLE 2 Chemical Requirements for Type 3 Bolts

Element	Heat Analysis, %	Product Analysis, %
Carbon		
Sizes 0.75 in. and smaller	0.20–0.53	0.19–0.55
Sizes larger than 0.75 in.	0.30–0.53	0.28–0.55
Manganese, min	0.40	0.37
Phosphorus, max	0.035	0.040
Sulfur, max	0.040	0.045
Copper	0.20–0.60	0.17–0.63
Chromium, min	0.45	0.42
Nickel, min	0.20	0.17
or		
Molybdenum, min	0.15	0.14

TABLE 3 Hardness Requirements for Bolts ½ to 1½ in. Nominal Size

Size, in.	Length, in.	Brinell		Rockwell C	
		min	max	min	max
½ to 1, incl.	Less than $2D^A$	311	352	33	39
	$2D^A$ and longer	...	352	...	39
Over 1 to 1½, incl.	Less than $3D^A$	311	352	33	39
	$3D^A$ and longer	...	352	...	39

[A] Heavy hex structural bolts 1 in. and smaller and shorter than 2D are subject only to minimum and maximum hardness. Heavy hex structural bolts larger than 1 through 1½, incl., in diameter and shorter than 3D are subject only to minimum and maximum hardness.

TABLE 4 Tensile Load Requirements for Full-Size Bolts

Bolt Size, Threads per Inch, and Series Designation	Stress Area,[A] in.²	Tensile Load,[B] lbf		Proof Load,[B] lbf	Alternative Proof Load,[B] lbf
		min	max	Length Measurement Method	Yield Strength Method
Column 1	Column 2	Column 3	Column 4	Column 5	Column 6
½-13 UNC	0.142	21 300	24 600	17 050	18 500
⅝-11 UNC	0.226	33 900	39 100	27 100	29 400
¾-10 UNC	0.334	50 100	57 800	40 100	43 400
⅞-9 UNC	0.462	69 300	79 950	55 450	60 100
1-8 UNC	0.606	90 900	104 850	72 700	78 800
1⅛-7 UNC	0.763	114 450	132 000	91 550	99 200
1¼-7 UNC	0.969	145 350	167 650	116 300	126 000
1⅜-6 UNC	1.155	173 250	199 850	138 600	150 200
1½-6 UNC	1.405	210 750	243 100	168 600	182 600

[A] The stress area is calculated as follows:

$$A_s = 0.7854 [D - (0.9743/n)]^2$$

where:
A_s = stress area, in.²
D = nominal bolt size, and
n = threads per inch.

[B] Loads tabulated and loads to be used for tests of full-size bolts larger than 1½ in. in diameter are based on the following:

Bolt Size	Column 3	Column 4	Column 5	Column 6
½ to 1½ in., incl	150 000 psi	173 000 psi	120 000 psi	130 000 psi

TABLE 5 Tensile Strength Requirements for Specimens Machined from Bolts

Bolt Size, in.	Tensile Strength, psi		Yield Strength (0.2 % offset), min, psi	Elongation in 2 in. or 50 mm, min, %	Reduction of Area, min, %
	min	max			
½ to 1½ in., incl	150 000	173 000	130 000	14	40

inspected, and all nonconforming bolts shall be removed and scrapped or destroyed.

10.2.3 Eddy current or liquid penetrant inspection shall be an acceptable substitute for the 100 % magnetic particle inspection when nonconforming bolts are found and 100 % inspection is required. On completion of the eddy current or liquid penetrant inspection, a random sample selected from each lot in accordance with Practice F1470 shall be re-examined by the magnetic particle method. In case of controversy, the magnetic particle test shall take precedence.

10.2.4 Magnetic particle indications of themselves shall not be cause for rejection. If in the opinion of the quality assurance representative the indications may be cause for rejection, a sample taken in accordance with Practice F1470 shall be examined by microscopic examination or removal by surface grinding to determine if the indicated discontinuities are within the specified limits.

11. Number of Tests and Retests

11.1 *Testing Responsibility*:

11.1.1 Each lot shall be tested by the manufacturer prior to shipment in accordance with the lot identification control quality assurance plan in 11.2-11.5.

11.1.2 When bolts are furnished by a source other than the manufacturer, the Responsible Party as defined in 16.1 shall be responsible for assuring all tests have been performed and the bolts comply with the requirements of this specification.

11.2 *Purpose of Lot Inspection*—The purpose of a lot inspection program shall be to ensure that each lot as represented by the samples tested conforms to the requirements of this specification. For such a plan to be fully effective, it is essential that secondary processors, distributors, and purchasers maintain the identification and integrity of each lot until the product is installed.

11.3 *Lot Method*—All bolts shall be processed in accordance with a lot identification-control quality assurance plan. The manufacturer, secondary processors, and distributors shall identify and maintain the integrity of each lot of bolts from raw-material selection through all processing operations and treatments to final packing and shipment. Each lot shall be assigned its own lot-identification number, each lot shall be tested, and the inspection test reports for each lot shall be retained.

11.4 *Lot Definition*—A lot shall be a quantity of uniquely identified heavy hex structural bolts of the same nominal size and length produced consecutively at the initial operation from a single mill heat of material and processed at one time, by the same process, in the same manner, so that statistical sampling is valid. The identity of the lot and lot integrity shall be maintained throughout all subsequent operations and packaging.

11.5 *Number of Tests*:

11.5.1 The minimum number of tests from each lot for the tests specified below shall be as follows:

Tests	Number of Tests in Accordance with
Hardness, tensile strength, proof load	Practice F1470
Surface discontinuities	Specification F788/F788M
Magnetic particle inspection	Specification F788/F788M
Dimensions and thread fit	ASME B18.2.6

11.5.2 For carburization and decarburization tests, not less than one sample unit per manufactured lot shall be tested for microhardness.

12. Test Methods

12.1 *Tensile, Proof Load, and Hardness*:

12.1.1 Tensile, proof load, and hardness tests shall be conducted in accordance with Test Methods F606.

12.1.2 Tensile strength shall be determined using the Wedge or Axial Tension Testing Method of Full Size Product Method or the Machined Test Specimens Method, depending on size and length as specified in 6.2.1-6.2.4. Fracture on full-size tests shall be in the body or threads of the bolt without a fracture at the junction of the head and body.

12.1.3 Proof load shall be determined using Method 1, Length Measurement, or Method 2, Yield Strength, at the option of the manufacturer.

12.2 *Carburization/Decarburization*—Tests shall be conducted in accordance with Test Method F2328 Hardness Method.

12.3 *Microhardness*—Tests shall be conducted in accordance with Test Method E384.

12.4 *Magnetic Particle*—Inspection shall be conducted in accordance with Section 10.

13. Inspection

13.1 If the inspection described in 13.2 is required by the purchaser, it shall be specified in the inquiry and contract or order.

13.2 The purchaser's representative shall have free entry to all parts of manufacturer's works or supplier's place of business that concern the manufacture of the material ordered. The manufacturer or supplier shall afford the purchaser's representative all reasonable facilities to satisfy him that the material is being furnished in accordance with this specification. All tests and inspections required by the specification that are requested by the purchaser's representative shall be made before shipment, and shall be conducted as not to interfere unnecessarily with the operation of the manufacturer's works or supplier's place of business.

14. Rejection and Rehearing

14.1 Disposition of nonconforming material shall be in accordance with Practice F1470 section titled "Disposition of Nonconforming Lots."

15. Certification

15.1 When specified on the purchase order, the manufacturer or supplier, whichever is the responsible party as defined in Section 16 shall furnish the purchaser a test report that includes the following:

15.1.1 Heat analysis, heat number, and a statement certifying that heats having bismuth, selenium, tellurium, or lead intentionally added were not used to produce the bolts;

15.1.2 Results of hardness, tensile, and proof load tests;

15.1.3 Results of magnetic particle inspection for longitudinal discontinuities and transverse cracks;

15.1.4 Results of tests and inspections for surface discontinuities including visual inspection for head bursts;

15.1.5 Results of carburization and decarburization tests;

15.1.6 Statement of compliance with dimensional and thread fit requirements;

15.1.7 Lot number and purchase order number;

15.1.8 Complete mailing address of responsible party; and

15.1.9 Title and signature of the individual assigned certification responsibility by the company officers.

15.2 Failure to include all the required information on the test report shall be cause for rejection.

16. Responsibility

16.1 The party responsible for the fastener shall be the organization that supplies the fastener to the purchaser.

17. Product Marking

17.1 *Manufacturer's Identification*—All Type 1 and Type 3 bolts shall be marked by the manufacturer with a unique identifier to identify the manufacturer or private label distributor, as appropriate.

17.2 *Grade Identification*:

17.2.1 Type 1 bolts shall be marked "A490M."

17.2.2 Type 3 bolts shall be marked "A490M" underlined.

17.3 *Marking Location and Methods*—All marking shall be located on the top of the bolt head and shall be either raised or depressed at the manufacturer's option.

17.4 *Acceptance Criteria*—Bolts that are not marked in accordance with these provisions shall be considered nonconforming and subject to rejection.

17.5 Type and manufacturer's or private label distributor's identification shall be separate and distinct. The two identifications shall preferably be in different locations and, when on the same level, shall be separated by at least two spaces.

18. Packaging and Package Marking

18.1 *Packaging:*

18.1.1 Unless otherwise specified, packaging shall be in accordance with Practice D3951.

18.1.2 When special packaging requirements are required, they shall be defined at the time of the inquiry and order.

18.2 *Package Marking*:

18.2.1 Each shipping unit shall include or be plainly marked with the following information:

18.2.1.1 ASTM designation and type,

18.2.1.2 Size,

18.2.1.3 Name and brand or trademark of the manufacturer,

18.2.1.4 Number of pieces,

18.2.1.5 Lot number,

18.2.1.6 Purchase order number, and

18.2.1.7 Country of origin.

19. Keywords

19.1 bolts; alloy steel; steel; structural; weathering steel

SUMMARY OF CHANGES

Committee F16 has identified the location of selected changes to this standard since the last issue (A490M–08b) that may impact the use of this standard. (Approved Nov. 1, 2009.)

(*1*) Revised 2.1, 7.2.2 and 12.2 covering Carburization and Decarburization requirements by removing SAE J121 and replacing with Test Method F2328.

(*2*) Deleted 2.3.

Committee F16 has identified the location of selected changes to this standard since the last issue (A490M–08a) that may impact the use of this standard. (Approved Dec. 1, 2008.)

(*1*) In 9.1 deleted Note 4 in regard to non injurious bursts.

Committee F16 has identified the location of selected changes to this standard since the last issue (A490M – 08) that may impact the use of this standard. (Approved Jan. 15th, 2008.)

(*1*) Revised 4.3 to add provision for specifying a Zinc/Aluminum corrosion protective coating conforming to Specification F1136, grade 3.

Committee F16 has identified the location of selected changes to this standard since the last issue (A490M – 06) that may impact the use of this standard. (Approved Jan. 1, 2008.)

(*1*) Deleted Table 6, Sample Sizes with Acceptance and Rejection Numbers for Inspection of Rejectable Longitudinal Discontinuities and Transverse Cracks.

(*2*) Deleted 10.3, Definitions.

(*3*) Revised 10.1.2, 10.2.1, 10.2.3 and 11.5.1 to use Specification F788/F788M and Practice F1470 for Magnetic Particle QA requirements.

ASTM International takes no position respecting the validity of any patent rights asserted in connection with any item mentioned in this standard. Users of this standard are expressly advised that determination of the validity of any such patent rights, and the risk of infringement of such rights, are entirely their own responsibility.

This standard is subject to revision at any time by the responsible technical committee and must be reviewed every five years and if not revised, either reapproved or withdrawn. Your comments are invited either for revision of this standard or for additional standards and should be addressed to ASTM International Headquarters. Your comments will receive careful consideration at a meeting of the responsible technical committee, which you may attend. If you feel that your comments have not received a fair hearing you should make your views known to the ASTM Committee on Standards, at the address shown below.

This standard is copyrighted by ASTM International, 100 Barr Harbor Drive, PO Box C700, West Conshohocken, PA 19428-2959, United States. Individual reprints (single or multiple copies) of this standard may be obtained by contacting ASTM at the above address or at 610-832-9585 (phone), 610-832-9555 (fax), or service@astm.org (e-mail); or through the ASTM website (www.astm.org). Permission rights to photocopy the standard may also be secured from the ASTM website (www.astm.org/COPYRIGHT/).

Designation: A496/A496M – 07

Standard Specification for
Steel Wire, Deformed, for Concrete Reinforcement[1]

This standard is issued under the fixed designation A496/A496M; the number immediately following the designation indicates the year of original adoption or, in the case of revision, the year of last revision. A number in parentheses indicates the year of last reapproval. A superscript epsilon (ε) indicates an editorial change since the last revision or reapproval.

This standard has been approved for use by agencies of the Department of Defense.

1. Scope*

1.1 This specification covers deformed steel wire which has been cold-worked by drawing, rolling, or both drawing and rolling, to be used as produced, or in fabricated form, for the reinforcement of concrete in sizes having nominal cross-sectional areas not less than 6.45 mm^2[0.01 in.^{2}].

1.2 Supplement S1 describes high-strength wire, which shall be furnished when specifically ordered. It shall be permissible to furnish high-strength wire in place of regular wire if mutually agreed to by the purchaser and manufacturer.

1.3 The values stated in either SI units or inch-pound units are to be regarded separately as standard. The values stated in each system may not be exact equivalents; therefore, each system shall be used independently of the other. Combining values from the two systems may result in non-conformance with the standard. (The inch-pound units are shown in brackets except in Table 1).

2. Referenced Documents

2.1 *ASTM Standards:*[2]
A370 Test Methods and Definitions for Mechanical Testing of Steel Products
A497/A497M Specification for Steel Welded Wire Reinforcement, Deformed, for Concrete
A700 Practices for Packaging, Marking, and Loading Methods for Steel Products for Shipment
E83 Practice for Verification and Classification of Extensometer Systems

2.2 *Military Standards:*[3]
MIL-STD-129 Marking for Shipment and Storage

2.3 *Federal Standard:*[3]
Fed. Std. No. 123 Marking for Shipments (Civil Agencies)

2.4 *ACI Standard:*[4]
ACI 318 Building Code Requirements for Structural Concrete

3. Terminology

3.1 *Definitions of Terms Specific to This Standard:*

3.1.1 *deformed steel wire for reinforcement—as used within the scope and intent of this specification*, shall mean any cold-worked, deformed steel wire intended for use as reinforcement in concrete construction, the wire surface having deformations that: (1) inhibit longitudinal movement of the wire in such construction; and (2) conform to the provisions of Section 5. It shall be permissible for the deformations to be raised or indented.

3.1.2 *size number—as used in this specification*, refers to the numerical designation of the wire as tabulated in Table 1 and Table 2 under the column headed Deformed Wire Size Number, or a number indicating the nominal cross-sectional area of the deformed wire in square millimeters [hundredths of a square inch].

4. Ordering Information

4.1 When deformed wire is ordered by size number, the dimensional requirements shall be as given in Table 1. When deformed wire is ordered to dimensions other than the sizes shown, the nominal dimensions shall be developed from the applicable unit mass per meter [weight per foot] of the section.

4.2 It shall be the responsibility of the purchaser to specify all requirements that are necessary for the manufacture and delivery of the wire under this specification. Such requirements to be considered include, but are not limited to, the following:
4.2.1 Quantity (mass [weight]),
4.2.2 Name of material (deformed steel wire for concrete reinforcement),
4.2.3 Wire diameter (see Table 1 and Table 2),
4.2.4 Yield strength measurement (see 8 and 13.3),
4.2.5 Packaging (see Section 17), and

---

[1] This specification is under the jurisdiction of ASTM A01 on Steel, Stainless Steel and Related Alloys and is the direct responsibility of A01.05 on Steel Reinforcement.
Current edition approved Sept. 1, 2007. Published October 2007. Originally approved in 1964. Last previous edition approved in 2005 as A496/A496M–05. DOI: 10.1520/A0496_A0496M-07.

[2] For referenced ASTM standards, visit the ASTM website, www.astm.org, or contact ASTM Customer Service at service@astm.org. For *Annual Book of ASTM Standards* volume information, refer to the standard's Document Summary page on the ASTM website.

[3] Available from Standardization Documents Order Desk, DODSSP, Bldg. 4, Section D, 700 Robbins Ave., Philadelphia, PA 19111-5098, http://www.dodssp.daps.mil.

[4] Available from American Concrete Institute (ACI), P.O. Box 9094, Farmington Hills, MI 48333-9094, http://www.aci-int.org.

A Summary of Changes section appears at the end of this standard.

TABLE 1 Dimensional Requirements for Deformed Wire for Concrete Reinforcement in SI Units

Nominal Dimensions						Deformation Requirements						
Deformed Wire Size[A,B,C]	[D in²] [× 100]	Unit Wt.		Diameter[D]		Cross-Sectional Area[2,E]		Perimeter		Spacing, Maximum		Spacing, Minimum		Min. Avg. Height of Deformations	
		kg/m	[lbs/ft.]	mm	[in.]	mm²	[in.²]	mm	[in.]	mm	[in.]	mm	[in.]	mm[F]	[in.]
MD 25	[D 3.9]	0.196	[0.132]	5.64	[0.222]	25	[0.039]	17.7	[0.698]	7.24	[0.285]	4.62	[0.182]	0.252	[0.010]
MD 30	[D 4.7]	0.235	[0.158]	6.18	[0.243]	30	[0.047]	19.4	[0.764]	7.24	[0.285]	4.62	[0.182]	0.279	[0.011]
MD 35	[D 5.4]	0.275	[0.185]	6.68	[0.263]	35	[0.054]	21.0	[0.826]	7.24	[0.285]	4.62	[0.182]	0.302	[0.012]
MD 40	[D 6.2]	0.314	[0.211]	7.14	[0.281]	40	[0.062]	22.4	[0.883]	7.24	[0.285]	4.62	[0.182]	0.320	[0.013]
MD 45	[D 7.0]	0.353	[0.237]	7.57	[0.298]	45	[0.070]	23.8	[0.936]	7.24	[0.285]	4.62	[0.182]	0.342	[0.014]
MD 50	[D 7.8]	0.392	[0.264]	7.98	[0.314]	50	[0.078]	25.1	[0.987]	7.24	[(0.285]	4.62	[0.182]	0.360	[0.014]
MD 55	[D 8.5]	0.432	[0.290]	8.37	[0.329]	55	[0.085]	26.3	[1.04]	7.24	[0.285]	4.62	[0.182]	0.378	[0.015]
MD 60	[D 9.3]	0.471	[0.316]	8.74	[0.344]	60	[0.093]	27.5	[1.08]	7.24	[0.285]	4.62	[0.182]	0.392	[0.015]
MD 65	[D 10.1]	0.510	[0.343]	9.10	[0.358]	65	[0.101]	28.6	[1.13]	7.24	[0.285]	4.62	[0.182]	0.455	[0.018]
MD 70	[D 10.9]	0.549	[0.369]	9.44	[0.372]	70	[0.109]	29.7	[1.17]	7.24	[0.285]	4.62	[0.182]	0.470	[0.018]
MD 80	[D 12.4]	0.628	[0.422]	10.1	[0.397]	80	[0.124]	31.7	[1.25]	7.24	[0.285]	4.62	[0.182]	0.505	[0.020]
MD 90	[D 14.0]	0.706	[0.475]	10.7	[0.421]	90	[0.140]	33.6	[1.32]	7.24	[0.285]	4.62	[0.182]	0.535	[0.021]
MD 100	[D 15.5]	0.785	[0.527]	11.3	[0.444]	100	[0.155]	35.4	[1.40]	7.24	[0.285]	4.62	[0.182]	0.565	[0.022]
MD 120	[D 18.6]	0.942	[0.633]	12.4	[0.487]	120	[0.186]	38.8	[1.53]	7.24	[0.285]	4.62	[0.182]	0.620	[0.024]
MD 130	[D 20.2]	1.02	[0.686]	12.9	[0.507]	130	[0.202]	40.4	[1.59]	7.24	[0.285]	4.62	[0.182]	0.645	[0.025]
MD 200	[D 31.0]	1.57	[1.05]	16.0	[0.628]	200	[0.310]	50.1	[1.97]	7.24	[0.285]	4.62	[0.182]	0.800	[0.031]
MD 290	[D 45.0]	2.28	[1.53]	19.2	[0.757]	290	[0.450]	60.4	[2.38]	7.24	[0.285]	4.62	[0.182]	0.961	[0.0379]

[A] The number following the prefix indicates the nominal cross-sectional area of the deformed wire in square millimeters.
[B] For sizes other than those shown above, the Size Number shall be the number of one hundredths of a square inch in the nominal area of the deformed wire cross section, prefixed by the letters MD.
[C] These sizes represent the most readily available sizes in the welded wire reinforcement industry. Other wire sizes are available and many manufacturers can produce them in 1-mm² [0.0015-in.²] increments.
[D] The nominal diameter of a deformed wire is equivalent to the diameter of a plain wire having the same mass per metre as the deformed wire.
[E] The cross-sectional area is based on the nominal diameter. The area in square millimetres [inches] is calculated by dividing the unit mass [weight] in kg/mm [lbs./in.] by 7×10⁻⁶ (mass of 1 mm³ of steel [0.2833 weight of 1 in.³ of steel]) or by dividing the unit mass [weight] in kg/m [lbs./ft.] by 0.007849 (mass of steel 1 mm square and 1 m long) [3.4 (weight of steel 1in. square and 1 ft. long)].
[F] The minimum average height of the deformations shall be determined from measurements made on not less than two typical deformations from each line of deformations on the wire. Measurements shall be made at the center of indentation as described in 7.7.

TABLE 2 Dimensional Requirements for Deformed Steel Wire for Concrete Reinforcement—US Customary Units Wire Sizes[A]

Deformed Wire Size Number[B,C]	Nominal Dimensions				Deformation Requirements		
	Unit Weight, lb/ft [kg/m]	Diameter, in. [mm][D]	Cross-sectional Area, in.²[mm²][E]	Perimeter, in. [mm]	Maximum Spacing, in. [mm]	Minimum Spacing, in. [mm]	Minimum Average Height of Deformations, in. [mm][F,G]
D-1	0.034 [0.051]	0.113 [2.87]	0.010 [6.45]	0.354 [9.00]	0.285 [7.24]	0.182 [4.62]	0.0045 [0.114]
D-2	0.068 [0.101]	0.160 [4.05]	0.020 [12.9]	0.501 [12.7]	0.285 [7.24]	0.182 [4.62]	0.0063 [0.160]
D-3	0.102 [0.152]	0.195 [4.96]	0.030 [19.4]	0.614 [15.6]	0.285 [7.24]	0.182 [4.62]	0.0078 [0.198]
D-4	0.136 [0.202]	0.226 [5.73]	0.040 [25.8]	0.709 [18.0]	0.285 [7.24]	0.182 [4.62]	0.0101 [0.257]
D-5	0.170 [0.253]	0.252 [6.41]	0.050 [32.3]	0.793 [20.1]	0.285 [7.24]	0.182 [4.62]	0.0113 [0.287]
D-6	0.204 [0.304]	0.276 [7.02]	0.060 [38.7]	0.868 [22.1]	0.285 [7.24]	0.182 [4.62]	0.0124 [0.315]
D-7	0.238 [0.354]	0.299 [7.58]	0.070 [45.2]	0.938 [23.8]	0.285 [7.24]	0.182 [4.62]	0.0134 [0.304]
D-8	0.272 [0.405]	0.319 [8.11]	0.080 [51.6]	1.00 [25.5]	0.285 [7.24]	0.182 [4.62]	0.0143 [0.363]
D-9	0.306 [0.455]	0.339 [8.60]	0.090 [58.1]	1.06 [27.0]	0.285 [7.24]	0.182 [4.62]	0.0152 [0.386]
D-10	0.340 [0.506]	0.357 [9.06]	0.100 [64.5]	1.12 [28.5]	0.285 [7.24]	0.182 [4.62]	0.0160 [0.406]
D-11	0.374 [0.557]	0.374 [9.51]	0.110 [71.0]	1.18 [29.9]	0.285 [7.24]	0.182 [4.62]	0.0187 [0.475]
D-12	0.408 [0.607]	0.391 [9.93]	0.120 [77.4]	1.23 [31.2]	0.285 [7.24]	0.182 [4.62]	0.0195 [0.495]
D-13	0.442 [0.658]	0.407 [10.3]	0.130 [83.9]	1.28 [32.5]	0.285 [7.24]	0.182 [4.62]	0.0203 [0.516]
D-14	0.476 [0.708]	0.422 [10.7]	0.140 [90.3]	1.33 [33.7]	0.285 [7.24]	0.182 [4.62]	0.0211 [0.536]
D-15	0.510 [0.759]	0.437 [11.1]	0.150 [96.8]	1.37 [34.9]	0.285 [7.24]	0.182 [4.62]	0.0218 [0.554]
D-16	0.544 [0.810]	0.451 [11.5]	0.160 [103]	1.42 [36.0]	0.285 [7.24]	0.182 [4.62]	0.0225 [0.572]
D-17	0.578 [0.860]	0.465 [11.8]	0.170 [110]	1.46 [37.1]	0.285 [7.24]	0.182 [4.62]	0.0232 [0.589]
D-18	0.612 [0.911]	0.479 [12.2]	0.180 [116]	1.50 [38.2]	0.285 [7.24]	0.182 [4.62]	0.0239 [0.607]
D-19	0.646 [0.961]	0.492 [12.5]	0.190 [122]	1.55 [39.2]	0.285 [7.24]	0.182 [4.62]	0.0245 [0.622]
D-20	0.680 [1.01]	0.505 [12.8]	0.200 [129]	1.59 [40.3]	0.285 [7.24]	0.182 [4.62]	0.0252 [0.604]
D-21	0.714 [1.06]	0.517 [13.1]	0.210 [135]	1.62 [41.3]	0.285 [7.24]	0.182 [4.62]	0.0259 [0.658]
D-22	0.748 [1.11]	0.529 [13.4]	0.220 [141]	1.66 [42.2]	0.285 [7.24]	0.182 [4.62]	0.0265 [0.673]
D-23	0.782 [1.16]	0.541 [13.7]	0.230 [148]	1.70 [43.2]	0.285 [7.24]	0.182 [4.62]	0.0271 [0.688]
D-24	0.816 [1.21]	0.553 [14.0]	0.240 [154]	1.74 [44.1]	0.285 [7.24]	0.182 [4.62]	0.0277 [0.704]
D-25	0.850 [1.26]	0.564 [14.3]	0.250 [161]	1.77 [45.0]	0.285 [7.24]	0.182 [4.62]	0.0282 [0.716]
D-26	0.884 [1.32]	0.575 [14.6]	0.260 [167]	1.81 [45.9]	0.285 [7.24]	0.182 [4.62]	0.0288 [0.732]
D-27	0.918 [1.37]	0.586 [14.9]	0.270 [174]	1.84 [46.8]	0.285 [7.24]	0.182 [4.62]	0.0293 [0.744]
D-28	0.952 [1.42]	0.597 [15.2]	0.280 [180]	1.88 [47.6]	0.285 [7.24]	0.182 [4.62]	0.0299 [0.759]
D-29	0.986 [1.47]	0.608 [15.4]	0.290 [187]	1.91 [48.5]	0.285 [7.24]	0.182 [4.62]	0.0304 [0.772]
D-30	1.02 [1.52]	0.618 [15.7]	0.300 [193]	1.94 [49.3]	0.285 [7.24]	0.182 [4.62]	0.0309 [0.785]
D-31	1.05 [1.57]	0.628 [16.0]	0.310 [200]	1.97 [50.1]	0.285 [7.24]	0.182 [4.62]	0.0314 [0.798]
D-45	1.53 [2.28]	0.757 [19.2]	0.450 [290]	2.38 [60.4]	0.285 [7.24]	0.182 [4.62]	0.0379 [0.961]

[A] In this table only, inch-pound units are regarded as standard and SI units are shown in brackets.
[B] The number following the prefix indicates the nominal cross-sectional area of the deformed wire in square inches [square millimeters].
[C] For sizes other than those shown above, the Size Number shall be the number of one hundredths of a square inch in the nominal area of the deformed wire cross section, prefixed by the letter D.
[D] The nominal diameter of a deformed wire is equivalent to the diameter of a plain wire having the same weight per foot as the deformed wire.
[E] The cross-sectional area is based on the nominal diameter. The area in square inches [millimeters] is calculated by dividing the unit weight [mass] in lbs./in. [kg/mm] by 0.2833 (weight of 1 in.³ of steel) [7×10^{-6}(mass of 1 mm³ of steel)], or by dividing the unit weight [mass] in lbs./ft. [kg/m] by 3.4 (weight of steel 1 in. square and 1 foot long) [0.007849 (mass of 1 mm square and 1 m long)].
[F] The minimum average height of the deformations shall be determined from measurements made on not less than two typical deformations from each line of deformations on the wire. Measurements shall be made at the center of indentation as described in 6.2.
[G] These sizes represent the most readily available sizes in the welded wire reinforcement industry. Other wire sizes are available and many manufacturers can produce them in 0.0015–in² [1–mm²] increments.

4.2.6 ASTM designation and year of issue.

4.2.7 Special requirements, if any. (See Supplement S1.)

Note 1—A typical ordering description is as follows: 50 000 lb deformed steel wire for concrete reinforcement, size No. MD80 [D12.4], in 800 kg [2000 lb] secured coils, to ASTM A496/A496M–___ .

5. Materials and Manufacture

5.1 The steel shall be made by one of the following processes: open-hearth, electric furnace, or basic oxygen.

5.2 The deformed steel wire shall be produced from rods or bars that have been hot rolled from billets.

6. Requirements

6.1 Deformations shall be spaced along the wire at a substantially uniform distance and shall be symmetrically dispersed around the perimeter of the section. The deformations on all longitudinal lines of the wire shall be similar in size and shape. A minimum of 25 % of the total surface area shall be deformed by measurable deformations.

6.2 Deformed wire shall have two or more lines of deformations.

6.3 The average longitudinal spacing of deformations shall be not less than 3.5 nor more than 5.5 deformations per 25.4 mm [1 in.] in each line of deformations on the wire.

6.4 The minimum average height of the center of typical deformations based on the nominal wire diameters shown in Table 1 and Table 2 shall be as follows:

Wire Sizes	Minimum Average Height of Deformations, Percent of Nominal Wire Diameter
MD20 [D3] and smaller	4
Larger than MD20 [D3] through MD65 [D10]	4 ½
Larger than MD65 [D10]	5

6.5 The deformations shall be placed with respect to the axis of the wire so that the included angle is not less than 45°; or if deformations are curvilinear, the angle formed by the transverse axis of the deformation and the wire axis shall be not less

TABLE 3 Tension Test Requirements

	MPa [psi] min
Tensile strength	585 [85 000]
Yield strength	515 [75 000]

TABLE 4 Tension Test Requirements (Material for Welded Wire Reinforcement)

	MPa [psi] min
Tensile strength	550 [80 000]
Yield strength	485 [70 000]

TABLE 5 Bend Test Requirements

Size Number of Wire	Bend Test
MD39 [D6] and smaller	Bend around a pin the diameter that is equal to twice the diameter of the specimen
Larger than MD39 [D6]	Bend around a pin the diameter that is equal to four times the diameter of the specimen

than 45°. Where the line of deformations forms an included angle with the axis of the wire from 45 to 70° inclusive, the deformations shall alternately reverse in direction on each side, or those on one side shall be reversed in direction from those on the opposite side. Where the included angle is greater than 70°, a reversal in direction is not required.

7. Dimensions

7.1 The average spacing of deformations shall be determined by dividing a measured length of the wire specimen by the number of individual deformations in any one row of deformations on any side of the wire specimen. A measured length of the wire specimen shall be considered the distance from a point on a deformation to a corresponding point on any other deformation in the same line of deformations on the wire.

7.2 The minimum average height of deformations shall be determined from measurements made on not less than two typical deformations from each line of deformations on the wire. Measurements shall be made at the center of indentations.

8. Mechanical Property Requirements

8.1 *Tension Tests*:

8.1.1 When tested as described in Test Methods and Definitions A370, the material, except as specified in 8.1.2 shall conform to the tensile property requirements in Table 3, based on the nominal area of wire.

8.1.2 When required by the purchaser, yield strength shall be determined using a Class B-1 extensometer as described in Practice E83. The yield strength shall be determined as described in Test Methods and Definitions A370 and an extension of 0.5 % of gage length. It shall be permissible to remove the extensometer after the yield strength has been determined. The wire shall meet the requirements of Table 3 or 4, whichever is applicable.

8.1.3 For material to be used in the fabrication of welded wire reinforcement, the tensile and yield strength properties shall conform to the requirements given in Table 4, based on the nominal area of the wire.

8.1.4 The material shall not exhibit a definite yield point as evidenced by a distinct drop of the beam or halt in the gage of the testing machine prior to reaching ultimate tensile load.

8.2 *Bend Test*—The bend test specimen shall withstand being bent at room temperature through 90° without cracking on the outside of the bent portion, as prescribed in Table 5.

9. Permissible Variation in Mass [Weight]

9.1 The permissible variation in mass [weight] of any deformed wire shall be ±6 % of its nominal weight. The theoretical masses [weights] shown in Table 1, or similar calculations on unlisted sizes, shall be used to establish the variation.

10. Workmanship, Finish, and Appearance

10.1 The wire shall be free of detrimental imperfections and shall have a workmanlike finish.

10.2 Rust, surface seams, or surface irregularities shall not be a cause for rejection provided the requirements of 10.3 are met, and the minimum dimensions and mechanical properties of a hand wire-brushed test specimen meet the requirements of this specification.

10.3 Wire intended for welded wire reinforcement shall be sufficiently free of rust and drawing lubricant, so as not to interfere with electric resistance welding.

11. Sampling

11.1 Test specimens for testing mechanical properties shall be full wire sections and shall be obtained from the ends of the wire product as drawn or rolled, or both drawn and rolled. The specimens shall be of sufficient length to perform testing described in 8.1 and 8.2.

11.2 Any test specimen exhibiting obvious isolated imperfections that are not representative of the product shall be discarded and another specimen substituted.

12. Number of Tests

12.1 One tension and one bend test shall be made from each 9000 kg [10 tons] or less of each size of wire or fraction thereof in a lot, or a total of seven samples, whichever is less. A lot shall consist of all the coils of a single size offered for delivery at the same time.

13. Inspection

13.1 The inspector representing the purchaser shall have free entry, at all times while work on the contract of the purchaser is being performed, to all parts of the manufacturer's facilities that concern the manufacture of the material ordered. The manufacturer shall afford the inspector all reasonable facilities to satisfy him that the material is being furnished in accordance with this specification.

13.2 Except for yield strength, all tests and inspections shall be made at the manufacturer's facilities prior to shipment, unless otherwise specified. Such tests shall be so conducted as not to interfere unnecessarily with the operation of the manufacturer's facilities.

13.3 The purchaser shall have the option to require a yield strength measurement to determine compliance with yield strength requirements in 8.1, and shall specify that the measurements be performed by the manufacturer at the manufacturer's facilities, a recognized laboratory, or the purchaser's representative at the manufacturer's facilities. Such measurements shall be conducted without unnecessarily interfering with manufacturing operations.

13.4 *For U.S. Government Procurement Only*—Except as otherwise specified in the contract, the contractor is responsible for the performance of all inspection and test requirements specified herein. Except as otherwise specified in the contract, the contractor shall have the option to use his own or any other suitable facilities for the performance of the inspection and test requirements specified herein, unless disapproved by the purchaser at the time of purchase. The purchaser shall have the right to perform any of the inspections and tests at the same frequency as set forth in this specification where such inspections are deemed necessary to ensure that material conforms to prescribed requirements.

14. Rejection

14.1 Material that shows detrimental imperfections subsequent to its acceptance at the manufacturer's facilities shall be rejected, and the manufacturer shall be notified.

14.2 Failure of any of the test specimens to comply with the requirements of this specification shall constitute grounds for rejection of the lot represented by the specimen.

14.3 Any rejection based on tests made in accordance with this specification shall be reported to the manufacturer within two weeks of the date of inspection or test. The material shall be adequately protected and correctly identified such that the manufacturer is able to make a proper investigation.

15. Rehearing

15.1 Rejected material shall be preserved for a period of not less than two weeks from the date of inspection, during which time the manufacturer shall have the option to make claim for a rehearing and retesting.

15.2 The manufacturer shall have the option to resubmit the rejected lot for re-inspection or retesting by inspecting or testing every coil for the property in which the specimen failed and sorting out non-conforming coils.

16. Certification

16.1 When specified in the purchase order or contract, the purchaser shall be furnished with the manufacturer's written certification that the material was manufactured, sampled, tested, and inspected in accordance with, and meets the requirements of, this specification. When specified in the purchase order or contract, a report of the test results shall be furnished. The certification shall include the specification number, year-date of issue and revision letter, if any.

16.2 A Material Test Report, Certificate of Inspection, or similar document printed from or used in electronic form from an electronic data interchange (EDI) transmission shall be regarded as having the same validity as a counterpart printed in the certifier's facility. The content of the EDI transmitted document must meet the requirements of the invoked ASTM standard(s) and conform to any existing EDI agreement between the purchaser and the manufacturer. Notwithstanding the absence of a signature, the organization submitting the EDI transmission is responsible for the content of the report.

NOTE 2—The industry definition as invoked here is: EDI is the computer-to-computer exchange of business information in a standard format such as ANSI ASC X12.

17. Packaging and Marking

17.1 The size of the wire, ASTM designation, and name or mark of the manufacturer shall be marked on a tag securely attached to each coil of wire.

17.2 Unless otherwise specified, packaging, marking, and loading for shipment shall be in accordance with Practices A700.

17.3 When specified in the contract or order, and for the direct procurement by or direct shipment to the U.S. government, marking for shipment, in addition to requirements specified in the contract or order, shall be in accordance with MIL-STD-129 for U.S. military agencies and in accordance with Fed. Std. No. 123 or U.S. government civil agencies.

18. Keywords

18.1 concrete reinforcement; deformations (indentations); steel wire

---

SUPPLEMENTARY REQUIREMENTS

S1. High-Strength Wire

S1.1 *Scope*—This supplement delineates only those details that are relative to high-strength wire and to the mechanical requirements for wire having properties generally as described in this specification.

NOTE S1.1—Building codes, for example, ACI 318 permit the use of reinforcement with a yield strength up to 550 MPa [80 000 psi]. For compatibility with the codes' design provisions for high-strength reinforcement, this supplement prescribes requirements for the mechanical properties of wire that exceed the minimum values for yield strength and tensile strength in Table 3 and Table 4 of this specification

S1.2 *Mechanical Property Requirements:*

S1.2.1 Minimum yield strength shall be specified in the purchase order in increments of 17.5 MPa [2500 psi]. When tested the yield strength shall be determined at an extension under load of 0.35 %.

NOTE S1.2—To conform to the limit on yield strength in building codes, the minimum yield strength specified in the purchase order should not be greater than 550 MPa [80 000 psi].

S1.2.2 Minimum tensile strength shall be 70 MPa [10 000 psi] greater than the minimum specified yield strength.

NOTE S1.3—A typical order entry line for minimum yield strength is,

"72 500 psi minimum yield strength" or "500 MPa minimum yield strength."

S1.3 *Certification*—Certification for material produced to this supplement shall include a report of the test results for yield strength, tensile strength, and bend tests. Frequency of testing shall conform to Section 12 of this specification and Section 12 of Specification A497/A497M as applicable.

SUMMARY OF CHANGES

Committee A01 has identified the location of selected changes to this standard since the last issue (A496/A496M–05) that may impact the use of this standard.

(*1*) Removed reference to MIL-STD-163.

ASTM International takes no position respecting the validity of any patent rights asserted in connection with any item mentioned in this standard. Users of this standard are expressly advised that determination of the validity of any such patent rights, and the risk of infringement of such rights, are entirely their own responsibility.

This standard is subject to revision at any time by the responsible technical committee and must be reviewed every five years and if not revised, either reapproved or withdrawn. Your comments are invited either for revision of this standard or for additional standards and should be addressed to ASTM International Headquarters. Your comments will receive careful consideration at a meeting of the responsible technical committee, which you may attend. If you feel that your comments have not received a fair hearing you should make your views known to the ASTM Committee on Standards, at the address shown below.

This standard is copyrighted by ASTM International, 100 Barr Harbor Drive, PO Box C700, West Conshohocken, PA 19428-2959, United States. Individual reprints (single or multiple copies) of this standard may be obtained by contacting ASTM at the above address or at 610-832-9585 (phone), 610-832-9555 (fax), or service@astm.org (e-mail); or through the ASTM website (www.astm.org).

Designation: A500/A500M – 09

Standard Specification for
Cold-Formed Welded and Seamless Carbon Steel Structural Tubing in Rounds and Shapes[1]

This standard is issued under the fixed designation A500/A500M; the number immediately following the designation indicates the year of original adoption or, in the case of revision, the year of last revision. A number in parentheses indicates the year of last reapproval. A superscript epsilon (ε) indicates an editorial change since the last revision or reapproval.

This standard has been approved for use by agencies of the Department of Defense.

1. Scope*

1.1 This specification covers cold-formed welded and seamless carbon steel round, square, rectangular, or special shape structural tubing for welded, riveted, or bolted construction of bridges and buildings, and for general structural purposes.

1.2 This tubing is produced in both welded and seamless sizes with a periphery of 64 in. [1630 mm] or less, and a specified wall thickness of 0.688 in. [18 mm] or less. Grade D requires heat treatment.

Note 1—Products manufactured to this specification may not be suitable for those applications such as dynamically loaded elements in welded structures, etc., where low-temperature notch-toughness properties may be important.

1.3 The values stated in either SI units or inch-pound units are to be regarded separately as standard. Within the text, the SI units are shown in brackets. The values stated in each system may not be exact equivalents; therefore, each system shall be used independently of the other. Combining values from the two systems may result in non-conformance with the standard. The inch-pound units shall apply unless the "M" designation of this specification is specified in the order.

1.4 The text of this specification contains notes and footnotes that provide explanatory material. Such notes and footnotes, excluding those in tables and figures, do not contain any mandatory requirements.

2. Referenced Documents

2.1 *ASTM Standards:*[2]
A370 Test Methods and Definitions for Mechanical Testing of Steel Products
A700 Practices for Packaging, Marking, and Loading Methods for Steel Products for Shipment
A751 Test Methods, Practices, and Terminology for Chemical Analysis of Steel Products
A941 Terminology Relating to Steel, Stainless Steel, Related Alloys, and Ferroalloys

2.2 *Military Standards:*
MIL-STD-129 Marking for Shipment and Storage[3]
MIL-STD-163 Steel Mill Products, Preparation for Shipment and Storage[3]

2.3 *Federal Standards:*
Fed. Std. No. 123 Marking for Shipment[3]
Fed. Std. No. 183 Continuous Identification Marking of Iron and Steel Products[3]

2.4 *AIAG Standard:*
B-1 Bar Code Symbology Standard[4]

3. Terminology

3.1 *Definitions*—For definitions of terms used in this specification, refer to Terminology A941.

4. Ordering Information

4.1 Orders for material under this specification shall contain information concerning as many of the following items as are required to describe the desired material adequately:
4.1.1 Quantity (feet [metres] or number of lengths),
4.1.2 Name of material (cold-formed tubing),
4.1.3 Method of manufacture (seamless or welded),
4.1.4 Grade (A, B, C, or D),
4.1.5 Size (outside diameter and wall thickness for round tubing, and outside dimensions and wall thickness for square and rectangular tubing),
4.1.6 Copper-containing steel (see Table 1), if applicable,
4.1.7 Length (random, multiple, specific; see 11.3),
4.1.8 End condition (see 16.3),
4.1.9 Burr removal (see 16.3),
4.1.10 Certification (see Section 18),
4.1.11 ASTM specification designation and year of issue,
4.1.12 End use,

---

[1] This specification is under the jurisdiction of ASTM Committee A01 on Steel, Stainless Steel and Related Alloys and is the direct responsibility of Subcommittee A01.09 on Carbon Steel Tubular Products.
Current edition approved Oct. 1, 2009. Published November 2009. Originally approved in 1964. Last previous edition approved in 2007 as A500/A500M – 07. DOI: 10.1520/A0500_A0500M-09.

[2] For referenced ASTM standards, visit the ASTM website, www.astm.org, or contact ASTM Customer Service at service@astm.org. For *Annual Book of ASTM Standards* volume information, refer to the standard's Document Summary page on the ASTM website.

[3] Available from Standardization Documents Order Desk, Bldg. 4 Section D, 700 Robbins Ave., Philadelphia, PA 19111-5094, Attn: NPODS.

[4] Available from Automotive Industry Action Group (AIAG), 26200 Lahser Rd., Suite 200, Southfield, MI 48033, http://www.aiag.org.

*A Summary of Changes section appears at the end of this standard.

TABLE 1 Chemical Requirements

Element	Grades A, B, and D		Grade C	
	Heat Analysis	Product Analysis	Heat Analysis	Product Analysis
Carbon, max[A]	0.26	0.30	0.23	0.27
Manganese, max[A]	1.35	1.40	1.35	1.40
Phosphorus, max	0.035	0.045	0.035	0.045
Sulfur, max	0.035	0.045	0.035	0.045
Copper, min[B]	0.20	0.18	0.20	0.18

[A] For each reduction of 0.01 percentage point below the specified maximum for carbon, an increase of 0.06 percentage point above the specified maximum for manganese is permitted, up to a maximum of 1.50 % by heat analysis and 1.60 % by product analysis.
[B] If copper-containing steel is specified in the purchase order.

4.1.13 Special requirements, and
4.1.14 Bar coding (see 19.3).

5. Process

5.1 The steel shall be made by one or more of the following processes: open-hearth, basic-oxygen, or electric-furnace.

5.2 When steels of different grades are sequentially strand cast, the steel producer shall identify the resultant transition material and remove it using an established procedure that positively separates the grades.

6. Manufacture

6.1 The tubing shall be made by a seamless or welding process.

6.2 Welded tubing shall be made from flat-rolled steel by the electric-resistance-welding process. The longitudinal butt joint of welded tubing shall be welded across its thickness in such a manner that the structural design strength of the tubing section is assured.

NOTE 2—Welded tubing is normally furnished without removal of the inside flash.

6.3 Except as required by 6.4, it shall be permissible for the tubing to be stress relieved or annealed.

6.4 Grade D tubing shall be heat treated at a temperature of at least 1100 °F [590 °C] for one hour per inch [25 mm] of thickness.

7. Heat Analysis

7.1 Each heat analysis shall conform to the requirements specified in Table 1 for heat analysis.

8. Product Analysis

8.1 The tubing shall be capable of conforming to the requirements specified in Table 1 for product analysis.

8.2 If product analyses are made, they shall be made using test specimens taken from two lengths of tubing from each lot of 500 lengths, or fraction thereof, or two pieces of flat-rolled stock from each lot of a corresponding quantity of flat-rolled stock. Methods and practices relating to chemical analysis shall be in accordance with Test Methods, Practices, and Terminology A751. Such product analyses shall conform to the requirements specified in Table 1 for product analysis.

8.3 If both product analyses representing a lot fail to conform to the specified requirements, the lot shall be rejected.

8.4 If only one product analysis representing a lot fails to conform to the specified requirements, product analyses shall be made using two additional test specimens taken from the lot. Both additional product analyses shall conform to the specified requirements or the lot shall be rejected.

9. Tensile Requirements

9.1 The material, as represented by the test specimen, shall conform to the requirements as to tensile properties prescribed in Table 2.

10. Flattening Test

10.1 The flattening test shall be made on round structural tubing. A flattening test is not required for shaped structural tubing.

10.2 For welded round structural tubing, a test specimen at least 4 in. [100 mm] in length shall be flattened cold between parallel plates in three steps, with the weld located 90° from the line of direction of force. During the first step, which is a test for ductility of the weld, no cracks or breaks on the inside or outside surfaces of the test specimen shall be present until the distance between the plates is less than two-thirds of the specified outside diameter of the tubing. For the second step, no cracks or breaks on the inside or outside parent metal surfaces of the test specimen, except as provided for in 10.5, shall be present until the distance between the plates is less than one-half of the specified outside diameter of the tubing. During the third step, which is a test for soundness, the flattening shall be continued until the test specimen breaks or the opposite walls of the test specimen meet. Evidence of

TABLE 2 Tensile Requirements

Round Structural Tubing				
	Grade A	Grade B	Grade C	Grade D
Tensile strength, min, psi [MPa]	45 000 [310]	58 000 [400]	62 000 [425]	58 000 [400]
Yield strength, min, psi [MPa]	33 000 [230]	42 000 [290]	46 000 [315]	36 000 [250]
Elongation in 2 in. [50 mm], min, %[D]	25[A]	23[B]	21[C]	23[B]

Shaped Structural Tubing				
	Grade A	Grade B	Grade C	Grade D
Tensile strength, min, psi [MPa]	45 000 [310]	58 000 [400]	62 000 [425]	58 000 [400]
Yield strength, min, psi [MPa]	39 000 [270]	46 000 [315]	50 000 [345]	36 000 [250]
Elongation in 2 in. [50 mm], min, %[D]	25[A]	23[B]	21[C]	23[B]

[A] Applies to specified wall thicknesses (t) equal to or greater than 0.120 in. [3.05 mm]. For lighter specified wall thicknesses, the minimum elongation values shall be calculated by the formula: percent elongation in 2 in. [50 mm] = $56t + 17.5$, rounded to the nearest percent.
[B] Applies to specified wall thicknesses (t) equal to or greater than 0.180 in. [4.57 mm]. For lighter specified wall thicknesses, the minimum elongation values shall be calculated by the formula: percent elongation in 2 in. [50 mm] = $61t + 12$, rounded to the nearest percent.
[C] Applies to specified wall thicknesses (t) equal to or greater than 0.120 in. [3.05 mm]. For lighter specified wall thicknesses, the minimum elongation values shall be by agreement with the manufacturer.
[D] The minimum elongation values specified apply only to tests performed prior to shipment of the tubing.

laminated or unsound material or of incomplete weld that is revealed during the entire flattening test shall be cause for rejection.

10.3 For seamless round structural tubing 2⅜ in. [60 mm] specified outside diameter and larger, a specimen not less than 2½ in. [65 mm] in length shall be flattened cold between parallel plates in two steps. During the first step, which is a test for ductility, no cracks or breaks on the inside or outside surfaces, except as provided for in 10.5, shall occur until the distance between the plates is less than the value of "H" calculated by the following equation:

$$H = (1 + e)t / (e + t/D) \quad (1)$$

where:
H = distance between flattening plates, in. [mm],
e = deformation per unit length (constant for a given grade of steel, 0.09 for Grade A, 0.07 for Grade B, and 0.06 for Grade C),
t = specified wall thickness of tubing, in. [mm], and
D = specified outside diameter of tubing, in. [mm].

During the second step, which is a test for soundness, the flattening shall be continued until the specimen breaks or the opposite walls of the specimen meet. Evidence of laminated or unsound material that is revealed during the entire flattening test shall be cause for rejection.

10.4 Surface imperfections not found in the test specimen before flattening, but revealed during the first step of the flattening test, shall be judged in accordance with Section 15.

10.5 When low D-to-t ratio tubulars are tested, because the strain imposed due to geometry is unreasonably high on the inside surface at the 6 and 12 o'clock locations, cracks at these locations shall not be cause for rejection if the D-to-t ratio is less than 10.

11. Permissible Variations in Dimensions

11.1 *Outside Dimensions*:

11.1.1 *Round Structural Tubing*—The outside diameter shall not vary more than ±0.5 %, rounded to the nearest 0.005 in. [0.1 mm], from the specified outside diameter for specified outside diameters 1.900 in. [48 mm] and smaller, and ± 0.75 %, rounded to the nearest 0.005 in. [0.1 mm], from the specified outside diameter for specified outside diameters 2.00 in. [5 cm] and larger. The outside diameter measurements shall be made at positions at least 2 in. [5 cm] from the ends of the tubing.

11.1.2 *Square and Rectangular Structural Tubing*—The outside dimensions, measured across the flats at positions at least 2 in. [5 cm] from the ends of the tubing, shall not vary from the specified outside dimensions by more than the applicable amount given in Table 3, which includes an allowance for convexity or concavity.

11.2 *Wall Thickness*—The minimum wall thickness at any point of measurement on the tubing shall be not more than 10 % less than the specified wall thickness. The maximum wall thickness, excluding the weld seam of welded tubing, shall be not more than 10 % greater than the specified wall thickness. For square and rectangular tubing, the wall thickness requirements shall apply only to the centers of the flats.

TABLE 3 Permissible Variations in Outside Flat Dimensions for Square and Rectangular Structural Tubing

Specified Outside Large Flat Dimension, in. [mm]	Permissible Variations Over and Under Specified Outside Flat Dimensions,[A] in. [mm]
2½ [65] or under	0.020 [0.5]
Over 2½ to 3½ [65 to 90], incl	0.025 [0.6]
Over 3½ to 5½ [90 to 140], incl	0.030 [0.8]
Over 5½ [140]	0.01 times large flat dimension

[A] The permissible variations include allowances for convexity and concavity. For rectangular tubing having a ratio of outside large to small flat dimension less than 1.5, and for square tubing, the permissible variations in small flat dimension shall be identical to the permissible variations in large flat dimension. For rectangular tubing having a ratio of outside large to small flat dimension in the range of 1.5 to 3.0 inclusive, the permissible variations in small flat dimension shall be 1.5 times the permissible variations in large flat dimension. For rectangular tubing having a ratio of outside large to small flat dimension greater than 3.0, the permissible variations in small flat dimension shall be 2.0 times the permissible variations in large flat dimension.

11.3 *Length*—Structural tubing is normally produced in random lengths 5 ft [1.5 m] and over, in multiple lengths, and in specific lengths. Refer to Section 4. When specific lengths are ordered, the length tolerance shall be in accordance with Table 4.

11.4 *Straightness*—The permissible variation for straightness of structural tubing shall be ⅛ in. times the number of feet [10 mm times the number of metres] of total length divided by 5.

11.5 *Squareness of Sides*—For square and rectangular structural tubing, adjacent sides shall be square (90°), with a permissible variation of ±2° max.

11.6 *Radius of Corners*—For square and rectangular structural tubing, the radius of each outside corner of the section shall not exceed three times the specified wall thickness.

11.7 *Twist*—For square and rectangular structural tubing, the permissible variations in twist shall be as given in Table 5. Twist shall be determined by holding one end of the tubing down on a flat surface plate, measuring the height that each corner on the bottom side of the tubing extends above the surface plate near the opposite ends of the tubing, and calculating the twist (the difference in heights of such corners), except that for heavier sections it shall be permissible to use a suitable measuring device to determine twist. Twist measurements shall not be taken within 2 in. [5 cm] of the ends of the tubing.

12. Special Shape Structural Tubing

12.1 The availability, dimensions, and tolerances of special shape structural tubing shall be subject to inquiry and negotiation with the manufacturer.

TABLE 4 Length Tolerances for Specific Lengths of Structural Tubing

	22 ft [6.5 m] and Under		Over 22 ft [6.5 m]	
	Over	Under	Over	Under
Length tolerance for specific lengths, in. [mm]	½ [13]	¼ [6]	¾ [19]	¼ [6]

TABLE 5 Permissible Variations in Twist for Square and Rectangular Structural Tubing

Specified Outside Large Flat Dimension, in. [mm]	Maximum Permissible Variations in Twist per 3 ft of Length [Twist per Metre of Length]	
	in.	[mm]
1½ [40] and under	0.050	[1.3]
Over 1½ to 2½ [40 to 65], incl	0.062	[1.6]
Over 2½ to 4 [65 to 100], incl	0.075	[1.9]
Over 4 to 6 [100 to 150], incl	0.087	[2.2]
Over 6 to 8 [150 to 200], incl	0.100	[2.5]
Over 8 [200]	0.112	[2.8]

13. Number of Tests

13.1 One tension test as specified in Section 15 shall be made from a length of tubing representing each lot.

13.2 The flattening test, as specified in Section 10, shall be made on one length of round tubing from each lot.

13.3 The term "lot" shall apply to all tubes of the same specified size that are produced from the same heat of steel.

14. Retests

14.1 If the results of the mechanical tests representing any lot fail to conform to the applicable requirements specified in Sections 9 and 10, the lot shall be rejected or retested using additional tubing of double the original number from the lot. The lot shall be acceptable if the results of all such retests representing the lot conform to the specified requirements.

14.2 If one or both of the retests specified in 14.1 fail to conform to the applicable requirements specified in Sections 9 and 10, the lot shall be rejected or, subsequent to the manufacturer heat treating, reworking, or otherwise eliminating the condition responsible for the failure, the lot shall be treated as a new lot and tested accordingly.

15. Test Methods

15.1 Tension test specimens shall conform to the applicable requirements of Test Methods and Definitions A370, Annex A2.

15.2 Tension test specimens shall be full–size longitudinal test specimens or longitudinal strip test specimens. For welded tubing, any longitudinal strip test specimens shall be taken from a location at least 90° from the weld and shall be prepared without flattening in the gage length. Longitudinal strip test specimens shall have all burrs removed. Tension test specimens shall not contain surface imperfections that would interfere with proper determination of the tensile properties.

15.3 The yield strength corresponding to an offset of 0.2 % of the gage length or to a total extension under load of 0.5 % of the gage length shall be determined.

16. Inspection

16.1 All tubing shall be inspected at the place of manufacture to ensure conformance to the requirements of this specification.

16.2 All tubing shall be free from defects and shall have a workmanlike finish.

16.2.1 Surface imperfections shall be classed as defects when their depth reduces the remaining wall thickness to less than 90 % of the specified wall thickness. It shall be permissible for defects having a depth not in excess of 33⅓ % of the specified wall thickness to be repaired by welding, subject to the following conditions:

16.2.1.1 The defect shall be completely removed by chipping or grinding to sound metal,

16.2.1.2 The repair weld shall be made using a low-hydrogen welding process, and

16.2.1.3 The projecting weld metal shall be removed to produce a workmanlike finish.

16.2.2 Surface imperfections such as handling marks, light die or roll marks, or shallow pits are not considered defects provided that the imperfections are removable within the specified limits on wall thickness. The removal of such surface imperfections is not required. Welded tubing shall be free of protruding metal on the outside surface of the weld seam.

16.3 Unless otherwise specified in the purchase order, structural tubing shall be furnished with square cut ends, with the burr held to a minimum. When so specified in the purchase order, the burr shall be removed on the outside diameter, inside diameter, or both.

17. Rejection

17.1 It shall be permissible for the purchaser to inspect tubing received from the manufacturer and reject any tubing that does not meet the requirements of this specification, based upon the inspection and test methods outlined herein. The purchaser shall notify the manufacturer of any tubing that has been rejected, and the disposition of such tubing shall be subject to agreement between the manufacturer and the purchaser.

17.2 It shall be permissible for the purchaser to set aside any tubing that is found in fabrication or installation within the scope of this specification to be unsuitable for the intended end use, based on the requirements of this specification. The purchaser shall notify the manufacturer of any tubing that has been set aside. Such tubing shall be subject to mutual investigation as to the nature and severity of the deficiency and the forming or installation, or both, conditions involved. The disposition of such tubing shall be subject to agreement between the manufacturer and the purchaser.

18. Certification

18.1 When specified in the purchase order or contract, the manufacturer shall furnish to the purchaser a certificate of compliance stating that the product was manufactured, sampled, tested, and inspected in accordance with this specification and any other requirements designated in the purchase order or contract, and was found to meet all such requirements. Certificates of compliance shall include the specification number and year of issue.

18.2 When specified in the purchase order or contract, the manufacturer shall furnish to the purchaser test reports for the product shipped that contain the heat analyses and the results of the tension tests required by this specification and the purchase order or contract. Test reports shall include the specification number and year of issue.

18.3 A signature or notarization is not required on certificates of compliance or test reports; however, the documents shall clearly identify the organization submitting them. Notwithstanding the absence of a signature, the organization submitting the document is responsible for its content.

18.4 A certificate of compliance or test report printed from, or used in electronic form from, an electronic data interchange (EDI) shall be regarded as having the same validity as a counterpart printed in the certifying organization's facility. The content of the EDI transmitted document shall conform to any existing EDI agreement between the purchaser and the manufacturer.

19. Product Marking

19.1 Except as noted in 19.2, each length of structural tubing shall be legibly marked to show the following information: manufacturer's name, brand, or trademark; the specification designation (year of issue not required); and grade letter.

19.2 For structural tubing having a specified outside diameter or large flat dimension of 4 in. [10 cm] or less, it shall be permissible for the information listed in 19.1 to be marked on a tag securely attached to each bundle.

19.3 *Bar Coding*—In addition to the requirements in 19.1 and 19.2, the manufacturer shall have the option of using bar coding as a supplementary identification method. When a specific bar coding system is specified in the purchase order, that system shall be used.

NOTE 3—In the absence of another bar coding system being specified in the purchase order, it is recommended that bar coding be consistent with AIAG Standard B-1.

20. Packing, Marking, and Loading

20.1 When specified in the purchase order, packaging, marking, and loading shall be in accordance with Practices A700.

21. Government Procurement

21.1 When specified in the contract, material shall be preserved, packaged and packed in accordance with the requirements of MIL-STD 163, with applicable levels being specified in the contract. Marking for shipment of such materials shall be in accordance with Federal Std. No. 123 for civil agencies and MIL-STD 129 or Federal Std. No. 183 if continuous marking is required.

21.2 *Inspection*—Unless otherwise specified in the contract, the manufacturer shall be responsible for the performance of all applicable inspection and test requirements specified herein. Except as otherwise specified in the contract, the manufacturer shall use its own or any other suitable facilities for the performance of such inspections and tests.

SUMMARY OF CHANGES

Committee A01 has identified the location of selected changes to this specification since the last issue, A500/A500M – 07, that may impact the use of this specification. (Approved October 1, 2009)

(*1*) Revised 1.2 to expand wall thickness range.

ASTM International takes no position respecting the validity of any patent rights asserted in connection with any item mentioned in this standard. Users of this standard are expressly advised that determination of the validity of any such patent rights, and the risk of infringement of such rights, are entirely their own responsibility.

This standard is subject to revision at any time by the responsible technical committee and must be reviewed every five years and if not revised, either reapproved or withdrawn. Your comments are invited either for revision of this standard or for additional standards and should be addressed to ASTM International Headquarters. Your comments will receive careful consideration at a meeting of the responsible technical committee, which you may attend. If you feel that your comments have not received a fair hearing you should make your views known to the ASTM Committee on Standards, at the address shown below.

This standard is copyrighted by ASTM International, 100 Barr Harbor Drive, PO Box C700, West Conshohocken, PA 19428-2959, United States. Individual reprints (single or multiple copies) of this standard may be obtained by contacting ASTM at the above address or at 610-832-9585 (phone), 610-832-9555 (fax), or service@astm.org (e-mail); or through the ASTM website (www.astm.org). Permission rights to photocopy the standard may also be secured from the ASTM website (www.astm.org/COPYRIGHT/).

Designation: A501 – 07

Standard Specification for
Hot-Formed Welded and Seamless Carbon Steel Structural Tubing[1]

This standard is issued under the fixed designation A501; the number immediately following the designation indicates the year of original adoption or, in the case of revision, the year of last revision. A number in parentheses indicates the year of last reapproval. A superscript epsilon (ε) indicates an editorial change since the last revision or reapproval.

This standard has been approved for use by agencies of the Department of Defense.

1. Scope*

1.1 This specification covers black and hot-dipped galvanized hot-formed welded and seamless carbon steel square, round, rectangular, or special shape structural tubing for welded, riveted, or bolted construction of bridges and buildings, and for general structural purposes.

1.2 Square and rectangular tubing is furnished in sizes 1 to 32 in. (25.4 to 813 mm) across flat sides with wall thicknesses 0.095 to 3.00 in. (2.41 to 76 mm), dependent upon size; round tubing is furnished in NPS ½ to NPS 24 (see Note 1) inclusive, with nominal (average) wall thicknesses 0.109 to 1.000 in. (2.77 to 25.40 mm), dependent upon size. Special shape tubing and tubing with other dimensions is permitted to be furnished, provided that such tubing complies with all other requirements of this specification.

NOTE 1—The dimensionless designator NPS (nominal pipe size) has been substituted in this standard for such traditional terms as "nominal diameter," "size," and "nominal size."

1.3 This specification covers the following grades:
1.3.1 Grade A — 36 000 psi (250 MPa) min yield strength.
1.3.2 Grade B — 50 000 psi (345 MPa) min yield strength.

1.4 An optional supplementary requirement is provided for Grade B and, when desired, shall be so stated on the order.

1.5 The following precautionary statement pertains only to the test method portion of this specification: *This standard does not purport to address all the safety concerns, if any, associated with its use. It is the responsibility of the user of this standard to establish appropriate safety and health practices and determine the applicability of regulatory limitations prior to use.*

1.6 The values stated in inch-pound units are to be regarded as standard. The values given in parentheses are mathematical conversions to SI units that are provided for information only and are not considered standard.

1.7 The text of this specification contains notes and footnotes that provide explanatory material. Such notes and footnotes, excluding those in tables and figures, do not contain any mandatory requirements.

2. Referenced Documents

2.1 *ASTM Standards:*[2]
A53/A53M Specification for Pipe, Steel, Black and Hot-Dipped, Zinc-Coated, Welded and Seamless
A370 Test Methods and Definitions for Mechanical Testing of Steel Products
A700 Practices for Packaging, Marking, and Loading Methods for Steel Products for Shipment
A751 Test Methods, Practices, and Terminology for Chemical Analysis of Steel Products
A941 Terminology Relating to Steel, Stainless Steel, Related Alloys, and Ferroalloys

2.2 *AIAG Standard:*[3]
B-1 Bar Code Symbology Standard

3. Terminology

3.1 *Definitions*—For definitions of terms used in this specification, refer to Terminology A941.

4. Ordering Information

4.1 Orders for material under this specification shall contain information concerning as many of the following items as are required to describe the desired material adequately:
4.1.1 Quantity (feet or number of lengths),
4.1.2 Name of material (hot-formed tubing),
4.1.3 Grade (A or B)
4.1.4 Method of manufacture (seamless or welded) (see Section 6),
4.1.5 Finish (black or galvanized),

---

[1] This specification is under the jurisdiction of ASTM Committee A01 on Steel, Stainless Steel and Related Alloys and is the direct responsibility of Subcommittee A01.09 on Carbon Steel Tubular Products.
Current edition approved March 1, 2007. Published April 2007. Originally approved in 1964. Last previous edition approved in 2005 as A501 – 01(2005). DOI: 10.1520/A0501-07.

[2] For referenced ASTM standards, visit the ASTM website, www.astm.org, or contact ASTM Customer Service at service@astm.org. For *Annual Book of ASTM Standards* volume information, refer to the standard's Document Summary page on the ASTM website.

[3] Available from Automotive Industry Action Group (AIAG), 26200 Lahser Rd., Suite 200, Southfield, MI 48033, http://www.aiag.org.

A Summary of Changes section appears at the end of this standard.

4.1.6 Size (outside diameter and calculated nominal wall thickness for round tubing and the outside dimensions and calculated nominal wall thickness for square and rectangular tubing (Section 11)),

4.1.7 Length (random, multiple, or specific; see 12.3),

4.1.8 End condition (see 17.3),

4.1.9 Burr removal (see 17.3),

4.1.10 Certification (see Section 19),

4.1.11 ASTM specification designation and year of issue,

4.1.12 End use,

4.1.13 Special requirements, and

4.1.14 Bar coding (see 20.3).

5. Process

5.1 The steel shall be made by one or more of the following processes: open-hearth, basic-oxygen, or electric-furnace.

5.2 When steels of different grades are sequentially strand cast, the steel producer shall identify the resultant transition material and remove it using an established procedure that positively separates the grades.

6. Manufacture

6.1 The tubing shall be made by one of the following processes: seamless; furnace-butt welding (continuous welding); electric-resistance welding or submerged arc welding followed by reheating throughout the cross section and hot forming by a reducing or shaping process, or both.

7. Heat Analysis

7.1 Each heat analysis shall conform to the requirements specified in Table 1 for heat analysis.

8. Product Analysis

8.1 The tubing shall be capable of conforming to the requirements specified in Table 1 for product analysis.

8.2 If product analyses are made, they shall be made using test specimens taken from two lengths of tubing from each lot of 500 lengths, or fraction thereof, or two pieces of flat-rolled stock from each lot of a corresponding quantity of flat-rolled stock. Methods and practices relating to chemical analysis shall be in accordance with Test Methods, Practices, and Terminology A751. Such product analyses shall conform to the requirements specified in Table 1 for product analysis.

8.3 If both product analyses representing a lot fail to conform to the specified requirements, the lot shall be rejected.

8.4 If only one product analysis representing a lot fails to conform to the specified requirements, product analyses shall be made using two additional test specimens taken from the lot. Both additional product analyses shall conform to the specified requirements or the lot shall be rejected.

9. Tensile Requirements

9.1 The material, as represented by the test specimen, shall conform to the requirements as to tensile properties prescribed in Table 2.

9.2 The yield strength corresponding to a permanent offset of 0.2 % of the gauge length of the specimen or to a total extension of 0.5 % of the gauge length under load shall be determined.

10. Charpy V-Notch Impact Test

10.1 The Charpy V-notch impact test applies to Grade B only and wall thickness greater than 0.312 in. (8 mm).

10.1.1 Charpy V-notch tests shall be made in accordance with Test Methods and Definitions A370

10.1.2 One Charpy V-notch impact test shall be made from a length of tubing representing each lot.

10.1.3 The test results of full-size longitudinal specimens shall meet an average value of 20 ft-lb at 0 °F (-18 °C).

11. Dimensions

11.1 *Square Structural Tubing*—The outside dimensions (across the flats), the weight per foot, and the calculated nominal wall thickness of common sizes of square structural tubing included in this specification are listed in Table 3.

11.2 *Rectangular Structural Tubing*—The outside dimensions (across the flats), the weight per foot, and the calculated nominal wall thickness of common sizes of rectangular structural tubing included in this specification are listed in Table 4.

11.3 *Round Structural Tubing*—The NPS and outside diameter dimensions, the weight per foot, and the calculated nominal wall thickness of common sizes of round structural tubing included in this specification are listed in Table 5.

11.4 *Special Shape Structural Tubing*—The dimensions and tolerances of special shape structural tubing are available by inquiry and negotiation with the manufacturer.

11.5 *Other Sizes*—The dimensional tolerances for hot-formed welded and seamless structural tubing manufactured in accordance with the requirements of this specification, but with ordered dimensions other than those listed in Table 3, Table 4,

TABLE 1 Chemical Requirements[A]

	Composition, %			
	Grade A		Grade B	
Element	Heat analysis	Product analysis	Heat analysis	Product analysis
Carbon, max	0.26	0.30	0.22[B]	0.26
Manganese, max	...	...	1.40[B]	1.45
Phosphorus, max	0.035	0.045	0.030	0.040
Sulfur, max	0.035	0.045	0.020	0.030
Copper, when copper steel is specified, min	0.20	0.18	0.20	0.18

[A] Where an ellipsis (...) appears in this table, there is no requirement.
[B] For each reduction of 0.01 percentage point below the specified maximum for carbon, an increase of 0.06 percentage point above the specified maximum for manganese is permitted, up to a maximum of 1.50 % by heat analysis and 1.60 % by product analysis.

TABLE 2 Tensile Requirements

	Grade A	Grade B
Tensile strength, min, psi (MPa)	58 000 (400)	70 000 (483)
Yield strength, min, psi (MPa)	36 000 (250)	50 000 (345)
Elongation in 2 in. (50.8 mm), min, %	23	23

and Table 5, shall be consistent with those given in this specification for similar sizes and type of product.

12. Permissible Variations in Dimensions of Square, Round, Rectangular, and Special Shape Structural Tubing

12.1 *Outside Dimensions*:

12.1.1 *Round Structural Tubing*—For round hot-formed structural tubing NPS 2 and over, the outside diameter shall not vary more than ±1 % from the specified outside diameter. For NPS 1½ and under, the outside diameter shall not vary more than 1/64 in. (0.40 mm) over or more than 1/32 in. (0.79 mm) under the specified outside diameter.

12.1.2 *Square, Rectangular, and Special Shape Structural Tubing*—The outside dimensions, measured across the flats at positions at least 2 in. (50.8 mm) from the ends of the tubing, shall not vary from the specified outside dimensions by more than the applicable amount given in Table 6, which includes an allowance for convexity or concavity.

12.2 *Weight*—The weight of the structural tubing shall be not more than 3.5 % under its theoretical weight, as calculated using its length and the applicable weight per unit length given in Table 3, Table 4, or Table 5.

12.3 *Length*—Structural tubing is commonly produced in random lengths of 16 to 22 ft. (4.9 to 6.7 m) or 32 to 44 ft. (9.8 to 13.4 m), in multiple lengths, and in specific lengths. When specific lengths are ordered, the permissible variations in length shall be as given in Table 7.

12.4 *Straightness*—The permissible variation for straightness of structural tubing shall be 1/8 in. times the number of feet (10.4 mm times the number of metres) of total length divided by five.

12.5 *Squareness of Sides*—For perpendicular and rectangular tubing, adjacent sides shall be square (90°), with a permissible variation of ±2°.

12.6 *Radius of Corners*—For square and rectangular structural tubing, the radius of each outside corner of the section shall not exceed three times the calculated nominal wall thickness.

12.7 *Twist*—For square, rectangular, and special shape structural tubing, the permissible variations in twist shall be as given in Table 8. Twist shall be determined by holding one end of the tubing down on a flat surface plate, measuring the height that each corner on the bottom side of the tubing extends above the surface plate near the opposite end of the tubing, and calculating the twist (the difference in the measured heights of such corners), except that for heavier sections it shall be permissible to use a suitable measuring device to determine twist. Twist measurements shall not be taken within 2 in. (50.8 mm) of the ends of the tubing.

13. Number of Tests

13.1 One tension test as specified in 15.2 shall be made from a length of tubing representing each lot.

13.2 The term "lot" shall apply to all tubes of the same specified size that are produced from the same heat of steel.

14. Retests

14.1 If the results of the mechanical tests representing any lot fail to conform to the applicable requirements specified in Sections 9 and 10, the lot shall be rejected or retested using additional tubing of double the original number from the lot. The lot shall be acceptable if the results of all such retests representing the lot conform to the specified requirements.

14.2 If one or both of the retests specified in 14.1 fail to conform to the applicable requirements specified in Sections 9 and 10, the lot shall be rejected or, subsequent to the manufacturer heat treating, reworking, or otherwise eliminating the condition responsible for the failure, the lot shall be treated as a new lot and tested accordingly.

15. Test Method

15.1 Tension test specimens shall conform to the applicable requirements of Test Methods and Definitions A370, Annex A2.

15.2 Tension test specimens shall be full-size longitudinal test specimens or longitudinal strip test specimens. For welded tubing, any longitudinal strip test specimens shall be taken from a location at least 90° from the weld and shall be prepared without flattening in the gauge length. Longitudinal strip test specimens shall have all burrs removed. Tension test specimens shall not contain surface imperfections that would interfere with proper determination of the tensile properties.

15.3 The yield strength corresponding to an offset of 0.2 % of the gauge length or to a total extension under load of 0.5 % of the gauge length shall be determined.

16. Galvanized Coatings

16.1 For structural tubing required to be hot-dipped galvanized, such coating shall comply with the requirements contained in Specification A53/A53M, except that the manufacturer shall additionally have the option of determining the coating weight using only the values obtained for the coating on the outside surface of the tubing.

17. Inspection

17.1 All tubing shall be inspected at the place of manufacture to ensure conformance with the requirements of this specification.

17.2 The structural tubing shall be free of defects and shall have a commercially smooth finish.

TABLE 3 Dimensions of Common Sizes of Square Structural Tubing

Size Given in Outside Dimensions Across Flat Sides, in. (mm)	Weight per Unit Length, lb/ft (kg/m)	Calculated Nominal Wall Thickness, in. (mm)
1 by 1 (25.4 by 25.4)	1.09 (1.62)	0.095 (2.41)
	1.41 (2.10)	0.133 (3.38)
2 by 2 (50.8 by 50.8)	2.69 (4.00)	0.110 (2.79)
	3.04 (4.52)	0.125 (3.18)
	3.65 (5.44)	0.154 (3.91)
	4.31 (6.41)	0.188 (4.78)
2½ by 2½ (63.5 by 63.5)	4.32 (6.43)	0.141 (3.58)
	5.59 (8.32)	0.188 (4.78)
	7.10 (10.56)	0.250 (6.35)
3 by 3 (76.2 by 76.2)	5.78 (8.60)	0.156 (3.96)
	6.86 (10.21)	0.188 (4.78)
	8.80 (13.09)	0.250 (6.35)
3½ by 3½ (88.9 by 88.9)	6.88 (10.24)	0.156 (3.96)
	8.14 (12.11)	0.188 (4.78)
	10.50 (15.62)	0.250 (6.35)
	12.69 (18.88)	0.312 (7.92)
4 by 4 (101.6 by 101.6)	9.31 (13.85)	0.188 (4.78)
	12.02 (17.89)	0.250 (6.35)
	14.52 (21.61)	0.312 (7.92)
	16.84 (25.06)	0.375 (9.52)
	20.88 (31.07)	0.500 (12.70)
5 by 5 (127.0 by 127.0)	11.86 (17.65)	0.188 (4.78)
	15.42 (22.94)	0.250 (6.35)
	18.77 (27.93)	0.312 (7.92)
	21.94 (32.65)	0.375 (9.52)
	27.68 (41.19)	0.500 (12.70)
6 by 6 (152.4 by 152.4)	14.41 (21.44)	0.188 (4.78)
	18.82 (28.00)	0.250 (6.35)
	23.02 (34.25)	0.312 (7.92)
	27.04 (40.28)	0.375 (9.52)
	34.48 (51.31)	0.500 (12.70)
7 by 7 (177.8 by 177.8)	16.85 (25.07)	0.188 (4.78)
	22.04 (32.80)	0.250 (6.35)
	26.99 (39.16)	0.312 (7.92)
	31.73 (47.21)	0.375 (9.52)
	40.55 (60.34)	0.500 (12.70)
8 by 8 (203.2 by 203.2)	25.44 (37.85)	0.250 (6.35)
	31.24 (46.49)	0.312 (7.92)
	36.83 (54.80)	0.375 (9.52)
	38.33 (57.03)	0.38 (9.65)
	47.35 (70.46)	0.500 (12.70)
	49.16 (73.15)	0.50 (12.70)
	56.98 (84.79)	0.625 (15.88)
	60.20 (89.57)	0.63 (16.00)
	65.73 (97.81)	0.750 (19.05)
10 by 10 (254.0 by 254.0)	32.23 (47.96)	0.250 (6.35)
	39.74 (59.13)	0.312 (7.92)
	47.03 (69.98)	0.375 (9.52)
	48.68 (72.43)	0.38 (9.65)
	60.95 (90.69)	0.500 (12.70)
	62.78 (93.41)	0.50 (12.70)
	73.98 (110.08)	0.625 (15.88)
	77.35 (115.10)	0.63 (16.00)
	86.13 (128.16)	0.750 (19.05)
	90.19 (134.19)	0.75 (19.05)
	107.79 (160.39)	1.000 (25.40)
12 by 12 (304.8 by 304.8)	76.39 (113.66)	0.50 (12.70)
	94.51 (140.62)	0.63 (16.00)
	110.61 (164.58)	0.75 (19.05)

TABLE 3 Continued

Size Given in Outside Dimensions Across Flat Sides, in. (mm)	Weight per Unit Length, lb/ft (kg/m)	Calculated Nominal Wall Thickness, in. (mm)
14 by 14 (355.6 by 355.6)	90.01 (133.92)	0.50 (12.70)
	111.66 (166.14)	0.63 (16.00)
	131.04 (194.97)	0.75 (19.05)
	140.49 (209.03)	0.81 (20.57)
	145.40 (216.35)	0.87 (22.00)
	162.18 (241.31)	0.98 (25.00)
16 by 16 (406.4 by 406.4)	103.62 (154.18)	0.50 (12.70)
	128.81 (191.66)	0.63 (16.00)
	162.52 (241.81)	0.81 (20.57)
	168.99 (251.44)	0.87 (22.00)
	188.98 (281.19)	0.98 (25.00)
	208.24 (309.84)	1.10 (28.00)
18 by 18 (457.2 by 457.2)	267.09 (397.40)	1.26 (32.00)
	294.62 (438.36)	1.42 (36.00)
	320.84 (477.38)	1.57 (40.00)
20 by 20 (508.0 by 508.0)	130.85 (194.70)	0.50 (12.70)
	163.12 (242.70)	0.63 (16.00)
	192.31 (286.13)	0.75 (19.05)
	206.66 (307.49)	0.81 (20.57)
	214.68 (319.42)	0.87 (22.00)
	240.67 (358.10)	0.98 (25.00)
	265.88 (395.60)	1.10 (28.00)
	298.26 (443.78)	1.26 (32.00)
	329.25 (489.88)	1.42 (36.00)
	358.83 (533.90)	1.57 (40.00)
	393.84 (585.99)	1.77 (45.00)
	426.66 (634.83)	1.97 (50.00)
22 by 22 (558.8 by 558.8)	177.48 (264.08)	0.63 (16.00)
	208.27 (309.88)	0.75 (19.00)
	238.27 (354.51)	0.87 (22.00)
	267.48 (397.98)	0.98 (25.00)
	295.90 (440.26)	1.10 (28.00)
	332.57 (494.83)	1.26 (32.00)
	367.84 (547.31)	1.42 (36.00)
	401.71 (597.70)	1.57 (40.00)
	442.08 (657.77)	1.77 (45.00)
	480.27 (714.58)	1.97 (50.00)
	516.26 (768.14)	2.17 (55.00)
24 by 24 (609.6 by 609.6)	194.64 (289.60)	0.63 (16.00)
	228.64 (340.19)	0.75 (19.00)
	261.85 (389.61)	0.87 (22.00)
	294.28 (437.85)	0.98 (25.00)
	325.92 (484.93)	1.10 (28.00)
	366.87 (545.87)	1.26 (32.00)
	406.43 (604.73)	1.42 (36.00)
	444.59 (661.51)	1.57 (40.00)
	490.32 (729.55)	1.77 (45.00)
	533.87 (794.34)	1.97 (50.00)
	575.22 (855.87)	2.17 (55.00)
	614.39 (914.15)	2.36 (60.00)
26 by 26 (660.4 by 660.4)	211.79 (315.12)	0.63 (16.00)
	249.01 (370.50)	0.75 (19.00)
	285.44 (424.70)	0.87 (22.00)
	321.08 (477.73)	0.98 (25.00)
	355.93 (529.59)	1.10 (28.00)
	401.18 (596.92)	1.26 (32.00)
	445.03 (662.16)	1.42 (36.00)
	487.48 (725.31)	1.57 (40.00)
	538.57 (801.33)	1.77 (45.00)
	587.47 (874.10)	1.97 (50.00)
	634.19 (943.61)	2.17 (55.00)
	678.72 (1009.86)	2.36 (60.00)
28 by 28 (711.2 by 711.2)	228.94 (340.64)	0.63 (16.00)
	269.38 (400.80)	0.75 (19.00)

TABLE 3 *Continued*

Size Given in Outside Dimensions Across Flat Sides, in. (mm)	Weight per Unit Length, lb/ft (kg/m)	Calculated Nominal Wall Thickness, in. (mm)
	309.02 (459.79)	0.87 (22.00)
	347.88 (517.61)	0.98 (25.00)
	385.95 (574.25)	1.10 (28.00)
	435.49 (647.96)	1.26 (32.00)
	483.62 (719.58)	1.42 (36.00)
	530.36 (789.12)	1.57 (40.00)
	586.81 (873.11)	1.77 (45.00)
	641.07 (953.85)	1.97 (50.00)
	693.15 (1031.34)	2.17 (55.00)
	743.04 (1105.57)	2.36 (60.00)
30 by 30 (762.0 by 762.0)	246.10 (366.17)	0.63 (16.00)
	289.75 (431.11)	0.75 (19.00)
	332.61 (494.88)	0.87 (22.00)
	374.68 (557.49)	0.98 (25.00)
	415.97 (618.92)	1.10 (28.00)
	469.79 (699.00)	1.26 (32.00)
	522.22 (777.00)	1.42 (36.00)
	573.24 (852.92)	1.57 (40.00)
	635.05 (944.89)	1.77 (45.00)
	694.68 (1033.61)	1.97 (50.00)
	752.12 (1119.07)	2.17 (55.00)
	807.36 (1201.28)	2.36 (60.00)

17.2.1 Surface imperfections shall be classed as defects when one or more of the following conditions exist:

17.2.1.1 The depth of the imperfections exceeds 15 % of the calculated nominal wall thickness.

17.2.1.2 The imperfections materially affect the appearance of the structural tubing.

17.2.1.3 At any location, the length of the imperfections, measured in the transverse direction, in combination with their depth materially reduce the total cross sectional area of the structural tubing.

17.2.2 It shall be permissible for defects having a depth not in excess of 33⅓ % of the calculated nominal wall thickness to be repaired by welding, subject to the following conditions:

17.2.2.1 The defect shall be completely removed by chipping or grinding to sound metal,

17.2.2.2 The repair weld shall be made using a low-hydrogen welding process, and

17.2.2.3 The projecting weld metal shall be removed to produce a workmanlike finish.

17.3 Unless otherwise specified in the purchase order, structural tubing shall be furnished with square cut ends. The burr shall be held to a minimum. When so specified in the purchase order, the burr shall be removed on the outside diameter, inside diameter, or both.

18. Rejection

18.1 It shall be permissible for the purchaser to inspect tubing received from the manufacturer and reject any tubing that does not meet the requirements of this specification, based upon the inspection and test methods outlined herein. The purchaser shall notify the manufacturer of any tubing that has been rejected, and the disposition of such tubing shall be subject to agreement between the manufacturer and the purchaser.

18.2 It shall be permissible for the purchaser to set aside any tubing that is found in fabrication or installation within the

TABLE 4 Dimensions of Common Sizes of Rectangular Structural Tubing

Size Given in Outside Dimensions Across Flat Sides, in. (mm)	Weight per Unit Length, lb/ft (kg/m)	Calculated Nominal Wall Thickness, in. (mm)
3 by 2 (76.2 by 50.8)	4.32 (6.43)	0.141 (3.58)
	5.59 (8.32)	0.188 (4.78)
	7.10 (10.56)	0.250 (6.35)
4 by 2 (101.6 by 50.8)	5.78 (8.60)	0.156 (3.96)
	6.86 (10.21)	0.188 (4.78)
	8.80 (13.09)	0.250 (6.35)
4 by 3 (101.6 by 76.2)	6.88 (10.24)	0.156 (3.96)
	8.14 (12.11)	0.188 (4.78)
	10.50 (15.62)	0.250 (6.35)
	12.69 (18.88)	0.312 (7.92)
5 by 3 (127.0 by 76.2)	9.31 (13.85)	0.188 (4.78)
	12.02 (17.89)	0.250 (6.35)
	14.52 (21.61)	0.312 (7.92)
	16.84 (25.06)	0.375 (9.52)
6 by 3 (152.4 by 76.2)	10.58 (15.74)	0.188 (4.78)
	13.72 (20.42)	0.250 (6.35)
	16.65 (24.78)	0.312 (7.92)
	19.39 (28.85)	0.375 (9.52)
6 by 4 (152.4 by 101.6)	11.86 (17.65)	0.188 (4.78)
	15.42 (22.94)	0.250 (6.35)
	18.77 (27.93)	0.312 (7.92)
	21.94 (32.65)	0.375 (9.52)
	27.68 (41.19)	0.500 (12.70)
7 by 5 (177.8 by 127.0)	14.41 (21.44)	0.188 (4.78)
	18.82 (28.00)	0.250 (6.35)
	23.02 (34.25)	0.312 (7.92)
	27.04 (40.28)	0.375 (9.52)
	34.48 (51.31)	0.500 (12.70)
8 by 4 (203.2 by 101.6)	14.41 (21.44)	0.188 (4.78)
	18.82 (28.00)	0.250 (6.35)
	23.02 (34.25)	0.312 (7.92)
	27.04 (40.28)	0.375 (9.52)
	34.48 (51.31)	0.500 (12.70)
8 by 6 (203.2 by 152.4)	16.85 (25.07)	0.188 (4.78)
	22.04 (32.80)	0.250 (6.35)
	26.99 (39.16)	0.312 (7.92)
	31.73 (47.21)	0.375 (9.52)
	33.20 (49.39)	0.38 (9.65)
	40.55 (60.34)	0.500 (12.70)
	42.41 (60.34)	0.500 (12.70)
10 by 4 (254.0 by 101.6)	42.35 (63.02)	0.50 (12.70)
	51.62 (76.81)	0.63 (16.00)
10 by 6 (254.0 by 152.4)	25.44 (37.85)	0.250 (6.35)
	31.24 (46.49)	0.312 (7.92)
	36.83 (54.80)	0.375 (9.52)
	38.33 (57.03)	0.38 (9.65)
	47.35 (70.46)	0.500 (12.70)
	49.16 (73.15)	0.50 (12.70)
	60.20 (89.57)	0.63 (16.00)
12 by 8 (304.8 by 203.2)	62.78 (93.41)	0.50 (12.70)
	77.35 (115.10)	0.63 (16.00)
16 by 8 (406.4 by 203.2)	76.39 (113.66)	0.50 (12.70)
	94.51 (140.62)	0.63 (16.00)
18 by 10 (457.2 by 254.0)	90.01 (133.92)	0.50 (12.70)
	111.66 (166.14)	0.63 (16.00)
20 by 12 (508.0 by 304.8)	103.62 (154.18)	0.50 (12.70)
	128.81 (191.66)	0.63 (16.00)

TABLE 4 Continued

Size Given in Outside Dimensions Across Flat Sides, in. (mm)	Weight per Unit Length, lb/ft (kg/m)	Calculated Nominal Wall Thickness, in. (mm)	
22 by 14 (550.0 by 350.0)	107.66 (160.19)	0.47	(12.00)
	140.75 (209.42)	0.63	(16.00)
	164.64 (244.97)	0.75	(19.00)
	187.75 (279.36)	0.87	(22.00)
	210.07 (312.57)	0.98	(25.00)
	231.61 (344.61)	1.10	(28.00)
24 by 16 (600.0 by 400.0)	120.32 (179.03)	0.47	(12.00)
	157.63 (234.54)	0.63	(16.00)
	184.69 (274.80)	0.75	(19.00)
	210.97 (313.90)	0.87	(22.00)
	236.45 (351.82)	0.98	(25.00)
	261.15 (388.57)	1.10	(28.00)
	292.86 (435.75)	1.26	(32.00)
	323.17 (480.84)	1.42	(36.00)
	352.08 (523.85)	1.57	(40.00)
26 by 18 (650.0 by 450.0)	174.51 (259.66)	0.63	(16.00)
	204.74 (304.63)	0.75	(19.00)
	234.18 (348.44)	0.87	(22.00)
	262.83 (391.07)	0.98	(25.00)
	290.70 (432.53)	1.10	(28.00)
	326.63 (485.99)	1.26	(32.00)
	361.15 (537.36)	1.42	(36.00)
	394.28 (586.65)	1.57	(40.00)
30 by 20 (750.0 by 500.0)	199.84 (297.34)	0.63	(16.00)
	234.81 (349.38)	0.75	(19.00)
	269.00 (400.25)	0.87	(22.00)
	302.40 (449.94)	0.98	(25.00)
	335.02 (498.47)	1.10	(28.00)
	377.27 (561.35)	1.26	(32.00)
	418.13 (622.14)	1.42	(36.00)
	457.59 (680.85)	1.57	(40.00)

TABLE 5 Dimensions of Common Sizes of Round Structural Tubing

NPS Designator	Outside Diameter, in. (mm)	Weight per Unit Length, lb/ft (kg/m)	Calculated Nominal Wall Thickness, in. (mm)
½	0.840 (21.3)	0.85 (1.27)	0.109 (2.77)
	0.840 (21.3)	1.09 (1.62)	0.147 (3.73)
¾	1.050 (26.7)	1.13 (1.69)	0.113 (2.87)
	1.050 (26.7)	1.47 (2.20)	0.154 (3.91)
1	1.315 (33.4)	1.34 (2.00)	0.104 (2.64)
	1.315 (33.4)	1.68 (2.50)	0.133 (3.38)
	1.315 (33.4)	2.17 (3.24)	0.179 (4.55)
1¼	1.660 (42.2)	1.82 (2.71)	0.110 (2.79)
	1.660 (42.2)	2.27 (3.39)	0.140 (3.56)
	1.660 (42.2)	3.00 (4.47)	0.191 (4.85)
1½	1.900 (48.3)	2.17 (3.25)	0.114 (2.90)
	1.900 (48.3)	2.72 (4.05)	0.145 (3.68)
	1.900 (48.3)	3.63 (5.41)	0.200 (5.08)
2	2.375 (60.3)	2.91 (4.33)	0.121 (3.07)
	2.375 (60.3)	3.65 (5.44)	0.154 (3.91)
	2.375 (60.3)	5.02 (7.48)	0.218 (5.54)
2½	2.875 (73.0)	4.53 (6.74)	0.156 (3.96)
	2.875 (73.0)	5.40 (8.04)	0.188 (4.78)
	2.875 (73.0)	5.79 (8.63)	0.203 (5.16)
	2.875 (73.0)	7.66 (11.41)	0.276 (7.01)
3	3.500 (88.9)	5.57 (8.29)	0.156 (3.96)
	3.500 (88.9)	6.65 (9.92)	0.188 (4.78)
	3.500 (88.9)	7.58 (11.29)	0.216 (5.49)
	3.500 (88.9)	10.25 (15.27)	0.300 (7.62)
3½	4.000 (101.6)	6.40 (9.53)	0.156 (3.96)
	4.000 (101.6)	7.65 (11.41)	0.188 (4.78)
	4.000 (101.6)	9.11 (13.57)	0.226 (5.74)
	4.000 (101.6)	12.50 (18.63)	0.318 (8.08)
4	4.500 (114.3)	7.24 (10.78)	0.156 (3.96)
	4.500 (114.3)	8.66 (12.91)	0.188 (4.78)
	4.500 (114.3)	10.01 (14.91)	0.219 (5.56)
	4.500 (114.3)	10.79 (16.07)	0.237 (6.02)
	4.500 (114.3)	14.98 (22.32)	0.337 (8.56)
5	5.563 (141.3)	14.62 (21.77)	0.258 (6.55)
	5.563 (141.3)	20.78 (30.97)	0.375 (9.53)
	5.563 (141.3)	38.55 (57.43)	0.750 (19.05)
6	6.625 (168.3)	18.97 (28.26)	0.280 (7.11)
	6.625 (168.3)	28.57 (42.56)	0.432 (10.97)
	6.625 (168.3)	53.16 (79.22)	0.864 (21.95)
8	8.625 (219.1)	28.55 (42.55)	0.322 (8.18)
	8.625 (219.1)	43.39 (64.64)	0.500 (12.70)
	8.625 (219.1)	72.42 (107.92)	0.875 (22.23)
10	10.750 (273.0)	40.48 (60.31)	0.365 (9.27)
	10.750 (273.0)	54.74 (81.55)	0.500 (12.70)
	10.750 (273.0)	104.13 (155.15)	1.000 (25.40)
12	12.750 (323.8)	49.56 (73.88)	0.375 (9.53)
	12.750 (323.8)	65.42 (97.46)	0.500 (12.70)
	12.750 (323.8)	125.49 (186.97)	1.000 (25.40)
14	14.000 (355.6)	54.57 (81.33)	0.375 (9.53)
	14.000 (355.6)	72.09 (107.39)	0.500 (12.70)
16	16.000 (406.4)	62.58 (93.27)	0.375 (9.53)
	16.000 (406.4)	82.77 (123.30)	0.500 (12.70)
18	18.000 (457.2)	70.59 (105.16)	0.375 (9.53)
	18.000 (457.2)	93.45 (139.15)	0.500 (12.70)
20	20.000 (508.0)	78.60 (117.15)	0.375 (9.53)
	20.000 (508.0)	104.13 (155.12)	0.500 (12.70)
24	24.000 (609.6)	94.62 (141.12)	0.375 (9.53)
	24.000 (609.6)	125.49 (187.06)	0.500 (12.70)

scope of this specification to be unsuitable for the intended end use, based on the requirements of this specification. The purchaser shall notify the manufacturer of any tubing that has been set aside. Such tubing shall be subject to mutual investigation as to the nature and severity of the deficiency and the forming or installation conditions, or both, involved. The disposition of such tubing shall be subject to agreement between the manufacturer and the purchaser.

19. Certification

19.1 When specified in the purchase order or contract, the manufacturer shall furnish to the purchaser a certificate of compliance stating that the product was manufactured, sampled, tested, and inspected in accordance with this specification and any other requirements designated in the purchase order or contract, and was found to meet all such requirements. Certificates of compliance shall include the specification number and year of issue.

19.2 When specified in the purchase order or contract, the manufacturer shall furnish to the purchaser test reports for the product shipped that contain the heat analyses and the results of the tension tests required by this specification and the purchase order or contract. Test reports shall include the specification number and year of issue.

19.3 A signature or notarization is not required on certificates of compliance or test reports; however, the documents shall clearly identify the organization submitting them. Notwithstanding the absence of a signature, the organization submitting the document is responsible for its content.

19.4 A certificate of compliance or test report printed from, or used in electronic form from, an electronic data interchange (EDI) shall be regarded as having the same validity as a counterpart printed in the certifying organization's facility. The

TABLE 6 Permissible Variations in Outside Flat Dimensions for Square, Rectangular, and Special Shape Structural Tubing

Specified Outside Large Flat Dimension, in. (mm)	Permissible Variations Over and Under Specified Outside Flat Dimensions,[A] in. (mm)
2½ (63.5) and under	0.020 (0.51)
Over 2½ to 3½ (63.5 to 88.9), incl	0.025 (0.64)
Over 3½ to 5½ (88.9 to 139.7), incl	0.030 (0.76)
Over 5½ (139.7) to 10 (254), incl	0.01 times large flat dimension
Over 10 (254)	0.02 times large flat dimension

[A] The permissible variations include allowances for convexity and concavity.

TABLE 7 Permissible Variations in Length for Specific Lengths of Structural Tubing

	22 ft (6.7 m) and Under		Over 22 to 44 ft (6.7 to 13.4 m), incl	
	Over	Under	Over	Under
Permissible variations in length for specific lengths, in. (mm)	½ (12.7)	¼ (6.4)	¾ (19.0)	¼ (6.4)

TABLE 8 Permissible Variations in Twist for Square, Rectangular, and Special Shape Structural Tubing

Specified Outside Large Flat Dimension, in. (mm)	Maximum Permissible Variations in Twist per 3 ft of Length (Twist per Metre of Length)	
	in.	mm
1½ (38.1) and under	0.050	1.39
Over 1½ to 2½ (38.1 to 63.5), incl	0.062	1.72
Over 2½ to 4 (63.5 to 101.6), incl	0.075	2.09
Over 4 to 6 (101.6 to 152.4), incl	0.087	2.42
Over 6 to 8 (152.4 to 203.2), incl	0.100	2.78
Over 8 (203.2)	0.112	3.11

content of the EDI transmitted document shall conform to any existing EDI agreement between the purchaser and the manufacturer.

20. Product Marking

20.1 Except as allowed by 20.2, each length of structural tubing shall be legibly marked by rolling, die-stamping, ink printing, or paint stenciling to show the following information: manufacturer's name, brand, or trademark; size; and the specification designation (year of issue not required).

20.2 For structural tubing having a specified outside diameter or large flat dimension less than 2 in. (50.8 mm), it shall be permissible for the information listed in 20.1 to be marked on a tag securely attached to each bundle.

20.3 *Bar Coding*—In addition to the requirements in 20.1 and 20.2, the manufacturer shall have the option of using bar coding as a supplementary identification method. When a specific bar coding system is specified in the purchase order, that system shall be used.

NOTE 2—In the absence of another bar coding system being specified in the purchase order, it is recommended that bar coding be consistent with AIAG Standard B-1.

21. Packaging, Marking, and Loading

21.1 When specified in the purchase order, packaging, marking, and loading shall be in accordance with Practices A700.

22. Keywords

22.1 steel tube; structural steel tubing

SUPPLEMENTARY REQUIREMENTS

The following supplementary requirement shall apply only when specified by the purchaser in the inquiry, contract, or order.

S1. Weld Line Integrity Evaluation

S1.1 The weld line integrity evaluation applies to Grade B only.

If NDT of the weld line is an express requirement of an order, 100 % of the weld line shall be subjected to nondestructive testing via eddy current or ultrasonic techniques. The technique, as well as acceptance criteria, shall be agreed upon by the purchaser and manufacturer, and specified on the order.

NOTE S1—Eddy current equipment usage is limited by its maximum thickness measurement capabilities.

A501 – 07

SUMMARY OF CHANGES

Committee A01 has identified the location of selected changes to this specification since the last issue, A501 – 01(2005), that may impact the use of this specification.

(*1*) Revised 1.2. Added new 1.3 and 1.4 and renumbered subsequent paragraphs.
(*2*) Added new 4.1.3 and renumbered subsequent paragraphs. Revised 6.1.
(*3*) Replaced old Section 10 Bend Test with new Section 10 Charpy V-Notch Impact Test.
(*4*) Added Supplementary Requirement.
(*5*) Deleted old Table 3 and renumbered subsequent tables. Revised Table 1, Table 2, Table 3, Table 4, and Table 6.

ASTM International takes no position respecting the validity of any patent rights asserted in connection with any item mentioned in this standard. Users of this standard are expressly advised that determination of the validity of any such patent rights, and the risk of infringement of such rights, are entirely their own responsibility.

This standard is subject to revision at any time by the responsible technical committee and must be reviewed every five years and if not revised, either reapproved or withdrawn. Your comments are invited either for revision of this standard or for additional standards and should be addressed to ASTM International Headquarters. Your comments will receive careful consideration at a meeting of the responsible technical committee, which you may attend. If you feel that your comments have not received a fair hearing you should make your views known to the ASTM Committee on Standards, at the address shown below.

This standard is copyrighted by ASTM International, 100 Barr Harbor Drive, PO Box C700, West Conshohocken, PA 19428-2959, United States. Individual reprints (single or multiple copies) of this standard may be obtained by contacting ASTM at the above address or at 610-832-9585 (phone), 610-832-9555 (fax), or service@astm.org (e-mail); or through the ASTM website (www.astm.org).

Designation: A514/A514M – 05 (Reapproved 2009)

Standard Specification for
High-Yield-Strength, Quenched and Tempered Alloy Steel Plate, Suitable for Welding[1]

This standard is issued under the fixed designation A514/A514M; the number immediately following the designation indicates the year of original adoption or, in the case of revision, the year of last revision. A number in parentheses indicates the year of last reapproval. A superscript epsilon (ε) indicates an editorial change since the last revision or reapproval.

This standard has been approved for use by agencies of the Department of Defense.

1. Scope

1.1 This specification covers quenched and tempered alloy steel plates of structural quality in thicknesses of 6 in. [150 mm] and under intended primarily for use in welded bridges and other structures.

NOTE 1—All grades are not available in a maximum thickness of 6 in. [150 mm]. See Table 1 for thicknesses available in each grade.

1.2 If the steel is to be welded, it is presupposed that a welding procedure suitable for the grade of steel and intended use or service will be utilized. See Appendix X 3 of Specification A6/A6M for information on weldability.

1.3 The values stated in either inch-pound units or SI units are to be regarded separately as standard. Within the text, the SI units are shown in brackets. The values stated in each system are not exact equivalents; therefore, each system is to be used independently of the other, without combining values in any way.

2. Referenced Documents

2.1 *ASTM Standards:*[2]

A6/A6M Specification for General Requirements for Rolled Structural Steel Bars, Plates, Shapes, and Sheet Piling

A370 Test Methods and Definitions for Mechanical Testing of Steel Products

3. General Requirements for Delivery

3.1 Plates furnished under this specification shall conform to the applicable requirements of the current edition of Specification A6/A6M unless a conflict exists in which case this specification shall prevail.

4. Materials and Manufacture

4.1 The steel shall be killed and conform to the requirements for fine austenitic grain size in Specification A6/A6M.

5. Heat Treatment

5.1 Except as allowed by 5.2, plates shall be heat treated to conform to the tensile and hardness requirements given in Table 2 by heating to not less than 1650°F [900°C], quenching in water or oil, and tempering at not less than 1150°F [620°C]. The heat-treatment temperatures shall be reported in the test report.

5.2 Plates ordered without the heat treatment specified in 5.1 shall be stress relieved by the manufacturer, and subsequent heat treatment of the plates to conform to 5.1 shall be the responsibility of the purchaser.

6. Chemical Composition

6.1 The heat analysis shall conform to the requirements given in Table 1.

6.2 The product analysis shall conform to the requirements given in Table 1, subject to the product analysis tolerances in Specification A6/A6M.

7. Mechanical Properties

7.1 *Tension Test*—The plates as represented by the tension test specimens shall conform to the tensile requirements given in Table 2.

7.2 *Hardness Test*—For plates 3/8 in. [10 mm] and under in thickness, a Brinell hardness test may be used instead of tension testing each plate, in which case a tension test shall be made from a corner of each of two plates per lot. A lot shall consist of plates from the same heat, thickness, prior condition, and scheduled heat treatment and shall not exceed 15 tons [15 Mg] in weight [mass]. A Brinell hardness test shall be made on each plate not tension tested and the results shall conform to the hardness requirements given in Table 2.

---

[1] This specification is under the jurisdiction of ASTM Committee A01 on Steel, Stainless Steel and Related Alloys and is the direct responsibility of Subcommittee A01.02 on Structural Steel for Bridges, Buildings, Rolling Stock and Ships.
Current edition approved Oct. 1, 2009. Published December 2009. Originally approved in 1964. Last previous edition approved in 2005 as A514/A514M – 05. DOI: 10.1520/A0514_A0514M-05R09.

[2] For referenced ASTM standards, visit the ASTM website, www.astm.org, or contact ASTM Customer Service at service@astm.org. For *Annual Book of ASTM Standards* volume information, refer to the standard's Document Summary page on the ASTM website.

A514/A514M − 05 (2009)

TABLE 1 Chemical Requirements (Heat Analysis)

Note 1—Where "..." appears in this table, there is no requirement.

	Chemical Composition, %							
	Grade A	Grade B	Grade E	Grade F	Grade H	Grade P	Grade Q	Grade S
	Maximum Thickness, in. [mm]							
Element	1¼ [32]	1¼ [32]	6 [150]	2½ [65]	2 [50]	6 [150]	6 [150]	2½ [65]
Carbon	0.15–0.21	0.12–0.21	0.12–0.20	0.10–0.20	0.12–0.21	0.12–0.21	0.14–0.21	0.11–0.21
Manganese	0.80–1.10	0.70–1.00	0.40–0.70	0.60–1.00	0.95–1.30	0.45–0.70	0.95–1.30	1.10–1.50
Phosphorus, max	0.035	0.035	0.035	0.035	0.035	0.035	0.035	0.035
Sulfur, max	0.035	0.035	0.035	0.035	0.035	0.035	0.035	0.020
Silicon	0.40–0.80	0.20–0.35	0.20–0.40	0.15–0.35	0.20–0.35	0.20–0.35	0.15–0.35	0.15–0.45
Nickel	...	...	...	0.70–1.00	0.30–0.70	1.20–1.50	1.20–1.50	...
Chromium	0.50–0.80	0.40–0.65	1.40–2.00	0.40–0.65	0.40–0.65	0.85–1.20	1.00–1.50	...
Molybdenum	0.18–0.28	0.15–0.25	0.40–0.60	0.40–0.60	0.20–0.30	0.45–0.60	0.40–0.60	0.10–0.60
Vanadium	...	0.03–0.08	[A]	0.03–0.08	0.03–0.08	...	0.03–0.08	0.06
Titanium	...	0.01–0.04	0.01–0.10	...	...	...	...	[B]
Zirconium	0.05–0.15[C]	...	...	...	...	...	...	...
Copper	...	...	...	0.15–0.50	...	...	...	...
Boron	0.0025 max	0.0005–0.005	0.001–0.005	0.0005–0.006	0.0005–0.005	0.001–0.005	...	0.001–0.005
Columbium, max	...	...	...	...	...	...	...	0.06

[A] May be substituted for part or all of titanium content on a one for one basis.
[B] Titanium may be present in levels up to 0.06 % to protect the boron additions.
[C] Zirconium may be replaced by cerium. When cerium is added, the cerium/sulfur ratio should be approximately 1.5 to 1, based upon heat analysis.

TABLE 2 Tensile and Hardness Requirements

Note 1— See the Orientation and Preparation subsections in the Tension Tests section of Specification A6/A6M.
Note 2—Where "..." appears in this table there is no requirement.

Thickness, in. [mm]	Tensile Strength, ksi [MPa]	Yield Strength, min[A], ksi [MPa]	Elongation in 2 in. [50 mm], min[BCD], %	Reduction of Area, min[BC], %	Brinell Hardness Number[E]
To ¾ [20], incl	110 to 130 [760 to 895]	100 [690]	18	40[F]	235 to 293 HBW
Over ¾ [20] to 2½ [65], incl	110 to 130 [760 to 895]	100 [690]	18	40[F], 50[G]	...
Over 2½ [65] to 6 [150], incl	100 to 130 [690 to 895]	90 [620]	16	50[G]	...

[A] Measured at 0.2 % offset or 0.5 % extension under load as described in the Determination of Tensile Properties section of Test Methods and Definitions A370.
[B] Elongation and reduction of area need not be determined for floor plates.
[C] For plates tested in the transverse direction, the elongation requirement is reduced by two percentage points and the reduction of area minimum requirement is reduced by five percentage points. See elongation requirement adjustments in the Tension Tests section of Specification A6/A6M.
[D] If measured on the Fig. 3 (Test Methods and Definitions A370) 1½-in. [40-mm] wide tension test specimen, the elongation is determined in a 2-in. [50-mm] gage length that includes the fracture and shows the greatest elongation.
[E] See 7.2.
[F] If measured on the Fig. 3 (Test Methods and Definitions A370) 1½-in. [40-mm] wide tension test specimen.
[G] If measured on the Fig. 4 (Test Methods and Definitions A370) ½-in. [12.5-mm] round tension test specimen.

8. Number of Tests

8.1 Except as allowed by 7.2, one tension test shall be taken from a corner of each plate as heat treated.

9. Retest

9.1 Plates that were subjected to Brinell hardness testing and failed to conform the specified hardness requirements may be subjected, at the manufacturer's option, to tension testing and shall be accepted if the results conform to the tensile requirements given in Table 2.

9.2 The manufacturer may re-heat treat plates that fail to meet the mechanical property requirements of this specification. All mechanical property tests shall be repeated after such heat treatment.

10. Test Specimens

10.1 If possible, all test specimens shall be cut from the plate in its heat-treated condition. If it is necessary to prepare test specimens from separate pieces, such pieces shall be full thickness, and shall be similarly and simultaneously heat treated with the plate. All such separate pieces shall be of such a size that the prepared test specimens are free of any variation in properties due to edge effects.

10.2 If specified in the purchase order, the test pieces shall be subjected to additional thermal treatments intended to simulate thermal treatments that subsequently might be done by the fabricator.

11. Keywords

11.1 alloy; bridges; high-yield-strength; plates; quenched; steel; structural steel; tempered; welded construction

SUPPLEMENTARY REQUIREMENTS

Supplementary requirements shall not apply unless specified in the purchase order or contract. Standardized supplementary requirements for use at the option of the purchaser are listed in Specification A6/A6M. Those that are considered suitable for use with this specification are listed by title:

S5. Charpy V-Notch Impact Test

S8. Ultrasonic Examination

Designation: A516/A516M – 06

Used in USDOE-NE Standards

Standard Specification for
Pressure Vessel Plates, Carbon Steel, for Moderate- and Lower-Temperature Service[1]

This standard is issued under the fixed designation A516/A516M; the number immediately following the designation indicates the year of original adoption or, in the case of revision, the year of last revision. A number in parentheses indicates the year of last reapproval. A superscript epsilon (ε) indicates an editorial change since the last revision or reapproval.

This standard has been approved for use by agencies of the Department of Defense.

1. Scope*

1.1 This specification[2] covers carbon steel plates intended primarily for service in welded pressure vessels where improved notch toughness is important.

1.2 Plates under this specification are available in four grades having different strength levels as follows:

Grade U.S. [SI]	Tensile Strength, ksi [MPa]
55 [380]	55–75 [380–515]
60 [415]	60–80 [415–550]
65 [450]	65–85 [450–585]
70 [485]	70–90 [485–620]

1.3 The maximum thickness of plates is limited only by the capacity of the composition to meet the specified mechanical property requirements; however, current practice normally limits the maximum thickness of plates furnished under this specification as follows:

Grade U.S. [SI]	Maximum Thickness, in. [mm]
55 [380]	12 [305]
60 [415]	8 [205]
65 [450]	8 [205]
70 [485]	8 [205]

1.4 For plates produced from coil and furnished without heat treatment or with stress relieving only, the additional requirements, including additional testing requirements and the reporting of additional test results of Specification A20/A20M apply.

1.5 The values stated in either inch-pound units or SI units are to be regarded separately as standard. Within the text, the SI units are shown in brackets. The values stated in each system are not exact equivalents; therefore, each system must be used independently of the other. Combining values from the two systems may result in nonconformance with the specification.

2. Referenced Documents

2.1 *ASTM Standards:*[3]

A20/A20M Specification for General Requirements for Steel Plates for Pressure Vessels

A435/A435M Specification for Straight-Beam Ultrasonic Examination of Steel Plates

A577/A577M Specification for Ultrasonic Angle-Beam Examination of Steel Plates

A578/A578M Specification for Straight-Beam Ultrasonic Examination of Rolled Steel Plates for Special Applications

3. General Requirements and Ordering Information

3.1 Plates supplied to this product specification shall conform to Specification A20/A20M, which outlines the testing and retesting methods and procedures, permissible variations in dimensions and mass, quality and repair of defects, marking, loading, and so forth.

3.2 Specification A20/A20M also establishes the rules for ordering information that should be complied with when purchasing plates to this specification.

3.3 In addition to the basic requirements of this specification, certain supplementary requirements are available where additional control, testing, or examination is required to meet end use requirements.

3.4 The purchaser is referred to the listed supplementary requirements in this specification and to the detailed requirements in Specification A20/A20M.

3.5 Coils are excluded from qualification to this specification until they are processed into finished plates. Plates produced from coil means plates that have been cut to individual lengths from coil. The processor directly controls, or

---

[1] This specification is under the jurisdiction of ASTM Committee A01 on Steel, Stainless Steel and Related Alloys and is the direct responsibility of Subcommittee A01.11 on Steel Plates for Boilers and Pressure Vessels.
Current edition approved March 1, 2006. Published March 2006. Originally approved in 1964. Last previous edition approved in 2005 as A516/A516M – 05ε1. DOI: 10.1520/A0516_A0516M-06.

[2] For ASME Boiler and Pressure Vessel Code applications, see related Specification SA-516/SA-516M in Section II of that Code.

[3] For referenced ASTM standards, visit the ASTM website, www.astm.org, or contact ASTM Customer Service at service@astm.org. For *Annual Book of ASTM Standards* volume information, refer to the standard's Document Summary page on the ASTM website.

*A Summary of Changes section appears at the end of this standard.

Copyright © ASTM International, 100 Barr Harbor Drive, PO Box C700, West Conshohocken, PA 19428-2959, United States.

is responsible for, the operations involved in the processing of coils into finished plates. Such operations include decoiling, leveling, cutting to length, testing, inspection, conditioning, heat treatment (if applicable), packaging, marking, loading for shipment, and certification.

NOTE 1—For plates produced from coil and furnished without heat treatment or with stress relieving only, three test results are reported for each qualifying coil. Additional requirements regarding plate produced from coil are described in Specification A20/A20M.

3.6 If the requirements of this specification are in conflict with the requirements of Specification A20/A20M, the requirements of this specification shall prevail.

4. Materials and Manufacture

4.1 *Steelmaking Practice*—The steel shall be killed and shall conform to the fine austenitic grain size requirement of Specification A20/A20M.

5. Heat Treatment

5.1 Plates 1.50 in. [40 mm] and under in thickness are normally supplied in the as-rolled condition. The plates may be ordered normalized or stress relieved, or both.

5.2 Plates over 1.50 in. [40 mm] in thickness shall be normalized.

5.3 When notch-toughness tests are required on plates 1½ in. [40 mm] and under in thickness, the plates shall be normalized unless otherwise specified by the purchaser.

5.4 If approved by the purchaser, cooling rates faster than those obtained by cooling in air are permissible for improvement of the toughness, provided the plates are subsequently tempered in the temperature range 1100 to 1300 °F [595 to 705 °C].

6. Chemical Composition

6.1 The steel shall conform to the chemical requirements given in Table 1 unless otherwise modified in accordance with Supplementary Requirement S17, Vacuum Carbon-Deoxidized Steel, in Specification A20/A20M.

7. Mechanical Properties

7.1 *Tension Test*—The plates, as represented by the tension test specimens, shall conform to the requirements given in Table 2.

8. Keywords

8.1 carbon steel; carbon steel plate; pressure containing parts; pressure vessel steels; steel plates for pressure vessels

TABLE 1 Chemical Requirements

Elements	Composition, %			
	Grade 55 [Grade 380]	Grade 60 [Grade 415]	Grade 65 [Grade 450]	Grade 70 [Grade 485]
Carbon, max[A,B]:				
½ in. [12.5 mm] and under	0.18	0.21	0.24	0.27
Over ½ in. to 2 in. [12.5 to 50 mm], incl	0.20	0.23	0.26	0.28
Over 2 in. to 4 in. [50 to 100 mm], incl	0.22	0.25	0.28	0.30
Over 4 to 8 in. [100 to 200 mm], incl	0.24	0.27	0.29	0.31
Over 8 in. [200 mm]	0.26	0.27	0.29	0.31
Manganese[B]:				
½ in. [12.5 mm] and under:				
Heat analysis	0.60–0.90	0.60–0.90[C]	0.85–1.20	0.85–1.20
Product analysis	0.55–0.98	0.55–0.98[C]	0.79–1.30	0.79–1.30
Over ½ in. [12.5 mm]:				
Heat analysis	0.60–1.20	0.85–1.20	0.85–1.20	0.85–1.20
Product analysis	0.55–1.30	0.79–1.30	0.79–1.30	0.79–1.30
Phosphorus, max[A]	0.035	0.035	0.035	0.035
Sulfur, max[A]	0.035	0.035	0.035	0.035
Silicon:				
Heat analysis	0.15–0.40	0.15–0.40	0.15–0.40	0.15–0.40
Product analysis	0.13–0.45	0.13–0.45	0.13–0.45	0.13–0.45

[A] Applies to both heat and product analyses.
[B] For each reduction of 0.01 percentage point below the specified maximum for carbon, an increase of 0.06 percentage point above the specified maximum for manganese is permitted, up to a maximum of 1.50 % by heat analysis and 1.60 % by product analysis.
[C] Grade 60 plates ½ in. [12.5 mm] and under in thickness may have 0.85–1.20 % manganese on heat analysis, and 0.79–1.30 % manganese on product analysis.

TABLE 2 Tensile Requirements

	Grade			
	55 [380]	60 [415]	65 [450]	70 [485]
Tensile strength, ksi [MPa]	55–75 [380–515]	60–80 [415–550]	65–85 [450–585]	70–90 [485–620]
Yield strength, min,[A] ksi [MPa]	30 [205]	32 [220]	35 [240]	38 [260]
Elongation in 8 in. [200 mm], min, %[B]	23	21	19	17
Elongation in 2 in. [50 mm], min, %[B]	27	25	23	21

[A] Determined by either the 0.2 % offset method or the 0.5 % extension-under-load method.
[B] See Specification A20/A20M for elongation adjustment.

SUPPLEMENTARY REQUIREMENTS

Supplementary requirements shall not apply unless specified in the purchase order.

A list of standardized supplementary requirements for use at the option of the purchaser is included in ASTM Specification A20/A20M. Those that are considered suitable for use with this specification are listed below by title.

S1. Vacuum Treatment,

S2. Product Analysis,

S3. Simulated Post-Weld Heat Treatment of Mechanical Test Coupons,

S4.1 Additional Tension Test,

S5. Charpy V-Notch Impact Test,

S6. Drop Weight Test (for Material 0.625 in. [16 mm] and over in Thickness),

S7. High-Temperature Tension Test,

S8. Ultrasonic Examination in accordance with Specification A435/A435M,

S9. Magnetic Particle Examination,

S11. Ultrasonic Examination in accordance with Specification A577/A577M,

S12. Ultrasonic Examination in accordance with Specification A578/A578M, and

S17. Vacuum Carbon-Deoxidized Steel.

ADDITIONAL SUPPLEMENTARY REQUIREMENTS

In addition, the following supplementary requirement is suitable for this application.

S54. Requirements for Carbon Steel Plate for Hydrofluoric Acid Alkylation Service

S54.1 Plates shall be provided in the normalized heat-treated condition.

S54.2 The maximum carbon equivalent shall be as follows:
Plate thickness less than or equal to 1 in. [25 mm]: CE maximum = 0.43
Plate thickness greater than 1 in. [25 mm]: CE maximum = 0.45

S54.3 Determine the carbon equivalent (CE) as follows:

$$CE = C + Mn/6 + (Cr + Mo + V)/5 + (Ni + Cu)/15$$

S54.4 Vanadium and niobium maximum content based on heat analysis shall be:
Maximum vanadium = 0.02 %
Maximum niobium = 0.02 %
Maximum vanadium plus niobium = 0.03 %
(Note: niobium = columbium)

S54.5 The maximum composition based on heat analysis of Ni + Cu shall be 0.15 %.

S54.6 The minimum C content based on heat analysis shall be 0.18 %. The maximum C content shall be as specified for the ordered grade.

S54.7 Welding consumables for repair welds shall be of the low-hydrogen type. E60XX electrodes shall not be used and the resulting weld chemistry shall meet the same chemistry requirements as the base metal.

S54.8 In addition to the requirements for product marking in the specification, an "HF-N" stamp or marking shall be provided on each plate to identify that the plate complies with this supplementary requirement.

A516/A516M – 06

SUMMARY OF CHANGES

Committee A01 has identified the location of selected changes to this standard since the last issue (A516/A516M– 05$^{\varepsilon1}$) that may impact the use of this standard. (Approved March 1, 2006.)

(1) Footnote C was added to Table 1.

Committee A01 has identified the location of selected changes to this standard since the last issue (A516/A516M– 03) that may impact the use of this standard. (Approved March 1, 2005.)

(1) Keywords were added.
(2) Supplementary Requirement S1, for plates for HF alkylation service, was added.
(3) Table 1 was corrected editorially.

ASTM International takes no position respecting the validity of any patent rights asserted in connection with any item mentioned in this standard. Users of this standard are expressly advised that determination of the validity of any such patent rights, and the risk of infringement of such rights, are entirely their own responsibility.

This standard is subject to revision at any time by the responsible technical committee and must be reviewed every five years and if not revised, either reapproved or withdrawn. Your comments are invited either for revision of this standard or for additional standards and should be addressed to ASTM International Headquarters. Your comments will receive careful consideration at a meeting of the responsible technical committee, which you may attend. If you feel that your comments have not received a fair hearing you should make your views known to the ASTM Committee on Standards, at the address shown below.

This standard is copyrighted by ASTM International, 100 Barr Harbor Drive, PO Box C700, West Conshohocken, PA 19428-2959, United States. Individual reprints (single or multiple copies) of this standard may be obtained by contacting ASTM at the above address or at 610-832-9585 (phone), 610-832-9555 (fax), or service@astm.org (e-mail); or through the ASTM website (www.astm.org).

Designation: A517/A517M – 06

Standard Specification for
Pressure Vessel Plates, Alloy Steel, High-Strength, Quenched and Tempered[1]

This standard is issued under the fixed designation A517/A517M; the number immediately following the designation indicates the year of original adoption or, in the case of revision, the year of last revision. A number in parentheses indicates the year of last reapproval. A superscript epsilon (ε) indicates an editorial change since the last revision or reapproval.

This standard has been approved for use by agencies of the Department of Defense.

1. Scope*

1.1 This specification[2] covers high-strength quenched and tempered alloy steel plates intended for use in fusion welded boilers and other pressure vessels.

1.2 This specification includes a number of grades as manufactured by different producers, but all having the same mechanical properties and general characteristics.

1.3 The maximum thickness of plates furnished under this specification shall be as follows:

Grade	Thickness
A, B	1.25 in. [32 mm]
H, S	2 in. [50 mm]
P	4 in. [100 mm]
F	2.50 in. [65 mm]
E, Q	6 in. [150 mm]

1.4 The values stated in either inch-pound units or SI units are to be regarded separately as standard. Within the text, the SI units are shown in brackets. The values stated in each system are not exact equivalents; therefore, each system is to be used independently of the other without combining values in any way.

2. Referenced Documents

2.1 *ASTM Standards:*[3]

A20/A20M Specification for General Requirements for Steel Plates for Pressure Vessels

A435/A435M Specification for Straight-Beam Ultrasonic Examination of Steel Plates

A577/A577M Specification for Ultrasonic Angle-Beam Examination of Steel Plates

A578/A578M Specification for Straight-Beam Ultrasonic Examination of Rolled Steel Plates for Special Applications

3. General Requirements and Ordering Information

3.1 Plates furnished to this material specification shall conform to Specification A20/A20M. These requirements outline the testing and retesting methods and procedures, permissible variations in dimensions, and mass, quality and repair of defects, marking, loading, etc.

3.2 Specification A20/A20M also establishes the rules for the ordering information which should be complied with when purchasing material to this specification.

3.3 In addition to the basic requirements of this specification, certain supplementary requirements are available when additional control, testing, or examination is required to meet end use requirements. These include:

3.3.1 Vacuum treatment,

3.3.2 Additional or special tension testing,

3.3.3 Impact testing, and

3.3.4 Nondestructive examination.

3.4 The purchaser is referred to the listed supplementary requirements in this specification and to the detailed requirements in Specification A20/A20M.

3.5 If the requirements of this specification are in conflict with the requirements of Specification A20/A20M, the requirements of this specification shall prevail.

4. Manufacture

4.1 *Steelmaking Practice*—The steel shall be killed and shall conform to the fine austenitic grain size requirement of Specification A20/A20M.

5. Heat Treatment

5.1 Except as allowed by section 5.2, the plates shall be heat treated by heating to not less than 1650°F [900°C], quenching in water or oil and tempering at not less than 1150°F [620°C].

---

[1] This specification is under the jurisdiction of ASTM Committee A01 on Steel, Stainless Steel and Related Alloys and is the direct responsibility of Subcommittee A01.11 on Steel Plates for Boilers and Pressure Vessels.
Current edition approved May 1, 2006. Published May 2006. Originally approved in 1964. Last previous edition approved in 2004 as A517/A517M – 93 (2004)[ε1]. DOI: 10.1520/A0517_A0517M-06.

[2] For ASME Boiler and Pressure Vessel Code applications, see related Specification SA-517/SA-517M in Section II of that Code.

[3] For referenced ASTM standards, visit the ASTM website, www.astm.org, or contact ASTM Customer Service at service@astm.org. For *Annual Book of ASTM Standards* volume information, refer to the standard's Document Summary page on the ASTM website.

*A Summary of Changes section appears at the end of this standard.

Copyright © ASTM International, 100 Barr Harbor Drive, PO Box C700, West Conshohocken, PA 19428-2959, United States.

A517/A517M – 06

TABLE 1 Chemical Requirements

Elements	Composition, %							
	Grade A	Grade B	Grade E	Grade F	Grade H	Grade P	Grade Q	Grade S
Carbon:								
Heat analysis	0.15–0.21	0.15–0.21	0.12–0.20	0.10–0.20	0.12–0.21	0.12–0.21	0.14–0.21	0.10–0.20
Product analysis	0.13–0.23	0.13–0.23	0.10–0.22	0.08–0.22	0.10–0.23	0.10–0.23	0.12–0.23	0.10–0.22
Manganese:								
Heat analysis	0.80–1.10	0.70–1.00	0.40–0.70	0.60–1.00	0.95–1.30	0.45–0.70	0.95–1.30	1.10–1.50
Product analysis	0.74–1.20	0.64–1.10	0.35–0.78	0.55–1.10	0.87–1.41	0.40–0.78	0.87–1.41	1.02–1.62
Phosphorus, max[A]	0.035	0.035	0.035	0.035	0.035	0.035	0.035	0.035
Sulfur, max[A]	0.035	0.035	0.035	0.035	0.035	0.035	0.035	0.035
Silicon:								
Heat analysis	0.40–0.80	0.15–0.35	0.10–0.40	0.15–0.35	0.15–0.35	0.20–0.35	0.15–0.35	0.15–0.40
Product analysis	0.34–0.86	0.13–0.37	0.08–0.45	0.13–0.37	0.13–0.37	0.18–0.37	0.13–0.37	0.13–0.45
Nickel:								
Heat analysis	...	...	...	0.70–1.00	0.30–0.70	1.20–1.50	1.20–1.50	...
Product analysis	...	...	...	0.67–1.03	0.27–0.73	1.15–1.55	1.15–1.55	...
Chromium:								
Heat analysis	0.50–0.80	0.40–0.65	1.40–2.00	0.40–0.65	0.40–0.65	0.85–1.20	1.00–1.50	...
Product analysis	0.46–0.84	0.36–0.69	1.34–2.06	0.36–0.69	0.36–0.69	0.79–1.26	0.94–1.56	...
Molybdenum:								
Heat analysis	0.18–0.28	0.15–0.25	0.40–0.60	0.40–0.60	0.20–0.30	0.45–0.60	0.40–0.60	0.10–0.35
Product analysis	0.15–0.31	0.12–0.28	0.36–0.64	0.36–0.64	0.17–0.33	0.41–0.64	0.36–0.64	0.10–0.38
Boron	0.0025 max	0.0005–0.005	0.001–0.005	0.0005–0.006	0.0005 min	0.001–0.005	...	...
Vanadium:								
Heat analysis	...	0.03–0.08	[B]	0.03–0.08	0.03–0.08	...	0.03–0.08	...
Product analysis	...	0.02–0.09	...	0.02–0.09	0.02–0.09	...	0.02–0.09	...
Titanium:								
Heat analysis	...	0.01–0.04	0.01–0.10	...	...	...	...	0.06
Product analysis	...	0.01–0.05	0.005–0.11	...	...	...	...	0.07
Zirconium:								
Heat analysis	0.05[C]–0.15	...	...	...	...	...	...	...
Product analysis	0.04–0.16	...	...	...	...	...	...	...
Copper:								
Heat analysis	...	...	...	0.15–0.50	...	...	...	...
Product analysis	...	...	...	0.12–0.53	...	...	...	...
Columbium, max								
Heat analysis	...	...	...	...	...	...	...	0.06
Product analysis	...	...	...	...	...	...	...	0.07

[A] Applied to both heat and product analyses.
[B] May be substituted for part or all of titanium content on a one for one basis.
[C] Zirconium may be replaced by cerium. When cerium is added, the cerium/sulfur ratio should be approximately 1.5 to 1, based on heat analysis.

5.2 Plates ordered without the heat treatment specified in section 5.1 shall be stress relieved by the manufacturer, and subsequent heat treatment of the plates to conform to section 5.1 shall be the responsibility of the purchaser.

6. Chemical Requirements

6.1 The steel shall conform to the chemical requirements shown in Table 1 unless otherwise modified in accordance with

A517/A517M – 06

TABLE 2 Tensile Requirements

	2.50 in. [65 mm] and Under	Over 2.50 to 6 in. [65 to 150 mm]
Tensile strength, ksi [MPa]	115–135 [795–930]	105–135 [725 to 930]
Yield strength, min, ksi [MPa]	100 [690]	90 [620]
Elongation in 2 in. [50 mm], min, %[A]	16	14
Reduction of area, min, %:		
Rectangular specimens	35	...
Round specimens	45	45

[A] See Specification A20/A20M for elongation adjustment.

Supplementary Requirement S17, Vacuum Carbon-Deoxidized Steel, in Specification A20/A20M for grades other than Grade A.

7. Mechanical Requirements

7.1 *Tension Tests*:

7.1.1 *Requirements*—The plates as represented by the tension-test specimens shall conform to the requirements given in Table 2.

7.1.2 *Test Methods*:

7.1.2.1 The yield strength may be determined by the 0.2 % offset method or by the total extension under load of 0.5 % method.

7.1.2.2 For plates ¾ in. [20 mm] and under in thickness, the test specimen shall be the 1½-in. [40-mm] wide rectangular-test specimen.

7.1.2.3 For plates over ¾ in. [20 mm], either the full thickness rectangular-test specimen or the ½-in. [12.5-mm] round-test specimen may be used.

7.1.2.4 When the 1½-in. [40-mm] wide rectangular-test specimen is used, the elongation is measured in a 2-in. or [50-mm] gage length which includes the fracture.

7.2 *Impact Properties Requirements*:

7.2.1 Transverse Charpy V-notch impact test specimens shall have a lateral expansion opposite the notch of not less than 0.015 in. [0.38 mm].

7.2.2 The test temperature shall be agreed upon between the manufacturer and the purchaser, but shall not be higher than 32°F [0°C].

8. Keywords

8.1 alloy steel; boilers; high-strength; impact tested; plates; pressure vessels; quenched; tempered

SUPPLEMENTARY REQUIREMENTS

Supplementary requirements shall not apply unless specified in the order.

A list of standardized supplementary requirements for use at the option of the purchaser are included in Specification A20/A20M. Several of those considered suitable for use with this specification are listed by title. Other tests may be performed by agreement between the supplier and the purchaser.

S1. Vacuum Treatment,

S2. Product Analysis,

S3. Simulated Post-Weld Heat Treatment of Mechanical Test Coupons,

S5. Charpy V-Notch Impact Test,

S6. Drop Weight Test (for Material 0.625 in. [16 mm] and over in Thickness),

S7. High-Temperature Tension Test,

S8. Ultrasonic Examination in accordance with Specification A435/A435M,

S9. Magnetic Particle Examination,

S11. Ultrasonic Examination in accordance with Specification A577/A577M,

S12. Ultrasonic Examination in accordance with Specification A578/A578M, and

S17. Vacuum Carbon-Deoxidized Steel.

A517/A517M – 06
SUMMARY OF CHANGES

Committee A01 has identified the location of selected changes to this standard since the last issue (A517/A517M – 93 (2004)$^{\varepsilon 1}$) that may impact the use of this standard.

(1) Revised Table 1.
(2) Revised section 1.3.
(3) Revised Section 5.
(4) Added keywords.
(5) Editorial changes made throughout.

ASTM International takes no position respecting the validity of any patent rights asserted in connection with any item mentioned in this standard. Users of this standard are expressly advised that determination of the validity of any such patent rights, and the risk of infringement of such rights, are entirely their own responsibility.

This standard is subject to revision at any time by the responsible technical committee and must be reviewed every five years and if not revised, either reapproved or withdrawn. Your comments are invited either for revision of this standard or for additional standards and should be addressed to ASTM International Headquarters. Your comments will receive careful consideration at a meeting of the responsible technical committee, which you may attend. If you feel that your comments have not received a fair hearing you should make your views known to the ASTM Committee on Standards, at the address shown below.

This standard is copyrighted by ASTM International, 100 Barr Harbor Drive, PO Box C700, West Conshohocken, PA 19428-2959, United States. Individual reprints (single or multiple copies) of this standard may be obtained by contacting ASTM at the above address or at 610-832-9585 (phone), 610-832-9555 (fax), or service@astm.org (e-mail); or through the ASTM website (www.astm.org).

Designation: A524 – 96 (Reapproved 2005)

Standard Specification for
Seamless Carbon Steel Pipe for Atmospheric and Lower Temperatures[1]

This standard is issued under the fixed designation A524; the number immediately following the designation indicates the year of original adoption or, in the case of revision, the year of last revision. A number in parentheses indicates the year of last reapproval. A superscript epsilon (ε) indicates an editorial change since the last revision or reapproval.

1. Scope

1.1 This specification[2] covers seamless carbon steel pipe intended primarily for service at atmospheric and lower temperatures, NPS ⅛ to 26 inclusive, with nominal (average) wall thickness as given in ANSI B36.10. Pipe having other dimensions may be furnished, provided such pipe complies with all other requirements of this specification. Pipe ordered to this specification shall be suitable both for welding, and for bending, flanging, and similar forming operations.

1.2 The values stated in inch-pound units are to be regarded as standard. The values given in parentheses are mathematical conversions to SI units that are provided for information only and are not considered standard.

NOTE 1—The dimensionless designator NPS (nominal pipe size) has been substituted in this standard for such traditional terms as "nominal diameter," "size," and "nominal size."

1.3 The following hazard caveat applies to the test methods portion, Section 16, only. *This standard does not purport to address all of the safety concerns, if any, associated with its use. It is the responsibility of the user of this standard to establish appropriate safety and health practices and determine the applicability of regulatory limitations prior to use.*

2. Referenced Documents

2.1 *ASTM Standards:*[3]
A530/A530M Specification for General Requirements for Specialized Carbon and Alloy Steel Pipe
E29 Practice for Using Significant Digits in Test Data to Determine Conformance with Specifications

2.2 *American National Standards Institute Standard:*
B36.10 Welded and Seamless Wrought Steel Pipe[4]

3. Ordering Information

3.1 Orders for material under this specification should include the following, as required, to describe the desired material adequately:

3.1.1 Quantity (feet or number of lengths),
3.1.2 Name of material (seamless carbon steel pipe),
3.1.3 Grade (Table 1 and Table 2),
3.1.4 Manufacture (hot finished or cold drawn),
3.1.5 Size (either nominal wall thickness and weight class or schedule number, or both, or outside diameter and nominal wall thickness, ANSI B36.10),
3.1.6 Length (17),
3.1.7 Optional requirements (Section 8 and Section 11 of Specification A530/A530M),
3.1.8 Test report required (Certification Section of Specification A530/A530M),
3.1.9 Specification designation,
3.1.10 End use of material, and
3.1.11 Special requirements.

4. General Requirements

4.1 Material furnished to this specification shall conform to the applicable requirements of the current edition of Specification A530/A530M unless otherwise provided herein.

5. Materials and Manufacture

5.1 *Process*:
5.1.1 The steel shall be killed steel made by one or more of the following processes: open-hearth, electric-furnace, or basic-oxygen.
5.1.2 The steel shall be made to fine grain practice.
5.1.3 Steel may be cast in ingots or may be strand cast. When steel of different grades are sequentially strand cast, identification of the resultant transition material is required. The producer shall remove the transition material by any established procedure that positively separates the grades.

---

[1] This specification is under the jurisdiction of ASTM Committee A01 on Steel, Stainless Steel and Related Alloys and is the direct responsibility of Subcommittee A01.09 on Carbon Steel Tubular Products.
Current edition approved Oct. 1, 2005. Published November 2005. Originally approved in 1965. Last previous edition approved in 2001 as A524 – 96 (2001). DOI: 10.1520/A0524-96R05.

[2] For ASME Boiler and Pressure Vessel Code Applications see related Specification SA-524 in Section II of that Code.

[3] For referenced ASTM standards, visit the ASTM website, www.astm.org, or contact ASTM Customer Service at service@astm.org. For *Annual Book of ASTM Standards* volume information, refer to the standard's Document Summary page on the ASTM website.

[4] Available from American National Standards Institute, 11 West 42nd St., 13th Floor, New York, NY 10036.

TABLE 1 Chemical Requirements

Element	Grades I and II, Composition, %
Carbon, max	0.21
Manganese	0.90–1.35
Phosphorus, max	0.035
Sulfur, max	0.035
Silicon	0.10–0.40

5.1.4 Pipe NPS 1½ and under may be either hot finished or cold drawn.

5.1.5 Unless otherwise specified, pipe NPS 2 and over shall be furnished hot finished. When agreed upon between the manufacturer and purchaser, cold-drawn pipe may be furnished.

5.2 *Heat Treatment*—All hot-finished and cold-drawn pipe shall be reheated to a temperature above 1550 °F (845°C) and followed by cooling in air or in the cooling chamber of a controlled atmosphere furnace.

6. Chemical Composition

6.1 The steel shall conform to the chemical requirements prescribed in Table 1.

7. Heat Analysis

7.1 An analysis of each heat of steel shall be made by the steel manufacturer to determine the percentages of the elements specified in Section 6. The chemical composition thus determined, or that determined from a product analysis made by the manufacturer, if the latter has not manufactured the steel, shall be reported to the purchaser or the purchaser's representative, and shall conform to the requirements specified in Section 6.

8. Product Analysis

8.1 At the request of the purchaser, analyses of two pipes from each lot (Note 2) shall be made by the manufacturer from the finished pipe. The chemical composition thus determined shall conform to the requirements specified in Section 6.

NOTE 2—A lot shall consist of 400 lengths, or fraction thereof, for each size NPS 2 up to but not including NPS 6, and of 200 lengths, or fraction thereof, for each size NPS 6 and over.

8.2 If the analysis of one of the tests specified in 8.1 does not conform to the requirements specified in 6, analyses shall be made on additional pipe of double the original number from the same lot, each of which shall conform to requirements specified.

9. Physical Properties

9.1 *Tensile Properties*—The material shall conform to the requirements as to tensile properties prescribed in Table 2.

9.2 *Bending Properties*:

9.2.1 For pipe NPS 2 and under, a sufficient length of pipe shall stand being bent cold through 90° around a cylindrical mandrel, the diameter of which is twelve times the nominal diameter of the pipe, without developing cracks. When ordered for close coiling, the pipe shall stand being bent cold through 180° around a cylindrical mandrel, the diameter of which is eight times the nominal diameter of the pipe, without failure.

9.2.2 For pipe whose diameter exceeds 25 in. (635 mm) and whose diameter to wall thickness ratio is 7.0 or less, bend test specimens shall be bent at room temperature through 180° without cracking on the outside of the bent portion. The inside diameter of the bend shall be 1 in. (25.4 mm). This test shall be in place of Section 10.

NOTE 3—Diameter to wall thickness ratio = specified outside diameter/ nominal wall thickness.
Example: For 28 in. diameter 5.000 in. thick pipe the diameter to wall thickness ratio = 28/5 = 5.6.

10. Flattening Test Requirements

10.1 For pipe over NPS 2, a section of pipe not less than 2½ in. (63.5 mm) in length shall be flattened cold between parallel plates until the opposite walls of the pipe meet. Flattening tests shall be in accordance with Specification A530/A530M, except that in the equation used to calculate the H value, the following e constants shall be used:

0.07 for Grade I
0.08 for Grade II

10.2 When low D-to-t ratio tubulars are tested, because the strain imposed due to geometry is unreasonably high on the inside surface at the 6 and 12 o'clock locations, cracks at these locations shall not be cause for rejection if the D-to-t ratio is less than ten.

11. Hydrostatic Test Requirements

11.1 Each length of pipe shall be subjected to the hydrostatic pressure, except as provided in 11.2.

11.2 When specified in the order, pipe may be furnished without hydrostatic testing and each length so furnished shall include with the mandatory marking the letters "NH."

11.3 When certification is required by the purchaser and the hydrostatic test has been omitted, the certification shall clearly state "Not Hydrostatically Tested," and the specification number and grade designation, as shown on the certification, shall be followed by the letters "NH."

12. Dimensions and Weights

12.1 The dimensions and weights of plain-end pipe are included in ANSI B36.10. Sizes and wall thicknesses most generally available are listed in Appendix X1.

13. Dimensions, Weight, and Permissible Variations

13.1 *Weight*—The weight of any length of pipe shall not vary more than 6.5 % over and 3.5 % under that specified for pipe of Schedule 120 and lighter nor more than 10 % over and 3.5 % under that specified for pipe heavier than Schedule 120. Unless otherwise agreed upon between the manufacturer and purchaser, pipe in sizes NPS 4 and smaller may be weighed in convenient lots; pipe in sizes larger than NPS 4 shall be weighed separately.

13.2 *Diameter*—Variations in outside diameter shall not exceed those specified in Table 3.

13.3 *Thickness*—The minimum wall thickness at any point shall not be more than 12.5 % under the nominal wall thickness specified.

NOTE 4—The minimum wall thickness on inspection is shown in Appendix X1.

TABLE 2 Tensile Requirements

	Wall Thicknesses			
	Grade I, 0.375 in. (9.52 mm) and under		Grade II, greater than 0.375 in. (9.52 mm)	
Tensile strength, psi (MPa)	60 000–85 000 (414–586)		55 000–80 000 (380–550)	
Yield strength, min, psi (MPa)	35 000 (240)		30 000 (205)	
	Longitudinal	Transverse	Longitudinal	Transverse
Elongation in 2 in. or 50 mm, min %:				
Basic minimum elongation for walls 5/16 in. (7.9 mm) and over in thickness, strip tests, and for all small sizes tested in full section	30	16.5	35	25
When standard round 2 in. or 50 mm gauge length test specimen is used for strip tests, a deduction for each 1/32 in. (0.8 mm) decrease in wall thickness below 5/16 in. (7.9 mm) from the basic minimum elongation of the following percentage	22 1.50[A]	12 1.00[A]	28 ...	20 ...

[A] The following table gives the computed minimum values:

Wall Thickness		Elongation in 2 in. or 50 mm, min, %	
		Grade I	
in.	mm	Longitudinal	Transverse
5/16 (0.312)	7.94	30.0	16.5
9/32 (0.281)	7.14	28.5	15.5
1/4 (0.250)	6.35	27.0	14.5
7/32 (0.219)	5.56	25.5	...
3/16 (0.188)	4.76	24.0	...
5/32 (0.156)	3.97	22.5	...
1/8 (0.125)	3.18	21.0	...
3/32 (0.094)	2.38	19.5	...
1/16 (0.062)	1.59	18.0	...

Note—The above table gives the computed minimum elongation values for each 1/32-in. (0.79-mm) decrease in wall thickness. Where the wall thickness lies between two values shown above, the minimum elongation value is determined by the following equation:

Grade	Direction of Test	Equation
I	transverse	$E = 32t + 6.50$
I	longitudinal	$E = 48t + 15.00$

where:
E = elongation in 2 in. or 50 mm in % and
t = actual thickness of specimen, in. (mm).

TABLE 3 Variations in Outside Diameter

NPS Designator	Permissible Variations in Outside Diameter, in. (mm)	
	Over	Under
1/8 to 1 1/2, incl	1/64 (0.4)	1/32 (0.8)
Over 1 1/2 to 4, incl	1/32 (0.8)	1/32 (0.8)
Over 4 to 8, incl	1/16 (1.6)	1/32 (0.8)
Over 8 to 18, incl	3/32 (2.4)	1/32 (0.8)
Over 18	1/8 (3.2)	1/32 (0.8)

14. Workmanship, Finish, and Appearance

14.1 The pipe manufacturer shall explore a sufficient number of visual surface imperfections to provide reasonable assurance that they have been properly evaluated with respect to depth. Exploration of all surface imperfections is not required but may be necessary to assure compliance with 14.2.

14.2 Surface imperfections that penetrate more than 12 1/2 % of the nominal wall thickness or encroach on the minimum wall thickness shall be considered defects. Pipe with such defects shall be given one of the following dispositions:

14.2.1 The defect may be removed by grinding provided that the remaining wall thickness is within specified limits.

14.2.2 Repaired in accordance with the repair welding provisions of 14.6.

14.2.3 The section of pipe containing the defect may be cut off within the limits of requirements on length.

14.2.4 Rejected.

14.3 To provide a workmanlike finish and basis for evaluating conformance with 14.2, the pipe manufacturer shall remove by grinding the following noninjurious imperfections:

14.3.1 Mechanical marks, abrasions (Note 5), and pits, any of which imperfections are deeper than 1/16 in. (1.58 mm).

NOTE 5—Marks and abrasions are defined as cable marks, dinges, guide marks, roll marks, ball scratches, scores, die marks, and the like.

14.3.2 Visual imperfections, commonly referred to as scabs, seams, laps, tears, or slivers, found by exploration in accordance with 14.1 to be deeper than 5 % of the nominal wall thickness.

14.4 At the purchaser's discretion, pipe shall be subject to rejection if surface imperfections acceptable under 14.2 are not scattered, but appear over a large area in excess of what is considered a workmanlike finish. Disposition of such pipe shall be a matter of agreement between the manufacturer and the purchaser.

14.5 When imperfections or defects are removed by grinding, a smooth curved surface shall be maintained, and the wall thickness shall not be decreased below that permitted by this specification. The outside diameter at the point of grinding may be reduced by the amount so removed.

14.5.1 Wall thickness measurements shall be made with a mechanical caliper or with a properly calibrated nondestructive testing device of appropriate accuracy. In case of dispute, the measurement determined by use of the mechanical caliper shall govern.

14.6 Weld repair shall be permitted only subject to the approval of the purchaser and in accordance with Specification A530/A530M.

14.7 The finished pipe shall be reasonably straight.

15. Number of Tests and Retests

15.1 One of either of the tests specified in 9.1 shall be made on one length of pipe from each lot (Note 2).

15.2 For pipe NPS 2 and under, the bend test specified in 9.2 shall be made on one pipe from each lot (Note 2). The bend tests specified in 9.2.2 shall be made on one end of each pipe.

15.3 The flattening test specified in 10 shall be made on one length of pipe from each lot (Note 2).

15.4 Retests shall be in accordance with Specification A530/A530M and as provided in 15.5 and 15.6.

15.5 If a specimen breaks in an inside or outside surface flaw, a retest shall be allowed.

15.6 Should a crop end of a finished pipe fail in the flattening test, one retest may be made from the broken end.

16. Test Specimens and Methods of Testing

16.1 Specimens cut either longitudinally or transversely shall be acceptable for the tension test.

16.2 Test specimens for the bend test specified in 9.2 and for the flattening tests specified in 10 shall consist of sections cut from a pipe. Specimens for flattening tests shall be smooth on the ends and free from burrs, except when made on crop ends.

16.3 Test specimens for the bend test specified in 9.2.2 shall be cut from one end of the pipe and, unless otherwise specified, shall be taken in a transverse direction. One test specimen shall be taken as close to the outer surface as possible and another from as close to the inner surface as possible. The specimens shall be either ½ by ½ in. (12.7 mm) in section or 1 by ½ in. (25.4 by 12.7 mm) in section with the corners rounded to a radius not over 1/16 in. (1.6 mm) and need not exceed 6 in. (152 mm) in length. The side of the samples placed in tension during the bend shall be the side closest to the inner and outer surface of the pipe respectively.

17. Lengths

17.1 Pipe lengths shall be in accordance with the following regular practice:

17.1.1 The lengths required shall be specified in the order, and

17.1.2 No jointers are permitted unless otherwise specified.

17.2 If definite lengths are not required, pipe may be ordered in single random lengths of 16 to 22 ft (4.9 to 6.7 m), with 5 % 12 to 16 ft (3.7 to 4.9 m), or in double random lengths with a minimum average of 35 ft (10.7 m) and a minimum length of 22 ft with 5 % 16 to 22 ft.

18. Rejection

18.1 Each length of pipe that develops injurious defects during shop working or application operations will be rejected, and the manufacturer shall be notified. No rejections under this or any other specifications shall be marked as specified in 19 for sale under this specification except where such pipe fails to comply with the weight requirements alone, in which case it may be sold under the weight specifications with which it does comply.

19. Product Marking

19.1 In addition to the marking prescribed in Specification A530/A530M, the marking shall include the hydrostatic test pressure when tested or the letters "NH" when not tested, the length and schedule number, and on pipe sizes larger than NPS 4 the weight shall be given. Length shall be marked in feet and tenths of a foot, or metres to two decimal places, depending on the units to which the material was ordered, or other marking subject to agreement.

19.2 *Bar Coding*—In addition to the requirements in 19.1, bar coding is acceptable as a supplemental identification method. The purchaser may specify in the order a specific bar coding system to be used.

APPENDIX

(Nonmandatory Information)

X1. DIMENSIONS AND WALL THICKNESSES

X1.1 Following are Tables X1.1 and X1.2, cited in the text of this standard.

A524 − 96 (2005)

TABLE X1.1 Table of Minimum Wall Thicknesses on Inspection for Nominal (Average) Pipe Wall Thickness

NOTE 1—The following equation, upon which this table is based, may be applied to calculate minimum wall thickness from nominal (average) wall thickness:

$$t_n \times 0.875 = t_m$$

where:
t_n = nominal (average) wall thickness, in. (mm), and
t_m = minimum wall thickness, in. (mm).

NOTE 2—The wall thickness is expressed to three decimal places, the fourth decimal place being carried forward or dropped, in accordance with Practice E29. This table is a master table covering wall thicknesses available in the purchase of different classifications of pipe, but it is not meant to imply that all of the walls listed therein are obtainable under this specification.

Nominal (Average) Thickness (t_n)		Minimum Thickness on Inspection (t_m)		Nominal (Average) Thickness (t_n)		Minimum Thickness on Inspection (t_m)		Nominal (Average) Thickness (t_n)		Minimum Thickness on Inspection (t_m)	
in.	mm	in.	mm	in.	mm	in.	mm	in.	mm	in.	mm
0.068	1.73	0.060	1.52	0.281	7.14	0.246	6.25	0.864	21.94	0.756	19.20
0.083	2.11	0.073	1.85	0.294	7.47	0.257	6.53	0.875	22.22	0.766	19.46
0.088	2.24	0.077	1.96	0.300	7.62	0.262	6.65	0.906	23.01	0.793	20.14
0.091	2.31	0.080	2.03	0.307	7.80	0.269	6.83	0.938	23.82	0.821	20.85
0.095	2.41	0.083	2.11	0.308	7.82	0.270	6.86	0.968	24.59	0.847	21.51
0.109	2.77	0.095	2.41	0.312	7.92	0.273	6.93	1.000	25.40	0.875	22.22
0.113	2.87	0.099	2.51	0.318	8.07	0.278	7.06	1.031	26.19	0.902	22.91
0.119	3.02	0.104	2.64	0.322	8.18	0.282	7.16	1.062	26.97	0.929	23.60
0.125	3.18	0.109	2.77	0.330	8.38	0.289	7.34	1.094	27.79	0.957	24.31
0.126	3.20	0.110	2.79	0.337	8.56	0.295	7.49	1.125	28.58	0.984	24.99
0.133	3.38	0.116	2.95	0.344	8.74	0.301	7.64	1.156	29.36	1.012	25.70
0.140	3.56	0.122	3.10	0.358	9.09	0.313	7.95	1.219	30.96	1.066	27.08
0.141	3.58	0.123	3.12	0.365	9.27	0.319	8.10	1.250	31.75	1.094	27.79
0.145	3.68	0.127	3.23	0.375	9.52	0.328	8.33	1.281	32.54	1.121	28.47
0.147	3.73	0.129	3.28	0.382	9.70	0.334	8.48	1.312	33.32	1.148	29.16
0.154	3.91	0.135	3.43	0.400	10.16	0.350	8.89	1.375	34.92	1.203	30.56
0.156	3.96	0.136	3.45	0.406	10.31	0.355	9.02	1.406	35.71	1.230	31.24
0.172	4.37	0.150	3.81	0.432	10.97	0.378	9.60	1.438	36.53	1.258	31.95
0.179	4.55	0.157	3.99	0.436	11.07	0.382	9.70	1.500	38.10	1.312	33.32
0.188	4.78	0.164	4.17	0.438	11.12	0.383	9.73	1.531	38.89	1.340	34.04
0.191	4.85	0.167	4.24	0.469	11.91	0.410	10.41	1.562	39.67	1.367	34.72
0.200	5.08	0.175	4.44	0.500	12.70	0.438	11.13	1.594	40.49	1.395	35.43
0.203	5.16	0.178	4.52	0.531	13.49	0.465	11.81	1.635	41.53	1.431	36.35
0.210	5.33	0.184	4.67	0.552	14.02	0.483	12.27	1.750	44.45	1.531	38.89
0.216	5.49	0.189	4.80	0.562	14.27	0.492	12.50	1.781	45.24	1.558	39.57
0.218	5.54	0.191	4.85	0.594	15.09	0.520	13.21	1.812	46.02	1.586	40.28
0.219	5.56	0.192	4.88	0.600	15.24	0.525	13.34	1.875	47.62	1.641	41.68
0.226	5.74	0.198	5.03	0.625	15.88	0.547	13.89	1.969	50.01	1.723	43.76
0.237	6.02	0.207	5.26	0.656	16.66	0.574	14.58	2.000	50.80	1.750	44.45
0.250	6.35	0.219	5.56	0.674	17.12	0.590	14.99	2.062	52.37	1.804	45.82
0.258	6.55	0.226	5.74	0.688	17.48	0.602	15.29	2.125	53.98	1.859	47.22
0.276	7.01	0.242	6.15	0.719	18.26	0.629	15.98	2.200	55.88	1.925	48.90
0.277	7.04	0.242	6.15	0.750	19.05	0.656	16.66	2.344	59.54	2.051	52.10
0.279	7.09	0.244	6.19	0.812	20.62	0.710	18.03	2.500	63.50	2.188	55.58
0.280	7.11	0.245	6.22	0.844	21.44	0.739	18.77				

TABLE X1.2 Dimensions, Weights and Test Pressures for Plain End Pipe
(As appears in American National Standard B36.10)

NPS Designator	Wall Thickness	Nominal Weight	Weight Class	Schedule No.	Test Pressure	
					Grade I	Grade II
	in. (mm)	lb/ft (kg/m)	psi (MPa)	psi (MPa)		
1/8	0.068 (1.73)	0.24 (0.36)	std	40	2500 (17.2)	...
	0.095 (2.41)	0.31 (0.46)	XS	80	2500 (17.2)	...
1/4	0.088 (2.24)	0.42 (0.63)	std	40	2500 (17.2)	...
	0.119 (3.02)	0.54 (0.80)	XS	80	2500 (17.2)	...
3/8	0.091 (2.31)	0.57 (0.85)	std	40	2500 (17.2)	...
	0.126 (3.20)	0.74 (1.10)	XS	80	2500 (17.2)	...
1/2	0.109 (2.77)	0.85 (1.27)	std	40	2500 (17.2)	...
	0.147 (3.73)	1.09 (1.62)	XS	80	2500 (17.2)	...
	0.294 (7.47)	1.71 (2.55)	XXS	...	2500 (17.2)	...
3/4	0.113 (2.87)	1.13 (1.68)	std	40	2500 (17.2)	...
	0.154 (3.91)	1.47 (2.19)	XS	80	2500 (17.2)	...
	0.308 (7.82)	2.44 (3.63)	XXS	...	2500 (17.2)	...

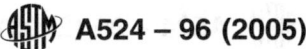

TABLE X1.2 Continued

NPS Designator	Wall Thickness in. (mm)	Nominal Weight lb/ft (kg/m)	Weight Class	Schedule No.	Test Pressure Grade I psi (MPa)	Test Pressure Grade II psi (MPa)
1	0.133 (3.38)	1.68 (2.50)	std	40	2500 (17.2)	...
	0.179 (4.55)	2.17 (3.23)	XS	80	2500 (17.2)	...
	0.358 (9.09)	3.66 (5.45)	XXS	...	2500 (17.2)	...
1¼	0.140 (3.56)	2.27 (3.38)	std	40	2500 (17.2)	...
	0.191 (4.85)	3.00 (4.47)	XS	80	2500 (17.2)	...
	0.382 (9.70)	5.21 (7.76)	XXS	...	...	2500 (17.2)
1½	0.145 (3.68)	2.72 (4.05)	std	40	2500 (17.2)	...
	0.200 (5.08)	3.63 (5.41)	XS	80	2500 (17.2)	...
	0.400 (10.16)	6.41 (9.55)	XXS	...	...	2500 (17.2)
2	0.154 (3.91)	3.65 (5.44)	std	40	2500 (17.2)	...
	0.218 (5.54)	5.02 (7.48)	XS	80	2500 (17.2)	...
	0.344 (8.74)	7.46 (11.12)	...	160	2500 (17.2)	...
	0.436 (11.07)	9.03 (13.45)	XXS	...	...	2500 (17.2)
2½	0.203 (5.16)	5.79 (8.62)	std	40	2500 (17.2)	...
	0.276 (7.01)	7.66 (11.41)	XS	80	2500 (17.2)	...
	0.375 (9.52)	10.01 (14.91)	...	160	2500 (17.2)	...
	0.552 (14.02)	13.70 (20.41)	XXS	...	...	2500 (17.2)
3	0.216 (5.49)	7.58 (11.29)	std	40	2500 (17.2)	...
	0.300 (7.62)	10.25 (15.27)	XS	80	2500 (17.2)	...
	0.438 (11.13)	14.32 (21.34)	...	160	...	2500 (17.2)
	0.600 (15.24)	18.58 (27.67)	XXS	...	...	2500 (17.2)
3½	0.226 (5.74)	9.11 (13.57)	std	40	2400 (16.5)	...
	0.318 (8.08)	12.51 (18.63)	XS	80	2800 (19.3)	...
4	0.237 (6.02)	10.79 (16.07)	std	40	2200 (15.2)	...
	0.337 (8.56)	14.98 (22.31)	XS	80	2800 (19.3)	...
	0.438 (11.13)	19.00 (28.30)	...	120	...	2800 (19.3)
	0.531 (13.49)	22.51 (33.53)	...	160	...	2800 (19.3)
	0.674 (17.12)	27.54 (41.02)	XXS	...	...	2800 (19.3)
5	0.258 (6.55)	14.62 (21.78)	std	40	1900 (13.1)	...
	0.375 (9.52)	20.78 (30.95)	XS	80	2800 (19.3)	...
	0.500 (12.70)	27.04 (40.28)	...	120	...	2800 (19.3)
	0.625 (15.88)	32.96 (49.09)	...	160	...	2800 (19.3)
	0.750 (19.05)	38.55 (57.42)	XXS	...	...	2800 (19.3)
6	0.280 (7.11)	18.97 (28.26)	std	40	1800 (12.4)	...
	0.432 (10.97)	28.57 (42.56)	XS	80	...	2300 (15.9)
	0.562 (14.27)	36.39 (54.20)	...	120	...	2800 (19.3)
	0.719 (18.26)	45.35 (67.55)	...	160	...	2800 (19.3)
	0.864 (21.95)	53.16 (79.68)	XXS	...	...	2800 (19.3)
8	0.250 (6.35)	22.36 (33.31)	...	20	1200 (8.3)	...
	0.277 (7.04)	24.70 (36.79)	...	30	1300 (9.0)	...
	0.322 (8.18)	28.55 (42.53)	std	40	1600 (11.0)	...
	0.406 (10.31)	35.64 (53.10)	...	60	...	1700 (11.7)
	0.500 (12.70)	43.39 (64.63)	XS	80	...	2100 (14.5)
	0.594 (15.09)	50.95 (75.92)	...	100	...	2500 (17.2)
	0.719 (18.26)	60.71 (90.43)	...	120	...	2800 (19.3)
	0.812 (20.62)	67.76 (100.96)	...	140	...	2800 (19.3)
	0.875 (22.22)	72.42 (107.87)	XXS	...	...	2800 (19.3)
	0.906 (23.01)	74.69 (111.29)	...	160	...	2800 (19.3)
10	0.250 (6.35)	28.04 (41.77)	...	20	1000 (6.9)	...
	0.279 (7.09)	31.20 (46.47)	...	...	1100 (7.6)	...
	0.307 (7.80)	34.24 (51.00)	...	30	1200 (8.3)	...
	0.365 (9.27)	40.48 (60.29)	std	40	1400 (9.7)	...
	0.500 (12.70)	54.74 (81.55)	XS	60	...	1700 (11.7)
	0.594 (15.09)	64.43 (96.00)	...	80	...	2000 (13.8)
	0.719 (18.26)	77.03 (114.74)	...	100	...	2400 (16.5)
	0.844 (21.44)	89.29 (133.04)	...	120	...	2800 (9.3)
	1.000 (25.40)	104.13 (155.15)	XXS	140	...	2800 (9.3)
	1.125 (28.58)	115.65 (172.32)	...	160	...	2800 (9.3)
12	0.250 (6.35)	33.38 (49.72)	...	20	800 (5.5)	...
	0.330 (8.38)	43.77 (65.20)	...	30	1100 (7.6)	...
	0.375 (9.52)	49.56 (73.82)	std	...	1200 (8.3)	...
	0.406 (10.31)	53.52 (79.74)	...	40	...	1100 (7.6)
	0.500 (12.70)	65.42 (97.44)	XS	...	...	1400 (9.7)
	0.562 (14.27)	73.15 (108.96)	...	60	...	1600 (11.0)
	0.688 (17.48)	88.63 (132.01)	...	80	...	1900 (13.1)
	0.844 (21.44)	107.32 (159.91)	...	100	...	2400 (16.5)
	1.000 (25.40)	125.49 (186.98)	XXS	120	...	2800 (19.3)
	1.125 (28.58)	139.68 (208.12)	...	140	...	2800 (19.3)
	1.312 (33.32)	160.27 (238.80)	...	160	...	2800 (19.3)
14	0.250 (6.35)	36.71 (54.68)	...	10	750 (5.2)	...

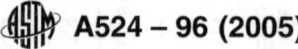

TABLE X1.2 *Continued*

NPS Designator	Wall Thickness in. (mm)	Nominal Weight lb/ft (kg/m)	Weight Class	Schedule No.	Test Pressure Grade I psi (MPa)	Test Pressure Grade II psi (MPa)
	0.312 (7.92)	45.61 (67.94)	...	20	950 (6.6)	...
	0.375 (9.52)	54.57 (81.28)	std	30	1100 (7.6)	...
	0.438 (11.13)	63.44 (94.49)	...	40	...	1100 (7.6)
	0.500 (12.70)	72.09 (107.38)	XS	...	...	1300 (9.0)
	0.594 (15.09)	85.05 (126.72)	...	60	...	1500 (10.3)
	0.750 (19.05)	106.13 (158.08)	...	80	...	1900 (13.1)
	0.938 (23.83)	130.85 (194.90)	...	100	...	2400 (16.5)
	1.094 (27.79)	150.79 (234.68)	...	120	...	2800 (19.3)
	1.250 (31.75)	170.22 (253.63)	...	140	...	2800 (19.3)
	1.406 (35.71)	189.11 (281.77)	...	160	...	2800 (19.3)
16	0.250 (6.35)	42.05 (62.63)	...	10	650 (4.5)	...
	0.312 (7.92)	52.27 (77.86)	...	20	800 (5.5)	...
	0.375 (9.52)	62.58 (93.21)	std	30	1000 (6.9)	...
	0.500 (12.70)	82.77 (123.29)	XS	40	...	1100 (7.6)
	0.656 (16.66)	107.50 (160.18)	...	60	...	1500 (10.3)
	0.844 (21.44)	136.62 (203.56)	...	80	...	1900 (13.1)
	1.031 (26.19)	164.82 (245.58)	...	100	...	2300 (15.9)
	1.219 (30.96)	192.43 (286.72)	...	120	...	2700 (18.6)
	1.438 (36.52)	223.64 (333.22)	...	140	...	2800 (19.3)
	1.594 (40.49)	245.25 (365.42)	...	160	...	2800 (19.3)
18	0.250 (6.35)	47.39 (70.59)	...	10	600 (4.1)	...
	0.312 (7.92)	58.94 (87.79)	...	20	750 (5.2)	...
	0.375 (9.52)	70.59 (105.14)	std	...	900 (6.2)	...
	0.438 (11.13)	82.15 (122.36)	...	30	...	900 (6.2)
	0.500 (12.70)	93.45 (139.19)	XS	...	...	1000 (6.9)
	0.562 (14.27)	104.67 (155.91)	...	40	...	1100 (7.6)
	0.750 (19.05)	138.17 (205.80)	...	60	...	1500 (10.3)
	0.938 (23.83)	170.92 (254.59)	...	80	...	1900 (13.1)
	1.156 (29.36)	207.96 (309.86)	...	100	...	2300 (15.9)
	1.375 (34.92)	244.14 (363.77)	...	120	...	2800 (19.3)
	1.562 (39.67)	274.22 (408.54)	...	140	...	2800 (19.3)
	1.781 (45.24)	308.50 (459.67)	...	160	...	2800 (19.3)
20	0.250 (6.35)	52.73 (78.54)	...	10	500 (3.4)	...
	0.375 (9.52)	78.60 (117.07)	std	20	800 (5.5)	...
	0.500 (12.70)	104.13 (155.10)	XS	30	...	900 (6.2)
	0.594 (15.09)	123.11 (183.43)	...	40	...	1100 (7.6)
	0.812 (20.62)	166.40 (247.85)	...	60	...	1500 (10.3)
	1.031 (26.19)	208.87 (311.22)	...	80	...	1900 (13.1)
	1.281 (32.54)	256.10 (381.59)	...	100	...	2300 (15.9)
	1.500 (38.10)	296.37 (441.59)	...	120	...	2700 (18.6)
	1.750 (44.45)	341.10 (508.24)	...	140	...	2800 (19.3)
	1.969 (50.01)	379.17 (564.96)	...	160	...	2800 (19.3)
24	0.250 (6.35)	63.41 (94.45)	...	10	450 (3.1)	...
	0.375 (9.52)	94.62 (140.94)	std	20	650 (4.5)	...
	0.500 (12.70)	125.49 (186.92)	XS	...	...	750 (5.2)
	0.562 (14.27)	140.68 (209.54)	...	30	...	850 (5.9)
	0.688 (17.48)	171.29 (255.14)	...	40	...	1000 (6.9)
	0.969 (24.61)	238.85 (355.89)	...	60	...	1500 (10.3)
	1.219 (30.96)	296.58 (441.90)	...	80	...	1800 (12.4)
	1.531 (38.89)	367.39 (547.41)	...	100	...	2300 (15.9)
	1.812 (46.02)	429.39 (639.79)	...	120	...	2700 (18.6)
	2.062 (52.37)	483.12 (719.85)	...	140	...	2800 (19.3)
	2.344 (59.64)	542.14 (807.79)	...	160	...	2800 (19.3)
26	0.250 (6.35)	68.75 (102.40)	...	...	400 (2.8)	...
	0.312 (7.92)	85.60 (127.50)	...	10	500 (3.4)	...
	0.375 (9.52)	102.63 (152.87)	std	...	610 (4.2)	...
	0.500 (12.70)	136.17 (202.83)	XS	20	...	690 (4.8)

ASTM International takes no position respecting the validity of any patent rights asserted in connection with any item mentioned in this standard. Users of this standard are expressly advised that determination of the validity of any such patent rights, and the risk of infringement of such rights, are entirely their own responsibility.

This standard is subject to revision at any time by the responsible technical committee and must be reviewed every five years and if not revised, either reapproved or withdrawn. Your comments are invited either for revision of this standard or for additional standards and should be addressed to ASTM International Headquarters. Your comments will receive careful consideration at a meeting of the responsible technical committee, which you may attend. If you feel that your comments have not received a fair hearing you should make your views known to the ASTM Committee on Standards, at the address shown below.

This standard is copyrighted by ASTM International, 100 Barr Harbor Drive, PO Box C700, West Conshohocken, PA 19428-2959, United States. Individual reprints (single or multiple copies) of this standard may be obtained by contacting ASTM at the above address or at 610-832-9585 (phone), 610-832-9555 (fax), or service@astm.org (e-mail); or through the ASTM website (www.astm.org).

Designation: A529/A529M − 05 (Reapproved 2009)

Standard Specification for
High-Strength Carbon-Manganese Steel of Structural Quality[1]

This standard is issued under the fixed designation A529/A529M; the number immediately following the designation indicates the year of original adoption or, in the case of revision, the year of last revision. A number in parentheses indicates the year of last reapproval. A superscript epsilon (ε) indicates an editorial change since the last revision or reapproval.

This standard has been approved for use by agencies of the Department of Defense.

1. Scope

1.1 This specification covers carbon-manganese steel shapes, plates, and bars of structural quality for use in riveted, bolted, or welded construction of buildings and for general structural purposes.

1.2 Material under this specification is available in two grades:

Grade	Yield Strength, ksi [MPa]	Thickness
50 [345]	50 [345]	Plates to 1 in. [25 mm] thick to 15 in. [380 mm] wide Bars to 3½ in. [90 mm] Shapes with flange or leg thickness to 1½ in. [40 mm] inclusive
55 [380]	55 [380]	Plates to 1 in. [25 mm] thick to 15 in. [380 mm] wide Bars to 3 in. [75 mm] Shapes with flange or leg thickness to 1½ in. [40 mm] inclusive

1.3 When the steel is to be welded, it is presupposed that a welding procedure suitable for the grade of steel and intended use or service will be utilized. See Appendix X3 of Specification A6/A6M for information on weldability.

1.4 The values stated in either inch-pound units or SI units are to be regarded as standard. Within the text, the SI units are shown in brackets. The values stated in each system are not exact equivalents; therefore, each system must be used independently of the other.

2. Referenced Documents

2.1 *ASTM Standards:*[2]

A6/A6M Specification for General Requirements for Rolled Structural Steel Bars, Plates, Shapes, and Sheet Piling

3. General Requirements for Delivery

3.1 Material furnished under this specification shall conform to the requirements of the current edition of Specification A6/A6M, for the ordered material, unless a conflict exists in which case this specification shall prevail.

4. Materials and Manufacture

4.1 The steel shall be killed, and such shall be affirmed in the test report by the inclusion of a statement of *killed steel*, a value of 0.10 % or more for silicon content, or a value of 0.015 % or more for total aluminum content.

5. Chemical Composition

5.1 *Heat Analysis*:

5.1.1 The heat analysis shall conform to the requirements prescribed in Table 1.

5.1.2 In addition to the elements specified in Table 1, test reports shall include for information the chemical analysis for copper, columbium, chromium, nickel, molybdenum, and vanadium. When the amount of copper, chromium, nickel, molybdenum, or silicon is less than 0.02 %, the analysis may be reported as "<0.02 %." When the amount of columbium or vanadium is less than 0.008 %, the analysis may be reported as "<0.008 %."

5.2 *Product Analysis*:

5.2.1 The steel shall conform on product analysis to the requirements of Table 1, subject to the product analysis tolerances in Specification A6/A6M.

6. Tension Test

6.1 The material as represented by the test specimen shall conform to the requirements as to the tensile properties prescribed in Table 2.

7. Keywords

7.1 bars; bolted construction; carbon; frames; metal building systems; plates; riveted construction; shapes; steel; structural steel; trusses; welded construction

---

[1] This specification is under the jurisdiction of ASTM Committee A01 on Steel, Stainless Steel and Related Alloys and is the direct responsibility of Subcommittee A01.02 on Structural Steel for Bridges, Buildings, Rolling Stock and Ships.
Current edition approved Oct. 1, 2009. Published December 2009. Originally approved in 1964. Last previous edition approved in 2005 as A529/A529M – 05. DOI: 10.1520/A0529_A0529M-05R09.

[2] For referenced ASTM standards, visit the ASTM website, www.astm.org, or contact ASTM Customer Service at service@astm.org. For *Annual Book of ASTM Standards* volume information, refer to the standard's Document Summary page on the ASTM website.

Copyright © ASTM International, 100 Barr Harbor Drive, PO Box C700, West Conshohocken, PA 19428-2959, United States.

TABLE 1 Chemical Requirements (Heat Analysis)

NOTE—A maximum of 1.50 % manganese is permissible, with an associated reduction of the carbon maximum of 0.01 percentage point for each 0.05 percentage point increase in manganese.

Element	Composition, % Grades 50 [345] and 55 [380]
Carbon, max	0.27
Manganese, max	1.35
Phosphorus, max	0.04
Sulfur, max	0.05
Silicon, max	0.40
Copper, min, when copper is specified	0.20

TABLE 2 Tensile Requirements[A]

	Grade 50 [345]		Grade 55 [380]	
	ksi	[MPa]	ksi	[MPa]
Tensile strength, min	70[B]	[485]	70	[485]
Tensile strength, max	100	[690]	100	[690]
Yield strength, min	50	[345]	55	[380]
Elongation in 8 in. [200 mm], min, %	18		17	
Elongation in 2 in. [50 mm], min, %	21		20	

[A] See the Orientation subsection in the Tension Tests section of Specification A6/A6M.
[B] Minimum tensile strength for Grade 50 shapes with flange or leg thickness to 1 ½ in. [40 mm] inclusive, shall be 65 ksi [450 MPa].

SUPPLEMENTARY REQUIREMENTS

Standardized supplementary requirements for use at the option of the purchaser are listed in Specification A6/A6M. Those that are considered suitable for use with this specification are listed by title:

S5. Charpy V-Notch Impact Test.

In addition, the following optional supplementary requirements are also suitable for use with this specification.

S32. Single Heat Bundles.

S32.1 Bundles containing shapes or bars shall be from a single heat of steel.

S78. Maximum Carbon Equivalent.

S78.1 This material shall be supplied with a maximum carbon equivalent value of 0.55 % or to a lower value specified in the purchase documents. This value will be based on heat analysis. The required chemical analysis as well as the carbon equivalent shall be reported.

S78.2 The carbon equivalent shall be calculated using the following formula:

$$CE = C + (Mn + Si)/6 + (Cu + Ni)/15 + (Cr + Mo + V + Cb)/5$$

S79. Maximum Tensile Strength.

S79.1 The maximum tensile strength shall be 90 ksi [620 MPa].

Designation: A537/A537M – 08

Standard Specification for
Pressure Vessel Plates, Heat-Treated, Carbon-Manganese-Silicon Steel[1]

This standard is issued under the fixed designation A537/A537M; the number immediately following the designation indicates the year of original adoption or, in the case of revision, the year of last revision. A number in parentheses indicates the year of last reapproval. A superscript epsilon (ε) indicates an editorial change since the last revision or reapproval.

This standard has been approved for use by agencies of the Department of Defense.

1. Scope*

1.1 This specification[2] covers heat-treated carbon-manganese-silicon steel plates intended for fusion welded pressure vessels and structures.

1.2 Plates furnished under this specification are available in the following three classes:

Class	Heat Treatment	Thickness,	Yield Strength, min, ksi [MPa]	Tensile Strength, min, ksi [MPa]
1	Normalized	2½ in. and under [65 mm and under]	50 [345]	70 [485]
		Over 2½ to 4 in. [Over 65 to 100 mm]	45 [310]	65 [450]
2	Quenched and tempered	2½ in. and under [65 mm and under]	60 [415]	80 [550]
		Over 2½ to 4 in. [Over 65 to 100 mm]	55 [380]	75 [515]
		Over 4 to 6 in. [Over 100 to 150 mm]	46 [315]	70 [485]
3	Quenched and tempered	2½ in. and under [65 mm and under]	55 [380]	80 [550]
		Over 2½ to 4 in. [Over 65 to 100 mm]	50 [345]	75 [515]
		Over 4 to 6 in. [Over 100 to 150 mm]	40 [275]	70 [485]

1.3 The maximum thickness of plates furnished under this specification is 4 in. [100 mm] for Class 1 and 6 in. [150 mm] for Class 2 and Class 3.

1.4 The values stated in either inch-pound units or SI units are to be regarded separately as standard. Within the text, the SI units are shown in brackets. The values stated in each system are not exact equivalents; therefore, each system is to be used independently of the other without combining values in any way.

2. Referenced Documents

2.1 *ASTM Standards:*[3]

A20/A20M Specification for General Requirements for Steel Plates for Pressure Vessels

A435/A435M Specification for Straight-Beam Ultrasonic Examination of Steel Plates

A577/A577M Specification for Ultrasonic Angle-Beam Examination of Steel Plates

A578/A578M Specification for Straight-Beam Ultrasonic Examination of Rolled Steel Plates for Special Applications

3. General Requirements and Ordering Information

3.1 Plates furnished supplied to this material specification shall conform to Specification A20/A20M. These requirements outline the testing and retesting methods and procedures; permissible variations in dimensions; and mass, quality, and repair of defects, marking, loading, etc.

3.2 Specification A20/A20M also establishes the rules for the ordering information which should be complied with when purchasing material to this specification.

3.3 In addition to the basic requirements of this specification, certain supplementary requirements are available when additional control, testing, or examination is required to meet end use requirements. These include:

3.3.1 Vacuum treatment,

3.3.2 Additional or special tension testing,

3.3.3 Impact testing, and

3.3.4 Nondestructive examination.

---

[1] This specification is under the jurisdiction of ASTM Committee A01 on Steel, Stainless Steel and Related Alloys and is the direct responsibility of Subcommittee A01.11 on Steel Plates for Boilers and Pressure Vessels.

Current edition approved Sept. 1, 2008. Published September 2008. Originally approved in 1965. Last previous edition approved in 2006 as A537/A537M – 06. DOI: 10.1520/A0537_A0537M-08.

[2] For ASME Boiler and Pressure Vessel Code applications, see related Specification SA-537/SA-537M in Section II of that Code.

[3] For referenced ASTM standards, visit the ASTM website, www.astm.org, or contact ASTM Customer Service at service@astm.org. For *Annual Book of ASTM Standards* volume information, refer to the standard's Document Summary page on the ASTM website.

---

*A Summary of Changes section appears at the end of this standard.

Copyright © ASTM International, 100 Barr Harbor Drive, PO Box C700, West Conshohocken, PA 19428-2959, United States.

3.4 The purchaser is referred to the listed supplementary requirements in this specification and to the detailed requirements in Specification A20/A20M.

3.5 If the requirements of this specification are in conflict with the requirements of Specification A20/A20M, the requirements of this specification shall prevail.

4. Manufacture

4.1 *Steelmaking Practice*—The steel shall be killed and conform to the fine austenitic grain size requirement of Specification A20/A20M.

5. Heat Treatment

5.1 All plates shall be thermally treated as follows:

5.1.1 Class 1 plates shall be normalized.

5.1.2 Class 2 and Class 3 plates shall be quenched and tempered. The tempering temperature for Class 2 plates shall not be less than 1100°F [595°C] and not less than 1150°F [620°C] for Class 3 plates.

6. Chemical Requirements

6.1 The steel shall conform to the chemical requirements shown in Table 1 unless otherwise modified in accordance with Supplementary Requirement S17, Vacuum Carbon-Deoxidized Steel, in Specification A20/A20M.

7. Mechanical Requirements

7.1 *Tension Tests*:

7.1.1 *Requirements*—The material as represented by the tension-test specimens shall conform to the requirements shown in Table 2.

7.1.2 For Class 2 and Class 3 plates with a nominal thickness of ¾ in. [20 mm] and under, the 1½-in. [40-mm] wide rectangular specimen may be used for the tension test, and the elongation may be determined in a 2-in. [50-mm] gage length that includes the fracture and that shows the greatest elongation.

8. Keywords

8.1 carbon steel plate; pressure containing parts; pressure vessel steels; steel plates for pressure vessel application

A537/A537M – 08

TABLE 1 Chemical Requirements

Element	Composition, %
Carbon, max[A]	0.24
Manganese:	
1½ in. [40 mm] and under in thickness:[B]	
Heat analysis	0.70–1.35
Product analysis	0.64–1.46
Over 1½ in. [40 mm] in thickness:	
Heat analysis	1.00–1.60
Product analysis	0.92–1.72
Phosphorus, max[A]	0.035
Sulfur, max[A]	0.035
Silicon:	
Heat analysis	0.15–0.50
Product analysis	0.13–0.55
Copper, max:	
Heat analysis	0.35
Product analysis	0.38
Nickel, max:[B]	
Heat analysis	0.25
Product analysis	0.28
Chromium, max:	
Heat analysis	0.25
Product analysis	0.29
Molybdenum, max:	
Heat analysis	0.08
Product analysis	0.09

[A] Applies to both heat and product analyses.

[B] Manganese may exceed 1.35 % on heat analysis, up to a maximum of 1.60 %, and nickel may exceed 0.25 % on heat analysis, up to a maximum of 0.50 %, provided the heat analysis carbon equivalent does not exceed 0.57 % when based upon the following equation:

$$CE = C + \frac{Mn}{6} + \frac{Cr + Mo + V}{5} + \frac{Ni + Cu}{15}$$

When this option is exercised, the manganese and nickel contents on product analysis shall not exceed the heat analysis content by more than 0.12 % and 0.03 %, respectively.

TABLE 2 Tensile Requirements

	Class 1	Class 2	Class 3
	ksi [MPa]	ksi [MPa]	ksi [MPa]
Tensile strength:			
2½ in. and under	70–90	80–100	80–100
[65 mm and under]	[485–620]	[550–690]	[550–690]
Over 2½ to 4 in., incl	65–85	75–95	75–95
[Over 65 to 100 mm, incl]	[450–585]	[515–655]	[515–655]
Over 4 to 6 in., incl	[A]	70–90	70–90
[Over 100 to 150 mm, incl]	[A]	[485–620]	[485–620]
Yield strength, min:			
2½ in. and under	50	60	55
[65 mm and under]	[345]	[415]	[380]
Over 2½ to 4 in., incl	45	55	50
[Over 65 to 100 mm, incl]	[310]	[380]	[345]
Over 4 in. to 6 in., incl	[A]	46	40
[Over 100 to 150 mm, incl]	[A]	[315]	[275]
Elongation in 2 in. [50 mm], min, %:[B]			
4 in. [100 mm] and under	22	22	22
Over 4 in. [100 mm]	[A]	20	20
Elongation in 8 in. [200 mm], min, %[B]	18	[C]	[C]

[A] Product is not available in this size range.

[B] See Specification A20/A20M for elongation adjustments.

[C] There is no requirement for elongation in 8 in.

A537/A537M – 08

SUPPLEMENTARY REQUIREMENTS

Supplementary requirements shall not apply unless specified in the order.

A list of standardized supplementary requirements for use at the option of the purchaser are included in Specification A20/A20M. Several of those considered suitable for use with this specification are listed by title. Other tests may be performed by agreement between the supplier and the purchaser.

S1. Vacuum Treatment,

S2. Product Analysis,

S3. Simulated Post-Weld Heat Treatment of Mechanical Test Coupons,

S4.1 Additional Tension Test,

S5. Charpy V-Notch Impact Test,

S6. Drop Weight Test (for Material 0.625 in. [16 mm] and over in Thickness),

S7. High-Temperature Tension Test,

S8. Ultrasonic Examination in accordance with Specification A435/A435M,

S9. Magnetic Particle Examination,

S11. Ultrasonic Examination in accordance with Specification A577/A577M,

S12. Ultrasonic Examination in accordance with Specification A578/A578M, and

S17. Vacuum Carbon-Deoxidized Steel.

SUMMARY OF CHANGES

Committee A01 has identified the location of selected changes to this standard since the last issue (A537/A537M – 06) that may impact the use of this standard. (Approved Sept. 1, 2008.)

(*1*) Revised 5.1.2.

ASTM International takes no position respecting the validity of any patent rights asserted in connection with any item mentioned in this standard. Users of this standard are expressly advised that determination of the validity of any such patent rights, and the risk of infringement of such rights, are entirely their own responsibility.

This standard is subject to revision at any time by the responsible technical committee and must be reviewed every five years and if not revised, either reapproved or withdrawn. Your comments are invited either for revision of this standard or for additional standards and should be addressed to ASTM International Headquarters. Your comments will receive careful consideration at a meeting of the responsible technical committee, which you may attend. If you feel that your comments have not received a fair hearing you should make your views known to the ASTM Committee on Standards, at the address shown below.

This standard is copyrighted by ASTM International, 100 Barr Harbor Drive, PO Box C700, West Conshohocken, PA 19428-2959, United States. Individual reprints (single or multiple copies) of this standard may be obtained by contacting ASTM at the above address or at 610-832-9585 (phone), 610-832-9555 (fax), or service@astm.org (e-mail); or through the ASTM website (www.astm.org).

Designation: A572/A572M – 07

American Association State
Highway and Transportation
Officals Standard
AASHTO No.: M223

Standard Specification for High-Strength Low-Alloy Columbium-Vanadium Structural Steel[1]

This standard is issued under the fixed designation A572/A572M; the number immediately following the designation indicates the year of original adoption or, in the case of revision, the year of last revision. A number in parentheses indicates the year of last reapproval. A superscript epsilon (ε) indicates an editorial change since the last revision or reapproval.

This standard has been approved for use by agencies of the Department of Defense.

1. Scope*

1.1 This specification covers five grades of high-strength low-alloy structural steel shapes, plates, sheet piling, and bars. Grades 42 [290], 50 [345], and 55 [380] are intended for riveted, bolted, or welded structures. Grades 60 [415] and 65 [450] are intended for riveted or bolted construction of bridges, or for riveted, bolted, or welded construction in other applications.

1.2 For applications, such as welded bridge construction, where notch toughness is important, notch toughness requirements are to be negotiated between the purchaser and the producer.

1.3 Specification A588/A588M shall not be substituted for Specification A572/A572M without agreement between the purchaser and the supplier.

1.4 The use of columbium, vanadium, titanium, nitrogen, or combinations thereof, within the limitations noted in Section 5, is required; the selection of type (1, 2, 3, or 5) is at the option of the producer, unless otherwise specified by the purchaser. (See Supplementary Requirement S90.)

1.5 The maximum thicknesses available in the grades and products covered by this specification are shown in Table 1.

1.6 When the steel is to be welded, a welding procedure suitable for the grade of steel and intended use or service is to be utilized. See Appendix X3 of Specification A6/A6M for information on weldability.

1.7 The values stated in either inch-pound units or SI units are to be regarded separately as standard. Within the text, the SI units are shown in brackets. The values stated in each system are not exact equivalents; therefore, each system is to be used independently of the other, without combining values in any way.

1.8 The text of this specification contains notes or footnotes, or both, that provide explanatory material. Such notes and footnotes, excluding those in tables and figures, do not contain any mandatory requirements.

1.9 For structural products produced from coil and furnished without heat treatment or with stress relieving only, the additional requirements, including additional testing requirements and the reporting of additional tests, of A6/A6M apply.

2. Referenced Documents

2.1 *ASTM Standards:*[2]

A6/A6M Specification for General Requirements for Rolled Structural Steel Bars, Plates, Shapes, and Sheet Piling
A36/A36M Specification for Carbon Structural Steel
A514/A514M Specification for High-Yield-Strength, Quenched and Tempered Alloy Steel Plate, Suitable for Welding
A588/A588M Specification for High-Strength Low-Alloy Structural Steel, up to 50 ksi [345 MPa] Minimum Yield Point, with Atmospheric Corrosion Resistance

3. General Requirements for Delivery

3.1 Structural products furnished under this specification shall conform to the requirements of the current edition of Specification A6/A6M, for the specific structural product ordered, unless a conflict exists in which case this specification shall prevail.

3.2 Coils are excluded from qualification to this specification until they are processed into a finished structural product. Structural products produced from coil means structural products that have been cut to individual lengths from a coil. The processor directly controls, or is responsible for, the operations involved in the processing of a coil into a finished structural product. Such operations include decoiling, leveling or

---

[1] This specification is under the jurisdiction of ASTM Committee A01 on Steel, Stainless Steel and Related Alloys and is the direct responsibility of Subcommittee A01.02 on Structural Steel for Bridges, Buildings, Rolling Stock and Ships.
Current edition approved March 1, 2007. Published March 2007. Originally approved in 1966. Last previous edition approved in 2006 as A572/A572M – 06. DOI: 10.1520/A0572_A0572M-07.

[2] For referenced ASTM standards, visit the ASTM website, www.astm.org, or contact ASTM Customer Service at service@astm.org. For *Annual Book of ASTM Standards* volume information, refer to the standard's Document Summary page on the ASTM website.

---

*A Summary of Changes section appears at the end of this standard.

Copyright © ASTM International, 100 Barr Harbor Drive, PO Box C700, West Conshohocken, PA 19428-2959, United States.

TABLE 1 Maximum Product Thickness or Size

Grade	Yield Point, min		Maximum Thickness or Size					
	ksi	[MPa]	Plates and Bars		Structural Shape Flange or Leg Thickness		Sheet Piling	Zees and Rolled Tees
			in.	[mm]	in.	[mm]		
42 [290][A]	42	[290]	6	[150]	all	all	all	all
50 [345][A]	50	[345]	4[B]	[100][B]	all	all	all	all
55 [380]	55	[380]	2	[50]	all	all	all	all
60 [415][A]	60	[415]	1¼[C]	[32][C]	2	[50]	all	all
65 [450]	65	[450]	1¼	[32]	2	[50]	not available	all

[A] In the above tabulation, Grades 42, 50, and 60 [290, 345, and 415], are the yield point levels most closely approximating a geometric progression pattern between 36 ksi [250 MPa], min, yield point steels covered by Specification A36/A36M and 100 ksi [690 MPa], min, yield strength steels covered by Specification A514/A514M.
[B] Round bars up to and including 11 in. [275 mm] in diameter are permitted.
[C] Round bars up to and including 3½ in. [90 mm] in diameter are permitted.

TABLE 2 Chemical Requirements[A]
(Heat Analysis)

Diameter, Thickness, or Distance Between Parallel Faces, in. [mm] Plates and Bars	Structural Shape Flange or Leg Thickness, in. [mm]	Grade	Carbon, max, %	Manganese,[B] max, %	Phosphorus, max, %	Sulfur, max, %	Silicon Plates to 1½ in. [40 mm] Thick, Shapes with Flange or Leg Thickness to 3 in. [75 mm] inclusive, Sheet Piling, Bars, Zees, and Rolled Tees[C]	Plates Over 1½ in. [40 mm] Thick and Shapes with Flange Thickness Over 3 in. [75 mm]
							max, %	range, %
6 [150]	all	42 [290]	0.21	1.35[D]	0.04	0.05	0.40	0.15–0.40
4 [100][E]	all	50 [345]	0.23	1.35[D]	0.04	0.05	0.40	0.15–0.40
2 [50][F]	all	55 [380]	0.25	1.35[D]	0.04	0.05	0.40	0.15–0.40
1¼ [32][F]	≤2 [50]	60 [415]	0.26	1.35[D]	0.04	0.05	0.40	[G]
>½ – 1¼ [13–32]	>1-2 [25-50]	65 [450]	0.23	1.65	0.04	0.05	0.40	[G]
≤½ [13][H]	≤ 1[H]	65 [450]	0.26	1.35	0.04	0.05	0.40	[G]

[A] Copper when specified shall have a minimum content of 0.20 % by heat analysis (0.18 % by product analysis).
[B] Manganese, minimum, by heat analysis of 0.80 % (0.75 % by product analysis) shall be required for all plates over ⅜ in. [10 mm] in thickness; a minimum of 0.50 % (0.45 % by product analysis) shall be required for plates ⅜ in. [10 mm] and less in thickness, and for all other products. The manganese to carbon ratio shall not be less than 2 to 1.
[C] Bars over 1½ in. [40 mm] in diameter, thickness, or distance between parallel faces shall be made by a killed steel practice.
[D] For each reduction of 0.01 percentage point below the specified carbon maximum, an increase of 0.06 percentage point manganese above the specified maximum is permitted, up to a maximum of 1.60 %.
[E] Round bars up to and including 11 in. [275 mm] in diameter are permitted.
[F] Round bars up to and including 3½ in. [90 mm] in diameter are permitted.
[G] The size and grade is not described in this specification.
[H] An alternative chemical requirement with a maximum carbon of 0.21 % and a maximum manganese of 1.65 % is permitted, with the balance of the elements as shown in Table 2.

straightening, hot-forming or cold-forming (if applicable), cutting to length, testing, inspection, conditioning, heat treatment (if applicable), packaging, marking, loading for shipment, and certification.

NOTE 1—For structural products produced from coil and furnished without heat treatment or with stress relieving only, two test results are to be reported for each qualifying coil. Additional requirements regarding structural products produced from coil are described in A6/A6M.

4. Materials and Manufacture

4.1 The steel shall be semi-killed or killed

5. Chemical Composition

5.1 The heat analysis shall conform to the requirements prescribed in Table 2 and Table 3.

5.2 The steel shall conform on product analysis to the requirements prescribed in Table 2 and Table 3, subject to the product analysis tolerances in Specification A6/A6M.

6. Mechanical Properties

6.1 *Tensile Properties*:

6.1.1 The material as represented by the test specimens shall conform to the tensile properties given in Table 4.

7. Test Reports

7.1 In addition to the Test Reports requirements in Specification A6/A6M, when Specification A588/A588M is substituted for Specification A572/A572M, the test report shall include the statement "Specification A588/A588M substituted" and the heat analysis of all elements required in Specification A588/A588M.

8. Keywords

8.1 bars; bolted construction; bridges; buildings; columbium-vanadium; high-strength; low-alloy; plates; riveted construction; shapes; sheet piling; steel; structural steel; welded construction

TABLE 3 Alloy Content

Type[A]	Elements	Heat Analysis, %
1	Columbium[B]	0.005–0.05[C]
2	Vanadium	0.01–0.15
3	Columbium[B]	0.005-0.05[C]
	Vanadium	0.01–0.15
	Columbium plus vanadium	0.02–0.15[D]
5	Titanium	0.006-0.04
	Nitrogen	0.003–0.015
	Vanadium	0.06 max

[A] Alloy content shall be in accordance with Type 1, 2, 3, or 5 and the contents of the applicable elements shall be reported on the test report.

[B] Columbium shall be restricted to the following thicknesses and sizes unless killed steel is furnished. Killed steel shall be confirmed by a statement of killed steel on the test report, or by a report on the presence of a sufficient quantity of a strong deoxidizing element, such as silicon at 0.10 % or higher, or aluminum at 0.015 % or higher. See table below.

[C] Product analysis limits = 0.004 to 0.06 %.

[D] Product analysis limits = 0.01 to 0.16 %.

Grades	Maximum Plate, Bar, Sheet Piling, Zees, and Rolled Tee Thicknesses, in. [mm]	Maximum Structural Shape Flange or Leg Thickness, in. [mm]
42, 50, and 55 [290, 345, and 380]	¾ [20]	1.5 [40]
60 and 65 [415 and 450]	½ [13]	1 [25]

TABLE 4 Tensile Requirements[A]

Grade	Yield Point, min		Tensile Strength, min		Minimum Elongation, %[B,C,D]	
	ksi	[MPa]	ksi	[MPa]	in 8 in. [200 mm]	in 2 in. [50 mm]
42 [290]	42	[290]	60	[415]	20	24
50 [345]	50	[345]	65	[450]	18	21
55 [380]	55	[380]	70	[485]	17	20
60 [415]	60	[415]	75	[520]	16	18
65 [450]	65	[450]	80	[550]	15	17

[A] See specimen Orientation under the Tension Tests section of Specification A6/A6M.

[B] Elongation not required to be determined for floor plate.

[C] For wide flange shapes over 426 lb/ft [634 kg/m], elongation in 2 in. [50 mm] of 19 % minimum applies.

[D] For plates wider than 24 in. [600 mm], the elongation requirement is reduced two percentage points for Grades 42, 50, and 55 [290, 345, and 380], and three percentage points for Grades 60 and 65 [415 and 450]. See elongation requirement adjustments in the Tension Tests section of Specification A6/A6M.

SUPPLEMENTARY REQUIREMENTS

Supplementary requirements shall not apply unless specified in the order or contract. Standardized supplementary requirements for use at the option of the purchaser are listed in Specification A6/A6M. Those that are considered suitable for use with this specification are listed by title:

S5. Charpy V-Notch Impact Test.
S18. Maximum Tensile Strength
S30. Charpy V-Notch Impact Test for Structural Shapes: Alternate Core Location
S32. Single Heat Bundles.

In addition, the following supplementary requirements are suitable for use:

S81. Tensile Strength

S81.1 For Grade 50 [345] steel of thicknesses ¾ in. [20 mm] and less, the tensile strength shall be a minimum of 70 ksi [485 MPa].

S90. Type

S90.1 The specific type of steel shall be as specified by the purchaser in the order or contract.

SUMMARY OF CHANGES

Committee A01 has identified the location of selected changes to this standard since the last issue (A572/A572M– 06) that may impact the use of this standard. (Approved March 1, 2007.)

(1) Added 1.3 and new Section 7.

Committee A01 has identified the location of selected changes to this standard since the last issue (A572/A572M– 04) that may impact the use of this standard. (Approved March 1, 2006.)

(1) The maximum permitted manganese was changed from 1.50 % to 1.60 % for grades 42, 50, 55, and 60 in Table 2.

ASTM International takes no position respecting the validity of any patent rights asserted in connection with any item mentioned in this standard. Users of this standard are expressly advised that determination of the validity of any such patent rights, and the risk of infringement of such rights, are entirely their own responsibility.

This standard is subject to revision at any time by the responsible technical committee and must be reviewed every five years and if not revised, either reapproved or withdrawn. Your comments are invited either for revision of this standard or for additional standards and should be addressed to ASTM International Headquarters. Your comments will receive careful consideration at a meeting of the responsible technical committee, which you may attend. If you feel that your comments have not received a fair hearing you should make your views known to the ASTM Committee on Standards, at the address shown below.

This standard is copyrighted by ASTM International, 100 Barr Harbor Drive, PO Box C700, West Conshohocken, PA 19428-2959, United States. Individual reprints (single or multiple copies) of this standard may be obtained by contacting ASTM at the above address or at 610-832-9585 (phone), 610-832-9555 (fax), or service@astm.org (e-mail); or through the ASTM website (www.astm.org).

Designation: A573/A573M – 05 (Reapproved 2009)

Standard Specification for
Structural Carbon Steel Plates of Improved Toughness[1]

This standard is issued under the fixed designation A573/A573M; the number immediately following the designation indicates the year of original adoption or, in the case of revision, the year of last revision. A number in parentheses indicates the year of last reapproval. A superscript epsilon (ε) indicates an editorial change since the last revision or reapproval.

1. Scope

1.1 This specification covers structural quality carbon-manganese-silicon steel plates in three tensile strength ranges intended primarily for service at atmospheric temperatures where improved notch toughness is important.

1.2 Plates covered by this specification are limited to a maximum thickness of 1.5 in. [40 mm].

1.3 If the steel is to be welded, it is presupposed that a welding procedure suitable for the grade of steel and intended use or service will be utilized. See Appendix X3 of Specification A6/A6M for information on weldability.

1.4 The values stated in either inch-pound units or SI units are to be regarded separately as standard. Within the text, the SI units are shown in brackets. The values stated in each system are not exact equivalents; therefore, each system is to be used independently of the other without combining values in any way.

2. Referenced Documents

2.1 *ASTM Standards:*[2]

A6/A6M Specification for General Requirements for Rolled Structural Steel Bars, Plates, Shapes, and Sheet Piling

3. General Requirements for Delivery

3.1 Plates furnished under this specification shall conform to the requirements of the current edition of Specification A6/A6M, unless a conflict exists in which case this specification shall prevail.

---

[1] This specification is under the jurisdiction of ASTM Committee A01 on Steel, Stainless Steel and Related Alloys and is the direct responsibility of Subcommittee A01.02 on Structural Steel for Bridges, Buildings, Rolling Stock and Ships.
Current edition approved Oct. 1, 2009. Published December 2009. Originally approved in 1966. Last previous edition approved in 2005 as A573/A573M – 05. DOI: 10.1520/A0573_A0573M-05R09.

[2] For referenced ASTM standards, visit the ASTM website, www.astm.org, or contact ASTM Customer Service at service@astm.org. For *Annual Book of ASTM Standards* volume information, refer to the standard's Document Summary page on the ASTM website.

TABLE 1 Chemical Requirements (Heat Analysis)

	Composition, %		
	Grade 58 [400]	Grade 65 [450]	Grade 70 [485]
Carbon, max:			
½ in. [13 mm] and under	0.23	0.24	0.27
Over ½ in. [13 mm] to 1½ in., [40 mm], incl	0.23	0.26	0.28
Manganese[A]	0.60–0.90	0.85–1.20	0.85–1.20
Phosphorus, max	0.035	0.035	0.035
Sulfur, max	0.04	0.04	0.04
Silicon	0.10–0.35	0.15–0.40	0.15–0.40

[A] For each reduction of 0.01 percentage point below the specified maximum for carbon, an increase of 0.06 percentage points above the specified maximum for manganese is permitted, up to a maximum of 1.50 % for Grades 58 and 65; and up to a maximum of 1.60 % for Grade 70.

4. Materials and Manufacture

4.1 The steel shall be made to fine grain practice.

5. Chemical Composition

5.1 The heat analysis shall conform to the requirements given in Table 1.

5.2 The product analysis shall conform to the requirements given in Table 1 subject to the product analysis tolerances in Specification A6/A6M.

6. Tension Test

6.1 The plates, as represented by the tension test specimens, shall conform to the tensile requirements given in Table 2.

7. Keywords

7.1 carbon steel; plates; structural steel; toughness; welded construction

TABLE 2 Tensile Requirements[A]

	Grade 58 [400]	Grade 65 [450]	Grade 70 [485]
Tensile strength, ksi [MPa]	58–71 [400–490]	65–77 [450–530]	70–90 [485–620]
Yield point, min, ksi [MPa]	32 [220]	35 [240]	42 [290]
Elongation in 8 in. [200 mm] min[B,C], %	21	20	18
Elongation in 2 in. [50 mm], min[B,C], %	24	23	21

[A] See the Orientation subsection in the Tension Tests section of Specification A6/A6M.
[B] Elongation need not be determined for floor plate.
[C] For plates wider than 24 in. [600 mm], the elongation requirement is reduced two percentage points. See the Elongation Requirement Adjustments subsection in the Tension Tests section of Specification A6/A6M.

SUPPLEMENTARY REQUIREMENTS

Standardized supplementary requirements for use at the option of the purchaser are listed in Specification A6/A6M. Supplementary requirements shall not apply unless specified in the purchase order or contract.

ASTM International takes no position respecting the validity of any patent rights asserted in connection with any item mentioned in this standard. Users of this standard are expressly advised that determination of the validity of any such patent rights, and the risk of infringement of such rights, are entirely their own responsibility.

This standard is subject to revision at any time by the responsible technical committee and must be reviewed every five years and if not revised, either reapproved or withdrawn. Your comments are invited either for revision of this standard or for additional standards and should be addressed to ASTM International Headquarters. Your comments will receive careful consideration at a meeting of the responsible technical committee, which you may attend. If you feel that your comments have not received a fair hearing you should make your views known to the ASTM Committee on Standards, at the address shown below.

This standard is copyrighted by ASTM International, 100 Barr Harbor Drive, PO Box C700, West Conshohocken, PA 19428-2959, United States. Individual reprints (single or multiple copies) of this standard may be obtained by contacting ASTM at the above address or at 610-832-9585 (phone), 610-832-9555 (fax), or service@astm.org (e-mail); or through the ASTM website (www.astm.org). Permission rights to photocopy the standard may also be secured from the ASTM website (www.astm.org/COPYRIGHT/).

Designation: A578/A578M – 07

Standard Specification for
Straight-Beam Ultrasonic Examination of Rolled Steel Plates for Special Applications[1]

This standard is issued under the fixed designation A578/A578M; the number immediately following the designation indicates the year of original adoption or, in the case of revision, the year of last revision. A number in parentheses indicates the year of last reapproval. A superscript epsilon (ε) indicates an editorial change since the last revision or reapproval.

This standard has been approved for use by agencies of the Department of Defense.

1. Scope*

1.1 This specification[2] covers the procedure and acceptance standards for straight-beam, pulse-echo, ultrasonic examination of rolled carbon and alloy steel plates, 3/8 in. [10 mm] in thickness and over, for special applications. The method will detect internal discontinuities parallel to the rolled surfaces. Three levels of acceptance standards are provided. Supplementary requirements are provided for alternative procedures.

1.2 Individuals performing examinations in accordance with this specification shall be qualified and certified in accordance with the requirements of the latest edition of ASNT SNT-TC-1A or an equivalent accepted standard. An equivalent standard is one which covers the qualification and certification of ultrasonic nondestructive examination candidates and which is acceptable to the purchaser.

1.3 The values stated in either SI units or inch-pound units are to be regarded separately as standard. The values stated in each system may not be exact equivalents; therefore, each system shall be used independently of the other. Combining values from the two systems may result in non-conformance with the standard.

1.4 *This standard does not purport to address all of the safety concerns, if any, associated with its use. It is the responsibility of the user of this standard to establish appropriate safety and health practices and determine the applicability of regulatory limitations prior to use.*

2. Referenced Documents

2.1 *ASTM Standards:*[3]
A263 Specification for Stainless Chromium Steel-Clad Plate
A264 Specification for Stainless Chromium-Nickel Steel-Clad Plate
A265 Specification for Nickel and Nickel-Base Alloy-Clad Steel Plate

2.2 *ANSI Standard:*[4]
B 46.1 Surface Texture

2.3 *ASNT Standard:*[5]
SNT-TC-1A

3. Ordering Information

3.1 The inquiry and order shall indicate the following:

3.1.1 Acceptance level requirements (Sections 7, 8, and 9). Acceptance Level B shall apply unless otherwise agreed to by purchaser and manufacturer.

3.1.2 Any additions to the provisions of this specification as prescribed in 5.2, 13.1, and Section 10.

3.1.3 Supplementary requirements, if any.

4. Apparatus

4.1 The amplitude linearity shall be checked by positioning the transducer over the depth resolution notch in the IIW or similar block so that the signal from the notch is approximately 30 % of the screen height, and the signal from one of the back

---

[1] This specification is under the jurisdiction of ASTM Committee A01 on Steel, Stainless Steel and Related Alloys and is the direct responsibility of Subcommittee A01.11 on Steel Plates for Boilers and Pressure Vessels.
Current edition approved Nov. 1, 2007. Published December 2007. Originally approved in 1967. Last previous edition approved in 2001 as A578/A578M–96(2001). DOI: 10.1520/A0578_A0578M-07.

[2] For ASME Boiler and Pressure Vessel Code applications, see related Specification SA-578/SA-578M in Section II of that Code.

[3] For referenced ASTM standards, visit the ASTM website, www.astm.org, or contact ASTM Customer Service at service@astm.org. For *Annual Book of ASTM Standards* volume information, refer to the standard's Document Summary page on the ASTM website.

[4] Available from American National Standards Institute (ANSI), 25 W. 43rd St., 4th Floor, New York, NY 10036, http://www.ansi.org.

[5] Available from American Society for Nondestructive Testing (ASNT), P.O. Box 28518, 1711 Arlingate Ln., Columbus, OH 43228-0518, http://www.asnt.org.

A Summary of Changes section appears at the end of this standard.

surfaces is approximately 60 % of the screen height (two times the height of the signal from the notch). A curve is then plotted showing the deviations from the above established 2:1 ratio that occurs as the amplitude of the signal from the notch is raised in increments of one scale division until the back reflection signal reaches full scale, and then is lowered in increments of one scale division until the notch signal reaches one scale division. At each increment the ratio of the two signals is determined. The ratios are plotted on the graph at the position corresponding to the larger signal. Between the limits of 20 and 80 % of the screen height, the ratio shall be within 10 % of 2:1. Instrument settings used during inspection shall not cause variation outside the 10 % limits established above.

4.2 The transducer shall be 1 or 1⅛ in. [25 or 30 mm] in diameter or 1 in. [25 mm] square.

4.3 Other search units may be used for evaluating and pinpointing indications.

5. Procedure

5.1 Perform the inspection in an area free of operations that interfere with proper performance of the test.

5.2 Unless otherwise specified, make the ultrasonic examination on either major surface of the plate.

5.3 The plate surface shall be sufficiently clean and smooth to maintain a first reflection from the opposite side of the plate at least 50 % of full scale during scanning. This may involve suitable means of scale removal at the manufacturer's option. Condition local rough surfaces by grinding. Restore any specified identification which is removed when grinding to achieve proper surface smoothness.

5.4 Perform the test by one of the following methods: direct contact, immersion, or liquid column coupling. Use a suitable couplant such as water, soluble oil, or glycerin. As a result of the test by this method, the surface of plates may be expected to have a residue of oil or rust or both.

5.5 A nominal test frequency of 2¼ MHz is recommended. When testing plates less than ¾ in. [20 mm] thick a frequency of 5 MHz may be necessary. Thickness, grain size or microstructure of the material and nature of the equipment or method may require a higher or lower test frequency. Use the transducers at their rated frequency. A clean, easily interpreted trace pattern should be produced during the examination.

5.6 *Scanning*:

5.6.1 Scanning shall be along continuous perpendicular grid lines on nominal 9-in. [225-mm] centers, or at the option of the manufacturer, shall be along continuous parallel paths, transverse to the major plate axis, on nominal 4-in. [100-mm] centers, or shall be along continuous parallel paths parallel to the major plate axis, on 3-in. [75-mm] or smaller centers. Measure the lines from the center or one corner of the plate with an additional path within 2 in. [50 mm] of all edges of the plate on the searching surface.

5.6.2 Conduct the general scanning with an instrument adjustment that will produce a first reflection from the opposite side of a sound area of the plate from 50 to 90 % of full scale. Minor sensitivity adjustments may be made to accommodate for surface roughness.

5.6.3 When a discontinuity condition is observed during general scanning adjust the instrument to produce a first reflection from the opposite side of a sound area of the plate of 75 ± 5 % of full scale. Maintain this instrument setting during evaluation of the discontinuity condition.

6. Recording

6.1 Record all discontinuities causing complete loss of back reflection.

6.2 For plates ¾ in. [20 mm] thick and over, record all indications with amplitudes equal to or greater than 50 % of the initial back reflection and accompanied by a 50 % loss of back reflection.

NOTE 1—Indications occurring midway between the initial pulse and the first back reflection may cause a second reflection at the location of the first back reflection. When this condition is observed it shall be investigated additionally by use of multiple back reflections.

6.3 Where grid scanning is performed and recordable conditions as in 6.1 and 6.2 are detected along a given grid line, the entire surface area of the squares adjacent to this indication shall be scanned. Where parallel path scanning is performed and recordable conditions as in 6.1 and 6.2 are detected, the entire surface area of a 9 by 9-in. [225 by 225-mm] square centered on this indication shall be scanned. The true boundaries where these conditions exist shall be established in either method by the following technique: Move the transducer away from the center of the discontinuity until the height of the back reflection and discontinuity indications are equal. Mark the plate at a point equivalent to the center of the transducer. Repeat the operation to establish the boundary.

7. Acceptance Standard—Level A

7.1 Any area where one or more discontinuities produce a continuous total loss of back reflection accompanied by continuous indications on the same plane (within 5 % of plate thickness) that cannot be encompassed within a circle whose diameter is 3 in. [75 mm] or ½ of the plate thickness, whichever is greater, is unacceptable.

8. Acceptance Standards—Level B

8.1 Any area where one or more discontinuities produce a continuous total loss of back reflection accompanied by continuous indications on the same plane (within 5 % of plate thickness) that cannot be encompassed within a circle whose diameter is 3 in. [75 mm] or ½ of the plate thickness, whichever is greater, is unacceptable.

8.2 In addition, two or more discontinuities smaller than described in 8.1 shall be unacceptable unless separated by a minimum distance equal to the greatest diameter of the larger discontinuity or unless they may be collectively encompassed by the circle described in 8.1.

9. Acceptance Standard—Level C

9.1 Any area where one or more discontinuities produce a continuous total loss of back reflection accompanied by continuous indications on the same plane (within 5 % of plate thickness) that cannot be encompassed within a 1-in. [25-mm] diameter circle is unacceptable.

10. Report

10.1 Unless otherwise agreed to by the purchaser and the manufacturer, the manufacturer shall report the following data:

10.1.1 All recordable indications listed in Section 6 on a sketch of the plate with sufficient data to relate the geometry and identity of the sketch to those of the plate.

10.1.2 Test parameters including: Make and model of instrument, test frequency, surface condition, transducer (type and frequency), and couplant.

10.1.3 Date of test.

11. Inspection

11.1 The inspector representing the purchaser shall have access at all times, while work on the contract of the purchaser is being performed, to all parts of the manufacturer's works that concern the ultrasonic testing of the material ordered. The manufacturer shall afford the inspector all reasonable facilities to satisfy him that the material is being furnished in accordance with this specification. All tests and inspections shall be made at the place of manufacture prior to shipment, unless otherwise specified, and shall be conducted without interfering unnecessarily with the manufacturer's operations.

12. Rehearing

12.1 The manufacturer reserves the right to discuss rejectable ultrasonically tested plate with the purchaser with the object of possible repair of the ultrasonically indicated defect before rejection of the plate.

13. Marking

13.1 Plates accepted according to this specification shall be identified by stenciling (stamping) "UT A578—A" on one corner for Level A, "UT A578—B" for Level B, and "UT A578—C" for Level C. The supplement number shall be added for each supplementary requirement ordered.

14. Keywords

14.1 nondestructive testing; pressure containing parts; pressure vessel steels; steel plate for pressure vessel applications; steel plates; ultrasonic examinations

SUPPLEMENTARY REQUIREMENTS

These supplementary requirements shall apply only when individually specified by the purchaser. When details of these requirements are not covered herein, they are subject to agreement between the manufacturer and the purchaser.

S1. Scanning

S1.1 Scanning shall be continuous over 100 % of the plate surface along parallel paths, transverse or parallel to the major plate axis, with not less than 10 % overlap between each path.

S2. Acceptance Standard

S2.1 Any recordable condition listed in Section 6 that (*1*) is continuous, (*2*) is on the same plane (within 5 % of the plate thickness), and (*3*) cannot be encompassed by a 3-in. [75-mm] diameter circle, is unacceptable. Two or more recordable conditions (see Section 5), that (*1*) are on the same plane (within 5 % of plate thickness), (*2*) individually can be encompassed by a 3-in. [75-mm] diameter circle, (*3*) are separated from each other by a distance less than the greatest dimension of the smaller indication, and (*4*) collectively cannot be encompassed by a 3-in. [75-mm] diameter circle, are unacceptable.

S2.2 An acceptance level more restrictive than Section 7 or 8 shall be used by agreement between the manufacturer and purchaser.

S3. Procedure

S3.1 The manufacturer shall provide a written procedure in accordance with this specification.

S4. Certification

S4.1 The manufacturer shall provide a written certification of the ultrasonic test operator's qualifications.

S5. Surface Finish

S5.1 The surface finish of the plate shall be conditioned to a maximum 125 μin. [3 μm] AA (see ANSI B 46.1) prior to test.

S6. Withdrawn

See Specifications A263, A264, and A265 for equivalent descriptions for clad quality level.

S7. Withdrawn

See Specifications A263, A264, and A265 for equivalent descriptions for clad quality level.

S8. Ultrasonic Examination Using Flat Bottom Hole Calibration (for Plates 4 in. [100 mm] Thick and Greater)

S8.1 Use the following calibration and recording procedures in place of 5.6.2, 5.6.3, and Section 6.

S8.2 The transducer shall be in accordance with 4.2.

S8.3 *Reference Reflectors*—The $T/4$, $T/2$, and $3T/4$ deep flat bottom holes shall be used to calibrate the equipment. The flat bottom hole diameter shall be in accordance with Table S8.1. The holes may be drilled in the plate to be examined if they can

TABLE S8.1 Calibration Hole Diameter as a Function of Plate Thickness (S8)

Plate Thickness, in. [mm]	4–6 [100–150]	>6–9 [>150–225]	>9–12 [>225–300]	>12–20 [>300–500]
Hole Diameter, in. [mm]	⅝ [16]	¾ [19]	⅞ [22]	1⅛ [29]

be located without interfering with the use of the plate, in a prolongation of the plate to be examined, or in a reference block of the same nominal composition, and thermal treatment as the plate to be examined. The surface of the reference block shall be no better to the unaided eye than the plate surface to be examined. The reference block shall be of the same nominal thickness (within 75 to 125 % or 1 in. [25 mm] of the examined plate, whichever is less) and shall have acoustical properties similar to the examined plate. Acoustical similarity is presumed when, without a change in instrument setting, comparison of the back reflection signals between the reference block and the examined plate shows a variation of 25 % or less.

S8.4 *Calibration Procedure:*

S8.4.1 Couple and position the search unit for maximum amplitudes from the reflectors at $T/4$, $T/2$, and $3T/4$. Set the instrument to produce a 75 ± 5 % of full scale indication from the reflector giving the highest amplitude.

S8.4.2 Without changing the instrument setting, couple and position the search unit over each of the holes and mark on the screen the maximum amplitude from each hole and each minimum remaining back reflection.

S8.4.3 Mark on the screen half the vertical distance from the sweep line to each maximum amplitude hole mark. Connect the maximum amplitude hole marks and extend the line through the thickness for the 100 % DAC (distance amplitude correction curve). Similarly connect and extend the half maximum amplitude marks for the 50 % DAC.

S8.5 *Recording:*

S8.5.1 Record all areas where the remaining back reflection is smaller than the highest of the minimum remaining back reflections found in S8.4.2.

S8.5.2 Record all areas where indications exceed 50 % DAC.

S8.5.3 Where recordable conditions listed in S8.5.1 and S8.5.2 are detected along a given grid line, continuously scan the entire surface area of the squares adjacent to the condition and record the boundaries or extent of each recordable condition.

S8.6 Scanning shall be in accordance with 5.6.

S8.7 The acceptance levels of Section 7 or 8 shall apply as specified by the purchaser except that the recordable condition shall be as given in S8.5.

S9. Ultrasonic Examination of Electroslag Remelted (ESR) and Vacuum-Arc Remelted (VAR) Plates, from 1 to 16 in. [25 to 400 mm] in Thickness, Using Flat-Bottom Hole Calibration and Distance-Amplitude Corrections

S9.1 The material to be examined must have a surface finish of 200 µin. [5 µm] as maximum for plates up to 8 in. [200 mm] thick, inclusive, and 250 µin. [6 µm] as maximum for plates over 8 to 16 in. [200 to 400 mm] thick.

S9.2 Use the following procedures in place of 5.6.1, 5.6.2, 5.6.3, and Section 6.

S9.3 The transducer shall be in accordance with 4.2.

S9.4 *Reference Reflectors*—The $T/4$, $T/2$, and $3T/4$ deep flat bottom holes shall be used to calibrate the equipment. The flat bottom hole diameter shall be in accordance with Table S9.1. The flat bottoms of the holes shall be within 1° of parallel to the examination surface. The holes may be drilled in the plate to be examined if they can be located without interfering with the use of the plate, in a prolongation of the plate to be examined, or in a reference block of the same nominal composition and thermal treatment as the plate to be examined. The surface of the reference block shall be no better to the unaided eye than the plate surface to be examined. The reference block shall be of the same nominal thickness (within 75 to 125 % or 1 in. [25 mm] of the examined plate, whichever is less) and shall have acoustical properties similar to the examined plate. Acoustical similarity is presumed when, without a change in instrument setting, comparison of the back reflection signals between the reference block and the examined plate shows a variation of 25 % or less.

S9.5 *Calibration Procedure:*

S9.5.1 Couple and position the search unit for maximum amplitudes from the reflectors at $T/4$, $T/2$, and $3T/4$. Set the instrument to produce a 75 ± 5 % of full-scale indication from the reflector giving the highest amplitude.

S9.5.2 Without changing the instrument setting, couple and position the search unit over each of the holes and mark on the screen the maximum amplitude from each of the holes.

S9.5.3 Mark on the screen half the vertical distances from the sweep line to each maximum amplitude hole mark. Connect the maximum amplitude hole marks and extend the line through the thickness for the 100 % DAC (distance amplitude correction curve). Similarly connect and extend the half maximim amplitude marks for the 50 % DAC.

S9.6 *Scanning*—Scanning shall cover 100 % of one major plate surface, with the search unit being indexed between each pass such that there is at least 15 % overlap of adjoining passes in order to assure adequate coverage for locating discontinuities.

S9.7 *Recording*—Record all areas where the back reflection drops below the 50 % DAC. If the drop in back reflection is not accompanied by other indications on the screen, recondition the surface in the area and reexamine ultrasonically. If the back reflection is still below 50 % DAC, the loss may be due to the metallurgical structure of the material being examined. The material shall be held for metallurgical review by the purchaser and manufacturer.

S9.8 *Acceptance Standards*—Any indication that exceeds the 100 % DAC shall be considered unacceptable. The manufacturer may reserve the right to discuss rejectable ultrasonically examined material with the purchaser, the object being the possible repair of the ultrasonically indicated defect before rejection of the plate.

TABLE S9.1 Calibration Hole Diameter as a Function of Plate Thickness (S9)

Plate Thickness, in. [mm]	1–4 [25–100]	>4–8 [>100–200]	>8–12 [>200–300]	>12–16 [>300–400]
Hole Diameter, in. [mm]	⅛ [3]	¼ [6]	⅜ [10]	½ [13]

A578/A578M – 07

SUMMARY OF CHANGES

Committee A01 has identified the location of selected changes to this standard since the last issue (A578/A578M–96(2001)) that may impact the use of this standard.

(*1*) Changed title to reflect deletion of clad reference.
(*2*) Deleted Supplementary Requirements S6 and S7.
(*3*) Expanded Supplementary Requirement S1 coverage to include overlap provision.
(*4*) Revised acceptance levels revised to clarify "same plane."

ASTM International takes no position respecting the validity of any patent rights asserted in connection with any item mentioned in this standard. Users of this standard are expressly advised that determination of the validity of any such patent rights, and the risk of infringement of such rights, are entirely their own responsibility.

This standard is subject to revision at any time by the responsible technical committee and must be reviewed every five years and if not revised, either reapproved or withdrawn. Your comments are invited either for revision of this standard or for additional standards and should be addressed to ASTM International Headquarters. Your comments will receive careful consideration at a meeting of the responsible technical committee, which you may attend. If you feel that your comments have not received a fair hearing you should make your views known to the ASTM Committee on Standards, at the address shown below.

This standard is copyrighted by ASTM International, 100 Barr Harbor Drive, PO Box C700, West Conshohocken, PA 19428-2959, United States. Individual reprints (single or multiple copies) of this standard may be obtained by contacting ASTM at the above address or at 610-832-9585 (phone), 610-832-9555 (fax), or service@astm.org (e-mail); or through the ASTM website (www.astm.org).

Designation: A588/A588M – 05

American Association State
Highway and Transportation Officials Standard
AASHTO No.: M 222

Standard Specification for
High-Strength Low-Alloy Structural Steel, up to 50 ksi [345 MPa] Minimum Yield Point, with Atmospheric Corrosion Resistance[1]

This standard is issued under the fixed designation A588/A588M; the number immediately following the designation indicates the year of original adoption or, in the case of revision, the year of last revision. A number in parentheses indicates the year of last reapproval. A superscript epsilon (ε) indicates an editorial change since the last revision or reapproval.

This standard has been approved for use by agencies of the Department of Defense.

1. Scope*

1.1 This specification covers high-strength low-alloy structural steel shapes, plates, and bars for welded, riveted, or bolted construction but intended primarily for use in welded bridges and buildings where savings in weight or added durability are important. The atmospheric corrosion resistance of this steel in most environments is substantially better than that of carbon structural steels with or without copper addition (see Note 1). When properly exposed to the atmosphere, this steel is suitable for many applications in the bare (unpainted) condition. This specification is limited to material up to 8 in. [200 mm] inclusive in thickness.

NOTE 1—For methods of estimating the atmospheric corrosion resistance of low-alloy steels, see Guide G101.

1.2 When the steel is to be welded, a welding procedure suitable for the grade of steel and intended use or service is to be utilized. See Appendix X3 of Specification A6/A6M for information on weldability.

1.3 The values stated in either inch-pound units or SI units are to be regarded separately as standard. Within the text, the SI units are shown in brackets. The values stated in each system are not exact equivalents; therefore, each system is to be used independently of the other, without combining values in any way.

1.4 The text of this specification contains notes, footnotes, or both, that provide explanatory material. Such notes and footnotes, excluding those in tables and figures, do not contain any mandatory requirements.

1.5 For structural products produced from coil and furnished without heat treatment or with stress relieving only, the additional requirements, including additional testing requirements and the reporting of additional test results, of Specification A6/A6M apply.

2. Referenced Documents

2.1 *ASTM Standards:*[2]

A6/A6M Specification for General Requirements for Rolled Structural Steel Bars, Plates, Shapes, and Sheet Piling

G101 Guide for Estimating the Atmospheric Corrosion Resistance of Low-Alloy Steels

3. General Requirements for Delivery

3.1 Structural products furnished under this specification shall conform to the requirements of the current edition of Specification A6/A6M, for the specific structural product ordered, unless a conflict exists, in which case this specification shall prevail.

3.2 Coils are excluded from qualification to this specification until they are processed into a finished structural product. Structural products produced from coil means structural products that have been cut to individual lengths from a coil. The processor directly controls, or is responsible for, the operations involved in the processing of a coil into a finished structural product. Such operations include decoiling, leveling or straightening, hot-forming or cold-forming (if applicable), cutting to length, testing, inspection, conditioning, heat treatment (if applicable), packaging, marking, loading for shipment, and certification.

NOTE 2—For structural products produced from coil and furnished without heat treatment or with stress relieving only, two test results are to be reported for each qualifying coil. Additional requirements regarding structural products produced from coil are described in Specification A6/A6M.

4. Materials and Manufacture

4.1 The steel shall be made to fine grain practice.

5. Chemical Composition

5.1 The heat analysis shall conform to the requirements prescribed in Table 1.

---

[1] This specification is under the jurisdiction of ASTM Committee A01 on Steel, Stainless Steel and Related Alloys and is the direct responsibility of Subcommittee A01.02 on Structural Steel for Bridges, Buildings, Rolling Stock and Ships.
Current edition approved April 1, 2005. Published April 2005. Originally approved in 1968. Last previous edition approved in 2004 as A588/A588M – 04. DOI: 10.1520/A0588_A0588M-05.

[2] For referenced ASTM standards, visit the ASTM website, www.astm.org, or contact ASTM Customer Service at service@astm.org. For *Annual Book of ASTM Standards* volume information, refer to the standard's Document Summary page on the ASTM website.

*A Summary of Changes section appears at the end of this standard.

Copyright © ASTM International, 100 Barr Harbor Drive, PO Box C700, West Conshohocken, PA 19428-2959, United States.

TABLE 1 Chemical Requirements (Heat Analysis)

NOTE 1—Where "..." appears in this table, there is no requirement.

Element	Composition, %			
	Grade A	Grade B	Grade C	Grade K
Carbon[A]	0.19 max	0.20 max	0.15 max	0.17 max
Manganese[A]	0.80–1.25	0.75–1.35	0.80–1.35	0.50–1.20
Phosphorus	0.04 max	0.04 max	0.04 max	0.04 max
Sulfur	0.05 max	0.05 max	0.05 max	0.05 max
Silicon	0.30–0.65	0.15–0.50	0.15–0.40	0.25–0.50
Nickel	0.40 max	0.50 max	0.25–0.50	0.40 max
Chromium	0.40–0.65	0.40–0.70	0.30–0.50	0.40–0.70
Molybdenum	...	...	...	0.10 max
Copper	0.25–0.40	0.20–0.40	0.20–0.50	0.30–0.50
Vanadium	0.02–0.10	0.01–0.10	0.01–0.10	...
Columbium	...	...	...	0.005–0.05[B]

[A]For each reduction of 0.01 percentage point below the specified maximum for carbon, an increase of 0.06 percentage point above the specified maximum for manganese is permitted, up to a maximum of 1.50 %.

[B]For plates under ½ in. [13 mm] in thickness, the minimum columbium is waived.

5.2 The steel shall conform on product analysis to the requirements prescribed in Table 1, subject to the product analysis tolerances in Specification A6/A6M.

5.3 The atmospheric corrosion-resistance index, calculated on the basis of the heat analysis of the steel, as described in Guide G101–Predictive Method Based on the Data of Larabee and Coburn, shall be 6.0 or higher.

NOTE 3—The user is cautioned that the Guide G101 predictive equation (Predictive Method Based on the Data of Larabee and Coburn) for calculation of an atmospheric corrosion-resistance index has only been verified for the composition limits stated in the guide.

5.4 When required, the manufacturer shall supply evidence of corrosion resistance satisfactory to the purchaser.

6. Tensile Requirements

6.1 The material as represented by the test specimens shall conform to the requirements for tensile properties prescribed in Table 2.

7. Keywords

7.1 atmospheric corrosion resistance; bars; bolted construction; bridges; buildings; durability; high-strength; low-alloy; plates; riveted construction; shapes; steel; structural steel; weight; welded construction

TABLE 2 Tensile Requirements[A]

NOTE 1—Where "..." appears in this table, there is no requirement.

	Plates and Bars			Structural Shapes
	For Thicknesses 4 in. [100 mm] and Under	For Thicknesses Over 4 in. [100 mm] to 5 in. [125 mm] incl	For Thicknesses Over 5 in. [125 mm] to 8 in. [200 mm] incl	All
Tensile strength, min, ksi [MPa]	70 [485]	67 [460]	63 [435]	70 [485]
Yield point, min, ksi [MPa]	50 [345]	46 [315]	42 [290]	50 [345]
Elongation in 8 in. [200 mm], min, %	18[B,C]	...	...	18[C]
Elongation in 2 in. [50 mm], min, %	21[B,C]	21[B,C]	21[B,C]	21[D]

[A]See specimen orientation under the Tension Tests section of Specification A6/A6M.

[B]Elongation not required to be determined for floor plate.

[C]For plates wider than 24 in. [600 mm], the elongation requirement is reduced two percentage points. See elongation requirement adjustments in the Tension Tests section of Specification A6/A6M.

[D]For wide flange shapes with flange thickness over 3 in. [75 mm], elongation in 2 in. [50 mm] of 18 % minimum applies.

SUPPLEMENTARY REQUIREMENTS

Supplementary requirements shall not apply unless specified in the order or contract. Standardized supplementary requirements for use at the option of the purchaser are listed in Specification A6/A6M. Those that are considered suitable for use with this specification are listed by title:

S2. Product Analysis,
S3. Simulated Post-Weld Heat Treatment of Mechanical Test Coupons,
S5. Charpy V-Notch Impact Test,
S6. Drop-Weight Test (for Material 0.625 in. [16 mm] and over in Thickness),
S8. Ultrasonic Examination,
S15. Reduction of Area Measurement,
S18. Maximum Tensile Strength,
S30. Charpy V-Notch Impact Test for Structural Shapes: Alternate Core Location, and
S32. Single Heat Bundles.

A588/A588M – 05
SUMMARY OF CHANGES

Committee A01 has identified the location of selected changes to this standard since the last issue (A588/A588M - 04) that may impact the use of this standard. (Approved April 1, 2005.)

(1) Revised the title.

Committee A01 has identified the location of selected changes to this standard since the last issue (A588/A588M - 03a) that may impact the use of this standard. (Approved April 1, 2004.)

(1) Added Supplementary Requirement S32.

ASTM International takes no position respecting the validity of any patent rights asserted in connection with any item mentioned in this standard. Users of this standard are expressly advised that determination of the validity of any such patent rights, and the risk of infringement of such rights, are entirely their own responsibility.

This standard is subject to revision at any time by the responsible technical committee and must be reviewed every five years and if not revised, either reapproved or withdrawn. Your comments are invited either for revision of this standard or for additional standards and should be addressed to ASTM International Headquarters. Your comments will receive careful consideration at a meeting of the responsible technical committee, which you may attend. If you feel that your comments have not received a fair hearing you should make your views known to the ASTM Committee on Standards, at the address shown below.

This standard is copyrighted by ASTM International, 100 Barr Harbor Drive, PO Box C700, West Conshohocken, PA 19428-2959, United States. Individual reprints (single or multiple copies) of this standard may be obtained by contacting ASTM at the above address or at 610-832-9585 (phone), 610-832-9555 (fax), or service@astm.org (e-mail); or through the ASTM website (www.astm.org).

Designation: A595/A595M – 06

Standard Specification for
Steel Tubes, Low-Carbon or High-Strength Low-Alloy, Tapered for Structural Use[1]

This standard is issued under the fixed designation A595/A595M; the number immediately following the designation indicates the year of original adoption or, in the case of revision, the year of last revision. A number in parentheses indicates the year of last reapproval. A superscript epsilon (ε) indicates an editorial change since the last revision or reapproval.

This standard has been approved for use by agencies of the Department of Defense.

1. Scope*

1.1 This specification covers three grades of seam-welded, round, tapered steel tubes for structural use. Grades A and B are of low-carbon steel or high-strength low-alloy steel composition and Grade C is of weather-resistant steel composition.

1.2 This tubing is produced in welded sizes in a range of diameters from 2 3/8 to 30 in. [60 to 762 mm] inclusive. Wall thicknesses range from 0.1046 to 0.375 in. [2.66 to 9.53 mm]. Tapers are subject to agreement with the manufacturer.

1.3 The values stated in either SI units or inch-pound units are to be regarded separately as standard. The values stated in each system may not be exact equivalents; therefore, each system shall be used indepedently of the other. Combining values from the two systems may result in non-conformance with the standard.

2. Referenced Documents

2.1 *ASTM Standards:*[2]

A370 Test Methods and Definitions for Mechanical Testing of Steel Products

A588/A588M Specification for High-Strength Low-Alloy Structural Steel, up to 50 ksi [345 MPa] Minimum Yield Point, with Atmospheric Corrosion Resistance

A606 Specification for Steel, Sheet and Strip, High-Strength, Low-Alloy, Hot-Rolled and Cold-Rolled, with Improved Atmospheric Corrosion Resistance

A751 Test Methods, Practices, and Terminology for Chemical Analysis of Steel Products

G101 Guide for Estimating the Atmospheric Corrosion Resistance of Low-Alloy Steels

3. Ordering Information

3.1 The inquiry and order should indicate the following:

3.1.1 Large and small diameters (in.) [mm], length (ft) [m], wall thickness (in.) [mm], taper (in./ft) [mm/m];

3.1.2 (see Table 1 and Table 2);

3.1.3 Extra test material requirements, if any; and

3.1.4 Supplementary requirements, if any.

4. General Requirements for Delivery

4.1 Required date of shipment or date of receipt, and

4.2 Special shipping instructions, if any.

5. Manufacture

5.1 Tube steel shall be hot-rolled aluminum-semikilled or fine-grained killed sheet or plate manufactured by one or more of the following processes: open-hearth, basic-oxygen, or electric-furnace.

5.2 Tubes shall be made from trapezoidal sheet or plate that is preformed and then seam welded. Tubes shall be brought to final size and properties by roll compressing cold on a hardened mandrel.

6. Chemical Composition

6.1 Steel shall conform to the requirements for chemical composition given in Tables 1 and 3. Chemical analysis shall be in accordance with Test Methods, Practices, and Terminology A751.

6.2 For Grade C material, the atmospheric corrosion-resistance index, calculated on the basis of the chemical composition of the steel, as described in Guide G101, shall be 6.0 or higher.

---

[1] This specification is under the jurisdiction of ASTM Committee A01 on Steel, Stainless Steel and Related Alloys and is the direct responsibility of Subcommittee A01.09 on Carbon Steel Tubular Products.
Current edition approved March 1, 2006. Published March 2006. Originally approved in 1969. Last previous edition approved in 2004 as A595 – 04a. DOI: 10.1520/A0595_A0595M-06.

[2] For referenced ASTM standards, visit the ASTM website, www.astm.org, or contact ASTM Customer Service at service@astm.org. For *Annual Book of ASTM Standards* volume information, refer to the standard's Document Summary page on the ASTM website.

*A Summary of Changes section appears at the end of this standard.

TABLE 1 Chemical Requirements

Composition by Heat Analysis, %

Elements	Grade A				Grade B				A 606	Grade C			
	Carbon Steel	HSLA SS	HSLAS Cl1	HSLAS Cl2	Carbon Steel	HSLA SS	HSLAS Cl1	HSLAS Cl2		A 588/A	A 588/B	A 588/C	A 588/K
Carbon	0.015–0.25	0.25 max	0.23 max	0.15 max	0.015–0.25	0.25 max	0.26 max	0.15 max	0.22 max	0.19 max	0.20 max	0.15 max	0.17 max
Manganese	0.30–0.90	1.35 max	1.35 max	1.35 max	0.40–1.35	1.35 max	1.50 max	1.50 max	1.25 max	0.80–1.25	0.75–1.35	0.80–1.35	0.50–1.20
Phosporous	0.035 max	0.035 max	0.04 max	0.04 max	0.035 max	0.035 max	0.04 max	0.04 max	[A]	0.04 max	0.04 max	0.04 max	0.04 max
Sulfur	0.035 max	0.04 max	0.04 max	0.04 max	0.035 max	0.04 max	0.04 max	0.04 max	0.04 max	0.05 max	0.05 max	0.05 max	0.05 max
Silicon	0.040 max[B]	0.040 max[B]	0.040 max[B]	0.040 max[B]	0.040 max[B]	0.040 max[B]	0.040 max[B]	0.040 max[B]	[A]	0.30–0.65	0.15–0.50	0.15–0.40	0.25–0.50
Copper[C,D]	...	0.20 max	0.20 max	0.20 max	...	0.20 max	0.20 max	0.20 max	[A]	0.25–0.40	0.20–0.40	0.20–0.50	0.30–0.50
Chromium[C,E]	...	0.15 max	0.15 max	0.15 max	...	0.15 max	0.15 max	0.15 max	[A]	0.40–0.65	0.40–0.70	0.30–0.50	0.40–0.70
Nickel[C]	...	0.20 max	0.20 max	0.20 max	...	0.20 max	0.20 max	0.20 max	[A]	0.40 max	0.50 max	0.25–0.50	0.40 max
Molybdenum[C,E]	...	0.06 max	0.06 max	0.06 max	...	0.06 max	0.06 max	0.06 max	...	...	[A]	[A]	0.10 max
Vanadium[F]	...	...	0.01 min	0.01 min	...	...	0.01 min	0.01 min	...	0.02–0.10	0.01–0.10	0.01–0.10	[A]
Columbium[F]	...	0.008 max	0.005 min	0.005 min	...	0.008 max	0.005 min	0.005 min	...	[A]	[A]	[A]	...
Nitrogen	...	0.008 max	[A]	[A]	...	0.008 max	[A]	[A]	...	...	...	...	0.005–0.05
Aluminum[B]	...	[A]	[A]	[A]	...	[A]	[A]	[A]	...	...	...	...	...

[A] There is no limit; however, the analysis shall be reported.
[B] Silicon or silicon in combination with aluminum must be sufficient to ensure uniform mechanical properties. Their sum shall be greater than or equal to 0.020 %.
[C] For HSLA steels the sum of copper, nickel, chromium, and molybdenum shall not exceed 0.50 % on heat analysis. When one of these elements are specified by the purchaser, the sum does not apply, in which case only the individual limits of the remaining elements shall apply.
[D] For HSLA steels when copper is specified, the copper limit is a minimum requirement. When copper steel is not specified, the copper limit is a maximum requirement.
[E] For SS steel the sum of chromium and molybdenum shall not exceed 0.16 % on heat analysis. When one or more of these elements are specified by the purchaser, the sum does not apply, in which case the individual limit on the remaining unspecified element shall apply.
[F] For HSLA steels vanadium and columbium minimums may be satisfied separately or by combining their values, in which event the sum shall exceed the combined minimums.

TABLE 2 Tensile Requirements

	Grade A	Grade B	Grade C
Yield point, min, ksi [MPa]	55 [380]	60 [410]	60 [410]
Ultimate tensile strength, min, ksi [MPa]	65 [450]	70 [480]	70 [480]
Elongation in 2 in. [50 mm], min %	23	21	21

NOTE 1—The user is cautioned that the Guide G101 predictive equation for calculation of an atmospheric corrosion-resistance index has been verified only for the composition limits stated in that guide.

6.3 When required by the purchase order, the manufacturer shall supply guidance concerning corrosion resistance that is satisfactory to the purchaser.

7. Mechanical Properties

7.1 *Tension Test*:

7.1.1 *Requirements*—The material, as represented by the test specimens, shall conform to the requirements as to tensile properties given in Table 2.

7.1.2 *Number of Tests*:

7.1.2.1 *For coil*—One or more tension tests as defined in Table 2 shall be made from the large end of one tube on each 100, or fewer, tubes produced from each coil in the applicable thickness class (see Table 4).

7.1.2.2 *For plate*—One or more tension tests as defined in Table 2 shall be made from the large end of one tube on a lot produced from a single heat of plate product of uniform thickness.

7.1.3 *Test Locations and Orientations*—Samples shall be taken at least 1 in. [25 mm] from the longitudinal seam weld.

7.1.4 *Test Method*:

7.1.4.1 Tension tests shall be made in accordance with Test Methods and Definitions A370. The yield strength corresponding to a permanent offset of 0.2 % of the gage length of the specimen or to a total extension of 0.5% of the gage length under load shall be determined in accordance with Test Methods and Definitions A370.

7.1.4.2 The ultimate tensile strength shall be determined in accordance with the Tensile Strength of Test Methods and Definitions A370.

7.1.5 Each test shall be identified as to the heat number of the basic material.

8. Dimensions and Tolerances

8.1 *Length*—The length shall be the specified length with a tolerance of +¾ in. [19 mm] or −¼ in. [6 mm].

8.2 *Diameter*—The outside diameter shall conform to the specified dimensions with a tolerance of ±1/16 in. [2 mm] as measured by girthing.

8.3 *Wall Thickness*—The tolerance for wall thickness exclusive of the weld area shall be +10 % or −5 % of the nominal wall thickness specified.

8.4 *Straightness*—The permissible variation for straightness of the tapered tube shall be 0.2 % or less of the total length.

9. Rework and Retreatment

9.1 In case any test fails to meet the requirements of Section 7, the manufacturer may elect to retreat, rework, or otherwise eliminate the condition responsible for failure to meet the specified requirements. Thereafter the material remaining from the respective class originally represented may be tested and shall comply with all requirements of this specification.

9.2 Imperfections in the outer surface, such as cracks, scabs, or excessive weld projections, shall be classed as injurious defects when their depth or projection exceeds 15 % of the wall thickness or when the imperfections materially affect the appearance of the tube.

9.2.1 Injurious defects having a depth not in excess of 33⅓ % of the specified wall thickness may be repaired by welding subject to the following conditions: (*1*) scabs shall be completely removed by chipping or grinding to sound metal, and (*2*) the repair weld shall be made using suitable electrodes.

9.2.2 Excessive projected weld metal shall be removed to produce a commercial finish.

10. Inspection

10.1 Inspection of material shall be made as agreed upon between the purchaser and the seller as part of the purchase contract.

11. Rejection and Rehearing

11.1 Each length of tubing received from the manufacturer may be inspected by the purchaser, and if it does not meet the requirements of this specification based on the inspection and test method as outlined in the specification, the length may be rejected and the manufacturer shall be notified. Disposition of rejected tubing shall be a matter of agreement between the manufacturer and the purchaser.

11.2 Tubing found in fabrication or in installation to be unsuitable for the intended use, under the scope and requirements of this specification, may be set aside and the manufacturer notified. Such tubing shall be subject to mutual investigation as to the nature and severity of the deficiency and the forming or installation, or both, conditions involved. Disposition shall be a matter for agreement.

12. Certification and Reports

12.1 Upon request of the purchaser in the contract or order, a manufacturer's certification that the material was manufactured and tested in accordance with this specification together with a report of the chemical and tension tests shall be furnished.

13. Product Marking

13.1 Each tapered tube shall be legibly marked by rolling, die stamping, ink printing, or paint stenciling to show the following information: thickness, taper, large diameter, small diameter, length, and the specification number, Grade A, B, or C.

13.2 *Bar Coding*—In addition to the requirements in 13.1, bar coding is acceptable as a supplemental identification method. The purchaser may specify in the order a specific bar coding system to be used.

14. Keywords

14.1 carbon steel tube; steel tube

TABLE 3 Chemical Requirements

Composition by Product Analysis, %

Elements	Grade A				Grade B				A 606	Grade C			
	Carbon Steel	HSLA SS	HSLAS Cl1	HSLAS Cl2	Carbon Steel	HSLA SS	HSLAS Cl1	HSLAS Cl2		A 588/A	A 588/B	A 588/C	A 588/K
Carbon	0.012–0.29	0.29 max	0.27 max	0.18 max	0.012–0.29	0.29 max	0.29 max	0.18 max	0.26 max	0.23 max	0.24 max	0.18 max	0.21 max
Manganese	0.26–0.94	1.40 max	1.40 max	1.40 max	0.35–1.40	1.40 max	1.40 max	1.40 max	1.3 max	0.72–1.35	0.67–1.45	0.72–1.45	0.42–1.30
Phosphorous	0.045 max	0.045 max	0.05 max	0.05 max	0.45 max	0.45 max	0.05 max	0.05 max	^{A}	0.05 max	0.05 max	0.05 max	0.05 max
Sulfur	0.045 max	0.05 max	0.05 max	0.05 max	0.45 max	0.05 max	0.05 max	0.05 max	0.06 max	0.06 max	0.06 max	0.06 max	0.06 max
Silicon	0.040 max^B	0.040 max^B	0.040 max^B	0.040 max^B	0.40 max^B	0.40 max^B	0.40 max^B	0.40 max^B	^{A}	0.25–0.70	0.15–0.60	0.13–0.43	0.20–0.55
Copper	...	0.22 max	0.22 max	0.22 max	...	0.22 max	0.22 max	0.22 max	^{A}	0.22–0.43	0.17–0.43	0.17–0.53	0.27–0.53
Chromium	...	0.19 max	0.19 max	0.19 max	...	0.19 max	0.19 max	0.19 max	^{A}	0.36–0.69	0.36–0.74	0.26–0.54	0.36–0.74
Nickel	...	0.23 max	0.23 max	0.23 max	...	0.23 max	0.23 max	0.23 max	^{A}	0.43 max	0.53 max	0.22–0.53	0.43 max
Molybdenum	...	0.07 max	0.07 max	0.07 max	...	0.07 max	0.07 max	0.07 max	...	...	...	...	...
Vanadium^C	...	0.018 max	0.00 min	0.00 min	...	0.018 max	0.00 min	0.00 min	...	0.01–0.11	0.00–0.11	0.00–0.11	0.11 max
Columbium^C	...	0.018 max	0.00 min	0.00 min	...	0.018 max	0.00 min	0.00 min	...	^{A}	^{A}	^{A}	^{A}
Nitrogen	...	^{A}	^{A}	^{A}	...	^{A}	^{A}	^{A}	^{A}	...	...	...	0.055 max
Aluminum^B	...	^{A}	^{A}	^{A}	...	^{A}	^{A}	^{A}	...	...	...	...	...

^{A}There is no limit; however, the analysis shall be reported.
^{B}Silicon or silicon in combination with aluminum must be sufficient to ensure uniform mechanical properties. Their sum shall be greater than or equal to 0.020 %.
^{C}For HSLA steels vanadium and columbium minimums may be satisfied separately or by combining their values, in which event the sum shall exceed the combined minimums.

TABLE 4 Thickness Class

Class	Thickness	
	in.	mm
1	0.1046 through 0.140	2.66 through 3.56
2	0.141 through 0.190	3.58 through 4.83
3	0.191 through 0.280	4.85 through 7.11
4	0.281 through 0.375	7.14 through 9.53

SUMMARY OF CHANGES

Committee A01 has identified the location of selected changes to this specification since the last issue, A595 – 04a, that may impact the use of this specification. (Approved March 1, 2006)

(1) Revised Sections 1, 3, 7, and 8 to include rationalized SI units, creating a combined standard.

(2) Revised the SI Yield point in Table 2 to 425 MPa, SI elongation to [50 mm], and elongation to whole numbers, not tenths of a percent.

ASTM International takes no position respecting the validity of any patent rights asserted in connection with any item mentioned in this standard. Users of this standard are expressly advised that determination of the validity of any such patent rights, and the risk of infringement of such rights, are entirely their own responsibility.

This standard is subject to revision at any time by the responsible technical committee and must be reviewed every five years and if not revised, either reapproved or withdrawn. Your comments are invited either for revision of this standard or for additional standards and should be addressed to ASTM International Headquarters. Your comments will receive careful consideration at a meeting of the responsible technical committee, which you may attend. If you feel that your comments have not received a fair hearing you should make your views known to the ASTM Committee on Standards, at the address shown below.

This standard is copyrighted by ASTM International, 100 Barr Harbor Drive, PO Box C700, West Conshohocken, PA 19428-2959, United States. Individual reprints (single or multiple copies) of this standard may be obtained by contacting ASTM at the above address or at 610-832-9585 (phone), 610-832-9555 (fax), or service@astm.org (e-mail); or through the ASTM website (www.astm.org).

Designation: A606/A606M – 09a

Standard Specification for
Steel, Sheet and Strip, High-Strength, Low-Alloy, Hot-Rolled and Cold-Rolled, with Improved Atmospheric Corrosion Resistance[1]

This standard is issued under the fixed designation A606/A606M; the number immediately following the designation indicates the year of original adoption or, in the case of revision, the year of last revision. A number in parentheses indicates the year of last reapproval. A superscript epsilon (ε) indicates an editorial change since the last revision or reapproval.

This standard has been approved for use by agencies of the Department of Defense.

1. Scope*

1.1 This specification covers high-strength, low-alloy, hot- and cold-rolled sheet and strip in cut lengths or coils, intended for use in structural and miscellaneous purposes, where savings in weight or added durability are important. These steels have enhanced atmospheric corrosion resistance and are supplied in two types: Type 2 contains 0.20 % minimum copper based on cast or heat analysis (0.18 % minimum Cu for product check). Type 4 contains additional alloying elements and provides a level of corrosion resistance substantially better than that of carbon steels with or without copper addition (Note 1). When properly exposed to the atmosphere, Type 4 steel can be used bare (unpainted) for many applications.

NOTE 1—For methods of establishing the atmospheric corrosion resistance of low-alloy steels, see Guide G101.

1.2 The values stated in either SI units or inch-pound units are to be regarded separately as standard. The values stated in each system may not be exact equivalents; therefore, each system shall be used independently of the other. Combining values from the two systems may result in non-conformance with the standard.

2. Referenced Documents

2.1 *ASTM Standards:*[2]

A109/A109M Specification for Steel, Strip, Carbon (0.25 Maximum Percent), Cold-Rolled

A568/A568M Specification for Steel, Sheet, Carbon, Structural, and High-Strength, Low-Alloy, Hot-Rolled and Cold-Rolled, General Requirements for

A749/A749M Specification for Steel, Strip, Carbon and High-Strength, Low-Alloy, Hot-Rolled, General Requirements for

G101 Guide for Estimating the Atmospheric Corrosion Resistance of Low-Alloy Steels

3. General Requirements for Delivery —

3.1 Material furnished under this specification shall conform to the applicable requirements of the current edition of Specification A568/A568M and the dimensional tolerance tables of Specification A109/A109M, unless otherwise provided herein.

4. Ordering Information

4.1 Orders for material under this specification shall include the following information, as required, to describe adequately the desired material:

4.1.1 ASTM specification number and date of issue, and type,

4.1.2 Name of material (high-strength low-alloy hot-rolled sheet or strip or high-strength low-alloy cold-rolled sheet or strip),

4.1.3 Condition (specify oiled or dry, as required),

4.1.4 Edges (must be specified for hot-rolled sheet or strip) (see 8.1),

4.1.5 Finish—Cold-rolled only (indicate exposed (E) or unexposed (U). Matte (dull) finish will be supplied unless otherwise specified), and

4.1.6 Dimensions (thickness, width, and whether cut lengths or coils).

NOTE 2—Not all producers are capable of meeting all of the limitations of the thickness tolerance tables in Specification A568/A568M. The purchaser should contact the producer regarding possible limitations prior to placing an order.

---

[1] This specification is under the jurisdiction of ASTM Committee A01 on Steel, Stainless Steel and Related Alloys and is the direct responsibility of Subcommittee A01.19 on Steel Sheet and Strip.

Current edition approved Nov. 1, 2009. Published December 2009. Originally approved in 1970. Last previous edition approved in 2009 as A606/A606M – 09. DOI: 10.1520/A0606_A0606M-09A.

[2] For referenced ASTM standards, visit the ASTM website, www.astm.org, or contact ASTM Customer Service at service@astm.org. For *Annual Book of ASTM Standards* volume information, refer to the standard's Document Summary page on the ASTM website.

*A Summary of Changes section appears at the end of this standard.

Copyright © ASTM International, 100 Barr Harbor Drive, PO Box C700, West Conshohocken, PA 19428-2959, United States.

4.1.7 Coil size (must include inside diameter, outside diameter, and maximum weight),

4.1.8 Application (show part identification and description),

4.1.9 Cast or heat (formerly ladle) analysis and mechanical properties report (if required) (see 10.1), and

4.1.10 Special requirements (if any).

4.1.10.1 When the purchaser requires thickness tolerances for 3/8 in. [10 mm] minimum edge distance (see Supplementary Requirement in Specification A568/A568M), this requirement shall be specified in the purchase order or contract.

NOTE 3—A typical ordering description is as follows: "ASTM A606–XX, Type 4 high-strength low-alloy hot-rolled sheet, dry, mill edge 0.106 by 48 by 96 in. for truck frame side members." Or, "ASTM A606M–XX, Type 4 high-strength low-alloy hot-rolled sheet, dry, mill edge, 2.7 by 1220 mm by coil for truck frame side members."

5. Materials and Manufacture

5.1 *Condition*—The material shall be furnished hot-rolled or cold-rolled as specified on the purchase order.

5.2 *Heat Treatment*— Unless otherwise specified, hot-rolled shall be furnished as rolled. When hot-rolled annealed or hot-rolled normalized material is required, it shall be specified on the purchase order.

6. Chemical Composition

6.1 The maximum limits of carbon, manganese, and sulfur shall be as prescribed in Table 1, unless otherwise agreed upon between the manufacturer and the purchaser.

6.2 The manufacturer shall use such alloying elements, combined with the carbon, manganese, and sulfur within the limits prescribed in Table 1 to satisfy the mechanical properties prescribed in Table 2 or Table 3. Such elements shall be included and reported in the specified heat or cast analysis. As indicated in 1.1, these steels have enhanced atmospheric corrosion resistance and are supplied in two types: Type 2 and Type 4. When requested, the producer shall supply acceptable evidence of corrosion resistance to the purchaser. For Type 2 steel, confirmation of the minimum copper content requirement of 1.1 shall be sufficient evidence of corrosion resistance. For Type 4 steel, the basis for this evidence can be a corrosion-resistance index calculated from the chemical composition of the steel in accordance with Guide G101. To comply with Specification A606, Type 4 steel shall have a minimum corrosion-resistance index of 6.0, based upon Guide G101— Predictive Method Based on the Data of Larabee and Coburn (see Note 4.)

TABLE 1 Chemical Requirements

	Composition, max, %	
	Cast or Heat (Formerly Ladle) Analysis	Product Check, or Verification Analysis
Carbon[A]	0.22	0.26
Manganese	1.25	1.30
Sulfur	0.04	0.06

[A] For compositions with a maximum carbon content of 0.15 % on heat or cast analysis, the maximum limit for manganese on heat or cast analysis may be increased to 1.40 % (with product analysis limits of 0.19 % carbon and 1.45 % manganese).

TABLE 2 Tensile Requirements[A] for Hot-Rolled Material

	As-Rolled	Annealed or Normalized
Tensile strength, min ksi (MPa)	70 [480]	65 [450]
Yield strength, min, ksi (MPa)	50 [340]	45 [310]
Elongation in 2 in. or 50 mm, min, %	22	22

[A] For coil products, testing by the producer is limited to the end of the coil. Mechanical properties throughout the coil shall comply with the minimum values specified.

TABLE 3 Tensile Requirements for Cold-Rolled Material

	Cut Lengths and Coils
Tensile strength, min, ksi (MPa)	65 [450]
Yield strength, min, ksi (MPa)	45 [310]
Elongation in 2 in. or 50 mm, min, %	22[A]

[A] 0.0448 in. [1.1 mm] and under in thickness—20 %.

NOTE 4—The user is cautioned that the Guide G101 predictive equation (Predictive Method Based on the Data of Larabee and Coburn) for calculation of an atmospheric corrosion index has been verified only for the composition limits stated in that guide.

6.3 When the steel is used in welded applications, welding procedure shall be suitable for the steel chemistry as described in 6.2 and the intended service.

7. Mechanical Property Requirements

7.1 *Tension Tests*:

7.1.1 *Requirements*—Material as represented by the test specimen shall conform to the tensile requirements specified in Table 2 (hot-rolled material) or in Table 3 (cold-rolled material).

7.1.2 *Number of Tests*—Two tensile tests shall be made from each heat or from each lot of 50 tons [45 000 kg]. When the amount of finished material from a heat or lot is less than 50 tons [45 000 kg], one test shall be made. When material rolled from one heat differs 0.050 in. [1.27 mm] or more in thickness, one tensile test shall be made from the thickest and thinnest material regardless of the weight represented.

7.1.3 *Location and Orientation*:

7.1.3.1 Tensile test specimens shall be taken at a point immediately adjacent to the material to be qualified.

7.1.3.2 Tensile test samples shall be taken from the full thickness of the sheet as rolled.

7.1.3.3 Tensile test specimens shall be taken from a location approximately halfway between the center of the sheet and the edge of the material as rolled.

7.1.3.4 Tensile test specimens shall be taken with the axis of the test specimen parallel to the rolling direction (longitudinal test).

7.1.4 *Test Method*—Yield strength shall be determined by either the 0.2 % offset method or by the 0.5 % extension under load method unless otherwise specified.

7.2 *Bending Properties*—The minimum forming radius (radii) that steel covered by this specification can be expected to sustain is listed in the Appendix X1 and is discussed in more detail in Specifications A568/A568M and A749/A749M. Where tighter bend radii are required, where curved or offset bends are involved, or where stretching or drawing are also a consideration, the producers should be consulted.

8. Workmanship, Finish, and Appearance

8.1 *Edges*:

8.1.1 *Hot-Rolled*—In the as-rolled condition the material has mill edges. Pickled or blast-cleaned material has cut edges. When required, as-rolled material may be specified to have cut edges. If mill edge material is required it must be specified.

8.1.2 *Cold-Rolled*—Cold-rolled material shall have cut edges only.

8.2 *Oiling*:

8.2.1 *Hot-Rolled*—Unless otherwise specified, hot-rolled as-rolled material shall be furnished dry, and hot-rolled pickled or blast-cleaned material shall be furnished oiled. When required, pickled or blast-cleaned material may be specified to be furnished dry, and as-rolled material may be specified to be furnished oiled.

8.2.2 *Cold-Rolled*—Unless otherwise specified, cold-rolled material shall be oiled. When required, cold-rolled material may be specified to be furnished dry, but is not recommended due to the increased possibility of rusting.

8.3 *Surface Finish*:

8.3.1 *Hot-Rolled*—Unless otherwise specified, hot-rolled material shall have an as-rolled, not pickled surface finish. When required, material may be specified to be pickled or blast-cleaned.

8.3.2 *Cold-Rolled*—Unless otherwise specified, cold-rolled material shall have a matte (dull) finish.

9. Retests and Disposition of Non-Conforming Material

9.1 Retests, conducted in accordance with the requirements of Section 11.1 of Specification A568/A568M, are permitted when an unsatisfactory test result is suspected to be the consequence of the test method procedure.

9.2 Disposition of non-conforming material shall be subject to the requirements of Section 11.2 of Specification A568/A568M.

10. Certification

10.1 When requested, the manufacturer shall furnish copies of a test report showing the results of the heat or cast analysis and mechanical property tests made to determine compliance with this specification. The report shall include the purchase order number, the ASTM designation number, and the heat or lot number correlating the test results with the material represented.

11. Keywords

11.1 alloy steel sheet; alloy steel strip; cold rolled steel sheet; cold rolled steel strip; high strength low alloy steel; hot rolled steel sheet; hot rolled steel strip; steel sheet; steel strip

APPENDIX

(Nonmandatory Information)

X1. BENDING PROPERTIES

TABLE X1.1 Suggested Minimum Inside Radius for Cold Bending[A]

NOTE 1—(t) equals a radius equivalent to the steel thickness.
NOTE 2—The suggested radii should be used as minimums for 90° bends in actual shop practice.

Grade	Minimum Inside Radius for Cold Bending
Hot Rolled or Cold Rolled	2½ t

[A] Material which does not perform satisfactorily, when fabricated in accordance with the above requirements, may be subject to rejection pending negotiation with the steel supplier.

SUMMARY OF CHANGES

Committee A01 has identified the location of selected changes to this standard since the last issue (A606 – 09) that may impact the use of this standard. (Approved November 1, 2009.)

(*1*) Section 4.1.6.1 deleted.
(*2*) Reversed order of 4.1.9 and 4.1.10 and added new section 4.1.10.1.

Committee A01 has identified the location of selected changes to this standard since the last issue (A606 – 04) that may impact the use of this standard. (Approved April 1, 2009.)

(*1*) 1.2 replaced.
(*2*) Note 3—Metric example added.
(*3*) 7.1.2—Metric values changed from "soft" to "hard."
(*4*) 9—Entire paragraph replaced with current wording on retesting, resampling, etc.
(*5*) Table 2—Metric values changed from "soft" to "hard."

(6) Table 3—Metric values changed from "soft" to "hard."

(7) Standard Designation changed to show dual standard.

ASTM International takes no position respecting the validity of any patent rights asserted in connection with any item mentioned in this standard. Users of this standard are expressly advised that determination of the validity of any such patent rights, and the risk of infringement of such rights, are entirely their own responsibility.

This standard is subject to revision at any time by the responsible technical committee and must be reviewed every five years and if not revised, either reapproved or withdrawn. Your comments are invited either for revision of this standard or for additional standards and should be addressed to ASTM International Headquarters. Your comments will receive careful consideration at a meeting of the responsible technical committee, which you may attend. If you feel that your comments have not received a fair hearing you should make your views known to the ASTM Committee on Standards, at the address shown below.

This standard is copyrighted by ASTM International, 100 Barr Harbor Drive, PO Box C700, West Conshohocken, PA 19428-2959, United States. Individual reprints (single or multiple copies) of this standard may be obtained by contacting ASTM at the above address or at 610-832-9585 (phone), 610-832-9555 (fax), or service@astm.org (e-mail); or through the ASTM website (www.astm.org). Permission rights to photocopy the standard may also be secured from the ASTM website (www.astm.org/COPYRIGHT/).

Designation: A618/A618M – 04

Standard Specification for Hot-Formed Welded and Seamless High-Strength Low-Alloy Structural Tubing[1]

This standard is issued under the fixed designation A618/A618M; the number immediately following the designation indicates the year of original adoption or, in the case of revision, the year of last revision. A number in parentheses indicates the year of last reapproval. A superscript epsilon (ε) indicates an editorial change since the last revision or reapproval.

This standard has been approved for use by agencies of the Department of Defense.

1. Scope*

1.1 This specification covers grades of hot-formed welded and seamless high-strength low-alloy square, rectangular, round, or special shape structural tubing for welded, riveted, or bolted construction of bridges and buildings and for general structural purposes. When the steel is used in welded construction, the welding procedure shall be suitable for the steel and the intended service.

1.2 Grade II has atmospheric corrosion resistance equivalent to that of carbon steel with copper (0.20 minimum Cu) Grades Ia and Ib have atmospheric corrosion resistance substantially better than that of Grade II (Note 1). When properly exposed to the atmosphere, Grades Ia and Ib can be used bare (unpainted) for many applications. When enhanced corrosion resistance is desired, Grade III, copper limits may be specified.

NOTE 1—For methods of estimating the atmospheric corrosion resistance of low alloy steels see Guide G101 or actual data.

1.3 The values stated in either SI units or inch-pound units are to be regarded separately as standard. Within the text, the SI units are shown in brackets. The values stated in each system may not be exact equivalents; therefore, each system shall be used independently of the other. Combining values from the two systems may result in non-conformance with the standard.

2. Referenced Documents

2.1 *ASTM Standards:*[2]

A370 Test Methods and Definitions for Mechanical Testing of Steel Products
A700 Practices for Packaging, Marking, and Loading Methods for Steel Products for Shipment
A751 Test Methods, Practices, and Terminology for Chemical Analysis of Steel Products
G101 Guide for Estimating the Atmospheric Corrosion Resistance of Low-Alloy Steels

3. Ordering Information

3.1 Orders for material under this specification should include the following as required to describe the material adequately:

3.1.1 Quantity (feet, metres, or number of lengths),
3.1.2 Grade (Table 1 and Table 2),
3.1.3 Material (round, square, or rectangular tubing),
3.1.4 Method of manufacture (seamless, buttwelded, or hot-stretch-reduced electric-resistance welded),
3.1.5 Size (outside diameter and nominal wall thickness for round tubing and the outside dimensions and calculated nominal wall thickness for square and rectangular tubing),
3.1.6 Length (specific or random, see 8.2),
3.1.7 End condition (see 9.2),
3.1.8 Burr removal (see 9.2),
3.1.9 Certification (see 12.1),
3.1.10 Specification designation (A618 or A618M, including yeardate),
3.1.11 End use, and
3.1.12 Special requirements.

4. Process

4.1 The steel shall be made by one or more of the following processes: open-hearth, basic-oxygen, or electric-furnace.

4.2 Steel may be cast in ingots or may be strand cast. When steels of different grades are sequentially strand cast, identification of the resultant transition material is required. The producer shall remove the transition material by any established procedure that positively separates the grades.

---

[1] This specification is under the jurisdiction of ASTM Committee A01 on Steel, Stainless Steel and Related Alloys and is the direct responsibility of Subcommittee A01.09 on Carbon Steel Tubular Products.
Current edition approved March 1, 2004. Published March 2004. Originally approved in 1968. Last previous edition approved in 2001 as A618–01. DOI: 10.1520/A0618_A0618M-04.

[2] For referenced ASTM standards, visit the ASTM website, www.astm.org, or contact ASTM Customer Service at service@astm.org. For *Annual Book of ASTM Standards* volume information, refer to the standard's Document Summary page on the ASTM website.

*A Summary of Changes section appears at the end of this standard.

Copyright © ASTM International, 100 Barr Harbor Drive, PO Box C700, West Conshohocken, PA 19428-2959, United States.

TABLE 1 Chemical Requirements

Element	Grade Ia		Grade Ib		Grade II		Grade III	
	Heat	Product	Heat	Product	Heat	Product	Heat	Product
Carbon, max	0.15	0.18	0.20	...	0.22	0.26	0.23[A]	0.27[A]
Manganese	1.00 max	1.04 max	1.35 max	1.40 max	0.85–1.25	1.30 max	1.35 max[A]	1.40 max[A]
Phosphorus, max	0.15	0.16	0.025	0.035	0.025	0.035	0.025	0.035
Sulfur, max	0.025	0.045	0.025	0.035	0.025	0.035	0.025	0.035
Silicon, max	...	...	...	...	0.30	0.33	0.30	0.35
Copper, min	0.20	0.18	0.20[B]	0.18[B]	0.20	0.18	...	...
Vanadium, min	...	...	...	...	0.02	0.01	0.02[C]	0.01

[A] For each reduction of 0.01 % C below the specified carbon maximum, an increase of 0.05 % manganese above the specified maximum will be permitted up to 1.45 % for the heat analysis and up to 1.50 % for the product analysis.
[B] If chromium and silicon contents are each 0.50 % min, then the copper minimums do not apply.
[C] For Grade III, columbium may be used in conformance with the following limits: 0.005 %, min (heat) and 0.004 %, min (product).

TABLE 2 Tensile Requirements

	Grades Ia, Ib, and II		Grade III	
	Walls ¾ in. [19.0 mm] and Under	Walls over ¾ to 1½ in. [19.0 to 38.0 mm], incl		
Tensile strength, min, ksi [MPa][A]	70 [485]	67 [460]	65	[450]
Yield strength, min, ksi [MPa][A]	50 [345]	46 [315]	50	[345]
Elongation in 2 in. or 50 mm, min, %	22	22	20	
Elongation in 8 in. or 200 mm, min, %	19	18	18	

[A] For Grade II, when the material is normalized, the minimum yield strength and minimum tensile strength required shall be reduced by 5 ksi [35 MPa].

5. Manufacture

5.1 The tubing shall be made by the seamless, furnace-buttwelded (continuous-welded), or hot-stretch-reduced electric-resistance-welded process.

6. Chemical Composition

6.1 When subjected to the heat and product analysis, respectively, the steel shall conform to the requirements prescribed in Table 1.

6.1.1 For Grades Ia and Ib, the choice and use of alloying elements, combined with carbon, manganese, and sulfur within the limits prescribed in Table 1 to give the mechanical properties prescribed in Table 2 and to provide the atmospheric corrosion resistance of 1.2, should be made by the manufacturer and included and reported in the heat analysis for information purposes only to identify the type of steel applied. For Grades Ia and Ib material, the atmospheric corrosion-resistance index, calculated on the basis of the chemical composition of the steel as described in Guide G101, shall be 6.0 or higher.

NOTE 2—The user is cautioned that the Guide G101 predictive equation for calculation of an atmospheric corrosion–resistance index has been verified only for the composition limits stated in that guide.

6.1.2 When Grade III is required for enhanced corrosion resistance, copper limits may be specified and the minimum content shall be 0.20 % by heat analysis and 0.18 % by product analysis.

6.2 *Heat Analysis*—An analysis of each heat of open-hearth, basic-oxygen, or electric-furnace steel shall be made by the manufacturer. This analysis shall be made from a test ingot taken during the pouring of the heat. The chemical composition thus determined shall conform to the requirements specified in Table 1 for heat analysis.

6.3 *Product Analysis*:

6.3.1 An analysis may be made by the purchaser from finished tubing manufactured in accordance with this specification, or an analysis may be made from flat-rolled stock from which the welded tubing is manufactured. When product analyses are made, two sample lengths from a lot of each 500 lengths, or fraction thereof, shall be selected. The specimens for chemical analysis shall be taken from the sample lengths in accordance with the applicable procedures of Test Methods, Practices, and Terminology A751. The chemical composition thus determined shall conform to the requirements specified in Table 1 for product analysis.

6.3.2 In the event the chemical composition of one of the sample lengths does not conform to the requirements shown in Table 1 for product analysis, an analysis of two additional lengths selected from the same lot shall be made, each of which shall conform to the requirements shown in Table 1 for product analysis, or the lot is subject to rejection.

7. *Mechanical Requirements* Mechanical Requirements

7.1 *Tensile Properties*:

7.1.1 The material, as represented by the test specimen, shall conform to the requirements prescribed in Table 2.

7.1.2 Elongation may be determined on a gage length of either 2 in. [50 mm] or 8 in. [200 mm] at the manufacturer's option.

7.1.3 For material under 5/16 in. [8.0 mm] in thickness, a deduction from the percentage elongation of 1.25 percentage points in 8 in. [200 mm] specified in Table 2 shall be made for each decrease of 1/32 in. [0.8 mm] of the specified thickness under 5/16 in. [8.0 mm].

7.2 *Bend Test*—The bend test specimen shall stand being bent cold through 180° without cracking on the outside of the bent portion, to an inside diameter which shall have a relation to the thickness of the specimen as prescribed in Table 3.

TABLE 3 Bend Test Requirements

Thickness of Material, in. [mm]	Ratio of Bend Diameter to Specimen Thickness
¾ [19.0] and under	1
Over ¾ to 1 [19.0 to 25.0], incl	1½
Over 1 [25.0]	2

7.3 *Number of Tests*—Two tension and two bend tests, as specified in 7.4.2, and 7.4.3, shall be made from tubing representing each heat. However, if tubing from one heat differs in the ordered nominal wall thickness, one tension test and one bend test shall be made from both the heaviest and lightest wall thicknesses processed.

7.4 *Test Specimens*:

7.4.1 The test specimens required by this specification shall conform to those described in the latest issue of Test Methods and Definitions A370.

7.4.2 The tension test specimen shall be taken longitudinally from a section of the finished tubing, at a location at least 90° from the weld in the case of welded tubing, and shall not be flattened between gage marks. If desired, the tension test may be made on the full section of the tubing; otherwise, a longitudinal strip test specimen shall be used as prescribed in Test Methods and Definitions A370, Annex A2. The specimens shall have all burrs removed and shall not contain surface imperfections that would interfere with the proper determination of the tensile properties of the metal.

7.4.3 The bend test specimen shall be taken longitudinally from the tubing, and shall represent the full wall thickness of material. The sides of the bend test specimen may have the corners rounded to a maximum radius of 1/16 in. [1.6 mm].

7.5 *Test Methods*:

7.5.1 The yield strength shall be determined in accordance with one of the alternatives described in Test Methods and Definitions A370.

7.5.2 The bend test shall be made on square or rectangular tubing manufactured in accordance with this specification.

7.6 *Retests*:

7.6.1 If the results of the mechanical tests representing any heat do not conform to a requirement, as specified in 7.1 and 7.2, retests may be made on additional tubing of double the original number from the same heat, each of which shall conform to the requirement specified, or the tubing represented by the test is subject to rejection.

7.6.2 In case of failure on retest to meet the requirements of 7.1 and 7.2, the manufacturer may elect to retreat, rework, or otherwise eliminate the condition responsible for failure to meet the specified requirements. Thereafter, the material remaining from the respective heat originally represented may be tested, and shall comply with all requirements of this specification.

8. Dimensions and Permissible Variations

8.1 The dimensions of square, rectangular, round, and special shape structural tubing to be ordered under this specification shall be subject to prior negotiation with the manufacturer. The dimensions agreed upon shall be indicated in the purchase order.

8.2 *Permissible Variations*:

8.2.1 *Outside Dimensions*:

8.2.1.1 For round tubing 2 in. [50 mm] and over in nominal diameter, the outside diameter shall not vary more than ±1 % from the specified outside diameter. For sizes 1½ in. [38 mm] and under, the outside diameter shall not vary more than 1/64 in. [0.4 mm] over and more than 1/32 in. [0.8 mm] under the specified outside diameter.

8.2.1.2 The specified dimensions, measured across the flats at positions at least 2 in. [50 mm] from either end of square and rectangular tubing and including an allowance for convexity and concavity, shall not exceed the plus and minus tolerance shown in Table 4.

8.2.2 *Mass*—The mass of structural tubing shall not be less than the specified value by more than 3.5 %. The mass tolerance shall be determined from individual lengths or for round tubing sizes 4½ in. [114 mm] in outside diameter and under and square and rectangular tubing having a periphery of 14 in. [356 mm] and under shall be determined from masses of the customary lifts produced by the mill. On round tubing sizes over 4½ in. [114 mm] in outside diameter and square and rectangular tubing having a periphery in excess of 14 in. [356 mm] the mass tolerance is applicable to the individual length.

8.2.3 *Length*—Structural tubing is commonly produced in random mill lengths of 16 to 22 ft [4.9 to 6.7 m] or 32 to 44 ft [9.8 to 13.4 m], in multiple lengths, and in definite cut lengths (Section 3). When cut lengths are specified for structural tubing, the length tolerances shall be in accordance with Table 5.

8.2.4 *Straightness*—The permissible variation for straightness of structural tubing shall be 1/8 in. times the number of feet of total length divided by 5 [2 mm times length in metres).

8.2.5 *Squareness of Sides*—For square or rectangular structural tubing, adjacent sides may deviate from 90° by a tolerance of ±2°, maximum.

8.2.6 *Radius of Corners*—For square or rectangular structural tubing, the radius of any outside corner of the section shall not exceed three times the specified wall thickness.

8.2.7 *Twist*:

8.2.7.1 The tolerance for twist, or variation with respect to axial alignment of the section for square, rectangular, or special shape structural tubing, shall be as prescribed in Table 6.

8.2.7.2 Twist is measured by holding down one end of a square or rectangular tube on a flat surface plate with the bottom side of the tube parallel to the surface plate, and noting the height that either corner at the opposite end of the bottom side of the tube extends above the surface plate. The difference in the height of the corners shall not exceed the values in Table 6.

9. Workmanship, Finish, and Appearance

9.1 The structural tubing shall be free of defects and shall have a commercially smooth finish.

9.1.1 Surface imperfections shall be classed as defects when their depth exceeds 15 % of the specified wall thickness and when the imperfections materially affected the appearance of the structural member, or when their length (measured in a

TABLE 4 Outside Dimension Tolerances for Square, Rectangular, and Special Shape Structural Tubing

Largest Outside Dimension Across Flats, in. [mm]	Tolerance ± in. [mm]
2½ [64] and under	0.020 [0.5]
Over 2½ to 3½ [64 to 89], incl	0.025 [0.6]
Over 3½ to 5½ [89 to 140], incl	0.030 [0.8]
Over 5½ [140]	1 %

TABLE 5 Cut Length Tolerances for Structural Tubing

	22 ft [6.7 m] and Under		Over 22 to 44 ft [6.7 to 13.4 m], incl	
	Over	Under	Over	Under
Length tolerance for specified cut lengths, in. [mm]	½ [13]	¼ [6]	¾ [19]	¼ [6]

TABLE 6 Twist Tolerances for Square, Rectangular, or Special Shape Structural Tubing

Specified Dimension of Longest Outside Side, in. [mm]	Maximum Twist per 3 ft of Length, in.	Maximum Twist per Metre of Length, mm
1½ [38] and under	0.050	1.4
Over 1½ to 2½ [38 to 64], incl	0.062	1.7
Over 2½ to 4 [64 to 102], incl	0.075	2.1
Over 4 to 6 [102 to 152], incl	0.087	2.4
Over 6 to 8 [152 to 203], incl	0.100	2.8
Over 8 [203]	0.112	3.1

transverse direction) and depth would materially reduce the total cross-sectional area at any location.

9.1.2 Defects having a depth not in excess of 33⅓ % of the wall thickness may be repaired by welding, subject to the following conditions:

9.1.2.1 The defect shall be completely removed by chipping or grinding to sound metal.

9.1.2.2 The repair weld shall be made using suitable coated electrodes.

9.1.2.3 The projecting weld metal shall be removed to produce a workmanlike finish.

9.2 The ends of structural tubing, unless otherwise specified, shall be finished square cut, and the burr held to a minimum. The burr can be removed on the outside diameter, inside diameter, or both, as a supplementary requirement. When the burrs are to be removed, it shall be specified in the purchase order.

10. Inspection

10.1 All tubing shall be subject to an inspection at the place of manufacture to assure conformance with the requirements of this specification.

11. Rejection

11.1 Each length of tubing received from the manufacturer may be inspected by the purchaser and, if it does not meet the requirements of this specification based on the inspection and test method as outlined in the specification, the length may be rejected and the manufacturer shall be notified. Disposition of rejected tubing shall be a matter of agreement between the manufacturer and the purchaser.

11.2 Tubing found in fabrication or in installation to be unsuitable for the intended use, under the scope and requirements of this specification, may be set aside and the manufacturer notified. Such tubing shall be subject to mutual investigation as to the nature and severity of the deficiency and the forming or installation, or both, conditions involved. Disposition shall be a matter for agreement.

12. Certification

12.1 Upon request of the purchaser in the contract or order, a manufacturer's certification that the material was manufactured and tested in accordance with this specification (including year of issue) together with a report of the chemical and tensile tests shall be furnished.

13. Packaging, Package Marking, and Loading

13.1 Except as noted in 13.2, each length of structural tubing shall be legibly marked by rolling, die stamping, ink printing, or paint stenciling to show the following information: manufacturer's name, brand, or trademark; size and wall thickness; steel grade; and the specification number (year of issue not required).

13.2 For structural tubing 1½ in. [38 mm] and under in nominal size or the greatest cross sectional dimension less than 2 in. [50 mm], the information listed in 10.1 may be marked on a tag securely attached to each bundle.

13.3 When specified in the order, contract, etc., packaging, marking, and loading shall be in accordance with the procedures of Practices A700.

13.4 *Bar Coding*—In addition to the requirements in 13.1, 13.2, and 13.3, bar coding is acceptable as a supplemental identification method. The purchaser may specify in the order a specific bar coding system to be used.

14. Keywords

14.1 high-strength low-alloy steel; seamless steel tube; steel tube; structural steel tubing; welded steel tubing

A618/A618M – 04

SUMMARY OF CHANGES

Committee A01 has identified the location of selected changes to this standard that have been incorporated since the last issue (A618–01) that may impact the use of this standard.

(1) Rationalized SI units have been added throughout the text and tables to create a combined standard.

ASTM International takes no position respecting the validity of any patent rights asserted in connection with any item mentioned in this standard. Users of this standard are expressly advised that determination of the validity of any such patent rights, and the risk of infringement of such rights, are entirely their own responsibility.

This standard is subject to revision at any time by the responsible technical committee and must be reviewed every five years and if not revised, either reapproved or withdrawn. Your comments are invited either for revision of this standard or for additional standards and should be addressed to ASTM International Headquarters. Your comments will receive careful consideration at a meeting of the responsible technical committee, which you may attend. If you feel that your comments have not received a fair hearing you should make your views known to the ASTM Committee on Standards, at the address shown below.

This standard is copyrighted by ASTM International, 100 Barr Harbor Drive, PO Box C700, West Conshohocken, PA 19428-2959, United States. Individual reprints (single or multiple copies) of this standard may be obtained by contacting ASTM at the above address or at 610-832-9585 (phone), 610-832-9555 (fax), or service@astm.org (e-mail); or through the ASTM website (www.astm.org).

Designation: A633/A633M – 01 (Reapproved 2006)

Standard Specification for
Normalized High-Strength Low-Alloy Structural Steel Plates[1]

This standard is issued under the fixed designation A633/A633M; the number immediately following the designation indicates the year of original adoption or, in the case of revision, the year of last revision. A number in parentheses indicates the year of last reapproval. A superscript epsilon (ε) indicates an editorial change since the last revision or reapproval.

This standard has been approved for use by agencies of the Department of Defense.

1. Scope

1.1 This specification covers normalized high-strength low-alloy structural steel plates for welded, riveted, or bolted construction.

1.2 This material is particularly suited for service at low ambient temperatures of −50°F [−45°C] and higher where notch toughness better than that expected in as-rolled material of a comparable strength level is desired.

1.3 Four grades, designated Grades A, C, D, and E (essentially former Specification A633 without a grade designation) are covered by this specification. Grade A provides a minimum yield point of 42 ksi [290 MPa] in thicknesses through 4 in. [100 mm], inclusive. Grades C and D provide a minimum yield point of 50 ksi [345 MPa] in thicknesses up to 2.50 in. [65 mm], inclusive and 46.0 ksi [315 MPa] in thicknesses over 2.50 in. to 4.0 in. [65 to 100 mm], inclusive. Grade E provides a minimum yield point of 60 ksi [415 MPa] in thicknesses up to 4.0 in. [100 mm], inclusive and 55 ksi [380 MPa] in thicknesses over 4 in. to 6 in. [100 to 150 mm], inclusive.

1.4 Current practice normally limits plates furnished under this specification to the maximum thicknesses shown in 1.3. The individual manufacturer should be consulted on size limitations for other product forms.

1.5 When the steel is to be welded, it is presupposed that a welding procedure suitable for the grade of steel and intended use or service will be utilized. See Appendix X3 no id found of Specification A6/A6M for information on weldability.

1.6 The values stated in either inch-pound units or SI units are to be regarded as standard. Within the text, the SI units are shown in brackets. The values stated in each system are not exact equivalents; therefore, each system must be used independently of the other. Combining values from the two systems may result in nonconformance with this specification.

2. Referenced Documents

2.1 *ASTM Standards:*[2]

A6/A6M Specification for General Requirements for Rolled Structural Steel Bars, Plates, Shapes, and Sheet Piling

3. General Requirements for Delivery

3.1 Material furnished under this specification shall conform to the requirements of the current edition of Specification A6/A6M, for the ordered material, unless a conflict exists in which case this specification shall prevail.

4. Manufacture

4.1 The requirements for fine austenitic grain size in Specification A6/A6M shall be met.

5. Heat Treatment

5.1 The material shall be normalized by heating to a suitable temperature which produces an austenitic structure, but not exceeding 1700°F [925°C], holding a sufficient time to attain uniform heat throughout the material and cooling in air.

5.1.1 Grade E material over 3 in. [75 mm] in thickness shall be double normalized.

5.2 If the purchaser elects to perform the required heat treatment, the material shall be accepted on the basis of mill tests made from test coupons heat treated in accordance with the purchase order requirements. If the test coupon heat treatment requirements are not indicated on the purchase order, the manufacturer shall heat treat the test coupons under conditions considered appropriate. The manufacturer shall inform the purchaser of the heat-treatment procedure followed in heat treating the test coupons at the mill.

6. Chemical Composition

6.1 The heat analysis shall conform to the chemical composition requirements listed in Table 1.

---

[1] This specification is under the jurisdiction of ASTM Committee A01 on Steel, Stainless Steel and Related Alloys and is the direct responsibility of Subcommittee A01.02 on Structural Steel for Bridges, Buildings, Rolling Stock and Ships.
Current edition approved March 1, 2006. Published March 2006. Originally approved in 1970. Last previous edition approved in 2001 as A633/A633M – 01. DOI: 10.1520/A0633_A0633M-01R06.

[2] For referenced ASTM standards, visit the ASTM website, www.astm.org, or contact ASTM Customer Service at service@astm.org. For *Annual Book of ASTM Standards* volume information, refer to the standard's Document Summary page on the ASTM website.

TABLE 1 Chemical Requirements

Note—Where "..." appears in this table, there is no requirement.

Element	Grade A, %	Grade C, %	Grade D, %	Grade E, %[A]
Carbon, max	0.18	0.20	0.20	0.22
Manganese:				
1 1/2 in. [40 mm] and under in thickness	1.00–1.35	1.15–1.50[B]	0.70–1.35	1.15–1.50
Over 1 1/2 in. to 4 in. [40 to 100 mm], incl	1.00–1.35	1.15–1.50[B]	1.00–1.60	1.15–1.50
Over 4 in. to 6 in. [100 to 150 mm], incl	[C]	[C]	[C]	1.15–1.50
Phosphorus, max	0.035	0.035	0.035	0.035
Sulfur, max	0.04	0.04	0.04	0.04
Silicon	0.15–0.50	0.15–0.50	0.15–0.50	0.15–0.50
Vanadium	...	...	...	0.04–0.11
Columbium	0.05 max	0.01–0.05	...	[D]
Nitrogen, max	...	...	...	0.03
Copper, max	...	...	0.35	...
Nickel, max	...	...	0.25	...
Chromium, max	...	...	0.25	...
Molybdenum, max	...	...	0.08	...

[A] For Grade E the minimum total aluminum content shall be 0.018 %, or the vanadium nitrogen ratio shall be 4:1 minimum.
[B] For Grade C manganese content may be increased to 1.60 % maximum provided the carbon content does not exceed 0.18 %.
[C] The size and grade is not described in this specification.
[D] Columbium may be present in the amount of 0.01 to 0.05 %.

TABLE 2 Tensile Requirements[A]

	Grade A	Grades C and D	Grade E
Yield point, min, ksi [MPa]:			
2.5 in. [65 mm] and under	42 [290]	50 [345]	60 [415]
Over 2.5 in. to 4 in. [65 to 100 mm], incl	42 [290]	46 [315]	60 [415]
Over 4 in. to 6 in. [100 to 150 mm], incl	[B]	[B]	55 [380]
Tensile strength, ksi [MPa]:			
2.5 in. [65 mm] and under	63 to 83 [430 to 570]	70 to 90 [485 to 620]	80 to 100 [550 to 690]
Over 2.5 in. to 4 in. [65 to 100 mm], incl	63 to 83 [430 to 570]	65 to 85 [450 to 590]	80 to 100 [550 to 690]
Over 4 in. to 6 in. [100 to 150 mm], incl	[B]	[B]	75 to 95 [515 to 655]
Elongation in 8 in. [200 mm], min, %[C]	18	18	18
Elongation in 2 in. [50 mm], min, %[C]	23	23	23

[A] See specimen Orientation under the Tension Tests of Specification A6/A6M.
[B] The size and grade is not described in this specification.
[C] For plates wider than 24 in. (610 mm), the elongation requirement is reduced two percentage points. See elongation requirement adjustments in the Tension Tests section of Specification A6/A6M.

6.2 The steel shall conform on product analysis to the requirements prescribed in Table 1, subject to the product analysis tolerances in Specification A6/A6M.

7. Tension Test

7.1 The material as represented by the test specimens shall conform to the requirements listed in Table 2.

8. Keywords

8.1 bolted construction; high-strength; low-alloy; low ambient temperatures; normalized; notch toughness; plates; riveted construction; steel; structural steel; welded construction

SUPPLEMENTARY REQUIREMENTS

Standardized supplementary requirements for use at the option of the purchaser are listed in Specification A6/A6M. Those that are considered suitable for use with this specification are listed by title:

S5. Charpy V-Notch Impact Test.
S23. Copper-Bearing Steel (for improved atmospheric corrosion resistance).

APPENDIX

(Nonmandatory Information)

X1. CHARPY V-NOTCH IMPACT TEST

X1.1 The values shown in Table X1.1 are included only as information as to the guarantees that are generally available. Mandatory conformance to any of the values listed is a matter for agreement between the purchaser and the manufacturer.

TABLE X1.1 Charpy V-Notch Impact Test Minimum Energy Values (Average of Three Specimens)

Test Temperature, °F [°C]	Longitudinal Specimens, ft·lbf [J]	Transverse Specimens, ft·lbf [J]
−75 [−60]	15 [20]	15 [20]
−60 [−50]	20 [27]	15 [20]
−50 [−45]	25 [34]	20 [27]
−40 [−40]	25 [34]	20 [27]
−30 [−35]	30 [41]	25 [34]
0 [−20]	40 [54]	30 [41]
32 [0]	45 [61]	30 [41]
75 [25]	50 [68]	30 [41]

ASTM International takes no position respecting the validity of any patent rights asserted in connection with any item mentioned in this standard. Users of this standard are expressly advised that determination of the validity of any such patent rights, and the risk of infringement of such rights, are entirely their own responsibility.

This standard is subject to revision at any time by the responsible technical committee and must be reviewed every five years and if not revised, either reapproved or withdrawn. Your comments are invited either for revision of this standard or for additional standards and should be addressed to ASTM International Headquarters. Your comments will receive careful consideration at a meeting of the responsible technical committee, which you may attend. If you feel that your comments have not received a fair hearing you should make your views known to the ASTM Committee on Standards, at the address shown below.

This standard is copyrighted by ASTM International, 100 Barr Harbor Drive, PO Box C700, West Conshohocken, PA 19428-2959, United States. Individual reprints (single or multiple copies) of this standard may be obtained by contacting ASTM at the above address or at 610-832-9585 (phone), 610-832-9555 (fax), or service@astm.org (e-mail); or through the ASTM website (www.astm.org).

Designation: A653/A653M – 09a

Standard Specification for
Steel Sheet, Zinc-Coated (Galvanized) or Zinc-Iron Alloy-Coated (Galvannealed) by the Hot-Dip Process[1]

This standard is issued under the fixed designation A653/A653M; the number immediately following the designation indicates the year of original adoption or, in the case of revision, the year of last revision. A number in parentheses indicates the year of last reapproval. A superscript epsilon (ε) indicates an editorial change since the last revision or reapproval.

1. Scope*

1.1 This specification covers steel sheet, zinc-coated (galvanized) or zinc-iron alloy-coated (galvannealed) by the hot-dip process in coils and cut lengths.

1.2 The product is produced in various zinc or zinc-iron alloy-coating weights [masses] or coating designations as shown in Table 1 and in Table S2.1.

1.3 Product furnished under this specification shall conform to the applicable requirements of the latest issue of Specification A924/A924M, unless otherwise provided herein.

1.4 The product is available in a number of designations, grades and classes in four general categories that are designed to be compatible with different application requirements.

1.4.1 Steels with mandatory chemical requirements and typical mechanical properties.

1.4.2 Steels with mandatory chemical requirements and mandatory mechanical properties.

1.4.3 Steels with mandatory chemical requirements and mandatory mechanical properties that are achieved through solid-solution or bake hardening.

1.5 This specification is applicable to orders in either inch-pound units (as A653) or SI units (as A653M). Values in inch-pound and SI units are not necessarily equivalent. Within the text, SI units are shown in brackets. Each system shall be used independently of the other.

1.6 Unless the order specifies the "M" designation (SI units), the product shall be furnished to inch-pound units.

1.7 The text of this specification references notes and footnotes that provide explanatory material. These notes and footnotes, excluding those in tables and figures, shall not be considered as requirements of this specification.

1.8 *This standard does not purport to address all of the safety concerns, if any, associated with its use. It is the responsibility of the user of this standard to establish appropriate safety and health practices and determine the applicability of regulatory limitations prior to use.*

2. Referenced Documents

2.1 *ASTM Standards:*[2]
A90/A90M Test Method for Weight [Mass] of Coating on Iron and Steel Articles with Zinc or Zinc-Alloy Coatings
A370 Test Methods and Definitions for Mechanical Testing of Steel Products
A568/A568M Specification for Steel, Sheet, Carbon, Structural, and High-Strength, Low-Alloy, Hot-Rolled and Cold-Rolled, General Requirements for
A902 Terminology Relating to Metallic Coated Steel Products
A924/A924M Specification for General Requirements for Steel Sheet, Metallic-Coated by the Hot-Dip Process
D7396 Guide for Preparation of New, Continuous Zinc-Coated (Galvanized) Steel Surfaces for Painting
E517 Test Method for Plastic Strain Ratio *r* for Sheet Metal
E646 Test Method for Tensile Strain-Hardening Exponents (*n* -Values) of Metallic Sheet Materials

2.2 *ISO Standard:*[3]
ISO 3575 Continuous Hot-Dip Zinc-Coated Carbon Steel of Commercial and Drawing Qualities
ISO 4998 Continuous Hot-Dip Zinc-Coated Carbon Steel of Structural Quality

3. Terminology

3.1 *Definitions*—See Terminology A902 for definitions of general terminology relating to metallic-coated hot-dip products.

3.2 *Definitions of Terms Specific to This Standard:*

3.2.1 *bake hardenable steel, n*—steel sheet in which a significant increase in yield strength is realized when moderate heat treatment, such as that used for paint baking, follows straining or cold working.

3.2.2 *differentially coated, n*—galvanized steel sheet having a specified "coating designation" on one surface and a significantly lighter specified "coating designation" on the other surface.

---

[1] This specification is under the jurisdiction of ASTM Committee A05 on Metallic-Coated Iron and Steel Products and is the direct responsibility of Subcommittee A05.11 on Sheet Specifications.
Current edition approved Dec. 1, 2009. Published December 2009. Originally approved in 1994. Last previous edition approved in 2009 as A653/A653M - 09. DOI: 10.1520/A0653_A0653M-09A.

[2] For referenced ASTM standards, visit the ASTM website, www.astm.org, or contact ASTM Customer Service at service@astm.org. For *Annual Book of ASTM Standards* volume information, refer to the standard's Document Summary page on the ASTM website.

[3] Available from American National Standards Institute (ANSI), 25 W. 43rd St., 4th Floor, New York, NY 10036, http://www.ansi.org.

*A Summary of Changes section appears at the end of this standard.

TABLE 1 Weight [Mass] of Coating Requirements[A,B,C]

NOTE 1— Use the information provided in 8.1.3 to obtain the approximate coating thickness from the coating weight [mass].

		Minimum Requirement[D]		
		Triple-Spot Test (TST)		Single-Spot Test (SST)
Inch-Pound Units				
Type	Coating Designation	TST Total Both Sides, oz/ft^2	TST One Side, oz/ft^2	SST Total Both Sides, oz/ft^2
Zinc	G01	no minimum	no minimum	no minimum
	G30	0.30	0.10	0.25
	G40	0.40	0.12	0.30
	G60	0.60	0.20	0.50
	G90	0.90	0.32	0.80
	G100	1.00	0.36	0.90
	G115	1.15	0.40	1.00
	G140	1.40	0.48	1.20
	G165	1.65	0.56	1.40
	G185	1.85	0.64	1.60
	G210	2.10	0.72	1.80
	G235	2.35	0.80	2.00
	G300	3.00	1.04	2.60
	G360	3.60	1.28	3.20
Zinc-iron alloy	A01	no minimum	no minimum	no minimum
	A25	0.25	0.08	0.20
	A40	0.40	0.12	0.30
	A60	0.60	0.20	0.50
SI Units				
Type	Coating Designation	TST Total Both Sides, g/m^2	TST One Side, g/m^2	SST Total Both Sides, g/m^2
Zinc	Z001	no minimum	no minimum	no minimum
	Z90	90	30	75
	Z120	120	36	90
	Z180	180	60	150
	Z275	275	94	235
	Z305	305	110	275
	Z350	350	120	300
	Z450	450	154	385
	Z500	500	170	425
	Z550	550	190	475
	Z600	600	204	510
	Z700	700	238	595
	Z900	900	316	790
	Z1100	1100	390	975
Zinc-iron alloy	ZF001	no minimum	no minimum	no minimum
	ZF75	75	24	60
	ZF120	120	36	90
	ZF180	180	60	150

[A] The coating designation is the term by which the minimum triple spot, total both sides coating weight [mass] is specified. Because of the many variables and changing conditions that are characteristic of continuous hot-dip coating lines, the zinc or zinc-iron alloy coating is not always evenly divided between the two surfaces of a coated sheet; nor is it always evenly distributed from edge to edge. However, the minimum triple-spot average coating weight (mass) on any one side shall not be less than 40 % of the single-spot requirement.
[B] As it is an established fact that the atmospheric corrosion resistance of zinc or zinc-iron alloy-coated sheet products is a direct function of coating thickness (weight (mass)), the selection of thinner (lighter) coating designations will result in almost linearly reduced corrosion performance of the coating. For example, heavier galvanized coatings perform adequately in bold atmospheric exposure whereas the lighter coatings are often further coated with paint or a similar barrier coating for increased corrosion resistance. Because of this relationship, products carrying the statement "meets ASTM A653/A653Mrequirements" should also specify the particular coating designation.
[C] International Standard, ISO 3575, continuous hot-dip zinc-coated carbon steel sheet contains Z100 and Z200 designations and does not specify a ZF75 coating.
[D] No minimum means that there are no established minimum requirements for triple- and single-spot tests.

3.2.2.1 *Discussion*—The single side relationship of either specified "coating designation" is the same as shown in the note of Table 1 regarding uniformity of coating.

3.2.3 *high strength low alloy steel, n*—a specific group of sheet steels whose strength is achieved through the use of microalloying elements such as columbium (niobium), vanadium, titanium, and molybdenum resulting in improved formability and weldability than is obtained from conventional carbon-manganese steels.

3.2.3.1 *Discussion*—Producers use one or a combination of microalloying elements to achieve the desired properties. The product is available in two designations, HSLAS and HSLAS-F. Both products are strengthened with microalloys, but HSLAS-F is further treated to achieve inclusion control.

3.2.4 *minimized spangle, n*—the finish produced on hot-dip zinc-coated steel sheet in which the grain pattern is visible to the unaided eye, and is typically smaller and less distinct than the pattern visible on regular spangle.

3.2.4.1 *Discussion*—This finish is produced by one of two methods: either (*1*) the zinc crystal growth has been started but arrested by special production practices during solidification of the zinc, or (*2*) the zinc crystal growth is inhibited by a combination of coating-bath chemistry plus cooling during solidification of the zinc. Minimized spangle is normally produced in coating designations G90 [Z275] and lighter.

3.2.5 *regular spangle, n*—the finish produced on hot-dip zinc-coated steel sheet in which there is a visible multifaceted zinc crystal structure.

3.2.5.1 *Discussion*—Solidification of the zinc coating is typically uncontrolled, which produces the variable grain size associated with this finish.

3.2.6 *spangle-free, n*—the uniform finish produced on hot-dip zinc-coated steel sheet in which the visual spangle pattern, especially the surface irregularities created by spangle formation, is not visible to the unaided eye.

3.2.6.1 *Discussion*—This finish is produced when the zinc crystal growth is inhibited by a combination of coating-bath chemistry, or cooling, or both during solidification of the zinc.

3.2.7 *solid-solution hardened steel or solution hardened steel, n*—steel sheet strengthened through additions of substitutional alloying elements such as Mn, P, or Si.

3.2.7.1 *Discussion*—Substitutional alloying elements such as Mn, P, and Si can occupy the same sites as iron atoms within the crystalline structure of steels. Strengthening arises as a result of the mismatch between the atomic sizes of these elements and that of iron.

3.2.8 *zinc-iron alloy, n*—a dull grey coating with no spangle pattern that is produced on hot-dip zinc-coated steel sheet.

3.2.8.1 *Discussion*—Zinc-iron alloy coating is composed entirely of inter-metallic alloys. It is typically produced by subjecting the hot-dip zinc-coated steel sheet to a thermal treatment after it emerges from the molten zinc bath. This type of coating is suitable for immediate painting without further treatment except normal cleaning (refer to Guide D7396). The lack of ductility of the alloy coating presents a potential for powdering, etc.

4. Classification

4.1 The material is available in several designations as follows:

4.1.1 Commercial steel (CS Types A, B, and C),
4.1.2 Forming steel (FS Types A and B),
4.1.3 Deep drawing steel (DDS Types A and C),
4.1.4 Extra deep drawing steel (EDDS),
4.1.5 Structural steel (SS),
4.1.6 High strength low alloy steel (HSLAS),
4.1.7 High strength low alloy steel with improved formability (HSLAS-F),
4.1.8 Solution hardened steel (SHS), and
4.1.9 Bake hardenable steel (BHS).

4.2 Structural steel, high strength low alloy steel, solution hardened steel, and bake hardenable steel are available in several grades based on mechanical properties. Structural Steel Grade 50 [340] is available in four classes based on tensile strength. Structural Steel Grade 80 [550] is available in three classes, based on chemistry.

4.3 The material is available as either zinc-coated or zinc-iron alloy-coated in several coating weights [masses] or coating designations as shown in Table 1 and in Table S2.1, and

4.3.1 The material is available with the same or different coating designations on each surface.

5. Ordering Information

5.1 Zinc-coated or zinc-iron alloy-coated sheet in coils and cut lengths is produced to thickness requirements expressed to 0.001 in. [0.01 mm]. The thickness of the sheet includes both the base metal and the coating.

5.2 Orders for product to this specification shall include the following information, as necessary, to adequately describe the desired product:

5.2.1 Name of product (steel sheet, zinc-coated (galvanized) or zinc-iron alloy-coated (galvannealed)),

5.2.2 Designation of sheet [CS (Types A, B, and C), FS (Types A and B), DDS (Types A and C), EDDS, SS, HSLAS, HSLAS-F, SHS, or BHS].

5.2.2.1 When a CS type is not specified, CS Type B will be furnished. When a FS type is not specified, FS Type B will be furnished. When a DDS type is not specified, DDS Type A will be furnished.

5.2.3 When a SS, HSLAS, HSLAS-F, SHS, or BHS designation is specified, state the grade, or class, or combination thereof.

5.2.4 ASTM designation number and year of issue, as A653 for inch-pound units or A653M for SI units.

5.2.5 Coating designation,
5.2.6 Chemically treated or not chemically treated,
5.2.7 Oiled or not oiled,
5.2.8 Minimized spangle (if required),
5.2.9 Extra smooth (if required),
5.2.10 Phosphatized (if required),
5.2.11 Dimensions (show thickness, minimum or nominal, width, flatness requirements, and length, (if cut lengths)).
5.2.12 Coil size requirements (specify maximum outside diameter (OD), acceptable inside diameter (ID), and maximum weight [mass]),
5.2.13 Packaging,
5.2.14 Certification, if required, heat analysis and mechanical property report,
5.2.15 Application (part identification and description), and
5.2.16 Special requirements (if any).

5.2.16.1 If required, the product may be ordered to a specified base metal thickness (see Supplementary Requirement S1.)

5.2.16.2 If required, the product may be ordered to a specified single spot/single side coating mass (see Supplementary Requirement S2.)

5.2.16.3 When the purchaser requires thickness tolerances for 3/8-in. [10-mm] minimum edge distance (see Supplementary Requirement in Specification A924/A924M), this requirement shall be specified in the purchase order or contract.

NOTE 1—Typical ordering descriptions are as follows: steel sheet, zinc-coated, commercial steel Type A, ASTM A653, Coating Designation G115, chemically treated, oiled, minimum 0.040 by 34 by 117 in., for stock tanks, or steel sheet, zinc-coated, high strength low alloy steel Grade 340, ASTM A653M, Coating Designation Z275, minimized spangle, not chemically treated, oiled, minimum 1.00 by 920 mm by coil, 1520-mm maximum OD, 600-mm ID, 10 000-kg maximum, for tractor inner fender.

NOTE 2—The purchaser should be aware that there are variations in manufacturing practices among the producers and therefore is advised to establish the producer's standard (or default) procedures for thickness tolerances.

6. Chemical Composition

6.1 *Base Metal*:

6.1.1 The heat analysis of the base metal shall conform to the requirements shown in Table 2 for CS (Types A, B, and C), FS (Types A and B), DDS (Types A and C), and EDDS, and Table 3 for SS, HSLAS, HSLAS-F, SHS, and BHS.

6.1.2 Each of the elements listed in Tables 2 and 3 shall be included in the report of heat analysis. When the amount of copper, nickel, chromium, or molybdenum is less than 0.02 %, report the analysis as either <0.02 % or the actual determined value. When the amount of vanadium, titanium, or columbium is less than 0.008 %, report the analysis as either <0.008 % or the actual determined value. When the amount of boron is less than 0.0005 %, report as <0.0005 % or the actual determined value.

6.1.3 See Specification A924/A924M for chemical analysis procedures and product analysis tolerances.

6.2 *Zinc Bath Analysis*—The bath metal used in continuous hot-dip galvanizing shall contain not less than 99 % zinc.

Note 3—To control alloy formation and promote adhesion of the zinc coating with the steel base metal, the molten coating metal composition normally contains a percentage of aluminum usually in the range from 0.05 to 0.25. This aluminum is purposely supplied to the molten coating bath, either as a specified ingredient in the zinc spelter or by the addition of a master alloy containing aluminum.

7. Mechanical Properties

7.1 Structural steel, high-strength low-alloy steel, high strength low alloy steel with improved formability, solution hardened steel, and bake hardenable steel shall conform to the mechanical property requirements in Table 4 for the grade, or class, or both.

7.1.1 Bake hardenable steel shall conform to bake hardening index requirements included in Table 4 for the grade specified. The method for measuring the bake hardening index is described in the Annex. Bake hardenable steel shall exhibit a minimum increase in yield strength of 4 ksi [25 MPa] as based on the upper yield point or of 3 ksi [20 MPa] as based on the lower yield stress, after a prestrained specimen has been exposed to a standard bake cycle (340°F [170°C] for 20 minutes).

TABLE 2 Chemical Requirements[A]

Designation	Composition, %—Heat Analysis Element, max (unless otherwise shown)													
	Carbon	Manganese	Phosphorus	Sulfur	Aluminum, min	Cu	Ni	Cr	Mo	V	Cb	Ti[B]	N	B
CS Type A[C,D,E]	0.10	0.60	0.030	0.035	...	0.25	0.20	0.15	0.06	0.008	0.008	0.025	...	...
CS Type B[F,C]	0.02 to 0.15	0.60	0.030	0.035	...	0.25	0.20	0.15	0.06	0.008	0.008	0.025	...	...
CS Type C[C,D,E]	0.08	0.60	0.100	0.035	...	0.25	0.20	0.15	0.06	0.008	0.008	0.025	...	...
FS Type A[C,G]	0.10	0.50	0.020	0.035	...	0.25	0.20	0.15	0.06	0.008	0.008	0.025	...	...
FS Type B[F,C]	0.02 to 0.10	0.50	0.020	0.030	...	0.25	0.20	0.15	0.06	0.008	0.008	0.025	...	...
DDS Type A[D,E]	0.06	0.50	0.020	0.025	0.01	0.25	0.20	0.15	0.06	0.008	0.008	0.025	...	...
DDS Type C[D,E]	0.02	0.50	0.020 to 0.100	0.025	0.01	0.25	0.20	0.15	0.06	0.10	0.10	0.15	...	...
EDDS[H]	0.02	0.40	0.020	0.020	0.01	0.25	0.20	0.15	0.06	0.10	0.10	0.15	...	...

[A] Where an ellipsis (...) appears in this table, there is no requirement, but the analysis shall be reported.
[B] For steels containing 0.02 % carbon or more, titanium is permitted at the producer's option, to the lesser of 3.4N + 1.5S or 0.025 %.
[C] When a deoxidized steel is required for the application, the purchaser has the option to order CS and FS to a minimum of 0.01 % total aluminum.
[D] Steel is permitted to be furnished as a vacuum degassed or chemically stabilized steel, or both, at the producer's option.
[E] For carbon levels less than or equal to 0.02 %, vanadium, columbium, or titanium, or combinations thereof are permitted to be used as stabilizing elements at the producer's option. In such cases, the applicable limit for vanadium and columbium shall be 0.10 % max and the limit for titanium shall be 0.15 % max.
[F] For CS and FS, specify Type B to avoid carbon levels below 0.02 %.
[G] Shall not be furnished as a stabilized steel.
[H] Shall be furnished as a stabilized steel.

TABLE 3 Chemical Requirements[A]

Designation	Composition, %—Heat Analysis Element, max (unless otherwise shown)													
	Carbon	Manganese	Phosphorus	Sulfur	Si	Al, min	Cu	Ni	Cr	Mo	V[B]	Cb[B]	Ti[B,C,D]	N
SS														
33 [230]	0.20	1.35	0.10	0.04			0.25	0.20	0.15	0.06	0.008	0.008	0.025	...
37 [255]	0.20	1.35	0.10	0.04			0.25	0.20	0.15	0.06	0.008	0.008	0.025	...
40 [275]	0.25	1.35	0.10	0.04			0.25	0.20	0.15	0.06	0.008	0.008	0.025	...
50 [340] Class 1, 2, and 4	0.25	1.35	0.20	0.04			0.25	0.20	0.15	0.06	0.008	0.008	0.025	...
50 [340] Class 3	0.25	1.35	0.04	0.04			0.25	0.20	0.15	0.06	0.008	0.008	0.025	...
55 [380]	0.25	1.35	0.04	0.04			0.25	0.20	0.15	0.06	0.008	0.008	0.025	...
60 [410]	0.25	1.35	0.04	0.04			0.25	0.20	0.15	0.06	0.008	0.008	0.025	...
70 [480]	0.25	1.35	0.04	0.04			0.25	0.20	0.15	0.06	0.008	0.008	0.025	...
80 [550] Class 1	0.20	1.35	0.04	0.04			0.25	0.20	0.15	0.06	0.008	0.015	0.025	...
80 [550] Class 2[E]	0.02	1.35	0.05	0.02			0.25	0.20	0.15	0.06	0.10	0.10	0.15	...
80 [550] Class 3	0.20	1.35	0.04	0.04			0.25	0.20	0.15	0.06	0.008	0.015	0.025	...
HSLAS[F]														

TABLE 3 Continued

Designation	Carbon	Manganese	Phosphorus	Sulfur	Si	Al, min	Cu	Ni	Cr	Mo	V[B]	Cb[B]	Ti[B,C,D]	N
40 [275]	0.20	1.20	...	0.035			...	0.20	0.15	0.16	0.01 min	0.005 min	0.01 min	...
50 [340]	0.20	1.20	...	0.035			0.20	0.20	0.15	0.16	0.01 min	0.005 min	0.01 min	...
55 [380] Class 1	0.25	1.35	...	0.035			0.20	0.20	0.15	0.16	0.01 min	0.005 min	0.01 min	...
55 [380] Class 2	0.15	1.20	...	0.035			0.20	0.20	0.15	0.16	0.01 min	0.005 min	0.01 min	...
60 [410]	0.20	1.35	...	0.035			0.20	0.20	0.15	0.16	0.01 min	0.005 min	0.01 min	...
70 [480]	0.20	1.65	...	0.035			0.20	0.20	0.15	0.16	0.01 min	0.005 min	0.01 min	...
80 [550]	0.20	1.65	...	0.035			0.20	0.20	0.15	0.16	0.01 min	0.005 min	0.01 min	...
HSLAS-F[F,G]														
40 [275]	0.15	1.20	...	0.035			...	0.20	0.15	0.16	0.01 min	0.005 min	0.01 min	...
50 [340]	0.15	1.20	...	0.035			0.20	0.20	0.15	0.16	0.01 min	0.005 min	0.01 min	...
55 [380] Class 1	0.20	1.35	...	0.035			0.20	0.20	0.15	0.16	0.01 min	0.005 min	0.01 min	...
55 [380] Class 2	0.15	1.20	...	0.035			0.20	0.20	0.15	0.16	0.01 min	0.005 min	0.01 min	...
60 [410]	0.15	1.20	...	0.035			0.20	0.20	0.15	0.16	0.01 min	0.005 min	0.01 min	...
70 [480]	0.15	1.65	...	0.035			0.20	0.20	0.15	0.16	0.01 min	0.005 min	0.01 min	...
80 [550]	0.15	1.65	...	0.035			0.20	0.20	0.15	0.16	0.01 min	0.005 min	0.01 min	...
SHS[D]	0.12	1.50	0.12	0.030	...	...	0.20	0.20	0.15	0.06	0.008	0.008	0.025	...
BHS[D]	0.12	1.50	0.12	0.030	...	...	0.20	0.20	0.15	0.06	0.008	0.008	0.025	...

[A] Where an ellipsis (...) appears in this table there is no requirement, but the analysis shall be reported.
[B] For carbon levels less than or equal to 0.02 %, vanadium, columbium, or titanium, or combinations thereof, are permitted to be used as stabilizing elements at the producer's option. In such cases, the applicable limit for vanadium and columbium shall be 0.10% max., and the limit for titanium shall be 0.15 % max.
[C] Titanium is permitted for SS steels at the producer's option, to the lesser of 3.4N +1.5S or 0.025 %.
[D] For steels containing more than 0.02 % carbon, titanium is permitted to the lesser of 3.4N + 1.5S or 0.025 %.
[E] Shall be furnished as a stabilized steel.
[F] HSLAS and HSLAS-F steels commonly contain the strengthening elements columbium, vanadium, and titanium added singly or in combination. The minimum requirements only apply to the microalloy elements selected for strengthening of the steel.
[G] HSLAS-F steel shall be treated to achieve inclusion control.

TABLE 4 Mechanical Requirements, Base Metal (Longitudinal)

Inch-Pound Units

Designation	Grade	Yield Strength, min, ksi	Tensile Strength, min, ksi[A]	Elongation in 2 in., min, %[A]	Bake Hardening Index, min, ksi Upper Yield/Lower Yield[A]
SS	33	33	45	20	...
	37	37	52	18	...
	40	40	55	16	...
	50 Class 1	50	65	12	...
	50 Class 2	50	...	12	...
	50 Class 3	50	70	12	...
	50 Class 4	50	60	12	...
	55	55	70	11	...
	60	60	70	10[B]	...
	70	70	80	9[B]	...
	80 Class 1[C]	80[D]	82	...	...
	80 Class 2[C,E]	80[D]	82	...	...
	80 Class 3	80[D]	82	3[F]	...
HSLAS	40	40	50[G]	22	...
	50	50	60[G]	20	...
	55 Class 1	55	70[G]	16	...
	55 Class 2	55	65[G]	18	...
	60	60	70[G]	16	...
	70	70	80[G]	12	...
	80	80	90[G]	10	...
HSLAS-F	40	40	50[G]	24	...
	50	50	60[G]	22	...

TABLE 4 Continued

Inch-Pound Units

Designation	Grade	Yield Strength, min, ksi	Tensile Strength, min, ksi[A]	Elongation in 2 in., min, %[A]	Bake Hardening Index, min, ksi Upper Yield/Lower Yield[A]
	55 Class 1	55	70[G]	18	...
	55 Class 2	55	65[G]	20	...
	60	60	70[G]	18	...
	70	70	80[G]	14	...
	80	80	90[G]	12	...
SHS	26	26	43	32	...
	31	31	46	30	...
	35	35	50	26	...
	41	41	53	24	...
	44	44	57	22	...
BHS	26	26	43	30	4 / 3
	31	31	46	28	4 / 3
	35	35	50	24	4 / 3
	41	41	53	22	4 / 3
	44	44	57	20	4 / 3

SI Units

Designation	Grade	Yield Strength, min, MPa	Tensile Strength, min, MPa[A]	Elongation in 50 mm, min, %[A]	Bake Hardening Index, min, MPa Upper Yield/Lower Yield[A]
SS	230	230	310	20	...
	255	255	360	18	...
	275	275	380	16	...
	340 Class 1	340	450	12	...
	340 Class 2	340	...	12	...
	340 Class 3	340	480	12	...
	340 Class 4	340	410	12	...
	380	380	480	11	...
	410	410	480	10[B]	...
	480	480	550	9[B]	...
	550 Class 1[C]	550[D]	570	...	...
	550 Class 2[C,E]	550[D]	570	...	...
	550 Class 3	550[D]	570	3[F]	...
HSLAS	275	275	340[G]	22	...
	340	340	410[G]	20	...
	380 Class 1	380	480[G]	16	...
	380 Class 2	380	450[G]	18	...
	410	410	480[G]	16	...
	480	480	550[G]	12	...
	550	550	620[G]	10	...
HSLAS-F	275	275	340[G]	24	...
	340	340	410[G]	22	...
	380 Class 1	380	480[G]	18	...
	380 Class 2	380	450[G]	20	...
	410	410	480[G]	18	...
	480	480	550[G]	14	...
	550	550	620[G]	12	...
SHS	180	180	300	32	...
	210	210	320	30	...
	240	240	340	26	...
	280	280	370	24	...
	300	300	390	22	...
BHS	180	180	300	30	25 / 20
	210	210	320	28	25 / 20
	240	240	340	24	25 / 20
	280	280	370	22	25 / 20
	300	300	390	20	25 / 20

[A] Where an ellipsis (. . .) appears in this table there is no requirement.
[B] For sheet thickness of 0.028 in. [0.71 mm] or thinner, the elongation requirement is reduced two percentage points for SS Grades 60 [410] and 70 [480].
[C] For sheet thickness of 0.028 in. [0.71 mm] or thinner, no tension test is required if the hardness result in Rockwell B85 or higher.
[D] As there is no discontinuous yield curve, the yield strength should be taken as the stress at 0.5 % elongation under load or 0.2 % offset.
[E] SS Grade 80 [550] Class 2 may exhibit different forming characteristics than Class 1, due to difference in chemistry.
[F] The purchaser should consult with the producer when ordering SS Grade 80 [550] Class 3 material in sheet thicknesses 0.028 in. [0.71 mm] or thinner regarding elongation and tension test requirements.
[G] If a higher tensile strength is required, the user should consult the producer.

7.2 The typical mechanical properties for CS (Types A, B, and C), FS (Types A and B), DDS (Types A and C), and EDDS sheet designations are listed in Table 5. These mechanical property values are nonmandatory. They are intended solely to provide the purchaser with as much information as possible to make an informed decision on the steel to be specified. Values outside of these ranges are to be expected.

7.3 When base metal mechanical properties are required, all tests shall be conducted in accordance with the methods specified in Specification A924/A924M.

7.4 *Bending Properties Minimum Cold Bending Radii*— Structural steel and high-strength low-alloy steel are commonly fabricated by cold bending. There are many interrelated factors that affect the ability of a steel to cold form over a given radius under shop conditions. These factors include thickness, strength level, degree of restraint, relationship to rolling direction, chemistry, and base metal microstructure. The table in Appendix X1 lists the suggested minimum inside radius for 90° cold bending for structural steel and high-strength low-alloy steel. They presuppose "hard way" bending (bend axis parallel to rolling direction) and reasonably good shop forming practices. Where possible, the use of larger radii or "easy way" bends are recommended for improved performance.

8. Coating Properties

8.1 *Coating Weight [Mass]*:

8.1.1 Coating weight [mass] shall conform to the requirements as shown in Table 1 for the specific coating designation, or

8.1.2 If required, the coating mass shall conform to the requirements as shown in Table S2.1 for the specific single spot/single side coating mass designation (single spot/single side designations are available only in SI units).

8.1.3 Use the following relationships to estimate the coating thickness from the coating weight [mass]:

8.1.3.1 1 oz/ft^2 coating weight = 1.7 mils coating thickness, and

8.1.3.2 7.14 g/m^2 coating mass = 1 μm coating thickness.

8.1.4 Use the following relationship to convert coating weight to coating mass:

8.1.4.1 1 oz/ft^2 coating weight = 305.15 g/m^2 coating mass.

8.2 *Coating Weight [Mass] Tests*:

8.2.1 Coating weight [mass] tests shall be performed in accordance with the requirements of Specification A924/A924M.

8.2.2 The referee method to be used shall be Test Method A90/A90M.

8.3 *Coating Bend Test*:

8.3.1 The bend test specimens of coated sheet designated by prefix "G" ["Z"] shall be capable of being bent through 180° in any direction without flaking of the coating on the outside of the bend only. The coating bend test inside diameter shall have a relation to the thickness of the specimen as shown in Table 6. Flaking of the coating within 0.25 in. [6 mm] of the edge of the bend specimen shall not be cause for rejection.

8.3.2 Because of the characteristics of zinc-iron alloy coatings designated by prefix "A" ["ZF"] as explained in the Discussion following 3.2.8, coating bend tests are not applicable.

9. Retests and Disposition of Non-Conforming Material

9.1 Retests, conducted in accordance with the requirements of the section on Retests and Disposition of Non-Conforming Material of Specification A924/A924M, are permitted when an unsatisfactory test result is suspected to be the consequence of the test method procedure.

9.2 Disposition of non-conforming material shall be subject to the requirements of 13.2 of Specification A924/A924M.

10. Dimensions and Permissible Variations

10.1 All dimensions and permissible variations shall comply with the requirements of Specification A924/A924M.

11. Keywords

11.1 alloyed coating; bake hardenable steel; high strength low alloy; minimized spangle coating; sheet steel; solution hardened steel; spangle; steel; steel sheet; structural steel; zinc; zinc coated (galvanized); zinc iron-alloy; zinc iron-alloy coated

TABLE 5 Typical Ranges of Mechanical Properties[A,B] **(Nonmandatory)**

Designation	(Longitudinal Direction)			r_m Value[C]	n Value[D]
	Yield Strength		Elongation in 2 in. [50 mm], %		
	ksi	[MPa]			
CS Type A	25/55	[170/380]	≥20	[E]	[E]
CS Type B	30/55	[205/380]	≥20	[E]	[E]
CS Type C	25/60	[170/410]	≥15	[E]	[E]
FS Types A and B	25/45	[170/310]	≥26	1.0/1.4	0.17/0.21
DDS Type A	20/35	[140/240]	≥32	1.4/1.8	0.19/0.24
DDS Type C	25/40	[170/280]	≥32	1.2/1.8	0.17/0.24
EDDS[F]	15/25	[105/170]	≥40	1.6/2.1	0.22/0.27

[A] The typical mechanical property values presented here are nonmandatory. They are intended solely to provide the purchaser with as much information as possible to make an informed decision on the steel to be specified. Values outside of these ranges are to be expected. The purchaser may negotiate with the supplier if a specific range or a more restrictive range is required for the application.
[B] These typical mechanical properties apply to the full range of steel sheet thicknesses. The yield strength tends to increase and some of the formability values tend to decrease as the sheet thickness decreases.
[C] r_m Value—Average plastic strain ratio as determined by Test Method E517.
[D] n Value—Strain-hardening exponent as determined by Test Method E646.
[E] No typical mechanical properties have been established.
[F] EDDS Sheet will be free from changes in mechanical properties over time, that is, nonaging.

TABLE 6 Coating Bend Test Requirements

Inch-Pound Units

Coating Designation[B]	Ratio of the Inside Bend Diameter to Thickness of the Specimen (Any Direction) CS, FS, DDS, EDDS, SHS, BHS			SS, Grade[A]		
	Sheet Thickness			33	37	40
	Through 0.039 in.	Over 0.039 through 0.079 in.	Over 0.079 in.			
G01	0	0	0	1½	2	2½
G30	0	0	0	1½	2	2½
G40	0	0	0	1½	2	2½
G60	0	0	0	1½	2	2½
G90	0	0	1	1½	2	2½
G100	0	0	1	1½	2	2½
G115	0	0	1	1½	2	2½
G140	1	1	2	2	2	2½
G165	2	2	2	2	2	2½
G185	2	2	2	2	2	2½
G210	2	2	2	2	2	2½
G235	2	2	3	3	3	3

Coating Designation	HSLAS[A]			HSLAS-F				
	40	50	60	40	50	60	70	80
G01	1½	1½	3	1	1	1	1½	1½
G30	1½	1½	3	1	1	1	1½	1½
G40	1½	1½	3	1	1	1	1½	1½
G60	1½	1½	3	1	1	1	1½	1½
G90	1½	1½	3	1	1	1	1½	1½
G100	1½	1½	3	1	1	1	1½	1½
G115	1½	1½	3	1	1	1	1½	1½

SI Units

Coating Designation[B]	Ratio of the Inside Bend Diameter to Thickness of the Specimen (Any Direction) CS, FS, DDS, EDDS, SHS, BHS			SS, Grade[C]		
	Sheet Thickness			230	255	275
	Through 1.0 mm	Over 1.0 mm through 2.0 m	Over 2.0 mm			
Z001	0	0	0	1½	2	2½
Z90	0	0	0	1½	2	2½
Z120	0	0	0	1½	2	2½
Z180	0	0	0	1½	2	2½
Z275	0	0	0	1½	2	2½
Z305	0	0	1	1½	2	2½
Z350	0	0	1	1½	2	2½
Z450	1	1	2	2	2	2½
Z500	2	2	2	2	2	2½
Z550	2	2	2	2	2	2½
Z600	2	2	2	2	2	2½
Z700	2	3	3	3	3	3

Coating Designation	HSLAS[C]			HSLAS-F				
	275	340	410	275	340	410	480	550
Z001	1½	1½	3	1	1	1	1½	1½
Z90	1½	1½	3	1	1	1	1½	1½
Z120	1½	1½	3	1	1	1	1½	1½
Z180	1½	1½	3	1	1	1	1½	1½
Z275	1½	1½	3	1	1	1	1½	1½
Z305	1½	1½	3	1	1	1	1½	1½
Z350	1½	1½	3	1	1	1	1½	1½

[A] SS Grades 50, 60, 70, and 80, HSLAS, and HSLAS-F Grades 70 and 80 are not subject to bend test requirements.
[B] If other coatings are required, the user should consult the producer for availability and suitable bend test requirements.
[C] SS Grades 340, 410, 480, and 550, HSLAS, and HSLAS-F Grades 480 and 550 are not subject to bend test requirements.

SUPPLEMENTARY REQUIREMENTS

The following standardized supplementary requirements are for use when desired by the purchaser. These additional requirements shall apply only when specified on the order.

S1. Base Metal Thickness

S1.1 The specified minimum thickness shall apply to the base metal only.

S1.2 The coating designation shown on the order indicates the coating to be applied to the specified minimum base metal thickness.

S1.3 The applicable tolerances for base metal thickness are shown in Tables 16 and Tables 17, Thickness Tolerance of Cold-Rolled Sheet (Carbon and High-Strength, Low-Alloy Steel), of Specification A568/A568M.

S2. Single Spot/Single Side Coating Mass

S2.1 The coating designation shown on the order indicates the coating mass to be applied to a single side. The order shall specify a coating mass designation from Table S2.1 for each surface. No inch pound designations are available, although for each SI coating mass designation in Table S2.1, corresponding inch-pound values are shown for information purposes

S2.2 The format for specifying the coating for each surface on the order shall be, for instance, 60G60G. In the case of differential coating masses, the thicker (heavier) coating mass side shall be specified first, for instance 90G45G.

TABLE S2.1 Mass of Coating Requirements—Single Spot/Single Side[A,B,C]

NOTE 1—Use the information provided in 8.1.3 to obtain the approximate coating thickness per side from the coating mass.
NOTE 2—As stated in 1.5, values in SI and inch-pound units are not necessarily equivalent.

Type	Single Spot/Single Side Coating Mass				
	SI Units			Inch-Pound Units (information only)	
	Coating Designation	Minimum, g/m²	Maximum, g/m²	Minimum, oz/ft²	Maximum, oz/ft²
Zinc	20G	20	70	0.07	0.23
	30G	30	80	0.10	0.26
	40G	40	90	0.12	0.29
	45G	45	95	0.15	0.31
	50G	50	100	0.16	0.33
	55G	55	105	0.18	0.34
	60G	60	110	0.20	0.36
	70G	70	120	0.23	0.40
	90G	90	160	0.30	0.62
	100G[D]	100	200	0.32	0.65
Zinc-Iron Alloy	40A	40	70	0.13	0.23
	45A	45	75	0.15	0.25
	50A	50	80	0.16	0.26

[A] The coating designation is the term by which the minimum single spot/single side coating mass is specified for each side.
[B] As it is an established fact that the atmospheric corrosion resistance of zinc or zinc-iron alloy-coated sheet products is a direct function of coating thickness (mass), the selection of thinner (lighter) coating designations will result in almost linearly reduced corrosion performance of the coating. For example, heavier galvanized coatings perform adequately in bold atmospheric exposure whereas the lighter coatings are often further coated with paint or a similar barrier coating for increased corrosion resistance. Because of this relationship, products carrying the statement "meets ASTM A653/A653Mrequirements" should also specify the particular coating designation.
[C] Ordering to single spot/single side spot coating weight designations allows for the possibility of receiving product with a higher average total coating mass on both surfaces than what might be expected on assumed equivalent product coated to the total both sides requirement of Table 1. The user should be aware that this may result in issues during forming and spot welding.
[D] It is permissible to order Coating Designation 100G as 98G. Historically, the conversion from 0.32 oz/ft² to g/m² resulted in a value of 98 g/m² which was subsequently rounded to 100 g/m². Both SI designations have the same limits.

ANNEX

(Mandatory Information)

A1. BAKE HARDENABLE STEELS

A1.1 Determination of Bake Hardening Index

A1.1.1 The bake hardening index (BHI) is determined by a two-step procedure using a standard longitudinal (rolling direction) tensile-test specimen, prepared in accordance with Test Methods A370. The test specimen is first strained in tension. The magnitude of this tensile "pre-strain" shall be 2 % (extension under load). The test specimen is then removed from the test machine and baked at a temperature of 340°F [170°C] for a period of 20 minutes. Referring to Fig. A1.1, the bake hardening index (BHI) of the material is calculated as follows:

$$BHI = B - A \quad (A1.1)$$

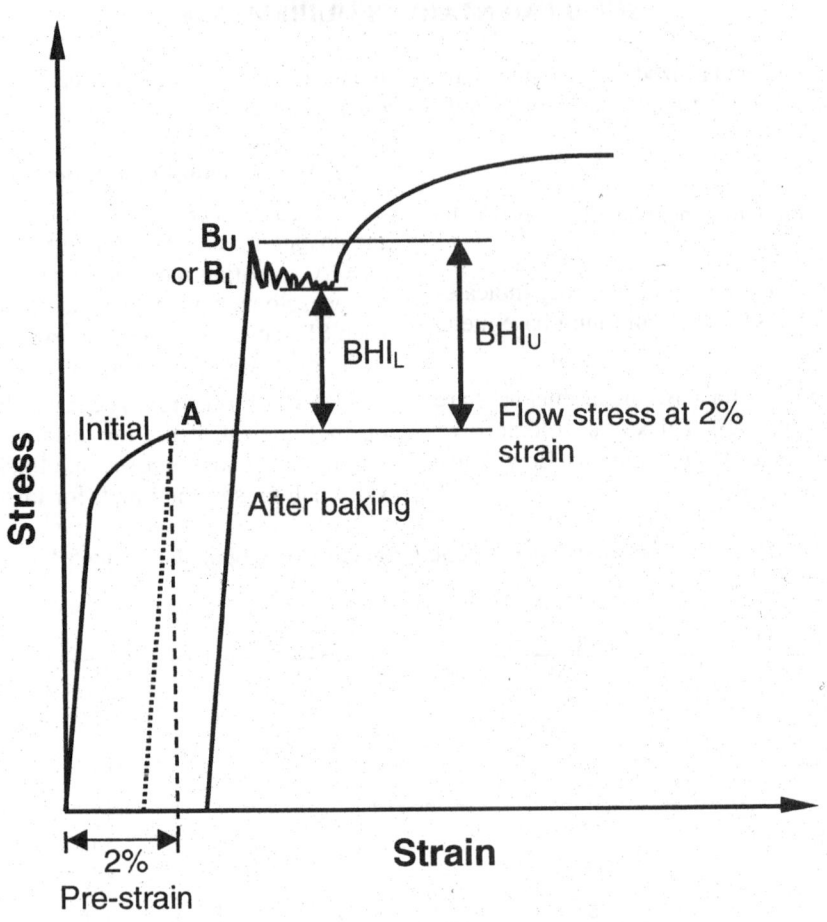

FIG. A1.1 Representation of Bake Hardening Index

where:
A = flow stress at 2 % extension under load
B = yield strength [upper yield strength (B_U) or lower yield stress (B_L)] after baking at 340°F [170°C] for 20 minutes.

A1.1.2 The original test specimen cross section (width and thickness) is used in the calculation of all engineering strengths in this test.

A1.1.3 The pre-straining of 2 % in tension is intended to simulate a modest degree of forming strain, while the subsequent baking is intended to simulate a paint-curing or similar treatment. In the production of actual parts, forming strains and baking treatments can differ from those employed here and, as a result, final properties can differ from the values obtained under these controlled conditions.

APPENDIXES

(Nonmandatory Information)

X1. BENDING PROPERTIES

X1.1 Table X1.1 lists suggested minimum inside radii for cold bending.

TABLE X1.1 Suggested Minimum Inside Radii for Cold Bending[A]

NOTE 1— (*t*) equals a radius equivalent to the steel thickness.
NOTE 2—The suggested radii should be used as minimums for 90° bends in actual shop practice.

Designation	Grade	Minimum Inside Radius for Cold Bending[B]
SS	33 [230]	1½ *t*
	37 [255]	2*t*
	40 [275]	2*t*
	50 [340] Class 1	not applicable
	50 [340] Class 2	not applicable
	50 [340] Class 3	not applicable
	50 [340] Class 4	not applicable
	55 [380]	not applicable
	60 [410]	not applicable
	70 [480]	not applicable
	80 [550] Class 1	not applicable
	80 [550] Class 2	not applicable
	80 [550] Class 3	not applicable
HSLAS	40 [275]	2*t*
	50 [340]	2½ *t*
	55 [380] Class 1	3*t*
	55 [380] Class 2	3*t*
	60 [410]	3*t*
	70 [480]	4*t*
	80 [550]	4½ *t*
HSLAS-F	40 [275]	1½ *t*
	50 [340]	2*t*
	55 [380] Class 1	2*t*
	55 [380] Class 2	2*t*
	60 [410]	2*t*
	70 [480]	3*t*
	80 [550]	3*t*
SHS	26 [180]	½ *t*
	31 [210]	1*t*
	35 [240]	1½ *t*
	41 [280]	2*t*
	44 [300]	2*t*
BHS	26 [180]	½ *t*
	31 [210]	1*t*
	35 [240]	1½ *t*
	41 [280]	2*t*
	44 [300]	2*t*

[A] Material that does not perform satisfactorily, when fabricated in accordance with the requirements in Table X1.1, may be subject to rejection pending negotiation with the steel supplier.
[B] Bending capability may be limited by coating designation.

X2. RATIONALE FOR CHANGES IN PRODUCT DESIGNATIONS

X2.1 Subcommittee A05.11 has revised the designations used to classify the various products available in each hot-dip coated specification. The previous "quality" designations have been replaced with designations and descriptions more closely related with product characteristics. Many of the former "quality" specifications described the steel only in terms of limited chemical composition, which in some cases was identical for two or more qualities. The former designations also did not reflect the availability of new steels which are the result of the use of new technologies such as vacuum degassing and steel ladle treatments.

X2.2 The former "quality" designators, defined in very broad qualitative terms, did not provide the user with all the information needed to select the appropriate steel for an application. The new designations are defined with technical information such as specific chemical composition limits and typical nonmandatory mechanical properties. These steel characteristics are important to users concerned with the weldability and formability of the coated steel products. The typical mechanical properties included in the new designation system are those indicated by the tension test. These properties are more predictive of steel formability than other tests such as the hardness test which may not compensate adequately for product variables such as substrate thickness and coating weight.

X2.3 The new designations also provide the user with the flexibility to restrict the steels applied on any order. For example, a user can restrict the application of ultra low carbon steels on an application through the selection of an appropriate "type" designator.

X2.4 There is a limited relationship between the former and current systems of designation. Some of the reasons for this limited relationship are: addition of steels not previously described in ASTM specifications, restrictions placed on ranges of chemical composition, the addition of typical mechanical properties, and the enhanced capability of steel producers to combine chemical composition and processing methods to achieve properties tailored to specific applications.

X2.5 The changes in designation are significant which may create transition issues that will have to be resolved. Continued dialogue between users and producers will have to be maintained to assist with the transition to the new system of designations. A user with concerns about the appropriate coated steel to order for a specific application should consult with a steel supplier or producer.

X3. RELATIONSHIP BETWEEN SPECIFICATIONS THAT DESCRIBE REQUIREMENTS FOR A COMMON PRODUCT

X3.1 ISO 3575 and ISO 4998 may be reviewed for comparison with this standard. The relationship between the standards may only be approximate; therefore, the respective documents should be consulted for actual requirements. Those who use these documents must determine which specifications address their needs.

X4. COATING MASS SELECTION BASED ON ATMOSPHERIC CORROSION RATES[4] FOR ZINC-COATED STEEL SHEET

X4.1 The proper selection of coating mass to meet a user's needs for zinc-coated steel sheet requires some knowledge about the relative corrosiveness of the environment in which the product will be used. The corrosion rate of the zinc coating varies widely depending upon many factors of the environment. For example, the time of wetness is an important issue that affects the corrosion rate. The presence of impurities such as chlorides, nitrates, and sulfates can also dramatically affect the rate of corrosion. Other issues such as the presence or absence of oxygen and the temperature of the environment are important determinants for predicting the "life of the product."

X4.2 The final performance requirements can also impact the minimum coating mass needed for a given application. For example, is the application an aesthetic one that requires no red rust. In this case, the time to failure is thus defined as the time for the onset of red rust (the time for the zinc coating to be consumed in a large enough area for rusting of the steel to be observed). Or, is the application one in which the time to failure is defined as the time when perforation of the steel sheet is observed? In this case, the thickness of the steel sheet as well as the thickness of the zinc coating impact the time to failure.

X4.3 No matter how one defines the "product life," there are data in the published literature to assist users once the environment and desired product life are determined.

X4.4 Although the corrosion rate can vary considerably depending on the environmental factors, it is well known that, in most instances, the life of the zinc coating is a linear function of coating mass for any specific environment. That means, to achieve twice the life for any specific application, the user should order twice the coating mass.

X4.4.1 *Examples*:

X4.4.1.1 A G60 coating mass will exhibit approximately twice the life of a G30 coating mass.

X4.4.1.2 A G90 coating mass will exhibit about 50 % longer life than a G60 coating mass.

X4.5 The following two reference books are excellent sources for additional and more detailed information on the corrosion behavior of zinc-coated steel sheet products:

X4.5.1 *Corrosion and Electrochemistry of Zinc*, X. Gregory Zhang, published by Plenum Press, 1996.

X4.5.2 *Corrosion Resistance of Zinc and Zinc Alloys*, Frank C. Porter, Published by Marcel Dekker, Inc., 1994

---

[4] Atmospheric corrosion rates do not apply to zinc-iron alloy coatings.

SUMMARY OF CHANGES

Committee A05 has identified the location of selected changes to this standard since the last issue (A653/A653M - 09) that may impact the use of this standard. (December 1, 2009)

(1) Revised 5.2.11.

(2) Added 5.2.16.3.

Committee A05 has identified the location of selected changes to this standard since the last issue (A653/A653M - 08) that may impact the use of this standard. (May 1, 2009)

(1) Added new Grade 80 Class 3 requirements to Tables 3, 4, and X1.1.

(2) Added new SS Grade 60 [410] and 70 [480] requirements to Tables 3, 4, 6, and X1.1.

(3) Changed reference from D2092 to D7396 in 2.1 and 3.2.8.1.

(4) Changed the order of the coating weight [mass] designations in Table 6 from descending to ascending.

(5) Added 20G coating designation to Table S2.1

Committee A05 has identified the location of selected changes to this standard since the last issue (A653/A653M - 07) that may impact the use of this standard. (July 15, 2008)

(1) Changed the order of the coating weight [mass] designations in Table 1 from descending to ascending, and changed Footnote A.

(2) Added Supplementary Requirement S2 that allows a purchaser to order product to single spot/single side coating designations. In support of this, revisions were made in 1.2, 4.3, and 8.1.1; and added 5.2.16.2 and 8.1.2, with previous 8.1.2 renumbered as 8.1.3.

(3) Added 8.1.4 added showing the formula for converting coating weight to coating mass.

(4) Replaced footnote H with footnotes D and E on DDS Type C in Table 2.

(5) In Table 3, revised SS GRD 33 to increase P level.

(6) Revised 10.1.

(7) Deleted Flatness Tables: 7 and 8.

ASTM International takes no position respecting the validity of any patent rights asserted in connection with any item mentioned in this standard. Users of this standard are expressly advised that determination of the validity of any such patent rights, and the risk of infringement of such rights, are entirely their own responsibility.

This standard is subject to revision at any time by the responsible technical committee and must be reviewed every five years and if not revised, either reapproved or withdrawn. Your comments are invited either for revision of this standard or for additional standards and should be addressed to ASTM International Headquarters. Your comments will receive careful consideration at a meeting of the responsible technical committee, which you may attend. If you feel that your comments have not received a fair hearing you should make your views known to the ASTM Committee on Standards, at the address shown below.

This standard is copyrighted by ASTM International, 100 Barr Harbor Drive, PO Box C700, West Conshohocken, PA 19428-2959, United States. Individual reprints (single or multiple copies) of this standard may be obtained by contacting ASTM at the above address or at 610-832-9585 (phone), 610-832-9555 (fax), or service@astm.org (e-mail); or through the ASTM website (www.astm.org). Permission rights to photocopy the standard may also be secured from the ASTM website (www.astm.org/COPYRIGHT/).

Designation: A656/A656M – 05ε2

Standard Specification for
Hot-Rolled Structural Steel, High-Strength Low-Alloy Plate with Improved Formability[1]

This standard is issued under the fixed designation A656/A656M; the number immediately following the designation indicates the year of original adoption or, in the case of revision, the year of last revision. A number in parentheses indicates the year of last reapproval. A superscript epsilon (ε) indicates an editorial change since the last revision or reapproval.

ε[1] NOTE—Table 1 was corrected editorially in February 2006.
ε[2] NOTE—The Scope (1.1) was corrected editorially in October 2007.

1. Scope*

1.1 This specification covers three types and four strength grades of high-strength low-alloy, hot rolled structural steel plate for use in truck frames, brackets, crane booms, rail cars, and similar applications. Steels that conform to this specification offer improved formability. These steels are normally furnished in the as-rolled condition. The type and strength grade furnished is as agreed upon between the manufacturer and the purchaser. The types and strength grades are shown in the tables.

1.2 The maximum thickness of plates shall be as follows:

Grade	Plate Thickness, max, in. [mm]
50	2 [50]
60	1½ [40]
70	1 [25]
80	¾ [20]

1.3 The values stated in either inch-pound units or SI units are to be regarded as standard. Within the text, the SI units are shown in brackets. The values stated in each system are not exact equivalents; therefore, each system must be used independently of the other. Combining values from the two systems may result in nonconformance with this specification. See Appendix X3 of Specification A6/A6M for information on weldability.

1.4 For plates produced from coil and furnished without heat treatment or with stress relieving only, the additional requirements, including additional testing requirements and the reporting of additional test results, of Specification A6/A6M apply.

2. Referenced Documents

2.1 *ASTM Standards:*[2]
A6/A6M Specification for General Requirements for Rolled Structural Steel Bars, Plates, Shapes, and Sheet Piling

3. General Requirements for Delivery

3.1 Plates furnished under this specification shall conform to the requirements of the current edition of Specification A6/A6M, for the specific plate ordered, unless a conflict exists, in which case this specification shall prevail.

3.2 Coils are excluded from qualification to this specification until they are processed into finished plates. Plates produced from coil means plates that have been cut to individual lengths from a coil. The processor directly controls, or is responsible for, the operations involved in the processing of a coil into finished plates. Such operations include decoiling, leveling, cutting to length, testing, inspection, conditioning, heat treatment (if applicable), packaging, marking, loading for shipment, and certification.

NOTE 1—For plates produced from coil and furnished without heat treatment or with stress relieving only, two test results are to be reported for each qualifying coil. Additional requirements regarding plate produced from coil are described in Specification A6/A6M.

4. Materials and Manufacture

4.1 The steel shall be made to fine grain practice.

5. Chemical Composition

5.1 Heat analyses shall conform to the chemical requirements given in Table 1. Dependent upon thickness, grade, and intended application, variations in the chemical composition are permitted within the limits given in Table 1 for the applicable type. Where it is of particular importance, the manufacturer should be consulted for specific chemical composition.

---

[1] This specification is under the jurisdiction of ASTM Committee A01 on Steel, Stainless Steel and Related Alloys and is the direct responsibility of Subcommittee A01.02 on Structural Steel for Bridges, Buildings, Rolling Stock and Ships.
Current edition approved Sept. 1, 2005. Published October 2005. Originally approved in 1972. Last previous edition approved in 2003 as A656/A656M–03. DOI: 10.1520/A0656_A0656M-05E02.

[2] For referenced ASTM standards, visit the ASTM website, www.astm.org, or contact ASTM Customer Service at service@astm.org. For *Annual Book of ASTM Standards* volume information, refer to the standard's Document Summary page on the ASTM website.

*A Summary of Changes section appears at the end of this standard.

A656/A656M − 05ε2

TABLE 1 Chemical Requirements

NOTE—An ellipsis (...) indicates that element is not defined for that Type.

Elements	Composition, %		
	Type 3	Type 7	Type 8
Carbon, max[A]	0.18	0.18	0.18
Manganese, max[A]	1.65	1.65	1.65
Phosphorus, max	0.025	0.025	0.025
Sulfur, max	0.035	0.035	0.035
Silicon, max	0.60	0.60	0.60
Vanadium, max	0.08	0.15[B]	0.15[C]†
Nitrogen, max	0.020	0.020	0.020
Columbium	0.008–0.10	0.10 max[B]	0.10 max[C]†
Titanium, max	...	...	0.15[C]

[A] For each reduction of 0.01 percentage point below the specified maximum for carbon, an increase of 0.06 percentage points above the specified maximum for manganese is permitted, up to a maximum of 1.75 % for Grades 50, 60, and 70; and up to a maximum of 1.90 % for Grade 80.
[B] The contents of columbium and vanadium shall additionally be in accordance with one of the following:
columbium 0.008-0.10 % with vanadium <0.008 %;
columbium <0.008 % with vanadium 0.008-0.15 %; or
columbium 0.008-0.10 % with vanadium 0.008-0.15 % and columbium plus vanadium not in excess of 0.20 %.
[C] The sum of Columbium, Vanadium, and Titanium shall be between 0.008 and 0.20 %.
† Footnote reference corrected editorially.

TABLE 2 Tensile Requirements[A]

	Grade 50 [345]	Grade 60 [415]	Grade 70 [485]	Grade 80 [550]
Yield point, min, ksi [MPa]	50 [345]	60 [415]	70 [485]	80 [550]
Tensile strength, min, ksi [MPa]	60 [415]	70 [485]	80 [550]	90 [620]
Elongation in 8 in. [200 mm], min, %[B]	20	17	14	12
Elongation in 2 in. [50 mm], min, %[B]	23	20	17	15

[A] See Specimen Orientation under the Tension Tests section of Specification A6/A6M.
[B] For plates wider than 24 in. [600 mm], the elongation requirement is reduced two percentage points for Grade 50 [345] and three percentage points for Grades 60, 70, and 80 [415, 485, and 550]. See Elongation Requirement Adjustments in the Tension Tests section of Specification A6/A6M.

5.2 *Product Analysis*—If a product analysis is made, it shall conform to the requirements given in Table 1, subject to the product analysis tolerances of Specification A6/A6M.

5.3 Where steel is to be welded, it is presupposed that a welding procedure suitable for the grade of steel and intended use or service will be utilized.

5.4 Unless specifically ordered, the type is at the discretion of the producer.

6. Tension Test

6.1 The plates as represented by the test specimens shall conform to the requirements given in Table 2.

7. Keywords

7.1 high-strength low-alloy steel; steel plates; structural applications

SUMMARY OF CHANGES

Committee A01 has identified the location of selected changes to this standard since the last issue (A656/A656M–03) that may impact the use of this standard.

(*1*) Table 1 was modified to include Type 8 chemistry and Titanium.

A656/A656M – 05ε2

ASTM International takes no position respecting the validity of any patent rights asserted in connection with any item mentioned in this standard. Users of this standard are expressly advised that determination of the validity of any such patent rights, and the risk of infringement of such rights, are entirely their own responsibility.

This standard is subject to revision at any time by the responsible technical committee and must be reviewed every five years and if not revised, either reapproved or withdrawn. Your comments are invited either for revision of this standard or for additional standards and should be addressed to ASTM International Headquarters. Your comments will receive careful consideration at a meeting of the responsible technical committee, which you may attend. If you feel that your comments have not received a fair hearing you should make your views known to the ASTM Committee on Standards, at the address shown below.

This standard is copyrighted by ASTM International, 100 Barr Harbor Drive, PO Box C700, West Conshohocken, PA 19428-2959, United States. Individual reprints (single or multiple copies) of this standard may be obtained by contacting ASTM at the above address or at 610-832-9585 (phone), 610-832-9555 (fax), or service@astm.org (e-mail); or through the ASTM website (www.astm.org).

Designation: A671 − 09

Standard Specification for
Electric-Fusion-Welded Steel Pipe for Atmospheric and Lower Temperatures[1]

This standard is issued under the fixed designation A671; the number immediately following the designation indicates the year of original adoption or, in the case of revision, the year of last revision. A number in parentheses indicates the year of last reapproval. A superscript epsilon (ε) indicates an editorial change since the last revision or reapproval.

1. Scope*

1.1 This specification[2] covers electric-fusion-welded steel pipe with filler metal added, fabricated from pressure vessel quality plate of several analyses and strength levels and suitable for high-pressure service at atmospheric and lower temperatures. Heat treatment may or may not be required to attain the desired properties or to comply with applicable code requirements. Supplementary requirements are provided for use when additional testing or examination is desired.

1.2 The specification nominally covers pipe 16 in. (405 mm) in outside diameter or larger and of ¼ in. (6.4 mm) wall thickness or greater. Pipe having other dimensions may be furnished provided it complies with all other requirements of this specification.

1.3 Several grades and classes of pipe are provided.

1.3.1 Grade designates the type of plate used as listed in 5.1.

1.3.2 Class designates the type of heat treatment performed during manufacture of the pipe, whether the weld is radiographically examined, and whether the pipe has been pressure tested as listed in 1.3.3.

1.3.3 Class designations are as follows (Note 1):

Class	Heat Treatment on Pipe	Radiography, see Section	Pressure Test, see:
10	none	none	none
11	none	9	none
12	none	9	8.3
13	none	none	8.3
20	stress relieved, see 5.3.1	none	none
21	stress relieved, see 5.3.1	9	none
22	stress relieved, see 5.3.1	9	8.3
23	stress relieved, see 5.3.1	none	8.3
30	normalized, see 5.3.2	none	none
31	normalized, see 5.3.2	9	none
32	normalized, see 5.3.2	9	8.3
33	normalized, see 5.3.2	none	8.3
40	normalized and tempered, see 5.3.3	none	none
41	normalized and tempered, see 5.3.3	9	none
42	normalized and tempered, see 5.3.3	9	8.3
43	normalized and tempered, see 5.3.3	none	8.3
50	quenched and tempered, see 5.3.4	none	none
51	quenched and tempered, see 5.3.4	9	none
52	quenched and tempered, see 5.3.4	9	8.3
53	quenched and tempered, see 5.3.4	none	8.3
60	normalized and precipitation heat treated	none	none
61	normalized and precipitation heat treated	9	none
62	normalized and precipitation heat treated	9	8.3
63	normalized and precipitation heat treated	none	8.3
70	quenched and precipitation heat treated	none	none
71	quenched and precipitation heat treated	9	none
72	quenched and precipitation heat treated	9	8.3
73	quenched and precipitation heat treated	none	8.3

NOTE 1—Selection of materials should be made with attention to temperature of service. For such guidance, Specification A20/A20M may be consulted.

1.4 The values stated in inch-pound units are to be regarded as standard. The values given in parentheses are mathematical conversions to SI units that are provided for information only and are not considered standard.

2. Referenced Documents

2.1 *ASTM Standards:*[3]

A20/A20M Specification for General Requirements for Steel Plates for Pressure Vessels

A370 Test Methods and Definitions for Mechanical Testing of Steel Products

A435/A435M Specification for Straight-Beam Ultrasonic Examination of Steel Plates

A530/A530M Specification for General Requirements for

---

[1] This specification is under the jurisdiction of ASTM Committee A01 on Steel, Stainless Steel and Related Alloys and is the direct responsibility of Subcommittee A01.09 on Carbon Steel Tubular Products.

Current edition approved Oct. 1, 2009. Published November 2009. Originally approved in 1972. Last previous edition approved in 2006 as A671 – 06. DOI: 10.1520/A0671-09.

[2] For ASME Boiler and Pressure Vessel Code applications see related Specification SA-671 in Section II of that Code.

[3] For referenced ASTM standards, visit the ASTM website, www.astm.org, or contact ASTM Customer Service at service@astm.org. For *Annual Book of ASTM Standards* volume information, refer to the standard's Document Summary page on the ASTM website.

*A Summary of Changes section appears at the end of this standard.

Specialized Carbon and Alloy Steel Pipe
A577/A577M Specification for Ultrasonic Angle-Beam Examination of Steel Plates
A578/A578M Specification for Straight-Beam Ultrasonic Examination of Rolled Steel Plates for Special Applications
E110 Test Method for Indentation Hardness of Metallic Materials by Portable Hardness Testers
E165 Practice for Liquid Penetrant Examination for General Industry
E709 Guide for Magnetic Particle Testing

2.2 *Plate Steels:*
A203/A203M Specification for Pressure Vessel Plates, Alloy Steel, Nickel
A285/A285M Specification for Pressure Vessel Plates, Carbon Steel, Low- and Intermediate-Tensile Strength
A299/A299M Specification for Pressure Vessel Plates, Carbon Steel, Manganese-Silicon
A353/A353M Specification for Pressure Vessel Plates, Alloy Steel, 9 Percent Nickel, Double-Normalized and Tempered
A515/A515M Specification for Pressure Vessel Plates, Carbon Steel, for Intermediate- and Higher-Temperature Service
A516/A516M Specification for Pressure Vessel Plates, Carbon Steel, for Moderate- and Lower-Temperature Service
A517/A517M Specification for Pressure Vessel Plates, Alloy Steel, High-Strength, Quenched and Tempered
A537/A537M Specification for Pressure Vessel Plates, Heat-Treated, Carbon-Manganese-Silicon Steel
A553/A553M Specification for Pressure Vessel Plates, Alloy Steel, Quenched and Tempered 8 and 9 % Nickel
A736/A736M Specification for Pressure Vessel Plates, Low-Carbon Age-Hardening Nickel-Copper-Chromium-Molybdenum-Columbium and Nickel-Copper-Manganese-Molybdenum-Columbium Alloy Steel

2.3 *ASME Boiler and Pressure Vessel Code:*[4]
Section II, Material Specifications
Section III, Nuclear Vessels
Section VIII, Unfired Pressure Vessels
Section IX, Welding Qualifications

3. Terminology

3.1 *Definitions of Terms Specific to This Standard:*

3.1.1 *lot*—a lot shall consist of 200 ft (61 m) or fraction thereof of pipe from the same heat of steel.

3.1.2 The description of a lot may be further restricted by the use of Supplementary Requirement S14.

4. Ordering Information

4.1 The inquiry and order for material under this specification should include the following information:

4.1.1 Quantity (feet, metres, or number of lengths),

4.1.2 Name of material (steel pipe, electric-fusionwelded),

4.1.3 Specification number,

4.1.4 Grade and class designations (see 1.3),

4.1.5 Size (inside or outside diameter, nominal or minimum wall thickness),

4.1.6 Length (specific or random),

4.1.7 End finish (11.4),

4.1.8 Purchase options, if any (see 5.2.3 and 11.3 of this specification. See also Specification A530/A530M),

4.1.9 Supplementary requirements, if any.

5. Materials and Manufacture

5.1 *Materials*—The steel plate material shall conform to the requirement of the applicable plate specification for the pipe grade ordered as listed in Table 1.

5.2 *Welding*:

5.2.1 The joints shall be double-welded, full-penetration welds made in accordance with procedures and by welders or welding operators qualified in accordance with the ASME Boiler and Pressure Vessel Code, Section IX.

5.2.2 The welds shall be made either manually or automatically by an electric process involving the deposition of filler metal.

5.2.3 As welded, the welded joint shall have positive reinforcement at the center of each side of the weld, but no more than ⅛ in. (3.2 mm). This reinforcement may be removed at the manufacturer's option or by agreement between the manufacturer and purchaser. The contour of the reinforcement

TABLE 1 Plate Specifications

Pipe Grade	Type of Steel	ASTM Specification No.	Grade
CA 55	plain carbon	A285/A285M	C
CB 60	plain carbon, killed	A515/A515M	60
CB 65	plain carbon, killed	A515/A515M	65
CB 70	plain carbon, killed	A515/A515M	70
CC 60	plain carbon, killed, fine grain	A516/A516M	60
CC 65	plain carbon, killed, fine grain	A516/A516M	65
CC 70	plain carbon, killed, fine grain	A516/A516M	70
CD 70	manganese-silicon, normalized	A537/A537M	1
CD 80	manganese-silicon, quenched and tempered	A537/A537M	2
CF 65	nickel steel	A203/A203M	A
CF 70	nickel steel	A203/A203M	B
CF 66	nickel steel	A203/A203M	D
CF 71	nickel steel	A203/A203M	E
CG 100	9 % nickel	A353/A353M	
CH 100	9 % nickel	A553/A553M	1
CJ 101	alloy steel, quenched and tempered	A517/A517M	A
CJ 102	alloy steel, quenched and tempered	A517/A517M	B
CJ 103	alloy steel, quenched and tempered	A517/A517M	C
CJ 104	alloy steel, quenched and tempered	A517/A517M	D
CJ 105	alloy steel, quenched and tempered	A517/A517M	E
CJ 106	alloy steel, quenched and tempered	A517/A517M	F
CJ 107	alloy steel, quenched and tempered	A517/A517M	G
CJ 108	alloy steel, quenched and tempered	A517/A517M	H
CJ 109	alloy steel, quenched and tempered	A517/A517M	J
CJ 110	alloy steel, quenched and tempered	A517/A517M	K
CJ 111	alloy steel, quenched and tempered	A517/A517M	L
CJ 112	alloy steel, quenched and tempered	A517/A517M	M
CJ 113	alloy steel, quenched and tempered	A517/A517M	P
CK 75	carbon-manganese-silicon	A299/A299M	
CP65	alloy steel, age hardening, normalized and precipitation heat treated	A736/A736M	2
CP75	alloy steel, age hardening, quenched and precipitation heat treated	A736/A736M	3

---
[4] Available from American Society of Mechanical Engineers (ASME), ASME International Headquarters, Three Park Ave., New York, NY 10016-5990, http://www.asme.org.

shall be smooth and the deposited metal shall be fused smoothly and uniformly into the plate surface.

5.2.4 When radiographic examination in accordance with 9.1 is to be used, the weld reinforcements shall be governed by the more restrictive provision UW–51 of Section VIII of the ASME Boiler and Pressure Vessel Code instead of 5.2.3 of this specification.

5.3 *Heat Treatment*—All classes other than 10, 11, 12, and 13 shall be heat treated in furnace controlled to ± 25 °F (± 14 °C) and equipped with a recording pyrometer so that heating records are available. Heat treating after forming and welding shall be to one of the following:

5.3.1 Classes 20, 21, 22, and 23 pipe shall be uniformly heated within the post-weld heat-treatment temperature range indicated in Table 2 for a minimum of 1 h/in. of thickness or for 1 h, whichever is greater.

5.3.2 Classes 30, 31, 32, and 33, pipe shall be uniformly heated to a temperature in the austenitizing range and not exceeding the maximum normalizing temperature indicated in Table 2 and subsequently cooled in air at room temperature.

5.3.3 Classes 40, 41, 42, and 43 pipe shall be normalized in accordance with 5.3.2. After normalizing, the pipe shall be reheated to the tempering temperature indicated in Table 2 as a minimum and held at temperature for a minimum of ½ h/in. of thickness or for ½ h, whichever is greater, and air cooled.

5.3.4 Classes 50, 51, 52, and 53 pipe shall be uniformly heated to a temperature in the austenitizing range, and not exceeding the maximum quenching temperature indicated in Table 2 and subsequently quenched in water or oil. After quenching, the pipe shall be reheated to the tempering temperature indicated in Table 2 as a minimum and held at that temperature for a minimum of ½ h/in. of thickness or for ½ h, whichever is greater, and air cooled.

5.3.5 Classes 60, 61, 62, and 63 pipe shall be normalized in accordance with 5.3.2. After normalizing, the pipe shall be precipitation heat treated in the range shown in Table 2 for a time to be determined by the manufacturer.

5.3.6 Classes 70, 71, 72, and 73 pipe shall be uniformly heated to a temperature in the austenitizing range, not exceeding the maximum quenching temperature indicated in Table 2, and subsequently quenched in water or oil. After quenching the pipe shall be reheated into the precipitation heat treating range indicated in Table 2 for a time to be determined by the manufacturer.

6. *General Requirements for Delivery* General Requirements for Delivery

6.1 Material furnished to this specification shall conform to the applicable requirements of the current edition of Specification A530/A530M unless otherwise provided herein.

TABLE 2 Heat Treatment Parameters

Pipe Grade[A]	ASTM Specification and Grade	Post-Weld Heat-Treatment Temperature Range °F (°C)	Normalizing Temperature, max, °F (°C)	Quenching Temperature, max, °F (°C)	Tempering Temperature, min, °F (°C)	Precipitation Heat Treatment Temperature Range °F (°C)
CA 55	A285/A285M (C)	1100–1250 (590–680)	1700 (925)	...	...	...
CB 60	A515/A515M (60)	1100–1250 (590–680)	1750 (950)	...	...	...
CB 65	A515/A515M (65)	1100–1250 (590–680)	1750 (950)	...	...	...
CB 70	A515/A515M	1100–1250 (590–680)	1750 (950)	...	...	...
CC 60	A516/A516M (60)	1100–1250 (590–680)[B]	1700 (925)	1650 (900)	1200 (650)[C]	...
CC 65	A516/A516M (65)	1100–1250 (590–680)[B]	1700 (925)	1650 (900)	1200 (650)	...
CC 70	A516/A516M (70)	1100–1250 (590–680)[B]	1700 (925)	1650 (900)	1200 (650)	...
CD 70	A537/A537M (1)	1100–1250 (590–680)	1700 (925)	...	...	...
CD 80	A537/A537M (2)	1100–1250 (590–680)[B]	...	1650 (900)	1100 (590)	...
CF 65	A203/A203M (A)	1100–1175 (590–635)	1750 (950)	...	...	...
CF 70	A203/A203M (B)	1100–1175 (590–635)	1750 (950)	...	...	...
CF 66	A203/A203M (D)	1100–1175 (590–635)	1750 (950)	...	...	...
CF 71	A203/A203M (E)	1100–1175 (590–635)	1750 (950)	...	...	...
CG 100	A353/A353M	1025–1085 (550–580)	1650 (900)	...	1050 (560)	...
CH 100	A553/A553M	1025–1085 (550–580)	1650 (900)	1700 (925)	1050 (560)	...
CJ 101	A517/A517M (A)	1000–1100 (540–590)	...	1725 (940)[D]	1150 (620)	...
CJ 102	A517/A517M (B)	1000–1100 (540–590)	...	1725 (940)[D]	1150 (620)	...
CJ 103	A517/A517M (C)	1000–1100 (540–590)	...	1725 (940)[D]	1150 (620)	...
CJ 104	A517/A517M (D)	1000–1100 (540–590)	...	1725 (940)[D]	1150 (620)	...
CJ 105	A517/A517M (E)	1000–1100 (540–590)	...	1725 (940)[D]	1150 (620)	...
CJ 106	A517/A517M (F)	1000–1100 (540–590)	...	1725 (940)[D]	1150 (620)	...
CJ 107	A517/A517M (G)	1000–1100 (540–590)	...	1725 (940)[D]	1150 (620)	...
CJ 108	A517/A517M (H)	1000–1100 (540–590)	...	1725 (940)[D]	1150 (620)	...
CJ 109	A517/A517M (J)	1000–1100 (540–590)	...	1725 (940)[D]	1150 (620)	...
CJ 110	A517/A517M (K)	1000–1100 (540–590)	...	1725 (940)[D]	1150 (620)	...
CJ 111	A517/A517M (L)	1000–1100 (540–590)	...	1725 (940)[D]	1150 (620)	...
CJ 112	A517/A517M (M)	1000–1100 (540–590)	...	1725 (940)[D]	1150 (620)	...
CJ 113	A517/A517M (P)	1000–1100 (540–590)	...	1725 (940)[D]	1150 (620)	...
CK 75	A299/A299M	1100–1250 (590–680)	1700 (925)	...	...	...
CP65	A736/A736M (2)	1000–1175 (540–635)	1725 (940)	...	...	1000–1200 (540–650)
CP75	A736/A736M (3)	1000–1175 (540–635)	...	1725 (940)	...	1000–1225 (540–665)

[A] Numbers indicate minimum tensile strength in ksi.
[B] In no case shall the post-weld heat-treatment temperature exceed the mill tempering temperature.
[C] Tempering range 1100 to 1300 (590 to 705), if accelerated cooling utilized per Specification A516/A516M.
[D] Per ASME Section VIII Specification A517/A517M specified 1650 (900) minimum quenching temperature.

7. Chemical Composition

7.1 *Product Analysis of Plate*—The pipe manufacturer shall make an analysis of each mill heat of plate material. The product analysis so determined shall meet the requirements of the plate specification to which the material was ordered.

7.2 *Product Analyses of Weld*—The pipe manufacturer shall make an analysis of finished deposited weld material from each 200 ft (61 m) or fraction thereof. Analyses shall conform to the welding procedure for deposited weld metal.

7.3 Analysis may be taken from the mechanical test specimens. The results of the analyses shall be reported to the purchaser.

8. Mechanical Requirements

8.1 *Tension Test*:

8.1.1 *Requirements*—Transverse tensile properties of the welded joint shall meet the minimum requirements for ultimate tensile strength of the specified plate material. In addition for Grades CD and CJ, when these are of Class 3x, 4x, or 5x, and Grade CP of Class 6x and 7x, the transverse tensile properties of the base plate shall be determined on specimens cut from the heat-treated pipe. These properties shall meet the mechanical test requirements of the plate specification.

8.1.2 *Number of Tests*—One test specimen of weld metal and one specimen of base metal, if required by 8.1.1, shall be made and tested to represent each lot of finished pipe.

8.1.3 *Test Specimen Location and Orientation*—The test specimens shall be taken transverse to the weld at the end of the finished pipe and may be flattened cold before final machining to size.

8.1.4 *Test Method*—The test specimen shall be made in accordance with QW-150 in Section IX of the ASME Boiler and Pressure Vessel Code. The test specimen shall be tested at room temperature in accordance with Test Methods and Definitions A370.

8.2 *Transverse Guided Weld Bend Test*:

8.2.1 *Requirements*—The bend test shall be acceptable if no cracks or other defects exceeding 1/8 in. (3.2 mm) in any direction are present in the weld metal or between the weld and the base metal after bending. Cracks that originate along the edges of the specimen during testing, and that are less than 1/4 in. (6.4 mm) measured in any direction shall not be considered.

8.2.2 *Number of Tests*—One test (two specimens) shall be made to represent each lot of finished pipe.

8.2.3 *Test Specimen Location and Orientation*—Two bend test specimens shall be taken transverse to the weld at the end of the finished pipe. As an alternative, by agreement between the purchaser and the manufacturer, the test specimens may be taken from a test plate of the same material as the pipe, the test plate being attached to the end of the cylinder and welded as a prolongation of the pipe longitudinal seam.

8.2.4 *Test Method*—The test requirements of A370, S9.1.7 shall be met. For wall thicknesses over 3/8 in. (9.5 mm) but less than 3/4 in. (19.0 mm) side-bend tests may be made instead of the face and root-bend tests. For wall thicknesses 3/4 in. and over both specimens shall be subjected to the side-bend test.

8.3 *Pressure Test*—Classes X2 and X3 pipe shall be tested in accordance with Specification A530/A530M, Section 6.

9. Radiographic Examination

9.1 The full length of each weld of Classes X1 and X2 shall be radiographically examined in accordance with and meet the requirements of ASME Boiler and Pressure Vessel Code, Section VIII, Paragraph UW–51.

9.2 Radiographic examination may be performed prior to heat treatment.

10. Rework

10.1 *Elimination of Surface Imperfections*—Unacceptable surface imperfections shall be removed by grinding or machining. The remaining thickness of the section shall be no less than the minimum specified in Section 11. The depression after grinding or machining shall be blended uniformly into the surrounding surface.

10.2 *Repair of Base Metal Defects by Welding*:

10.2.1 The manufacturer may repair, by welding, base metal where defects have been removed, provided the depth of the repair cavity as prepared for welding does not exceed 1/3 of the nominal thickness and the requirements of 10.2.2, 10.2.3, 10.2.4, 10.2.5 and 10.2.6 are met. Base metal defects in excess of these may be repaired with prior approval of the customer.

10.2.2 The defect shall be removed by suitable mechanical or thermal cutting or gouging methods and the cavity prepared for repair welding.

10.2.3 The welding procedure and welders or welding operators are to be qualified in accordance with Section IX of the ASME Boiler and Pressure Vessel Code.

10.2.4 The full length of the repaired pipe shall be heat treated after repair in accordance with the requirements of the pipe class specified.

10.2.5 Each repair weld of a defect where the cavity, prepared for welding, has a depth exceeding the lesser of 3/8 in. (9.5 mm) or 10 % of the nominal thickness shall be examined by radiography in accordance with the methods and the acceptance standards of Section 9.

10.2.6 The repair surface shall be blended uniformly into the surrounding base metal surface and examined and accepted in accordance with Supplementary Requirements S6 or S8.

10.3 *Repair of Weld Metal Defects by Welding*:

10.3.1 The manufacturer may repair weld metal defects if he meets the requirements of 10.2.3, 10.2.4, 10.3.2, 10.3.3 and 10.4.

10.3.2 The defect shall be removed by suitable mechanical or thermal cutting or gouging methods and the repair cavity examined and accepted in accordance with Supplementary Requirements S7 or S9.

10.3.3 The weld repair shall be blended uniformly into the surrounding metal surfaces and examined and accepted in accordance with 9.1 and with Supplementary Requirements S7 or S9.

10.4 *Retest*—Each length of repaired pipe of a class requiring a pressure test shall be hydrostatically tested following repair.

11. Dimensions, Mass and Permissible Variations

11.1 The wall thickness and weight for welded pipe furnished to this specification shall be governed by the requirements of the specification to which the manufacturer ordered the plate.

11.2 Permissible variations in dimensions at any point in a length of pipe shall not exceed the following:

11.2.1 *Outside Diameter*—Based on circumferential measurement ± 0.5 % of the specified outside diameter.

11.2.2 *Out-of-Roundness*—Difference between major and minor outside diameters, 1 %.

11.2.3 *Alignment*—Using a 10-ft (3-m) straight edge placed so that both ends are in contact with the pipe, 1/8 in. (3.2 mm).

11.2.4 *Thickness*—The minimum wall thickness at any point in the pipe shall not be more than 0.01 in. (0.25 mm) under the specified nominal thickness.

11.3 Circumferential welded joints of the same quality as the longitudinal joints shall be permitted by agreement between the manufacturer and the purchaser.

11.4 Lengths with unmachined ends shall be within –0, +1/2 in. (–0, +13 mm) of that specified. Lengths with machined ends shall be as agreed between the manufacturer and the purchaser.

12. Workmanship, Finish, and Appearance

12.1 The finished pipe shall be free of injurious defects and shall have a workmanlike finish. This requirement is to mean the same as the identical requirement that appears in Specification A20/A20M with respect to steel plate surface finish.

13. Product Marking

13.1 In addition to the marking provision of Specification A530/A530M, class marking in accordance with 1.3.3 shall follow the grade marking, for example, CC 70–10.

13.2 *Bar Coding*—In addition to the requirements in 13.1, bar coding is acceptable as a supplemental identification method. The purchaser may specify in the order a specific bar coding system to be used.

SUPPLEMENTARY REQUIREMENTS

One or more of the following supplementary requirements shall be applied only when specified by the purchaser in the inquiry, contract, or order. Details of these supplementary requirements shall be agreed upon in writing by the manufacturer and purchaser. Supplementary requirements shall in no way negate any requirement of the specification itself.

S1. Tension and Bend Tests

S1.1 Tension tests in accordance with 8.1 and bend tests in accordance with 8.2 shall be made on specimens representing each length of pipe.

S2. Charpy V-Notch Test

S2.1 *Requirements*—The acceptable test energies for material shown in Specification A20/A20M shall conform to the energy values shown in Specification A20/A20M.

S2.1.1 Materials not listed in Specification A20/A20M shall be in accordance with the purchase order requirements.

S2.2 *Number of Specimens*—Each test shall consist of at least three specimens.

S2.2.1 One base metal test shall be made from one pipe length per heat-treat charge per nominal wall thickness. For pipe from Classes 10, 11, 12, and 13, one base metal test shall be made per heat per size and per wall thickness.

S2.2.2 One weld-metal test shall be made in accordance with UG–84 of Section VIII of the ASME Boiler and Pressure Vessel Code.

S2.2.3 One heat-affected-zone test shall be made in accordance with UG–84 of Section VIII of the ASME Boiler and Pressure Vessel Code.

S2.3 *Test Specimen Location and Orientation:*

S2.3.1 Specimens for base-metal tests in Grades CA, CB, and CC in the as rolled stress relieved or normalized condition (classes of the 10, 20, 30, and 40 series) shall be taken so that the longitudinal axis of the specimen is parallel to the longitudinal axis of the pipe.

S2.3.2 Base-metal specimens of quench and tempered pipe, when the quenching and tempering follows the welding operation, shall be taken in accordance with the provision of N330 of Section III of the ASME Boiler and Pressure Vessel Code.

S2.4 *Test Method*—The specimen shall be Charpy-V Type A in accordance with Test Methods and Definitions A370. The specimens shall be tested in accordance with Test Methods and Definitions A370. Unless otherwise indicated by the purchaser, the test temperature shall be as given in Specification A20/A20M for those base materials covered by Specification A20/A20M. For materials not covered by Specification A20/A20M the test temperature shall be 10 °F (–12 °C) unless otherwise stated in the purchase order.

S3. Hardness Test

S3.1 Hardness tests shall be made in accordance with Test Methods and Definitions A370 or Test Method E110 across the welded joint of both ends of each length of pipe. In addition, hardness tests shall be made to include the heat-affected zone if so required by the purchaser. The maximum acceptable hardness shall be as agreed upon between the manufacturer and the purchaser.

S3.2 As an alternative to the heat-affected zone hardness, by agreement between the manufacturer and purchaser, maximum heat-affected zone hardness may be specified for the procedure test results.

S4. Product Analysis

S4.1 Product analyses in accordance with 7.1 shall be made on each 500 ft (152 m) of pipe of fraction thereof, or alternatively, on each length of pipe as designated in the order.

S5. Metallography

S5.1 The manufacturer shall furnish one photomicrograph to show the microstructure at 100× magnification of the weld metal or base metal of the pipe in the as-finished condition. The purchaser shall state in the order: the material, base metal or weld, and the number and locations of tests to be made. This test is for information only.

S6. Magnetic Particle Examination of Base Metal

S6.1 All accessible surfaces of the pipe shall be examined in accordance with Guide E709. Accessible is defined as: All outside surfaces, all inside surfaces of pipe 24 in. (610 mm) in diameter and greater, and inside surfaces of pipe less than 24 in. in diameter for a distance of 1 pipe diameter from the ends.

S6.2 *Acceptance Standards*—The following relevant indications are unacceptable:

S6.2.1 Any linear indications greater than 1/16 in. (1.6 mm) long for materials less than 5/8 in (15.9 mm) thick; greater than 1/8 in. (3.2 mm) long for materials from 5/8 in. thick to under 2 in. (51 mm) thick; and greater than 3/16 in. (4.8 mm) long for materials 2 in. thick or greater.

S6.2.2 Rounded indications with dimensions greater than 1/8 in. (3.2 mm) for thicknesses less than 5/8 in. (15.9 mm), and greater than 3/16 in. (4.8 mm) for thicknesses 5/8 in. and greater.

S6.2.3 Four or more indications in any line separated by 1/16 in. (1.6 mm) or less edge-to-edge.

S6.2.4 Ten or more indications in any 6 in.^{2} (39 cm^2) of surface with the major dimension of this area not to exceed 6 in. (152 mm) when it is taken in the most unfavorable orientation relative to the indications being evaluated.

S7. Magnetic Particle Examination of Weld Metal

S7.1 All accessible welds shall be examined in accordance with Guide E709. Accessible is defined as: All outside surfaces, all inside surfaces of pipe 24 in. (610 mm) in diameter and greater, and inside surfaces of pipe less than 24 in. in diameter for a distance of one pipe diameter from the ends.

S7.2 *Acceptance Criteria*—The following relevant indications are unacceptable:

S7.2.1 Any cracks and linear indications.

S7.2.2 Rounded indications with dimensions greater than 3/16 in. (4.8 mm).

S7.2.3 Four or more indications in any line separated by 1/16 in. (1.6 mm) or less edge-to-edge.

S7.2.4 Ten or more indications in any 6 in.^{2} (39 cm^2) of surface with the major dimension of this area not to exceed 6 in. (152 mm) when it is taken in the most unfavorable orientation relative to the indications being evaluated.

S8. Liquid Penetrant Examination of Base Metal

S8.1 All accessible surfaces of the pipe shall be examined in accordance with Test Method E165. Accessible is as defined in S7.1.

S8.2 The acceptance criteria shall be in accordance with S6.2.

S9. Liquid Penetrant Examination of Weld Metal

S9.1 All accessible surfaces of the pipe shall be examined in accordance with Test Method E165. Accessible is as defined in S7.1.

S9.2 The acceptance criteria shall be in accordance with S7.2.

S10. Straight Beam Ultrasonic Examination of Flat Plate—UT 1

S10.1 The plate shall be examined and accepted in accordance with Specification A435/A435M except that 100 % of one surface shall be scanned by moving the search unit in parallel paths with not less than 10 % overlap.

S11. Straight Beam Ultrasonic Examination of Flat Plate—UT 2

S11.1 The plate shall be examined in accordance with Specification A578/A578M except that 100 % of one surface shall be scanned and the acceptance criteria shall be as follows:

S11.1.1 Any area, where one or more discontinuities produce a continuous total loss of back reflection accompanied by continuous indications on the same plane that cannot be encompassed within a circle whose diameter is 3 in. (76.2 mm) or one half of the plate thickness, whichever is greater, is unacceptable.

S11.1.2 In addition, two or more discontinuities on the same plane and having the same characteristics but smaller than described above shall be unacceptable unless separated by a minimum distance equal to the largest diameter of the larger discontinuity or unless they may be collectively encompassed by the circle described above.

S12. Angle Beam Ultrasonic Examination (Plate Less than 2 in. (50.8 mm) Thick)—UT 3

S12.1 The plate shall be examined in accordance with Specification A577/A577M except that the calibration notch shall be vee shaped and the acceptance criteria shall be as follows: Any area showing one or more reflections producing indications whose amplitude exceeds that of the calibration notch is unacceptable.

S13. Repair Welding

S13.1 Repair of base metal defects by welding shall be done only with customer approval.

S14. Description of Term

S14.1 *lot*—all pipe of the same mill heat of plate material and wall thickness (within ±1/4 in. (6.4 mm)) heat treated in one furnace charge. For pipe that is not heat treated or that is heat treated in a continuous furnace, a lot shall consist of each 200 ft (61 m) or fraction thereof of all pipe of the same mill heat of plate material and wall thickness (within ±1/4 in. (6.4 mm)), subjected to the same heat treatment. For pipe heat treated in a batch-type furnace that is automatically controlled within a 50 °F (28 °C) range and is equipped with recording pyrometers so that heating records are available, a lot shall be defined the same as for continuous furnaces.

A671 – 09

SUMMARY OF CHANGES

Committee A01 has identified the location of selected changes to this specification since the last issue, A671 – 06, that may impact the use of this specification. (Approved October 1, 2009)

(*1*) Added quenching temperature for CH 100 in Table 2.

ASTM International takes no position respecting the validity of any patent rights asserted in connection with any item mentioned in this standard. Users of this standard are expressly advised that determination of the validity of any such patent rights, and the risk of infringement of such rights, are entirely their own responsibility.

This standard is subject to revision at any time by the responsible technical committee and must be reviewed every five years and if not revised, either reapproved or withdrawn. Your comments are invited either for revision of this standard or for additional standards and should be addressed to ASTM International Headquarters. Your comments will receive careful consideration at a meeting of the responsible technical committee, which you may attend. If you feel that your comments have not received a fair hearing you should make your views known to the ASTM Committee on Standards, at the address shown below.

This standard is copyrighted by ASTM International, 100 Barr Harbor Drive, PO Box C700, West Conshohocken, PA 19428-2959, United States. Individual reprints (single or multiple copies) of this standard may be obtained by contacting ASTM at the above address or at 610-832-9585 (phone), 610-832-9555 (fax), or service@astm.org (e-mail); or through the ASTM website (www.astm.org). Permission rights to photocopy the standard may also be secured from the ASTM website (www.astm.org/COPYRIGHT/).

Designation: A673/A673M – 07

American Association State Highway
and Transportation Officials
Standard AASHTO No.: T 243

Standard Specification for
Sampling Procedure for Impact Testing of Structural Steel[1]

This standard is issued under the fixed designation A673/A673M; the number immediately following the designation indicates the year of original adoption or, in the case of revision, the year of last revision. A number in parentheses indicates the year of last reapproval. A superscript epsilon (ε) indicates an editorial change since the last revision or reapproval.

This standard has been approved for use by agencies of the Department of Defense.

1. Scope*

1.1 This specification covers the procedure for longitudinal Charpy V-notch testing of structural steel and contains two frequencies of testing. The impact properties of steel can vary within the same heat and piece, be it as rolled, control rolled, or heat treated. The purchaser should, therefore, be aware that testing of one plate, bar, or shape does not provide assurance all plates, bars, or shapes of the same heat as processed will be identical in toughness with the product tested. Normalizing or quenching and tempering the product will reduce the degree of variation.

1.2 This specification is intended to supplement specifications for structural steel when so specified.

1.3 This specification does not necessarily apply to all product specifications; therefore, the manufacturer or processor should be consulted for energy absorption levels and minimum testing temperatures that can be expected or supplied.

1.4 Two frequencies of testing (P and H) are prescribed.

1.5 The values stated in either inch-pound units or SI units are to be regarded as standard. Within the text, the SI units are shown in brackets. The values stated in each system are not exact equivalents; therefore, each system must be used independently of the other. Combining values from the two systems may result in nonconformance with this specification.

2. Referenced Documents

2.1 *ASTM Standards:*[2]

A6/A6M Specification for General Requirements for Rolled Structural Steel Bars, Plates, Shapes, and Sheet Piling

A370 Test Methods and Definitions for Mechanical Testing of Steel Products

3. Ordering Information

3.1 The inquiry and order shall indicate the following:

3.1.1 Frequency of testing, (P) or (H),

3.1.2 Test temperature (see 4.5 and 4.6),

3.1.3 Minimum average absorbed energy value (see 4.1 and 4.6),

3.1.4 Transverse impact test orientation for plate widths over 24 in. [600 mm], if desired (see 4.2.2),

3.1.5 Alternate core location (see 4.3), if applicable, and

3.1.6 Condition (as-rolled, stress relieved, normalized, normalized and stress relieved, or quenched and tempered).

4. Tests

4.1 Impact testing shall be in accordance with Test Methods and Definitions A370. An impact test shall consist of testing three specimens taken from a single test coupon or test location, the average result of which shall be not less than the minimum average absorbed energy specified in the purchase order, which in no case shall be less than 7 ft·lbf [10 J] for full size specimens.

4.2 Except as allowed by 4.3, specimens for plates and bars shall be taken from a location adjacent to the location specified for the tension test specimen, and specimens for shapes shall be taken from a location at an end of the shape at a point one third the distance from the outer edge of the flange or leg to the web or heel of the shape (see Fig. 1 and Fig. 2). For plates produced from coils, three impact tests shall be taken from the product of each coil or qualifying coil (see Section 5); one test coupon shall be obtained from a location adjacent to the location specified for each of the two required tension tests (see Specification A6/A6M) and the third test coupon shall be obtained from a location immediately after the last plate produced to the qualifying specification.

4.2.1 Except as allowed by 4.2.2, the longitudinal axis of each specimen shall be parallel to the final direction of rolling of the plate or parallel to the major axis of the shape.

---

[1] This specification is under the jurisdiction of ASTM Committee A01 on Steel, Stainless Steel and Related Alloys and is the direct responsibility of Subcommittee A01.02 on Structural Steel for Bridges, Buildings, Rolling Stock and Ships.
Current edition approved March 1, 2007. Published March 2007. Originally approved in 1972. Last previous edition approved in 2005 as A673/A673M – 05a. DOI: 10.1520/A0673_A0673M-07.

[2] For referenced ASTM standards, visit the ASTM website, www.astm.org, or contact ASTM Customer Service at service@astm.org. For *Annual Book of ASTM Standards* volume information, refer to the standard's Document Summary page on the ASTM website.

*A Summary of Changes section appears at the end of this standard.

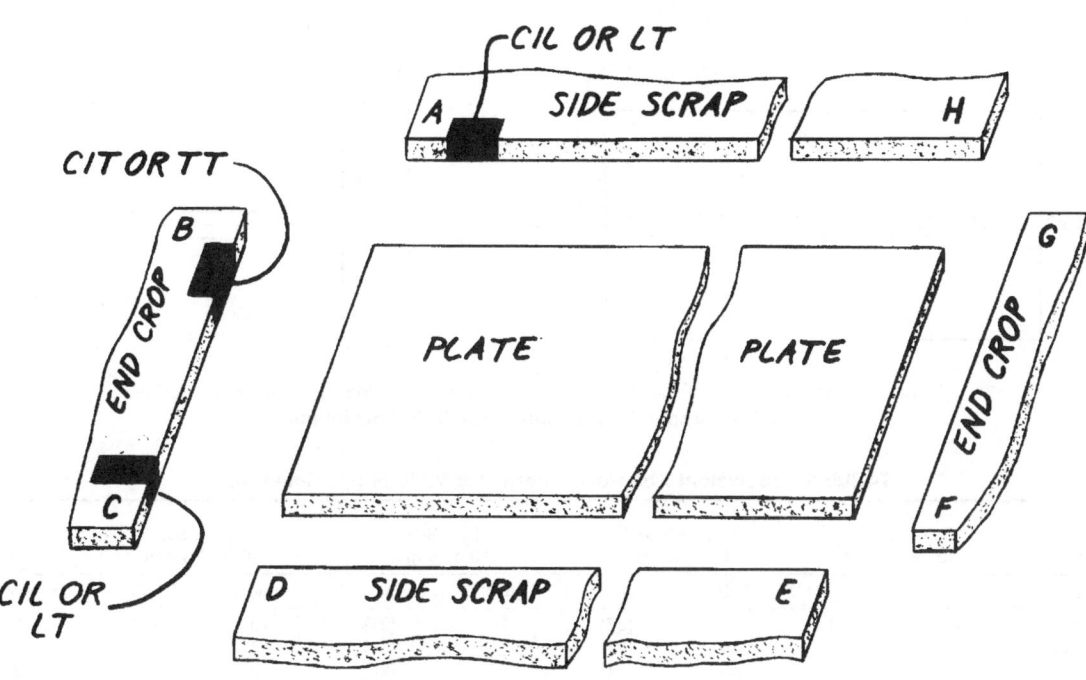

NOTE 1—*LT* (Longitudinal tensile test) For widths through 24 in. [600 mm], may be taken at any location, *A* through *H*.
NOTE 2—*TT* (Transverse tensile test) For widths over 24 in. [600 mm], may be taken at location *B*, *C*, *F*, or *G*.
NOTE 3—*CIL* (Charpy impact longitudinal) May be taken at any location, *A* through *H*.
NOTE 4—*CIT* (Charpy impact transverse) For widths over 24 in. [600 mm], may be taken at location *B*, *C*, *F*, or *G*.
FIG. 1 Plate Test Location

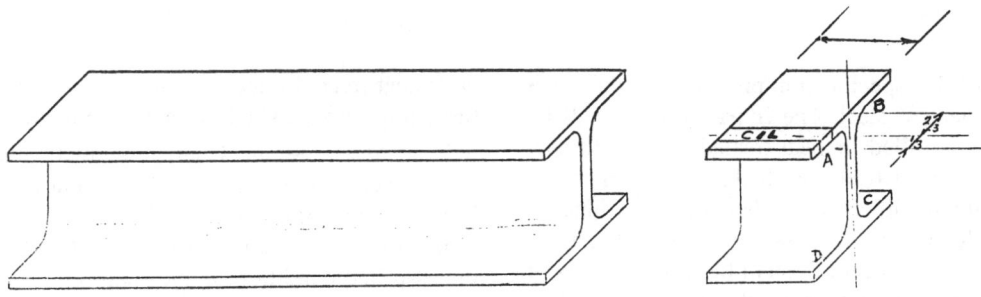

NOTE 1—*CIL* = Charpy impact longitudinal.
NOTE 2—Test coupon for impact specimens may be taken from locations *A*, *B*, *C*, or *D* as shown laid out at location *A*.
FIG. 2 Shape Test Location

4.2.2 If specified in the purchase order, for plate widths over 24 in. [600 mm], the longitudinal axis of each specimen shall be transverse to the final direction of rolling of the plate.

4.2.3 The longitudinal axis of each specimen shall be located midway between the surface and the center of the product thickness, and the length of the notch shall be perpendicular to the rolled surface of the product.

4.3 For shapes with a flange thickness equal to or greater than 1 ½ in. [38.1 mm], where alternate core location testing is specified in the purchase order, the longitudinal axis of each specimen shall be located midway between the inner flange surface and the center of the flange thickness at the intersection with the web mid-thickness (see Fig. 3).

4.4 The absorbed energy values obtained for subsize specimens shall not be less than the applicable values given in Table 1, which are proportional to the absorbed energy values required for full-size specimens.

4.5 Except as allowed by 4.6, the test temperature shall be as specified in the purchase order.

4.6 The manufacturer shall have the option of using a lower test temperature than is specified in the purchase order, provided that the absorbed energy values specified in the purchase order are met.

4.7 The actual test temperature used shall be reported with the test results.

5. Frequency of Testing

5.1 *Frequency (H) Heat Testing for Plates, Shapes, and Bars*—One impact test (a set of three specimens) shall be made for each 50 tons [45 Mg] of the same type of product subject

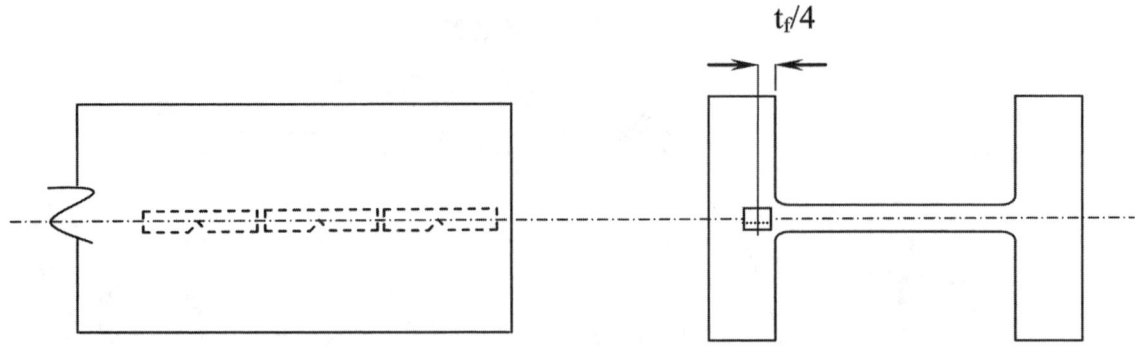

NOTE—The notch for any of the specimens can be on either side of the beam centerline.
FIG. 3 Alternate Core Location for CVN Specimens

TABLE 1 Equivalent Absorbed Energy for Various Specimen Sizes

Full Size, 10 by 10 mm		3/4 Size, 10 by 7.5 mm		2/3 Size, 10 by 6.7 mm		1/2 Size, 10 by 5 mm		1/3 Size, 10 by 3.3 mm		1/4 Size, 10 by 2.5 mm	
ft·lbf	[J]	ft·lbf	[J]	ft·lbf	[J]	ft·lbf	[J]	ft·lbf	[J]	ft·lbf	[J]
40	[54]	30	[41]	27	[37]	20	[27]	13	[18]	10	[14]
35	[48]	26	[35]	23	[31]	18	[24]	12	[16]	9	[12]
30	[41]	22	[30]	20	[27]	15	[20]	10	[14]	8	[11]
25	[34]	19	[26]	17	[23]	12	[16]	8	[11]	6	[8]
20	[27]	15	[20]	13	[18]	10	[14]	7	[10]	5	[7]
16	[22]	12	[16]	11	[15]	8	[11]	5	[7]	4	[5]
15	[20]	11	[15]	10	[14]	8	[11]	5	[7]	4	[5]
13	[18]	10	[14]	9	[12]	6	[8]	4	[5]	3	[4]
12	[16]	9	[12]	8	[11]	6	[8]	4	[5]	3	[4]
10	[14]	8	[11]	7	[10]	5	[7]	3	[4]	2	[3]
7	[10]	5	[7]	5	[7]	4	[5]	2	[3]	2	[3]

to the requirements of this specification produced on the same mill from the same heat of steel. The impact test(s) shall be taken from different as-rolled or heat-treated pieces. Impact specimens shall be selected from the thickest material rolled subject to the following modifications: When material rolled up to 2 in. [50 mm] inclusive in thickness differs ⅜ in. [10 mm] or more in thickness, one impact test shall be made from both the thickest and thinnest material rolled. When material rolled over 2 in. [50 mm] in thickness differs 1 in. [25 mm] or more in thickness, one impact test shall be made from both the thickest and thinnest material rolled that is more than 2 in. [50 mm] in thickness. If insufficient pieces of the thickest or thinnest material are produced to permit compliance with the above, then testing may proceed to the next nearest thickness available. When plates are produced from coils, three impact tests shall be taken from each qualifying coil (see 4.2). One such coil shall be tested for each 50 tons [45 Mg] of the same product produced on the same mill from the same heat of steel. When material from one heat differs ¹⁄₁₆ in. [2 mm] or more in thickness, tests shall be made from both the thickest and thinnest material rolled regardless of the number of coils represented.

5.2 *Frequency (P) Piece Testing*:

5.2.1 *Plates*—One Charpy V-notch impact test (a set of three specimens) shall be made from each plate-as-rolled except for material that has been heat treated by quenching and tempering, in which case specimens shall be selected from each heat-treated plate. When plates are produced from coils, three impact tests shall be taken from each coil.

5.2.2 *Shapes*—One Charpy V-notch impact test (a set of three specimens) shall be made from at least each 15 tons [15 Mg] or each single length of 15 tons [15 Mg] or more, of the same nominal shape size, excluding length, from each heat in the as-rolled condition. If the shapes are heat treated, one test shall be taken from each heat of each furnace lot. For shapes heat treated in a continuous furnace, a lot shall not exceed 15 tons [15 Mg].

5.2.3 *Bars*—One Charpy V-notch impact test (a set of three specimens) shall be made for each 5 tons [5 Mg] of the same heat and same diameter or thickness if the material is furnished as rolled or is heat treated in a continuous-type furnace. For material heat treated in a noncontinuous furnace, one test shall be taken from each heat of the same bar diameter or thickness for each furnace charge.

6. Heat Treatment

6.1 The material shall be heat treated when specified on the purchase order.

6.2 When the plates are to be supplied in the as-rolled condition the manufacturer or processor (see Specification A6/A6M) has the option to heat treat the plates by normalizing or stress relieving or normalizing and stress relieving to meet the desired toughness properties.

6.3 When the fabricator elects to perform the required heat treatment or fabricates by hot forming instead of heat treating, the plates shall be accepted on the basis of tests made on full-thickness specimens heat treated in accordance with the purchaser's order requirements. If the heat treatment temperatures are not indicated on the purchase order, the manufacturer or processor shall heat treat the specimens under conditions considered appropriate for grain refinement and to meet the toughness requirements. The plate manufacturer or processor shall inform the purchaser of the procedure followed in treating the specimens.

7. Retests

7.1 If more than one individual test value is below the specified minimum average value, or if individual test value is below the greater of 5 ft·lbf [7 J] and two thirds of the specified minimum average value, a retest of three additional specimens shall be made, and each individual test value for such retest shall be equal to or greater than the specified minimum average value.

7.2 If the required energy values are not obtained upon retest, the material may at the option of the manufacturer or processor be heat treated in the case of as-rolled material or reheat treated in the case of heat-treated material.

7.3 After heat treatment or reheat treatment a set of three specimens shall be tested and qualified in the same manner for the original material.

7.4 If the impact test fails for the thickest product tested when testing to frequency (H) that material shall be rejected and the next thickest material tested to qualify the heat in accordance with 4.1. At the option of the manufacturer or processor retests may be made on the rejected material in which case each piece shall be accepted or rejected on the basis of the results of its own test.

8. Test Reports

8.1 Test reports for each heat supplied are required when specified by the purchase order.

8.1.1 Test reports shall show the results of each test required by the specification. However, for (H) frequency, only one test need be reported when the amount of material from a shipment is less than 50 tons [45 Mg] or two tests when the amount of material from a shipment is 50 tons [45 Mg] or more.

8.1.2 The thickness of the product tested may not necessarily be the same as an individual ordered thickness when (H) heat testing is ordered. Tests from material thicknesses in accordance with 5.1 and encompassing the thicknesses in a shipment shall be sufficient for qualifying the material in the shipment. These test thicknesses may or may not be within previously tested and shipped thicknesses from the same heat.

8.1.3 For plates produced from coils, all three test results shall be reported for each qualifying coil. If only half or less of a coil is utilized, then only one test from the outer lap and one from the innermost portion shipped need be reported.

8.1.4 For plates produced from coils, both the manufacturer and processor shall be identified on the test report.

8.1.5 A signature is not required on the test report. However, the document shall clearly identify the organization submitting the report. Notwithstanding the absence of a signature, the organization submitting the report is responsible for the content of the report.

8.1.6 When finished material is supplied to a purchase order specifying ASTM A673/A 673M, the organization supplying that material shall provide the purchaser with a copy of the original manufacturer's test report.

9. Keywords

9.1 charpy V-notch; impact; sampling procedure; steel; structural steel; testing

APPENDIX

(Nonmandatory Information)

X1. VARIATION IN CHARPY V-NOTCH TESTS

X1.1 A survey of the variation to be expected in Charpy V-notch test results obtained from three common fine grain plate steels was conducted by the American Iron and Steel Institute (AISI).[3] The results of the survey are contained in a Contributions to the Metallurgy of Steel entitled, "The Variations of Charpy V-Notch Impact Test Properties in Steel Plates," (SU/24), published January 1979. The survey data consists of test values obtained from six locations in addition to the locations shown in Fig. 1 of this specification. The plate conditions tested involved as-rolled, normalized, and quench and tempered. Sufficient full-size specimens were taken from each sample so that three longitudinal and three transverse specimens could be broken at three test temperatures defined for each grade. The data is presented in tables of probability that impact properties at other than the official location which may differ from those of the reported test location. Additional data of the same type, but utilizing samples from thicker plates, was published by AISI as SU/27[4]. Another survey sponsored by the AISI, entitled "Statistical Analysis of Structural Plate Mechanical Properties" was published in January 2003[5]. That study analyzed the impact properties of more modern higher strength as-rolled structural plate steels.

---

[3] Originally published by the American Iron and Steel Institute. Available from ASTM Headquarters as PCN:29-000390-02.

[4] "The Variations in Charpy V-Notch Impact Properties in Steel Plates" originally published by the American Iron and Steel Institute, July 1989.

[5] Available from AISI directly at http://www.steel.org/infrastructure/bridges/index.html.

A673/A673M – 07

SUMMARY OF CHANGES

Committee A01 has identified the location of selected changes to this standard since the last version (A673/A673M– 05a) that may impact the use of this standard. (Approved March. 1, 2007.)

(1) Revised Fig. 3.

ASTM International takes no position respecting the validity of any patent rights asserted in connection with any item mentioned in this standard. Users of this standard are expressly advised that determination of the validity of any such patent rights, and the risk of infringement of such rights, are entirely their own responsibility.

This standard is subject to revision at any time by the responsible technical committee and must be reviewed every five years and if not revised, either reapproved or withdrawn. Your comments are invited either for revision of this standard or for additional standards and should be addressed to ASTM International Headquarters. Your comments will receive careful consideration at a meeting of the responsible technical committee, which you may attend. If you feel that your comments have not received a fair hearing you should make your views known to the ASTM Committee on Standards, at the address shown below.

This standard is copyrighted by ASTM International, 100 Barr Harbor Drive, PO Box C700, West Conshohocken, PA 19428-2959, United States. Individual reprints (single or multiple copies) of this standard may be obtained by contacting ASTM at the above address or at 610-832-9585 (phone), 610-832-9555 (fax), or service@astm.org (e-mail); or through the ASTM website (www.astm.org).

Designation: A678/A678M − 05 (Reapproved 2009)

Standard Specification for
Quenched-and-Tempered Carbon and High-Strength Low-Alloy Structural Steel Plates[1]

This standard is issued under the fixed designation A678/A678M; the number immediately following the designation indicates the year of original adoption or, in the case of revision, the year of last revision. A number in parentheses indicates the year of last reapproval. A superscript epsilon (ε) indicates an editorial change since the last revision or reapproval.

1. Scope

1.1 This specification covers quenched-and-tempered carbon steel and high-strength low-alloy steel plates of structural quality for welded, riveted, or bolted construction.

1.2 If the steel is to be welded, it is presupposed that a welding procedure suitable for the grade of steel and intended use or service will be used.

1.3 Plates under this specification are available in four grades as follows:

Grade	Yield Strength, min, ksi [MPa]	Tensile Strength, ksi [MPa]	Maximum Thickness, in. [mm]
A	50 [345]	70–90 [485–620]	1½ [40]
B	60 [415]	80–100 [550–690]	2½ [65]
C	[A]	[A]	2 [50]
D	75 [515]	90–110 [620–760]	3 [75]

[A] Varies with thickness. See Table 1.

1.4 The values stated in either inch-pound units or SI units are to be regarded separately as standard. Within the text, the SI units are shown in brackets. The values stated in each system are not exact equivalents; therefore, each system is to be used independently of the other.

2. Referenced Documents

2.1 *ASTM Standards:*[2]

A6/A6M Specification for General Requirements for Rolled Structural Steel Bars, Plates, Shapes, and Sheet Piling

A370 Test Methods and Definitions for Mechanical Testing of Steel Products

3. General Requirements for Delivery

3.1 Plates furnished under this specification shall conform to the applicable requirements of the current edition of Specification A6/A6M unless a conflict exists, in which case this specification shall prevail.

4. Materials and Manufacture

4.1 The requirements for fine austenitic grain size in Specification A6/A6M shall be met.

5. Heat Treatment

5.1 The plates shall be heat treated by heating to a temperature that produces an austenitic structure, but not exceeding 1700°F [925°C], holding a sufficient time to attain uniform heat throughout the material, quenching in a suitable medium, and tempering at not less than 1100°F [593°C]. The heat-treatment temperatures shall be reported in the test report.

6. Chemical Composition

6.1 The heat analysis shall conform to the requirements given in Table 2 for the applicable grade.

6.2 The product analysis shall conform to the requirements given in Table 2, subject to the product analysis tolerances in Specification A6/A6M.

7. Tension Test

7.1 The plates as represented by the test specimens shall conform to the requirements given in Table 1 specified for the applicable grade.

7.2 *Number of Tests*—One tension test shall be taken from a corner of each plate as heat treated.

8. Keywords

8.1 bolted construction; carbon; high-strength; low-alloy; plates; quenched; steel; structural steel; tempered; welded construction

---

[1] This specification is under the jurisdiction of ASTM Committee A01 on Steel, Stainless Steel and Related Alloys and is the direct responsibility of Subcommittee A01.02 on Structural Steel for Bridges, Buildings, Rolling Stock and Ships.

Current edition approved Oct. 1, 2009. Published December 2009. Originally approved in 1973. Last previous edition approved in 2005 as A678/A678M − 05. DOI: 10.1520/A0678_A0678M-05R09.

[2] For referenced ASTM standards, visit the ASTM website, www.astm.org, or contact ASTM Customer Service at service@astm.org. For *Annual Book of ASTM Standards* volume information, refer to the standard's Document Summary page on the ASTM website.

TABLE 1 Mechanical Requirements[A]

	Grade A	Grade B	Grade C	Grade D
Yield strength[B], min. ksi [MPa]				
To ¾ in. [20 mm], incl	50 [345]	60 [415]	75 [515]	75 [515]
Over ¾ to 1½ in. [20 to 40 mm], incl	50 [345]	60 [415]	70 [485]	75 [515]
Over 1½ to 2 in. [40 to 50 mm], incl	C	60 [415]	65 [450]	75 [515]
Over 2 to 2½ in. [50 to 65 mm], incl	C	60 [415]	C	75 [515]
Over 2½ to 3 in. [65 to 75 mm], incl	C	C	C	75 [515]
Tensile strength, ksi [MPa]				
To ¾ in. [20 mm], incl	70–90 [485–620]	80–100 [550–690]	95–115 [655–790]	90–110 [620–760]
Over ¾ to 1½ in. [20 to 40 mm], incl	70–90 [485–620]	80–100 [550–690]	90–110 [620–760]	90–110 [620–760]
Over 1½ to 2 in. [40 to 50 mm], incl	C	80–100 [550–690]	85–105 [585–720]	90–110 [620–760]
Over 2 to 2½ in. [50 to 65 mm], incl	C	80–100 [550–690]	C	90–110 [620–760]
Over 2½ to 3 in. [65 to 75 mm], incl	C	C	C	90–110 [620–760]
Elongation in 2 in. [50 mm], min, %[D,E]	22	22	19	18

[A] See Specimen Orientation under the Tension Tests section of Specification A6/A6M.
[B] Measured at 0.2 % offset or 0.5 % extension under load.
[C] The size and grade is not described in this specification.
[D] For thickness of ¾ in. [20 mm] and under, measured on 1½-in. [40-mm] wide full thickness rectangular specimen as shown in Fig. 3 of Test Methods and Definitions A370. The elongation is measured in a 2-in. [50-mm] gage length that includes the fracture and which shows the greatest elongation.
[E] For plates wider than 24 in. [600 mm], the elongation requirement is reduced two percentage points. See elongation requirement adjustments in the Tension Tests section of Specification A6/A6M.

TABLE 2 Chemical Requirements

NOTE 1—Small amounts of alloying elements may be present, but shall not exceed the following amounts: Cu-0.35; Ni-0.25; Cr-0.25; Mo-0.08.
NOTE 2—Where "..." appears in this table there is no requirement.

Element	Composition, %			
	Grade A[A]	Grade B[A]	Grade C[A]	Grade D[A]
Carbon, max	0.16	0.20	0.22	0.22
Manganese				
1½ in. [40 mm] and under in thickness	0.90–1.50	0.70–1.35	1.00–1.60	1.15–1.50
Over 1½ to 2½ in. [40 to 65 mm], incl	B	1.00–1.60	1.00–1.60	1.15–1.50
Over 2½ to 3 in. [65 to 75 mm], incl	B	B	B	1.15–1.50
Phosphorus, max	0.035	0.035	0.035	0.035
Sulfur, max	0.04	0.04	0.04	0.04
Silicon	0.15–0.50	0.15–0.50	0.20–0.50	0.15–0.50
Vanadium	...	...	...	0.04–0.11
Columbium	...	...	...	C
Nitrogen, max	...	...	...	0.03
Copper, min, if specified	0.20	0.20	0.20	0.20

[A] Boron may be added only by agreement between the manufacturer and the purchaser.
[B] This size and grade is not described in this specification.
[C] Columbium may be present in the amount of 0.01 to 0.05 %.

SUPPLEMENTARY REQUIREMENTS

Supplementary requirements shall not apply unless specified in the purchase order or contract. Standardized supplementary requirements for use at the option of the purchaser are listed in Specification A6/A6M. Those that are considered suitable for use with this specification are listed in this section by title.

S5. Charpy V-Notch Impact Test

ASTM International takes no position respecting the validity of any patent rights asserted in connection with any item mentioned in this standard. Users of this standard are expressly advised that determination of the validity of any such patent rights, and the risk of infringement of such rights, are entirely their own responsibility.

This standard is subject to revision at any time by the responsible technical committee and must be reviewed every five years and if not revised, either reapproved or withdrawn. Your comments are invited either for revision of this standard or for additional standards and should be addressed to ASTM International Headquarters. Your comments will receive careful consideration at a meeting of the responsible technical committee, which you may attend. If you feel that your comments have not received a fair hearing you should make your views known to the ASTM Committee on Standards, at the address shown below.

This standard is copyrighted by ASTM International, 100 Barr Harbor Drive, PO Box C700, West Conshohocken, PA 19428-2959, United States. Individual reprints (single or multiple copies) of this standard may be obtained by contacting ASTM at the above address or at 610-832-9585 (phone), 610-832-9555 (fax), or service@astm.org (e-mail); or through the ASTM website (www.astm.org). Permission rights to photocopy the standard may also be secured from the ASTM website (www.astm.org/COPYRIGHT/).

Designation: A709/A709M – 09a

Standard Specification for
Structural Steel for Bridges[1]

This standard is issued under the fixed designation A709/A709M; the number immediately following the designation indicates the year of original adoption or, in the case of revision, the year of last revision. A number in parentheses indicates the year of last reapproval. A superscript epsilon (ε) indicates an editorial change since the last revision or reapproval.

1. Scope*

1.1 This specification covers carbon and high-strength low-alloy steel structural shapes, plates, and bars and quenched and tempered alloy steel for structural plates intended for use in bridges. Seven grades are available in four yield strength levels as follows:

Grade U.S. [SI]	Yield Strength, ksi [MPa]
36 [250]	36 [250]
50 [345]	50 [345]
50S [345S]	50 [345]
50W [345W]	50 [345]
HPS 50W [HPS 345W]	50 [345]
HPS 70W [HPS 485W]	70 [485]
HPS 100W [HPS 690W]	100 [690]

1.1.1 Grades 36 [250], 50 [345], 50S [345S], and 50W [345W] are also included in Specifications A36/A36M, A572/A572M, A992/A992M, and A588/A588M, respectively. When the supplementary requirements of this specification are specified, they exceed the requirements of Specifications A36/A36M, A572/A572M, A992/A992M, and A588/A588M.

1.1.2 Grades 50W [345W], HPS 50W [HPS 345W], HPS 70W [HPS 485W], and HPS 100W [HPS 690W] have enhanced atmospheric corrosion resistance (see 13.1.2). Product availability is shown in Table 1.

1.2 Grade HPS 70W [HPS 485W] or HPS 100W [HPS 690W] shall not be substituted for Grades 36 [250], 50 [345], 50S [345S], 50W [345W], or HPS 50W [HPS 345W]. Grade 50W [345W], or HPS 50W [HPS 345W] shall not be substituted for Grades 36 [250], 50 [345] or 50S [345S] without agreement between the purchaser and the supplier.

1.3 When the steel is to be welded, it is presupposed that a welding procedure suitable for the grade of steel and intended use or service will be utilized. See Appendix X3 of Specification A6/A6M for information on weldability.

1.4 For structural products to be used as tension components requiring notch toughness testing, standardized requirements are provided in this standard, and they are based upon American Association of State Highway and Transportation Officials (AASHTO) requirements for both fracture critical and non-fracture critical members.

1.5 Supplementary requirements are available but shall apply only if specified in the purchase order.

1.6 The values stated in either SI units or inch-pound units are to be regarded separately as standard. The values stated in each system may not be exact equivalents; therefore, each system shall be used independently of the other. Combining values from the two systems may result in non-conformance with the standard.

1.7 For structural products produced from coil and furnished without heat treatment or with stress relieving only, the additional requirements, including additional testing requirements and the reporting of additional test results, of Specification A6/A6M apply.

2. Referenced Documents

2.1 *ASTM Standards:*[2]

A6/A6M Specification for General Requirements for Rolled Structural Steel Bars, Plates, Shapes, and Sheet Piling
A36/A36M Specification for Carbon Structural Steel
A370 Test Methods and Definitions for Mechanical Testing of Steel Products
A572/A572M Specification for High-Strength Low-Alloy Columbium-Vanadium Structural Steel
A588/A588M Specification for High-Strength Low-Alloy Structural Steel, up to 50 ksi [345 MPa] Minimum Yield Point, with Atmospheric Corrosion Resistance
A673/A673M Specification for Sampling Procedure for Impact Testing of Structural Steel
A992/A992M Specification for Structural Steel Shapes
G101 Guide for Estimating the Atmospheric Corrosion Resistance of Low-Alloy Steels

3. Terminology

3.1 *Definitions of Terms Specific to This Standard:*

3.1.1 *fracture critical member*—a main load-carrying tension member or tension component of a bending member

---

[1] This specification is under the jurisdiction of ASTM Committee A01 on Steel, Stainless Steel and Related Alloys and is the direct responsibility of Subcommittee A01.02 on Structural Steel for Bridges, Buildings, Rolling Stock and Ships.
Current edition approved Oct. 1, 2009. Published October 2009. Originally approved in 1974. Last previous edition approved in 2009 as A709/A709M – 09. DOI: 10.1520/A0709_A0709M-09A.

[2] For referenced ASTM standards, visit the ASTM website, www.astm.org, or contact ASTM Customer Service at service@astm.org. For *Annual Book of ASTM Standards* volume information, refer to the standard's Document Summary page on the ASTM website.

*A Summary of Changes section appears at the end of this standard.

Copyright © ASTM International, 100 Barr Harbor Drive, PO Box C700, West Conshohocken, PA 19428-2959, United States.

TABLE 1 Tensile and Hardness Requirements[A]

NOTE 1—Where "..." appears in this table, there is no requirement.

Grade	Plate Thickness, in. [mm]	Structural Shape Flange or Leg Thickness, in. [mm]	Yield Point or Yield Strength,[B] ksi [MPa]	Tensile Strength, ksi [MPa]	Minimum Elongation, %				Reduction of Area[C,D] min, %
					Plates and Bars[C,E]		Shapes[E]		
					8 in. or 200 mm	2 in. or 50 mm	8 in. or 200 mm	2 in. or 50 mm	
36 [250]	to 4 [100], incl	to 3 in. [75 mm], incl	36 [250] min	58–80 [400–550]	20	23	20	21	...
		over 3 in. [75 mm]	36 [250] min	58 [400] min	...	...	20	19	...
50 [345]	to 4 [100], incl	all	50 [345] min	65 [450] min	18	21	18	21[F]	...
50S [345S]	G	all	50–65 [345–450][H,I]	65 [450][H] min	...	...	18	21	...
50W [345W] and HPS 50W [HPS 345W]	to 4 [100], incl	all	50 [345] min	70 [485] min	18	21	18	21[J]	...
HPS 70W [HPS 485 W]	to 4 [100], incl	G	70 [485] min[B]	85–110 [585–760]	...	19[K]	...	...	...
HPS 100W [HPS 690W]	to 2½ [65], incl	G	100 [690] min[B]	110–130 [760–895]	...	18[K]	...	...	L
	over 2½ to 4 [65 to 100], incl[M]	G	90 [620] min[B]	100–130 [690–895]	...	16[K]	...	...	L

[A] See specimen orientation and preparation subsection in the Tension Tests section of Specification A6/A6M.
[B] Measured at 0.2 % offset or 0.5 % extension under load as described in Section 13 of Test Methods A370.
[C] Elongation and reduction of area not required to be determined for floor plates.
[D] For plates wider than 24 in. [600 mm], the reduction of area requirement, where applicable, is reduced by five percentage points.
[E] For plates wider than 24 in. [600 mm], the elongation requirement is reduced by two percentage points. See elongation requirement adjustments in the Tension Tests section of Specification A6/A6M.
[F] Elongation in 2 in. or 50 mm: 19 % for shapes with flange thickness over 3 in. [75 mm].
[G] Not applicable.
[H] The yield to tensile ratio shall be 0.87 or less for shapes that are tested from the web location; for all other shapes, the requirement is 0.85.
[I] A maximum yield strength of 70 ksi [480 MPa] is permitted for structural shapes that are required to be tested from the web location.
[J] For wide flange shapes with flange thickness over 3 in. [75 mm], elongation in 2 in. or 50 mm of 18 % minimum applies.
[K] If measured on the Fig. 3 (Test Methods A370) 1½-in. [40-mm] wide specimen, the elongation is determined in a 2-in. or 50-mm gage length that includes the fracture and shows the greatest elongation.
[L] 40 % minimum applies if measured on the Fig 3 (Test Methods A370) 1½-in. [40-mm] wide specimen; 50 % minimum applies if measured on the Fig. 4 (Test Methods A370) ½-in. [12.5-mm] round specimen.
[M] Not applicable to Fracture Critical Tension Components (see Table 9).

TABLE 2 Grade 36 [250] Chemical Requirements (Heat Analysis)

NOTE 1—Where "..." appears in this table there is no requirement. The heat analysis for manganese shall be determined and reported as described in the Heat Analysis section of Specification A6/A6M.

Product Thickness, in. (mm)	Shapes[A] All	Plates[B]				Bars[B]		
		To ¾ [20], incl	Over ¾ to 1½ [20 to 40], incl	Over 1½ to 2½ [40 to 65], incl	Over 2½ to 4 [65 to 100], incl	To ¾ [20], incl	Over ¾ to 1½ [20 to 40], incl	Over 1½ to 4 [100], incl
Carbon, max, %	0.26	0.25	0.25	0.26	0.27	0.26	0.27	0.28
Manganese, %	...	...	0.80–1.20	0.80–1.20	0.85–1.20	...	0.60–0.90	0.60–0.90
Phosphorus, max, %	0.04	0.04	0.04	0.04	0.04	0.04	0.04	0.04
Sulfur, max, %	0.05	0.05	0.05	0.05	0.05	0.05	0.05	0.05
Silicon, %	0.40 max	0.40 max	0.40 max	0.15–0.40	0.15–0.40	0.40 max	0.40 max	0.40 max
Copper, min, % when copper steel is specified	0.20	0.20	0.20	0.20	0.20	0.20	0.20	0.20

[A] Manganese content of 0.85 to 1.35 % and silicon content of 0.15 to 0.40 % is required for shapes with flange thickness over 3 in. [75 mm].
[B] For each reduction of 0.01 % below the specified carbon maximum, an increase of 0.06 % manganese above the specified maximum will be permitted up to a maximum of 1.35 %.

whose failure would be expected to cause collapse of a structure or bridge without multiple, redundant load paths.

3.1.2 *main load-carrying member*—a steel member designed to carry primary design loads, including dead, live, impact, and other loads.

3.1.3 *non-fracture critical member*—a main load-carrying member whose failure would not be expected to cause collapse of a structure or bridge with multiple, redundant load paths.

3.1.4 *non-tension component*—a steel member that is not in tension under any design loading.

3.1.5 *secondary member*—a steel member used for aligning and bracing of main load-carrying members, or for attaching utilities, signs, or other items to them, but not to directly support primary design loads

TABLE 3 Grade 50 [345] Chemical Requirements[A] (Heat Analysis)

Maximum Diameter, Thickness, or Distance Between Parallel Faces, in. [mm]	Carbon, max, %	Manganese,[B] max, %	Phosphorus, max, %	Sulfur, max, %	Silicon[C] Plates to 1½-in. [40-mm] Thick, Shapes with flange or leg thickness to 3 in. [75 mm] inclusive, Sheet Piling, Bars, Zees, and Rolled Tees, max, %[D]	Silicon[C] Plates Over 1½-in. [40-mm] Thick and Shapes with flange thickness over 3 in. [75 mm], %	Columbium, Vanadium and Nitrogen
4 [100]	0.23	1.35	0.04	0.05	0.40	0.15–0.40	[E]

[A] Copper when specified shall have a minimum content of 0.20 % by heat analysis (0.18 % by product analysis).
[B] Manganese, minimum by heat analysis of 0.80 % (0.75 % by product analysis) shall be required for all plates over ⅜ in. [10 mm] in thickness; a minimum of 0.50 % (0.45 % by product analysis) shall be required for plates ⅜ in. [10 mm] and less in thickness, and for all other products. The manganese to carbon ratio shall not be less than 2 to 1. For each reduction of 0.01 percentage point below the specified carbon maximum, an increase of 0.06 percentage point manganese above the specified maximum is permitted, up to a maximum of 1.60 %.
[C] Silicon content in excess of 0.40 % by heat analysis must be negotiated.
[D] Bars over 1½ in. [40 mm] in diameter, thickness, or distance between parallel faces, shall be made by a killed steel practice.
[E] Alloy content shall be in accordance with Type 1, 2, 3, or 5 and the contents of the applicable elements shall be reported on the test report.

Type	Elements	Heat Analysis, %
1	Columbium[A]	0.005–0.05[B]
2	Vanadium	0.01–0.15
3	Columbium[A]	0.005–0.05[B]
	Vanadium	0.01–0.15
	Columbium plus vanadium	0.02–0.15[C]
5	Titanium	0.006–0.04
	Nitrogen	0.003–0.015
	Vanadium	0.06 max

[A] Columbium shall be restricted to Grade 50 [345] plate, bar, zee, and rolled tee thickness of ¾ in. [20 mm] max, and to shapes with flange or leg thickness to 1½ in. [40 mm] inclusive unless killed steel is furnished. Killed steel shall be confirmed by a statement of killed steel on the test report, or by a report of the presence of a sufficient quantity of a strong deoxidizing element, such as silicon at 0.10 % or higher, or aluminum at 0.015 % or higher.
[B] Product analysis limits = 0.004 to 0.06 %.
[C] Product analysis limits = 0.01 to 0.16 %.

TABLE 4 Grade 50W [345 W] Chemical Requirements (Heat Analysis)

NOTE 1—Types A, B, and C are equivalent to Specification A588/A588M Grades A, B, and C, respectively.

Element	Composition, %[A]		
	Type A	Type B	Type C
Carbon[B]	0.19 max	0.20 max	0.15 max
Manganese[B]	0.80–1.25	0.75–1.35	0.80–1.35
Phosphorus	0.04 max	0.04 max	0.04 max
Sulfur	0.05 max	0.05 max	0.05 max
Silicon	0.30–0.65	0.15–0.50	0.15–0.40
Nickel	0.40 max	0.50 max	0.25–0.50
Chromium	0.40–0.65	0.40–0.70	0.30–0.50
Copper	0.25–0.40	0.20–0.40	0.20–0.50
Vanadium	0.02–0.10	0.01–0.10	0.01–0.10

[A] Weldability data for these types have been qualified by FHWA for use in bridge construction.
[B] For each reduction of 0.01 percentage point below the specified maximum for carbon, an increase of 0.06 percentage point above the specified maximum for manganese is permitted, up to a maximum of 1.50 %.

3.1.6 *tension component*—a part or element of a fracture critical or non-fracture critical member that is in tension under various design loadings.

TABLE 5 Grades HPS 50W [HPS 345W] and HPS 70W [HPS 485 W], and HPS 100W [HPS 690W] Chemical Requirements (Heat Analysis)

NOTE 1—Where "..." appears in this table, there is no requirement.

Element	Composition, %	
	Grades HPS 50W [HPS 345W], HPS 70W [HPS 485W]	Grade HPS 100W [HPS 690W]
Carbon	0.11 max	0.08 max
Manganese		
2.5 in. [65 mm] and under	1.10–1.35	0.95–1.50
Over 2.5 in. [65 mm]	1.10–1.50	0.95–1.50
Phosphorus	0.020 max	0.015 max
Sulfur[A]	0.006 max	0.006 max
Silicon	0.30–0.50	0.15–0.35
Copper	0.25–0.40	0.90–1.20
Nickel	0.25–0.40	0.65–0.90
Chromium	0.45–0.70	0.40–0.65
Molybdenum	0.02–0.08	0.40–0.65
Vanadium	0.04–0.08	0.04–0.08
Columbium (Niobium)	...	0.01–0.03
Aluminum	0.010–0.040	0.020–0.050
Nitrogen	0.015 max	0.015 max

[A] The steel shall be calcium treated for sulfide shape control.

TABLE 6 Grade 50S [345S] Chemical Requirements (Heat Analysis)

Element	Composition, %
Carbon, max	0.23
Manganese	0.50 to 1.60[A]
Silicon, max	0.40
Vanadium, max	0.15[B]
Columbium, max	0.05[B]
Phosphorus, max	0.035
Sulfur, max	0.045
Copper, max	0.60
Nickel, max	0.45
Chromium, max	0.35
Molybdenum, max	0.15

[A] Provided that the ratio of manganese to sulfur is not less than 20 to 1, the minimum limit for manganese for shapes with flange or leg thickness not exceeding 1 in. [25 mm] shall be 0.30 %.

[B] The sum of columbium and vanadium shall not exceed 0.15 %.

TABLE 7 Relationship Between Impact Testing Temperature Zones and Minimum Service Temperature

Zone	Minimum Service Temperature, °F [°C]
1	0 [−18]
2	below 0 to −30 [−18 to −34]
3	below −30 to −60 [−34 to −51]

4. Ordering Requirements

4.1 In addition to the items listed in the ordering information section of Specification A6/A6M, the following items should be considered if applicable:

4.1.1 Type of component (tension or non-tension, fracture critical or non-fracture critical) (see Section 10).

4.2 Impact testing temperature zone (see Table 7).

5. General Requirements for Delivery

5.1 Structural products furnished under this specification shall conform to the requirements of the current edition of Specification A6/A6M, for the specific structural product ordered, unless a conflict exists in which case this specification shall prevail.

5.2 Coils are excluded from qualification to this specification until they are processed into a finished structural product. Structural products produced from coil means structural products that have been cut to individual lengths from a coil. The processor directly controls, or is responsible for, the operations involved in the processing of a coil into a finished structural product. Such operations include decoiling, leveling or straightening, hot-forming or cold-forming (if applicable), cutting to length, testing, inspection, conditioning, heat treatment (if applicable), packaging, marking, loading for shipment, and certification.

Note 1—For structural products produced from coil and furnished without heat treatment or with stress relieving only, two test results are to be reported for each qualifying coil. Additional requirements regarding structural products produced from coil are described in Specification A6/A6M.

6. Materials and Manufacture

6.1 For Grades 36 [250] and 50 [345], the steel shall be semi-killed or killed.

6.2 For Grades 50W [345W], HPS 50W [HPS 345W], and HPS 70W [HPS 485W], the steel shall be made to fine grain practice.

6.3 For Grade 50S [345S], the steel shall be killed and such shall be affirmed in the test report by a statement of *killed steel*, a value of 0.10 % or more for the silicon content, or a value of 0.015 % or more for the total aluminum content.

6.4 For Grade 50S [345S], the steelmaking practice used shall be one that produces steel having a nitrogen content not greater than 0.015 % and includes the addition of one or more nitrogen-binding elements, or one that produces steel having a nitrogen content of not greater than 0.012 % (with or without the addition of nitrogen-binding elements). The nitrogen content need not be reported, regardless of which steelmaking practice was used.

6.5 For Grades HPS 50W [HPS 345W], HPS 70W [HPS 485W], and HPS 100W [HPS 690W], the steel shall be made using a low-hydrogen practice, such as vacuum degassing during steel making; controlled soaking of the ingots, slabs; controlled slow cooling of the ingots, slabs, or plates, or a combination thereof.

6.6 For Grade HPS 100W [HPS 690W], the requirements for fine austenitic grain size in Specification A6/A6M shall be met.

6.7 Grades HPS 50W [HPS 345W] and HPS 70W [HPS 485W] shall be furnished in one of the following conditions: as-rolled, control-rolled, thermo-mechanical control processed (TMCP) with or without accelerated cooling, or quenched and tempered.

7. Heat Treatment

7.1 For quenched and tempered Grades HPS 50W [HPS 345W] and HPS 70W [HPS 485W], the heat treatment shall be performed by the manufacturer and shall consist of heating the steel to not less than 1650°F [900°C], quenching it in water or oil, and tempering it at not less than 1100°F [590°C]. The heat-treating temperatures shall be reported on the test certificates.

7.2 For Grade HPS 100W [HPS 690W], the heat treatment shall be performed by the manufacturer and shall consist of heating the steel to a temperature in the range from 1600 to 1700°F [870 to 925°C], quenching it in water, and tempering it at not less than 1050°F [565°C] for a time to be determined by the manufacturer. The heat-treating temperatures shall be reported on the test certificates.

8. Chemical Requirements

8.1 The heat analysis shall conform to the requirements for the specified grade, as given in Tables 2-6.

8.2 For Grade 50S [345S], in addition to the elements listed in Table 6, test reports shall include, for information, the chemical analysis for tin. Where the amount of tin is less than 0.02 %, it shall be permissible for the analysis to be reported as <0.02 %.

8.3 For Grade 50S [345S], the maximum permissible carbon equivalent value shall be 0.47 % for structural shapes with flange thickness over 2 in. [50 mm], and 0.45 % for other structural shapes. The carbon equivalent shall be based on heat analysis. The required chemical analysis as well as the carbon equivalent shall be reported. The carbon equivalent shall be calculated using the following formula:

$$CE = C + \frac{Mn}{6} + \frac{(Cr + Mo + V)}{5} + \frac{(Ni + Cu)}{15} \quad (1)$$

9. Tensile Requirements

9.1 The material as represented by test specimens, except as specified in 9.2, shall conform to the requirements for tensile properties given in Table 1.

9.2 For Grade 36 [250], shapes less than 1 in.² [645 mm²] in cross section and bars, other than flats, less than ½ in. [12.5 mm] in thickness or diameter need not be subjected to tension tests by the manufacturer.

10. Impact Testing Requirements

10.1 *Non-Fracture-Critical, T, Tension Components*—Structural products ordered for use as tension components of non-fracture-critical members shall be impact tested in accordance with Specification A673/A673M and as given in Table 8. The test results shall meet the requirements given in Table 8.

10.2 *Fracture-Critical, F, Tension Components*—Structural products ordered for use as tension components of fracture-critical members shall be impact tested in accordance with Specification A673/A673M and as given in Table 9. The test results shall meet the requirements given in Table 9.

10.3 Steel grades ordered for use without suffix T or F as listed in 9.1 and 9.2 do not require impact testing and shall be used as non-tension components or secondary members only.

11. Test Specimens and Number of Tension Tests

11.1 For Grades 36 [250], 50 [345], and 50W [345W], and non-quenched and tempered Grades HPS 50W [HPS 345W] and HPS 70W [HPS 485W], location and condition, number of tests, and preparation of test specimens shall meet the requirements of Specification A6/A6M.

11.2 The following requirements, which are in addition to those of Specification A6/A6M, shall apply only to Grade HPS 100W [HPS 690W] and quenched and tempered Grades HPS 50W [HPS 345W] and HPS 70W [HPS 485W].

11.2.1 When possible, all test specimens shall be cut from the plate in its heat-treated condition. If it is necessary to prepare test specimens from separate pieces, all of these pieces shall be full thickness, and shall be similarly and simultaneously heat treated with the material. All such separate pieces shall be of such size that the prepared test specimens are free of any variation in properties due to edge effects.

11.2.2 After final heat treatment of the plates, one tension test specimen shall be taken from a corner of each plate as heat treated.

Note 2—The term "plate" identifies the "plate as heat treated."

12. Retests

12.1 Grades 36 [250], 50 [345], 50S [345S], and 50W [345W], and non-quenched and tempered HPS 50W [HPS 345W] and HPS 70W [HPS 485W] shall be retested in accordance with Specification A6/A6M.

12.2 The manufacturer may reheat treat quenched and tempered plates that fail to meet the mechanical property requirements of this specification. All mechanical property tests shall be repeated when the material is resubmitted for inspection.

13. Atmospheric Corrosion Resistance

13.1 Steels meeting this specification provide two levels of atmospheric corrosion resistance:

13.1.1 Steel grades without suffix provide a level of atmospheric corrosion resistance typical of carbon or alloy steel without copper.

13.1.2 The steel for Grades 50W [345W], HPS 50W [HPS 345W], and HPS 70W [HPS 485W] shall have an atmospheric corrosion resistance index of 6.0 or higher, calculated from the heat analysis in accordance with Guide G101, Predictive Method Based on the Data of Larabee and Coburn (see Note 3). When properly exposed to the atmosphere, these steels can

TABLE 8 Non-Fracture Critical Tension Component Impact Test Requirements

Grade	Thickness, in. [mm]	Minimum Average Energy, ft·lbf [J]		
		Zone 1	Zone 2	Zone 3
36T [250T][A]	to 4 [100] incl	15 [20] at 70°F [21°C]	15 [20] at 40°F [4°C]	15 [20] at 10°F [−12°C]
50T [345T][A,B], 50ST [345ST][A,B], 50WT [345WT][A,B]	to 2 [50] incl over 2 to 4 [50 to 100] incl	15 [20] at 70°F [21°C] 20 [27] at 70°F [21°C]	15 [20] at 40°F [4°C] 20 [27] at 40°F [4°C]	15 [20] at 10°F [−12°C] 20 [27] at 10°F [−12°C]
HPS 50WT [HPS 345WT][A,B]	to 4 [100] incl	20 [27] at 10°F [−12°C]	20 [27] at 10°F [−12°C]	20 [27] at 10°F [−12°C]
HPS 70WT [HPS 485WT][C,D]	to 4 [100] incl	25 [34] at −10°F [−23°C]	25 [34] at −10°F [−23°C]	25 [34] at −10°F [−23°C]
HPS 100WT [HPS 690WT][C]	to 2½ [65] incl over 2½ to 4 [65 to 100] incl	25 [34] at −30°F [−34°C] 35 [48] at −30°F [−34°C]	25 [34] at −30°F [−34°C] 35 [48] at −30°F [−34°C]	25 [34] at −30°F [−34°C] 35 [48] at −30°F [−34°C]

[A] The CVN-impact testing shall be at "H" frequency in accordance with Specification A673/A673M.
[B] If the yield point of the structural product exceeds 65 ksi [450 MPa], the testing temperature for the minimum average energy required shall be reduced by 15°F [8°C] for each increment of 10 ksi [70 MPa] above 65 ksi [450 MPa]. The yield point is the value given in the test report.
[C] The CVN-impact testing shall be at "P" frequency in accordance with Specification A673/A673M.
[D] If the yield strength of the structural product exceeds 85 ksi [585 MPa], the testing temperature for the minimum average energy required shall be reduced by 15°F [8°C] for each increment of 10 ksi [70 MPa] above 85 ksi [585 MPa]. The yield strength is the value given in the test report.

TABLE 9 Fracture Critical Tension Component Impact Test Requirements

Grade	Thickness, in. [mm]	Minimum Test Value Energy,[A] ft·lbf [J]	Minimum Average Energy[A], ft·lbf [J]		
			Zone 1	Zone 2	Zone 3
36F [250F]	to 4 [100], incl	20 [27]	25 [34] at 70°F [21°C]	25 [34] at 40°F [4°C]	25 [34] at 10°F [−12°C]
50F [345F][B], 50SF [345SF][B], 50WF [345WF][B]	to 2 [50], incl over 2 to 4 [50 to 100], incl	20 [27] 24 [33]	25 [34] at 70°F [21°C] 30 [41] at 70°F [21°C]	25 [34] at 40°F [4°C] 30 [41] at 40°F [4°C]	25 [34] at 10°F [−12°C] 30 [41] at 10°F [−12°C]
HPS 50WF [HPS 345WF][B]	to 4 [100], incl	24 [33]	30 [41] at 10°F [−12°C]	30 [41] at 10°F [−12°C]	30 [41] at 10°F [−12°C]
HPS 70WF [HPS 485WF][C]	to 4 [100], incl	28 [38]	35 [48] at −10°F [−23°C]	35 [48] at −10°F [−23°C]	35 [48] at −10°F [−23°C]
HPS 100WF [HPS 690WF]	to 2½ [65], incl over 2½ to 4 [65 to 100], incl	28 [38] [D]	35 [48] at −30°F [−34°C] Not permitted	35 [48] at −30°F [−34°C] Not permitted	35 [48] at −30°F [−34°C] Not permitted

[A] The CVN-impact testing shall be at "P" frequency in accordance with Specification A673/A673M except for plates of Grades 36F [250F], 50F [345F], 50WF [345WF], HPS 50WF [HPS 345WF], and HPS 70WF [HPS 485WF], for which the sampling shall be as follows:
  (1) As-rolled (including control-rolled and TMCP) plates shall be sampled at each end of each plate-as-rolled.
  (2) Normalized plates shall be sampled at one end of each plate, as heat treated.
  (3) Quenched and tempered plates shall be sampled at each end of each plate, as heat treated.
[B] If the yield point of the structural product exceeds 65 ksi [450 MPa], the testing temperature for the minimum average energy and minimum test value energy required shall be reduced by 15°F [8°C] for each increment of 10 ksi [70 MPa] above 65 ksi [450 MPa]. The yield point is the value given in the test report.
[C] If the yield strength of the structural product exceeds 85 ksi [585 MPa], the testing temperature for the minimum average energy and minimum test value energy required shall be reduced by 15°F [8°C] for each increment of 10 ksi [70 MPa] above 85 ksi [585 MPa]. The yield strength is the value given in the test report.
[D] Not applicable.

be used bare (unpainted) for many applications. The steel for Grade HPS 100W [HPS 690W] provides an improved level of atmospheric corrosion resistance over alloy steel without copper.

NOTE 3—For methods of estimating the atmospheric corrosion resistance of low-alloy steels, see Guide G101. The user is cautioned that the Guide G101 predictive equation (Predictive Method Based on the Data of Larabee and Coburn) for calculation of an atmospheric corrosion resistance index has only been verified for the composition limits stated in that guide.

14. Marking

14.1 In addition to the marking requirements of Specification A6/A6M, the structural product shall be marked as follows:

14.1.1 For Grade 50W [345W], the composition type shall be included.

14.1.2 For structural products that conform to the requirements of 10.1, the letter T and the applicable zone number (1, 2, or 3) shall follow the grade designation.

14.1.3 For structural products that conform to the requirements of 10.2, the letter F and the applicable zone number (1, 2, or 3) shall follow the grade designation.

15. Keywords

15.1 alloy; atmospheric corrosion resistance; bars; bridges; carbon; fracture-critical; high-strength; low-alloy; non-fracture critical; plates; quenched; shapes; steel; structural steel; tempered

SUPPLEMENTARY REQUIREMENTS

Supplementary requirements shall not apply unless specified in the purchase order or contract. Standardized supplementary requirements for use at the option of the purchaser are listed in Specification A6/A6M. Those that are considered suitable for use with this specification are listed by title:

S8. Ultrasonic Examination

S5.1 Refer to S8 of Specification A6/A6M.

S32. Single Heat Bundles

S32.1 Bundles containing shapes or bars shall be from a single heat of steel.

S60. Frequency of Tension Tests

S60.1 Tension testing that is additional to the tension testing required by Specification A6/A6M shall be made, as follows:

S60.1.1 *Plate*—One tension test shall be made using a test specimen taken from each as-rolled or as-heat treated plate.

S60.1.2 *Structural Shapes*—One tension test shall be made using a test specimen taken from each 5 tons [5 Mg] of material produced on the same mill of the same nominal size, excluding length, from each heat of steel. For single pieces that weigh more than 5 tons [5 Mg] individually, each piece shall be tested. If shapes are heat treated, one test shall be made on specimens taken from each heat of the same nominal size, excluding length, in each furnace lot.

S60.1.3 *Bars*—One tension test shall be made using a test specimen taken from each 5 tons [5 Mg] of the same heat and same diameter or thickness if the material is furnished as-rolled or is heat treated in a continuous-type furnace. For material heat treated in other than a continuous-type furnace, one test shall be taken from each heat of the same bar diameter or thickness for each furnace charge.

S92. Atmospheric Corrosion Resistance

S92.1 When specified, the material manufacturer shall supply to the purchaser evidence of atmospheric corrosion resistance satisfactory to the purchaser.

S92.2 Refer to S23 of Specification A6/A6M (applicable only to Grades 36 [250] and 50 [345]).

S93. Limitation on Weld Repair (Fracture Critical Material Only)

S93.1 Weld repair of the base metal by the material manufacturer or supplier is not permitted.

ADDITIONAL SUPPLEMENTARY REQUIREMENTS

Standardized supplementary requirements for use at the option of the purchaser are listed in Specification A6/A6M as follows:

S18. Maximum Tensile Strength (Grades 50 [345], 50S [345S], 50W [345W], and HPS 50W [HPS 345W]).

SUMMARY OF CHANGES

Committee A01 has identified the location of selected changes to this standard since the last issue (A709/A709M – 09) that may impact the use of this standard. (Approved Oct. 1, 2009.)

(1) Revised Table 1, Table 5, and Table 9.

Committee A01 has identified the location of selected changes to this standard since the last issue (A709/A709M – 08) that may impact the use of this standard. (Approved April 1, 2009.)

(1) Deleted Section 11 and Table 5 and renumbered subsequent sections and tables.

(2) Revised Sections 1, 2, 5-8, and 11-14.

(3) Revised Table 1, Table 8, and Table 9.

Committee A01 has identified the location of selected changes to this standard since the last issue (A709/A709M – 07) that may impact the use of this standard. (Approved Oct. 1, 2008.)

(1) Revised Table 8 and Table 9.

ASTM International takes no position respecting the validity of any patent rights asserted in connection with any item mentioned in this standard. Users of this standard are expressly advised that determination of the validity of any such patent rights, and the risk of infringement of such rights, are entirely their own responsibility.

This standard is subject to revision at any time by the responsible technical committee and must be reviewed every five years and if not revised, either reapproved or withdrawn. Your comments are invited either for revision of this standard or for additional standards and should be addressed to ASTM International Headquarters. Your comments will receive careful consideration at a meeting of the responsible technical committee, which you may attend. If you feel that your comments have not received a fair hearing you should make your views known to the ASTM Committee on Standards, at the address shown below.

This standard is copyrighted by ASTM International, 100 Barr Harbor Drive, PO Box C700, West Conshohocken, PA 19428-2959, United States. Individual reprints (single or multiple copies) of this standard may be obtained by contacting ASTM at the above address or at 610-832-9585 (phone), 610-832-9555 (fax), or service@astm.org (e-mail); or through the ASTM website (www.astm.org).

Designation: A710/A710M – 02 (Reapproved 2007)

Standard Specification for
Precipitation–Strengthened Low-Carbon Nickel-Copper-Chromium-Molybdenum-Columbium Alloy Structural Steel Plates[1]

This standard is issued under the fixed designation A710/A710M; the number immediately following the designation indicates the year of original adoption or, in the case of revision, the year of last revision. A number in parentheses indicates the year of last reapproval. A superscript epsilon (ε) indicates an editorial change since the last revision or reapproval.

This standard has been approved for use by agencies of the Department of Defense.

1. Scope

1.1 This specification covers low-carbon precipitation—strengthened nickel - copper - chromium - molybdenum - columbium alloy steel plates for general applications. The alloys in this specification are strengthened by precipitation in various temperature ranges. Precipitation strengthening can occur upon air cooling after hot rolling, during normalizing, and by another heat treatment. These grades are not intended for use in applications above 900°F [540°C].

1.2 Two grades, each with three classes, are provided as follows:

Grade and Class	Condition
Grade A, Class 1	as-rolled and precipitation heat treated
Grade A, Class 2	normalized and precipitation heat treated
Grade A, Class 3	quenched and precipitation heat treated
Grade B, Class 1	as-rolled
Grade B, Class 2	normalized
Grade B, Class 3	normalized and precipitation heat treated

1.3 Grade A provides minimum yield strength levels ranging from 50 to 85 ksi [345 to 585 MPa], depending on thickness and condition.

1.4 Grade A, Class 1, plates are limited to a maximum thickness of ¾ in. [20 mm]. The maximum thickness of Grade A, Classes 2 and 3, is limited only by the capacity of the composition to meet the specified mechanical property requirements; however, current practice normally limits the maximum thickness to 8 in. [200 mm].

1.5 Mandatory notch toughness requirements are specified for Grade A, Class 1.

1.6 Grade B provides minimum yield strength levels ranging from 70 to 75 ksi [480 to 515 MPa], depending on thickness and condition.

1.7 Grade B plates are limited to a maximum thickness of 2 in. [50 mm].

1.8 Mandatory notch toughness requirements are specified for the three classes of Grade B.

1.9 When the steel is to be welded, it is presupposed that a welding procedure suitable for the grade of steel and intended use or service will be utilized. See Appendix X3 of Specification A6/A6M for information on weldability.

1.10 The values stated in either inch-pound units or SI units are to be regarded separately as standard. Within the text, the SI units are shown in brackets. The values stated in each system are not exact equivalents; therefore, each system must be used independently of the other. Combining values from the two systems may result in nonconformance with the specification.

2. Referenced Documents

2.1 *ASTM Standards:*[2]

A6/A6M Specification for General Requirements for Rolled Structural Steel Bars, Plates, Shapes, and Sheet Piling

A673/A673M Specification for Sampling Procedure for Impact Testing of Structural Steel

3. Terminology

3.1 *Definitions of Terms Specific to This Standard:*

3.1.1 *precipitation heat treatment*—a sub-critical temperature thermal treatment performed to cause precipitation of submicroscopical constituents, etc., so as to result in enhancement of some desirable property.

3.1.2 *precipitation strengthening*—the precipitation of submicroscopic and/or microscopic constituents of an alloy at various temperatures, which results in the alteration of certain properties.

---

[1] This specification is under the jurisdiction of ASTM Committee A01 on Steel, Stainless Steel and Related Alloys and is the direct responsibility of Subcommittee A01.02 on Structural Steel for Bridges, Buildings, Rolling Stock and Ships.
Current edition approved March 1, 2007. Published March 2007. Originally approved in 1974. Last previous edition approved in 2002 as A710/A710M – 02. DOI: 10.1520/A0710_A0710M-02R07.

[2] For referenced ASTM standards, visit the ASTM website, www.astm.org, or contact ASTM Customer Service at service@astm.org. For *Annual Book of ASTM Standards* volume information, refer to the standard's Document Summary page on the ASTM website.

3.1.3 *soak*—to hold at temperature after the material has attained the temperature throughout.

4. General Requirements for Delivery

4.1 Material furnished under this specification shall conform to the requirements of the current edition of Specification A6/A6M, for the ordered material, unless a conflict exists in which case this specification shall prevail.

5. Materials and Manufacture

5.1 The steel shall be made to fine grain practice.

6. Heat Treatment

6.1 Grade A, Class 1 material shall be precipitation heat treated in the temperature range from 1000 to 1300°F [540 to 705°C] for a time to be determined by the material manufacturer.

6.2 Grade A, Class 2 material shall be normalized at a temperature in the range from 1600 to 1700°F [870 to 925°C] and then precipitation heat treated at a temperature in the range from 1000 to 1300°F [540 to 705°C] for a time to be determined by the material manufacturer.

6.3 Grade A, Class 3 material shall be quenched in water or oil from a temperature in the range from 1600 to 1700°F [870 to 925°C] and then precipitation heat treated at a temperature in the range from 1000 to 1300°F [540 to 705°C] for a time to be determined by the material manufacturer.

6.4 Grade B, Class 1 shall be hot-rolled.

6.5 Grade B, Class 2 shall be normalized after hot rolling by reheating to 1600 to 1700°F [870 to 925°C], and then cooled in still air.

6.6 Grade B, Class 3, shall be normalized at 1600 to 1700°F [870 to 925°C], and then precipitation heat treated at 1000 to 1300°F [540 to 705°C] for a time to be determined by the material manufacturer. One hour at a specified temperature is generally considered as a maximum.

6.7 If the purchaser elects to perform the thermal (heat) treatment, the material shall be accepted on the basis of mill tests from test coupons heat treated in accordance with the purchase order requirements. If the test coupon heat treatment requirements are not indicated on the purchase order, the manufacturer shall heat treat the test coupons under conditions he considers appropriate. The manufacturer shall inform the purchaser of the procedure followed in thermally treating the test coupons at the mill.

7. Chemical Composition

7.1 The heat analysis shall conform to the requirements as to chemical composition prescribed in Table 1.

7.2 The steel shall conform on product analysis to the requirements prescribed in Table 1, subject to the product analysis tolerance in Specification A6/A6M for alloy steels.

8. Tension Test

8.1 The material, as represented by the test specimens, shall conform to the requirements specified in Table 2.

8.2 *Number of Tests*—One tension test shall be taken from a corner of each plate as heat treated for each class of material. For plates 3/8 in. [10 mm] and under in thickness, a tension test shall be made from a corner of each of two plates per lot. A lot shall consist of plates from the same heat and thickness, same prior condition and scheduled heat treatment, and shall not exceed 15 tons [13.6 Mg] in weight. Plates wider than 24 in. [610 mm] shall be tested in the transverse direction and are subject to the modifications for elongation contained in footnote[D] of Table 2.

9. Notch Toughness Requirements

9.1 *Notch Toughness Tests—Grade A, Class 1*:

9.1.1 Notch toughness tests shall be made in accordance with Test Frequency H of Specification A673/A673M. Upon agreement, transverse tests may be specified instead of the longitudinal tests specified in Specification A673/A673M (plates only).

9.1.2 The test results shall meet a minimum average value of 20 ft·lbf [27 J] at −50°F [−45°C] for longitudinal specimens. For transverse specimens, the test results shall meet a minimum average value of 15 ft·lbf [20 J] at −50°F [−45°C]. By agreement, a test temperature lower than −50°F [−45°C] may be used.

9.2 *Notch Toughness Tests—Grade B*:

9.3 Notch toughness tests shall be made in accordance with Test Frequency H of Specification A673/A673M.

TABLE 1 Chemical Requirements

Element	Composition, %	
	Grade A	Grade B
Carbon	0.07 max	0.03–0.09
Manganese	0.40–0.70	0.45–1.30
Phosphorus, max	0.025	0.025
Sulfur, max	0.025	0.025
Silicon	0.40 max	0.30–0.50
Nickel	0.70–1.00	0.80–1.00
Chromium	0.60–0.90	0.30 max
Molybdenum	0.15–0.25	0.25 max
Copper	1.00–1.30	1.25–1.50
Columbium	0.02 min	0.02–0.06
Titanium	...	0.01–0.03

NOTE: Where an ellipsis (...) appears in the table, there is no requirement.

TABLE 2 Tensile Requirements[A]

	Grade A Class 1	Grade A Class 2	Grade A Class 3	Grade B Class 1	Grade B Class 2	Grade B Class 3
Yield strength,[B] min, ksi [MPa]						
¼ in. [6.5 mm] and under	85 [585]	65 [450]	80 [550]	70 [485]	70 [485]	70 [515]
Over ¼ in. to ⁵⁄₁₆ in. [6.5 to 8 mm], incl	85 [585]	65 [450]	80 [550]	70 [485]	70 [485]	70 [515]
Over ⁵⁄₁₆ in. to ⅜ in. [8 to 10 mm], incl	80 [550]	65 [450]	80 [550]	70 [485]	70 [485]	70 [515]
Over ⅜ in. to ½ in. [10 to 12.5 mm], incl	80 [550]	65 [450]	80 [550]	70 [485]	70 [485]	70 [515]
Over ½ in. to ¾ in. [12.5 to 20 mm], incl	80 [550]	65 [450]	80 [550]	70 [485]	70 [485]	70 [515]
Over ¾ in. to 1 in. [20 to 25 mm], incl	...	65 [450]	80 [550]	70 [485]	70 [485]	70 [485]
Over 1 in. to 1¼ in. [25 to 30 mm], incl	...	60 [415]	80 [550]	...	70 [485]	70 [485]
Over 1¼ in. to 2 in. [30 to 50 mm], incl	...	60 [415]	75 [515]	...	70 [485]	70 [485]
Over 2 in. to 4 in. [50 to 100 mm], incl	...	55 [380]	65 [450]	...	...	...
Over 4 in. [100 mm]	...	50 [345]	60 [415]	...	...	...
Tensile strength, min, ksi [MPa]						
¼ in. [6.5 mm] and under	90 [620]	72 [495]	85 [585]	80 [550]	80 [550]	80 [585]
Over ¼ in. to ⁵⁄₁₆ in. [6.5 to 8 mm], incl	90 [620]	72 [495]	85 [585]	80 [550]	80 [550]	80 [585]
Over ⁵⁄₁₆ in. to ⅜ in. [8 to 10 mm], incl	90 [620]	72 [495]	85 [585]	80 [550]	80 [550]	80 [585]
Over ⅜ in. to ½ in. [10 to 12.5 mm], incl	90 [620]	72 [495]	85 [585]	80 [550]	80 [550]	80 [585]
Over ½ in. to ¾ in. [12.5 to 20 mm], incl	90 [620]	72 [495]	85 [585]	80 [550]	80 [550]	80 [585]
Over ¾ in. to 1 in. [20 to 25 mm], incl	...	72 [495]	85 [585]	80 [550]	80 [550]	80 [550]
Over 1 in. to 1¼ in. [25 to 30 mm], incl	...	72 [495]	85 [585]	...	80 [550]	80 [550]
Over 1¼ in. to 2 in. [30 to 50 mm], incl	...	72 [495]	85 [585]	...	80 [550]	80 [550]
Over 2 in. to 4 in. [50 to 100 mm], incl	...	65 [450]	75 [515]	...	...	...
Over 4 in. [100 mm]	...	60 [415]	70 [485]	...	...	...
Elongation in 2 in. or 50 mm, min, %[C][D]	20	20	20	20	20	20

NOTE: Where ellipses (...) appear in the table, there is no requirement.

[A] For plates wider than 24 in. [600 mm], the test specimen is taken in the transverse direction. See Specification A6/A6M, Tension Tests requirements.
[B] 0.2 % offset or 0.5 % extension-under-load.
[C] For thickness of ¾ in. [20 mm] and under, measured on 1½ in. [40 mm] wide full thickness rectangular specimen as shown in Fig. 4 of Test Methods and Definitions A370. The elongation is measured in a 2-in. [50-mm] gage length which includes the fracture and which shows the greatest elongation.
[D] For plates wider than 24 in. [600 mm], the elongation requirement is reduced two percentage points.

9.4 The test results for Grade B, Class 1, and Grade B, Class 3, shall have a minimum average notch toughness of 35 ft·lbf [47 J] at -10°F [-23°C] in the longitudinal direction of the plate. For Grade B, Class 2, test results shall have a minumum average notch toughness of 20 ft·lbf [27 J] at -50°F [-45°C] in the longitudinal direction of the plate.

10. Retreatment

10.1 Thermally treated material that fails to meet the mechanical property requirements may be reheat-treated. All required tests shall be repeated when material is resubmitted for inspection.

11. Keywords

11.1 alloy; carbon; chromium; columbium; copper; general applications; low carbon; manganese; molybdenum; nickel; plates; precipitation heat treatment; precipitation strengthening; steel; structural steel

SUPPLEMENTARY REQUIREMENTS

Supplementary requirements shall not apply unless specified in the purchase order or contract. Standardized supplementary requirements for use at the option of the purchaser are listed in Specification A6/A6M. In addition, the following supplementary requirement is also suitable for use with this specification:

S94. Notch Toughness—Grade A, Classes 2 and 3

S94.1 Notch toughness tests on longitudinal specimens shall be made in accordance with Test Frequency H of Specification A673/A673M.

S94.2 *Grade A, Class 2*—The test results shall meet a minimum average value of 50 ft·lbf [69 J] at −50°F [−45°C].

S94.3 *Grade A, Class 3*—The test results shall meet a minimum average value of 50 ft·lbf [69 J] at −80°F [−60°C].

ASTM International takes no position respecting the validity of any patent rights asserted in connection with any item mentioned in this standard. Users of this standard are expressly advised that determination of the validity of any such patent rights, and the risk of infringement of such rights, are entirely their own responsibility.

This standard is subject to revision at any time by the responsible technical committee and must be reviewed every five years and if not revised, either reapproved or withdrawn. Your comments are invited either for revision of this standard or for additional standards and should be addressed to ASTM International Headquarters. Your comments will receive careful consideration at a meeting of the responsible technical committee, which you may attend. If you feel that your comments have not received a fair hearing you should make your views known to the ASTM Committee on Standards, at the address shown below.

This standard is copyrighted by ASTM International, 100 Barr Harbor Drive, PO Box C700, West Conshohocken, PA 19428-2959, United States. Individual reprints (single or multiple copies) of this standard may be obtained by contacting ASTM at the above address or at 610-832-9585 (phone), 610-832-9555 (fax), or service@astm.org (e-mail); or through the ASTM website (www.astm.org).

Standard Specification for Through-Thickness Tension Testing of Steel Plates for Special Applications[1]

This standard is issued under the fixed designation A770/A770M; the number immediately following the designation indicates the year of original adoption or, in the case of revision, the year of last revision. A number in parentheses indicates the year of last reapproval. A superscript epsilon (ε) indicates an editorial change since the last revision or reapproval.

This standard has been approved for use by agencies of the Department of Defense.

1. Scope

1.1 This specification[2] covers the procedures and acceptance standards for the determination of reduction of area using a tension test specimen whose axis is perpendicular to the rolled surfaces of steel plates 1 in. [25 mm] and greater in thickness. The principal purpose of the testing is to provide a measure of the resistance of a steel plate to lamellar tearing. (See Appendix X1.)

1.2 The values stated in either inch-pound units or SI units are to be regarded as standard. Within the text, the SI units are shown in brackets. The values stated in each system are not exact equivalents; therefore, each system must be used independently of the other. Combining values from the two systems may result in nonconformance with the specification.

1.3 This specification is expressed in both inch-pound and SI units. However, unless the order specifies the applicable "M" specification designation (SI units), the material shall be furnished to inch-pound units.

2. Referenced Documents

2.1 *ASTM Standards:*[3]

A370 Test Methods and Definitions for Mechanical Testing of Steel Products

3. Ordering Information

3.1 The inquiry and order shall include the following, if required:

3.1.1 Supplementary requirements that are available to meet end use requirements (see S1 through S5).

3.1.2 Special requirements agreed upon between the manufacturer and the purchaser.

4. Tension Tests

4.1 *Number of Tests*:

4.1.1 Two tests shall be required from each plate-as-rolled, except for plates subjected to heat treatment by quenching and tempering. Two tests shall be required from each quenched-and-tempered plate. The tests shall be representative of the plate in its final condition.

4.1.2 When plates are furnished by the manufacturer in an unheat-treated condition and qualified by heat-treated specimens (including normalized, normalized and tempered, and quenched and tempered), two tests shall be required from each plate-as-rolled.

NOTE 1—The term "plate-as-rolled" refers to the unit plate rolled from a slab or directly from an ingot. It does not refer to the condition of the plate.

4.2 *Location of Test Coupons*—Take one test coupon at each end of each plate as defined in 4.1. Take the test coupons from the center of the plate width.

4.3 *Orientation of Test Specimens*—The longitudinal axis of the reduced section of the test specimens shall be perpendicular to the rolled surface of the plate.

4.4 *Preparation of Test Specimens*:

4.4.1 *Welded Prolongations*—When required, join welded prolongations to the surface(s) of the plate being tested. The joining method used shall be one which results in a minimal heat-affected zone in the portion of the plate to be tested. Shielded metal arc, friction, stud, or electron-beam welding methods have proven to be suitable.

4.4.2 *Standard Test Specimens*:

4.4.2.1 Three types of standard round tension test specimens are shown in Fig. 1 and Table 1. For Types 1 and 2 specimens, locate the center of the length of the reduced section at the approximate mid-point of the plate thickness. For Type 3 specimens, locate the weld fusion line of one plate surface within ¼ in. [6 mm] of one end of the reduced section.

4.4.2.2 For plates from 1 in. [25 mm] to 1¼ in. [32 mm] inclusive in thickness, use either the 0.350-in. [8.75-mm] Type 1 specimen or the 0.500-in. [12.5-mm] Type 2 specimen.

4.4.2.3 For plates over 1¼ in. to 2 in. [50 mm] inclusive in thickness, use the 0.500-in. [12.5-mm] Type 2 specimen.

---

[1] This specification is under the jurisdiction of ASTM Committee A01 on Steel, Stainless Steel and Related Alloys and is the direct responsibility of Subcommittee A01.11 on Steel Plates for Boilers and Pressure Vessels.
Current edition approved Nov. 1, 2007. Published March 2008. Originally approved in 1980. Last previous edition approved in 2003 as A770/A770M – 03. DOI: 10.1520/A0770_A0770M-03R07.

[2] For ASME Boiler and Pressure Vessel Code applications, see related Specification SA-770/SA-770M in Section II of that Code.

[3] For referenced ASTM standards, visit the ASTM website, www.astm.org, or contact ASTM Customer Service at service@astm.org. For *Annual Book of ASTM Standards* volume information, refer to the standard's Document Summary page on the ASTM website.

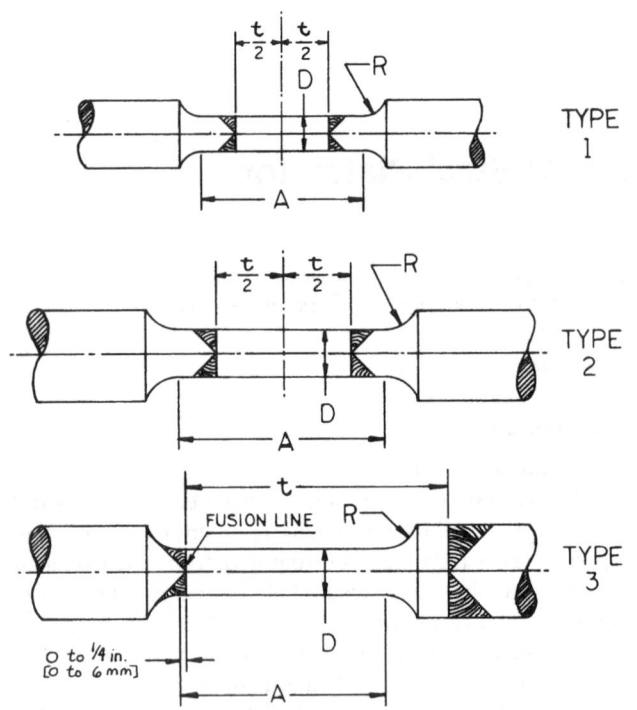

NOTE—For Type 3 only one welded prolongation may be needed, depending upon plate thickness.

FIG. 1 Standard Round Tension Test Specimens

TABLE 1 Schedule of Standard Test Specimens, Inches [Millimetres][A]

	Specimen Type		
	1	2	3
Plate thickness (t)	$1 \leq t \leq 1\frac{1}{4}$	$1 < t \leq 2$	$2 < t$
Diameter (D)	0.350 [8.75]	0.500 [12.5]	0.500 [12.5]
Radius, minimum (R)	¼ [6]	⅜ [10]	⅜ [10]
Length of reduced section (A)	1¾ [45]	2¼ [60]	2¼ [60]

[A] See Test Methods and Definitions A370 (Fig. 5 for further details and Fig. 6 for various types of ends).

4.4.2.4 For plates greater than 2 in. [50 mm] in thickness, use the Type 3 specimen.

4.4.3 *Alternative Test Specimens*—The alternative test specimens in Fig. 2 and Table 2 may be used in place of the standard specimens in Fig. 1 and Table 1.

4.4.3.1 For plates over 2 in. [50 mm] in thickness, Type A or Type B specimens may be used. The Type A specimen provides a reduced section length greater than the plate thickness. The Type B specimen provides a reduced section length of 2¼ in. [57 mm] with its center at the mid-thickness of the plate. Over a minimum plate thickness determined by the specimen end configuration, no welded prolongations may be needed for the Type B specimen. For plates over 4¼ in. [108 mm] in thickness, the Type C specimen may be used. For plates over 6 in. [150 mm] in thickness, a series of two or more Type A or Type C specimens with reduced sections of 4 in. [100 mm] or less may be used to cover the full thickness of the plate. The number of tests required will depend upon the thickness of the plate being tested and the reduced section length selected.

4.4.3.2 For plates over 1 in. [25 mm] in thickness, a series of button-head specimens shown in Fig. 2 and Table 2 may be

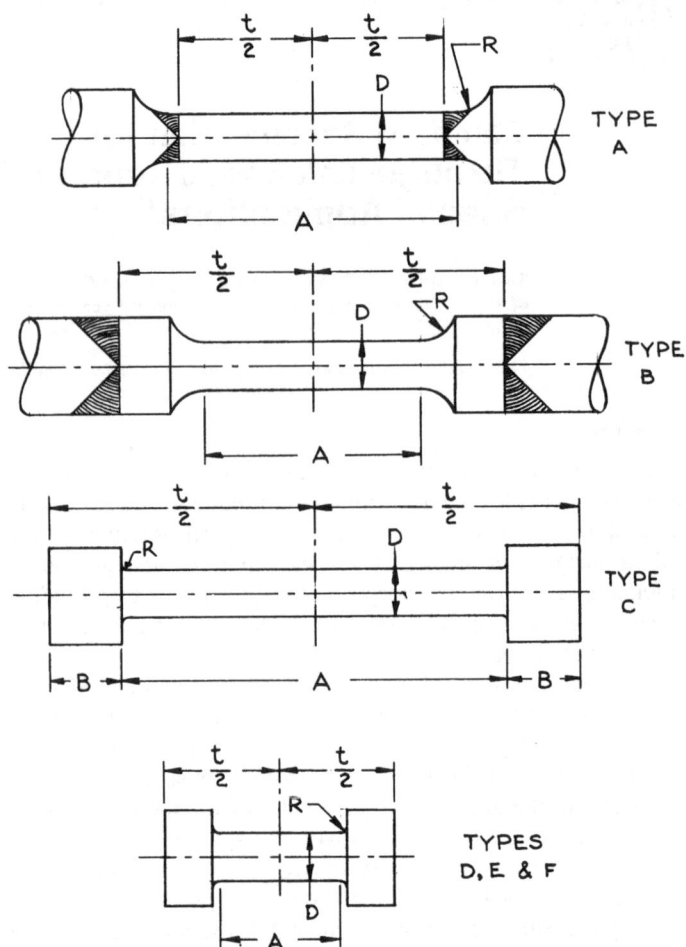

FIG. 2 Alternative Tension Test Specimens

used. The test specimen type to be used, Type D, Type E, or Type F, is determined by the nominal plate thickness as described in Table 2. A series of two or more Type F specimens may be used to cover the full thickness of the plate. The length of the reduced section (A), as shown in Fig. 2 and specified in Table 2, is the length of the reduced section excluding the machined radius (R). Within the plate thickness dimension specified for each test specimen type, either the button-head thickness, the reduced section length, or the machined radius may be varied. In all cases, the minimum length of the reduced section must be as specified in Table 2 to maintain a minimum length to diameter ratio (see Appendix X2.2).

5. Acceptance Standards

5.1 Each tension test shall have a minimum reduction of area no less than 20 %. If the reduction of area of both tests is less than 20 %, no retest shall be permitted. If the reduction of area of one of the two tests from a plate is less than 20 %, one retest of two additional specimens taken from a location adjacent to the specimen that failed may be made, and both of these additional specimens shall have a reduction of area of 20 % or more.

5.2 Failures occurring in the prolongations, the weld, or in the fusion line shall be considered as a "no-test," and an additional specimen shall be tested.

TABLE 2 Schedule of Alternative Test Specimens, Inches [Millimetres]

	Specimen Type					
	A[A]	B[A]	C[B]	D	E	F
Plate thickness (t)	2 < t [50 < t]	2 < t [50 < t]	4¼ < t [108 < t]	1 ≤ t ≤ 1¾ [25 ≤ t ≤ 45]	1¾ < t ≤ 2½ [45 ≤ t ≤ 64]	2½ < t [64 < t]
Diameter (D)	0.500 [12.5]	0.500 [12.5]	0.500 [12.5]	0.250 [6.25][C] ± 0.005 [0.10]	0.350 [8.75][C] ± 0.007 [0.18]	0.500 [12.5][C] ± 10.010 [0.25]
Radius, min (R)	⅜ [10]	⅜ [10]	1/16 [2]	optional	optional	optional
Length of reduced section, min (A)	t + ¼ min [t + 6]	2¼ [60]	t − 1½ [t − 38]	0.625 [16]	0.875 [22]	1.250 [32]

[A] See Test Methods and Definitions A370 (Fig. 5 for further details and Fig. 6 for various types of ends).
[B] See Test Methods and Definitions A370 (Fig. 6, specimen 3 for further details).
[C] The reduced section may have a gradual taper from the ends toward the center, with the ends not more than 1 % larger in diameter than the center (controlling dimension).

6. Marking

6.1 Plates accepted in accordance with this specification shall be identified by stamping or stenciling ZT adjacent to the marking required by the applicable product specification.

7. Keywords

7.1 lamellar tearing; special steel-making processes; steel plate; through-thickness tension testing

SUPPLEMENTARY REQUIREMENTS

These requirements apply only when specified by the purchaser.

S1. Tensile Strength Requirements

S1.1 Tensile strength shall conform to a minimum value which is subject to agreement between the manufacturer and purchaser.

S2. Yield Strength Requirements

S2.1 Yield strength, for plates 2 in. [50 mm] and over in thickness, shall conform to a minimum value which is subject to agreement between the manufacturer and purchaser.

S3. Reduction of Area Requirements

S3.1 A minimum reduction of area limit higher than that in 5.1 may be specified subject to agreement between the manufacturer and purchaser.

S4. Number of Tests

S4.1 A greater number of tests than indicated in 4.1 may be specified subject to agreement between the manufacturer and purchaser.

S5. Location of Test Coupons

S5.1 Test coupons from locations in addition to those specified in 4.2 may be specified subject to agreement between the manufacturer and purchaser.

APPENDIXES

(Nonmandatory Information)

X1. LAMELLAR TEARING ADJACENT TO WELDS

X1.1 Introduction

X1.1.1 Lamellar tearing is a particular type of cracking that occurs under the weld of a steel plate weldment. It is generally caused by strain induced in the thickness direction resulting from shrinkage of the weld deposit and by the restraint imposed by the components that comprise the weldment. High restraint increases the possibility of lamellar tearing. However, lamellar tearing is not solely confined to highly restrained weldments. Lamellar tearing may also result from loads on the plate surface.

X1.2 Characteristics of Lamellar Tearing

X1.2.1 Lamellar tearing normally occurs in susceptible material underneath the weld, in a direction generally parallel to the plate surface and often slightly outside the heat-affected zone. Lamellar tearing generally has a step-like appearance consisting of "terraces" (cracks running parallel to the plate surface) and "walls" (cracks which connect the individual terraces). The tearing may remain completely subsurface or appear at plate edges or at weld toes.

X1.3 Inclusions

X1.3.1 The step-like cracking characteristic of lamellar tearing is usually considered to result from small elongated nonmetallic inclusions that are normally present in the steel. Strains in the through-thickness direction can cause individual inclusions to fractures or decohere from the surrounding steel matrix, thus initiating a void. Further strain can cause the remaining metallic ligaments to shear or rupture, resulting in the step-like fracture appearance.

X1.3.2 A high or concentrated inclusion content in the steel produces planar regions of poor ductility parallel to the steel surface. On the other hand, a reduction in the magnitude and concentration of these inclusions to a low level tends to preclude any easy fracture path along the low ductility inclusions and the steel exhibits improved ductility in a through-thickness direction.

X1.3.3 The extent of nonmetallic inclusions depends on the type of steel. In silicon semikilled or fully killed steels, these inclusions are primarily oxides (present as silicates) and sulfides (present as manganese sulfides). For aluminum-silicon killed steels, these inclusions are primarily sulfides (manganese sulfides). To improve the through-thickness ductility and thus the resistance of the steel to lamellar tearing, it is necessary to reduce the level of the nonmetallic inclusions. To provide a high resistance to lamellar tearing may require the use of special steel-making processes that can reduce the oxygen and sulfur contents in the steel to very low levels.

X1.4 Steel Manufacturing Processes

X1.4.1 Special steel-making processes are available for improving the through-thickness ductility. The more common processes, used singly or in combination, are: (*1*) low sulfur practices; (*2*) inclusion shape control; (*3*) electroslag or vacuum arc remelting; and (*4*) vacuum degassing. The steel-making processes are not all intended for the same purpose, but will improve the through-thickness ductility to various degrees depending on the process used.

X1.5 Through-Thickness Ductility Requirements

X1.5.1 Susceptibility to lamellar tearing depends on many factors (for example, restraint, welding conditions, etc.) and, consequently a specific through-thickness ductility requirement does not provide a guarantee against lamellar tearing. The most widely accepted method of measuring the material ductility factor of susceptibility to lamellar tearing is the reduction of area of a round tension test specimen oriented perpendicular to the rolled surface of a plate.

X2. TESTING PARAMETERS AFFECTING REDUCTION OF AREA VALUES

X2.1 Variability of Through-Thickness Properties

X2.1.1 Through-thickness tension test results, and in particular the reduction of area determination as provided for in this specification, are subject to substantially greater scatter than would normally be expected from standard tension tests of a plate in the longitudinal or transverse direction. This scatter of test results is due in part to the inherent variability of the distribution of the nonmetallic inclusions discussed in X1.3. For example, those nonmetallic inclusions that form during the solidification phase of the steelmaking process tend to occur with a higher frequency in the area of final solidification.

X2.1.2 Test specimen design may also have an effect on the test results. Some of these factors are discussed in X2.2. Operator technique will also be a factor in increasing scatter, particularly in the measurement of the final diameter of the test specimen. Because of the effect of inclusions on the fracture process, the appearance of the final fracture may be quite different than the classical cup-cone fractures common to longitudinal and transverse tension testing. For those materials with approximately 20 % reduction of area, the final diameter measurement may require a substantial amount of judgment on the part of the test operator.

X2.1.3 In view of the potential variability of the through-thickness reduction of area test results, it is recognized that two tests per plate are not sufficient to fully characterize the through-thickness ductility of that plate. The number of tests and test positions have not been established that would provide a good estimate of both the mean and the variability of through-thickness tensile reduction of values of a plate. Therefore, an average value requirement is not included in this specification. The intent of this specification is to qualify a plate according to the described testing procedures using only a minimum value requirement. The potential variability of the test results also increases the possibility that subsequent testing of a steel plate qualified according to this specification may produce results that do not meet the specified acceptance standard.

X2.2 Effects of Test Specimen Design

X2.2.1 Two main factors considered in the selection of test specimen geometry were the diameter and the slenderness ratio. It is generally accepted that there is a diameter effect on reduction-of-area values such that a smaller diameter specimen generally yields a higher average reduction in area value. It is also accepted that smaller diameter test specimens will tend to give greater variability to the resulting reduction in area values. Because these relationships between the test specimen diameter and the average and variability of the test result have not been satisfactorily quantified at this time, the same minimum requirement has been applied to all test specimen diameters.

X2.2.2 The slenderness ratio (reduced section length/ reduced section diameter) is known to affect the reduction in area values when below a minimum value. This minimum value may be from 1.5 to 2.5, depending on the material. Below this minimum value, the reduction at the failure point in the reduced section is restrained by the larger cross section away from the reduced section. A minimum slenderness ratio of 2 was selected for the standard Type 2 specimen to allow a 0.500-in. [12.5-mm] diameter specimen to be used on a 1-in. [25-mm] plate. A minimum slenderness ratio of 2.5 was selected for the collar-button specimens (Types D, E, and F) to ensure that this effect is minimized for these test specimens.

ASTM International takes no position respecting the validity of any patent rights asserted in connection with any item mentioned in this standard. Users of this standard are expressly advised that determination of the validity of any such patent rights, and the risk of infringement of such rights, are entirely their own responsibility.

This standard is subject to revision at any time by the responsible technical committee and must be reviewed every five years and if not revised, either reapproved or withdrawn. Your comments are invited either for revision of this standard or for additional standards and should be addressed to ASTM International Headquarters. Your comments will receive careful consideration at a meeting of the responsible technical committee, which you may attend. If you feel that your comments have not received a fair hearing you should make your views known to the ASTM Committee on Standards, at the address shown below.

This standard is copyrighted by ASTM International, 100 Barr Harbor Drive, PO Box C700, West Conshohocken, PA 19428-2959, United States. Individual reprints (single or multiple copies) of this standard may be obtained by contacting ASTM at the above address or at 610-832-9585 (phone), 610-832-9555 (fax), or service@astm.org (e-mail); or through the ASTM website (www.astm.org).

Designation: A852/A852M – 03 (Reapproved 2007)

Standard Specification for
Quenched and Tempered Low-Alloy Structural Steel Plate with 70 ksi [485 MPa] Minimum Yield Strength to 4 in. [100 mm] Thick[1]

This standard is issued under the fixed designation A852/A852M; the number immediately following the designation indicates the year of original adoption or, in the case of revision, the year of last revision. A number in parentheses indicates the year of last reapproval. A superscript epsilon (ε) indicates an editorial change since the last revision or reapproval.

1. Scope*

1.1 This specification covers quenched and tempered high-strength low-alloy structural steel plates for welded, riveted, or bolted construction. It is intended primarily for use in welded bridges and buildings where savings in weight, added durability, and good notch toughness are important. The atmospheric corrosion resistance of this steel in most environments is substantially better than that of carbon structural steels with or without copper addition (see Note 1). When properly exposed to the atmosphere, this steel can be used bare (unpainted) for many applications. Welding technique is of fundamental importance, and it is presupposed that the welding procedure will be suitable for the steel and the intended service. This specification is limited to material up to 4 in. [100 mm], inclusive, in thickness. See Appendix X3 of Specification A6/A6M for information on weldability.

NOTE 1—For methods of estimating the atmospheric corrosion resistance of low-alloy steels, see Guide G101.

1.2 Plates produced under this specification are impact tested at a temperature not higher than 50°F [10°C].

1.3 The values stated in inch-pound units or SI units are to be regarded separately as standard. Within the text, the SI units are shown in brackets. The values stated in each item are not exact equivalents. Therefore, each system must be used independently of the other. Combining values from the two systems may result in nonconformance with this specification.

2. Referenced Documents

2.1 *ASTM Standards:*[2]

A6/A6M Specification for General Requirements for Rolled Structural Steel Bars, Plates, Shapes, and Sheet Piling

A370 Test Methods and Definitions for Mechanical Testing of Steel Products

A673/A673M Specification for Sampling Procedure for Impact Testing of Structural Steel

G101 Guide for Estimating the Atmospheric Corrosion Resistance of Low-Alloy Steels

3. General Requirements for Delivery

3.1 Material furnished under this specification shall conform to the requirements of the current edition of Specification A6/A6M, for the ordered material, unless a conflict exists in which case this specification shall prevail.

4. Materials and Manufacture

4.1 The steel shall be made to fine grain practice, and the fine austenitic grain size requirements of Specification A6/A6M shall be met.

5. Heat Treatment

5.1 The material shall be heat treated by the manufacturer by heating to a temperature that produces an austenitic structure, but not less than 1650°F [900°C], holding a sufficient time to attain uniform heat throughout the material, quenching in a

---

[1] This specification is under the jurisdiction of ASTM Committee A01 on Steel, Stainless Steel and Related Alloys and is the direct responsibility of Subcommittee A01.02 on Structural Steel for Bridges, Buildings, Rolling Stock and Ships.
Current edition approved Sept. 1, 2007. Published September 2007. Originally approved in 1985. Last previous edition approved in 2003 as A852/A852M – 03. DOI: 10.1520/A0852_A0852M-03R07.

[2] For referenced ASTM standards, visit the ASTM website, www.astm.org, or contact ASTM Customer Service at service@astm.org. For *Annual Book of ASTM Standards* volume information, refer to the standard's Document Summary page on the ASTM website.

A Summary of Changes section appears at the end of this standard.

Copyright © ASTM International, 100 Barr Harbor Drive, PO Box C700, West Conshohocken, PA 19428-2959, United States.

TABLE 1 Chemical Composition Requirements

Element	Composition, %
Carbon[A]	0.19 max
Manganese[A]	0.80–1.35
Phosphorus	0.035 max
Sulfur	0.04 max
Silicon	0.20–0.65
Nickel	0.50 max
Chromium	0.40–0.70
Copper	0.20–0.40
Vanadium	0.02–0.10

[A] For each reduction of 0.01 percentage point below the specified maximum for carbon, an increase of 0.06 percentage point above the specified maximum for manganese is permitted, up to a maximum of 1.50 %.

TABLE 2 Tensile Requirements[A]

Yield strength, min, ksi [MPa][B]	70 [485]
Tensile strength, range, ksi [MPa]	90–110 [620–760]
Elongation in 2 in. [50 mm], min, %[C,D]	19

[A] See Specimen Orientation under the Tension Tests section of Specification A6/A6M.
[B] Measured at 0.2 % offset or 0.5 % extension underload.
[C] For thicknesses of ¾ in. [19 mm] and under, measured on 1½-in. [40-mm] wide full thickness rectangular specimen as shown in Fig. 3 of Test Methods A370, the elongation is measured in a 2-in. [50 mm] gage length that includes the fracture and shows the greatest elongation.
[D] For plates wider than 24 in. [600 mm], the elongation requirement is reduced two percentage points. See Elongation Requirement Adjustments in the Tension Tests section of Specification A6/A6M.

suitable medium, and tempering at not less than 1100°F [593°C]. Heat treating temperatures shall be reported on the test certificates.

6. Chemical Composition

6.1 The heat analysis shall conform to the requirement prescribed in Table 1.

6.2 When product analysis is required, the steel shall conform on product analysis to the requirements prescribed in Table 1, subject to the product analysis tolerances in Specification A6/A6M.

6.3 The atmospheric corrosion-resistance index calculated on the basis of the heat analysis of the steel, as described in Guide G101–Predictive Method Based on the Data of Larabee and Coburn, shall be 6.0 or higher.

NOTE 2—The user is cautioned that the Guide G101 (Predictive Method Based on the Data of Larabee and Coburn) for calculation of an atmospheric corrosion-resistance index has only been verified for the composition limits stated in that guide.

6.4 When required, the manufacturer shall supply evidence of corrosion resistance satisfactory to the purchaser.

7. Tensile Test Requirements

7.1 *Tension Tests*—The material as represented by the test specimens shall conform to the requirements specified in Table 2.

7.2 *Number of Tests*—One tension test shall be taken from a corner of each plate as heat treated. Plates wider than 24 in. [610 mm] shall be tested in the transverse direction and are subject to the modifications for elongation contained in Footnote C of Table 2.

8. Impact Test Requirements

8.1 Longitudinal Charpy V-notch impact tests shall be made in accordance with Test Frequency H of Specification A673/A673M. By agreement, Charpy V-notch impact tests may be made in accordance with Test Frequency P of Specification A673/A673M.

8.2 The tests results shall meet an average minimum value of 20 ft·lbf [27 J] at 50°F [10°C]. By agreement, a test temperature lower than 50°F [10°C] or an energy level greater than 20 ft·lbf, or both, may be specified.

9. Retest

9.1 The manufacturer may reheat treat plates that fail to meet the mechanical property requirements of this specification. All mechanical property tests shall be repeated when material is reheat treated.

SUPPLEMENTARY REQUIREMENTS

Supplementary requirements shall not apply unless specified in the purchase order or contract. Standardized supplementary requirements for use at the option of the purchaser are listed in Specification A6/A6M. Those that are considered suitable for use with this specification are listed by title.

S1. Vacuum Treatment,
S2. Product Analysis,
S3. Simulated Post-Weld Heat Treatment of Mechanical Test Coupons,
S6. Drop Weight Test (for Material 0.625 in. [16 mm] and Over in Thickness), and
S8. Ultrasonic Examination.

A852/A852M – 03 (2007)

SUMMARY OF CHANGES

Committee A01 has identified the location of the following changes to this standard since A852/A852M – 01 that may impact the use of this standard.

(1) Table 1 has been revised.

ASTM International takes no position respecting the validity of any patent rights asserted in connection with any item mentioned in this standard. Users of this standard are expressly advised that determination of the validity of any such patent rights, and the risk of infringement of such rights, are entirely their own responsibility.

This standard is subject to revision at any time by the responsible technical committee and must be reviewed every five years and if not revised, either reapproved or withdrawn. Your comments are invited either for revision of this standard or for additional standards and should be addressed to ASTM International Headquarters. Your comments will receive careful consideration at a meeting of the responsible technical committee, which you may attend. If you feel that your comments have not received a fair hearing you should make your views known to the ASTM Committee on Standards, at the address shown below.

This standard is copyrighted by ASTM International, 100 Barr Harbor Drive, PO Box C700, West Conshohocken, PA 19428-2959, United States. Individual reprints (single or multiple copies) of this standard may be obtained by contacting ASTM at the above address or at 610-832-9585 (phone), 610-832-9555 (fax), or service@astm.org (e-mail); or through the ASTM website (www.astm.org).

Designation: A871/A871M − 03 (Reapproved 2007)

Standard Specification for
High-Strength Low-Alloy Structural Steel Plate With Atmospheric Corrosion Resistance[1]

This standard is issued under the fixed designation A871/A871M; the number immediately following the designation indicates the year of original adoption or, in the case of revision, the year of last revision. A number in parentheses indicates the year of last reapproval. A superscript epsilon (ε) indicates an editorial change since the last revision or reapproval.

1. Scope

1.1 This specification covers high-strength low-alloy steel plate intended for use in tubular structures and poles or in other suitable applications. Two grades, 60 and 65, may be provided as-rolled, normalized or quenched and tempered as required to meet the specified mechanical requirements.

1.2 The atmospheric corrosion resistance of this steel in most environments is substantially better than that of carbon structural steels with or without copper addition (see Note 1). When properly exposed to the atmosphere, this steel can be used bare (unpainted) for many applications.

Note 1—For methods of estimating the atmospheric corrosion resistance of low-alloy steels, see Guide G101.

1.3 When the steel is to be welded, it is presupposed that welding procedures suitable for the grade of steel and intended use or service will be utilized. See Appendix X3 of Specification A6/A6M for information on weldability.

1.4 Supplementary requirements in accordance with Specification A6/A6M are available, but shall apply only when specified by the purchaser at time of ordering.

1.5 The values stated in either inch-pound units or SI units are to be regarded as standard. Within the text, the SI units are shown in brackets. The values stated in each system are not exact equivalents; therefore, each system must be used independently of the other. Combining values from the two systems may result in nonconformance with the specification.

2. Referenced Documents

2.1 *ASTM Standards:*[2]
A6/A6M Specification for General Requirements for Rolled Structural Steel Bars, Plates, Shapes, and Sheet Piling
A370 Test Methods and Definitions for Mechanical Testing of Steel Products
A673/A673M Specification for Sampling Procedure for Impact Testing of Structural Steel
G101 Guide for Estimating the Atmospheric Corrosion Resistance of Low-Alloy Steels

3. General Requirements for Delivery

3.1 Material furnished under this specification shall conform to the requirements of the current edition of Specification A6/A6M, for the ordered material, unless a conflict exists in which case this specification shall prevail.

4. Materials and Manufacture

4.1 The steel shall be made to fine grain practice.

5. Heat Treatment

5.1 Grade 65 in thicknesses of 3/16 to 3/4 in. [5 to 20 mm] and Grade 60 in thicknesses of 3/16 to 1 3/8 in. [5 to 35 mm] are normally furnished in the as-rolled condition. The manufacturer has the option to heat treat this material to meet the mechanical requirements of Section 7. Quenched and tempered material shall be heat treated by heating to not less than 1650°F [900°C], holding a sufficient time to attain uniform heat throughout the material, quenching in a suitable medium, and tempering at not less than 1100°F [595°C]. Heat treating temperatures shall be reported on the test certificates.

5.2 The maximum thickness of plates is limited only by the capacity of the composition to meet the specified mechanical requirements. The individual manufacturer shall be contacted to determine the actual maximum thickness for each grade and heat treatment method.

6. Chemical Requirements

6.1 The heat analysis shall conform to the chemical requirements of Table 1.

6.2 The steel shall conform on product analysis to the chemical requirements of Table 1, subject to the product analysis tolerances in Specification A6/A6M.

---

[1] This specification is under the jurisdiction of ASTM Committee A01 on Steel, Stainless Steel and Related Alloys and is the direct responsibility of Subcommittee A01.02 on Structural Steel for Bridges, Buildings, Rolling Stock and Ships.
Current edition approved Sept. 1, 2007. Published September 2007. Originally approved in 1987. Last previous edition approved in 2003 as A871/A871M – 03. DOI: 10.1520/A0871_A0871M-03R07.

[2] For referenced ASTM standards, visit the ASTM website, www.astm.org, or contact ASTM Customer Service at service@astm.org. For *Annual Book of ASTM Standards* volume information, refer to the standard's Document Summary page on the ASTM website.

Copyright © ASTM International, 100 Barr Harbor Drive, PO Box C700, West Conshohocken, PA 19428-2959, United States.

TABLE 1 Chemical Requirements (Heat Analysis)

Element	Composition, %			
	Type I	Type II	Type III	Type IV
Carbon[A]	0.19 max	0.20 max	0.15 max	0.17 max
Manganese[A]	0.80–1.35	0.75–1.35	0.80–1.35	0.50–1.20
Phosphorus	0.04 max	0.04 max	0.04 max	0.04 max
Sulfur	0.05 max	0.05 max	0.05 max	0.05 max
Silicon	0.30–0.65	0.15–0.50	0.15–0.40	0.25–0.50
Nickel	0.40 max	0.50 max	0.25–0.50	0.40 max
Chromium	0.40–0.70	0.40–0.70	0.30–0.50	0.40–0.70
Molybdenum	...	...	...	0.10 max
Copper	0.25–0.40	0.20–0.40	0.20–0.50	0.30–0.50
Vanadium	0.02–0.10	0.01–0.10	0.01–0.10	...
Columbium	...	...	...	0.005–0.05[B]

[A] For each reduction of 0.01 percentage point below the specified maximum for carbon, an increase of 0.06 percentage point above the specified maximum for manganese is permitted, up to a maximum of 1.50 %.

[B] For plates under ½ in. [13 mm] in thickness, the minimum columbium is waived.

6.3 The atmospheric corrosion-resistance index, calculated on the basis of the heat analysis for the steel, as described in Guide G101—Predictive Method Based on the Data of Larabee and Coburn, shall be 6.0 or higher.

NOTE 2—The user is cautioned that the Guide G101 (Predictive Method Based on the Data of Larabee and Coburn) for calculation of an atmospheric corrosion-resistance index has only been verified for the composition limits stated in that guide.

6.4 When required, the manufacturer shall supply evidence of corrosion resistance satisfactory to the purchaser.

7. Mechanical Requirements

7.1 *Tension Tests*:

7.1.1 The steel as represented by the tension test specimens shall conform to the tensile requirements of Table 2.

7.1.2 For adjustments in Table 2 percentage elongation requirements for material thickness under 0.312 in. [8 mm] and over 3.5 in. [90 mm], see Specification A6/A6M.

TABLE 2 Tensile Requirements[A]

Grade	Yield Strength[B] min. ksi [MPa]	Tensile Strength min. ksi [MPa]	Minimum Elongation, %[C]	
			In 8 in. [200 mm]	In 2 in. [50 mm]
60	60 [415]	75 [520]	16	18
65	65 [450]	80 [550]	15	17

[A] For plates wider than 24 in. [600 mm], the test specimen is taken in the transverse direction. See 11.2 of Specification A6/A6M.

[B] Measured at 0.2 % offset or 0.5 % extension under load as described in Section 13 on yield strength of Test Methods A370.

[C] For plates wider than 24 in. [600 mm], the elongation requirement is reduced three percentage points.

7.2 *Charpy V-Notch Impact Tests*:

7.2.1 The steel, as represented by the Charpy V-Notch test, shall conform to the impact test requirements of Table 3.

7.2.2 If more stringent impact requirements are required, they shall be negotiated between the purchaser and the manufacturer.

8. Test Specimens and Number of Tests

8.1 The purchaser shall indicate on the purchase order the frequency of Charpy V-Notch Impact Testing, as provided for in Specification A673/A673M. If the purchase order does not specify the frequency, "H" testing frequency shall be supplied.

9. Keywords

9.1 as-rolled; atmospheric corrosion resistance; high-strength; low-alloy; normalized; plate; poles; quenched; steel; structural steel; tempered; tubular structures

TABLE 3 Charpy V-Notch Impact Test Requirements

Plate Thickness in [mm]	Absorbed Energy ft-lb [J]	Temperature °F [°C]
Up to ½ [12] incl	15 [20]	0 [−18]
Over ½ [12]	15 [20]	−20 [−29]

SUPPLEMENTARY REQUIREMENTS

Supplementary requirements shall not apply unless specified in the purchase order or contract. Standardized supplementary requirements for use at the option of the purchaser are listed in Specification A6/A6M.

ASTM International takes no position respecting the validity of any patent rights asserted in connection with any item mentioned in this standard. Users of this standard are expressly advised that determination of the validity of any such patent rights, and the risk of infringement of such rights, are entirely their own responsibility.

This standard is subject to revision at any time by the responsible technical committee and must be reviewed every five years and if not revised, either reapproved or withdrawn. Your comments are invited either for revision of this standard or for additional standards and should be addressed to ASTM International Headquarters. Your comments will receive careful consideration at a meeting of the responsible technical committee, which you may attend. If you feel that your comments have not received a fair hearing you should make your views known to the ASTM Committee on Standards, at the address shown below.

This standard is copyrighted by ASTM International, 100 Barr Harbor Drive, PO Box C700, West Conshohocken, PA 19428-2959, United States. Individual reprints (single or multiple copies) of this standard may be obtained by contacting ASTM at the above address or at 610-832-9585 (phone), 610-832-9555 (fax), or service@astm.org (e-mail); or through the ASTM website (www.astm.org).

Designation: A913/A913M − 07

Standard Specification for
High-Strength Low-Alloy Steel Shapes of Structural Quality, Produced by Quenching and Self-Tempering Process (QST)[1]

This standard is issued under the fixed designation A913/A913M; the number immediately following the designation indicates the year of original adoption or, in the case of revision, the year of last revision. A number in parentheses indicates the year of last reapproval. A superscript epsilon (ε) indicates an editorial change since the last revision or reapproval.

1. Scope*

1.1 This specification covers high-strength low-alloy structural steel shapes in Grades 50 [345], 60 [415], 65 [450] and 70 [485], produced by the quenching and self-tempering process (QST).[2] The shapes are intended for riveted, bolted or welded construction of bridges, buildings and other structures.

1.2 The QST process consists of in line heat treatment and cooling rate controls which result in mechanical properties in the finished condition that are equivalent to those attained using heat treating processes which entail reheating after rolling. A description of the QST process is given in Appendix X1.

1.3 Due to the inherent characteristics of the QST process, the shapes shall not be formed and post weld heat treated at temperatures exceeding 1100°F [600°C].

1.4 When the steel is to be welded, it is presupposed that a welding procedure suitable for the grade of steel and intended use or service will be utilized. See Appendix X3 of Specification A6/A6M for information on weldability.

1.5 The values stated in either inch-pound units or SI units are to be regarded separately as standard. Within the text, the SI units are shown in brackets. The values stated in each system are not exact equivalents; therefore, each system must be used independently of the other. Combining values from the two systems may result in nonconformance with this specification.

2. Referenced Documents

2.1 *ASTM Standards:*[3]
A6/A6M Specification for General Requirements for Rolled Structural Steel Bars, Plates, Shapes, and Sheet Piling
A673/A673M Specification for Sampling Procedure for Impact Testing of Structural Steel
A898/A898M Specification for Straight Beam Ultrasonic Examination of Rolled Steel Structural Shapes

3. General Requirements for Delivery

3.1 Material furnished under this specification shall conform to the applicable requirements of the current edition of Specification A6/A6M.

4. Materials and Manufacture

4.1 The shapes shall be produced by the quenching and self-tempering process (QST). Self-tempering temperature shall be a minimum of 1100°F [600°C] and the self-tempering temperature for the material represented shall be reported on the mill test report. See Appendix X1 for Process Description.

4.2 For grades 60 [415], 65 [450], and 70 [485], the requirements for fine austenitic grain size in Specification A6/A6M shall be met.

5. Chemical Composition

5.1 The chemical analysis of the heat shall conform to the requirements prescribed in Table 1.

5.2 The steel shall conform on product analysis to the requirements prescribed in Table 1 subject to the product analysis tolerances in Specification A6/A6M.

6. Mechanical Properties

6.1 *Tensile Properties*—The material as represented by the test specimens shall conform to the tensile properties given in Table 2.

---

[1] This specification is under the jurisdiction of ASTM Committee A01 on Steel, Stainless Steel and Related Alloys and is the direct responsibility of Subcommittee A01.02 on Structural Steel for Bridges, Buildings, Rolling Stock and Ships.
Current edition approved Nov. 1, 2007. Published November 2007. Originally approved in 1993. Last previous edition approved in 2004 as A913/A913M – 04. DOI: 10.1520/A0913_A0913M-07.

[2] The quenching and self-tempering process (QST) and the used apparatus are covered by patents held by the Centre de Recherches Métallurgiques (CRM)—Rue Ernest Solvay, 11, B4000, Liège (Belgium). Interested parties are invited to submit information regarding the identification of acceptable alternatives to these patented items to the Committee on Standards, ASTM Headquarters, 100 Barr Harbor Drive, West Conshohocken, PA 19428-2959. Comments will receive careful consideration at the meeting of the responsible technical committee, which any interested party may attend.

[3] For referenced ASTM standards, visit the ASTM website, www.astm.org, or contact ASTM Customer Service at service@astm.org. For *Annual Book of ASTM Standards* volume information, refer to the standard's Document Summary page on the ASTM website.

*A Summary of Changes section appears at the end of this standard.

Copyright © ASTM International, 100 Barr Harbor Drive, PO Box C700, West Conshohocken, PA 19428-2959, United States.

TABLE 1 Chemical Requirements (Heat Analysis)

Element	Maximum content in %			
	Grade 50 [345]	Grade 60 [415]	Grade 65 [450]	Grade 70 [485]
Carbon	0.12	0.14	0.16	0.16
Manganese	1.60	1.60	1.60	1.60
Phosphorus	0.040	0.030	0.030	0.040
Sulfur	0.030	0.030	0.030	0.030
Silicon	0.40	0.40	0.40	0.40
Copper	0.45	0.35	0.35	0.45
Nickel	0.25	0.25	0.25	0.25
Chromium	0.25	0.25	0.25	0.25
Molybdenum	0.07	0.07	0.07	0.07
Columbium	0.05	0.04	0.05	0.05
Vanadium	0.06	0.06	0.06	0.09

TABLE 2 Tensile Requirements

Grade	Yield Point, min.		Tensile Strength, min.		Elongation, min	
	ksi	[MPa]	ksi	[MPa]	8 in. [200 mm], %	2 in. [50 mm], %
50 [345]	50	[345]	65	[450]	18	21
60 [415]	60	[415]	75	[520]	16	18
65 [450]	65	[450]	80	[550]	15	17
70 [485]	70	[485]	90	[620]	14	16

6.2 Charpy V-notch tests shall be made in accordance with Specification A673/A673M, Frequency II:

6.2.1 The test results of full-size specimens shall meet an average value of 40 ft-lbf [54 J] at 70°F [21°C].

6.2.1.1 Test reports for every heat supplied are required.

6.2.2 Charpy V-notch test requirements exceeding the value specified in 6.2.1 or lower test temperatures are subject to agreement between the purchaser and the producer.

7. Maximum Carbon Equivalent Requirement

7.1 The carbon equivalent on heat analysis shall not exceed the limits listed in this section. The chemical analysis (heat analysis) of the elements that appear in the carbon equivalent formula and the actual carbon equivalent shall be reported.

Carbon equivalent limits
Grade 50 [345]: 0.38 %
Grade 60 [415]: 0.40 %
Grade 65 [450]: 0.43 %
Grade 70 [485]: 0.45 %

7.2 Calculate the carbon equivalent using the following equation:

$$CE = C + Mn/6 + (Cr + Mo + V)/5 + (Cu + Ni)/15$$

8. Keywords

8.1 high-strength low-alloy steel; QST; quenching and self-tempering process; steel shapes; structural shapes; structural steel

---

SUPPLEMENTARY REQUIREMENTS

Supplementary requirements shall not apply unless specified in the purchase order or contract. Standardized supplementary requirements for use at the option of the purchaser are listed in Specification A6/A6M. Those that are considered suitable for use with this specification are listed by title:

S1. Vacuum Treatment.
S2. Product Analysis.
S3. Simulated Post-Weld Heat Treatment of Mechanical Test Coupons.
S5. Charpy V-Notch Impact Test.
S18. Maximum Tensile Strength.
S30. Charpy V-Notch Impact Test for Structural Shapes: Alternate Core Location.

ADDITIONAL SUPPLEMENTARY REQUIREMENTS

In addition, the following special supplementary requirements are also suitable for use with this specification:

S4. *Additional Tension Test:*
S4.1 One tension test shall be made per ingot or per bloom. The results obtained and the actual self-tempering temperature for the ingot or bloom represented shall be reported on the mill test report when such tests are required by the order.

S8. *Ultrasonic Examination:*
S8.1 Ultrasonic Examination in accordance with Specification A898/A898M.

S32. *Single Heat Bundles:*
S32.1 Bundles containing shapes or bars shall be from a single heat of steel.

S75. *Maximum Yield Point to Tensile Strength Ratio—Grade 50 [345]:*
S75.1 The maximum yield point shall be 65 ksi. [450].
S75.2 The maximum yield to tensile strength ratio shall be 0.85.

S77. *Reduced Sulfur—Grade 65 [450]:*
S77.1 The Grade 65 [450] shall be furnished with a maximum sulfur of 0.010 %. This may be desirable in material subjected to high through-thickness stresses.

APPENDIX

(Nonmandatory Information)

X1. QUENCHING AND SELF-TEMPERING PROCESS (QST)

X1.1 *Introduction*—The quenching and self-tempering process, commonly referred to as "QST," has evolved from the "thermo-mechanical control processes" (TMCP) that have been known and used for a number of years. QST, which is a variation of TMCP, produces fine-grained steel by a combination of chemical composition and integrated controls of manufacturing processes from ingot or bloom reheating to in-line interrupted quenching and self-tempering, thereby achieving the specified mechanical properties in the required product thicknesses.

X1.2 *Outline of QST*—Given in Fig. X1.1.

X1.2.1 Quenching and self-tempering (QST)-steels of fine grain size are manufactured by producing tempered martensite and varying the pearlite or bainite, or both, volume fraction through time quenching (interrupted quenching in which the duration of holding in the quenching medium (water) is controlled) in the temperature region above Ar3 (the temperature at which austenite begins to transform to ferrite during cooling) to a minimum self-tempering temperature (STT, that is, the maximum surface temperature after quenching) of 1100°F [600°C] in order to meet the specified requirements. Time quenching can be performed after conventional rolling processes (AR) and (CR) or after thermomechanical controlled rolling (TMCR).

X1.2.2 The selection of the rolling process to be used is made by the producer depending upon the chemical composition, the product thickness, and the required properties.

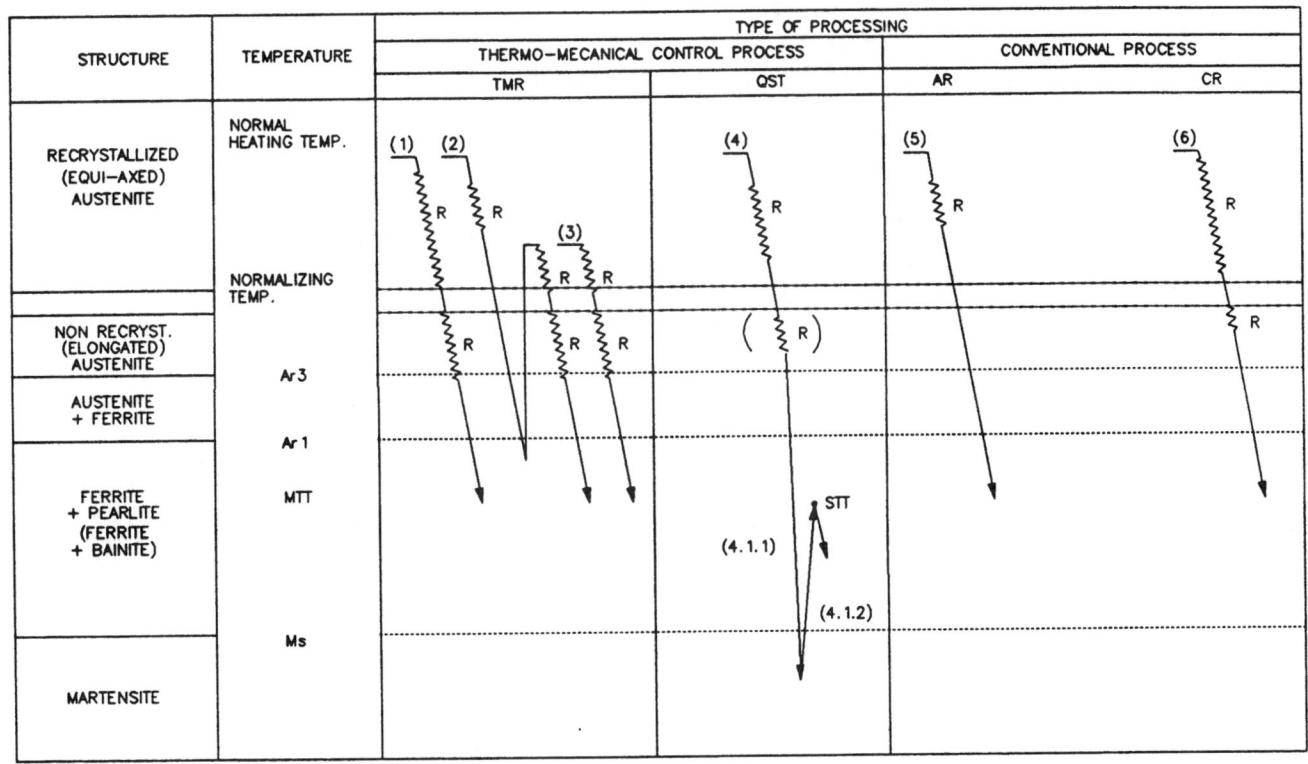

FIG. X1.1 Schematic Diagrams of Thermo-Mechanical Control and Conventional Process

A913/A913M – 07

SUMMARY OF CHANGES

Committee A01 has identified the location of selected changes to this standard since the last issue (A913/A913M–04) that may impact the use of this standard.

(1) Added 6.2.1.1.

ASTM International takes no position respecting the validity of any patent rights asserted in connection with any item mentioned in this standard. Users of this standard are expressly advised that determination of the validity of any such patent rights, and the risk of infringement of such rights, are entirely their own responsibility.

This standard is subject to revision at any time by the responsible technical committee and must be reviewed every five years and if not revised, either reapproved or withdrawn. Your comments are invited either for revision of this standard or for additional standards and should be addressed to ASTM International Headquarters. Your comments will receive careful consideration at a meeting of the responsible technical committee, which you may attend. If you feel that your comments have not received a fair hearing you should make your views known to the ASTM Committee on Standards, at the address shown below.

This standard is copyrighted by ASTM International, 100 Barr Harbor Drive, PO Box C700, West Conshohocken, PA 19428-2959, United States. Individual reprints (single or multiple copies) of this standard may be obtained by contacting ASTM at the above address or at 610-832-9585 (phone), 610-832-9555 (fax), or service@astm.org (e-mail); or through the ASTM website (www.astm.org).

Designation: A992/A992M – 06a

Standard Specification for
Structural Steel Shapes[1]

This standard is issued under the fixed designation A992/A992M; the number immediately following the designation indicates the year of original adoption or, in the case of revision, the year of last revision. A number in parentheses indicates the year of last reapproval. A superscript epsilon (ε) indicates an editorial change since the last revision or reapproval.

1. Scope*

1.1 This specification covers rolled steel structural shapes for use in building framing or bridges, or for general structural purposes.

1.2 Supplementary requirements are provided for use where additional testing or additional restrictions are required by the purchaser. Such requirements apply only when specified in the purchase order.

1.3 When the steel is to be welded, a welding procedure suitable for the grade of steel and intended use or service is to be utilized. See Appendix X3 of Specification A6/A6M for information on weldability.

1.4 The values stated in either inch-pound units or SI units are to be regarded separately as standard. Within the text, the SI units are shown in brackets. The values stated in each system are not exact equivalents; therefore, each system is to be used independently of the other without combining values in any way.

1.5 The text of this specification contains notes or footnotes, or both, that provide explanatory material; such notes and footnotes, excluding those in tables and figures, do not contain any mandatory requirements.

2. Referenced Documents

2.1 *ASTM Standards:*[2]
A6/A6M Specification for General Requirements for Rolled Structural Steel Bars, Plates, Shapes, and Sheet Piling
A370 Test Methods and Definitions for Mechanical Testing of Steel Products
A673/A673M Specification for Sampling Procedure for Impact Testing of Structural Steel
E112 Test Methods for Determining Average Grain Size

3. General Requirements For Delivery

3.1 Material furnished under this specification shall conform to the requirements of the current edition of Specification A6/A6M for the ordered material, unless a conflict exists, in which case this specification shall prevail.

4. Materials and Manufacture

4.1 The steel shall be killed, and such shall be affirmed in the test report by a statement of *killed steel*, a value of 0.10 % or more for the silicon content, or a value of 0.015 % or more for the total aluminum content.

4.2 The steelmaking practice used shall be one that produces steel having a nitrogen content not greater than 0.015 % and includes the addition of one or more nitrogen-binding elements, or one that produces steel having a nitrogen content not greater than 0.012 % (with or without the addition of nitrogen-binding elements). The nitrogen content need not be reported, regardless of which steelmaking practice was used.

5. Chemical Composition

5.1 The heat analysis shall conform to the requirements in Table 1.

5.2 In addition to the elements listed in Table 1, test reports shall include, for information, the chemical analysis for tin. Where the amount of tin is less than 0.02 %, it shall be permissible for the analysis to be reported as "< 0.02 %."

5.3 The steel shall conform on product analysis to the requirements prescribed in Table 1, subject to the product analysis tolerances in Specification A6/A6M.

5.4 The maximum permissible carbon equivalent value shall be 0.47 % for shapes with flange thickness over 2 in. [50 mm], and 0.45 % in other shapes. The carbon equivalent value shall be based on heat analysis. The required chemical analysis as well as the carbon equivalent shall be reported. The carbon equivalent shall be calculated using the following formula:

*A Summary of Changes section appears at the end of this standard.

---

[1] This specification is under the jurisdiction of ASTM Committee A01 on Steel, Stainless Steel, and Related Alloys and is the direct responsibility of Subcommittee A01.02 on Structural Steel for Bridges, Buildings, Rolling Stock, and Ships.
Current edition approved July 1, 2006. Published July 2006. Originally approved in 1998. Last previous edition approved in 2006 as A992/A992M– 06. DOI: 10.1520/A0992_A0992M-06A.

[2] For referenced ASTM standards, visit the ASTM website, www.astm.org, or contact ASTM Customer Service at service@astm.org. For *Annual Book of ASTM Standards* volume information, refer to the standard's Document Summary page on the ASTM website.

TABLE 1 Chemical Requirements (Heat Analysis)

Element	Composition, %
Carbon, max	0.23
Manganese,	0.50 to 1.60[A]
Silicon, max	0.40
Vanadium, max	0.15[B]
Columbium, max	0.05[B]
Phosphorus, max	0.035
Sulfur, max	0.045
Copper, max	0.60
Nickel, max	0.45
Chromium, max	0.35
Molybdenum, max	0.15

[A] Provided that the ratio of manganese to sulfur is not less than 20 to 1, the minimum limit for manganese for shapes with flange or leg thickness not exceeding 1 in. [25 mm] shall be 0.30 %.

[B] The sum of columbium and vanadium shall not exceed 0.15 %.

$$CE = C + (Mn)/6 + (Cr + Mo + V)/5 + (Ni + Cu)/15 \quad (1)$$

6. Tensile Requirements

6.1 The material as represented by the test specimens shall conform to the requirements for tensile properties prescribed in Table 2.

TABLE 2 Tensile Requirements

Tensile strength, min ksi [MPa]	65 [450]
Yield point, ksi [MPa]	50 to 65 [345 to 450][A]
Yield to tensile ratio, max	0.85[B]
Elongation in 8 in. [200 mm], min, %[C]	18
Elongation in 2 in. [50 mm], min, %[C]	21

[A] A maximum yield strength of 70 ksi [480 MPa] is permitted for structural shapes that are required to be tested from the web location.

[B] A maximum ratio of 0.87 is permitted for structural shapes that are tested from the web location.

[C] See elongation requirement adjustments under the Tension Tests section of Specification A6/A6M.

7. Keywords

7.1 bridges; building framing; shapes; structural steel; welded construction

SUPPLEMENTARY REQUIREMENTS

Supplementary requirements shall apply only if specified in the purchase order or contract.

Standardized supplementary requirements for use at the option of the purchaser are listed in Specification A6/A6M. Those that are considered suitable for use with this specification are listed by title:

S1. Vacuum Treatment
S2. Product Analysis
S5. Charpy V-Notch Impact Test
S8. Ultrasonic Examination
S30. Charpy V-Notch Impact Test for Structural Shapes: Alternate Core Location

In addition, the following optional supplementary requirements are suitable for use with this specification.

S32. Single Heat Bundles.

S32.1 Bundles containing shapes or bars shall be from a single heat of steel.

S79. Maximum Tensile Strength

S79.1 The maximum tensile strength shall be 90 ksi [620 MPa].

SUMMARY OF CHANGES

Committee A01 has identified the location of selected changes to this standard since the last issue (A992/A992M– 06) that may impact the use of this standard. (Approved July 1, 2006.)

(1) The value for maximum manganese in Table 1 was changed from 1.50 % to 1.60 %.

Committee A01 has identified the location of selected changes to this standard since the last issue (A992/A992M– 04a) that may impact the use of this standard. (Approved March 15, 2006.)

(1) Table 2 was modified to allow a maximum yield-to-tensile ratio of 0.87 for structural shapes that are tested from the web location.

(2) Table 2 was modified to change the maximum yield strength from 65 ksi to 70 ksi [450 MPa to 480 MPa] for sections tensile tested in the web.

Committee A01 has identified the location of selected changes to this standard since the last issue (A992/A992M– 04) that may impact the use of this standard. (Approved Sept. 1, 2004.)

(1) Numeric value of vanadium maximum increased from 0.11 to 0.15 in Table 1.

ASTM International takes no position respecting the validity of any patent rights asserted in connection with any item mentioned in this standard. Users of this standard are expressly advised that determination of the validity of any such patent rights, and the risk of infringement of such rights, are entirely their own responsibility.

This standard is subject to revision at any time by the responsible technical committee and must be reviewed every five years and if not revised, either reapproved or withdrawn. Your comments are invited either for revision of this standard or for additional standards and should be addressed to ASTM International Headquarters. Your comments will receive careful consideration at a meeting of the responsible technical committee, which you may attend. If you feel that your comments have not received a fair hearing you should make your views known to the ASTM Committee on Standards, at the address shown below.

This standard is copyrighted by ASTM International, 100 Barr Harbor Drive, PO Box C700, West Conshohocken, PA 19428-2959, United States. Individual reprints (single or multiple copies) of this standard may be obtained by contacting ASTM at the above address or at 610-832-9585 (phone), 610-832-9555 (fax), or service@astm.org (e-mail); or through the ASTM website (www.astm.org).

Designation: A1008/A1008M – 09a

Standard Specification for
Steel, Sheet, Cold-Rolled, Carbon, Structural, High-Strength Low-Alloy, High-Strength Low-Alloy with Improved Formability, Solution Hardened, and Bake Hardenable[1]

This standard is issued under the fixed designation A1008/A1008M; the number immediately following the designation indicates the year of original adoption or, in the case of revision, the year of last revision. A number in parentheses indicates the year of last reapproval. A superscript epsilon (ε) indicates an editorial change since the last revision or reapproval.

1. Scope*

1.1 This specification covers cold-rolled, carbon, structural, high-strength low-alloy, high-strength low-alloy with improved formability, solution hardened, and bake hardenable steel sheet, in coils and cut lengths.

1.2 Cold rolled steel sheet is available in the designations as listed in 4.1.

1.3 This specification does not apply to steel strip as described in Specification A109/A109M.

1.4 The values stated in either SI units or inch-pound units are to be regarded separately as standard. The values stated in each system may not be exact equivalents; therefore, each system shall be used independently of the other. Combining values from the two systems may result in non-conformance with the standard.

2. Referenced Documents

2.1 *ASTM Standards:*[2]
A109/A109M Specification for Steel, Strip, Carbon (0.25 Maximum Percent), Cold-Rolled
A366/A366M Specification for Commercial Steel (CS) Sheet, Carbon (0.15 Maximum Percent) Cold-Rolled[3]
A370 Test Methods and Definitions for Mechanical Testing of Steel Products
A568/A568M Specification for Steel, Sheet, Carbon, Structural, and High-Strength, Low-Alloy, Hot-Rolled and Cold-Rolled, General Requirements for
A620/A620M Specification for Drawing Steel (DS), Sheet, Carbon, Cold-Rolled[3]
A941 Terminology Relating to Steel, Stainless Steel, Related Alloys, and Ferroalloys
E18 Test Methods for Rockwell Hardness of Metallic Materials
E517 Test Method for Plastic Strain Ratio *r* for Sheet Metal
E646 Test Method for Tensile Strain-Hardening Exponents (*n* -Values) of Metallic Sheet Materials

3. Terminology

3.1 *Definitions of Terms Specific to This Standard:*

3.1.1 For definitions of other terms used in this specification, refer to Terminology A941.

3.1.2 *aging*—loss of ductility with an increase in hardness, yield strength, and tensile strength that occurs when steel that has been slightly cold worked (such as by temper rolling) is stored for some time.

3.1.2.1 *Discussion*—Aging increases the tendency of a steel to exhibit stretcher strains and fluting.

3.1.3 *bake hardenable steel*—steel in which significant aging is realized when moderate heat treatment, such as that used for paint baking, follows straining or cold working.

3.1.4 *inclusion control, n*—the process of reducing the volume fraction of inclusions or modifying the shape of inclusions to improve formability, weldability, and machinability.

3.1.4.1 *Discussion*—Inclusions, especially those elongated during the rolling process, create the conditions for initiating and/or propagating cracks when the material is stretched or bent during the manufacture of a part. The adverse effects of inclusions are minimized by reducing the content of inclusions in the steel and/or by altering the shape of inclusions through

---

[1] This specification is under the jurisdiction of ASTM Committee A01 on Steel, Stainless Steel and Related Alloys and is the direct responsibility of Subcommittee A01.19 on Steel Sheet and Strip.
Current edition approved Nov. 1, 2009. Published December 2009. Originally approved in 2000. Last previous edition approved in 2009 as A1008/A1008M – 09. DOI: 10.1520/A1008_A1008M-09A.
[2] For referenced ASTM standards, visit the ASTM website, www.astm.org, or contact ASTM Customer Service at service@astm.org. For *Annual Book of ASTM Standards* volume information, refer to the standard's Document Summary page on the ASTM website.
[3] Withdrawn. The last approved version of this historical standard is referenced on www.astm.org.

*A Summary of Changes section appears at the end of this standard.

Copyright © ASTM International, 100 Barr Harbor Drive, PO Box C700, West Conshohocken, PA 19428-2959, United States.

the use of additions during the steelmaking process that change the elongated shape of the inclusions to less harmful small, well dispersed globular inclusions.

3.1.5 *solid-solution hardened steel or solution hardened steel*—steel strengthened through additions of elements, such as Mn, P, or Si, that can be dissolved within the crystalline structure of steels.

3.1.5.1 *Discussion*—Alloying elements that form a solid-solution with iron provide strengthening as a result of local distortions in atomic arrangements, which arise as a result of the mismatch between the atomic sizes of such elements and that of iron.

3.1.6 *stabilization*—addition of one or more nitride- or carbide-forming elements, or both, such as titanium and columbium, to control the level of the interstitial elements of carbon and nitrogen in the steel.

3.1.6.1 *Discussion*—Stabilizing improves formability and increases resistance to aging.

3.1.7 *vacuum degassing*—process of refining liquid steel in which the liquid is exposed to a vacuum as part of a special technique for removing impurities or for decarburizing the steel.

4. Classification

4.1 Cold-rolled steel sheet is available in the following designations:

4.1.1 Commercial Steel (CS Types A, B, and C),

4.1.2 Drawing Steel (DS Types A and B),

NOTE 1—CS Type B and DS Type B describe the most common product previously included, respectively, in Specifications A366/A366M and A620/A620M.

4.1.3 Deep Drawing Steel (DDS),

4.1.4 Extra Deep Drawing Steel (EDDS),

4.1.5 Structural Steel (SS grades 25[170], 30[205], 33[230] Types 1 and 2, 40[275] Types 1 and 2, 50[340], 60[410], 70[480], and 80[550]).

4.1.6 High-Strength Low-Alloy Steel (HSLAS, in classes 1 and 2, in grades 45[310], 50[340]. 55[380], 60[410], 65[450], and 70[480] in Classes 1 and 2), and

4.1.7 High-Strength Low-Alloy Steel with Improved Formability (HSLAS-F grades 50[340], 60[410], 70[480], and 80[550]).

4.1.7.1 HSLAS-F steel has improved formability when compared to HSLAS. The steel is fully deoxidized, made to fine grain practice and includes microalloying elements such as columbium, vanadium, zirconium, etc. The steel shall be treated to achieve inclusion control.

4.1.8 Solution hardened steel (SHS), and

4.1.9 Bake hardenable steel (BHS).

4.2 When required for HSLAS and HSLAS-F steels, limitations on the use of one or more of the microalloy elements shall be specified on the order.

4.3 Cold-rolled steel sheet is supplied for either exposed or unexposed applications. Within the latter category, cold-rolled sheet is specified either "temper rolled" or "annealed last." For details on processing, attributes and limitations, and inspection standards, refer to Specification A568/A568M.

5. Ordering Information

5.1 It is the purchaser's responsibility to specify in the purchase order all ordering information necessary to describe the required material. Examples of such information include, but are not limited to, the following:

5.1.1 ASTM specification number and year of issue;

5.1.2 Name of material and designation (cold-rolled steel sheet) (include grade, type, and class, as appropriate, for CS, DS, DDS, EDDS, SS, HSLAS, HSLAS-F, SHS, or BHS) (see 4.1);

5.1.2.1 When a type is not specified for CS or DS, Type B will be furnished (see 4.1);

5.1.2.2 When a class is not specified for HSLAS, Class 1 will be furnished (see 4.1);

5.1.2.3 When a type is not specified for SS 33 [230] and SS 40 [275], Type 1 will be furnished (see 4.1);

5.1.3 Classification (either exposed, unexposed, temper rolled, or annealed last) (see 4.3);

5.1.4 Finish (see 9.1);

5.1.5 Oiled or not oiled, as required (see 9.2);

5.1.6 Dimensions (thickness, width, and whether cut lengths or coils);

NOTE 2—Not all producers are capable of meeting all the limitations of the thickness tolerance tables in Specification A568/A568M. The purchaser should contact the producer regarding possible limitations prior to placing an order.

5.1.7 Coil size (must include inside diameter, outside diameter, and maximum weight);

5.1.8 Copper bearing steel (if required);

5.1.9 Quantity;

5.1.10 Application (part identification and description);

5.1.11 A report of heat analysis will be supplied, if requested, for CS, DS, DDS, and EDDS. For materials with required mechanical properties, SS, HSLAS, HSLAS-F, SHS, and BHS, a report is required of heat analysis and mechanical properties as determined by the tension test, and

5.1.12 Special requirements (if any).

5.1.12.1 When the purchaser requires thickness tolerances for 3/8 in. [10 mm] minimum edge distance (see Supplementary Requirement in Specification A568/A568M), this requirement shall be specified in the purchase order or contract.

NOTE 3—A typical ordering description is as follows: ASTM A 1008-XX, cold rolled steel sheet, CS Type A, exposed, matte finish, oiled, 0.035 by 30 in. by coil, ID 24 in., OD 48 in., max weight 15 000 lbs, 100 000 lb, for part No. 4560, Door Panel.
or:
ASTM A 1008M-XX, cold-rolled steel sheet, SS grade 275, unexposed, matte finish, oiled, 0.88 mm by 760 mm by 2440 mm, 10 000 kg, for shelf bracket.

6. General Requirements for Delivery

6.1 Material furnished under this specification shall conform to the applicable requirements of the current edition of Specification A568/A568M unless otherwise provided herein.

7. Chemical Composition

7.1 The heat analysis of the steel shall conform to the chemical composition requirements of the appropriate designation shown in Table 1 for CS, DS, DDS, and EDDS and in Table 2 for SS, HSLAS, HSLAS-F, SHS, and BHS.

7.2 Each of the elements listed in Table 1 and Table 2 shall be included in the report of the heat analysis. When the amount of copper, nickel, chromium, or molybdenum is less than 0.02 %, report the analysis as <0.02 % or the actual determined value. When the amount of vanadium, columbium, or titanium is less than 0.008 %, report the analysis as <0.008 % or the actual determined value. When the amount of boron is less than 0.0005 %, report the analysis as <0.0005 % or the actual determined value.

7.3 Sheet steel grades defined by this specification are suitable for welding if appropriate welding conditions are selected. For certain welding processes, if more restrictive composition limits are desirable, they shall be specified at the time of inquiry and confirmed at the time of ordering.

8. Mechanical Properties

8.1 *CS, DS, DDS, and EDDS*:

8.1.1 Typical nonmandatory mechanical properties for CS, DS, DDS and EDDS are shown in Table 3.

8.1.2 The material shall be capable of being bent, at room temperature, in any direction through 180° flat on itself without cracking on the outside of the bent portion (see Section 14 of Test Methods and Definitions A370). The bend test is not a requirement of delivery. However, if testing is performed by the purchaser, material not conforming to the requirement shall be subject to rejection.

8.1.3 Sheet of these designations except for EDDS are subject to aging dependent upon processing factors such as the method of annealing (continuous annealing or box annealing), and chemical composition. For additional information on aging, see Appendix X1 of Specification A568/A568M.

8.1.4 EDDS steel is stabilized to be nonaging and so is not subject to stretcher strains and fluting. Other steels are processed to be nonaging; please consult your supplier.

8.2 *SS, HSLAS, HSLAS-F, SHS, and BHS*:

8.2.1 The available strength grades for SS, HSLAS and HSLAS-F are shown in Table 4.

8.2.2 The available strength grades for SHS and BHS are shown in Table 5.

8.2.3 *Tension Tests*:

8.2.3.1 *Requirements*—Material as represented by the test specimen shall conform to the mechanical property requirements specified in Table 4. These requirements do not apply to the uncropped ends of unprocessed coils.

8.2.3.2 *Number of Tests*—Two tension tests shall be made from each heat or from each 50 tons [45 000 kg]. When the amount of finished material from a heat is less than 50 tons [45 000 kg], one test shall be made. When material rolled from heat differs 0.050 in. [1.27 mm] or more in thickness, one tension test shall be made from the thickest and thinnest material regardless of the weight represented.

8.2.3.3 Tension test specimens shall be taken at a point immediately adjacent to the material to be qualified.

8.2.3.4 Tension test specimens shall be taken from the full thickness of the sheet.

8.2.3.5 Tension test specimens shall be taken from a location approximately halfway between the center of the sheet and the edge of the material as rolled.

8.2.3.6 Tension test samples shall be taken with the lengthwise axis of the test specimen parallel to the rolling direction (longitudinal test).

8.2.3.7 *Test Method*—Yield strength shall be determined by either the 0.2 % offset method or the 0.5 % extension under load method unless otherwise specified.

8.2.3.8 Bake hardenable steel shall conform to bake hardening index requirements included in Table 5 for the grade specified. The method for measuring the bake hardening index is described in Annex A1. Bake hardenable steel shall exhibit a minimum increase in yield strength of 4 ksi [25 MPa] as based on the upper yield point or 3 ksi [20 MPa] as based on the lower yield stress, after a prestrained specimen has been exposed to a standard bake cycle (340°F [170°C]) for 20 min.

TABLE 1 Chemical Composition[A]
For Cold Rolled Steel Sheet Designations CS, DS, DDS, and EDDS

Designation	% Heat Analysis, Element Maximum Unless Otherwise Shown														
	C	Mn	P	S	Al	Si	Cu	Ni	Cr[B]	Mo	V	Cb	Ti[C]	N	B
CS Type A[D,E,F,G]	0.10	0.60	0.030	0.035	...	...	0.20[H]	0.20	0.15	0.06	0.008	0.008	0.025	...	...
CS Type B[D]	0.02 to 0.15	0.60	0.030	0.035	...	...	0.20[H]	0.20	0.15	0.06	0.008	0.008	0.025	...	...
CS Type C[D,E,F,G]	0.08	0.60	0.10	0.035	...	...	0.20[H]	0.20	0.15	0.06	0.008	0.008	0.025	...	...
DS Type A[E,I]	0.08	0.50	0.020	0.030	0.01 min	...	0.20	0.20	0.15	0.06	0.008	0.008	0.025	...	...
DS Type B	0.02 to 0.08	0.50	0.020	0.030	0.02 min	...	0.20	0.20	0.15	0.06	0.008	0.008	0.025	...	...
DDS[F,G]	0.06	0.50	0.020	0.025	0.01 min	...	0.20	0.20	0.15	0.06	0.008	0.008	0.025	...	...
EDDS[J]	0.02	0.40	0.020	0.020	0.01 min	...	0.10	0.10	0.15	0.03	0.10	0.10	0.15	...	...

[A] Where an ellipsis (. . .) appears in the table, there is no requirement, but the analysis result shall be reported.
[B] Chromium is permitted, at the producer's option, to 0.25 % maximum when the carbon content is less than or equal to 0.05 %.
[C] For steels containing 0.02 % or more carbon, titanium is permitted at the producer's option, to the lesser of 3.4N + 1.5S or 0.025 %.
[D] When an aluminum deoxidized steel is required for the application, it is permissible to order Commercial Steel (CS) to a minimum of 0.01 % total aluminum.
[E] Specify Type B to avoid carbon levels below 0.02 %.
[F] It is permissible to furnish as a vacuum degassed or chemically stabilized steel, or both, at the producer's option.
[G] For carbon levels less than or equal to 0.02 %, it is permissible to use vanadium, columbium or titanium, or a combination thereof, as stabilizing elements at the producer's option. In such cases, the applicable limit for vanadium or columbium shall be 0.10 % max. and the limit on titanium shall be 0.15 % max.
[H] When copper steel is specified, the copper limit is a minimum requirement. When copper steel is not specified, the copper limit is a maximum requirement.
[I] If produced utilizing a continuous anneal process, stabilized steel is permissible at the producer's option, and Footnotes F and G apply.
[J] Shall be furnished as a vacuum degassed and stabilized steel.

TABLE 2 Chemical Composition[A]
For Cold Rolled Steel Sheet Designations SS, HSLAS, HSLAS-F, SHS, and BHS

Designation	C	Mn	P	S	Al	Si	Cu[B]	Ni	Cr	Mo	V	Cb	Ti	N
SS:[C]														
Grade 25 [170]	0.20	0.60	0.035	0.035	...	...	0.20	0.20	0.15	0.06	0.008	0.008	0.025	...
Grade 30 [205]	0.20	0.60	0.035	0.035	...	...	0.20	0.20	0.15	0.06	0.008	0.008	0.025	...
Grade 33 [230] Type 1	0.20	0.60	0.035	0.035	...	...	0.20	0.20	0.15	0.06	0.008	0.008	0.025	...
Grade 33 [230] Type 2	0.15	0.60	0.20	0.035	...	...	0.20	0.20	0.15	0.06	0.008	0.008	0.025	...
Grade 40 [275] Type 1	0.20	1.35	0.035	0.035	...	...	0.20	0.20	0.15	0.06	0.008	0.008	0.025	...
Grade 40 [275] Type 2	0.15	0.60	0.20	0.035	...	...	0.20	0.20	0.15	0.06	0.008	0.008	0.025	...
Grade 50 [340]	0.20	1.35	0.035	0.035	...	...	0.20	0.20	0.15	0.06	0.008	0.008	0.025	...
Grade 60 [410]	0.20	1.35	0.035	0.035	...	...	0.20	0.20	0.15	0.06	0.008	0.008	0.025	...
Grade 70 [480]	0.20	1.35	0.035	0.035	...	...	0.20	0.20	0.15	0.06	0.008	0.008	0.025	...
Grade 80 [550]	0.20	1.35	0.035	0.035	...	...	0.20	0.20	0.15	0.06	0.008	0.008	0.025	...
HSLAS:[D]														
Grade 45 [310] Class 1	0.22	1.65	0.04	0.04	...	...	0.20	0.20	0.15	0.06	0.005 min	0.005 min	0.005 min	...
Grade 45 [310] Class 2	0.15	1.65	0.04	0.04	...	...	0.20	0.20	0.15	0.06	0.005 min	0.005 min	0.005 min	...
Grade 50 [340] Class 1	0.23	1.65	0.04	0.04	...	...	0.20	0.20	0.15	0.06	0.005 min	0.005 min	0.005 min	...
Grade 50 [340] Class 2	0.15	1.65	0.04	0.04	...	...	0.20	0.20	0.15	0.06	0.005 min	0.005 min	0.005 min	...
Grade 55 [380] Class 1	0.25	1.65	0.04	0.04	...	...	0.20	0.20	0.15	0.06	0.005 min	0.005 min	0.005 min	...
Grade 55 [380] Class 2	0.15	1.65	0.04	0.04	...	...	0.20	0.20	0.15	0.06	0.005 min	0.005 min	0.005 min	...
Grade 60 [410] Class 1	0.26	1.65	0.04	0.04	...	...	0.20	0.20	0.15	0.06	0.005 min	0.005 min	0.005 min	...
Grade 60 [410] Class 2	0.15	1.65	0.04	0.04	...	...	0.20	0.20	0.15	0.06	0.005 min	0.005 min	0.005 min	...
Grade 65 [450] Class 1	0.26	1.65	0.04	0.04	...	...	0.20	0.20	0.15	0.06	0.005 min	0.005 min	0.005 min	E
Grade 65 [450] Class 2	0.15	1.65	0.04	0.04	...	...	0.20	0.20	0.15	0.06	0.005 min	0.005 min	0.005 min	E
Grade 70 [480] Class 1	0.26	1.65	0.04	0.04	...	...	0.20	0.20	0.15	0.16	0.005 min	0.005 min	0.005 min	E
Grade 70 [480] Class 2	0.15	1.65	0.04	0.04	...	...	0.20	0.20	0.15	0.16	0.005 min	0.005 min	0.005 min	E
HSLAS-F:[D]														
Grade 50 [340] and 60 [410]	0.15	1.65	0.020	0.025	...	...	0.20	0.20	0.15	0.06	0.005 min	0.005 min	0.005 min	E
Grade 70 [480] and 80 [550]	0.15	1.65	0.020	0.025	...	...	0.20	0.20	0.15	0.16	0.005 min	0.005 min	0.005 min	E
SHS[F]	0.12	1.50	0.12	0.030	...	...	0.20	0.20	0.15	0.06	0.008	0.008	0.008	...
BHS[F]	0.12	1.50	0.12	0.030	...	...	0.20	0.20	0.15	0.06	0.008	0.008	0.008	...

[A] Where an ellipsis (. . .) appears in the table, there is no requirement but, the analysis shall be reported.
[B] When copper is specified, the copper limit is a minimum requirement. When copper steel is not specified, the copper limit is a maximum requirement.
[C] Titanium is permitted for SS designations, at the producer's option, to the lesser of 3.4N + 1.5S or 0.025 %.
[D] HSLAS and HSLAS-F steels contain the strengthening elements columbium (niobium), vanadium, titanium, and molybdenum added singly or in combination. The minimum requirements only apply to the microalloy elements selected for strengthening of the steel.
[E] The purchaser has the option of restricting the nitrogen content. It should be noted that, depending on the microalloying scheme (for example, use of vanadium) of the producer, nitrogen may be a deliberate addition. Consideration should be made for the use of nitrogen binding elements (for example, vanadium, titanium).
[F] For carbon levels less than or equal to 0.02 % vanadium, columbium, or titanium, or a combination thereof, are permitted to be used as stabilizing elements at the producer's option. In such cases, the applicable limit for vanadium and columbium shall be 0.10 % max., and the limit for titanium shall be 0.15 % max.

TABLE 3 Typical Ranges of Mechanical Properties[A]
(Nonmandatory)[B]
For Cold Rolled Steel Sheet Designations CS, DS, DDS and EDDS

Designation	Yield Strength[C]		Elongation in 2 in. [50 mm] %[C]	r_m Value[D]	n-Value[E]
	ksi	MPa			
CS Types A, B, and C	20 to 40	[140 to 275]	≥30	F	F
DS Types A and B	22 to 35	[150 to 240]	≥36	1.3 to 1.7	0.17 to 0.22
DDS	17 to 29	[115 to 200]	≥38	1.4 to 1.8	0.20 to 0.25
EDDS	15 to 25	[105 to 170]	≥40	1.7 to 2.1	0.23 to 0.27

[A] These typical mechanical properties apply to the full range of steel sheet thicknesses. The yield strength tends to increase, the elongation decreases and some of the formability values tend to decrease as the sheet thickness decreases.
[B] The typical mechanical property values presented here are nonmandatory. They are provided to assist the purchaser in specifying a suitable steel for a given application. Values outside of these ranges are to be expected.
[C] Yield Strength and elongation are measured in the longitudinal direction in accordance with Test Methods and Definitions A370.
[D] Average plastic strain ratio (r_m value) as determined by Test Method E517.
[E] The strain hardening exponent (n-value) as determined by Test Method E646.
[F] No typical properties have been established.

8.2.4 *Bending Properties*:

8.2.4.1 The suggested minimum inside radii for cold bending are listed in Appendix X1 and is discussed in more detail in Specification A568/A568M (Section 6). Where a tighter bend radius is required, where curved or offset bends are involved, or where stretching or drawing are also a consideration, the producer shall be consulted.

9. Finish and Appearance

9.1 *Surface Finish*:

TABLE 4 Mechanical Property Requirements[A]
For Cold Rolled Steel Sheet Designations SS, HSLAS, and HSLAS-F

Designation	Yield Strength, min		Tensile Strength, min		Elongation in 2 in. or 50 mm, min, %
	ksi	[MPa]	ksi	[MPa]	
SS:					
Grade 25 [170]	25	[170]	42	[290]	26
Grade 30 [205]	30	[205]	45	[310]	24
Grade 33 [230] Types 1 and 2	33	[230]	48	[330]	22
Grade 40 [275] Types 1 and 2	40	[275]	52	[360]	20
Grade 50 [340]	50	[340]	65	[450]	18
Grade 60 [410]	60	[410]	75	[520]	12
Grade 70 [480]	70	[480]	85	[585]	6
Grade 80 [550]	80[B]	[550]	82	[565]	[C]
HSLAS:					
Grade 45 [310] Class 1	45	[310]	60	[410]	22
Grade 45 [310] Class 2	45	[310]	55	[380]	22
Grade 50 [340] Class 1	50	[340]	65	[450]	20
Grade 50 [340] Class 2	50	[340]	60	[410]	20
Grade 55 [380] Class 1	55	[380]	70	[480]	18
Grade 55 [380] Class 2	55	[380]	65	[450]	18
Grade 60 [410] Class 1	60	[410]	75	[520]	16
Grade 60 [410] Class 2	60	[410]	70	[480]	16
Grade 65 [450] Class 1	65	[450]	80	[550]	15
Grade 65 [450] Class 2	65	[450]	75	[520]	15
Grade 70 [480] Class 1	70	[480]	85	[585]	14
Grade 70 [480] Class 2	70	[480]	80	[550]	14
HSLAS-F:					
Grade 50 [340]	50	[340]	60	[410]	22
Grade 60 [410]	60	[410]	70	[480]	18
Grade 70 [480]	70	[480]	80	[550]	16
Grade 80 [550]	80	[550]	90	[620]	14

[A] For coil products, testing by the producer is limited to the end of the coil. Mechanical properties throughout the coil shall comply with the minimum values specified.
[B] On this full-hard product, the yield strength approaches the tensile strength and since there is no halt in the gage or drop in the beam, the yield point shall be taken as the yield stress at 0.5 % extension under load.
[C] There is no requirement for elongation in 2 in. for SS Grade 80.

TABLE 5 Mechanical Property Requirements[A,B]
For Cold Rolled Steel Sheet Designations SHS and BHS

Designation	Yield Strength, min		Tensile Strength, min		Elongation in 2 in. or 50 mm, min., %	Bake Hardening Index, min Upper Yield/Lower Yield	
	ksi	[MPa]	ksi	[MPa]		ksi	[MPa]
SHS:							
Grade 26 [180]	26	[180]	43	[300]	32	...	...
Grade 31 [210]	31	[210]	46	[320]	30	...	...
Grade 35 [240]	35	[240]	50	[340]	26	...	...
Grade 41 [280]	41	[280]	53	[370]	24	...	...
Grade 44 [300]	44	[300]	57	[390]	22	...	...
BHS:							
Grade 26 [180]	26	[180]	43	[300]	30	4/3	25/20
Grade 31 [210]	31	[210]	46	[320]	28	4/3	25/20
Grade 35 [240]	35	[240]	50	[340]	24	4/3	25/20
Grade 41 [280]	41	[280]	53	[370]	22	4/3	25/20
Grade 44 [300]	44	[300]	57	[390]	20	4/3	25/20

[A] Where an ellipsis (. . .) appears in the table, there is no requirement.
[B] For coil products, testing by the producer is limited to the end of the coil. Mechanical properties throughout the coil shall comply with the minimum values specified.

9.1.1 Unless otherwise specified, the sheet shall have a matte finish. When required, specify the appropriate surface texture and condition. For additional information, see the Finish and Condition section of Specification A568/A568M.

For additional information see "Finish and Condition" section of Specification A568/A568M.

9.2 *Oiling*:

9.2.1 Unless otherwise specified, the sheet shall be oiled.

9.2.2 When required, specify the sheet to be furnished not oiled (dry).

10. Retests and Disposition of Non-Conforming Material

10.1 Retests, conducted with the requirements of Section 11.1 of Specification A568/A568M, are permitted when an unsatisfactory test result is suspected to be the consequence of the test method procedure.

10.2 Disposition of non-conforming material shall be subject to the requirements of Section 11.2 of Specification A568/A568M.

11. Certification

11.1 A report of heat analysis shall be supplied, if requested, for CS, DS, DDS, and EDDS steels. For material with required mechanical properties, SS, HSLAS, HSLAS-F, SHS, and BHS, a report is required of heat analysis and mechanical properties as determined by the tension test.

11.2 The report shall include the purchase order number, the ASTM designation number and year date, product designation, grade, type or class, as applicable, the heat number, and as required, heat analysis and mechanical properties as indicated by the tension test.

11.3 A signature is not required on the test report. However, the document shall clearly identify the organization submitting the report. Notwithstanding the absence of a signature, the organization submitting the report is responsible for the content of the report.

11.4 A Material Test Report, Certificate of Inspection, or similar document printed from or used in electronic form from an electronic data interchange (EDI) transmission shall be regarded as having the same validity as a counterpart printed in the certifier's facility. The content of the EDI transmitted document must meet the requirements of the invoked ASTM standard(s) and conform to any existing EDI agreement between the purchaser and the supplier. Notwithstanding the absence of a signature, the organization submitting the EDI transmission is responsible for the content of the report.

12. Product Marking

12.1 In addition to the requirements of Specification A568/A568M, each lift or coil shall be marked with the designation shown on the order (CS (Type A, B, or C), DS (Type A or B), DDS, EDDS, SS, HSLAS, HSLAS-F, SHS, or BHS). The designation shall be legibly stenciled on the top of each lift or shown on a tag attached to each coil or shipping unit.

13. Keywords

13.1 bake hardenable steel; bake hardening index; carbon steel sheet; cold-rolled steel sheet; commercial steel; deep drawing steel; drawing steel; extra deep drawing steel; high-strength low-alloy steel; high-strength low-alloy steel with improved formability; solution hardened steel; steel sheet; structural steel

---

ANNEX

(Mandatory Information)

A1. BAKE HARDENABLE STEELS

A1.1 Determination of Bake Hardening Index

A1.1.1 The bake hardening index (BHI) is determined by a two-step procedure using a standard longitudinal (rolling direction) tensile-test specimen, prepared in accordance with Test Methods and Definitions A370. The test specimen is first strained in tension. The magnitude of this tensile "pre-strain" shall be 2 % (extension under load). The test specimen is then removed from the test machine and baked at a temperature of 340°F [170°C] for a period of 20 min. Referring to Fig. A1.1, the bake hardening index (BHI) of the material is calculated as follows:

$$BHI = B - A \quad (A1.1)$$

where:
A = flow stress at 2 % extension under load, and
B = yield strength [upper yield strength (B_U) or lower yield strength (B_L)] after baking at 340°F [170°C] for 20 min.

A1.1.2 The original test specimen cross section (width and thickness) is used in the calculation of all engineering strengths in this test.

A1.1.3 The pre-straining of 2 % in tension is intended to simulate a modest degree of forming strain, while the subsequent baking is intended to simulate a paint-curing or similar treatment. In the production of actual parts, forming strains and baking treatments can differ from those employed here, and as a result, final properties can differ from the values obtained under these controlled conditions.

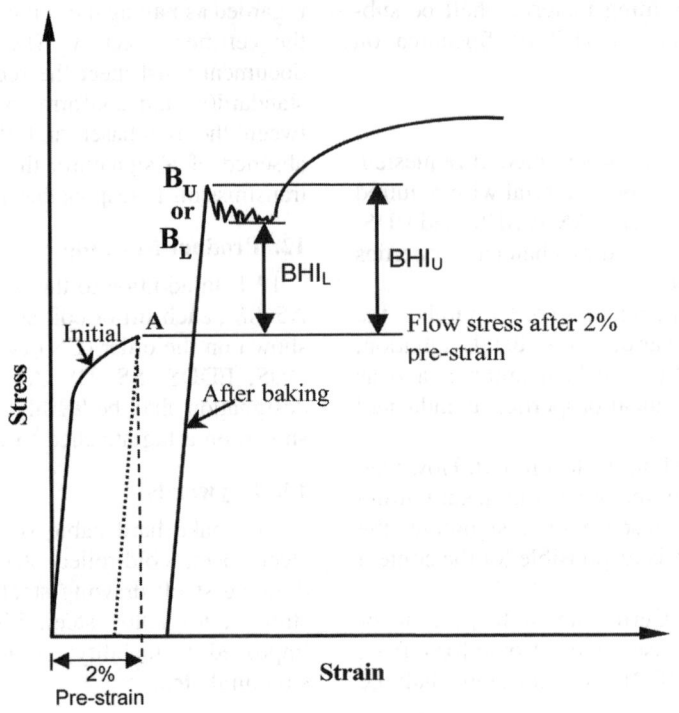

FIG. A1.1 Representation of Bake Hardening Index

APPENDIXES

(Nonmandatory Information)

X1. BENDING PROPERTIES

TABLE X1.1 Suggested Minimum Inside Radius for Cold Bending

NOTE 1—(*t*) Equals a radius equivalent to the steel thickness.
NOTE 2—The suggested radius should be used as a minimum for 90° bends in actual shop practice
NOTE 3—Material which does not perform satisfactorily, when fabricated in accordance with the requirements, may be subject to rejection pending negotiation with the steel supplier.

Designation	Grade	Minimum Inside Radius for Cold Bending	
Structural Steel	25 [170]	½ t	
	30 [205]	1 t	
	33 [230]	1½ t	
	40 [275]	2 t	
	50 [340]	2½ t	
	60 [410]	3 t	
	70 [480]	4 t	
	80 [550]	not applicable	
High-Strength Low-Alloy Steel		Class 1	Class 2
	45[310]	1½ t	1½ t
	50[340]	2 t	1½ t
	55[380]	2 t	2 t
	60[410]	2½ t	2 t
	65[450]	3 t	2½ t
	70[480]	3½ t	3 t
High-Strength Low-Alloy Steel with Improved Formability	50[340]	1 t	
	60[410]	1½ t	
	70[480]	2 t	
	80[550]	2 t	
Solution Hardened Steel	26 [180]	½ t	
	31 [210]	1 t	
	35 [240]	1½ t	
	41 [280]	2 t	
	44 [300]	2 t	
Bake Hardenable Steel	26 [180]	½ t	
	31 [210]	1 t	
	35 [240]	1½ t	
	41 [280]	2 t	
	44 [300]	2 t	

X2. RELATED ISO STANDARDS

The ISO standards listed below may be reviewed for comparison with this ASTM standard. The relationship between the standards may only be approximate; therefore, the respective standards should be consulted for actual requirements. Those who use these documents must determine which specifications address their needs.

ISO 3574 Cold-Reduced Carbon Steel Sheet of Commercial and Drawing Qualities

ISO 4997 Cold-Reduced Steel Sheet of Structural Quality

ISO 13887 Cold-Reduced Steel Sheet of Higher Strength with Improved Formability

X3. HARDNESS PROPERTIES

TABLE X3.1 Typical Hardness Values

Note 1—The hardness values shown are at the time of shipment.
Note 2—Test for hardness shall be conducted in accordance with the requirements of Test Methods E18.
Note 3—The hardness values are Rockwell B scale as measured or converted from the appropriate Rockwell scales.
Note 4—The typical hardness values apply to the full range of steel sheet thickness. Hardness tends to increase as the steel sheet thickness decreases.
Note 5—Hardness testing is commonly used to assess the relative formability of various designations of uncoated steel sheet. This assessment done by many users is recognized to be only an approximation of the relative formability and therefore cannot be used as a specification requirement.

Designation	Hardness-Rockwell B Scale
CS Type A	70 or less
CS Type B	70 or less
CS Type C	70 or less
DS Type A	60 or less
DS Type B	60 or less
DDS	55 or less
EDDS	45 or less

SUMMARY OF CHANGES

Committee A01 has identified the location of selected changes to this standard since the last issue (A1008/A1008M – 09) that may impact the use of this standard. (Approved November 1, 2009.)

(1) Section 5.1.6.1 deleted.
(2) Reversed order of 5.1.11 and 5.1.12 and added new section 5.1.12.1.

Committee A01 has identified the location of selected changes to this standard since the last issue (A1008/A1008M – 08a) that may impact the use of this standard. (Approved May 1, 2009.)

(1) Corrected SI equivalence values for SS Grade 60 and SS Grade 70 tensile strength in Table 4.

Committee A01 has identified the location of selected changes to this standard since the last issue (A1008/A1008M – 08) that may impact the use of this standard. (Approved October 1, 2008.)

(1) Composition (Mn levels in SS grades) changed in Table 2.

ASTM International takes no position respecting the validity of any patent rights asserted in connection with any item mentioned in this standard. Users of this standard are expressly advised that determination of the validity of any such patent rights, and the risk of infringement of such rights, are entirely their own responsibility.

This standard is subject to revision at any time by the responsible technical committee and must be reviewed every five years and if not revised, either reapproved or withdrawn. Your comments are invited either for revision of this standard or for additional standards and should be addressed to ASTM International Headquarters. Your comments will receive careful consideration at a meeting of the responsible technical committee, which you may attend. If you feel that your comments have not received a fair hearing you should make your views known to the ASTM Committee on Standards, at the address shown below.

This standard is copyrighted by ASTM International, 100 Barr Harbor Drive, PO Box C700, West Conshohocken, PA 19428-2959, United States. Individual reprints (single or multiple copies) of this standard may be obtained by contacting ASTM at the above address or at 610-832-9585 (phone), 610-832-9555 (fax), or service@astm.org (e-mail); or through the ASTM website (www.astm.org). Permission rights to photocopy the standard may also be secured from the ASTM website (www.astm.org/COPYRIGHT/).

Designation: A1011/A1011M – 09b

Standard Specification for
Steel, Sheet and Strip, Hot-Rolled, Carbon, Structural, High-Strength Low-Alloy, High-Strength Low-Alloy with Improved Formability, and Ultra-High Strength[1]

This standard is issued under the fixed designation A1011/A1011M; the number immediately following the designation indicates the year of original adoption or, in the case of revision, the year of last revision. A number in parentheses indicates the year of last reapproval. A superscript epsilon (ε) indicates an editorial change since the last revision or reapproval.

1. Scope*

1.1 This specification covers hot-rolled, carbon, structural, high-strength low-alloy, high-strength low-alloy with improved formability, and ultra-high strength steel sheet and strip, in coils and cut lengths.

1.2 Hot rolled steel sheet and strip is available in the designations as listed in 4.1.

1.3 This specification is not applicable to the steel covered by Specification A635/A635M.

1.4 The values stated in either SI units or inch-pound units are to be regarded separately as standard. The values stated in each system may not be exact equivalents; therefore, each system shall be used independently of the other. Combining values from the two systems may result in non-conformance with the standard.

2. Referenced Documents

2.1 *ASTM Standards:*[2]
A370 Test Methods and Definitions for Mechanical Testing of Steel Products
A568/A568M Specification for Steel, Sheet, Carbon, Structural, and High-Strength, Low-Alloy, Hot-Rolled and Cold-Rolled, General Requirements for
A569/A569M Specification for Steel, Carbon (0.15 Maximum, Percent), Hot-Rolled Sheet and Strip Commercial[3]
A622/A622M Specification for Drawing Steel (DS), Sheet and Strip, Carbon, Hot-Rolled[3]
A635/A635M Specification for Steel, Sheet and Strip, Heavy-Thickness Coils, Hot-Rolled, Alloy, Carbon, Structural, High-Strength Low-Alloy, and High-Strength Low-Alloy with Improved Formability, General Requirements for
A749/A749M Specification for Steel, Strip, Carbon and High-Strength, Low-Alloy, Hot-Rolled, General Requirements for
A941 Terminology Relating to Steel, Stainless Steel, Related Alloys, and Ferroalloys
E18 Test Methods for Rockwell Hardness of Metallic Materials

3. Terminology

3.1 *Definitions*—For definitions of other terms used in this specification refer to Terminology A941.

3.2 *Definitions of Terms Specific to This Standard:*

3.2.1 *aging*—loss of ductility with an increase in hardness, yield strength, and tensile strength that occurs when steel, which has been slightly cold worked (such as by temper rolling) is stored for some time.

3.2.1.1 *Discussion*—Aging also increases the tendency toward stretcher strains and fluting.

3.2.2 *inclusion control, n*—the process of reducing the volume fraction of inclusions or modifying the shape of inclusions to improve formability, weldability, and machinability.

3.2.2.1 *Discussion*—Inclusions, especially those elongated during the rolling process, create the conditions for initiating or propagating cracks when the material is stretched or bent during the manufacture of a part (or both). The adverse effects of inclusions are minimized by reducing the content of inclusions in the steel or by altering the shape of inclusions through the use of additions during the steelmaking process that change the elongated shape of the inclusions to less harmful small, well dispersed globular inclusions (or both).

3.2.3 *stabilization*—addition of one or more nitride or carbide forming elements, or both, such as titanium and columbium, to control the level of the interstitial elements carbon and nitrogen in the steel.

---

[1] This specification is under the jurisdiction of ASTM Committee A01 on Steel, Stainless Steel and Related Alloys and is the direct responsibility of Subcommittee A01.19 on Steel Sheet and Strip.
Current edition approved Nov. 1, 2009. Published December 2009. Originally approved in 2000. Last previous edition approved in 2009 as A1011/A1011M – 09a. DOI: 10.1520/A1011_A1011M-09B.
[2] For referenced ASTM standards, visit the ASTM website, www.astm.org, or contact ASTM Customer Service at service@astm.org. For *Annual Book of ASTM Standards* volume information, refer to the standard's Document Summary page on the ASTM website.
[3] Withdrawn. The last approved version of this historical standard is referenced on www.astm.org.

*A Summary of Changes section appears at the end of this standard.

3.2.3.1 *Discussion*—Stabilization improves formability and increases resistance to aging.

3.2.4 *vacuum degassing*—process of refining liquid steel in which the liquid is exposed to a vacuum as part of a special technique for removing impurities or for decarburizing the steel.

4. Classification

4.1 Hot-rolled steel sheet and steel strip is available in the following designations:

4.1.1 Commercial Steel (CS Types A, B, C, and D),

4.1.2 Drawing Steel (DS Types A and B),

NOTE 1—CS Type B and DS Type B describe the most common product previously included, respectively, in Specifications A569/A569M and A622/A622M.

4.1.3 Structural Steel (SS grades 30[205], 33[230], 36[250] Types 1 and 2, 40[275], 45[310], 50[340], 55[380], 60[410], 70[480], and 80[550]),

4.1.4 High-Strength Low-Alloy Steel (HSLAS, classes 1 and 2, in grades 45[310], 50[340], 55[380], 60[410], 65[450], and 70[480]).

4.1.5 High-Strength Low-Alloy Steel with Improved Formability (HSLAS-F grades 50[340], 60[410], 70[480], and 80[550]).

4.1.5.1 HSLAS-F steel has improved formability when compared to HSLAS. The steel is fully deoxidized, made to a fine grain practice, and includes microalloying elements such as columbium, vanadium, and zirconium. The steel shall be treated to achieve inclusion control.

4.1.6 Ultra-High Strength (UHSS Types 1 and 2, in Grades 90 [620] and 100 [690]).

4.1.6.1 UHSS steel has increased strength compared with HSLAS-F. The steel is killed and made to a fine ferritic grain practice, and includes microalloying elements such as columbium (niobium), titanium, vanadium, molybdenum, and so forth. The steel shall be treated to achieve inclusion control. The material is intended for miscellaneous applications where higher strength, savings in weight, and weldability are important. Atmospheric corrosion resistance of these steels is equivalent to plain carbon steels. With copper specified, the atmospheric corrosion resistance is somewhat enhanced.

4.1.7 When required for HSLAS, HSLAS-F, and UHSS steels, limitations on the use of one or more of the microalloy elements shall be specified on the order.

5. Ordering Information

5.1 It is the purchaser's responsibility to specify in the purchase order all ordering information necessary to describe the required material. Examples of such information include, but are not limited to, the following:

5.1.1 ASTM specification number and year of issue,

5.1.2 Name of material and designation (hot-rolled steel sheet) (include grade, type and class, as appropriate, for CS, DS, SS, HSLAS, HSLAS-F, and UHSS) (see 4.1),

5.1.2.1 When a type is not specified for CS or DS, Type B will be furnished (see 4.1),

5.1.2.2 When a class is not specified for HSLAS, Class 1 will be furnished (see 4.1),

5.1.2.3 When a type is not specified for SS Grade 36, Type 1 will be furnished (see 4.1),

5.1.2.4 When a type is not specified for UHSS, Type 1 shall be furnished (see 4.1).

5.1.3 Finish (see 9.1)

5.1.4 Type of edge (see 9.3),

5.1.5 Oiled or not oiled, as required (see 9.2),

5.1.6 Dimensions (thickness, width, and whether cut lengths or coils),

NOTE 2—Not all producers are capable of meeting all the limitations of the thickness tolerance tables in Specifications A568/A568M and A749/A749M. The purchaser should contact the producer prior to placing an order.

5.1.7 Coil size (inside diameter, outside diameter, and maximum weight),

5.1.8 Copper bearing steel (if required),

5.1.9 Quantity,

5.1.10 Application (part identification and description),

5.1.11 A report of heat analysis will be supplied, if requested, for CS and DS. For materials with required mechanical properties, SS, HSLAS, HSLAS-F, and UHSS, a report is required of heat analysis and mechanical properties as determined by the tension test, and

5.1.12 Special requirements (if any).

5.1.12.1 When the purchaser requires thickness tolerances for 3/8 in. [10 mm] minimum edge distance (see Supplementary Requirement in Specification A568/A568M), this requirement shall be specified in the purchase order or contract.

NOTE 3—A typical ordering description is as follows: ASTM A1011-XX, hot rolled steel sheet, CS Type A, pickled and oiled, cut edge, 0.075 by 36 by 96 in., 100 000 lb, for part no. 6310, for shelf bracket.
or:
ASTM A1011M-XX, hot rolled steel sheet, CS Type B, pickled and oiled, cut edge, 3.7 by 117 mm by coil, ID 600 mm, OD 1500 mm, max weight 10 000 kg, 50 000 kg, for upper control arm.

6. General Requirements for Delivery

6.1 Material furnished under this specification shall conform to the applicable requirements of the current edition of Specification A568/A568M for sheets and Specification A749/A749M for strip, unless otherwise provided for herein.

7. Chemical Composition

7.1 The heat analysis of the steel shall conform to the chemical composition requirements of the appropriate designation shown in Table 1 for CS and DS and Table 2 for SS, HSLAS, HSLAS-F, and UHSS.

7.2 Each of the elements listed in Tables 1 and 2 shall be included in the report of the heat analysis. When the amount of copper, nickel, chromium, or molybdenum is less than 0.02 %, report the analysis as <0.02 % or the actual determined value. When the amount of vanadium, columbium, or titanium is less than 0.008 %, report the analysis as <0.008 % or the actual determined value. When the amount of boron is less than 0.0005 %, report the analysis as <0.0005 % or the actual determined value.

7.3 Sheet steel grades defined by this specification are suitable for welding if appropriate welding conditions are selected. For certain welding processes, if more restrictive

TABLE 1 Chemical Composition[A]
For Hot Rolled Steel Sheet and Strip Designations CS and DS

	\multicolumn{14}{c}{Composition, % Heat Analysis Element maximum unless otherwise shown}
	C	Mn	P	S	Al	Si	Cu	Ni	Cr[B]	Mo	V	Cb	Ti[C]	N	B
CS Type A[D,E,F,G]	0.10	0.60	0.030	0.035	...	...	0.20[H]	0.20	0.15	0.06	0.008	0.008	0.025	...	...
CS Type B[F]	0.02 to 0.15	0.60	0.030	0.035	...	...	0.20[H]	0.20	0.15	0.06	0.008	0.008	0.025	...	...
CS Type C[D,E,F]	0.08	0.60	0.10	0.035	...	...	0.20[H]	0.20	0.15	0.06	0.008	0.008	0.025	...	...
CS Type D[F]	0.10	0.70	0.030	0.035	...	...	0.20[H]	0.20	0.15	0.06	0.008	0.008	0.008	...	...
DS Type A[D,E,G]	0.08	0.50	0.020	0.030	0.01 min	...	0.20	0.20	0.15	0.06	0.008	0.008	0.025	...	...
DS Type B	0.02 to 0.08	0.50	0.020	0.030	0.01 min	...	0.20	0.20	0.15	0.06	0.008	0.008	0.025	...	...

[A] Where an ellipsis (...) appears in the table, there is no specified limit, but the analysis shall be reported.
[B] Chromium is permitted, at the producer's option, to 0.25 % maximum when the carbon content is less than or equal to 0.05 %.
[C] For steels containing 0.02 % carbon or more, titanium is permitted at the producer's option, to the lesser of 3.4N + 1.5S or 0.025 %.
[D] Specify Type B to avoid carbon levels below 0.02 %.
[E] For carbon levels less than or equal to 0.02 %, it is permissible to use vanadium, columbium, or titanium, or combinations thereof, as stabilizing elements at the producer's option. In such case, the limits for these elements are 0.10 % for vanadium or columbium and 0.15 % for titanium.
[F] When an aluminum deoxidized steel is required, it is permissible to order a minimum of 0.01 % total aluminum.
[G] It is permissible to furnish as a vacuum degassed or chemically stabilized steel, or both, at producer's option.
[H] When copper steel is specified, the copper limit is a minimum requirement. When copper steel is not specified, the copper limit is a maximum requirement.

TABLE 2 Chemical Composition[A]
For Hot Rolled Steel Sheet and Strip Designations SS, HSLAS, HSLAS-F, and UHSS

Designation	C	Mn	P	S	Al	Si	Cu[B]	Ni	Cr	Mo	V	Cb	Ti	N
SS:[C]														
Grade 30 [205]	0.25	0.90	0.035	0.04	...	...	0.20	0.20	0.15	0.06	0.008	0.008	0.025	...
Grade 33 [230]	0.25	0.90	0.035	0.04	...	...	0.20	0.20	0.15	0.06	0.008	0.008	0.025	...
Grade 36 [250] Type 1	0.25	0.90	0.035	0.04	...	...	0.20	0.20	0.15	0.06	0.008	0.008	0.025	...
Grade 36 [250] Type 2[D]	0.25	1.35	0.035	0.04	...	...	0.20	0.20	0.15	0.06	0.008	0.008	0.025	...
Grade 40 [275]	0.25	0.90	0.035	0.04	...	...	0.20	0.20	0.15	0.06	0.008	0.008	0.025	...
Grade 45 [310][D]	0.25	1.35	0.035	0.04	...	...	0.20	0.20	0.15	0.06	0.008	0.008	0.025	...
Grade 50 [340][D]	0.25	1.35	0.035	0.04	...	...	0.20	0.20	0.15	0.06	0.008	0.008	0.025	...
Grade 55 [380][D]	0.25	1.35	0.035	0.04	...	...	0.20	0.20	0.15	0.06	0.008	0.008	0.025	...
Grade 60 [410]	0.25	1.35	0.035	0.04	...	...	0.20	0.20	0.15	0.06	0.008	0.008	0.025	...
Grade 70 [480]	0.25	1.35	0.035	0.04	...	...	0.20	0.20	0.15	0.06	0.008	0.008	0.025	...
Grade 80 [550]	0.25	1.35	0.035	0.04	...	...	0.20	0.20	0.15	0.06	0.008	0.008	0.025	...
HSLAS:[E]														
Grade 45 [310] Class 1[D]	0.22	1.35	0.04	0.04	...	...	0.20	0.20	0.15	0.06	0.005 min	0.005 min	0.005 min	...
Grade 45 [310] Class 2	0.15	1.35	0.04	0.04	...	...	0.20	0.20	0.15	0.06	0.005 min	0.005 min	0.005 min	...
Grade 50 [340] Class 1[D]	0.23	1.35	0.04	0.04	...	...	0.20	0.20	0.15	0.06	0.005 min	0.005 min	0.005 min	...
Grade 50 [340] Class 2	0.15	1.35	0.04	0.04	...	...	0.20	0.20	0.15	0.06	0.005 min	0.005 min	0.005 min	...
Grade 55 [380] Class 1[D]	0.25	1.35	0.04	0.04	...	...	0.20	0.20	0.15	0.06	0.005 min	0.005 min	0.005 min	...
Grade 55 [380] Class 2	0.15	1.35	0.04	0.04	...	...	0.20	0.20	0.15	0.06	0.005 min	0.005 min	0.005 min	...
Grade 60 [410] Class 1	0.26	1.50	0.04	0.04	...	...	0.20	0.20	0.15	0.06	0.005 min	0.005 min	0.005 min	...
Grade 60 [410] Class 2	0.15	1.50	0.04	0.04	...	...	0.20	0.20	0.15	0.06	0.005 min	0.005 min	0.005 min	...
Grade 65 [450] Class 1	0.26	1.50	0.04	0.04	...	...	0.20	0.20	0.15	0.06	0.005 min	0.005 min	0.005 min	[F]
Grade 65 [450] Class 2	0.15	1.50	0.04	0.04	...	...	0.20	0.20	0.15	0.06	0.005 min	0.005 min	0.005 min	[F]
Grade 70 [480] Class 1	0.26	1.65	0.04	0.04	...	...	0.20	0.20	0.15	0.16	0.005 min	0.005 min	0.005 min	[F]
Grade 70 [480] Class 2	0.15	1.65	0.04	0.04	...	...	0.20	0.20	0.15	0.16	0.005 min	0.005 min	0.005 min	[F]
HSLAS-F:[E]														
Grade 50 [340] and 60 [410]	0.15	1.65	0.020	0.025	...	...	0.20	0.20	0.15	0.06	0.005 min	0.005 min	0.005 min	[F]
Grade 70 [480] and 80 [550]	0.15	1.65	0.020	0.025	...	...	0.20	0.20	0.15	0.16	0.005 min	0.005 min	0.005 min	[F]
UHSS:[E]														
Grade 90 [620] and 100 [690] Type 1	0.15	2.00	0.020	0.025	...	...	0.20	0.20	0.15	0.40	0.005 min	0.005 min	0.005 min	[F]
Grade 90 [620] and 100 [690] Type 2	0.15	2.00	0.020	0.025	...	...	0.60	0.50	0.30	0.40	0.005 min	0.005 min	0.005 min	[F]

[A] Where an ellipsis (...) appears in the table, there is no requirement but the analysis shall be reported.
[B] When copper is specified, a minimum of 0.20 % is required. When copper steel is not specified, the copper limit is a maximum requirement.
[C] Titanium is permitted for SS designations, at the producer's option, to the lesser of 3.4N + 1.5S or 0.025 %.
[D] For each reduction of 0.01 % below the specified carbon maximum, an increase of 0.06 % manganese above the specified maximum will be permitted up to a maximum of 1.50 %.
[E] HSLAS, HSLAS-F, and UHSS steels contain the strengthening elements columbium (niobium), vanadium, titanium, and molybdenum added singly or in combination. The minimum requirements only apply to the microalloy elements selected for strengthening of the steel.
[F] The purchaser has the option of restricting the nitrogen content. It should be noted that, depending on the microalloying scheme (for example, use of vanadium) of the producer, nitrogen may be a deliberate addition. Consideration should be made for the use of nitrogen binding elements (for example, vanadium, titanium).

composition limits are desirable, they shall be specified at the time of inquiry and confirmed at the time of ordering.

8. Mechanical Properties

8.1 *CS and DS*:

8.1.1 Typical, nonmandatory mechanical properties for CS and DS are found in Table 3.

TABLE 3 Typical Ranges of Mechanical Properties[A] (Nonmandatory)[B]
For Hot-Rolled Steel Sheet and Strip Designations CS and DS

Designation	Yield Strength[C]		Elongation in 2 in. [50 mm]%[C]
	ksi	MPa	
CS Types A, B, C, and D	30 to 50	[205 to 340]	≥25
DS Types A and B	30 to 45	[205 to 310]	≥28

[A] The yield strength tends to increase and the elongation tends to decrease as the sheet thickness decreases. These properties represent those typical of material in the thickness range of 0.100 to 0.150 in. [2.5 to 3.5 mm] for CS Types A, B, and DS Types A and B and in the thickness ranges of 0.060 to 0.075 in. [1.5 to 1.9 mm] for CS Type D.

[B] The typical mechanical property values presented here are nonmandatory. They are provided to assist the purchaser in specifying a suitable steel for a given application. Values outside these ranges are to be expected.

[C] Yield strength and elongation are measured in the longitudinal direction in accordance with Test Methods and Definitions A370.

8.1.2 The material shall be capable of being bent at room temperature in any direction through 180° flat on itself without cracking on the outside of the bent portion (see the section on bend test in Test Methods and Definitions A370). The bend test is not a requirement of delivery. However, if testing is performed by the purchaser, material not conforming to the requirement shall be subject to rejection.

8.2 *SS, HSLAS, HSLAS-F, and UHSS*:

8.2.1 The available grades and corresponding mechanical properties for SS, HSLAS, HSLAS-F, and UHSS are shown in Table 4.

8.2.2 *Tension Tests*:

8.2.2.1 *Requirements*—Material as represented by the test specimen shall conform to the mechanical property requirements specified in Table 4. These requirements do not apply to the uncropped ends of unprocessed coils.

8.2.2.2 *Number of Tests*—Two tension tests shall be made from each heat or from each 50 tons [45 000 kg]. When the amount of finished material from a heat is less than 50 tons [45 000 kg], one tension test shall be made. When material rolled from one heat differs 0.050 in. [1.27 mm] or more in thickness, one tension test shall be made from the thickest and thinnest material regardless of the weight represented.

8.2.2.3 Tension test specimens shall be taken at a point immediately adjacent to the material to be qualified.

TABLE 4 Mechanical Property Requirements[A]
For Hot Rolled Steel Sheet and Strip Designations SS, HSLAS, HSLAS-F, and UHSS

Designation	Yield Strength	Tensile Strength[B]	Elongation in 2 in. [50 mm] min, % for Thicknesses:			Elongation in 8 in. [200 mm], % for Thickness:
	ksi [MPa] min	ksi [MPa] min or range	Under 0.230 [6.0 mm] to 0.097 [2.5 mm]	Under 0.097 [2.5 mm] to 0.064 [1.6 mm]	Under 0.064 [1.6 mm] to 0.025 [0.65 mm]	Under 0.230 [6.0 mm]
SS:						
Grade 30 [205]	30 [205]	49 [340]	25	24	21	19
Grade 33 [230]	33 [230]	52 [360]	23	22	18	18
Grade 36 [250] Type 1	36 [250]	53 [365]	22	21	17	17
Grade 36 [250] Type 2	36 [250]	58-80 [400-550]	21	20	16	16
Grade 40 [275]	40 [275]	55 [380]	21	20	15	16
Grade 45 [310]	45 [310]	60 [410]	19	18	13	14
Grade 50 [340]	50 [340]	65 [450]	17	16	11	12
Grade 55 [380]	55 [380]	70 [480]	15	14	9	10
Grade 60 [410]	60 [410]	75 [520]	14	13	8	9
Grade 70 [480]	70 [480]	85 [585]	13	12	7	8
Grade 80 [550]	80 [550]	95 [620]	12	11	6	7
HSLAS:			Over 0.097 in. [2.5 mm]	Up to 0.097 in. [2.5 mm]		
Grade 45 [310] Class 1	45 [310]	60 [410]	25	23		...
Grade 45 [310] Class 2	45 [310]	55 [380]	25	23		...
Grade 50 [340] Class 1	50 [340]	65 [450]	22	20		...
Grade 50 [340] Class 2	50 [340]	60 [410]	22	20		...
Grade 55 [380] Class 1	55 [380]	70 [480]	20	18		...
Grade 55 [380] Class 2	55 [380]	65 [450]	20	18		...
Grade 60 [410] Class 1	60 [410]	75 [520]	18	16		...
Grade 60 [410] Class 2	60 [410]	70 [480]	18	16		...
Grade 65 [450] Class 1	65 [450]	80 [550]	16	14		...
Grade 65 [450] Class 2	65 [450]	75 [520]	16	14		...
Grade 70 [480] Class 1	70 [480]	85 [585]	14	12		...
Grade 70 [480] Class 2	70 [480]	80 [550]	14	12		...
HSLAS-F:						
Grade 50 [340]	50 [340]	60 [410]	24	22		...
Grade 60 [410]	60 [410]	70 [480]	22	20		...
Grade 70 [480]	70 [480]	80 [550]	20	18		...
Grade 80 [550]	80 [550]	90 [620]	18	16		...
UHSS:						
Grade 90 [620] Types 1 and 2	90 [620]	100 [690]	16	14		...
Grade 100 [690] Types 1 and 2	100 [690]	110 [760]	14	12		...

[A] For coil products, testing by the producer is limited to the end of the coil. Mechanical properties throughout the coil shall comply with the minimum values specified.

[B] A minimum and maximum tensile strength has been specified for SS36 Type 2.

8.2.2.4 Tension test specimens shall be taken from the full thickness of the sheet as-rolled.

8.2.2.5 Tension test specimens shall be taken from a location approximately halfway between the center of sheet and the edge of the material as-rolled.

8.2.2.6 Tension test specimens shall be taken with the lengthwise axis of the test specimen parallel to the rolling direction (longitudinal test).

8.2.2.7 *Test Method*—Yield strength shall be determined by either the 0.2 % offset method or the 0.5 % extension under load method unless otherwise specified.

8.2.3 *Bending Properties*:

8.2.3.1 The suggested minimum inside radii for cold bending are listed in Appendix X1 and is discussed in more detail in Specifications A568/A568M (6.6) and A749/A749M (7.6). Where a tighter bend radius is required, where curved or offset bends are involved, or where stretching or drawing are also a consideration, the producer shall be consulted.

9. Finish and Appearance

9.1 *Surface Finish*:

9.1.1 Unless otherwise specified, the material shall be furnished as rolled, that is, without removing the hot-rolled oxide or scale.

9.1.2 When required, it is permissible to specify that the material be pickled or blast cleaned (descaled).

9.2 *Oiling*:

9.2.1 Unless otherwise specified, as-rolled material shall be furnished not oiled (that is, dry), and pickled or blast cleaned material shall be furnished oiled.

9.3 *Edges*:

9.3.1 Steel sheet is available with mill edge or cut edge.

9.3.2 Steel strip is available with mill edge or cut edge.

10. Retests and Disposition of Non-Conforming Material

10.1 Retests, conducted in accordance with the requirements of Section 11.1 of Specification A568/A568M, are permitted when an unsatisfactorily test result is suspected to be the consequence of the test method procedure.

10.2 Disposition of non-conforming material shall be subject to the requirements of Section 11.2 of Specification A568/A568M.

11. Certification

11.1 A report of heat analysis shall be supplied, if requested, for CS and DS steels. For material with required mechanical properties, SS, HSLAS, HSLAS-F, and UHSS a report is required of heat analysis and mechanical properties as determined by the tension test.

11.2 The report shall include the purchase order number; the ASTM designation number and year date; product designation; grade; type or class, as applicable; the heat number; and as required, heat analysis and mechanical properties as indicated by the tension test.

11.3 A signature is not required on the test report. However, the document shall clearly identify the organization submitting the report. Notwithstanding the absence of a signature, the organization submitting the report is responsible for the content of the report.

11.4 A Material Test Report, Certificate of Inspection, or similar document printed from or used in electronic form from an electronic data interchange (EDI) transmission shall be regarded as having the same validity as a counterpart printed in the certifier's facility. The content of the EDI transmitted document must meet the requirements of the invoked ASTM standard and the purchaser and the supplier. Notwithstanding the absence of a signature, the organization submitting the EDI transmission is responsible for the content of the report.

12. Product Marking

12.1 In addition to the requirements of Specification A568/A568M for sheet and Specification A749/A749M for strip, each lift or coil shall be marked with the designation shown on the order {CS (Type A, B, or C), DS (Type A or B), SS (Grade and for SS36, Type), HSLAS (Grade and Class), HSLAS-F (Grade), or UHSS (Type and Grade)}. The designation shall be legibly stenciled on the top of each lift or shown on a tag attached to each coil or shipping unit.

13. Keywords

13.1 carbon steel sheet; carbon steel strip; commercial steel; drawing steel; high strength-low alloy steel; high strength-low alloy steel with improved formability; hot-rolled steel sheet; hot-rolled steel strip; steel sheet; steel strip; structural steel; ultra-high strength steel

---

APPENDIXES

(Nonmandatory Information)

X1. BENDING PROPERTIES

TABLE X1.1 Suggested Minimum Inside Radius for Cold Bending

NOTE 1—(t) Equals a radius equivalent to the steel thickness.
NOTE 2—The suggested radius should be used as a minimum for 90° bends in actual shop practice.
NOTE 3—Material which does not perform satisfactorily, when fabricated in accordance with the above requirements, may be subject to rejection pending negotiation with the steel supplier.

Designation	Grade	Minimum Inside Radius for Cold Bending	
Structural Steel			
	30[205]	1 t	
	33[230]	1 t	
	36[250] Type 1	1½ t	
	36[250] Type 2	2 t	
	40[275]	2 t	
	45[310]	2 t	
	50[340]	2½ t	
	55[380]	3 t	
	60[410]	3½ t	
	70[480]	4 t	
	80[550]	4 t	
High-Strength Low-Alloy Steel		Class 1	Class 2
	45[310]	1½ t	1½ t
	50[340]	2 t	1½ t
	55[380]	2 t	2 t
	60[410]	2½ t	2 t
	65[450]	3 t	2½ t
	70[480]	3½ t	3 t
High-Strength Low-Alloy Steel with Improved Formability			
	50[340]	1 t	
	60[410]	1½ t	
	70[480]	2 t	
	80[550]	2 t	
Ultra-High Strength Steel Types 1 and 2			
	90 [620]	2½ t	
	100 [690]	2½ t	

X2. RELATED ISO STANDARDS

The ISO standards listed below may be reviewed for comparison with this ASTM standard. The relationship between the standards may only be approximate; therefore, the respective standards should be consulted for actual requirements. Those who use these documents must determine which specifications address their needs.

ISO 3573 Hot-rolled Carbon Steel Sheet of Commercial and Drawing Qualities

ISO 4995 Hot-rolled Steel Sheet of Structural Quality

ISO 4996 Hot-rolled Steel Sheet of High Yield Stress Structural Quality

ISO 5951 Hot-rolled Steel Sheet of Higher Yield Strength with Improved Formability

ISO 6316 Hot-rolled Carbon Steel Strip of Structural Quality

ISO 6317 Hot-rolled Carbon Steel Strip of Commercial and Drawing Qualities

X3. HARDNESS PROPERTIES

X3.1 Table X3.1 lists the typical hardness values.

TABLE X3.1 Typical Hardness Values

NOTE 1—The hardness values shown are at the time of shipment.
NOTE 2—Tests for hardness shall be conducted in accordance with the requirements of Test Methods E18.
NOTE 3—The hardness values are Rockwell B scale as measured or converted from the appropriate Rockwell scales.
NOTE 4—The typical hardness values apply to the full range of steel sheet thickness. Hardness tends to increase as the steel sheet thickness decreases.
NOTE 5—Hardness testing is commonly used to assess the relative formability of various designations of uncoated steel sheet. This assessment done by many users is recognized to be only an approximation of the relative formability and therefore cannot be used as a specification requirement.

Designation	Hardness-Rockwell B Scale
CS Type A	75 or less
CS Type B	75 or less
CS Type C	75 or less
DS	65 or less

SUMMARY OF CHANGES

Committee A01 has identified the location of selected changes to this standard since the last issue (A1011/A1011M – 09a) that may impact the use of this standard. (Approved November 1, 2009.)

(1) Section 5.1.6.1 deleted.
(2) Reversed order of 5.1.11 and 5.1.12 and added new section 5.1.12.1.

Committee A01 has identified the location of selected changes to this standard since the last issue (A1011/A1011M – 09) that may impact the use of this standard. (Approved May 1, 2009.)

(1) Corrected SI equivalence values for SS Grade 60 and SS Grade 70 tensile strength in Table 4.

Committee A01 has identified the location of selected changes to this standard since the last issue (A1011/A1011M – 08) that may impact the use of this standard. (Approved April 1, 2009.)

(1) Revised Section 8.1.2.

ASTM International takes no position respecting the validity of any patent rights asserted in connection with any item mentioned in this standard. Users of this standard are expressly advised that determination of the validity of any such patent rights, and the risk of infringement of such rights, are entirely their own responsibility.

This standard is subject to revision at any time by the responsible technical committee and must be reviewed every five years and if not revised, either reapproved or withdrawn. Your comments are invited either for revision of this standard or for additional standards and should be addressed to ASTM International Headquarters. Your comments will receive careful consideration at a meeting of the responsible technical committee, which you may attend. If you feel that your comments have not received a fair hearing you should make your views known to the ASTM Committee on Standards, at the address shown below.

This standard is copyrighted by ASTM International, 100 Barr Harbor Drive, PO Box C700, West Conshohocken, PA 19428-2959, United States. Individual reprints (single or multiple copies) of this standard may be obtained by contacting ASTM at the above address or at 610-832-9585 (phone), 610-832-9555 (fax), or service@astm.org (e-mail); or through the ASTM website (www.astm.org). Permission rights to photocopy the standard may also be secured from the ASTM website (www.astm.org/COPYRIGHT/).

Designation: A1018/A1018M – 09

Standard Specification for
Steel, Sheet and Strip, Heavy-Thickness Coils, Hot-Rolled, Carbon, Commercial, Drawing, Structural, High-Strength Low-Alloy, High-Strength Low-Alloy with Improved Formability, and Ultra-High Strength[1]

This standard is issued under the fixed designation A1018/A1018M; the number immediately following the designation indicates the year of original adoption or, in the case of revision, the year of last revision. A number in parentheses indicates the year of last reapproval. A superscript epsilon (ε) indicates an editorial change since the last revision or reapproval.

1. Scope*

1.1 This specification covers hot-rolled, heavy-thickness coils beyond the size limits of Specification A1011/A1011M.

1.2 The product is available in six designations: Commercial Steel, Drawing Steel, Structural Steel, High-Strength Low-Alloy Steel, High-Strength Low-Alloy Steel with Improved Formability, and Ultra-High Strength Steel.

1.3 This material is available only in coils described as follows:

Product	Size Limits, Coils Only	
	Width, in. [mm]	Thickness, in. [mm]
Strip	Over 8 to 12, incl [Over 200 to 300]	0.230 to 1.000, incl [From 6.0 through 25]
Sheet	Over 12 [Over 300]	0.230 to 1.000, incl [From 6.0 through 25]

NOTE 1—The changes in width limits with the publication of A635/A635M – 06a result in a change in tensile testing direction for material from 0.180 in. [4.5 mm] to 0.230 in. exclusive [6.0 mm exclusive] over 48 in. [1200 mm] wide as that material is now covered by Specification A568/A568M – 06a. The purchaser is advised to discuss this change with the supplier.

1.4 Sheet and strip in coils of sizes noted in 1.3 are covered by this specification only with the following provisions:

1.4.1 The material is to be fed directly from coils into a blanking press, drawing or forming operation, tube mill, rolling mill, or sheared or slit into blanks for subsequent drawing or forming.

1.4.2 The material is not to be converted into steel plates for structural or pressure vessel use unless tested in complete accordance with the appropriate sections of Specifications A6/A6M (plates provided from coils) or A20/A20M (plates produced from coils). Plate converted from coils is no longer governed by this sheet steel specification and since this material is now a plate, the requirements of the appropriate plate specification shall apply, except in cases where there is a conflict between the requirements of the plate specification and this specification. In these cases, the more restrictive limits of either specification shall apply.

1.4.3 The dimensional tolerances of Specification A635/A635M are applicable to material produced to this specification.

1.4.4 Not all strength levels are available in all thicknesses. The user should consult the producer for appropriate size limitations.

1.5 The values stated in either SI units or inch-pound units are to be regarded separately as standard. The values stated in each system may not be exact equivalents; therefore, each system shall be used independently of the other. Combining values from the two systems may result in non-conformance with the standard.

2. Referenced Documents

2.1 *ASTM Standards:*[2]

A6/A6M Specification for General Requirements for Rolled Structural Steel Bars, Plates, Shapes, and Sheet Piling

A20/A20M Specification for General Requirements for Steel Plates for Pressure Vessels

A370 Test Methods and Definitions for Mechanical Testing of Steel Products

A568/A568M Specification for Steel, Sheet, Carbon, Structural, and High-Strength, Low-Alloy, Hot-Rolled and Cold-Rolled, General Requirements for

A572/A572M Specification for High-Strength Low-Alloy Columbium-Vanadium Structural Steel

A635/A635M Specification for Steel, Sheet and Strip,

---

[1] This specification is under the jurisdiction of ASTM Committee A01 on Steel, Stainless Steel and Related Alloys and is the direct responsibility of Subcommittee A01.19 on Steel Sheet and Strip.
Current edition approved Nov. 1, 2009. Published December 2009. Originally approved in 2001. Last previous edition approved in 2008 as A1018/A1018M – 08a. DOI: 10.1520/A1018_A1018M-09.

[2] For referenced ASTM standards, visit the ASTM website, www.astm.org, or contact ASTM Customer Service at service@astm.org. For *Annual Book of ASTM Standards* volume information, refer to the standard's Document Summary page on the ASTM website.

*A Summary of Changes section appears at the end of this standard.

Copyright © ASTM International, 100 Barr Harbor Drive, PO Box C700, West Conshohocken, PA 19428-2959, United States.

Heavy-Thickness Coils, Hot-Rolled, Alloy, Carbon, Structural, High-Strength Low-Alloy, and High-Strength Low-Alloy with Improved Formability, General Requirements for

A751 Test Methods, Practices, and Terminology for Chemical Analysis of Steel Products

A941 Terminology Relating to Steel, Stainless Steel, Related Alloys, and Ferroalloys

A1011/A1011M Specification for Steel, Sheet and Strip, Hot-Rolled, Carbon, Structural, High-Strength Low-Alloy, High-Strength Low-Alloy with Improved Formability, and Ultra-High Strength

E29 Practice for Using Significant Digits in Test Data to Determine Conformance with Specifications

G101 Guide for Estimating the Atmospheric Corrosion Resistance of Low-Alloy Steels

3. Terminology

3.1 *Definitions*—For definitions of other terms used in this specification refer to Terminology A941.

3.2 Definitions of Terms Specific to this Standard

3.2.1 *inclusion control, n*—the process of reducing the volume fraction of inclusions or modifying the shape of inclusions to improve formability, weldability, and machinability.

3.2.1.1 *Discussion*—Inclusions, especially those elongated during the rolling process, create the conditions for initiating and/or propagating cracks when the material is stretched or bent during the manufacture of a part. The adverse effects of inclusions are minimized by reducing the content of inclusions in the steel and/or by altering the shape of inclusions through the use of additions during the steelmaking process that change the elongated shape of the inclusions to less harmful small, well dispersed globular inclusions.

4. General Requirements for Delivery

4.1 Material furnished under this specification shall conform to the applicable requirements of the current edition of Specification A635/A635M, unless otherwise provided herein.

5. Classification

5.1 Heavy thickness coils are available in the following designations:

5.1.1 *Commercial Steel (CS Types A and B, and Standard Steel Designations)*

5.1.2 *Drawing Steel (DS Types A and B, and Standard Steel Designations)*

5.1.3 *Structural Steel*—(SS Grades 30[205], 33[230], 36[250] Types 1 and 2, and 40[275]).

5.1.4 *High-Strength Low-Alloy Steel*—(HSLAS Grades 45[310], 50[340], 55[380], 60[410], 65[450], 70[480]) in Classes 1 and 2.

5.1.4.1 This material is intended for miscellaneous applications where greater strength and savings in weight are important. The material is available in two classes. They are similar in strength level, except that Class 2 offers improved weldability and more formability than Class 1. Atmospheric corrosion resistance of these steels is equivalent to plain carbon steels. With copper specified, the atmospheric corrosion is somewhat enhanced.

5.1.5 *High-Strength Low-Alloy Steel with Improved Formability*—(HSLAS-F Grades 50[340], 60[410], 70[480], 80[550]).

5.1.5.1 This material has improved formability when compared with HSLAS. The steel is killed and made to a fine ferritic grain practice and includes microalloying elements such as columbium, titanium, vanadium, zirconium, etc. The steel shall be treated to achieve inclusion control. The material is intended for miscellaneous applications where higher strength, savings in weight, improved formability, and weldability are important. Atmospheric corrosion resistance of these steels is equivalent to plain carbon steels. With copper specified, the atmospheric corrosion resistance is somewhat enhanced.

NOTE 2—For methods of establishing the atmospheric corrosion resistance of low-alloy steels, see Guide G101.

5.1.6 *Ultra-High Strength Steel*—(UHSS Grades 90 [620] and 100 [690], Types 1 and 2).

5.1.6.1 This material has increased strength compared with HSLAS-F. The steel is killed and made to a fine ferritic grain practice, and includes microalloying elements such as columbium (niobium), titanium, vanadium, molybdenum, and so forth. The steel shall be treated to achieve inclusion control. The material is intended for miscellaneous applications where higher strength, savings in weight, and weldability are important. Atmospheric corrosion resistance of these steels is equivalent to plain carbon steels. With copper specified, the atmospheric corrosion resistance is somewhat enhanced.

5.1.7 When required for HSLAS, HSLAS-F, and UHSS steels, limitations on the use of one or more of the microalloy elements shall be specified on the order.

5.2 The limits for copper, chromium, nickel, and molybdenum are available in two levels, Limits A and Limits B (see Table 1).

6. Ordering Information

6.1 Orders for material under this specification shall include the following information, as required, to describe adequately the desired material.

6.1.1 ASTM specification number and year of issue.

6.1.2 Name of material and designation (hot-rolled steel sheet or hot-rolled strip) (include grade and, as appropriate, type and class for CS, DS, SS, HSLAS, HSLAS-F, and UHSS) (see 5.1).

6.1.2.1 For CS and DS, when a type is not specified, Type B will be furnished.

6.1.2.2 For SS Grade 36, when a type is not specified, Type 1 will be furnished (see 5.1).

6.1.2.3 For UHSS, when a type is not specified, Type 1 shall be furnished.

6.1.2.4 For HSLAS, when a class is not specified, Class 1 will be furnished (see 5.1),

6.1.3 Copper bearing, (if required),

6.1.4 For SS, HSLAS, and HSLAS-F and selected CS steels, specify the limits for chemical requirements listed in

TABLE 1 Chemical Requirements: Cu, Ni, Cr and Mo for Commercial Steels, Structural Steels, High-Strength Low-Alloy Steels, and High-Strength Low-Alloy Steels with Improved Formability

Designation	Limits	Cu[A,B]	Ni[B]	Cr[B,C]	Mo[B,C]
% Heat Analysis, Element Maximum Unless Otherwise Shown					
CS:					
Grades 1015, 1016, 1017, 1018, 1019, 1020, 1021, 1022, 1023, 1524	A	0.20	0.20	0.15	0.06
	B	0.40	0.40	0.30	0.12
SS:					
All grades	A	0.20	0.20	0.15	0.06
	B	0.40	0.40	0.30	0.12
HSLAS:					
All grades and classes except for Grade 70 [480]	A	0.20	0.20	0.15	0.06
	B	0.40	0.40	0.30	0.12
Grade 70 [480] Class 1 and Class 2	A	0.20	0.20	0.15	0.16
	B	0.40	0.40	0.30	0.16
HSLAS-F:					
Grades 50 [340] and 60 [410]	A	0.20	0.20	0.15	0.06
	B	0.40	0.40	0.30	0.12
Grade 70 [480] and 80 [550]	A	0.20	0.20	0.15	0.16
	B	0.40	0.40	0.30	0.16

[A] When copper bearing steel is specified, the minimum limit for copper is 0.20 %. When copper bearing steel is not specified, the maximum limit for copper is as shown in the table.

[B] For Limits B steels, the sum of copper, nickel, chromium and molybdenum shall not exceed 1.00 % on heat analysis. When one or more of these elements are specified by the purchaser, the sum does not apply; in which case, only the individual limits on the remaining elements shall apply.

[C] For Limits B steels, the sum of chromium and molybdenum shall not exceed 0.32 % on heat analysis. When one or more of these elements are specified, the sum does not apply; in which case, only the individual limits on the remaining elements shall apply.

Table 1 (elements Cu, Cr, Ni, and Mo). When Limits A or Limits B is not specified, Limits A shall be furnished.

6.1.5 *Condition*—Material in accordance with this specification is furnished in the hot rolled condition. Pickled (or blast cleaned) must be specified if required. Material ordered as pickled (or blast cleaned) will be oiled unless ordered dry,

6.1.6 Type of edge must be specified for hot rolled sheet coils and strip coils, either mill edge or cut edge (sheet), mill edge or slit edge (strip),

6.1.7 Dimensions (decimal thickness and width of material),

NOTE 3—Not all producers are capable of meeting all the limitations of the thickness tolerance tables in Specification A635/A635M. The purchaser should contact the producer regarding possible limitations prior to placing an order.

6.1.8 Coils size and weight requirements (must include inside diameter (ID), outside diameter (OD), and maximum weight,

6.1.9 Quantity (weight),

6.1.10 Application (part identification and description). Orders for conversion to plate shall include reference to the applicable ASTM plate specification.

6.1.11 A report is required of heat analysis and mechanical properties as determined by the tension test, and

6.1.12 Special requirements (if any).

6.1.12.1 When the purchaser requires a limit on "carbon equivalent" (see Supplementary Requirement S1), this requirement shall be specified in the purchase order or contract.

6.1.12.2 When the purchaser requires thickness tolerances for 3/8 in. [10 mm] minimum edge distance (see Supplementary Requirement in Specification A635/A635M), this requirement shall be specified in the purchase order or contract.

NOTE 4—A typical ordering description is as follows: (inch pound units) ASTM A1018/A1018M: Grade 50, High-Strength, Low-Alloy Steel, Class 2, Limits B, hot-rolled sheet coils, pickled and oiled, cut edge, 0.500 by 40 in. by coil; ID 24 in., OD 72 in., maximum; coil weight 40 000 lb., maximum; 200 000 lb. for roll forming shapes; (SI units) ASTM A1018/A1018M: Grade 340, High-Strength Low-Alloy Steel, Class 2, Limits B, hot-rolled sheet coils, pickled and oiled, cut edge; 10 mm by 900 mm by coil; ID 600 mm, OD 1800 mm, maximum; coil weight 18 000 kg maximum; 90 000 kg for roll forming shapes. For conversion to plate: (inch-pound units) ASTM A1018/A1018M: Grade 50, High-Strength Low-Alloy Steel, Class 1, Limits A, hot-rolled sheet coils, as rolled, mill edge, 0.500 in. by 50 in. by coil, ID 24 in., OD 72 in., maximum; coil weight 40 000 lb., maximum; 200 000 lb. for conversion to plate, Specification A572/A572M Grade 50; (SI units) ASTM A1018/A1018M: Grade 340, Structural Steel, hot-rolled sheet coils, as rolled, mill edge; 10 mm by 1000 mm by coil; ID 600 mm, OD 1800 mm, maximum; coil weight 18 000 kg maximum; 100 000 kg for conversion to plate, Specification A572/A572M Grade 340.

7. Chemical Composition

7.1 The heat analysis of commercial steel and drawing steel shall conform to the requirements of Table 2, and where appropriate, Table 3.

7.2 The heat analysis of structural steel, high-strength low-alloy steel, high-strength low-alloy steel with improved formability, and ultra-high strength steel shall conform to the requirements of Table 2 and Table 3.

7.3 Table 1 describes the heat analysis requirements for two sets of limits (A and B) for the elements copper, chromium, nickel, and molybdenum. The required set of limits (A and B) for these elements shall be specified on the order.

7.4 Chemical analysis shall be conducted in accordance with Test Methods, Practices, and Terminology A751.

7.5 Each of the elements listed in Tables 2 and 4 shall be included in the report of the heat analysis. When the amount of copper, nickel, chromium, or molybdenum is less than 0.02 %, report the analysis as <0.02 % or the actual value. When the amount of columbium, titanium, or vanadium is less than

TABLE 2 Chemical Requirements[A]
Commercial and Drawing Steels

% Heat Analysis, Element Maximum Unless Otherwise Shown

Designation	C	Mn	P	S	Al	Si	Cu	Ni	Cr	Mo	V	Cb	Ti[B]	N	B
Commercial Steels (CS)															
CS Type A	0.10	0.60	0.030	0.035	...	...	0.20[C]	0.20	0.15	0.06	0.008	0.008	0.025	...	...
CS Type B	0.02 to 0.15	0.60	0.030	0.035	...	...	0.20[C]	0.20	0.15	0.06	0.008	0.008	0.025	...	...
1007	0.02 to 0.10	0.50	0.030	0.035	...	...	0.20[C]	0.20	0.15	0.06	0.008	0.008	0.025	...	...
1008	0.10	0.50	0.030	0.035	...	...	0.20[C]	0.20	0.15	0.06	0.008	0.008	0.025	...	...
1009	0.15	0.60	0.030	0.035	...	...	0.20[C]	0.20	0.15	0.06	0.008	0.008	0.025	...	...
1010	0.08 to 0.13	0.30 to 0.60	0.030	0.035	...	...	0.20[C]	0.20	0.15	0.06	0.008	0.008	0.025	...	...
1012	0.10 to 0.15	0.30 to 0.60	0.030	0.035	...	...	0.20[C]	0.20	0.15	0.06	0.008	0.008	0.025	...	...
1015	0.13 to 0.18	0.30 to 0.60	0.030	0.035	...	...					0.008	0.008	0.025	...	...
1016	0.13 to 0.18	0.60 to 0.90	0.030	0.035	...	...					0.008	0.008	0.025	...	...
1017	0.15 to 0.20	0.30 to 0.60	0.030	0.035	...	...					0.008	0.008	0.025	...	...
1018	0.15 to 0.20	0.60 to 0.90	0.030	0.035	...	...	See Table 1 for limits of copper, chromium, nickel, and molybdenum				0.008	0.008	0.025	...	...
1019	0.15 to 0.20	0.70 to 1.00	0.030	0.035	...	...					0.008	0.008	0.025	...	...
1020	0.18 to 0.23	0.30 to 0.60	0.030	0.035	...	...					0.008	0.008	0.025	...	...
1021	0.18 to 0.23	0.60 to 0.90	0.030	0.035	...	...					0.008	0.008	0.025	...	...
1022	0.18 to 0.23	0.70 to 1.00	0.030	0.035	...	...					0.008	0.008	0.025	...	...
1023	0.20 to 0.25	0.30 to 0.60	0.030	0.035	...	...					0.008	0.008	0.025	...	...
1524	0.19 to 0.25	1.35 to 1.65	0.030	0.035	...	...					0.008	0.008	0.025	...	...
Drawing Steels (DS)															
DS Type A	0.08	0.50	0.020	0.030	0.01 min	...	0.20	0.20	0.15	0.06	0.008	0.008	0.025	...	...
DS Type B	0.02 to 0.08	0.50	0.020	0.030	0.01 min	...	0.20	0.20	0.15	0.06	0.008	0.008	0.025	...	...
1006	0.08	0.45	0.030	0.035	0.01 min	...	0.20	0.20	0.15	0.06	0.008	0.008	0.025	...	...
1006A	0.02 to 0.08	0.45	0.030	0.035	0.01 min	...	0.20	0.20	0.15	0.06	0.008	0.008	0.025	...	...

[A] Where an ellipsis (. . .) appears in the table, there is no requirement, but the analysis shall be reported.
[B] Titanium is permitted at the producer's option, to the lesser of 3.4N + 1.5S or 0.025 %.
[C] When copper steel is specified, the copper limit is a minimum requirement. When copper steel is not specified, the copper limit is a maximum requirement.

TABLE 3 Chemical Requirements[A]
Structural Steels, High-Strength Low-Alloy Steels, and High-Strength Low-Alloy Steels with Improved Formability

% Heat Analysis, Element Maximum Unless Otherwise Shown

Designation	C	Mn	P	S	Al	Si	V	Cb	Ti	N
SS[B]										
Grade 30 [205]	0.25	1.50	0.035	0.04	...	...	0.008	0.008	0.025	0.014
Grade 33 [230]	0.25	1.50	0.035	0.04	...	...	0.008	0.008	0.025	0.014
Grade 36 [250] Type 1	0.25	1.50	0.035	0.04	...	...	0.008	0.008	0.025	0.014
Grade 36 [250] Type 2	0.25	...[C]	0.035	0.04	...	...	0.008	0.008	0.025	0.014
Grade 40 [275]	0.25	1.50	0.035	0.04	...	...	0.008	0.008	0.025	0.014
HSLAS[D]										
Grade 45 [310] Class 1	0.22	1.50	0.04	0.04	...	...	0.005 min	0.005 min	0.005 min	...
Grade 45 [310] Class 2	0.15	1.50	0.04	0.04	...	...	0.005 min	0.005 min	0.005 min	[E]
Grade 50 [340] Class 1	0.23	1.50	0.04	0.04	...	...	0.005 min	0.005 min	0.005 min	...
Grade 50 [340] Class 2	0.15	1.50	0.04	0.04	...	...	0.005 min	0.005 min	0.005 min	[E]
Grade 55 [380] Class 1	0.25	1.50	0.04	0.04	...	...	0.005 min	0.005 min	0.005 min	...
Grade 55 [380] Class 2	0.15	1.50	0.04	0.04	...	...	0.005 min	0.005 min	0.005 min	[E]
Grade 60 [410] Class 1	0.26	1.50	0.04	0.04	...	...	0.005 min	0.005 min	0.005 min	...
Grade 60 [410] Class 2	0.15	1.50	0.04	0.04	...	...	0.005 min	0.005 min	0.005 min	[E]
Grade 65 [450] Class 1	0.26	1.50	0.04	0.04	...	...	0.005 min	0.005 min	0.005 min	[E]
Grade 65 [450] Class 2	0.15	1.50	0.04	0.04	...	...	0.005 min	0.005 min	0.005 min	[E]
Grade 70 [480] Class 1	0.26	1.65	0.04	0.04	...	...	0.005 min	0.005 min	0.005 min	[E]
Grade 70 [480] Class 2	0.15	1.65	0.04	0.04	...	...	0.005 min	0.005 min	0.005 min	[E]
HSLAS-F[D]										
Grade 50 [340]	0.15	1.65	0.025	0.035	...	...	0.005 min	0.005 min	0.005 min	[E]
Grade 60 [410]	0.15	1.65	0.025	0.035	...	...	0.005 min	0.005 min	0.005 min	[E]
Grade 70 [480]	0.15	1.65	0.025	0.035	...	...	0.005 min	0.005 min	0.005 min	[E]
Grade 80 [550]	0.15	1.65	0.025	0.035	...	...	0.005 min	0.005 min	0.005 min	[E]

[A] An ellipsis (. . .) indicates that no limits have been set for that element. See Table 1 for requirements for Cu, Ni, Cr, and Mo.
[B] Titanium is permitted for SS designations, at the producer's option, to the lesser of 3.4N + 1.5S or 0.025 %.
[C] For product greater than 0.75 in. [20 mm] in thickness, the manganese requirement is 0.80 to 1.20 %. For each reduction of 0.01 % below the specified carbon maximum, an increase of 0.06 % manganese above the specified maximum will be permitted up to a maximum of 1.35 %.
[D] HSLAS and HSLAS-F steels contain the strengthening elements columbium, vanadium, and titanium added singly or in combination. The minimum requirements only apply to the microalloy elements selected for strengthening of the steel.
[E] The purchaser has the option of restricting the nitrogen content. It should be noted that, depending on the microalloying scheme (for example, use of vanadium) of the producer, nitrogen may be a deliberate addition. Consideration should be made for the use of nitrogen binding elements (for example, vanadium, titanium).

0.008 %, report the analysis as <0.008 % or the actual deter-

**TABLE 4 Chemical Requirements
Ultra-High Strength Steels**

Designation	% Heat Analysis, Element Maximum Unless Otherwise Shown											
	C	Mn	P	S	Cu[A]	Ni	Cr	Mo	V[B]	Cb[B]	Ti[B]	N
UHSS												
Grade 90 [620] Type 1	0.15	2.00	0.020	0.025	0.20	0.20	0.15	0.40	0.005 min	0.005 min	0.005 min	[C]
Grade 90 [620] Type 2	0.15	2.00	0.020	0.025	0.60	0.50	0.30	0.40	0.005 min	0.005 min	0.005 min	[C]
Grade 100 [690] Type 1	0.15	2.00	0.020	0.025	0.20	0.20	0.15	0.40	0.005 min	0.005 min	0.005 min	[C]
Grade 100 [690] Type 2	0.15	2.00	0.020	0.025	0.60	0.50	0.30	0.40	0.005 min	0.005 min	0.005 min	[C]

[A] When copper steel is specified, a minimum of 0.20 % is required. When copper steel is not specified, the copper limit is a maximum requirement.
[B] UHSS steels contain the strengthening elements columbium (niobium), vanadium, and titanium added singly or in combination. The minimum requirements only apply to the microalloy elements selected for strengthening of the steel.
[C] The purchaser has the option of restricting the nitrogen content. It should be noted that, depending on the microalloying scheme (for example, use of vanadium) of the producer, nitrogen may be a deliberate addition. Consideration should be made for the use of nitrogen binding elements (for example, vanadium, titanium).

mined value. When the amount of boron is less than 0.0005 %, report the analysis as <0.0005 % or the actual determined value.

7.6 For Structural Steel (SS) the addition of microalloying elements, including columbium, vanadium, or titanium, as well as nitrogen, as strength enhancers is prohibited.

7.7 Sheet steel grades defined by this specification are suitable for welding if appropriate welding conditions are selected. For certain welding processes, more restrictive composition limits may be desirable and should be requested at the time of inquiry and ordering.

8. Mechanical Properties

8.1 Test specimen preparation and mechanical testing shall be in accordance with Test Methods and Definitions A370.

8.2 *Tensile Properties*—The material, structural steel, high-strength low-alloy steel, high-strength low-alloy steel with improved formability, and ultra-high strength steel, as represented by the test specimens shall conform to the mechanical property requirements as stated in Table 5. These requirements do not apply to the uncropped ends of unprocessed coils.

TABLE 5 Mechanical Property Requirements[A] For Hot Rolled Heavy Thickness Coils

Designations SS, HSLAS, and HSLAS-F				
Designation	Yield Strength min ksi [MPa]	Tensile Strength min[B] ksi [MPa]	Elongation in 2 in. [50 mm], min, % for thicknesses to 1 in. [25 mm] incl.	Elongation in 8 in. [200 mm], min, % for thicknesses 0.180 in. [4.5 mm] to 1 in. [25 mm] incl.
SS				
Grade 30 [205]	30 [205]	49 [340]	22	17
Grade 33 [230]	33 [230]	52 [360]	22	16
Grade 36 [250] Type 1	36 [250]	53 [365]	21	15
Grade 36 [250] Type 2	36 [250]	58 to 80 [400 to 550]	21	18
Grade 40 [275]	40 [275]	55 [380]	19	14
HSLAS				
Grade 45 [310] Class 1	45 [310]	60 [410]	22	17
Grade 45 [310] Class 2	45 [310]	55 [380]	22	17
Grade 50 [340] Class 1	50 [340]	65 [450]	20	16
Grade 50 [340] Class 2	50 [340]	60 [410]	20	16
Grade 55 [380] Class 1	55 [380]	70 [480]	18	15
Grade 55 [380] Class 2	55 [380]	65 [450]	18	15
Grade 60 [410] Class 1	60 [410]	75 [520]	16	14
Grade 60 [410] Class 2	60 [410]	70 [480]	16	14
Grade 65 [450] Class 1	65 [450]	80 [550]	14	12
Grade 65 [450] Class 2	65 [450]	75 [520]	14	12
Grade 70 [480] Class 1	70 [480]	85 [590]	12	10
Grade 70 [480] Class 2	70 [480]	80 [550]	12	10
HSLAS-F				
Grade 50 [340]	50 [340]	60 [410]	22	16
Grade 60 [410]	60 [410]	70 [480]	16	14
Grade 70 [480]	70 [480]	80 [550]	12	10
Grade 80 [550]	80 [550]	90 [620]	12	10
UHSS				
Grade 90 [620] Types 1 and 2	90 [620]	100 [690]	10	8
Grade 100 [690] Types 1 and 2	100 [690]	110 [760]	10	8

[A] For coil products, testing by the producer is limited to the end of the coil. Mechanical properties throughout the coil shall comply with the minimum values specified.
[B] A minimum and maximum tensile strength are specified for SS Grade 36 Type 2.

8.3 *Tension Test Specimen Location and Orientation*—Tension test specimens shall be taken sufficiently far from the as hot-rolled coil ends so that the sample is representative of material which received the designed processing. The test shall be taken approximately midway between the center and edge of the material as rolled. For coils wider than 24 in. [600 mm], Tension test specimens shall be taken such that the longitudinal axis of the specimens is perpendicular to the direction of rolling (transverse test). For coils through 24 in. [600 mm] in width, tension test specimens shall be taken such that longitudinal axis of the specimen is parallel to the direction of rolling (longitudinal test).

8.4 *Tension Tests*—Two tension tests shall be conducted from each heat or each of 50 tons [45 Mg]. When the amount of finished material from a heat is less than 50 tons [45 Mg], only one tension test shall be conducted. When material rolled from one heat differs 0.050 in. [1.3 mm] or more in thickness, one tension test shall be conducted from both the thickest and the thinnest material rolled regardless of the weight represented.

8.5 To determine conformance with this specification, a test value should be rounded to the nearest 1 ksi [7 Mpa] of tensile strength and yield point, and to the nearest unit in the right-hand place of figures used in expressing the limiting value for other places in accordance with the rounding off methods given in Practice E29.

8.6 Structural steel, high-strength low-alloy steel, high-strength low-alloy steel with improved formability, and ultra-high strength steel covered by this specification are commonly fabricated by cold bending. There are many interrelated factors that affect the ability of a given steel to cold form over a given radius under shop conditions. These factors include thickness, strength level, degree of restraint, relationship to rolling direction, chemistry, and microstructure. The producer shall be consulted concerning the recommended minimum inside radius and bending direction. Where possible, a larger radius or "easy way" bending (with the bend axis perpendicular to rolling direction), or both, are recommended for improved performance.

8.7 Fabricators must be aware that cracks may initiate upon bending a sheared or burned edge. This is not considered a fault of the steel, but is rather a function of the induced cold work or heat affected zone.

9. Workmanship, Finish, and Appearance

9.1 *Edges*—The normal edge condition in heavy-thickness coils is mill edge. If cut edge is required, it must be specified.

9.2 *Oiling*—Unless otherwise specified, hot-rolled as-rolled material shall be furnished dry, and hot rolled pickled or blast cleaned material shall be furnished oiled. When required, it is permissible to specify pickled or blast cleaned material be furnished dry, or that as-rolled material be furnished oiled.

9.3 *Surface Finish*—Unless otherwise specified, hot-rolled material shall have an as-rolled, not pickled surface finish. When required, material shall be specified to be pickled or blast-cleaned.

10. Retests and Disposition of Non-Conforming Material

10.1 Retests, conducted in accordance with the requirements of Section 10.1 of Specification A635/A635M, are permitted when an unsatisfactory test result is suspected to be the consequence of the test method procedure.

10.2 Disposition of non-conforming material shall be subject to the requirements of Section 10.2 of Specification A635/A635M.

11. Certification

11.1 A report of heat analysis shall be supplied, if requested, for CS and DS steels. For material with required mechanical properties, SS, HSLAS, HSLAS-F, and UHSS, a report is required of heat analysis and mechanical properties as determined by the tension test.

12. Keywords

12.1 carbon steel sheet; carbon steel strip; commercial steel; drawing steel; heavy-thickness coils; high-strength low-alloy steel; hot-rolled steel sheet; hot-rolled steel strip; improved formability; steel sheet; steel strip; structural applications; ultra-high strength steel

---

SUPPLEMENTARY REQUIREMENTS

The following standardized supplementary requirement is for use when desired by the purchaser. These additional requirements shall apply only when specified on the order.

S1. Carbon Equivalent

S1.1 When a purchaser places limits on the carbon equivalent (CE), the CE value shall be calculated in accordance with the following formula:

$$CE = \% \text{ carbon} + \frac{\% \text{ manganese}}{6} + \frac{(\% \text{ chromium} + \% \text{ molybdenum} + \% \text{ vanadium})}{5} + \frac{(\% \text{ nickel} + \% \text{ copper})}{15}$$

APPENDIX

(Nonmandatory Information)

X1. BENDING PROPERTIES—STRUCTURAL STEEL, HIGH-STRENGTH LOW-ALLOY STEEL, HIGH-STRENGTH LOW-ALLOY STEEL WITH IMPROVED FORMABILITY, AND ULTRA-HIGH STRENGTH STEEL

TABLE X1.1 Suggested Minimum Inside Radius for Cold Bending

NOTE 1—(t) equals a radius equivalent to the steel thickness
NOTE 2—The suggested radius should be used as a minimum for 90° bend in actual shop practice.
NOTE 3—Material that does not perform satisfactorily, when fabricated in accordance with the above requirements, may be subject to rejection pending negotiation with the steel supplier.

Designation	Grade	Minimum Inside Radius for Cold Bending	
Structural Steel			
	30 [205]	1t	
	33 [230]	1t	
	36 [250] Type 1	1½t	
	36 [250] Type 2	2t	
	40 [275]	2t	
High-Strength Low-Alloy Steel		Class 1	Class 2
	45 [310]	1½t	1½t
	50 [340]	2t	1½t
	55 [380]	2t	2t
	60 [410]	2½t	2t
	65 [450]	3t	2½t
	70 [480]	3½t	3t
High-Strength Low-Alloy Steel with Improved Formability			
	50 [340]	1t	
	60 [410]	1½t	
	70 [480]	2t	
	80 [550]	2t	
Ultra-High Strength Steel Types 1 and 2			
	90 [620]	2½ t	
	100 [690]	2½ t	

SUMMARY OF CHANGES

Committee A01 has identified the location of selected changes to this standard since the last issue (A1018/A1018M – 08a) that may impact the use of this standard. (Approved November 1, 2009.)

(1) Section 6.1.7.1 deleted
(2) Reversed order of 6.1.11 and 6.1.12, revised original 6.1.11.1 to remove permissive language, and added new section 6.1.12.2

Committee A01 has identified the location of selected changes to this standard since the last issue (A1018/A1018M – 08) that may impact the use of this standard. (Approved October 1, 2008)

(1) Revised maximum limits for phosphorus and sulfur of DS Type A and Type B steels in Table 2.

ASTM International takes no position respecting the validity of any patent rights asserted in connection with any item mentioned in this standard. Users of this standard are expressly advised that determination of the validity of any such patent rights, and the risk of infringement of such rights, are entirely their own responsibility.

This standard is subject to revision at any time by the responsible technical committee and must be reviewed every five years and if not revised, either reapproved or withdrawn. Your comments are invited either for revision of this standard or for additional standards and should be addressed to ASTM International Headquarters. Your comments will receive careful consideration at a meeting of the responsible technical committee, which you may attend. If you feel that your comments have not received a fair hearing you should make your views known to the ASTM Committee on Standards, at the address shown below.

This standard is copyrighted by ASTM International, 100 Barr Harbor Drive, PO Box C700, West Conshohocken, PA 19428-2959, United States. Individual reprints (single or multiple copies) of this standard may be obtained by contacting ASTM at the above address or at 610-832-9585 (phone), 610-832-9555 (fax), or service@astm.org (e-mail); or through the ASTM website (www.astm.org). Permission rights to photocopy the standard may also be secured from the ASTM website (www.astm.org/COPYRIGHT/).

Standard Specification for Structural Steel with Low Yield to Tensile Ratio for Use in Buildings[1]

This standard is issued under the fixed designation A1043/A1043M; the number immediately following the designation indicates the year of original adoption or, in the case of revision, the year of last revision. A number in parentheses indicates the year of last reapproval. A superscript epsilon (ε) indicates an editorial change since the last revision or reapproval.

1. Scope

1.1 This specification covers two grades, 36 [250] and 50 [345] of rolled steel structural shapes and plates with low yield to tensile ratio for use in building framing or for general structural purposes.

1.2 All shape profiles with a flange width of 6 in. [150 mm] and greater described in Specification A6/A6M Annex A2 and plates up to and including 5 in. [125 mm] thick are included in this specification.

1.3 Supplementary requirements are provided for use where additional testing or additional restrictions are required by the purchaser. Such requirements apply only when specified in the purchase order.

1.4 When the steel is to be welded, a welding procedure suitable for the grade of steel and intended use or service is to be utilized. See Appendix X3 of Specification A6/A6M for information on weldability.

1.5 The text of this specification contains notes or footnotes, or both, that provide explanatory material; such notes and footnotes, excluding those in tables and figures, do not contain any mandatory requirements.

1.6 The values stated in either inch-pound units or SI units are to be regarded separately as standard. Within the text, the SI units are shown in brackets. The values stated in each system are not exact equivalents; therefore, each system is to be used independently of the other without combining values in any way.

2. Referenced Documents

2.1 *ASTM Standards:*[2]

A6/A6M Specification for General Requirements for Rolled Structural Steel Bars, Plates, Shapes, and Sheet Piling

A370 Test Methods and Definitions for Mechanical Testing of Steel Products

A673/A673M Specification for Sampling Procedure for Impact Testing of Structural Steel

A770/A770M Specification for Through-Thickness Tension Testing of Steel Plates for Special Applications

3. General Requirements for Delivery

3.1 Product furnished under this specification shall conform to the requirements of the current edition of Specification A6/A6M for the ordered product, unless a conflict exists, in which case this specification shall prevail.

4. Materials and Manufacture

4.1 The steel shall be killed, and such shall be affirmed in the test report by a statement of *killed steel*, a value of 0.10 % or more for the silicon content, a value of 0.015 % or more for the total aluminum content, or a value of 0.006 % or more for the titanium content.

5. Chemical Composition

5.1 The heat analysis shall conform to the requirements given in Table 1.

5.2 The steel shall conform on product analysis to the requirements given in Table 1, subject to the product analysis tolerances in Specification A6/A6M.

---

[1] This specification is under the jurisdiction of ASTM Committee A01 on Steel, Stainless Steel and Related Alloys and is the direct responsibility of Subcommittee A01.02 on Structural Steel for Bridges, Buildings, Rolling Stock and Ships.
Current edition approved Oct. 1, 2009. Published December 2009. Originally approved in 2005. Last previous edition approved in 2005 as A1043/A1043M – 05. DOI: 10.1520/A1043_A1043M-05R09.

[2] For referenced ASTM standards, visit the ASTM website, www.astm.org, or contact ASTM Customer Service at service@astm.org. For *Annual Book of ASTM Standards* volume information, refer to the standard's Document Summary page on the ASTM website.

TABLE 1 Chemical Requirements (Heat Analysis)

Element		Composition, %	
		Grade 36 [250]	Grade 50 [345]
Carbon, max	2 in. [50 mm] and less in thickness	0.20	0.20
	Over 2 in. [50 mm] in thickness	0.22	0.22
Manganese		0.50–1.40	0.50–1.60
Phosphorus, max		0.035	0.035
Sulfur, max		0.045	0.045
Silicon, max		0.35	0.55
Nitrogen, max		0.012	0.012
Nickel, max		0.45	0.45
Chromium, max		0.35	0.35
Molybdenum, max		0.15	0.15
Copper, max		0.50	0.50
Vanadium, max		0.15[A]	0.15[A]
Columbium, max		0.05[A]	0.05[A]
Titanium, max		0.03	0.03

[A] The sum of columbium and vanadium shall not exceed 0.15 %.

5.3 The maximum permissible carbon equivalent values shall be:

Grade	Thickness of Plate and Shape Flange	CE Value max %
36 [250]	All	0.37
50 [345]	≤ 2 in. [50 mm]	0.45
50 [345]	> 2 in. [50 mm]	0.47

5.3.1 The carbon equivalent value shall be based upon heat analysis. The required chemical analysis as well as the carbon equivalent shall be reported. The carbon equivalent shall be calculated using the following formula:

$$CE = C + (Mn)/6 + (Cr + Mo + V)/5 + (Ni + Cu)/15 \quad (1)$$

6. Tensile Requirements

6.1 The product as represented by the test specimens shall conform to the requirements for tensile properties given in Table 2.

TABLE 2 Tensile Requirements[A]

	Grade	
	36 [250]	50 [345]
Tensile strength, min, ksi [MPa]	58 [400]	65 [450]
Yield point, ksi [MPa]	36 to 52 [250 to 360]	50 to 65 [345 to 450]
Yield to tensile ratio, max	0.80	0.80
Elongation in 8 in. [200 mm], min %[B]	20	18
Elongation in 2 in. [50 mm], min %[B]	23	21

[A] See Specimen Orientation under the Tension Tests section of Specification A6/A6M.
[B] For plates wider than 24 in. [600 mm] the elongation requirement is reduced two percentage points. See Elongation Requirement Adjustments in the Tension Tests section of Specification A6/A6M.

7. Charpy Impact Requirements

7.1 Charpy V-notch tests shall be conducted in accordance with Specification A673/A673M, frequency H. The test results for full-size test specimens shall conform to the following minimum average value for Grades 36 [250] and 50 [345]:

40 ft·lbf [54 J] at 70°F [21°C]

or a lower test temperature as specified in the purchase order.

7.2 For shapes with a flange thickness equal to or greater than 1½ in. [38.1 mm] the test specimens shall be taken from the Alternate Core Location as defined in Specification A673/A673M.

8. Keywords

8.1 building framing; low yield to tensile ratio; plates; shapes; structural steel; welded construction

---

SUPPLEMENTARY REQUIREMENTS

Supplementary requirements shall not apply unless specified in the purchase order or contract. Standardized supplementary requirements for use at the option of the purchaser are listed in Specification A6/A6M. Those that are considered suitable for use with this specification are listed by title.

S1. Vacuum Treatment,
S2. Product Analysis, and
S8. Ultrasonic Examination.

ASTM A1043/A1043M − 05 (2009)

ADDITIONAL SUPPLEMENTARY REQUIREMENTS

In addition, the following optional supplementary requirements are suitable for use with this specification.

S71. Through-thickness Tension Tests

S71.1 Through-thickness tension test shall be made in accordance with the requirements of Specification A770/A770M except for test frequency. The test frequency shall be the same as the tension test frequency.

S72. P_{cm} Carbon Equivalent Limit

S72.1 The P_{cm} carbon equivalent for each heat, based upon heat analysis, shall not exceed 0.26 % for grade 36 [250] or 0.29 % for Grade 50 [345], calculated using the following formula:

$$P_{cm} = C + Si/30 + (Mn + Cu + Cr)/20 + Ni/60 + Mo/15 + V/10 + 5B \%$$

S73. Grade 50 Plates Restricted Carbon Equivalent Limit

S73.1 The plates shall be produced by the thermo-mechanical control process (TMCP) or shall be quenched and tempered.

S73.2 The carbon equivalent for each heat, based upon heat analysis, shall not exceed 0.38 % for plates 2 in. [50 mm] or under in thickness and 0.40 % for plates over 2 in. [50 mm] in thickness, calculated using the following formula:

$$CE = C + Mn/6 + (Cr + Mo + V)/5 + (Cu + Ni)/15 \%$$

S73.3 For plates produced by the thermo-mechanical control process, in addition to the marking required by Specification A6/A6M, the letters "TMC" shall be marked following the specification designation mark.

S74. Grade 50 Plates Restricted P_{cm} Carbon Equivalent Limit

S74.1 The plates shall be produced by the thermo-mechanical control process (TMCP) or shall be quenched and tempered.

S74.2 The P_{cm} carbon equivalent for each heat, based upon heat analysis, shall not exceed 0.24 % for plates 2 in. [50 mm] or under in thickness and 0.26 % for plates over 2 in. [50 mm] in thickness, calculated using the following formula:

$$P_{cm} = C + Si/30 + (Mn + Cu + Cr)/20 + Ni/60 + Mo/15 + V/10 + 5B \%$$

S74.3 For plates produced by the thermo-mechanical control process, in addition to the marking required by Specification A6/A6M, the letters "TMC" shall be marked following the specification designation mark.

ASTM International takes no position respecting the validity of any patent rights asserted in connection with any item mentioned in this standard. Users of this standard are expressly advised that determination of the validity of any such patent rights, and the risk of infringement of such rights, are entirely their own responsibility.

This standard is subject to revision at any time by the responsible technical committee and must be reviewed every five years and if not revised, either reapproved or withdrawn. Your comments are invited either for revision of this standard or for additional standards and should be addressed to ASTM International Headquarters. Your comments will receive careful consideration at a meeting of the responsible technical committee, which you may attend. If you feel that your comments have not received a fair hearing you should make your views known to the ASTM Committee on Standards, at the address shown below.

This standard is copyrighted by ASTM International, 100 Barr Harbor Drive, PO Box C700, West Conshohocken, PA 19428-2959, United States. Individual reprints (single or multiple copies) of this standard may be obtained by contacting ASTM at the above address or at 610-832-9585 (phone), 610-832-9555 (fax), or service@astm.org (e-mail); or through the ASTM website (www.astm.org). Permission rights to photocopy the standard may also be secured from the ASTM website (www.astm.org/COPYRIGHT/).

Designation: E23 − 07a$^{\varepsilon 1}$

An American National Standard

Standard Test Methods for Notched Bar Impact Testing of Metallic Materials[1]

This standard is issued under the fixed designation E23; the number immediately following the designation indicates the year of original adoption or, in the case of revision, the year of last revision. A number in parentheses indicates the year of last reapproval. A superscript epsilon (ε) indicates an editorial change since the last revision or reapproval.

This standard has been approved for use by agencies of the Department of Defense.

ε^1 NOTE—Editorial changes made throughout in September 2007.

1. Scope

1.1 These test methods describe notched-bar impact testing of metallic materials by the Charpy (simple-beam) test and the Izod (cantilever-beam) test. They give the requirements for: test specimens, test procedures, test reports, test machines (see Annex A1) verifying Charpy impact machines (see Annex A2), optional test specimen configurations (see Annex A3), pre-cracking Charpy V-notch specimens (see Annex A4), designation of test specimen orientation (see Annex A5), and determining the percent of shear fracture on the surface of broken impact specimens (see Annex A6). In addition, information is provided on the significance of notched-bar impact testing (see Appendix X1), methods of measuring the center of strike (see Appendix X2).

1.2 These test methods do not address the problems associated with impact testing at temperatures below −196 °C (−320 °F, 77 K).

1.3 The values stated in SI units are to be regarded as the standard. Inch-pound units are provided for information only.

1.4 *This standard does not purport to address all of the safety concerns, if any, associated with its use. It is the responsibility of the user of this standard to establish appropriate safety and health practices and determine the applicability of regulatory limitations prior to use.* Specific precautionary statements are given in Section 5.

2. Referenced Documents

2.1 *ASTM Standards:*[2]
B925 Practices for Production and Preparation of Powder Metallurgy (PM) Test Specimens
E177 Practice for Use of the Terms Precision and Bias in ASTM Test Methods
E399 Test Method for Linear-Elastic Plane-Strain Fracture Toughness K_{Ic} of Metallic Materials
E604 Test Method for Dynamic Tear Testing of Metallic Materials
E691 Practice for Conducting an Interlaboratory Study to Determine the Precision of a Test Method
E1313 Guide for Recommended Formats for Data Records Used in Computerization of Mechanical Test Data for Metals (Discontinued 2000)[3]

3. Summary of Test Method

3.1 The essential features of an impact test are: a suitable specimen (specimens of several different types are recognized), a set of anvils, and specimen supports on which the test specimen is placed to receive the blow of the moving mass, a moving mass that has sufficient energy to break the specimen placed in its path, and a device for measuring the energy absorbed by the broken specimen.

4. Significance and Use

4.1 These test methods of impact testing relate specifically to the behavior of metal when subjected to a single application of a force resulting in multi-axial stresses associated with a notch, coupled with high rates of loading and in some cases with high or low temperatures. For some materials and temperatures the results of impact tests on notched specimens, when correlated with service experience, have been found to predict the likelihood of brittle fracture accurately. Further information on significance appears in Appendix X1.

5. Precautions in Operation of Machine

5.1 Safety precautions should be taken to protect personnel from the swinging pendulum, flying broken specimens, and hazards associated with specimen warming and cooling media.

6. Apparatus

6.1 *General Requirements*:
6.1.1 The testing machine shall be a pendulum type of rigid construction.

---
[1] These test methods are under the jurisdiction of ASTM Committee E28 on Mechanical Testing and are the direct responsibility of Subcommittee E28.07 on Impact Testing.
Current edition approved June 1, 2007. Published July 2007. Originally approved in 1933. Last previous edition approved 2007 as E23 – 07. DOI: 10.1520/E0023-07AE01.
[2] For referenced ASTM standards, visit the ASTM website, www.astm.org, or contact ASTM Customer Service at service@astm.org. For *Annual Book of ASTM Standards* volume information, refer to the standard's Document Summary page on the ASTM website.
[3] Withdrawn. The last approved version of this historical standard is referenced on www.astm.org.

Copyright © ASTM International, 100 Barr Harbor Drive, PO Box C700, West Conshohocken, PA 19428-2959, United States.

6.1.2 The testing machine shall be designed and built to conform with the requirements given in Annex A1.

6.2 *Inspection and Verification*

6.2.1 Inspection procedures to verify impact machines directly are provided in A2.2 and A2.3. The items listed in A2.2 must be inspected annually.

6.2.2 The procedures to verify Charpy V-notch machines indirectly, using verification specimens, are given in A2.4. Charpy impact machines must be verified directly and indirectly annually.

7. Test Specimens

7.1 *Configuration and Orientation*:

7.1.1 Specimens shall be taken from the material as specified by the applicable specification. Specimen orientation should be designated according to the terminology given in Annex A5.

7.1.2 The type of specimen chosen depends largely upon the characteristics of the material to be tested. A given specimen may not be equally satisfactory for soft nonferrous metals and hardened steels; therefore, many types of specimens are recognized. In general, sharper and deeper notches are required to distinguish differences in very ductile materials or when using low testing velocities.

7.1.3 The specimens shown in Figs. 1 and 2 are those most widely used and most generally satisfactory. They are particularly suitable for ferrous metals, excepting cast iron.[4]

7.1.4 The specimen commonly found suitable for die-cast alloys is shown in Fig. 3.

7.1.5 The specimens commonly found suitable for Powder Metallurgy (P/M) materials are shown in Figs. 4 and 5. P/M impact test specimens shall be produced following the procedure in Practice B925. The impact test results of these materials are affected by specimen orientation. Therefore,

---

[4] Report of Subcommittee XV on Impact Testing of Committee A-3 on Cast Iron, Proceedings, ASTM, Vol 33 Part 1, 1933.

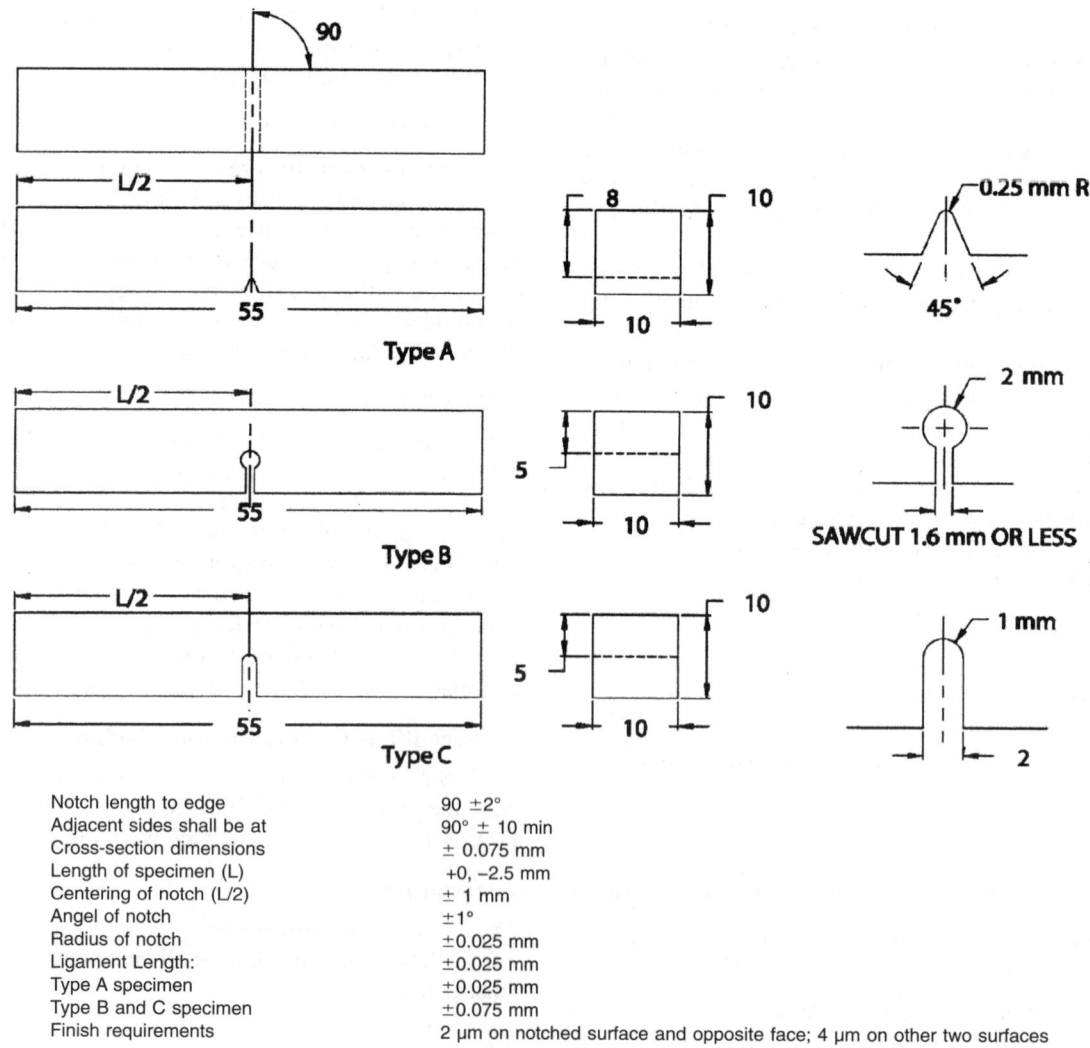

Notch length to edge	90 ±2°
Adjacent sides shall be at	90° ± 10 min
Cross-section dimensions	± 0.075 mm
Length of specimen (L)	+0, −2.5 mm
Centering of notch (L/2)	± 1 mm
Angel of notch	±1°
Radius of notch	±0.025 mm
Ligament Length:	±0.025 mm
Type A specimen	±0.025 mm
Type B and C specimen	±0.075 mm
Finish requirements	2 µm on notched surface and opposite face; 4 µm on other two surfaces

FIG. 1 Charpy (Simple-Beam) Impact Test Specimens, Types A, B, and C

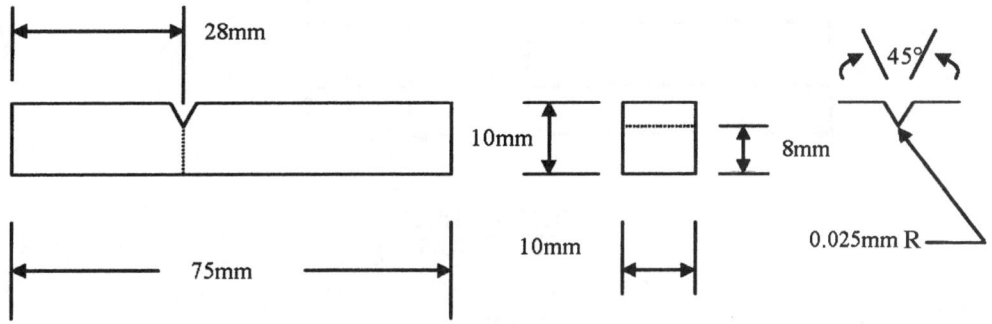

NOTE 1—Permissible variations shall be as follows:

Notch length to edge	90 ±2°
Cross-section dimensions	±0.025 mm
Length of specimen	+0, −2.5 mm
Angle of notch	±1°
Radius of notch	±0.025 mm
Ligament Length	±0.025 mm
Adjacent sides shall be at	90° ± 10 min
Finish requirements	2 µm on notched surface and opposite face; 4 µm on other two surfaces

FIG. 2 Izod (Cantilever-Beam) Impact Test Specimen, Type D

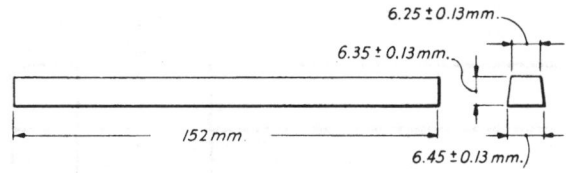

NOTE 1—Two Izod specimens may be cut from this bar.
NOTE 2—Blow shall be struck on narrowest face.

FIG. 3 Izod Impact Test Bar for Die Castings Alloys

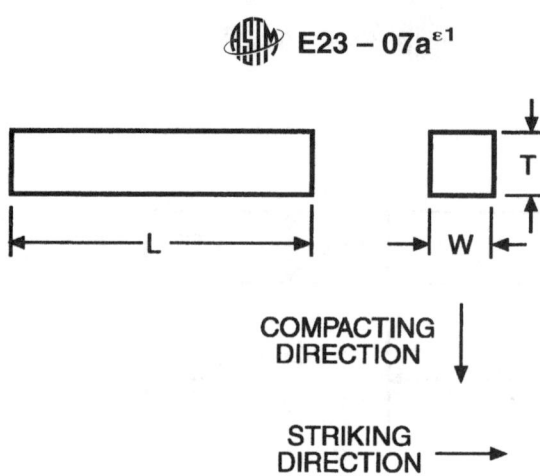

Dimensions		
	mm	in.
L- Overall Length	55.0 ± 1.0	2.16 ± 0.04
W-Width	10.00 ± 0.13	0.394 ± 0.005
T-Thickness	10.00 ± 0.13	0.394 ± 0.005

Note 1—Adjacent sides shall be 90°± 10 min.

FIG. 4 Unnotched Charpy (Simple Beam) Impact Test Specimen for P/M Structural Materials

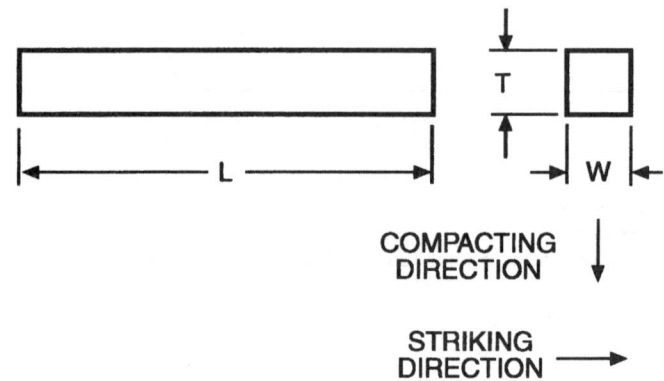

Dimensions		
	mm	in.
L- Overall Length	75.0 ± 1.5†	2.95 ± 0.06
W-Width	10.00 ± 0.13	0.394 ± 0.005
T-Thickness	10.00 ± 0.13	0.394 ± 0.005

Note 1—Adjacent sides shall be 90°± 10 min.
† Editorially corrected in August 2007.

FIG. 5 Izod (Cantilever-Beam) Impact Test Specimen for P/M Structural Materials

unless otherwise specified, the position of the specimen in the machine shall be such that the pendulum will strike a surface that is parallel to the compacting direction. For P/M materials the impact test results are reported as unnotched absorbed impact energy.

7.1.6 Sub-size and supplementary specimen recommendations are given in Annex A3.

7.2 *Specimen Machining*:

7.2.1 When heat-treated materials are being evaluated, the specimen shall be finish machined, including notching, after the final heat treatment, unless it can be demonstrated that the impact properties of specimens machined before heat treatment are identical to those machined after heat treatment.

7.2.2 Notches shall be smoothly machined but polishing has proven generally unnecessary. However, since variations in notch dimensions will seriously affect the results of the tests, adhering to the tolerances given in Fig. 1 is necessary (Appendix X1.2 illustrates the effects from varying notch dimensions on Type A specimens). In keyhole specimens, the round hole shall be carefully drilled with a slow feed rate. The slot may be cut by any feasible method, but care must be exercised in cutting the slot to ensure that the surface of the drilled hole opposite the slot is not damaged.

7.2.3 Identification marks shall only be placed in the following locations on specimens: either of the 10-mm square ends; the side of the specimen that faces up when the specimen is positioned in the anvils (see Note 1); or the side of the specimen opposite the notch. No markings, on any side of the specimen, shall be within 15 mm of the center line of the notch. An electrostatic pencil may be used for identification purposes, but caution must be taken to avoid excessive heat.

NOTE 1—Careful consideration should be given before placing identification marks on the side of the specimen to be placed up when positioned in the anvils. If the test operator is not careful, the specimen may be placed in the machine with the identification marking resting on the specimen supports. Under these circumstances, the absorbed energy value obtained may be unreliable.

8. Procedure

8.1 *Preparation of the Apparatus*:

8.1.1 Perform a routine procedure for checking impact machines at the beginning of each day, each shift, or just prior to testing on a machine used intermittently. It is recommended that the results of these routine checks be kept in a log book for the machine. After the testing machine has been ascertained to comply with Annex A1 and Annex A2, carry out the routine check as follows:

8.1.1.1 Visually examine the striker and anvils for obvious damage and wear.

8.1.1.2 Check the zero position of the machine by using the following procedure: raise the pendulum to the latched position, move the pointer to near the maximum capacity of the range being used, release the pendulum, and read the indicated value. The pointer should indicate zero on machines reading directly in energy. On machines reading in degrees, the reading should correspond to zero on the conversion chart furnished by the machine manufacturer.

NOTE 2—On machines that do not compensate for windage and friction losses, the pointer will not indicate zero. In this case, the indicated values, when converted to energy, shall be corrected for frictional losses that are assumed to be proportional to the arc of swing.

8.1.1.3 To ensure that friction and windage losses are within allowable tolerances, the following procedure is recommended: raise the pendulum to the latched position, move the pointer to the negative side of zero, release the pendulum and allow it to cycle five times (a forward and a backward swing together count as one swing), prior to the sixth forward swing, set the pointer to between 5 and 10 % of the scale capacity of the dial, after the sixth forward swing (eleven half swings), record the value indicated by the pointer, convert the reading to energy (if necessary), divide it by 11 (half swings), then divide by the maximum scale value being used and multiply it by 100 to get the percent friction. The result, friction and windage loss, shall not exceed 0.4 % of scale range capacity being tested and should not change by more than 10 % of friction measurements previously made on the machine. If the friction and windage loss value does exceed 0.4 % or is significantly different from previous measurements, check the indicating mechanism, the latch height, and the bearings for wear and damage. However, if the machine has not been used recently, let the pendulum swing for 50 to 100 cycles, and repeat the friction test before undertaking repairs to the machine.

8.2 *Test Temperature Considerations*:

8.2.1 The temperature of testing affects the impact properties of most materials. For materials with a body centered cubic structure, a transition in fracture mode occurs over a temperature range that depends on the chemical composition and microstructure of the material. Test temperatures may be chosen to characterize material behavior at fixed values, or over a range of temperatures to characterize the transition region, lower shelf, or upper shelf behavior, or all of these. The choice of test temperature is the responsibility of the user of this test method and will depend on the specific application. For tests performed at room temperature, a temperature of 20 ± 5°C (68 ± 9°F) is recommended.

8.2.2 The temperature of a specimen can change significantly during the interval it is removed from the temperature conditioning environment, transferred to the impact machine, and the fracture event is completed (see Note 5). When using a heating or cooling medium near its boiling point, use data from the references in Note 5 or calibration data with thermocouples to confirm that the specimen is within the stated temperature tolerances when the striker contacts the specimen. If excessive adiabatic heating is expected, monitor the specimen temperature near the notch during fracture.

8.2.3 Verify temperature-measuring equipment at least every six months. If liquid-in-glass thermometers are used, an initial verification shall be sufficient, however, the device shall be inspected for problems, such as the separation of liquid, at least twice annually.

8.2.4 Hold the specimen at the desired temperature within ± 1 °C (± 2 °F) in the temperature conditioning environment (see 8.2.4.1 and 8.2.4.2). Any method of heating or cooling or transferring the specimen to the anvils may be used provided the temperature of the specimen immediately prior to fracture is essentially the same as the holding temperature (see Note 5). The maximum change in the temperature of the specimen allowed for the interval between the temperature conditioning treatment and impact is not specified here, because it is dependent on the material being tested and the application. The user of nontraditional or lesser used temperature conditioning and transfer methods (or sample sizes) shall show that the temperature change for the specimen prior to impact is comparable to or less than the temperature change for a standard size specimen of the same material that has been thermally conditioned in a commonly used medium (oil, air, nitrogen, acetone, methanol), and transferred for impact within 5 seconds (see Note 5). Three temperature conditioning and transfer methods used in the past are: liquid bath thermal conditioning and transfer to the specimen supports with centering tongs; furnace thermal conditioning and robotic transfer to the specimen supports; placement of the specimen on the supports followed by in situ heating and cooling.

8.2.4.1 For liquid bath cooling or heating use a suitable container, which has a grid or another type of specimen positioning fixture. Cover the specimens, when immersed, with at least 25 mm (1 in.) of the liquid, and position so that the notch area is not closer than 25 mm (1 in.) to the sides or bottom of the container, and no part of the specimen is in contact with the container. Place the device used to measure the temperature of the bath in the center of a group of the specimens. Agitate the bath and hold at the desired temperature within ± 1°C (± 2°F). Thermally condition the specimens for at least 5 min before testing, unless a shorter thermal conditioning time can be shown to be valid by measurements with thermocouples. Leave the mechanism (tongs, for example) used to handle the specimens in the bath for at least 5 min before testing, and return the mechanism to the bath between tests.

8.2.4.2 When using a gas medium, position the specimens so that the gas circulates around them and hold the gas at the desired temperature within ± 1°C (± 2°F) for at least 30 min. Leave the mechanism used to remove the specimen from the medium in the medium except when handling the specimens.

NOTE 3—Temperatures up to +260°C (+500°F) may be obtained with certain oils, but "flash-point" temperatures must be carefully observed.

NOTE 4—For testing at temperatures down to –196°C (–320 °F, 77 °K), standard testing procedures have been found to be adequate for most metals.

NOTE 5—A study has shown that a specimen heated to 100 C in water can cool 10 C in the 5 s allowed for transfer to the specimen supports **(1)**[5]. Other studies, using cooling media that are above their boiling points at room temperature have also shown large changes in specimen temperature during the transfer of specimens to the machine anvils. In addition, some materials change temperature dramatically during impact testing at cryogenic temperatures due to adiabatic heating **(2)**.

---

[5] The boldface numbers given in parentheses refer to a list of references at the end of the text.

8.3 *Charpy Test Procedure*:

8.3.1 The Charpy test procedure may be summarized as follows: the test specimen is thermally conditioned and positioned on the specimen supports against the anvils; the pendulum is released without vibration, and the specimen is impacted by the striker. Information is obtained from the machine and from the broken specimen.

8.3.2 To position a test specimen in the machine, it is recommended that self-centering tongs similar to those shown in Fig. 6 be used (see A1.10.1). The tongs illustrated in Fig. 6 are for centering V-notch specimens. If keyhole specimens are used, modification of the tong design may be necessary. If an end-centering device is used, caution must be taken to ensure that low-energy high-strength specimens will not rebound off this device into the pendulum and cause erroneously high recorded values. Many such devices are permanent fixtures of machines, and if the clearance between the end of a specimen in the test position and the centering device is not approximately 13 mm (0.5 in.), the broken specimens may rebound into the pendulum.

8.3.3 To conduct the test, prepare the machine by raising the pendulum to the latched position, set the energy indicator at the maximum scale reading, or initialize the digital display, or both, position the specimen on the anvils, and release the pendulum. If a liquid bath or gas medium is being used for

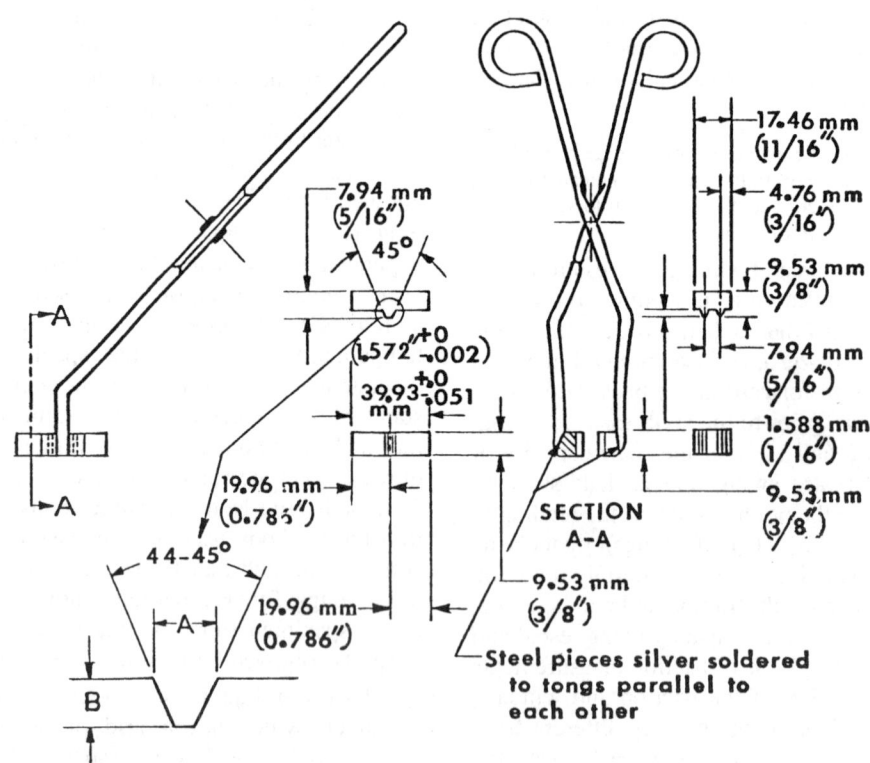

NOTE 1—Unless otherwise shown, permissible variation shall be ±1 mm (0.04 in.).

Specimen Depth, mm (in.)	Base Width (A), mm (in.)	Height (B), mm (in.)
10 (0.394)	1.60 to 1.70 (0.063 to 0.067)	1.52 to 1.65 (0.060 to 0.065)
5 (0.197)	0.74 to 0.80 (0.029 to 0.033)	0.69 to 0.81 (0.027 to 0.032)
3 (0.118)	0.45 to 0.51 (0.016 to 0.020)	0.36 to 0.48 (0.014 to 0.019)

FIG. 6 Centering Tongs for V-Notch Charpy Specimens

thermal conditioning, perform the following sequence in less than 5 s (for standard 10 × 10 × 55 mm (0.394 × 0.394 × 2.165 in.) specimens, see 8.2.4). Remove the test specimen from its cooling (or heating) medium with centering tongs that have been temperature conditioned with the test specimen, place the specimen in the test position, and, release the pendulum smoothly. If a test specimen has been removed from the temperature conditioning bath and it is questionable that the test can be conducted within the 5 s time frame, return the specimen to the bath for the time required in 8.2 before testing.

8.3.3.1 If a fractured impact specimen does not separate into two pieces, report it as unbroken (see 9.2.2 for separation instructions). Unbroken specimens with absorbed energies of less than 80 % of the machine capacity may be averaged with values from broken specimens. If the individual values are not listed, report the percent of unbroken specimens with the average. If the absorbed energy exceeds 80 % of the machine capacity and the specimen passes completely between the anvils, report the value as approximate (see 10.1) do not average it with other values. If an unbroken specimen does not pass between the machine anvils, (for example, it stops the pendulum), the result shall be reported as exceeding the machine capacity. A specimen shall never be struck more than once.

8.3.3.2 If a specimen jams in the machine, disregard the results and check the machine thoroughly for damage or misalignment, which would affect its calibration.

8.3.3.3 To prevent recording an erroneous value, caused by jarring the indicator when locking the pendulum in its upright (ready) position, read the value for each test from the indicator prior to locking the pendulum for the next test.

8.4 *Izod Test Procedure*:

8.4.1 The Izod test procedure may be summarized as follows: the test specimen is positioned in the specimen-holding fixture and the pendulum is released without vibration. Information is obtained from the machine and from the broken specimen. The details are described as follows:

8.4.2 Testing at temperatures other than room temperature is difficult because the specimen-holding fixture for Izod specimens is often part of the base of the machine and cannot be readily cooled (or heated). Consequently, Izod testing is not recommended at other than room temperature.

8.4.3 Clamp the specimen firmly in the support vise so that the centerline of the notch is in the plane of the top of the vise within 0.125 mm (0.005 in.). Set the energy indicator at the maximum scale reading, and release the pendulum smoothly. Sections 8.3.3.1-8.3.3.3 inclusively, also apply when testing Izod specimens.

9. Information Obtainable from Impact Tests

9.1 *The absorbed energy* shall be taken as the difference between the energy in the striking member at the instant of impact with the specimen and the energy remaining after breaking the specimen. This value is determined by the machine's scale reading which has been corrected for windage and friction losses.

NOTE 6—Alternative means for energy measurement are acceptable provided the accuracy of such methods can be demonstrated. Methods used in the past include optical encoders and strain gaged strikers.

9.2 *Lateral expansion measurement* methods must take into account the fact that the fracture path seldom bisects the point of maximum expansion on both sides of a specimen. One half of a broken specimen may include the maximum expansion for both sides, one side only, or neither. Therefore, the expansion on each side of each specimen half must be measured relative to the plane defined by the undeformed portion on the side of the specimen, as shown in Fig. 7. For example, if A_1 is greater than A_2, and A_3 is less than A_4, then the lateral expansion is the sum of $A_1 + A_4$.

9.2.1 Before making any expansion measurements, it is essential that the two specimen halves are visually examined for burrs that may have formed during impact testing; if the burrs will influence the lateral expansion measurements, they must be removed (by rubbing on emery cloth or any other suitable method), making sure that the protrusions to be measured are not rubbed during the removal of the burr. Then, examine each fracture surface to ascertain that the protrusions have not been damaged by contacting an anvil, a machine mounting surface, etc. Lateral expansion shall not be measured on a specimen with this type of damage.

9.2.2 Lateral expansion measurements shall be reported as follows. The lateral expansion of an unbroken specimen can be reported as broken if the specimen can be separated by pushing the hinged halves together once and then pulling them apart without further fatiguing the specimen, and the lateral expansion measured for the unbroken specimen (prior to bending) is equal to or greater than that measured for the separated halves. In the case where a specimen cannot be separated into two halves, the lateral expansion can be measured as long as the shear lips can be accessed without interference from the hinged ligament that has been deformed during testing. The specimen should be reported as unbroken.

9.2.3 Lateral expansion may be measured easily by using a gage like the one shown in Fig. 8 (assembly and details shown in Fig. 9). Using this type of gage the measurement is made with the following procedure: orient the specimen halves so that the compression sides are facing each another, take one half of the fractured specimen and press it against the anvil and

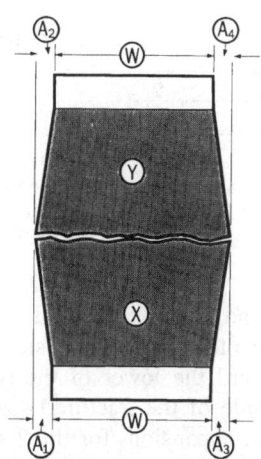

FIG. 7 Halves of Broken Charpy V-Notch Impact Specimen Illustrating the Measurement of Lateral Expansion, Dimensions A_1, A_2, A_3, A_4 and Original Width, Dimension W

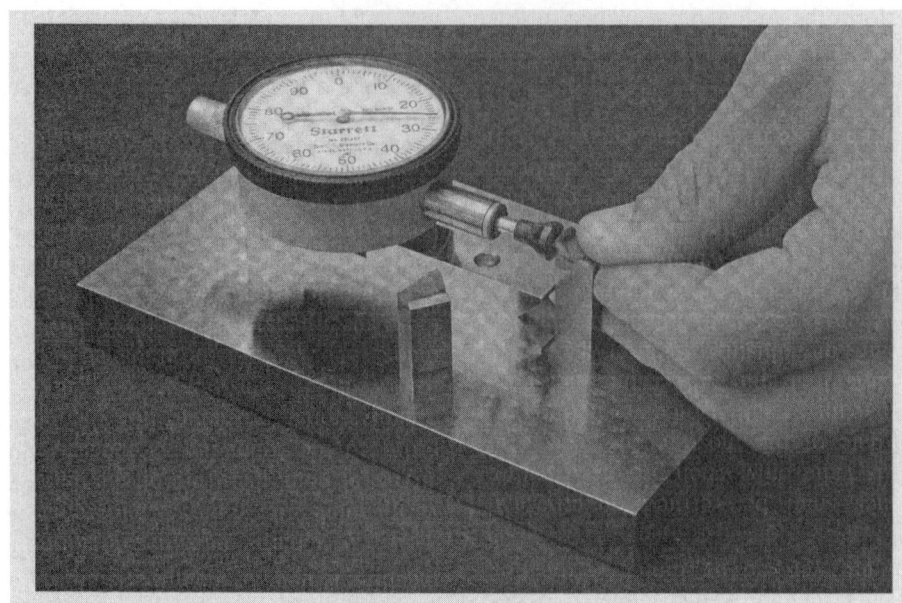

FIG. 8 Lateral Expansion Gage for Charpy Impact Specimens

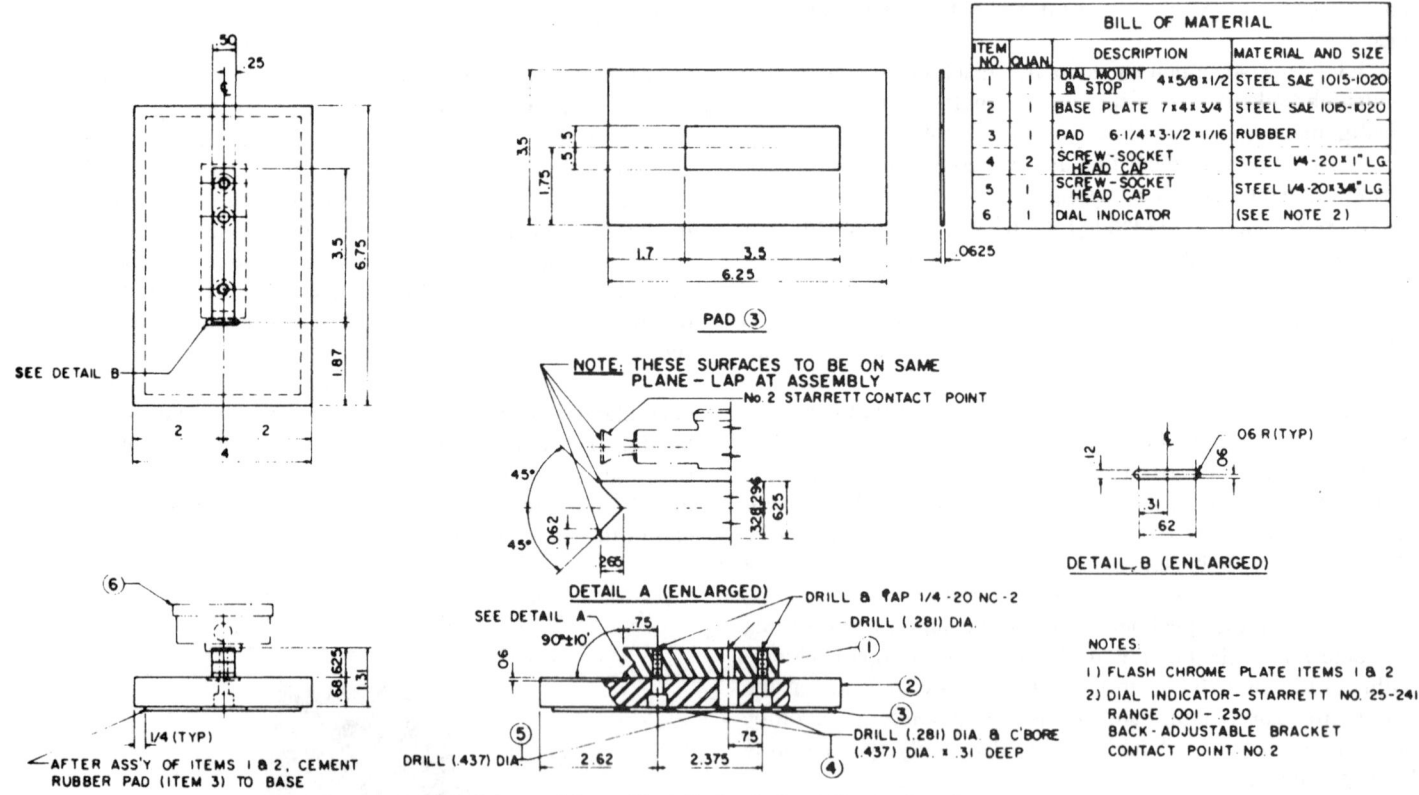

FIG. 9 Assembly and Details for Lateral Expansion Gage

dial gage plunger and record the reading, make a similar measurement on the other half (same side) of the fractured specimen and disregard the lower of the two values, do the same for the other side of the fractured specimen, report the sum of the maximum expansions for the 2 sides as the lateral expansion for the specimen.

9.3 *The percentage of shear fracture* on the fracture surfaces of impact specimens may be determined using a variety of methods. The acceptable methods are defined in Annex A6. For each method, the user must distinguish between regions formed by ductile stable crack growth mechanisms, and regions formed by brittle fast crack propagation (unstable crack growth mechanisms). The typical zones of fracture appearance are shown in Fig. 10, where the "flat fracture" region is the region in which unstable crack growth occurs on a microsecond time scale.

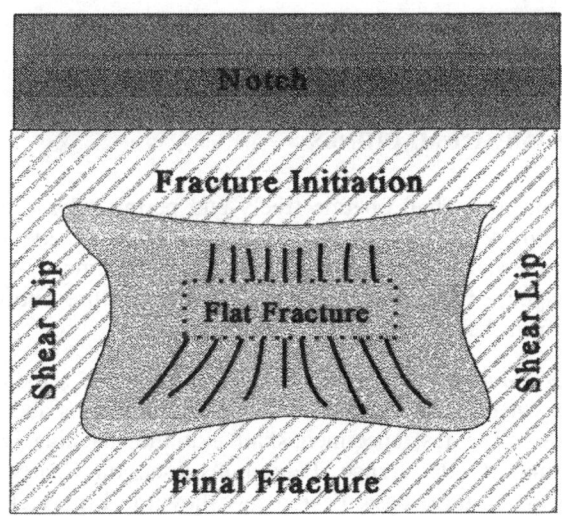

NOTE 1—The shear of ductile fracture regions on the fracture surface include the fracture initiation region, the two shear lips, and the region of final fracture. The flat or radial fracture region is a region of less ductile unstable crack growth.

FIG. 10 Determination of Percent Shear Fracture

The percent shear area on the fracture surface of a Charpy impact specimen is typically calculated as the difference between the total fractured area and the area of flat fracture. The measurement methods described here provide estimates for the area of the macroscopically flat fracture region (directly or indirectly), but do not consider details of the fracture mode for this " flat" region of unstable fracture. The flat fracture region could be 100 percent cleavage, a mixture of cleavage and ductile-dimple fracture morphologies, or other combinations of ductile-brittle fracture morphologies. Estimates of ductility within the unstable crack growth region are beyond the scope of these methods.

10. Report

10.1 *Absorbed energy values above 80 %* of the scale range are inaccurate and shall be reported as approximate. Ideally an impact test would be conducted at a constant impact velocity. In a pendulum-type test, the velocity decreases as the fracture progresses. For specimens that have impact energies approaching 80 % of the capacity of the pendulum, the velocity of the pendulum decreases (to about 45 % of the initial velocity) during fracture to the point that accurate impact energies are no longer obtained.

10.2 *For commercial acceptance testing*, report the following information (for each specimen tested):

10.2.1 Specimen type (and size if not the full-size specimen),

10.2.2 Test temperature,
10.2.3 Absorbed energy, and
10.2.4 Any other contractual requirements.

10.3 *For other than commercial acceptance testing* the following information is often reported in addition to the information in 10.2:

10.3.1 Lateral expansion,
10.3.2 Unbroken specimens,
10.3.3 Fracture appearance (% shear, See Note A6.1),
10.3.4 Specimen orientation, and
10.3.5 Specimen location.

NOTE 7—A recommended format for computerization of notched bar impact test data is available in Practice E1313.

NOTE 8—When the test temperature is specified as room temperature, report the actual temperature.

11. Precision and Bias

11.1 *An Interlaboratory study* used CVN specimens of low energy and of high energy to find sources of variation in the CVN absorbed energy. Data from 29 laboratories were included with each laboratory testing one set of five specimens of each energy level. Except being limited to only two energy levels (by availability of reference specimens), Practice E691 was followed for the design and analysis of the data, the details are given in ASTM Research Report NO. E28-1014.[6]

11.2 *Precision—The Precision* information given below (in units of J and ft·lbf) is for the average CVN absorbed energy of five test determinations at each laboratory for each material.

Material	Low Energy		High Energy	
	J	ft-lbf	J	ft-lbf
Absorbed Energy	15.9	11.7	96.2	71.0
95 % Repeatability Limit	2.4	1.7	8.3	6.1
95 % Reproducibility Limits	2.7	2.0	9.2	6.8

The terms repeatability and reproducibility limit are used as defined in Practice E177. The respective standard deviations among test results may be obtained by dividing the above limits by 2.8.

11.3 *Bias*— Bias cannot be defined for CVN absorbed energy. The physical simplicity of the pendulum design is complicated by complex energy loss mechanisms within the machine and the specimen. Therefore, there is no absolute standard to which the measured values can be compared.

12. Keywords

12.1 Charpy test; fracture appearance; Izod test; impact test; notched specimens; pendulum machine

---

[6] Supporting data have been filed at ASTM Headquarters and may be obtained by requesting Research Report E28–1014.

ANNEXES

(Mandatory Information)

A1. GENERAL REQUIREMENTS FOR IMPACT MACHINES

A1.1 *The machine frame* shall be equipped with a bubble level or a machined surface suitable for establishing levelness of the axis of pendulum bearings or, alternatively, the levelness of the axis of rotation of the pendulum may be measured directly. The machine shall be level to within 3:1000 and securely bolted to a concrete floor not less than 150 mm (6 in.) thick or, when this is not practical, the machine shall be bolted to a foundation having a mass not less than 40 times that of the pendulum. The bolts shall be tightened as specified by the machine manufacturer.

A1.2 *A scale or digital display*, graduated in degrees or energy, on which readings can be estimated in increments of 0.25 % of the energy range or less shall be furnished for the machine.

A1.2.1 The scales and digital displays may be compensated for windage and pendulum friction. The error in the scale reading at any point shall not exceed 0.2 % of the range or 0.4 % of the reading, whichever is larger. (See A2.3.8.)

A1.3 *The total friction and windage losses* of the machine during the swing in the striking direction shall not exceed 0.75 % of the scale range capacity, and pendulum energy loss from friction in the indicating mechanism shall not exceed 0.25 % of scale range capacity. See A2.3.8 for friction and windage loss calculations.

A1.4 *The position of the pendulum*, when hanging freely, shall be such that the striker is within 2.5 mm (0.10 in.) from the test specimen. When the indicator has been positioned to read zero energy in a free swing, it shall read within 0.2 % of scale range when the striker of the pendulum is held against the test specimen. The plane of swing of the pendulum shall be perpendicular to the transverse axis of the Charpy specimen anvils or Izod vise within 3:1000.

A1.5 *Transverse play of the pendulum* at the striker shall not exceed 0.75 mm (0.030 in.) under a transverse force of 4 % of the effective weight of the pendulum applied at the center of strike. Radial play of the pendulum bearings shall not exceed 0.075 mm (0.003 in.).

A1.6 *The impact velocity* (tangential velocity) of the pendulum at the center of the strike shall not be less than 3 nor more than 6 m/s (not less than 10 nor more than 20 ft/s).

A1.7 *The height of the center of strike* in the latched position, above its free hanging position, shall be within 0.4 % of the range capacity divided by the supporting force, measured as described in A2.3.5.1 If windage and friction are compensated for by increasing the height of drop, the height of drop may be increased by not more than 1 %.

A1.8 *The mechanism for releasing the pendulum* from its initial position shall operate freely and permit release of the pendulum without initial impulse, retardation, or side vibration. If the same lever used to release the pendulum is also used to engage the brake, means shall be provided for preventing the brake from being accidentally engaged.

A1.9 *Specimen clearance* is needed to ensure satisfactory results when testing materials of different strengths and compositions. The test specimen shall exit the machine with a minimum of interference. Pendulums used on Charpy machines are of three basic designs, as shown in Fig. A1.1.

A1.9.1 When using a C-type pendulum or a compound pendulum, the broken specimen will not rebound into the pendulum and slow it down if the clearance at the end of the specimen is at least 13 mm (0.5 in.) or if the specimen is deflected out of the machine by some arrangement such as that shown in Fig. A1.1.

A1.9.2 When using the U-type pendulum, means shall be provided to prevent the broken specimen from rebounding against the pendulum (see Fig. A1.1). In most U-type pendulum machines, steel shrouds should be designed and installed to the following requirements: (*a*) have a thickness of approximately 1.5 mm (0.06 in.), (*b*) have a minimum hardness of 45 HRC, (*c*) have a radius of less than 1.5 mm (0.06 in.) at the underside corners, and (*d*) be so positioned that the clearance between them and the pendulum overhang (both top and sides) does not exceed 1.5 mm (0.06 in.).

NOTE A1.1—In machines where the opening within the pendulum permits clearance between the ends of a specimen (resting on the specimen supports) and the shrouds, and this clearance is at least 13 mm (0.5 in.), the requirements (*a*) and (*d*) need not apply.

A1.10 *Charpy Apparatus*:

A1.10.1 Means shall be provided (see Fig. A1.2) to locate and support the test specimen against two anvil blocks in such a position that the center of the notch can be located within 0.25 mm (0.010 in.) of the midpoint between the anvils (see 8.3.2).

A1.10.2 The supports and striker shall be of the forms and dimensions shown in Fig. A1.2. Other dimensions of the pendulum and supports should be such as to minimize interference between the pendulum and broken specimens.

A1.10.3 The center line of the striker shall advance in the plane that is within 0.40 mm (0.016 in.) of the midpoint between the supporting edges of the anvils. The striker shall be perpendicular to the longitudinal axis of the specimen within 5:1000. The striker shall be parallel within 1:1000 to the face of a perfectly square test specimen held against the anvils.

A1.11 *Izod Apparatus*:

A1.11.1 Means shall be provided (see Fig. A1.3) for clamping the specimen in such a position that the face of the specimen is parallel to the striker within 1:1000. The edges of

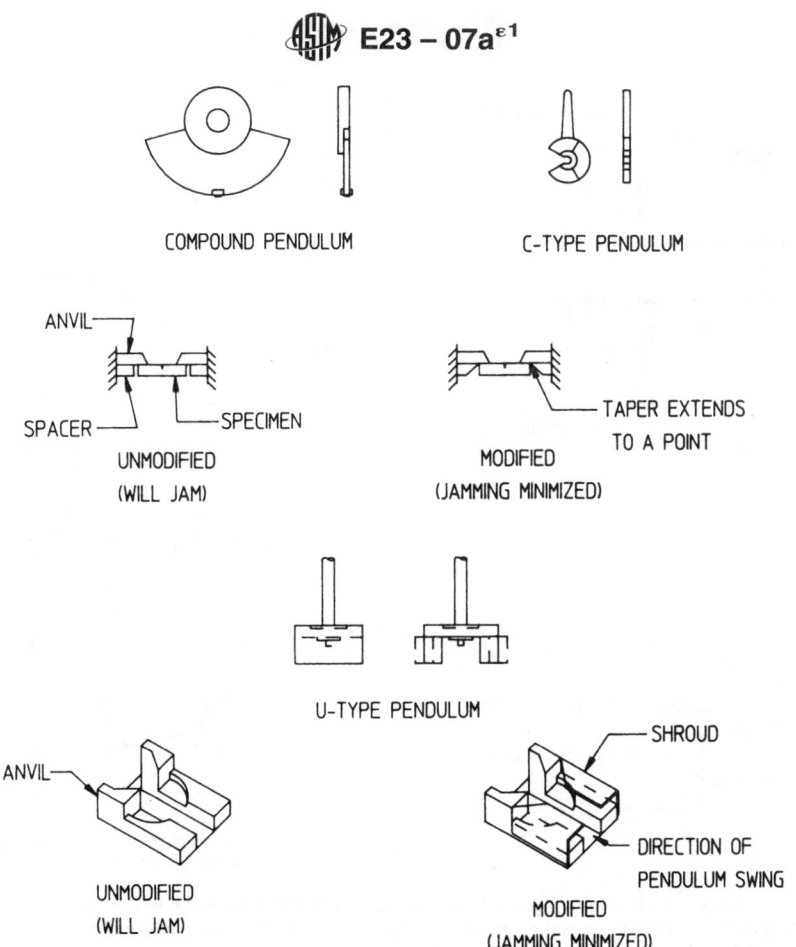

FIG. A1.1 Typical Pendulums and Anvils for Charpy Machines, Shown with Modifications to Minimize Jamming

the clamping surfaces shall be sharp angles of 90 ± 1° with radii less than 0.40 mm (0.016 in.). The clamping surfaces shall be smooth with a 2 μm (63 μin.) finish or better, and shall clamp the specimen firmly at the notch with the clamping force applied in the direction of impact. For rectangular specimens, the clamping surfaces shall be flat and parallel within 0.025 mm (0.001in.). For cylindrical specimens, the clamping surfaces shall be contoured to match the specimen and each surface shall contact a minimum of π/2 rad (90°) of the specimen circumference.

A1.11.2 The dimensions of the striker and its position relative to the specimen clamps shall be as shown in Fig. A1.3.

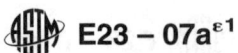

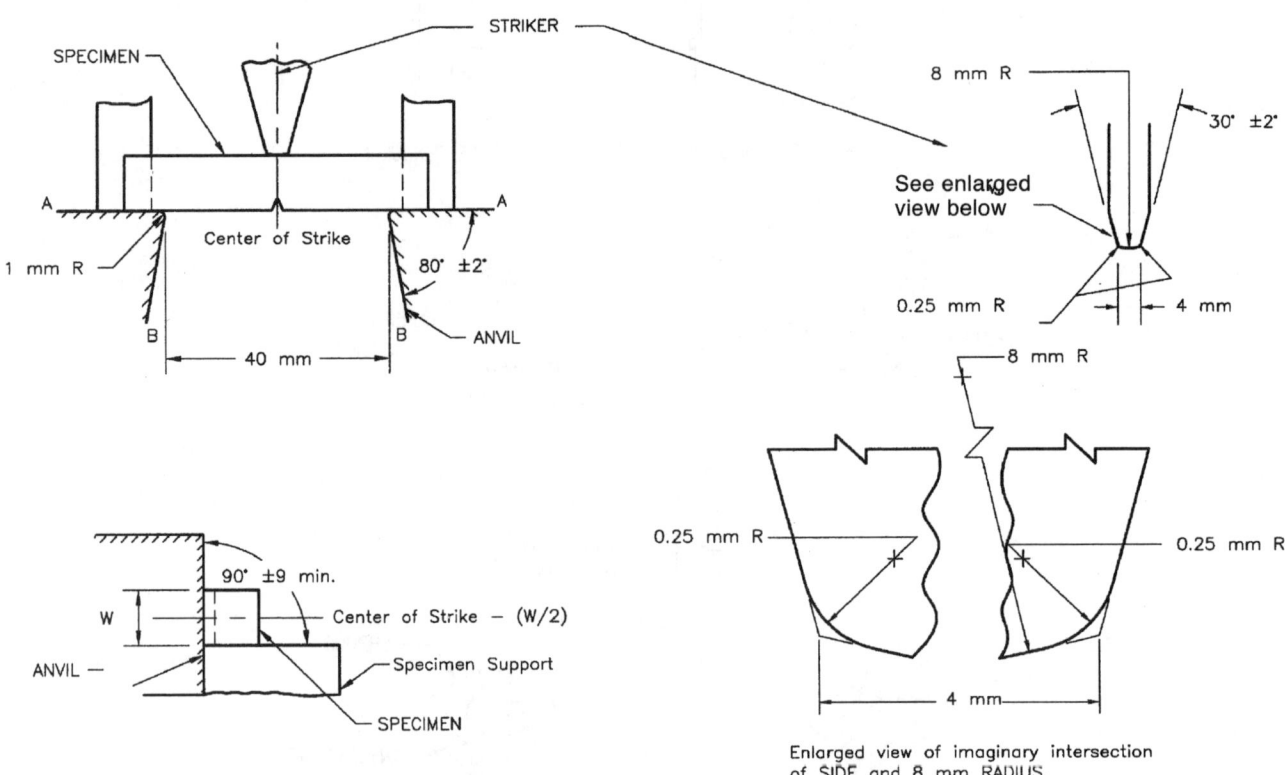

Note 1—Anvils shall be manufactured with a surface finish of 0.1 μm or better on surfaces A and B above the anvil supports when mounted on the machine.
Note 2— Striker shall be manufactured with a surface finish of 0.1 μm or better along the front radius and along both sides.
Note 3—All dimensional tolerances shall be ±0.05 mm unless otherwise specified.

FIG. A1.2 Charpy Striker

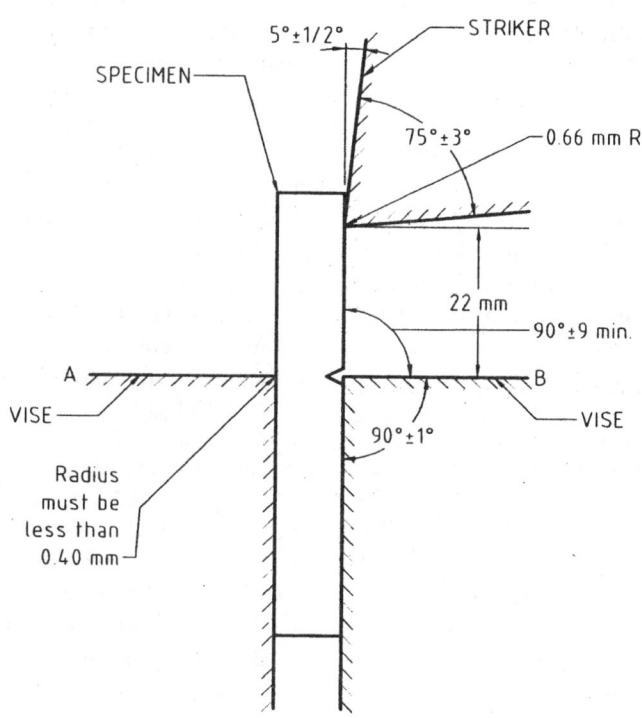

NOTE 1—All dimensional tolerances shall be ±0.05 mm unless otherwise specified.
NOTE 2—The clamping surfaces of A and B shall be flat and parallel within 0.025 mm.
NOTE 3— Surface finish on striker and vise shall be 2 μm.
NOTE 4—Striker width must be greater than that of the specimen being tested.

FIG. A1.3 Izod (Cantilever-Beam) Impact Test

A2. VERIFICATION OF PENDULUM IMPACT MACHINES

A2.1 *The verification of impact machines has two parts:* direct verification, which consists of inspecting the machine to ensure that the requirements of this annex and Annex A1 are met, and indirect verification, which entails the testing of verification specimens.

A2.1.1 Izod machines are verified by direct verification annually.

A2.1.2 Charpy machines shall be verified directly and indirectly annually. Data is valid only when produced within 365 days following the date of the most recent successful verification test. Charpy machines shall also be verified immediately after replacing parts that may affect the measured energy, after making repairs or adjustments, after they have been moved, or whenever there is reason to doubt the accuracy of the results, without regard to the time interval. These restrictions include cases where parts, which may affect the measured energy, are removed from the machine and then reinstalled without modification (with the exception of when the striker or anvils are removed to permit use of a different striker or set of anvils and then are reinstalled, see A2.1.3). It is not intended that parts not subjected to wear (such as pendulum and scale linearity) are to be directly verified each year unless a problem is evident. Only the items cited in A2.2 are required to be inspected annually. Other parts of the machine shall be directly verified at least once, when the machine is new, or when parts are replaced.

A2.1.3 Charpy machines do not require immediate indirect verification after removal and replacement of the striker or anvils, or both, that were on the machine when it was verified provided the following safeguards are implemented: *(1)* an organizational procedure for the change is developed and followed, *(2)* high-strength low-energy quality control specimens, (See A2.4.1.1 for guidance in breaking energy range for these specimens), are tested prior to removal and immediately after installation of the previously verified striker or anvils, or both within the 365 day verification period, *(3)* the results of the before and after tests of the quality control specimens are within 1.4 Joules (1.0 ft-lbf) of each other, *(4)* the results of the comparisons are kept in a log book, and *(5)* before reattachment, the striker and anvils are visually inspected for wear and dimensionally verified to assure that they meet the required tolerances of Fig. A1.2. The use of certified impact verification specimens is not required and internal quality control specimens are permitted.

A2.2 *Direct Verification of Parts Requiring Annual Inspection*:

A2.2.1 Inspect the specimen supports, anvils, and striker and replace any of these parts that show signs of wear. A straight edge or radius gage can be used to discern differences between the used and unused portions of these parts to help identify a worn condition (see Note A2.1).

NOTE A2.1—To measure the anvil or striker radii, the recommended procedure is to make a replica (casting) of the region of interest and measure cross sections of the replica. This can be done with the anvils and striker in place on the machine or removed from the machine. Make a dam with cardboard and tape surrounding the region of interest, then pour a low-shrinkage casting compound into the dam (silicon rubber casting compounds work well). Allow the casting to cure, remove the dam, and slice cross sections through the region of interest with a razor. Use these cross sections to make radii measurements on optical comparators or other instruments.

A2.2.2 Ensure the bolts that attach the anvils and striker to the machine are tightened to the manufacture's specifications.

A2.2.3 Verify that the shrouds, if applicable, are properly installed (see A1.9.2).

A2.2.4 The pendulum release mechanism, which releases the pendulum from its initial position, shall comply with A1.8.

A2.2.5 Check the level of the machine in both directions (see A1.1).

A2.2.6 Check that the foundation bolts are tightened to the manufacturer's specifications.

NOTE A2.2—Expansion bolts or fasteners with driven in inserts shall not be used for foundations. These fasteners will work loose and/or tighten up against the bottom of the machine indicating a false high torque value when the bolts are tightened.

A2.2.7 Check the indicator zero and the friction loss of the machine as described in 8.1.

A2.3 *Direct Verification of Parts to be Verified at Least Once*:

A2.3.1 Charpy anvils and supports or Izod vises shall conform to the dimensions shown in Fig. A1.2 or Fig. A1.3.

NOTE A2.3—The impact machine will be inaccurate to the extent that some energy is used in deformation or movement of its component parts or of the machine as a whole; this energy will be registered as used in fracturing the specimen.

A2.3.2 The striker shall conform to the dimensions shown in Fig. A1.2 or Fig. A1.3. The mounting surfaces must be clean and free of defects that would prevent a good fit. Check that the striker complies with A1.10.3 (for Charpy tests) or A1.11.1 (for Izod tests).

A2.3.3 The pendulum alignment shall comply with A1.4 and A1.5. If the side play in the pendulum or the radial play in the bearings exceeds the specified limits, adjust or replace the bearings.

A2.3.4 *Determine the Center of Strike*—For Charpy machines the center of strike of the pendulum is determined using a half-width specimen (10 × 5 × 55 mm) in the test position. With the striker in contact with the specimen, a line marked along the top edge of the specimen on the striker will indicate the center of strike. For Izod machines, the center of strike may be considered to be the contact line when the pendulum is brought into contact with a specimen in the normal testing position.

A2.3.5 *Determine the Potential Energy*—The following procedure shall be used when the center of strike of the pendulum is coincident with the radial line from the centerline of the pendulum bearings (herein called the axis of rotation) to the center of gravity (see Appendix X2). If the center of strike is more than 1.0 mm (0.04 in.) from this line, suitable corrections in elevation of the center of strike must be made in A2.3.8.1 and A2.3.9, so that elevations set or measured correspond to what they would be if the center of strike were on this line. The potential energy of the system is equal to the height from which the pendulum falls, as determined in A2.3.5.2, times the supporting force, as determined in A2.3.5.1

A2.3.5.1 To measure the supporting force, support the pendulum horizontally to within 15:1000 with two supports, one at the bearings (or center of rotation) and the other at the center of strike on the striker (see Fig. A2.1). Then arrange the support at the striker to react upon some suitable weighing device such as a platform scale or balance, and determine the weight to within 0.4 %. Take care to minimize friction at either point of support. Make contact with the striker through a round rod crossing the center of strike. The supporting force is the scale reading minus the weights of the supporting rod and any shims that may be used to maintain the pendulum in a horizontal position.

A2.3.5.2 Determine the height of pendulum drop for compliance with the requirement of A1.7. On Charpy machines determine the height from the top edge of a half-width (or center of a full-width) specimen to the elevated position of the center of strike to 0.1 %. On Izod machines determine the height from a distance 22.66 mm (0.892 in.) above the vise to the release position of the center of strike to 0.1 %. The height may be determined by direct measurement of the elevation of the center of strike or by calculation from the change in angle of the pendulum using the following formulas (see Fig. A2.1):

$$h = S(1 - cos(\beta)) \tag{A2.1}$$

$$h_1 = S(1 - cos(\alpha)) \tag{A2.2}$$

where
h = initial elevation of the striker, m (ft),
S = length of the pendulum distance to the center of strike, m (ft),
β = angle of fall,
h_1 = height of rise, m (ft), and
α = angle of rise.

A2.3.6 Determine the impact velocity, $[v]$, of the machine, neglecting friction, by means of the following equation:

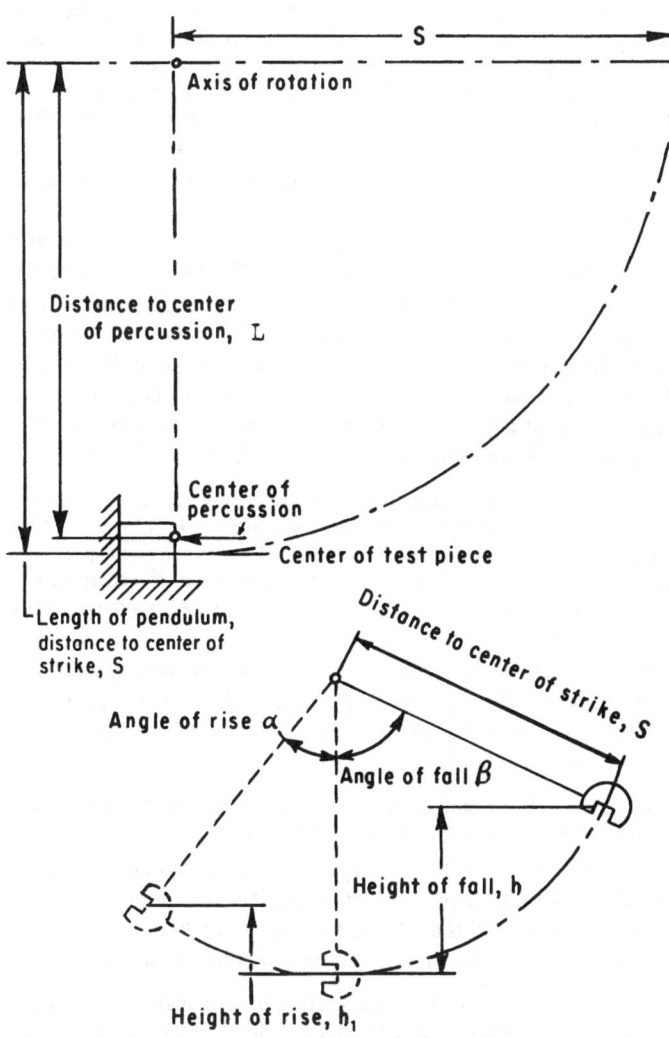

FIG. A2.1 Dimensions for Calculations

$$v = \sqrt{2gh} \quad (A2.3)$$

where:
v = velocity, m/s (ft/s),
g = acceleration of gravity, 9.81 m/s² (32.2 ft/s²), and
h = initial elevation of the striker, m (ft).

A2.3.7 The center of percussion shall be at a point within 1 % of the distance from the axis of rotation to the center of strike in the specimen, to ensure that minimum force is transmitted to the point of rotation. Determine the location of the center of percussion as follows:

A2.3.7.1 Using a stop watch or some other suitable time-measuring device, capable of measuring time to within 0.2 s, swing the pendulum through a total angle not greater than 15° and record the time for 100 complete cycles (to and fro). The period of the pendulum then, is the time for 100 cycles divided by 100.

A2.3.7.2 Determine the center of percussion by means of the following equation:

$$L = \frac{gp^2}{4\pi^2} \quad (A2.4)$$

where:
L = distance from the axis to the center of percussion, m (ft),
g = local gravitational acceleration (accuracy of one part in one thousand), m/s² (ft/s²),
π = 3.1416, and
p = period of a complete swing (to and fro), s.

A2.3.8 *Determination of the Friction Losses*—The energy loss from friction and windage of the pendulum and friction in the recording mechanism, if not corrected, will be included in the energy loss attributed to breaking the specimen and can result in erroneously high measurements of absorbed energy. For machines recording in degrees, frictional losses are usually not compensated for by the machine manufacturer, whereas in machines recording directly in energy, they are usually compensated for by increasing the starting height of the pendulum. Determine energy losses from friction as follows:

A2.3.8.1 Without a specimen in the machine, and with the indicator at the maximum energy reading, release the pendulum from its starting position and record the energy value indicated. This value should indicate zero energy if frictional losses have been corrected by the manufacturer. Now raise the pendulum slowly until it just contacts the indicator at the value obtained in the free swing. Secure the pendulum at this height and determine the vertical distance from the center of strike to the top of a half-width specimen positioned on the specimen rest supports within 0.1 % (see A2.3.5). Determine the supporting force as in A2.3.5.1 and multiply by this vertical distance. The difference in this value and the initial potential energy is the total energy loss in the pendulum and indicator combined. Without resetting the pointer, repeatedly release the pendulum from its initial position until the pointer shows no further movement. The energy loss determined by the final position of the pointer is that due to the pendulum alone. The frictional loss in the indicator alone is then the difference between the combined indicator and pendulum losses and those due to the pendulum alone.

A2.3.9 The indicating mechanism accuracy shall be checked to ensure that it is recording accurately over the entire range (see A1.2.1). Check it at graduation marks corresponding to approximately 0, 10, 20, 30, 50, and 70 % of each range. With the striker marked to indicate the center of strike, lift the pendulum and set it in a position where the indicator reads, for example, 13 J (10 ft·lbf). Secure the pendulum at this height and determine the vertical distance from the center of strike to the top of a half-width specimen positioned on the specimen supports within 0.1 % (see A2.3.5). Determine the residual energy by multiplying the height of the center of strike by the supporting force, as described in A2.3.5.1. Increase this value by the total frictional and windage losses for a free swing (see A2.3.8.1) multiplied by the ratio of the angle of swing of the pendulum from the latch to the energy value being evaluated to the angle of swing of the pendulum from the latch to the zero energy reading. Subtract the sum of the residual energy and proportional frictional and windage loss from the potential energy at the latched position (see A2.3.5). The indicator shall agree with the energy calculated within the limits of A1.2.1. Make similar calculations at other points of the scale. The indicating mechanism shall not overshoot or drop back with the pendulum. Make test swings from various heights to check visually the operation of the pointer over several portions of the scale.

Note A2.4—Indicators that indicate in degrees shall be checked using the above procedure. Degree readings from the scale shall be converted to energy readings using the conversion formula or table normally used in testing. In this way the formula or table can also be checked for windage and friction corrections.

A2.4 *Indirect Verification*:

A2.4.1 Indirect verification requires the testing of specimens with certified values to verify the accuracy of Charpy impact machines.

A2.4.1.1 Verification specimens with certified values are produced at low (13 to 20 J), high (88 to 136 J), and super-high (176 to 244 J) energy levels. To meet the verification requirements, the average value determined for a set of verification specimens at each energy level tested shall correspond to the certified values of the verification specimens within 1.4 J (1.0 ft·lbf) or 5.0 %, whichever is greater.

A2.4.1.2 Other sources of verification specimens[7] may be used provided their reference value has been established on the three reference machines owned, maintained, and operated by NIST in Boulder, CO.

A2.4.2 The verified range of a Charpy impact machine is described with reference to the lowest and highest energy specimens tested on the machine. These values are determined from tests on sets of verification specimens at two or more levels of absorbed energy, except in the case where a Charpy machine has a maximum capacity that is too low for two energy levels to be tested. In this case, one level of absorbed energy can be used for indirect verification.

A2.4.3 Determine the usable range of the impact testing machine prior to testing verification specimens. The usable range of an impact machine is dependent upon the resolution of the scale or readout device at the low end and the capacity of the machine at the high end.

A2.4.3.1 The resolution of the scale or readout device establishes the lower limit of the usable range for the machine. The lower limit is equal to 25 times the resolution of the scale or readout device at 15 J (11 ft-lbf).

Note A2.5—On analog scales, the resolution is the smallest change in energy that can be discerned on the scale. This is usually ¼ to ⅕ of the difference between 2 adjacent marks on the scale at the 15 J (11 ft-lbf) energy level.

Note A2.6—Digital readouts usually incorporate devices, such as digital encoders, with a fixed discrete angular resolution. The resolution of these types of readout devices is the smallest change in energy that can be consistently measured at 15 J. The resolution of these types of devices is usually not a change in the last digit shown on the display because resolution is a function of the angular position of the pendulum and changes throughout the swing. For devices which incorporate a verification mode in which a live readout of absorbed energy is available, the pendulum may be moved slowly in the area of 15 J to observe the smallest change in the readout device (the resolution).

A2.4.3.2 The upper limit of the usable range of the machine is equal to 80 % of the capacity of the machine.

A2.4.4 Only verification specimens that are within the usable range of the impact machine shall be tested. To verify the machine over its full usable range, test the lowest and highest energy levels of verification specimens commercially available that are within the machines' usable range. If the ratio of the highest and lowest certified values tested is greater than four, testing of a third set of intermediate energy specimens is required (if the specimens are commercially available).

Note A2.7—Use the upper bound of the energy range given for the low, high, and super-high verification specimens (20, 136, and 245 J respectively) to determine the highest energy level verification specimens that can be tested. Alternately, use the lower bound of the energy range given for the verification specimens to determine the minimum energy level for testing.

A2.4.4.1 If the low energy verification specimens were not tested (tested only high and super-high), the lower limit of the verified range shall be one half the energy of the lowest energy verification set tested.

Note A2.8—For example, if the certified value of the high energy specimens tested was 100 J, the lower limit would be 50 J.

A2.4.4.2 If the highest energy verification specimens available for a given Charpy machine capacity have not been tested, the upper value of the verified range shall be 1.5 times the certified value of the highest energy specimens tested.

Note A2.9—For example, if the machine being tested has a maximum capacity of 325 J (240 ft-lbf) and only low and high energy verification specimens were tested, the upper bound of the verified range would be 150 J (100 J * 1.5 = 150 J), assuming that the high energy samples tested had a certified value of 100 J. To verify this machine over its full range, low, high, and super-high verification specimens would have to be tested, because super-high verification specimens can be tested on a machine with a 325 J capacity (80 % of 325 J is 260 J, and the certified value of super-high specimens never exceed 260 J). See Table A2.1.

TABLE A2.1 Verified Ranges for Various Machine Capacities and Verification Specimens Tested[A]

Machine Capacity J	Resolution J	Usable Range J	Verification Specimens Tested			Verified Range J
			Low	High	Super-high	
80	0.10	2.5 to 64	X	...	...	2.5 to 64
160	0.20	5.0 to 128	X	X	...	5.0 to 128
325	0.25	6.25 to 260	X	X	X	6.25 to 260
400	0.30	7.5 to 320	...	X	X	50 to 320
400	0.15	3.75 to 320	X	X	...	3.75 to 150
400	0.15	3.75 to 320	X	X	X	3.75 to 320

[A] In these examples, the high energy verification specimens are assumed to have a certified value of 100 J.

---

[7] Some sources for verification specimens maybe listed in the ASTM International Equipment Directory, www.astm.org.

A3. ADDITIONAL IMPACT TEST SPECIMEN CONFIGURATIONS

A3.1 *Sub-Size Specimen*—When the amount of material available does not permit making the standard impact test specimens shown in Figs. 1 and 2, smaller specimens may be used, but the results obtained on different sizes of specimens cannot be compared directly (X1.3). When Charpy specimens other than the standard are necessary or specified, it is recommended that they be selected from Fig. A3.1.

A3.2 *Supplementary Specimens*—For economy in preparation of test specimens, special specimens of round or rectangular cross section are sometimes used for cantilever beam test. These are shown as Specimens X, Y, and Z in Figs. A3.2 and A3.3. Specimen Z is sometimes called the Philpot specimen, after the name of the original designer. For hard materials, the machining of the flat surface struck by the pendulum is sometimes omitted. Types Y and Z require a different vise from that shown in Fig. A1.3, each half of the vise having a semi-cylindrical recess that closely fits the clamped portion of the specimen. As previously stated, the results cannot be reliably compared with those obtained using specimens of other sizes or shapes.

On sub-size specimens the length, notch angle, and notch radius are constant (see Fig. 1); depth (*D*), notch depth (*N*), and width (*W*) vary as indicated below.

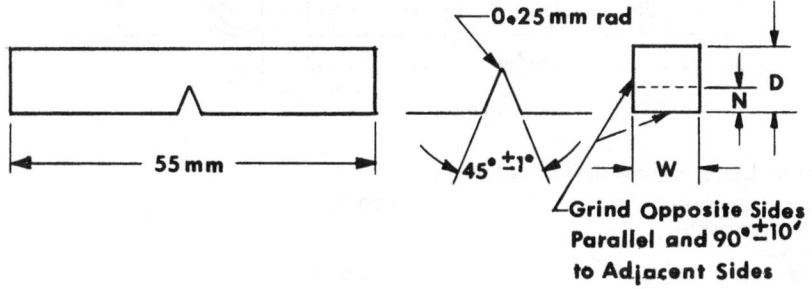

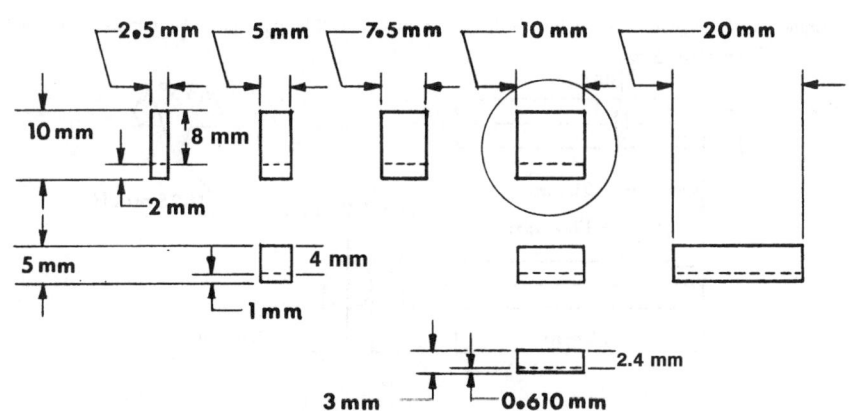

Note 1—Circled specimen is the standard specimen (see Fig. 1).
Note 2—Permissible variations shall be as follows:

Cross-section dimensions	±1 % or ±0.075 mm, whichever is smaller
Radius of notch	±0.025 mm
Ligament length	±0.025 mm
Finish requirements	2 µm on notched surface and opposite face; 4 µm on other two surfaces

FIG. A3.1 Non-Standard Charpy (Simple-Beam) (Type A) Impact Test Specimens

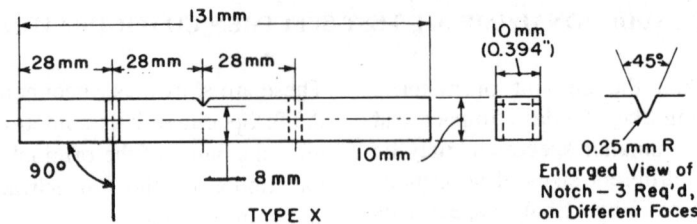

NOTE 1—Permissible variations for type X specimens shall be as follows:

    Notch length to edge   90± 2°
    Adjacent sides shall be at   90°± 10 min
    Ligament length of Type X specimen ±0.025 mm

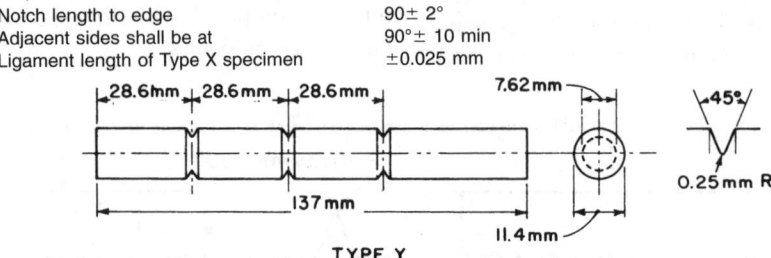

NOTE 2—Permissible variations for both specimens shall be as follows:

    Cross-section dimensions   ±0.025 mm
    Lengthwise dimensions   +0, −2.5 mm
    Angle of notch   ±1°
    Radius of notch   ±0.025 mm
    Notch diameter of Type Y specimen ±0.025 mm

FIG. A3.2 Izod (Cantilever-Beam) Impact Test Specimens, Types X and Y

The flat shall be parallel to the longitudinal centerline of the specimen and shall be parallel to the bottom of the notch within 2:1000.

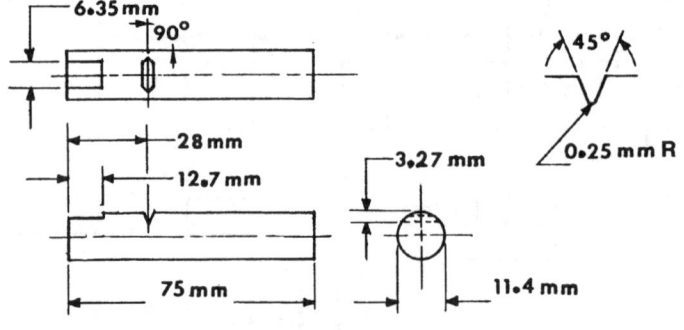

TYPE Z

NOTE 1—Permissible variations shall be as follows:

    Notch length to longitudinal centerline 90± 2°
    Cross-section dimensions   ±0.025 mm
    Length of specimen   +0, −2.5 mm
    Angle of notch   ±1°
    Radius of notch   ±0.025 mm
    Notch depth   ±0.025 mm

FIG. A3.3 Izod (Cantilever-Beam) Impact Test Specimen (Philpot), Type Z

A4. PRECRACKING CHARPY V-NOTCH IMPACT SPECIMENS

A4.1 Scope

A4.1.1 This annex describes the procedure for the fatigue precracking of standard Charpy V-notch (CVN) impact specimens. The annex provides information on applications of precracked Charpy impact testing and fatigue-precracking procedures.

A4.2 Significance and Use

A4.2.1 Section 4 also applies to precracked Charpy V-notch impact specimens.

A4.2.2 It has been found that fatigue-precracked CVN specimens generally result in better correlations with other impact toughness tests such as Test Method E604 and with fracture toughness tests such as Test Method E399 than the standard V-notch specimens (3,4,5,6,7,8). Also, the sharper notch yields more conservative estimations of the notched impact toughness and the transition temperature of the material (9,10).

A4.3 Apparatus

A4.3.1 The equipment for fatigue cracking shall be such that the stress distribution is symmetrical through the specimen thickness; otherwise, the crack will not grow uniformly. The stress distribution shall also be symmetrical about the plane of the prospective crack; otherwise the crack will deviate unduly from that plane and the test result will be significantly affected.

A4.3.2 The recommended fixture to be used is shown in Fig. A4.1. The nominal span between support rollers shall be $4D \pm 0.2D$, where D is the depth of the specimen. The diameter of the rollers shall be between $D/2$ and D. The radius of the ram shall be between $D/8$ and D. This fixture is designed to minimize frictional effects by allowing the support rollers to rotate and move apart slightly as the specimen is loaded, thus permitting rolling contact. The rollers are initially positioned against stops that set the span length and are held in place by low-tension springs (such as rubber bands). Fixtures, rolls, and ram should be made of high hardness (greater than 40 HRC) steels.

A4.4 Test Specimens

A4.4.1 The dimensions of the precracked Charpy specimen are essentially those of type-A shown in Fig. 1. The notch depth plus the fatigue crack extension length shall be designated as N as shown in Fig. A4.2. When the amount of material available does not permit making the standard impact test specimen, smaller specimens may be made by reducing the

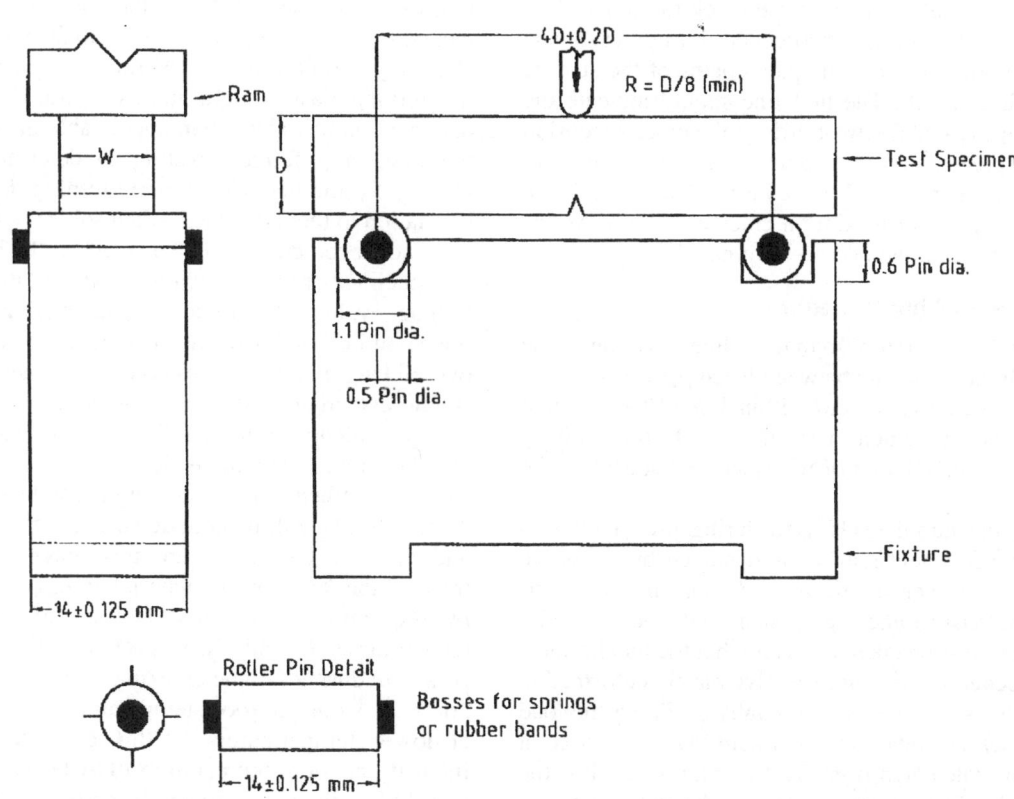

FIG. A4.1 Fatigue Precracking-Fixture Design

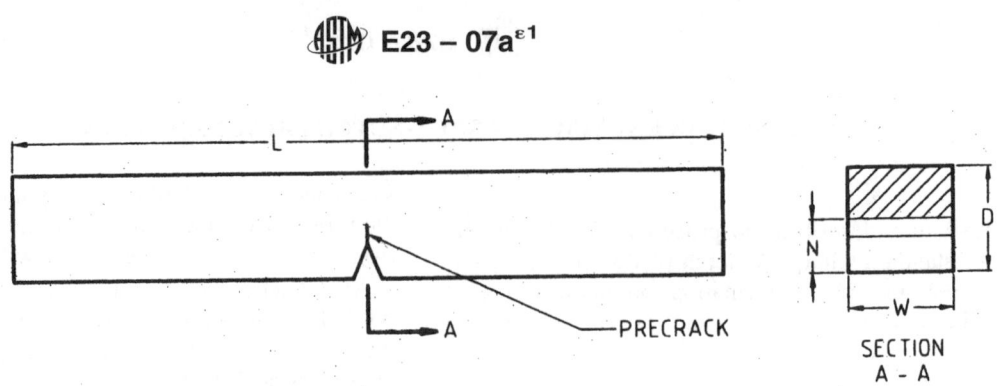

L = Length
D = Depth
W = Width
N = Notch depth plus fatigue crack extension length

FIG. A4.2 Charpy (Simple-Beam, type A) Impact Test Specimen

width; but the results obtained on different sizes of specimens cannot be compared directly (see X1.3).

A4.4.2 The fatigue precracking is to be done with the material in the same heat-treated condition as that in which it will be impact tested. No intermediate treatments between fatigue precracking and testing are allowed.

A4.4.3 Because of the relatively blunt machined V-notch in the Charpy impact specimen, fatigue crack initiation can be difficult. Early crack initiation can be promoted by pressing or milling a sharper radius into the V-notch. Care must be taken to ensure that excessive deformation at the crack tip is avoided.

A4.4.4 It is advisable to mark two pencil lines on each side of the specimen normal to the anticipated paths of the surface traces of the fatigue crack. The first line should indicate the point at which approximately two-thirds of the crack extension has been accomplished. At this point, the stress intensity applied to the specimen should be reduced. The second line should indicate the point of maximum crack extension. At this point, fatigue precracking should be terminated.

A4.5 Fatigue Precracking Procedure

A4.5.1 Set up the test fixture so that the line of action of the applied load shall pass midway between the support roll center within 1 mm. Measure the span to within 1 % of the nominal length. Locate the specimen with the crack tip midway between the rolls within 1 mm of the span, and square to the roll axes within 2°.

A4.5.2 Select the initial loads used during precracking so that the remaining ligament remains undamaged by excessive plasticity. If the load cycle is maintained constant, the maximum K (stress intensity) and the K range will increase with crack length; care must be taken to ensure that the maximum K value is not exceeded to prevent excessive plastic deformation at the crack tip. This is done by continually shedding the load as the fatigue crack extends. The maximum load to be used at any instant can be calculated from Eq A4.1 and A4.2 while the minimum load should be kept at 10 % of the maximum. Eq A4.1 relates the maximum load to a stress intensity (K) value for the material that will ensure an acceptable plastic-zone size at the crack tip. It is also advisable to check this maximum load to ensure that it is below the limit load for the material using Eq A4.2. When the most advanced crack trace has almost reached the first scribed line corresponding to approximately two-thirds of the final crack length, reduce the maximum load so that 0.6 K_{max} is not exceeded.

A4.5.3 Fatigue cycling is begun, usually with a sinusoidal waveform and near to the highest practical frequency. There is no known marked frequency effect on fatigue precrack formation up to at least 100 Hz in the absence of adverse environments; however, frequencies of 15 to 30 Hz are typically used. Carefully monitor the crack growth optically. A low-power magnifying glass is useful in this regard. If crack growth is not observed on one side when appreciable growth is observed on the first, stop fatigue cycling to determine the cause and remedy for the behavior. Simply turning the specimen around in relation to the fixture will often solve the problem. When the most advanced crack trace has reached the halfway mark, turn the specimen around in relation to the fixture and complete the fatigue cycling. Continue fatigue cycling until the surface traces on both sides of the specimen indicate that the desired overall length of notch plus crack is reached. The fatigue crack should extend at least 1 mm beyond the tip of the V-notch but no more than 3 mm. A fatigue crack extension of approximately 2 mm is recommended.

A4.5.4 When fatigue cracking is conducted at a temperature T_1 and testing will be conducted at a different temperature T_2, and $T_1 > T_2$, the maximum stress intensity must not exceed 60 % of the K_{max} of the material at temperature T_1 multiplied by the ratio of the yield stresses of the material at the temperatures T_1 and T_2, respectively. Control of the plastic-zone size during fatigue cracking is important when the fatigue cracking is done at room temperature and the test is conducted at lower temperatures. In this case, the maximum stress intensity at room temperature must be kept to low values so that the plastic-zone size corresponding to the maximum stress intensity at low temperatures is smaller.

A4.6 Calculation

A4.6.1 Specimens shall be precracked in fatigue at load values that will not exceed a maximum stress intensity, K_{max}. or three-point bend specimens use:

$$P_{max} = [K_{max}*W*D^{3/2}] / [S*f(N/D)] \quad (A4.1)$$

where:
- P_{max} = maximum load to be applied during precracking,
- K_{max} = maximum stress intensity = $\sigma_{ys}*(2*\pi*r_y)^{1/2}$, where r_y = is the radius of the induced plastic zone size which should be less than or equal to 0.5 mm,
- D = specimen depth,
- W = specimen width,
- S = span, and
- $f(N/D)$ = geometrical factor (see Table A4.1).

A4.6.2 See the appropriate section of Test Method E399 for the $f(N/D)$ calculation. Table A4.1 contains calculated values for $f(N/D)$ for CVN precracking. Eq A4.2 should be used to ensure that the loads used in fatigue cracking are well below the calculated limit load for the material.

$$P_L = (4/3)*[D*(D-N)^2*\sigma_{ys}]/S \quad (A4.2)$$

where:
- P_L = limit load for the material.

A4.7 Crack Length Measurement

A4.7.1 After fracture, measure the initial notch plus fatigue crack length, N, to the nearest 1 % at the following three positions: at the center of the crack front and midway between the center and the intersection of the crack front with the specimen surfaces. Use the average of these three measurements as the crack length.

A4.7.2 If the difference between any two of the crack length measurements exceeds 10 % of the average, or if part of the crack front is closer to the machine notch root than 5 % of the average, the specimen should be discarded. Also, if the length of either surface trace of the crack is less than 80 % of the average crack length, the specimen should be discarded.

A4.8 Report

A4.8.1 Report the following information for each specimen tested: type of specimen used (and size if not the standard size), test temperatures, and energy absorption. Report the average precrack length in addition to these Test Method E23 requirements.

A4.8.2 The following information may be provided as supplementary information: lateral expansion, fracture appearance, and also, it would probably be useful to report energy absorption normalized in some manner.

TABLE A4.1 Calculations of f(N/D)

N (mm)	D (mm)	N/D	f(N/D)
2.00	10.00	0.20	1.17
2.10	10.00	0.21	1.21
2.20	10.00	0.22	1.24
2.30	10.00	0.23	1.27
2.40	10.00	0.24	1.31
2.50	10.00	0.25	1.34
2.60	10.00	0.26	1.37
2.70	10.00	0.27	1.41
2.80	10.00	0.28	1.45
2.90	10.00	0.29	1.48
3.00	10.00	0.30	1.52
3.10	10.00	0.31	1.56
3.20	10.00	0.32	1.60
3.30	10.00	0.33	1.64
3.40	10.00	0.34	1.69
3.50	10.00	0.35	1.73
3.60	10.00	0.36	1.78
3.70	10.00	0.37	1.83
3.80	10.00	0.38	1.88
3.90	10.00	0.39	1.93
4.00	10.00	0.40	1.98
4.10	10.00	0.41	2.04
4.20	10.00	0.42	2.10
4.30	10.00	0.43	2.16
4.40	10.00	0.44	2.22
4.50	10.00	0.45	2.29
4.60	10.00	0.46	2.35
4.70	10.00	0.47	2.43
4.80	10.00	0.48	2.50
4.90	10.00	0.49	2.58
5.00	10.00	0.50	2.66

A5. SPECIMEN ORIENTATION

A5.1 Designation of Specimen Axis:

A5.1.1 The L-axis is coincident with the main direction of grain flow due to processing. This axis is usually referred to as the longitudinal direction (see Fig. A5.1, Fig. A5.2, and Fig. A5.3).

A5.1.2 The S-axis is coincident with the direction of the main working force. This axis is usually referred to as the short-transverse-direction.

A5.1.3 The T-axis is normal to the L- and S-axies. This axis is usually referred to as the transverse direction.

A5.1.4 Specimens parallel to the surface of wrought products, processed with the same degree of homogenous deformation along the L- and T axies may be called T specimens.

A5.1.5 Specimens normal to the uniform grain flow of wrought products (or grain growth in cast products), whose grain flow is exclusively in one direction, so that T- and S specimens are equivalent, may be called S specimens.

A5.2 Designation of Notch Orientation:

A5.2.1 The notch orientation is designated by the direction in which fracture propagates. This letter is separated from the specimen-axis designation by a hyphen. In unique cases (Fig. A5.3), when fracture propagates across two planes, two letters are required to designate notch orientation.

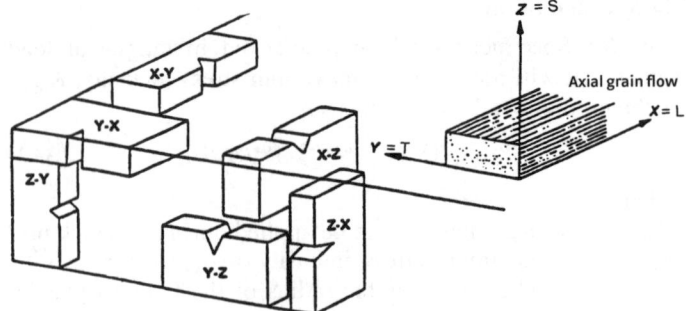

FIG. A5.1 Fracture Planes Along Principal Axes

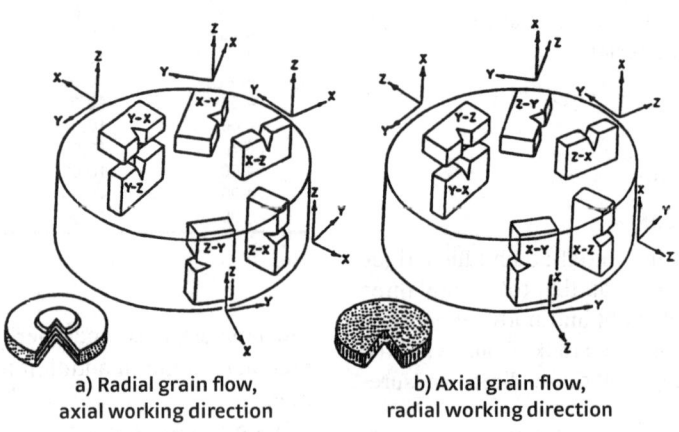

a) Radial grain flow, axial working direction

b) Axial grain flow, radial working direction

FIG. A5.2 Fracture Planes—Cylindrical Sections

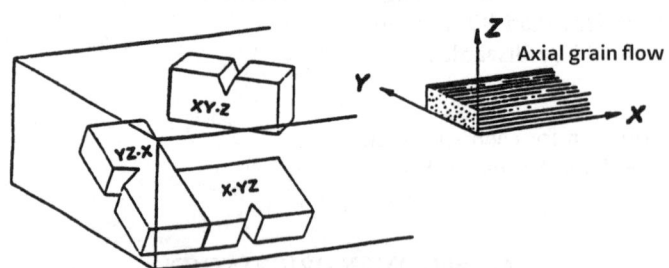

FIG. A5.3 Fracture Planes not Along Principal Axes

A6. DETERMINATION OF THE PROPORTION OF SHEAR FRACTURE SURFACE

A6.1 These fracture-appearance methods are based on the concept that 100% shear (ductile) fracture occurs above the transition-temperature range and cleavage (brittle) fracture occurs below the range. This concept appears to be appropriate, at least for body-centered-cubic iron-based alloys that undergo a distinct ductile to brittle transition, but interpretation is complicated in materials that exhibit mixed mode fracture during unstable crack growth. In the transition-temperature range, fracture is initiated at the root of the notch by fibrous tearing. A short distance from the notch, unstable crack growth occurs as the fracture mechanism changes to cleavage or mixed mode mechanism, which often results in distinct radial markings in the central portion of the specimen (indicative of fast, unstable fracture). After several microseconds the unstable crack growth arrests. Final fracture occurs at the remaining ligament and at the sides of the specimen in a ductile manner.

As shear-lips are formed at the sides of the specimen, the plastic hinge at the remaining ligament ruptures. In the ideal case, a "picture frame" of fibrous (ductile) fracture surrounds a relatively flat area of cleavage (brittle) fracture.

The five methods used below may be used to determine the percentage of ductile fracture on the surface of impact specimens. It is recommended that the user qualitatively characterize the fracture mode of the flat fracture zone, and provide a description of how the shear measurements were made. The accuracy of the methods are grouped in order of increasing precision. In the case where a specimen does not separate into two halves during the impact test and the fracture occurs without any evidence of cleavage (brittle) fracture, the percent shear fracture can be considered to be 100% and the specimen should be reported as unbroken.

NOTE A6.1—Round robin data (five U.S. companies, 1990) estimates of the percent shear for five quenched and tempered 8219 steels and four microalloyed 1040 steels indicated the following: (1) results using method A6.1.1 systematically underestimated the percent shear (compared with method A6.1.4), (2) the error using method A6.1.2 was random and, (3) The typical variation in independent measurements using method A6.1.4 was on the order of 5 to 10 % for microalloyed 1040 steels.

A6.1.1 Measure the length and width of the flat fracture region of the fracture surface, as shown in Fig. 10, and determine the percent shear from either Table A6.1 or Table A6.2 depending on the units of measurement.

A6.1.2 Compare the appearance of the fracture of the specimen with a fracture appearance chart such as that shown in Fig. A6.1.

A6.1.3 Magnify the fracture surface and compare it to a precalibrated overlay chart or measure the percent shear fracture by means of a planimeter.

A6.1.4 Photograph the fracture surface at a suitable magnification and measure the percent shear fracture by means of a planimeter.

A6.1.5 Capture a digital image of the fracture surface and measure the percent shear fracture using image analysis software.

TABLE A6.1 Percent Shear for Measurements Made in Millimetres

NOTE 1—100 % shear is to be reported when either A or B is zero.

Dimension B, mm	Dimension A, mm																	
	1.0	1.5	2.0	2.5	3.0	3.5	4.0	4.5	5.0	5.5	6.0	6.5	7.0	7.5	8.0	8.5	9.0	9.5	10
1.0	99	98	98	97	96	96	95	94	94	93	92	92	91	91	90	89	89	88	88
1.5	98	97	96	95	94	93	92	92	91	90	89	88	87	86	85	84	83	82	81
2.0	98	96	95	94	92	91	90	89	88	86	85	84	82	81	80	79	77	76	75
2.5	97	95	94	92	91	89	88	86	84	83	81	80	78	77	75	73	72	70	69
3.0	96	94	92	91	89	87	85	83	81	79	77	76	74	72	70	68	66	64	62
3.5	96	93	91	89	87	85	82	80	78	76	74	72	69	67	65	63	61	58	56
4.0	95	92	90	88	85	82	80	77	75	72	70	67	65	62	60	57	55	52	50
4.5	94	92	89	86	83	80	77	75	72	69	66	63	61	58	55	52	49	46	44
5.0	94	91	88	85	81	78	75	72	69	66	62	59	56	53	50	47	44	41	37
5.5	93	90	86	83	79	76	72	69	66	62	59	55	52	48	45	42	38	35	31
6.0	92	89	85	81	77	74	70	66	62	59	55	51	47	44	40	36	33	29	25
6.5	92	88	84	80	76	72	67	63	59	55	51	47	43	39	35	31	27	23	19
7.0	91	87	82	78	74	69	65	61	56	52	47	43	39	34	30	26	21	17	12
7.5	91	86	81	77	72	67	62	58	53	48	44	39	34	30	25	20	16	11	6
8.0	90	85	80	75	70	65	60	55	50	45	40	35	30	25	20	15	10	5	0

TABLE A6.2 Percent Shear for Measurements Made in Inches

NOTE 1—100 % shear is to be reported when either A or B is zero.

Dimension B, in.	Dimension A, in.																	
	0.05	0.10	0.12	0.14	0.16	0.18	0.20	0.22	0.24	0.26	0.28	0.30	0.32	0.34	0.36	0.38	0.40
0.05	98	96	95	94	94	93	92	91	90	90	89	88	87	86	85	85	84
0.10	96	92	90	89	87	85	84	82	81	79	77	76	74	73	71	69	68
0.12	95	90	88	86	85	83	81	79	77	75	73	71	69	67	65	63	61
0.14	94	89	86	84	82	80	77	75	73	71	68	66	64	62	59	57	55
0.16	94	87	85	82	79	77	74	72	69	67	64	61	59	56	53	51	48
0.18	93	85	83	80	77	74	72	68	65	62	59	56	54	51	48	45	42
0.20	92	84	81	77	74	72	68	65	61	58	55	52	48	45	42	39	36
0.22	91	82	79	75	72	68	65	61	57	54	50	47	43	40	36	33	29
0.24	90	81	77	73	69	65	61	57	54	50	46	42	38	34	30	27	23
0.26	90	79	75	71	67	62	58	54	50	46	41	37	33	29	25	20	16
0.28	89	77	73	68	64	59	55	50	46	41	37	32	28	23	18	14	10
0.30	88	76	71	66	61	56	52	47	42	37	32	27	23	18	13	9	3
0.31	88	75	70	65	60	55	50	45	40	35	30	25	20	18	10	5	0

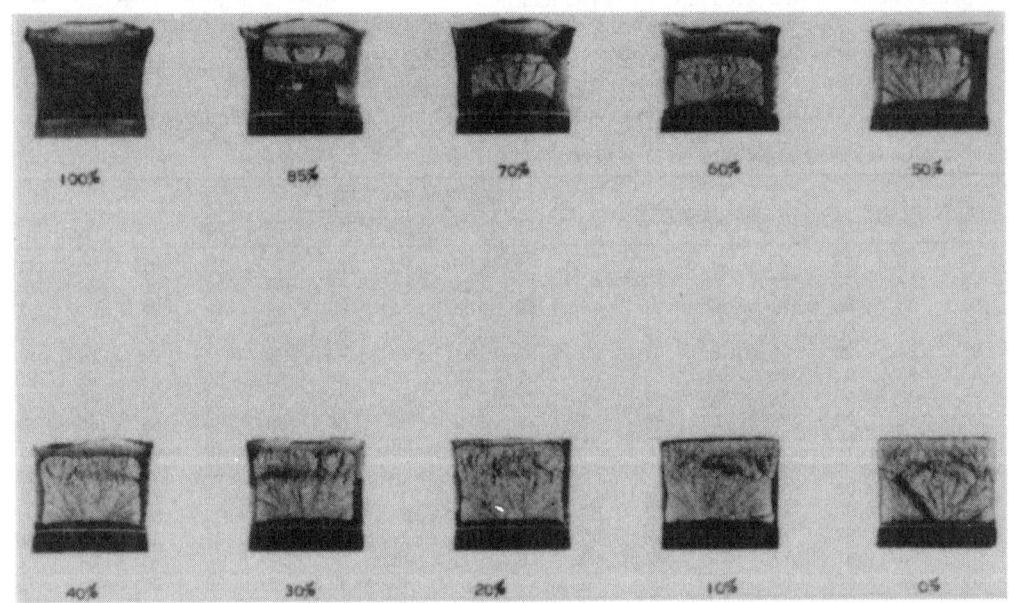

(*a*) Fracture Appearance Charts and Percent Shear Fracture Comparator

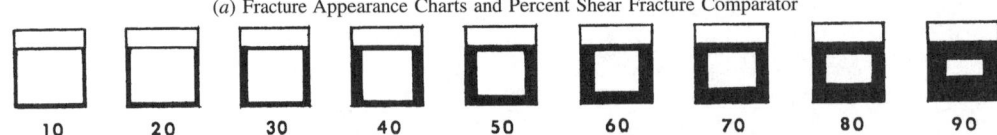

(*b*) Guide for Estimating Fracture Appearance

FIG. A6.1 Fracture Appearance

APPENDIXES

(Nonmandatory Information)

X1. NOTES ON SIGNIFICANCE OF NOTCHED-BAR IMPACT TESTING

X1.1 *Notch Behavior:*

X1.1.1 The Charpy V-notch (CVN) impact test has been used extensively in mechanical testing of steel products, in research, and in procurement specifications for over three decades. Where correlations with fracture mechanics parameters are available, it is possible to specify CVN toughness values that would ensure elastic-plastic or plastic behavior for fracture of fatigue cracked specimens subjected to minimum operating temperatures and maximum in service rates of loading.

X1.1.2 The notch behavior of the face-centered cubic metals and alloys, a large group of nonferrous materials and the austenitic steels can be judged from their common tensile properties. If they are brittle in tension, they will be brittle when notched, while if they are ductile in tension they will be ductile when notched, except for unusually sharp or deep notches (much more severe than the standard Charpy or Izod specimens). Even low temperatures do not alter this characteristic of these materials. In contrast, the behavior of the ferritic steels under notch conditions cannot be predicted from their properties as revealed by the tension test. For the study of these materials the Charpy and Izod type tests are accordingly very useful. Some metals that display normal ductility in the tension test may nevertheless break in brittle fashion when tested or when used in the notched condition. Notched conditions include constraints to deformation in directions perpendicular to the major stress, or multi axial stresses, and stress concentrations. It is in this field that the Charpy and Izod tests prove useful for determining the susceptibility of a steel to notch-brittle behavior though they cannot be directly used to appraise the serviceability of a structure.

X1.2 *Notch Effect*:

X1.2.1 The notch results in a combination of multi axial stresses associated with restraints to deformation in directions perpendicular to the major stress, and a stress concentration at the base of the notch. A severely notched condition is generally not desirable, and it becomes of real concern in those cases in which it initiates a sudden and complete failure of the brittle type. Some metals can be deformed in a ductile manner even down to very low temperatures, while others may crack. This difference in behavior can be best understood by considering the cohesive strength of a material (or the property that holds it together) and its relation to the yield point. In cases of brittle fracture, the cohesive strength is exceeded before significant plastic deformation occurs and the fracture appears crystalline. In cases of the ductile or shear type of failure, considerable deformation precedes the final fracture and the broken surface appears fibrous instead of crystalline. In intermediate cases, the fracture comes after a moderate amount of deformation and is part crystalline and part fibrous in appearance.

X1.2.2 When a notched bar is loaded, there is a normal stress across the base of the notch which tends to initiate fracture. The property that keeps it from cleaving, or holds it together, is the "cohesive strength". The bar fractures when the normal stress exceeds the cohesive strength. When this occurs without the bar deforming it is the condition for brittle fracture.

X1.2.3 In testing, though not in service because of side effects, it happens more commonly that plastic deformation precedes fracture. In addition to the normal stress, the applied load also sets up shear stresses which are about 45° to the normal stress. The elastic behavior terminates as soon as the shear stress exceeds the shear strength of the material and deformation or plastic yielding sets in. This is the condition for ductile failure.

X1.2.4 This behavior, whether brittle or ductile, depends on whether the normal stress exceeds the cohesive strength before the shear stress exceeds the shear strength. Several important facts of notch behavior follow from this. If the notch is made sharper or more drastic, the normal stress at the root of the notch will be increased in relation to the shear stress and the bar will be more prone to brittle fracture (see Table X1.1). Also, as the speed of deformation increases, the shear strength increases and the likelihood of brittle fracture increases. On the other hand, by raising the temperature, leaving the notch and

TABLE X1.1 Effect of Varying Notch Dimensions on Standard Specimens

	High-Energy Specimens, J (ft·lbf)	Medium-Energy Specimens, J (ft·lbf)	Low-Energy Specimens, J (ft·lbf)
Specimen with standard dimensions	103.0 ± 5.2 (76.0 ± 3.8)	60.3 ± 3.0 (44.5 ± 2.2)	16.9 ± 1.4 (12.5 ± 1.0)
Depth of notch, 2.13 mm (0.084 in.)[A]	97.9 (72.2)	56.0 (41.3)	15.5 (11.4)
Depth of notch, 2.04 mm (0.0805 in.)[A]	101.8 (75.1)	57.2 (42.2)	16.8 (12.4)
Depth of notch, 1.97 mm (0.0775 in.)[A]	104.1 (76.8)	61.4 (45.3)	17.2 (12.7)
Depth of notch, 1.88 mm (0.074 in.)[A]	107.9 (79.6)	62.4 (46.0)	17.4 (12.8)
Radius at base of notch 0.13 mm (0.005 in.)[B]	98.0 (72.3)	56.5 (41.7)	14.6 (10.8)
Radius at base of notch 0.38 mm (0.015 in.)[B]	108.5 (80.0)	64.3 (47.4)	21.4 (15.8)

[A] Standard 2.0 ± 0.025 mm (0.079 ± 0.001 in.).
[B] Standard 0.25 ± 0.025 mm (0.010 ± 0.001 in.).

the speed of deformation the same, the shear strength is lowered and ductile behavior is promoted, leading to shear failure.

X1.2.5 Variations in notch dimensions will seriously affect the results of the tests. Tests on E4340 steel specimens[5] have shown the effect of dimensional variations on Charpy results (see Table X1.1).

X1.3 *Size Effect*:

X1.3.1 Increasing either the width or the depth of the specimen tends to increase the volume of metal subject to distortion, and by this factor tends to increase the energy absorption when breaking the specimen. However, any increase in size, particularly in width, also tends to increase the degree of constraint and by tending to induce brittle fracture, may decrease the amount of energy absorbed. Where a standard-size specimen is on the verge of brittle fracture, this is particularly true, and a double width specimen may actually require less energy for rupture than one of standard width.

X1.3.2 In studies of such effects where the size of the material precludes the use of the standard specimen, for example when the material is 6.35-mm (0.25-in.) plate, subsize specimens are used. Such specimens (Fig. A3.1) are based on the Type A specimen of Fig. 1.

X1.3.3 General correlation between the energy values obtained with specimens of different size or shape is not feasible, but limited correlations may be established for specification purposes on the basis of special studies of particular materials and particular specimens. On the other hand, in a study of the relative effect of process variations, evaluation by use of some arbitrarily selected specimen with some chosen notch will in most instances place the methods in their proper order.

X1.4 *Temperature Effect*:

X1.4.1 The testing conditions also affect the notch behavior. So pronounced is the effect of temperature on the behavior of steel when notched that comparisons are frequently made by examining specimen fractures and by plotting energy value and fracture appearance versus temperature from tests of notched bars at a series of temperatures. When the test temperature has been carried low enough to start cleavage fracture, there may be an extremely sharp drop in absorbed energy or there may be a relatively gradual falling off toward the lower temperatures. This drop in energy value starts when a specimen begins to exhibit some crystalline appearance in the fracture. The transition temperature at which this embrittling effect takes place varies considerably with the size of the part or test specimen and with the notch geometry.

X1.5 *Testing Machine*:

X1.5.1 The testing machine itself must be sufficiently rigid or tests on high-strength low-energy materials will result in excessive elastic energy losses either upward through the pendulum shaft or downward through the base of the machine. If the anvil supports, the striker, or the machine foundation bolts are not securely fastened, tests on ductile materials in the range from 108 J (80 ft·lbf) may actually indicate values in excess of 122 to 136 J (90 to 100 ft·lbf)

X1.5.2 A problem peculiar to Charpy-type tests occurs when high-strength, low-energy specimens are tested at low temperatures. These specimens may not leave the machine in the direction of the pendulum swing but rather in a sidewise direction. To ensure that the broken halves of the specimens do not rebound off some component of the machine and contact the pendulum before it completes its swing, modifications may be necessary in older model machines. These modifications differ with machine design. Nevertheless the basic problem is the same in that provisions must be made to prevent rebounding of the fractured specimens into any part of the swinging pendulum. Where design permits, the broken specimens may be deflected out of the sides of the machine and yet in other designs it may be necessary to contain the broken specimens within a certain area until the pendulum passes through the anvils. Some low-energy high-strength steel specimens leave impact machines at speeds in excess of 15.2 m/s (50 ft/s) although they were struck by a pendulum traveling at speeds approximately 5.2 m/s (17 ft/s). If the force exerted on the pendulum by the broken specimens is sufficient, the pendulum will slow down and erroneously high energy values will result. This problem accounts for many of the inconsistencies in Charpy results reported by various investigators within the 14 to 34 J (10 to 25 ft-lb) range. Figure A1.1 illustrates a modification found to be satisfactory in minimizing jamming.

X1.6 *Velocity of Straining*:

X1.6.1 Velocity of straining is likewise a variable that affects the notch behavior of steel. The impact test shows somewhat higher energy absorption values than the static tests above the transition temperature and yet, in some instances, the reverse is true below the transition temperature.

X1.7 *Correlation with Service*:

X1.7.1 While Charpy or Izod tests may not directly predict the ductile or brittle behavior of steel as commonly used in large masses or as components of large structures, these tests can be used as acceptance tests or tests of identity for different lots of the same steel or in choosing between different steels, when correlation with reliable service behavior has been established. It may be necessary to make the tests at properly chosen temperatures other than room temperature. In this, the service temperature or the transition temperature of full-scale specimens does not give the desired transition temperatures for Charpy or Izod tests since the size and notch geometry may be so different. Chemical analysis, tension, and hardness tests may not indicate the influence of some of the important processing factors that affect susceptibility to brittle fracture nor do they comprehend the effect of low temperatures in inducing brittle behavior.

X2. SUGGESTED METHODS OF MEASUREMENT

X2.1 *Position of the Center of Strike Relative to the Center of Gravity:*

X2.1.1 Since the center of strike can only be marked on an assembled machine, only the methods applicable to an assembled machine are described as follows:

X2.1.1.1 The fundamental fact on which all the methods are based is that when the friction forces are negligible, the center of gravity is vertically below the axis of rotation of a pendulum supported by the bearings only, (herein referred to as a free hanging pendulum). Paragraph A1.3 limits the friction forces in impact machines to a negligible value. The required measurements may be made using specialized instruments such as transits, clinometers, or cathometers. However, simple instruments have been used as described in the following to make measurements of sufficient accuracy.

X2.1.1.2 Suspend a plumb bob from the frame. The plumb line should appear visually to be in the plane of swing of the striking edge.

X2.1.1.3 Place a massive object on the base close to the latch side of the pendulum. Adjust the position of this object so that when back lighted, a minimal gap is visible between it and the pendulum. (See Fig. X2.1.)

X2.1.1.4 With a scale or depth gage pressed lightly against the striking edge at the center of strike, measure the horizontal distance between the plumb line and striking edge. (The dimension B in Fig. X2.1.)

X2.1.1.5 Similarly, measure the distance in a horizontal plane through the axis of rotation from the plumb line to the clamp block or enlarged end of the pendulum stem. (Dimension A in Fig. X2.1.)

X2.1.1.6 Use a depth gage to measure the radial distance from the surface contacted in measuring A to a machined surface of the shaft which connects the pendulum to the bearings in the machine frame. (Dimension C in Fig. X2.1.)

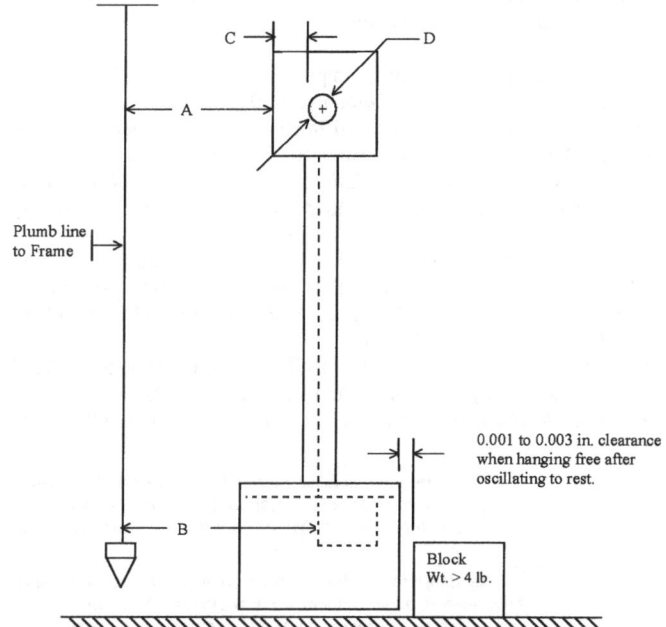

FIG. X2.1 Measurement of Deviation of Center of Strike from Vertical Plane through Axis of Rotation when Pendulum is Hanging Free

X2.1.1.7 Use an outside caliper or micrometer to measure the diameter of the shaft at the same location contacted in measuring C. (Dimension D in Fig. X2.1.)

X2.1.1.8 Substitute the measured dimensions in the equation

$$X = A + C + D/2 - B \tag{X2.1}$$

where:
X = deviation of the center of strike from a line from the center of rotation through the center of gravity.

X3. INSTRUCTIONS FOR TESTING NON-STANDARD SPECIMENS

X3.1 When testing non-standard size specimens (see Fig. A3.1), the specimen support height should be changed to ensure that the center of strike is maintained (see A2.3.4 and A2.3.7 for instructions). To comply with this change, new specimen supports can be manufactured or shims may be added to the specimen supports in a secure manner so that they do not interfere with the test.

X3.2 In order to maintain the center of strike requirements, the following procedure should be used when testing a non-standard specimen. The height of the specimen supports should be changed to ensure that the centerline of the non-standard specimen will coincide with the centerline of the standard specimen. Higher specimen supports should be used when testing a sub-size specimen and lower specimen supports should be used when testing an oversized specimen.

X3.3 Determine the nominal height of the non-standard specimen. When testing sub-size specimens, subtract this value from the standard height specimens (10 mm). Divide this value by two. This amount shall be added to the standard specimen support height. For oversize specimens, the result of the subtraction is a negative number. Therefore, the thickness of the supports shall be reduced by the amount calculated.

REFERENCES

(1) Nanstad, R. K., Swain, R. L. and Berggren, R. G., "Influence of Thermal Conditioning Media on Charpy Specimen Test Temperature," *Charpy Impact Test: Factors and Variables, ASTM STP 1072*, ASTM, 1990.

(2) Tobler R. L. Et al.," Charpy Impact Tests Near Absolute Zero," *Journal of Testing and Evaluation*, Vol 19, 1 1992.

(3) Wullaert, R. A., Ireland, D. R., and Tetelman, A. S., "Radiation Effects on the Metallurgical Fracture Parameters and Fracture Toughness of Pressure Vessel Steels," *Irradiation Effects on Structural Alloys for Nuclear Reactor Applications, ASTM STP 484*, ASTM, 1970, pp. 20–41.

(4) Sovak, J. F., "Correlation of Data from Standard and Precracked Charpy Specimens with Fracture Toughness Data for HY-130, A517-F, and HY-80 Steel," *Journal of Testing and Evaluation*, JTEVA, Vol 10, No. 3, May 1982, pp. 102–114.

(5) Succop, G. and Brown, W. F., Jr., "Estimation of K_{Ic} from Slow Bend Precracked Charpy Specimen Strength Ratios," *Developments in Fracture Mechanics Test Methods Standardization, ASTM STP 632*, W. F. Brown, Jr., and J. G. Kaufman, Eds., ASTM, 1977, pp. 179–192.

(6) Tauscher, S., "The Correlation of Fracture Toughness with Charpy V-notch Impact Test Data," Army Armament Research and Development Command, Technical Report ARLCB-TR-81012, 1981.

(7) Wullaert, R. A., Ireland, D. R., and A. S. Tetelman, "Use of the Precracked Charpy Specimen in Fracture Toughness Testing," *Fracture Prevention and Control*, pp. 255–282.

(8) Barsom, J. M. and Rolfe, S. T., "Correlations Between K_{Ic} and Charpy V-notch Test Results in the Transition-Temperature Range," *Impact Testing of Metals, ASTM STP 466*, ASTM, 1970, pp. 281–302.

(9) Mikalac, S., Vassilaros, M. G., and H. C. Rogers, "Precracking and Strain Rate Effects on HSLA-100 Steel Charpy Specimens," *Charpy Impact Test: Factors and Variables, ASTM STP 1072*, J. M. Holt, Ed., ASTM, 1990.

(10) Sharkey, R. L. and Stone, D. H., "A Comparison of Charpy V-notch, Dynamic Tear, and Precracked Charpy Impact Transition-Temperature Curves for AAR Grades of Cast Steel."

ASTM International takes no position respecting the validity of any patent rights asserted in connection with any item mentioned in this standard. Users of this standard are expressly advised that determination of the validity of any such patent rights, and the risk of infringement of such rights, are entirely their own responsibility.

This standard is subject to revision at any time by the responsible technical committee and must be reviewed every five years and if not revised, either reapproved or withdrawn. Your comments are invited either for revision of this standard or for additional standards and should be addressed to ASTM International Headquarters. Your comments will receive careful consideration at a meeting of the responsible technical committee, which you may attend. If you feel that your comments have not received a fair hearing you should make your views known to the ASTM Committee on Standards, at the address shown below.

This standard is copyrighted by ASTM International, 100 Barr Harbor Drive, PO Box C700, West Conshohocken, PA 19428-2959, United States. Individual reprints (single or multiple copies) of this standard may be obtained by contacting ASTM at the above address or at 610-832-9585 (phone), 610-832-9555 (fax), or service@astm.org (e-mail); or through the ASTM website (www.astm.org).

Designation: E92 − 82 (Reapproved 2003)$^{\varepsilon 2}$

Standard Test Method for
Vickers Hardness of Metallic Materials[1]

This standard is issued under the fixed designation E92; the number immediately following the designation indicates the year of original adoption or, in the case of revision, the year of last revision. A number in parentheses indicates the year of last reapproval. A superscript epsilon (ε) indicates an editorial change since the last revision or reapproval.

This standard has been approved for use by agencies of the Department of Defense.

ε^1 NOTE—Section 3.2 was editorially updated in June 2003.
ε^2 NOTE—Table 3 was editorially corrected in June 2004.

1. Scope

1.1 This test method covers the determination of the Vickers hardness of metallic materials, using applied forces of 1 kgf to 120 kgf,[2] the verification of Vickers hardness testing machines (Part B), and the calibration of standardized hardness test blocks (Part C). Two general classes of standard tests are recognized:

1.1.1 *Verification, Laboratory, or Referee Tests*, where a high degree of accuracy is required.

1.1.2 *Routine Tests*, where a somewhat lower degree of accuracy is permissible.

1.2 *This standard does not purport to address all of the safety concerns, if any, associated with its use. It is the responsibility of the user of this standard to establish appropriate safety and health practices and determine the applicability of regulatory limitations prior to use.*

2. Referenced Documents

2.1 *ASTM Standards:* [3]
E4 Practices for Force Verification of Testing Machines [3]
E140 Hardness Conversion Tables for Metals Relationship Among Brinell Hardness, Vickers Hardness, Rockwell Hardness, Superficial Hardness, Knoop Hardness, and Scleroscope Hardness [3]

E384 Test Method for Microindentation Hardness of Materials

3. Terminology

3.1 *calibration*—determination of the values of the significant parameters by comparison with values indicated by a reference instrument or by a set of reference standards.

3.2 *verification*—confirmation by examination and provision of evidence that an instrument, material, reference, or standard is in conformance with a specification.

3.3 *Vickers hardness number, HV*—a number related to the applied force and the surface area of the permanent impression made by a square-based pyramidal diamond indenter having included face angles of 136° (see Fig. 1 and Table 1), computed from the equation:

$$\text{HV} = 2P \sin(\alpha/2)/d^2 = 1.8544 P/d^2 \quad (1)$$

where:
P = force, kgf,
d = mean diagonal of impression, mm, and
α = face angle of diamond = 136°.

3.4 *Vickers hardness test*—an indentation hardness test using calibrated machines to force a square-based pyramidal diamond indenter having specified face angles, under a predetermined force, into the surface of the material under test and to measure the diagonals of the resulting impression after removal of the force.

3.4.1 Vickers hardness tests are made at test forces of 1 kgf to 120 kgf.

3.4.2 For practical purposes the Vickers hardness number is constant when a square-based diamond pyramid with a face angle of 136° is used with applied forces of 5 kgf and higher.

---

[1] This test method is under the jurisdiction of ASTM Committee E28 on Mechanical Testing and is the direct responsibility of Subcommittee E28.06 on Indentation Hardness Testing.
Current edition approved Jan. 10, 2003. Published April 2003. Originally approved in 1952. Last previous edition approved in 1997 as E92 – 82 (1997)$^{\varepsilon 3}$. DOI: 10.1520/E0092-82R03E02.

[2] A procedure covering Vickers tests using applied forces of 1 gf to 1000 gf (1 kgf) may be found in Test Method E384, Test Method for Microindentation Hardness of Materials, appearing in the *Annual Book of ASTM Standards*, Vol 03.01.

[3] *Annual Book of ASTM Standards*, Vol 03.01.

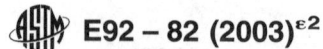

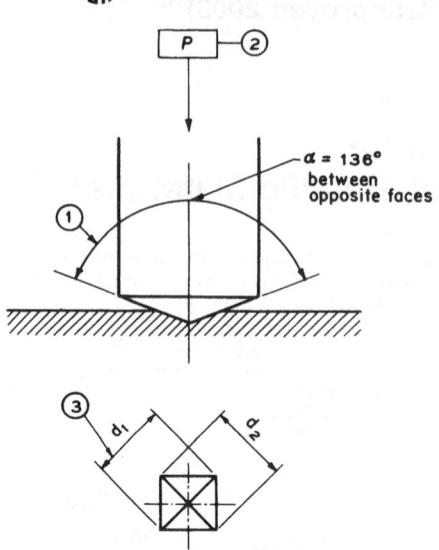

FIG. 1 Vickers Hardness Test (see Table 1)

TABLE 1 Symbols and Designations Associated with Fig. 1

Number	Symbol	Designation
1	...	Angle at the vertex of the pyramidal indenter (136°)
2	P	Test force in kilograms-force
3	d	Arithmetic mean of the two diagonals d^1 and d^2

TABLE 2 Vickers Hardness Numbers
(Diamond, 136° Face Angle, force of 1 kgf)

Diagonal of Impression, mm	Vickers Hardness Number for Diagonal Measured to 0.0001 mm									
	0.0000	0.0001	0.0002	0.0003	0.0004	0.0005	0.0006	0.0007	0.0008	0.0009
0.005	74 170	71 290	68 580	66 020	63 590	61 300	59 130	57 080	55 120	53 270
0.006	51 510	49 840	48 240	46 720	45 270	43 890	42 570	41 310	40 100	38 950
0.007	37 840	36 790	35 770	34 800	33 860	32 970	32 100	31 280	30 480	29 710
0.008	28 970	28 260	27 580	26 920	26 280	25 670	25 070	24 500	23 950	23 410
0.009	22 890	22 390	21 910	21 440	20 990	20 550	20 120	19 710	19 310	18 920
0.010	18 540	18 180	17 820	17 480	17 140	16 820	16 500	16 200	15 900	15 610
0.011	15 330	15 050	14 780	14 520	14 270	14 020	13 780	13 550	13 320	13 090
0.012	12 880	12 670	12 460	12 260	12 060	11 870	11 680	11 500	11 320	11 140
0.013	10 970	10 810	10 640	10 480	10 330	10 170	10 030	9 880	9 737	9 598
0.014	9 461	9 327	9 196	9 068	8 943	8 820	8 699	8 581	8 466	8 353
0.015	8 242	8 133	8 026	7 922	7 819	7 718	7 620	7 523	7 428	7 335
0.016	7 244	7 154	7 066	6 979	6 895	6 811	6 729	6 649	6 570	6 493
0.017	6 416	6 342	6 268	6 196	6 125	6 055	5 986	5 919	5 853	5 787
0.018	5 723	5 660	5 598	5 537	5 477	5 418	5 360	5 303	5 247	5 191
0.019	5 137	5 083	5 030	4 978	4 927	4 877	4 827	4 778	4 730	4 683
0.020	4 636	4 590	4 545	4 500	4 456	4 413	4 370	4 328	4 286	4 245
0.021	4 205	4 165	4 126	4 087	4 049	4 012	3 975	3 938	3 902	3 866
0.022	3 831	3 797	3 763	3 729	3 696	3 663	3 631	3 599	3 567	3 536
0.023	3 505	3 475	3 445	3 416	3 387	3 358	3 329	3 301	3 274	3 246
0.024	3 219	3 193	3 166	3 140	3 115	3 089	3 064	3 039	3 015	2 991
0.025	2 967	2 943	2 920	2 897	2 874	2 852	2 830	2 808	2 786	2 764
0.026	2 743	2 722	2 701	2 681	2 661	2 641	2 621	2 601	2 582	2 563
0.027	2 544	2 525	2 506	2 488	2 470	2 452	2 434	2 417	2 399	2 382
0.028	2 365	2 348	2 332	2 315	2 299	2 283	2 267	2 251	2 236	2 220
0.029	2 205	2 190	2 175	2 160	2 145	2 131	2 116	2 102	2 088	2 074
0.030	2 060	2 047	2 033	2 020	2 007	1 993	1 980	1 968	1 955	1 942
0.031	1 930	1 917	1 905	1 893	1 881	1 869	1 857	1 845	1 834	1 822
0.032	1 811	1 800	1 788	1 777	1 766	1 756	1 745	1 734	1 724	1 713
0.033	1 703	1 693	1 682	1 672	1 662	1 652	1 643	1 633	1 623	1 614

ASTM E92 – 82 (2003)ε2

TABLE 2 *Continued*

Diagonal of Impression, mm	Vickers Hardness Number for Diagonal Measured to 0.0001 mm									
	0.0000	0.0001	0.0002	0.0003	0.0004	0.0005	0.0006	0.0007	0.0008	0.0009
0.034	1 604	1 595	1 585	1 576	1 567	1 558	1 549	1 540	1 531	1 522
0.035	1 514	1 505	1 497	1 488	1 480	1 471	1 463	1 455	1 447	1 439
0.036	1 431	1 423	1 415	1 407	1 400	1 392	1 384	1 377	1 369	1 362
0.037	1 355	1 347	1 340	1 333	1 326	1 319	1 312	1 305	1 298	1 291
0.038	1 284	1 277	1 271	1 264	1 258	1 251	1 245	1 238	1 232	1 225
0.039	1 219	1 213	1 207	1 201	1 195	1 189	1 183	1 177	1 171	1 165
0.040	1 159	1 153	1 147	1 142	1 136	1 131	1 125	1 119	1 114	1 109
0.041	1 103	1 098	1 092	1 087	1 082	1 077	1 072	1 066	1 061	1 056
0.042	1 051	1 046	1 041	1 036	1 031	1 027	1 022	1 017	1 012	1 008
0.043	1 003	998	994	989	985	980	975	971	967	962
0.044	958	953	949	945	941	936	932	928	924	920
0.045	916	912	908	904	900	896	892	888	884	880
0.046	876	873	869	865	861	858	854	850	847	843
0.047	839	836	832	829	825	822	818	815	812	808
0.048	805	802	798	795	792	788	785	782	779	775
0.049	772	769	766	763	760	757	754	751	748	745
0.050	742	739	736	733	730	727	724	721	719	716
0.051	713	710	707	705	702	699	696	694	691	688
0.052	686	683	681	678	675	673	670	668	665	663
0.053	660	658	655	653	650	648	645	643	641	638
0.054	636	634	631	629	627	624	622	620	617	615
0.055	613	611	609	606	604	602	600	598	596	593
0.056	591	589	587	585	583	581	579	577	575	573
0.057	571	569	567	565	563	561	559	557	555	553
0.058	551	549	547	546	544	542	540	538	536	535
0.059	533	531	529	527	526	524	522	520	519	516.8
0.060	515.1	513.4	511.7	510.0	508.3	506.6	505.0	503.3	501.6	500.0
0.061	498.4	496.7	495.1	493.5	491.9	490.3	488.7	487.1	485.5	484.0
0.062	482.4	480.9	479.3	477.8	476.2	474.7	473.2	471.7	470.2	468.7
0.063	467.2	465.7	464.3	462.8	461.3	459.9	458.4	457.0	455.6	454.1
0.064	452.7	451.3	449.9	448.5	447.1	445.7	444.4	443.0	441.6	440.3
0.065	438.9	437.6	436.2	434.9	433.6	432.2	430.9	429.6	428.3	427.0
0.066	425.7	424.4	423.1	421.9	420.6	419.3	418.1	416.8	415.6	414.3
0.067	413.1	411.9	410.6	409.4	408.2	407.0	405.8	404.6	403.4	402.2
0.068	401.0	399.9	398.7	397.5	396.6	395.2	394.0	392.9	391.8	390.6
0.069	389.5	388.4	387.2	386.1	385.0	383.9	382.8	381.7	380.6	379.5
0.070	378.4	377.4	376.3	375.2	374.2	373.1	372.0	371.0	369.9	368.9
0.071	367.9	366.8	365.8	364.8	363.7	362.7	361.7	360.7	359.7	358.7
0.072	357.7	356.7	355.7	354.7	353.8	352.8	351.8	350.9	349.9	348.9
0.073	348.0	347.0	346.1	345.1	344.2	343.3	342.3	341.4	340.5	339.6
0.074	338.6	337.7	336.8	335.9	335.0	334.1	333.2	332.3	331.4	330.5
0.075	329.7	328.8	327.9	327.0	326.2	325.3	324.5	323.6	322.7	321.9
0.076	321.0	320.2	319.4	318.5	317.7	316.9	316.0	315.2	314.4	313.6
0.077	312.8	312.0	311.1	310.3	309.5	308.7	307.9	307.2	306.4	305.6
0.078	304.8	304.0	303.2	302.5	301.7	300.9	300.2	299.4	298.6	297.9
0.079	297.1	296.4	295.6	294.9	294.1	293.4	292.7	291.9	291.2	290.5
0.080	289.7	289.0	288.3	287.6	286.9	286.2	285.4	284.7	284.0	283.3
0.081	282.6	281.9	281.2	280.6	279.9	279.2	278.5	277.8	277.1	276.5
0.082	275.8	275.1	274.4	273.8	273.1	272.4	271.8	271.1	270.5	269.8
0.083	269.2	268.5	267.9	267.2	266.6	266.0	265.3	264.7	264.1	263.4
0.084	262.8	262.2	261.6	260.9	260.3	259.7	259.1	258.5	257.9	257.3
0.085	256.7	256.1	255.5	254.9	254.3	253.7	253.1	252.5	251.9	251.3
0.086	250.7	250.1	249.6	249.0	248.4	247.8	247.3	246.7	246.1	245.6
0.087	245.0	244.4	243.9	243.3	242.8	242.2	241.6	241.1	240.6	240.0
0.088	239.5	238.9	238.4	237.8	237.3	236.8	236.2	235.7	235.2	234.6
0.089	234.1	233.6	233.1	232.5	232.0	231.5	231.0	230.5	230.0	229.4
0.090	228.9	228.4	227.9	227.4	226.9	226.4	225.9	225.4	224.9	224.4
0.091	223.9	223.4	222.9	222.5	222.0	221.5	221.0	220.5	220.0	219.6
0.092	219.1	218.6	218.1	217.7	217.1	216.7	216.3	215.8	215.3	214.9
0.093	214.4	213.9	213.5	213.0	212.6	212.1	211.7	211.2	210.8	210.3

TABLE 2 Continued

Diagonal of Impression, mm	Vickers Hardness Number for Diagonal Measured to 0.0001 mm									
	0.0000	0.0001	0.0002	0.0003	0.0004	0.0005	0.0006	0.0007	0.0008	0.0009
0.094	209.9	209.4	209.0	208.5	208.1	207.6	207.2	206.8	206.3	205.9
0.095	205.5	205.0	204.6	204.2	203.8	203.3	202.9	202.5	202.1	201.6
0.096	201.2	200.8	200.4	200.0	199.5	199.1	198.7	198.3	197.9	197.5
0.097	197.1	196.7	196.3	195.9	195.5	195.1	194.7	194.3	193.9	193.5
0.098	193.1	192.7	192.3	191.9	191.5	191.1	190.7	190.4	190.0	189.6
0.099	189.2	188.8	188.4	188.1	187.7	187.3	186.9	186.6	186.2	185.5

At lower test forces the Vickers hardness may be force-dependent. In Table 2 are given the Vickers hardness numbers for a test force of 1 kgf. For obtaining hardness numbers when other test forces are used, the Vickers hardness number obtained from Table 2 is multiplied by the test force in kilograms-force (Table 3).

NOTE 1—The Vickers hardness number is followed by the symbol HV with a suffix number denoting the force and second suffix number indicating the duration of forceing when the latter differs from 10 to 15 s, which is the normal force time. *Example:*

440 HV 30 = Vickers hardness of 440 measured under a force of 30 kgf applied for 10 to 15 s.

440 HV 30/20 = Vickers hardness of 440 measured under a force of 30 kgf applied for 20 s.

A. GENERAL DESCRIPTION AND TEST PROCEDURE FOR VICKERS HARDNESS TESTS

4. Apparatus

4.1 *Testing Machine*—Equipment for Vickers hardness testing usually consists of a testing machine which supports the specimen and permits the indenter and the specimen to be brought into contact gradually and smoothly under a predetermined force, which is applied for a fixed period of time. The design of the machine should be such that no rocking or lateral movement of the indenter or specimen is permitted while the force is being applied or removed. A measuring microscope is usually mounted on the machine in such a manner that the impression in the specimen may be readily located in the optical field.

4.2 *Indenter*:

4.2.1 The indenter shall be a highly polished, pointed, square-based pyramidal diamond with face angles of 136° ± 30 min.

4.2.2 All four faces of the indenter shall be equally inclined to the axis of the indenter (within ±30 min) and meet at a sharp point, that is, the line of junction between opposite faces shall not be more than 0.001 mm in length as shown in Fig. 2.

4.2.3 The diamond should be examined periodically and if it is loose in the mounting material, chipped, or cracked, it should be discarded or reconditioned.

TABLE 3 Decimal Point Finder for Use with Table 2
An example of determination of hardness numbers follows the table.

Diagonal Length, mm	Vickers Hardness (HV), 1-kgf Force
0.005	74 200
0.006	51 500
0.007	37 800
0.008	29 000
0.009	22 900
0.010	18 540
0.020	4 640
0.030	2 060
0.040	1 159
0.050	742
0.060	515
0.070	378
0.080	290
0.090	229
0.100	185.4
0.200	46.4
0.300	20.6
0.400	11.6
0.500	7.42
0.600	5.15
0.700	3.78
0.800	2.90
0.900	2.29
1.000	1.85
1.100	1.53
1.200	1.29
1.300	1.10
1.400	0.946
1.500	0.824
1.600	0.724
1.700	0.642
1.800	0.572
1.900	0.514
2.000	0.464

Example—Using a 50-kgf test force, the average measured diagonal length = 0.644 mm.
In Table 2 read:

  HV = 447 at 0.0644-mm diagonal length at 1-kgf force.

Using Table 3 determine:

  HV = 4.47 at 0.644-mm diagonal length at 1-kgf force.

  50 × 4.47 = 224 HV for 50-kg test force.

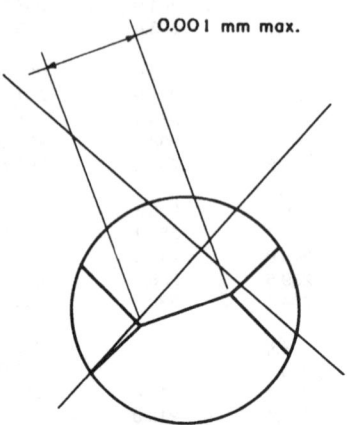

FIG. 2 Junction of Indenter Faces

NOTE 2—The condition of the point of the indenter is of considerable importance where the test force is light and the impression is small. It is recommended that the point be periodically checked by examining an impression made in a polished steel block. Under a magnification of 600× or more, using a vertical illuminator, any chipping or rounding of the point can be detected and the extent of the defect measured with a filar micrometer. It is recommended that a diamond pyramid indenter should not be used for tests in which the maximum length of such a defect exceeds 5 % of the length of the impression diagonal.

4.3 *Measuring Microscope*—The divisions of the micrometer scale of the measuring microscope or other measuring device shall be so constructed that the length of the diagonals of an impression in a properly surface-finished specimen (see section 5.1.2) can be measured to within ±0.0005 mm or ±0.5 %, whichever is larger.

5. Test Specimen

5.1 The Vickers hardness test is adaptable to a wide variety of test specimens ranging from large bars and rolled sections to minute pieces in metallographic mounts. In general the backs of the specimens shall be so finished or the specimens shall be so clamped that there is no possibility of their rocking or shifting under the test force. The specimens shall also conform to the requirements given in the following 5.1.1, 5.1.2, and 5.1.3.

5.1.1 *Thickness*—The thickness of the test specimen shall be such that no bulge or marking showing the effect of the force appears on the side of the specimen opposite the impression. In any event the thickness of the specimen shall be at least one and one half times the length of the diagonal. When laminated material is tested, the thickness of the individual component being tested shall be used for the thickness-diagonal length relationship.

5.1.2 *Finish*—The surface of the specimen should be so prepared that the ends of the diagonals are clearly defined and can be read with precision of ±0.0005 mm or ±0.5 % of the length of the diagonals, whichever is larger. Care should be taken in specimen preparation to avoid tempering during grinding, or work-hardening the surface during polishing.

5.1.3 *Alignment*—The specimen should be so prepared or mounted that the surface is normal to the axis of the indenter within ±1° of angle. This can readily be accomplished by surface grinding (or otherwise machining) the opposite side of the specimen to parallelism with the side to be tested.

5.1.4 *Radius of Curvature*—Until further investigative work is accomplished to determine the effect of the radius of curvature on readings, due caution should be used in interpreting or accepting the results of tests made on cylindrical surfaces.

NOTE 3—A method recommended by the International Organization for Standardization for correcting Vickers hardness readings taken on spherical or cylindrical surfaces is given in Table 4, Table 5, and Table 6.

NOTE 4—These tables give correction factors to be applied to Vickers hardness values obtained when tests are made on spherical or cylindrical surfaces. The correction factors are tabulated in terms of the ratio of the mean diagonal d of the indentation to the diameter D of the sphere or cylinder. Examples of the use of these tables are:

TABLE 4 Correction Factors for Use in Vickers Hardness Tests Made on Spherical Surfaces

Convex Surface		Concave Surface	
d/D[A]	Correction Factor	d/D[A]	Correction Factor
0.004	0.995	0.004	1.005
0.009	0.990	0.008	1.010
0.013	0.985	0.012	1.015
0.018	0.980	0.016	1.020
0.023	0.975	0.020	1.025
0.028	0.970	0.024	1.030
0.033	0.965	0.028	1.035
0.038	0.960	0.031	1.040
0.043	0.955	0.035	1.045
0.049	0.950	0.038	1.050
0.055	0.945	0.041	1.055
0.061	0.940	0.045	1.060
0.067	0.935	0.048	1.065
0.073	0.930	0.051	1.070
0.079	0.925	0.054	1.075
0.086	0.920	0.057	1.080
0.093	0.915	0.060	1.085
0.100	0.910	0.063	1.090
0.107	0.905	0.066	1.095
0.114	0.900	0.069	1.100
0.122	0.895	0.071	1.105
0.130	0.890	0.074	1.110
0.139	0.885	0.077	1.115
0.147	0.880	0.079	1.200
0.156	0.875	0.082	1.125
0.165	0.870	0.084	1.130
0.175	0.865	0.087	1.135
0.185	0.860	0.089	1.140
0.195	0.855	0.091	1.145
0.206	0.850	0.094	1.150

[A] D = diameter of sphere.
d = mean diagonal of impression in millimeters.

Example 1. Convex Sphere:
Diameter of sphere, D	= 10 mm
Load	= 10 kgf
Mean diagonal of impression, d	= 0.150 mm
d/D = 0.150/10 = 0.015	
From Tables 2 and 3, HV	= 824
From Table 4, by interpolation, correction factor	= 0.983
Hardness of sphere = 824 × 0.983	= 810 HV 10

Example 2. Concave Cylinder, One Diagonal Parallel to Axis:
Diameter of cylinder, D	= 5 mm
Load	= 30 kgf
Mean diagonal of impression, d	= 0.415 mm
d/D = 0.415/5 = 0.083	
From Tables 2 and 3, HV	= 323
From Table 6, correction factor	= 1.075
Hardness of cylinder = 323 × 1.075	= 347 HV 30.

6. Verification of Apparatus

6.1 The hardness testing machine shall be verified as specified in Part B.

6.1.1 Two acceptable methods of verifying Vickers hardness testing machines are given in Part B.

TABLE 5 Correction Factors for Use in Vickers Hardness Tests Made on Cylindrical Surfaces
(Diagonals at 45° to the axis)

Convex Surface		Concave Surface	
d/D[A]	Correction Factor	d/D[A]	Correction Factor
0.009	0.995	0.009	1.005
0.017	0.990	0.017	1.020
0.026	0.985	0.025	1.015
0.035	0.980	0.034	1.020
0.044	0.975	0.042	1.025
0.053	0.970	0.050	1.030
0.062	0.965	0.058	1.035
0.071	0.960	0.066	1.040
0.081	0.955	0.074	1.045
0.090	0.950	0.082	1.050
0.100	0.945	0.089	1.055
0.109	0.940	0.097	1.060
0.119	0.935	0.104	1.065
0.129	0.930	0.112	1.070
0.139	0.925	0.119	1.075
0.149	0.920	0.127	1.080
0.159	0.915	0.134	1.085
0.169	0.910	0.141	1.090
0.179	0.905	0.148	1.095
0.189	0.900	0.155	1.100
0.200	0.895	0.162	1.105
		0.169	1.110
		0.176	1.115
		0.183	1.120
		0.189	1.125
		0.196	1.130
		0.203	1.135
		0.209	1.140
		0.216	1.140
		0.222	1.150

[A] D = diameter of cylinder.
d = mean diagonal of impression in millimeters.

TABLE 6 Correction Factors for Use in Vickers Hardness Tests Made on Cylindrical Surfaces
(One diagonal parallel to axis)

Convex Surface		Concave Surface	
d/D[A]	Correction Factor	d/D[A]	Correction Factor
0.009	0.995	0.048	1.035
0.019	0.990	0.053	1.040
0.029	0.985	0.058	1.045
0.041	0.980	0.063	1.050
0.054	0.975	0.067	1.055
0.068	0.970	0.071	1.060
0.085	0.965	0.076	1.065
0.104	0.960	0.079	1.070
0.126	0.955	0.083	1.075
0.153	0.950	0.087	1.080
0.189	0.945	0.090	1.085
0.243	0.940	0.093	1.090

Concave Surface

d/D[A]	Correction Factor
0.097	1.095
0.100	1.100
0.103	1.105
0.105	1.110
0.108	1.115
0.111	1.120
0.008	1.005
0.016	1.020
0.023	1.015
0.030	1.020
0.036	1.025
0.042	1.030
0.113	1.125
0.116	1.130
0.118	1.135
0.120	1.140
0.123	1.145
0.125	1.150

[A] D = diameter of cylinder.
d = mean diagonal of impression in millimeters.

7. Procedure

7.1 *Magnitude of Test Force*—Test forces of 1 kgf to 120 kgf may be used, depending on the requirements of the test. Although tests on homogeneous materials indicate that the Vickers hardness number is nearly independent of the test force, this condition will not be present in cases where there is a hardness gradient from the specimen surface to the interior of the specimen. The magnitude of the test force should therefore be stated in the test report (Section 11).

7.2 *Application of Test Force*—Apply the test force and release smoothly without shock or vibration. The time of application of the full test force shall be 10 to 15 s, unless otherwise specified.

7.3 *Spacing of Indentations*—The center of the impression shall not be closer to any edge of the test specimen or to another impression than a distance equal to two and one half times the length of diagonal of the impression. When laminated material is tested, a bond surface shall be considered as an edge for spacing of indentation calculations.

8. Measurement of Impression

8.1 Both diagonals of the impression shall be measured and their mean value used as a basis for calculation of the Vickers hardness number. It is recommended that the measurement be made with the impression centered as nearly as possible in the field of the microscope.

8.2 In the case of anisotropic materials, for example materials that have been heavily cold worked, there may be a difference between the lengths of the two diagonals of the impression. In such cases, the test specimen should be reoriented so that the diagonals of a new impression are approximately of equal length.

9. Accuracy

9.1 The accuracy of the Vickers hardness method is a function of the accuracies of the test force, indenter, and measuring device. The condition of the test and support surfaces and support of the test piece during application of the test force also affect accuracy. Under optimum conditions of these factors the accuracy that can be expected is the equivalent of 4 % of the Vickers hardness number of the standardized reference hardness test blocks (see section 18.2). Under less than ideal conditions the reduction in accuracy, when required, can be established empirically by employing statistical methods.

10. Conversion to Other Hardness Scales or Tensile Strength Values

10.1 There is no general method for converting accurately Vickers hardness numbers to other hardness scales or tensile strength values. Such conversions are, at best, approximations and therefore should be avoided, except for special cases where a reliable basis for the approximate conversions has been obtained by comparison tests.

NOTE 5—Standard E140 gives approximate conversion values for specific materials such as steel, nickel and high-nickel alloys, and cartridge brass.

11. Report

11.1 The report shall include the following information:
11.1.1 The Vickers hardness number,
11.1.2 The test force used (see 3.4.2, Note 1), and
11.1.3 The force application time, if other than 10 to 15 s (see 3.4.2, Note 1).

12. Precision and Bias

12.1 Due to the wide variety of materials tested by this method and the possible variations in test specimens, the precision of this method has not been established. The accepted practice is to utilize the information in 9.1 when establishing hardness tolerances for specific applications. The precision of this method, whether involving a single operator, multiple operators, or multiple laboratories, can be established by employing statistical methods.

B. VERIFICATION OF VICKERS HARDNESS TESTING MACHINES

13. Scope

13.1 Part B covers two procedures for the verification of Vickers hardness testing machines and a procedure that is recommended for use to confirm that the machine has not become maladjusted in the intervals between the periodical routine checks. The two methods of verification are:
13.1.1 Separate verification of force application, indenter, and measuring microscope.
13.1.2 Vertification by standardized test block method.
13.2 The first procedure (13.1.1) is mandatory for new and rebuilt machines.
13.3 The second procedure (13.1.2) shall be used for verifying machines in service.

14. General Requirements

14.1 Before a Vickers hardness testing machine is verified the machine shall be examined to ensure that:
14.1.1 The machine is properly set up.
14.1.2 The indenter holder is mounted normally in the plunger.
14.1.3 The force can be applied and removed without shock or vibration in such a manner that the readings are not influenced.
14.2 If the measuring device is integral with the machine, the machine shall be examined to ensure that:
14.2.1 The change from forceing to measuring does not influence the readings.
14.2.2 The method of illumination does not affect the readings.
14.2.3 The center of the impression is in the center of the field of view.

15. Verification

15.1 *Separate Verification of Force Application, Indenter, and Measuring Microscope*:

15.1.1 *force Application*—The applied force shall be checked by the use of dead weights and proving levers, or by an elastic calibration device or springs in the manner described in Practices E4. Such dead weights or other forceing devices shall be accurate to ±0.2 %. Vickers hardness testing machines shall be verified at a minimum of three applied forces including the test force specified. A minimum of three readings should be taken at each force. A Vickers hardness testing machine is acceptable for use over a forceing range within which the machine error does not exceed ±1 %.

15.1.2 *Indenter*—The form of the diamond indenter shall be verified by direct measurement of its shape or by measurements of its projection on a screen. The angle between opposite faces of the pyramid shall be 136° ± 30 min. All four faces shall be equally inclined to the axis of the pyramid within ±30 min. The four faces of indenters used for laboratory, or routine tests, shall meet at a point no more than 0.001 mm in length (see Fig. 2). The four faces of indenters used in calibrating standardized hardness test blocks, shall meet at a point in which the line of junction between opposite faces is no more than 0.0005 mm in length (see Fig. 3). The quadrilateral that would be formed by the intersection of the four faces with a plane perpendicular to the axis of the indenter shall have angles of 90° ± 12 min.

15.1.3 *Measuring Microscope*—The measuring microscope or other device for measuring the diagonals of the impression shall be calibrated against an accurately ruled line scale (stage micrometer). The errors of the line scale shall not exceed 0.05 μm (0.00005 mm) or 0.05 % of any interval, whichever is greater. The measuring microscope shall be calibrated throughout its range of use and a calibration factor chosen such that the error shall not exceed ±0.5 %. It may be necessary to divide the complete range of the micrometer microscope into several subranges, each having its own factor.

15.2 *Verification by Standardized Test Block Method*:

15.2.1 A Vickers hardness testing machine used only for routine testing may be checked by making a series of impressions on standardized hardness test blocks (Part C).

15.2.2 A minimum of five Vickers hardness readings shall be taken on at least three blocks having different levels of

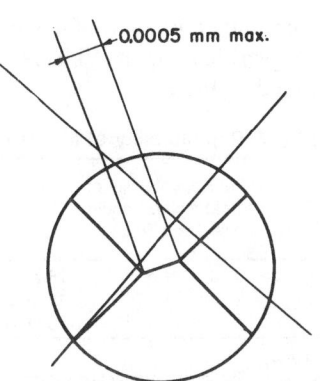

FIG. 3 Junction of Indenter Faces

hardness using a test force or forces as specified by the user with the test force applied for 12 s.

15.2.3 Vickers hardness testing machines shall be considered verified if the mean diagonal for five hardness impressions meets the requirements of 17.2.

16. Procedure for Periodic Checks by the User

16.1 Verification by the standardized test block method (15.2.2) is too lengthy for daily use. Instead the following is recommended:

16.1.1 Make at least one routine check each day that the testing machine is used.

16.1.2 Before making the check, verify that the zero reading of the measuring apparatus is correctly adjusted.

16.1.3 Make at least five hardness readings on a standardized hardness test block on the scale and at the hardness level at which the machine is being used. If the values fall within the range of the standardized hardness test block the machine may be regarded as satisfactory; if not the machine should be verified as described in 15.2.2.

17. Repeatability and Error

17.1 *Repeatability*:

17.1.1 For each standardized block, let $d_1, d_2, \cdots d_5$ be the arithmetic means of the two diagonals of the indentations, arranged in increasing order of magnitude.

17.1.2 The repeatability of the machine under the particular verification conditions is expressed by the quantity $d_5 - d_1$.

17.2 *Error*:

17.2.1 The error of the machine under the particular verification conditions is expressed by the quantity $\bar{d} - d$, where $\bar{d} = (d_1 + d_2 + \cdots d_5)/5$, and d is the reported mean diagonal of impressions on the standardized hardness test block.

18. Assessment of Verification

18.1 *Repeatability*—The repeatability of the machine verified is considered satisfactory if it satisfies the conditions given in Table 7.

18.2 *Error*—The mean diagonal for five impressions should not differ from the mean diagonal corresponding to the Vickers hardness of the standardized test block by more than 2 % or 0.5 µm (0.0005 mm), whichever is greater.

C. CALIBRATION OF STANDARD HARDNESS TEST BLOCKS FOR VICKERS HARDNESS MACHINES

19. Scope

19.1 Part C covers the calibration of standardized hardness test blocks for the verification of Vickers hardness testing machines as described in Part B.

TABLE 7 Repeatability of Machines

Range of Standardized Hardness of Test Blocks	The Repeatability of the Machine Should be Less Than:	Examples of Equivalents in Hardness Units
100 to 240, incl	4 % of d [A,B]	8 at 100 HV; 16 at 200 HV
Over 240 to 600, incl	3 % of d [A,B]	18 at 300 HV; 36 at 600 HV
Over 600	2 % of d [A,B]	28 at 700 HV

[A] $d = (d_1 + d_2 + ... + d_5)/5$.
[B] In all cases the repeatability is the percentage given or 1 µm (0.001 mm), whichever is the greater.

20. Manufacture

20.1 Each metal block to be standardized shall be not less than ¼ in. (6 mm) in thickness.

20.2 Each block shall be specially prepared and heat treated to give the necessary homogeneity and stability of structure.

20.3 Each block, if of steel, shall be demagnetized by the manufacturer and maintained demagnetized by the user.

20.4 The lower surface of the test block shall have a fine ground finish.

20.5 The test (upper) surface shall be polished and free of scratches which would interfere with measurements of the diagonals of the impression.

20.5.1 The mean surface roughness height rating shall not exceed 4 µin. (0.0001 mm) center line average.

20.6 To ensure that no material is subsequently removed from the test surface of the standardized test block, an official mark or the thickness at the time of calibration shall be marked on the test surface to an accuracy of ±0.005 in.(±0.1 mm).

21. Standardizing Procedure

21.1 The standardized hardness test blocks shall be calibrated on a Vickers hardness testing machine verified in accordance with the requirements of 13.1.1.

21.2 The mechanism that controls the application of force should either:

21.2.1 Employ a device such as a spring to reduce the velocity of indentation of the indenter during the period of indentation, or

21.2.2 Employ a device to maintain a constant velocity of indentation of the indenter.

21.3 The full test force shall be applied for 12 s.

22. Number of Indentations

22.1 At least five and preferably ten randomly distributed indentations shall be made on each test block.

23. Measurement of the Diagonals of the Indentation

23.1 The illuminating system of the measuring microscope shall be adjusted to give uniform intensity over the field of view and maximum contrast between the indentation and the undistributed surface of the block.

23.2 The measuring microscope shall be graduated to read 0.001 mm with estimates made to the nearest ±0.0002 mm.

23.3 The measuring microscope shall be checked by a stage micrometer, or by other suitable means, to ensure that the difference between readings corresponding to any two divisions of the instrument is correct within ± 0.0005 mm.

23.4 It is recommended that each indentation be measured by two observers.

24. Repeatability

24.1 Let $d_1, d_2, \cdots d_n$ be the mean values of the measured diagonals as determined by one observer, arranged in increasing order of magnitude.

24.2 The repeatability of the hardness readings on the block is defined as $(d_{10} - d_1)$, when ten readings have been made or $1.32 (d_5 - d_1)$ when five readings are taken on the block.

25. Uniformity of Hardness

25.1 Unless the repeatability of hardness readings as measured by the mean diagonals of five or ten impressions is within the limits given in Table 8, the block cannot be regarded as sufficiently uniform for standardization purposes.

TABLE 8 Repeatability of Hardness Readings

Range of Standardized Hardness of Test Block	The Repeatability of the Test Block Readings Shall be Less Than:
100 to 240, incl	3 % of d [A,B]
Over 240 to 600, incl	2 % of d [A,B]
Over 600	1.5 % of d [A,B]

[A] $d = (d_1 + d_2 + ... + d_n)/n$.
[B] In all cases the repeatability is the percentage given or 1 μm (0.001 mm), whichever is the greater.

26. Marking

26.1 Each block shall be marked with the following:

26.1.1 Arithmetic mean of the hardness values found in the standardization test (see also 3.4.2, Note 1).

26.1.2 The name or mark of the supplier,

26.1.3 The serial number of the block, and

26.1.4 The thickness of the test block or an official mark on the top surface (see section 19.6).

NOTE 6—All of the markings except the official mark or thickness should be placed on the side of the block, the markings being upright when the test surface is the upper face.

27. Keywords

27.1 metallic; Vickers hardness

ASTM International takes no position respecting the validity of any patent rights asserted in connection with any item mentioned in this standard. Users of this standard are expressly advised that determination of the validity of any such patent rights, and the risk of infringement of such rights, are entirely their own responsibility.

This standard is subject to revision at any time by the responsible technical committee and must be reviewed every five years and if not revised, either reapproved or withdrawn. Your comments are invited either for revision of this standard or for additional standards and should be addressed to ASTM International Headquarters. Your comments will receive careful consideration at a meeting of the responsible technical committee, which you may attend. If you feel that your comments have not received a fair hearing you should make your views known to the ASTM Committee on Standards, at the address shown below.

This standard is copyrighted by ASTM International, 100 Barr Harbor Drive, PO Box C700, West Conshohocken, PA 19428-2959, United States. Individual reprints (single or multiple copies) of this standard may be obtained by contacting ASTM at the above address or at 610-832-9585 (phone), 610-832-9555 (fax), or service@astm.org (e-mail); or through the ASTM website (www.astm.org).

Designation: E94 – 04

Standard Guide for Radiographic Examination[1]

This standard is issued under the fixed designation E94; the number immediately following the designation indicates the year of original adoption or, in the case of revision, the year of last revision. A number in parentheses indicates the year of last reapproval. A superscript epsilon (ε) indicates an editorial change since the last revision or reapproval.

1. Scope

1.1 This guide[2] covers satisfactory X-ray and gamma-ray radiographic examination as applied to industrial radiographic film recording. It includes statements about preferred practice without discussing the technical background which justifies the preference. A bibliography of several textbooks and standard documents of other societies is included for additional information on the subject.

1.2 This guide covers types of materials to be examined; radiographic examination techniques and production methods; radiographic film selection, processing, viewing, and storage; maintenance of inspection records; and a list of available reference radiograph documents.

Note 1—Further information is contained in Guide E999, Practice E1025, Test Methods E1030, and E1032.

1.3 *Interpretation and Acceptance Standards*—Interpretation and acceptance standards are not covered by this guide, beyond listing the available reference radiograph documents for castings and welds. Designation of accept - reject standards is recognized to be within the cognizance of product specifications and generally a matter of contractual agreement between producer and purchaser.

1.4 *Safety Practices*—Problems of personnel protection against X rays and gamma rays are not covered by this document. For information on this important aspect of radiography, reference should be made to the current document of the National Committee on Radiation Protection and Measurement, Federal Register, U.S. Energy Research and Development Administration, National Bureau of Standards, and to state and local regulations, if such exist. For specific radiation safety information refer to NIST Handbook ANSI 43.3, 21 CFR 1020.40, and 29 CFR 1910.1096 or state regulations for agreement states.

1.5 *This standard does not purport to address all of the safety problems, if any, associated with its use. It is the responsibility of the user of this standard to establish appropriate safety and health practices and determine the applicability of regulatory limitations prior to use.* (See 1.4.)

1.6 If an NDT agency is used, the agency shall be qualified in accordance with Practice E543.

2. Referenced Documents

2.1 *ASTM Standards:*[3]

E543 Specification for Agencies Performing Nondestructive Testing

E746 Practice for Determining Relative Image Quality Response of Industrial Radiographic Imaging Systems

E747 Practice for Design, Manufacture and Material Grouping Classification of Wire Image Quality Indicators (IQI) Used for Radiology

E801 Practice for Controlling Quality of Radiological Examination of Electronic Devices

E999 Guide for Controlling the Quality of Industrial Radiographic Film Processing

E1025 Practice for Design, Manufacture, and Material Grouping Classification of Hole-Type Image Quality Indicators (IQI) Used for Radiology

E1030 Test Method for Radiographic Examination of Metallic Castings

E1032 Test Method for Radiographic Examination of Weldments

E1079 Practice for Calibration of Transmission Densitometers

E1254 Guide for Storage of Radiographs and Unexposed Industrial Radiographic Films

E1316 Terminology for Nondestructive Examinations

E1390 Specification for Illuminators Used for Viewing Industrial Radiographs

E1735 Test Method for Determining Relative Image Quality of Industrial Radiographic Film Exposed to X-Radiation from 4 to 25 MeV

E1742 Practice for Radiographic Examination

E1815 Test Method for Classification of Film Systems for Industrial Radiography

2.2 *ANSI Standards:*

PH1.41 Specifications for Photographic Film for Archival

---

[1] This guide is under the jurisdiction of ASTM Committee E07 on Nondestructive Testing and is the direct responsibility of Subcommittee E07.01 on Radiology (X and Gamma) Method.
Current edition approved Jan. 1, 2004. Published February 2004. Originally approved in 1952. Last previous edition approved in 2000 as E94 - 00. DOI: 10.1520/E0094-04.

[2] For ASME Boiler and Pressure Vessel Code applications see related Guide SE-94 in Section V of that Code.

[3] For referenced ASTM standards, visit the ASTM website, www.astm.org, or contact ASTM Customer Service at service@astm.org. For *Annual Book of ASTM Standards* volume information, refer to the standard's Document Summary page on the ASTM website.

Copyright © ASTM International, 100 Barr Harbor Drive, PO Box C700, West Conshohocken, PA 19428-2959, United States.

Records, Silver-Gelatin Type, on Polyester Base[4]

PH2.22 Methods for Determining Safety Times of Photographic Darkroom Illumination[4]

PH4.8 Methylene Blue Method for Measuring Thiosulfate and Silver Densitometric Method for Measuring Residual Chemicals in Films, Plates, and Papers[4]

T9.1 Imaging Media (Film)—Silver-Gelatin Type Specifications for Stability[4]

T9.2 Imaging Media—Photographic Process Film Plate and Paper Filing Enclosures and Storage Containers[4]

2.3 *Federal Standards:*

Title 21, Code of Federal Regulations (CFR) 1020.40, Safety Requirements of Cabinet X-Ray Systems[5]

Title 29, Code of Federal Regulations (CFR) 1910.96, Ionizing Radiation (X-Rays, RF, etc.)[5]

2.4 *Other Document:*

NBS Handbook ANSI N43.3 General Radiation Safety Installations Using NonMedical X-Ray and Sealed Gamma Sources up to 10 MeV[6]

3. Terminology

3.1 *Definitions*—For definitions of terms used in this guide, refer to Terminology E1316.

4. Significance and Use

4.1 Within the present state of the radiographic art, this guide is generally applicable to available materials, processes, and techniques where industrial radiographic films are used as the recording media.

4.2 *Limitations*—This guide does not take into consideration special benefits and limitations resulting from the use of nonfilm recording media or readouts such as paper, tapes, xeroradiography, fluoroscopy, and electronic image intensification devices. Although reference is made to documents that may be used in the identification and grading, where applicable, of representative discontinuities in common metal castings and welds, no attempt has been made to set standards of acceptance for any material or production process. Radiography will be consistent in sensitivity and resolution only if the effect of all details of techniques, such as geometry, film, filtration, viewing, etc., is obtained and maintained.

5. Quality of Radiographs

5.1 To obtain quality radiographs, it is necessary to consider as a minimum the following list of items. Detailed information on each item is further described in this guide.

5.1.1 Radiation source (X-ray or gamma),

5.1.2 Voltage selection (X-ray),

5.1.3 Source size (X-ray or gamma),

5.1.4 Ways and means to eliminate scattered radiation,

5.1.5 Film system class,

5.1.6 Source to film distance,

5.1.7 Image quality indicators (IQI's),

5.1.8 Screens and filters,

5.1.9 Geometry of part or component configuration,

5.1.10 Identification and location markers, and

5.1.11 Radiographic quality level.

6. Radiographic Quality Level

6.1 Information on the design and manufacture of image quality indicators (IQI's) can be found in Practices E747, E801, E1025, and E1742.

6.2 The quality level usually required for radiography is 2 % (2-2T when using hole type IQI) unless a higher or lower quality is agreed upon between the purchaser and the supplier. At the 2 % subject contrast level, three quality levels of inspection, 2-1T, 2-2T, and 2-4T, are available through the design and application of the IQI (Practice E1025, Table 1). Other levels of inspection are available in Practice E1025 Table 1. The level of inspection specified should be based on the service requirements of the product. Great care should be taken in specifying quality levels 2-1T, 1-1T, and 1-2T by first determining that these quality levels can be maintained in production radiography.

NOTE 2—The first number of the quality level designation refers to IQI thickness expressed as a percentage of specimen thickness; the second number refers to the diameter of the IQI hole that must be visible on the radiograph, expressed as a multiple of penetrameter thickness, T.

6.3 If IQI's of material radiographically similar to that being examined are not available, IQI's of the required dimensions but of a lower-absorption material may be used.

6.4 The quality level required using wire IQI's shall be equivalent to the 2-2T level of Practice E1025 unless a higher or lower quality level is agreed upon between purchaser and supplier. Table 4 of Practice E747 gives a list of various hole-type IQI's and the diameter of the wires of corresponding EPS with the applicable 1T, 2T, and 4T holes in the plaque IQI. Appendix X1 of Practice E747 gives the equation for calculating other equivalencies, if needed.

7. Energy Selection

7.1 X-ray energy affects image quality. In general, the lower the energy of the source utilized the higher the achievable radiographic contrast, however, other variables such as geometry and scatter conditions may override the potential advantage of higher contrast. For a particular energy, a range of thicknesses which are a multiple of the half value layer, may be radiographed to an acceptable quality level utilizing a particular X-ray machine or gamma ray source. In all cases the specified IQI (penetrameter) quality level must be shown on the radiograph. In general, satisfactory results can normally be obtained for X-ray energies between 100 kV to 500 kV in a range between 2.5 to 10 half value layers (HVL) of material thickness (see Table 1). This range may be extended by as much as a factor of 2 in some situations for X-ray energies in the 1 to 25 MV range primarily because of reduced scatter.

8. Radiographic Equivalence Factors

8.1 The radiographic equivalence factor of a material is that factor by which the thickness of the material must be multiplied to give the thickness of a "standard" material (often steel)

---

[4] Available from American National Standards Institute (ANSI), 25 W. 43rd St., 4th Floor, New York, NY 10036.

[5] Available from U.S. Government Printing Office Superintendent of Documents, 732 N. Capitol St., NW, Mail Stop: SDE, Washington, DC 20401.

[6] Available from National Technical Information Service (NTIS), U.S. Department of Commerce, 5285 Port Royal Rd., Springfield, VA 22161.

TABLE 1 Typical Steel HVL Thickness in Inches (mm) for Common Energies

Energy	Thickness, Inches (mm)
120 kV	0.10 (2.5)
150 kV	0.14 (3.6)
200 kV	0.20 (5.1)
250 kV	0.25 (6.4)
400 kV (Ir 192)	0.35 (8.9)
1 MV	0.57 (14.5)
2 MV (Co 60)	0.80 (20.3)
4 MV	1.00 (25.4)
6 MV	1.15 (29.2)
10 MV	1.25 (31.8)
16 MV and higher	1.30 (33.0)

which has the same absorption. Radiographic equivalence factors of several of the more common metals are given in Table 2, with steel arbitrarily assigned a factor of 1.0. The factors may be used:

8.1.1 To determine the practical thickness limits for radiation sources for materials other than steel, and

8.1.2 To determine exposure factors for one metal from exposure techniques for other metals.

9. Film

9.1 Various industrial radiographic film are available to meet the needs of production radiographic work. However, definite rules on the selection of film are difficult to formulate because the choice depends on individual user requirements. Some user requirements are as follows: radiographic quality levels, exposure times, and various cost factors. Several methods are available for assessing image quality levels (see Test Method E746, and Practices E747 and E801). Information about specific products can be obtained from the manufacturers.

9.2 Various industrial radiographic films are manufactured to meet quality level and production needs. Test Method E1815 provides a method for film manufacturer classification of film systems. A film system consist of the film and associated film processing system. Users may obtain a classification table from the film manufacturer for the film system used in production radiography. A choice of film class can be made as provided in Test Method E1815. Additional specific details regarding classification of film systems is provided in Test Method E1815. ANSI Standards PH1.41, PH4.8, T9.1, and T9.2 provide specific details and requirements for film manufacturing.

10. Filters

10.1 *Definition*—Filters are uniform layers of material placed between the radiation source and the film.

10.2 *Purpose*—The purpose of filters is to absorb the softer components of the primary radiation, thus resulting in one or several of the following practical advantages:

10.2.1 Decreasing scattered radiation, thus increasing contrast.

10.2.2 Decreasing undercutting, thus increasing contrast.

10.2.3 Decreasing contrast of parts of varying thickness.

10.3 *Location*—Usually the filter will be placed in one of the following two locations:

10.3.1 As close as possible to the radiation source, which minimizes the size of the filter and also the contribution of the filter itself to scattered radiation to the film.

10.3.2 Between the specimen and the film in order to absorb preferentially the scattered radiation from the specimen. It should be noted that lead foil and other metallic screens (see 13.1) fulfill this function.

10.4 *Thickness and Filter Material*— The thickness and material of the filter will vary depending upon the following:

10.4.1 The material radiographed.

10.4.2 Thickness of the material radiographed.

10.4.3 Variation of thickness of the material radiographed.

10.4.4 Energy spectrum of the radiation used.

10.4.5 The improvement desired (increasing or decreasing contrast). Filter thickness and material can be calculated or determined empirically.

11. Masking

11.1 Masking or blocking (surrounding specimens or covering thin sections with an absorptive material) is helpful in reducing scattered radiation. Such a material can also be used to equalize the absorption of different sections, but the loss of detail may be high in the thinner sections.

12. Back-Scatter Protection

12.1 Effects of back-scattered radiation can be reduced by confining the radiation beam to the smallest practical cross

TABLE 2 Approximate Radiographic Equivalence Factors for Several Metals (Relative to Steel)

Metal	Energy Level									
	100 kV	150 kV	220 kV	250 kV	400 kV	1 MV	2 MV	4 to 25 MV	^{192}Ir	^{60}Co
Magnesium	0.05	0.05	0.08							
Aluminum	0.08	0.12	0.18						0.35	0.35
Aluminum alloy	0.10	0.14	0.18						0.35	0.35
Titanium		0.54	0.54		0.71	0.9	0.9	0.9	0.9	0.9
Iron/all steels	1.0	1.0	1.0	1.0	1.0	1.0	1.0	1.0	1.0	1.0
Copper	1.5	1.6	1.4	1.4	1.4	1.1	1.1	1.2	1.1	1.1
Zinc		1.4	1.3		1.3			1.2	1.1	1.0
Brass		1.4	1.3		1.3	1.2	1.1	1.0	1.1	1.0
Inconel X		1.4	1.3		1.3	1.3	1.3	1.3	1.3	1.3
Monel	1.7		1.2							
Zirconium	2.4	2.3	2.0	1.7	1.5	1.0	1.0	1.0	1.2	1.0
Lead	14.0	14.0	12.0			5.0	2.5	2.7	4.0	2.3
Hafnium			14.0	12.0	9.0	3.0				
Uranium			20.0	16.0	12.0	4.0		3.9	12.6	3.4

section and by placing lead behind the film. In some cases either or both the back lead screen and the lead contained in the back of the cassette or film holder will furnish adequate protection against back-scattered radiation. In other instances, this must be supplemented by additional lead shielding behind the cassette or film holder.

12.2 If there is any question about the adequacy of protection from back-scattered radiation, a characteristic symbol (frequently a 1/8-in. (3.2-mm) thick letter *B*) should be attached to the back of the cassette or film holder, and a radiograph made in the normal manner. If the image of this symbol appears on the radiograph as a lighter density than background, it is an indication that protection against back-scattered radiation is insufficient and that additional precautions must be taken.

13. Screens

13.1 *Metallic Foil Screens*:

13.1.1 Lead foil screens are commonly used in direct contact with the films, and, depending upon their thickness, and composition of the specimen material, will exhibit an intensifying action at as low as 90 kV. In addition, any screen used in front of the film acts as a filter (Section 10) to preferentially absorb scattered radiation arising from the specimen, thus improving radiographic quality. The selection of lead screen thickness, or for that matter, any metallic screen thickness, is subject to the same considerations as outlined in 10.4. Lead screens lessen the scatter reaching the film regardless of whether the screens permit a decrease or necessitate an increase in the radiographic exposure. To avoid image unsharpness due to screens, there should be intimate contact between the lead screen and the film during exposure.

13.1.2 Lead foil screens of appropriate thickness should be used whenever they improve radiographic quality or penetrameter sensitivity or both. The thickness of the front lead screens should be selected with care to avoid excessive filtration in the radiography of thin or light alloy materials, particularly at the lower kilovoltages. In general, there is no exposure advantage to the use of 0.005 in. in front and back lead screens below 125 kV in the radiography of 1/4-in. (6.35-mm) or lesser thickness steel. As the kilovoltage is increased to penetrate thicker sections of steel, however, there is a significant exposure advantage. In addition to intensifying action, the back lead screens are used as protection against back-scattered radiation (see Section 12) and their thickness is only important for this function. As exposure energy is increased to penetrate greater thicknesses of a given subject material, it is customary to increase lead screen thickness. For radiography using radioactive sources, the minimum thickness of the front lead screen should be 0.005 in. (0.13 mm) for iridium-192, and 0.010 in. (0.25 mm) for cobalt-60.

13.2 *Other Metallic Screen Materials*:

13.2.1 Lead oxide screens perform in a similar manner to lead foil screens except that their equivalence in lead foil thickness approximates 0.0005 in. (0.013 mm).

13.2.2 Copper screens have somewhat less absorption and intensification than lead screens, but may provide somewhat better radiographic sensitivity with higher energy above 1 MV.

13.2.3 Gold, tantalum, or other heavy metal screens may be used in cases where lead cannot be used.

13.3 *Fluorescent Screens*—Fluorescent screens may be used as required providing the required image quality is achieved. Proper selection of the fluorescent screen is required to minimize image unsharpness. Technical information about specific fluorescent screen products can be obtained from the manufacturers. Good film-screen contact and screen cleanliness are required for successful use of fluorescent screens. Additional information on the use of fluorescent screens is provided in Appendix X1.

13.4 *Screen Care*—All screens should be handled carefully to avoid dents and scratches, dirt, or grease on active surfaces. Grease and lint may be removed from lead screens with a solvent. Fluorescent screens should be cleaned in accordance with the recommendations of the manufacturer. Screens showing evidence of physical damage should be discarded.

14. Radiographic Image Quality

14.1 *Radiographic image quality* is a qualitative term used to describe the capability of a radiograph to show flaws in the area under examination. There are three fundamental components of radiographic image quality as shown in Fig. 1. Each component is an important attribute when considering a specific radiographic technique or application and will be briefly discussed below.

14.2 *Radiographic contrast* between two areas of a radiograph is the difference between the film densities of those areas. The degree of radiographic contrast is dependent upon both subject contrast and film contrast as illustrated in Fig. 1.

14.2.1 *Subject contrast* is the ratio of X-ray or gamma-ray intensities transmitted by two selected portions of a specimen. Subject contrast is dependent upon the nature of the specimen (material type and thickness), the energy (spectral composition, hardness or wavelengths) of the radiation used and the intensity and distribution of scattered radiation. It is independent of time, milliamperage or source strength (curies), source distance and the characteristics of the film system.

14.2.2 *Film contrast* refers to the slope (steepness) of the film system characteristic curve. Film contrast is dependent upon the type of film, the processing it receives and the amount of film density. It also depends upon whether the film was exposed with lead screens (or without) or with fluorescent screens. Film contrast is independent, for most practical purposes, of the wavelength and distribution of the radiation reaching the film and, hence is independent of subject contrast. For further information, consult Test Method E1815.

14.3 *Film system granularity* is the objective measurement of the local density variations that produce the sensation of graininess on the radiographic film (for example, measured with a densitometer with a small aperture of ≤ 0.0039 in. (0.1 mm)). Graininess is the subjective perception of a mottled random pattern apparent to a viewer who sees small local density variations in an area of overall uniform density (that is, the visual impression of irregularity of silver deposit in a processed radiograph). The degree of granularity will not affect the overall spatial radiographic resolution (expressed in line pairs per mm, etc.) of the resultant image and is usually independent of exposure geometry arrangements. Granularity

Radiographic Image Quality					
Radiographic Contrast		Film System Granularity	Radiographic Definition	
Subject Contrast	Film Contrast		Inherent Unsharpness	Geometric Unsharpness
Affected by: • Absorption differences in specimen (thickness, composition, density) • Radiation wavelength • Scattered radiation Reduced or enhanced by: • Masks and diaphragms • Filters • Lead screens • Potter-Bucky diaphragms	Affected by: • Type of film • Degree of development (type of developer, time, temperature and activity of developer, degree of agitation) • Film density • Type of screens (that is, fluorescent, lead or none)	• Grain size and distribution within the film emulsion • Processing conditions (type and activity of developer, temperature of developer, etc.) • Type of screens (that is, fluorescent, lead or none) • Radiation quality (that is, energy level, filtration, etc. • Exposure quanta (that is, intensity, dose, etc.)	Affected by: • Degree of screen-film contact • Total film thickness • Single or double emulsion coatings • Radiation quality • Type and thickness of screens (fluorescent, lead or none)	Affected by: • Focal spot or source physical size • Source-to-film distance • Specimen-to-film distance • Abruptness of thickness changes in specimen • Motion of specimen or radiation source

FIG. 1 Variables of Radiographic Image Quality

is affected by the applied screens, screen-film contact and film processing conditions. For further information on detailed perceptibility, consult Test Method E1815.

14.4 *Radiographic definition* refers to the sharpness of the image (both the image outline as well as image detail). Radiographic definition is dependent upon the inherent unsharpness of the film system and the geometry of the radiographic exposure arrangement (geometric unsharpness) as illustrated in Fig. 1.

14.4.1 *Inherent unsharpness* (U_i) is the degree of visible detail resulting from geometrical aspects within the film-screen system, that is, screen-film contact, screen thickness, total thickness of the film emulsions, whether single or double-coated emulsions, quality of radiation used (wavelengths, etc.) and the type of screen. Inherent unsharpness is independent of exposure geometry arrangements.

14.4.2 *Geometric unsharpness* (U_g) determines the degree of visible detail resultant from an "in-focus" exposure arrangement consisting of the source-to-film-distance, object-to-film-distance and focal spot size. Fig. 2(a) illustrates these conditions. Geometric unsharpness is given by the equation:

$$U_g = Ft/d_o \qquad (1)$$

where:
U_g = geometric unsharpness,
F = maximum projected dimension of radiation source,
t = distance from source side of specimen to film, and
d_o = source-object distance.

NOTE 3—d_o and t must be in the same units of measure; the units of U_g will be in the same units as F.

NOTE 4—A nomogram for the determination of U_g is given in Fig. 3 (inch-pound units). Fig. 4 represents a nomogram in metric units.
Example:
Given:
Source-object distance (d_o) = 40 in.,
Source size (F) = 500 mils, and
Source side of specimen to film distance (t) = 1.5 in.
Draw a straight line (dashed in Fig. 3) between 500 mils on the F scale and 1.5 in. on the t scale. Note the point on intersection (P) of this line with the pivot line. Draw a straight line (solid in Fig. 3) from 40 in. on the d_o scale through point P and extend to the U_g scale. Intersection of this line with the U_g scale gives geometrical unsharpness in mils, which in the example is 19 mils.

Inasmuch as the source size, F, is usually fixed for a given radiation source, the value of U_g is essentially controlled by the simple d_o/t ratio.

Geometric unsharpness (U_g) can have a significant effect on the quality of the radiograph; therefore source-to-film-distance (SFD) selection is important. The geometric unsharpness (U_g) equation, Eq 1, is for information and guidance and provides a means for determining geometric unsharpness values. The amount or degree of unsharpness should be minimized when establishing the radiographic technique.

15. Radiographic Distortion

15.1 The radiographic image of an object or feature within an object may be larger or smaller than the object or feature itself, because the penumbra of the shadow is rarely visible in a radiograph. Therefore, the image will be larger if the object or feature is larger than the source of radiation, and smaller if object or feature is smaller than the source. The degree of reduction or enlargement will depend on the source-to-object

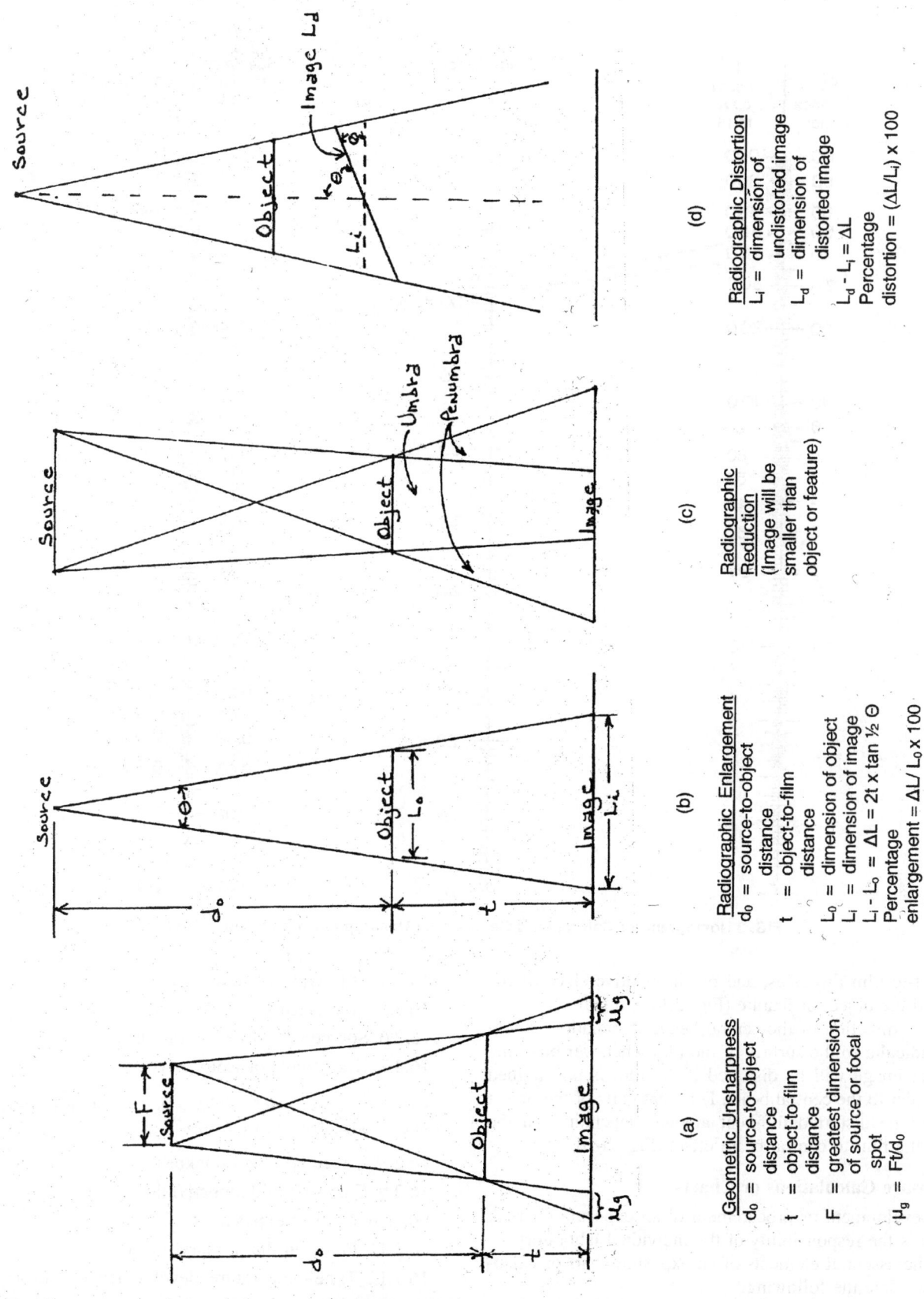

FIG. 2 Effects of Object-Film Geometry

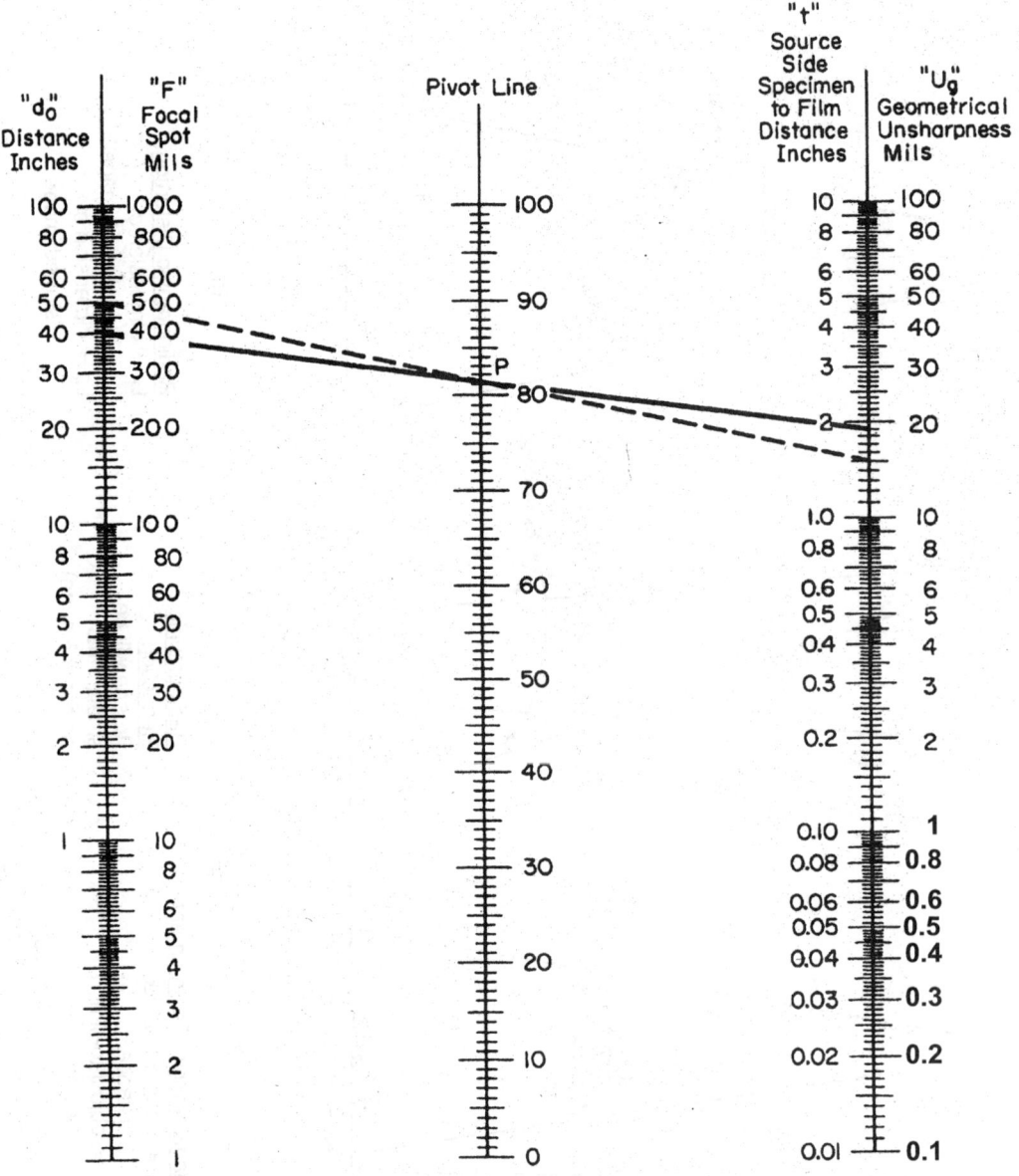

FIG. 3 Nomogram for Determining Geometrical Unsharpness (Inch-Pound Units)

and object-to-film distances, and on the relative sizes of the source and the object or feature (Fig. 2(b) and (c)).

15.2 The direction of the central beam of radiation should be perpendicular to the surface of the film whenever possible. The object image will be distorted if the film is not aligned perpendicular to the central beam. Different parts of the object image will be distorted different amount depending on the extent of the film to central beam offset (Fig. 2(d)).

16. Exposure Calculations or Charts

16.1 Development or procurement of an exposure chart or calculator is the responsibility of the individual laboratory.

16.2 The essential elements of an exposure chart or calculator must relate the following:

16.2.1 Source or machine,
16.2.2 Material type,
16.2.3 Material thickness,
16.2.4 Film type (relative speed),
16.2.5 Film density, (see Note 5),
16.2.6 Source or source to film distance,
16.2.7 Kilovoltage or isotope type,

NOTE 5—For detailed information on film density and density measurement calibration, see Practice E1079.

16.2.8 Screen type and thickness,
16.2.9 Curies or milliampere/minutes,
16.2.10 Time of exposure,
16.2.11 Filter (in the primary beam),
16.2.12 Time-temperature development for hand processing; access time for automatic processing; time-temperature development for dry processing, and
16.2.13 Processing chemistry brand name, if applicable.

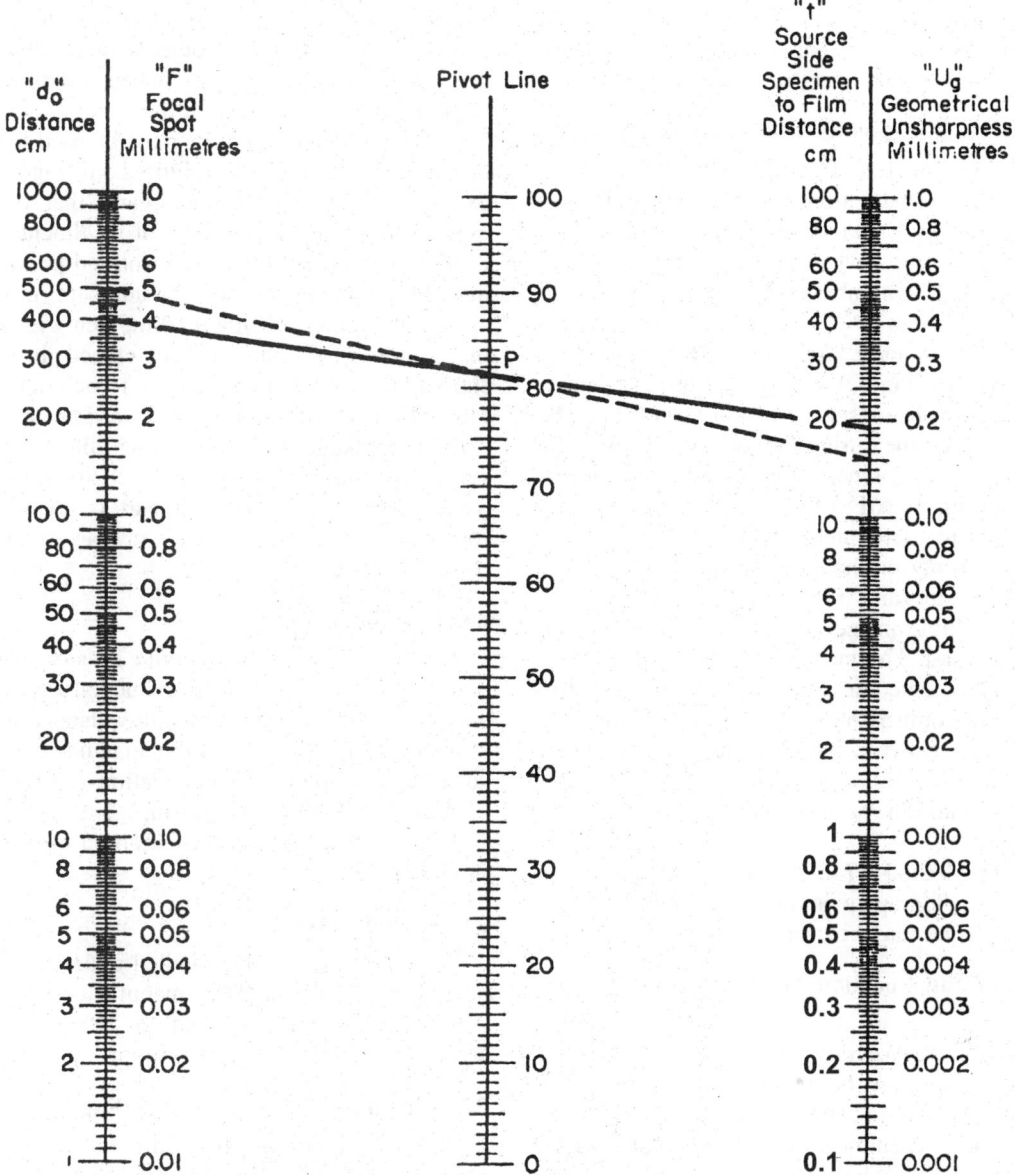

FIG. 4 Nomogram for Determining Geometrical Unsharpness (Metric Units)

16.3 The essential elements listed in 16.2 will be accurate for isotopes of the same type, but will vary with X-ray equipment of the same kilovoltage and milliampere rating.

16.4 Exposure charts should be developed for each X-ray machine and corrected each time a major component is replaced, such as the X-ray tube or high-voltage transformer.

16.5 The exposure chart should be corrected when the processing chemicals are changed to a different manufacturer's brand or the time-temperature relationship of the processor may be adjusted to suit the exposure chart. The exposure chart, when using a dry processing method, should be corrected based upon the time-temperature changes of the processor.

17. Technique File

17.1 It is recommended that a radiographic technique log or record containing the essential elements be maintained.

17.2 The radiographic technique log or record should contain the following:

17.2.1 Description, photo, or sketch of the test object illustrating marker layout, source placement, and film location.
17.2.2 Material type and thickness,
17.2.3 Source to film distance,
17.2.4 Film type,
17.2.5 Film density, (see Note 5),
17.2.6 Screen type and thickness,
17.2.7 Isotope or X-ray machine identification,
17.2.8 Curie or milliampere minutes,
17.2.9 IQI and shim thickness,
17.2.10 Special masking or filters,
17.2.11 Collimator or field limitation device,
17.2.12 Processing method, and
17.2.13 View or location.

17.3 The recommendations of 17.2 are not mandatory, but are essential in reducing the overall cost of radiography, and serve as a communication link between the radiographic interpreter and the radiographic operator.

18. Penetrameters (Image Quality Indicators)

18.1 Practices E747, E801, E1025, and E1742 should be consulted for detailed information on the design, manufacture and material grouping of IQI's. Practice E801 addresses IQI's for examination of electronic devices and provides additional details for positioning IQI's, number of IQI's required, and so forth.

18.2 Test Methods E746 and E1735 should be consulted for detailed information regarding IQI's which are used for determining relative image quality response of industrial film. The IQI's can also be used for measuring the image quality of the radiographic system or any component of the systems equivalent penetrameter sensitivity (EPS) performance.

18.2.1 An example for determining and EPS performance evaluation of several X-ray machines is as follows:

18.2.1.1 Keep the film and film processing parameters constant, and take multiple image quality exposures with all machines being evaluated. The machines should be set for a prescribed exposure as stated in the standard and the film density equalized. By comparison of the resultant films, the relative EPS variations between the machines can be determined.

18.2.2 Exposure condition variables may also be studied using this plaque.

18.2.3 While Test Method E746 plaque can be useful in quantifying relative radiographic image quality, these other applications of the plaque may be useful.

19. Identification of and Location Markers on Radiographs

19.1 *Identification of Radiographs*:

19.1.1 Each radiograph must be identified uniquely so that there is a permanent correlation between the part radiographed and the film. The type of identification and method by which identification is achieved shall be as agreed upon between the customer and inspector.

19.1.2 The minimum identification should at least include the following: the radiographic facility's identification and name, the date, part number and serial number, if used, for unmistakable identification of radiographs with the specimen. The letter *R* should be used to designate a radiograph of a repair area, and may include − 1, − 2, etc., for the number of repair.

19.2 *Location Markers*:

19.2.1 Location markers (that is, lead or high-atomic number metals or letters that are to appear as images on the radiographic film) should be placed on the part being examined, whenever practical, and not on the cassette. Their exact locations should also be marked on the surface of the part being radiographed, thus permitting the area of interest to be located accurately on the part, and they should remain on the part during radiographic inspection. Their exact location may be permanently marked in accordance with the customer's requirements.

19.2.2 Location markers are also used in assisting the radiographic interpreter in marking off defective areas of components, castings, or defects in weldments; also, sorting good and rejectable items when more than one item is radiographed on the same film.

19.2.3 Sufficient markers must be used to provide evidence on the radiograph that the required coverage of the object being examined has been obtained, and that overlap is evident, especially during radiography of weldments and castings.

19.2.4 Parts that must be identified permanently may have the serial numbers or section numbers, or both, stamped or written upon them with a marking pen with a special indelible ink, engraved, die stamped, or etched. In any case, the part should be marked in an area not to be removed in subsequent fabrication. If die stamps are used, caution is required to prevent breakage or future fatigue failure. The lowest stressed surface of the part should be used for this stamping. Where marking or stamping of the part is not permitted for some reason, a marked reference drawing or shooting sketch is recommended.

20. Storage of Film

20.1 Unexposed films should be stored in such a manner that they are protected from the effects of light, pressure, excessive heat, excessive humidity, damaging fumes or vapors, or penetrating radiation. Film manufacturers should be consulted for detailed recommendations on film storage. Storage of film should be on a "first in," "first out" basis.

20.2 More detailed information on film storage is provided in Guide E1254.

21. Safelight Test

21.1 Films should be handled under safelight conditions in accordance with the film manufacturer's recommendations. ANSI PH2.22 can be used to determine the adequacy of safelight conditions in a darkroom.

22. Cleanliness and Film Handling

22.1 Cleanliness is one of the most important requirements for good radiography. Cassettes and screens must be kept clean, not only because dirt retained may cause exposure or processing artifacts in the radiographs, but because such dirt may also be transferred to the loading bench, and subsequently to other film or screens.

22.2 The surface of the loading bench must be kept clean. Where manual processing is used cleanliness will be promoted by arranging the darkroom with processing facilities on one side and film-handling facilities on the other. The darkroom will then have a wet side and a dry side and the chance of chemical contamination of the loading bench will be relatively slight.

22.3 Films should be handled only at their edges, and with dry, clean hands to avoid finger marks on film surfaces.

22.4 Sharp bending, excessive pressure, and rough handling of any kind must be avoided.

23. Film Processing, General

23.1 To produce a satisfactory radiograph, the care used in making the exposure *must* be followed by equal care in processing. The most careful radiographic techniques can be nullified by incorrect or improper darkroom procedures.

23.2 Sections 24-26 provide general information for film processing. Detailed information on film processing is provided in Guide E999.

24. Automatic Processing

24.1 *Automatic Processing*—The essence of the automatic processing system is control. The processor maintains the chemical solutions at the proper temperature, agitates and replenishes the solutions automatically, and transports the films mechanically at a carefully controlled speed throughout the processing cycle. Film characteristics must be compatible with processing conditions. It is, therefore, essential that the recommendations of the film, processor, and chemical manufacturers be followed.

24.2 *Automatic Processing, Dry*—The essence of dry automatic processing is the precise control of development time and temperature which results in reproducibility of radiographic density. Film characteristics must be compatible with processing conditions. It is, therefore, essential that the recommendations of the film and processor manufacturers be followed.

25. Manual Processing

25.1 Film and chemical manufacturers should be consulted for detailed recommendations on manual film processing. This section outlines the steps for one acceptable method of manual processing.

25.2 *Preparation*—No more film should be processed than can be accommodated with a minimum separation of ½ in. (12.7 mm). Hangers are loaded and solutions stirred before starting development.

25.3 *Start of Development*—Start the timer and place the films into the developer tank. Separate to a minimum distance of ½ in. (12.7 mm) and agitate in two directions for about 15 s.

25.4 *Development*—Normal development is 5 to 8 min at 68°F (20°C). Longer development time generally yields faster film speed and slightly more contrast. The manufacturer's recommendation should be followed in choosing a development time. When the temperature is higher or lower, development time must be changed. Again, consult manufacturer-recommended development time versus temperature charts. Other recommendations of the manufacturer to be followed are replenishment rates, renewal of solutions, and other specific instructions.

25.5 *Agitation*—Shake the film horizontally and vertically, ideally for a few seconds each minute during development. This will help film develop evenly.

25.6 *Stop Bath or Rinse*—After development is complete, the activity of developer remaining in the emulsion should be neutralized by an acid stop bath or, if this is not possible, by rinsing with vigorous agitation in clear water. Follow the film manufacturer's recommendation of stop bath composition (or length of alternative rinse), time immersed, and life of bath.

25.7 *Fixing*—The films must not touch one another in the fixer. Agitate the hangers vertically for about 10 s and again at the end of the first minute, to ensure uniform and rapid fixation. Keep them in the fixer until fixation is complete (that is, at least twice the clearing time), but not more than 15 min in relatively fresh fixer. Frequent agitation will shorten the time of fixation.

25.8 *Fixer Neutralizing*—The use of a hypo eliminator or fixer neutralizer between fixation and washing may be advantageous. These materials permit a reduction of both time and amount of water necessary for adequate washing. The recommendations of the manufacturers as to preparation, use, and useful life of the baths should be observed rigorously.

25.9 *Washing*—The washing efficiency is a function of wash water, its temperature, and flow, and the film being washed. Generally, washing is very slow below 60°F (16°C). When washing at temperatures above 85°F (30°C), care should be exercised not to leave films in the water too long. The films should be washed in batches without contamination from new film brought over from the fixer. If pressed for capacity, as more films are put in the wash, partially washed film should be moved in the direction of the inlet.

25.9.1 The cascade method of washing uses less water and gives better washing for the same length of time. Divide the wash tank into two sections (may be two tanks). Put the films from the fixer in the outlet section. After partial washing, move the batch of film to the inlet section. This completes the wash in fresh water.

25.9.2 For specific washing recommendations, consult the film manufacturer.

25.10 *Wetting Agent*—Dip the film for approximately 30 s in a wetting agent. This makes water drain evenly off film which facilitates quick, even drying.

25.11 *Residual Fixer Concentrations*— If the fixing chemicals are not removed adequately from the film, they will in time cause staining or fading of the developed image. Residual fixer concentrations permissible depend upon whether the films are to be kept for commercial purposes (3 to 10 years) or must be of archival quality. Archival quality processing is desirable for all radiographs whenever average relative humidity and temperature are likely to be excessive, as is the case in tropical and subtropical climates. The method of determining residual fixer concentrations may be ascertained by reference to ANSI PH4.8, PH1.28, and PH1.41.

25.12 *Drying*—Drying is a function of (*1*) film (base and emulsion); (*2*) processing (hardness of emulsion after washing, use of wetting agent); and (*3*) drying air (temperature, humidity, flow). Manual drying can vary from still air drying at ambient temperature to as high as 140°F (60°C) with air circulated by a fan. Film manufacturers should again be contacted for recommended drying conditions. Take precaution to tighten film on hangers, so that it cannot touch in the dryer. Too hot a drying temperature at low humidity can result in uneven drying and should be avoided.

26. Testing Developer

26.1 It is desirable to monitor the activity of the radiographic developing solution. This can be done by periodic development of film strips exposed under carefully controlled conditions, to a graded series of radiation intensities or time, or by using a commercially available strip carefully controlled for film speed and latent image fading.

27. Viewing Radiographs

27.1 Guide E1390 provides detailed information on requirements for illuminators. The following sections provide general information to be considered for use of illuminators.

27.2 *Transmission*—The illuminator must provide light of an intensity that will illuminate the average density areas of the radiographs without glare and it must diffuse the light evenly over the viewing area. Commercial fluorescent illuminators are satisfactory for radiographs of moderate density; however, high light intensity illuminators are available for densities up to 3.5 or 4.0. Masks should be available to exclude any extraneous light from the eyes of the viewer when viewing radiographs smaller than the viewing port or to cover low-density areas.

27.3 *Reflection*—Radiographs on a translucent or opaque backing may be viewed by reflected light. It is recommended that the radiograph be viewed under diffuse lighting conditions to prevent excess glare. Optical magnification can be used in certain instances to enhance the interpretation of the image.

28. Viewing Room

28.1 Subdued lighting, rather than total darkness, is preferable in the viewing room. The brightness of the surroundings should be about the same as the area of interest in the radiograph. Room illumination must be so arranged that there are no reflections from the surface of the film under examination.

29. Storage of Processed Radiographs

29.1 Guide E1254 provides detailed information on controls and maintenance for storage of radiographs and unexposed film. The following sections provide general information for storage of radiographs.

29.2 Envelopes having an edge seam, rather than a center seam, and joined with a nonhygroscopic adhesive, are preferred, since occasional staining and fading of the image is caused by certain adhesives used in the manufacture of envelopes (see ANSI PH1.53).

30. Records

30.1 It is recommended that an inspection log (a log may consist of a card file, punched card system, a book, or other record) constituting a record of each job performed, be maintained. This record should comprise, initially, a job number (which should appear also on the films), the identification of the parts, material or area radiographed, the date the films are exposed, and a complete record of the radiographic procedure, in sufficient detail so that any radiographic techniques may be duplicated readily. If calibration data, or other records such as card files or procedures, are used to determine the procedure, the log need refer only to the appropriate data or other record. Subsequently, the interpreter's findings and disposition (acceptance or rejection), if any, and his initials, should also be entered for each job.

31. Reports

31.1 When written reports of radiographic examinations are required, they should include the following, plus such other items as may be agreed upon:

31.1.1 Identification of parts, material, or area.

31.1.2 Radiographic job number.

31.1.3 Findings and disposition, if any. This information can be obtained directly from the log.

32. Identification of Completed Work

32.1 Whenever radiography is an inspective (rather than investigative) operation whereby material is accepted or rejected, all parts and material that have been accepted should be marked permanently, if possible, with a characteristic identifying symbol which will indicate to subsequent or final examiners the fact of radiographic acceptance.

32.2 Whenever possible, the completed radiographs should be kept on file for reference. The custody of radiographs and the length of time they are preserved should be agreed upon between the contracting parties.

33. Keywords

33.1 exposure calculations; film system; gamma-ray; image quality indicator (IQI); radiograph; radiographic examination; radiographic quality level; technique file; X-ray

---

APPENDIX

(Nonmandatory Information)

X1. USE OF FLUORESCENT SCREENS

X1.1 *Description*—Fluorescent intensifying screens have a cardboard or plastic support coated with a uniform layer of inorganic phosphor (crystalline substance). The support and phosphor are held together by a radiotransparent binding material. Fluorescent screens derive their name from the fact that their phosphor crystals "fluoresce" (emit visible light) when struck by X or gamma radiation. Some phosphors like calcium tungstate ($CaWO_4$) give off blue light while others known as rare earth emit light green.

X1.2 *Purpose and Film Types*—Fluorescent screen exposures are usually much shorter than those made without screens or with lead intensifying screens, because radiographic films generally are more responsive to visible light than to direct X-radiation, gamma radiation, and electrons.

X1.2.1 Films fall into one of two categories: non-screen type film having moderate light response, and screen type film specifically sensitized to have a very high blue or green light response. Fluorescent screens can reduce conventional exposures by as much as 150 times, depending on film type.

X1.3 *Image Quality and Use*—The image quality associated with fluorescent screen exposures is a function of sharpness, mottle, and contrast. Screen sharpness depends on phosphor crystal size, thickness of the crystal layer, and the reflective base coating. Each crystal emits light relative to its size and in all directions thus producing a relative degree of image unsharpness. To minimize this unsharpness, screen to film contact should be as intimate as possible. Mottle adversely affects image quality in two ways. First, a "quantum" mottle is dependent upon the amount of X or gamma radiation actually absorbed by the fluorescent screen, that is, faster screen/film systems lead to greater mottle and poorer image quality. A "structural" mottle, which is a function of crystal size, crystal uniformity, and layer thickness, is minimized by using screens having small, evenly spaced crystals in a thin crystalline layer. Fluorescent screens are highly sensitive to longer wavelength scattered radiation. Consequently, to maximize contrast when this non-image forming radiation is excessive, fluorometallic intensifying screens or fluorescent screens backed by lead screens of appropriate thickness are recommended. Screen technology has seen significant advances in recent years, and today's fluorescent screens have smaller crystal size, more uniform crystal packing, and reduced phosphor thickness. This translates into greater screen/film speed with reduced unsharpness and mottle. These improvements can represent some meaningful benefits for industrial radiography, as indicated by the three examples as follows:

X1.3.1 *Reduced Exposure (Increased Productivity)*—There are instances when prohibitively long exposure times make conventional radiography impractical. An example is the inspection of thick, high atomic number materials with low curie isotopes. Depending on many variables, exposure time may be reduced by factors ranging from 2× to 105× when the appropriate fluorescent screen/film combination is used.

X1.3.2 *Improved Safety Conditions (Field Sites)*—Because fluorescent screens provide reduced exposure, the length of time that non-radiation workers must evacuate a radiographic inspection site can be reduced significantly.

X1.3.3 *Extended Equipment Capability*—Utilizing the speed advantage of fluorescent screens by translating it into reduced energy level. An example is that a 150 kV X-ray tube may do the job of a 300 kV tube, or that iridium 192 may be used in applications normally requiring cobalt 60. It is possible for overall image quality to be better at the lower kV with fluorescent screens than at a higher energy level using lead screens.

BIBLIOGRAPHY ON INDUSTRIAL RADIOGRAPHY

For conciseness, this bibliography has been limited to books and specifically to books in English published after 1950.

(1) Clark, G. L., *Applied X-Rays,* 4th ed., McGraw Hill Book Co., Inc., New York, 1955.
(2) Clauser, H. R., *Practical Radiography for Industry,* Reinhold Publishing Corp., New York, 1952.
(3) Hogarth, C. A., and Blitz, J. (Editors), *Techniques of Nondestructive Testing,* Butte Worth and Co., Ltd., London, 1960.
(4) McMaster, R. C. (Editor), *Nondestructive Testing Handbook,* The Ronald Press, New York, 1960.
(5) Morgan, R. H., and Corrigan, K. E. (Editors), *Handbook of Radiology,* The Year Book Publishers, Inc., Chicago, 1955.
(6) Reed, M. E., *Cobalt-60 Radiography in Industry,* Tracer-lab, Inc., Boston, 1954.
(7) Robertson, J. K., *Radiology Physics,* 3rd ed., D. Van Nostrand Company, New York, 1956.
(8) Weyl, C., and Warren, S. R., *Radiologic Physics,* 2nd ed., Charles C. Thomas, Springfield, IL, 1951.
(9) Wilshire, W. J. (Editor), *A Further Handbook of Industrial Radiology,* Edward Arnold and Company, London, 1957.
(10) McGonnagle, W. J., *Nondestructive Testing,* McGraw Hill Book Co., Inc., New York, 1961.
(11) *Handbook on Radiography,* Revised edition, Atomic Energy of Canada Ltd. Ottawa, Ont., 1950.
(12) *Papers on Radiography, ASTM STP 96,* ASTM, 1950.
(13) *Symposium on the Role of Nondestructive Testing in the Economics of Production, ASTM STP 112,* ASTM, 1951.
(14) *Radioisotope Technique,* Vol II, H. M. Stationery Office, London, 1952.
(15) *Symposium on Nondestructive Testing, ASTM STP 145,* ASTM, 1953.
(16) *Memorandum on Gamma-Ray Sources for Radiography,* Revised edition, Institute of Physics, London, 1954.
(17) *Papers on Nondestructive Testing,* see *Proceedings,* ASTM, Vol 54, 1954.
(18) *Radiography in Modern Industry* (3rd edition), Eastman Kodak Co., Rochester, NY, 1969.
(19) *Symposium on Nondestructive Tests in the Field of Nuclear Energy, ASTM STP 223,* ASTM, 1958.
(20) *Radiographer's Reference* (3rd edition), E. I. du Pont de Nemours & Co., Inc., Wilmington, DE, 1974 (or latest revision).

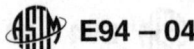

ASTM International takes no position respecting the validity of any patent rights asserted in connection with any item mentioned in this standard. Users of this standard are expressly advised that determination of the validity of any such patent rights, and the risk of infringement of such rights, are entirely their own responsibility.

This standard is subject to revision at any time by the responsible technical committee and must be reviewed every five years and if not revised, either reapproved or withdrawn. Your comments are invited either for revision of this standard or for additional standards and should be addressed to ASTM International Headquarters. Your comments will receive careful consideration at a meeting of the responsible technical committee, which you may attend. If you feel that your comments have not received a fair hearing you should make your views known to the ASTM Committee on Standards, at the address shown below.

This standard is copyrighted by ASTM International, 100 Barr Harbor Drive, PO Box C700, West Conshohocken, PA 19428-2959, United States. Individual reprints (single or multiple copies) of this standard may be obtained by contacting ASTM at the above address or at 610-832-9585 (phone), 610-832-9555 (fax), or service@astm.org (e-mail); or through the ASTM website (www.astm.org).

Designation: E140 − 07

Standard Hardness Conversion Tables for Metals Relationship Among Brinell Hardness, Vickers Hardness, Rockwell Hardness, Superficial Hardness, Knoop Hardness, and Scleroscope Hardness[1]

This standard is issued under the fixed designation E140; the number immediately following the designation indicates the year of original adoption or, in the case of revision, the year of last revision. A number in parentheses indicates the year of last reapproval. A superscript epsilon (ε) indicates an editorial change since the last revision or reapproval.

This standard has been approved for use by agencies of the Department of Defense.

1. Scope*

1.1 Conversion Table 1 presents data in the Rockwell C hardness range on the relationship among Brinell hardness, Vickers hardness, Rockwell hardness, Rockwell superficial hardness, Knoop hardness, and Scleroscope hardness of non-austenitic steels including carbon, alloy, and tool steels in the as-forged, annealed, normalized, and quenched and tempered conditions provided that they are homogeneous.

1.2 Conversion Table 2 presents data in the Rockwell B hardness range on the relationship among Brinell hardness, Vickers hardness, Rockwell hardness, Rockwell superficial hardness, Knoop hardness, and Scleroscope hardness of non-austenitic steels including carbon, alloy, and tool steels in the as-forged, annealed, normalized, and quenched and tempered conditions provided that they are homogeneous.

1.3 Conversion Table 3 presents data on the relationship among Brinell hardness, Vickers hardness, Rockwell hardness, Rockwell superficial hardness, and Knoop hardness of nickel and high-nickel alloys (nickel content over 50 %). These hardness conversion relationships are intended to apply particularly to the following: nickel-aluminum-silicon specimens finished to commercial mill standards for hardness testing, covering the entire range of these alloys from their annealed to their heavily cold-worked or age-hardened conditions, including their intermediate conditions.

1.4 Conversion Table 4 presents data on the relationship among Brinell hardness, Vickers hardness, Rockwell hardness, and Rockwell superficial hardness of cartridge brass.

1.5 Conversion Table 5 presents data on the relationship between Brinell hardness and Rockwell B hardness of austenitic stainless steel plate in the annealed condition.

1.6 Conversion Table 6 presents data on the relationship between Rockwell hardness and Rockwell superficial hardness of austenitic stainless steel sheet.

1.7 Conversion Table 7 presents data on the relationship among Brinell hardness, Vickers hardness, Rockwell hardness, Rockwell superficial hardness, and Knoop hardness of copper.

1.8 Conversion Table 8 presents data on the relationship among Brinell hardness, Rockwell hardness, and Vickers hardness of alloyed white iron.

1.9 Conversion Table 9 presents data on the relationship among Brinell hardness, Vickers hardness, Rockwell hardness, and Rockwell superficial hardness of wrought aluminum products.

1.10 Many of the conversion values presented herein were obtained from computer-generated curves of actual test data. Most Rockwell hardness numbers are presented to the nearest 0.1 or 0.5 hardness number to permit accurate reproduction of these curves. Since all converted hardness values must be considered approximate, however, all converted Rockwell hardness numbers shall be rounded to the nearest whole number in accordance with Practice E29.

1.11 Appendix X1-Appendix X9 contain equations developed from the data in Tables 1-9, respectively, to convert from one hardness scale to another. Since all converted hardness values must be considered approximate, however, all converted hardness numbers shall be rounded in accordance with Practice E29.

1.12 Conversion of hardness values should be used only when it is impossible to test the material under the conditions specified, and when conversion is made it should be done with discretion and under controlled conditions. Each type of hardness test is subject to certain errors, but if precautions are carefully observed, the reliability of hardness readings made on instruments of the indentation type will be found comparable. Differences in sensitivity within the range of a given hardness scale (for example, Rockwell B) may be greater than between two different scales or types of instruments. The conversion

---

[1] These conversion tables are under the jurisdiction of ASTM Committee E28 on Mechanical Testing and are the direct responsibility of Subcommittee E28.06 on Indentation Hardness Testing.
Current edition approved Jan. 1, 2007. Published January 2007. Originally approved in 1958. Last previous edition approved in 2005 as E140 – 05ε1. DOI: 10.1520/E0140-07.

*A Summary of Changes section appears at the end of this standard.

Copyright © ASTM International, 100 Barr Harbor Drive, PO Box C700, West Conshohocken, PA 19428-2959, United States.

values, whether from the tables or calculated from the equations, are only approximate and may be inaccurate for specific application.

2. Referenced Documents

2.1 *ASTM Standards:*[2]

E10 Test Method for Brinell Hardness of Metallic Materials

E18 Test Methods for Rockwell Hardness of Metallic Materials

E29 Practice for Using Significant Digits in Test Data to Determine Conformance with Specifications

E92 Test Method for Vickers Hardness of Metallic Materials

E384 Test Method for Microindentation Hardness of Materials

E448 Practice for Scleroscope Hardness Testing of Metallic Materials

3. Methods for Hardness Determinations

3.1 The hardness readings used with these conversion tables shall be determined in accordance with one of the following ASTM test methods:

3.1.1 *Vickers Hardness*—Test Method E92.

3.1.2 *Brinell Hardness*—Test Method E10.

3.1.3 *Rockwell Hardness*—Test Method E18 Scales A, B, C, D, E, F, G, H, K, 15-N, 30-N, 45-N, 15-T, 30-T, 45-T, 15-W.

3.1.4 *Knoop Hardness*—Test Method E384.

3.1.5 *Scleroscope*[3] *Hardness*—Practice E448.

NOTE 1—The comparative hardness test done to generate the conversion tables in this standard were preformed in past years using ASTM test methods in effect at the time of testing. In some cases, the standards have changed in ways that could affect the final results. For example, currently both the Rockwell and Brinell hardness standards (Test Method E10 and E18, respectively) allow or require the use of tungsten carbide ball indenters; however, all of the ball scale Rockwell hardness tests (HRB, HR30T, etc.) and most of the Brinell hardness tests preformed to develop these tables used hardened steel ball indenters. The use of tungsten carbide balls will produce slightly different hardness results than steel balls. Therefore, the user is cautioned to consider these differences and to keep in mind the approximate nature of these conversions when applying them to the results of tests using tungsten carbide balls.

4. Apparatus and Reference Standards

4.1 The apparatus and reference standards shall conform to the description in Test Methods E92, E10, E18, E384, and Practice E448.

---

[2] For referenced ASTM standards, visit the ASTM website, www.astm.org, or contact ASTM Customer Service at service@astm.org. For *Annual Book of ASTM Standards* volume information, refer to the standard's Document Summary page on the ASTM website.

[3] Registered trademark of the Shore Instrument and Manufacturing Co., Inc.

5. Principle of Method of Conversion

5.1 Tests have proved that even the most reliable data cannot be fitted to a single conversion relationship for all metals. Indentation hardness is not a single fundamental property but a combination of properties, and the contribution of each to the hardness number varies with the type of test. The modulus of elasticity has been shown to influence conversions at high hardness levels; and at low hardness levels conversions between hardness scales measuring depth and those measuring diameter are likewise influenced by differences in the modulus of elasticity. Therefore separate conversion tables are necessary for different materials.

NOTE 2—Hardness conversion values for other metals based on comparative test on similar materials having similar mechanical properties will be added to this standard as the need arises.

6. Significance and Use

6.1 The conversion values given in the tables, or calculated by the equations given in the appendixes, should only be considered valid for the specific materials indicated. This is because conversions can be affected by several factors, including the material alloy, grain structure, heat treatment, etc.

6.2 Since the various types of hardness tests do not all measure the same combination of material properties, conversion from one hardness scale to another is only an approximate process. Because of the wide range of variation among different materials, it is not possible to state confidence limits for the errors in using a conversion chart. Even in the case of a table established for a single material, such as the table for cartridge brass, some error is involved depending on composition and methods of processing.

6.3 Because of their approximate nature, conversion tables must be regarded as only an estimate of comparative values. It is recommended that hardness conversions be applied primarily to values such as specification limits, which are established by agreement or mandate, and that the conversion of test data be avoided whenever possible (see Note 1).

7. Reporting of Hardness Numbers

7.1 When reporting converted hardness numbers the measured hardness and test scale shall be indicated in parentheses as in the following example:

$$353\ HBW\ (38\ HRC) \tag{1}$$

8. Keywords

8.1 conversion; hardness scale; metallic

TABLE 1 Approximate Hardness Conversion Numbers for Non-Austenitic Steels (Rockwell C Hardness Range)[A, B]

Rockwell C Hardness Number 150 kgf (HRC)	Vickers Hardness Number (HV)	Brinell Hardness Number[C]		Knoop Hardness, Number 500-gf and Over (HK)	Rockwell Hardness Number		Rockwell Superficial Hardness Number			Scleroscope Hardness Number[D]	Rockwell C Hardness Number 150 kgf (HRC)
		10-mm Standard Ball, 3000-kgf (HBS)	10-mm Carbide Ball, 3000-kgf (HBW)		A Scale, 60-kgf (HRA)	D Scale, 100-kgf (HRD)	15-N Scale, 15-kgf (HR 15-N)	30-N Scale, 30-kgf (HR 30-N)	45-N Scale, 45-kgf (HR 45-N)		
68	940	...	...	920	85.6	76.9	93.2	84.4	75.4	97.3	68
67	900	...	...	895	85.0	76.1	92.9	83.6	74.2	95.0	67
66	865	...	...	870	84.5	75.4	92.5	82.8	73.3	92.7	66
65	832	...	(739)	846	83.9	74.5	92.2	81.9	72.0	90.6	65
64	800	...	(722)	822	83.4	73.8	91.8	81.1	71.0	88.5	64
63	772	...	(705)	799	82.8	73.0	91.4	80.1	69.9	86.5	63
62	746	...	(688)	776	82.3	72.2	91.1	79.3	68.8	84.5	62
61	720	...	(670)	754	81.8	71.5	90.7	78.4	67.7	82.6	61
60	697	...	(654)	732	81.2	70.7	90.2	77.5	66.6	80.8	60
59	674	...	634	710	80.7	69.9	89.8	76.6	65.5	79.0	59
58	653	...	615	690	80.1	69.2	89.3	75.7	64.3	77.3	58
57	633	...	595	670	79.6	68.5	88.9	74.8	63.2	75.6	57
56	613	...	577	650	79.0	67.7	88.3	73.9	62.0	74.0	56
55	595	...	560	630	78.5	66.9	87.9	73.0	60.9	72.4	55
54	577	...	543	612	78.0	66.1	87.4	72.0	59.8	70.9	54
53	560	...	525	594	77.4	65.4	86.9	71.2	58.6	69.4	53
52	544	(500)	512	576	76.8	64.6	86.4	70.2	57.4	67.9	52
51	528	(487)	496	558	76.3	63.8	85.9	69.4	56.1	66.5	51
50	513	(475)	481	542	75.9	63.1	85.5	68.5	55.0	65.1	50
49	498	(464)	469	526	75.2	62.1	85.0	67.6	53.8	63.7	49
48	484	451	455	510	74.7	61.4	84.5	66.7	52.5	62.4	48
47	471	442	443	495	74.1	60.8	83.9	65.8	51.4	61.1	47
46	458	432	432	480	73.6	60.0	83.5	64.8	50.3	59.8	46
45	446	421	421	466	73.1	59.2	83.0	64.0	49.0	58.5	45
44	434	409	409	452	72.5	58.5	82.5	63.1	47.8	57.3	44
43	423	400	400	438	72.0	57.7	82.0	62.2	46.7	56.1	43
42	412	390	390	426	71.5	56.9	81.5	61.3	45.5	54.9	42
41	402	381	381	414	70.9	56.2	80.9	60.4	44.3	53.7	41
40	392	371	371	402	70.4	55.4	80.4	59.5	43.1	52.6	40
39	382	362	362	391	69.9	54.6	79.9	58.6	41.9	51.5	39
38	372	353	353	380	69.4	53.8	79.4	57.7	40.8	50.4	38
37	363	344	344	370	68.9	53.1	78.8	56.8	39.6	49.3	37
36	354	336	336	360	68.4	52.3	78.3	55.9	38.4	48.2	36
35	345	327	327	351	67.9	51.5	77.7	55.0	37.2	47.1	35
34	336	319	319	342	67.4	50.8	77.2	54.2	36.1	46.1	34
33	327	311	311	334	66.8	50.0	76.6	53.3	34.9	45.1	33
32	318	301	301	326	66.3	49.2	76.1	52.1	33.7	44.1	32
31	310	294	294	318	65.8	48.4	75.6	51.3	32.5	43.1	31
30	302	286	286	311	65.3	47.7	75.0	50.4	31.3	42.2	30
29	294	279	279	304	64.8	47.0	74.5	49.5	30.1	41.3	29
28	286	271	271	297	64.3	46.1	73.9	48.6	28.9	40.4	28
27	279	264	264	290	63.8	45.2	73.3	47.7	27.8	39.5	27
26	272	258	258	284	63.3	44.6	72.8	46.8	26.7	38.7	26
25	266	253	253	278	62.8	43.8	72.2	45.9	25.5	37.8	25
24	260	247	247	272	62.4	43.1	71.6	45.0	24.3	37.0	24
23	254	243	243	266	62.0	42.1	71.0	44.0	23.1	36.3	23
22	248	237	237	261	61.5	41.6	70.5	43.2	22.0	35.5	22
21	243	231	231	256	61.0	40.9	69.9	42.3	20.7	34.8	21
20	238	226	226	251	60.5	40.1	69.4	41.5	19.6	34.2	20

[A] In the table headings, force refers to total test forces.
[B] Appendix X1 contains equations converting determined hardness scale numbers to Rockwell C hardness numbers for non-austenitic steels. Refer to 1.11 before using conversion equations.
[C] The Brinell hardness numbers in parentheses are outside the range recommended for Brinell hardness testing in 8.1 of Test Method E10.
[D] These Scleroscope hardness conversions are based on Vickers—Scleroscope hardness relationships developed from Vickers hardness data provided by the National Bureau of Standards for 13 steel reference blocks, Scleroscope hardness values obtained on these blocks by the Shore Instrument and Mfg. Co., Inc., the Roll Manufacturers Institute, and members of this institute, and also on hardness conversions previously published by the American Society for Metals and the Roll Manufacturers Institute.

TABLE 2 Approximate Hardness Conversion Numbers for Non-Austenitic Steels (Rockwell B Hardness Range)[A,B]

| Rockwell B Hardness Number, 100-kgf (HRB) | Vickers Hardness Number (HV) | Brinell Hardness Number, 3000-kgf (HBS) | Knoop Hardness Number, 500-gf, and Over (HK) | Rockwell A Hardness Number, 60-kgf (HRA) | Rockwell F Hardness Number, 60-kgf (HRF) | Rockwell Superficial Hardness Number | | | Rockwell B Hardness Number, 100-kgf (HRB) |
						15-T Scale, 15-kgf (HR 15-T)	30-T Scale, 30-kgf (HR 30-T)	45-T Scale, 45-kgf (HR 45-T)	
100	240	240	251	61.5	...	93.1	83.1	72.9	100
99	234	234	246	60.9	...	92.8	82.5	71.9	99
98	228	228	241	60.2	...	92.5	81.8	70.9	98
97	222	222	236	59.5	...	92.1	81.1	69.9	97
96	216	216	231	58.9	...	91.8	80.4	68.9	96
95	210	210	226	58.3	...	91.5	79.8	67.9	95
94	205	205	221	57.6	...	91.2	79.1	66.9	94
93	200	200	216	57.0	...	90.8	78.4	65.9	93
92	195	195	211	56.4	...	90.5	77.8	64.8	92
91	190	190	206	55.8	...	90.2	77.1	63.8	91
90	185	185	201	55.2	...	89.9	76.4	62.8	90
89	180	180	196	54.6	...	89.5	75.8	61.8	89
88	176	176	192	54.0	...	89.2	75.1	60.8	88
87	172	172	188	53.4	...	88.9	74.4	59.8	87
86	169	169	184	52.8	...	88.6	73.8	58.8	86
85	165	165	180	52.3	...	88.2	73.1	57.8	85
84	162	162	176	51.7	...	87.9	72.4	56.8	84
83	159	159	173	51.1	...	87.6	71.8	55.8	83
82	156	156	170	50.6	...	87.3	71.1	54.8	82
81	153	153	167	50.0	...	86.9	70.4	53.8	81
80	150	150	164	49.5	...	86.6	69.7	52.8	80
79	147	147	161	48.9	...	86.3	69.1	51.8	79
78	144	144	158	48.4	...	86.0	68.4	50.8	78
77	141	141	155	47.9	...	85.6	67.7	49.8	77
76	139	139	152	47.3	...	85.3	67.1	48.8	76
75	137	137	150	46.8	99.6	85.0	66.4	47.8	75
74	135	135	147	46.3	99.1	84.7	65.7	46.8	74
73	132	132	145	45.8	98.5	84.3	65.1	45.8	73
72	130	130	143	45.3	98.0	84.0	64.4	44.8	72
71	127	127	141	44.8	97.4	83.7	63.7	43.8	71
70	125	125	139	44.3	96.8	83.4	63.1	42.8	70
69	123	123	137	43.8	96.2	83.0	62.4	41.8	69
68	121	121	135	43.3	95.6	82.7	61.7	40.8	68
67	119	119	133	42.8	95.1	82.4	61.0	39.8	67
66	117	117	131	42.3	94.5	82.1	60.4	38.7	66
65	116	116	129	41.8	93.9	81.8	59.7	37.7	65
64	114	114	127	41.4	93.4	81.4	59.0	36.7	64
63	112	112	125	40.9	92.8	81.1	58.4	35.7	63
62	110	110	124	40.4	92.2	80.8	57.7	34.7	62
61	108	108	122	40.0	91.7	80.5	57.0	33.7	61
60	107	107	120	39.5	91.1	80.1	56.4	32.7	60

TABLE 2 *Continued*

Rockwell B Hardness Number, 100-kgf, (HRB)	Vickers Hardness Number (HV)	Brinell Hardness Number, 3000-kgf, 10-mm Ball	Knoop Hardness Number, 500-gf and Over	Rockwell A Hardness Number, 60-kgf, Diamond Penetrator	Rockwell F Hardness Number, 60-kgf, 1/16-in. (1.588-mm) Ball	Rockwell Superficial Hardness Number			Rockwell B Hardness Number, 100-kgf, 1/16-in. (1.588-mm) Ball
						15-T Scale, 15-kgf, 1/16-in. (1.588-mm) Ball	30-T Scale, 30-kgf, 1/16-in. (1.588-mm) Ball	45-T Scale, 45-kgf, 1/16-in. (1.588-mm) Ball	
59	106	106	118	39.0	90.5	79.8	55.7	31.7	59
58	104	104	117	38.6	90.0	79.5	55.0	30.7	58
57	103	103	115	38.1	89.4	79.2	54.4	29.7	57
56	101	101	114	37.7	88.8	78.8	53.7	28.7	56
55	100	100	112	37.2	88.2	78.5	53.0	27.7	55
54	...	...	111	36.8	87.7	78.2	52.4	26.7	54
53	...	...	110	36.3	87.1	77.9	51.7	25.7	53
52	...	...	109	35.9	86.5	77.5	51.0	24.7	52
51	...	...	108	35.5	86.0	77.2	50.3	23.7	51
50	...	...	107	35.0	85.4	76.9	49.7	22.7	50
49	...	...	106	34.6	84.8	76.6	49.0	21.7	49
48	...	...	105	34.1	84.3	76.2	48.3	20.7	48
47	...	...	104	33.7	83.7	75.9	47.7	19.7	47
46	...	...	103	33.3	83.1	75.6	47.0	18.7	46
45	...	...	102	32.9	82.6	75.3	46.3	17.7	45
44	...	...	101	32.4	82.0	74.9	45.7	16.7	44
43	...	...	100	32.0	81.4	74.6	45.0	15.7	43
42	...	...	99	31.6	80.8	74.3	44.3	14.7	42
41	...	...	98	31.2	80.3	74.0	43.7	13.6	41
40	...	...	97	30.7	79.7	73.6	43.0	12.6	40
39	...	...	96	30.3	79.1	73.3	42.3	11.6	39
38	...	...	95	29.9	78.6	73.0	41.6	10.6	38
37	...	...	94	29.5	78.0	72.7	41.0	9.6	37
36	...	...	93	29.1	77.4	72.3	40.3	8.6	36
35	...	...	92	28.7	76.9	72.0	39.6	7.6	35
34	...	...	91	28.2	76.3	71.7	39.0	6.6	34
33	...	...	90	27.8	75.7	71.4	38.3	5.6	33
32	...	...	89	27.4	75.2	71.0	37.6	4.6	32
31	...	...	88	27.0	74.6	70.7	37.0	3.6	31
30	...	...	87	26.6	74.0	70.4	36.3	2.6	30

[A] In table headings, kgf refers to total test force.
[B] Appendix X2 contains equations converting determined hardness numbers to Rockwell B hardness numbers for non-austenitic steels. Refer to 1.11 before using conversion equations.

TABLE 3 Approximate Hardness Conversion Numbers for Nickel and High-Nickel Alloys[A,B,C]

NOTE 1—See Supplement to Table 3.

NOTE 2—The use of hardness scales for hardness values shown in parentheses is not recommended since they are beyond the ranges recommended for accuracy. Such values are shown for comparative purposes only, where comparisons may be desired and the recommended machine and scale are not available.

Vickers Hardness Number	Brinell Hardness Number	Rockwell Hardness Number									Rockwell Superficial Hardness Number				
Vickers Indenter 1, 5, 10, 30-kgf (HV)	10-mm Standard Ball, 3000-kgf (HBS)	A Scale 60-kgf Diamond Penetrator (HRA)	B Scale 100-kgf 1/16-in. (1.588-mm) Ball (HRB)	C Scale 150-kgf Diamond Penetrator (HRC)	D Scale 100-kgf Diamond Penetrator (HRD)	E Scale 100-kgf 1/8-in. (3.175-mm) Ball (HRE)	F Scale 60-kgf 1/16-in. (1.588-mm) Ball (HRF)	G Scale 150-kgf 1/16-in. (1.588-mm) Ball (HRG)	K Scale 150-kgf 1/8-in. (3.175-mm) Ball (HRK)	15-N Scale 15-kgf Superficial Diamond Penetrator (HR 15-N)	30-N Scale 30-kgf Superficial Diamond Penetrator (HR 30-N)	45-N Scale 45-kgf Superficial Diamond Penetrator (HR 45-N)	15-T Scale 15-kgf 1/16-in. (1.588-mm) Ball (HR 15-T)	30-T Scale 30-kgf 1/16-in. (1.588-mm) Ball (HR 30-T)	45-T Scale 45-kgf 1/16-in. (1.588-mm) Ball (HR 45-T)
513	(479)	75.5	...	50.0	63.0	...	...	...	...	85.5	68.0	54.5	...	...	...
481	450	74.5	...	48.0	61.5	...	...	...	...	84.5	66.5	52.5	...	...	...
452	425	73.5	...	46.0	60.0	...	...	...	...	83.5	64.5	50.0	...	...	...
427	403	72.5	...	44.0	58.5	...	...	...	...	82.5	63.0	47.5	...	...	...
404	382	71.5	...	42.0	57.0	...	...	...	...	81.5	61.0	45.5	...	...	...
382	363	70.5	...	40.0	55.5	...	...	...	...	80.5	59.5	43.0	...	...	...
362	346	69.5	...	38.0	54.0	...	...	...	...	79.5	58.0	41.0	...	...	...
344	329	68.5	...	36.0	52.5	...	...	...	...	78.5	56.0	38.5	...	...	...
326	313	67.5	...	34.0	50.5	...	...	...	...	77.5	54.5	36.0	...	...	...
309	298	66.5	(106)	32.0	49.5	...	(116.5)	94.0	...	76.5	52.5	34.0	94.5	85.5	77.0
285	275	64.5	(104)	28.5	46.5	...	(115.5)	91.0	...	75.0	49.5	30.0	94.0	84.5	75.0
266	258	63.0	(102)	25.5	44.5	...	(114.5)	87.5	...	73.5	47.0	26.5	93.0	83.0	73.0
248	241	61.5	100	22.5	42.0	...	(113.0)	84.5	...	72.0	44.5	23.0	92.5	81.5	71.0
234	228	60.5	98	20.0	40.0	...	(112.0)	81.5	...	70.5	42.0	20.0	92.0	80.5	69.0
220	215	59.0	96	(17.0)	38.0	...	(111.0)	78.5	100.0	69.0	39.5	17.0	91.0	79.0	67.0
209	204	57.5	94	(14.5)	36.0	...	(110.0)	75.5	98.0	68.0	37.5	14.0	90.5	77.5	65.0
198	194	56.5	92	(12.0)	34.0	...	(108.5)	72.0	96.5	66.5	35.5	11.0	89.5	76.0	63.0
188	184	55.0	90	(9.0)	32.0	(108.5)	(107.5)	69.0	94.5	65.0	32.5	7.5	89.0	75.0	61.0
179	176	53.5	88	(6.5)	30.0	(107.0)	(106.5)	65.5	93.0	64.0	30.5	5.0	88.0	73.5	59.5
171	168	52.5	86	(4.0)	28.0	(106.0)	(105.0)	62.5	91.0	62.5	28.5	2.0	87.5	72.0	57.5
164	161	51.5	84	(2.0)	26.5	(104.5)	(104.0)	59.5	89.0	61.5	26.5	(-0.5)	87.0	70.5	55.5
157	155	50.0	82	...	24.5	(103.0)	(103.0)	56.5	87.5	...	...	...	86.0	69.5	53.5
151	149	49.0	80	...	22.5	(102.0)	(101.5)	53.0	85.5	...	...	...	85.5	68.0	51.5
145	144	47.5	78	...	21.0	(100.5)	(100.5)	50.0	83.5	...	...	...	84.5	66.5	49.5
140	139	46.5	76	...	(19.0)	99.5	99.5	47.0	82.0	...	...	...	84.0	65.5	47.5
135	134	45.5	74	...	(17.5)	98.0	98.5	43.5	80.0	...	...	...	83.0	64.0	45.5
130	129	44.0	72	...	(16.0)	97.0	97.0	40.5	78.0	...	...	...	82.5	62.5	43.5
126	125	43.0	70	...	(14.5)	95.5	96.0	37.5	76.5	...	...	...	82.0	61.0	41.5
122	121	42.0	68	...	(13.0)	94.5	95.0	34.5	74.5	...	...	...	81.0	60.0	39.5
119	118	41.0	66	...	(11.5)	93.0	93.5	31.0	72.5	...	...	...	80.5	58.5	37.5
115	114	40.0	64	...	(10.0)	91.5	92.5	...	71.0	...	...	...	79.5	57.0	35.5
112	111	39.0	62	...	(8.0)	90.5	91.5	...	69.0	...	...	...	79.0	56.0	33.5
108	108	...	60	...	...	89.0	90.0	...	67.5	...	...	...	78.5	54.5	31.5
106	106	...	58	...	...	88.0	89.0	...	65.5	...	...	...	78.0	53.0	29.5
103	103	...	56	...	...	86.5	88.0	...	63.5	...	...	...	77.5	51.5	27.5
100	100	...	54	...	...	85.5	87.0	...	62.0	...	...	...	77.0	50.5	25.5
98	98	...	52	...	...	84.0	85.5	...	60.0	...	...	...	76.0	49.0	23.5
95	95	...	50	...	...	83.0	84.5	...	58.0	...	...	...	74.5	47.5	21.5

TABLE 3 Continued

Vickers Hardness Number	Brinell Hardness Number	Rockwell Hardness Number								Rockwell Superficial Hardness Number					
Vickers Indenter 1, 5, 10, 30-kgf (HV)	10-mm Standard Ball, 3000-kgf (HBS)	A Scale 60-kgf Diamond Penetrator (HRA)	B Scale 100-kgf 1/16-in. (1.588-mm) Ball (HRB)	C Scale 150-kgf Diamond Penetrator (HRC)	D Scale 100-kgf Diamond Penetrator (HRD)	E Scale 100-kgf 1/8-in. (3.175-mm) Ball (HRE)	F Scale 60-kgf 1/16-in. (1.588-mm) Ball (HRF)	G Scale 150-kgf 1/16-in. (1.588-mm) Ball (HRG)	K Scale 150-kgf 1/8-in. (3.175-mm) Ball (HRK)	15-N Scale 15-kgf Superficial Diamond Penetrator (HR 15-N)	30-N Scale 30-kgf Superficial Diamond Penetrator (HR 30-N)	45-N Scale 45-kgf Superficial Diamond Penetrator (HR 45-N)	15-T Scale 15-kgf 1/16-in. (1.588-mm) Ball (HR 15-T)	30-T Scale 30-kgf 1/16-in. (1.588-mm) Ball (HR 30-T)	45-T Scale 45-kgf 1/16-in. (1.588-mm) Ball (HR 45-T)
93	93	...	48	...	...	81.5	83.5	...	56.5	...	...	...	74.0	46.5	19.5
91	91	...	46	...	...	80.5	82.0	...	54.5	...	...	...	73.5	45.0	17.0
89	89	...	44	...	...	79.0	81.0	...	52.5	...	...	...	72.5	43.5	14.5
87	87	...	42	...	...	78.0	80.0	...	51.0	...	...	...	72.0	42.0	12.5
85	85	...	40	...	...	76.5	79.0	...	49.0	...	...	...	71.0	41.0	10.0
83	83	...	38	...	...	75.0	77.5	...	47.0	...	...	...	70.5	39.5	7.5
81	81	...	36	...	...	74.0	76.5	...	45.5	...	...	...	70.0	38.0	5.5
79	79	...	34	...	...	72.5	75.5	...	43.5	...	...	...	69.0	36.5	3.0
78	78	...	32	...	...	71.5	74.0	...	42.0	...	...	...	68.5	35.5	1.0
77	77	...	30	...	...	70.0	73.0	...	40.0	...	...	...	67.5	34.0	(−1.5)

TABLE 3 *Continued*

Vickers Hardness Number	Knoop Hardness Number
Vickers Indenter 1,5,10,30-kgf (HV)	Knoop Indenter 500 and 1000-gf (HK)
382	436
362	413
344	392
326	372
309	352
285	325
266	304
248	283
234	267
220	251
209	239
198	226
188	215
179	204
171	195
164	187
157	179
151	173
145	166
140	160
135	154
130	149
126	144
122	140
119	136

[A] In table headings, kgf or gf refers to total test force.
[B] Appendix X3 contains equations converting determined hardness scale numbers to Vickers hardness numbers for nickel and high-nickel alloys. Refer to 1.11 before using conversion equations.
[C] Note that in Table 5 of Test Method E10 (appears in the *Annual Book of ASTM Standards*, Vol 03.01), the use of a 3000-kgf force is recommended (but not mandatory) for material in the hardness range from 96 to 600 HV, and a 1500-kgf force is recommended (but not mandatory) for material in the hardness range from 48 to 300 HV. These recommendations are designed to limit impression diameters to the range from 2.50 to 6.0 mm. The Brinell hardness numbers in this conversion table are based on tests using a 3000-kgf force. When the 1500-kgf force is used for the softer nickel and high-nickel alloys, these conversion relationships do not apply.

TABLE 4 Approximate Hardness Conversion Numbers for Cartridge Brass (70 % Copper 30 % Zinc Alloy)[A,B]

Vickers Hardness Number (HV)	Rockwell Hardness Number		Rockwell Superficial Hardness Number			Brinell Hardness Number
	B Scale, 100-kgf, 1/16-in. (1.588-mm) Ball (HRB)	F Scale, 60-kgf 1/16-in. (1.588-mm) Ball (HRF)	15-T Scale, 15-kgf, 1/16-in. (1.588-mm) Ball (HR 15-T)	30-T Scale, 30-kgf, 1/16-in. (1.588-mm) Ball (HR 30-T)	45-T Scale, 45-kgf, 1/16-in. (1.588-mm) Ball (HR 45-T)	500-kgf, 10-mm Ball (HBS)
196	93.5	110.0	90.0	77.5	66.0	169
194	...	109.5	...	...	65.5	167
192	93.0	...	...	77.0	65.0	166
190	92.5	109.0	...	76.5	64.5	164
188	92.0	...	89.5	...	64.0	162
186	91.5	108.5	...	76.0	63.5	161
184	91.0	...	...	75.5	63.0	159
182	90.5	108.0	89.0	...	62.5	157
180	90.0	107.5	...	75.0	62.0	156
178	89.0	...	...	74.5	61.5	154
176	88.5	107.0	...	...	61.0	152
174	88.0	...	88.5	74.0	60.5	150
172	87.5	106.5	...	73.5	60.0	149
170	87.0	...	...	...	59.5	147
168	86.0	106.0	88.0	73.0	59.0	146
166	85.5	...	...	72.5	58.5	144
164	85.0	105.5	...	72.0	58.0	142
162	84.0	105.0	87.5	...	57.5	141
160	83.5	...	...	71.5	56.5	139
158	83.0	104.5	...	71.0	56.0	138
156	82.0	104.0	87.0	70.5	55.5	136
154	81.5	103.5	...	70.0	54.5	135
152	80.5	103.0	...	...	54.0	133
150	80.0	...	86.5	69.5	53.5	131
148	79.0	102.5	...	69.0	53.0	129
146	78.0	102.0	...	68.5	52.5	128
144	77.5	101.5	86.0	68.0	51.5	126
142	77.0	101.0	...	67.5	51.0	124
140	76.0	100.5	85.5	67.0	50.0	122
138	75.0	100.0	...	66.5	49.0	121
136	74.5	99.5	85.0	66.0	48.0	120
134	73.5	99.0	...	65.5	47.5	118
132	73.0	98.5	84.5	65.0	46.5	116
130	72.0	98.0	84.0	64.5	45.5	114
128	71.0	97.5	...	63.5	45.0	113
126	70.0	97.0	83.5	63.0	44.0	112
124	69.0	96.5	...	62.5	43.0	110
122	68.0	96.0	83.0	62.0	42.0	108
120	67.0	95.5	...	61.0	41.0	106
118	66.0	95.0	82.5	60.5	40.0	105
116	65.0	94.5	82.0	60.0	39.0	103
114	64.0	94.0	81.5	59.5	38.0	101
112	63.0	93.0	81.0	58.5	37.0	99
110	62.0	92.6	80.5	58.0	35.5	97
108	61.0	92.0	...	57.0	34.5	95
106	59.5	91.2	80.0	56.0	33.0	94
104	58.0	90.5	79.5	55.0	32.0	92
102	57.0	89.8	79.0	54.5	30.5	90
100	56.0	89.0	78.5	53.5	29.5	88
98	54.0	88.0	78.0	52.5	28.0	86
96	53.0	87.2	77.5	51.5	26.5	85
94	51.0	86.3	77.0	50.5	24.5	83
92	49.5	85.4	76.5	49.0	23.0	82
90	47.5	84.4	75.5	48.0	21.0	80
88	46.0	83.5	75.0	47.0	19.0	79
86	44.0	82.3	74.5	45.5	17.0	77
84	42.0	81.2	73.5	44.0	14.5	76
82	40.0	80.0	73.0	43.0	12.5	74
80	37.5	78.6	72.0	41.0	10.0	72
78	35.0	77.4	71.5	39.5	7.5	70
76	32.5	76.0	70.5	38.0	4.5	68
74	30.0	74.8	70.0	36.0	1.0	66
72	27.5	73.2	69.0	34.0	...	64
70	24.5	71.8	68.0	32.0	...	63
68	21.5	70.0	67.0	30.0	...	62
66	18.5	68.5	66.0	28.0	...	61
64	15.5	66.8	65.0	25.5	...	59
62	12.5	65.0	63.5	23.0	...	57

TABLE 4 Continued

Vickers Hardness Number (HV)	Rockwell Hardness Number		Rockwell Superficial Hardness Number			Brinell Hardness Number
	B Scale, 100-kgf, 1/16-in. (1.588-mm) Ball (HRB)	F Scale, 60-kgf 1/16-in. (1.588-mm) Ball (HRF)	15-T Scale, 15-kgf, 1/16-in. (1.588-mm) Ball (HR 15-T)	30-T Scale, 30-kgf, 1/16-in. (1.588-mm) Ball (HR 30-T)	45-T Scale, 45-kgf, 1/16-in. (1.588-mm) Ball (HR 45-T)	500-kgf, 10-mm Ball (HBS)
60	10.0	62.5	62.5	...	...	55
58	...	61.0	61.0	18.0	...	53
56	...	58.8	60.0	15.0	...	52
54	...	56.5	58.5	12.0	...	50
52	...	53.5	57.0	...	...	48
50	...	50.5	55.5	...	...	47
49	...	49.0	54.5	...	...	46
48	...	47.0	53.5	...	...	45
47	...	45.0	...	...	...	44
46	...	43.0	...	...	...	43
45	...	40.0	...	...	...	42

[A] In table headings, kgf or gf refers to total test force.
[B] Appendix X4 contains equations converting determined hardness scale numbers to Vickers hardness numbers for cartridge brass. Refer to 1.11 before using conversion equations.

TABLE 5 Approximate Brinell-Rockwell B Hardness Conversion Numbers for Austenitic Stainless Steel Plate in Annealed Condition[A,B]

Rockwell Hardness Number, B Scale (100-kgf, 1/16-in. (1.588-mm) ball) (HRB)	Brinell Hardness Number (3000-kgf, 10-mm ball) (HBS)
100	256
99	248
98	240
97	233
96	226
95	219
94	213
93	207
92	202
91	197
90	192
89	187
88	183
87	178
86	174
85	170
84	167
83	163
82	160
81	156
80	153
79	150
78	147
77	144
76	142
75	139
74	137
73	135
72	132
71	130
70	128
69	126
68	124
67	122
66	120
65	118
64	116
63	114
62	113
61	111
60	110

[A] In table headings, kgf or gf refers to total test force.
[B] Appendix X5 contains an equation converting determined Brinell hardness numbers to Rockwell B hardness numbers for austenitic steel plate in the annealed condition. Refer to 1.11 before using this conversion equation.

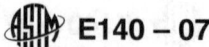

TABLE 6 Approximate Rockwell Hardness Conversion Numbers for Austenitic Stainless Steel Sheet[A,B]

NOTE 1—These conversions are based on interlaboratory tests conducted on the following grades: Types 201, 202, 301, 302, 304, 304L, 305, 316, 316L, 321, and 347. Tempers ranged from annealed to extra hard for Type 301, with a smaller range of tempers for the other types. Test coupon thicknesses ranged from approximately 0.1 in. (2.5 mm) to 0.050 in. (1.27 mm).

Rockwell Hardness Number			Rockwell Superficial Hardness Number		
C Scale, 150-kgf, Diamond Penetrator (HRC)	A Scale, 60-kgf, Diamond Penetrator (HRA)	15-N Scale, 15-kgf, Superficial Diamond Penetrator (HR 15-N)	30-N Scale, 30-kgf, Superficial Diamond Penetrator (HR 30-N)	45-N Scale, 45-kgf, Superficial Diamond Penetrator (HR 45-N)	
48	74.4	84.1	66.2	52.1	
47	73.9	83.6	65.3	50.9	
46	73.4	83.1	64.5	49.8	
45	72.9	82.6	63.6	48.7	
44	72.4	82.1	62.7	47.5	
43	71.9	81.6	61.8	46.4	
42	71.4	81.0	61.0	45.2	
41	70.9	80.5	60.1	44.1	
40	70.4	80.0	59.2	43.0	
39	69.9	79.5	58.4	41.8	
38	69.3	79.0	57.5	40.7	
37	68.8	78.5	56.6	39.6	
36	68.3	78.0	55.7	38.4	
35	67.8	77.5	54.9	37.3	
34	67.3	77.0	54.0	36.1	
33	66.8	76.5	53.1	35.0	
32	66.3	75.9	52.3	33.9	
31	65.8	75.4	51.4	32.7	
30	65.3	74.9	50.5	31.6	
29	64.8	74.4	49.6	30.4	
28	64.3	73.9	48.8	29.3	
27	63.8	73.4	47.9	28.2	
26	63.3	72.9	47.0	27.0	
25	62.8	72.4	46.2	25.9	
24	62.3	71.9	45.3	24.8	
23	61.8	71.3	44.4	23.6	
22	61.3	70.8	43.5	22.5	
21	60.8	70.3	42.7	21.3	
20	60.3	69.8	41.8	20.2	
B Scale, 100-kgf, 1/16-in. (1.588-mm) Ball (HRB)	A Scale, 60-kgf, Diamond Penetrator (HRA)	F Scale, 60-kgf, 1/16-in. (1.588-mm) Ball[C] (HRF)	15-T Scale, 15-kgf, 1/16-in. (1.588-mm) Ball (HR 15-T)	30-T Scale, 30-kgf, 1/16-in. (1.588-mm) Ball (HR 30-T)	45-T Scale, 45-kgf, 1/16-in. (1.588-mm) Ball (HR 45-T)
100	61.5	(113.9)	91.5	80.4	70.2
99	60.9	(113.2)	91.2	79.7	69.2
98	60.3	(112.5)	90.8	79.0	68.2
97	59.7	(111.8)	90.4	78.3	67.2
96	59.1	(111.1)	90.1	77.7	66.1
95	58.5	(110.5)	89.7	77.0	65.1
94	58.0	(109.8)	89.3	76.3	64.1
93	57.4	(109.1)	88.9	75.6	63.1
92	56.8	(108.4)	88.6	74.9	62.1
91	56.2	(107.8)	88.2	74.2	61.1
90	55.6	(107.1)	87.8	73.5	60.1
89	55.0	(106.4)	87.5	72.8	59.0
88	54.5	(105.7)	87.1	72.1	58.0
87	53.9	(105.0)	86.7	71.4	57.0
86	53.3	(104.4)	86.4	70.7	56.0
85	52.7	(103.7)	86.0	70.0	55.0
84	52.1	(103.0)	85.6	69.3	54.0
83	51.5	(102.3)	85.2	68.6	52.9
82	50.9	(101.7)	84.9	67.9	51.9
81	50.4	(101.0)	84.5	67.2	50.9
80	49.8	(100.3)	84.1	66.5	49.9
79	49.2	99.6	83.8	65.8	48.9
78	48.6	99.0	83.4	65.1	47.9
77	48.0	98.3	83.0	64.4	46.8
76	47.4	97.6	82.6	63.7	45.8
75	46.9	96.9	82.3	63.0	44.8
74	46.3	96.2	81.9	62.4	43.8
73	45.7	95.6	81.5	61.7	42.8
72	45.1	94.9	81.2	61.0	41.8
71	44.5	94.2	80.8	60.3	40.7
70	43.9	93.5	80.4	59.6	39.7

TABLE 6 *Continued*

B Scale, 100-kgf, 1/16-in. (1.588-mm) Ball (HRB)	A Scale, 60-kgf, Diamond Penetrator (HRA)	F Scale, 60-kgf, 1/16-in. (1.588-mm) Ball[C] (HRF)	15-T Scale, 15-kgf, 1/16-in. (1.588-mm) Ball (HR 15-T)	30-T Scale, 30-kgf, 1/16-in. (1.588-mm) Ball (HR 30-T)	45-T Scale, 45-kgf, 1/16-in. (1.588-mm) Ball (HR 45-T)
69	43.3	92.8	80.1	58.9	38.7
68	42.8	92.2	79.7	58.2	37.7
67	42.2	91.5	79.3	57.5	36.7
66	41.6	90.8	78.9	56.8	35.7
65	41.0	90.1	78.6	56.1	34.7
64	40.4	89.5	78.2	55.4	33.6
63	39.8	88.8	77.8	54.7	32.6
62	39.3	88.1	77.5	54.0	31.6
61	38.7	87.4	77.1	53.3	30.6
60	38.1	86.8	76.7	52.6	29.6
Standard deviation[C]	1.44	2.75	2.29	1.67	1.57

[A] In table headings, kgf or gf refers to total test force.
[B] Appendix X6 contains equations converting determined hardness numbers to Rockwell C and Rockwell B hardness numbers for austenitic stainless steel sheet. Refer to 1.11 before using conversion equations.
[C] Observed standard deviation of the interlaboratory test data about the indicated conversion line.

TABLE 7 Approximate Hardness Conversion Numbers for Copper, No. 102 to 142 Inclusive[A,B]

Vickers Hardness Number		Knoop Hardness Number		Rockwell Superficial Hardness Number			Rockwell Hardness Number		Rockwell Superficial Hardness Number			Brinell Hardness Number	
1-kgf (HV)	100-gf (HV)	1-kgf (HK)	500-gf (HK)	15-T Scale, 15-kgf 1/16-in. (1.588-mm) Ball (HR 15-T)	15-T Scale, 15-kgf 1/16-in. (1.588-mm) Ball (HR 15-T)	30-T Scale, 30-kgf 1/16-in. (1.588-mm) Ball (HR 30-T)	B Scale, 100-kgf 1/16-in. (1.588-mm) Ball (HRB)	F Scale, 60-kgf 1/16-in. (1.588-mm) Ball (HRF)	15-T Scale, 15-kgf 1/16-in. (1.588-mm) Ball (HR 15-T)	30-T Scale, 30-kgf 1/16-in. (1.588-mm) Ball (HR 30-T)	45-T Scale, 45-kgf 1/16-in. (1.588-mm) Ball (HR 45-T)	500-kgf, 10-mm Diameter Ball (HBS)	20-kgf 2-mm Diameter Ball (HBS)
				0.010-in. (0.25-mm) Strip	0.020-in. (0.51-mm) Strip	0.040-in. (1.02-mm) Strip and Greater						0.080-in. (2.03-mm) Strip	0.040-in. (1.02-mm) Strip
130	127.0	138.7	133.8	...	85.0	...	67.0	99.0	...	69.5	49.0	...	119.0
128	125.2	136.8	132.1	83.0	84.5	...	66.0	98.0	87.0	68.5	48.0	...	117.5
126	123.6	134.9	130.4	...	84.0	...	65.0	97.0	...	67.5	46.5	120.0	115.0
124	121.9	133.0	128.7	82.5	83.5	...	64.0	96.0	86.0	66.5	45.0	117.5	113.0
122	121.1	131.0	127.0	...	83.0	...	62.5	95.5	85.5	66.0	44.0	115.0	111.0
120	118.5	129.0	125.2	82.0	82.5	...	61.0	95.0	...	65.0	42.5	112.0	109.0
118	116.8	127.1	123.5	81.5	...	...	59.5	94.0	85.0	64.0	41.0	110.0	107.5
116	115.0	125.1	121.7	...	82.0	...	58.5	93.0	...	63.0	40.0	107.0	105.5
114	113.5	123.2	119.9	81.0	81.5	...	57.0	92.5	84.5	62.0	38.5	105.0	103.5
112	111.8	121.4	118.1	80.5	81.0	...	55.0	91.5	...	61.0	37.0	102.0	102.0
110	109.9	119.5	116.3	80.0	...	...	53.5	91.0	84.0	60.0	36.0	99.5	100.0
108	108.3	117.5	114.5	...	80.5	...	52.0	90.5	83.5	59.0	34.5	97.0	98.0
106	106.6	115.6	112.6	79.5	80.0	...	50.0	89.5	...	58.0	33.0	94.5	96.0
104	104.9	113.5	110.1	79.0	79.5	...	48.0	88.5	83.0	57.0	32.0	92.0	94.0
102	103.2	111.5	108.0	78.5	79.0	...	46.5	87.5	82.5	56.0	30.0	89.5	92.0
100	101.5	109.4	106.0	78.0	78.0	...	44.5	87.0	82.0	55.0	28.5	87.0	90.0
98	99.8	107.3	104.0	77.5	77.5	...	42.0	85.5	81.0	53.5	26.5	84.5	88.0
96	98.0	105.3	102.1	77.0	77.0	...	40.0	84.5	80.5	52.0	25.5	82.0	86.5
94	96.4	103.2	100.0	76.5	76.5	...	38.0	83.0	80.0	51.0	23.0	79.5	85.0
92	94.7	101.0	98.0	76.0	75.5	...	35.5	82.0	79.0	49.0	21.0	77.0	83.0
90	93.0	98.9	96.0	75.5	75.0	...	33.0	81.0	78.0	47.5	19.0	74.5	81.0
88	91.2	96.9	94.0	75.0	74.5	...	30.5	79.5	77.0	46.0	16.5	...	79.0
86	89.7	95.5	92.0	74.5	73.5	...	28.0	78.0	76.0	44.0	14.0	...	77.0
84	87.9	92.3	90.0	74.0	73.0	...	25.5	76.5	75.0	43.0	12.0	...	75.0
82	86.1	90.1	87.9	73.5	72.0	...	23.0	74.5	74.5	41.0	9.5	...	73.0
80	84.5	87.9	86.0	72.5	71.0	...	20.0	73.0	73.5	39.5	7.0	...	71.5
78	82.8	85.7	84.0	72.0	70.0	...	17.0	71.0	72.5	37.5	5.0	...	69.5
76	81.0	83.5	81.9	71.5	69.5	...	14.5	69.0	71.5	36.0	2.0	...	67.5
74	79.2	81.1	79.9	71.0	68.5	...	11.5	67.5	70.0	34.0	...	...	66.0
72	77.6	78.9	78.7	70.0	67.5	...	8.5	66.0	69.0	32.0	...	...	64.0
70	75.8	76.8	76.6	69.5	66.5	...	5.0	64.0	67.5	30.0	...	...	62.0
68	74.3	74.1	74.4	69.0	65.5	...	2.0	62.0	66.0	28.0	...	...	60.5
66	72.6	71.9	71.9	68.0	64.5	...	...	60.0	64.5	25.5	...	...	58.5
64	70.9	69.5	70.0	67.5	63.5	...	...	58.0	63.5	23.5	...	...	57.0
62	69.1	67.0	67.9	66.5	62.0	...	...	56.0	61.0	21.0	...	...	55.0
60	67.5	64.6	65.9	66.0	61.0	...	...	54.0	59.0	18.0	...	...	53.0
58	65.8	62.0	63.8	65.0	60.0	...	...	51.5	57.0	15.5	...	...	51.5
56	64.0	59.8	61.8	64.5	58.5	...	...	49.0	55.0	13.0	...	...	49.5
54	62.3	57.4	59.5	63.5	57.5	...	...	47.0	53.0	10.0	...	...	48.0
52	60.7	55.0	57.2	63.0	56.0	...	...	44.0	51.5	7.5	...	...	46.5
50	58.9	52.8	55.0	62.0	55.0	...	...	41.5	49.5	4.5	...	...	44.5
48	57.3	50.3	52.7	61.0	53.5	...	...	39.0	47.5	1.5	...	...	42.0
46	55.8	48.0	50.2	60.5	52.0	...	...	36.0	45.0	...	...	...	41.0
44	53.9	45.9	47.8	59.5	51.0	...	...	33.5	43.0	...	...	...	...
42	52.2	43.7	45.2	58.5	49.5	...	...	30.5	41.0	...	...	...	...
40	51.3	40.2	42.8	57.5	48.0	...	...	28.0	38.5	...	...	...	...

[A] In table headings, kgf or gf refers to total test force.
[B] Appendix X7 contains equations converting determined hardness scale numbers to Vickers hardness numbers for copper, numbers 102 to 142 inclusive. Refer to 1.11 before using conversion equations.

TABLE 8 Approximate Hardness Conversion Numbers for Alloyed White Irons [A, B, C]

Vickers Hardness, HV 50	Brinell Hardness, HBW	Rockwell C Hardness, HRC	Vickers Hardness, HV 50	Brinell[D] Hardness, HBW	Rockwell C Hardness, HRC
1000	(903)[E]	70	680	621	57
980	(886)	69	660	604	56
960	(868)	68	640	586	55
940	(850)	68	620	569	54
920	(833)	67	600	551	53
900	(815)	66	580	533	52
880	(798)	66	560	516	51
860	(780)	65	540	498	50
840	(762)	64	520	481	48
820	(745)	63	500	463	47
800	(727)	62	480	445	45
780	(710)	62	460	428	44
760	(692)	61	440	410	42
740	(674)	60	420	393	40
720	(657)	59	400	375	38
700	(639)	58	380	357	35

[A] Data were generated in an interlaboratory comparison program conducted by American Foundrymen's Society Special Irons Subcommittee, 5-D. Supporting data available on loan from ASTM Headquarters. Request E28-1003.
[B] In table headings, kgf or gf refers to total test force.
[C] Appendix X8 contains equations converting determined hardness scale numbers to Vickers hardness numbers for alloyed white irons. Refer to 1.11 before using conversion equations.
[D] Ten-millimetre tungsten carbide ball.
[E] Brinell hardness numbers in parentheses are above the maximum hardness recommended by Test Method E10 and are presented for information only.

TABLE 9 Approximate Hardness Conversion Numbers for Wrought Aluminum Products [A, B, C]

Brinell Hardness Number 500-kgf, (10-mm Ball) (HBS)	Vickers Hardness Number 15-kgf, (HV)	Rockwell Hardness Number			Rockwell Superficial Hardness Number		
		B Scale 100-kgf, 1/16-in. Ball (HRB)	E Scale 100-kgf, 1/8-in. Ball (HRE)	H Scale 60-kgf, 1/8-in. Ball (HRH)	15-T Scale 15-kgf, 1/16-in. Ball (HR 15-T)	30-T Scale 30-kgf, 1/16-in. Ball (HR 30-T)	15-W Scale 15-kgf, 1/8-in. Ball (HR 15-W)
160	189	91	...	...	89	77	95
155	183	90	...	...	89	76	95
150	177	89	...	...	89	75	94
145	171	87	...	...	88	74	94
140	165	86	...	...	88	73	94
135	159	84	...	...	87	71	93
130	153	81	...	...	87	70	93
125	147	79	...	...	86	68	92
120	141	76	...	...	86	67	92
115	135	72	101	...	86	65	91
110	129	69	100	...	85	63	91
105	123	65	99	...	84	61	91
100	117	60	98	...	83	59	90
95	111	56	96	...	82	57	90
90	105	51	94	108	81	54	89
85	98	46	91	107	80	52	89
80	92	40	88	106	78	50	88
75	86	34	84	104	76	47	87
70	80	28	80	102	74	44	86
65	74	...	75	100	72	...	85
60	68	...	70	97	70	...	83
55	62	...	65	94	67	...	82
50	56	...	59	91	64	...	80
45	50	...	53	87	62	...	79
40	44	...	46	83	59	...	77

[A] Data were generated in an interlaboratory test program conducted by ASTM Subcommittee E28.06. Supporting data available from ASTM Headquarters. Request E28-1005.
[B] In table headings, kgf or gf refers to total test force.
[C] Appendix X9 contains equations converting determined hardness scale numbers to Brinell numbers for wrought aluminum products. Refer to 1.11 before using conversion equations.

APPENDIXES

(Nonmandatory Information)

X1. HARDNESS CONVERSION EQUATIONS FOR NON-AUSTENITIC STEELS (DETERMINED HARDNESS SCALE NUMBERS TO ROCKWELL C HARDNESS NUMBERS)

X1.1 The following equations were generated from the specific hardness numbers contained in Table 1 and should not be used for converting numbers outside of the defined hardness range. Due to inherent inaccuracies in the conversion process, the converted number should be rounded to the nearest whole number in accordance with Practice E29.

X1.1.1 From Vickers hardness to Rockwell C hardness:

$$HRC = +3.14900E+01 + 7.96683E-02(HV) - 3.55432E-05(HV)^2 - 6.72816E+03(HV)^{-1}$$
$$R^2 = 0.9999 \tag{X1.1}$$

X1.1.2 From Brinell hardness (10-mm diameter steel ball, 3000-kgf force) to Rockwell C hardness:

$$HRC = +8.35260E+01 - 8.68203E-02(HBS) + 1.44229E-04(HBS)^2 - 1.15905E+04(HBS)^{-1}$$
$$R^2 = 0.9998 \tag{X1.2}$$

X1.1.3 From Brinell hardness (10-mm diameter tungsten carbide ball, 3000-kgf force) to Rockwell C hardness:

$$HRC = +1.81673E+01 + 1.20388E-01(HBW) - 6.94388E-05(HBW)^2 - 4.88327E+03(HBW)^{-1}$$
$$R^2 = 0.9998 \tag{X1.3}$$

X1.1.4 From Knoop hardness (500-gf force and greater) to Rockwell C hardness:

$$HRC = +6.43102E+01 + 7.59497E-03(HK_{500-1000}) + 1.13729E-05(HK_{500-1000})^2 - 1.17515E+04(HK500-1000)^{-1}$$
$$R^2 = 1.0000 \tag{X1.4}$$

X1.1.5 From Rockwell A hardness to Rockwell C hardness:

$$HRC = -1.25501E+02 + 2.76747E+00(HRA) - 5.94178E-03(HRA)^2$$
$$R^2 = 0.9999 \tag{X1.5}$$

X1.1.6 From Rockwell D hardness to Rockwell C hardness:

$$HRC = -3.20806E+01 + 1.30193E+00(HRD)$$
$$R^2 = 1.0000 \tag{X1.6}$$

X1.1.7 From Rockwell 15N hardness to Rockwell C hardness:

$$HRC = -3.74666E+02 + 1.27582E+01(HR15N) - 1.48317E-01(HR15N)^2 + 6.68816E-04(HR15N)^3$$
$$R^2 = 0.9999 \tag{X1.7}$$

X1.1.8 From Rockwell 30N hardness to Rockwell C hardness:

$$HRC = -2.60390E+01 + 1.11079E+00(HR30N)$$
$$R^2 = 1.0000 \tag{X1.8}$$

X1.1.9 From Rockwell 45N hardness to Rockwell C hardness:

$$HRC = +3.18978E+00 + 8.54135E-01(HR45N)$$
$$R^2 = 0.9999 \tag{X1.9}$$

X1.1.10 From Scleroscope hardness to Rockwell C hardness:

$$HRC = +1.14708E+01 + 9.61667E-01(HSc) - 3.15195E-03(HSc)^2 - 6.97208E+02(HSc)^{-1}$$
$$R^2 = 1.0000 \tag{X1.10}$$

X2. HARDNESS CONVERSION EQUATIONS FOR NON-AUSTENITIC STEELS (DETERMINED HARDNESS SCALE NUMBERS TO ROCKWELL B HARDNESS NUMBERS)

X2.1 The following equations were generated from the specific hardness numbers contained in Table 2 and should not be used for converting numbers outside of the defined hardness range. Due to inherent inaccuracies in the conversion process, the converted number should be rounded to the nearest whole number in accordance with Practice E29.

X2.1.1 From Vickers hardness to Rockwell B hardness:

$$HRB = +1.14665E+02 + 8.82795E-02(HV) - 1.41855E-04(HV)^2 - 6.69528E+03(HV)^{-1}$$
$$R^2 = 0.9998 \tag{X2.1}$$

X2.1.2 From Brinell hardness (10-mm diameter steel ball, 3000-kgf force) to Rockwell B hardness:

$$HRB = +1.14665E+02 + 8.82795E-02(HBS) - 1.41855E-04(HBS)^2 - 6.69528E+03(HBS)^{-1}$$
$$R^2 = 0.9998 \tag{X2.2}$$

X2.1.3 From Knoop hardness (500-gf force and greater) to Rockwell B hardness:

$$HRB = +1.75357E+02 - 2.37706E-01(HK_{500-1000}) + 4.56743E-04(HK_{500-1000})^2 - 1.12480E+04(HK_{500-1000})^{-1}$$
$$R^2 = 0.9996 \tag{X2.3}$$

X2.1.4 From Rockwell A hardness to Rockwell B hardness:

$$HRB = -4.82350E+01 + 3.33354E+00(HRA) - 1.50107E-02(HRA)^2$$
$$R^2 = 1.0000 \tag{X2.4}$$

X2.1.5 From Rockwell F hardness to Rockwell B hardness:

$$HRB = -9.99816E+01 + 1.75617E+00(HRF)$$
$$R^2 = 1.0000 \tag{X2.5}$$

X2.1.6 From Rockwell 15T hardness to Rockwell B hardness:

$$\text{HRB} = -1.86934\text{E}+02 + 3.08173\text{E}+00(\text{HR15T})$$
$$R^2 = 1.0000 \tag{X2.6}$$

X2.1.7 From Rockwell 30T hardness to Rockwell B hardness:

$$\text{HRB} = -2.42568\text{E}+01 + 1.49484\text{E}+00(\text{HR30T})$$
$$R^2 = 1.0000 \tag{X2.7}$$

X2.1.8 From Rockwell 45T hardness to Rockwell B hardness:

$$\text{HRB} = +2.74135\text{E}+01 + 9.95874\text{E}-01(\text{HR45T})$$
$$R^2 = 1.0000 \tag{X2.8}$$

X3. HARDNESS CONVERSION EQUATIONS FOR NICKEL AND HIGH-NICKEL ALLOYS (DETERMINED HARDNESS SCALE NUMBERS TO VICKERS HARDNESS NUMBERS)

X3.1 The following equations were generated from the specific hardness numbers contained in Table 3 and should not be used for converting numbers outside of the defined hardness range. Due to inherent inaccuracies in the conversion process, the converted number should be rounded to the nearest whole number in accordance with Practice E29.

X3.1.1 From Brinell hardness (10-mm diameter steel ball, 3000-kgf force) to Vickers hardness (1.5, 10, and 30-kgf forces):

$$\text{HV } 1.5,10,30 = +8.52592\text{E}-02 + 9.82889\text{E}-01(\text{HBS}) + 1.89707\text{E}-04(\text{HBS})^2$$
$$R^2 = 1.0000 \tag{X3.1}$$

X3.1.2 From Rockwell A hardness to Vickers hardness (1.5, 10, and 30-kgf forces):

$$(\text{HV } 1.5,10,30)^{-1} = +2.13852\text{E}-02 - 3.84341\text{E}-04(\text{HRA}) + 1.67455\text{E}-06(\text{HRA})^2$$
$$R^2 = 0.9998 \tag{X3.2}$$

X3.1.3 From Rockwell B hardness to Vickers hardness (1.5, 10, and 30-kgf forces):

$$(\text{HV } 1.5, 10, 30)^{-1} = +1.69552\text{E}-02 - 1.29200\text{E}-04(\text{HRB})$$
$$R^2 = 0.9999 \tag{X3.3}$$

X3.1.4 From Rockwell C hardness to Vickers hardness (1.5, 10, and 30-kgf forces):

$$(\text{HV } 1.5, 10, 30)^{-1} = +6.24553\text{E}-03 - 1.08014\text{E}-04(\text{HRC}) + 4.32021\text{E}-07(\text{HRC})^2$$
$$R^2 = 0.9995 \tag{X3.4}$$

X3.1.5 From Rockwell D hardness to Vickers hardness (1.5, 10, and 30-kgf forces):

$$(\text{HV } 1.5, 10, 30)^{-1} = +1.04408\text{E}-02 - 1.86498\text{E}-04(\text{HRD}) + 8.16952\text{E}-07(\text{HRD})^2$$
$$R^2 = 0.9998 \tag{X3.5}$$

X3.1.6 From Rockwell E hardness to Vickers hardness (1.5, 10, and 30-kgf forces):

$$(\text{HV } 1.5, 10, 30)^{-1} = +2.72286\text{E}-02 - 2.01993\text{E}-04(\text{HRE})$$
$$R^2 = 0.9994 \tag{X3.6}$$

X3.1.7 From Rockwell F hardness to Vickers hardness (1.5, 10, and 30-kgf forces):

$$(\text{HV } 1.5, 10, 30)^{-1} = +2.94130\text{E}-02 - 2.23861\text{E}-04(\text{HRF})$$
$$R^2 = 0.9991 \tag{X3.7}$$

X3.1.8 From Rockwell G hardness to Vickers hardness (1.5, 10, and 30-kgf forces):

$$(\text{HV } 1.5, 10, 30)^{-1} = +1.10239\text{E}-02 - 8.27628\text{E}-05(\text{HRG})$$
$$R^2 = 0.9999 \tag{X3.8}$$

X3.1.9 From Rockwell K hardness to Vickers hardness (1.5, 10, and 30-kgf forces):

$$(\text{HV } 1.5, 10, 30)^{-1} = +1.87458\text{E}-02 - 1.41851\text{E}-04(\text{HRK})$$
$$R^2 = 0.9998 \tag{X3.9}$$

X3.1.10 From Rockwell 15N hardness to Vickers hardness (1.5, 10, and 30-kgf forces):

$$(\text{HV } 1.5, 10, 30)^{-1} = +2.59838\text{E}-02 - 4.31479\text{E}-04(\text{HR15N}) + 1.75469\text{E}-06(\text{HR15N})^2$$
$$R^2 = 0.9998 \tag{X3.10}$$

X3.1.11 From Rockwell 30N hardness to Vickers hardness (1.5, 10, and 30-kgf forces):

$$(\text{HV } 1.5, 10, 30)^{-1} = +9.85078\text{E}-03 - 1.58346\text{E}-04(\text{HR30N}) + 6.16727\text{E}-07(\text{HR30N})^2$$
$$R^2 = 0.9997 \tag{X3.11}$$

X3.1.12 From Rockwell 45N hardness to Vickers hardness (1.5, 10, and 30-kgf forces):

$$(\text{HV } 1.5, 10, 30)^{-1} = +6.03882\text{E}-03 - 9.51201\text{E}-05(\text{HR45N}) + 3.63345\text{E}-07(\text{HR45N})^2$$
$$R^2 = 0.9998 \tag{X3.12}$$

X3.1.13 From Rockwell 15T hardness to Vickers hardness (1.5, 10, and 30-kgf forces):

$$(\text{HV } 1.5, 10, 30)^{-1} = +3.71482\text{E}-02 - 3.49957\text{E}-04(\text{HR15T}) - 8.92693\text{E}-08(\text{HR15T})^2$$
$$R^2 = 0.9996 \tag{X3.13}$$

X3.1.14 From Rockwell 30T hardness to Vickers hardness (1.5, 10, and 30-kgf forces):

$$(\text{HV } 1.5, 10, 30)^{-1} = +1.94133\text{E}-02 - 1.85296\text{E}-04(\text{HR30T}) - 4.01798\text{E}-08(\text{HR30T})^2$$
$$R^2 = 0.9998 \tag{X3.14}$$

X3.1.15 From Rockwell 45T hardness to Vickers hardness (1.5, 10, and 30-kgf forces):

$$(\text{HV } 1.5, 10, 30)^{-1} = +1.29736\text{E}-02 - 1.14693\text{E}-04(\text{HR45T}) - 1.61879\text{E}-07(\text{HR45T})^2$$

$R^2 = 0.9998$ (X3.15)

X3.1.16 From Knoop hardness (500 and 1000-gf forces) to Vickers hardness (1.5, 10, and 30-kgf forces):

HV 1.5, 10, 30 $= -5.08687\mathrm{E}{-}01 + 8.78046\mathrm{E}{-}01(\mathrm{HK}_{500,1000})$
$R^2 = 1.0000$ (X3.16)

X4. HARDNESS CONVERSION EQUATIONS FOR CARTRIDGE BRASS
(DETERMINED HARDNESS SCALE NUMBERS TO VICKERS HARDNESS NUMBERS)

X4.1 The following equations were generated from the specific hardness numbers contained in Table 4 and should not be used for converting numbers outside of the defined hardness range. Due to inherent inaccuracies in the conversion process, the converted number should be rounded to the nearest whole number in accordance with Practice E29.

X4.1.1 From Rockwell B hardness to Vickers hardness:

$(\mathrm{HV})^{-1} = +1.77793\mathrm{E}{-}02 - 1.31112\mathrm{E}{-}04(\mathrm{HRB}) - 3.77903\mathrm{E}{-}07(\mathrm{HRB})^2 + 3.55271\mathrm{E}{-}09(\mathrm{HRB})^3$
$R^2 = 0.9996$ (X4.1)

X4.1.2 From Rockwell F hardness to Vickers hardness:

$(\mathrm{HV})^{-1} = +2.95966\mathrm{E}{-}02 - 1.03725\mathrm{E}{-}04(\mathrm{HRF}) - 2.31669\mathrm{E}{-}06(\mathrm{HRF})^2 + 1.12203\mathrm{E}{-}08(\mathrm{HRF})^3$
$R^2 = 0.9998$ (X4.2)

X4.1.3 From Rockwell 15T hardness to Vickers hardness:

$(\mathrm{HV})^{-1} = +7.65595\mathrm{E}{-}02 - 1.79133\mathrm{E}{-}03(\mathrm{HR15T}) + 1.84105\mathrm{E}{-}05(\mathrm{HR15T})^2 - 8.14318\mathrm{E}{-}08(\mathrm{HR15T})^3$
$R^2 = 0.9998$ (X4.3)

X4.1.4 From Rockwell 30T hardness to Vickers hardness:

$(\mathrm{HV})^{-1} = +2.08924\mathrm{E}{-}02 - 2.03448\mathrm{E}{-}04(\mathrm{HR30T}) - 2.80441\mathrm{E}{-}08(\mathrm{HR30T})^2 + 1.33185\mathrm{E}{-}10(\mathrm{HR30T})^3$
$R^2 = 0.9998$ (X4.4)

X4.1.5 From Rockwell 45T hardness to Vickers hardness:

$(\mathrm{HV})^{-1} = +1.36295\mathrm{E}{-}02 - 1.03553\mathrm{E}{-}04(\mathrm{HR45T}) - 9.70546\mathrm{E}{-}07(\mathrm{HR45T})^2 + 8.77834\mathrm{E}{-}09(\mathrm{HR45T})^3$
$R^2 = 0.9999$ (X4.5)

X4.1.6 From Brinell hardness (10-mm diameter steel ball, 500-kgf force) to Vickers hardness:

HV $= -5.60725\mathrm{E}{+}00 + 1.19007\mathrm{E}{+}00(\mathrm{HBS}\ 10/500/15)$
$R^2 = 0.9998$ (X4.6)

X5. HARDNESS CONVERSION EQUATION FOR ANNEALED AUSTENITIC STAINLESS STEEL PLATE
(DETERMINED BRINELL HARDNESS NUMBERS TO ROCKWELL B HARDNESS NUMBERS)

X5.1 The following equation was generated from the specific hardness numbers contained in Table 5 and should not be used for converting numbers outside of the defined hardness range. Due to inherent inaccuracies in the conversion process, the converted number should be rounded to the nearest whole number in accordance with Practice E29.

X5.1.1 From Brinell hardness (10-mm steel diameter ball, 3000-kgf force) to Rockwell B hardness:

HRB $= +1.29998\mathrm{E}{+}02 - 7.66860\mathrm{E}{+}03(\mathrm{HBS})^{-1}$
$R^2 = 0.9999$ (X5.1)

X6. HARDNESS CONVERSION EQUATIONS FOR AUSTENITIC STAINLESS STEEL SHEET
(DETERMINED HARDNESS SCALE NUMBERS TO ROCKWELL C OR ROCKWELL B HARDNESS NUMBERS

X6.1 The following equations were generated from the specific hardness numbers contained in Table 6 and should not be used for converting numbers outside of the defined hardness range. Due to inherent inaccuracies in the conversion process, the converted number should be rounded to the nearest whole number in accordance with Practice E29.

X6.1.1 From Rockwell A hardness to Rockwell C hardness:

HRC $= -9.94148\mathrm{E}{+}01 + 1.98137\mathrm{E}{+}00(\mathrm{HRA})$
$R^2 = 1.0000$ (X6.1)

X6.1.2 From Rockwell 15N hardness to Rockwell C hardness:

HRC $= -1.16608\mathrm{E}{+}02 + 1.95692\mathrm{E}{+}00(\mathrm{HR15N})$
$R^2 = 1.0000$ (X6.2)

X6.1.3 From Rockwell 30N hardness to Rockwell C hardness:

HRC $= -2.79663\mathrm{E}{+}01 + 1.14752\mathrm{E}{+}00(\mathrm{HR30N})$
$R^2 = 1.0000$ (X6.3)

X6.1.4 From Rockwell 45N hardness to Rockwell C hardness:

HRC $= +2.25782\mathrm{E}{+}00 + 8.78362\mathrm{E}{-}01(\mathrm{HR45N})$
$R^2 = 1.0000$ (X6.4)

X6.1.5 From Rockwell A hardness to Rockwell B hardness:

HRB = − 5.16024E+00 + 1.71080E+00(HRA)
R^2 = 1.0000   (X6.5)

X6.1.6 From Rockwell F hardness to Rockwell B hardness:

HRB = − 6.79918E+01 + 1.47539E+00(HRF)
R^2 = 0.9999   (X6.6)

X6.1.7 From Rockwell 15T hardness to Rockwell B hardness:

HRB = − 1.47089E+02 + 2.69928E+00(HR15T)
R^2 = 1.0000   (X6.7)

X6.1.8 From Rockwell 30T hardness to Rockwell B hardness:

HRB = − 1.56777E+01 + 1.43818E+00(HR30T)
R^2 = 1.0000   (X6.8)

X6.1.9 From Rockwell 45T hardness to Rockwell B hardness:

HRB = + 3.08896E+01 + 9.84321E−01(HR45T)
R^2 = 1.0000   (X6.9)

X7. HARDNESS CONVERSION EQUATIONS FOR COPPER, NOS. 102 TO 142 INCLUSIVE (DETERMINED HARDNESS SCALE NUMBERS TO VICKERS HARDNESS NUMBERS)

X7.1 The following equations were generated from the specific hardness numbers contained in Table 7 and should not be used for converting numbers outside of the defined hardness range. Due to inherent inaccuracies in the conversion process, the converted number should be rounded to the nearest whole number in accordance with Practice E29.

X7.1.1 From Vickers hardness (100-gf force) to Vickers hardness (1-kgf force):

HV 1 = − 1.94066E+01 + 1.17624E+00(HV_{100})
R^2 = 0.9999   (X7.1)

X7.1.2 From Knoop hardness (1-kgf force) to Vickers hardness (1-kgf force):

HV 1 = + 1.1858E+01 + 6.42195E−01(HK_{1000}) + 1.50709E−03(HK_{1000})^2
R^2 = 0.9999   (X7.2)

X7.1.3 From Knoop hardness (500 gf force) to Vickers hardness (1-kgf force):

HV 1 = + 4.04249E+00 + 7.73167E−01(HK_{500}) + 1.22866E−03(HK_{500})^2
R^2 = 0.9998   (X7.3)

X7.1.4 From Rockwell 15T hardness to Vickers hardness (1-kgf force) for 0.010-in. (0.25-mm) strip:

(HV 1)$^{-1}$ = + 3.37918E−01 − 1.15500E−02(HR15T) + 1.40059E−04(HR15T)^2 − 5.88157E−07(HR15T)^3
R^2 = 0.9997   (X7.4)

X7.1.5 From Rockwell 15T hardness to Vickers hardness (1-kgf force) for 0.020-in. (0.51-mm) strip:

(HV 1)$^{-1}$ = + 1.25038E−01 − 3.80747E−03(HR15T) + 4.54150E−05(HR15T)^2 − 1.98661E−07(HR15T)^3
R^2 = 0.9997   (X7.5)

X7.1.6 From Rockwell B hardness to Vickers hardness (1-kgf force) for 0.040-in. (1.02-mm) and greater strip:

(HV 1)$^{-1}$ = + 1.49881E−02 − 1.39326E−04(HRB) + 8.82686E−07(HRB)^2 − 6.30498E−09(HRB)^3
R^2 = 0.9999   (X7.6)

X7.1.7 From Rockwell F hardness to Vickers hardness (1-kgf force) for 0.040-in. (1.02-mm) and greater strip:

(HV 1)$^{-1}$ = + 4.03378E−02 − 7.12218E−04(HRF) + 6.46922E−06(HRF)^2 − 2.64942E−08(HRF)^3
R^2 = 0.9998   (X7.7)

X7.1.8 From Rockwell 15T hardness to Vickers hardness (1-kgf force) for 0.040-in. (1.02-mm) and greater strip:

(HV 1)$^{-1}$ = + 6.91162E−02 − 1.89938E−03(HR15T) + 2.43142E−05(HR15T)^2 − 1.21657E−07(HR15T)^3
R^2 = 0.9994   (X7.8)

X7.1.9 From Rockwell 30T hardness to Vickers hardness (1-kgf force) for 0.040-in. (1.02-mm) and greater strip:

(HV 1)$^{-1}$ = + 2.12081E−02 − 2.79029E−04(HR30T) + 1.85833E−06(HR30T)^2 − 9.41015E−09(HR30T)^3
R^2 = 0.9999   (X7.9)

X7.1.10 From Rockwell 45T hardness to Vickers hardness (1-kgf force) for 0.040-in. (1.02-mm) and greater strip:

(HV 1)$^{-1}$ = + 1.33602E−02 − 1.16936E−04(HR45T) − 2.02801E−07(HR45T)^2 + 4.40268E−09(HR45T)^3
R^2 = 0.9995   (X7.10)

X7.1.11 From Brinell hardness (10-mm diameter steel ball, 500-kgf force) to Vickers hardness (1-kgf force) for 0.080- in. (2.03-mm) strip:

HV 1 = + 2.77693E + 01 + 8.62358E−01(HBS 10/500/15) − 3.66858E−04(HBS 10/500/15)^2
R^2 = 0.9999   (X7.11)

X7.1.12 From Brinell hardness (2-mm diameter steel ball, 20-kgf force) to Vickers hardness (1-kgf force) for 0.040 in. (1.02-mm) strip:

HV 1 = − 1.01087E+00 + 1.18352E+00(HBS 2/20/15) − 7.02625E−04(HBS 2/20/15)^2
R^2 = 0.9999   (X7.12)

X8. HARDNESS CONVERSION EQUATIONS FOR ALLOYED WHITE IRON (DETERMINED HARDNESS SCALE NUMBERS TO VICKERS HARDNESS NUMBERS)

X8.1 The following equations were generated from the specific hardness numbers contained in Table 8 and should not be used for converting numbers outside of the defined hardness range. Due to inherent inaccuracies in the conversion process, the converted number should be rounded to the nearest whole number in accordance with Practice E29.

X8.1.1 From Brinell hardness (10-mm diameter tungsten carbide ball, 3000-kgf force) to Vickers hardness (50-kgf force):

$$\text{HV } 50 = -2.61008E+01 + 1.13635E+00(\text{HBW})$$
$$R^2 = 1.0000 \tag{X8.1}$$

X8.1.2 From Rockwell C hardness to Vickers hardness (50-kgf force):

$$\text{HV } 50 = +5.72753E+02 - 1.71996E+01(\text{HRC}) + 3.33893E-01(\text{HRC})^2$$
$$R^2 = 0.9991 \tag{X8.2}$$

X9. HARDNESS CONVERSION EQUATIONS FOR WROUGHT ALUMINUM PRODUCTS (DETERMINED HARDNESS SCALE NUMBERS TO BRINELL HARDNESS NUMBERS)

X9.1 The following equations were generated from the specific hardness numbers contained in Table 9 and should not be used for converting numbers outside of the defined hardness range. Due to inherent inaccuracies in the conversion process, the converted number should be rounded to the nearest whole number in accordance with Practice E29.

X9.1.1 From Vickers hardness (15-kgf force) to Brinell hardness (10-mm diameter steel ball, 500-kgf force):

$$\text{HBS } 10/500/15 = +3.76211E+00 + 8.25368E-01(\text{HV } 15)$$
$$R^2 = 1.0000 \tag{X9.1}$$

X9.1.2 From Rockwell B hardness to Brinell hardness (10-mm diameter steel ball, 500-kgf force):

$$(\text{HBS } 10/500/15)^{-1} = +2.09261E-02 - 3.13747E-04(\text{HRB}) + 3.24720E-06(\text{HRB})^2 - 1.71476E-08(\text{HRB})^3$$
$$R^2 = 0.9995 \tag{X9.2}$$

X9.1.3 From Rockwell E hardness to Brinell hardness (10-mm diameter steel ball, 500-kgf force):

$$(\text{HBS } 10/500/15)^{-1} = +6.91185E-02 - 1.57873E-03(\text{HRE}) + 1.66991E-05(\text{HRE})^2 - 6.90196E-08(\text{HRE})^3$$
$$R^2 = 0.9994 \tag{X9.3}$$

X9.1.4 From Rockwell H hardness to Brinell hardness (10-mm diameter steel ball, 500-kgf force):

$$(\text{HBS } 10/500/15)^{-1} = +4.00460E-01 - 1.06615E-02(\text{HRH}) + 1.02525E-04(\text{HRH})^2 - 3.44242E-07(\text{HRH})^3$$
$$R^2 = 0.9995 \tag{X9.4}$$

X9.1.5 From Rockwell 15T hardness to Brinell hardness (10-mm diameter steel ball, 500-kgf force):

$$(\text{HBS } 10/500/15)^{-1} = +3.35165E-01 - 1.16197E-02(\text{HR15T}) + 1.44778E-04(\text{HR15T})^2 - 6.26187E-07(\text{HR15T})^3$$
$$R^2 = 0.9988 \tag{X9.5}$$

X9.1.6 From Rockwell 30T hardness to Brinell hardness (10-mm diameter steel ball, 500-kgf force):

$$(\text{HBS } 10/500/15)^{-1} = +4.68610E-02 - 1.24964E-03(\text{HR30T}) + 1.45528E-05(\text{HR30T})^2 - 6.71417E-08(\text{HR30T})^3$$
$$R^2 = 0.9994 \tag{X9.6}$$

X9.1.7 From Rockwell 15W hardness to Brinell hardness (10-mm diameter steel ball, 500-kgf force):

$$(\text{HBS } 10/500/15) = -7.10127E+03 + 2.71267E+02(\text{HR15W}) - 3.46213E+00(\text{HR15W})^2 + 1.48551E-02(\text{HR15W})^3$$
$$R^2 = 0.9924 \tag{X9.7}$$

X10. EFFECT OF STRAIN HARDENING ON HARDNESS CONVERSION RELATIONSHIPS

X10.1 For ferrous and nonferrous metals softer than 240 HB, a single set of hardness conversion relationships inevitably introduces large errors because of the wide difference that may exist in the amount of cold working before testing, as well as the amount that occurs during the test itself. This dependence on strain-hardening characteristics can be demonstrated by the Rockwell scales 15-T, 30-T, 45-T, F, and B, in which forces ranging from 15 to 100 kgf are applied on a ¹⁄₁₆-in. (1.588-mm) diameter ball indenter. As higher forces are used, the increased strain raises the hardness by an amount that depends on the pretest capacity of the metal for strain hardening. An annealed metal of high capacity for strain hardening will harden much more in the test than will a cold-worked metal. For example, an annealed iron and a cold-rolled aluminum alloy may have hardnesses of 71 and 72 HR 15T, respectively. The hardnesses are 31 HRB for the soft annealed iron and 7 HRB for the cold-rolled aluminum alloy.

X10.2 On the other hand, if materials have Brinell or Rockwell hardness values that are approximately equal in the annealed state as well as after heavy cold deformation, these materials will have similar hardness conversion relationships

for all degrees of strain hardening. This is true of yellow brasses and low-carbon steels and irons. The limiting conditions can usually be identified by the appearance of the hardness indentations themselves. Soft annealed metals have characteristic "sinking" type indentation contours when indenters of the ball type are used. On the other hand, heavily cold-worked metals have sharp "ridging" type indentations. While annealed metals are being progressively cold worked, the indentation contours pass through a "flat" stage in which the lip of the indentation is neither round nor sharply ridged. It is necessary to base hardness conversions on comparative tests of similar materials that also have very similar mechanical properties.

SUMMARY OF CHANGES

Committee E28 has identified the location of selected changes to this standard since the last issue E104-05$^{\varepsilon 1}$ that may impact the use of this standard. (Approved Jan. 1, 2007)

(1) Note 1 was added.

(2) Reference to Note 1 was added to 6.3.

ASTM International takes no position respecting the validity of any patent rights asserted in connection with any item mentioned in this standard. Users of this standard are expressly advised that determination of the validity of any such patent rights, and the risk of infringement of such rights, are entirely their own responsibility.

This standard is subject to revision at any time by the responsible technical committee and must be reviewed every five years and if not revised, either reapproved or withdrawn. Your comments are invited either for revision of this standard or for additional standards and should be addressed to ASTM International Headquarters. Your comments will receive careful consideration at a meeting of the responsible technical committee, which you may attend. If you feel that your comments have not received a fair hearing you should make your views known to the ASTM Committee on Standards, at the address shown below.

This standard is copyrighted by ASTM International, 100 Barr Harbor Drive, PO Box C700, West Conshohocken, PA 19428-2959, United States. Individual reprints (single or multiple copies) of this standard may be obtained by contacting ASTM at the above address or at 610-832-9585 (phone), 610-832-9555 (fax), or service@astm.org (e-mail); or through the ASTM website (www.astm.org). Permission rights to photocopy the standard may also be secured from the ASTM website (www.astm.org/COPYRIGHT/).

Designation: E165 – 09

Standard Practice for
Liquid Penetrant Examination for General Industry[1]

This standard is issued under the fixed designation E165; the number immediately following the designation indicates the year of original adoption or, in the case of revision, the year of last revision. A number in parentheses indicates the year of last reapproval. A superscript epsilon (ε) indicates an editorial change since the last revision or reapproval.

1. Scope

1.1 This practice[2] covers procedures for penetrant examination of materials. Penetrant testing is a nondestructive testing method for detecting discontinuities that are open to the surface such as cracks, seams, laps, cold shuts, shrinkage, laminations, through leaks, or lack of fusion and is applicable to in-process, final, and maintenance testing. It can be effectively used in the examination of nonporous, metallic materials, ferrous and nonferrous metals, and of nonmetallic materials such as nonporous glazed or fully densified ceramics, as well as certain nonporous plastics, and glass.

1.2 This practice also provides a reference:

1.2.1 By which a liquid penetrant examination process recommended or required by individual organizations can be reviewed to ascertain its applicability and completeness.

1.2.2 For use in the preparation of process specifications and procedures dealing with the liquid penetrant testing of parts and materials. Agreement by the customer requesting penetrant inspection is strongly recommended. All areas of this practice may be open to agreement between the cognizant engineering organization and the supplier, or specific direction from the cognizant engineering organization.

1.2.3 For use in the organization of facilities and personnel concerned with liquid penetrant testing.

1.3 This practice does not indicate or suggest criteria for evaluation of the indications obtained by penetrant testing. It should be pointed out, however, that after indications have been found, they must be interpreted or classified and then evaluated. For this purpose there must be a separate code, standard, or a specific agreement to define the type, size, location, and direction of indications considered acceptable, and those considered unacceptable.

1.4 The values stated in inch-pound units are to be regarded as the standard. SI units are provided for information only.

1.5 *This standard does not purport to address all of the safety concerns, if any, associated with its use. It is the responsibility of the user of this standard to establish appropriate safety and health practices and determine the applicability of regulatory limitations prior to use.*

2. Referenced Documents

2.1 *ASTM Standards:*[3]

D129 Test Method for Sulfur in Petroleum Products (General Bomb Method)

E516 Practice for Testing Thermal Conductivity Detectors Used in Gas Chromatography

D808 Test Method for Chlorine in New and Used Petroleum Products (Bomb Method)

D1193 Specification for Reagent Water

D1552 Test Method for Sulfur in Petroleum Products (High-Temperature Method)

D4327 Test Method for Anions in Water by Chemically Suppressed Ion Chromatography

E433 Reference Photographs for Liquid Penetrant Inspection

E543 Specification for Agencies Performing Nondestructive Testing

E1208 Test Method for Fluorescent Liquid Penetrant Examination Using the Lipophilic Post-Emulsification Process

E1209 Test Method for Fluorescent Liquid Penetrant Examination Using the Water-Washable Process

E1210 Test Method for Fluorescent Liquid Penetrant Examination Using the Hydrophilic Post-Emulsification Process

E1219 Test Method for Fluorescent Liquid Penetrant Examination Using the Solvent-Removable Process

E1220 Test Method for Visible Penetrant Examination Using Solvent-Removable Process

E1316 Terminology for Nondestructive Examinations

E1417 Practice for Liquid Penetrant Testing

E1418 Test Method for Visible Penetrant Examination Using the Water-Washable Process

2.2 *ASNT Document:*[4]

---

[1] This practice is under the jurisdiction of ASTM Committee E07 on Nondestructive Testing and is the direct responsibility of Subcommittee E07.03 on Liquid Penetrant and Magnetic Particle Methods.
Current edition approved July 1, 2009. Published July 2009. Originally approved in 1960. Last previous edition approved in 2002 as E165 - 02. DOI: 10.1520/E0165-09.

[2] For ASME Boiler and Pressure Vessel Code applications see related Recommended Test Method SE-165 in the Code.

[3] For referenced ASTM standards, visit the ASTM website, www.astm.org, or contact ASTM Customer Service at service@astm.org. For *Annual Book of ASTM Standards* volume information, refer to the standard's Document Summary page on the ASTM website.

[4] Available from American Society for Nondestructive Testing (ASNT), P.O. Box 28518, 1711 Arlingate Ln., Columbus, OH 43228-0518, http://www.asnt.org.

Copyright © ASTM International, 100 Barr Harbor Drive, PO Box C700, West Conshohocken, PA 19428-2959, United States.

SNT-TC-1A Recommended Practice for Nondestructive Testing Personnel Qualification and Certification
ANSI/ASNT CP-189 Standard for Qualification and Certification of Nondestructive Testing Personnel

2.3 *Military Standard:*
MIL-STD-410 Nondestructive Testing Personnel Qualification and Certification[5]

2.4 *APHA Standard:*
429 Method for the Examination of Water and Wastewater[6]

2.5 *AIA Standard:*
NAS-410 Certification and Qualification of Nondestructive Test Personnel[7]

2.6 *SAE Standards:*[8]
AMS 2644 Inspection Material, Penetrant
QPL-AMS-2644 Qualified Products of Inspection Materials, Penetrant

3. Terminology

3.1 The definitions relating to liquid penetrant examination, which appear in Terminology E1316, shall apply to the terms used in this practice.

4. Summary of Practice

4.1 Liquid penetrant may consist of visible or fluorescent material. The liquid penetrant is applied evenly over the surface being examined and allowed to enter open discontinuities. After a suitable dwell time, the excess surface penetrant is removed. A developer is applied to draw the entrapped penetrant out of the discontinuity and stain the developer. The test surface is then examined to determine the presence or absence of indications.

NOTE 1—The developer may be omitted by agreement between the contracting parties.

NOTE 2—Fluorescent penetrant examination shall not follow a visible penetrant examination unless the procedure has been qualified in accordance with 10.2, because visible dyes may cause deterioration or quenching of fluorescent dyes.

4.2 Processing parameters, such as surface precleaning, penetrant dwell time and excess penetrant removal methods, are dependent on the specific materials used, the nature of the part under examination, (that is, size, shape, surface condition, alloy) and type of discontinuities expected.

5. Significance and Use

5.1 Liquid penetrant testing methods indicate the presence, location and, to a limited extent, the nature and magnitude of the detected discontinuities. Each of the various penetrant methods has been designed for specific uses such as critical service items, volume of parts, portability or localized areas of examination. The method selected will depend accordingly on the design and service requirements of the parts or materials being tested.

6. Classification of Penetrant Materials and Methods

6.1 Liquid penetrant examination methods and types are classified in accordance with MIL-I-25135 and AMS 2644 as listed in Table 1.

6.2 *Fluorescent Penetrant Testing (Type 1)*—Fluorescent penetrant testing utilizes penetrants that fluoresce brilliantly when excited by black light (UVA). The sensitivity of fluorescent penetrants depends on their ability to be retained in the various size discontinuities during processing, and then to bleed out into the developer coating and produce indications that will fluoresce. Fluorescent indications are many times brighter than their surroundings when viewed under appropriate black light illumination.

6.3 *Visible Penetrant Testing (Type 2)*—Visible penetrant testing uses a penetrant that can be seen in visible light. The penetrant is usually red, so that resultant indications produce a definite contrast with the white background of the developer. Visible penetrant indications must be viewed under adequate white light.

7. Materials

7.1 *Liquid Penetrant Testing Materials* consist of fluorescent or visible penetrants, emulsifiers (oil-base and water-base), removers (water and solvent), and developers (dry powder, aqueous and nonaqueous). A family of liquid penetrant examination materials consists of the applicable penetrant and emulsifier, as recommended by the manufacturer. Any liquid penetrant, remover and developer listed in QPL-25135/QPL-AMS2644 can be used, regardless of the manufacturer. Intermixing of penetrants and emulsifiers from different manufacturers is prohibited.

NOTE 3—Refer to 9.1 for special requirements for sulfur, halogen and alkali metal content.

NOTE 4—While approved penetrant materials will not adversely affect common metallic materials, some plastics or rubbers may be swollen or stained by certain penetrants.

7.2 *Penetrants*:

7.2.1 *Post-Emulsifiable Penetrants* are insoluble in water and cannot be removed with water rinsing alone. They are formulated to be selectively removed from the surface using a separate emulsifier. Properly applied and given a proper emulsification time, the emulsifier combines with the excess surface penetrant to form a water-washable mixture, which can

TABLE 1 Classification of Penetrant Examination Types and Methods

Type I—Fluorescent Penetrant Examination
Method A—Water-washable (see Test Method E1209)
Method B—Post-emulsifiable, lipophilic (see Test Method E1208)
Method C—Solvent removable (see Test Method E1219)
Method D—Post-emulsifiable, hydrophilic (see Test Method E1210)
Type II—Visible Penetrant Examination
Method A—Water-washable (see Test Method E1418)
Method C—Solvent removable (see Test Method E1220)

---

[5] Available from Standardization Documents Order Desk, DODSSP, Bldg. 4, Section D, 700 Robbins Ave., Philadelphia, PA 19111-5098, http://www.dodssp.daps.mil.

[6] Available from American Public Health Association, Publication Office, 1015 Fifteenth Street, NW, Washington, DC 20005.

[7] Available from Aerospace Industries Association of America, Inc. (AIA), 1000 Wilson Blvd., Suite 1700, Arlington, VA 22209-3928, http://www.aia-aerospace.org.

[8] Available from Society of Automotive Engineers (SAE), 400 Commonwealth Dr., Warrendale, PA 15096-0001, http://www.sae.org.

be rinsed from the surface, leaving the surface free of excessive fluorescent background. Proper emulsification time must be experimentally established and maintained to ensure that over-emulsification does not result in loss of indications.

7.2.2 *Water-Washable Penetrants* are formulated to be directly water-washable from the surface of the test part, after a suitable penetrant dwell time. Because the emulsifier is "built-in," water-washable penetrants can be washed out of discontinuities if the rinsing step is too long or too vigorous. It is therefore extremely important to exercise proper control in the removal of excess surface penetrant to ensure against overwashing. Some penetrants are less resistant to overwashing than others, so caution should be exercised.

7.2.3 *Solvent-Removable Penetrants* are formulated so that excess surface penetrant can be removed by wiping until most of the penetrant has been removed. The remaining traces should be removed with the solvent remover (see 8.6.4). To prevent removal of penetrant from discontinuities, care should be taken to avoid the use of excess solvent. Flushing the surface with solvent to remove the excess penetrant is prohibited as the penetrant indications could easily be washed away.

7.3 *Emulsifiers*:

7.3.1 *Lipophilic Emulsifiers* are oil-miscible liquids used to emulsify the post-emulsified penetrant on the surface of the part, rendering it water-washable. The individual characteristics of the emulsifier and penetrant, and the geometry/surface roughness of the part material contribute to determining the emulsification time.

7.3.2 *Hydrophilic Emulsifiers* are water-miscible liquids used to emulsify the excess post-emulsified penetrant on the surface of the part, rendering it water-washable. These water-base emulsifiers (detergent-type removers) are supplied as concentrates to be diluted with water and used as a dip or spray. The concentration, use and maintenance shall be in accordance with manufacturer's recommendations.

7.3.2.1 Hydrophilic emulsifiers function by displacing the excess penetrant film from the surface of the part through detergent action. The force of the water spray or air/mechanical agitation in an open dip tank provides the scrubbing action while the detergent displaces the film of penetrant from the part surface. The individual characteristics of the emulsifier and penetrant, and the geometry and surface roughness of the part material contribute to determining the emulsification time. Emulsification concentration shall be monitored weekly using a suitable refractometer.

7.4 *Solvent Removers*—Solvent removers function by dissolving the penetrant, making it possible to wipe the surface clean and free of excess penetrant.

7.5 *Developers*—Developers form a translucent or white absorptive coating that aids in bringing the penetrant out of surface discontinuities through blotting action, thus increasing the visibility of the indications.

7.5.1 *Dry Powder Developers*—Dry powder developers are used as supplied, that is, free-flowing, non-caking powder (see 8.8.1). Care should be taken not to contaminate the developer with fluorescent penetrant, as the contaminated developer specks can appear as penetrant indications.

7.5.2 *Aqueous Developers*—Aqueous developers are normally supplied as dry powder particles to be either suspended (water suspendable) or dissolved (water soluble) in water. The concentration, use and maintenance shall be in accordance with manufacturer's recommendations. Water soluble developers shall not be used with Type 2 penetrants or Type 1, Method A penetrants.

NOTE 5—Aqueous developers may cause stripping of indications if not properly applied and controlled. The procedure should be qualified in accordance with 10.2.

7.5.3 *Nonaqueous Wet Developers*—Nonaqueous wet developers are supplied as suspensions of developer particles in a nonaqueous solvent carrier ready for use as supplied. Nonaqueous, wet developers are sprayed on to form a thin coating on the surface of the part when dried. This thin coating serves as the developing medium.

NOTE 6—This type of developer is intended for application by spray only.

7.5.4 *Liquid Film Developers* are solutions or colloidal suspensions of resins/polymer in a suitable carrier. These developers will form a transparent or translucent coating on the surface of the part. Certain types of film developer may be stripped from the part and retained for record purposes (see 8.8.4).

8. Procedure

8.1 The following processing parameters apply to both fluorescent and visible penetrant testing methods.

8.2 *Temperature Limits*—The temperature of the penetrant materials and the surface of the part to be processed shall be between 40° and 125°F (4° and 52°C) or the procedure must be qualified at the temperature used as described in 10.2.

8.3 *Examination Sequence*—Final penetrant examination shall be performed after the completion of all operations that could cause surface-connected discontinuities or operations that could expose discontinuities not previously open to the surface. Such operations include, but are not limited to, grinding, welding, straightening, machining, and heat treating. Satisfactory inspection results can usually be obtained on surfaces in the as-welded, as-rolled, as-cast, as-forged, or ceramics in the densified condition.

8.3.1 *Surface Treatment*—Final penetrant examination may be performed prior to treatments that can smear the surface but not by themselves cause surface discontinuities. Such treatments include, but are not limited to, vapor blasting, deburring, sanding, buffing, sandblasting, or lapping. Performance of final penetrant examination after such surface treatments necessitates that the part(s) be etched to remove smeared metal from the surface prior to testing unless otherwise agreed by the contracting parties. Note that final penetrant examination shall always precede surface peening.

NOTE 7—Sand or shot blasting can close discontinuities so extreme care should be taken to avoid masking discontinuities. Under certain circumstances, however, grit blasting with certain air pressures and/or mediums may be acceptable without subsequent etching when agreed by the contracting parties.

NOTE 8—Surface preparation of structural or electronic ceramics for penetrant testing by grinding, sand blasting and etching is not recommended because of the potential for damage.

8.4 *Precleaning*—The success of any penetrant examination procedure is greatly dependent upon the surrounding surface and discontinuity being free of any contaminant (solid or liquid) that might interfere with the penetrant process. All parts or areas of parts to be examined must be clean and dry before the penetrant is applied. If only a section of a part, such as a weld, including the heat affected zone is to be examined, all contaminants shall be removed from the area being examined as defined by the contracting parties. "Clean" is intended to mean that the surface must be free of rust, scale, welding flux, weld spatter, grease, paint, oily films, dirt, and so forth, that might interfere with the penetrant process. All of these contaminants can prevent the penetrant from entering discontinuities (see Annex on Cleaning of Parts and Materials).

8.4.1 *Drying after Cleaning*—It is essential that the surface of parts be thoroughly dry after cleaning, since any liquid residue will hinder the entrance of the penetrant. Drying may be accomplished by warming the parts in drying ovens, with infrared lamps, forced hot air, or exposure to ambient temperature.

NOTE 9—Residues from cleaning processes such as strong alkalies, pickling solutions and chromates, in particular, may adversely react with the penetrant and reduce its sensitivity and performance.

8.5 *Penetrant Application*—After the part has been cleaned, dried, and is within the specified temperature range, the penetrant is applied to the surface to be examined so that the entire part or area under examination is completely covered with penetrant. Application methods include dipping, brushing, flooding, or spraying. Small parts are quite often placed in suitable baskets and dipped into a tank of penetrant. On larger parts, and those with complex geometries, penetrant can be applied effectively by brushing or spraying. Both conventional and electrostatic spray guns are effective means of applying liquid penetrants to the part surfaces. Not all penetrant materials are suitable for electrostatic spray applications, so tests should be conducted prior to use. Electrostatic spray application can eliminate excess liquid build-up of penetrant on the part, minimize overspray, and minimize the amount of penetrant entering hollow-cored passages which might serve as penetrant reservoirs, causing severe bleedout problems during examination. Aerosol sprays are conveniently portable and suitable for local application.

NOTE 10—With spray applications, it is important that there be proper ventilation. This is generally accomplished through the use of a properly designed spray booth and exhaust system.

8.5.1 *Penetrant Dwell Time*—After application, allow excess penetrant to drain from the part (care should be taken to prevent pools of penetrant from forming on the part), while allowing for proper penetrant dwell time (see Table 2). The length of time the penetrant must remain on the part to allow proper penetration should be as recommended by the penetrant manufacturer. Table 2, however, provides a guide for selection of penetrant dwell times for a variety of materials, forms, and types of discontinuities. Unless otherwise specified, the dwell time shall not exceed the maximum recommended by the manufacturer.

8.6 *Penetrant Removal*

8.6.1 *Water Washable (Method A)*:

8.6.1.1 *Removal of Water Washable Penetrant*—After the required penetrant dwell time, the excess penetrant on the surface being examined must be removed with water. It can be removed manually with a coarse spray or wiping the part surface with a dampened rag, automatic or semi-automatic water-spray equipment, or by water immersion. For immersion rinsing, parts are completely immersed in the water bath with air or mechanical agitation.

(*a*) The temperature of the water shall be maintained within the range of 50° to 100°F (10° to 38°C).

(*b*) Spray-rinse water pressure shall not exceed 40 psi (275 kPa). When hydro-air pressure spray guns are used, the air pressure should not exceed 25 psi (172 kPa).

NOTE 11—Overwashing should be avoided. Excessive washing can cause penetrant to be washed out of discontinuities. With fluorescent penetrant methods perform the rinsing operation under black light so that it can be determined when the surface penetrant has been adequately removed.

8.6.1.2 *Removal by Wiping (Method C)*—After the required penetrant dwell time, the excess penetrant is removed by wiping with a dry, clean, lint-free cloth/towel. Then use a clean lint-free cloth/towel lightly moistened with water or solvent to

TABLE 2 Recommended Minimum Dwell Times

Material	Form	Type of Discontinuity	Dwell Times[A] (minutes)	
			Penetrant[B]	Developer[C]
Aluminum, magnesium, steel, brass and bronze, titanium and high-temperature alloys	castings and welds	cold shuts, porosity, lack of fusion, cracks (all forms)	5	10
	wrought materials—extrusions, forgings, plate	laps, cracks (all forms)	10	10
Carbide-tipped tools		lack of fusion, porosity, cracks	5	10
Plastic	all forms	cracks	5	10
Glass	all forms	cracks	5	10
Ceramic	all forms	cracks, porosity	5	10

[A] For temperature range from 50° to 125°F (10° to 52°C). For temperatures between 40° and 50°F (4.4° and 10°C), recommend a minimum dwell time of 20 minutes.
[B] Maximum penetrant dwell time in accordance with 8.5.1.
[C] Development time begins as soon as wet developer coating has dried on surface of parts (recommended minimum). Maximum development time in accordance with 8.8.5.

remove the remaining traces of surface penetrant as determined by examination under black light for fluorescent methods and visible light for visible methods.

8.6.2 *Lipophilic Emulsification (Method B)*:

8.6.2.1 *Application of Lipophilic Emulsifier*—After the required penetrant dwell time, the excess penetrant on the part must be emulsified by immersing or flooding the parts with the required emulsifier (the emulsifier combines with the excess surface penetrant and makes the mixture removable by water rinsing). Lipophilic emulsifier shall not be applied by spray or brush and the part or emulsifier shall not be agitated while being immersed. After application of the emulsifier, the parts shall be drained and positioned in a manner that prevents the emulsifier from pooling on the part(s).

8.6.2.2 *Emulsification Time*—The emulsification time begins as soon as the emulsifier is applied. The length of time that the emulsifier is allowed to remain on a part and in contact with the penetrant is dependent on the type of emulsifier employed and the surface roughness. Nominal emulsification time should be as recommended by the manufacturer. The actual emulsification time must be determined experimentally for each specific application. The surface finish (roughness) of the part is a significant factor in the selection of and in the emulsification time of an emulsifier. Contact time shall be kept to the minimum time to obtain an acceptable background and shall not exceed three minutes.

8.6.2.3 *Post Rinsing*—Effective post rinsing of the emulsified penetrant from the surface can be accomplished using either manual, semi-automated, or automated water immersion or spray equipment or combinations thereof.

8.6.2.4 *Immersion*—For immersion post rinsing, parts are completely immersed in the water bath with air or mechanical agitation. The amount of time the part is in the bath should be the minimum required to remove the emulsified penetrant. In addition, the temperature range of the water should be 50 to 100°F (10 to 38°C). Any necessary touch-up rinse after an immersion rinse shall meet the requirements of 8.6.2.5.

8.6.2.5 *Spray Post Rinsing*—Effective post rinsing following emulsification can also be accomplished by either manual or automatic water spray rinsing. The water temperature shall be between 50 and 100°F (10 and 38°C). The water spray pressure shall not exceed 40 psi (275 kPa) when manual spray guns are used. When hydro-air pressure spray guns are used, the air pressure should not exceed 25 psi (172 kPa).

8.6.2.6 *Rinse Effectiveness*—If the emulsification and final rinse step is not effective, as evidenced by excessive residual surface penetrant after emulsification and rinsing; thoroughly reclean and completely reprocess the part.

8.6.3 *Hydrophilic Emulsification (Method D)*:

8.6.3.1 *Application of Hydrophilic Remover*—Following the required penetrant dwell time, the parts may be prerinsed with water prior to the application of hydrophilic emulsifier. This prerinse allows for the removal of excess surface penetrant from the parts prior to emulsification so as to minimize penetrant contamination in the hydrophilic emulsifier bath, thereby extending its life. It is not necessary to prerinse a part if a spray application of emulsifier is used.

8.6.3.2 *Prerinsing Controls*—Effective prerinsing is accomplished by manual, semi-automated, or automated water spray rinsing of the part(s). The water spray pressure shall not exceed 40 psi (275 kPa) when manual or hydro air spray guns are used. When hydro-air pressure spray guns are used, the air pressure shall not exceed 25 psi (172 kPa). Water free of contaminants that could clog spray nozzles or leave a residue on the part(s) is recommended.

8.6.3.3 *Application of Emulsifier*—The residual surface penetrant on part(s) must be emulsified by immersing the part(s) in an agitated hydrophilic emulsifier bath or by spraying the part(s) with water/emulsifier solutions thereby rendering the remaining residual surface penetrant water-washable for the final rinse station. The emulsification time begins as soon as the emulsifier is applied. The length of time that the emulsifier is allowed to remain on a part and in contact with the penetrant is dependent on the type of emulsifier employed and the surface roughness. The emulsification time should be determined experimentally for each specific application. The surface finish (roughness of the part is a significant factor in determining the emulsification time necessary for an emulsifier. Contact emulsification time should be kept to the least possible time consistent with an acceptable background and shall not exceed two minutes.

8.6.3.4 *Immersion*—For immersion application, parts shall be completely immersed in the emulsifier bath. The hydrophilic emulsifier concentration shall be as recommended by the manufacturer and the bath or part shall be gently agitated by air or mechanically throughout the cycle. The minimum time to obtain an acceptable background shall be used, but the dwell time shall not be more than two minutes unless approved by the contracting parties.

8.6.3.5 *Spray Application*—For spray applications, all part surfaces should be evenly and uniformly sprayed with a water/emulsifier solution to effectively emulsify the residual penetrant on part surfaces to render it water-washable. The concentration of the emulsifier for spray application should be in accordance with the manufacturer's recommendations, but it shall not exceed 5 %. The water spray pressure should be less than 40 psi (275 kpa). Contact with the emulsifier shall be kept to the minimum time to obtain an acceptable background and shall not exceed two minutes. The water temperature shall be maintained between 50 and 100°F (10 and 38°C).

8.6.3.6 *Post-Rinsing of Hydrophilic Emulsified Penetrants*—Effective post-rinsing of emulsified penetrant from the surface can be accomplished using either manual or automated water spray, water immersion, or combinations thereof. The total rinse time shall not exceed two minutes regardless of the number of rinse methods used.

8.6.3.7 *Immersion Post-Rinsing*—If an agitated immersion rinse is used, the amount of time the part(s) is (are) in the bath shall be the minimum required to remove the emulsified penetrant and shall not exceed two minutes. In addition, the temperature range of the water shall be within 50 and 100°F (10 and 38°C). Be aware that a touch-up rinse may be necessary after immersion rinse, but the total wash time still shall not exceed two minutes.

8.6.3.8 *Spray Post-Rinsing*—Effective post-rinsing following emulsification can also be accomplished by manual, semi-automatic, or automatic water spray. The water spray pressure shall not exceed 40 psi (275 kPa) when manual or hydro air spray guns are used. When hydro-air pressure spray guns are used, the air pressure shall not exceed 25 psi (172 kPa). The water temperature shall be between 50 and 100°F (10 and 38°C). The spray rinse time shall be less than two minutes, unless otherwise specified.

8.6.3.9 *Rinse Effectiveness*—If the emulsification and final rinse steps are not effective, as evidenced by excessive residual surface penetrant after emulsification and rinsing, thoroughly reclean, and completely reprocess the part.

8.6.4 *Removal of Solvent-Removable Penetrant (Method C)*—After the required penetrant dwell time, the excess penetrant is removed by wiping with a dry, clean, lint-free cloth/towel. Then use a clean, lint-free cloth/towel lightly moistened with solvent remover to remove the remaining traces of surface penetrant. Gentle wiping must be used to avoid removing penetrant from any discontinuity. On smooth surfaces, an alternate method of removal can be done by wiping with a clean, dry cloth. Flushing the surface with solvent following the application of the penetrant and prior to developing is prohibited.

8.7 *Drying*—Regardless of the type and method of penetrant used, drying the surface of the part(s) is necessary prior to applying dry or nonaqueous developers or following the application of the aqueous developer. Drying time will vary with the type of drying used and the size, nature, geometry, and number of parts being processed.

8.7.1 *Drying Parameters*—Components shall be air dried at room temperature or in a drying oven. Room temperature drying can be aided by the use of fans. Oven temperatures shall not exceed 160°F (71°C). Drying time shall only be that necessary to adequately dry the part. Components shall be removed from the oven after drying. Components should not be placed in the oven with pooled water or pooled aqueous solutions/suspensions.

8.8 *Developer Application*—There are various modes of effective application of the various types of developers such as dusting, immersing, flooding or spraying. The developer form, the part size, configuration, and surface roughness will influence the choice of developer application.

8.8.1 *Dry Powder Developer (Form A)*—Dry powder developers shall be applied after the part is dry in such a manner as to ensure complete coverage of the area of interest. Parts can be immersed in a container of dry developer or in a fluid bed of dry developer. They can also be dusted with the powder developer through a hand powder bulb or a conventional or electrostatic powder gun. It is common and effective to apply dry powder in an enclosed dust chamber, which creates an effective and controlled dust cloud. Other means suited to the size and geometry of the specimen may be used, provided the powder is applied evenly over the entire surface being examined. Excess developer powder may be removed by shaking or tapping the part, or by blowing with low-pressure dry, clean, compressed air not exceeding 5 psi (34 kPa). Dry developers shall not be used with Type II penetrant.

8.8.2 *Aqueous Developers (Forms B and C)*—Water soluble developers (Form B) are prohibited for use with Type 2 penetrants or Type 1, Method A penetrants. Water suspendable developers (Form C) can be used with both Type 1 and Type 2 penetrants. Aqueous developers shall be applied to the part immediately after the excess penetrant has been removed and prior to drying. Aqueous developers shall be prepared and maintained in accordance with the manufacturer's instructions and applied in such a manner as to ensure complete, even, part coverage. Aqueous developers may be applied by spraying, flowing, or immersing the part in a prepared developer bath. Immerse the parts only long enough to coat all of the part surfaces with the developer since indications may leach out if the parts are left in the bath too long. After the parts are removed from the developer bath, allow the parts to drain. Drain all excess developer from recesses and trapped sections to eliminate pooling of developer, which can obscure discontinuities. Dry the parts in accordance with 8.7. The dried developer coating appears as a translucent or white coating on the part.

8.8.3 *Nonaqueous Wet Developers (Forms D and E)*—After the excess penetrant has been removed and the surface has been dried, apply nonaqueous wet developer by spraying in such a manner as to ensure complete part coverage with a thin, even film of developer. The developer shall be applied in a manner appropriate to the type of penetrant being used. For visible dye, the developer must be applied thickly enough to provide a contrasting background. For fluorescent dye, the developer must be applied thinly to produce a translucent covering. Dipping or flooding parts with nonaqueous developers is prohibited, because the solvent action of these types of developers can flush or dissolve the penetrant from within the discontinuities.

NOTE 12—The vapors from the volatile solvent carrier in the developer may be hazardous. Proper ventilation should be provided at all times, but especially when the developer is applied inside a closed area.

8.8.4 *Liquid Film Developers*—Apply by spraying as recommended by the manufacturer. Spray parts in such a manner as to ensure complete part coverage of the area being examined with a thin, even film of developer.

8.8.5 *Developing Time*—The length of time the developer is to remain on the part prior to inspection shall be not less than ten minutes. Developing time begins immediately after the application of dry powder developer or as soon as the wet (aqueous or nonaqueous) developer coating is dry (that is, the water or solvent carrier has evaporated to dryness). The maximum permitted developing times shall be four hours for dry powder developer (Form A), two hours for aqueous developer (Forms B and C), and one hour for nonaqueous developer (Forms D and E).

8.9 *Inspection*—After the applicable development time, perform inspection of the parts under visible or ultraviolet light as appropriate. It may be helpful to observe the bleed out during the development time as an aid in interpreting indications.

8.9.1 *Ultraviolet Light Examination*—Examine parts tested with Type 1 fluorescent penetrant under black light in a darkened area. Ambient light shall not exceed 2 fc (21.5 lx).

The measurement shall be made with a suitable visible light sensor at the inspection surface.

NOTE 13—Because the fluorescent constituents in the penetrant will eventually fade with direct exposure to ultraviolet lights, direct exposure of the part under test to ultraviolet light should be minimized when not removing excess penetrant or evaluating indications.

8.9.1.1 *Black Light Level Control*—Black lights shall provide a minimum light intensity of 1000 µW/cm^2, at a distance of 15 in. (38.1 cm). The intensity shall be checked daily to ensure the required output. Reflectors and filters shall also be checked daily for cleanliness and integrity. Cracked or broken ultraviolet filters shall be replaced immediately. Since a drop in line voltage can cause decreased black light output with consequent inconsistent performance, a constant-voltage transformer should be used when there is evidence of voltage fluctuation.

NOTE 14—Certain high-intensity black lights may emit unacceptable amounts of visible light, which can cause fluorescent indications to disappear. Care should be taken to only use bulbs suitable for fluorescent penetrant testing purposes.

8.9.1.2 *Black Light Warm-Up*—Unless otherwise specified by the manufacturer, allow the black light to warm up for a minimum of five minutes prior to use or measurement of its intensity.

8.9.1.3 *Visual Adaptation*—Personnel examining parts after penetrant processing shall be in the darkened area for at least one minute before examining parts. Longer times may be necessary under some circumstances. Photochromic or tinted lenses shall not be worn during the processing and examination of parts.

8.9.2 *Visible Light Examination*—Inspect parts tested with Type 2 visible penetrant under either natural or artificial visible light. Proper illumination is required to ensure adequate sensitivity of the examination. A minimum light intensity at the examination surface of 100 fc (1076 lx) is required.

8.9.3 *Housekeeping*—Keep the examination area free of interfering debris, including fluorescent residues and objects.

8.9.4 *Indication Verification*—For Type 1 inspections only, it is common practice to verify indications by wiping the indication with a solvent-dampened swab or brush, allowing the area to dry, and redeveloping the area. Redevelopment time shall be a minimum of ten minutes, except nonaqueous redevelopment time should be a minimum of three minutes. If the indication does not reappear, the original indication may be considered false. This procedure may be performed up to two times for any given original indication.

8.9.5 *Evaluation*—All indications found during inspection shall be evaluated in accordance with acceptance criteria as specified. Reference Photographs of indications are noted in E433).

8.10 *Post Cleaning*—Post cleaning is necessary when residual penetrant or developer could interfere with subsequent processing or with service requirements. It is particularly important where residual penetrant testing materials might combine with other factors in service to produce corrosion and prior to vapor degreasing or heat treating the part as these processes can bake the developer onto the part. A suitable technique, such as a simple water rinse, water spray, machine wash, solvent soak, or ultrasonic cleaning may be employed (see Annex A1 for further information on post cleaning). It is recommended that if developer removal is necessary, it should be carried out as promptly as possible after examination so that the developer does not adhere to the part.

9. Special Requirements

9.1 *Impurities*:

9.1.1 When using penetrant materials on austenitic stainless steels, titanium, nickel-base or other high-temperature alloys, the need to restrict certain impurities such as sulfur, halogens and alkali metals must be considered. These impurities may cause embrittlement or corrosion, particularly at elevated temperatures. Any such evaluation shall also include consideration of the form in which the impurities are present. Some penetrant materials contain significant amounts of these impurities in the form of volatile organic solvents that normally evaporate quickly and usually do not cause problems. Other materials may contain impurities, which are not volatile and may react with the part, particularly in the presence of moisture or elevated temperatures.

9.1.2 Because volatile solvents leave the surface quickly without reaction under normal examination procedures, penetrant materials are normally subjected to an evaporation procedure to remove the solvents before the materials are analyzed for impurities. The residue from this procedure is then analyzed in accordance with Test Method D1552 or Test Method D129 decomposition followed by Test Method E516, Method B (Turbidimetric Method) for sulfur. The residue may also be analyzed by Test Method D808 or Annex A2 on Methods for Measuring Total Chlorine Content in Combustible Liquid Penetrant Materials (for halogens other than fluorine) and Annex A3 on Method for Measuring Total Fluorine Content in Combustible Liquid Penetration Materials (for fluorine). An alternative procedure, Annex A4 on Determination of Anions by Ion Chromatography, provides a single instrumental technique for rapid sequential measurement of common anions such as chloride, fluoride, and sulfate. Alkali metals in the residue are determined by flame photometry, atomic absorption spectrophotometry, or ion chromatography (see ASTM D4327).

NOTE 15—Some current standards require impurity levels of sulfur and halogens to not exceed 1 % of any one suspect element. This level, however, may be unacceptable for some applications, so the actual maximum acceptable impurity level must be decided between supplier and user on a case by case basis.

9.2 *Elevated-Temperature Testing*—Where penetrant testing is performed on parts that must be maintained at elevated temperature during examination, special penetrant materials and processing techniques may be required. Such examination requires qualification in accordance with 10.2 and the manufacturer's recommendations shall be observed.

10. Qualification and Requalification

10.1 *Personnel Qualification*—When required by the customer, all penetrant testing personnel shall be qualified/certified in accordance with a written procedure conforming to the applicable edition of recommended Practice SNT-TC-1A, ANSI/ASNT CP-189, NAS-410, or MIL-STD-410.

10.2 *Procedure Qualification*—Qualification of procedures using times, conditions, or materials differing from those specified in this general practice or for new materials may be performed by any of several methods and should be agreed upon by the contracting parties. A test piece containing one or more discontinuities of the smallest relevant size is generally used. When agreed upon by the contracting parties, the test piece may contain real or simulated discontinuities, providing it displays the characteristics of the discontinuities encountered in product examination.

10.2.1 Requalification of the procedure to be used may be required when a change is made to the procedure or when material substitution is made.

10.3 *Nondestructive Testing Agency Qualification*—If a nondestructive testing agency as described in Practice E543 is used to perform the examination, the agency should meet the requirements of Practice E543.

10.4 *Requalification* may be required when a change or substitution is made in the type of penetrant materials or in the procedure (see 10.2).

11. Keywords

11.1 fluorescent liquid penetrant testing; hydrophilic emulsification; lipophilic emulsification; liquid penetrant testing; nondestructive testing; solvent removable; visible liquid penetrant testing; water-washable; post-emulsified; black light; ultraviolet light; visible light

ANNEXES

(Mandatory Information)

A1. CLEANING OF PARTS AND MATERIALS

A1.1 Choice of Cleaning Method

A1.1.1 The choice of a suitable cleaning method is based on such factors as: (*1*) type of contaminant to be removed since no one method removes all contaminants equally well; (*2*) effect of the cleaning method on the parts; (*3*) practicality of the cleaning method for the part (for example, a large part cannot be put into a small degreaser or ultrasonic cleaner); and (*4*) specific cleaning requirements of the purchaser. The following cleaning methods are recommended:

A1.1.1.1 *Detergent Cleaning*—Detergent cleaners are non-flammable water-soluble compounds containing specially selected surfactants for wetting, penetrating, emulsifying, and saponifying various types of soils, such as grease and oily films, cutting and machining fluids, and unpigmented drawing compounds, etc. Detergent cleaners may be alkaline, neutral, or acidic in nature, but must be noncorrosive to the item being inspected. The cleaning properties of detergent solutions facilitate complete removal of soils and contamination from the surface and void areas, thus preparing them to absorb the penetrant. Cleaning time should be as recommended by the manufacturer of the cleaning compound.

A1.1.1.2 *Solvent Cleaning*—There are a variety of solvent cleaners that can be effectively utilized to dissolve such soils as grease and oily films, waxes and sealants, paints, and in general, organic matter. These solvents should be residue-free, especially when used as a hand-wipe solvent or as a dip-tank degreasing solvent. Solvent cleaners are not recommended for the removal of rust and scale, welding flux and spatter, and in general, inorganic soils. Some cleaning solvents are flammable and can be toxic. Observe all manufacturers' instructions and precautionary notes.

A1.1.1.3 *Vapor Degreasing*—Vapor degreasing is a preferred method of removing oil or grease-type soils from the surface of parts and from open discontinuities. It will not remove inorganic-type soils (dirt, corrosion, salts, etc.), and may not remove resinous soils (plastic coatings, varnish, paint, etc.). Because of the short contact time, degreasing may not completely clean out deep discontinuities and a subsequent solvent soak is recommended.

A1.1.1.4 *Alkaline Cleaning*:

(*a*) Alkaline cleaners are nonflammable water solutions containing specially selected detergents for wetting, penetrating, emulsifying, and saponifying various types of soils. Hot alkaline solutions are also used for rust removal and descaling to remove oxide scale which can mask surface discontinuities. Alkaline cleaner compounds must be used in accordance with the manufacturers' recommendations. Parts cleaned by the alkaline cleaning process must be rinsed completely free of cleaner and thoroughly dried prior to the penetrant testing process (part temperature at the time of penetrant application shall not exceed 125°F (52°C).

(*b*) Steam cleaning is a modification of the hot-tank alkaline cleaning method, which can be used for preparation of large, unwieldy parts. It will remove inorganic soils and many organic soils from the surface of parts, but may not reach to the bottom of deep discontinuities, and a subsequent solvent soak is recommended.

A1.1.1.5 *Ultrasonic Cleaning*—This method adds ultrasonic agitation to solvent or detergent cleaning to improve cleaning efficiency and decrease cleaning time. It should be used with water and detergent if the soil to be removed is inorganic (rust, dirt, salts, corrosion products, etc.), and with organic solvent if the soil to be removed is organic (grease and oily films, etc.). After ultrasonic cleaning, parts must be rinsed completely free of cleaner, thoroughly dried, and cooled to at least 125°F (52°C), before application of penetrant.

A1.1.1.6 *Paint Removal*—Paint films can be effectively removed by bond release solvent paint remover or disintegrating-type hot-tank alkaline paint strippers. In most cases, the paint film must be completely removed to expose the surface of the metal. Solvent-type paint removers can be of the high-viscosity thickened type for spray or brush application or can be of low viscosity two-layer type for dip-tank application. Both types of solvent paint removers are generally used at ambient temperatures, as received. Hot-tank alkaline strippers should be used in accordance with the manufacturer's instructions. After paint removal, the parts must be thoroughly rinsed to remove all contamination from the void openings, thoroughly dried, and cooled to at least 125°F (52°C) before application of penetrant.

A1.1.1.7 *Mechanical Cleaning and Surface Conditioning*—Metal-removing processes such as filing, buffing, scraping, mechanical milling, drilling, reaming, grinding, liquid honing, sanding, lathe cutting, tumble or vibratory deburring, and abrasive blasting, including abrasives such as glass beads, sand, aluminum oxide, ligno-cellulose pellets, metallic shot, etc., are often used to remove such soils as carbon, rust and scale, and foundry adhering sands, as well as to deburr or produce a desired cosmetic effect on the part. *These processes may decrease the effectiveness of the penetrant testing by smearing or peening over metal surfaces and filling discontinuities open to the surface, especially for soft metals such as aluminum, titanium, magnesium, and beryllium alloy.*

A1.1.1.8 *Acid Etching*—Inhibited acid solutions (pickling solutions) are routinely used for descaling part surfaces. Descaling is necessary to remove oxide scale, which can mask surface discontinuities and prevent penetrant from entering. Acid solutions/etchants are also used routinely to remove smeared metal that peens over surface discontinuities. Such etchants should be used in accordance with the manufacturers' recommendations.

NOTE A1.1—Etched parts and materials should be rinsed completely free of etchants, the surface neutralized and thoroughly dried by heat prior to application of penetrants. Acids and chromates can adversely affect the fluorescence of fluorescent materials.

NOTE A1.2—Whenever there is a possibility of hydrogen embrittlement as a result of acid solution/etching, the part should be baked at a suitable temperature for an appropriate time to remove the hydrogen before further processing. After baking, the part shall be cooled to a temperature below 125°F (52°C) before applying penetrants.

A1.1.1.9 *Air Firing of Ceramics*—Heating of a ceramic part in a clean, oxidizing atmosphere is an effective way of removing moisture or light organic soil or both. The maximum temperature that will not cause degradation of the properties of the ceramic should be used.

A1.2 Post Cleaning

A1.2.1 *Removal of Developer*—Dry powder developer can be effectively removed with an air blow-off (free of oil) or it can be removed with water rinsing. Wet developer coatings can be removed effectively by water rinsing or water rinsing with detergent either by hand or with a mechanical assist (scrub brushing, machine washing, etc.). The soluble developer coatings simply dissolve off of the part with a water rinse.

A1.2.2 Residual penetrant may be removed through solvent action. Solvent soaking (15 min minimum), and ultrasonic solvent cleaning (3 min minimum) techniques are recommended. In some cases, it is desirable to vapor degrease, then follow with a solvent soak. The actual time required in the vapor degreaser and solvent soak will depend on the nature of the part and should be determined experimentally.

A2. METHODS FOR MEASURING TOTAL CHLORINE CONTENT IN COMBUSTIBLE LIQUID PENETRANT MATERIALS

A2.1 Scope and Application

A2.1.1 These methods cover the determination of chlorine in combustible liquid penetrant materials, liquid or solid. Its range of applicability is 0.001 to 5 % using either of the alternative titrimetric procedures. The procedures assume that bromine or iodine will not be present. If these elements are present, they will be detected and reported as chlorine. The full amount of these elements will not be reported. Chromate interferes with the procedures, causing low or nonexistent end points. The method is applicable only to materials that are totally combustible.

A2.2 Summary of Methods

A2.2.1 The sample is oxidized by combustion in a bomb containing oxygen under pressure (see A2.2.1.1). The chlorine compounds thus liberated are absorbed in a sodium carbonate solution and the amount of chloride present is determined titrimetrically either against silver nitrate with the end point detected potentiometrically (Method A) or coulometrically with the end point detected by current flow increase (Method B).

A2.2.1.1 *Safety*—Strict adherence to all of the provisions prescribed hereinafter ensures against explosive rupture of the bomb, or a blow-out, provided the bomb is of proper design and construction and in good mechanical condition. It is desirable, however, that the bomb be enclosed in a shield of steel plate at least ½ in. (12.7 mm) thick, or equivalent protection be provided against unforeseeable contingencies.

A2.3 Apparatus

A2.3.1 *Bomb*, having a capacity of not less than 300 mL, so constructed that it will not leak during the test, and that quantitative recovery of the liquids from the bomb may be readily achieved. The inner surface of the bomb may be made of stainless steel or any other material that will not be affected by the combustion process or products. Materials used in the bomb assembly, such as the head gasket and leadwire insulation, shall be resistant to heat and chemical action, and shall not undergo any reaction that will affect the chlorine content of the liquid in the bomb.

A2.3.2 *Sample Cup*, platinum, 24 mm in outside diameter at the bottom, 27 mm in outside diameter at the top, 12 mm in height outside and weighing 10 to 11 g, opaque fused silica, wide-form with an outside diameter of 29 mm at the top, a height of 19 mm, and a 5-mL capacity (Note 1), or nickel (Kawin capsule form), top diameter of 28 mm, 15 mm in height, and 5-mL capacity.

NOTE A2.1—Fused silica crucibles are much more economical and longer-lasting than platinum. After each use, they should be scrubbed out with fine, wet emery cloth, heated to dull red heat over a burner, soaked in hot water for 1 h, then dried and stored in a desiccator before reuse.

A2.3.3 *Firing Wire*, platinum, approximately No. 26 B & S gage.

A2.3.4 *Ignition Circuit* (Note A2.2), capable of supplying sufficient current to ignite the nylon thread or cotton wicking without melting the wire.

NOTE A2.2—The switch in the ignition circuit should be of a type that remains open, except when held in closed position by the operator.

A2.3.5 *Nylon Sewing Thread*, or *Cotton Wicking*, white.

A2.4 Purity of Reagents

A2.4.1 Reagent grade chemicals shall be used in all tests. Unless otherwise indicated, it is intended that all reagents shall conform to the specifications of the Committee on Analytical Reagents of the American Chemical Society, where such specifications are available.[9] Other grades may be used provided it is first ascertained that the reagent is of sufficiently high purity to permit its use without lessening the accuracy of the determination.

A2.4.2 Unless otherwise indicated, references to water shall be understood to mean referee grade reagent water conforming to Specification D1193.

A2.5 Decomposition

A2.5.1 *Reagents and Materials*:

A2.5.1.1 *Oxygen*, free of combustible material and halogen compounds, available at a pressure of 40 atm (4.05 MPa).

A2.5.1.2 *Sodium Carbonate Solution (50 g Na_2CO_3/L)*—Dissolve 50 g of anhydrous Na_2CO_3 or 58.5 g of $Na_2CO_3 \cdot {}_2O$) or 135 g of $Na_2CO_3 \cdot 10H_2O$ in water and dilute to 1 L.

A2.5.1.3 *White Oil*, refined.

A2.5.2 *Procedure*:

A2.5.2.1 *Preparation of Bomb and Sample*—Cut a piece of firing wire approximately 100 mm in length. Coil the middle section (about 20 mm) and attach the free ends to the terminals. Arrange the coil so that it will be above and to one side of the sample cup. Place 5 mL of Na_2CO_3 solution in the bomb (Note A2.3), place the cover on the bomb and vigorously shake for 15 s to distribute the solution over the inside of the bomb. Open the bomb, place the sample-filled sample cup in the terminal holder, and insert a short length of thread between the firing wire and the sample. Use of a sample weight containing over 20 mg of chlorine may cause corrosion of the bomb. The sample weight should not exceed 0.4 g if the expected chlorine content is 2.5 % or above. If the sample is solid, not more than 0.2 g should be used. Use 0.8 g of white oil with solid samples. If white oil will be used (Note A2.4), add it to the sample cup by means of a dropper at this time (see Note A2.5 and Note A2.6).

NOTE A2.3—After repeated use of the bomb for chlorine determination, a film may be noticed on the inner surface. This dullness should be removed by periodic polishing of the bomb. A satisfactory method for doing this is to rotate the bomb in a lathe at about 300 rpm and polish the inside surface with Grit No. 2/0 or equivalent paper coated with a light machine oil to prevent cutting, and then with a paste of grit-free chromic oxide and water. This procedure will remove all but very deep pits and put a high polish on the surface. Before using the bomb, it should be washed with soap and water to remove oil or paste left from the polishing operation. Bombs with porous or pitted surfaces should never be used because of the tendency to retain chlorine from sample to sample. It is recommended to not use more than 1 g total of sample and white oil or other chlorine-free combustible material.

NOTE A2.4—If the sample is not readily miscible with white oil, some other nonvolatile, chlorine-free combustible diluent may be employed in place of white oil. However, the combined weight of sample and nonvolatile diluent shall not exceed 1 g. Some solid additives are relatively insoluble, but may be satisfactorily burned when covered with a layer of white oil.

NOTE A2.5—The practice of running alternately samples high and low in chlorine content should be avoided whenever possible. It is difficult to rinse the last traces of chlorine from the walls of the bomb and the tendency for residual chlorine to carry over from sample to sample has been observed in a number of laboratories. When a sample high in chlorine has preceded one low in chlorine content, the test on the low-chlorine sample should be repeated and one or both of the low values thus obtained should be considered suspect if they do not agree within the limits of repeatability of this method.

A2.5.2.2 *Addition of Oxygen*—Place the sample cup in position and arrange the nylon thread, or wisp of cotton so that the end dips into the sample. Assemble the bomb and tighten the cover securely. Admit oxygen (see Note A2.6) slowly (to avoid blowing the sample from the cup) until a pressure is reached as indicated in Table A2.1.

NOTE A2.6—It is recommended to not add oxygen or ignite the sample if the bomb has been jarred, dropped, or tilted.

A2.5.2.3 *Combustion*—Immerse the bomb in a cold-water bath. Connect the terminals to the open electrical circuit. Close the circuit to ignite the sample. Remove the bomb from the bath after immersion for at least ten minutes. Release the pressure at a slow, uniform rate such that the operation requires not less than 1 min. Open the bomb and examine the contents. If traces of unburned oil or sooty deposits are found, discard the determination, and thoroughly clean the bomb before again putting it in use (Note A2.3).

TABLE A2.1 Gauge Pressures

Capacity of Bomb, mL	Gauge Pressure, atm (MPa)	
	min[A]	max
300 to 350	38 (3.85)	40 (4.05)
350 to 400	35 (3.55)	37 (3.75)
400 to 450	30 (3.04)	32 (3.24)
450 to 500	27 (2.74)	29 (2.94)

[A] The minimum pressures are specified to provide sufficient oxygen for complete combustion and the maximum pressures present a safety requirement.

---

[9] *Reagent Chemicals, American Chemical Society Specifications*, American Chemical Society, Washington, DC. For suggestions on the testing of reagents not listed by the American Chemical Society, see *Analar Standards for Laboratory Chemicals*, BDH Ltd., Poole, Dorset, U.K., and the *United States Pharmacopeia and National Formulary*, U.S. Pharmaceutical Convention, Inc. (USPC), Rockville, MD.

A2.6 Analysis, Method A, Potentiometric Titration Procedure

A2.6.1 *Apparatus*:
A2.6.1.1 *Silver Billet Electrode*.
A2.6.1.2 *Glass Electrode*, pH measurement type.
A2.6.1.3 *Buret*, 25-mL capacity, 0.05-mL graduations.
A2.6.1.4 *Millivolt Meter*, or expanded scale pH meter capable of measuring 0 to 220 mV.

NOTE A2.7—An automatic titrator is highly recommended in place of items A2.6.1.3 and A2.6.1.4. Repeatability and sensitivity of the method are much enhanced by the automatic equipment while much tedious effort is avoided.

A2.6.2 *Reagents and Materials*:
A2.6.2.1 *Acetone*, chlorine-free.
A2.6.2.2 *Methanol*, chlorine-free.
A2.6.2.3 *Silver Nitrate Solution (0.0282 N)*—Dissolve 4.7910 ± 0.0005 g of silver nitrate ($AgNO_3$) in water and dilute to 1 L.
A2.6.2.4 *Sodium Chloride Solution (0.0282 N)*—Dry a few grams of sodium chloride (NaCl) for 2 h at 130 to 150°C, weigh out 1.6480 ± 0.0005 g of the dried NaCl, dissolve in water, and dilute to 1 L.
A2.6.2.5 *Sulfuric Acid (1 + 2)*—Mix 1 volume of concentrated sulfuric acid (H_2SO_4, sp. gr 1.84) with 2 volumes of water.

A2.6.3 *Collection of Chlorine Solution*—Remove the sample cup with clean forceps and place in a 400-mL beaker. Wash down the walls of the bomb shell with a fine stream of methanol from a wash bottle, and pour the washings into the beaker. Rinse any residue into the beaker. Next, rinse the bomb cover and terminals into the beaker. Finally, rinse both inside and outside of the sample crucible into the beaker. Washings should equal but not exceed 100 mL. Add methanol to make 100 mL.

A2.6.4 *Determination of Chlorine*—Add 5 mL of H_2SO_4 (1:2) to acidify the solution (solution should be acid to litmus and clear of white Na_2CO_3 precipitate). Add 100 mL of acetone. Place the electrodes in the solution, start the stirrer (if mechanical stirrer is to be used), and begin titration. If titration is manual, set the pH meter on the expanded millivolt scale and note the reading. Add exactly 0.1 mL of $AgNO_3$ solution from the buret. Allow a few seconds stirring; then record the new millivolt reading. Subtract the second reading from the first. Continue the titration, noting each amount of $AgNO_3$ solution and the amount of difference between the present reading and the last reading. Continue adding 0.1-mL increments, making readings and determining differences between readings until a maximum difference between readings is obtained. The total amount of $AgNO_3$ solution required to produce this maximum differential is the end point. Automatic titrators continuously stir the sample, add titrant, measure the potential difference, calculate the differential, and plot the differential on a chart. The maximum differential is taken as the end point.

NOTE A2.8—For maximum sensitivity, 0.00282 N $AgNO_3$ solution may be used with the automatic titrator. This dilute reagent should not be used with large samples or where chlorine content may be over 0.1 % since these tests will cause end points of 10 mL or higher. The large amount of water used in such titrations reduces the differential between readings, making the end point very difficult to detect. For chlorine contents over 1 % in samples of 0.8 g or larger, 0.282 N $AgNO_3$ solution will be required to avoid exceeding the 10-mL water dilution limit.

A2.6.5 *Blank*—Make blank determinations with the amount of white oil used but omitting the sample. (Liquid samples normally require only 0.15 to 0.25 g of white oil while solids require 0.7 to 0.8 g.) Follow normal procedure, making two or three test runs to be sure the results are within the limits of repeatability for the test. Repeat this blank procedure whenever new batches of reagents or white oil are used. The purpose of the blank run is to measure the chlorine in the white oil, the reagents, and that introduced by contamination.

A2.6.6 *Standardization*—Silver nitrate solutions are not permanently stable, so the true activity should be checked when the solution is first made up and then periodically during the life of the solution. This is done by titration of a known NaCl solution as follows: Prepare a mixture of the amounts of the chemicals (Na_2CO_3 solution, H_2SO_4 solution, acetone, and methanol) specified for the test. Pipet in 5.0 mL of 0.0282-N NaCl solution and titrate to the end point. Prepare and titrate a similar mixture of all the chemicals except the NaCl solution, thus obtaining a reagent blank reading. Calculate the normality of the $AgNO_3$ solution as follows:

$$N_{AgNO3} = \frac{5.0 \times N_{NaCl}}{V_A - V_B} \quad (A2.1)$$

where:
N_{AgNO3} = normality of the $AgNO_3$ solution,
N_{NaCl} = normality of the NaCl solution,
V_A = millilitres of $AgNO_3$ solution used for the titration including the NaCl solution, and
V_B = millilitres of $AgNO_3$ solution used for the titration of the reagents only.

A2.6.7 *Calculation*—Calculate the chlorine content of the sample as follows:

$$\text{Chlorine, weight \%} = \frac{(V_S - V_B) \times N \times 3.545}{W} \quad (A2.2)$$

where:
V_S = millilitres of $AgNO_3$ solution used by the sample,
V_B = millilitres of $AgNO_3$ solution used by the blank,
N = normality of the $AgNO_3$ solution, and
W = grams of sample used.

A2.6.8 *Precision and Accuracy*:
A2.6.8.1 The following criteria should be used for judging the acceptability of results:
A2.6.8.1.1 *Repeatability*—Results by the same analyst should not be considered suspect unless they differ by more than 0.006 % or 10.5 % of the value determined, whichever is higher.
A2.6.8.1.2 *Reproducibility*—Results by different laboratories should not be considered suspect unless they differ by more than 0.013 % or 21.3 % of the value detected, whichever is higher.
A2.6.8.1.3 *Accuracy*—The average recovery of the method is 86 % to 89 % of the actual amount present.

A2.7 Analysis, Method B, Coulometric Titration

A2.7.1 *Apparatus*:
A2.7.1.1 *Coulometric Chloride Titrator*.

A2.7.1.2 *Beakers*, two, 100-mL, or glazed crucibles (preferably with 1½ in.-outside diameter bottom).
A2.7.1.3 *Refrigerator*.
A2.7.2 *Reagents*:
A2.7.2.1 *Acetic Acid, Glacial*.
A2.7.2.2 *Dry Gelatin Mixture*. [10]
A2.7.2.3 *Nitric Acid*.
A2.7.2.4 *Sodium Chloride Solution*—100 meq C/1. Dry a quantity of NaCl for 2 h at 130 to 150°C. Weigh out 5.8440 ± 0.0005 g of dried NaCl in a closed container, dissolve in water, and dilute to 1 L.
A2.7.3 *Reagent Preparation*:

NOTE A2.9—The normal reagent preparation process has been slightly changed, due to the interference from the 50 mL of water required to wash the bomb. This modified process eliminates the interference and does not alter the quality of the titration.

A2.7.3.1 *Gelatin Solution*—A typical preparation is: Add approximately 1 L of hot distilled or deionized water to the 6.2 g of dry gelatin mixture contained in one vial supplied by the equipment manufacturer. Gently heat with continuous mixing until the gelatin is completely dissolved.
A2.7.3.2 Divide into aliquots each sufficient for one day's analyses. (Thirty millilitres is enough for approximately eleven titrations.) Keep the remainder in a refrigerator, but do not freeze. The solution will keep for about six months in the refrigerator. When ready to use, immerse the day's aliquot in hot water to liquefy the gelatin.
A2.7.3.3 *Glacial Acetic Acid-Nitric Acid Solution*—A typical ratio is 12.5 to 1 (12.5 parts CH_3COOH to 1 part HNO_3).
A2.7.3.4 Mix enough gelatin solution and of acetic acid-nitric acid mixture for one titration. (A typical mixture is 2.5 mL of gelatin solution and 5.4 mL of acetic-nitric acid mixture.)

NOTE A2.10—The solution may be premixed in a larger quantity for convenience, but may not be useable after 24 h.

A2.7.3.5 Run at least three blank values and take an average according to the operating manual of the titrator. Determine separate blanks for both five drops of mineral oil and 20 drops of mineral oil.
A2.7.4 *Titration*:
A2.7.4.1 Weigh to the nearest 0.1 g and record the weight of the 100-mL beaker.

A2.7.4.2 Remove the sample crucible from the cover assembly support ring using a clean forceps, and, using a wash bottle, rinse both the inside and the outside with water into the 100-mL beaker.
A2.7.4.3 Empty the bomb shell into the 100-mL beaker. Wash down the sides of the bomb shell with water, using a wash bottle.
A2.7.4.4 Remove the cover assembly from the cover assembly support, and, using the wash bottle, rinse the under side, the platinum wire, and the terminals into the same 100-mL beaker. The total amount of washings should be 50 ± 1 g.
A2.7.4.5 Add specified amounts of gelatin mixture and acetic acid-nitric acid mixture, or gelatin mix-acetic acid-nitric acid mixture, if this was premixed, into the 100-mL beaker that contains the 50 g of washings including the decomposed sample.
A2.7.4.6 Titrate using a coulometric titrimeter, according to operating manual procedure.
A2.7.5 *Calculations*—Calculate the chloride ion concentration in the sample as follows:

$$\text{Chlorine, weight \%} = \frac{(P - B) \times M}{W} \quad (A2.3)$$

where:
P = counter reading obtained with the sample,
B = average counter reading obtained with average of the three blank readings,
M = standardization constant. This is dependent on the instrument range setting in use and the reading obtained with a known amount of the 100 meq of Cl per litre of solution, and
W = weight of sample used, g.

A2.7.6 *Precision and Accuracy*:
A2.7.6.1 Duplicate results by the same operator can be expected to exhibit the following relative standard deviations:

Approximate % Chlorine	RSD, %
1.0 and above	0.10
0.1	2.5
0.003	5.9

A2.7.6.2 The method can be expected to report values that vary from the true value by the following amounts:

0.1 % chlorine and above	±2%
0.001 to 0.01 % chlorine	±9%.

A2.7.6.3 If bromine is present, 36.5 % of the true amount will be reported. If iodine is present, 20.7 % of the true amount will be reported. Fluorine will not be detected.

---

[10] May be purchased from the equipment supplier. A typical mixture consists of 6 g of gelatin powder, 0.1 g of thymol blue, water-soluble, and 0.1 g of thymol, reagent grade, crystal.

A3. METHOD FOR MEASURING TOTAL FLUORINE CONTENT IN COMBUSTIBLE LIQUID PENETRANT MATERIALS

A3.1 Scope and Application

A3.1.1 This method covers the determination of fluorine in combustible liquid penetrant materials, liquid or solid, that do not contain appreciable amounts of interfering elements, or have any insoluble residue after combustion. Its range of applicability is 1 to 200 000 ppm.

A3.1.2 The measure of the fluorine content employs the fluoride selective ion electrode.

A3.2 Summary of Method

A3.2.1 The sample is oxidized by combustion in a bomb containing oxygen under pressure (see A3.2.1.1). The fluorine compounds thus liberated are absorbed in a sodium citrate solution and the amount of fluorine present is determined potentiometrically through the use of a fluoride selective ion electrode.

A3.2.1.1 *Safety*—Strict adherence to all of the provisions prescribed hereinafter ensures against explosive rupture of the bomb, or a blow-out, provided the bomb is of proper design and construction and in good mechanical condition. It is desirable, however, that the bomb be enclosed in a shield of steel plate at least ½ in. (12.7 mm) thick, or equivalent protection be provided against unforeseeable contingencies.

A3.3 Interferences

A3.3.1 Silicon, calcium, aluminum, magnesium, and other metals forming precipitates with fluoride ion will interfere if they are present in sufficient concentration to exceed the solubility of their respective fluorides. Insoluble residue after combustion will entrain fluorine even if otherwise soluble.

A3.4 Apparatus

A3.4.1 *Bomb*, having a capacity of not less than 300 mL, so constructed that it will not leak during the test, and that quantitative recovery of the liquids from the bomb may be readily achieved. The inner surface of the bomb may be made of stainless steel or any other material that will not be affected by the combustion process or products. Materials used in the bomb assembly, such as the head gasket and leadwire insulation, shall be resistant to heat and chemical action, and shall not undergo any reaction that will affect the fluorine content of the liquid in the bomb.

A3.4.2 *Sample Cup*, nickel, 20 mm in outside diameter at the bottom, 28 mm in outside diameter at the top, and 16 mm in height; or platinum, 24 mm in outside diameter at the bottom, 27 mm in outside diameter at the top, 12 mm in height, and weighing 10 to 11 g.

A3.4.3 *Firing Wire*, platinum, approximately No. 26 B & S gage.

A3.4.4 *Ignition Circuit* (Note A3.1), capable of supplying sufficient current to ignite the nylon thread or cotton wicking without melting the wire.

NOTE A3.1—**Caution:** The switch in the ignition circuit should be of a type that remains open, except when held in closed position by the operator.

A3.4.5 *Nylon Sewing Thread*, or *Cotton Wicking*, white.

A3.4.6 *Funnel*, polypropylene (Note A3.2).

A3.4.7 *Volumetric Flask*, polypropylene, 100-mL (Note A3.2).

A3.4.8 *Beaker*, polypropylene, 150-mL (Note A3.2).

A3.4.9 *Pipet*, 100-µL, Eppendorf-type (Note A3.2).

A3.4.10 *Magnetic Stirrer* and TFE-coated magnetic stirring bar.

A3.4.11 *Fluoride Specific Ion Electrode* and suitable reference electrode.

A3.4.12 *Millivolt Meter* capable of measuring to 0.1 mV.

NOTE A3.2—Glassware should never be used to handle a fluoride solution as it will remove fluoride ions from solution or on subsequent use carry fluoride ion from a concentrated solution to one more dilute.

A3.5 Reagents

A3.5.1 *Purity of Reagents*—Reagent grade chemicals shall be used in all tests. Unless otherwise indicated, it is intended that all reagents shall conform to the specifications of the Committee on Analytical Reagents of the American Chemical Society, where such specifications are available.[9] Other grades may be used, provided it is first ascertained that the reagent is of sufficiently high purity to permit its use without lessening the accuracy of the determination.

A3.5.2 *Purity of Water*—Unless otherwise indicated, all references to water shall be understood to mean Type I reagent water conforming to Specification D1193.

A3.5.3 *Fluoride Solution, Stock (2000 ppm)*—Dissolve 4.4200 ± 0.0005 g of predried (at 130 to 150°C for 1 h, then cooled in a desiccator) sodium fluoride in distilled water and dilute to 1 L.

A3.5.4 *Oxygen*, free of combustible material and halogen compounds, available at a pressure of 40 atm (4.05 MPa).

A3.5.5 *Sodium Citrate Solution*—Dissolve 27 g of sodium citrate dihydrate in water and dilute to 1 L.

A3.5.6 *Sodium Hydroxide Solution (5 N)*—Dissolve 200 g of sodium hydroxide (NaOH) pellets in water and dilute to 1 L; store in a polyethylene container.

A3.5.7 *Wash Solution (Modified TISAB, Total Ionic Strength Adjustment Buffer)*—To 300 mL of distilled water, add 32 mL of glacial acetic acid, 6.6 g of sodium citrate dihydrate, and 32.15 g of sodium chloride. Stir to dissolve and then adjust the pH to 5.3 using 5 *N* NaOH solution. Cool and dilute to 1 L.

A3.5.8 *White Oil*, refined.

A3.6 Decomposition Procedure

A3.6.1 *Preparation of Bomb and Sample*—Cut a piece of firing wire approximately 100 mm in length. Coil the middle section (about 20 mm) and attach the free ends to the terminals. Arrange the coil so that it will be above and to one side of the sample cup. Place 10 mL of sodium citrate solution in the bomb, place the cover on the bomb, and vigorously shake for 15 s to distribute the solution over the inside of the bomb. Open the bomb, place the sample-filled sample cup in the terminal holder, and insert a short length of thread between the firing wire and the sample. The sample weight used should not exceed 1 g. If the sample is a solid, add a few drops of white oil at this time to ensure ignition of the sample.

NOTE A3.3—Use of sample weights containing over 20 mg of chlorine may cause corrosion of the bomb. To avoid this it is recommended that for samples containing over 2 % chlorine, the sample weight be based on the following table:

Chlorine Content, %	Sample weight, g	White Oil weight, g
2 to 5	0.4	0.4
5 to 10	0.2	0.6
10 to 20	0.1	0.7
20 to 50	0.05	0.7

Do not use more than 1 g total of sample and white oil or other fluorine-free combustible material.

A3.6.2 *Addition of Oxygen*—Place the sample cup in position and arrange the nylon thread, or wisp of cotton so that the end dips into the sample. Assemble the bomb and tighten the cover securely. Admit oxygen (see Note A3.4) slowly (to avoid blowing the sample from the cup) until a pressure is reached as indicated in Table A3.1.

NOTE A3.4—**Caution:** It is recommended to not add oxygen or ignite the sample if the bomb has been jarred, dropped, or tilted.

A3.6.3 *Combustion*—Immerse the bomb in a cold-water bath. Connect the terminals to the open electrical circuit. Close the circuit to ignite the sample. Remove the bomb from the bath after immersion for at least 10 min. Release the pressure at a slow, uniform rate such that the operation requires not less than 1 min. Open the bomb and examine the contents. If traces of unburned oil or sooty deposits are found, discard the determination, and thoroughly clean the bomb before again putting it in use.

A3.6.4 *Collection of Fluorine Solution*—Remove the sample cup with clean forceps and rinse with wash solution into a 100-mL volumetric flask. Rinse the walls of the bomb shell with a fine stream of wash solution from a wash bottle,

TABLE A3.1 Gauge Pressures

Capacity of Bomb, mL	Gauge Pressure atm (MPa)	
	min[A]	max
300 to 350	38	40
350 to 400	35	37
400 to 450	30	32
450 to 500	27	29

[A] The minimum pressures are specified to provide sufficient oxygen for complete combustion and the maximum pressures present a safety requirement.

and add the washings to the flask. Next, rinse the bomb cover and terminals into the volumetric flask. Finally, add wash solution to bring the contents of the flask to the line.

A3.7 Procedure

A3.7.1 Ascertain the slope (millivolts per ten-fold change in concentration) of the electrode as described by the manufacturer.

A3.7.2 Obtain a blank solution by performing the procedure without a sample.

A3.7.3 Immerse the fluoride and reference electrodes in solutions and obtain the equilibrium reading to 0.1 mV. (The condition of the electrode determines the length of time necessary to reach equilibrium. This may be as little as 5 min or as much as 20 min.)

A3.7.4 Add 100 µL of stock fluoride solution and obtain the reading after the same length of time necessary for A3.7.3.

A3.8 Calculation

A3.8.1 Calculate the fluorine content of the sample as follows:

$$\text{Fluorine, ppm} = \frac{\left[\dfrac{2 \times 10^{-4}}{10^{\Delta E_1/S} - 1} - \dfrac{2 \times 10^{-4}}{10^{\Delta E_2/S} - 1}\right]}{W} \times 10^6 \quad (A3.1)$$

where:
ΔE_1 = millivolt change in sample solution on addition of 100 µL of stock fluoride solution,
ΔE_2 = millivolt change in blank solution on addition of 100 µL of the stock fluoride solution,
S = slope of fluoride electrode as determined in A3.7.1, and
W = grams of sample.

A3.9 Precision and Bias

A3.9.1 *Repeatability*—The results of two determinations by the same analyst should not be considered suspect unless they differ by more than 1.1 ppm (0.00011 %) or 8.0 % of the amount detected, whichever is greater.

A3.9.2 *Reproducibility*—The results of two determinations by different laboratories should not be considered suspect unless they differ by 6.7 ppm or 129.0 % of the amount detected, whichever is greater.

A3.9.3 *Bias*—The average recovery of the method is 62 to 64 % of the amount actually present although 83 to 85 % recoveries can be expected with proper technique.

A4. DETERMINATION OF ANIONS BY ION CHROMATOGRAPHY WITH CONDUCTIVITY MEASUREMENT

A4.1 Scope and Application

A4.1.1 This method is condensed from ASTM procedures and APHA Method 429 and optimized for the analysis of detrimental substances in organic based materials. It provides a single instrumental technique for rapid, sequential measurement of common anions such as bromide, chloride, fluoride, nitrate, nitrite, phosphate, and sulfate.

A4.2 Summary of Method

A4.2.1 The material must be put in the form of an aqueous solution before analysis can be attempted. The sample is oxidized by combustion in a bomb containing oxygen under pressure. The products liberated are absorbed in the eluant present in the bomb at the time of ignition. This solution is washed from the bomb, filtered, and diluted to a known volume.

A4.2.1.1 A filtered aliquot of sample is injected into a stream of carbonate-bicarbonate eluant and passed through a series of ion exchangers. The anions of interest are separated on the basis of their relative affinities for a low capacity, strongly basic anion exchanger (guard and separator column). The separated anions are directed onto a strongly acidic cation exchanger (suppressor column) where they are converted to their highly conductive acid form and the carbonate-bicarbonate eluant is converted to weakly conductive carbonic acid. The separated anions in their acid form are measured by conductivity. They are identified on the basis of retention time as compared to standards. Quantitation is by measurement of peak area or peak height. Blanks are prepared and analyzed in a similar fashion.

A4.2.2 *Interferences*—Any substance that has a retention time coinciding with that of any anion to be determined will interfere. For example, relatively high concentrations of low-molecular-weight organic acids interfere with the determination of chloride and fluoride. A high concentration of any one ion also interferes with the resolution of others. Sample dilution overcomes many interferences. To resolve uncertainties of identification or quantitation use the method of known additions. Spurious peaks may result from contaminants in reagent water, glassware, or sample processing apparatus. Because small sample volumes are used, scrupulously avoid contamination.

A4.2.3 *Minimum Detectable Concentration*—The minimum detectable concentration of an anion is a function of sample size and conductivity scale used. Generally, minimum detectable concentrations are in the range of 0.05 mg/L for F^- and 0.1 mg/L for Br^-, Cl^-, NO_3^-, NO_2^-, PO_4^{3-}, and SO_4^{2-} with a 100-µL sample loop and a 10-µmho full-scale setting on the conductivity detector. Similar values may be achieved by using a higher scale setting and an electronic integrator.

A4.3 Apparatus

A4.3.1 *Bomb*, having a capacity of not less than 300 mL, so constructed that it will not leak during the test, and that quantitative recovery of the liquids from the bomb may be readily achieved. The inner surface of the bomb may be made of stainless steel or any other material that will not be affected by the combustion process or products. Materials used in the bomb assembly, such as the head gasket and leadwire insulation, shall be resistant to heat and chemical action, and shall not undergo any reaction that will affect the chlorine content of the liquid in the bomb.

A4.3.2 *Sample Cup*, platinum, 24 mm in outside diameter at the bottom, 27 mm in outside diameter at the top, 12 mm in height outside, and weighing 10 to 11 g; opaque fused silica, wide-form with an outside diameter of 29 mm at the top, a height of 19 mm, and a 5-mL capacity (Note A4.1), or nickel (Kawin capsule form), top diameter of 28 mm, 15 mm in height, and 5-mL capacity.

NOTE A4.1—Fused silica crucibles are much more economical and longer lasting than platinum. After each use, they should be scrubbed out with fine, wet emery cloth, heated to dull red heat over a burner, soaked in hot water for 1 h then dried and stored in a desiccator before reuse.

A4.3.3 *Firing Wire*, platinum, approximately No. 26 B and S gage.

A4.3.4 *Ignition Circuit* (Note A4.2), capable of supplying sufficient current to ignite the nylon thread or cotton wicking without melting the wire.

NOTE A4.2—The switch in the ignition circuit should be of a type that remains open, except when held in closed position by the operator.

A4.3.5 *Nylon Sewing Thread*, or *Cotton Wicking*, white.

A4.3.6 *Ion Chromatograph*, including an injection valve, a sample loop, guard, separator, and suppressor columns, a temperature-compensated small-volume conductivity cell (6 µL or less), and a strip chart recorder capable of full-scale response of 2 s or less. An electronic peak integrator is optional. The ion chromatograph shall be capable of delivering 2 to 5 mL eluant/min at a pressure of 1400 to 6900 kPa.

A4.3.7 *Anion Separator Column*, with styrene divinylbenzene-based low-capacity pellicular anion-exchange resin capable of resolving Br^-, Cl^-, F^-, NO_3^-, NO_2^-, PO_4^{3-}, and SO_4^{2-}; 4 × 250 mm.

A4.3.8 *Guard Column*, identical to separator column except 4 × 50 mm, to protect separator column from fouling by particulates or organics.

A4.3.9 *Suppressor Column*, high-capacity cation-exchange resin capable of converting eluant and separated anions to their acid forms.

A4.3.10 *Syringe*, minimum capacity of 2 mL and equipped with a male pressure fitting.

A4.4 Reagents

A4.4.1 *Purity of Reagents*—Reagent grade chemicals shall be used in all tests. Unless otherwise indicated, it is intended that all reagents shall conform to the specifications of the Committee on Analytical Reagents of the American Chemical Society, where such specifications are available.[9] Other grades may be used provided it is first ascertained that the reagent has sufficiently high purity to permit its use without lessening the accuracy of the determination.

A4.4.2 *Deionized or Distilled Water*, free from interferences at the minimum detection limit of each constituent and filtered through a 0.2-µm membrane filter to avoid plugging columns.

A4.4.3 *Eluant Solution*, sodium bicarbonate-sodium carbonate, 0.003M $NaHCO_3$ 0.0024M Na_2CO_3: dissolve 1.008 g $NaHCO_3$ and 1.0176 g Na_2CO_3 in water and dilute to 4 L.

A4.4.4 *Regenerant Solution 1*, H_2SO_4, 1 N, use this regenerant when suppressor is not a continuously regenerated one.

A4.4.5 *Regenerant Solution 2*, H_2SO_4, 0.025 N, dilute 2.8 mL conc H_2SO_4 to 4 L or 100 mL regenerant solution 1 to 4 L. Use this regenerant with continuous regeneration fiber suppressor system.

A4.4.6 *Standard Anion Solutions*, 1000 mg/L, prepare a series of standard anion solutions by weighing the indicated amount of salt, dried to a constant weight at 105°C, to 1000 mL. Store in plastic bottles in a refrigerator; these solutions are stable for at least one month.

Anion	Salt	Amount, g/L
Cl^-	NaCl	1.6485
F^-	NaF	2.2100
Br^-	NaBr	1.2876
NO_3^-	$NaNO_3$	1.3707
NO_2^-	$NaNO_2$	1.4998
PO_4^{3-}	KH_2PO_4	1.4330
SO_4^{2-}	K_2SO_4	1.8141

A4.4.7 *Combined Working Standard Solution, High Range*—Combine 10 mL of the Cl^-, F^-, NO_3^-, NO_2^-, and PO_4^{3-} standard anion solutions, 1 mL of the Br^-, and 100 mL of the SO_4^{2-} standard solutions, dilute to 1000 mL, and store in a plastic bottle protected from light; contains 10 mg/L each of Cl^-, F^-, NO_3^-, NO_2^-, and PO_4^{3-}, 1 mg Br^-/L, and 100 mg SO_4^{2-}/L. Prepare fresh daily.

A4.4.8 *Combined Working Standard Solution, Low Range*—Dilute 100 mL combined working standard solution, high range, to 1000 mL and store in a plastic bottle protected from light; contains 1.0 mg/L each Cl^-, F^-, NO_3^-, NO_2^-, and PO_4^{3-}, 0.1 mg Br^-/L, and 10 mg SO_4^{2-}/L. Prepare fresh daily.

A4.4.9 *Alternative Combined Working Standard Solutions*—Prepare appropriate combinations according to anion concentration to be determined. If NO_2^- and PO_4^{3-} are not included, the combined working standard is stable for one month.

A4.5 Decomposition Procedure

A4.5.1 *Preparation of Bomb and Sample*—Cut a piece of firing wire approximately 100 mm in length. Coil the middle section (about 20 mm) and attach the free ends to the terminals. Arrange the coil so that it will be above and to one side of the sample cup. Place 5 mL of $Na_2CO_3/NaHCO_3$ solution in the bomb, place the cover on the bomb, and vigorously shake for 15 s to distribute the solution over the inside of the bomb. Open the bomb, place the sample-filled sample cup in the terminal holder, and insert a short length of thread between the firing wire and the sample. The sample weight used should not exceed 1 g. If the sample is a solid, add a few drops of white oil at this time to ensure ignition of the sample.

NOTE A4.3—Use of sample weights containing over 20 mg of chlorine may cause corrosion of the bomb. To avoid this it is recommended that for samples containing over 2 % chlorine, the sample weight be based on the following:

Chlorine content, %	Sample weight, g	White Oil weight, g
2 to 5	0.4	0.4
5 to 10	0.2	0.6
10 to 20	0.1	0.7
20 to 50	0.05	0.7

CAUTION: Do not use more than 1 g total of sample and white oil or other fluorine-free combustible material.

A4.5.2 *Addition of Oxygen*—Place the sample cup in position and arrange the nylon thread, or wisp of cotton so that the end dips into the sample. Assemble the bomb and tighten the cover securely. Admit oxygen (see Note A4.4) slowly (to avoid blowing the sample from the cup) until a pressure is reached as indicated in Table A4.1.

NOTE A4.4—It is recommended to not add oxygen or ignite the sample if the bomb has been jarred, dropped, or tilted.

A4.5.3 *Combustion*—Immerse the bomb in a cold-water bath. Connect the terminals to the open electrical circuit. Close the circuit to ignite the sample. Remove the bomb from the bath after immersion for at least 10 min. Release the pressure at a slow, uniform rate such that the operation requires not less than 1 min. Open the bomb and examine the contents. If traces of unburned oil or sooty deposits are found, discard the determination, and thoroughly clean the bomb before again putting it in use.

A4.5.4 *Collection of Solution*—Remove the sample cup with clean forceps and rinse with deionized water and filter the washings into a 100-mL volumetric flask. Rinse the walls of the bomb shell with a fine stream of deionized water from a wash bottle, and add the washings through the filter paper to the flask. Next, rinse the bomb cover and terminals and add the washings through the filter into the volumetric flask. Finally, add deionized water to bring the contents of the flask to the line. Use aliquots of this solution for the ion chromatography (IC) analysis.

A4.6 Procedure

A4.6.1 *System Equilibration*—Turn on ion chromatograph and adjust eluant flow rate to approximate the separation achieved in Fig. A4.1 (2 to 3 mL/min). Adjust detector to desired setting (usually 10 µmho) and let system come to equilibrium (15 to 20 min). A stable base line indicates equilibrium conditions. Adjust detector offset to zero-out eluant conductivity; with the fiber suppressor adjust the regeneration flow rate to maintain stability, usually 2.5 to 3 mL/min.

A4.6.1.1 Set up the ion chromatograph in accordance with the manufacturer's instructions.

A4.6.2 *Calibration*—Inject standards containing a single anion or a mixture and determine approximate retention times. Observed times vary with conditions but if standard eluant and anion separator column are used, retention always is in the order F^-, Cl^-, NO_2^-, PO_4^{3-}, Br^-, NO_3^-, and SO_4^{2-}. Inject at least three different concentrations for each anion to be measured and construct a calibration curve by plotting peak height or area against concentration on linear graph paper. Recalibrate whenever the detector setting is changed. With a system requiring suppressor regeneration, NO_2^- interaction with the suppressor may lead to erroneous NO_2^- results; make this determination only when the suppressor is at the same stage of exhaustion as during standardization or recalibrate

TABLE A4.1 Gage Pressures

Capacity of Bomb, mL	Gage Pressures, atm	
	mm[A]	max
300 to 350	38	40
350 to 400	35	37
400 to 450	30	32
450 to 500	27	29

[A] The minimum pressures are specified to provide sufficient oxygen for complete combustion and the maximum pressures present a safety requirement.

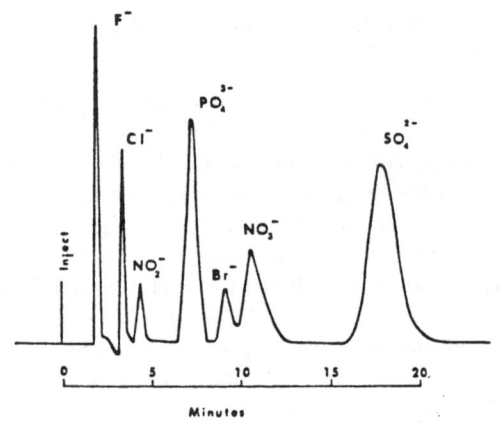

FIG. A4.1 Typical Anion Profile

frequently. In this type of system the water dip (see NoteNote A4.4) may shift slightly during suppressor exhaustion and with a fast run column this may lead to slight interference for F^- or Cl^-. To eliminate this interference, analyze standards that bracket the expected result or eliminate the water dip by diluting the sample with eluant or by adding concentrated eluant to the sample to give the same HCO_3^-/CO_3^{2-} concentration as in the eluant. If sample adjustments are made, adjust standards and blanks identically.

NOTE A4.5—Water dip occurs because water conductivity in sample is less than eluant conductivity (eluant is diluted by water).

A4.6.2.1 If linearity is established for a given detector setting, it is acceptable to calibrate with a single standard. Record the peak height or area and retention time to permit calculation of the calibration factor, F.

A4.6.3 *Sample Analysis*—Remove sample particulates, if necessary, by filtering through a prewashed 0.2-μm-pore-diam membrane filter. Using a prewashed syringe of 1 to 10 mL capacity equipped with a male luer fitting inject sample or standard. Inject enough sample to flush sample loop several times: for 0.1 mL sample loop inject at least 1 mL. Switch ion chromatograph from load to inject mode and record peak heights and retention times on strip chart recorder. After the last peak (SO_4^{2-}) has appeared and the conductivity signal has returned to base line, another sample can be injected.

A4.6.4 *Regeneration*—For systems without fiber suppressor regenerate with 1 N H_2SO_4 in accordance with the manufacturer's instructions when the conductivity base line exceeds 300 μmho when the suppressor column is on line.

A4.7 Calculation

A4.7.1 Calculate concentration of each anion, in mg/L, by referring to the appropriate calibration curve. Alternatively, when the response is shown to be linear, use the following equation:

$$C = H \times F \times D \qquad (A4.1)$$

where:
C = mg anion/L,
H = peak height or area,
F = response factor – concentration of standard/height (or area) of standard, and
D = dilution factor for those samples requiring dilution.

A4.8 Precision and Bias

A4.8.1 Samples of reagent water to which were added the common anions were analyzed in 15 laboratories with the results shown in Table A4.2.

TABLE A4.2 Precision and Accuracy Observed for Anions at Various Concentration Levels in Reagent Water

Anion	Amount Added, mg/L	Amount Found, mg/L	Overall Precision, mg/L	Single-Operator Precision, mg/L	Significant Bias 95 % Level
F^-	0.48	0.49	0.05	0.03	No
F^-	4.84	4.64	0.52	0.46	No
Cl^-	0.76	0.86	0.38	0.11	No
Cl^-	17	17.2	0.82	0.43	No
Cl^-	455	471	46	13	No
NO_2^-	0.45	0.09	0.09	0.04	Yes, neg
NO_2^-	21.8	19.4	1.9	1.3	Yes, neg
Br^-	0.25	0.25	0.04	0.02	No
Br^-	13.7	12.9	1.0	0.6	No
PO_4^{3-}	0.18	0.10	0.06	0.03	Yes, neg
PO_4^{3-}	0.49	0.34	0.15	0.17	Yes, neg
NO_3^-	0.50	0.33	0.16	0.03	No
NO_3^-	15.1	14.8	1.15	0.9	No
SO_4^{2-}	0.51	0.52	0.07	0.03	No
SO_4^{2-}	43.7	43.5	2.5	2.2	No

ASTM International takes no position respecting the validity of any patent rights asserted in connection with any item mentioned in this standard. Users of this standard are expressly advised that determination of the validity of any such patent rights, and the risk of infringement of such rights, are entirely their own responsibility.

This standard is subject to revision at any time by the responsible technical committee and must be reviewed every five years and if not revised, either reapproved or withdrawn. Your comments are invited either for revision of this standard or for additional standards and should be addressed to ASTM International Headquarters. Your comments will receive careful consideration at a meeting of the responsible technical committee, which you may attend. If you feel that your comments have not received a fair hearing you should make your views known to the ASTM Committee on Standards, at the address shown below.

This standard is copyrighted by ASTM International, 100 Barr Harbor Drive, PO Box C700, West Conshohocken, PA 19428-2959, United States. Individual reprints (single or multiple copies) of this standard may be obtained by contacting ASTM at the above address or at 610-832-9585 (phone), 610-832-9555 (fax), or service@astm.org (e-mail); or through the ASTM website (www.astm.org).

Designation: E709 − 08

Standard Guide for
Magnetic Particle Testing[1]

This standard is issued under the fixed designation E709; the number immediately following the designation indicates the year of original adoption or, in the case of revision, the year of last revision. A number in parentheses indicates the year of last reapproval. A superscript epsilon (ε) indicates an editorial change since the last revision or reapproval.

This standard has been approved for use by agencies of the Department of Defense.

1. Scope

1.1 This guide[2] describes techniques for both dry and wet magnetic particle testing, a nondestructive method for detecting cracks and other discontinuities at or near the surface in ferromagnetic materials. Magnetic particle testing may be applied to raw material, semifinished material (billets, blooms, castings, and forgings), finished material and welds, regardless of heat treatment or lack thereof. It is useful for preventive maintenance testing.

1.1.1 This guide is intended as a reference to aid in the preparation of specifications/standards, procedures and techniques.

1.2 This guide is also a reference that may be used as follows:

1.2.1 To establish a means by which magnetic particle testing, procedures recommended or required by individual organizations, can be reviewed to evaluate their applicability and completeness.

1.2.2 To aid in the organization of the facilities and personnel concerned in magnetic particle testing.

1.2.3 To aid in the preparation of procedures dealing with the examination of materials and parts. This guide describes magnetic particle testing techniques that are recommended for a great variety of sizes and shapes of ferromagnetic materials and widely varying examination requirements. Since there are many acceptable differences in both procedure and technique, the explicit requirements should be covered by a written procedure (see Section 21).

1.3 This guide does not indicate, suggest, or specify acceptance standards for parts/pieces examined by these techniques. It should be pointed out, however, that after indications have been produced, they must be interpreted or classified and then evaluated. For this purpose there should be a separate code, specification, or a specific agreement to define the type, size, location, degree of alignment and spacing, area concentration, and orientation of indications that are unacceptable in a specific part versus those which need not be removed before part acceptance. Conditions where rework or repair is not permitted should be specified.

1.4 This guide describes the use of the following magnetic particle method techniques.

1.4.1 Dry magnetic powder (see 8.4),

1.4.2 Wet magnetic particle (see 8.5),

1.4.3 Magnetic slurry/paint magnetic particle (see 8.5.7), and

1.4.4 Polymer magnetic particle (see 8.5.8).

1.5 *Personnel Qualification*—Personnel performing examinations in accordance with this guide should be qualified and certified in accordance with ASNT Recommended Practice No. SNT-TC-1A, ANSI/ASNT Standard CP-189, NAS 410, or as specified in the contract or purchase order.

1.6 *Nondestructive Testing Agency*—If a nondestructive testing agency as described in Practice E543 is used to perform the examination, the nondestructive testing agency should meet the requirements of Practice E543.

1.7 The numerical values shown in inch-pound units are to be regarded as the standard. SI units are provided for information only.

1.8 *This standard does not purport to address all of the safety concerns, if any, associated with its use. It is the responsibility of the user of this standard to establish appropriate safety and health practices and determine the applicability of regulatory limitations prior to use.*

2. Referenced Documents

2.1 *ASTM Standards:*[3]

A275/A275M Practice for Magnetic Particle Examination of Steel Forgings

A456/A456M Specification for Magnetic Particle Examination of Large Crankshaft Forgings

D93 Test Methods for Flash Point by Pensky-Martens Closed Cup Tester

D445 Test Method for Kinematic Viscosity of Transparent

---

[1] This guide is under the jurisdiction of ASTM Committee E07 on Nondestructive Testing and is the direct responsibility of Subcommittee E07.03 on Liquid Penetrant and Magnetic Particle Methods.
Current edition approved Feb. 15, 2008. Published April 2008. Originally approved in 1980. Last previous edition approved in 2001 as E709 - 01. DOI: 10.1520/E0709-08.
[2] For ASME Boiler and Pressure Vessel Code Applications see related Guide SE-709 in Section II of that Code.

[3] For referenced ASTM standards, visit the ASTM website, www.astm.org, or contact ASTM Customer Service at service@astm.org. For *Annual Book of ASTM Standards* volume information, refer to the standard's Document Summary page on the ASTM website.

Copyright © ASTM International, 100 Barr Harbor Drive, PO Box C700, West Conshohocken, PA 19428-2959, United States.

and Opaque Liquids (and Calculation of Dynamic Viscosity)

E165 Practice for Liquid Penetrant Examination for General Industry

E543 Specification for Agencies Performing Nondestructive Testing

E1316 Terminology for Nondestructive Examinations

E1444 Practice for Magnetic Particle Testing

E2297 Guide for Use of UV-A and Visible Light Sources and Meters used in the Liquid Penetrant and Magnetic Particle Methods

2.2 *Society of Automotive Engineers (SAE): Aerospace Materials Specifications:*[4]

AMS 2300 Premium Aircraft Quality Steel Cleanliness Magnetic Particle Inspection Procedure

AMS 2301 Aircraft Quality Steel Cleanliness Magnetic Particle Inspection Procedure

AMS 2303 Aircraft Quality Steel Cleanliness Martensitic Corrosion Resistant Steels Magnetic Particle Inspection Procedure

AMS 2641 Vehicle Magnetic Particle Inspection

AMS 3040 Magnetic Particles, Non-fluorescent, Dry Method

AMS 3041 Magnetic Particles, Non-fluorescent, Wet Method, Oil Vehicle, Ready to Use

AMS 3042 Magnetic Particles, Non-fluorescent, Wet Method, Dry Powder

AMS 3043 Magnetic Particles, Non-fluorescent, Oil Vehicle, Aerosol Packaged

AMS 3044 Magnetic Particles, Fluorescent, Wet Method, Dry Powder

AMS 3045 Magnetic Particles, Non-fluorescent, Wet Method, Oil Vehicle, Ready to Use

AMS 3046 Magnetic Particles, Non-fluorescent, Wet Method, Oil Vehicle, Aerosol Packaged

AMS 5062 Steel, Low Carbon Bars, Forgings, Tubing, Sheet, Strip, and Plate 0.25 Carbon, Maximum

AMS 5355 Investment Castings

AMS-I-83387 Inspection Process, Magnetic Rubber

AS 4792 Water Conditioning Agents for Aqueous Magnetic Particle Inspection

AS 5282 Tool Steel Ring Standard for Magnetic Particle Inspection

AS 5371 Reference Standards Notched Shims for Magnetic Particle Inspection

2.3 *American Society for Nondestructive Testing:*[5]

SNT-TC-1A Personnel Qualification and Certification in Nondestructive Testing

CP-189 ASNT Qualification and Certification of Nondestructive Testing Personnel

2.4 *Federal Standards:*[6]

A-A-59230 Fluid, Magnetic Particle Inspection, Suspension

FED-STD 313 Material Safety Data Sheets Preparation and the Submission of

2.5 *OSHA Document:*[7]

29CFR 1910.1200 Hazard Communication

2.6 *AIA Documents:*[8]

NAS 410 Nondestructive Testing Personnel Qualification and Certification

3. Terminology

3.1 For definitions of terms used in the practice, refer to Terminology E1316

4. Summary of Guide

4.1 *Principle*—The magnetic particle method is based on establishing a magnetic field with high flux density in a ferromagnetic material. The flux lines must spread out when they pass through non-ferromagnetic material such as air in a discontinuity or an inclusion. Because flux lines can not cross, this spreading action may force some of the flux lines out of the material (flux leakage). Flux leakage is also caused by reduction in ferromagnetic material (cross-sectional change), a sharp dimensional change, or the end of the part. If the flux leakage is strong enough, fine magnetic particles will be held in place and an accumulation of particles will be visible under the proper lighting conditions. While there are variations in the magnetic particle method, they all are dependent on this principle, that magnetic particles will be retained at the locations of magnetic flux leakage. The amount of flux leakage at discontinuities depends primarily on the following factors; flux density in the material, and size, orientation, and proximity to the surface of a discontinuity. With longitudinal fields, all of the flux lines must complete their loops though air and an excessively strong magnetic field may interfere with examination near the flux entry and exit points due to the high flux-density present at these points.

4.2 *Method*—While this practice permits and describes many variables in equipment, materials, and procedures, there are three steps essential to the method:

4.2.1 The part must be magnetized.

4.2.2 Magnetic particles of the type designated in the contract/purchase order/specification should be applied while the part is magnetized or immediately thereafter.

4.2.3 Any accumulation of magnetic particles must be observed, interpreted, and evaluated.

4.3 *Magnetization*:

4.3.1 *Ways to Magnetize*—A ferromagnetic material can be magnetized either by passing an electric current through the material or by placing the material within a magnetic field originated by an external source. The entire mass or a portion of the mass can be magnetized as dictated by size and equipment capacity or need. As previously noted, in order to be detectable, the discontinuity must interrupt the normal path of the magnetic field lines. If a discontinuity is open to the

---

[4] Available from Society of Automotive Engineers (SAE), 400 Commonwealth Dr., Warrendale, PA 15096-0001, http://www.sae.org.

[5] Available from American Society for Nondestructive Testing (ASNT), P.O. Box 28518, 1711 Arlingate Ln., Columbus, OH 43228-0518, http://www.asnt.org.

[6] Available from Standardization Documents Order Desk, DODSSP, Bldg. 4, Section D, 700 Robbins Ave., Philadelphia, PA 19111-5098, http://www.dodssp.daps.mil.

[7] Available from Occupational Safety and Health Administration (OSHA), 200 Constitution Ave., NW, Washington, DC 20210, http://www.osha.gov.

[8] Available from Aerospace Industries Association of America, Inc. (AIA), 1000 Wilson Blvd., Suite 1700, Arlington, VA 22209-3928, http://www.aia-aerospace.org.

surface, the flux leakage attracting the particles will be at the maximum value for that particular discontinuity. When that same discontinuity is below the surface, flux leakage evident on the surface will be a lesser value.

4.3.2 *Field Direction*—If a discontinuity is oriented parallel to the magnetic field lines, it may be essentially undetectable. Therefore, since discontinuities may occur in any orientation, it may be necessary to magnetize the part or the area of interest twice or more sequentially in different directions by the same method or a combination of different methods (see Section 13) to induce magnetic field lines in a suitable direction in which to perform an adequate examination.

4.3.3 *Field Strength*—The magnetic field must be of sufficient strength to indicate those discontinuities which are unacceptable, yet must not be so strong that an excess of local particle accumulation masks relevant indications (see Section 14).

4.4 *Types of Magnetic Particles and Their Use*—There are various types of magnetic particles available for use in magnetic particle testing. They are available as dry powders (fluorescent and nonfluorescent) ready for use as supplied (see 8.4), powder concentrates (fluorescent and nonfluorescent) for dispersion in water or suspending in light petroleum distillates (see 8.5), magnetic slurries/paints (see 8.5.7), and magnetic polymer dispersions (see 8.5.8).

4.5 *Evaluation of Indications*—When the material to be examined has been properly magnetized, the magnetic particles have been properly applied, and the excess particles properly removed, there will be accumulations of magnetic particles remaining at the points of flux leakage. These accumulations show the distortion of the magnetic field and are called indications. Without disturbing the particles, the indications must be examined, classified, compared with the acceptance standards, and a decision made concerning the disposition of the material that contains the indication.

4.6 *Typical Magnetic Particle Indications*:

4.6.1 *Surface Discontinuities*—Surface discontinuities, with few exceptions, produce sharp, distinct patterns (see Annex A1).

4.6.2 *Near-surface discontinuities*—Near-surface discontinuities produce less distinct indications than those open to the surface. The patterns tend to be broad, rather than sharp, and the particles are less tightly held (see Annex A1).

5. Significance and Use

5.1 The magnetic particle method of nondestructive testing indicates the presence of surface and near-surface discontinuities in materials that can be magnetized (ferromagnetic). This method can be used for production examination of parts/components or structures and for field applications where portability of equipment and accessibility to the area to be examined are factors. The ability of the method to find small discontinuities can be enhanced by using fluorescent particles suspended in a suitable vehicle and by introducing a magnetic field of the proper strength whose orientation is as close as possible to 90° to the direction of the suspected discontinuity (see 4.3.2). A smoother surface or a pulsed current improves mobility of the magnetic particles under the influence of the magnetic field to collect on the surface where magnetic flux leakage occurs.

6. Equipment

6.1 *Types*—There are a number of types of equipment available for magnetizing ferromagnetic parts and components. With the exception of a permanent magnet, all equipment requires a power source capable of delivering the required current levels to produce the magnetic field. The current used dictates the sizes of cables and the capability of relays, switching contacts, meters and rectifier if the power source is alternating current.

6.2 *Portability*—Portability, which includes the ability to hand carry the equipment, can be obtained from yokes, portable coils with power supplies, and capacitor discharge power supplies with cables. Generally, portable coils provide high magnetizing forces by using higher numbers of turns to compensate for their lower current flow. Capacitor discharge units use high current storage capacity and provide these high current levels for only a very short duration.

6.3 *Yokes*—Yokes are usually C-shaped electromagnets which induce a magnetic field between the poles (legs) and are used for local magnetization (Fig. 1). Many portable yokes

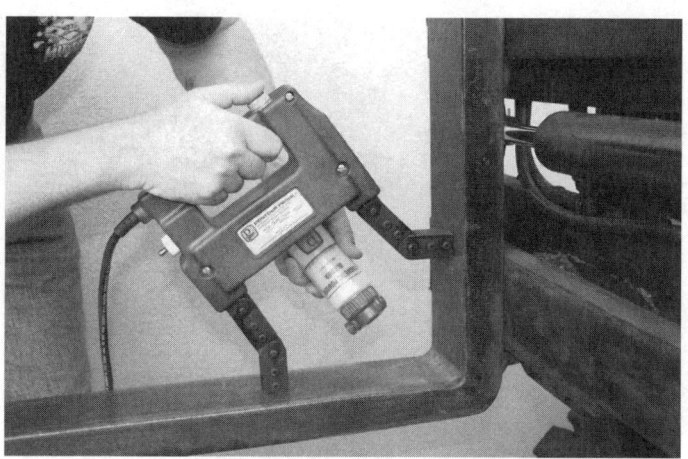

FIG. 1 Yoke Method of Part Magnetization

have articulated legs (poles) that allow the legs to be adjusted to contact irregular surfaces or two surfaces that join at an angle.

6.3.1 *Permanent Magnets*—Permanent magnets are available but their use may be restricted for many applications. This restriction may be due to application impracticality, or due to the specifications governing the examination. Permanent magnets can lose their magnetic field generating capacity by being partially demagnetized by a stronger flux field, being damaged, or dropped. In addition, the particle mobility created by AC current or HW current pulsations produced by electromagnetic yokes are not present. Particles, steel filings, chips, and scale clinging to the poles can create a housekeeping problem.

6.4 *Prods*—Prods are used for local magnetizations, see Fig. 2. The prod tips that contact the piece should be aluminum, copper braid, or copper pads rather than solid copper. With solid copper tips, accidental arcing during prod placement or removal can cause copper penetration into the surface which may result in metallurgical damage (softening, hardening, cracking, etc.). Open-circuit voltages should not exceed 25 V.

6.4.1 *Remote Control Switch*—A remote-control switch, which may be built into the prod handles, should be provided to permit the current to be turned on after the prods have been properly placed and to turn it off before the prods are removed in order to prevent arcing (arc burns).

6.5 *Bench Unit*—A typical bench type unit is shown in Fig. 3. The unit normally is furnished with a head/tailstock combination along with a fixed coil (see Fig. 4).

6.6 *Black Light*—The black light must be capable of developing a peak wavelength output at or near 365 nm with an intensity at the examination surface that satisfies 7.1.2. Suitable filters are used to remove the extraneous visible light and any harmful UV radiation emitted by the black light bulb. Some high intensity black light bulbs may emit unacceptable amounts of blue light that may cause indications to become invisible due to the increase in surface background. Refer to E2297 for more detail. When using a mercury vapor bulb a change in line voltage greater than $\pm 10\%$ can cause a change in black light output with consequent inconsistent performance. A constant voltage transformer may be used where there is evidence of voltage changes greater than 10 %.

6.7 *Equipment Verification*—See Section 20.

7. Examination Area

7.1 *Light Intensity for Examination*—Magnetic indications found using nonfluorescent particles are examined under visible light. Indications found using fluorescent particles must be examined under black (ultraviolet) light. This requires a darkened area with accompanying control of the visible light intensity.

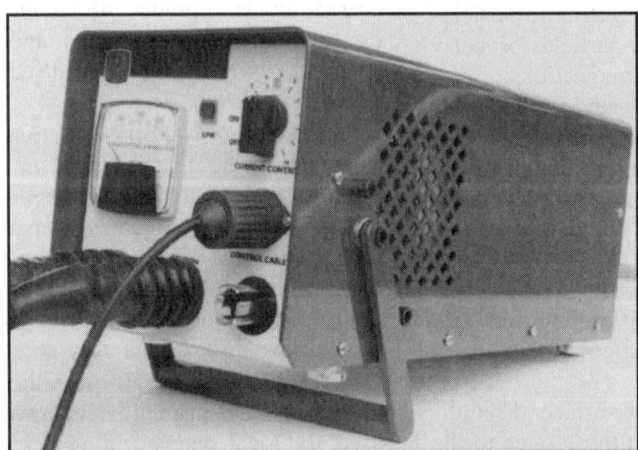
Typical portable power pack for prods

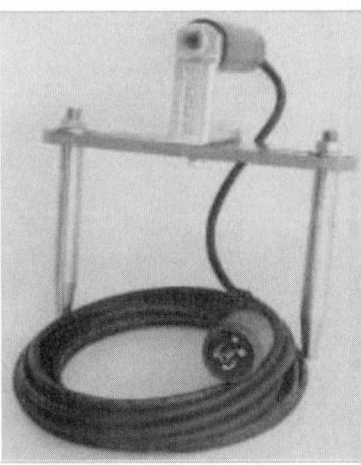
Typical Single Prod Set

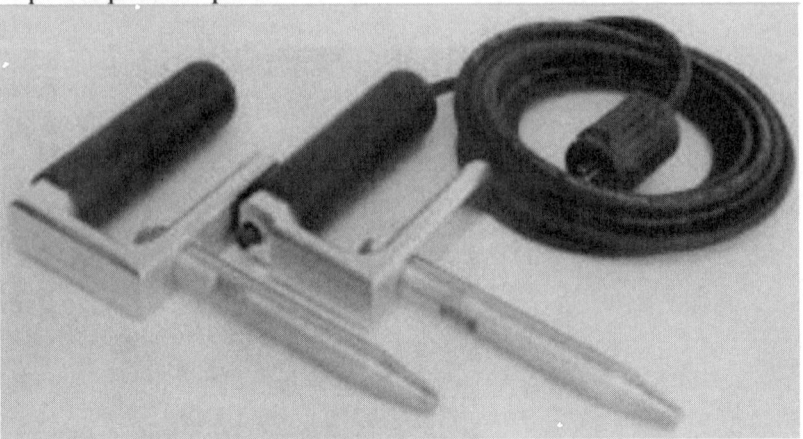
Typical Double Prod Set

FIG. 2

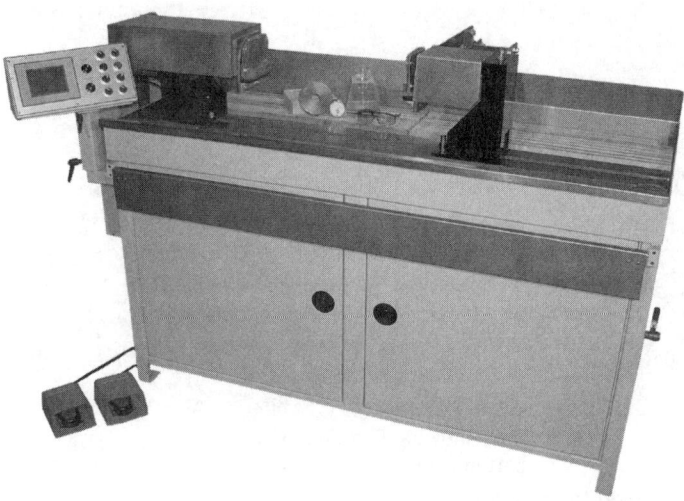

FIG. 3 Bench Unit

7.1.1 *Visible Light Intensity*—The intensity of the visible light at the surface of the part/work piece undergoing nonfluorescent particle examination is recommended to be a minimum of 100 foot candles (1076 lux). The intensity of ambient visible light in the darkened area where fluorescent magnetic particle testing is performed is recommended to not exceed 2 foot candles (21.5 lux).

7.1.1.1 *Field Examinations*—For some field examinations using nonfluorescent particles, visible light intensities as low as 50 foot candles (538 lux) may be used when agreed on by the contracting agency.

7.1.2 *Black (Ultraviolet) Light*:

7.1.2.1 *Black Light Intensity*—The black light intensity at the examination surface is recommended to not be less than 1000 µW/cm^2 when measured with a suitable black light meter.

7.1.2.2 *Black Light Warm-up*—When using a mercury vapor bulb, allow the black light to warm up for a minimum of five minutes prior to its use or measurement of the intensity of the ultraviolet light emitted.

7.1.3 *Dark Area Eye Adaptation*—The generally accepted practice is that an inspector be in the darkened area at least one (1) minute so that his/her eyes will adapt to dark viewing prior to examining parts under UV illumination. (**Warning**—Photochromic or permanently tinted lenses should not be worn during examination.)

7.2 *Housekeeping*—The examination area should be kept free of interfering debris. If fluorescent materials are involved, the area should also be kept free of fluorescent objects not related to the part/piece being examined.

8. Magnetic Particle Materials

8.1 *Magnetic Particle Properties*:

8.1.1 *Dry Particle Properties*—AMS 3040 describes the generally accepted properties of dry method particles.

8.1.2 *Wet Particle Properties*—The following documents describe the generally accepted properties of wet method particles in their various forms:

AMS 3041 Magnetic Particles, Non-fluorescent, Wet Method, Oil Vehicle, Ready to Use

AMS 3042 Magnetic Particles, Non-fluorescent, Wet Method, Dry Powder

AMS 3043 Magnetic Particles, Non-fluorescent, Oil Vehicle, Aerosol Packaged

AMS 3044 Magnetic Particles, Fluorescent, Wet Method, Dry Powder

AMS 3045 Magnetic Particles, Non-fluorescent, Wet Method, Oil Vehicle, Ready to Use

AMS 3046 Magnetic Particles, Non-fluorescent, Wet Method, Oil Vehicle, Aerosol Packaged

8.1.3 *Suspension Vehicle*—The suspension vehicle for wet-method examination may be either a light oil distillate fluid (refer to AMS 2641 or A-A-52930) or a conditioned water vehicle (refer to AS 4792).

8.2 *Particle Types*—The particles used in either dry or wet magnetic particle testing techniques are basically finely divided ferromagnetic materials which have been treated to impart color (fluorescent and nonfluorescent) in order to make them highly visible (contrasting) against the background of the surface being examined. The particles are designed for use either as a free flowing dry powder or for suspension at a given concentration in a suitable liquid medium.

8.3 *Particle Characteristics*—The magnetic particles must have high permeability to allow ease of magnetizing and attraction to the site of the flux leakage and low retentivity so they will not be attracted (magnetic agglomeration) to each other. Control of particle size and shape is required to obtain consistent results. The particles should be nontoxic, free from rust, grease, paint, dirt, and other deleterious materials that might interfere with their use; see 20.5 and 20.6. Both dry and wet particles are considered safe when used in accordance with the manufacturer's instructions. They generally afford a very low hazard potential with regard to flammability and toxicity.

8.4 *Dry Particles*—Dry magnetic powders are designed to be used as supplied and are applied by spraying or dusting directly onto the surface of the part being examined. They are generally used on an expendable basis because of the requirement to maintain particle size and control possible contamination. Reuse is not a normal practice. Dry powders may also be used under extreme environmental conditions. They are not affected by cold; therefore examination can be carried out at temperatures that would thicken or freeze wet baths. They are also heat resistant; some powders may be usable at temperatures up to 600°F (315°C). Some colored, organic coatings applied to dry particles to improve contrast lose their color at temperatures this high, making the contrast less effective. Fluorescent dry particles cannot be used at this high a temperature; the manufacturer should be contacted for the temperature limitations.

8.4.1 *Advantages*—The dry magnetic particle technique is generally superior to the wet technique for detection of near-surface discontinuities on parts with a gross indication size. Refer to 8.5.1: (a) for large objects when using portable equipment for local magnetization; (b) superior particle mobility is obtained for relatively deep-seated flaws using half-wave rectified current as the magnetizing source; (c) ease of removal.

Pinion gear in coil

Conception showing flux distribution with part in the bottom of the coil

FIG. 4 Bench Fixed Coil and Field Distribution

8.4.2 *Disadvantages*—The dry magnetic particle technique; (a) cannot be used in confined areas without proper safety breathing apparatus; (b) can be difficult to use in overhead magnetizing positions; (c) does not always leave evidence of complete coverage of part surface as with the wet technique; (d) is likely to have lower production rates than the wet technique; and (e) is difficult to adapt to any type of automatic system.

8.4.3 *Nonfluorescent Colors*—Although dry magnetic particle powder can be almost any color, the most frequently employed colors are light gray, black, red, or yellow. The choice is generally based on maximum contrast with the surface to be examined. The examination is done under visible light.

8.4.4 *Fluorescent*—Fluorescent dry magnetic particles are also available, but are not in general use primarily because of their higher cost and use limitations. They require a black light source and a darkened work area. These requirements are not often available in the field-type locations where dry magnetic particle examinations are especially suitable.

8.4.5 *Dual Colors*—Dual-colored particles are available that are readily detectable in visible light and also display fluorescence when viewed under ultra-violet light or a combination visible and ultra-violet light. Use in accordance with the manufacturer's recommendations.

8.5 *Wet Particle Systems*—Wet magnetic particles are designed to be suspended in a vehicle such as water or light petroleum distillate at a given concentration for application to the examination surface by flowing, spraying, or pouring. They are available in both fluorescent and nonfluorescent concentrates. In some cases the particles are premixed with the suspending vehicle by the supplier, but usually the particles are supplied as a dry concentrate or paste concentrate which is mixed with the distillate or water by the user. The suspensions are normally used in wet horizontal magnetic particle equipment in which the suspension is retained in a reservoir and recirculated for continuous use. The suspension may also be used on an expendable basis dispensed from an aerosol or other suitable dispensers.

8.5.1 *Primary Use*—Because the particles used are smaller, wet method techniques are generally used to locate smaller discontinuities than the dry method is used for. The liquid vehicles used may not perform satisfactorily when their viscosity exceeds 5cSt (5 mm^2/s) at the operating temperature. If the suspension vehicle is a hydrocarbon, its flash point limits the top temperature of usage. Mixing equipment for bulk reservoirs or manual agitation for portable dispensers is usually required to keep wet method particles uniformly in suspension.

8.5.2 *Where Used*—The wet fluorescent method usually is performed indoors or in areas where shelter and ambient light level can be controlled and where proper application equipment is available.

8.5.3 *Color*—The color chosen for any given examination should be one that best contrasts with the test surface. Because contrast is invariably higher with fluorescent materials, these are utilized in most wet process examinations. Fluorescent wet method particles normally glow a bright yellow-green when viewed under black light, although other colors are available. Non-fluorescent particles are usually black or reddish brown, although other colors are available. Dual-colored particles are available that are readily detectable in visible light and also display fluorescence when viewed under ultra-violet light or a combination visible and ultra-violet light. Refer to 8.5.5.

8.5.4 *Suspension Vehicles*—Generally the particles are suspended in a light petroleum (low-viscosity) distillate or conditioned water. (If sulfur or chlorine limits are specified, use Test Methods E165, Annex A2 or A4 to determine their values.

8.5.4.1 *Petroleum Distillates*—Low-viscosity light petroleum distillates vehicles (AMS 2641 Type 1 or equal) are ideal for suspending both fluorescent and nonfluorescent magnetic particles and are commonly employed.

(1) Advantages—Two significant advantages for the use of petroleum distillate vehicles are: (a) the magnetic particles are suspended and dispersed in petroleum distillate vehicles without the use of conditioning agents; and (b) the petroleum distillate vehicles provide a measure of corrosion protection to parts and the equipment used.

(2) Disadvantages—Principal disadvantages are flammability, fumes, and availability. It is essential, therefore, to select and maintain readily available sources of supply of petroleum distillate vehicles that have as high a flash point as practicable

to avoid possible flammability problems and provide a work area with proper ventilation.

(3) Characteristics—Petroleum distillate vehicles to be used in wet magnetic particle testing should possess the following: (*a*) viscosity should not exceed 3.0 cSt (3 mm^2/s) at 100°F (38°C) and not more than 5.0 cSt (5 mm^2/s) at the lowest temperature at which the vehicle will be used; when verified in accordance with Test Method D445, in order not to impede particle mobility (see 20.7.3), (*b*) minimum flash point, when verified in accordance with Test Methods D93, should be 200°F (93°C) in order to minimize fire hazards (see 20.7.4), (*c*) odorless; not objectionable to user, (*d*) low inherent fluorescence if used with fluorescent particles; that is, it should not interfere significantly with the fluorescent particle indications (see 20.6.4.1), and (*e*) nonreactive; should not degrade suspended particles.

8.5.4.2 *Water Vehicles with Conditioning Agents*—Water may be used as a suspension vehicle for wet magnetic particles provided suitable conditioning agents are added which provide proper wet dispersing, in addition to corrosion protection for the parts being examined and the equipment in use. Plain water does not disperse some types of magnetic particles, does not wet all surfaces, and is corrosive to parts and equipment. On the other hand, conditioned water suspensions of magnetic particles are safer to use since they are nonflammable. The selection and concentration of the conditioning agent should be as recommended by the particle manufacturer. The following are recommended properties for water vehicles containing conditioning agents for use with wet magnetic particle testing:

(1) Wetting Characteristics—The vehicle should have good wetting characteristics; that is, wet the surface to be examined, give even, complete coverage without evidence of dewetting the examination surface. The surface tension (coverage) should be observed under both black light and visible light. Smooth examination surfaces require that a greater percentage of wetting agent be added than is required for rough surface. Nonionic wetting agents are recommended (see 20.7.5).

(2) Suspension Characteristics—Impart good dispersability; that is, thoroughly disperse the magnetic particles without evidence of particle agglomeration.

(3) Foaming—Minimize foaming; that is, it should not produce excessive foam which would interfere with indication formation or cause particles to form scum with the foam.

(4) Corrosiveness—It should not corrode parts to be examined or the equipment in which it is used.

(5) Viscosity Limit—The viscosity of the conditioned water should not exceed a maximum viscosity of 3 cSt (3 mm^2/s) at 100°F (38°C) (see 20.7.3).

(6) Fluorescence—The conditioned water should not produce excessive fluorescence if intended for use with fluorescent particles.

(7) Nonreactiveness—The conditioned water should not cause deterioration of the suspended magnetic particles.

(8) Water pH—The pH of the conditioned water should not be less than 7.0 or exceed 10.5.

(9) Odor—The conditioned water should be essentially odorless.

8.5.5 *Concentration of Wet Magnetic Particle Suspension*—The initial bath concentration of suspended magnetic particles should be as specified or as recommended by the manufacturer and should be checked by settling volume measurements and maintained at the specified concentration on a daily basis. If the concentration is not maintained properly, examination results can vary greatly. The concentration of dual-colored particles in the wet-method bath suspension may be adjusted to best perform in the desired lighting environment. Higher particle concentration is recommended for visible light areas and lower particle concentration is recommended for ultraviolet light areas. Use in accordance with the particle manufacturer's recommendations.

8.5.6 *Application of Wet Magnetic Particles* (see 15.2).

8.5.7 *Magnetic Slurry/Paint Systems*—Another type of examination vehicle is the magnetic slurry/paint type consisting of a heavy oil in which flake-like particles are suspended. The material is normally applied by brush before the part is magnetized. Because of the high viscosity, the material does not rapidly run off surfaces, facilitating the examination of vertical or overhead surfaces. The vehicles may be combustible, but the fire hazard is very low. Other hazards are very similar to those of the oil and water vehicles previously described.

8.5.8 *Polymer-Based Systems*—The vehicle used in the magnetic polymer is basically a liquid polymer which disperses the magnetic particles and which cures to an elastic solid in a given period of time, forming fixed indications. Viscosity limits of standard wet technique vehicles do not apply. Care should be exercised in handling these polymer materials. Use in accordance with manufacturer's instructions and precautions. This technique is particularly applicable to examination areas of limited visual accessibility, such as bolt holes.

9. Part Preparation

9.1 *General*—The surface of the part to be examined should be essentially clean, dry, and free of contaminants such as dirt, oil, grease, loose rust, loose mill sand, loose mill scale, lint, thick paint, welding flux/slag, and weld splatter that might restrict particle movement. See 15.1.2 about applying dry particles to a damp/wet surface. When examining a local area, such as a weld, the areas adjacent to the surface to be examined, as agreed by the contracting parties, must also be cleaned to the extent necessary to permit detection of indications.

9.1.1 *Nonconductive Coatings*—Thin nonconductive coatings, such as paint in the order of 0.02 to 0.05 mm (1 or 2 mil) will not normally interfere with the formation of indications, but they must be removed at all points where electrical contact is to be made for direct magnetization. Indirect magnetization does not require electrical contact with the part/piece. See Section 12.2. If a nonconducting coating/plating is left on the area to be examined that has a thickness greater than 0.05 mm (2 mil), it must be demonstrated that unacceptable discontinuities can be detected through the maximum thickness applied.

9.1.2 *Conductive Coatings*—A conductive coating (such as chrome plating and heavy mill scale on wrought products resulting from hot forming operations) can mask discontinuities. As with nonconductive coatings, it must be demonstrated that the unacceptable discontinuities can be detected through the coating.

9.1.3 *Residual Magnetic Fields*—If the part/piece holds a residual magnetic field from a previous magnetization that will interfere with the examination, the part must be demagnetized. See Section 18.

9.2 *Cleaning Examination Surface*—Cleaning of the examination surface may be accomplished by detergents, organic solvents, or mechanical means. As-welded, as-rolled, as-cast, or as-forged surfaces are generally satisfactory, but if the surface is unusually nonuniform, as with burned-in sand, a very rough weld deposit, or scale, interpretation may be difficult because of mechanical entrapment of the magnetic particles. In case of doubt, any questionable area should be recleaned and reexamined (see 9.1).

9.2.1 *Plugging and Masking Small Holes and Openings*—Unless prohibited by the purchaser, small openings and oil holes leading to obscure passages or cavities can be plugged or masked with a suitable nonabrasive material which is readily removed. In the case of engine parts, the material must be soluble in oil. Effective masking must be used to protect components that may be damaged by contact with the particles or particle suspension.

10. Sequence of Operations

10.1 *Sequencing Particle Application and Establishing Magnetic Flux Field*—The sequence of operation in magnetic particle examination applies to the relationship between the timing and application of particles and establishing the magnetizing flux field. Two basic techniques apply, that is, continuous (see 10.1.1 and 10.1.2) and residual (see 10.1.3), both of which are commonly employed in industry.

10.1.1 *Continuous Magnetization*—Continuous magnetization is employed for most applications utilizing either dry or wet particles and will provide higher magnetic field strengths, to aid indication formation better, than residual magentic fields. The continuous method must be used when performing multi-directional magnetization. The sequence of operation for the dry and the wet continuous magnetization techniques are significantly different and are discussed separately in 10.1.1.1 and 10.1.1.2.

10.1.1.1 *Dry Continuous Magnetization Technique*—Unlike a wet suspension, dry particles lose most of their mobility when they contact the surface of a part. Therefore, it is imperative that the part/area of interest be under the influence of the applied magnetic field while the particles are still airborne and free to be attracted to leakage fields. This dictates that the flow of magnetizing current be initiated prior to the application of dry magnetic particles and terminated after the application of powder has been completed and any excess has been blown off. Magnetizing with HW current and AC current provide additional particle mobility on the surface of the part. Examination with dry particles is usually carried out in conjunction with prod-type or yoke localized magnetizations, and buildup of indications is observed as the particles are being applied.

10.1.1.2 *Wet Continuous Magnetization Technique*—The wet continuous magnetization technique involves bathing the part with the examination medium to provide an abundant source of suspended particles on the surface of the part and terminating the bath application immediately prior to cutting off of the magnetizing current. The duration of the magnetizing current is typically on the order of ½ s for each magnetizing pulse (shot), with two or more shots given to the part.

10.1.1.3 *Polymer or Slurry Continuous Magnetization Technique*—Prolonged or repeated periods of magnetization are often necessary for polymer- or slurry-base suspensions because of slower inherent magnetic particle mobility in the high-viscosity suspension vehicles.

10.1.2 *True Continuous Magnetization Technique*—In this technique, the magnetizing current is sustained throughout both the processing and examination of the part.

10.1.3 *Residual Magnetization Techniques*:

10.1.3.1 *Residual Magnetization*—In this technique, the examination medium is applied after the magnetizing force has been discontinued. It can be used only if the material being examined has relatively high retentivity so the residual leakage field will be of sufficient strength to attract and hold the particles and produce indications. This technique may be advantageous for integration with production or handling requirements or when higher than residual field strengths are not required to achieve satisfactory results. When inducing circular fields and longitudinal fields of long pieces, residual fields are normally sufficient to meet magnetizing requirements consistent with the requirements of Section 14. The residual method has found wide use examining pipe and tubular goods. For magnetization requirements of oilfield tubulars, refer to Appendix X1. Unless demonstrations with typical parts indicate that the residual field has sufficient strength to produce relevant indications of discontinuities (see 20.8) when the field is in proper orientation, the continuous method should be used.

11. Types of Magnetizing Currents

11.1 *Basic Current Types*—The four basic types of current used in magnetic particle testing to establish part magnetization are alternating current (AC), half-wave rectified current (HW), full-wave rectified current (FW), and for a special application, DC.

11.1.1 *Alternating Current (AC)*—Part magnetization with alternating current is preferred for those applications where examination requirements call for the detection of discontinuities, such as fatigue cracks, that are open to the surface to which the magnetizing force is applied. Associated with AC is a "skin effect" that confines the magnetic field at or near to the surface of a part. In contrast, both HW current and FW current produce a magnetic field having penetrating capabilities proportional to the amount of applied current, which should be used when near-surface or inside surface discontinuities are of concern.

11.1.2 *Half-Wave Rectified Current (HW)*—Half-wave current is frequently used in conjunction with wet and dry particles because the current pulses provide more mobility to the particles. This waveform is used with prods, yokes, mobile and bench units. Half-wave rectified current is used to achieve depth of penetration for detection of typical discontinuities found in weldments, forgings, and ferrous castings. As with AC for magnetization, single-phase current is utilized and the average value measured as "magnetizing current."

11.1.3 *Full-Wave Rectified Current (FW)*—Full-wave current may utilize single- or three-phase current. Three-phase current has the advantage of lower line amperage draws, whereas single-phase equipment is less expensive. Full-wave rectified current is commonly used when the residual method is to be employed. Because particle movement, either dry or wet is noticeably less, precautions must be taken to ensure that sufficient time is allowed for formation of indications.

11.1.4 *Direct Current (DC)*—A bank of batteries, full-wave rectified AC filtered through capacitors or a DC generator produce direct magnetizing current. They have largely given way to half-wave rectified or full-wave rectified DC except for a few specialized applications, primarily because of broad application advantages when using other types of equipment.

11.1.5 *Capacitor Discharge (CD) Current*—A bank of capacitors are used to store energy and when triggered the energy reaches high amperage with a very short duration (normally less than 25 milliseconds). Because of the short pulse duration the current requirements are affected by the amount of material to be magnetized as well as the applied amperage. The capacitor discharge technique is widely used to establish a residual magnetic field in tubing, casing, line pipe, and drill pipe. For specific requirements, see Appendix X1.

12. Part Magnetization Techniques

12.1 *Examination Coverage*—All examinations should be conducted with sufficient area overlap to assure the required coverage at the specified sensitivity has been obtained.

12.2 *Direct and Indirect Magnetization*—A part can be magnetized either directly or indirectly. For direct magnetization the magnetizing current is passed directly through the part creating a magnetic field oriented 90 degrees to current flow in the part. With indirect magnetization techniques a magnetic field is induced in the part, which can create a circular/toroidal, longitudinal, or multidirectional magnetic field in the part. The techniques described in 20.8 for verifying that the magnetic fields have the anticipated direction and strength should be employed. This is especially important when using multidirectional techniques to examine complex shapes.

12.3 *Choosing Magnetization Technique*—The choice of direct or indirect magnetization will depend on such factors as size, configuration, or ease of processing. Table 1 compares the advantages and limitations of the various methods of part magnetization.

TABLE 1 Advantages and Limitations of the Various Ways of Magnetizing a Part

Magnetizing Technique and Material Form	Advantages	Limitations
I. Direct Contact Part Magnetization (see 12.3.1)		
Head/Tailstock Contact Solid, relatively small parts (castings, forgings, machined pieces) that can be processed on a horizontal wet unit	1. Fast, easy technique.	1. Possibility of arc burns if poor contact conditions exist.
	2. Circular magnetic field surrounds current path.	2. Long parts should be examined in sections to facilitate bath application without resorting to an overly long current shot.
	3. Good sensitivity to surface and near-surface discontinuities.	
	4. Simple as well as relatively complex parts can usually be easily processed with one or more shots.	
	5. Complete magnetic path is conducive to maximizing residual characteristics of material.	
Large castings and forgings	1. Large surface areas can be processed and examined in relatively short time.	1. High amperage requirements (16 000 to 20 000 A) dictate costly DC power supply.
Cylindrical parts such as tubing, pipe, hollow shafts, etc.	1. Entire length can be circularly magnetized by contacting, end to end.	1. Effective field limited to outside surface and cannot be used for inside diameter examination.
		2. Ends must be conductive to electrical contacts and capable of carrying required current without excessive heat. Cannot be used on oilfield tubulars because of possibility of arc burns.
Long solid parts such as billets, bars, shafts, etc.	1. Entire length can be circularly magnetized by contacting, end to end.	1. Output voltage requirements increase as the part length increases, due to greater value of the impedance and/or resistance as the cables and part length grows.
	2. Current requirements are independent of length.	2. Ends must be conductive to electrical contact and capable of carrying required current without excessive heat.
	3. No end loss.	
Prods: Welds	1. Circular field can be selectively directed to weld area by prod placement.	1. Only small area can be examined at one time.
	2. In conjunction with half-wave rectified alternating current and dry powder, provides excellent sensitivity to subsurface discontinuities as well as surface type.	2. Arc burns due to poor contact.
	3. Flexible, in that prods, cables, and power packs can be brought to examination site.	3. Surface must be dry when dry powder is being used.

TABLE 1 *Continued*

Magnetizing Technique and Material Form	Advantages	Limitations
Large castings or forgings	1. Entire surface area can be examined in small increments using nominal current values. 2. Circular field can be concentrated in specific areas that historically are prone to discontinuities. 3. Equipment can be brought to the location of parts that are difficult to move. 4. In conjunction with half-wave rectified alternating current and dry powder, provides excellent sensitivity to near surface subsurface type discontinuities that are difficult to locate by other methods.	4. Prod spacing must be in accordance with the magnetizing current level. 1. Coverage of large surface area require a multiplicity of shots that can be very time-consuming. 2. Possibility of arc burns due to poor contact. Surface should be dry when dry powder is being used. 3. Large power packs (over 6000A) often require a large capacity voltage source to operate. 4. When using HW current or FW current on retentive materials, it is often necessary that the power pack be equipped with a reversing DC demagnetizing option.
II. Indirect Part Magnetization (see 12.3.2) Central Conductor		

Miscellaneous parts having holes through which a conductor can be placed such as:
  Bearing race
  Hollow cylinder
  Gear
  Large nut
	1. When used properly, no electrical contact is made with the part and possibility of arc burns eliminated.
	2. Circumferentially directed magnetic field is generated in all surfaces, surrounding the conductor (inside diameter, faces, etc.).
	3. Ideal for those cases where the residual method is applicable.
	4. Light weight parts can be supported by the central conductor.
	5. Smaller central conductor and multiple coil wraps may be used to reduce current requirements.
		1. Size of conductor must be ample to carry required current.
		2. Larger diameters require repeated magnetization with conductor against inside diameter and rotation of part between processes. Where continuous magnetization technique is being employed, examination is required after each magnetization step.

Large clevis
Pipe coupling, casing/tubing
Tubular type parts such as:
  Pipe/Casting
  Tubing
  Hollow shaft
	1. When used properly, no electrical contact is made with the part and possibility of arc burns eliminated.
	2. Inside diameter as well as outside diameter examination.
	3. Entire length of part circularly magnetized.
		1. Outside surface sensitivity may be somewhat less than that obtained on the inside surface for large diameter and extremely heavy wall sections.

Large valve bodies and similar parts	1. Provides good sensitivity for detection of discontinuities located on internal surfaces.	1. Outside surface sensitivity may be somewhat less than that obtained on the inside diameter for heavy wall sections.
Coil/Cable Wrap Miscellaneous medium-sized parts where the length predominates such as a crankshaft	1. All generally longitudinal surfaces are longitudinally magnetized to effectively locate transverse discontinuities.	1. Length may dictate multiple shot as coil is repositioned. 2. Longitudinal magnetization of complex parts with upsets such as crankshafts will lead to dead spots where the magnetic field is cancelled out. Care must be taken to assure magnetization of all areas in perpendicular directions.
Large castings, forgings, or shafting	1. Longitudinal field easily attained by means of cable wrapping.	1. Multiple magnetization may be required due to configuration of part.
Miscellaneous small parts	1. Easy and fast, especially where residual magnetization is applicable. 2. No electrical contact. 3. Relatively complex parts can usually be processed with same ease as those with simple cross section.	1. L/D (length/diameter) ratio important consideration in determining adequacy of ampere-turns. 2. Effective L/D ratio can be altered by utilizing pieces of similar cross-sectional area. 3. Use smaller coil for more intense field. 4. Sensitivity diminishes at ends of part due to general leakage field pattern. 5. Quick break desirable to minimize end effect on short parts with low L/D ratio.
Induced Current Fixtures		
Examination of ring-shaped part for circumferential-type discontinuities.	1. No electrical contact. 2. All surface of part subjected to toroidal-type magnetic field. 3. Single process for 100 % coverage. 4. Can be automated.	1. Laminated core required through ring. 2. Type of magnetizing current must be compatible with method. 3. Other conductors encircling field must be avoided. 4. Large diameters require special consideration.
Ball examination	1. No electrical contact. 2. 100 % coverage for discontinuities in any direction with three-step process and proper orientation between steps.	1. For small-diameter balls, limited to residual magnetization.

TABLE 1 Continued

Magnetizing Technique and Material Form	Advantages	Limitations
Disks and gears	3. Can be automated. 1. No electrical contact. 2. Good sensitivity at or near periphery or rim. 3. Sensitivity in various areas can be varied by core or pole-piece selection.	1. 100 % coverage may require two-step process with core or pole-piece variation, or both. 2. Type of magnetizing current must be compatible with part geometry.
Yokes: Examination of large surface areas for surface-type discontinuities.	1. No electrical contact. 2. Highly portable. 3. Can locate discontinuities in any direction with proper orientation.	1. Time consuming. 2. Must be systematically repositioned in view of random discontinuity orientation.
Miscellaneous parts requiring examination of localized areas.	1. No electrical contact. 2. Good sensitivity to direct surface discontinuities. 3. Highly portable. 4. Wet or dry technique. 5. Alternating-current type can also serve as demagnetizer in some instances.	1. Must be properly positioned relative to orientation of discontinuities. 2. Relatively good contact must be established between part and poles. 3. Complex part geometry may cause difficulty. 4. Poor sensitivity to subsurface-type discontinuities except in isolated areas.

12.3.1 *Direct Contact Magnetization*—For direct magnetization, physical contact must be made between the ferromagnetic part and the current carrying electrodes connected to the power source. Both localized area magnetization and overall part magnetization are direct contact means of part magnetization, and can be achieved through the use of prods, head and tailstock, clamps, and magnetic leeches.

12.3.2 *Localized Area Magnetization*:

12.3.2.1 *Prod Technique*—The prod electrodes are first pressed firmly against the part under examination (see Fig. 2). The magnetizing current is then passed through the prods and into the area of the part in contact with the prods. This establishes a circular magnetic field in the part around and between each prod electrode, sufficient to carry out a local magnetic particle examination (see Fig. 2). (**Warning**—Extreme care should be taken to maintain clean prod tips, to minimize heating at the point of contact and to prevent arc burns and local overheating on the surface being examined since these may cause adverse effects on material properties. Arc burns may cause metallurgical damage; if the tips are solid copper, copper penetration into the part may occur. Prods should not be used on machined surfaces or on aerospace component parts.)

(1) Unrectified AC limits the prod technique to the detection of surface discontinuities. Half-wave rectified AC is most desirable since it will detect both surface and near-surface discontinuities. The prod technique generally utilizes dry magnetic particle materials due to better particle mobility. Wet magnetic particles are not generally used with the prod technique because of potential electrical and flammability hazards.

(2) Proper prod examination requires a second placement with the prods rotated approximately 90° from the first placement to assure that all existing discontinuities are revealed. Depending on the surface coverage requirements, overlap between successive prod placements may be necessary. On large surfaces, it is good practice to layout a grid for prod/yoke placement.

12.3.2.2 *Manual Clamp/Magnetic Leech Technique*—Local areas of complex components may be magnetized by electrical contacts manually clamped or attached with magnetic leeches to the part (Fig. 5). As with prods, sufficient overlap may be necessary if examination of the contact location is required.

12.3.2.3 *Overall Magnetization*:

(1) Head and Tailstock Contact—Parts may be clamped between two electrodes (such as a head and tailstock of

FIG. 5 Direct Contact Magnetization through Magnetic Leech Clamp of Part

horizontal wet magnetic particle equipment) and the magnetizing current applied directly through the part (Fig. 6). The size and shape of the part will determine whether both field directions can be obtained with such equipment.

(2) Clamps—The magnetizing current may be applied to the part under examination by clamping (Fig. 7) the current carrying electrodes to the part, producing a circular magnetic field.

(3) Multidirectional Magnetization Technique—With suitable circuitry, it is possible to produce a multidirectional (oscillating) field in a part by selectively switching the magnetic field within the part between electrode contacts/clamps positioned approximately 90° apart or by using a combination of switched direct and indirect methods, such as contact and coil. This permits building up indications in all possible directions and may be considered the equivalent of magnetizing in two or more directions (Fig. 8). On some complex shapes as many as 16 to 20 steps may be required with conventional equipment. With multidirectional magnetization, it is usually possible to reduce the magnetizing steps required by more than half. In many instances, the number of steps may be reduced to one. It is essential that the wet continuous method, be used and that the magnetic field direction and relative intensity be determined by AS 5371 shims as described in Appendix X2 or with an identical part with discontinuities in all areas of interest.

12.3.3 *Indirect Magnetization*—Indirect part magnetization involves the use of a preformed coil, cable wrap, yoke, or a central conductor to induce a magnetic field. Coil, cable wrap, and yoke magnetization are referred to as longitudinal magnetization in the part (see 13.4).

12.3.3.1 *Coil and Cable Magnetization*—When coil (Fig. 4) or cable wrap (Fig. 9) techniques are used, the magnetizing force is proportional to ampere turns (see X3.2.2.1).

12.3.3.2 *Central Conductor, Induced Current Magnetization*—Indirect circular magnetization of hollow pieces/parts can be performed by passing the magnetizing current through a central conductor (Fig. 10(a) and Fig. 10(b)) or cable used as a central conductor or through an induced current fixture (Fig. 8(A)). Central conductors may be solid or hollow and are ideally made from non-ferrous material. Ferrous central conductors will function as well, but will generate substantial heat due to magnetic domain movement and a reduced magnetic field outside the conductor when compared to a non-ferrous conductor. Additionally, when using ferromagnetic conductors, the inspector must be made aware of the possibility of magnetic writing. When using a bench-type unit, the distance along the part circumference, which may be effectively examined should be taken as approximately four times the diameter of the central conductor, as illustrated in Fig. 10 (b). The entire circumference should be examined by rotating the part on the conductor, allowing for approximately a 10 % magnetic field overlap. Central conductors are widely used in magnetic particle examination to provide:

(1) A circular field on both the inside surface and outside surface of tubular pieces that cannot be duplicated by the direct current technique.

(2) A non-contact means of part magnetization virtually eliminating the possibility of arc burning the material, as can be the case with current flow through contacts, such as prods or clamps.

(3) Substantial processing advantages over direct contact techniques on ring-shaped parts.

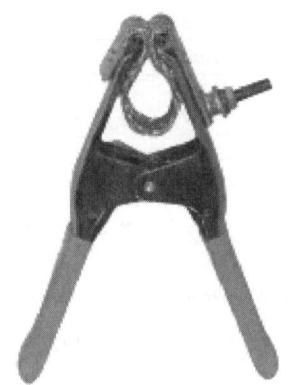

FIG. 7 Spring Loaded Contact Clamp

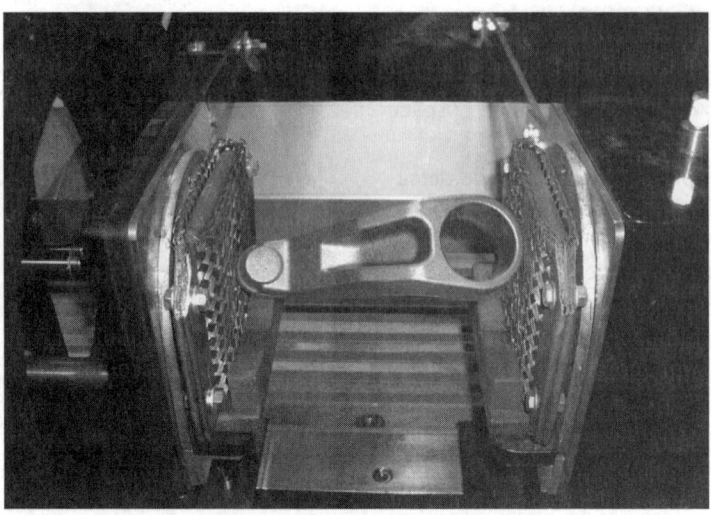

FIG. 6 Direct Contact Shot

Figure A - Typical 2-vector multidirectional wet horizontal using laminated core

Figure B - Typical 2-vector multidirectional wet horizontal

FIG. 8 Multidirectional Magnetic Particle Units

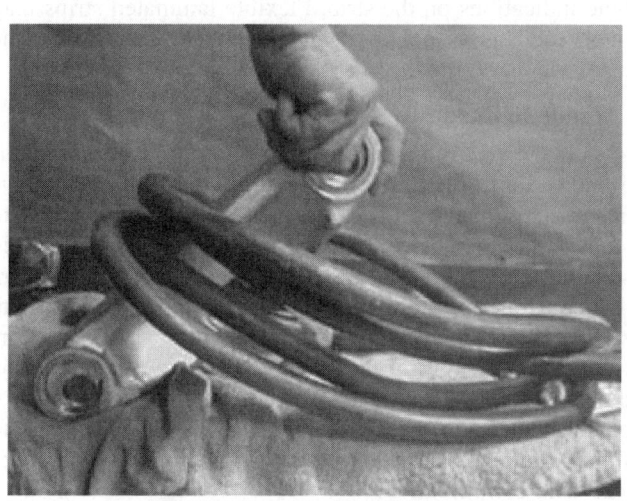

FIG. 9 Cable Wrap Magnetization

(4) In general it is not important for the central conductor to be centered because the flux lines follow the path of least resistance through the ferromagnetic material. On large diameter materials the central conductor should be within 6 in. of the center. The resulting field is concentric relative to the axis of the piece and is maximum at the inside surface.

12.3.3.3 *Yoke Magnetization*—A magnetic field can be induced into a part by means of an electromagnet (see Fig. 1), where the part or a portion thereof becomes the magnetic path between the poles (acts as a keeper) and discontinuities preferentially transverse to the alignment of the pole pieces are indicated. Most yokes are energized by an input of AC and produce a magnetizing field of AC, half-wave DC, or full-wave DC. A permanent magnet can also introduce a magnetic field in the part, but its use is restricted (see 6.3.1).

13. Direction of Magnetic Fields

13.1 *Discontinuity Orientation vs. Magnetic Field Direction*—Since indications are not normally obtained when discontinuities are parallel to the magnetic field, and since indications may occur in various or unknown directions in a part, each part must be magnetized in at least two directions approximately at right angles to each other as noted in 4.3.2. On some parts circular magnetization may be used in two or more directions, while on others both circular and longitudinal magnetization are used to achieve the same result. For purposes of demagnetization verification, circular magnetism normally precedes longitudinal magnetization. A multidirectional field can also be employed to achieve part magnetization in more than one direction.

13.2 *Circular Magnetization*—Circular magnetization (Fig. 11) is the term used when electric current is passed through a part, or by use of a central conductor (see 12.3.3.2) through a central opening in the part, inducing a magnetic field at right angles to the current flow. Circular fields normally produce strong residual fields, but are not measurable because the flux is contained within the part.

13.3 *Transverse Magnetization*—Transverse magnetization is the term used when the magnetic field is established across the part and the lines of flux complete their loop outside the part. Placing a yoke across a bar normal to the bar axis would produce a transverse field.

13.4 *Toroidal Magnetization*—When magnetizing a part with a toroidal shape, such as a solid wheel or the disk with a center opening, an induced field that is radial to the disk is most useful for the detection of discontinuities in a circumferential direction. In such applications this field may be more effective than multiple shots across the periphery, but requires special equipment.

13.5 *Longitudinal Magnetization*—Longitudinal magnetization (Fig. 12) is the term used when a magnetic field is generated by an electric current passing through a multiturn, which encloses the part or section of the part to be examined.

13.6 *Multidirectional Magnetization*—Multidirectional magnetization may be used to fulfill the requirement for magnetization in two directions if it is demonstrated that it is effective in all areas of interest. Examine parts in accordance with 20.8.2 or shims manufactured to the requirements of

FIG. 10 Central Bar Conductors

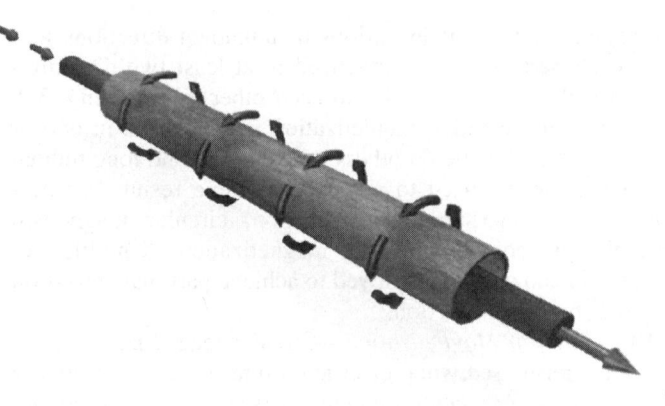

FIG. 11 Circular Magnetism

AS 5371, or as otherwise approved by the Level 3 and the Cognizant Engineering Organization, may be used to verify field direction, strength, and balance in multidirectional magnetization. Balance of the field intensity is critical. The field intensity should be balanced in all directions. The particle application must be timed so that the magnetization levels reach full value in all directions, while the particles are mobile on the surface under examination.

13.6.1 When actual parts with known defects are used, the number and orientation(s) of the defects (for example, axial, longitudinal, circumferential, etc.) should be noted. The magnetic field intensity can be considered as being properly balanced when all noted defects can be readily identified with particle indications.

13.7 *Flexible Laminated Strips for Magnetic Particle Testing*

13.7.1 Flexible laminated strips as described in Annex A2 of E1444 may be used to ensure proper field direction during magnetic particle examination. The longitudinal axis of the strip should be placed perpendicular to the direction of the magnetic field of interest in order to generate the strongest particle indications on the strip. Flexible laminated strips may only be used as a tool to demonstrate the direction of the external magnetic field.

14. Magnetic Field Strength

14.1 *Magnetizing Field Strengths*—To produce interpretable indications, the magnetic field in the part must have sufficient strength and proper orientation. For the indications to be consistent, this field strength must be controlled within reasonable limits, usually ±25 % on single vector equipment and when using multi-directional equipment, the field strength must be controlled much closer, often within ±5 %. Factors that affect the strength of the field are the size, shape, section thickness, material of the part/piece, and the technique of magnetization. Since these factors vary widely, it is difficult to establish rigid rules for magnetic field strengths for every conceivable configuration.

14.2 *Establishing Field Strengths*—Sufficient magnetic field strength can be established by:

14.2.1 *Known Discontinuities*—Experiments with similar/identical parts having known discontinuities in all areas of interest.

14.2.2 *Artificial Discontinuities*—Verification of indications derived from AS 5371 shims (see Appendix X2) taped or glued defect side in contact with the part under examination is an effective means of verifying field strength when using the continuous method.

14.2.3 *Hall-effect Meter Tangential Field Strengths*—A minimum tangential applied field strength of 30 G (2.4 kAM^{-1}) should be adequate when using single vector equipment. Stronger field strengths are allowed, but it must not be so strong that it causes the masking of relevant indications by nonrelevant accumulations of magnetic particles. Due to the complex number of variables, the use of Gaussmeters should not be the sole source of determining an acceptable field on multi-directional techniques.

14.2.3.1 *Circular Magnetism Hall-effect Meter Measurement*—On a part with consistent diameter or thickness, the transverse probe may be placed anywhere along the

FIG. 12 Longitudinal Magnetism

length of the part as the tangential circular field is consistent across the length. The transverse probe should be positioned upright such that the circular field is normal to the major dimension of the Hall-effect sensor and within 5° of perpendicularity to the part. More than one measurement should be taken to ensure consistent readings. On parts with more than one diameter/thickness, multiple measurements should be taken to ensure a minimum measurement of 30 gauss on all areas to be examined. Measurement is made of the applied field, that is, during the magnetizing shot, not the residual flux field.

14.2.3.2 *Longitudinal Magnetism Hall-effect Meter Measurement*—On a part with consistent diameter or thickness, the probe may be placed anywhere along the length of the part, except near the poles as the tangential longitudinal field is consistent across the length, except at the poles. Measurement near the poles will yield a skewed reading due to detection of the normal flux field at each pole. Also, measurement near any geometry change that would produce a non-relevant flux leakage should be avoided. The probe should be positioned within 5° of perpendicularity to the part and such that the longitudinal field is normal to the major dimension of the Hall-effect sensor. More than one measurement should be taken to ensure consistent readings. The Hall-effect probe may be placed within the coil or outside the vicinity of the coil if the part is longer than the width of the coil. On parts with more than one diameter/thickness, multiple measurements should be taken to ensure a minimum measurement of 30 gauss on all areas to be examined. Measurement is made of the applied field, that is, during the magnetizing shot, not the residual flux field.

14.2.4 *Using Empirical Formulas*—Appendix X3 details the use of empirical formulas for determining field strength. Amperages derived from empirical formulas should be verified with a Hall-effect gaussmeter or AS 5371 shims.

14.3 *Localized Magnetization:*:

14.3.1 *Using Prods*—When using prods on material ¾ in. (19 mm) in thickness or less, it is recommended to use 90 to 115 A/in. of prod spacing (3.5 to 4.5 A/mm). For material greater than ¾ in. (19 mm) in thickness, it is recommended to use 100 to 125 A/in. of prod spacing. Prod spacing is recommended to be not less than 2 in. (50 mm) or greater than 8 in. (200 mm). The effective width of the magnetizing field when using prods is one fourth of the prod spacing on each side of a line through the prod centers.

14.3.2 *Using Yokes*—The field strength of a yoke (or a permanent magnet) can be empirically determined by measuring its lifting power (see 20.3.7). If a Hall-effect probe is used, it shall be placed on the surface midway between the poles.

15. Application of Dry and Wet Magnetic Particles

15.1 *Dry Magnetic Particles*:

15.1.1 *Magnetic Fields for Dry Particles*—Dry magnetic powders are generally applied with the continuous magnetizing techniques. When utilizing AC, the current must be on before application of the dry powder and remain on through the examination phase. With Half-wave rectified AC or yoke DC magnetization, a current duration of at least ½ s should be used. The current duration should be short enough to prevent any damage from overheating or from other causes. It should be noted that AC and half-wave rectified DC impart better particle mobility to the powder than DC or full-wave rectified AC. Dry magnetic powders are widely used for magnetic particle examination of large parts as well as on localized areas such as welds. Dry magnetic particles are widely used for oil field applications and are frequently used in conjunction with capacitor discharge style equipment and the residual method.

15.1.2 *Dry Powder Application*—It is recommended that dry powders be applied in such a manner that a light uniform, dust-like coating settles upon the surface of the part/piece while it is being magnetized. Dry particles must not be applied to a damp surface; they will have limited mobility. Neither should they be applied where there is excessive wind. The preferred application technique suspends the particles in air in such a manner that they reach the part surface being magnetized in a uniform cloud with a minimum of force. Usually, specially designed powder blowers and hand powder applicators are employed (see Fig. 1). Dry particles should not be applied by pouring, throwing, or spreading with the fingers.

15.1.3 *Excess Powder Removal*—Care is needed in both the application and removal of excess dry powder. Removal of excess powder is generally done while the magnetizing current is present and care must be exercised to prevent the removal of particles attracted by a leakage field, which may prove to be a relevant indication.

15.1.4 *Near-surface Discontinuities Powder Patterns*—In order to recognize the broad, fuzzy, weakly held powder patterns produced by near-surface discontinuities, it is essential to observe carefully the formation of indications while the powder is being applied and also while the excess is being removed. Sufficient time for indication formation and examination should be allowed between successive magnetization cycles.

15.2 *Wet Particle Application*—Wet magnetic particles, fluorescent or nonfluorescent, suspended in a vehicle at a recommended concentration may be applied either by spraying or flowing over the areas to be examined during the application of the magnetizing field current (continuous technique) or after turning off the current (residual technique). Proper sequencing of operation (part magnetization and timing of bath application) is essential to indication formation and retention. For the continuous technique multiple current shots should be applied. The last shot should be applied after the particle flow has been diverted and while the particle bath is still on the part. A single shot may be sufficient. Care should be taken to prevent damage to a part due to overheating or other causes. Since fine or weakly held indications on highly finished or polished surfaces may be washed away or obliterated, care must be taken to prevent high-velocity flow over critical surfaces and to cut off the bath application before removing the magnetizing force. Discontinuity detection may benefit from an extended drain time of several seconds before actual examination.

15.3 *Magnetic Slurry/Paints*—Magnetic slurry/paints are applied to the part with a brush before or during part magnetization. Indications appear as a dark line against a light silvery background. Magnetic slurry is ideal for overhead or underwater magnetic particle examination.

15.4 *Magnetic Polymers*—Magnetic polymers are applied to the part to be examined as a liquid polymer suspension. The part is then magnetized, the polymer is allowed to cure, and the elastic coating is removed from the examination surface for interpretation and evaluation. Care must be exercised to ensure that magnetization is completed within the active migration period of the polymer which is usually about 10 min. This method is particularly applicable to areas of limited visual access such as bolt holes. Detailed application and use instructions of the manufacturer should be followed for optimum results.

15.5 *White Background and Black Oxide*—A thin white background is applied by aerosol to provide a thin (≤ 2 mil), smooth, high contrast background prior to magnetization and particle application. After background has dried, magnetization and particle application follow normal procedures. The high contrast between the white background and black particles provides high sensitivity in white light conditions. Detailed application and use instructions of the manufacturer should be followed for optimum results.

16. Interpretation of Indications

16.1 *Valid Indications*—All valid indications formed by magnetic particle examination are the result of magnetic leakage fields. Indications may be relevant (16.1.1), nonrelevant (16.1.2), or false (16.1.3).

16.1.1 *Relevant Indications*—Relevant indications are produced by leakage fields which are the result of discontinuities. Relevant indications require evaluation with regard to the acceptance standards agreed upon between the manufacturer/test agency and the purchaser (see Annex A1).

16.1.2 *Nonrelevant Indications*—Nonrelevant indications can occur singly or in patterns as a result of leakage fields created by conditions that require no evaluation such as changes in section (like keyways and drilled holes), inherent material properties (like the edge of a bimetallic weld), magnetic writing, etc.

16.1.3 *False Indications*—False indications are not the result of magnetic forces. Examples are particles held mechanically or by gravity in shallow depressions or particles held by rust or scale on the surface.

17. Recording of Indications

17.1 *Means of Recording*—When required by a written procedure, permanent records of the location, type, direction, length(s), and spacing(s) of indications may be made by one or more of the following means.

17.1.1 *Sketches*—Sketching the indication(s) and their locations.

17.1.2 *Transfer (Dry Powder Only)*—Covering the indication(s) with transparent adhesive-backed tape, removing the tape with the magnetic particle indication(s) adhering to it, and placing it on paper or other appropriate background material indicating locations.

17.1.3 *Strippable Film (Dry Powder Only)*—Covering the indication(s) with a spray-on strippable film that fixes the indication(s) in place. When the film is stripped from the part, the magnetic particle indication(s) adhere to it.

17.1.4 *Photographing*—Photographing the indications themselves, the tape, or the strippable film reproductions of the indications.

17.1.5 *Written Records*—Recording the location, length, orientation, and number of indications.

17.1.5.1 *Defect or Indication Sizing Accuracy*—For situations where defect or indication size limits are specified by the acceptance criteria, measurement equipment should be selected with an accuracy being precise enough to determine compliance. For example, to verify maximum defect length does not exceed 0.150 in. (3.81 mm) a measuring device accurate to ±0.010 in. (0.254 mm) could be used by reducing the allowable limit too 0.140 in. (3.56 mm), but using a measuring device accurate to ±0.150 in. (3.81 mm) or one with 0.100 in. (2.54 mm) increments is not accurate enough.

17.1.5.2 For situations where no defect or indication tolerances are specified (for example, reporting the length of a crack when the acceptance criteria is "No cracks allowed") the crack length should not be reported with more precision than the resolution of the measurement equipment allows. For example, when using a measuring device accurate to ±0.010 in. (0.254 mm) report the crack length in 0.010 in. (0.254 mm) increments.

17.1.5.3 Some contracts may require better than the minimum measurement accuracy needed to determine compliance. These situations are generally limited to critical direct measurement of deliverable product features, rather than examination parameter checks. For example, an accuracy ratio of 2 to 1 may be specified for measurement of defects or product geometry, which means an instrument with a calibrated accuracy of ±0.005 in. (0.127 mm) would be needed for verifying or reporting dimensions to the nearest ±0.010 in. (0.254 mm).

17.2 *Accompanying Information*—A record of the procedure parameters listed below as applicable should accompany the examination results:

17.2.1 *Method Used*—Magnetic particle method (dry, wet, fluorescent, etc.).

17.2.2 *Magnetizing Technique*—Magnetizing technique (continuous, true-continuous, residual).

17.2.3 *Current Type*—Magnetizing current (AC, half-wave rectified or full-wave rectified AC, etc.).

17.2.4 *Field Direction*—Direction of magnetic field (prod placement, cable wrap sequence, etc.).

17.2.5 *Field Strength*—Magnetic current strength (ampere turns, amperes per inch (millimetre) of prod spacing, lifting force, etc.).

18. Demagnetization

18.1 *Applicability*—All ferromagnetic material will retain some residual magnetism, the strength of which is dependent on the retentivity of the part. Residual magnetism does not affect the mechanical properties of the part. However, a residual field may cause chips, filing, scale, etc. to adhere to the surface affecting subsequent machining operations, painting, or plating. Additionally, if the part will be used in locations near sensitive instruments, high residual fields could affect the operation of these instruments. Furthermore, a strong residual magnetic field in a part to be welded or electroplated could interfere with welding or plating process. Residual fields may also interfere with later magnetic particle examination. Demagnetization is required only if specified in the drawings, specification, or purchase order. When required, an acceptable level of residual magnetization and the measuring method should also be specified. See 18.3.

18.2 *Demagnetization Methods*—The ease of demagnetization is dependent on the coercive force of the metal. High retentivity is not necessarily related to high coercive force in that the strength of the residual field is not always an indicator of ease of demagnetizing. In general, demagnetization is accomplished by subjecting the part to a field equal to or greater than that used to magnetize the part and in nearly the same direction, then continuously reversing the field direction while gradually decreasing it to zero.

18.2.1 *Withdrawal from Alternating Current Coil*—The fastest and most simple technique is to pass the part through a high intensity alternating current coil and then slowly withdraw the part from the field of the coil. A coil of 5000 to 10,000 ampere turns is recommended. Line frequency is usually from 50 to 60 Hz alternating current. The piece should enter the coil from a 12-in. (300-mm) distance and move through it steadily and slowly until the piece is at least 36 in. (900 mm) beyond the coil. Care should be exercised to ensure that the part is entirely removed from the influence of the coil before the demagnetizing force is discontinued, otherwise the demagnetizer may have the reverse effect and actually remagnetize the part. This should be repeated as necessary to reduce the residual field to an acceptable level. See 18.3. Small parts of complex figuration can be rotated and tumbled while passing through the field of the coil. Use of this technique may not be effective on large parts in which the alternating magnetic current field is insufficient to penetrate.

18.2.2 *Decreasing Alternating Current*—An alternative technique for part demagnetization is subjecting the part to the alternating magnetic field while gradually reducing its strength to a desired level.

18.2.3 *Demagnetizing with Yokes*—Alternating current yokes may be used for local demagnetization by placing the poles on the surface, moving them around the area, and slowly withdrawing the yoke while it is still energized.

18.2.4 *Reversing Direct Current*—The part to be demagnetized is subjected to consecutive steps of reversed and reduced direct current magnetization to a desired level. (This is the most effective process of demagnetizing large parts in which the alternating current field has insufficient penetration to remove the internal residual magnetization.) This technique requires special equipment for reversing the current while simultaneously reducing it in small increments.

18.3 *Extent of Demagnetization*—The effectiveness of the demagnetizing operation can be indicated by the use of appropriate magnetic field indicators. (**Warning**—A part may retain a strong residual field after having been circularly magnetized and exhibit little or no external evidence of this field. Therefore, the circular magnetization should be conducted before longitudinal magnetization if complete demagnetization is required. If a sacrificial part is available, in the case of a part such as a bearing race that has been circularly magnetized, it is often advisable to section one side of it and measure the remaining leakage field in order to check the demagnetizing process.)

18.3.1 After demagnetization, measurable residual fields should not exceed a value agreed upon or as specified on the engineering drawing or in the contract, purchase order, or specification.

19. Post Examination Cleaning

19.1 *Particle Removal*—Post-examination cleaning is necessary where magnetic particle material(s) could interfere with subsequent processing or with service requirements. Demagnetization should always precede particle removal. The purchaser should specify when post-examination cleaning is needed and the extent required.

19.2 *Means of Particle Removal*—Typical post-examination cleaning techniques employed are: (a) the use of compressed air to blow off unwanted dry magnetic particles; (b) drying of wet particles and subsequent removal by brushing or with compressed air; (c) removal of wet particles by flushing with solvent; and (d) other suitable post-examination cleaning techniques may be used if they will not interfere with subsequent requirements.

20. Process Controls

20.1 *Contributing Factors*—The overall performance of a magnetic particle testing system is dependent upon the following:

20.1.1 Operator capability, if a manual operation is involved.

20.1.2 Control of process steps.

20.1.3 The particles or suspension, or both.

20.1.4 The equipment.

20.1.5 Visible light level.

20.1.6 Black light monitoring where applicable.

20.1.7 Magnetic field strength.

20.1.8 Field direction or orientation.

20.1.9 Residual field strength.

20.1.10 These factors should all be controlled individually.

20.2 *Maintenance and Calibration of Equipment*—The magnetic particle equipment employed should be maintained in proper working order at all times. The frequency of verification calibration, usually every six months, see Table 2, or whenever a malfunction is suspected, should be specified in the written procedures of the nondestructive testing facility. Records of the checks and results provide useful information for quality control purposes and should be maintained. In addition, any or all of the checks described should be performed whenever a malfunction of the system is suspected. Calibration checks should be conducted in accordance with the specifications or documents that are applicable.

20.2.1 *Equipment Calibration*—It is good practice that all calibrated equipment be traceable to the job it was used on. This facilitates possible re-examination or evaluation should a piece of equipment be found not working properly.

20.2.2 Some examination procedures may require equipment calibration or operational checks, but no accuracy requirement is specified, for that equipment, by the contractually specified magnetic particle examination procedure (for example, ASTM E1444 light meters and gaussmeter accuracy), however the accuracy of the measuring device should be reasonably suited for the situation with the resolution of the equipment being precise enough to determine compliance.

20.2.3 Equipment that meets an accuracy requirement specified by the contractually specified magnetic particle examination procedure (for example, ASTM E1444 ammeter accuracy of ±10 % or 50 amperes, or a timer control ±0.01 second) should be considered adequate, with no additional accuracy or uncertainty determination needed.

20.2.4 Measurement equipment that the contractually specified magnetic particle inspection procedure does not specifically require to be calibrated or meet a specified accuracy (for example, timers, shop air pressure gauge, etc.) should be maintained in good working order and have measurement resolution reasonably suited for the intended use.

20.3 *Equipment Checks*—The following checks are recommended for ensuring the accuracy of magnetic particle magnetizing equipment.

20.3.1 *Ammeter Accuracy*—The equipment meter readings should be compared to those of a control check meter incorporating a shunt or current transformer connected to monitor the output current. The accuracy of the entire control check meter arrangement should be verified at six-month intervals or as agreed upon between the purchaser and supplier by a means traceable to the National Institute of Standards and Technology (NIST). Comparative readings should be taken at a minimum of three output levels encompassing the usable range. The equipment meter reading should not deviate by more than ±10 % of full scale relative to the actual current values as shown by the check meter. (**Warning**—When measuring half-wave DC, the direct current reading of a conventional DC check meter reading should be doubled.)

20.3.2 *Timer Control Check*—On equipment utilizing a timer to control the duration of the current flow, the timer should be checked for accuracy as specified in Table 2 or whenever a malfunction is suspected. The timer should be calibrated to within ±0.1 seconds using a suitable electronic timer.

20.3.3 *Magnetic Field Quick Break Check*—On equipment that has a quick break feature, the functioning of this circuit should be checked and verified. This check may be performed using a suitable oscilloscope or a simple test device usually available from the manufacturer. Normally, only the fixed coil is checked for quick break functionality. Headstocks would need to be checked only if cables are attached to the headstocks to form a coil wrap. On electronic power packs or machines, failure to achieve indication of a "quick break" would indicate that a malfunction exists in the energizing circuit.

20.3.4 *Equipment Current Output Check*—To ensure the continued accuracy of the equipment, ammeter readings at each transformer tap should be made with a calibrated ammeter-shunt combination. This accessory is placed in series with the contacts. The equipment shunt should not be used to check the machine of which it is a part. For infinite current control units (non-tap switch), settings at 500-A intervals should be used. On uni-directional equipment, variations exceeding ±10 % from the equipment ammeter readings

TABLE 2 Recommended Verification Intervals

Item	Maximum Time Between Verifications[A]	Reference Paragraphs
Lighting:		
Visible light intensity	1 week	7.1.1
Black light intensity	1 day	7.1.2
Ambient visible light intensity	1 week[A]	7.1.1
System performance using test piece or ring specimen of Fig. 13	1 day	20.8
Wet particle concentration	8 h, or every shift change	20.6
Wet particle contamination	1 week	20.6.4
Water break check	1 day	20.7.5
Equipment calibration/check:		
Ammeter accuracy	6 months	20.3.1
Timer control	6 months	20.3.2
Quick break	6 months	20.3.3
Hall-effect gaussmeter	6 months	20.3.6
Dead weight check	6 months	20.3.7
Light meter checks	6 months	20.4

[A] Note—The maximum time between verifications may be extended when substantiated by actual technical stablity/reliability data.

indicate the equipment needs service or repair. On multi-vector equipment, variations exceeding ±5 % from the equipment ammeter readings indicate the equipment needs service or repair.

20.3.5 *Internal Short Circuit Check*—Magnetic particle equipment should be checked periodically for internal short circuiting. With the headstocks set for maximum amperage output, any deflection of the ammeter when the current is activated with no conductor between the contacts is an indication of an internal short circuit and must be repaired prior to use.

20.3.6 *Hall-effect Meters*—Depending upon the manufacturer, meters are normally accurate for use with full-wave DC only. Hall-effect meter readings for HW and AC current applications should be correlated to the results of the application of AS 5371 shims. Hall-effect gaussmeters should be calibrated every six months in accordance with the manufacturer's instructions.

NOTE 1—When used with SCR controlled equipment, the Gaussmeter's accuracy is dependant upon the actual circuit design of each model meter and results may vary.

20.3.7 *Electromagnetic Yoke Lifting Force Check*—The magnetizing force of a yoke (or a permanent magnet) should be checked by determining its lifting power on a steel plate. See Table 3. The lifting force relates to the electromagnetic strength of the yoke.

20.3.8 *Powder Blower*—The performance of powder blowers used to apply the dry magnetic particles should be checked at routine intervals or whenever a malfunction is suspected. The check should be made on a representative examination part. The blower should coat the area under evaluation with a light, uniform dust-like coating of dry magnetic particles and have sufficient force to remove the excess particles without disturbing those particles that are evidence of indications. Necessary adjustments to the blower's flow rate or air velocity should be made in accordance with the manufacturer's recommendations.

20.4 *Examination Area Light Level Control*:

20.4.1 *Visible Light Intensity*—Light intensity in the examination area should be checked at specified intervals with the designated light meter at the surface of the parts being examined. See Table 2.

20.4.2 *Black (ultraviolet) Light Intensity*—Black light intensity and wavelength should be checked at the specified intervals but not to exceed one-week intervals and whenever a bulb is changed. Reflectors and filters should be cleaned daily and checked for integrity. See Table 2. Cracked or broken UV filters should be replaced immediately. Defective bulbs which radiate UV energy must also be replaced before further use.

20.5 *Dry Particle Quality Control Checks*—In order to assure uniform and consistent performance from the dry magnetic powder selected for use, it is advisable that all incoming powders be certified or checked for conformance with quality control standards established between the user and supplier.

20.5.1 *Contamination*:

20.5.1.1 *Degradation Factors*—Dry magnetic particles are generally very rugged and perform with a high degree of consistency over a wide process envelope. Their performance, however, is susceptible to degradation from such contaminants as moisture, grease, oil, rust and mill scale particles, nonmagnetic particles such as foundry sand, and excessive heat. These contaminants will usually manifest themselves in the form of particle color change and particle agglomeration, the degree of which will determine further use of the powder. Over-heated dry particles can lose their color, thereby reducing the color contrast with the part and thus hinder part examination. Particle agglomeration can reduce particle mobility during processing, and large particle agglomerates may not be retained at an indication. Dry particles should not be recycled as fractionation, the subsequent depletion of finer particles from the aggregate powder composition, degrades the quality of the particles.

20.5.1.2 *Ensuring Particle Quality*—To ensure against deleterious effects from possible contaminants, it is recommended that a routine performance check be conducted (see 20.8.3).

20.6 *Wet Particle Quality Control Checks*—The following checks for wet magnetic particle suspensions should be conducted at startup and at regular intervals to assure consistent performance. See Table 2. Since bath contamination will occur as the bath is used, monitoring the working bath at regular intervals is essential.

20.6.1 *Determining Bath Concentration*—Bath concentration and sometimes bath contamination are determined by measuring its settling volume through the use of a pear-shaped centrifuge tube with a 1-mL stem (0.05-mL divisions) for fluorescent particle suspensions or a 1.5-mL stem (0.1-mL divisions) for nonfluorescent suspensions. (See Appendix X5.) Before sampling, the suspension should be run through the recirculating system for at least 30 min to ensure thorough mixing of all particles which could have settled on the sump screen and along the sides or bottom of the tank. Take a 100-mL portion of the suspension from the hose or nozzle into a clean, non-fluorescing centrifuge tube, demagnetize and allow it to settle for approximately 60 min with petroleum distillate suspensions or 30 min with water-based suspensions before reading. These times are average times based upon the most commonly used products; actual times should be adjusted so that the particles have substantially settled out of suspension. The volume settling out at the bottom of the tube is indicative of the particle concentration in the bath.

20.6.2 *Sample Interpretation*—If the bath concentration is low in particle content, add a sufficient amount of particle materials to obtain the desired concentration; if the suspension is high in particle content, add sufficient vehicle to obtain the desired concentration. If the settled particles appear to be loose agglomerates rather than a solid layer, take a second sample. If still agglomerated, the particles may have become magnetized; replace the suspension.

TABLE 3 Minimum Yoke Lifting Force

Type Current	Yoke Pole Leg Spacing	
	2 to 4 in. (50 to 100 mm)	4 to 6 in. (100 to 150 mm)
AC	10 lb (45 N)	
DC	30 lb (135 N)	50 lb (225 N)

20.6.3 *Settling Volumes*—For fluorescent particles, the recommended settling volume (see 15.2) is from 0.1 to 0.4 mL in a 100-mL bath sample and from 1.2 to 2.4 mL per 100 mL of vehicle for non-fluorescent particles, unless otherwise approved by the Cognizant Engineering Organization (CEO). Refer to appropriate AMS document (3041, 3042, 3043, 3044, 3045, and/or 3046). For dual-colored particles, the recommended settling volume should be determined by the performance requirements and lighting environment of a given application as recommended by the manufacturer. See 8.5.5.

20.6.4 *Bath Contamination*—Both fluorescent and nonfluorescent suspensions should be checked periodically for contaminants such as dirt, scale, oil, lint, loose fluorescent pigment, water (in the case of oil suspensions), and particle agglomerates which can adversely affect the performance of the magnetic particle examination process. See Table 2.

20.6.4.1 *Carrier Contamination*—For fluorescent baths, the liquid directly above the precipitate should be evaluated with black light. The liquid will have a little fluorescence. Its color can be compared with a freshly made-up sample using the same materials or with an unused sample from the original bath that was retained for this purpose. If the "used" sample is noticeably more fluorescent than the comparison standard, the bath should be replaced.

20.6.4.2 *Particle Contamination*—The graduated portion of the tube should be evaluated under black light if the bath is fluorescent and under visible light (for both fluorescent and nonfluorescent particles) for striations or bands, differences in color or appearance. Bands or striations may indicate contamination. If the total volume of the contaminates, including bands or striations exceeds 30 % of the volume of magnetic particles, or if the liquid is noticeably fluorescent (see 20.6.4.1), the bath should be replaced.

20.6.5 *Particle Durability*—The durability of both the fluorescent and nonfluorescent magnetic particles in suspension should be checked periodically to ensure that the particles have not degraded due to chemical attack from the suspending oil or conditioned water vehicles or mechanically degraded by the rotational forces of the recirculating pump in a wet horizontal magnetic particle unit. Fluorescent magnetic particle breakdown in particular can result in a decrease in sensitivity and an increase in nonmagnetic fluorescent background. Lost fluorescent pigment can produce false indications that can interfere with the examination process.

20.6.6 *Fluorescent Brightness*—It is important that the brightness of fluorescent magnetic particle powder be maintained at the established level so that indication and background brightness can be kept at a relatively constant level. Variations in contrast can noticeably affect examination results. Lack of adequate contrast is generally caused by:

20.6.6.1 An increase in contamination level of the vehicle increasing background fluorescence, or

20.6.6.2 Loss of vehicle because of evaporation, increasing concentration, or

20.6.6.3 Degradation of fluorescent particles. A change in contrast ratio can be observed by using a verification ring specimen with an etched surface.

20.6.7 *System Performance*—Failure to find a known discontinuity in a part or obtain the specified indications on the test ring (see 20.8.4) indicates a need for changing of the entire bath. If a part was used, it must have been completely demagnetized and cleaned so that no fluorescent background can be detected when viewed under black light with a surface intensity of at least 1000 $\mu W/cm^2$. If any background is noted that interferes with either detection or interpretation, the bath should be drained and a new suspension made.

20.6.8 *Magnetic Stripe Cards*—The encoded pattern on the magnetic stripes of magnetic stripe cards may serve as a verification piece for the evaluation of particle sensitivity. Particles are attracted to magnetic gradients formed when the stripe has been encoded. See Appendix X4 for further information.

20.7 *Bath Characteristics Control*:

20.7.1 *Oil Bath Fluids*—Properties of oil-bath fluids are described in AMS 2641 or A-A–59230.

20.7.2 *Water Bath Fluids*—Properties of conditioned water-bath fluids are described in AS 4792.

20.7.3 *Viscosity*—The recommended viscosity of the suspension is not to exceed 5 mm^2/s (5.0 cSt), at any temperature at which the bath may be used, when verified in accordance with Test Method D445.

20.7.4 *Flash Point*—The recommended flash point of wet magnetic particle light petroleum distillate suspension is a minimum of 200°F (93°C); use Test Method D93.

20.7.5 *Water Break Check for Conditioned Water Vehicles*—Properly conditioned water will provide proper wetting, particle dispersion, and corrosion protection. The water break check should be performed by flooding a part, similar in surface finish to those under examination, with suspension, and then noting the appearance of the surface of the part after the flooding is stopped. If the film of suspension is continuous and even all over the part, sufficient wetting agent is present. If the film of suspension breaks, exposing bare surfaces of the part, and the suspension forms many separate droplets on the surface, more wetting agent is needed or the part has not been sufficiently cleaned. When using the fluorescent method, this check should be performed under both blacklight and whitelight.

20.7.6 *pH of Conditioned Water Vehicles*—The recommended pH of the conditioned water bath is between 7.0 and 10.5 as determined by a suitable pH meter or special pH paper.

20.8 *Verifying System Performance*—System performance checks must be conducted in accordance with a written procedure so that the verification is performed in the same manner each time.

20.8.1 *Production Verification Parts with Discontinuities*—A practical way to evaluate the performance and sensitivity of the dry or wet magnetic particles or overall system performance, or both, is to use representative verification parts with known discontinuities of the type and severity normally encountered during actual production examination. However, the usefulness of such parts is limited because the orientation and magnitude of the discontinuities cannot be controlled. The use of flawed parts with gross discontinuities is not recommended. (**Warning**—If such parts are used, they must be thoroughly demagnetized and cleaned after each use.)

20.8.2 *Fabricated Test Parts with Discontinuities*—Often, production verification parts with known discontinuities of the type and severity needed for evaluation are not available. As an alternative, fabricated verification specimens with discontinuities of varying degree and severity can be used to provide an indication of the effectiveness of the dry or wet magnetic particle examination process. If such parts are used, they should be thoroughly demagnetized and cleaned after each use.

20.8.3 *Test Plate*—A magnetic particle system performance verification plate, such as shown in Fig. 13 is useful for checking the overall performance of wet or dry techniques using prods and yokes. Recommended minimum dimensions are ten inches per side and nominal thickness of one inch. Discontinuities can be formed by controlled heating/cooling, EDM notches, artificial discontinuities in accordance with 14.2.2 or other means. (**Warning**—Notches should be filled flush to the surface with a nonconducting material, such as epoxy, to prevent the mechanical holding of the indicating medium.)

20.8.4 *Test Ring Specimen*—A verification (Ketos) ring specimen may also be used in evaluating and comparing the overall performance and sensitivity of both dry and wet, fluorescent and non-fluorescent magnetic particle techniques using a central conductor magnetization technique. Refer to Practice E1444, Appendix X1.

20.8.4.1 *Using the Test Ring*—If using the verification ring, place a conductor with a diameter between 1 and 1.25 in. (25 and 31 mm) through the center of the ring. Center the ring on the length of the conductor. Magnetize the ring circularly by passing the current through the conductor as described in Appendix X1 of Practice E1444. Gently apply particles to the surface of the ring while the current is flowing. Examine the ring within one minute after current application. The number of hole indications visible should meet or exceed those as specified in Appendix X1 of Practice E1444.

20.8.5 *Magnetic Field Indicators*:

20.8.5.1 *"Pie" Field Indicator*—The magnetic field indicator shown in Fig. 14 relies on the slots between the pie shaped segments to show the presence and the approximate direction of the external magnetic field. Because "pie" field indicators are constructed of highly permeable material with 100 % through wall flaws, indications do not mean that suitable field strength is present for the location of relevant indications in the part under examination. The "pie" field indicator is used with

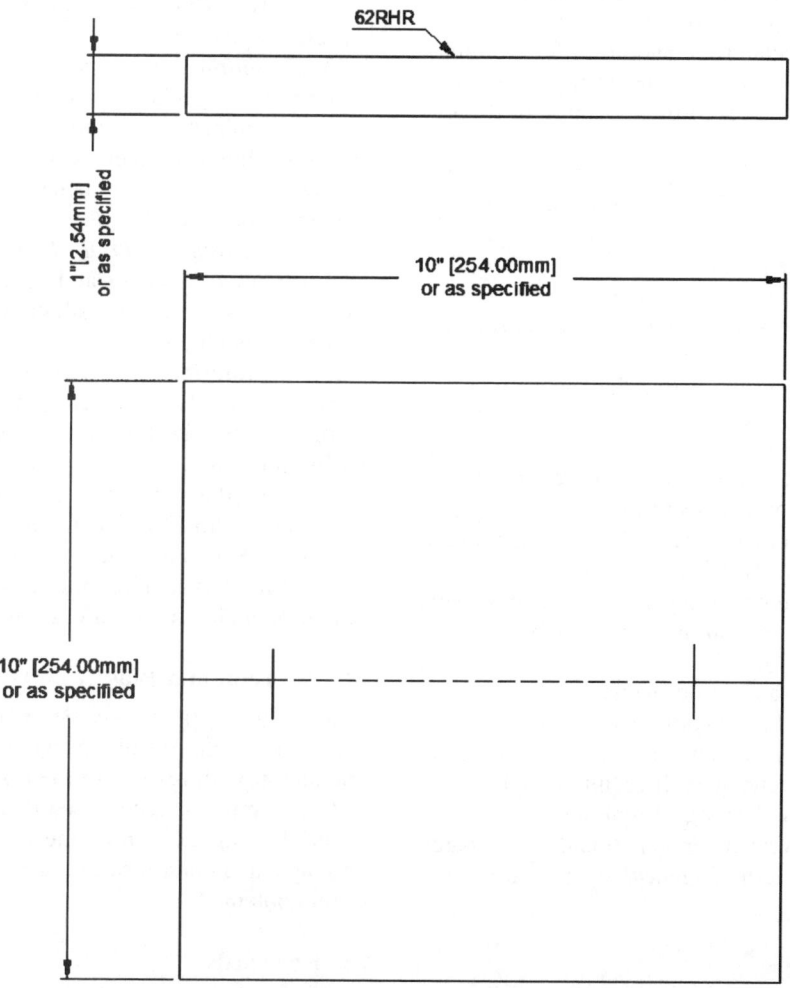

FIG. 13 Sample of a Magnetic Particle Performance Verification Plate. Defects are formed and located in accordance with plate manufacturers' specifications.

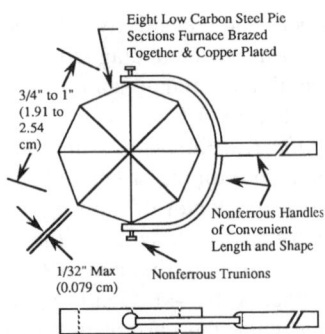

FIG. 14 Magnetic Field Indicator

the magnetic particles applied across the copper face of the indicator (the slots are against the piece) simultaneously with the magnetizing force. Typical "pie" field indicators show a clear indication in a five gauss external field. These devices are generally used as instructional aids.

20.8.5.2 *Slotted Shims*—Several types of slotted shims exist. Refer to AS 5371 and to illustrations in Appendix X2.

21. Procedures

21.1 When specified a procedure should be written for all magnetic particle examinations and should include as a minimum the following information. A sketch is usually used for illustrating part geometry, techniques, and areas for examination. This sketch may also be used for recording location of magnetic field indicators and for recording location of discontinuities.

21.1.1 Area to be examined (entire part or specific area),

21.1.2 Type of magnetic particle material (dry or wet, visible or fluorescent),

21.1.3 Magnetic particle equipment,

21.1.4 Part surface preparation requirements,

21.1.5 Magnetizing process (continuous, true-continuous, residual),

21.1.6 Magnetizing current (alternating, half-wave rectified AC, full-wave rectified AC, direct),

21.1.7 Means of establishing part magnetization (direct-prods, head/tailstock contact or cable wrap, indirect-coil/cable wrap, yoke, central conductor, and so forth),

21.1.8 Direction of magnetic field (circular or longitudinal),

21.1.9 System performance/sensitivity checks,

21.1.10 Magnetic field strength (ampere turns, field density, magnetizing force, and number and duration of application of magnetizing current),

21.1.11 Application of examination media,

21.1.12 Interpretation and evaluation of indications,

21.1.13 Type of records including accept/reject criteria,

21.1.14 Demagnetizing techniques, if required, and

21.1.15 Post-examination cleaning, if required,

21.2 *Written Reports*—Written reports should be prepared as agreed upon between the testing agency/department and the purchaser/user.

22. Acceptance Standards

22.1 The acceptability of parts examined by this method is not specified herein. Acceptance standards are a matter of agreement between the manufacturer and the purchaser and should be stated in a referenced contract, specification, or code.

23. Safety

23.1 Those involved with hands-on magnetic particle examination exposure to hazards include:

23.1.1 *Electric Shock and Burns*—Electric short circuits can cause shock and particularly burns from the high amperages at relatively low voltages that are used. Equipment handling water suspensions should have good electrical grounds.

23.1.2 *Flying Particles*—Magnetic particles, particularly the dry ones, dirt, foundry sand, rust, and mill scale can enter the eyes and ears when they are blown off the part when applying them to a vertical or overhead surface or when cleaning an examined surface with compressed air. Dry particles are easy to inhale and the use of a dust respirator is recommended.

23.1.3 *Falls*—A fall from a scaffold or ladder if working on a large structure in the field or shop.

23.1.4 *Fire*—Ignition of a petroleum distillate bath.

23.1.5 *Environment*—Doing magnetic particle examination where flammable vapors are present as in a petrochemical plant or oil refinery. Underwater work has its own set of hazards and should be addressed independently.

23.1.6 *Wet Floors*—Slipping on a floor wetted with a particle suspension.

23.1.7 *Shifting or Dropping of Large Components*—Large components, especially those on temporary supports can shift during examination or fall while being lifted. In addition, operators should be alert to the possibility of injury to body members being caught beneath a sling/chain or between head/tail stock and the piece.

23.1.8 *Ultraviolet Light Exposure*—Ultraviolet light can adversely affect the eyes and skin. Safety goggles designed to absorb UV wavelength radiation are suggested where high intensity blacklight is used.

23.1.9 *Materials and Concentrates*—The safe handling of magnetic particles and concentrates are governed by the supplier's Material Safety Data Sheets (MSDS). The MSDS conforming to 29 CFR 1910.1200 or equivalent must be provided by the supplier to any user and must be prepared in accordance with FED-STD-313.

23.1.10 *Equipment Hazards*—Because of the large breadth of equipment available, unique safety hazards may exist and should be addressed on a case by case basis.

24. Precision and Bias

24.1 The methodology described in the practice will produce repeatable results provided the field has the proper orientation with respect to the discontinuities being sought.

24.2 It must be recognized that the surface condition of the material being examined, the material's magnetic properties, its shape, and control of the factors listed in 20.1 influence the results obtained.

25. Keywords

25.1 dye; evaluation; examination; fluorescent; inspection; magnetic particle; nondestructive; testing

ANNEX

(Mandatory Information)

A1. TYPICAL MAGNETIC PARTICLE INDICATIONS

A1.1 Surface discontinuities with few exceptions produce sharp and distinct magnetic particle indications. Near-surface discontinuities on the other hand produce less distinct or fuzzy magnetic particle indications in comparison to surface discontinuities; the magnetic particle indications are broad rather than sharp and the particles are less tightly held.

A1.2 *Wet Method:*

A1.2.1 *Fluorescent*—Indications of surface cracks, surface indications, and an indication of a near surface discontinuity are shown in Figs. A1.1-A1.6.

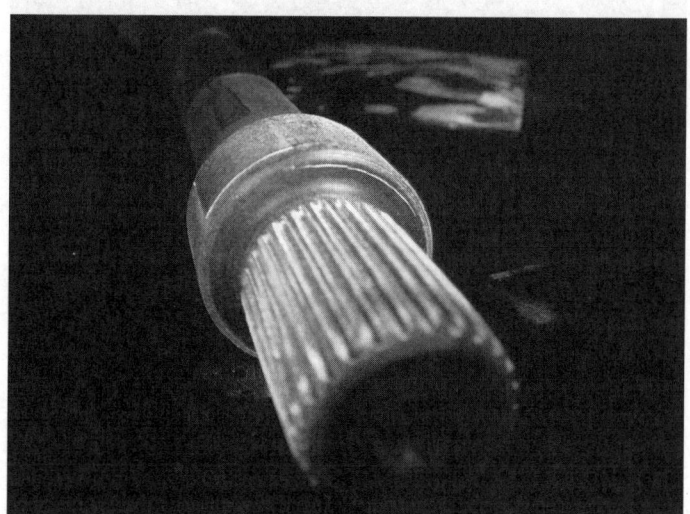

FIG. A1.1 Axle with Circumferential Crack in Shoulder

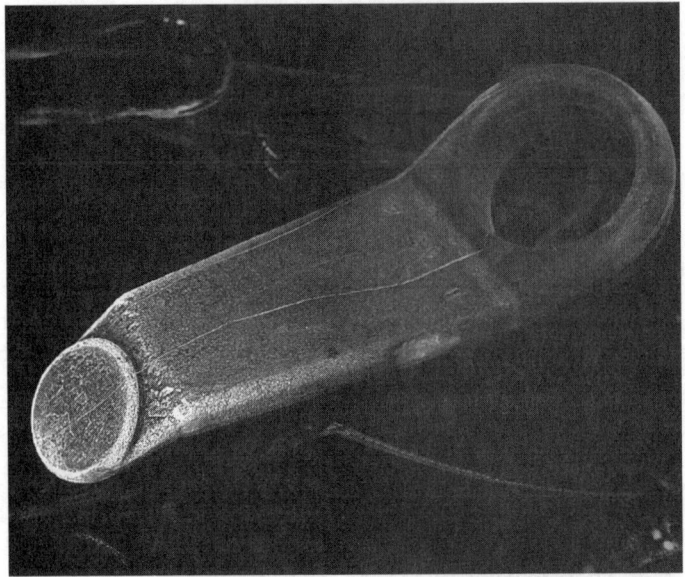

FIG. A1.2 Arm with Two Longitudinal Indications

FIG. A1.3 Hub with Both Radial and Longitudinal Indications

A1.2.2 *Nonfluorescent*—Indications of surface cracks are shown in Figs. A1.7-A1.16.

A1.3 *Dry Method*—Indications of surface cracks are shown in Figs. A1.17-A1.23.

A1.4 Nonrelevant indications are shown in Figs. A1.24-A1.26.

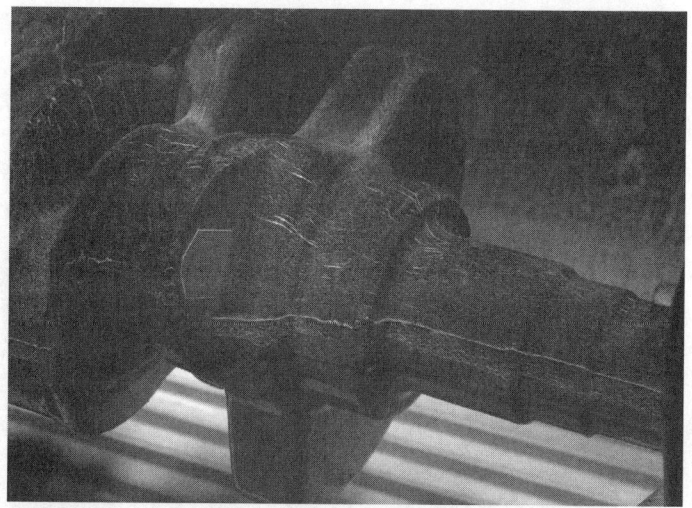

FIG. A1.4 Crankshaft with Various Longitudinal Indications

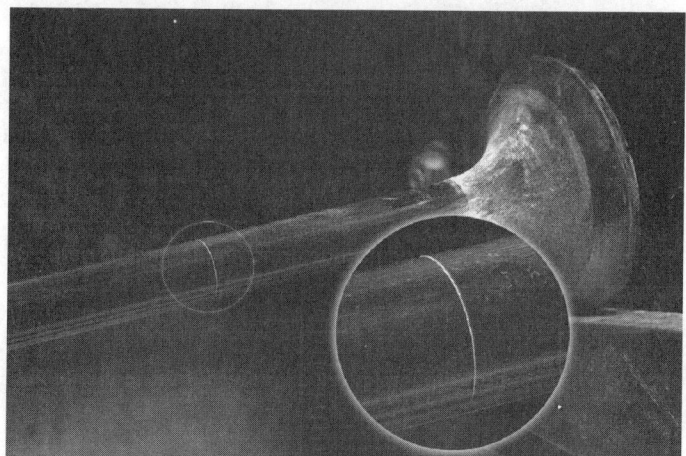

FIG. A1.5 Valve with Indication on the Stem

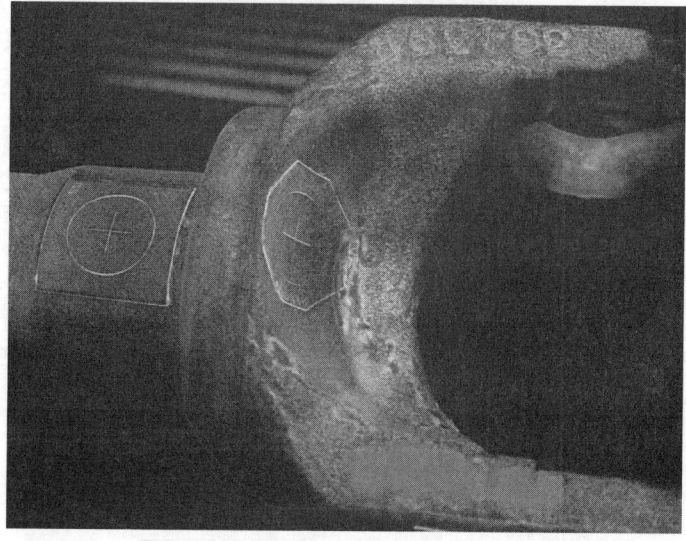

FIG. A1.6 Yoke Showing Balanced QQIs

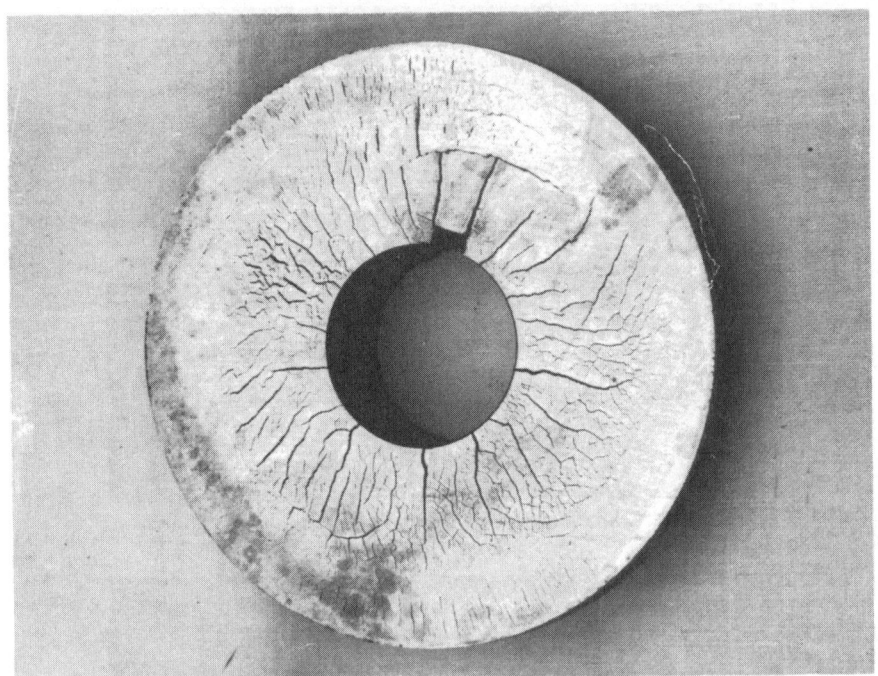

FIG. A1.7 Indications of Surface Cracking (Produced by Central Conductor Magnetization DC Continuous)

FIG. A1.8 Indications of Surface Cracking (Produced by Circular Direct Magnetization DC Continuous)

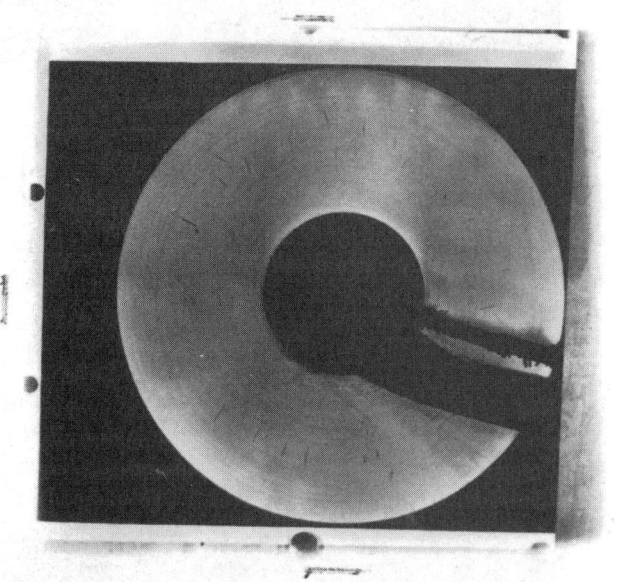

FIG. A1.9 Indications of Surface Cracks (Produced by Central Conductor Magnetization DC Continuous)

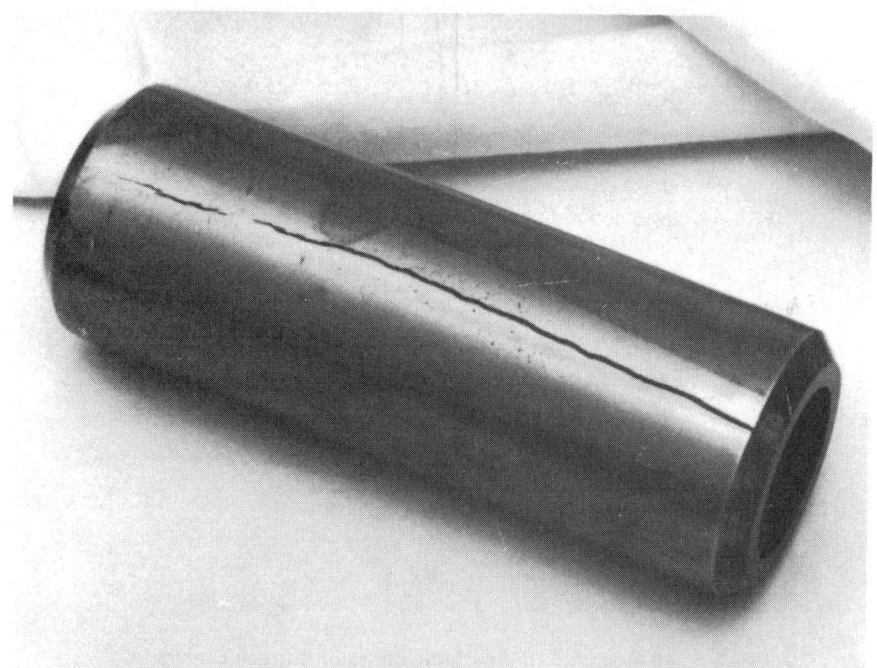

FIG. A1.10 Indications of Surface Cracks (Produced by Circular Indirect Magnetization DC)

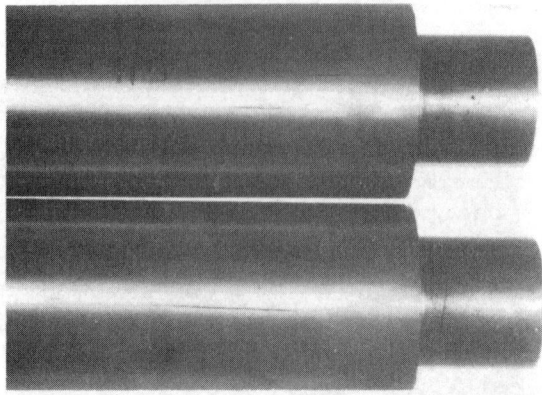

FIG. A1.11 Indications of a Near-Surface Discontinuity (Produced by Circular Direct Magnetization AC Continuous)

FIG. A1.12 Indications of Near-Surface Indications (Produced by Circular Direct Magnetization AC Continuous)

FIG. A1.13 Magnetic Rubber Indications of Surface Cracks in Aircraft Fastener Holes (Produced by Yoke Magnetization DC Continuous)

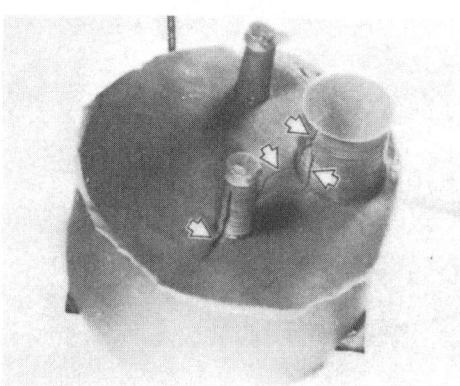

FIG. A1.14 Magnetic Rubber Indications of Surface Cracks in Aircraft Fastener Holes (Produced by Yoke Magnetization DC Continuous)

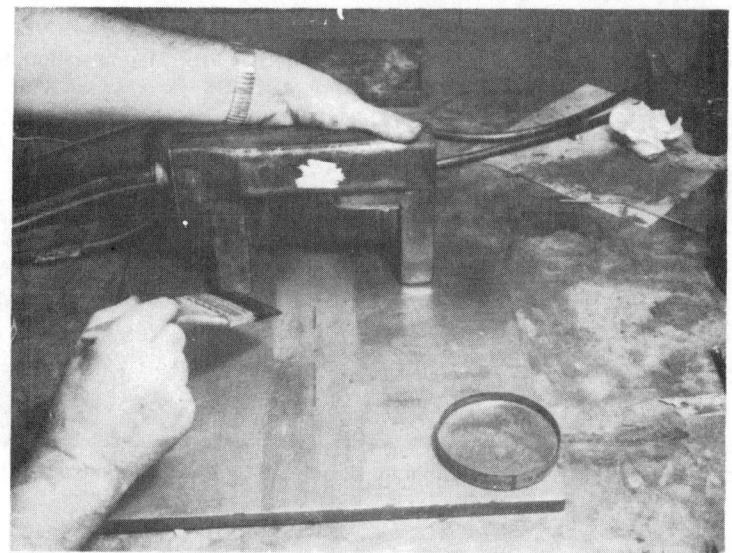

FIG. A1.15 Magnetic Slurry Indications of Surface Cracks in Weldment (Produced by Yoke Magnetization, AC Continuous)

FIG. A1.16 Magnetic Slurry Indications of Surface Cracks (Produced by Yoke Magnetization, AC Continuous)

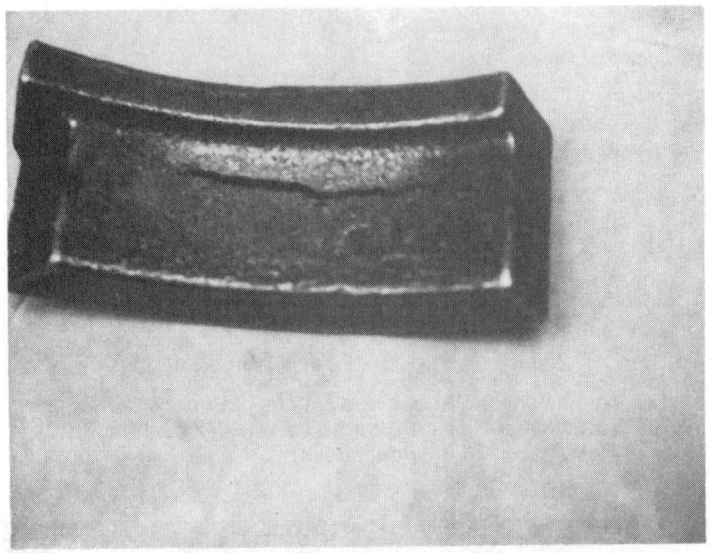

FIG. A1.17 Indications of a Near-Surface Discontinuity (Produced by Prod Magnetization, HWDC Continuous)

FIG. A1.18 Indications of a Near-Surface Discontinuity (Produced by Prod Magnetization, HWDC Continuous)

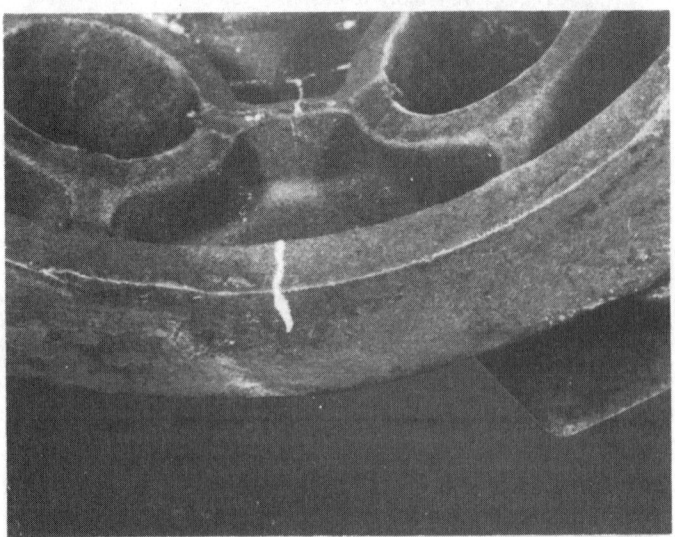

FIG. A1.19 Indication of Surface Cracks (Produced by Circular Indirect Magnetization, AC Continuous)

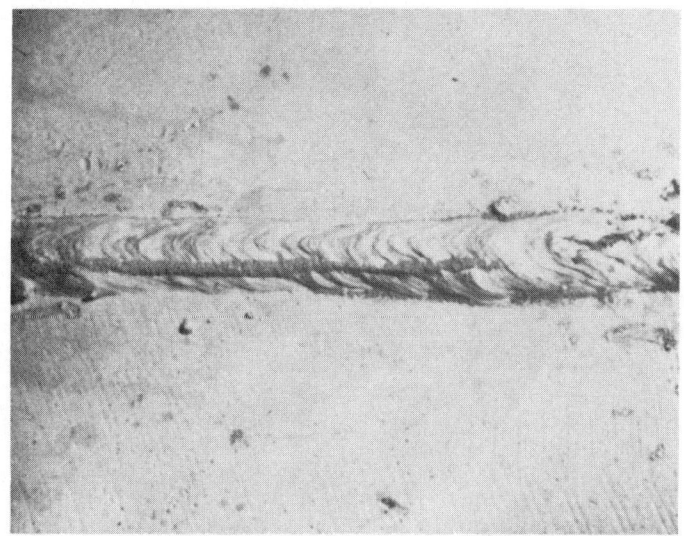

FIG. A1.20 Indication of Surface Cracks (Produced by Prod Magnetization, AC Continuous)

FIG. A1.21 Indications of Surface Cracks (Produced by Prod Magnetization, DC Continuous)

FIG. A1.22 Indications of Surface Cracks (Produced by Circular Direct Magnetization, AC Continuous)

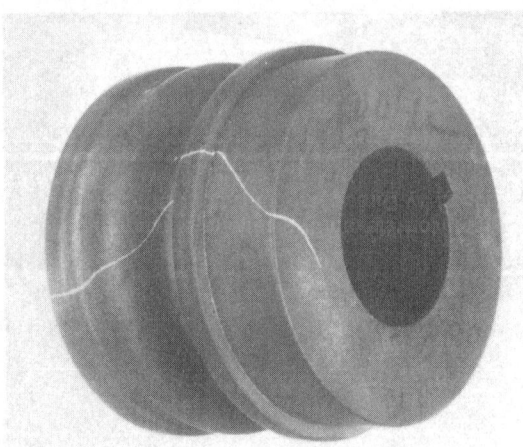

FIG. A1.23 Indications of Surface Cracks (Produced by Central Conductor Magnetization, AC Continuous)

FIG. A1.24 Nonrelevant Indications of Magnetic Writing (Produced by Direct Magnetization, DC Continuous)

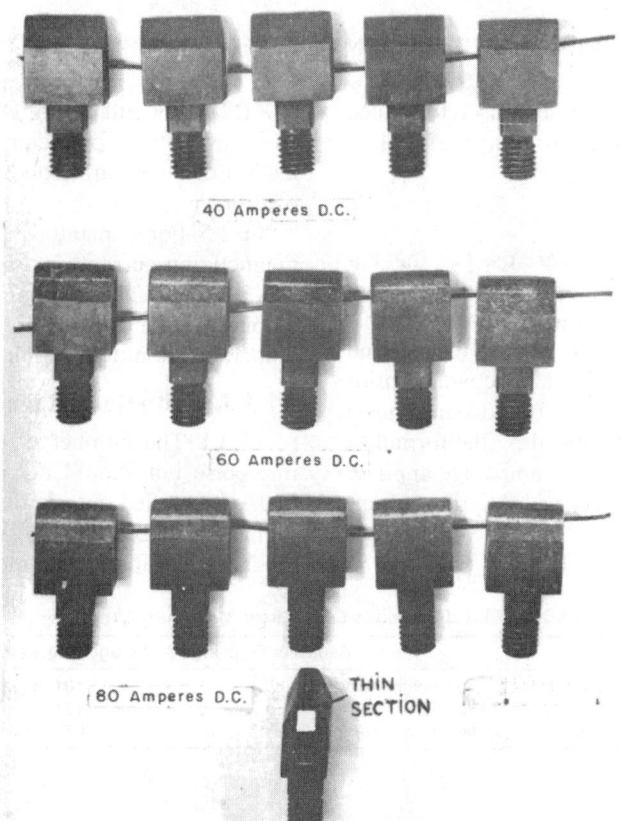

FIG. A1.25 Nonrelevant Indications Due to Change in Section on a Small Part (Produced by Indirect, Circular Magnetization, DC Continuous)

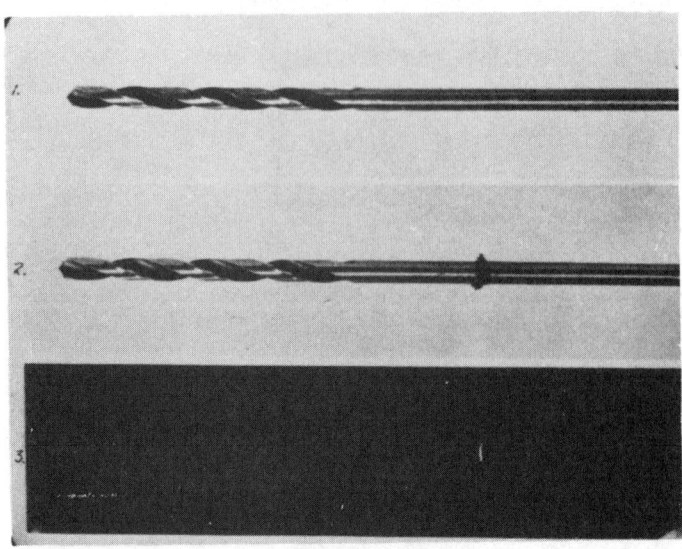

FIG. A1.26 Nonrelevant Indications of Junction Between Dissimilar Materials (Produced by Coil DC Residual Magnetization)

APPENDIXES

(Nonmandatory Information)

X1. MAGNETIZATION OF OILFIELD TUBULARS

X1.1 The following requirements should be used to induce residual magnetic fields in oilfield tubulars (tubing, casing, line pipe, and drill pipe).

X1.2 Circular Magnetism

X1.2.1 When capacitor-discharge units are used as magnetizing sources, the oilfield tubulars should be insulated from metal racks and adjacent oilfield tubulars to prevent arc burns.

X1.2.2 Partial demagnetization might occur in a magnetized length of oilfield tubulars if it is not sufficiently separated prior to magnetizing the next adjacent length. The distance used should be at least 36 inches or as determined by the formula I (0.006), whichever is greater, where I is the amperage applied.

X1.2.3 For battery or three-phase rectified-AC power supplies, a minimum magnetizing current of 300 Amps/in of specified outside diameter should be used.

X1.2.4 For full circumference inspection of material with a specified outside diameter of 16 inches and smaller, centralization of the central conductor is not required during magnetization.

X1.2.5 For capacitor-discharge units, see Table X1.1 for magnetizing current requirements.

X1.2.6 The above requirements have been demonstrated by empirical data and do not require verification, however, the amperage should be monitored during current application.

X1.3 Longitudinal Magnetization

X1.3.1 The number of coil turns and current required are imprecise but should not be less than 500 ampere-turns per inch of specified outside diameter. The current should be set as high as possible, but not so high as to cause furring of dry magnetic particles or immobility of wet magnetic particles.

TABLE X1.1 Capacitor Discharge Minimum Current

Number of Pulses	Capacitor Discharge Amperage Requirements	
Single	240 times specified weight per foot in lb/ft	161 times specified weight per metre in kg/m
Double	180 times specified weight per foot in lb/ft	121 times specified weight per metre in kg/m
Triple	145 times specified weight per foot in lb/ft	97 times specified weight per metre in kg/m

X2. REFERENCE STANDARD NOTCHED SHIMS FOR MAGNETIC PARTICLE TESTING IN ACCORDANCE WITH AS 5371

X2.1 The following standard flawed shims are typically used to establish proper field direction and ensure adequate field strength during technique development in magnetic particle examination. The shims of Fig. X2.1 may be used to ensure the establishment and balance of fields in the multidirectional magnetization method.

X2.1.1 The shims are available in two thicknesses, 0.002 in. (0.05 mm) and 0.004 in. (0.10 mm). Thinner shims are used when the thicker shims cannot conform to the part surface in the area of interest.

X2.1.2 The shims are available in two sizes, 0.75 in. (19 mm) square for Figs. X2.1 and X2.2 and 0.79 in. (20 mm) square of Fig. X2.3. The shims of Fig. X2.3 are cut, by the user, into four 0.395 in. (10 mm) square shims for use in restricted areas.

X2.1.3 Shims should be low carbon steel, AMS 5062 or equivalent.

X2.1.4 Shims should be used as specified in AS 5371. Shims are placed in the area(s) of interest with notches toward the surface of the part being examined. Use enough shims or place the shims in multiple areas to ensure proper field directions and strengths are obtained.

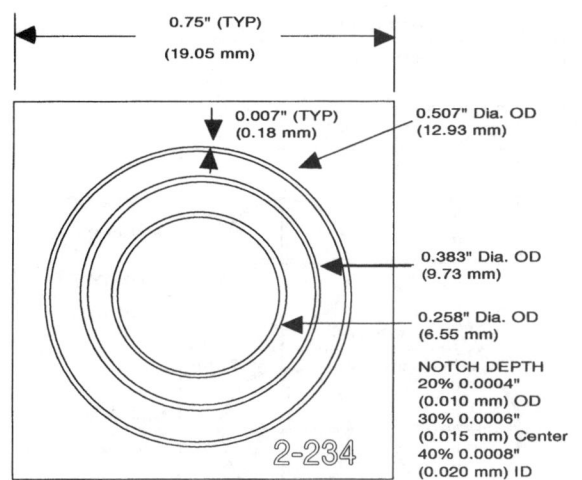 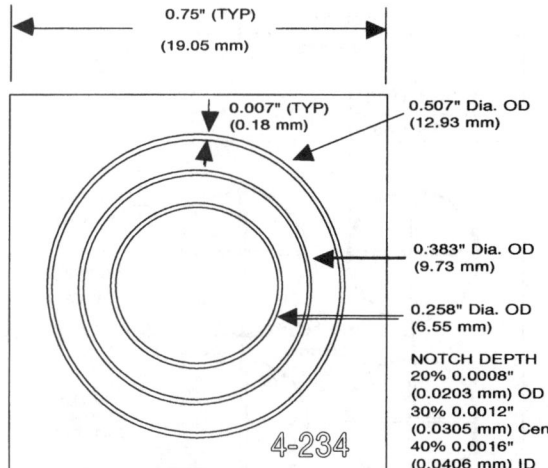

FIG. X2.1 Shim Thicknesses for Shim Types 3C2–234 and 3C4–234

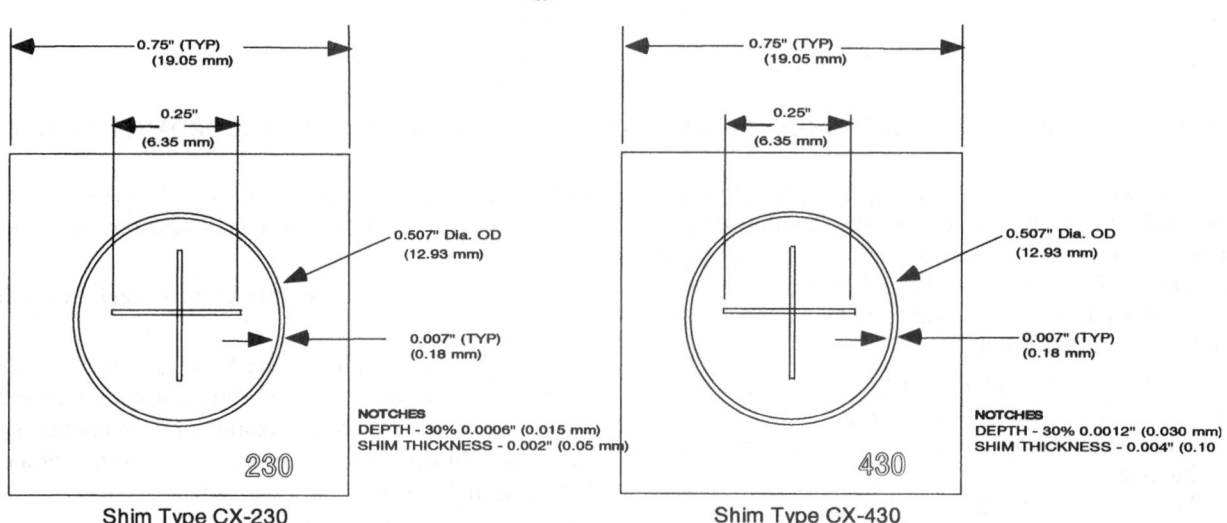

FIG. X2.2 Shim Types CX-230 and CX-430

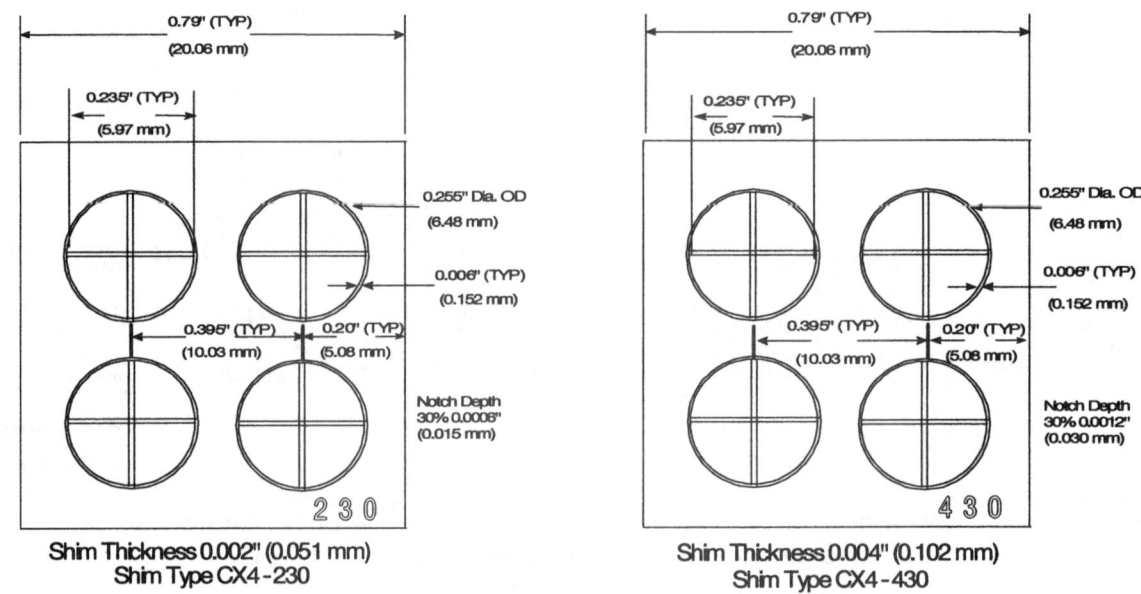

FIG. X2.3 Shim Thicknesses for Shim Types CX4–230 and CX4–430

X3. EMPIRICAL FORMULAS

X3.1 This appendix has empirical formulas for establishing magnetic field strengths; they are rules of thumb. As such, they must be used with judgment. Their use may lead to:

X3.1.1 Over magnetization, which causes excessive particle background that makes interpretation more difficult if not impossible.

X3.1.2 Poor coverage.

X3.1.3 Poor choice of examination geometries.

X3.1.4 A combination of the above.

X3.2 *Guidelines for Establishing Magnetic Fields*—The following guidelines can be effectively applied for establishing proper levels of circular and longitudinal magnetization using empirical formulas.

X3.2.1 *Circular Magnetization*

Magnetic Field Strength:

X3.2.1.1 *Direct Circular Magnetization*

When magnetizing by passing current directly through the part the nominal current should generally be 300–800 A/in. of part diameter (12 to 32 A/mm). The diameter of the part should be taken as the greatest distance between any two points on the outside circumference of the part. Currents will normally be 500 A/in. (20 A/mm) or lower, with the higher currents up to 800 A/in. (32 A/mm) being used to examine for inclusions or to examine low-permeability alloys. Amperages of less than 300 A/in. may be used when part configuration dictates and approval is obtained from the Level III and the Cognizant Engineering Organization. The field strengths generated through the use of empirical formulas should be verified with a Hall effect gaussmeter or AS 5371 shims.

X3.2.1.2 *Central Conductor Induced Magnetization*

When using offset central conductors the conductor passing through the inside of the part is placed against an inside wall of the part. The current should be from 12 A per mm of part diameter to 32 A per mm of part diameter (300 to 800 A/in.). The diameter of the part should be taken as the largest distance between any two points on the outside circumference of the part. Generally, currents will be 500 A/in. (20 A per mm) or lower with the higher currents (up to 800 A/in.) being used to examine for inclusions or to examine low permeability alloys such as precipitation-hardening steels. For examinations used to locate inclusions in precipitation-hardening steels even higher currents, up to 1000 A/in. (40 A per mm) may be used. The distance along the part circumference, which may be effectively examined should be taken as approximately four times the diameter of the central conductor, as illustrated in Fig. 10(b). The entire circumference should be examined by rotating the part on the conductor, allowing for approximately a 10 % magnetic field overlap. Less overlap, different current levels, and larger effective regions (up to 360°) may be used if the presence of suitable field levels is verified.

X3.2.2 *Air-Core Coil Longitudinal Magnetization*

Longitudinal part magnetization is produced by passing a current through a multi-turn coil encircling the part, or section of the part to be examined. A magnetic field is produced parallel to the axis of the coil. The unit of measurement is ampere turns (NI) (the actual amperage multiplied by the number of turns in the encircling coil or cable). The effective is variable and is a function of the fill factor and field extends on either side of the coil. The effective distance can easily be determined by use of a Gauss (Tesla) meter to identify where the flux lines are leaving to complete their return loop. Long parts should be examined in sections that do not exceed this length. There are four empirical longitudinal magnetization formulas employed for using encircling coils, the formula to be used depending on the fill factor. The formulas are included for historical continuity only. If used its use should be limited to simple shaped parts. It would be quicker and more accurate to use a Gauss (Tesla) meter, lay its probe on the part and measure the field rather than to calculate using the formulas.

X3.2.2.1 *Low Fill-Factor Coils*

In this case, the cross-sectional area of the fixed encircling coil greatly exceeds the cross-sectional area of the part (less than 10 % coil inside diameter). For proper part magnetization, such parts should be placed well within the coils and close to the inside wall of the coil. With this low fill-factor, adequate field strength for eccentrically positioned parts with a length-over-diameter ratio (L/D) between 3 and 15 is calculated from the following equations:[9]

(1) Parts with Low Fill-Factor Positioned Close to Inside Wall of Coil:

$$NI = K/(L/D) \; (\pm 10\,\%) \quad (X3.1)$$

where:
- N = number of turns in the coil,
- I = coil current to be used, amperes (A),
- K = 45 000 (empirically derived constant),
- L = part, length, in., (see Note),
- D = part diameter, in.; for hollow parts, see X3.2.2.4, and
- NI = ampere turns.

For example, a part 15 in. (38.1 cm) long with 5-in. (12.7-cm) outside diameter has an L/D ratio of 15/5 or 3. Accordingly, the ampere turn requirement ($NI = 45\,000/3$) to provide adequate field strength in the part would be 15 000 ampere turns. If a five-turn coil or cable is used, the coil amperage requirements would be ($I = 15\,000/5$) = 3000 A ($\pm 10\,\%$). A 500 turn coil would require 30 A ($\pm 10\,\%$).

(2) Parts with a Low Fill-Factor Positioned in the Center of the Coil:

$$NI = KR/\{(6L/D) - 5\}(\pm 10\,\%) \quad (X3.2)$$

where:
- N = number of turns in the coil,
- I = coil current to be used, A,
- K = 43 000 (empirically derived constant),
- R = coil radius, in.,
- L = part length, in. (see Note),
- D = part diameter, in., for hollow parts (see X3.2.2.4), and
- NI = ampere turns.

For example, a part 15 in. (38.1 cm) long with 5-in. (12.7-cm) outside diameter has a L/D ratio of 15/5 or 3. If a five-turn 12-in. diameter (6-in. radius) (30.8-cm diameter (15.4-cm radius)) coil or cable is used, (*1*) the ampere turns requirement would be as follows:

$$NI = \frac{(43\,000 \times 6)}{((6 \times 3) - 5)} \; or \; 19\,846$$

and (*2*) the coil amperage requirement would be as follows:

$$\frac{19\,846}{5} \; or \; 3\,969\,A\;(\pm 10\,\%)$$

X3.2.2.2 *Intermediate Fill-Factor Coils*

When the cross section of the coil is greater than twice and less than ten times the cross section of the part being examined:

$$NI = (NI)_{hf}(10 - Y) + (NI)_{lf}(Y - 2)/8 \quad (X3.3)$$

where:
- NI_{hf} = value of NI calculated for high fill-factor coils using Eq X3.3,
- NI_{lf} = value of NI calculated for low fill-factor coils using Eq X3.1 or Eq X3.2, and
- Y = ratio of the cross-sectional area of the coil to the cross section of the part. For example, if the coil has an inside diameter of 10 in. (25.4 cm) and part (a bar) has an outside diameter of 5 in. (12.2 cm)

$$Y = (\pi(5)^2)/(\pi(2.5)^2) = 4$$

X3.2.2.3 *High Fill-Factor Coils*

---

[9] These equations are included for historical continuity only. It is faster to buy a Tesla meter, lay the probe on the part and measure the field strength than calculating using the equations.

In this case, when fixed coils or cable wraps are used and the cross-sectional area of the coil is less than twice the cross-sectional area (including hollow portions) of the part, the coil has a high fill-factor.

(1) For Parts Within a High Fill-Factor Positioned Coil and for Parts with an L/D ratio equal to or greater than 3:

$$NI = \frac{K}{\{(L/D) + 2\}} (\pm 10\,\%)$$

where:
N = number of turns in the coil or cable wrap,
I = coil current, A,
K = 35 000 (empirically derived constant),
L = part length, in.,
D = part diameter, in., and
NI = ampere turns.

For example, the application of Eq X3.3 can be illustrated as follows: a part 10 in. (25.4 cm) long-with 2-in. (5.08-cm) outside diameter would have an L/D ratio of 5 and an ampere turn requirements of $NI = 35\,000/(5 + 2)$ or 5000 ($\pm 10\,\%$) ampere turns. If a five-turn coil or cable wrap is employed, the amperage requirement is 5000/5 or 1000 A ($\pm 10\,\%$).

Note X3.1—For *L/D* ratios less than 3, a pole piece (ferromagnetic material approximately the same diameter as part) should be used to effectively increase the *L/D* ratio or utilize an alternative magnetization method such as induced current. For *L/D* ratios greater than 15, a maximum *L/D* value of 15 should be used for all formulas cited above.

X3.2.2.4 *L/D Ratio for a Hollow Piece*

When calculating the *L/D* ratio for a hollow piece, *D* should be replaced with an effective diameter D_{eff} calculated using:

$$D_{eff} = 2[(A_t - A_h)/\pi]^{1/2}$$

where:
A_t = total cross-sectional area of the part, and
A_h = cross-sectional area of the hollow portion(s) of the part.

$$D_{eff} = [(OD)^2 - (ID)^2]^{1/2}$$

where:
OD = outside diameter of the cylinder, and
ID = inside diameter of the cylinder.

X4. DEVICES FOR EVALUATION OF MAGNETIC PARTICLE EXAMINATION MATERIALS

X4.1 Scope

X4.1.1 The purpose of this appendix is to describe the capabilities and use of various devices that may be utilized to monitor and evaluate the performance of materials and systems for magnetic particle examination.

X4.2 Magnetic Stripe Cards. The magnetically encoded pattern in magnetic stripes, as on cards used for personal banking, identification and other purposes, can serve as a tool to evaluate magnetic particle examination materials. Particles are attracted to the magnetic gradients formed in the stripe when the stripe has been magnetically encoded with a pattern of flux reversals. The encoding of the stripe can be controlled to provide gradients of varying magnitude. Particles can be evaluated for sensitivity when observed to see how small a gradient can generate a particle indication.

X4.3 Characteristics

X4.3.1 Magnetic stripe cards should be made in accordance with ISO 7810– Identification Cards— Physical Characteristics.

X4.3.2 The stripe may be made of either low-coercivity (lo-co) or high-coercivity (hi-co) material, as designated by the manufacturer.

X4.3.3 A constant encoding pattern, decaying encoding pattern, reverse decaying pattern or other pattern may be encoded into the stripe. See Fig. X4.1 photograph of fluorescent particle indications of decaying and reverse decaying encoding patterns.

X4.4 Use of the Magnetic Stripe Card for Magnetic Particle Material Evaluation

X4.4.1 *Wet Method Materials*—Wet method materials may be poured, sprayed or otherwise applied to the stripe, as they would be used for MPI. Excess bath should be allowed to flow away from the stripe. The stripe should be observed under suitable illumination (See Section 7) for the formation of particle indications. Observations should be noted as to the quantity of particle indications and the clarity thereof.

Note X4.1—Dark colored non-fluorescent particles may be more readily observed with the use of a white contrast paint applied over the stripe prior to particle evaluation. Particle indications may also be observed and/or permanently recorded per Section 17 (Paragraph 17.1.2 can apply to wet method powder after the fluid has been allowed to evaporate.).

X4.4.2 *Dry Method Materials*—Dry method materials should be poured, dusted, blown or otherwise applied to the stripe, as they would be used for MPI. Excess powder should be removed with a gentle blowing action. The stripe should be observed under suitable illumination (See Section 7) for the formation of particle indications. Observations should be noted as to the quantity of particle indications and the clarity therof.

X4.4.3 *Recording of Indications*—Recorded particle indications (See 17.1.2) may serve as material documentation records and standards for material performance. Other material, or the same material at a later time, can be compared at any time to the recorded standard.

X4.5 *Loss of Indications on the Stripe*—There are several circumstances where particle indications may not be visible on

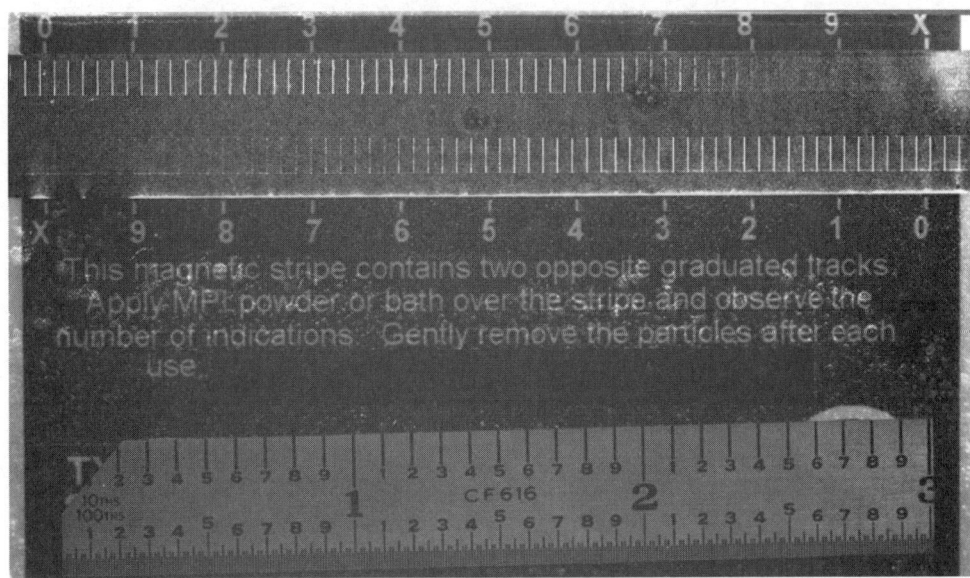

FIG. X4.1 Example of fluorescent particle indications of a decaying encoding pattern (top track) and a reverse-decaying pattern (bottom track) on the magnetic stripe of a magnetic stripe card.

the magnetic stripe. When indications are not visible the subject particles should not be used for examination unless otherwise verified as being acceptable.

X4.5.1 *Concentration*—The subject wet method particles may not have a sufficient level of concentration. In this case, increase the concentration level of the bath and re-perform the check until the particles demonstrate suitable performance.

X4.5.2 *Sensitivity*—The subject particles may not provide necessary sensitivity. In this case, replace the material with a suitably sensitive material and re-perform the check until the particles demonstrate suitable performance.

X4.5.3 *Erasure*—The stripe has become magnetically erased. In this case, no discernible particle indication will appear. In this case, repeat the check with another card and/or sensitivity check until the particles demonstrate suitable performance. Either destroy the card with the de-encoded stripe or report it to the manufacturer and follow the manufacturer's recommendations.

X4.6 Precautions

X4.6.1 *Preparation*—The surface of the stripe must be clean of any fluid or foreign matter prior to the application of the MPI material. The encoded stripe should not be re-magnetized in any manner prior to use or de-magnetized in any manner following its use.

X4.6.2 *Storage*—The surface of the stripe should be cleaned of remaining fluid and particles after the observations of the MPI material have been made. When not in use, the card should be stored away from excessive heat and strong magnetic fields.

X5. CENTRIFUGE TUBES

X5.1 Centrifuge tubes should be pear-shaped, made from thoroughly annealed glass, and conform to the dimensions given in Figs. X5.1 and X5.2 as applicable. The graduations, numbered as shown, should be clear and distinct.

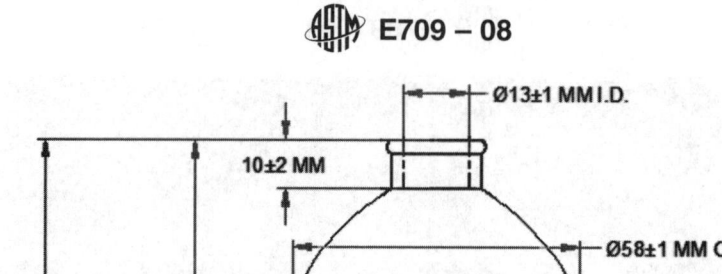
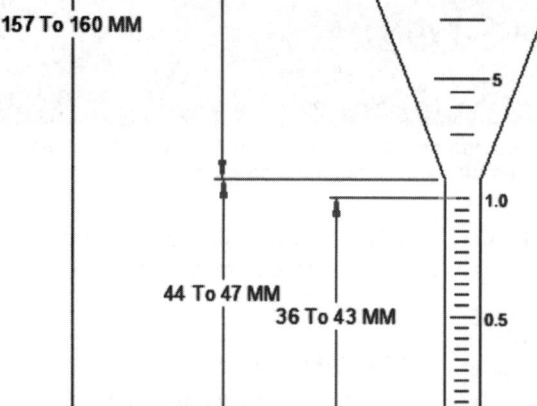

FIG. X5.1 Pear Shaped Centrifuge Tube – Fluorescent Bath

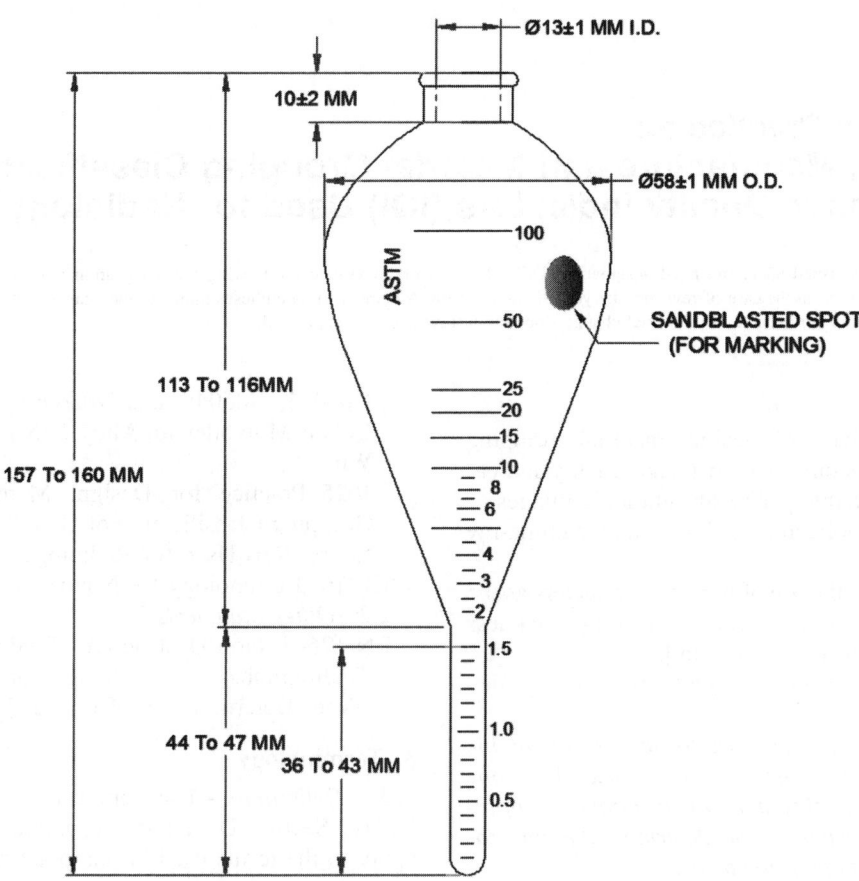

FIG. X5.2 Pear Shaped Centrifuge Tube – Non-Fluorescent Bath

ASTM International takes no position respecting the validity of any patent rights asserted in connection with any item mentioned in this standard. Users of this standard are expressly advised that determination of the validity of any such patent rights, and the risk of infringement of such rights, are entirely their own responsibility.

This standard is subject to revision at any time by the responsible technical committee and must be reviewed every five years and if not revised, either reapproved or withdrawn. Your comments are invited either for revision of this standard or for additional standards and should be addressed to ASTM International Headquarters. Your comments will receive careful consideration at a meeting of the responsible technical committee, which you may attend. If you feel that your comments have not received a fair hearing you should make your views known to the ASTM Committee on Standards, at the address shown below.

This standard is copyrighted by ASTM International, 100 Barr Harbor Drive, PO Box C700, West Conshohocken, PA 19428-2959, United States. Individual reprints (single or multiple copies) of this standard may be obtained by contacting ASTM at the above address or at 610-832-9585 (phone), 610-832-9555 (fax), or service@astm.org (e-mail); or through the ASTM website (www.astm.org).

Designation: E747 – 04

Standard Practice for
Design, Manufacture and Material Grouping Classification of Wire Image Quality Indicators (IQI) Used for Radiology[1]

This standard is issued under the fixed designation E747; the number immediately following the designation indicates the year of original adoption or, in the case of revision, the year of last revision. A number in parentheses indicates the year of last reapproval. A superscript epsilon (ε) indicates an editorial change since the last revision or reapproval.

1. Scope

1.1 This practice[2] covers the design, material grouping classification, and manufacture of wire image quality indicators (IQI) used to indicate the quality of radiologic images.

1.2 This practice is applicable to X-ray and gamma-ray radiology.

1.3 This practice covers the use of wire penetrameters as the controlling image quality indicator for the material thickness range from 6.4 to 152 mm [0.25 to 6.0 in.].

1.4 The values stated in inch-pound units are to be regarded as standard.

1.5 *This standard does not purport to address all of the safety concerns, if any, associated with its use. It is the responsibility of the user of this standard to establish appropriate safety and health practices and determine the applicability of regulatory limitations prior to use.*

2. Referenced Documents

2.1 *ASTM Standards:*[3]
B139/B139M Specification for Phosphor Bronze Rod, Bar, and Shapes
B150M Specification for Aluminum Bronze, Rod, Bar, and Shapes [Metric][4]
B161 Specification for Nickel Seamless Pipe and Tube
B164 Specification for Nickel-Copper Alloy Rod, Bar, and Wire
B166 Specification for Nickel-Chromium-Iron Alloys (UNS N06600, N06601, N06603, N06690, N06693, N06025, N06045, and N06696)* and Nickel-Chromium-Cobalt-Molybdenum Alloy (UNS N06617) Rod, Bar, and Wire
E1025 Practice for Design, Manufacture, and Material Grouping Classification of Hole-Type Image Quality Indicators (IQI) Used for Radiology
E1316 Terminology for Nondestructive Examinations

2.2 *Other Standards:*[5]
EN 426–1 Non-Destructive Testing—Image Quality of Radiographs-Part 1: Image Quality Indicators (Wire-Type)-Determination of Image Quality Value

3. Terminology

3.1 *Definitions*—The definitions of terms in Terminology E1316, Section D, relating to gamma and X-radiology, shall apply to the terms used in this practice.

4. Wire IQI Requirements

4.1 The quality of all levels of examination shall be determined by a set of wires conforming to the following requirements:

4.1.1 Wires shall be fabricated from materials or alloys identified or listed in accordance with 7.2. Other materials may be used in accordance with 7.3.

4.1.2 The IQI consists of sets of wires arranged in order of increasing diameter. The diameter sizes specified in Table 1 are established from a consecutive series of numbers taken in general from the ISO/R 10 series. The IQI shall be fabricated in accordance with the requirements specified in Figs. 1-8 and Tables 1-3. IQIs previously manufactured to the requirements of Annex A1 may be used as an alternate provided all other requirements of this practice are met.

4.1.3 Image quality indicator (IQI) designs other than those shown in Figs. 1-8 and Annex A1 are permitted by contractual agreement. If an IQI set as listed in Table 1 or Annex A1 is modified in size, it must contain the grade number, set identity, and essential wire. It must also contain two additional wires

---

[1] This practice is under the jurisdiction of ASTM Committee E07 on Nondestructive Testing and is the direct responsibility of Subcommittee E07.01 on Radiographic Practice and Penetrameters.
Current edition approved January 1, 2004. Published February 2004. Originally approved in 1980. Last previous edition approved in 1997 as E747 - 97. DOI: 10.1520/E0747-04.

[2] For ASME Boiler and Pressure Vessel Code applications see related Practice SE-747 in Section II of that Code.

[3] For referenced ASTM standards, visit the ASTM website, www.astm.org, or contact ASTM Customer Service at service@astm.org. For *Annual Book of ASTM Standards* volume information, refer to the standard's Document Summary page on the ASTM website.

[4] Withdrawn. The last approved version of this historical standard is referenced on www.astm.org.

[5] Available from American National Standards Institute (ANSI), 25 W. 43rd St., 4th Floor, New York, NY 10036.

TABLE 1 Wire IQI Sizes and Wire Identity Numbers

SET A		SET B	
Wire Diameter in. [mm]	Wire Identity	Wire Diameter in. [mm]	Wire Identity
0.0032 [0.08][A]	1	0.010 [0.25]	6
0.004 [0.1]	2	0.013 [0.33]	7
0.005 [0.13]	3	0.016 [0.4]	8
0.0063 [0.16]	4	0.020 [0.51]	9
0.008 [0.2]	5	0.025 [0.64]	10
0.010 [0.25]	6	0.032 [0.81]	11

SET C		SET D	
Wire Diameter in. [mm]	Wire Identity	Wire Diameter in. [mm]	Wire Identity
0.032 [0.81]	11	0.10 [2.5]	16
0.040 [1.02]	12	0.126 [3.2]	17
0.050 [1.27]	13	0.160 [4.06]	18
0.063 [1.6]	14	0.20 [5.1]	19
0.080 [2.03]	15	0.25 [6.4]	20
0.100 [2.5]	16	0.32 [8]	21

[A] The 0.0032 wire may be used to establish a special quality level as agreed upon between the purchaser and the supplier.

that are the next size larger and the next size smaller as specified in the applicable set listed in Table 1.

4.1.4 Each set must be identified using letters and numbers made of industrial grade lead or of a material of similar radiographic density. Identification shall be as shown on Figs. 1-8 or Annex A1, unless otherwise specified by contractual agreement.

4.1.5 European standard EN 462-1 contains similar provisions (with nominal differences-see table A1.1) for wire image quality indicators as this standard (E747). International users of these type IQI standards who prefer the use of EN 462-1 for their particular applications should specify such alternate provisions within separate contractual arrangements from this standard.

5. Image Quality Indicator (IQI) Procurement

5.1 When selecting IQI's for procurement, the following factors should be considered:

5.1.1 Determine the alloy group(s) of the material to be examined.

5.1.2 Determine the thickness or thickness range of the material(s) to be examined.

5.1.3 Select the applicable IQI's that represent the required IQI thickness(s) and alloy(s).

6. Image Quality Levels

6.1 The quality level required using wire penetrameters shall be equivalent to the 2-2T level of Practice E1025 for hole-type IQI's unless a higher or lower quality level is agreed upon between purchaser and supplier. Table 4 provides a list of various hole-type IQI's and the diameter of wires of corresponding equivalent penetrameter sensitivity (EPS) with the applicable 1T, 2T, and 4T holes in the IQI. This table can be used for determining 1T, 2T, and 4T quality levels. Appendix X1 gives the equation for calculating other equivalencies if needed.

6.2 In specifying quality levels, the contract, purchase order, product specification, or drawing should clearly indicate the thickness of material to which the quality level applies. Careful consideration of required quality levels is particularly important.

7. Material Groups

7.1 *General*:

7.1.1 Materials have been designated in eight groups based on their radiographic absorption characteristics: groups 03, 02, and 01 for light metals and groups 1 through 5 for heavy metals.

7.1.2 The light metal groups, magnesium (Mg), aluminum (Al), and titanium (Ti) are identified 03, 02, and 01 respectively, for their predominant alloying constituent. The materials are listed in order of increasing radiation absorption.

7.1.3 The heavy metal groups, steel, copper-base, nickel-base, and kindred alloys are identified 1 through 5. The materials increase in radiation absorption with increasing numerical designation.

7.1.4 Common trade names or alloy designations have been used for clarification of the pertinent materials.

7.1.5 The materials from which the IQI for the group are to be made are designated in each case and these IQI's are applicable for all materials listed in that group. In addition, any group IQI may be used for any material with a higher group number, provided the applicable quality level is maintained.

7.2 *Materials Groups*:

7.2.1 *Materials Group 01*:

7.2.1.1 Image quality indicators (IQI's) shall be made of titanium or titanium shall be the predominant alloying constituent.

7.2.1.2 Use on all alloys of which titanium is the predominant alloying constituent.

7.2.2 *Materials Group 02*:

7.2.2.1 Image quality indicators (IQI's) shall be made of aluminum or aluminum shall be the predominant alloying constituent.

7.2.2.2 Use on all alloys of which aluminum is the predominant alloying constituent.

7.2.3 *Materials Group 03*:

7.2.3.1 Image quality indicators (IQI's) shall be made of magnesium or magnesium shall be the predominant alloying constituent.

7.2.3.2 Use on all alloys of which magnesium is the predominant alloying constituent.

7.2.4 *Materials Group 1*:

7.2.4.1 Image quality indicators (IQI's) shall be made of carbon steel or Type 300 series stainless steel.

7.2.4.2 Use on all carbon steel, low-alloy steels, stainless steels, and manganese-nickel-aluminum bronze (Superston).[6]

7.2.5 *Materials Group 2*:

7.2.5.1 Image quality indicators (IQI's) shall be made of aluminum bronze (Alloy No. 623 of Specification B150M) or equivalent, or nickel-aluminum bronze (Alloy No. 630 of Specification B150M) or equivalent.

7.2.5.2 Use on all aluminum bronzes and all nickel-aluminum bronzes.

---

[6] Superston is a registered trademark of Superston Corp., Jersey City, NJ.

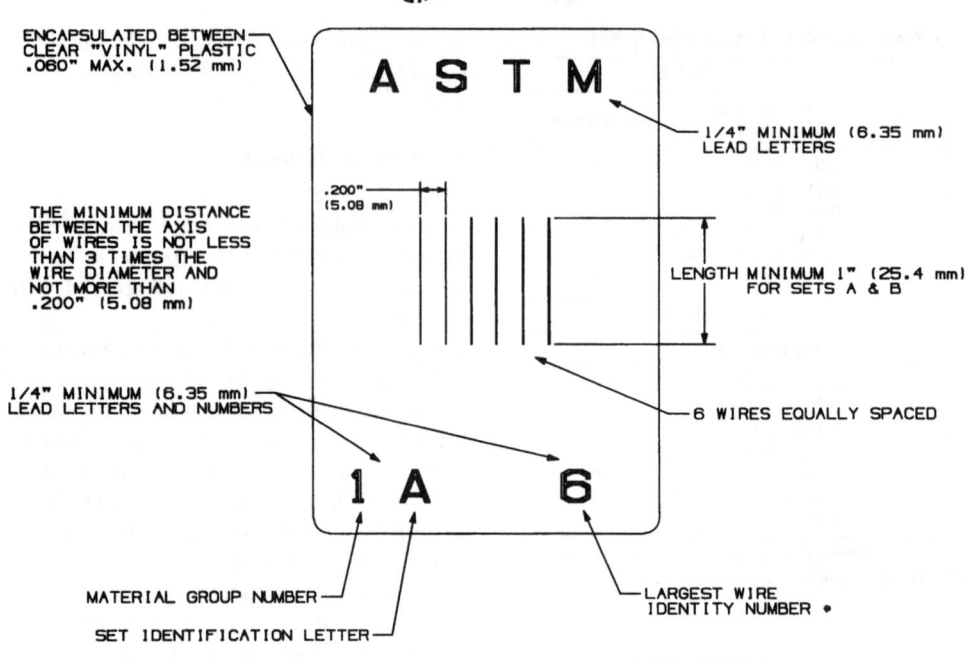

FIG. 1 Set A/Alternate 1

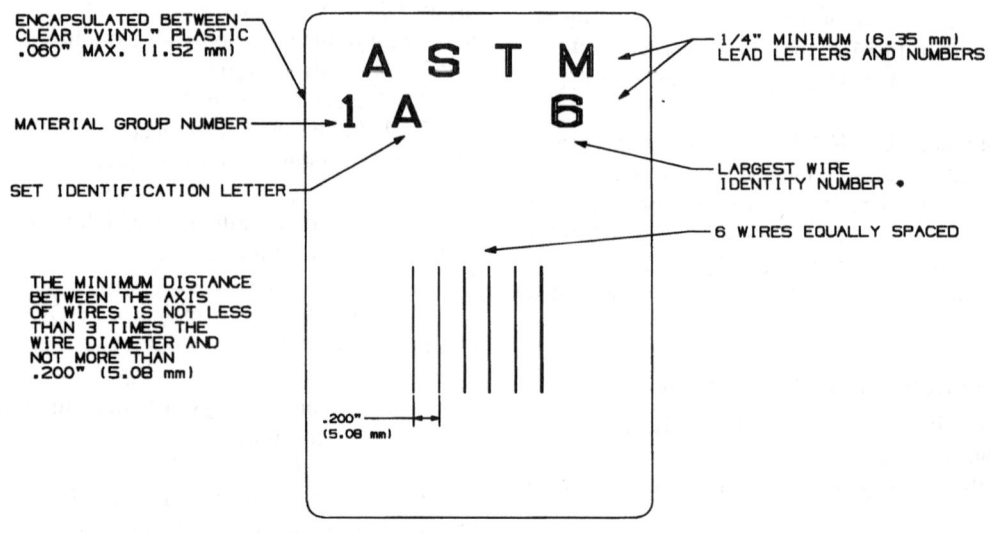

FIG. 2 Set A/Alternate 2

7.2.6 *Materials Group 3*:

7.2.6.1 Image quality indicators (IQI's) shall be made of nickel-chromium-iron alloy (UNS No. N06600) (Inconel).[7] (See Specification B166).

7.2.6.2 Use on nickel-chromium-iron alloy and 18 % nickel-maraging steel.

7.2.7 *Materials Group 4*:

7.2.7.1 Image quality indicators (IQI's) shall be made of 70 to 30 nickel-copper alloy (Monel)[8] (Class A or B of Specification B164) or equivalent, or 70 to 30 copper-nickel alloy (Alloy G of Specification B161) or equivalent.

7.2.7.2 Use on nickel, copper, all nickel-copper series, or copper-nickel series of alloys, and all brasses (copper-zinc alloys). Group 4 IQI's may include the leaded brasses since leaded brass increases in attenuation with increase in lead content. This would be equivalent to using a lower group IQI.

7.2.8 *Materials Group 5*:

7.2.8.1 Image quality indicators (IQI's) shall be made of tin bronze (Alloy D of Specification B139/B139M).

7.2.8.2 Use on tin bronzes including gun-metal and valve bronze, or leaded-tin bronze of higher lead content than valve bronze. Group 5 IQI's may include bronze of higher lead

---

[7] Inconel is a registered trademark of The International Nickel Co., Inc., Huntington, WV 25720.

[8] Monel is a registered trademark of The International Nickel Co., Inc., Huntington,, WV 25720.

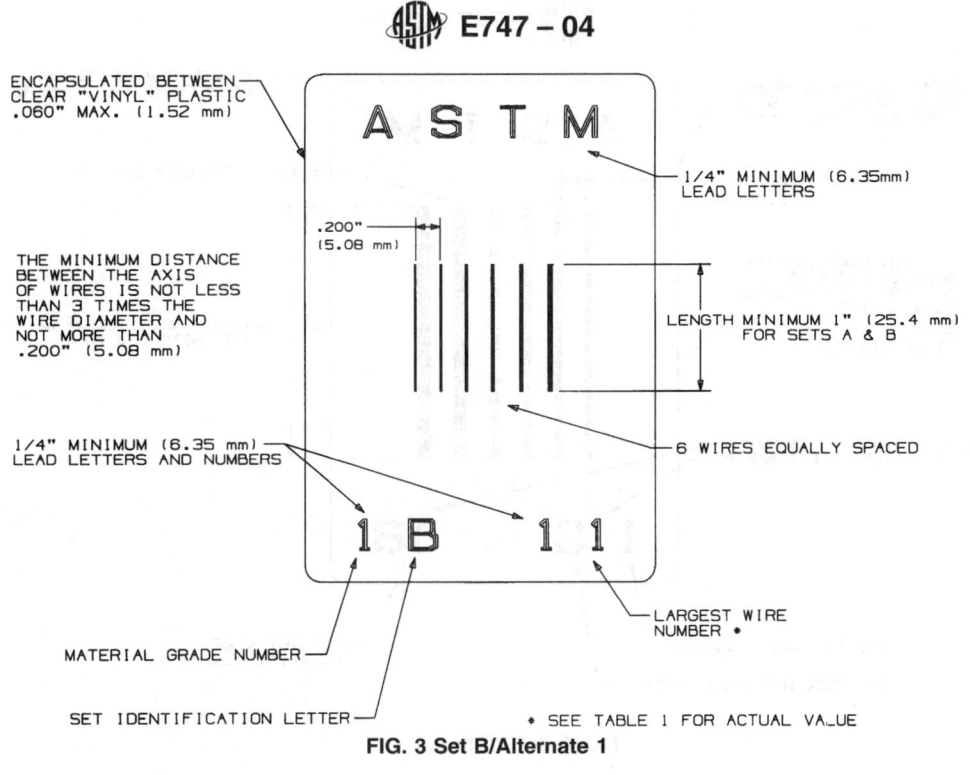

FIG. 3 Set B/Alternate 1

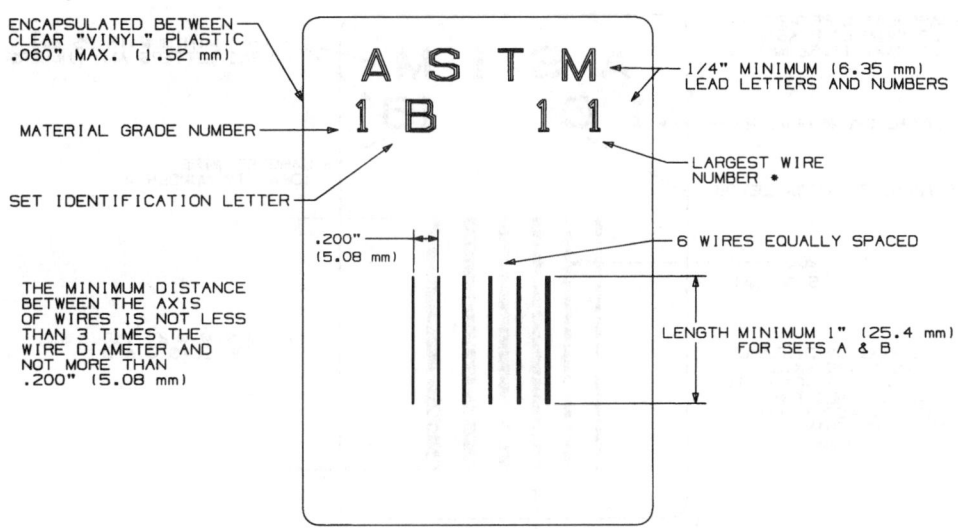

FIG. 4 Set B/Alternate 2

content since leaded bronze increases in attenuation with increase in lead content. This would be equivalent to using a lower group IQI.

NOTE 1—In developing the eight listed materials groups, a number of other trade names or other nominal alloy designations were evaluated. For the purpose of making this practice as useful as possible, these materials are listed and categorized, by group, as follows:

(1) Group 2—Haynes Alloy IN-100.[9]

(2) Group 3—Haynes Alloy No. 713C, Hastelloy D[10], G.E. Alloy SEL, Haynes Stellite Alloy No. 21, GMR-235 Alloy, Haynes Alloy No. 93, Inconel X[7], Inconel 718, and Haynes Stellite Alloy No. S-816.

(3) Group 4—Hastelloy Alloy F, Hastelloy Alloy X, and Multimeter Alloy Rene 41.

(4) Group 5—Alloys in order of increasing attenuation: Hastelloy Alloy B, Hastelloy Alloy C, Haynes Stellite Alloy No. 31, Thetaloy, Haynes Stellite No. 3, Haynes Alloy No. 25. Image quality indicators

---

[9] All Haynes alloys are registered trademarks of Union Carbide Corp., New York, NY.

[10] All Hastelloys and Haynes Stellite alloys are registered trademarks of the Cabot Corp., Boston, MA.

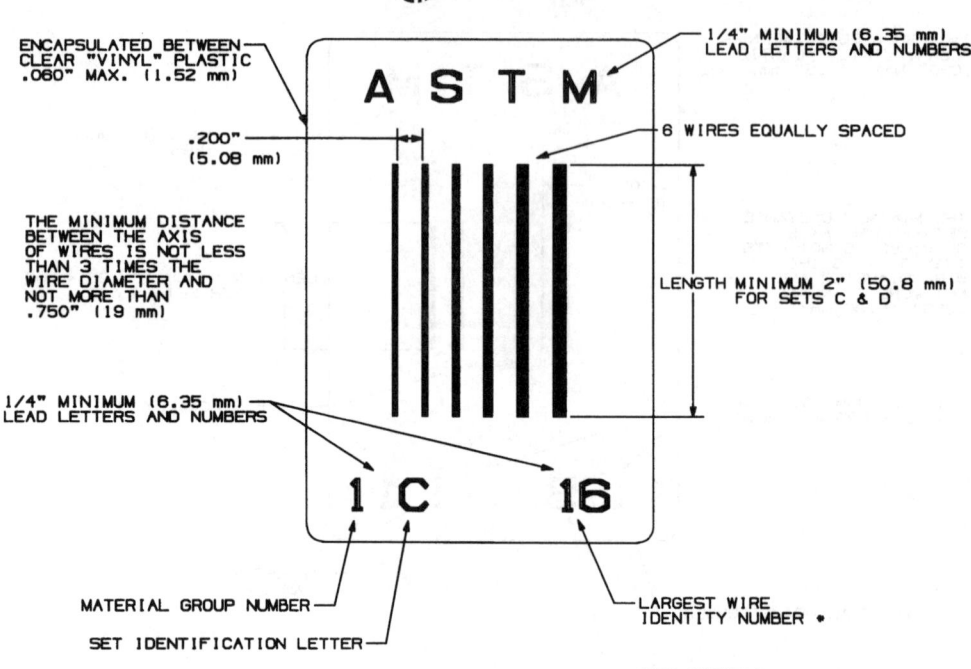

FIG. 5 Set C/Alternate 1

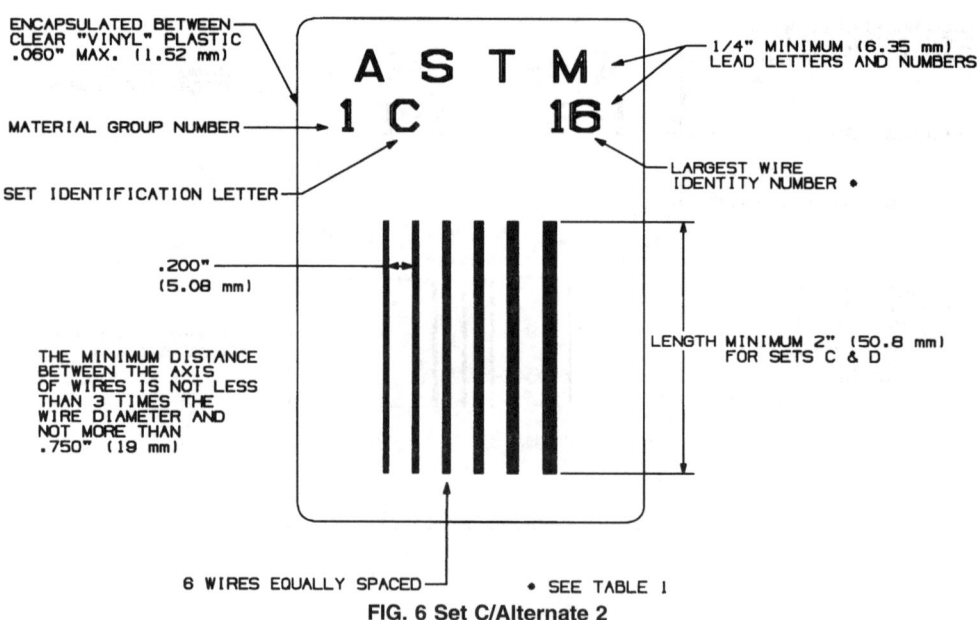

FIG. 6 Set C/Alternate 2

(IQI's) of any of these materials are considered applicable for the materials that follow it.

NOTE 2—The committee formulating these recommendations recommend other materials may be added to the materials groups listed as the need arises or as more information is gained, or that additional materials groups may be added.

7.3 *Method for Other Materials*:

7.3.1 For materials not herein covered, IQI's of the same materials, or any other material, may be used if the following requirements are met. Two blocks of equal thickness, one of the material to be examined (production material) and one of the IQI material, shall be radiographed on one film by one exposure at the lowest energy level to be used for production. Transmission densitometer measurements of the radiographic image of each material shall be made. The density of each image shall be between 2.0 and 4.0. If the image density of the IQI material is within 1.00 to 1.15 times (−0 % to + 15 %) the image density of the production material, IQI's made of that IQI material may be used in radiography of that production material. The percentage figure is based on the radiographic density of the IQI material.

7.3.2 It shall always be permissible to use IQI's of similar composition as the material being examined.

8. Image Quality Indicator (IQI) Certification

8.1 Documents shall be provided by the IQI manufacturer attesting to the following:

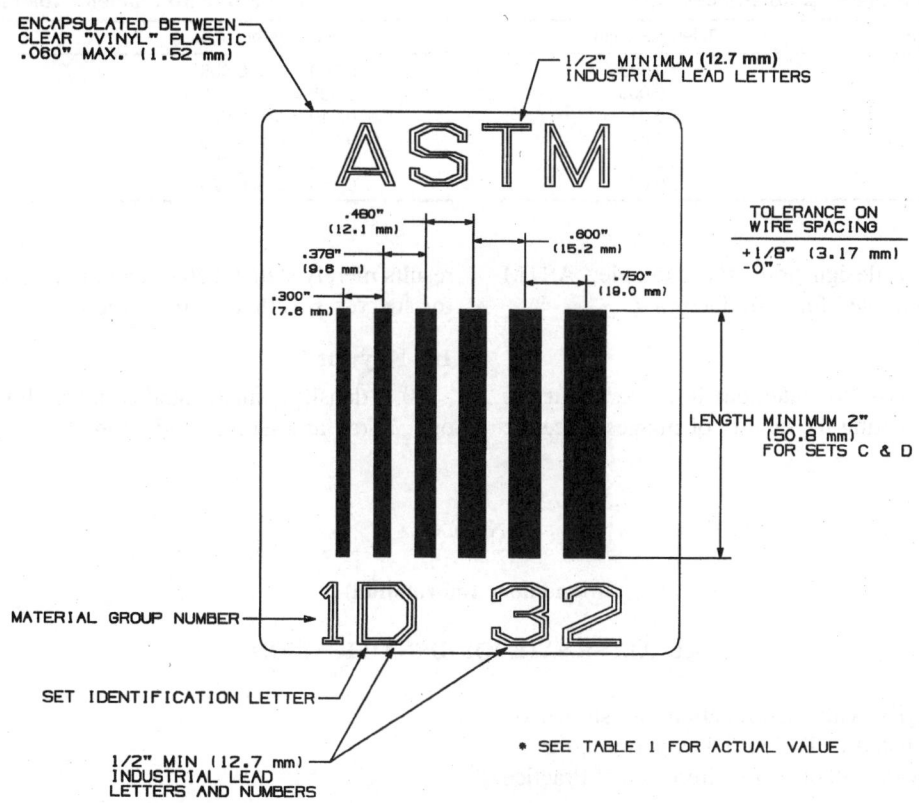

FIG. 7 Set D/Alternate 1

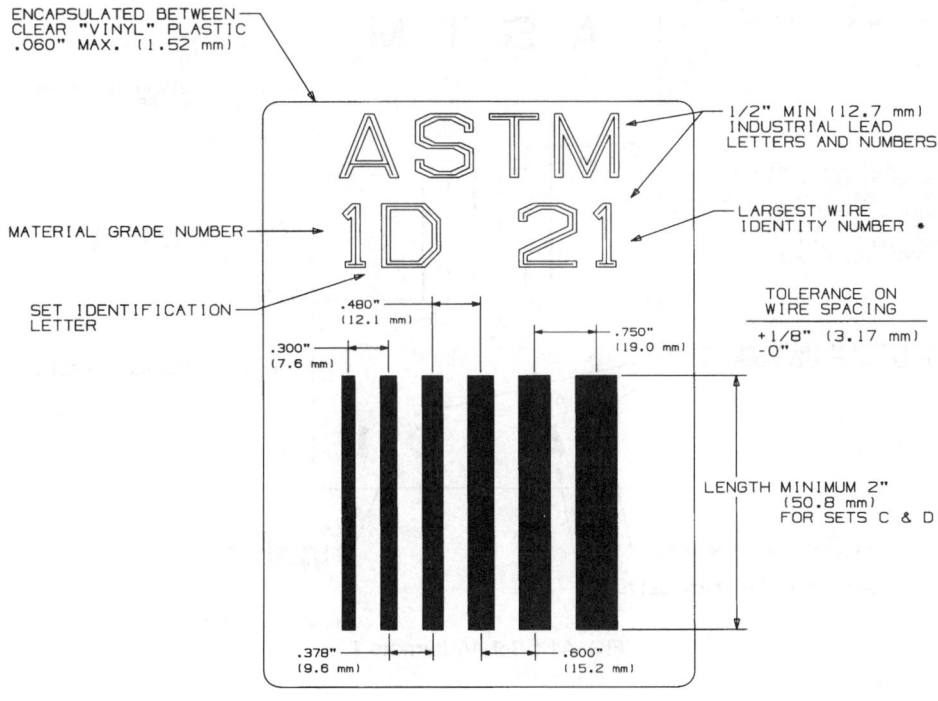

FIG. 8 Set D/Alternate 2

8.1.1 IQI identification alternate, if used.

8.1.2 Material type.

8.1.3 Conformance to specified tolerances for dimensional values.

TABLE 2 Wire Diameter Tolerances, mm	
Wire Diameter [d], mm	Tolerance, mm
$0.000 < d \leq 0.125$	±0.0025
$0.125 < d \leq 0.25$	±0.005
$0.25 < d \leq 0.5$	±0.01
$0.50 < d \leq 1.6$	±0.02
$1.6 < d \leq 4$	±0.03
$4.0 < d \leq 8$	±0.05

TABLE 3 Wire Diameter Tolerances, in.	
Wire Diameter (d), in.	Tolerance, in.
$0.000 < d \leq 0.005$	±0.0001
$0.005 < d \leq 0.010$	±0.0002
$0.010 < d \leq 0.020$	±0.0004
$0.020 < d \leq 0.063$	±0.0008
$0.063 < d \leq 0.160$	±0.0012
$0.160 < d \leq 0.320$	±0.0020

8.1.4 ASTM standard designation, for example, ASTM E747—(year designation) used for manufacturing.

9. Precision and Bias

9.1 *Precision and Bias*—No statement is made about the precision or bias for indicating the quality of images since the results merely state whether there is conformance to the criteria for success specified in this practice.

10. Keywords

10.1 density; image quality level; IQI; radiologic; radiology; X-ray and gamma radiation

ANNEX

(Mandatory Information)

A1. ALTERNATE IQI IDENTIFICATION

A1.1 The use of IQI's with identifications as shown on Figs. A1.1-A1.9 and as listed in Table A1.1 is permitted as an acceptable alternate provided all other requirements of Practice E747 are satisfied.

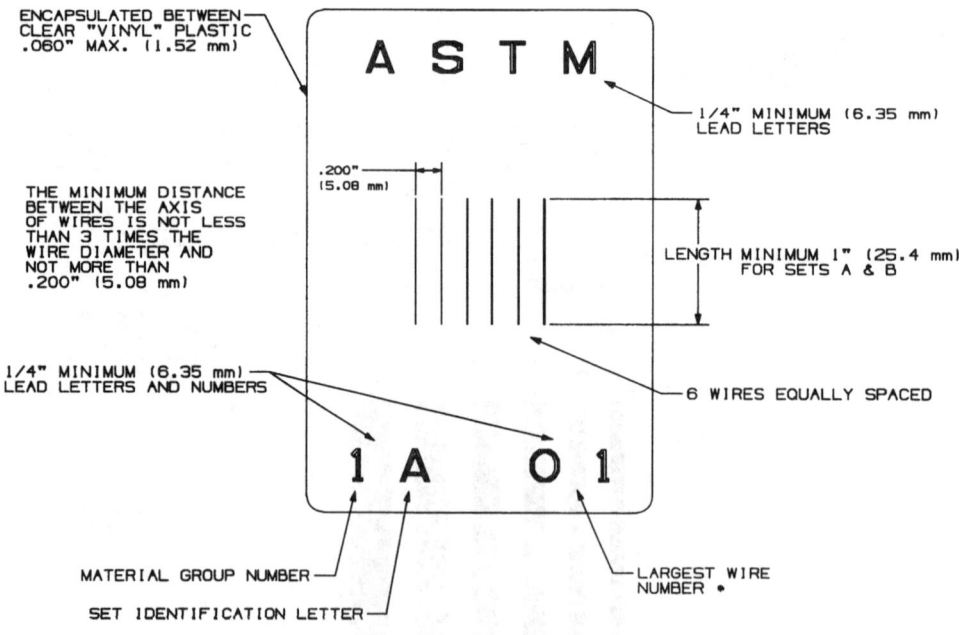

FIG. A1.1 Set A/Alternate 1

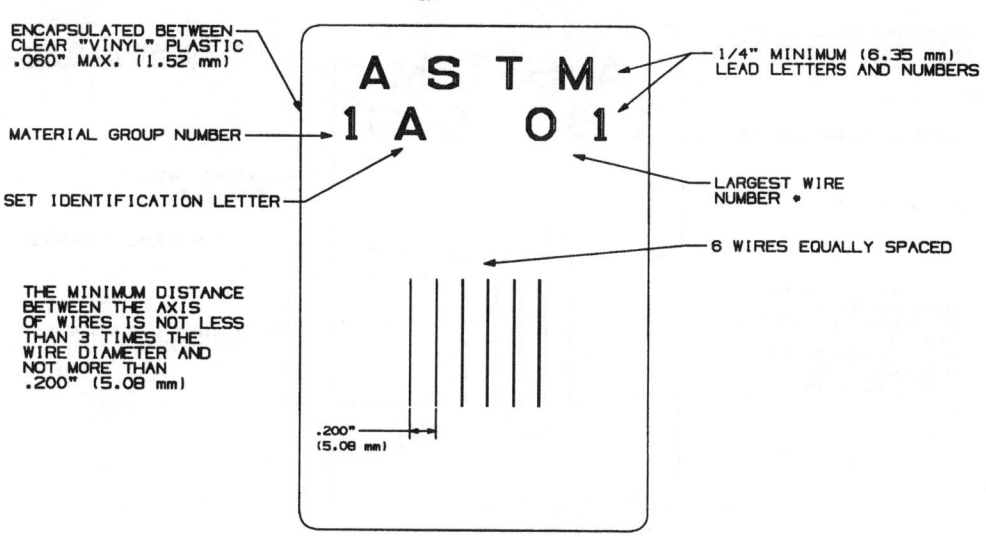

FIG. A1.2 Set A/Alternate 2

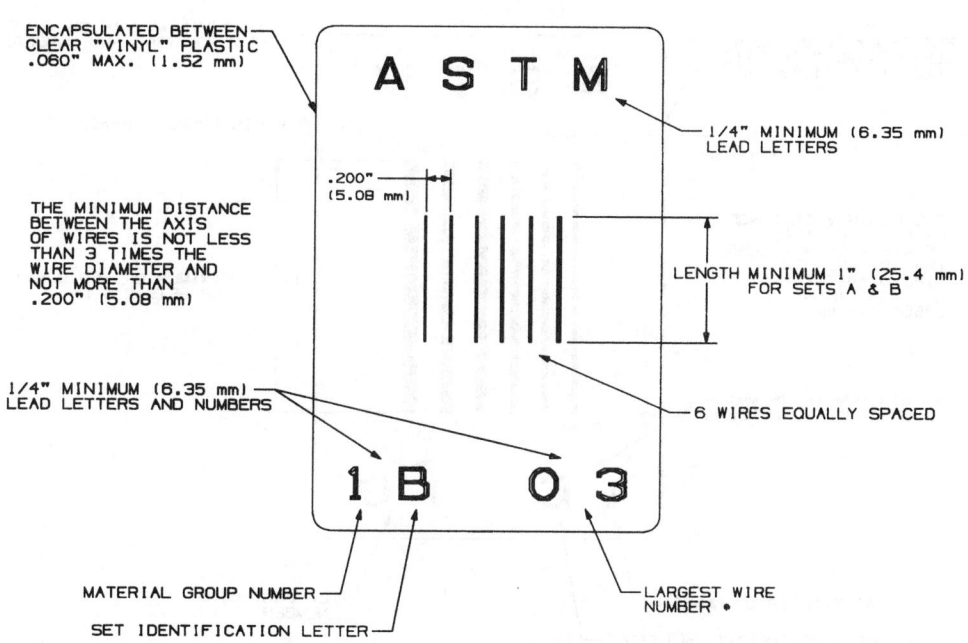

FIG. A1.3 Set B/Alternate 1

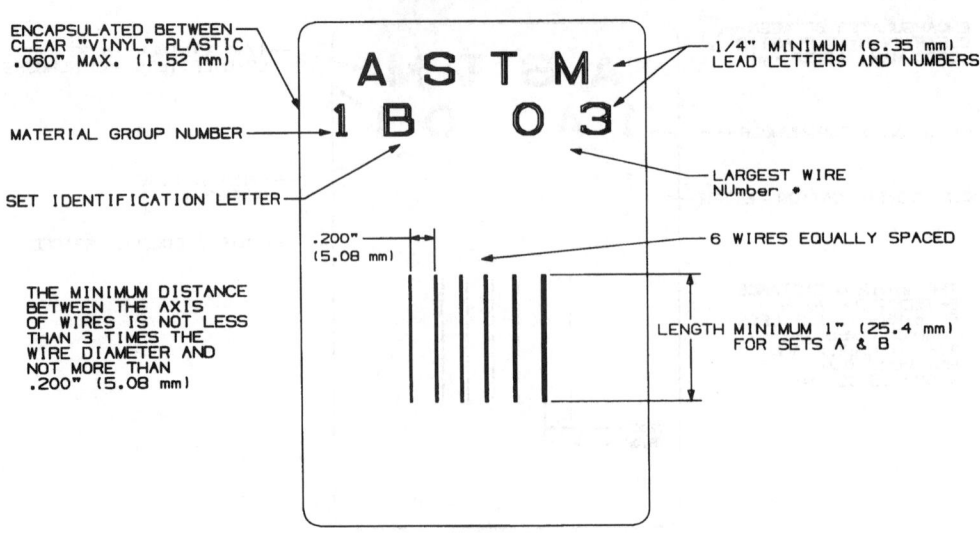

FIG. A1.4 Set B/Alternate 2

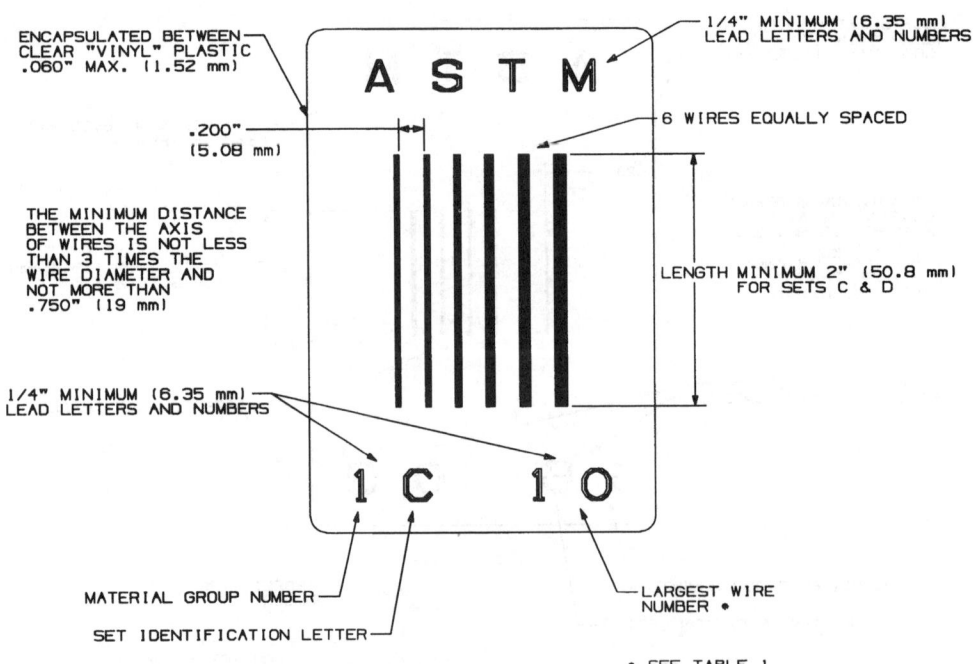

FIG. A1.5 Set C/Alternate 1

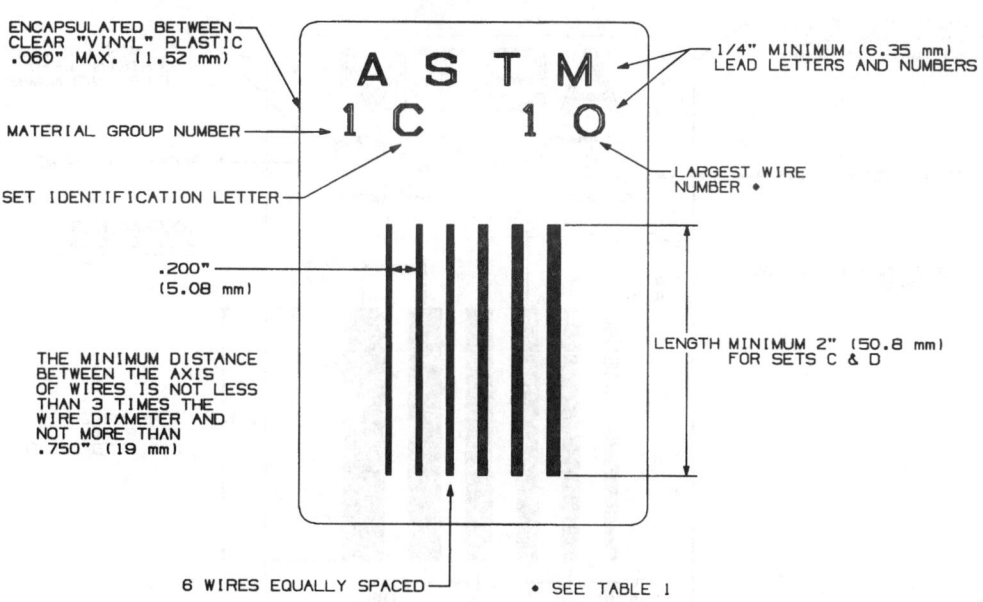

FIG. A1.6 Set C/Alternate 2

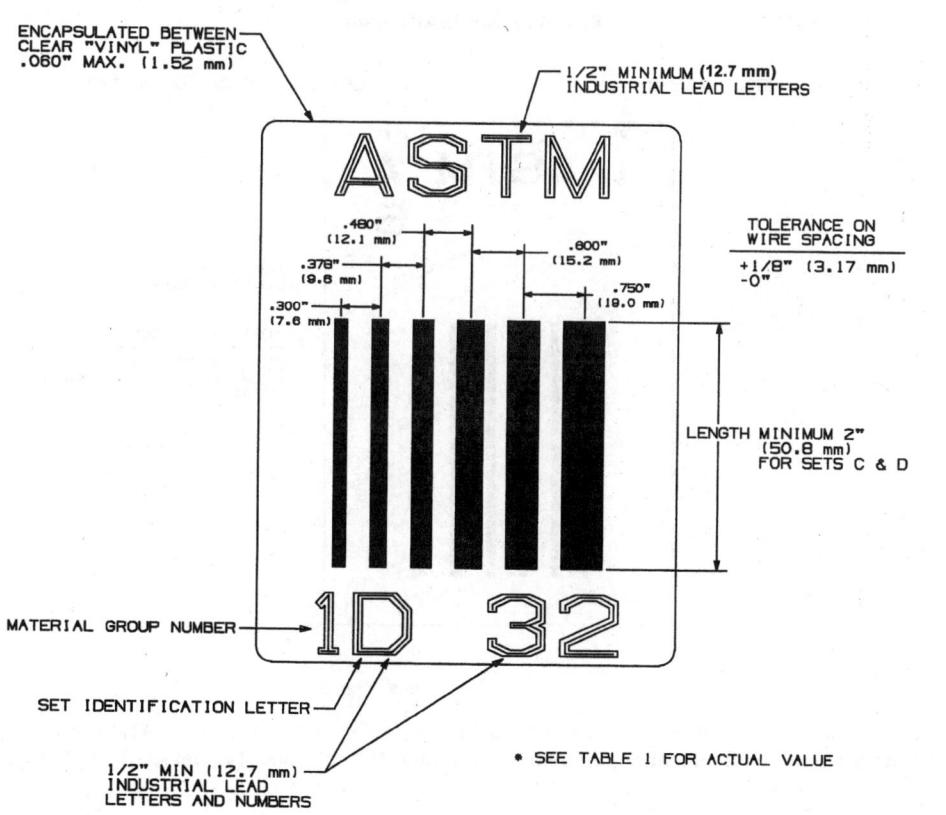

FIG. A1.7 Set D/Alternate 1

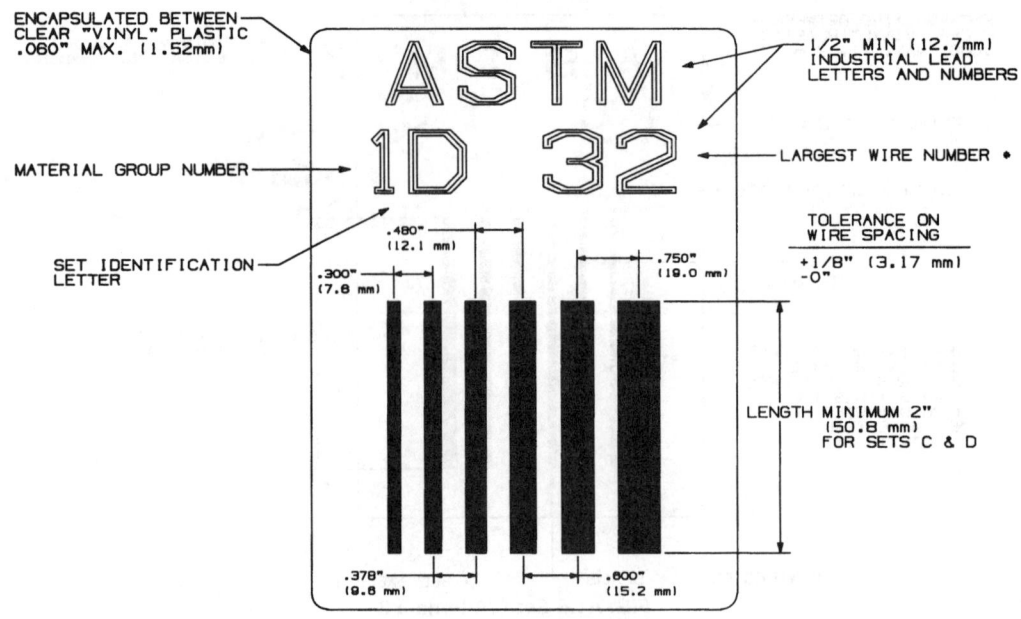

FIG. A1.8 Set D/Alternate 2

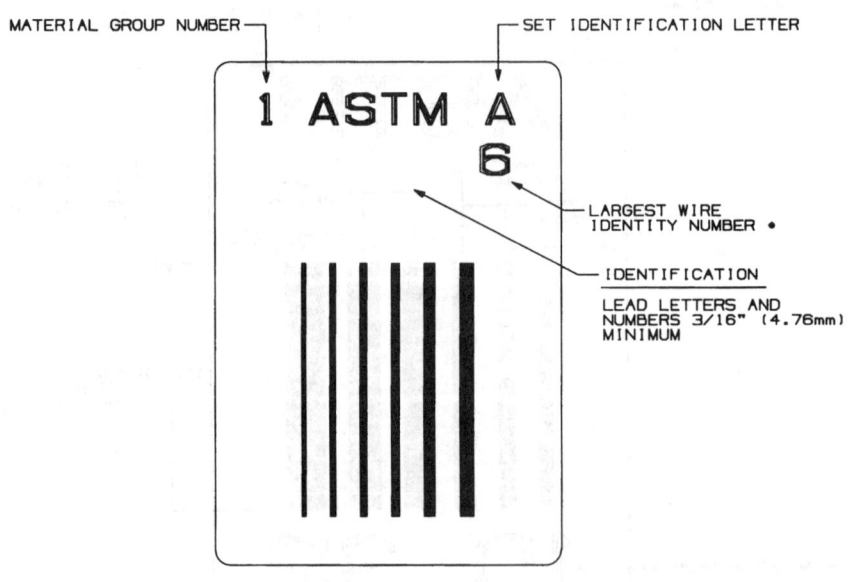

Note 1—All other IQI requirements as shown on Figs. 1-8 or Figs. A1.1-A1.8 apply.
FIG. A1.9 Alternate Identification Locations and Letter, Number Size-Typical All Sets (A, B, C, D)

TABLE A1.1 Penetrameter Sizes
Wire Diameter in. [mm]

SET A	ASTM Wire Identity	CEN Alternate Wire No. EN 462–1[A]	SET B	ASTM Wire Identity	CEN Alternate Wire No. EN 462–1[A]
0.0032[0.08]	1	W 17	0.010[0.25]	6	W 12
0.0040[0.1]	2	W 16	0.013[0.33]	7	W 11
0.0050[0.13]	3	W 15	0.016[0.41]	8	W 10
0.0063[0.16]	4	W 14	0.020[0.51]	9	W 9
0.0080[0.2]	5	W 13	0.025[0.64]	10	W 8
0.010[0.25]	6	W 12	0.032[0.81]	11	W 7

SET C	ASTM Wire Identity	CEN Alternate Wire No. EN 462–1[A]	SET D	ASTM Wire Identity	CEN Alternate Wire No. EN 462–1[A]
0.032[0.81]	11	W 7	0.100[2.5]	16	W 2
0.040[1.02]	12	W 6	0.126[3.2]	17	W 1
0.050[1.27]	13	W 5	0.160[4.06]	18	...
0.063[1.6]	14	W 4	0.20[5.1]	19	...
0.080[2.03]	15	W 3	0.25[6.4]	20	...
0.100[2.50]	16	W 2	0.32[8.1]	21	...

[A] As governed under provisions of paragraph 4.1.5 of this practice.

TABLE 4 Wire Sizes Equivalent to Corresponding 1T, 2T, and 4T Holes in Various Hole Type Plaques

Plaque Thickness, in. [mm]	Plaque IQI Identification Number	Diameter of wire with EPS of hole in plaque, in. [mm][A]		
		1T	2T	4T
0.005 [0.13]	5		0.0038 [0.09]	0.006 [0.15]
0.006 [0.16]	6		0.004 [0.10]	0.0067 [0.18]
0.008 [0.20]	8	0.0032 [0.08]	0.005 [0.13]	0.008 [0.20]
0.009 [0.23]	9	0.0035 [0.09]	0.0056 [0.14]	0.009 [0.23]
0.010 [0.25]	10	0.004 [0.10]	0.006 [0.15]	0.010 [0.25]
0.012 [0.30]	12	0.005 [0.13]	0.008 [0.20]	0.012 [0.28]
0.015 [0.38]	15	0.0065 [0.16]	0.010 [0.25]	0.016 [0.41]
0.017 [0.43]	17	0.0076 [0.19]	0.012 [0.28]	0.020 [0.51]
0.020 [0.51]	20	0.010 [0.25]	0.015 [0.38]	0.025 [0.63]
0.025 [0.64]	25	0.013 [0.33]	0.020 [0.51]	0.032 [0.81]
0.030 [0.76]	30	0.016 [0.41]	0.025 [0.63]	0.040 [1.02]
0.035 [0.89]	35	0.020 [0.51]	0.032 [0.81]	0.050 [1.27]
0.040 [1.02]	40	0.025 [0.63]	0.040 [0.02]	0.063 [1.57]
0.050 [1.27]	50	0.032 [0.81]	0.050 [1.27]	0.080 [2.03]
0.060 [1.52]	60	0.040 [1.02]	0.063 [1.57]	0.100 [2.54]
0.070 [1.78]	70	0.050 [1.27]	0.080 [2.03]	0.126 [3.20]
0.080 [2.03]	80	0.063 [1.57]	0.100 [2.54]	0.160 [4.06]
0.100 [2.50]	100	0.080 [2.03]	0.126 [3.20]	0.200 [5.08]
0.120 [3.05]	120	0.100 [2.54]	0.160 [4.06]	0.250 [6.35]
0.140 [3.56]	140	0.126 [3.20]	0.200 [5.08]	0.320 [8.13]
0.160 [4.06]	160	0.160 [4.06]	0.250 [6.35]	
0.200 [5.08]	200	0.200 [5.08]	0.320 [8.13]	
0.240 [6.10]	240	0.250 [6.35]		
0.280 [7.11]	280	0.320 [8.13]		

[A] Minimum plaque hole sizes were used as defined within Practice E1025.

APPENDIX

(Nonmandatory Information)

X1. CALCULATING OTHER EQUIVALENTS

X1.1 The equation to determine the equivalencies between wire and (hole type) IQI's is as follows:

$$F^3 d^3 l = T^2 H^2 (\pi/4)$$

where:
F = form factor for wire, 0.79,
d = wire diameter, in. [mm],
l = effective length of wire, 0.3 in. [7.6 mm],
T = plaque thickness, in. [mm], and
H = diameter of hole, in. [mm].

X1.2 It should be noted that the wire and plaque (hole type) IQI sensitivities cannot be related by a fixed constant.

X1.3 Figs. X1.1 and X1.2 are conversion charts for hole type IQI's containing 1T and 2T holes to wires. The sensitivities are given as a percentage of the specimen thickness.

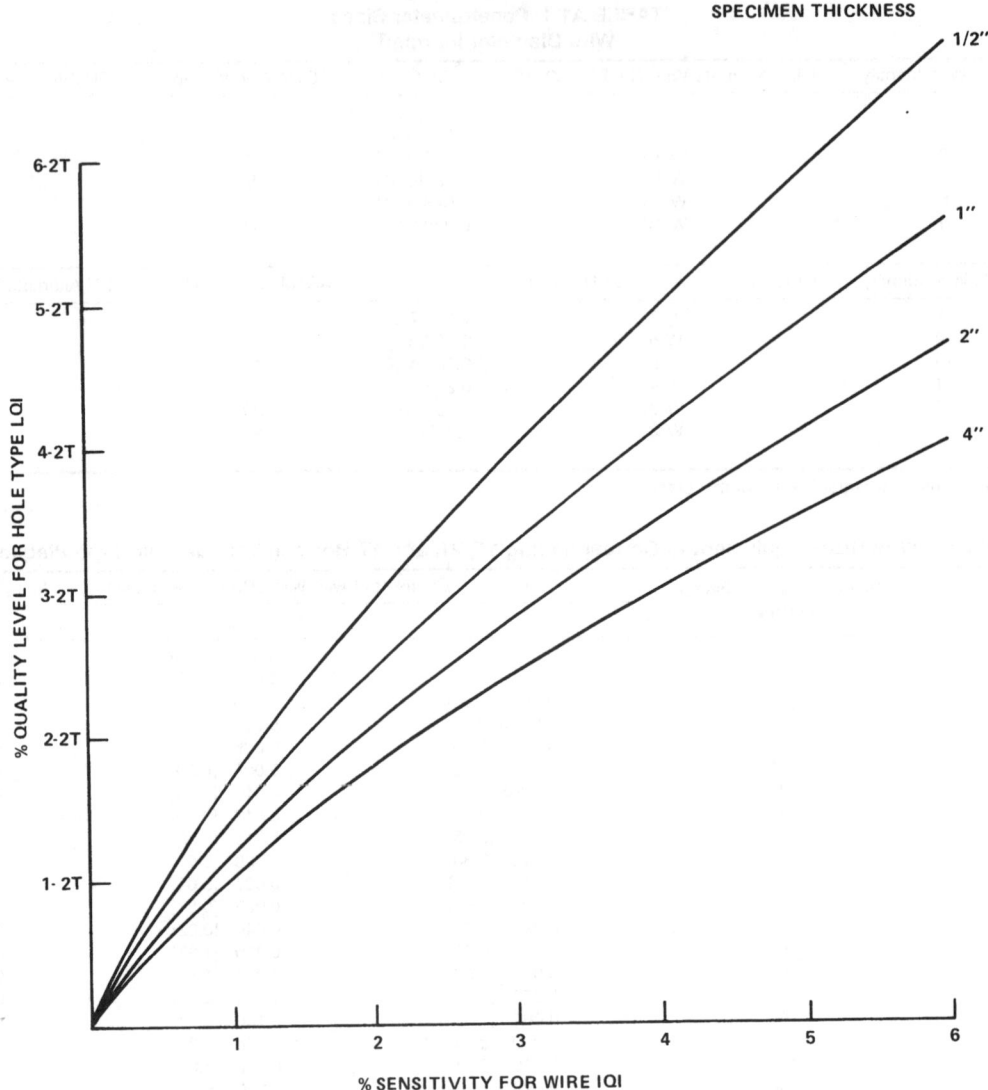

FIG. X1.1 Conversion Chart for 2-T Quality Level Holes to % Wire Sensitivity

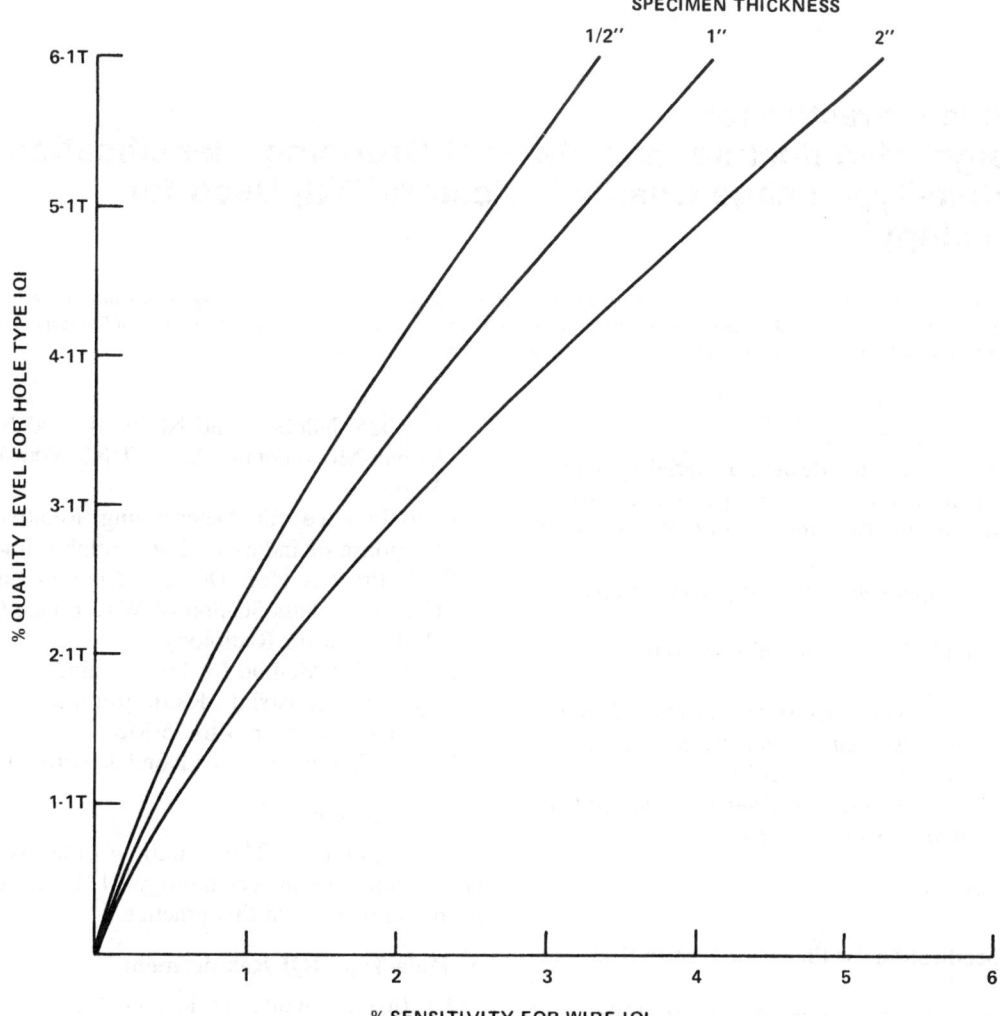

FIG. X1.2 Conversion Chart for 1-T Quality Level Holes to % Wire Sensitivity

ASTM International takes no position respecting the validity of any patent rights asserted in connection with any item mentioned in this standard. Users of this standard are expressly advised that determination of the validity of any such patent rights, and the risk of infringement of such rights, are entirely their own responsibility.

This standard is subject to revision at any time by the responsible technical committee and must be reviewed every five years and if not revised, either reapproved or withdrawn. Your comments are invited either for revision of this standard or for additional standards and should be addressed to ASTM International Headquarters. Your comments will receive careful consideration at a meeting of the responsible technical committee, which you may attend. If you feel that your comments have not received a fair hearing you should make your views known to the ASTM Committee on Standards, at the address shown below.

This standard is copyrighted by ASTM International, 100 Barr Harbor Drive, PO Box C700, West Conshohocken, PA 19428-2959, United States. Individual reprints (single or multiple copies) of this standard may be obtained by contacting ASTM at the above address or at 610-832-9585 (phone), 610-832-9555 (fax), or service@astm.org (e-mail); or through the ASTM website (www.astm.org).

Designation: E1025 − 05

Standard Practice for
Design, Manufacture, and Material Grouping Classification of Hole-Type Image Quality Indicators (IQI) Used for Radiology[1]

This standard is issued under the fixed designation E1025; the number immediately following the designation indicates the year of original adoption or, in the case of revision, the year of last revision. A number in parentheses indicates the year of last reapproval. A superscript epsilon (ε) indicates an editorial change since the last revision or reapproval.

1. Scope

1.1 This practice[2] covers the design, material grouping classification, and manufacture of hole-type image quality indicators (IQI) used to indicate the quality of radiologic images.

1.2 This practice is applicable to X-ray and gamma-ray radiology.

1.3 The values stated in inch-pound units are to be regarded as standard.

1.4 *This standard does not purport to address all of the safety concerns, if any, associated with its use. It is the responsibility of the user of this standard to establish appropriate safety and health practices and determine the applicability of regulatory limitations prior to use.*

2. Referenced Documents

2.1 *ASTM Standards:*[3]

B139/B139M Specification for Phosphor Bronze Rod, Bar, and Shapes

B150M Specification for Aluminum Bronze, Rod, Bar, and Shapes [Metric][4]

B161 Specification for Nickel Seamless Pipe and Tube

B164 Specification for Nickel-Copper Alloy Rod, Bar, and Wire

B166 Specification for Nickel-Chromium-Iron Alloys (UNS N06600, N06601, N06603, N06690, N06693, N06025, N06045, and N06696)* and Nickel-Chromium-Cobalt-Molybdenum Alloy (UNS N06617) Rod, Bar, and Wire

E746 Practice for Determining Relative Image Quality Response of Industrial Radiographic Imaging Systems

E747 Practice for Design, Manufacture and Material Grouping Classification of Wire Image Quality Indicators (IQI) Used for Radiology

E1735 Test Method for Determining Relative Image Quality of Industrial Radiographic Film Exposed to X-Radiation from 4 to 25 MeV

E1316 Terminology for Nondestructive Examinations

3. Terminology

3.1 *Definitions*—The definitions of terms relating to gamma and X-radiology in Terminology E1316, Section D, shall apply to the terms used in this practice.

4. Hole-Type IQI Requirements

4.1 Image quality indicators (IQIs) used to determine radiologic-image quality levels shall conform to the following requirements.

4.1.1 *Standard Hole-Type IQIs*:

4.1.1.1 Image quality indicators (IQIs) shall be fabricated from materials or alloys identified or listed in accordance with 7.3. Other materials may be used in accordance with 7.4.

4.1.1.2 Image quality indicators (IQIs) shall dimensionally conform to the requirements of Fig. 1.

4.1.1.3 Both the rectangular and the circular IQI shall be identified with number(s) made of lead or a material of similar radiation opacity. The number shall be bonded to the rectangular IQI's and shall be placed adjacent to circular IQI's to provide identification of the IQI on the image. The identification numbers shall indicate the thickness of the IQI in thousandths of an inch, that is, a number 10 IQI is 0.010 in. thick, a number 100 IQI is 0.100 in. thick, etc. Additional identification requirements are provided in 7.2.

4.1.1.4 Alloy-group identification shall be in accordance with Fig. 2. Rectangular IQI's shall be notched. Image quality indicators (IQI's) shall be vibrotooled or etched as specified.

---

[1] This practice is under the jurisdiction of ASTM Committee E07 on Nondestructive Testing and is the direct responsibility of Subcommittee E07.01 on Radiographic Practice and Penetrameters.
Current edition approved June 1, 2005. Published June 2005. Originally approved in 1984. Last previous edition approved in 1998 as E1025 - 98. DOI: 10.1520/E1025-05.

[2] For ASME Boiler and Pressure Vessel Code applications see related Practice SE-1025 in Section II of that Code.

[3] For referenced ASTM standards, visit the ASTM website, www.astm.org, or contact ASTM Customer Service at service@astm.org. For *Annual Book of ASTM Standards* volume information, refer to the standard's Document Summary page on the ASTM website.

[4] Withdrawn. The last approved version of this historical standard is referenced on www.astm.org.

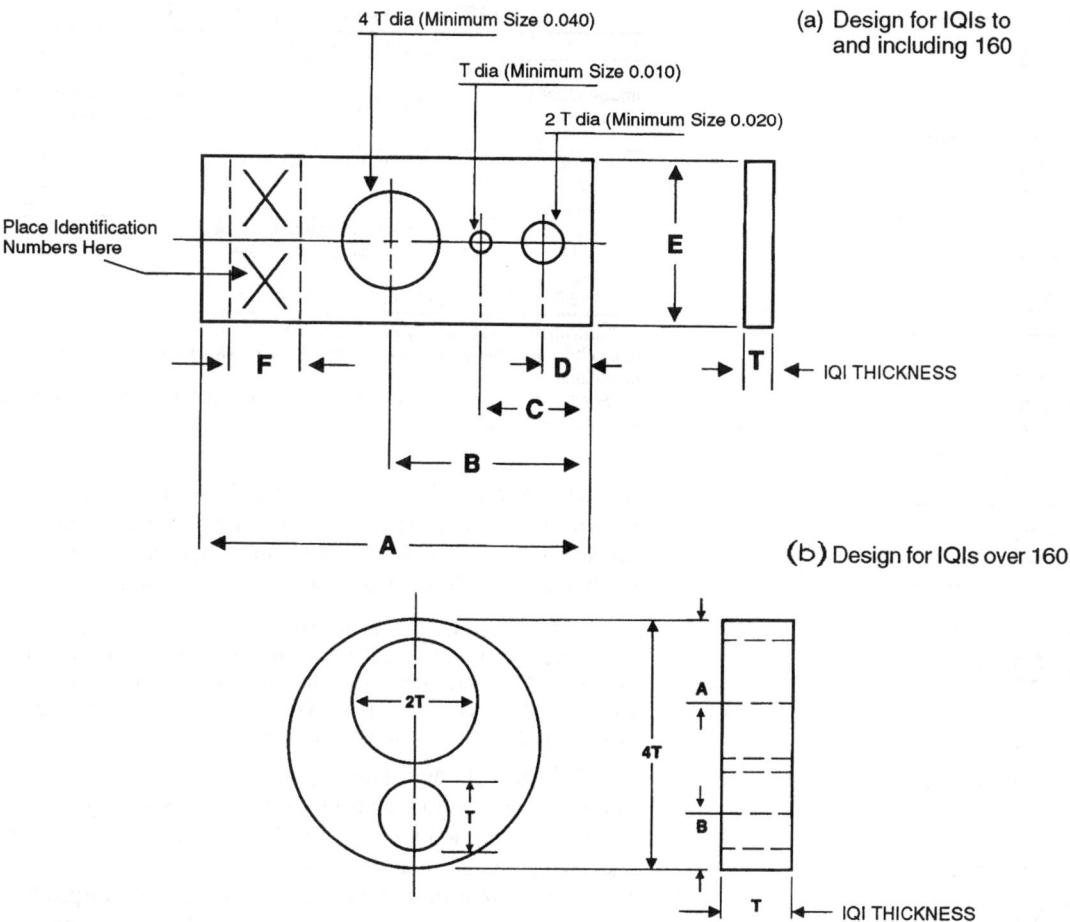

NOTE 1—All dimensions in inches (Note 6).
NOTE 2—Tolerances for IQI thickness and hole diameter.
NOTE 3—XX identification number equals T in .001 inches.
NOTE 4—IQIs No. 1 through 9 are not 1T, 2T, and 4T.
NOTE 5—Holes shall be true and normal to the IQI. Do not chamfer.
NOTE 6—To convert inch dimensions to metric, multiply by 25.4.

Identification Number T (Note 3)	A	B	C	D	E	F	Tolerances (Note 2)
1–4	1.500 ±0.015	0.750 ±0.015	0.438 ±0.015	0.250 ±0.015	0.500 ±0.015	0.250 ±0.030	±10%
5–20	1.500 ±0.015	0.750 ±0.015	0.438 ±0.015	0.250 ±0.015	0.500 ±0.015	0.250 ±0.030	±0.0005
21–50	...	...	...	...	...	...	±0.0025
Over 50–160	2.250 ±0.030	1.375 ±0.030	0.750 ±0.030	0.375 ±0.030	1.000 ±0.030	0.375 ±0.030	±0.005
Over 160	1.330T ±0.005	0.830T ±0.005	...	...	...	...	±0.010

FIG. 1 IQI Design

4.1.2 *Modified Hole-Type IQI*:
4.1.2.1 The rectangular IQI may be modified in length and width as necessary for special applications, provided the hole size(s) and IQI thickness conform to Fig. 1.
4.1.2.2 The IQI's shall be identified as specified in 4.1.1.3, except that the identification numbers may be placed adjacent to the IQI if placement on the IQI is impractical.

4.1.2.3 When modified IQI's are used, details of the modification shall be documented in the records accompanying the examination results.

5. IQI Procurement

5.1 When selecting IQI's for procurement, the following factors should be considered:

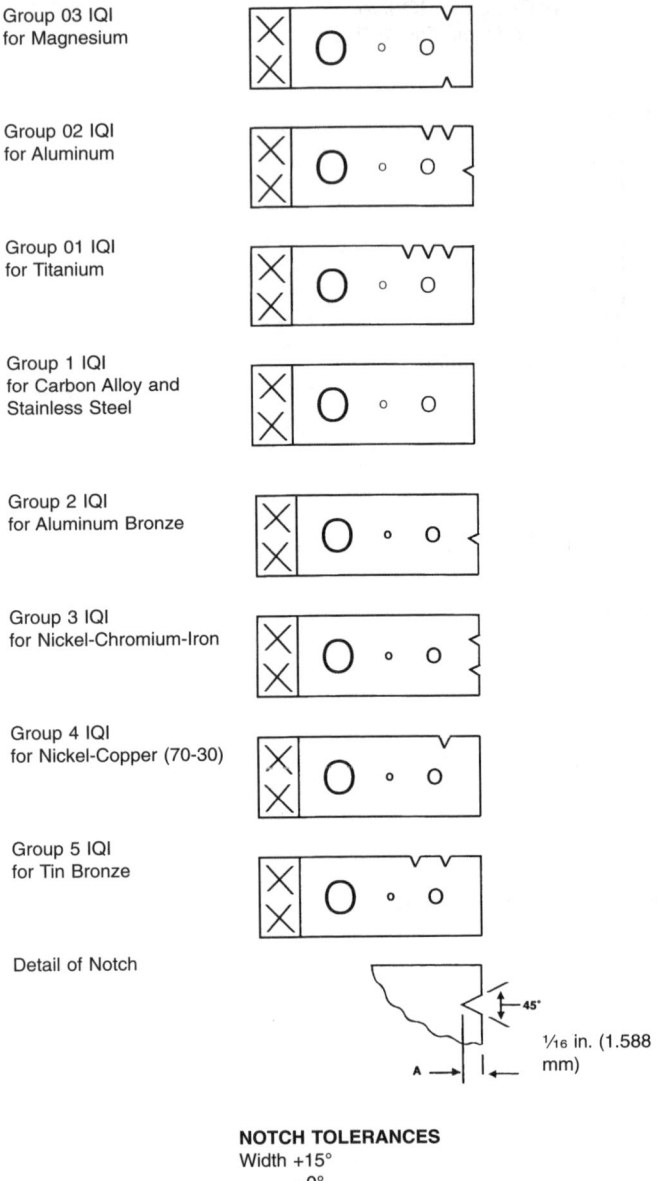

FIG. 2 Rectangular IQI Notch Identification and Material Grouping

TABLE 1 Typical Image Quality Levels

Image Quality Levels	IQI Thickness	Minimum Preceptible Hole Diameter	Equivalent IQI Sensitivity, %[A]
Standard Image Quality Levels			
2-1T	1/50 (2 %) of Specimen Thickness	1T	1.4
2-2T[B]		2T	2.0
2-4T		4T	2.8
Special Image Quality Levels			
1-1T	1/100 (1 %) of Specimen Thickness	1T	0.7
1-2T		2T	1
4-2T	1/25 (4 %) of Specimen Thickness	2T	4

[A] Equivalent IQI sensitivity is that thickness of the IQI, expressed as a percentage of the part thickness, in which the 2T hole would be visible under the same conditions.

[B] For Level 2-2T Radiologic—The 2T hole in an IQI, 1/50 (2 %) of the specimen thickness, is visible.

diameter of the hole and is expressed as a multiple of the IQI thickness, T. The image quality level 2-2T means that the IQI thickness T is 2 % of the specimen thickness and that the diameter of the IQI imaged hole is 2 × the IQI thickness.

NOTE 2—Image Quality Indicators (IQI's) less than number 10 have hole sizes 0.010, 0.020, and 0.040 in. diameter regardless of the IQI thickness. Therefore, IQI's less than number 10 do not represent the quality levels specified in 6.1 and Table 1. The equivalent sensitivity can be computed from data furnished in Appendix X1.

6.2 Typical image quality level designations are shown in Table 1. The level of inspection specified should be based on service requirements of the product. Care should be taken in specifying image quality levels 2-1T, 1-1T, and 1-2T by first determining that these levels can be maintained in production.

6.3 In specifying image quality levels, the contract, purchase order, product specification, or drawing should state the proper two-part expression and clearly indicate the thickness of the metal to which the level refers. In place of a designated two–part expression, the IQI number and minimum discernible hole size shall be specified.

6.4 Appendix X1 of this practice provides methods for determining equivalent penetrameter sensitivity (EPS) in percent. Under certain conditions (as described within the purchaser-supplier agreement), EPS may be useful in relating a discernible hole size of the IQI thickness with the section thickness radiographed for establishing an overall technical image quality equivalency. This is not an alternative IQI provision for the originally specified IQI requirement of this practice, but may be a useful tool for establishing technical image equivalency on a case basis need with specific customer approvals.

6.5 Practice E747 contains provisions for wire IQI's that use varying length and diameter wires to affect image quality requirements. The requirements of Practice E747 are different from this standard; however, Practice E747 (see Table 4) contains provisions whereby wire sizes equivalent to corresponding 1T, 2T and 4T holes for various plaque thicknesses are provided. Appendix X1 of Practice E747 also provides methods for determining equivalencies between wire and hole type IQI's. This is not an alternative IQI provision for the originally specified IQI requirements of this practice, but may be useful for establishing technical image equivalency on a case basis need with specific customer approvals.

5.1.1 Determine the alloy group(s) of the material to be examined.

5.1.2 Determine the thickness or thickness range of the material(s) to be examined.

5.1.3 Select the applicable IQI's that represent the required IQI thickness and alloy(s).

NOTE 1—This practice does not recommend or suggest specific IQI sets to be procured. Section 5 is an aid in selecting IQI's based on specific needs.

6. Image Quality Levels

6.1 Image quality levels are designated by a two part expression X-YT. The first part of the expression X refers to the IQI thickness expressed as a percentage of the specimen thickness. The second part of the expression YT refers to the

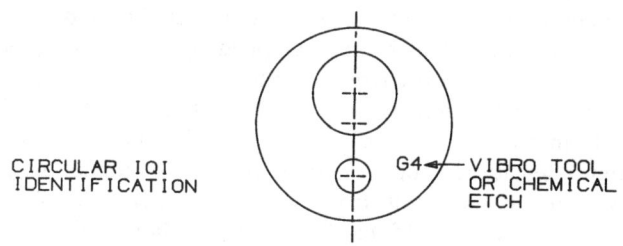

FIG. 3 Circular IQI Identification

6.6 Test Methods E746 and E1735 provide additional tools for determining relative image quality response of industrial radiographic film systems when exposed to energy levels described within those test methods. Both of these test methods use the "equivalent penetrameter sensitivity" (EPS) concept to provide statistical image quality information that allows the film system or other exposure components to be assessed on a relative basis. These test methods are not alternative IQI provisions for the originally specified IQI requirements of this practice, but may be useful on a case basis with specific customer approvals, for establishing technical image equivalency of certain aspects of the radiographic imaging process.

7. Material Groups

7.1 *General*:

7.1.1 Materials have been designated in eight groups based on their radiation absorption characteristics: Groups 03, 02, and 01 for light metals and Groups 1 through 5 for heavy metals.

7.1.2 The light metal groups, magnesium (Mg), aluminum (Al), and titanium (Ti) are identified 03, 02, and 01 respectively for their predominant alloying constituent. The materials are listed in order of increasing radiation absorption.

7.1.3 The heavy metal groups, steel, copper base, nickel base, and kindred alloys are identified 1 through 5. The materials increase in radiation absorption with increasing numerical designation.

NOTE 3—These groups were established experimentally at 180 kV on ¾-in. (19-mm) thick specimens. They apply from 125 kV to the multivolt range.

7.1.4 Common trade names or alloy designations have been used for clarification of the pertinent materials.

7.1.5 The materials from which the IQI for the group are to be made are designated in each case, and these IQI's are applicable for all materials listed in that group. In addition, any group IQI may be used for any material with a higher group number, provided the applicable quality level is maintained.

7.2 *Identification System*:

7.2.1 A notching system has been designated for the eight groups of IQI's and is shown in Fig. 2.

7.2.2 For circular IQI's, a group designation shall be vibrotooled or chemically etched on the IQI to identify it by using the letter "G" followed by the group number, that is, G4 for a Group 4 IQI. For identification of the group on the image, corresponding lead characters shall be placed adjacent to the circular IQI, just as is done with the lead numbers identifying the thickness. The identification is shown in Fig. 3.

7.3 *Materials Groups*:

7.3.1 *Materials Group* 03:

7.3.1.1 Image quality indicators (IQI's) shall be made of magnesium or magnesium shall be the predominant alloying constituent.

7.3.1.2 Use on all alloys of which magnesium is the predominant alloying constituent.

7.3.2 *Materials Group* 02:

7.3.2.1 Image quality indicators (IQI's) shall be made of aluminum or aluminum shall be the predominant alloying constituent.

7.3.2.2 Use on all alloys of which aluminum is the predominant alloying constituent.

7.3.3 *Materials Group* 01:

7.3.3.1 Image quality indicators (IQI's) shall be made of titanium or titanium shall be the predominant alloying constituent.

7.3.3.2 Use on all alloys of which titanium is the predominant alloying constituent.

7.3.4 *Materials Group* 1:

7.3.4.1 Image quality indicators (IQI's) shall be made of carbon steel or Type 300 series stainless steel.

7.3.4.2 Use on all carbon steel, all low-alloy steels, all stainless steels, manganese-nickel-aluminum bronze (Superston).[5]

7.3.5 *Materials Group* 2:

7.3.5.1 Image quality indicators (IQI's) shall be made of aluminum bronze (Alloy No. 623, of Specification B150M or equivalent, or nickel-aluminum bronze (Alloy No. 630 of Specification B150M) or equivalent.

7.3.5.2 Use on all aluminum bronzes and all nickel-aluminum bronzes.

7.3.6 *Materials Group* 3:

7.3.6.1 Image quality indicators (IQI's) shall be made of nickel-chromium-iron alloy (UNS No. NO6600) (Inconel).[6] (See Specification B166.)

7.3.6.2 Use on nickel-chromium-iron alloy and 18 % nickel-maraging steel.

7.3.7 *Materials Group* 4:

7.3.7.1 Image quality indicators (IQI's) shall be made of 70 to 30 nickel-copper alloy (Monel)[7] (Class A or B of Specification B164) or equivalent, or 70 to 30 copper-nickel alloy, (Alloy G of Specification B161) or equivalent.

7.3.7.2 Use on nickel, copper, all nickel-copper series, or copper-nickel series of alloys, and all brasses (copper-zinc alloys). Group 4 IQI's may be used on the leaded brasses, since leaded brass increases in attenuation with increase in lead content. This would be equivalent to using a lower group IQI.

7.3.8 *Materials Group* 5:

7.3.8.1 Image quality indicators (IQI's) shall be made of tin bronze (Alloy D of Specification B139/B139M).

7.3.8.2 Use on tin bronzes including gun-metal and valve bronze, leaded-tin bronze of higher lead content than valve bronze. Group 5 IQI's may be used on bronze of higher lead

---

[5] Superston is a registered trademark of Superston Corp., Jersey City, NJ.
[6] Inconel is a registered trademark of The International Nickel Co., Inc., Huntington, WV 25720.
[7] Monel is a registered trademark of The International Nickel Co., Inc., Huntington, WV 25720.

content since leaded bronze increases in attenuation with increase in lead content. This would be equivalent to using a lower group IQI.

NOTE 4—In developing the eight listed materials groups, a number of other trade names or other nominal alloy designations were evaluated. For the purpose of making this practice as useful as possible, these materials are listed and categorized, by group, as follows:
(1) Group 2—Haynes Alloy IN-100.[8]
(2) Group 3—Haynes Alloy No. 713C, Hastelloy D,[9] G.E. Alloy SEL, Haynes Stellite Alloy No. 21,[9] GMR-235 Alloy, Haynes Alloy No. 93, Inconel X,[6] Inconel 718, and Haynes Stellite Alloy NO. S-816.
(3) Group 4—Hastelloy Alloy F, Hastelloy Alloy X, and Multimeter Alloy Rene 41.
(4) Group 5—Alloys in order of increasing attenuation: Hastelloy Alloy B, Hastelloy Alloy C, Haynes Stellite Alloy No. 31, Thetaloy, Haynes Stellite No. 3, Haynes Alloy No. 25. IQIs of any of these materials are considered applicable for the materials that follow it.

NOTE 5—The committee formulating these recommendations, recommended other materials may be added to the materials groups listed as the need arises or as more information is gained, or that additional materials groups may be added.

7.4 *Radiographic Method for Other Materials*:

7.4.1 For materials not herein covered, IQI's of the same materials, or any other material, may be used if the following requirements are met. Two blocks of equal thickness, one of the material to be examined (production material) and one of the IQI material, shall be radiographed on one film by one exposure at the lowest energy level to be used for production. Transmission densitometer readings for both materials shall be read from the film and shall be between 2.0 and 4.0 (radiographic) density for both materials. If the radiographic image density of the material from which the IQI's are to be fabricated is within +15 to −0 % of the radiographic image density of the production material, the IQI material may be used to fabricate IQI's for examination of the production material. The percentage figure is based on the radiographic density of the IQI material.

7.4.2 It shall always be permissible to use IQI's of similar composition as the material being examined.

8. IQI Certification

8.1 Records shall be available that attest to the conformance of the material type, grouping (notches), and dimensional tolerances of the IQI's specified by this practice.

9. Precision and Bias

9.1 *Precision and Bias*—No statement is made about the precision or bias for indicating the quality of radiographs since the results merely state whether there is conformance to the criteria for success specified in this practice.

10. Keywords

10.1 density; image quality level; IQI; radiologic; radiology; X-ray and gamma radiation

---

[8] All Haynes alloys are registered trademarks of Union Carbide Corp., New York, NY.

[9] All Hastelloys and Haynes Stellite alloys are registered trademarks of Cabot Corp., Boston, MA.

APPENDIX

(Nonmandatory Information)

X1. EQUIVALENT IQI (PENETRAMETER) SENSITIVITY (EPS)[9]

X1.1 To find the equivalent IQI sensitivity (percent), the hole size (diameter in inches), of the IQI thickness (inches), for a section thickness (inches), the following computations may be used:

where:
$$\alpha = \frac{100}{X}\sqrt{\frac{TH}{2}},$$

α = equivalent IQI sensitivity, %,
X = section thickness to be examined, in.,
T = IQI Thickness, in., and
H = hole diameter, in.

X1.2 Alternate method for determining EPS using Fig. X1.1 Nomograph:

---

[10] O'Connor, D. T., and Criscuolo, E. L., "The Quality of Radiographic Inspection," *ASTM Bulletin*, ASTM, Vol 213, 1956, p. 52.

Example:
Given:
$$X = 0.5 \text{ in.,} \quad (X1.1)$$
$$T = 0.005 \text{ in., and}$$
$$H = 0.0625 \text{ in.}$$

Solution:
$$A = \frac{100T}{X} = \frac{100 \times 0.005}{0.5} = 1.0\,\% \quad (X1.2)$$

$$B = \frac{100H}{X} = \frac{100 \times 0.0625}{0.5} = 12.5\,\% \quad (X1.3)$$

X1.3 Proceed to the nomograph (Fig. X1.1) and draw a line joining the 1.0 % Value A and the 12.5 % Value B and look on the center percent scale where the line crosses it and read the answer—2.5 %. Thus under the given conditions, equivalent IQI (penetrameter) sensitivity (EPS) is 2.5 %.

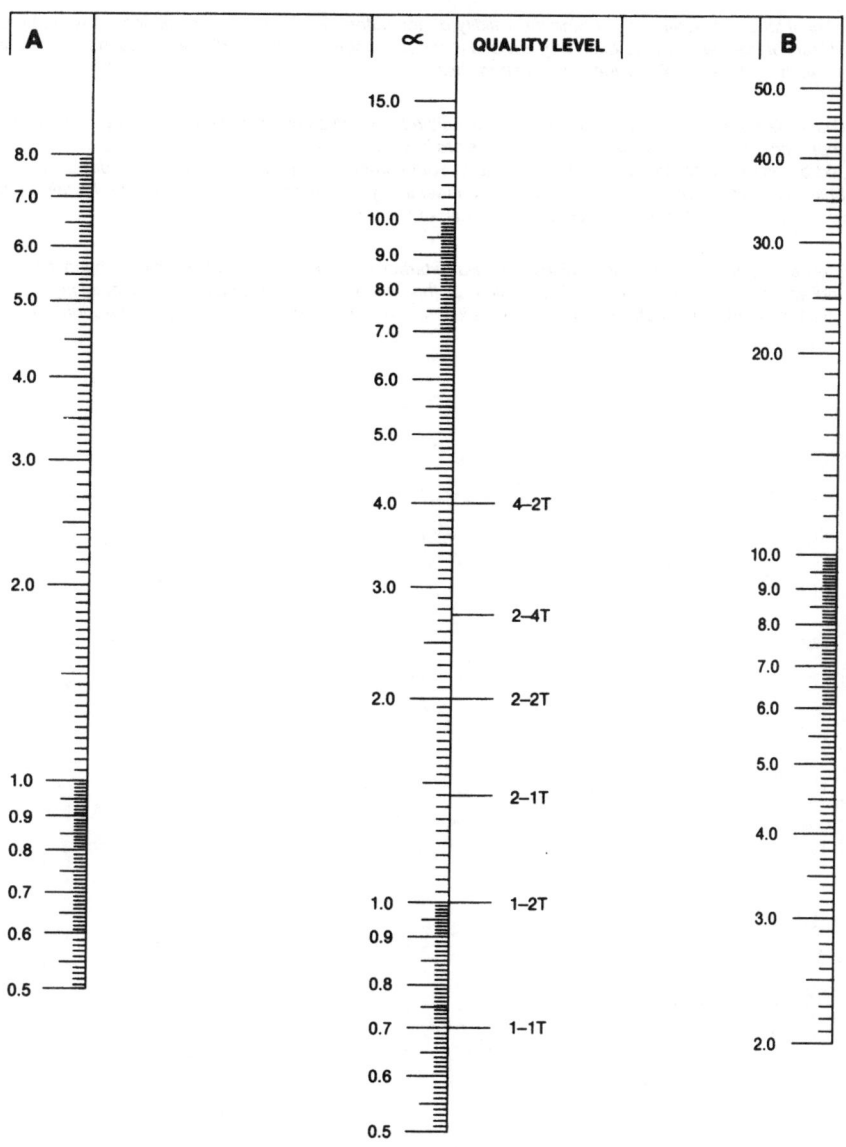

Definitions:

A equals the visible IQI (penetrameter) plaque thickness (*T*) expressed as a percentage of the section (object) thickness to be radiographed in (inches).

B equals the diameter of the smallest IQI (penetrameter) hole (*H*) for which the image is visibly expressed as a percentage of the section (object) thickness to be radiographed in (inches).

NOTE 1—The nomograph is used for computing equivalent IQI sensitivity from T (T equals penetrameter thickness) inches and H (H equals hole diameter) inches. Draw a straight line joining the values on any two scales, and look on the third scale where the line crosses and read the answer. Due to normal reproduction methods in producing the nomograph, some small error (that is, less than 5 %) may occur. If more accurate results are required, the formula in Appendix X1 should be used.

FIG. X1.1 Equivalent IQI (Penetrameter) Sensitivity Nomograph

ASTM International takes no position respecting the validity of any patent rights asserted in connection with any item mentioned in this standard. Users of this standard are expressly advised that determination of the validity of any such patent rights, and the risk of infringement of such rights, are entirely their own responsibility.

This standard is subject to revision at any time by the responsible technical committee and must be reviewed every five years and if not revised, either reapproved or withdrawn. Your comments are invited either for revision of this standard or for additional standards and should be addressed to ASTM International Headquarters. Your comments will receive careful consideration at a meeting of the responsible technical committee, which you may attend. If you feel that your comments have not received a fair hearing you should make your views known to the ASTM Committee on Standards, at the address shown below.

This standard is copyrighted by ASTM International, 100 Barr Harbor Drive, PO Box C700, West Conshohocken, PA 19428-2959, United States. Individual reprints (single or multiple copies) of this standard may be obtained by contacting ASTM at the above address or at 610-832-9585 (phone), 610-832-9555 (fax), or service@astm.org (e-mail); or through the ASTM website (www.astm.org).

Designation: E1032 − 06

Standard Test Method for Radiographic Examination of Weldments[1]

This standard is issued under the fixed designation E1032; the number immediately following the designation indicates the year of original adoption or, in the case of revision, the year of last revision. A number in parentheses indicates the year of last reapproval. A superscript epsilon (ε) indicates an editorial change since the last revision or reapproval.

1. Scope

1.1 This test method provides a uniform procedure for radiographic examination of weldments using industrial radiographic film. Requirements expressed in this method are intended to control the quality of the radiographic images and are not intended for controlling acceptability or quality of welds.

1.2 The radiographic extent, the quality level, and the acceptance criteria to be applied shall be specified in the contract, purchase order, product specification, or drawings.

1.3 The radiographic techniques stated herein provide adequate assurance for defect detectability; however, it is recognized that, for special applications, specific techniques using more or less stringent requirements may be required than those specified. In these cases, the use of alternative radiographic techniques shall be as agreed upon between purchaser and supplier (also see Section 4).

1.4 The values stated in inch-pound units are to be regarded as standard.

1.5 *This standard does not purport to address all of the safety concerns, if any, associated with its use. It is the responsibility of the user of this standard to establish appropriate safety and health practices and determine the applicability of regulatory limitations prior to use.* (For more specific safety precautionary information, see Section 7.)

2. Referenced Documents

2.1 *ASTM Standards:*[2]
E94 Guide for Radiographic Examination
E242 Reference Radiographs for Appearances of Radiographic Images as Certain Parameters Are Changed
E390 Reference Radiographs for Steel Fusion Welds
E543 Specification for Agencies Performing Nondestructive Testing
E747 Practice for Design, Manufacture and Material Grouping Classification of Wire Image Quality Indicators (IQI) Used for Radiology
E999 Guide for Controlling the Quality of Industrial Radiographic Film Processing
E1025 Practice for Design, Manufacture, and Material Grouping Classification of Hole-Type Image Quality Indicators (IQI) Used for Radiology
E1079 Practice for Calibration of Transmission Densitometers
E1254 Guide for Storage of Radiographs and Unexposed Industrial Radiographic Films
E1316 Terminology for Nondestructive Examinations
E1815 Test Method for Classification of Film Systems for Industrial Radiography

2.2 *ASNT Standards:*[3]
Recommended Practice No. SNT-TC-1A Personnel Qualification and Certification in Nondestructive Testing
ANSI/ASNT-CP-189 Standard for Qualification and Certification of Nondestructive Testing Personnel

2.3 *Other Standards:*
NAS 410 National Aerospace Standard Certification and Qualification of Nondestructive Test Personnel[4]
EN 444 Nondestructive Testing—General Principles for Radiographic Examination of Metallic Materials by X and Gamma Rays—Basic Rules[5]
ISO 5579 Nondestructive Testing—Radiographic Examination of Metallic Materials by X and Gamma Rays—Basic Rules[5]

3. Terminology

3.1 *Definitions*—For definitions of terms used in this test method, see Terminology E1316.

4. Basis of Application

4.1 *Personnel Qualification*—Nondestructive testing (NDT) personnel shall be qualified in accordance with a nationally recognized NDT personnel qualification practice or standard such as ANSI/ASNT-CP-189, SNT-TC-1A, NAS 410 or a

---

[1] This test method is under the jurisdiction of ASTM Committee E07 on Nondestructive Testing and is the direct responsibility of Subcommittee E07.01 on Radiology (X and Gamma) Method.
Current edition approved July 1, 2006. Published July 2006. Originally approved in 1985. Last previous edition approved in 2001 as E1032 - 01. DOI: 10.1520/E1032-06.

[2] For referenced ASTM standards, visit the ASTM website, www.astm.org, or contact ASTM Customer Service at service@astm.org. For *Annual Book of ASTM Standards* volume information, refer to the standard's Document Summary page on the ASTM website.

[3] Available from American Society for Nondestructive Testing (ANST), 1711 Arlingate Lane, P.O. Box 28518, Columbus, OH 43228-0518.

[4] Available from Aerospace Industries Association of America, Inc. (AIA), 1250 Eye St., NW, Washington, DC 20005.

[5] Available from American National Standards Institute (ANSI), 25 W. 43rd St., 4th Floor, New York, NY 10036.

similar document. The practice or standard used and its applicable revision shall be specified in the contractual agreement between the using parties.

4.2 *Qualification of Nondestructive Agencies*—If specified in the contractual agreement, NDT agencies shall be qualified and evaluated in accordance with Practice E543. The applicable edition of Practice E543 shall be specified in the contractual agreement.

4.3 *Time of Examination*—The time of examination shall be in accordance with 8.1 unless otherwise specified.

4.4 *Procedures*—The procedures to be utilized shall be as described in 7.1.

4.5 *Extent of Examination*—The extent of the examination shall be in accordance with 7.2.

4.6 *Reporting Criteria/Acceptance Criteria*—Reporting criteria of the examination results shall be in accordance with Section 11.

4.7 *Reexamination of Repaired or Reworked Items*—Reexamination of repaired or reworked items is not addressed in this test method and if required shall be specified in the contractual agreement.

4.8 *Radiographic Quality Level*—The radiographic quality level shall be in accordance with 7.4.

5. Materials

5.1 *Film Systems*—Only film systems having cognizant engineering organization (CEO) approval or meeting the requirements of test method E1815 shall be used to meet the requirements of this standard.

6. Apparatus

6.1 *Radiation Source (X-Ray or Gamma-Ray)*—Selection of the appropriate source is dependent upon variables regarding the weld being examined (material composition and thickness). The suitability of the source shall be demonstrated by attainment of the required IQI sensitivity and compliance with all other requirements stipulated herein (film density and area of interest density tolerances, etc.).

6.2 *Film Holders and Cassettes*—Film holders and cassettes shall be light tight and shall be handled properly to reduce the likelihood that they may be damaged. They may be flexible vinyl, plastic, or other durable material, or they may be made from metallic materials. In the event that light leaks into the film holder and produces images on the radiograph, the radiograph need not be rejected unless the images encroach on the radiographic area of interest. If the film holder exhibits light leaks, it shall be repaired before reuse or discarded. Film holders and cassettes should be routinely examined to minimize the likelihood of light leaks.

6.3 *Intensifying Screens*:

6.3.1 *Lead-Foil Screens*:

6.3.1.1 Intensifying screens of the lead-foil type are generally used for production radiography. Lead-foil screens shall be of the same approximate dimensions as the film being used and shall be in direct contact with the film during exposure.

6.3.1.2 Unless otherwise specified in the purchaser-supplier agreement, the lead-foil screens shown in Table 1 shall be used, except as provided within the tabular notes below it.

6.3.2 *Fluorescent, Fluorometallic, or Other Metallic Screens*—Such screens may be used with CEO approval as described under 5.1; however, they must be capable of demonstrating the required IQI sensitivity. Fluorescent or fluorometallic screens may cause limitations in image quality (see Guide E94, Appendix X1).

6.3.3 *Screen Care*:

6.3.3.1 All screens should be handled carefully to avoid dents, scratches, grease, or dirt on active surfaces. Screens that render nonrelevant indications on radiographs shall be visually examined and discarded if physical damage is observed.

6.3.3.2 Screens, with or without backing, shall be free of dust, dirt, oxidation, or any other foreign material that render undesirable nonrelevant images on the film.

6.3.3.3 *Other Screens*—European Standard CEN EN 444 contains similar provisions for intensifying screens as in this test method. International users of these type screens who prefer the use of CEN EN 444 or ISO 5579 for their particular applications should specify such alternative provisions within separate contractual arrangements from this test method.

6.4 *Filters*—Filters shall be used whenever the contrast reductions caused by low energy, scattered radiation, or the extent of undercut (edge burn-off) occurring on production radiographs is of significant magnitude to cause difficulty in meeting the quality level or radiographic coverage requirements stipulated by the job order or contract (see Guide E94).

6.5 *Masking*—Masking material may improve radiographic quality (see Guide E94).

6.6 *IQI's (Penetrameters)*—Unless otherwise specified by the applicable job order or contract, only those IQI's that comply with the design and identification requirements specified in Practice E1025 or Practice E747 shall be used.

TABLE 1 Lead-Foil Screens

KeV Range	Front Screen[A]	Back Screen Minimum
0 to 150 KeV[B]	0.000 to 0.001 in. [0 to 0.025 mm]	0.005 in. [0.127 mm][C]
150 to 200 KeV; Ir 192; Se 75	0.000 to 0.005 in. [0 to 0.127 mm]	0.005 in. [0.127 mm]
200 KeV to 2 MeV; Co 60	0.005 to 0.010 in. [0.126 to 0.254 mm]	0.010 in. [0.254 mm]
2 to 4 MeV	0.010 to 0.020 in. [0.254 to 0.508 mm]	0.010 in. [0.254 mm]
4 to 10 MeV	0.010 to 0.030 in. [0.254 to 0.762 mm]	0.010 in. [0.254 mm]
10 to 25 MeV	0.010 to 0.050 in. [0.254 to 1.27 mm]	0.010 in. [0.254 mm]

[A]The lead screen thickness listed for the various voltage ranges are recommended thicknesses and not required thicknesses. Other thicknesses and materials may be used provided the required radiographic quality level, contrast, and density are achieved.
[B]Prepacked film with lead screens may be used from 80 to 150 KeV. No lead screens are recommended below 80 KeV. Prepacked film may be used at higher energy levels provided the contrast, density, radiographic quality level, and backscatter requirements are achieved. Additional intermediate lead screens may be used for reduction of scattered radiation at higher energies.
[C]No back screen is required provided the backscatter requirements of 8.5 are met.

6.7 *Shims, Separate Blocks, or Like Sections*—Shims, separate blocks, or like sections made of the same or radiographically similar materials (as defined in Practice E1025) may be used to facilitate IQI positioning. There is no restriction on shim or separate block maximum thickness, provided the IQI and area-of-interest density variation requirements of 8.8.2 are met. The like section should be geometrically similar to the object being radiographed.

6.8 *Radiographic Location and Identification Markers*—Lead numbers and letters are used to designate the part number and location number. The size and thickness of the markers shall depend on the ability of the radiographic technique to discern the markers on the radiograph. As a general rule, markers 1/16 in. thick will suffice for most low energy (less than 1 MeV) X ray and Iridium 192 radiography; for higher energy radiography it may be necessary to use markers that are thicker (1/8 in. thick or more).

6.9 *Radiographic Density Measurement Apparatus*—Either a transmission densitometer or a step-wedge comparison film shall be used for judging film-density requirements. Step-wedge comparison films or densitometers calibration, or both, shall be verified by comparison with a calibrated step-wedge film traceable to the National Institute of Standards and Technology. Where applicable, a film digitization and analysis system may be substituted for a transmission densitometer provided the film digitization and analysis system has been calibrated and verified by comparison with a calibrated step-wedge film traceable to the National Institute of Standards and Technology. Densitometers shall be calibrated in accordance with Practice E1079.

7. Requirements

7.1 *Procedure Requirement*—Unless otherwise specified by the applicable job order or contract, radiographic examination shall be performed in accordance with a written procedure. Specific requirements regarding the preparation and approval of the written procedures shall be dictated by purchaser and supplier agreement. The production procedure shall address all applicable portions of this document and shall be available for review during interpretation of the radiographs.

7.2 *Radiographic Coverage*—Unless otherwise specified by purchaser and supplier agreement, the extent of radiographic coverage shall include 100 % of the volume of the weld.

7.3 *Radiographic Film Quality*—All radiographs shall be free of mechanical, chemical, handling-related, or other blemishes which could mask or be confused with the image of any discontinuity in the area of interest on the radiograph. If any doubt exists as to the true nature of an indication exhibited by the film, the radiograph shall be rejected and the view retaken.

NOTE 1—Digital image enhancement techniques applied to scanned radiographic images have, in some cases, shown the ability to resolve doubts regarding the true nature of indications shown in the original radiograph. Where applicable, these techniques may be used in an effort to resolve questions regarding the nature of the indication.

7.4 *Radiographic Quality Level*—Radiographic quality level shall be determined upon agreement between the purchaser and supplier and shall be specified in the applicable job order or contract.

7.5 *Acceptance Level*—Accept and reject levels shall be stipulated by the applicable contract, job order, drawing, or other purchaser and supplier agreement.

7.6 *Radiographic Density Limitations*—The density through the body of the IQI and area of interest shall be 1.5 to 4.0 for single film viewing and 2.0 to 4.0 for composite viewing.

7.7 *Film Handling*:

7.7.1 *Darkroom Facilities*—Darkroom facilities should be kept clean and as dust-free as practical. Safe-lights should be those recommended by film manufacturers for the radiographic materials used and should be positioned in accordance with the manufacturer's recommendations. All darkroom equipment and materials should be capable of producing radiographs that are suitable for interpretation.

7.7.2 *Film Processing*—Radiographic film processing should be controlled in accordance with Guide E999.

7.7.3 *Film-Viewing Facilities*—Viewing facilities shall provide subdued background lighting of an intensity that will not cause troublesome reflection, shadows, or glare on the radiograph. The viewing light shall be of sufficient intensity to view densities up to 4.0 and be appropriately controlled so that the optimum intensity for single or superimposed viewing of radiographs may be selected.

7.7.4 *Storage of Radiographs*—When storage is required by the applicable job order or contract, the radiographs should be stored in an area with sufficient environmental control to preclude image deterioration or other damage. The radiograph storage duration and location shall be as agreed upon between purchaser and supplier. (See Guide E1254 for storage information.)

8. Procedure

8.1 *Time of Examination*—Unless otherwise specified by the applicable job order or contract, radiography may be performed prior to heat treatment.

8.2 *Surface Preparation*—Unless otherwise agreed upon, remove the weld ripples or weld-surface irregularities on both the inside (where accessible) and outside by any suitable process so that the image of the irregularities cannot mask, or be confused with, the image of any discontinuity. Interpretation can be optimized if surface irregularities are removed such that the image of the irregularities is not discernible on the radiograph.

8.3 *Source to Film Distance*—Unless otherwise specified in the applicable job order or contract, geometric unsharpness (U_g) shall not exceed the following:

Material Thickness	Ug Maximum
Under 1 in. [25.4 mm]	0.010 in. [0.254 mm]
1 through 2 in. [25.4 through 50.8 mm]	0.020 in. [0.508 mm]
Over 2 through 3 in. [Over 50.8 through 76.2 mm]	0.030 in. [0.762 mm]
Over 3 through 4 in. [Over 76.2 through 101.6 mm]	0.040 in. [1.016 mm]
Greater than 4 in. [Greater than 101.6 mm]	0.070 in. [1.778 mm]

Geometric unsharpness values shall be determined (calculated) as specified by the formula in Guide E94.

8.4 *Direction of the Radiation*—Direct the central beam of radiation perpendicularly toward the center of the effective area of the film or to a plane tangent to the center of the film, to the maximum extent possible, except for double-wall exposure—double-wall viewing elliptical-projection techniques, as described in 8.14.2.

8.5 *Back-Scattered Radiation Protection*:

8.5.1 Back-scattered radiation (radiation reflected from surfaces behind the film, (that is, walls, floors, etc.) serves to reduce radiographic contrast and may produce undesirable effects on radiographic quality. A ⅛-in. lead sheet, placed behind the film, generally furnishes adequate protection against back-scattered radiation.

8.5.2 To detect back-scattered radiation, position a lead letter *B* (approximately ⅛ in. thick by ½ in. high) on the rear side of the film holder. If a light image of the lead letter *B* appears on the radiograph, it indicates that more back-scatter protection is necessary. The appearance of a dark image of the lead letter *B* should be disregarded, unless the dark image could mask or be confused with rejectable weld defects.

8.6 *IQI Selection*—The thickness on which the IQI is based is the single-wall thickness plus actual reinforcement thickness up to the maximum allowed. Backing strips or rings are not considered as part of the weld or reinforcement thickness in IQI selection. For any thickness, a thinner IQI may be used, provided all other requirements for radiography are met.

8.7 *IQI Placement*:

8.7.1 Place the IQIs on the source side adjacent to the weld being radiographed. Where the weld metal is not radiographically similar to the base material or where geometry precludes placement, the IQI may be placed over the weld.

8.7.2 *Film Side IQI*—In those cases where the physical placement of the IQI on the source side is not possible, the IQI may be placed on the film side. The applicable job order or contract shall specify the applicable film-side quality level. Place a lead letter *F* adjacent to the IQI for identification.

8.8 *Separate Block*—When configuration or size prevents placing the IQI on the object being radiographed, a shim or separate block or like section conforming to the requirements of 6.7 may be used, provided the following conditions are met:

8.8.1 The IQI shall be no closer to the film than the source side of the object being radiographed (unless otherwise specified).

8.8.2 The radiographic density measured through the body of the IQI on the shim, separate block, or like section shall not exceed the density measured in the area of interest by more than 15 %. The penetrameter density may be lighter than the area of interest density, provided the specified quality level is obtained and the density requirements of 7.6 are met.

8.8.3 The shim, separate block, or like section shall be placed as close as possible to the object being radiographed.

8.8.4 The shim, separate block, or like section dimensions shall exceed the IQI dimensions such that the outline of at least three sides of the IQI image shall be visible on the radiograph.

8.9 *Number of IQIs*:

8.9.1 One IQI shall represent an area within which radiographic densities are not less than 15 % from the density measured through the body of the IQI. At least one IQI per radiograph, exposed simultaneously with the specimen, shall be used except as noted in 8.9.2 and 8.9.3.

8.9.2 When film density, in the area of interest, is lower by more than 15 % of that measured through the body of the IQI, two IQIs used in the following manner will be satisfactory: (*1*) if one IQI shows an acceptable sensitivity at the most dense portion of the radiograph and (*2*) the second IQI shows an acceptable sensitivity at the least dense portion of the radiograph. These two IQIs will then serve to qualify the radiograph technique.

8.9.3 For cylindrical vessels or flat components where one or more film holders and cassettes are used for an exposure, at least one IQI image shall appear on each radiograph, except where the source is placed on the axis of the object and a complete circumference or portion of the circumference radiographed with a single exposure. In which case, at least three IQIs shall be placed approximately equidistant apart. When the source is placed on the axis of the circumference and a portion of that circumference (four or more continuous film locations) is radiographed during a single exposure, at least three IQIs placed approximately equidistant apart shall be used. Otherwise, at least one IQI image shall appear on each radiograph. Where portions of longitudinal welds adjoining the circumferential weld are being examined simultaneously with the circumferential weld, additional IQIs shall be placed on the longitudinal welds at the ends of the sections most remote from the position of the source used to radiograph the circumferential weld.

8.9.4 Qualifying radiographs, on which one or more IQIs were imaged during exposure, shall always be retained as part of record to validate required IQI sensitivity and placement.

NOTE 2—For parts of irregular geometry or widely varying thickness, it may be necessary to radiograph the first unit of a given design to determine proper placement of IQI for subsequent radiography.

8.10 *Shim Utilization*—When a weld reinforcement or backing ring and strip is not removed, place a shim of material which is radiographically similar to the backing ring and strip under the IQI to provide approximately the same thickness of material under the IQI as the average thickness of the weld reinforcement plus the wall thickness and backing ring and strip. There is no restriction on shim thickness, provided the IQI and area-of-interest density variation requirements are met.

8.10.1 *Shim Dimensions and Location*—The shim dimension and location shall exceed the IQI dimensions by at least 1/8 in. on at least three sides. At least three sides of the IQI shall be discernible in accordance with 8.8.4 except that only the two ends of the IQI need to be discernible when located on piping less than 1-in. nominal pipe size. The shim shall be placed so as not to overlap the weld image including the backing strip or ring.

8.10.2 *Shim Image Film Density*—The film density of the shim image shall not be greater than 15 % more than the lightest film density of the area of interest. It may be less dense than the lightest film density of the area of interest.

8.11 *Location Markers*—Location markers shall be placed outside the weld area. The radiographic image of the location markers for the coordination of the part with the film shall appear on the film without interfering with the interpretation and with such an arrangement that it is evident that complete coverage was obtained.

8.11.1 *Double-Wall Technique*—When using a technique in which radiation passes through two walls and the welds in both walls are viewed for acceptance, and the entire image of the object being radiographed is shown on the radiograph, only one location marker is required in the radiograph.

8.11.2 *Series of Radiographs*—For welds that require a series of radiographs to cover the full length or circumference of the weld, the complete set of location markers must be applied at one time, wherever possible. A reference or zero position for each series must be identified on the component. A known feature on the object (for example, keyway, nozzle, and axis line) may also be used for establishment of a zero position; indicate this feature on the radiographic record.

8.11.3 *Similar Welds*—On similar type welds on a single component, the sequence and spacing of the location markers must conform to a uniform system that shall be positively identified in the radiographic procedure or interpretation records. In addition, reference points on the component will be shown on the sketch to indicate the direction of the numbering system.

8.12 *Radiograph Identification*—A system of positive identification of the film shall be provided. As a minimum, the following shall appear on the radiograph: the name or symbol of the company performing radiography, the date, and the weld identification number traceable to part and contract. Subsequent radiographs made by reasons of a repaired area shall be identified with the letter *R*.

8.13 *Multiple-Film Techniques*—Film techniques with two or more films of equal or different speeds in the same cassette are allowed, provided prescribed quality level and density requirements stipulated herein are met.

8.14 *Radiographic Techniques*:

8.14.1 *Single-Wall Technique*—Except as provided in 8.14.2 and 8.14.3, radiography shall be performed using a technique in which the radiation passes through only one wall.

8.14.2 *Double-Wall Technique for Circumferential Welds*—For circumferential welds 3½ in. (OD) outside diameter or less, a technique may be used in which the radiation passes through both walls and both walls are viewed for acceptance on the same film. Unless otherwise specified, either elliptical or superimposed projections may be used.

8.14.2.1 For elliptical projections, where the weld is not superimposed, at least two views separated by 90° shall be required.

8.14.2.2 Where design or access restricts a practical technique from obtaining 90° separation of views, agreement between contracting parties must specify necessary weld coverage.

8.14.2.3 For superimposed projections a minimum of three views is required at approximately 0°, 60°, and 120°.

8.14.2.4 For circumferential welds greater than 3½ in. outside diameter (OD), a technique shall be used in which only single-wall viewing is performed. Sufficient exposures shall be taken to ensure complete coverage.

8.14.3 For radiographic techniques which prevent single-wall exposures due to restricted access, such as jacketed pipe or ship hull, technique should be agreed upon in advance between the purchaser and supplier. It should be recognized that IQI sensitivities based on single-wall thickness may not be obtainable under some conditions.

9. Safety

9.1 Radiographic procedures shall comply with applicable city, state, and federal regulations.

10. Radiograph Evaluation

10.1 *Film Quality*— Verify that the radiograph meets the quality requirements specified in 7.3, 7.4, 7.6, 8.5.2, 8.8, and 8.9.

10.2 *Film Evaluation*— Determine the acceptance or rejection of the weldment by comparing the radiographic image to the agreed upon acceptance criteria (see 7.5).

10.3 *Reference Radiographs*— Graded reference radiographs showing typical indications of various welding defects in graded levels of severity are useful tools for specifying and evaluating acceptance criteria for weld radiographs. Since severity levels are typically provided for each of a variety of flaw types, the acceptance criteria may specify different severity levels for the different conditions. For optimal utility the reference radiographs should be representative of the part to be radiographed in both material and section thickness. Reference radiographs for steel fusion welds are available in Reference Radiographs E390. Additional reference radiographs which illustrate the effects of modifying certain radiographic parameters, particularly energy and screen combinations, are available in Reference Radiographs E242.

11. Records

11.1 The following radiographic records shall be maintained as agreed upon between purchaser and supplier:

11.1.1 Radiographic Standard Shooting Sketch.

11.1.2 Weld Repair Documentation.

11.1.3 Film.

11.1.4 Film interpretation record shall contain as a minimum the following information:

11.1.4.1 Disposition of each radiograph (acceptable or rejectable).

11.1.4.2 If rejectable, cause for rejection (slag, crack, porosity, etc.).

11.1.4.3 Surface indication verified by visual examination (grinding marks, weld ripple, spatter, etc.).

11.1.4.4 Signature of the film interpreter, including certification level.

12. Keywords

12.1 gamma ray; nondestructive testing; radiographic examination; radiography; weldments; X-ray

ASTM International takes no position respecting the validity of any patent rights asserted in connection with any item mentioned in this standard. Users of this standard are expressly advised that determination of the validity of any such patent rights, and the risk of infringement of such rights, are entirely their own responsibility.

This standard is subject to revision at any time by the responsible technical committee and must be reviewed every five years and if not revised, either reapproved or withdrawn. Your comments are invited either for revision of this standard or for additional standards and should be addressed to ASTM International Headquarters. Your comments will receive careful consideration at a meeting of the responsible technical committee, which you may attend. If you feel that your comments have not received a fair hearing you should make your views known to the ASTM Committee on Standards, at the address shown below.

This standard is copyrighted by ASTM International, 100 Barr Harbor Drive, PO Box C700, West Conshohocken, PA 19428-2959, United States. Individual reprints (single or multiple copies) of this standard may be obtained by contacting ASTM at the above address or at 610-832-9585 (phone), 610-832-9555 (fax), or service@astm.org (e-mail); or through the ASTM website (www.astm.org).

RELATED MATERIAL

 Index

ASTM Standards for Welding

This index covers the standards and related material appearing in this volume. The boldface references represent the ASTM designations.

Alphabetization in the index is letter-for-letter, with no consideration to punctuation or word division. Initial prepositions of (indented) subentries are ignored for alphabetization.

In the preparation of indexes, every attempt has been made to index standards on three levels: (1) by main subject, using general and specific search terms; (2) by tests or other significant sections of ASTM standards; and (3) by cross-references to locate main subject entry terms. *(See also* references are abbreviated as *Sa* and appear under main entry terms.) The following examples illustrate ASTM's method of indexing.

INDEX TERMS FOR SPECIFICATIONS

Steel pipe—carbon steel
Seamless and Welded Steel Pipe for Low-Temperature Service, Specification for, **A333/A333M**

Steel tube—ferritic stainless steel
Seamless and Welded Ferritic/Austenitic Stainless Steel Tubing for General Service, Specification for, **A789/A789M**

INDEX TERMS FOR TESTS

Gold—electrodeposited coating
Installing Corrugated Aluminum Structural Plate Pipe for Culverts and Sewers, Practice for, **B789/B789M**

Effective elastic parameter (E_{eff})
Determining the Effective Elastic Parameter for X-Ray Diffraction Measurements of Residual Stress, Test Method for, **E1426**

CROSS-REFERENCES

Water—drinking
 See **Drinking water**

PDB (pressure design basis)
 See **Pressure design basis (PDB)**
 Sa **Pressure testing**

Index, ASTM Standards for Welding

A

Acid cleaning
   See **Cleaning agents/processes**
Acoustic emission (AE) testing
   See **Nondestructive evaluation (NDE)**
Airtight doors—specifications
   See **Shipbuilding steel materials—specifications**
Alkali-ion diode halogen leak testing
   See **Leak testing**
Alloy steel bars—specifications
   High-Strength Low-Alloy Columbium-Vanadium Structural Steel, Specification for, **A572/A572M**
   Steel Bars, Carbon and Alloy, Hot-Wrought, General Requirements for, Specification for, **A29/A29M**
   Structural Steel for Bridges, Specification for, **A709/A709M**
Alloy steel bolting materials—specifications
   Structural Bolts, Alloy Steel, Heat Treated, 150 ksi Minimum Tensile Strength, Specification for, **A490**
Alloy steel pipe—specifications
   Seamless and Welded Steel Pipe for Low-Temperature Service, Specification for, **A333/A333M**
Alloy steel plate—specifications
   High-Strength Low-Alloy Steel Shapes of Structural Quality, Produced by Quenching and Self-Tempering Process (QST), Specification for, **A913/A913M**
   High-Yield-Strength, Quenched and Tempered Alloy Steel Plate, Suitable for Welding, Specification for, **A514/A514M**
   Hot-Rolled Structural Steel, High-Strength Low-Alloy Plate with Improved Formability, Specification for, **A656/A656M**
   Pressure Vessel Plates, Alloy Steel, High-Strength, Quenched and Tempered, Specification for, **A517/A517M**
   Quenched and Tempered Low-Alloy Structural Steel Plate with 70 ksi [485 MPa] Minimum Yield Strength to 4 in. [100 mm] Thick, Specification for, **A852/A852M**
   Straight-Beam Ultrasonic Examination of Rolled Steel Plates for Special Applications, Specification for, **A578/A578M**
   Straight-Beam Ultrasonic Examination of Steel Plates, Specification for, **A435/A435M**
   Structural Steel for Bridges, Specification for, **A709/A709M**
Alloy steel sheet/strip—specifications
   Sa **Steel sheet/strip**
   Steel, Sheet and Strip, High-Strength, Low-Alloy, Hot-Rolled and Cold-Rolled, with Improved Atmospheric Corrosion Resistance, Specification for, **A606/A606M**
Alloy steel tube—specifications
   Sa **Steel tube**
   Hot-Formed Welded and Seamless High-Strength Low-Alloy Structural Tubing, Specification for, **A618/A618M**
   Seamless and Welded Carbon and Alloy-Steel Tubes for Low-Temperature Service, Specification for, **A334/A334M**
Anions content
   Liquid Penetrant Examination for General Industry, Practice for, **E165**
Architectural materials/applications—specifications
   See **Building materials/applications—specifications**
Arc-welded steel pipe/tube
   See **Electric-fusion-welded steel pipe**
Arc-welded tube
   See **Welding/welds**
ASTM equivalent penetrameter sensitivity
   See **Penetrameters**
Atmospheric corrosion materials/applications—specifications
   High-Strength Low-Alloy Structural Steel Plate With Atmospheric Corrosion Resistance, Specification for, **A871/A871M**
   High-Strength Low-Alloy Structural Steel, up to 50 ksi [345 MPa] Minimum Yield Point, with Atmospheric Corrosion Resistance, Specification for, **A588/A588M**
   Steel, Sheet and Strip, High-Strength, Low-Alloy, Hot-Rolled and Cold-Rolled, with Improved Atmospheric Corrosion Resistance, Specification for, **A606/A606M**
   Structural Steel for Bridges, Specification for, **A709/A709M**
Atmospheric steel pipe
   Electric-Fusion-Welded Steel Pipe for Atmospheric and Lower Temperatures, Specification for, **A671**
   Seamless Carbon Steel Pipe for Atmospheric and Lower Temperatures, Specification for, **A524**
Austenitic stainless steel tube—specifications
   See **Steel tube**
Automotive steel materials—structural
   Hot-Rolled Structural Steel, High-Strength Low-Alloy Plate with Improved Formability, Specification for, **A656/A656M**
Axle steel bars
   See **Alloy steel bars—specifications**

B

Bars/forging/forging stock—specifications
   See **Alloy steel bars—specifications**
Basic-oxygen steel
   High-Strength Low-Alloy Structural Steel, Specification for, **A242/A242M**
   High-Yield-Strength, Quenched and Tempered Alloy Steel Plate, Suitable for Welding, Specification for, **A514/A514M**
   Quenched and Tempered Low-Alloy Structural Steel Plate with 70 ksi [485 MPa] Minimum Yield Strength to 4 in. [100 mm] Thick, Specification for, **A852/A852M**
   Structural Carbon Steel Plates of Improved Toughness, Specification for, **A573/A573M**
Batter piles
   See **Piles**
Bend testing—metallic materials
   Mechanical Testing of Steel Products, Test Methods and Definitions for, **A370**
Billets
   Magnetic Particle Testing, Guide for, **E709**
Black steel pipe—specifications
   Pipe, Steel, Black and Hot-Dipped, Zinc-Coated, Welded and Seamless, Specification for, **A53/A53M**
Blooms
   Magnetic Particle Testing, Guide for, **E709**
Bolted construction materials/applications—specifications
   Carbon Structural Steel, Specification for, **A36/A36M**
   High-Strength Carbon-Manganese Steel of Structural Quality, Specification for, **A529/A529M**
   High-Strength Low-Alloy Columbium-Vanadium Structural Steel, Specification for, **A572/A572M**
   High-Strength Low-Alloy Structural Steel, up to 50 ksi [345 MPa] Minimum Yield Point, with Atmospheric Corrosion Resistance, Specification for, **A588/A588M**
   Normalized High-Strength Low-Alloy Structural Steel Plates, Specification for, **A633/A633M**
   Quenched-and-Tempered Carbon and High-Strength Low-Alloy Structural Steel Plates, Specification for, **A678/A678M**
   Structural Bolts, Alloy Steel, Heat Treated, 150 ksi Minimum Tensile Strength, Specification for, **A490**
   Structural Bolts, Steel, Heat Treated, 120/105 ksi Minimum Tensile Strength, Specification for, Δ **A325**
Bolting materials
   See **Steel bolting materials**
   See **Steel bolting materials—specifications**
Bridge/structural materials
   Structural Steel Shapes, Specification for, **A992A/A992AM**
Bridge/structural materials—specifications
   Carbon Structural Steel, Specification for, **A36/A36M**
   General Requirements for Rolled Structural Steel Bars, Plates, Shapes, and Sheet Piling, Specification for, **A6/A6M**

Index, ASTM Standards for Welding

High-Strength Low-Alloy Columbium-Vanadium Structural Steel, Specification for, **A572/A572M**

High-Strength Low-Alloy Structural Steel, up to 50 ksi [345 MPa] Minimum Yield Point, with Atmospheric Corrosion Resistance, Specification for, **A588/A588M**

High-Yield-Strength, Quenched and Tempered Alloy Steel Plate, Suitable for Welding, Specification for, **A514/A514M**

Structural Steel for Bridges, Specification for, **A709/A709M**

Brinell hardness
Mechanical Testing of Steel Products, Test Methods and Definitions for, **A370**

Standard Hardness Conversion Tables for Metals Relationship Among Brinell Hardness, Vickers Hardness, Rockwell Hardness, Superficial Hardness, Knoop Hardness, and Scleroscope Hardness, **E140**

Building design and construction—specifications
Structural Steel Shapes, Specification for, **A992A/A992AM**

Building framing
Structural Steel Shapes, Specification for, **A992A/A992AM**
Structural Steel with Low Yield to Tensile Ratio for Use in Buildings, Specification for, **A1043/A1043M**

Building materials/applications—specifications
Carbon Structural Steel, Specification for, **A36/A36M**
Structural Steel Shapes, Specification for, **A992A/A992AM**

Building steel materials—structural
See **Bridge/structural materials**
See **Generator materials**
See **Pressure vessel steel**
See **Steel bolting materials—specifications**
See **Structural steel (SS)—specifications**
See **Structural steel (SS) plate—specifications**

Bulkheads
See **Shipbuilding steel materials—specifications**

Butterfly valves
See **Shipbuilding steel materials—specifications**

C

Capped steel
General Requirements for Rolled Structural Steel Bars, Plates, Shapes, and Sheet Piling, Specification for, **A6/A6M**

Carbon-manganese steel—specifications
High-Strength Carbon-Manganese Steel of Structural Quality, Specification for, **A529/A529M**
Pressure Vessel Plates, Heat-Treated, Carbon-Manganese-Silicon Steel, Specification for, **A537/A537M**

Carbon steel bars/shapes—specifications
Carbon Structural Steel, Specification for, **A36/A36M**
High-Strength Carbon-Manganese Steel of Structural Quality, Specification for, **A529/A529M**
Steel Bar, Carbon and Alloy, Cold-Finished, Specification for, **A108**
Steel Bars, Carbon and Alloy, Hot-Wrought, General Requirements for, Specification for, **A29/A29M**
Structural Steel for Bridges, Specification for, **A709/A709M**

Carbon steel pipe—specifications
Seamless Carbon Steel Pipe for Atmospheric and Lower Temperatures, Specification for, **A524**
Seamless Carbon Steel Pipe for High-Temperature Service, Specification for, **A106/A106M**
Seamless and Welded Steel Pipe for Low-Temperature Service, Specification for, **A333/A333M**

Carbon steel plate—specifications
Carbon Structural Steel, Specification for, **A36/A36M**
High-Strength Carbon-Manganese Steel of Structural Quality, Specification for, **A529/A529M**
Precipitation–Strengthened Low-Carbon Nickel-Copper-Chromium-Molybdenum-Columbium Alloy Structural Steel Plates, Specification for, **A710/A710M**

Pressure Vessel Plates, Carbon Steel, for Moderate- and Lower-Temperature Service, Specification for, **A516/A516M**
Pressure Vessel Plates, Heat-Treated, Carbon-Manganese-Silicon Steel, Specification for, **A537/A537M**
Quenched-and-Tempered Carbon and High-Strength Low-Alloy Structural Steel Plates, Specification for, **A678/A678M**
Straight-Beam Ultrasonic Examination of Steel Plates, Specification for, **A435/A435M**
Structural Carbon Steel Plates of Improved Toughness, Specification for, **A573/A573M**
Structural Steel for Bridges, Specification for, **A709/A709M**

Carbon steel sheet—specifications
High-Strength Low-Alloy Columbium-Vanadium Structural Steel, Specification for, **A572/A572M**
Steel, Sheet, Cold-Rolled, Carbon, Structural, High-Strength Low-Alloy, High-Strength Low-Alloy with Improved Formability, Solution Hardened, and Bake Hardenable, Specification for, **A1008/A1008M**

Carbon steel sheet/strip—specifications
Steel, Sheet and Strip, Hot-Rolled, Carbon, Structural, High-Strength Low-Alloy, High-Strength Low-Alloy with Improved Formability, and Ultra-High Strength, Specification for, **A1011/A1011M**
Steel, Strip, Carbon (0.25 Maximum Percent), Cold-Rolled, Specification for, **A109/A109M**

Carbon steel tube—specifications
Cold-Formed Welded and Seamless Carbon Steel Structural Tubing in Rounds and Shapes, Specification for, **A500/A500M**
Hot-Formed Welded and Seamless Carbon Steel Structural Tubing, Specification for, **A501**
Seamless and Welded Carbon and Alloy-Steel Tubes for Low-Temperature Service, Specification for, **A334/A334M**
Steel Tubes, Low-Carbon or High-Strength Low-Alloy, Tapered for Structural Use, Specification for, **A595/A595M**

Castings
Magnetic Particle Testing, Guide for, **E709**

Cast-in-place concrete piles
Welded and Seamless Steel Pipe Piles, Specification for, **A252**

Centrifugally cast steel pipe—specifications
See **Steel pipe**

Charpy impact test
Mechanical Testing of Steel Products, Test Methods and Definitions for, **A370**
Notched Bar Impact Testing of Metallic Materials, Test Methods for, **E23**
Sampling Procedure for Impact Testing of Structural Steel, Specification for, **A673/A673M**

Chlorine content
Liquid Penetrant Examination for General Industry, Practice for, **E165**

Chromium steel plate—specifications
Precipitation–Strengthened Low-Carbon Nickel-Copper-Chromium-Molybdenum-Columbium Alloy Structural Steel Plates, Specification for, **A710/A710M**

Circular bead weldment
See **Welding/welds**

Circular magnetization
See **Magnetic particle inspection**

Classification
Design, Manufacture and Material Grouping Classification of Wire Image Quality Indicators (IQI) Used for Radiology, Practice for, **E747**
Design, Manufacture, and Material Grouping Classification of Hole-Type Image Quality Indicators (IQI) Used for Radiology, Practice for, **E1025**

Cleaning agents/processes
Liquid Penetrant Examination for General Industry, Practice for, **E165**

Index, ASTM Standards for Welding

Coated sheet products
Coated sheet products
  See **Zinc-coated steel sheet—specifications**
Cofferdams
  See **Structural steel (SS) piles—specifications**
Cold cleaning operations
  See **Cleaning agents/processes**
Cold-drawn steel tube—specifications
  See **Steel tube**
Cold-drawn steel wire—specifications
  See **Steel wire**
Cold-finished steel bars—specifications
  Steel Bar, Carbon and Alloy, Cold-Finished, Specification for, **A108**
  Steel Bars, Carbon and Alloy, Hot-Wrought, General Requirements for, Specification for, **A29/A29M**
Cold-formed steel sheet—specifications
  See **Structural steel (SS) piles—specifications**
Cold-formed steel tube—specifications
  Cold-Formed Welded and Seamless Carbon Steel Structural Tubing in Rounds and Shapes, Specification for, **A500/A500M**
Cold-rolled steel sheet/strip—specifications
  Sa **Steel sheet**
  Steel, Sheet and Strip, High-Strength, Low-Alloy, Hot-Rolled and Cold-Rolled, with Improved Atmospheric Corrosion Resistance, Specification for, **A606/A606M**
  Steel, Sheet, Cold-Rolled, Carbon, Structural, High-Strength Low-Alloy, High-Strength Low-Alloy with Improved Formability, Solution Hardened, and Bake Hardenable, Specification for, **A1008/A1008M**
  Steel, Strip, Carbon (0.25 Maximum Percent), Cold-Rolled, Specification for, **A109/A109M**
Columbium alloy steel—specifications
  High-Strength Low-Alloy Columbium-Vanadium Structural Steel, Specification for, **A572/A572M**
  Precipitation–Strengthened Low-Carbon Nickel-Copper-Chromium-Molybdenum-Columbium Alloy Structural Steel Plates, Specification for, **A710/A710M**
  Steel, Sheet and Strip, Heavy-Thickness Coils, Hot-Rolled, Carbon, Commercial, Drawing, Structural, High-Strength Low-Alloy, High-Strength Low-Alloy with Improved Formability, and Ultra-High Strength, Specification for, **A1018/A1018M**
Combustible liquid penetrant materials
  Liquid Penetrant Examination for General Industry, Practice for, **E165**
Commercial steel (CS) sheet/strip—specifications
  Sa **Steel sheet**
  Steel, Sheet and Strip, Hot-Rolled, Carbon, Structural, High-Strength Low-Alloy, High-Strength Low-Alloy with Improved Formability, and Ultra-High Strength, Specification for, **A1011/A1011M**
  Steel, Sheet, Cold-Rolled, Carbon, Structural, High-Strength Low-Alloy, High-Strength Low-Alloy with Improved Formability, Solution Hardened, and Bake Hardenable, Specification for, **A1008/A1008M**
Common requirements (steel)
  See **General delivery requirements—steel**
Composition analysis—metals/alloys
  Steel Sheet, Zinc-Coated (Galvanized) or Zinc-Iron Alloy-Coated (Galvannealed) by the Hot-Dip Process, Specification for, **A653/A653M**
Concrete reinforcement—specifications
  Steel Wire, Deformed, for Concrete Reinforcement, Specification for, **A496/A496M**
  Welded and Seamless Steel Pipe Piles, Specification for, **A252**
Continuous magnetization technique
  See **Magnetic particle inspection**
Continuous welding (CW)
  See **Welded steel materials/applications—specifications**

Conversion units/factors
  Standard Hardness Conversion Tables for Metals Relationship Among Brinell Hardness, Vickers Hardness, Rockwell Hardness, Superficial Hardness, Knoop Hardness, and Scleroscope Hardness, **E140**
Copper structural materials/applications—specifications
  Precipitation–Strengthened Low-Carbon Nickel-Copper-Chromium-Molybdenum-Columbium Alloy Structural Steel Plates, Specification for, **A710/A710M**
Corrosion-resistant steel—specifications
  High-Strength Low-Alloy Structural Steel, Specification for, **A242/A242M**
Corrosive service applications—bars/rods/shapes
  High-Strength Low-Alloy Structural Steel, Specification for, **A242/A242M**
Corrosive service applications—pipe (steel)
  See **Steel pipe**
Corrosive service applications—plate/sheet/strip
  High-Strength Low-Alloy Structural Steel Plate With Atmospheric Corrosion Resistance, Specification for, **A871/A871M**
Cracking
  Magnetic Particle Testing, Guide for, **E709**
Crankshafts
  See **Shipbuilding steel materials—specifications**

D

DDS (deep drawing steel) sheet
  See **Steel sheet**
Deck treads (steel plate for)
  See **Rolled steel—specifications**
Deep drawing steel
  Steel, Sheet, Cold-Rolled, Carbon, Structural, High-Strength Low-Alloy, High-Strength Low-Alloy with Improved Formability, Solution Hardened, and Bake Hardenable, Specification for, **A1008/A1008M**
Deep foundation units
  See **Piles**
Defects—metals/alloys
  Magnetic Particle Testing, Guide for, **E709**
Deformed steel wire—specifications
  Sa **Steel wire (concrete reinforcement applications)—specifications**
  Steel Wire, Deformed, for Concrete Reinforcement, Specification for, **A496/A496M**
Design—radiology
  Design, Manufacture and Material Grouping Classification of Wire Image Quality Indicators (IQI) Used for Radiology, Practice for, **E747**
  Design, Manufacture, and Material Grouping Classification of Hole-Type Image Quality Indicators (IQI) Used for Radiology, Practice for, **E1025**
Discontinuities
  See **Fluorescent liquid penetrant testing**
  See **Magnetic particle inspection**
  See **Penetrant inspection**
  See **Radiographic examination**
Discontinuities—metals/alloys
  Magnetic Particle Testing, Guide for, **E709**
Discontinuities—steel
  Straight-Beam Ultrasonic Examination of Rolled Steel Plates for Special Applications, Specification for, **A578/A578M**
  Straight-Beam Ultrasonic Examination of Steel Plates, Specification for, **A435/A435M**
Drawing steel (DS) sheet/strip—specifications
  Sa **Steel sheet**
  Steel, Sheet and Strip, Hot-Rolled, Carbon, Structural, High-Strength Low-Alloy, High-Strength Low-Alloy with Im-

proved Formability, and Ultra-High Strength, Specification for, **A1011/A1011M**
Steel, Sheet, Cold-Rolled, Carbon, Structural, High-Strength Low-Alloy, High-Strength Low-Alloy with Improved Formability, Solution Hardened, and Bake Hardenable, Specification for, **A1008/A1008M**

Dry film thickness
See **Thickness**

Dry magnetic particle inspection
Sa **Magnetic particle inspection**
Magnetic Particle Testing, Guide for, **E709**

Dual column ion chromatography
See **Ion chromatography (IC)**

Durability
High-Strength Low-Alloy Structural Steel, Specification for, **A242/A242M**

E

Electric-furnace steel
High-Yield-Strength, Quenched and Tempered Alloy Steel Plate, Suitable for Welding, Specification for, **A514/A514M**
Quenched and Tempered Low-Alloy Structural Steel Plate with 70 ksi [485 MPa] Minimum Yield Strength to 4 in. [100 mm] Thick, Specification for, **A852/A852M**

Electric-fusion-welded steel pipe
Electric-Fusion (Arc)-Welded Steel Pipe (NPS 4 and Over), Specification for, **A139/A139M**
Electric-Fusion-Welded Steel Pipe for Atmospheric and Lower Temperatures, Specification for, **A671**
Welded and Seamless Steel Pipe Piles, Specification for, **A252**

Electric-resistance-welded (ERW) steel pipe—specifications
Welded and Seamless Steel Pipe Piles, Specification for, **A252**

Electrodeposited Zn coatings—specifications
See **Zinc electrodeposited coatings—specifications**

Electronically suppressed ion chromatography
See **Ion chromatography (IC)**

Elongation—metallic materials
Mechanical Testing of Steel Products, Test Methods and Definitions for, **A370**

ERW line pipe
See **Electric-resistance-welded (ERW) steel pipe—specifications**

Extra-deep drawing steel
Steel, Sheet, Cold-Rolled, Carbon, Structural, High-Strength Low-Alloy, High-Strength Low-Alloy with Improved Formability, Solution Hardened, and Bake Hardenable, Specification for, **A1008/A1008M**

Eyebolts (steel)
See **Steel bolting materials**

F

Fasteners (metal)
See **Steel bolting materials**

Fasteners (metal)—specifications
Structural Bolts, Alloy Steel, Heat Treated, 150 ksi Minimum Tensile Strength, Specification for, **A490**

Fatigue cracking analysis
Notched Bar Impact Testing of Metallic Materials, Test Methods for, **E23**

Ferromagnetic material/testing
Magnetic Particle Testing, Guide for, **E709**

Fiber-reinforced concrete
See **Concrete reinforcement—specifications**

Fine-grain steel
High-Strength, Quenched and Tempered Alloy Steel Plate, Suitable for Welding, Specification for, **A514/A514M**
Hot-Rolled Structural Steel, High-Strength Low-Alloy Plate with Improved Formability, Specification for, **A656/A656M**
Quenched and Tempered Low-Alloy Structural Steel Plate with 70 ksi [485 MPa] Minimum Yield Strength to 4 in. [100 mm] Thick, Specification for, **A852/A852M**
Quenched-and-Tempered Carbon and High-Strength Low-Alloy Structural Steel Plates, Specification for, **A678/A678M**

Fine wire
See **Steel wire**

Flat bar steel
See **Alloy steel bars—specifications**

Flat weldment
Sa **Welding/welds**
Radiographic Examination of Weldments, Test Method for, **E1032**

Floor plate
See **Rolled steel—specifications**

Fluorescent liquid penetrant testing
Sa **Penetrant inspection**
Liquid Penetrant Examination for General Industry, Practice for, **E165**

Fluorine content
Liquid Penetrant Examination for General Industry, Practice for, **E165**

Flux leak testing
See **Leak testing**

Forgings
Magnetic Particle Testing, Guide for, **E709**
Mechanical Testing of Steel Products, Test Methods and Definitions for, **A370**

Formability (in steel specifications)
See **Improved formability steel**

Foundation piles
See **Piles**

Fracture appearance transition temperature (FATT)
Mechanical Testing of Steel Products, Test Methods and Definitions for, **A370**

Framework
High-Strength Carbon-Manganese Steel of Structural Quality, Specification for, **A529/A529M**

Fusion-welded materials/applications—specifications
Pressure Vessel Plates, Heat-Treated, Carbon-Manganese-Silicon Steel, Specification for, **A537/A537M**
Structural Carbon Steel Plates of Improved Toughness, Specification for, **A573/A573M**

G

Gages—ships
See **Shipbuilding steel materials—specifications**

Galvanized steel bolts
See **Steel bolting materials—specifications**

Galvannealed steel sheet—specifications
Steel Sheet, Zinc-Coated (Galvanized) or Zinc-Iron Alloy-Coated (Galvannealed) by the Hot-Dip Process, Specification for, **A653/A653M**

Gamma radiation
Design, Manufacture and Material Grouping Classification of Wire Image Quality Indicators (IQI) Used for Radiology, Practice for, **E747**
Design, Manufacture, and Material Grouping Classification of Hole-Type Image Quality Indicators (IQI) Used for Radiology, Practice for, **E1025**

Gas-filled pressure vessels
See **Pressure vessel steel**

Gas pressure systems/applications—specifications
Electric-Fusion (Arc)-Welded Steel Pipe (NPS 4 and Over), Specification for, **A139/A139M**

Gastight doors
See **Shipbuilding steel materials—specifications**

Index, ASTM Standards for Welding

General delivery requirements—steel

General delivery requirements—steel
General Requirements for Rolled Structural Steel Bars, Plates, Shapes, and Sheet Piling, Specification for, **A6/A6M**
Steel Bars, Carbon and Alloy, Hot-Wrought, General Requirements for, Specification for, **A29/A29M**

Generator materials
Mechanical Testing of Steel Products, Test Methods and Definitions for, **A370**

Grain size
See **Nondestructive evaluation (NDE)**

Groupings
General Requirements for Rolled Structural Steel Bars, Plates, Shapes, and Sheet Piling, Specification for, **A6/A6M**

H

Hardness (indentation)—metallic materials
Mechanical Testing of Steel Products, Test Methods and Definitions for, **A370**
Standard Hardness Conversion Tables for Metals Relationship Among Brinell Hardness, Vickers Hardness, Rockwell Hardness, Superficial Hardness, Knoop Hardness, and Scleroscope Hardness, **E140**

Hardness tests—Brinell
See **Brinell hardness**

Hardness tests—Knoop (HK)
See **Knoop hardness (HK) number**

Hardness tests—metals
See **Metallic hardness**

Hardness tests—Rockwell
See **Rockwell hardness**

Hardness tests—Vickers (HV)
See **Vickers hardness (HV)**

Heat-treated steel bolts
See **Steel bolting materials**

Heat-treated steel sheet and strip
See **Steel sheet/strip**

Heavy-thickness steel coils—specifications
Steel, Sheet and Strip, Heavy-Thickness Coils, Hot-Rolled, Carbon, Commercial, Drawing, Structural, High-Strength Low-Alloy, High-Strength Low-Alloy with Improved Formability, and Ultra-High Strength, Specification for, **A1018/A1018M**

High pressure service applications—specifications
Metal-Arc-Welded Steel Pipe for Use With High-Pressure Transmission Systems, Specification for, **A381**

High-strength low-alloy (HSLA) steel
See **HSLA (high-strength low-alloy) steel—specifications**

High-strength structural steel—specifications
High-Strength Carbon-Manganese Steel of Structural Quality, Specification for, **A529/A529M**
Pressure Vessel Plates, Alloy Steel, High-Strength, Quenched and Tempered, Specification for, **A517/A517M**
Structural Bolts, Steel, Heat Treated, 120/105 ksi Minimum Tensile Strength, Specification for, Δ **A325**
Structural Steel for Ships, Specification for, **A131/A131M**

High-temperature service applications—steel bolting applications
See **Steel pipe**

High-temperature service applications—steel pipe
Seamless Carbon Steel Pipe for High-Temperature Service, Specification for, **A106/A106M**

High-yield-strength steel plate
Sa **Alloy steel plate—specifications**
High-Yield-Strength, Quenched and Tempered Alloy Steel Plate, Suitable for Welding, Specification for, **A514/A514M**

Hot-dip (galvanized) coatings—specifications
Steel Sheet, Zinc-Coated (Galvanized) or Zinc-Iron Alloy-Coated (Galvannealed) by the Hot-Dip Process, Specification for, **A653/A653M**

Hot-dip processed coatings—specifications
Steel Sheet, Zinc-Coated (Galvanized) or Zinc-Iron Alloy-Coated (Galvannealed) by the Hot-Dip Process, Specification for, **A653/A653M**

Hot forging
See **Forgings**

Hot-formed seamless/welded steel tubing—specifications
Hot-Formed Welded and Seamless Carbon Steel Structural Tubing, Specification for, **A501**
Hot-Formed Welded and Seamless High-Strength Low-Alloy Structural Tubing, Specification for, **A618/A618M**

Hot-rolled stainless steel bars—specifications
See **Alloy steel bars—specifications**

Hot-rolled steel sheet/strip—specifications
Sa **Steel sheet/strip**
Steel, Sheet and Strip, High-Strength, Low-Alloy, Hot-Rolled and Cold-Rolled, with Improved Atmospheric Corrosion Resistance, Specification for, **A606/A606M**
Steel, Sheet and Strip, Hot-Rolled, Carbon, Structural, High-Strength Low-Alloy, High-Strength Low-Alloy with Improved Formability, and Ultra-High Strength, Specification for, **A1011/A1011M**
Steel, Sheet, Cold-Rolled, Carbon, Structural, High-Strength Low-Alloy, High-Strength Low-Alloy with Improved Formability, Solution Hardened, and Bake Hardenable, Specification for, **A1008/A1008M**

Hot wrought steel bars—specifications
Steel Bars, Carbon and Alloy, Hot-Wrought, General Requirements for, Specification for, **A29/A29M**

H-piles
See **Structural steel (SS) piles—specifications**

HSLA (high-strength low-alloy) steel—specifications
Sa **Structural steel (SS) plate—specifications**
High-Strength Low-Alloy Columbium-Vanadium Structural Steel, Specification for, **A572/A572M**
High-Strength Low-Alloy Structural Steel, Specification for, **A242/A242M**
High-Strength Low-Alloy Structural Steel Plate With Atmospheric Corrosion Resistance, Specification for, **A871/A871M**
Hot-Rolled Structural Steel, High-Strength Low-Alloy Plate with Improved Formability, Specification for, **A656/A656M**
Steel Sheet, Zinc-Coated (Galvanized) or Zinc-Iron Alloy-Coated (Galvannealed) by the Hot-Dip Process, Specification for, **A653/A653M**

HSLA (high-strength low-alloy) steel bars—specifications
High-Strength Low-Alloy Structural Steel, up to 50 ksi [345 MPa] Minimum Yield Point, with Atmospheric Corrosion Resistance, Specification for, **A588/A588M**
Structural Steel for Bridges, Specification for, **A709/A709M**

HSLA (high-strength low-alloy) steel plate—specifications
High-Strength Low-Alloy Structural Steel, up to 50 ksi [345 MPa] Minimum Yield Point, with Atmospheric Corrosion Resistance, Specification for, **A588/A588M**
Normalized High-Strength Low-Alloy Structural Steel Plates, Specification for, **A633/A633M**
Quenched-and-Tempered Carbon and High-Strength Low-Alloy Structural Steel Plates, Specification for, **A678/A678M**
Structural Steel for Bridges, Specification for, **A709/A709M**

HSLA (high-strength low-alloy) steel sheet/strip—specifications
Steel, Sheet and Strip, Heavy-Thickness Coils, Hot-Rolled, Carbon, Commercial, Drawing, Structural, High-Strength Low-Alloy, High-Strength Low-Alloy with Improved Formability, and Ultra-High Strength, Specification for, **A1018/A1018M**
Steel, Sheet and Strip, High-Strength, Low-Alloy, Hot-Rolled and Cold-Rolled, with Improved Atmospheric Corrosion Resistance, Specification for, **A606/A606M**
Steel, Sheet and Strip, Hot-Rolled, Carbon, Structural, High-Strength Low-Alloy, High-Strength Low-Alloy with Im-

Index, ASTM Standards for Welding

proved Formability, and Ultra-High Strength, Specification for, **A1011/A1011M**

Steel, Sheet, Cold-Rolled, Carbon, Structural, High-Strength Low-Alloy, High-Strength Low-Alloy with Improved Formability, Solution Hardened, and Bake Hardenable, Specification for, **A1008/A1008M**

HSLA (high-strength low-alloy) steel sheet/tube—specifications
Hot-Formed Welded and Seamless High-Strength Low-Alloy Structural Tubing, Specification for, **A618/A618M**

Hydrophilic emulsifiers
Liquid Penetrant Examination for General Industry, Practice for, **E165**

I

Image analysis—radiographic testing
Design, Manufacture and Material Grouping Classification of Wire Image Quality Indicators (IQI) Used for Radiology, Practice for, **E747**

Design, Manufacture, and Material Grouping Classification of Hole-Type Image Quality Indicators (IQI) Used for Radiology, Practice for, **E1025**

Image quality indicator
See **IQI (image quality indicators)**

Impact resistance
Mechanical Testing of Steel Products, Test Methods and Definitions for, **A370**

Notched Bar Impact Testing of Metallic Materials, Test Methods for, **E23**

Impact testing—Charpy
Mechanical Testing of Steel Products, Test Methods and Definitions for, **A370**

Notched Bar Impact Testing of Metallic Materials, Test Methods for, **E23**

Sampling Procedure for Impact Testing of Structural Steel, Specification for, **A673/A673M**

Improved formability steel
Hot-Rolled Structural Steel, High-Strength Low-Alloy Plate with Improved Formability, Specification for, **A656/A656M**

Steel, Sheet and Strip, Heavy-Thickness Coils, Hot-Rolled, Carbon, Commercial, Drawing, Structural, High-Strength Low-Alloy, High-Strength Low-Alloy with Improved Formability, and Ultra-High Strength, Specification for, **A1018/A1018M**

Steel, Sheet and Strip, Hot-Rolled, Carbon, Structural, High-Strength Low-Alloy, High-Strength Low-Alloy with Improved Formability, and Ultra-High Strength, Specification for, **A1011/A1011M**

Steel, Sheet, Cold-Rolled, Carbon, Structural, High-Strength Low-Alloy, High-Strength Low-Alloy with Improved Formability, Solution Hardened, and Bake Hardenable, Specification for, **A1008/A1008M**

Inclusions
Magnetic Particle Testing, Guide for, **E709**

Indentation hardness (of metallic materials)
See **Hardness (indentation)—metallic materials**

Indirect verification
Notched Bar Impact Testing of Metallic Materials, Test Methods for, **E23**

Industrial radiographic film processing/testing
See **Radiographic examination**

Internal threaded fasteners
See **Steel bolting materials**

Ion chromatography (IC)
Liquid Penetrant Examination for General Industry, Practice for, **E165**

IQI (image quality indicators)
Design, Manufacture and Material Grouping Classification of Wire Image Quality Indicators (IQI) Used for Radiology, Practice for, **E747**

Design, Manufacture, and Material Grouping Classification of Hole-Type Image Quality Indicators (IQI) Used for Radiology, Practice for, **E1025**

Iron bolting materials
See **Steel bolting materials**

Iron pipe/fittings—specifications
See **Steel pipe**

Izod impact testing
Notched Bar Impact Testing of Metallic Materials, Test Methods for, **E23**

J

Joint steel bars—specifications
See **Alloy steel bars—specifications**

K

Killed steel
General Requirements for Rolled Structural Steel Bars, Plates, Shapes, and Sheet Piling, Specification for, **A6/A6M**

Quenched-and-Tempered Carbon and High-Strength Low-Alloy Structural Steel Plates, Specification for, **A678/A678M**

Knoop hardness (HK) number
Standard Hardness Conversion Tables for Metals Relationship Among Brinell Hardness, Vickers Hardness, Rockwell Hardness, Superficial Hardness, Knoop Hardness, and Scleroscope Hardness, **E140**

L

Laps
Magnetic Particle Testing, Guide for, **E709**

Leak testing
Liquid Penetrant Examination for General Industry, Practice for, **E165**

Length of inclusions
See **Inclusions**

Light-gaged steel sheet piling
See **Structural steel (SS) piles—specifications**

Light gage steel sheet piling
See **Structural steel (SS) piles—specifications**

Lipophilic emulsification
Liquid Penetrant Examination for General Industry, Practice for, **E165**

Liquid conveyance
Electric-Fusion (Arc)-Welded Steel Pipe (NPS 4 and Over), Specification for, **A139/A139M**

Liquid penetration
Liquid Penetrant Examination for General Industry, Practice for, **E165**

Load-carrying piles
Welded and Seamless Steel Pipe Piles, Specification for, **A252**

Long bar steel
See **Alloy steel bars—specifications**

Longitudinal magnetization
See **Magnetic particle inspection**

Long-terne steel sheet
See **Steel sheet**

Low-alloy/carbon steel plate
See **Carbon steel plate—specifications**

Low-alloy high-strength steel
See **HSLA (high-strength low-alloy) steel—specifications**

Low-alloy steel—specifications
Quenched and Tempered Low-Alloy Structural Steel Plate with 70 ksi [485 MPa] Minimum Yield Strength to 4 in. [100 mm] Thick, Specification for, **A852/A852M**

Index, ASTM Standards for Welding

Low-carbon steel plate—specifications
Low-carbon steel plate—specifications
    Sa **Carbon steel plate—specifications**
    Precipitation–Strengthened Low-Carbon Nickel-Copper-Chromium-Molybdenum-Columbium Alloy Structural Steel Plates, Specification for, **A710/A710M**
    Steel Tubes, Low-Carbon or High-Strength Low-Alloy, Tapered for Structural Use, Specification for, **A595/A595M**

Low-temperature service applications—steel
    Electric-Fusion-Welded Steel Pipe for Atmospheric and Lower Temperatures, Specification for, **A671**
    Seamless Carbon Steel Pipe for Atmospheric and Lower Temperatures, Specification for, **A524**
    Seamless and Welded Carbon and Alloy-Steel Tubes for Low-Temperature Service, Specification for, **A334/A334M**
    Seamless and Welded Steel Pipe for Low-Temperature Service, Specification for, **A333/A333M**

Low-temperature service applications—steel plate
    Pressure Vessel Plates, Carbon Steel, for Moderate- and Lower-Temperature Service, Specification for, **A516/A516M**

Low yield to tensile ratio
    Structural Steel with Low Yield to Tensile Ratio for Use in Buildings, Specification for, **A1043/A1043M**

M

Machinery (for ships)
    See **Shipbuilding steel materials—specifications**
Magnetic particle inspection
    Sa **Nondestructive evaluation (NDE)**
    Magnetic Particle Testing, Guide for, **E709**
Manganese alloy steel plate—specifications
    Structural Carbon Steel Plates of Improved Toughness, Specification for, **A573/A573M**
Marine systems/subsystems/equipment
    See **Shipbuilding steel materials—specifications**
Marine systems/subsystems/equipment (steel)—specifications
    General Requirements for Rolled Structural Steel Bars, Plates, Shapes, and Sheet Piling, Specification for, **A6/A6M**
Mass transit
    See **Railroad steel materials**
Mechanical analysis/testing
    Mechanical Testing of Steel Products, Test Methods and Definitions for, **A370**
Merchant quality steel bars
    See **Alloy steel bars—specifications**
Metal-arc-welded steel pipe
    Sa **Electric-resistance-welded (ERW) steel pipe—specifications**
    Metal-Arc-Welded Steel Pipe for Use With High-Pressure Transmission Systems, Specification for, **A381**
Metal buildings
    High-Strength Carbon-Manganese Steel of Structural Quality, Specification for, **A529/A529M**
Metal fasteners
    See **Fasteners (metal)—specifications**
Metallic hardness
    Standard Hardness Conversion Tables for Metals Relationship Among Brinell Hardness, Vickers Hardness, Rockwell Hardness, Superficial Hardness, Knoop Hardness, and Scleroscope Hardness, **E140**
    Vickers Hardness of Metallic Materials, Test Method for, **E92**
Metallurgical structure
    General Requirements for Rolled Structural Steel Bars, Plates, Shapes, and Sheet Piling, Specification for, **A6/A6M**
Metals and metallic materials
    Magnetic Particle Testing, Guide for, **E709**
    Vickers Hardness of Metallic Materials, Test Method for, **E92**

Mill edge
    General Requirements for Rolled Structural Steel Bars, Plates, Shapes, and Sheet Piling, Specification for, **A6/A6M**
Minimized spangle coating
    Steel Sheet, Zinc-Coated (Galvanized) or Zinc-Iron Alloy-Coated (Galvannealed) by the Hot-Dip Process, Specification for, **A653/A653M**
Moderate-temperature service applications—steel plate
    Pressure Vessel Plates, Carbon Steel, for Moderate- and Lower-Temperature Service, Specification for, **A516/A516M**
Molybdenum alloys—specifications
    Precipitation–Strengthened Low-Carbon Nickel-Copper-Chromium-Molybdenum-Columbium Alloy Structural Steel Plates, Specification for, **A710/A710M**

N

NDT laboratories
    See **Nondestructive evaluation (NDE)**
Negative buoyancy pipe
    See **Steel pipe**
Neutron radiographic testing
    See **Radiographic examination**
Nickel alloy steel plate—specifications
    Precipitation–Strengthened Low-Carbon Nickel-Copper-Chromium-Molybdenum-Columbium Alloy Structural Steel Plates, Specification for, **A710/A710M**
Nickel-copper alloy steel—specifications
    Precipitation–Strengthened Low-Carbon Nickel-Copper-Chromium-Molybdenum-Columbium Alloy Structural Steel Plates, Specification for, **A710/A710M**
Nipples for steel pipe
    See **Steel pipe**
Nondestructive evaluation (NDE)
    Liquid Penetrant Examination for General Industry, Practice for, **E165**
Nondestructive evaluation (NDE)—magnetic particle inspection
    See **Magnetic particle inspection**
Nondestructive evaluation (NDE)—penetrant inspection
    See **Penetrant inspection**
Nondestructive evaluation (NDE)—radiographic
    See **Radiographic examination**
Nonferrous bolting materials
    See **Steel bolting materials**
Nonferrous metals/alloys
    See **Metals and metallic materials**
Nonpressure piping—specifications
    See **Pressure vessel steel**
Normalized steel
    Normalized High-Strength Low-Alloy Structural Steel Plates, Specification for, **A633/A633M**
Notched-bar impact
    Notched Bar Impact Testing of Metallic Materials, Test Methods for, **E23**
Notch testing
    Notched Bar Impact Testing of Metallic Materials, Test Methods for, **E23**
Notch toughness
    Normalized High-Strength Low-Alloy Structural Steel Plates, Specification for, **A633/A633M**

O

Oersted conversion units
    See **Conversion units/factors**
Oily penetrants
    See **Penetrant inspection**

Index, ASTM Standards for Welding

Open-hearth steel—specifications
  High-Strength Carbon-Manganese Steel of Structural Quality, Specification for, **A529/A529M**
  High-Yield-Strength, Quenched and Tempered Alloy Steel Plate, Suitable for Welding, Specification for, **A514/A514M**
  Precipitation–Strengthened Low-Carbon Nickel-Copper-Chromium-Molybdenum-Columbium Alloy Structural Steel Plates, Specification for, **A710/A710M**
  Quenched and Tempered Low-Alloy Structural Steel Plate with 70 ksi [485 MPa] Minimum Yield Strength to 4 in. [100 mm] Thick, Specification for, **A852/A852M**

Ordinary-strength steel
  Structural Steel for Ships, Specification for, **A131/A131M**

P

Pendulum test
  Notched Bar Impact Testing of Metallic Materials, Test Methods for, **E23**

Penetrameters
  Design, Manufacture and Material Grouping Classification of Wire Image Quality Indicators (IQI) Used for Radiology, Practice for, **E747**
  Radiographic Examination of Weldments, Test Method for, **E1032**

Penetrant inspection
  Sa **Nondestructive evaluation (NDE)**
  Liquid Penetrant Examination for General Industry, Practice for, **E165**

Piles
  General Requirements for Rolled Structural Steel Bars, Plates, Shapes, and Sheet Piling, Specification for, **A6/A6M**
  High-Strength Low-Alloy Columbium-Vanadium Structural Steel, Specification for, **A572/A572M**
  Welded and Seamless Steel Pipe Piles, Specification for, **A252**

Pipe piles
  Welded and Seamless Steel Pipe Piles, Specification for, **A252**

Plain end steel pipe—specifications
  See **Black steel pipe—specifications**

Power generating facilities
  See **Generator materials**

Precipitation-hardening metals/alloys—specifications
  Precipitation–Strengthened Low-Carbon Nickel-Copper-Chromium-Molybdenum-Columbium Alloy Structural Steel Plates, Specification for, **A710/A710M**

Precracking Charpy V-notch (CVN) impact specimens
  Notched Bar Impact Testing of Metallic Materials, Test Methods for, **E23**

Pressure vessel steel
  Sa **Steel bolting materials**
  Mechanical Testing of Steel Products, Test Methods and Definitions for, **A370**

Pressure vessel steel pipe—specifications
  Metal-Arc-Welded Steel Pipe for Use With High-Pressure Transmission Systems, Specification for, **A381**

Pressure vessel steel plate—specifications
  Straight-Beam Ultrasonic Examination of Rolled Steel Plates for Special Applications, Specification for, **A578/A578M**

Product analysis specifications/tolerances
  General Requirements for Rolled Structural Steel Bars, Plates, Shapes, and Sheet Piling, Specification for, **A6/A6M**

Propellers
  See **Shipbuilding steel materials—specifications**

Q

QST process
  High-Strength Low-Alloy Steel Shapes of Structural Quality, Produced by Quenching and Self-Tempering Process (QST),
  Specification for, **A913/A913M**

Quality control (QC)—radiographic testing
  Design, Manufacture and Material Grouping Classification of Wire Image Quality Indicators (IQI) Used for Radiology, Practice for, **E747**

Quenched and tempered steels (specifications)
  High-Strength Low-Alloy Steel Shapes of Structural Quality, Produced by Quenching and Self-Tempering Process (QST), Specification for, **A913/A913M**
  High-Yield-Strength, Quenched and Tempered Alloy Steel Plate, Suitable for Welding, Specification for, **A514/A514M**
  Pressure Vessel Plates, Alloy Steel, High-Strength, Quenched and Tempered, Specification for, **A517/A517M**
  Quenched and Tempered Low-Alloy Structural Steel Plate with 70 ksi [485 MPa] Minimum Yield Strength to 4 in. [100 mm] Thick, Specification for, **A852/A852M**
  Quenched-and-Tempered Carbon and High-Strength Low-Alloy Structural Steel Plates, Specification for, **A678/A678M**
  Structural Steel for Bridges, Specification for, **A709/A709M**

R

Radiographic examination
  Design, Manufacture and Material Grouping Classification of Wire Image Quality Indicators (IQI) Used for Radiology, Practice for, **E747**
  Design, Manufacture, and Material Grouping Classification of Hole-Type Image Quality Indicators (IQI) Used for Radiology, Practice for, **E1025**
  Radiographic Examination, Guide for, **E94**
  Radiographic Examination of Weldments, Test Method for, **E1032**

Radiological examination
  Design, Manufacture and Material Grouping Classification of Wire Image Quality Indicators (IQI) Used for Radiology, Practice for, **E747**
  Design, Manufacture, and Material Grouping Classification of Hole-Type Image Quality Indicators (IQI) Used for Radiology, Practice for, **E1025**

Railroad steel materials
  Carbon Structural Steel, Specification for, **A36/A36M**
  Hot-Rolled Structural Steel, High-Strength Low-Alloy Plate with Improved Formability, Specification for, **A656/A656M**
  Mechanical Testing of Steel Products, Test Methods and Definitions for, **A370**

Rapid indentation hardness testing
  See **Hardness (indentation)—metallic materials**

Reinforcement (concrete)
  See **Concrete reinforcement—specifications**

Reinforcing steel/concrete—specifications
  See **Concrete reinforcement—specifications**

Rimmed structural steel bars
  General Requirements for Rolled Structural Steel Bars, Plates, Shapes, and Sheet Piling, Specification for, **A6/A6M**

Riveted construction—specifications
  Carbon Structural Steel, Specification for, **A36/A36M**
  High-Strength Carbon-Manganese Steel of Structural Quality, Specification for, **A529/A529M**
  High-Strength Low-Alloy Columbium-Vanadium Structural Steel, Specification for, **A572/A572M**
  High-Strength Low-Alloy Structural Steel, Specification for, **A242/A242M**
  High-Strength Low-Alloy Structural Steel, up to 50 ksi [345 MPa] Minimum Yield Point, with Atmospheric Corrosion Resistance, Specification for, **A588/A588M**
  Normalized High-Strength Low-Alloy Structural Steel Plates, Specification for, **A633/A633M**
  Quenched-and-Tempered Carbon and High-Strength Low-Alloy Structural Steel Plates, Specification for, **A678/A678M**
  Structural Steel for Ships, Specification for, **A131/A131M**

Riveted construction—specifications
  Specification for, **A913/A913M**

Index, ASTM Standards for Welding

Rockwell hardness

Rockwell hardness
   Mechanical Testing of Steel Products, Test Methods and Definitions for, **A370**
   Standard Hardness Conversion Tables for Metals Relationship Among Brinell Hardness, Vickers Hardness, Rockwell Hardness, Superficial Hardness, Knoop Hardness, and Scleroscope Hardness, **E140**

Rolled steel—specifications
   General Requirements for Rolled Structural Steel Bars, Plates, Shapes, and Sheet Piling, Specification for, **A6/A6M**
   Structural Steel Shapes, Specification for, **A992A/A992AM**

Roofing materials/applications—bridges
   See **Bridge/structural materials**

Rotor forgings
   See **Generator materials**

Round bar steel
   See **Alloy steel bars—specifications**

S

Sampling metals/alloys
   Sampling Procedure for Impact Testing of Structural Steel, Specification for, **A673/A673M**

Sampling steel
   See **Steel sampling**

Seamless steel pipe—specifications
   Pipe, Steel, Black and Hot-Dipped, Zinc-Coated, Welded and Seamless, Specification for, **A53/A53M**
   Seamless Carbon Steel Pipe for Atmospheric and Lower Temperatures, Specification for, **A524**
   Seamless and Welded Steel Pipe for Low-Temperature Service, Specification for, **A333/A333M**

Seamless steel tube—specifications
   Hot-Formed Welded and Seamless High-Strength Low-Alloy Structural Tubing, Specification for, **A618/A618M**

Sea walls
   See **Structural steel (SS) piles—specifications**

Semi-killed steel
   General Requirements for Rolled Structural Steel Bars, Plates, Shapes, and Sheet Piling, Specification for, **A6/A6M**

Seven-wire strand (tendon)
   Mechanical Testing of Steel Products, Test Methods and Definitions for, **A370**

Sheared edge
   General Requirements for Rolled Structural Steel Bars, Plates, Shapes, and Sheet Piling, Specification for, **A6/A6M**

Sheet piling
   See **Structural steel (SS) piles—specifications**

Sheet steel
   See **Steel sheet/strip**

Shipbuilding steel materials—specifications
   General Requirements for Rolled Structural Steel Bars, Plates, Shapes, and Sheet Piling, Specification for, **A6/A6M**
   Mechanical Testing of Steel Products, Test Methods and Definitions for, **A370**

Shipbuilding steel materials (plate)
   See **Alloy steel plate—specifications**

Shipbuilding steel materials (structural)
   Structural Steel for Ships, Specification for, **A131/A131M**

Single-reduced tin mill black plate
   See **Tin mill products**

Slush castings
   See **Castings**

Solvent removable penetrants
   Sa **Penetrant inspection**
   Liquid Penetrant Examination for General Industry, Practice for, **E165**

Spangle
   Steel Sheet, Zinc-Coated (Galvanized) or Zinc-Iron Alloy-Coated (Galvannealed) by the Hot-Dip Process, Specification for, **A653/A653M**

Specifications—bridge/structural materials
   See **Bridge/structural materials—specifications**

Specifications—metals/alloys (copper)
   See **Copper structural materials/applications—specifications**

Specifications—quenched and tempered steels
   See **Quenched and tempered steels (specifications)**

Specifications—shipbuilding materials/applications
   See **Shipbuilding steel materials—specifications**

Stainless steel—structural
   See **Structural steel (SS)**

Stainless steel bars/billets—specifications
   See **Alloy steel bars—specifications**

Stainless steel pipe—specifications
   Seamless and Welded Steel Pipe for Low-Temperature Service, Specification for, **A333/A333M**

Steam turbine materials/applications
   See **Shipbuilding steel materials—specifications**

Steel
   Sa **Steel sheet**
   Sa **Steel sheet/strip**
   Sa **Steel wire**
   Magnetic Particle Testing, Guide for, **E709**
   Mechanical Testing of Steel Products, Test Methods and Definitions for, **A370**

Steel—carbon steel
   See **Carbon steel bars/shapes—specifications**
   See **Carbon steel pipe—specifications**
   See **Carbon steel plate—specifications**
   See **Carbon steel sheet—specifications**

Steel bars
   Mechanical Testing of Steel Products, Test Methods and Definitions for, **A370**

Steel bars—specifications
   High-Strength Low-Alloy Structural Steel, Specification for, **A242/A242M**
   High-Strength Low-Alloy Structural Steel, up to 50 ksi [345 MPa] Minimum Yield Point, with Atmospheric Corrosion Resistance, Specification for, **A588/A588M**
   Steel Bars, Carbon and Alloy, Hot-Wrought, General Requirements for, Specification for, **A29/A29M**

Steel bars and shapes—specifications
   General Requirements for Rolled Structural Steel Bars, Plates, Shapes, and Sheet Piling, Specification for, **A6/A6M**
   High-Strength Carbon-Manganese Steel of Structural Quality, Specification for, **A529/A529M**
   High-Strength Low-Alloy Steel Shapes of Structural Quality, Produced by Quenching and Self-Tempering Process (QST), Specification for, **A913/A913M**

Steel bars (carbon steel)
   See **Carbon steel bars/shapes—specifications**

Steel bars/shapes/wires
   Carbon Structural Steel, Specification for, **A36/A36M**

Steel bars (structural)
   See **Structural steel (SS) bars—specifications**

Steel bolting materials
   Mechanical Testing of Steel Products, Test Methods and Definitions for, **A370**

Steel bolting materials—specifications
   Structural Bolts, Alloy Steel, Heat Treated, 150 ksi Minimum Tensile Strength, Specification for, **A490**
   Structural Bolts, Steel, Heat Treated, 120/105 ksi Minimum Tensile Strength, Specification for, Δ **A325**

Steel bolting materials (alloy steel)
   See **Alloy steel bolting materials—specifications**

Index, ASTM Standards for Welding

Steel chain
  Mechanical Testing of Steel Products, Test Methods and Definitions for, **A370**

Steel fiber-reinforced concrete/shotcrete
  See **Concrete reinforcement—specifications**

Steel hull
  See **Shipbuilding steel materials (structural)**

Steel line pipe
  See **Black steel pipe—specifications**

Steel pipe
  Sa **Carbon steel pipe—specifications**
  Mechanical Testing of Steel Products, Test Methods and Definitions for, **A370**

Steel pipe—specifications
  Metal-Arc-Welded Steel Pipe for Use With High-Pressure Transmission Systems, Specification for, **A381**
  Welded and Seamless Steel Pipe Piles, Specification for, **A252**

Steel pipe (black)
  See **Black steel pipe—specifications**

Steel pipe (electric-fusion-welded)
  See **Electric-fusion-welded steel pipe**

Steel pipe (electric-resistance-welded (ERW))
  See **Electric-resistance-welded (ERW) steel pipe—specifications**

Steel pipe (high-temperature/pressure service)
  See **High-temperature service applications—steel pipe**

Steel pipe (zinc-coated)
  See **Zinc-coated steel pipe—specifications**

Steel plate—specifications
  General Requirements for Rolled Structural Steel Bars, Plates, Shapes, and Sheet Piling, Specification for, **A6/A6M**
  Through-Thickness Tension Testing of Steel Plates for Special Applications, Specification for, **A770/A770M**

Steel plate (ultrasonic examination)
  Straight-Beam Ultrasonic Examination of Rolled Steel Plates for Special Applications, Specification for, **A578/A578M**
  Straight-Beam Ultrasonic Examination of Steel Plates, Specification for, **A435/A435M**

Steel plate (alloy steel)
  See **Alloy steel plate—specifications**

Steel plate (carbon steel)
  See **Carbon steel plate—specifications**

Steel plate (high-strength/low alloy (HSLA))
  See **HSLA (high-strength low-alloy) steel—specifications**

Steel plate (manganese alloy)
  See **Manganese alloy steel plate—specifications**

Steel plate (nickel alloy)
  See **Nickel alloy steel plate—specifications**

Steel plate (pressure vessels)
  See **Pressure vessel steel plate—specifications**

Steel plates/shapes/bars—specifications
  General Requirements for Rolled Structural Steel Bars, Plates, Shapes, and Sheet Piling, Specification for, **A6/A6M**
  High-Strength Low-Alloy Steel Shapes of Structural Quality, Produced by Quenching and Self-Tempering Process (QST), Specification for, **A913/A913M**

Steel plate (structural)
  See **Structural steel (SS) plate—specifications**

Steel reinforcing bars
  See **Structural steel (SS) bars—specifications**

Steel sampling
  Sampling Procedure for Impact Testing of Structural Steel, Specification for, **A673/A673M**

Steel shapes
  See **Structural steel (SS) shapes—specifications**

Steel sheet
  Mechanical Testing of Steel Products, Test Methods and Definitions for, **A370**

Steel sheet (alloy steel)
  See **Alloy steel sheet/strip—specifications**

Steel sheet (carbon steel)
  See **Carbon steel sheet—specifications**

Steel sheet (high-strength/low-alloy)
  See **HSLA (high-strength low-alloy) steel sheet/strip—specifications**

Steel sheet piling—specifications
  Sa **Structural steel (SS) piles—specifications**
  High-Strength Low-Alloy Columbium-Vanadium Structural Steel, Specification for, **A572/A572M**

Steel sheet/strip
  Mechanical Testing of Steel Products, Test Methods and Definitions for, **A370**

Steel sheet/strip—alloy steel
  See **Alloy steel sheet/strip—specifications**

Steel sheet/strip—carbon (cold-rolled)
  See **Carbon steel sheet/strip—specifications**

Steel sheet/strip—carbon steel
  Sa **Carbon steel sheet/strip—specifications**
  Steel, Sheet and Strip, Heavy-Thickness Coils, Hot-Rolled, Carbon, Commercial, Drawing, Structural, High-Strength Low-Alloy, High-Strength Low-Alloy with Improved Formability, and Ultra-High Strength, Specification for, **A1018/A1018M**

Steel sheet/strip—structural
  See **Structural steel (SS) sheet/strip—specifications**

Steel sheet/strip/plate
  Mechanical Testing of Steel Products, Test Methods and Definitions for, **A370**

Steel sheet (zinc/zinc alloy-coated)
  See **Zinc-coated steel sheet—specifications**

Steel strip—carbon steel
  See **Carbon steel sheet/strip—specifications**

Steel structural shapes
  See **Structural steel (SS)**

Steel tube
  Mechanical Testing of Steel Products, Test Methods and Definitions for, **A370**

Steel tube (alloy steel)
  See **Alloy steel tube—specifications**

Steel tube (structural)
  See **Structural steel (SS) tube—specifications**

Steel valves
  Mechanical Testing of Steel Products, Test Methods and Definitions for, **A370**

Steel wire
  Mechanical Testing of Steel Products, Test Methods and Definitions for, **A370**

Steel wire (concrete reinforcement applications)—specifications
  Sa **Concrete reinforcement—specifications**
  Steel Wire, Deformed, for Concrete Reinforcement, Specification for, **A496/A496M**

Straight-beam ultrasonic testing
  Sa **Ultrasonic testing—steel**
  Straight-Beam Ultrasonic Examination of Rolled Steel Plates for Special Applications, Specification for, **A578/A578M**

Stress-relieved steel bars
  See **Alloy steel bars—specifications**

Structural shipbuilding materials/applications
  Structural Steel for Ships, Specification for, **A131/A131M**

Structural steel (SS)
  Sa **Bridge/structural materials**
  Sa **Generator materials**
  Sa **Pressure vessel steel**
  Sa **Shipbuilding steel materials—specifications**
  Sa **Structural steel (SS) bars—specifications**
  Structural Steel with Low Yield to Tensile Ratio for Use in Buildings, Specification for, **A1043/A1043M**

Index, ASTM Standards for Welding

Structural steel (SS)—specifications
Structural steel (SS)—specifications
  High-Strength Carbon-Manganese Steel of Structural Quality, Specification for, **A529/A529M**
  High-Strength Low-Alloy Steel Shapes of Structural Quality, Produced by Quenching and Self-Tempering Process (QST), Specification for, **A913/A913M**
  Sampling Procedure for Impact Testing of Structural Steel, Specification for, **A673/A673M**
Structural steel (SS) bars—specifications
  General Requirements for Rolled Structural Steel Bars, Plates, Shapes, and Sheet Piling, Specification for, **A6/A6M**
  High-Strength Carbon-Manganese Steel of Structural Quality, Specification for, **A529/A529M**
  High-Strength Low-Alloy Columbium-Vanadium Structural Steel, Specification for, **A572/A572M**
  High-Strength Low-Alloy Structural Steel, Specification for, **A242/A242M**
  High-Strength Low-Alloy Structural Steel, up to 50 ksi [345 MPa] Minimum Yield Point, with Atmospheric Corrosion Resistance, Specification for, **A588/A588M**
  Normalized High-Strength Low-Alloy Structural Steel Plates, Specification for, **A633/A633M**
  Structural Steel for Bridges, Specification for, **A709/A709M**
  Structural Steel for Ships, Specification for, **A131/A131M**
Structural steel (SS) bolting materials—specifications
  Structural Bolts, Alloy Steel, Heat Treated, 150 ksi Minimum Tensile Strength, Specification for, **A490**
  Structural Bolts, Steel, Heat Treated, 120/105 ksi Minimum Tensile Strength, Specification for, Δ **A325**
Structural steel (SS) piles—specifications
  General Requirements for Rolled Structural Steel Bars, Plates, Shapes, and Sheet Piling, Specification for, **A6/A6M**
  High-Strength Low-Alloy Columbium-Vanadium Structural Steel, Specification for, **A572/A572M**
  Welded and Seamless Steel Pipe Piles, Specification for, **A252**
Structural steel (SS) plate—specifications
  Carbon Structural Steel, Specification for, **A36/A36M**
  General Requirements for Rolled Structural Steel Bars, Plates, Shapes, and Sheet Piling, Specification for, **A6/A6M**
  High-Strength Carbon-Manganese Steel of Structural Quality, Specification for, **A529/A529M**
  High-Strength Low-Alloy Columbium-Vanadium Structural Steel, Specification for, **A572/A572M**
  High-Strength Low-Alloy Structural Steel, Specification for, **A242/A242M**
  High-Strength Low-Alloy Structural Steel Plate With Atmospheric Corrosion Resistance, Specification for, **A871/A871M**
  High-Yield-Strength, Quenched and Tempered Alloy Steel Plate, Suitable for Welding, Specification for, **A514/A514M**
  Hot-Rolled Structural Steel, High-Strength Low-Alloy Plate with Improved Formability, Specification for, **A656/A656M**
  Normalized High-Strength Low-Alloy Structural Steel Plates, Specification for, **A633/A633M**
  Precipitation–Strengthened Low-Carbon Nickel-Copper-Chromium-Molybdenum-Columbium Alloy Structural Steel Plates, Specification for, **A710/A710M**
  Quenched and Tempered Low-Alloy Structural Steel Plate with 70 ksi [485 MPa] Minimum Yield Strength to 4 in. [100 mm] Thick, Specification for, **A852/A852M**
  Quenched-and-Tempered Carbon and High-Strength Low-Alloy Structural Steel Plates, Specification for, **A678/A678M**
  Structural Carbon Steel Plates of Improved Toughness, Specification for, **A573/A573M**
  Structural Steel for Bridges, Specification for, **A709/A709M**
  Structural Steel for Ships, Specification for, **A131/A131M**
Structural steel (SS) shapes—specifications
  Carbon Structural Steel, Specification for, **A36/A36M**
  General Requirements for Rolled Structural Steel Bars, Plates, Shapes, and Sheet Piling, Specification for, **A6/A6M**
  High-Strength Carbon-Manganese Steel of Structural Quality, Specification for, **A529/A529M**
  High-Strength Low-Alloy Columbium-Vanadium Structural Steel, Specification for, **A572/A572M**
  High-Strength Low-Alloy Steel Shapes of Structural Quality, Produced by Quenching and Self-Tempering Process (QST), Specification for, **A913/A913M**
  High-Strength Low-Alloy Structural Steel, Specification for, **A242/A242M**
  Normalized High-Strength Low-Alloy Structural Steel Plates, Specification for, **A633/A633M**
  Structural Steel Shapes, Specification for, **A992A/A992AM**
  Structural Steel for Bridges, Specification for, **A709/A709M**
  Structural Steel for Ships, Specification for, **A131/A131M**
Structural steel (SS) sheet/strip—specifications
  General Requirements for Rolled Structural Steel Bars, Plates, Shapes, and Sheet Piling, Specification for, **A6/A6M**
  Steel, Sheet and Strip, Hot-Rolled, Carbon, Structural, High-Strength Low-Alloy, High-Strength Low-Alloy with Improved Formability, and Ultra-High Strength, Specification for, **A1011/A1011M**
  Steel, Sheet, Cold-Rolled, Carbon, Structural, High-Strength Low-Alloy, High-Strength Low-Alloy with Improved Formability, Solution Hardened, and Bake Hardenable, Specification for, **A1008/A1008M**
Structural steel (SS) tube—specifications
  Cold-Formed Welded and Seamless Carbon Steel Structural Tubing in Rounds and Shapes, Specification for, **A500/A500M**
  Hot-Formed Welded and Seamless Carbon Steel Structural Tubing, Specification for, **A501**
  Hot-Formed Welded and Seamless High-Strength Low-Alloy Structural Tubing, Specification for, **A618/A618M**
  Steel Tubes, Low-Carbon or High-Strength Low-Alloy, Tapered for Structural Use, Specification for, **A595/A595M**
Structural steel (SS) wire
  See **Steel wire**
Superficial Rockwell hardness
  See **Rockwell hardness**
Superstructures—ships
  See **Shipbuilding steel materials—specifications**
Surface analysis—metals/alloys
  Magnetic Particle Testing, Guide for, **E709**
  Straight-Beam Ultrasonic Examination of Rolled Steel Plates for Special Applications, Specification for, **A578/A578M**
  Straight-Beam Ultrasonic Examination of Steel Plates, Specification for, **A435/A435M**

T

Tables (measurement)
  Standard Hardness Conversion Tables for Metals Relationship Among Brinell Hardness, Vickers Hardness, Rockwell Hardness, Superficial Hardness, Knoop Hardness, and Scleroscope Hardness, **E140**
Tapered structural steel tube
  Sa **Structural steel (SS) tube—specifications**
  Steel Tubes, Low-Carbon or High-Strength Low-Alloy, Tapered for Structural Use, Specification for, **A595/A595M**
Temperature tests—metals/alloys
  Mechanical Testing of Steel Products, Test Methods and Definitions for, **A370**
Tempered structural steel—specifications
  Sa **Quenched and tempered steels (specifications)**
  High-Strength Low-Alloy Structural Steel Plate With Atmospheric Corrosion Resistance, Specification for, **A871/A871M**
  Quenched and Tempered Low-Alloy Structural Steel Plate with 70 ksi [485 MPa] Minimum Yield Strength to 4 in. [100 mm] Thick, Specification for, **A852/A852M**
  Structural Steel for Bridges, Specification for, **A709/A709M**

Index, ASTM Standards for Welding

Tensile properties/testing—metallic materials
General Requirements for Rolled Structural Steel Bars, Plates, Shapes, and Sheet Piling, Specification for, **A6/A6M**

Tensile properties/testing—steel
Mechanical Testing of Steel Products, Test Methods and Definitions for, **A370**
Through-Thickness Tension Testing of Steel Plates for Special Applications, Specification for, **A770/A770M**

Tensile property classification
General Requirements for Rolled Structural Steel Bars, Plates, Shapes, and Sheet Piling, Specification for, **A6/A6M**

Thermal neutron radiographic testing
See **Radiographic examination**

Thickness
Through-Thickness Tension Testing of Steel Plates for Special Applications, Specification for, **A770/A770M**

Thin-walled pressure vessels
See **Pressure vessel steel**

Through-thickness testing
Through-Thickness Tension Testing of Steel Plates for Special Applications, Specification for, **A770/A770M**

Thrusters
See **Shipbuilding steel materials—specifications**

Tin-coated steel sheet
See **Steel sheet**

Tin mill products
Mechanical Testing of Steel Products, Test Methods and Definitions for, **A370**

Total chlorine
Sa **Chlorine content**
Liquid Penetrant Examination for General Industry, Practice for, **E165**

Total fluorine
Liquid Penetrant Examination for General Industry, Practice for, **E165**

Transmission systems
Metal-Arc-Welded Steel Pipe for Use With High-Pressure Transmission Systems, Specification for, **A381**

Trucks
Hot-Rolled Structural Steel, High-Strength Low-Alloy Plate with Improved Formability, Specification for, **A656/A656M**

True continuous magnetization technique
See **Magnetic particle inspection**

Trusses
High-Strength Carbon-Manganese Steel of Structural Quality, Specification for, **A529/A529M**

U

Ultrasonic testing
See **Radiographic examination**

Ultrasonic testing—steel
Straight-Beam Ultrasonic Examination of Rolled Steel Plates for Special Applications, Specification for, **A578/A578M**
Straight-Beam Ultrasonic Examination of Steel Plates, Specification for, **A435/A435M**

Unexposed radiographic film
See **Radiographic examination**

Universal mill edge
General Requirements for Rolled Structural Steel Bars, Plates, Shapes, and Sheet Piling, Specification for, **A6/A6M**

V

Vanadium alloy steel
Steel, Sheet and Strip, Heavy-Thickness Coils, Hot-Rolled, Carbon, Commercial, Drawing, Structural, High-Strength Low-Alloy, High-Strength Low-Alloy with Improved Formability, and Ultra-High Strength, Specification for, **A1018/A1018M**

Vapor transmission
Electric-Fusion (Arc)-Welded Steel Pipe (NPS 4 and Over), Specification for, **A139/A139M**

Vickers hardness (HV)
Vickers Hardness of Metallic Materials, Test Method for, **E92**

Visible liquid penetrant testing
Sa **Liquid penetration**
Sa **Penetrant inspection**
Liquid Penetrant Examination for General Industry, Practice for, **E165**

Visual examination—steel
Straight-Beam Ultrasonic Examination of Rolled Steel Plates for Special Applications, Specification for, **A578/A578M**
Straight-Beam Ultrasonic Examination of Steel Plates, Specification for, **A435/A435M**

W

Washers/washer assemblies—specifications
See **Steel bolting materials**

Watertight doors
See **Shipbuilding steel materials—specifications**

Water washable penetrants
Sa **Penetrant inspection**
Liquid Penetrant Examination for General Industry, Practice for, **E165**

Water well pipe
See **Carbon steel pipe—specifications**

Weight saving steel
High-Strength Low-Alloy Structural Steel, Specification for, **A242/A242M**

Welded construction
Structural Steel with Low Yield to Tensile Ratio for Use in Buildings, Specification for, **A1043/A1043M**

Welded steel bars/shapes—specifications
Carbon Structural Steel, Specification for, **A36/A36M**
High-Strength Low-Alloy Columbium-Vanadium Structural Steel, Specification for, **A572/A572M**
High-Strength Low-Alloy Structural Steel, Specification for, **A242/A242M**

Welded steel materials/applications—specifications
Sa **Structural steel (SS)**
High-Strength Carbon-Manganese Steel of Structural Quality, Specification for, **A529/A529M**
Structural Steel Shapes, Specification for, **A992A/A992AM**

Welded steel pipe—specifications
Electric-Fusion (Arc)-Welded Steel Pipe (NPS 4 and Over), Specification for, **A139/A139M**
Metal-Arc-Welded Steel Pipe for Use With High-Pressure Transmission Systems, Specification for, **A381**
Pipe, Steel, Black and Hot-Dipped, Zinc-Coated, Welded and Seamless, Specification for, **A53/A53M**
Seamless and Welded Steel Pipe for Low-Temperature Service, Specification for, **A333/A333M**
Welded and Seamless Steel Pipe Piles, Specification for, **A252**

Welded steel plate/sheet/strip—specifications
Carbon Structural Steel, Specification for, **A36/A36M**
High-Strength Low-Alloy Columbium-Vanadium Structural Steel, Specification for, **A572/A572M**
High-Strength Low-Alloy Structural Steel, Specification for, **A242/A242M**
High-Yield-Strength, Quenched and Tempered Alloy Steel Plate, Suitable for Welding, Specification for, **A514/A514M**
Normalized High-Strength Low-Alloy Structural Steel Plates, Specification for, **A633/A633M**
Pressure Vessel Plates, Carbon Steel, for Moderate- and Lower-Temperature Service, Specification for, **A516/A516M**
Quenched-and-Tempered Carbon and High-Strength Low-Alloy Structural Steel Plates, Specification for, **A678/A678M**

Index, ASTM Standards for Welding

Welded steel tube—specifications

Welded steel tube—specifications
    Cold-Formed Welded and Seamless Carbon Steel Structural Tubing in Rounds and Shapes, Specification for, **A500/A500M**
    Hot-Formed Welded and Seamless Carbon Steel Structural Tubing, Specification for, **A501**
    Hot-Formed Welded and Seamless High-Strength Low-Alloy Structural Tubing, Specification for, **A618/A618M**
    Seamless and Welded Carbon and Alloy-Steel Tubes for Low-Temperature Service, Specification for, **A334/A334M**
    Steel Tubes, Low-Carbon or High-Strength Low-Alloy, Tapered for Structural Use, Specification for, **A595/A595M**

Welded steel wire—specifications
    See **Steel wire (concrete reinforcement applications)—specifications**

Welding/welds
    Magnetic Particle Testing, Guide for, **E709**

Welding/welds (nondestructive testing)
    Radiographic Examination of Weldments, Test Method for, **E1032**

Welding/welds (pressure vessels)
    See **Pressure vessel steel**

Wet magnetic particle examination
    Sa **Magnetic particle inspection**
    Magnetic Particle Testing, Guide for, **E709**

Winding steel wire
    See **Steel wire**

"W" shapes
    Structural Steel Shapes, Specification for, **A992A/A992AM**

X

X-irradiation
    Design, Manufacture and Material Grouping Classification of Wire Image Quality Indicators (IQI) Used for Radiology, Practice for, **E747**
    Design, Manufacture, and Material Grouping Classification of Hole-Type Image Quality Indicators (IQI) Used for Radiology, Practice for, **E1025**

X-ray monitoring/processing
    Design, Manufacture and Material Grouping Classification of Wire Image Quality Indicators (IQI) Used for Radiology, Practice for, **E747**
    Design, Manufacture, and Material Grouping Classification of Hole-Type Image Quality Indicators (IQI) Used for Radiology, Practice for, **E1025**

Y

Yield strength and yield point
    Mechanical Testing of Steel Products, Test Methods and Definitions for, **A370**
    Quenched and Tempered Low-Alloy Structural Steel Plate with 70 ksi [485 MPa] Minimum Yield Strength to 4 in. [100 mm] Thick, Specification for, **A852/A852M**

Z

Zinc-coated steel pipe—specifications
    Pipe, Steel, Black and Hot-Dipped, Zinc-Coated, Welded and Seamless, Specification for, **A53/A53M**

Zinc-coated steel sheet—specifications
    Steel Sheet, Zinc-Coated (Galvanized) or Zinc-Iron Alloy-Coated (Galvannealed) by the Hot-Dip Process, Specification for, **A653/A653M**

Zinc electrodeposited coatings—specifications
    Steel Sheet, Zinc-Coated (Galvanized) or Zinc-Iron Alloy-Coated (Galvannealed) by the Hot-Dip Process, Specification for, **A653/A653M**